# College Algebra
## GRAPHS AND MODELS

EDITION 5

# College Algebra

## GRAPHS AND MODELS

**MARVIN L. BITTINGER**
Indiana University Purdue University Indianapolis

**JUDITH A. BEECHER**
Indiana University Purdue University Indianapolis

**DAVID J. ELLENBOGEN**
Community College of Vermont

**JUDITH A. PENNA**
Indiana University Purdue University Indianapolis

**PEARSON**

Boston   Columbus   Indianapolis   New York   San Francisco   Upper Saddle River
Amsterdam   Cape Town   Dubai   London   Madrid   Milan   Munich   Paris   Montréal   Toronto
Delhi   Mexico City   São Paulo   Sydney   Hong Kong   Seoul   Singapore   Taipei   Tokyo

| | |
|---|---|
| *Editor-in-Chief* | Anne Kelly |
| *Acquisitions Editor* | Kathryn O'Connor |
| *Senior Content Editor* | Joanne Dill |
| *Editorial Assistant* | Judith Garber |
| *Senior Managing Editor* | Karen Wernholm |
| *Senior Production Project Manager* | Kathleen A. Manley |
| *Digital Assets Manager* | Marianne Groth |
| *Media Producer* | Aimee Thorne |
| *Software Development* | Kristina Evans, MathXL; Mary Durnwald, TestGen |
| *Marketing Manager* | Peggy Sue Lucas |
| *Marketing Assistant* | Justine Goulart |
| *Senior Author Support/ Technology Specialist* | Joe Vetere |
| *Rights and Permissions Advisor* | Michael Joyce |
| *Image Manager* | Rachel Youdelman |
| *Procurement Manager* | Evelyn Beaton |
| *Procurement Specialist* | Debbie Rossi |
| *Media Procurement Specialist* | Ginny Michaud |
| *Associate Director of Design* | Andrea Nix |
| *Art Director/Cover Designer* | Beth Paquin |
| *Text Design, Art Editing, and Photo Research* | The Davis Group, Inc. |
| *Editorial and Production Coordination* | Martha Morong/Quadrata, Inc. |
| *Composition* | PreMediaGlobal |
| *Illustrations* | Network Graphics, William Melvin, and PreMediaGlobal |
| *Cover Image* | Marek Chalupnik/Shutterstock |

For permission to use copyrighted material, grateful acknowledgment is made to the copyright holders on page 706, which is hereby made part of this copyright page.

All screenshots of graphs from the Appcylon graphing calculator app are reprinted with permission from Appcylon LLC.

Many of the designations used by manufacturers and sellers to distinguish their products are claimed as trademarks. Where those designations appear in this book, and Pearson Education was aware of a trademark claim, the designations have been printed in initial caps or all caps.

**Library of Congress Cataloging-in-Publication Data**
College algebra : graphs and models/Marvin L. Bittinger [... et al.].
  —5th ed.
   p. cm.
  Includes index.
  ISBN-13: 978-0-321-78395-0
  ISBN-10: 0-321-78395-6
1. Algebra—Textbooks. 2. Algebra—Graphic methods—Textbooks.
3. Function algebras—Textbooks. I. Bittinger, Marvin L.
  QA154.3.C65   2013
  512—dc22             2011011992

2 3 4 5 6 7 8 9 10—DOW—15 14 13 12

**PEARSON**

www.pearsonhighered.com

ISBN-10: 0-321-78395-6
ISBN-13: 978-0-321-78395-0

# Contents

v

## 2  More on Functions                                159

## 7     Conic Sections                                                          577

## 8     Sequences, Series, and Combinatorics                                     627

This *College Algebra* textbook is known for enabling students to "see the math" through its

- focus on visualization,
- early introduction to functions,
- integration of technology, and
- connections between mathematical concepts and the real world.

With the new edition, we continue to innovate by incorporating more ongoing review to help students develop understanding and study effectively. We have new Mid-Chapter Mixed Review exercise sets to give students an extra opportunity to master concepts and new Study Guides that provide built-in tools to help prepare for tests.

MyMathLab® has been expanded so that the online content is even more integrated with the text's approach, with the addition of Vocabulary Reinforcement, Synthesis, and Mid-Chapter Mixed Review exercises from the text as well as example-based videos.

Our overarching goal is to provide students with a learning experience that will not only lead to success in this course but also prepare them to be successful in the mathematics courses they take in the future.

## Content Changes to the Fourth Edition

- The material on solving formulas for a specified letter has been moved from Chapter 1 in the Fourth Edition to Chapter R in the Fifth Edition so that it can be reviewed before it is used to solve a linear equation for one of the variables in Chapter 1.
- Symmetry and transformations are now presented in two sections rather than one.

## Emphasis on Functions

Functions are the core of this course and should be presented as a thread that runs throughout the course rather than as an isolated topic. We introduce functions in Chapter 1, whereas many traditional college algebra textbooks cover equation-solving in Chapter 1. Our approach of introducing students to a relatively new concept at the beginning of the course, rather than requiring them to begin with a review of material that was previously covered in intermediate algebra, immediately engages them and serves to help them avoid the temptation not to study early in the course because "I already know this."

The concept of a function can be challenging for students. By repeatedly exposing them to the language, notation, and use of functions, demonstrating visually how functions relate to equations and graphs, and also showing how functions can be used to model real data, we hope to ensure that students not only become comfortable with functions but also come to understand and appreciate them. You will see this emphasis on functions woven throughout the other themes that follow.

**Classify the Function Exercises** With a focus on conceptual understanding, students are asked periodically to identify a number of functions by their type (linear, quadratic, rational, and so on). As students progress through the text, the variety of functions with which they are familiar increases and these exercises become more challenging. The "classifying the function" exercises appear with the review exercises in the Skill Maintenance portion of an exercise set. (See pp. 344 and 442.)

## ■ Visual Emphasis

Our early introduction of functions allows graphs to be used to provide a visual aspect to solving equations and inequalities. For example, we are able to show students both algebraically and visually that the solutions of a quadratic equation $ax^2 + bx + c = 0$ are the zeros of the quadratic function $f(x) = ax^2 + bx + c$ as well as the first coordinates of the $x$-intercepts of the graph of that function. This makes it possible for students, particularly visual learners, to gain a quick understanding of these concepts. (See pp. 251, 255, 299, 366, and 429.)

**Visualizing the Graph**    Appearing at least once in every chapter, this feature provides students with an opportunity to match an equation with its graph by focusing on the characteristics of the equation and the corresponding attributes of the graph. (See pp. 209, 269, and 361.) In addition to this full-page feature, many of the exercise sets include exercises in which the student is asked to match an equation with its graph or to find an equation of a function from its graph. (See pp. 211, 212, 310, and 412.) In MyMathLab, animated Visualizing the Graph features for each chapter allow students to interact with graphs on a whole new level.

**Side-by-Side Examples**    Many examples are presented in a side-by-side, two-column format in which the algebraic solution of an equation appears in the left column and a graphical solution appears in the right column. (See pp. 244, 368–369, and 446–447.) This enables students to visualize and comprehend the connections among the solutions of an equation, the zeros of a function, and the $x$-intercepts of the graph of a function.

**Integrated Technology**    In order to increase students' understanding of the course content through a visual means, we integrate graphing calculator technology throughout. The use of the graphing calculator is woven throughout the text's exposition, exercise sets, and testing program without sacrificing algebraic skills. Graphing calculator technology is included in order to enhance—not replace—students' mathematical skills, and to alleviate the tedium associated with certain procedures. (See pp. 249, 357–358, and 443.) In addition, each copy of the student text is bundled with a complimentary graphing calculator manual created by author Beverly Fusfield. A GCM icon GCM indicates the correlation of the manual to the main text content.

*New!*    **Interactive Figures**    These new figures help bring mathematical concepts to life. They are included in MyMathLab as both a teaching tool and a learning tool. Used as a lecture tool, the figures help engage students more fully and save the time that would otherwise be spent drawing figures by hand. Questions pertaining to each interactive figure are assignable in MyMathLab and reinforce active learning, critical thinking, and conceptual learning. Interactive Figure icons appear throughout the text. (See pp. 82 and 177.)

*New!*    **App for the iPhone**    The advent of the iPhone and other sophisticated mobile phones has made available many inexpensive mathematics applications. One useful app is Appcylon LLC Graphing Calculator (which can be purchased at the Apps Store). Among the features of this app are the ability to graph most of the functions we encounter in this book and find the exact $(x, y)$ coordinates for zeros, intersections, and maximum and minimum values using the TRACE mode. We illustrate these features by including occasional app windows throughout. (See pp. 162, 244, 359, and 448.)

## ■ Making Connections

**Zeros, Solutions, and x-Intercepts**    We find that when students understand the connections among the real zeros of a function, the solutions of its associated equation, and the first coordinates of the $x$-intercepts of its graph, a door opens to a new level of mathematical comprehension that increases the probability of success in this course. We emphasize zeros, solutions, and $x$-intercepts throughout the text by using consistent, precise terminology and including exceptional graphics. Seeing this theme repeated in different contexts leads to a better understanding and retention of these concepts. (See pp. 244 and 255.)

**Connecting the Concepts**   This feature highlights the importance of connecting concepts. When students are presented with concepts in visual form—using graphs, an outline, or a chart—rather than merely in paragraphs of text, comprehension is streamlined and retention is enhanced. The visual aspect of this feature invites students to stop and check their understanding of how concepts work together in one section or in several sections. This check in turn enhances student performance on homework assignments and exams. (See pp. 133, 255, and 330.)

**Annotated Examples**   We have included over 770 Annotated Examples designed to fully prepare the student to work the exercises. Learning is carefully guided with the use of numerous color-coded art pieces and step-by-step annotations. Substitutions and annotations are highlighted in red for emphasis. (See pp. 247 and 437–438.)

**Now Try Exercises**   Now Try Exercises are found after nearly every example. This feature encourages active learning by asking students to do an exercise in the exercise set that is similar to the example the student has just read. (See pp. 246, 351, and 407.)

**Synthesis Exercises**   These exercises appear at the end of each exercise set and encourage critical thinking by requiring students to synthesize concepts from several sections or to take a concept a step further than in the general exercises. For the Fifth Edition, these exercises are assignable in MyMathLab. (See pp. 332–333, 416, and 469–470.)

**Real-Data Applications and Regression**   We encourage students to see and interpret the mathematics that appears every day in the world around them. Throughout the writing process, we conducted an energetic search for real-data applications, and the result is a variety of examples and exercises that connect the mathematical content with everyday life. Most of these applications feature source lines and many include charts and graphs. Many are drawn from the fields of health, business and economics, life and physical sciences, social science, and areas of general interest such as sports and travel. (See pp. 96–97, 257–258, 413–414, and 463–466.)

In Chapter 1, we introduce the use of regression or curve fitting to model data with linear functions, and continue this visual theme with quadratic, cubic, quartic, exponential, logarithmic, and logistic functions. (See pp. 115–116, 308–309, and 461–462.)

## ▪ Ongoing Review

The most significant addition to the Fifth Edition is the new ongoing review features that have been integrated throughout to help students reinforce their understanding and improve their success in the course.

*New!*   **Mid-Chapter Mixed Review**   This new review reinforces understanding of the mathematical concepts and skills covered in the first half of the chapter before students move on to new material in the second half of the chapter. Each review begins with at least three True/False exercises that require students to consider the concepts they have studied and also contains exercises that drill the skills from all prior sections of the chapter. These exercises are assignable in MyMathLab. (See pp. 190–191 and 433–434.)

**Collaborative Discussion and Writing Exercises** appear in the Mid-Chapter Mixed Review as well. These exercises can be discussed in small groups or by the class as a whole to encourage students to talk about the key mathematical concepts in the chapter. They can also be assigned to individual students to give them an opportunity to write about mathematics. (See pp. 274 and 334.)

A section reference is provided for each exercise in the Mid-Chapter Mixed Review. This tells the student which section to refer to if help is needed to work the exercise. Answers to all exercises in the Mid-Chapter Mixed Review are given at the back of the book.

**New!**

**Study Guide**    This new feature is found at the beginning of the **Summary and Review** near the end of each chapter. Presented in a two-column format organized by section, this feature gives key concepts and terms in the left column and a worked-out example in the right column. It provides students with a concise and effective review of the chapter that is a solid basis for studying for a test. In MyMathLab, these Study Guides are accompanied by narrated examples to reinforce the key concepts and ideas. (See pp. 287–292 and 470–477.)

**Exercise Sets**    There are over 5850 exercises in this text. The exercise sets are enhanced with real-data applications and source lines, detailed art pieces, tables, graphs, and photographs. In addition to the exercises that provide students with concepts presented in the section, the exercise sets feature the following elements to provide ongoing review of topics presented earlier:

- **Skill Maintenance Exercises.** These exercises provide an ongoing review of concepts previously presented in the course, enhancing students' retention of these concepts. These exercises include **Vocabulary Reinforcement**, described below, and **Classifying the Function** exercises, described earlier in the section "Emphasis on Functions." Answers to all Skill Maintenance exercises appear in the answer section at the back of the book, along with a section reference that directs students quickly and efficiently to the appropriate section of the text if they need help with an exercise. (See pp. 198, 282, 364–365, and 432.)
- **Enhanced Vocabulary Reinforcement Exercises.** This feature checks and reviews students' understanding of the vocabulary introduced throughout the text. It appears once in every chapter, in the Skill Maintenance portion of an exercise set, and is intended to provide a continuing review of the terms that students must know in order to be able to communicate effectively in the language of mathematics. (See pp. 220–221, 286, and 364–365.) These are now assignable in MyMathLab and can serve as reading quizzes.
- **Enhanced Synthesis Exercises.** These exercises are described under the Making Connections heading and are also assignable in MyMathLab.

**Review Exercises**    These exercises in the **Summary and Review** supplement the Study Guide by providing a thorough and comprehensive review of the skills taught in the chapter. A group of true/false exercises appears first, followed by a large number of exercises that drill the skills and concepts taught in the chapter. In addition, three multiple-choice exercises, one of which involves identifying the graph of a function, are included in the Review Exercises for every chapter with the exception of Chapter R, the review chapter. Each Review Exercise is accompanied by a section reference that, as in the Mid-Chapter Mixed Review, directs students to the section in which the material being reviewed can be found. Collaborative Discussion and Writing exercises are also included. These exercises are described under the Mid-Chapter Mixed Review heading on p. xiii. (See pp. 293–295 and 477–481.)

**Chapter Test**    The test at the end of each chapter allows students to test themselves and target areas that need further study before taking the in-class test. Each Chapter Test includes a multiple-choice exercise involving identifying the graph of a function (except in Chapter R, in which graphing has not yet been introduced). Answers to all questions in the Chapter Tests appear in the answer section at the back of the book, along with corresponding section references. (See pp. 295–296 and 481–482.)

**Review Icons**    Placed next to the concept that a student is currently studying, a review icon references a section of the text in which the student can find and review topics on which the current concept is built. (See pp. 346 and 392.)

# Supplements

## Student Supplements

### Graphing Calculator Manual
ISBN: 0-321-79099-5; 978-0-321-79099-6

- By Beverly Fusfield
- Contains keystroke level instruction for the Texas Instruments TI-84 Plus using the new MathPrint OS
- Teaches students how to use a graphing calculator using actual examples and exercises from the main text
- Mirrors the topic order in the main text to provide a just-in-time mode of instruction
- Automatically ships with each new copy of the text

### Student's Solutions Manual
ISBN: 0-321-79125-8; 978-0-321-79125-2

- By Judith A. Penna
- Contains completely worked-out solutions with step-by-step annotations for all the odd-numbered exercises in the exercise sets, the Mid-Chapter Mixed Review exercises, and Chapter Review exercises, as well as solutions for all the Chapter Test exercises

### Video Resources with Optional Subtitles

- Complete set of digitized videos in both streaming and Podcast formats for student use at home or on campus are available within their MyMathLab course
- Ideal for distance learning or supplemental instruction
- Features authors Judy Beecher and Judy Penna working through and explaining examples in the text

## Instructor Supplements

### Annotated Instructor's Edition
ISBN: 0-321-79061-8; 978-0-321-79061-3

- Includes all the answers to the exercise sets, usually right on the page where the exercises appear
- Readily accessible answers help both new and experienced instructors prepare for class efficiently
- *New!* Sample homework assignments are now indicated by a blue underline within each end-of-section exercise set and may be assigned in MyMathLab

### Instructor's Solutions Manual (Download only)

- By Judith A. Penna
- Contains worked-out solutions to all exercises in the exercise sets and solutions for all Mid-Chapter Mixed Review exercises, Chapter Review exercises, and Chapter Test exercises
- Available for download through www.pearsonhighered.com/irc or inside your MyMathLab course

### Online Test Bank (Download only)

- By Laurie Hurley
- Contains four free-response test forms for each chapter following the same format and having the same level of difficulty as the tests in the main text, plus two multiple-choice test forms for each chapter
- Provides six forms of the final examination, four with free-response questions and two with multiple-choice questions
- Available for download through www.pearsonhighered.com/irc or inside your MyMathLab course

### TestGen®

- Enables instructors to build, edit, print, and administer tests
- Features a computerized bank of questions developed to cover all text objectives
- Available for download through www.pearsonhighered.com/irc or inside your MyMathLab course

### Pearson Adjunct Support Center

- Offers consultation on suggested syllabi, helpful tips for using the textbook support package, assistance with content, and advice on classroom strategies
- Available Sunday through Thursday evenings from 5 P.M. to midnight EST
- e-mail: AdjunctSupport@pearson.com; Telephone: 1-800-435-4084; fax: 1-877-262-9774

# ■ Media Supplements

**MyMathLab® Online Course (Access code required.)**  MyMathLab delivers **proven results** in helping individual students succeed.

- MyMathLab has a consistently positive impact on the quality of learning in higher education math instruction. MyMathLab can be successfully implemented in any environment—lab-based, hybrid, fully online, traditional—and demonstrates the quantifiable difference that integrated usage has on student retention, subsequent success, and overall achievement.

- MyMathLab's comprehensive online gradebook automatically tracks your students' results on tests, on quizzes, on homework, and in the study plan. You can use the gradebook to quickly intervene if your students have trouble, or to provide positive feedback on a job well done. The data within MyMathLab is easily exported to a variety of spreadsheet programs, such as Microsoft Excel. You can determine which points of data you want to export and then analyze the results to determine success.

MyMathLab provides **engaging experiences** that personalize, stimulate, and measure learning for each student.

- **Exercises:** The homework and practice exercises in MyMathLab are correlated to the exercises in the textbook, and they regenerate algorithmically to give students unlimited opportunity for practice and mastery. The software offers immediate, helpful feedback when students enter incorrect answers. This includes:

  **Vocabulary exercises** that can serve as reading quizzes.

  **Mid-Chapter Mixed Reviews** are new to the text and are assignable online, helping students to reinforce their understanding of the concepts.

  **Synthesis exercises** are now assignable online, testing students' ability to answer questions that cover multiple concepts.

  **Sample homework assignments** are now indicated by a blue underline within each end-of-section exercise set and may be assigned in MyMathLab.

  **Interactive figures** are new to the text and are indicated by icons. There are homework exercises that pertain to each interactive figure. They are assignable in MyMathLab and available in the Multimedia Library.

- **Multimedia Learning Aids:** Exercises include guided solutions, sample problems, animations, videos, and eText clips for extra help at point-of-use. This includes **example-based videos**, created by the authors themselves, to walk students through the detailed solution process for key examples in the textbook. Videos are closed-captioned in English.

- **Expert Tutoring:** Although many students describe MyMathLab as "like having your own personal tutor," students using MyMathLab do have access to live tutoring from Pearson, from qualified math and statistics instructors who provide tutoring sessions for students via MyMathLab.

- **Study Guides** are new to the text. In MyMathLab, these Study Guides are accompanied by narrated examples to reinforce the key concepts and ideas.

And, MyMathLab comes from a **trusted partner** with educational expertise and an eye to the future.

• Knowing that you are using a Pearson product means knowing that you are using quality content. This means that our eTexts are accurate, that our assessment tools work, and that our questions are error-free. And whether you are just getting started with MyMathLab or have a question along the way, we are here to help you learn about our technologies and how to incorporate them into your course.

To learn more about how MyMathLab combines proven learning applications with powerful assessment, visit **www.mymathlab.com** or contact your Pearson representative.

**MathXL® Online Course (Access code required.)**   MathXL® is a powerful online homework, tutorial, and assessment system that accompanies Pearson Education's textbooks in mathematics and statistics.

With MathXL, instructors can:

• Create, edit, and assign online homework and tests using algorithmically generated exercises correlated at the objective level to the textbook.
• Create and assign their own online exercises and import TestGen tests for added flexibility.
• Maintain records of all student work tracked in MathXL's online gradebook.

With MathXL, students can:

• Take chapter tests in MathXL and receive personalized study plans and/or personalized homework assignments based on their test results.
• Use the study plan and/or the homework to link directly to tutorial exercises for the objectives they need to study.
• Access supplemental animations and video clips directly from selected exercises.

MathXL is available to qualified adopters. For more information, visit our website at www.mathxl.com, or contact your Pearson representative.

**Videos**   The video lectures for this text are available within MyMathLab in both streaming and Video Podcast format, making it easy and convenient for students to watch the videos from a computer either at home or on campus. The videos feature authors Judy Beecher and Judy Penna, who present Example Solutions. Example Solutions walk students through the detailed solution process for the examples in the textbook. The format provides distance-learning students with comprehensive video instruction, but also allows students needing less review to watch instruction on a specific skill or procedure. The videos have optional text subtitles, which can be easily turned on or off for individual student needs. Subtitles are available in English and Spanish.

**PowerPoints**   PowerPoint Lecture Slides feature presentations written and designed specifically for this text. These lecture slides provide an outline to use in a lecture setting, presenting definitions, figures, and key examples from the text. They are available online within MyMathLab or from the Instructor Resource Center at www.pearsonhighered.com/irc.

# Acknowledgments

We wish to express our heartfelt thanks to a number of people who have contributed in special ways to the development of this textbook. Our editor, Kathryn O'Connor, encouraged and supported our vision. We are very appreciative of the marketing insight provided by Peggy Lucas, our marketing manager, and of the support that we received from the entire Pearson team, including Kathy Manley, senior production project manager, Joanne Dill, senior content editor, Judith Garber, editorial assistant, and Justine Goulart, marketing assistant. We also thank Aimee Thorne, media producer, for her creative work on the media products that accompany this text and Tom Atwater, for accuracy checking the videos. And we are immensely grateful to Martha Morong for her editorial and production services, and to Geri Davis for her text design and art editing, and for the endless hours of hard work they have done to make this a book of which we are proud. We also thank Laurie Hurley, Holly Martinez, and Julie Stephenson for their meticulous accuracy checking and proofreading of the text.

The following reviewers made invaluable contributions to the development of the Fifth Edition and we thank them for that:

Gerald Allen, *Howard College*
Robin Ayers, *Western Kentucky University*
Heidi Barrett, *Arapahoe Community College*
George Behr, *Coastline Community College*
Kimberly Bennekin, *Georgia Perimeter College*
Marc Campbell, *Daytona Beach Community College*
\* Mark A. Crawford, Jr., *Waubonsee Community College*
Brad Feldser, *Kennesaw State University*
Homa Ghaussi-Mujtaba, *Lansing Community College*
Bob Gravelle, *Colorado Technical University, Colorado Springs*
Judy Hayes, *Lake-Sumter Community College*
Michelle Hollis, *Bowling Green Community College*
\* Glenn Jablonski, *Triton College*
Bridgette Jacob, *Onondaga Community College*
Symon Kimitei, *Kennesaw State University*
Deanna Kindhart, *Illinois Central College*
\* Pamela Krompak, *Owens Community College*
\* Laud Kwaku, *Owens Community College*
\* Carol A. Lucas, *University of Central Oklahoma*
Daniel Olson, *Purdue University, North Central*
\* Cloyd A. Payne, *Owens Community College*
Randy K. Ross, *Morehead State University*
\* Daniel Russow, *Arizona Western College*
Brian Schworm, *Morehead State University*
Judith Staver, *Florida Community College at Jacksonville, South Campus*
\* Laura Taylor, *Cape Fear Community College*
\* Jean Hunt Thorton, *Western Kentucky University*
\* Pat Velicky, *Florence–Darlington Technical College*
Douglas Windham, *Tallahassee Community College*
\* Weicheng Xuan, *Arizona Western College*

\*  *reviewers of the Fifth Edition*

## GUIDE TO SUCCESS

Success can be planned. Combine goals and good study habits to create a plan for success that works for you. The following list contains study tips that your authors consider most helpful.

### SKILLS FOR SUCCESS

- **Set goals and expect success.** Approach this class experience with a positive attitude.
- **Communicate with your instructor** by learning the office hours and making a point to visit when you need extra help. Many instructors welcome student e-mails.
- **Take your text with you to class and lab.** Each section in the text is designed with headings and boxed information that provide an outline for easy reference.
- **Ask questions in class, lab, and tutoring sessions.** Instructors encourage them, and other students probably have the same questions.
- **Begin each homework assignment as soon as possible.** Then if you have difficulty, you have the time to access supplementary resources.
- **Carefully read the instructions** before working homework exercises **and include all steps.**
- **Form a study group** with fellow students. Verbalizing questions about topics that you do not understand can clarify the material for you.
- After each quiz or test, **write out corrected step-by step solutions** to all missed questions. They will provide a valuable study guide for the mid-term exam and the final exam.
- **MyMathLab has numerous tools to help you succeed.** Use MyMathLab to create a personalized study plan and practice skills with sample quizzes and tests.
- **Knowing math vocabulary is an important step toward success.** Review vocabulary with Vocabulary Reinforcement exercises in the text and in MyMathLab.
- **Learning with the Interactive Figures increases understanding and retention of the concepts.** An index of these interactive figures is on the following page.
- If you miss a lecture, **watch the video in the Multimedia Library** of MyMathLab that explains the concepts you missed.

In writing this textbook, we challenged ourselves to do everything possible to help you learn the concepts and skills contained between its covers so that you will be successful in this course and in the mathematics courses you take in the future. We realize that your time is both valuable and limited, so we communicate in a highly visual way that allows you to learn quickly and efficiently. We are confident that, if you invest an adequate amount of time in the learning process, this text will be of great value to you. We wish you a positive learning experience.

Marv Bittinger
Judy Beecher
David Ellenbogen
Judy Penna

# INTERACTIVE FIGURES

New Interactive Figures enable you to manipulate figures to bring math concepts to life. Icons throughout the text direct you to the MyMathLab® course whenever an Interactive Figure is available.

# College Algebra
## GRAPHS AND MODELS

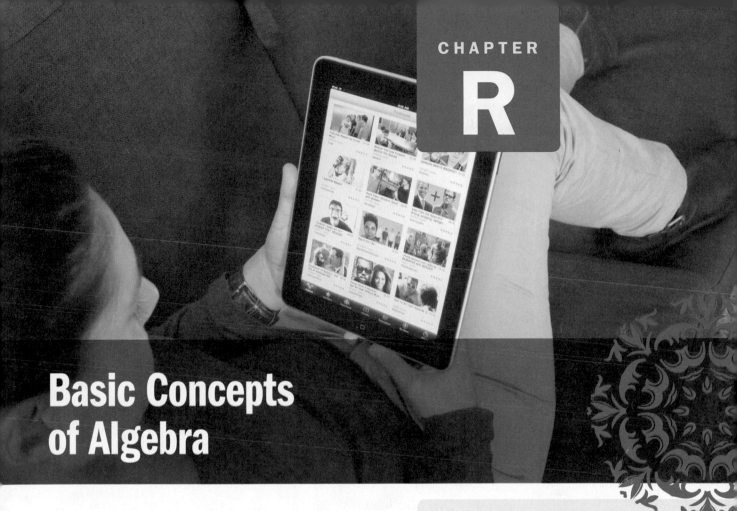

# CHAPTER R

# Basic Concepts of Algebra

## APPLICATION

In the first two months after having been uploaded to a popular video-sharing Web site, a video showing a baby laughing hysterically at ripping paper received approximately 12,630,000 views. Convert the number 12,630,000 to scientific notation.

**This problem appears as Example 5 in Section R.2.**

## The Real-Number System

- Identify various kinds of real numbers.
- Use interval notation to write a set of numbers.
- Identify the properties of real numbers.
- Find the absolute value of a real number.

### ■ Real Numbers

In applications of algebraic concepts, we use real numbers to represent quantities such as distance, time, speed, area, profit, loss, and temperature. Some frequently used sets of real numbers and the relationships among them are shown below.

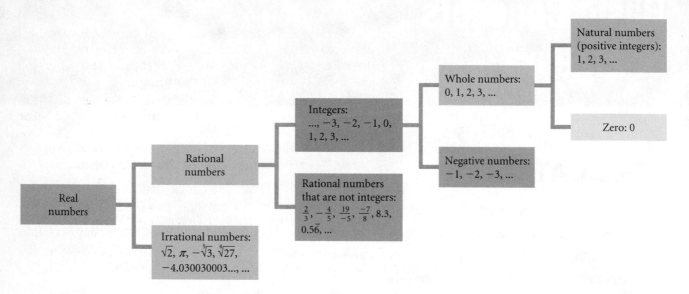

Numbers that can be expressed in the form $p/q$, where $p$ and $q$ are integers and $q \neq 0$, are **rational numbers**. Decimal notation for rational numbers either *terminates* (ends) or *repeats*. Each of the following is a rational number.

a)  0                                         $0 = \dfrac{0}{a}$ for any nonzero integer $a$

b)  $-7$                                      $-7 = \dfrac{-7}{1}, \text{or} \dfrac{7}{-1}$

c)  $\dfrac{1}{4} = 0.25$                      **Terminating decimal**

d)  $-\dfrac{5}{11} = -0.454545\ldots = -0.\overline{45}$    **Repeating decimal**

e)  $\dfrac{5}{6} = 0.8333\ldots = 0.8\overline{3}$    **Repeating decimal**

The real numbers that are not rational are **irrational numbers**. Decimal notation for irrational numbers neither terminates nor repeats. Each of the following is an irrational number.

**a)**  $\pi = 3.1415926535\ldots$     There is no repeating block of digits.
$\left(\frac{22}{7}\text{ and }3.14\text{ are rational }approximations\text{ of the irrational number }\pi.\right)$

**b)**  $\sqrt{2} = 1.414213562\ldots$     There is no repeating block of digits.

**c)**  $-6.12122122212222\ldots$     Although there is a pattern, there is no repeating block of digits.

The set of all rational numbers combined with the set of all irrational numbers gives us the set of **real numbers**. The real numbers are modeled using a **number line**, as shown below.

Each point on the line represents a real number, and every real number is represented by a point on the line.

$$-2.9 \qquad -\tfrac{3}{5} \qquad \sqrt{3} \qquad \pi \qquad \tfrac{17}{4}$$

$$-5\ \ -4\ \ -3\ \ -2\ \ -1\ \ \ 0\ \ \ 1\ \ \ 2\ \ \ 3\ \ \ 4\ \ \ 5$$

The order of the real numbers can be determined from the number line. If a number $a$ is to the left of a number $b$, then $a$ **is less than** $b$ $(a < b)$. Similarly, $a$ **is greater than** $b$ $(a > b)$ if $a$ is to the right of $b$ on the number line. For example, we see from the number line above that $-2.9 < -\frac{3}{5}$, because $-2.9$ is to the left of $-\frac{3}{5}$. Also, $\frac{17}{4} > \sqrt{3}$, because $\frac{17}{4}$ is to the right of $\sqrt{3}$.

The statement $a \leq b$, read "$a$ is less than or equal to $b$," is true if either $a < b$ is true or $a = b$ is true.

The symbol $\in$ is used to indicate that a member, or **element**, belongs to a set. Thus if we let $\mathbb{Q}$ represent the set of rational numbers, we can see from the diagram on p. 2 that $0.5\overline{6} \in \mathbb{Q}$. We can also write $\sqrt{2} \notin \mathbb{Q}$ to indicate that $\sqrt{2}$ is *not* an element of the set of rational numbers.

When *all* the elements of one set are elements of a second set, we say that the first set is a **subset** of the second set. The symbol $\subseteq$ is used to denote this. For instance, if we let $\mathbb{R}$ represent the set of real numbers, we can see from the diagram that $\mathbb{Q} \subseteq \mathbb{R}$ (read "$\mathbb{Q}$ is a subset of $\mathbb{R}$").

## ▪ Interval Notation

Sets of real numbers can be expressed using **interval notation**. For example, for real numbers $a$ and $b$ such that $a < b$, the **open interval** $(a, b)$ is the set of real numbers between, but not including, $a$ and $b$. That is,

$$(a, b) = \{x \mid a < x < b\}.$$

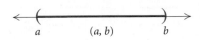

The points $a$ and $b$ are **endpoints** of the interval. The parentheses indicate that the endpoints are not included in the interval.

Some intervals extend without bound in one or both directions. The interval $[a, \infty)$, for example, begins at $a$ and extends to the right without bound. That is,

$$[a, \infty) = \{x \mid x \geq a\}.$$

The bracket indicates that $a$ is included in the interval.

The various types of intervals are listed below.

## Intervals: Types, Notation, and Graphs

| TYPE | INTERVAL NOTATION | SET NOTATION | GRAPH |
|---|---|---|---|
| Open | $(a, b)$ | $\{x \mid a < x < b\}$ | |
| Closed | $[a, b]$ | $\{x \mid a \leq x \leq b\}$ | |
| Half-open | $[a, b)$ | $\{x \mid a \leq x < b\}$ | |
| Half-open | $(a, b]$ | $\{x \mid a < x \leq b\}$ | |
| Open | $(a, \infty)$ | $\{x \mid x > a\}$ | |
| Half-open | $[a, \infty)$ | $\{x \mid x \geq a\}$ | |
| Open | $(-\infty, b)$ | $\{x \mid x < b\}$ | |
| Half-open | $(-\infty, b]$ | $\{x \mid x \leq b\}$ | |

The interval $(-\infty, \infty)$, graphed below, names the set of all real numbers, $\mathbb{R}$.

**EXAMPLE 1**  Write interval notation for each set and graph the set.

**a)** $\{x \mid -4 < x < 5\}$         **b)** $\{x \mid x \geq 1.7\}$

**c)** $\{x \mid -5 < x \leq -2\}$        **d)** $\{x \mid x < \sqrt{5}\}$

*Solution*

**a)** $\{x \mid -4 < x < 5\} = (-4, 5);$

$-5\ -4\ -3\ -2\ -1\ \ 0\ \ 1\ \ 2\ \ 3\ \ 4\ \ 5$

**b)** $\{x \mid x \geq 1.7\} = [1.7, \infty);$

$-5\ -4\ -3\ -2\ -1\ \ 0\ \ 1\ \ 2\ \ 3\ \ 4\ \ 5$

**c)** $\{x \mid -5 < x \leq -2\} = (-5, -2];$

$-5\ -4\ -3\ -2\ -1\ \ 0\ \ 1\ \ 2\ \ 3\ \ 4\ \ 5$

**d)** $\{x \mid x < \sqrt{5}\} = (-\infty, \sqrt{5});$

$-5\ -4\ -3\ -2\ -1\ \ 0\ \ 1\ \ 2\ \ 3\ \ 4\ \ 5$

**Now Try Exercises 13 and 15.**

## ▪ Properties of the Real Numbers

The following properties can be used to manipulate algebraic expressions as well as real numbers.

---

**PROPERTIES OF THE REAL NUMBERS**

For any real numbers $a$, $b$, and $c$:

| | |
|---|---|
| $a + b = b + a$ and $ab = ba$ | Commutative properties of addition and multiplication |
| $a + (b + c) = (a + b) + c$ and $a(bc) = (ab)c$ | Associative properties of addition and multiplication |
| $a + 0 = 0 + a = a$ | Additive identity property |
| $-a + a = a + (-a) = 0$ | Additive inverse property |
| $a \cdot 1 = 1 \cdot a = a$ | Multiplicative identity property |
| $a \cdot \dfrac{1}{a} = \dfrac{1}{a} \cdot a = 1 \, (a \neq 0)$ | Multiplicative inverse property |
| $a(b + c) = ab + ac$ | Distributive property |

---

Note that the distributive property is also true for subtraction since $a(b - c) = a[b + (-c)] = ab + a(-c) = ab - ac$.

**EXAMPLE 2**  State the property being illustrated in each sentence.

**a)** $8 \cdot 5 = 5 \cdot 8$
**b)** $5 + (m + n) = (5 + m) + n$
**c)** $14 + (-14) = 0$
**d)** $6 \cdot 1 = 1 \cdot 6 = 6$
**e)** $2(a - b) = 2a - 2b$

*Solution*

| SENTENCE | PROPERTY |
|---|---|
| **a)** $8 \cdot 5 = 5 \cdot 8$ | Commutative property of multiplication: $ab = ba$ |
| **b)** $5 + (m + n) = (5 + m) + n$ | Associative property of addition: $a + (b + c) = (a + b) + c$ |
| **c)** $14 + (-14) = 0$ | Additive inverse property: $a + (-a) = 0$ |
| **d)** $6 \cdot 1 = 1 \cdot 6 = 6$ | Multiplicative identity property: $a \cdot 1 = 1 \cdot a = a$ |
| **e)** $2(a - b) = 2a - 2b$ | Distributive property: $a(b + c) = ab + ac$ |

**Now Try Exercises 49 and 55.**

■ **Absolute Value**

The number line can be used to provide a geometric interpretation of *absolute value*. The **absolute value** of a number $a$, denoted $|a|$, is its distance from 0 on the number line. For example, $|-5| = 5$, because the distance of $-5$ from 0 is 5. Similarly, $\left|\frac{3}{4}\right| = \frac{3}{4}$, because the distance of $\frac{3}{4}$ from 0 is $\frac{3}{4}$.

---

**ABSOLUTE VALUE**

For any real number $a$,

$$|a| = \begin{cases} a, & \text{if } a \geq 0, \\ -a, & \text{if } a < 0. \end{cases}$$

When $a$ is nonnegative, the absolute value of $a$ is $a$. When $a$ is negative, the absolute value of $a$ is the opposite, or additive inverse, of $a$. Thus, $|a|$ is never negative; that is, for any real number $a$, $|a| \geq 0$.

---

Absolute value can be used to find the distance between two points on the number line.

$|a - b| = |b - a|$

---

**DISTANCE BETWEEN TWO POINTS ON THE NUMBER LINE**

For any real numbers $a$ and $b$, the **distance between $a$ and $b$** is $|a - b|$, or equivalently, $|b - a|$.

---

GCM

**EXAMPLE 3** Find the distance between $-2$ and 3.

*Solution* The distance is

$$|-2 - 3| = |-5| = 5, \quad \text{or equivalently,}$$
$$|3 - (-2)| = |3 + 2| = |5| = 5.$$

We can also use the absolute-value operation on a graphing calculator to find the distance between two points. On many graphing calculators, absolute value is denoted "abs" and is found in the MATH NUM menu and also in the CATALOG.

| | |
|---|---|
| $\lvert -2-3 \rvert$ | |
| | 5 |
| $\lvert 3-(-2) \rvert$ | |
| | 5 |

**Now Try Exercise 69.**

## R.1 Exercise Set

In Exercises 1–10, consider the numbers

$$\tfrac{2}{3},\ 6,\ \sqrt{3},\ -2.45,\ \sqrt[6]{26},\ 18.\overline{4},\ -11,\ \sqrt[3]{27},$$
$$5\tfrac{1}{6},\ 7.151551555\ldots,\ -\sqrt{35},\ \sqrt[5]{3},\ -\tfrac{8}{7},\ 0,\ \sqrt{16}.$$

**1.** Which are rational numbers?

**2.** Which are natural numbers?

**3.** Which are irrational numbers?

**4.** Which are integers?

**5.** Which are whole numbers?

**6.** Which are real numbers?

**7.** Which are integers but not natural numbers?

**8.** Which are integers but not whole numbers?

**9.** Which are rational numbers but not integers?

**10.** Which are real numbers but not integers?

*Write interval notation. Then graph the interval.*

**11.** $\{x \mid -5 \le x \le 5\}$      **12.** $\{x \mid -2 < x < 2\}$

**13.** $\{x \mid -3 < x \le -1\}$      **14.** $\{x \mid 4 \le x < 6\}$

**15.** $\{x \mid x \le -2\}$      **16.** $\{x \mid x > -5\}$

**17.** $\{x \mid x > 3.8\}$      **18.** $\{x \mid x \ge \sqrt{3}\}$

**19.** $\{x \mid 7 < x\}$      **20.** $\{x \mid -3 > x\}$

*Write interval notation for the graph.*

**21.**

**22.**

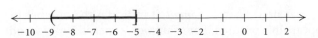

**23.**

**24.**

**25.**

**26.**

**27.**

**28.**

In Exercises 29–46, the following notation is used:
$\mathbb{N}$ = the set of natural numbers, $\mathbb{W}$ = the set of whole numbers, $\mathbb{Z}$ = the set of integers, $\mathbb{Q}$ = the set of rational numbers, $\mathbb{I}$ = the set of irrational numbers, and $\mathbb{R}$ = the set of real numbers. Classify the statement as true or false.

**29.** $6 \in \mathbb{N}$      **30.** $0 \notin \mathbb{N}$

**31.** $3.2 \in \mathbb{Z}$      **32.** $-10.\overline{1} \in \mathbb{R}$

**33.** $-\dfrac{11}{5} \in \mathbb{Q}$      **34.** $-\sqrt{6} \in \mathbb{Q}$

**35.** $\sqrt{11} \notin \mathbb{R}$      **36.** $-1 \in \mathbb{W}$

**37.** $24 \notin \mathbb{W}$      **38.** $1 \in \mathbb{Z}$

**39.** $1.089 \notin \mathbb{I}$      **40.** $\mathbb{N} \subseteq \mathbb{W}$

**41.** $\mathbb{W} \subseteq \mathbb{Z}$      **42.** $\mathbb{Z} \subseteq \mathbb{N}$

**43.** $\mathbb{Q} \subseteq \mathbb{R}$      **44.** $\mathbb{Z} \subseteq \mathbb{Q}$

**45.** $\mathbb{R} \subseteq \mathbb{Z}$      **46.** $\mathbb{Q} \subseteq \mathbb{I}$

*Name the property illustrated by the sentence.*

**47.** $3 + y = y + 3$

**48.** $6(xz) = (6x)z$

**49.** $-3 \cdot 1 = -3$

**50.** $4(y - z) = 4y - 4z$

**51.** $5 \cdot x = x \cdot 5$

**52.** $7 + (x + y) = (7 + x) + y$

**53.** $2(a + b) = (a + b)2$

**54.** $-11 + 11 = 0$

**55.** $-6(m + n) = -6(n + m)$

**56.** $t + 0 = t$

**57.** $8 \cdot \frac{1}{8} = 1$

**58.** $9x + 9y = 9(x + y)$

*Simplify.*

**59.** $|-8.15|$          **60.** $|-14.7|$

**61.** $|295|$          **62.** $|-93|$

**63.** $|-\sqrt{97}|$          **64.** $\left|\frac{12}{19}\right|$

**65.** $|0|$          **66.** $|15|$

**67.** $\left|\frac{5}{4}\right|$          **68.** $|-\sqrt{3}|$

*Find the distance between the given pair of points on the number line.*

**69.** $-8, 14$          **70.** $-5.2, 0$

**71.** $-9, -3$          **72.** $\frac{15}{8}, \frac{23}{12}$

**73.** $6.7, 12.1$          **74.** $-15, -6$

**75.** $-\frac{3}{4}, \frac{15}{8}$          **76.** $-3.4, 10.2$

**77.** $-7, 0$          **78.** $3, 19$

## Synthesis

*To the student and the instructor: The Synthesis exercises found at the end of every exercise set challenge students to combine concepts or skills studied in that section or in preceding parts of the text.*

*Between any two (different) real numbers there are many other real numbers. Find each of the following. Answers may vary.*

**79.** An irrational number between 0.124 and 0.125

**80.** A rational number between $-\sqrt{2.01}$ and $-\sqrt{2}$

**81.** A rational number between $-\frac{1}{101}$ and $-\frac{1}{100}$

**82.** An irrational number between $\sqrt{5.99}$ and $\sqrt{6}$

**83.** The hypotenuse of an isosceles right triangle with legs of length 1 unit can be used to "measure" a value for $\sqrt{2}$ by using the Pythagorean theorem, as shown.

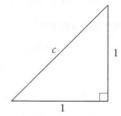

$$c^2 = 1^2 + 1^2$$
$$c^2 = 2$$
$$c = \sqrt{2}$$

Draw a right triangle that could be used to "measure" $\sqrt{10}$ units.

---

## R.2   Integer Exponents, Scientific Notation, and Order of Operations

- Simplify expressions with integer exponents.
- Solve problems using scientific notation.
- Use the rules for order of operations.

### ■ Integers as Exponents

When a positive integer is used as an *exponent*, it indicates the number of times that a factor appears in a product. For example, $7^3$ means $7 \cdot 7 \cdot 7$ and $5^1$ means 5.

For any *positive* integer $n$,

$$a^n = \underbrace{a \cdot a \cdot a \cdots a,}_{n \text{ factors}}$$

where $a$ is the **base** and $n$ is the **exponent**.

Zero and negative-integer exponents are defined as follows.

For any *nonzero* real number $a$ and any integer $m$,

$$a^0 = 1 \quad \text{and} \quad a^{-m} = \frac{1}{a^m}.$$

**EXAMPLE 1** Simplify each of the following.

**a)** $6^0$                                    **b)** $(-3.4)^0$

*Solution*

**a)** $6^0 = 1$                         **b)** $(-3.4)^0 = 1$

> **Now Try Exercise 7.**

**EXAMPLE 2** Write each of the following with positive exponents.

**a)** $4^{-5}$             **b)** $\dfrac{1}{(0.82)^{-7}}$           **c)** $\dfrac{x^{-3}}{y^{-8}}$

*Solution*

**a)** $4^{-5} = \dfrac{1}{4^5}$

**b)** $\dfrac{1}{(0.82)^{-7}} = (0.82)^{-(-7)} = (0.82)^7$

**c)** $\dfrac{x^{-3}}{y^{-8}} = x^{-3} \cdot \dfrac{1}{y^{-8}} = \dfrac{1}{x^3} \cdot y^8 = \dfrac{y^8}{x^3}$

> **Now Try Exercise 1.**

The results in Example 2 can be generalized as follows.

For any nonzero numbers $a$ and $b$ and any integers $m$ and $n$,

$$\frac{a^{-m}}{b^{-n}} = \frac{b^n}{a^m}.$$

(A factor can be moved to the other side of the fraction bar if the sign of the exponent is changed.)

**EXAMPLE 3**    Write an equivalent expression without negative exponents:

$$\frac{x^{-3}y^{-8}}{z^{-10}}.$$

*Solution*    Since each exponent is negative, we move each factor to the other side of the fraction bar and change the sign of each exponent:

$$\frac{x^{-3}y^{-8}}{z^{-10}} = \frac{z^{10}}{x^3 y^8}.$$

**Now Try Exercise 5.**

The following properties of exponents can be used to simplify expressions.

## PROPERTIES OF EXPONENTS

For any real numbers $a$ and $b$ and any integers $m$ and $n$, assuming 0 is not raised to a nonpositive power:

| | |
|---|---|
| $a^m \cdot a^n = a^{m+n}$ | Product rule |
| $\dfrac{a^m}{a^n} = a^{m-n} \ (a \neq 0)$ | Quotient rule |
| $(a^m)^n = a^{mn}$ | Power rule |
| $(ab)^m = a^m b^m$ | Raising a product to a power |
| $\left(\dfrac{a}{b}\right)^m = \dfrac{a^m}{b^m} \ (b \neq 0)$ | Raising a quotient to a power |

**EXAMPLE 4**    Simplify each of the following.

**a)** $y^{-5} \cdot y^3$     **b)** $\dfrac{48x^{12}}{16x^4}$     **c)** $(t^{-3})^5$

**d)** $(2s^{-2})^5$     **e)** $\left(\dfrac{45x^{-4}y^2}{9z^{-8}}\right)^{-3}$

*Solution*

**a)** $y^{-5} \cdot y^3 = y^{-5+3} = y^{-2}$, or $\dfrac{1}{y^2}$

**b)** $\dfrac{48x^{12}}{16x^4} = \dfrac{48}{16}x^{12-4} = 3x^8$

**c)** $(t^{-3})^5 = t^{-3 \cdot 5} = t^{-15}$, or $\dfrac{1}{t^{15}}$

**d)** $(2s^{-2})^5 = 2^5(s^{-2})^5 = 32s^{-10}$, or $\dfrac{32}{s^{10}}$

**e)** $\left(\dfrac{45x^{-4}y^2}{9z^{-8}}\right)^{-3} = \left(\dfrac{5x^{-4}y^2}{z^{-8}}\right)^{-3} = \dfrac{5^{-3}x^{12}y^{-6}}{z^{24}} = \dfrac{x^{12}}{5^3 y^6 z^{24}}$, or $\dfrac{x^{12}}{125 y^6 z^{24}}$

**Now Try Exercises 15 and 39.**

## ■ Scientific Notation

We can use scientific notation to name both very large and very small positive numbers and to perform computations.

> **SCIENTIFIC NOTATION**
>
> **Scientific notation** for a number is an expression of the type
>
> $$N \times 10^m,$$
>
> where $1 \leq N < 10$, $N$ is in decimal notation, and $m$ is an integer.

Keep in mind that in scientific notation positive exponents are used for numbers greater than or equal to 10 and negative exponents for numbers between 0 and 1.

**EXAMPLE 5** *Views of Video.* In the first two months after having been uploaded to a popular video-sharing Web site, a video showing a baby laughing hysterically at ripping paper received approximately 12,630,000 views (*Source*: www.youtube.com). Convert the number 12,630,000 to scientific notation.

*Solution* We want the decimal point to be positioned between the 1 and the 2, so we move it 7 places to the left. Since the number to be converted is greater than 10, the exponent must be positive. Thus we have

$$12,630,000 = 1.263 \times 10^7.$$

**Now Try Exercise 51.**

**EXAMPLE 6** *Volume Conversion.* One cubic inch is approximately equal to 0.000016 m³. Convert the number 0.000016 to scientific notation.

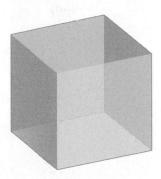

*Solution* We want the decimal point to be positioned between the 1 and the 6, so we move it 5 places to the right. Since the number to be converted is between 0 and 1, the exponent must be negative. Thus we have

$$0.000016 = 1.6 \times 10^{-5}.$$

**Now Try Exercise 59.**

**EXAMPLE 7** Convert each of the following to decimal notation.

**a)** $7.632 \times 10^{-4}$ **b)** $9.4 \times 10^5$

*Solution*

**a)** The exponent is negative, so the number is between 0 and 1. We move the decimal point 4 places to the left.

$$7.632 \times 10^{-4} = 0.0007632$$

**b)** The exponent is positive, so the number is greater than 10. We move the decimal point 5 places to the right.

$$9.4 \times 10^5 = 940,000$$

> **Now Try Exercises 61 and 63.**

GCM **EXAMPLE 8** *Distance to Neptune.* The distance from Earth to the sun is defined as **1 astronomical unit,** or AU. It is about 93 million mi. The average distance from Earth to Neptune is 30.1 AUs (*Source*: Astronomy Department, Cornell University). How many miles is it from Earth to Neptune? Express your answer in scientific notation.

*Solution* First, we convert each number to scientific notation:

$$93 \text{ million} = 93,000,000 = 9.3 \times 10^7;$$
$$30.1 = 3.01 \times 10^1, \text{ or } 3.01 \times 10.$$

Now we multiply:

$$(9.3 \times 10^7)(3.01 \times 10) = (9.3 \times 3.01) \times (10^7 \times 10)$$
$$= 27.993 \times 10^8 \qquad \text{This is not scientific notation because } 27.993 > 10.$$
$$= (2.7993 \times 10) \times 10^8$$
$$= 2.7993 \times (10 \times 10^8)$$
$$= 2.7993 \times 10^9 \text{ mi.} \qquad \text{Writing scientific notation}$$

> **Now Try Exercise 79.**

Most calculators make use of scientific notation. For example, the number 48,000,000,000,000 might be expressed as shown on the left below. The computation in Example 8 can be performed on a calculator as shown on the right below.

| 4.8E13 | 9.3E7*3.01E1 |
| | 2.7993E9 |

## ■ Order of Operations

Recall that to simplify the expression $3 + 4 \cdot 5$, first we multiply 4 and 5 to get 20 and then we add 3 to get 23. Mathematicians have agreed on the following procedure, or rules for order of operations.

> ## RULES FOR ORDER OF OPERATIONS
>
> 1. Do all calculations within grouping symbols before operations outside. When nested grouping symbols are present, work from the inside out.
> 2. Evaluate all exponential expressions.
> 3. Do all multiplications and divisions in order from left to right.
> 4. Do all additions and subtractions in order from left to right.

GCM **EXAMPLE 9** Calculate each of the following.

**a)** $8(5 - 3)^3 - 20$

**b)** $\dfrac{10 \div (8 - 6) + 9 \cdot 4}{2^5 + 3^2}$

*Solution*

**a)**
$$
\begin{aligned}
8(5 - 3)^3 - 20 &= 8 \cdot 2^3 - 20 && \text{Doing the calculation within parentheses} \\
&= 8 \cdot 8 - 20 && \text{Evaluating the exponential expression} \\
&= 64 - 20 && \text{Multiplying} \\
&= 44 && \text{Subtracting}
\end{aligned}
$$

**b)**
$$
\frac{10 \div (8 - 6) + 9 \cdot 4}{2^5 + 3^2} = \frac{10 \div 2 + 9 \cdot 4}{32 + 9}
$$
$$
= \frac{5 + 36}{41} = \frac{41}{41} = 1
$$

Note that fraction bars act as grouping symbols. That is, the given expression is equivalent to $\left[ 10 \div (8 - 6) + 9 \cdot 4 \right] \div (2^5 + 3^2)$.

We can also enter these computations on a graphing calculator as shown below.

```
8(5−3)³−20
                          44
(10/(8−6)+9∗4)/(2⁵+3²)
                           1
```

**Now Try Exercises 87 and 91.**

To confirm that it is essential to include parentheses around the numerator and around the denominator when the computation in Example 9(b) is entered in a calculator, enter the computation without using these parentheses. Note that the result is 15.125 rather than 1.

GCM **EXAMPLE 10** *Compound Interest.* If a principal $P$ is invested at an interest rate $r$, compounded $n$ times per year, in $t$ years it will grow to an amount $A$ given by

$$
A = P\left(1 + \frac{r}{n}\right)^{nt}.
$$

Suppose that $1250 is invested at 4.6% interest, compounded quarterly. How much is in the account at the end of 8 years?

**Solution**    We have $P = 1250$, $r = 4.6\%$, or $0.046$, $n = 4$, and $t = 8$. Substituting, we find that the amount in the account at the end of 8 years is given by

$$A = 1250\left(1 + \frac{0.046}{4}\right)^{4 \cdot 8}.$$

Next, we evaluate this expression:

| | | |
|---|---|---|
| $A$ | $= 1250(1 + 0.0115)^{4 \cdot 8}$ | Dividing inside the parentheses |
| | $= 1250(1.0115)^{4 \cdot 8}$ | Adding inside the parentheses |
| | $= 1250(1.0115)^{32}$ | Multiplying in the exponent |
| | $\approx 1250(1.441811175)$ | Evaluating the exponential expression |
| | $\approx 1802.263969$ | Multiplying |
| | $\approx 1802.26.$ | Rounding to the nearest cent |

The amount in the account at the end of 8 years is $1802.26. The TVM SOLVER option in the FINANCE APP on a graphing calculator can be used to do this computation.

**Now Try Exercise 93.**

```
N=32
I%=4.6
PV=-1250
PMT=0
FV=1802.263969
P/Y=4
C/Y=4
PMT: END BEGIN
```

## R.2    Exercise Set

*Write an equivalent expression without negative exponents.*

**1.** $3^{-7}$

**2.** $\dfrac{1}{(5.9)^{-4}}$

**3.** $\dfrac{x^{-5}}{y^{-4}}$

**4.** $\dfrac{a^{-2}}{b^{-8}}$

**5.** $\dfrac{m^{-1}n^{-12}}{t^{-6}}$

**6.** $\dfrac{x^{-9}y^{-17}}{z^{-11}}$

*Simplify.*

**7.** $23^0$

**8.** $\left(-\dfrac{2}{5}\right)^0$

**9.** $z^0 \cdot z^7$

**10.** $x^{10} \cdot x^0$

**11.** $5^8 \cdot 5^{-6}$

**12.** $6^2 \cdot 6^{-7}$

**13.** $m^{-5} \cdot m^5$

**14.** $n^9 \cdot n^{-9}$

**15.** $y^3 \cdot y^{-7}$

**16.** $b^{-4} \cdot b^{12}$

**17.** $(x + 3)^4(x + 3)^{-2}$

**18.** $(y - 1)^{-1}(y - 1)^5$

**19.** $3^{-3} \cdot 3^8 \cdot 3$

**20.** $6^7 \cdot 6^{-10} \cdot 6^2$

**21.** $2x^3 \cdot 3x^2$

**22.** $3y^4 \cdot 4y^3$

**23.** $(-3a^{-5})(5a^{-7})$

**24.** $(-6b^{-4})(2b^{-7})$

**25.** $(6x^{-3}y^5)(-7x^2y^{-9})$

**26.** $(8ab^7)(-7a^{-5}b^2)$

**27.** $(2x)^4(3x)^3$

**28.** $(4y)^2(3y)^3$

**29.** $(-2n)^3(5n)^2$

**30.** $(2x)^5(3x)^2$

**31.** $\dfrac{y^{35}}{y^{31}}$

**32.** $\dfrac{x^{26}}{x^{13}}$

**33.** $\dfrac{b^{-7}}{b^{12}}$

**34.** $\dfrac{a^{-18}}{a^{-13}}$

**35.** $\dfrac{x^2y^{-2}}{x^{-1}y}$

**36.** $\dfrac{x^3y^{-3}}{x^{-1}y^2}$

**37.** $\dfrac{32x^{-4}y^3}{4x^{-5}y^8}$

**38.** $\dfrac{20a^5b^{-2}}{5a^7b^{-3}}$

**39.** $(2x^2y)^4$

**40.** $(3ab^5)^3$

**41.** $(-2x^3)^5$

**42.** $(-3x^2)^4$

**43.** $(-5c^{-1}d^{-2})^{-2}$

**44.** $(-4x^{-5}z^{-2})^{-3}$

**45.** $(3m^4)^3(2m^{-5})^4$

**46.** $(4n^{-1})^2(2n^3)^3$

**47.** $\left(\dfrac{2x^{-3}y^7}{z^{-1}}\right)^3$

**48.** $\left(\dfrac{3x^5y^{-8}}{z^{-2}}\right)^4$

**49.** $\left(\dfrac{24a^{10}b^{-8}c^7}{12a^6b^{-3}c^5}\right)^{-5}$

**50.** $\left(\dfrac{125p^{12}q^{-14}r^{22}}{25p^8q^6r^{-15}}\right)^{-4}$

*Convert to scientific notation.*

**51.** 16,500,000

**52.** 359,000

**53.** 0.000000437

**54.** 0.0056

**55.** 234,600,000,000

**56.** 8,904,000,000

**57.** 0.00104

**58.** 0.00000000514

**59.** *Mass of a Neutron.*  The mass of a neutron is about 0.00000000000000000000000000167 kg.

**60.** *Tech Security Spending.*  It is estimated that $37,800,000,000 will be spent on information technology security systems in 2013 (*Source:* IDC).

*Convert to decimal notation.*

**61.** $7.6 \times 10^5$

**62.** $3.4 \times 10^{-6}$

**63.** $1.09 \times 10^{-7}$

**64.** $5.87 \times 10^8$

**65.** $3.496 \times 10^{10}$

**66.** $8.409 \times 10^{11}$

**67.** $5.41 \times 10^{-8}$

**68.** $6.27 \times 10^{-10}$

**69.** The amount of solid waste generated in the United States in a recent year was $2.319 \times 10^8$ tons (*Source:* Franklin Associates, Ltd.).

**70.** The mass of a proton is about $1.67 \times 10^{-24}$ g.

*Compute. Write the answer using scientific notation.*

**71.** $(4.2 \times 10^7)(3.2 \times 10^{-2})$

**72.** $(8.3 \times 10^{-15})(7.7 \times 10^4)$

**73.** $(2.6 \times 10^{-18})(8.5 \times 10^7)$

**74.** $(6.4 \times 10^{12})(3.7 \times 10^{-5})$

**75.** $\dfrac{6.4 \times 10^{-7}}{8.0 \times 10^6}$

**76.** $\dfrac{1.1 \times 10^{-40}}{2.0 \times 10^{-71}}$

**77.** $\dfrac{1.8 \times 10^{-3}}{7.2 \times 10^{-9}}$

**78.** $\dfrac{1.3 \times 10^4}{5.2 \times 10^{10}}$

*Solve. Write the answer using scientific notation.*

**79.** *Trash on U.S. Roadways.*  It is estimated that there were 51.2 billion pieces of trash on 76 million mi of U.S. roadways in a recent year (*Source:* Keep America Beautiful). On average, how many pieces of trash were on each mile of roadway?

**80.** *Nanowires.*  A **nanometer** is 0.000000001 m. Scientists have developed optical nanowires to transmit light waves short distances. A nanowire with a diameter of 360 nanometers has been used in experiments on the transmission of light (*Source: The New York Times,* January 29, 2004). Find the diameter of such a wire, in meters.

**81.** *Population Density.*  The tiny country of Monaco has an area of 0.75 mi$^2$. It is estimated that the population of Monaco will be 38,000 in 2050. Find the number of people per square mile in 2050.

**82.** *Chesapeake Bay Bridge-Tunnel.*  The 17.6-mi long Chesapeake Bay Bridge-Tunnel was completed in 1964. Construction costs were $210 million. Find the average cost per mile.

83. *Distance to a Star.* The nearest star, Alpha Centauri C, is about 4.22 light-years from Earth. One **light-year** is the distance that light travels in 1 year and is about $5.88 \times 10^{12}$ mi. How many miles is it from Earth to Alpha Centauri C?

84. *Parsecs.* One **parsec** is about 3.26 light-years and 1 light-year is about $5.88 \times 10^{12}$ miles. Find the number of miles in 1 parsec.

85. *Nuclear Disintegration.* One gram of radium produces 37 billion disintegrations per second. How many disintegrations are produced in 1 hour?

86. *Length of Earth's Orbit.* The average distance from Earth to the sun is 93 million mi. About how far does Earth travel in a yearly orbit? (Assume a circular orbit.)

*Calculate.*

87. $5 \cdot 3 + 8 \cdot 3^2 + 4(6-2)$

88. $5[3 - 8 \cdot 3^2 + 4 \cdot 6 - 2]$

89. $16 \div 4 \cdot 4 \div 2 \cdot 256$

90. $2^6 \cdot 2^{-3} \div 2^{10} \div 2^{-8}$

91. $\dfrac{4(8-6)^2 - 4 \cdot 3 + 2 \cdot 8}{3^1 + 19^0}$

92. $\dfrac{[4(8-6)^2 + 4](3 - 2 \cdot 8)}{2^2(2^3 + 5)}$

*Compound Interest.* Use the compound interest formula from Example 10 for Exercises 93–96. Round to the nearest cent.

93. Suppose that \$3225 is invested at 3.1%, compounded semiannually. How much is in the account at the end of 4 years?

94. Suppose that \$7550 is invested at 2.8%, compounded semiannually. How much is in the account at the end of 5 years?

95. Suppose that \$4100 is invested at 2.3%, compounded quarterly. How much is in the account at the end of 6 years?

96. Suppose that \$4875 is invested at 1.8%, compounded quarterly. How much is in the account at the end of 9 years?

## Synthesis

*Savings Plan.* The formula

$$S = P \left[ \frac{\left(1 + \dfrac{r}{12}\right)^{12 \cdot t} - 1}{\dfrac{r}{12}} \right]$$

*gives the amount S accumulated in a savings plan when a deposit of P dollars is made each month for t years in an account with interest rate r, compounded monthly. Use this formula for Exercises 97–100.*

97. Kathryn deposits \$250 in a retirement account each month beginning at age 40. If the investment earns 5% interest, compounded monthly, how much will have accumulated in the account when she retires 27 years later?

98. Charles deposits \$100 in a retirement account each month beginning at age 25. If the investment earns 4% interest, compounded monthly, how much will have accumulated in the account when he retires at age 65?

99. Sue and Richard want to establish a college fund for their daughter that will have accumulated \$120,000 at the end of 18 years. If they can count on an interest rate of 3%, compounded monthly, how much should they deposit each month in order to accomplish this?

**100.** Francisco wants to have $200,000 accumulated in a retirement account by age 70. If he starts making monthly deposits to the plan at age 30 and can count on an interest rate of 4.5%, compounded monthly, how much should he deposit each month in order to accomplish this?

*Simplify. Assume that all exponents are integers, all denominators are nonzero, and zero is not raised to a nonpositive power.*

**101.** $(x^t \cdot x^{3t})^2$

**102.** $(x^y \cdot x^{-y})^3$

**103.** $(t^{a+x} \cdot t^{x-a})^4$

**104.** $(m^{x-b} \cdot n^{x+b})^x (m^b n^{-b})^x$

**105.** $\left[ \dfrac{(3x^a y^b)^3}{(-3x^a y^b)^2} \right]^2$

**106.** $\left[ \left( \dfrac{x^r}{y^t} \right)^2 \left( \dfrac{x^{2r}}{y^{4t}} \right)^{-2} \right]^{-3}$

---

## R.3  Addition, Subtraction, and Multiplication of Polynomials

- Identify the terms, the coefficients, and the degree of a polynomial.
- Add, subtract, and multiply polynomials.

### Polynomials

Polynomials are a type of algebraic expression that you will often encounter in your study of algebra. Some examples of polynomials are

$$3x - 4y, \qquad 5y^3 - \tfrac{7}{3}y^2 + 3y - 2, \qquad -2.3a^4, \quad \text{and} \quad z^6 - \sqrt{5}.$$

All but the first are polynomials in one variable.

---

**POLYNOMIALS IN ONE VARIABLE**

A **polynomial in one variable** is any expression of the type

$$a_n x^n + a_{n-1} x^{n-1} + \cdots + a_2 x^2 + a_1 x + a_0,$$

where $n$ is a nonnegative integer and $a_n, \ldots, a_0$ are real numbers, called **coefficients**. The parts of a polynomial separated by plus signs are called **terms**. The **leading coefficient** is $a_n$, and the **constant term** is $a_0$. If $a_n \neq 0$, the **degree** of the polynomial is $n$. The polynomial is said to be written in **descending order**, because the exponents decrease from left to right.

---

**EXAMPLE 1**   Identify the terms of the polynomial

$$2x^4 - 7.5x^3 + x - 12.$$

*Solution*   Writing plus signs between the terms, we have

$$2x^4 - 7.5x^3 + x - 12 = 2x^4 + (-7.5x^3) + x + (-12),$$

so the terms are

$$2x^4, \qquad -7.5x^3, \qquad x, \quad \text{and} \quad -12.$$

A polynomial consisting of only a nonzero constant term, like 23, has degree 0. It is agreed that the polynomial consisting only of 0 has *no* degree.

**EXAMPLE 2**   Find the degree of each polynomial.

**a)** $2x^3 - 9$          **b)** $y^2 - \frac{3}{2} + 5y^4$          **c)** 7

*Solution*

| POLYNOMIAL | DEGREE |
|---|---|
| **a)** $2x^3 - 9$ | 3 |
| **b)** $y^2 - \frac{3}{2} + 5y^4 = 5y^4 + y^2 - \frac{3}{2}$ | 4 |
| **c)** $7 = 7x^0$ | 0 |

**Now Try Exercise 1.**

Algebraic expressions like $3ab^3 - 8$ and $5x^4y^2 - 3x^3y^8 + 7xy^2 + 6$ are **polynomials in several variables**. The **degree of a term** is the sum of the exponents of the variables in that term. The **degree of a polynomial** is the degree of the term of highest degree.

**EXAMPLE 3**   Find the degree of the polynomial
$$7ab^3 - 11a^2b^4 + 8.$$

*Solution*   The degrees of the terms of $7ab^3 - 11a^2b^4 + 8$ are 4, 6, and 0, respectively, so the degree of the polynomial is 6.

**Now Try Exercise 3.**

A polynomial with just one term, like $-9y^6$, is a **monomial**. If a polynomial has two terms, like $x^2 + 4$, it is a **binomial**. A polynomial with three terms, like $4x^2 - 4xy + 1$, is a **trinomial**.

Expressions like

$$2x^2 - 5x + \frac{3}{x}, \quad 9 - \sqrt{x}, \quad \text{and} \quad \frac{x + 1}{x^4 + 5}$$

are not polynomials, because they cannot be written in the form $a_n x^n + a_{n-1}x^{n-1} + \cdots + a_1 x + a_0$, where the exponents are all nonnegative integers and the coefficients are all real numbers.

## Addition and Subtraction

If two terms of an expression have the same variables raised to the same powers, they are called **like terms**, or **similar terms**. We can **combine**, or **collect, like terms** using the distributive property. For example, $3y^2$ and $5y^2$ are like terms and

$$3y^2 + 5y^2 = (3 + 5)y^2$$
$$= 8y^2.$$

We add or subtract polynomials by combining like terms.

**EXAMPLE 4**  Add or subtract each of the following.

**a)** $(-5x^3 + 3x^2 - x) + (12x^3 - 7x^2 + 3)$

**b)** $(6x^2y^3 - 9xy) - (5x^2y^3 - 4xy)$

*Solution*

**a)** $(-5x^3 + 3x^2 - x) + (12x^3 - 7x^2 + 3)$

$= (-5x^3 + 12x^3) + (3x^2 - 7x^2) - x + 3$  Rearranging using commutative and associative properties

$= (-5 + 12)x^3 + (3 - 7)x^2 - x + 3$  Using the distributive property

$= 7x^3 - 4x^2 - x + 3$

**b)**  We can subtract by adding an opposite:

$(6x^2y^3 - 9xy) - (5x^2y^3 - 4xy)$

$= (6x^2y^3 - 9xy) + (-5x^2y^3 + 4xy)$  Adding the opposite of $5x^2y^3 - 4xy$

$= 6x^2y^3 - 9xy - 5x^2y^3 + 4xy$

$= x^2y^3 - 5xy.$  Combining like terms

**Now Try Exercises 5 and 9.**

## ▪ Multiplication

To multiply monomials, we first multiply their coefficients, and then we multiply their variables.

**EXAMPLE 5**  Multiply each of the following.

**a)** $(-2x^3)(5x^4)$ **b)** $(3yz^2)(8y^3z^5)$

*Solution*

**a)** $(-2x^3)(5x^4) = (-2 \cdot 5)(x^3 \cdot x^4) = -10x^7$

**b)** $(3yz^2)(8y^3z^5) = (3 \cdot 8)(y \cdot y^3)(z^2 \cdot z^5) = 24y^4z^7$

**Now Try Exercises 13 and 15.**

Multiplication of polynomials other than monomials is based on the distributive property—for example,

$(x + 4)(x + 3) = x(x + 3) + 4(x + 3)$  Using the distributive property

$= x^2 + 3x + 4x + 12$  Using the distributive property two more times

$= x^2 + 7x + 12.$  Combining like terms

In general, to multiply two polynomials, we multiply each term of one by each term of the other and then add the products.

**EXAMPLE 6** Multiply: $(4x^4y - 7x^2y + 3y)(2y - 3x^2y)$.

*Solution* We have

$(4x^4y - 7x^2y + 3y)(2y - 3x^2y)$

$= 4x^4y(2y - 3x^2y) - 7x^2y(2y - 3x^2y) + 3y(2y - 3x^2y)$

Using the distributive property

$= 8x^4y^2 - 12x^6y^2 - 14x^2y^2 + 21x^4y^2 + 6y^2 - 9x^2y^2$

Using the distributive property three more times

$= 29x^4y^2 - 12x^6y^2 - 23x^2y^2 + 6y^2.$     Combining like terms

We can also use columns to organize our work, aligning like terms under each other in the products.

$$\begin{array}{r} 4x^4y - 7x^2y + 3y \\ 2y - 3x^2y \\ \hline -12x^6y^2 + 21x^4y^2 - 9x^2y^2 \\ 8x^4y^2 - 14x^2y^2 + 6y^2 \\ \hline -12x^6y^2 + 29x^4y^2 - 23x^2y^2 + 6y^2 \end{array}$$

Multiplying by $-3x^2y$
Multiplying by $2y$
Adding

**Now Try Exercise 17.**

We can find the product of two binomials by multiplying the **First** terms, then the **Outer** terms, then the **Inner** terms, then the **Last** terms. Then we combine like terms, if possible. This procedure is sometimes called **FOIL**.

**EXAMPLE 7** Multiply: $(2x - 7)(3x + 4)$.

*Solution* We have

$$(2x - 7)(3x + 4) = 6x^2 + 8x - 21x - 28$$
$$= 6x^2 - 13x - 28.$$

**Now Try Exercise 19.**

We can use FOIL to find some special products.

**SPECIAL PRODUCTS OF BINOMIALS**

$(A + B)^2 = A^2 + 2AB + B^2$     Square of a sum

$(A - B)^2 = A^2 - 2AB + B^2$     Square of a difference

$(A + B)(A - B) = A^2 - B^2$     Product of a sum and a difference

**EXAMPLE 8** Multiply each of the following.

**a)** $(4x + 1)^2$      **b)** $(3y^2 - 2)^2$      **c)** $(x^2 + 3y)(x^2 - 3y)$

*Solution*

a) $(4x + 1)^2 = (4x)^2 + 2 \cdot 4x \cdot 1 + 1^2 = 16x^2 + 8x + 1$

b) $(3y^2 - 2)^2 = (3y^2)^2 - 2 \cdot 3y^2 \cdot 2 + 2^2 = 9y^4 - 12y^2 + 4$

c) $(x^2 + 3y)(x^2 - 3y) = (x^2)^2 - (3y)^2 = x^4 - 9y^2$

> **Now Try Exercises 27 and 37.**

Division of polynomials is discussed in Section 4.3.

## R.3 Exercise Set

*Determine the terms and the degree of the polynomial.*

**1.** $7x^3 - 4x^2 + 8x + 5$

**2.** $-3n^4 - 6n^3 + n^2 + 2n - 1$

**3.** $3a^4b - 7a^3b^3 + 5ab - 2$

**4.** $6p^3q^2 - p^2q^4 - 3pq^2 + 5$

*Perform the indicated operations.*

**5.** $(3ab^2 - 4a^2b - 2ab + 6) + (-ab^2 - 5a^2b + 8ab + 4)$

**6.** $(-6m^2n + 3mn^2 - 5mn + 2) + (4m^2n + 2mn^2 - 6mn - 9)$

**7.** $(2x + 3y + z - 7) + (4x - 2y - z + 8) + (-3x + y - 2z - 4)$

**8.** $(2x^2 + 12xy - 11) + (6x^2 - 2x + 4) + (-x^2 - y - 2)$

**9.** $(3x^2 - 2x - x^3 + 2) - (5x^2 - 8x - x^3 + 4)$

**10.** $(5x^2 + 4xy - 3y^2 + 2) - (9x^2 - 4xy + 2y^2 - 1)$

**11.** $(x^4 - 3x^2 + 4x) - (3x^3 + x^2 - 5x + 3)$

**12.** $(2x^4 - 3x^2 + 7x) - (5x^3 + 2x^2 - 3x + 5)$

**13.** $(3a^2)(-7a^4)$

**14.** $(8y^5)(9y)$

**15.** $(6xy^3)(9x^4y^2)$

**16.** $(-5m^4n^2)(6m^2n^3)$

**17.** $(a - b)(2a^3 - ab + 3b^2)$

**18.** $(n + 1)(n^2 - 6n - 4)$

**19.** $(y - 3)(y + 5)$

**20.** $(z + 4)(z - 2)$

**21.** $(x + 6)(x + 3)$

**22.** $(a - 8)(a - 1)$

**23.** $(2a + 3)(a + 5)$

**24.** $(3b + 1)(b - 2)$

**25.** $(2x + 3y)(2x + y)$

**26.** $(2a - 3b)(2a - b)$

**27.** $(x + 3)^2$

**28.** $(z + 6)^2$

**29.** $(y - 5)^2$

**30.** $(x - 4)^2$

**31.** $(5x - 3)^2$

**32.** $(3x - 2)^2$

**33.** $(2x + 3y)^2$

**34.** $(5x + 2y)^2$

**35.** $(2x^2 - 3y)^2$

**36.** $(4x^2 - 5y)^2$

**37.** $(n + 6)(n - 6)$

**38.** $(m + 1)(m - 1)$

**39.** $(3y + 4)(3y - 4)$

**40.** $(2x - 7)(2x + 7)$

**41.** $(3x - 2y)(3x + 2y)$

**42.** $(3x + 5y)(3x - 5y)$

**43.** $(2x + 3y + 4)(2x + 3y - 4)$

**44.** $(5x + 2y + 3)(5x + 2y - 3)$

**45.** $(x + 1)(x - 1)(x^2 + 1)$

**46.** $(y - 2)(y + 2)(y^2 + 4)$

## Synthesis

*Multiply. Assume that all exponents are natural numbers.*

**47.** $(a^n + b^n)(a^n - b^n)$

**48.** $(t^a + 4)(t^a - 7)$

**49.** $(a^n + b^n)^2$

**50.** $(x^{3m} - t^{5n})^2$

**51.** $(x - 1)(x^2 + x + 1)(x^3 + 1)$

**52.** $\left[ (2x - 1)^2 - 1 \right]^2$

**53.** $(x^{a-b})^{a+b}$

**54.** $(t^{m+n})^{m+n} \cdot (t^{m-n})^{m-n}$

**55.** $(a + b + c)^2$

## R.4 Factoring

- Factor polynomials by removing a common factor.
- Factor polynomials by grouping.
- Factor trinomials of the type $x^2 + bx + c$.
- Factor trinomials of the type $ax^2 + bx + c$, $a \neq 1$, using the FOIL method and the grouping method.
- Factor special products of polynomials.

To factor a polynomial, we do the reverse of multiplying; that is, we find an equivalent expression that is written as a product.

### Terms with Common Factors

When a polynomial is to be factored, we should always look first to factor out a factor that is common to all the terms using the distributive property. We generally look for the constant common factor with the largest absolute value and for variables with the largest exponent common to all the terms. In this sense, we factor out the "largest" common factor.

**EXAMPLE 1**    Factor each of the following.

**a)** $15 + 10x - 5x^2$                    **b)** $12x^2y^2 - 20x^3y$

*Solution*

**a)** $15 + 10x - 5x^2 = 5 \cdot 3 + 5 \cdot 2x - 5 \cdot x^2 = 5(3 + 2x - x^2)$

We can always check a factorization by multiplying:

$$5(3 + 2x - x^2) = 15 + 10x - 5x^2.$$

**b)** There are several factors common to the terms of $12x^2y^2 - 20x^3y$, but $4x^2y$ is the "largest" of these.

$$\begin{aligned} 12x^2y^2 - 20x^3y &= 4x^2y \cdot 3y - 4x^2y \cdot 5x \\ &= 4x^2y(3y - 5x) \end{aligned}$$

**Now Try Exercise 3.**

### Factoring by Grouping

In some polynomials, pairs of terms have a common binomial factor that can be removed in a process called **factoring by grouping**.

**EXAMPLE 2** Factor: $x^3 + 3x^2 - 5x - 15$.

*Solution* We have

$x^3 + 3x^2 - 5x - 15 = (x^3 + 3x^2) + (-5x - 15)$     Grouping. Each group of terms has a common factor.

$= x^2(x + 3) - 5(x + 3)$     Factoring a common factor out of each group

$= (x + 3)(x^2 - 5).$     Factoring out the common binomial factor

> **Now Try Exercise 9.**

## ▪ Trinomials of the Type $x^2 + bx + c$

Some trinomials can be factored into the product of two binomials. To factor a trinomial of the form $x^2 + bx + c$, we look for binomial factors of the form

$$(x + p)(x + q),$$

where $p \cdot q = c$ and $p + q = b$. That is, we look for two numbers $p$ and $q$ whose sum is the coefficient of the middle term of the polynomial, $b$, and whose product is the constant term, $c$.

When we factor any polynomial, we should always check first to determine whether there is a factor common to all the terms. If there is, we factor it out first.

**EXAMPLE 3** Factor: $x^2 + 5x + 6$.

*Solution* First, we look for a common factor. There is none. Next, we look for two numbers whose product is 6 and whose sum is 5. Since the constant term, 6, and the coefficient of the middle term, 5, are both positive, we look for a factorization of 6 in which both factors are positive.

| Pairs of Factors | Sums of Factors |
|:---:|:---:|
| 1, 6 | 7 |
| 2, 3 | 5 |

The numbers we need are 2 and 3.

The factorization is $(x + 2)(x + 3)$. We have

$$x^2 + 5x + 6 = (x + 2)(x + 3).$$

We can check this by multiplying:

$$(x + 2)(x + 3) = x^2 + 3x + 2x + 6 = x^2 + 5x + 6.$$

> **Now Try Exercise 21.**

**EXAMPLE 4** Factor: $x^4 - 6x^3 + 8x^2$.

*Solution* First, we look for a common factor. Each term has a factor of $x^2$, so we factor it out first:

$$x^4 - 6x^3 + 8x^2 = x^2(x^2 - 6x + 8).$$

Now we consider the trinomial $x^2 - 6x + 8$. We look for two numbers whose product is 8 and whose sum is $-6$. Since the constant term, 8, is positive and the coefficient of the middle term, $-6$, is negative, we look for a factorization of 8 in which both factors are negative.

| Pairs of Factors | Sums of Factors |
|:---:|:---:|
| $-1, -8$ | $-9$ |
| $-2, -4$ | $-6$ ← |

The numbers we need are $-2$ and $-4$.

The factorization of $x^2 - 6x + 8$ is $(x - 2)(x - 4)$. We must also include the common factor that we factored out earlier. Thus we have

$$x^4 - 6x^3 + 8x^2 = x^2(x - 2)(x - 4).$$

**Now Try Exercise 31.**

**EXAMPLE 5** Factor: $x^2 + xy - 12y^2$.

*Solution* Having checked for a common factor and finding none, we think much the same way we would if we were factoring $x^2 + x - 12$. We look for factors of $-12$ whose sum is the coefficient of the $xy$-term, 1. Since the last term, $-12y^2$, is negative, one factor will be positive and the other will be negative.

| Pairs of Factors | Sums of Factors |
|:---:|:---:|
| $-1, \quad 12$ | $11$ |
| $1, -12$ | $-11$ |
| $-2, \quad 6$ | $4$ |
| $2, -6$ | $-4$ |
| $-3, \quad 4$ | $1$ ← |
| $3, -4$ | $-1$ |

The numbers we need are $-3$ and 4.

We might have observed at the outset that since the sum of the factors must be 1, a positive number, we need consider only pairs of factors for which the positive factor has the greater absolute value. Thus only the pairs $-1$ and 12, $-2$ and 6, $-3$ and 4 need have been considered.

The factorization is

$$(x - 3y)(x + 4y).$$

**Now Try Exercise 25.**

## ◼ Trinomials of the Type $ax^2 + bx + c,\ a \neq 1$

We consider two methods for factoring trinomials of the type $ax^2 + bx + c$, $a \neq 1$.

### The FOIL Method

We first consider the **FOIL method** for factoring trinomials of the type $ax^2 + bx + c$, $a \neq 1$. Consider the following multiplication.

$$
\begin{array}{cccc}
\textbf{F} & \textbf{O} & \textbf{I} & \textbf{L} \\
\downarrow & \downarrow & \downarrow & \downarrow
\end{array}
$$

$$(3x + 2)(4x + 5) = 12x^2 + \underbrace{15x + 8x}_{} + 10$$

$$= 12x^2 + \quad 23x \quad + 10$$

To factor $12x^2 + 23x + 10$, we must reverse what we just did. We look for two binomials whose product is this trinomial. The product of the First terms must be $12x^2$. The product of the Outer terms plus the product of the Inner terms must be $23x$. The product of the Last terms must be 10. We know from the preceding discussion that the answer is $(3x + 2)(4x + 5)$. In general, however, finding such an answer involves trial and error. We use the following method.

---

To factor trinomials of the type $ax^2 + bx + c,\ a \neq 1$, using the **FOIL method**:

1. Factor out the largest common factor.
2. Find two First terms whose product is $ax^2$:

$$(\boxed{\phantom{x}}\,x +\quad)(\boxed{\phantom{x}}\,x +\quad) = ax^2 + bx + c.$$

$$\text{FOIL}$$

3. Find two Last terms whose product is $c$:

$$(\,x + \boxed{\phantom{x}})(\,x + \boxed{\phantom{x}}) = ax^2 + bx + c.$$

$$\text{FOIL}$$

4. Repeat steps (2) and (3) until a combination is found for which the sum of the Outer product and the Inner product is $bx$:

$$(\boxed{\phantom{x}}\,x + \boxed{\phantom{x}})(\boxed{\phantom{x}}\,x + \boxed{\phantom{x}}) = ax^2 + bx + c.$$

$$\text{I} \qquad\qquad \text{FOIL}$$

$$\text{O}$$

---

**EXAMPLE 6**  Factor: $3x^2 - 10x - 8$.

*Solution*

1. There is no common factor (other than 1 or $-1$).
2. Factor the first term, $3x^2$. The only possibility (with positive integer coefficients) is $3x \cdot x$. The factorization, if it exists, must be of the form $(3x + \boxed{\phantom{x}})(x + \boxed{\phantom{x}})$.

3. Next, factor the constant term, $-8$. The possibilities are $-8(1)$, $8(-1)$, $-2(4)$, and $2(-4)$. The factors can be written in the opposite order as well: $1(-8)$, $-1(8)$, $4(-2)$, and $-4(2)$.

4. Find a pair of factors for which the sum of the outer product and the inner product is the middle term, $-10x$. Each possibility should be checked by multiplying. Some trials show that the desired factorization is $(3x + 2)(x - 4)$.

> **Now Try Exercise 35.**

## The Grouping Method

The second method for factoring trinomials of the type $ax^2 + bx + c$, $a \neq 1$, is known as the **grouping method**, or the *ac*-**method**.

---

To factor $ax^2 + bx + c$, $a \neq 1$, using the **grouping method**:

1. Factor out the largest common factor.
2. Multiply the leading coefficient $a$ and the constant $c$.
3. Try to factor the product $ac$ so that the sum of the factors is $b$. That is, find integers $p$ and $q$ such that $pq = ac$ and $p + q = b$.
4. Split the middle term. That is, write it as a sum using the factors found in step (3).
5. Factor by grouping.

---

**EXAMPLE 7** Factor: $12x^3 + 10x^2 - 8x$.

*Solution*

1. Factor out the largest common factor, $2x$:

$$12x^3 + 10x^2 - 8x = 2x(6x^2 + 5x - 4).$$

2. Now consider $6x^2 + 5x - 4$. Multiply the leading coefficient, 6, and the constant, $-4$: $6(-4) = -24$.

3. Try to factor $-24$ so that the sum of the factors is the coefficient of the middle term, 5.

| Pairs of Factors | Sums of Factors |
|:---:|:---:|
| 1, −24 | −23 |
| −1, 24 | 23 |
| 2, −12 | −10 |
| −2, 12 | 10 |
| 3, −8 | −5 |
| −3, 8 | 5 ← |
| 4, −6 | −2 |
| −4, 6 | 2 |

$-3 \cdot 8 = -24; \ -3 + 8 = 5$

4. Split the middle term using the numbers found in step (3):

$$5x = -3x + 8x.$$

**5.** Finally, factor by grouping:

$$6x^2 + 5x - 4 = 6x^2 - 3x + 8x - 4$$
$$= 3x(2x - 1) + 4(2x - 1)$$
$$= (2x - 1)(3x + 4).$$

Be sure to include the common factor to get the complete factorization of the original trinomial:

$$12x^3 + 10x^2 - 8x = 2x(2x - 1)(3x + 4).$$

> **Now Try Exercise 45.**

## ■ Special Factorizations

We reverse the equation $(A + B)(A - B) = A^2 - B^2$ to factor a **difference of squares**.

$$A^2 - B^2 = (A + B)(A - B)$$

**EXAMPLE 8**   Factor each of the following completely.

**a)** $x^2 - 16$         **b)** $9a^2 - 25$         **c)** $6x^4 - 6y^4$

*Solution*

**a)** $x^2 - 16 = x^2 - 4^2 = (x + 4)(x - 4)$

**b)** $9a^2 - 25 = (3a)^2 - 5^2 = (3a + 5)(3a - 5)$

**c)** $6x^4 - 6y^4 = 6(x^4 - y^4)$
$$= 6[(x^2)^2 - (y^2)^2]$$
$$= 6(x^2 + y^2)(x^2 - y^2) \qquad x^2 - y^2 \text{ can be factored further.}$$
$$= 6(x^2 + y^2)(x + y)(x - y) \qquad \text{Because none of these factors can be factored further, we have } \textit{factored completely.}$$

> **Now Try Exercise 49.**

The rules for squaring binomials can be reversed to factor trinomials that are squares of binomials:

$$A^2 + 2AB + B^2 = (A + B)^2;$$
$$A^2 - 2AB + B^2 = (A - B)^2.$$

**EXAMPLE 9**   Factor each of the following.

**a)** $x^2 + 8x + 16$                    **b)** $25y^2 - 30y + 9$

*Solution*

$$A^2 + 2 \cdot A \cdot B + B^2 = (A + B)^2$$

**a)** $x^2 + 8x + 16 = x^2 + 2 \cdot x \cdot 4 + 4^2 = (x + 4)^2$

$$A^2 - 2 \cdot A \cdot B + B^2 = (A - B)^2$$

**b)** $25y^2 - 30y + 9 = (5y)^2 - 2 \cdot 5y \cdot 3 + 3^2 = (5y - 3)^2$

**Now Try Exercises 57 and 59.**

We can use the following rules to factor a **sum** or a **difference of cubes**:

$$A^3 + B^3 = (A + B)(A^2 - AB + B^2);$$
$$A^3 - B^3 = (A - B)(A^2 + AB + B^2).$$

These rules can be verified by multiplying.

**EXAMPLE 10**   Factor each of the following.

**a)** $x^3 + 27$ 

**b)** $16y^3 - 250$

*Solution*

**a)** $x^3 + 27 = x^3 + 3^3 = (x + 3)(x^2 - 3x + 9)$

**b)** $16y^3 - 250 = 2(8y^3 - 125)$

$$= 2\left[(2y)^3 - 5^3\right] = 2(2y - 5)(4y^2 + 10y + 25)$$

**Now Try Exercises 67 and 69.**

Not all polynomials can be factored into polynomials with integer coefficients. An example is $x^2 - x + 7$. There are no integer factors of 7 whose sum is $-1$. In such a case, we say that the polynomial is "not factorable," or **prime**.

## CONNECTING THE CONCEPTS

### A Strategy for Factoring

**A.** Always factor out the largest common factor first.

**B.** Look at the number of terms.

*Two terms:* Try factoring as a difference of squares first. Next, try factoring as a sum or a difference of cubes. There is no rule for factoring a *sum* of squares.

*Three terms:* Try factoring as the square of a binomial. Next, try using the FOIL method or the grouping method for factoring a trinomial.

*Four or more terms:* Try factoring by grouping and factoring out a common binomial factor.

**C.** Always *factor completely*. If a factor with more than one term can itself be factored further, do so.

# R.4  Exercise Set

*Factor out the largest common factor.*

**1.** $3x + 18$

**2.** $5y - 20$

**3.** $2z^3 - 8z^2$

**4.** $12m^2 + 3m^6$

**5.** $4a^2 - 12a + 16$

**6.** $6n^2 + 24n - 18$

**7.** $a(b - 2) + c(b - 2)$

**8.** $a(x^2 - 3) - 2(x^2 - 3)$

*Factor by grouping.*

**9.** $3x^3 - x^2 + 18x - 6$

**10.** $x^3 + 3x^2 + 6x + 18$

**11.** $y^3 - y^2 + 2y - 2$

**12.** $y^3 - y^2 + 3y - 3$

**13.** $24x^3 - 36x^2 + 72x - 108$

**14.** $5a^3 - 10a^2 + 25a - 50$

**15.** $x^3 - x^2 - 5x + 5$

**16.** $t^3 + 6t^2 - 2t - 12$

**17.** $a^3 - 3a^2 - 2a + 6$

**18.** $x^3 - x^2 - 6x + 6$

*Factor the trinomial.*

**19.** $w^2 - 7w + 10$

**20.** $p^2 + 6p + 8$

**21.** $x^2 + 6x + 5$

**22.** $x^2 - 8x + 12$

**23.** $t^2 + 8t + 15$

**24.** $y^2 + 12y + 27$

**25.** $x^2 - 6xy - 27y^2$

**26.** $t^2 - 2t - 15$

**27.** $2n^2 - 20n - 48$

**28.** $2a^2 - 2ab - 24b^2$

**29.** $y^2 - 4y - 21$

**30.** $m^2 - m - 90$

**31.** $y^4 - 9y^3 + 14y^2$

**32.** $3z^3 - 21z^2 + 18z$

**33.** $2x^3 - 2x^2y - 24xy^2$

**34.** $a^3b - 9a^2b^2 + 20ab^3$

**35.** $2n^2 + 9n - 56$

**36.** $3y^2 + 7y - 20$

**37.** $12x^2 + 11x + 2$

**38.** $6x^2 - 7x - 20$

**39.** $4x^2 + 15x + 9$

**40.** $2y^2 + 7y + 6$

**41.** $2y^2 + y - 6$

**42.** $20p^2 - 23p + 6$

**43.** $6a^2 - 29ab + 28b^2$

**44.** $10m^2 + 7mn - 12n^2$

**45.** $12a^2 - 4a - 16$

**46.** $12a^2 - 14a - 20$

*Factor the difference of squares.*

**47.** $z^2 - 81$

**48.** $m^2 - 4$

**49.** $16x^2 - 9$

**50.** $4z^2 - 81$

**51.** $6x^2 - 6y^2$

**52.** $8a^2 - 8b^2$

**53.** $4xy^4 - 4xz^2$

**54.** $5x^2y - 5yz^4$

**55.** $7pq^4 - 7py^4$

**56.** $25ab^4 - 25az^4$

*Factor the square of the binomial.*

**57.** $x^2 + 12x + 36$

**58.** $y^2 - 6y + 9$

**59.** $9z^2 - 12z + 4$

**60.** $4z^2 + 12z + 9$

**61.** $1 - 8x + 16x^2$

**62.** $1 + 10x + 25x^2$

**63.** $a^3 + 24a^2 + 144a$

**64.** $y^3 - 18y^2 + 81y$

**65.** $4p^2 - 8pq + 4q^2$

**66.** $5a^2 - 10ab + 5b^2$

*Factor the sum or the difference of cubes.*

**67.** $x^3 + 64$

**68.** $y^3 - 8$

**69.** $m^3 - 216$

**70.** $n^3 + 1$

**71.** $8t^3 + 8$

**72.** $2y^3 - 128$

**73.** $3a^5 - 24a^2$

**74.** $250z^4 - 2z$

**75.** $t^6 + 1$

**76.** $27x^6 - 8$

*Factor completely.*

**77.** $18a^2b - 15ab^2$

**78.** $4x^2y + 12xy^2$

**79.** $x^3 - 4x^2 + 5x - 20$

**80.** $z^3 + 3z^2 - 3z - 9$

**81.** $8x^2 - 32$

**82.** $6y^2 - 6$

**83.** $4y^2 - 5$

**84.** $16x^2 - 7$

**85.** $m^2 - 9n^2$

**86.** $25t^2 - 16$

**87.** $x^2 + 9x + 20$

**88.** $y^2 + y - 6$

**89.** $y^2 - 6y + 5$

**90.** $x^2 - 4x - 21$

**91.** $2a^2 + 9a + 4$

**92.** $3b^2 - b - 2$

**93.** $6x^2 + 7x - 3$

**94.** $8x^2 + 2x - 15$

**95.** $y^2 - 18y + 81$

**96.** $n^2 + 2n + 1$

**97.** $9z^2 - 24z + 16$

**98.** $4z^2 + 20z + 25$

**99.** $x^2y^2 - 14xy + 49$

**100.** $x^2y^2 - 16xy + 64$

**101.** $4ax^2 + 20ax - 56a$

**102.** $21x^2y + 2xy - 8y$

**103.** $3z^3 - 24$      **104.** $4t^3 + 108$

**105.** $16a^7b + 54ab^7$      **106.** $24a^2x^4 - 375a^8x$

**107.** $y^3 - 3y^2 - 4y + 12$

**108.** $p^3 - 2p^2 - 9p + 18$

**109.** $x^3 - x^2 + x - 1$      **110.** $x^3 - x^2 - x + 1$

**111.** $5m^4 - 20$      **112.** $2x^2 - 288$

**113.** $2x^3 + 6x^2 - 8x - 24$

**114.** $3x^3 + 6x^2 - 27x - 54$

**115.** $4c^2 - 4cd + d^2$      **116.** $9a^2 - 6ab + b^2$

**117.** $m^6 + 8m^3 - 20$      **118.** $x^4 - 37x^2 + 36$

**119.** $p - 64p^4$      **120.** $125a - 8a^4$

## Synthesis

*Factor.*

**121.** $y^4 - 84 + 5y^2$      **122.** $11x^2 + x^4 - 80$

**123.** $y^2 - \frac{8}{49} + \frac{2}{7}y$      **124.** $t^2 - \frac{27}{100} + \frac{3}{5}t$

**125.** $x^2 + 3x + \frac{9}{4}$      **126.** $x^2 - 5x + \frac{25}{4}$

**127.** $x^2 - x + \frac{1}{4}$      **128.** $x^2 - \frac{2}{3}x + \frac{1}{9}$

**129.** $(x + h)^3 - x^3$      **130.** $(x + 0.01)^2 - x^2$

**131.** $(y - 4)^2 + 5(y - 4) - 24$

**132.** $6(2p + q)^2 - 5(2p + q) - 25$

*Factor. Assume that variables in exponents represent natural numbers.*

**133.** $x^{2n} + 5x^n - 24$      **134.** $4x^{2n} - 4x^n - 3$

**135.** $x^2 + ax + bx + ab$

**136.** $bdy^2 + ady + bcy + ac$

**137.** $25y^{2m} - (x^{2n} - 2x^n + 1)$

**138.** $x^{6a} - t^{3b}$

**139.** $(y - 1)^4 - (y - 1)^2$

**140.** $x^6 - 2x^5 + x^4 - x^2 + 2x - 1$

---

# R.5

## The Basics of Equation Solving

- Solve linear equations.
- Solve quadratic equations.
- Solve a formula for a given letter.

An **equation** is a statement that two expressions are equal. To **solve** an equation in one variable is to find all the values of the variable that make the equation true. Each of these numbers is a **solution** of the equation. The set of all solutions of an equation is its **solution set**. Equations that have the same solution set are called **equivalent equations**.

### ■ Linear Equations and Quadratic Equations

A **linear equation in one variable** is an equation that is equivalent to one of the form $ax + b = 0$, where $a$ and $b$ are real numbers and $a \neq 0$.

A **quadratic equation** is an equation that is equivalent to one of the form $ax^2 + bx + c = 0$, where $a$, $b$, and $c$ are real numbers and $a \neq 0$.

The following principles allow us to solve many linear equations and quadratic equations.

---

**EQUATION-SOLVING PRINCIPLES**

For any real numbers $a$, $b$, and $c$,

**The Addition Principle:** If $a = b$ is true, then $a + c = b + c$ is true.

**The Multiplication Principle:** If $a = b$ is true, then $ac = bc$ is true.

**The Principle of Zero Products:** If $ab = 0$ is true, then $a = 0$ or $b = 0$, and if $a = 0$ or $b = 0$, then $ab = 0$.

**The Principle of Square Roots:** If $x^2 = k$, then $x = \sqrt{k}$ or $x = -\sqrt{k}$.

---

Let's first consider a linear equation. We will use the addition and multiplication principles to solve it.

**EXAMPLE 1** Solve: $2x + 3 = 1 - 6(x - 1)$.

**Solution** We begin by using the distributive property to remove the parentheses.

$$2x + 3 = 1 - 6(x - 1)$$
$$2x + 3 = 1 - 6x + 6 \qquad \text{Using the distributive property}$$
$$2x + 3 = 7 - 6x \qquad \text{Combining like terms on the right}$$
$$8x + 3 = 7 \qquad \text{Using the addition principle to add } 6x \text{ on both sides}$$
$$8x = 4 \qquad \text{Using the addition principle to add } -3, \text{ or subtract 3, on both sides}$$
$$x = \frac{4}{8} \qquad \text{Using the multiplication principle to multiply by } \frac{1}{8}, \text{ or divide by 8, on both sides}$$
$$x = \frac{1}{2} \qquad \text{Simplifying}$$

We check the result in the original equation.

Check: 
$$\begin{array}{c|c} \multicolumn{2}{c}{2x + 3 = 1 - 6(x - 1)} \\ \hline 2 \cdot \frac{1}{2} + 3 \ ? \ 1 - 6\left(\frac{1}{2} - 1\right) & \text{Substituting } \frac{1}{2} \text{ for } x \\ 1 + 3 \ \bigg| \ 1 - 6\left(-\frac{1}{2}\right) & \\ 4 \ \bigg| \ 1 + 3 & \\ 4 \ \bigg| \ 4 & \text{TRUE} \end{array}$$

The solution is $\frac{1}{2}$.

**Now Try Exercise 13.**

Now we consider a quadratic equation that can be solved using the principle of zero products.

**EXAMPLE 2** Solve: $x^2 - 3x = 4$.

**Solution** First, we write the equation with 0 on one side.

$$x^2 - 3x = 4$$
$$x^2 - 3x - 4 = 0 \qquad \text{Subtracting 4 on both sides}$$
$$(x + 1)(x - 4) = 0 \qquad \text{Factoring}$$
$$x + 1 = 0 \quad or \quad x - 4 = 0 \qquad \text{Using the principle of}$$
$$\text{zero products}$$
$$x = -1 \quad or \qquad x = 4$$

*Check:*    For $-1$:                       For 4:

$$\begin{array}{c|c} x^2 - 3x = 4 \\ \hline (-1)^2 - 3(-1) \ ? \ 4 \\ 1 + 3 \\ 4 \ \bigm| \ 4 \quad \text{TRUE} \end{array} \qquad \begin{array}{c|c} x^2 - 3x = 4 \\ \hline 4^2 - 3 \cdot 4 \ ? \ 4 \\ 16 - 12 \\ 4 \ \bigm| \ 4 \quad \text{TRUE} \end{array}$$

The solutions are $-1$ and 4.                       ▮ **Now Try Exercise 35.**

The principle of square roots can be used to solve some quadratic equations, as we see in the next example.

**EXAMPLE 3** Solve: $3x^2 - 6 = 0$.

**Solution** We will use the principle of square roots:

$$3x^2 - 6 = 0$$
$$3x^2 = 6 \qquad \text{Adding 6 on both sides}$$
$$x^2 = 2 \qquad \text{Dividing by 3 on both sides to isolate } x^2$$
$$x = \sqrt{2} \quad or \quad x = -\sqrt{2}. \qquad \text{Using the principle of square roots}$$

Both numbers check. The solutions are $\sqrt{2}$ and $-\sqrt{2}$, or $\pm\sqrt{2}$ (read "plus or minus $\sqrt{2}$").                       ▮ **Now Try Exercise 53.**

## ▪ Formulas

A **formula** is an equation that can be used to *model* a situation. For example, the formula $P = 2l + 2w$ in Example 4 gives the perimeter of a rectangle with length $l$ and width $w$.

The equation-solving principles presented earlier can be used to solve a formula for a given letter.

**EXAMPLE 4** Solve: $P = 2l + 2w$ for $l$.

**Solution** We have

$$P = 2l + 2w \qquad \text{We want to isolate } l.$$
$$P - 2w = 2l \qquad \text{Subtracting 2w on both sides}$$
$$\frac{P - 2w}{2} = l. \qquad \text{Dividing by 2 on both sides}$$

▮ **Now Try Exercise 63.**

The formula $l = \dfrac{P - 2w}{2}$ can be used to determine the length of a rectangle if we are given its perimeter and its width.

**EXAMPLE 5** The formula $A = P + Prt$ gives the amount $A$ to which a principal of $P$ dollars will grow when invested at simple interest rate $r$ for $t$ years. Solve the formula for $P$.

*Solution* We have

$$A = P + Prt \qquad \text{We want to isolate } P.$$
$$A = P(1 + rt) \qquad \text{Factoring}$$

$$\frac{A}{1 + rt} = \frac{P(1 + rt)}{1 + rt} \qquad \text{Dividing by } 1 + rt \text{ on both sides}$$

$$\frac{A}{1 + rt} = P. \qquad \qquad \boxed{\textbf{Now Try Exercise 77.}}$$

The formula $P = \dfrac{A}{1 + rt}$ can be used to determine how much should be invested at simple interest rate $r$ in order to have $A$ dollars $t$ years later.

**EXAMPLE 6** Solve: $A = \frac{1}{2}h(b_1 + b_2)$ for $b_1$.

*Solution* We have

$$A = \frac{1}{2}h(b_1 + b_2) \qquad \text{Formula for the area of a trapezoid}$$

$$2A = h(b_1 + b_2) \qquad \text{Multiplying by 2}$$
$$2A = hb_1 + hb_2 \qquad \text{Removing parentheses}$$
$$2A - hb_2 = hb_1 \qquad \text{Subtracting } hb_2$$
$$\frac{2A - hb_2}{h} = b_1. \qquad \text{Dividing by } h$$

$$\boxed{\textbf{Now Try Exercise 65.}}$$

## R.5  Exercise Set

*Solve.*

**1.** $x - 5 = 7$

**2.** $y + 3 = 4$

**3.** $3x + 4 = -8$

**4.** $5x - 7 = 23$

**5.** $5y - 12 = 3$

**6.** $6x + 23 = 5$

**7.** $6x - 15 = 45$

**8.** $4x - 7 = 81$

**9.** $5x - 10 = 45$

**10.** $6x - 7 = 11$

**11.** $9t + 4 = -5$

**12.** $5x + 7 = -13$

**13.** $8x + 48 = 3x - 12$

**14.** $15x + 40 = 8x - 9$

**15.** $7y - 1 = 23 - 5y$

**16.** $3x - 15 = 15 - 3x$

**17.** $3x - 4 = 5 + 12x$

**18.** $9t - 4 = 14 + 15t$

**19.** $5 - 4a = a - 13$

**20.** $6 - 7x = x - 14$

**21.** $3m - 7 = -13 + m$

**22.** $5x - 8 = 2x - 8$

**23.** $11 - 3x = 5x + 3$

**24.** $20 - 4y = 10 - 6y$

**25.** $2(x + 7) = 5x + 14$

**26.** $3(y + 4) = 8y$

**27.** $24 = 5(2t + 5)$    **28.** $9 = 4(3y - 2)$

**29.** $5y - 4(2y - 10) = 25$

**30.** $8x - 2(3x - 5) = 40$

**31.** $7(3x + 6) = 11 - (x + 2)$

**32.** $9(2x + 8) = 20 - (x + 5)$

**33.** $4(3y - 1) - 6 = 5(y + 2)$

**34.** $3(2n - 5) - 7 = 4(n - 9)$

**35.** $x^2 + 3x - 28 = 0$    **36.** $y^2 - 4y - 45 = 0$

**37.** $x^2 + 5x = 0$    **38.** $t^2 + 6t = 0$

**39.** $y^2 + 6y + 9 = 0$    **40.** $n^2 + 4n + 4 = 0$

**41.** $x^2 + 100 = 20x$    **42.** $y^2 + 25 = 10y$

**43.** $x^2 - 4x - 32 = 0$    **44.** $t^2 + 12t + 27 = 0$

**45.** $3y^2 + 8y + 4 = 0$    **46.** $9y^2 + 15y + 4 = 0$

**47.** $12z^2 + z = 6$    **48.** $6x^2 - 7x = 10$

**49.** $12a^2 - 28 = 5a$    **50.** $21n^2 - 10 = n$

**51.** $14 = x(x - 5)$    **52.** $24 = x(x - 2)$

**53.** $x^2 - 36 = 0$    **54.** $y^2 - 81 = 0$

**55.** $z^2 = 144$    **56.** $t^2 = 25$

**57.** $2x^2 - 20 = 0$    **58.** $3y^2 - 15 = 0$

**59.** $6z^2 - 18 = 0$    **60.** $5x^2 - 75 = 0$

*Solve.*

**61.** $A = \frac{1}{2}bh$, for $b$
(Area of a triangle)

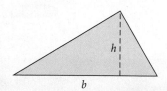

**62.** $A = \pi r^2$, for $\pi$
(Area of a circle)

**63.** $P = 2l + 2w$, for $w$
(Perimeter of a rectangle)

**64.** $A = P + Prt$, for $r$
(Simple interest)

**65.** $A = \frac{1}{2}h(b_1 + b_2)$, for $b_2$
(Area of a trapezoid)

**66.** $A = \frac{1}{2}h(b_1 + b_2)$, for $h$

**67.** $V = \frac{4}{3}\pi r^3$, for $\pi$
(Volume of a sphere)

**68.** $V = \frac{4}{3}\pi r^3$, for $r^3$

**69.** $F = \frac{9}{5}C + 32$, for $C$
(Temperature conversion)

**70.** $Ax + By = C$, for $y$
(Standard linear equation)

**71.** $Ax + By = C$, for $A$

**72.** $2w + 2h + l = p$, for $w$

**73.** $2w + 2h + l = p$, for $h$

**74.** $3x + 4y = 12$, for $y$

**75.** $2x - 3y = 6$, for $y$

**76.** $T = \frac{3}{10}(I - 12{,}000)$, for $I$

**77.** $a = b + bcd$, for $b$

**78.** $q = p - np$, for $p$

**79.** $z = xy - xy^2$, for $x$

**80.** $st = t - 4$, for $t$

## Synthesis

*Solve.*

**81.** $3[5 - 3(4 - t)] - 2 = 5[3(5t - 4) + 8] - 26$

**82.** $6[4(8 - y) - 5(9 + 3y)] - 21 = -7[3(7 + 4y) - 4]$

**83.** $x - \{3x - [2x - (5x - (7x - 1))]\} = x + 7$

**84.** $23 - 2[4 + 3(x - 1)] + 5[x - 2(x + 3)] = 7\{x - 2[5 - (2x + 3)]\}$

**85.** $(5x^2 + 6x)(12x^2 - 5x - 2) = 0$

**86.** $(3x^2 + 7x - 20)(x^2 - 4x) = 0$

**87.** $3x^3 + 6x^2 - 27x - 54 = 0$

**88.** $2x^3 + 6x^2 = 8x + 24$

**R.6**

# Rational Expressions

- Determine the domain of a rational expression.
- Simplify rational expressions.
- Multiply, divide, add, and subtract rational expressions.
- Simplify complex rational expressions.

A **rational expression** is the quotient of two polynomials. For example,

$$\frac{3}{5}, \quad \frac{2}{x-3}, \quad \text{and} \quad \frac{x^2-4}{x^2-4x-5}$$

are rational expressions.

## ■ The Domain of a Rational Expression

The **domain** of an algebraic expression is the set of all real numbers for which the expression is defined. Since division by 0 is not defined, any number that makes the denominator 0 is not in the domain of a rational expression.

**EXAMPLE 1**    Find the domain of each of the following.

a) $\dfrac{2}{x-3}$
    b) $\dfrac{x^2-4}{x^2-4x-5}$

**Solution**

a) We solve the equation $x - 3 = 0$ to determine the numbers that are *not* in the domain:

$$x - 3 = 0$$
$$x = 3. \qquad \text{Adding 3 on both sides}$$

Since the denominator is 0 when $x = 3$, the domain of $2/(x - 3)$ is the set of all real numbers except 3.

b) We solve the equation $x^2 - 4x - 5 = 0$ to find the numbers that are not in the domain:

$$x^2 - 4x - 5 = 0$$
$$(x + 1)(x - 5) = 0 \qquad \text{Factoring}$$
$$x + 1 = 0 \quad or \quad x - 5 = 0 \qquad \text{Using the principle of zero products}$$
$$x = -1 \quad or \qquad x = 5.$$

Since the denominator is 0 when $x = -1$ or $x = 5$, the domain is the set of all real numbers except $-1$ and 5.

**Now Try Exercise 3.**

We can describe the domains found in Example 1 using *set-builder notation*. For example, we write "The set of all real numbers $x$ such that $x$ is not equal to 3"as

$$\{x \mid x \text{ is a real number } and\ x \neq 3\}.$$

Similarly, we write "The set of all real numbers $x$ such that $x$ is not equal to $-1$ and $x$ is not equal to 5" as

$$\{x \mid x \text{ is a real number } and\ x \neq -1\ and\ x \neq 5\}.$$

## ■ Simplifying, Multiplying, and Dividing Rational Expressions

To simplify rational expressions, we use the fact that

$$\frac{a \cdot c}{b \cdot c} = \frac{a}{b} \cdot \frac{c}{c} = \frac{a}{b} \cdot 1 = \frac{a}{b}.$$

**EXAMPLE 2**  Simplify: $\dfrac{9x^2 + 6x - 3}{12x^2 - 12}$.

*Solution*

$$\begin{aligned}
\frac{9x^2 + 6x - 3}{12x^2 - 12} &= \frac{3(3x^2 + 2x - 1)}{12(x^2 - 1)} \\
&= \frac{3(x + 1)(3x - 1)}{3 \cdot 4(x + 1)(x - 1)}
\end{aligned}$$

$\left.\phantom{\begin{aligned}a\\b\end{aligned}}\right\}$ Factoring the numerator and the denominator

$$= \frac{3(x + 1)}{3(x + 1)} \cdot \frac{3x - 1}{4(x - 1)}$$

Factoring the rational expression

$$= 1 \cdot \frac{3x - 1}{4(x - 1)}$$

$\dfrac{3(x + 1)}{3(x + 1)} = 1$

$$= \frac{3x - 1}{4(x - 1)}$$

Removing a factor of 1

**Now Try Exercise 9.**

Canceling is a shortcut that is often used to remove a factor of 1.

**EXAMPLE 3**  Simplify each of the following.

a) $\dfrac{4x^3 + 16x^2}{2x^3 + 6x^2 - 8x}$

b) $\dfrac{2 - x}{x^2 + x - 6}$

*Solution*

a)
$$\begin{aligned}
\frac{4x^3 + 16x^2}{2x^3 + 6x^2 - 8x} &= \frac{2 \cdot 2 \cdot x \cdot x(x + 4)}{2 \cdot x(x + 4)(x - 1)}
\end{aligned}$$

Factoring the numerator and the denominator

$$= \frac{2 \cdot 2 \cdot x \cdot x(x + 4)}{2 \cdot x(x + 4)(x - 1)}$$

Removing a factor of 1:
$\dfrac{2x(x + 4)}{2x(x + 4)} = 1$

$$= \frac{2x}{x - 1}$$

**b)** $\dfrac{2-x}{x^2+x-6} = \dfrac{2-x}{(x+3)(x-2)}$   Factoring the denominator

$$= \dfrac{-1(x-2)}{(x+3)(x-2)} \qquad 2 - x = -1(x-2)$$

$$= \dfrac{-1(x\cancel{-2})}{(x+3)(x\cancel{-2})} \qquad \text{Removing a factor of 1: } \dfrac{x-2}{x-2} = 1$$

$$= \dfrac{-1}{x+3}, \text{ or } -\dfrac{1}{x+3}$$

**Now Try Exercises 11 and 15.**

In Example 3(b), we saw that

$$\dfrac{2-x}{x^2+x-6} \quad \text{and} \quad -\dfrac{1}{x+3}$$

are **equivalent expressions**. This means that they have the same value for all numbers that are in *both* domains. Note that $-3$ is not in the domain of *either* expression, whereas 2 is in the domain of $-1/(x+3)$ but not in the domain of $(2-x)/(x^2+x-6)$ and thus is not in the domain of *both* expressions.

To multiply rational expressions, we multiply numerators and multiply denominators and, if possible, simplify the result. To divide rational expressions, we multiply the dividend by the reciprocal of the divisor and, if possible, simplify the result; that is,

$$\dfrac{a}{b} \cdot \dfrac{c}{d} = \dfrac{ac}{bd} \quad \text{and} \quad \dfrac{a}{b} \div \dfrac{c}{d} = \dfrac{a}{b} \cdot \dfrac{d}{c} = \dfrac{ad}{bc}.$$

**EXAMPLE 4**   Multiply or divide and simplify each of the following.

**a)** $\dfrac{a^2-4}{16a} \cdot \dfrac{20a^2}{a+2}$ 

**b)** $\dfrac{x+4}{2x^2-6x} \cdot \dfrac{x^4-9x^2}{x^2+2x-8}$

**c)** $\dfrac{x-2}{12} \div \dfrac{x^2-4x+4}{3x^3+15x^2}$ 

**d)** $\dfrac{y^3-1}{y^2-1} \div \dfrac{y^2+y+1}{y^2+2y+1}$

*Solution*

**a)** $\dfrac{a^2-4}{16a} \cdot \dfrac{20a^2}{a+2} = \dfrac{(a^2-4)(20a^2)}{16a(a+2)}$   Multiplying the numerators and the denominators

$$= \dfrac{\cancel{(a+2)}(a-2) \cdot \cancel{4} \cdot 5 \cdot a \cdot \cancel{a}}{\cancel{4} \cdot 4 \cdot \cancel{a} \cdot \cancel{(a+2)}}$$   Factoring and removing a factor of 1

$$= \dfrac{5a(a-2)}{4}$$

**b)** $\dfrac{x+4}{2x^2-6x} \cdot \dfrac{x^4-9x^2}{x^2+2x-8}$

$$= \frac{(x+4)(x^4-9x^2)}{(2x^2-6x)(x^2+2x-8)} \qquad \text{Multiplying the numerators and the denominators}$$

$$= \frac{(x+4)(x^2)(x^2-9)}{2x(x-3)(x+4)(x-2)} \qquad \text{Factoring}$$

$$= \frac{(x+4) \cdot x \cdot x \cdot (x+3)(x-3)}{2 \cdot x \cdot (x-3)(x+4)(x-2)} \qquad \text{Factoring further and removing a factor of 1}$$

$$= \frac{x(x+3)}{2(x-2)}$$

**c)** $\dfrac{x-2}{12} \div \dfrac{x^2-4x+4}{3x^3+15x^2}$

$$= \frac{x-2}{12} \cdot \frac{3x^3+15x^2}{x^2-4x+4} \qquad \text{Multiplying by the reciprocal of the divisor}$$

$$= \frac{(x-2)(3x^3+15x^2)}{12(x^2-4x+4)}$$

$$= \frac{(x-2)(3)(x^2)(x+5)}{3 \cdot 4(x-2)(x-2)} \qquad \text{Factoring and removing a factor of 1}$$

$$= \frac{x^2(x+5)}{4(x-2)}$$

**d)** $\dfrac{y^3-1}{y^2-1} \div \dfrac{y^2+y+1}{y^2+2y+1}$

$$= \frac{y^3-1}{y^2-1} \cdot \frac{y^2+2y+1}{y^2+y+1} \qquad \text{Multiplying by the reciprocal of the divisor}$$

$$= \frac{(y^3-1)(y^2+2y+1)}{(y^2-1)(y^2+y+1)}$$

$$= \frac{(y-1)(y^2+y+1)(y+1)(y+1)}{(y+1)(y-1)(y^2+y+1)} \qquad \text{Factoring and removing a factor of 1}$$

$$= y+1 \qquad \boxed{\text{Now Try Exercises 19 and 25.}}$$

## ■ Adding and Subtracting Rational Expressions

When rational expressions have the same denominator, we can add or subtract by adding or subtracting the numerators and retaining the common denominator. If the denominators differ, we must find equivalent rational expressions that have a common denominator. In general, it is most efficient to find the **least common denominator (LCD)** of the expressions.

To find the least common denominator of rational expressions, factor each denominator and form the product that uses each factor the greatest number of times it occurs in any factorization.

**EXAMPLE 5** Add or subtract and simplify each of the following.

a) $\dfrac{x^2 - 4x + 4}{2x^2 - 3x + 1} + \dfrac{x + 4}{2x - 2}$

b) $\dfrac{x - 2}{x^2 - 25} - \dfrac{x + 5}{x^2 - 5x}$

c) $\dfrac{x}{x^2 + 11x + 30} - \dfrac{5}{x^2 + 9x + 20}$

*Solution*

a) $\dfrac{x^2 - 4x + 4}{2x^2 - 3x + 1} + \dfrac{x + 4}{2x - 2}$

$= \dfrac{x^2 - 4x + 4}{(2x - 1)(x - 1)} + \dfrac{x + 4}{2(x - 1)}$   Factoring the denominators

The LCD is $(2x - 1)(x - 1)(2)$, or $2(2x - 1)(x - 1)$.

$= \dfrac{x^2 - 4x + 4}{(2x - 1)(x - 1)} \cdot \dfrac{2}{2} + \dfrac{x + 4}{2(x - 1)} \cdot \dfrac{2x - 1}{2x - 1}$   Multiplying each term by 1 to get the LCD

$= \dfrac{2x^2 - 8x + 8}{(2x - 1)(x - 1)(2)} + \dfrac{2x^2 + 7x - 4}{2(x - 1)(2x - 1)}$

$= \dfrac{4x^2 - x + 4}{2(2x - 1)(x - 1)}$   Adding the numerators and keeping the common denominator. We cannot simplify.

b) $\dfrac{x - 2}{x^2 - 25} - \dfrac{x + 5}{x^2 - 5x}$

$= \dfrac{x - 2}{(x + 5)(x - 5)} - \dfrac{x + 5}{x(x - 5)}$   Factoring the denominators

The LCD is $(x + 5)(x - 5)(x)$, or $x(x + 5)(x - 5)$.

$= \dfrac{x - 2}{(x + 5)(x - 5)} \cdot \dfrac{x}{x} - \dfrac{x + 5}{x(x - 5)} \cdot \dfrac{x + 5}{x + 5}$   Multiplying each term by 1 to get the LCD

$= \dfrac{x^2 - 2x}{x(x + 5)(x - 5)} - \dfrac{x^2 + 10x + 25}{x(x + 5)(x - 5)}$

$= \dfrac{x^2 - 2x - (x^2 + 10x + 25)}{x(x + 5)(x - 5)}$   Subtracting the numerators and keeping the common denominator

$= \dfrac{x^2 - 2x - x^2 - 10x - 25}{x(x + 5)(x - 5)}$   Removing parentheses

$= \dfrac{-12x - 25}{x(x + 5)(x - 5)}$

c) $\dfrac{x}{x^2 + 11x + 30} - \dfrac{5}{x^2 + 9x + 20}$

$= \dfrac{x}{(x + 5)(x + 6)} - \dfrac{5}{(x + 5)(x + 4)}$   Factoring the denominators

The LCD is $(x + 5)(x + 6)(x + 4)$.

$= \dfrac{x}{(x + 5)(x + 6)} \cdot \dfrac{x + 4}{x + 4} - \dfrac{5}{(x + 5)(x + 4)} \cdot \dfrac{x + 6}{x + 6}$

Multiplying each term by 1 to get the LCD

$= \dfrac{x^2 + 4x}{(x + 5)(x + 6)(x + 4)} - \dfrac{5x + 30}{(x + 5)(x + 4)(x + 6)}$

$= \dfrac{x^2 + 4x - 5x - 30}{(x + 5)(x + 6)(x + 4)}$   Be sure to change the sign of *every* term in the numerator of the expression being subtracted: $-(5x + 30) = -5x - 30$

$= \dfrac{x^2 - x - 30}{(x + 5)(x + 6)(x + 4)}$

$= \dfrac{(x + 5)(x - 6)}{(x + 5)(x + 6)(x + 4)}$   Factoring and removing a factor of 1: $\dfrac{x + 5}{x + 5} = 1$

$= \dfrac{x - 6}{(x + 6)(x + 4)}$   **Now Try Exercises 37 and 39.**

## ◼ Complex Rational Expressions

A **complex rational expression** has rational expressions in its numerator or its denominator or both.

> To simplify a complex rational expression:
>
> *Method 1.* Find the LCD of all the denominators within the complex rational expression. Then multiply by 1 using the LCD as the numerator and the denominator of the expression for 1.
>
> *Method 2.* First add or subtract, if necessary, to get a single rational expression in the numerator and in the denominator. Then divide by multiplying by the reciprocal of the denominator.

**EXAMPLE 6**   Simplify: $\dfrac{\dfrac{1}{a} + \dfrac{1}{b}}{\dfrac{1}{a^3} + \dfrac{1}{b^3}}$.

*Solution*

**Method 1.** The LCD of the four rational expressions in the numerator and the denominator is $a^3b^3$.

$$\frac{\dfrac{1}{a} + \dfrac{1}{b}}{\dfrac{1}{a^3} + \dfrac{1}{b^3}} = \frac{\dfrac{1}{a} + \dfrac{1}{b}}{\dfrac{1}{a^3} + \dfrac{1}{b^3}} \cdot \frac{a^3b^3}{a^3b^3} \qquad \text{Multiplying by 1 using } \frac{a^3b^3}{a^3b^3}$$

$$= \frac{\left(\dfrac{1}{a} + \dfrac{1}{b}\right)(a^3b^3)}{\left(\dfrac{1}{a^3} + \dfrac{1}{b^3}\right)(a^3b^3)}$$

$$= \frac{a^2b^3 + a^3b^2}{b^3 + a^3}$$

$$= \frac{a^2b^2(b + a)}{(b + a)(b^2 - ba + a^2)} \qquad \text{Factoring and removing a}$$
$$\text{factor of 1: } \frac{b + a}{b + a} = 1$$

$$= \frac{a^2b^2}{b^2 - ba + a^2}$$

**Method 2.** We add in the numerator and in the denominator.

$$\frac{\dfrac{1}{a} + \dfrac{1}{b}}{\dfrac{1}{a^3} + \dfrac{1}{b^3}} = \frac{\dfrac{1}{a} \cdot \dfrac{b}{b} + \dfrac{1}{b} \cdot \dfrac{a}{a}}{\dfrac{1}{a^3} \cdot \dfrac{b^3}{b^3} + \dfrac{1}{b^3} \cdot \dfrac{a^3}{a^3}} \quad \leftarrow \text{The LCD is } ab.$$
$$\leftarrow \text{The LCD is } a^3b^3.$$

$$= \frac{\dfrac{b}{ab} + \dfrac{a}{ab}}{\dfrac{b^3}{a^3b^3} + \dfrac{a^3}{a^3b^3}}$$

$$= \frac{\dfrac{b + a}{ab}}{\dfrac{b^3 + a^3}{a^3b^3}} \qquad \begin{array}{l}\text{We have a single rational expression}\\ \text{in both the numerator and the}\\ \text{denominator.}\end{array}$$

$$= \frac{b + a}{ab} \cdot \frac{a^3b^3}{b^3 + a^3} \qquad \begin{array}{l}\text{Multiplying by the reciprocal of}\\ \text{the denominator}\end{array}$$

$$= \frac{(b + a)(a)(b)(a^2b^2)}{(a)(b)(b + a)(b^2 - ba + a^2)}$$

$$= \frac{a^2b^2}{b^2 - ba + a^2}$$

**Now Try Exercise 57.**

## R.6 Exercise Set

*Find the domain of the rational expression.*

**1.** $-\dfrac{5}{3}$

**2.** $\dfrac{4}{7 - x}$

**3.** $\dfrac{3x - 3}{x(x - 1)}$

**4.** $\dfrac{15x - 10}{2x(3x - 2)}$

**5.** $\dfrac{x + 5}{x^2 + 4x - 5}$

**6.** $\dfrac{(x^2 - 4)(x + 1)}{(x + 2)(x^2 - 1)}$

**7.** $\dfrac{7x^2 - 28x + 28}{(x^2 - 4)(x^2 + 3x - 10)}$

**8.** $\dfrac{7x^2 + 11x - 6}{x(x^2 - x - 6)}$

*Simplify.*

**9.** $\dfrac{x^2 - 4}{x^2 - 4x + 4}$

**10.** $\dfrac{x^2 + 2x - 3}{x^2 - 9}$

**11.** $\dfrac{x^3 - 6x^2 + 9x}{x^3 - 3x^2}$

**12.** $\dfrac{y^5 - 5y^4 + 4y^3}{y^3 - 6y^2 + 8y}$

**13.** $\dfrac{6y^2 + 12y - 48}{3y^2 - 9y + 6}$

**14.** $\dfrac{2x^2 - 20x + 50}{10x^2 - 30x - 100}$

**15.** $\dfrac{4 - x}{x^2 + 4x - 32}$

**16.** $\dfrac{6 - x}{x^2 - 36}$

*Multiply or divide and, if possible, simplify.*

**17.** $\dfrac{r - s}{r + s} \cdot \dfrac{r^2 - s^2}{(r - s)^2}$

**18.** $\dfrac{x^2 - y^2}{(x - y)^2} \cdot \dfrac{1}{x + y}$

**19.** $\dfrac{x^2 + 2x - 35}{3x^3 - 2x^2} \cdot \dfrac{9x^3 - 4x}{7x + 49}$

**20.** $\dfrac{x^2 - 2x - 35}{2x^3 - 3x^2} \cdot \dfrac{4x^3 - 9x}{7x - 49}$

**21.** $\dfrac{4x^2 + 9x + 2}{x^2 + x - 2} \cdot \dfrac{x^2 - 1}{3x^2 + x - 2}$

**22.** $\dfrac{2a^2 + 5a - 3}{a^2 + 8a + 15} \cdot \dfrac{a^2 + 4a - 5}{3a^2 - a - 2}$

**23.** $\dfrac{m^2 - n^2}{r + s} \div \dfrac{m - n}{r + s}$

**24.** $\dfrac{a^2 - b^2}{x - y} \div \dfrac{a + b}{x - y}$

**25.** $\dfrac{3x + 12}{2x - 8} \div \dfrac{(x + 4)^2}{(x - 4)^2}$

**26.** $\dfrac{a^2 - a - 2}{a^2 - a - 6} \div \dfrac{a^2 - 2a}{2a + a^2}$

**27.** $\dfrac{x^2 - y^2}{x^3 - y^3} \cdot \dfrac{x^2 + xy + y^2}{x^2 + 2xy + y^2}$

**28.** $\dfrac{c^3 + 8}{c^2 - 4} \div \dfrac{c^2 - 2c + 4}{c^2 - 4c + 4}$

**29.** $\dfrac{(x - y)^2 - z^2}{(x + y)^2 - z^2} \div \dfrac{x - y + z}{x + y - z}$

**30.** $\dfrac{(a + b)^2 - 9}{(a - b)^2 - 9} \cdot \dfrac{a - b - 3}{a + b + 3}$

*Add or subtract and, if possible, simplify.*

**31.** $\dfrac{7}{5x} + \dfrac{3}{5x}$

**32.** $\dfrac{7}{12y} - \dfrac{1}{12y}$

**33.** $\dfrac{4}{3a + 4} + \dfrac{3a}{3a + 4}$

**34.** $\dfrac{a - 3b}{a + b} + \dfrac{a + 5b}{a + b}$

**35.** $\dfrac{5}{4z} - \dfrac{3}{8z}$

**36.** $\dfrac{12}{x^2 y} + \dfrac{5}{xy^2}$

**37.** $\dfrac{3}{x + 2} + \dfrac{2}{x^2 - 4}$

**38.** $\dfrac{5}{a - 3} - \dfrac{2}{a^2 - 9}$

**39.** $\dfrac{y}{y^2 - y - 20} - \dfrac{2}{y + 4}$

**40.** $\dfrac{6}{y^2 + 6y + 9} - \dfrac{5}{y + 3}$

**41.** $\dfrac{3}{x + y} + \dfrac{x - 5y}{x^2 - y^2}$

**42.** $\dfrac{a^2 + 1}{a^2 - 1} - \dfrac{a - 1}{a + 1}$

**43.** $\dfrac{y}{y - 1} + \dfrac{2}{1 - y}$

(*Note:* $1 - y = -1(y - 1)$.)

**44.** $\dfrac{a}{a - b} + \dfrac{b}{b - a}$

(*Note:* $b - a = -1(a - b)$.)

**45.** $\dfrac{x}{2x - 3y} - \dfrac{y}{3y - 2x}$

**46.** $\dfrac{3a}{3a - 2b} - \dfrac{2a}{2b - 3a}$

**47.** $\dfrac{9x + 2}{3x^2 - 2x - 8} + \dfrac{7}{3x^2 + x - 4}$

**48.** $\dfrac{3y}{y^2 - 7y + 10} - \dfrac{2y}{y^2 - 8y + 15}$

**49.** $\dfrac{5a}{a - b} + \dfrac{ab}{a^2 - b^2} + \dfrac{4b}{a + b}$

**50.** $\dfrac{6a}{a - b} - \dfrac{3b}{b - a} + \dfrac{5}{a^2 - b^2}$

**51.** $\dfrac{7}{x + 2} - \dfrac{x + 8}{4 - x^2} + \dfrac{3x - 2}{4 - 4x + x^2}$

**52.** $\dfrac{6}{x + 3} - \dfrac{x + 4}{9 - x^2} + \dfrac{2x - 3}{9 - 6x + x^2}$

**53.** $\dfrac{1}{x + 1} + \dfrac{x}{2 - x} + \dfrac{x^2 + 2}{x^2 - x - 2}$

**54.** $\dfrac{x - 1}{x - 2} - \dfrac{x + 1}{x + 2} - \dfrac{x - 6}{4 - x^2}$

*Simplify.*

**55.** $\dfrac{\dfrac{a - b}{b}}{\dfrac{a^2 - b^2}{ab}}$

**56.** $\dfrac{\dfrac{x^2 - y^2}{xy}}{\dfrac{x - y}{y}}$

**57.** $\dfrac{\dfrac{x}{y} - \dfrac{y}{x}}{\dfrac{1}{y} + \dfrac{1}{x}}$

**58.** $\dfrac{\dfrac{a}{b} - \dfrac{b}{a}}{\dfrac{1}{a} - \dfrac{1}{b}}$

**59.** $\dfrac{c + \dfrac{8}{c^2}}{1 + \dfrac{2}{c}}$

**60.** $\dfrac{a - \dfrac{a}{b}}{b - \dfrac{b}{a}}$

**61.** $\dfrac{x^2 + xy + y^2}{\dfrac{x^2}{y} - \dfrac{y^2}{x}}$

**62.** $\dfrac{\dfrac{a^2}{b} + \dfrac{b^2}{a}}{a^2 - ab + b^2}$

**63.** $\dfrac{a - a^{-1}}{a + a^{-1}}$

**64.** $\dfrac{x^{-1} + y^{-1}}{x^{-3} + y^{-3}}$

**65.** $\dfrac{\dfrac{1}{x - 3} + \dfrac{2}{x + 3}}{\dfrac{3}{x - 1} - \dfrac{4}{x + 2}}$

**66.** $\dfrac{\dfrac{5}{x + 1} - \dfrac{3}{x - 2}}{\dfrac{1}{x - 5} + \dfrac{2}{x + 2}}$

**67.** $\dfrac{\dfrac{a}{1 - a} + \dfrac{1 + a}{a}}{\dfrac{1 - a}{a} + \dfrac{a}{1 + a}}$

**68.** $\dfrac{\dfrac{1 - x}{x} + \dfrac{x}{1 + x}}{\dfrac{1 + x}{x} + \dfrac{x}{1 - x}}$

**69.** $\dfrac{\dfrac{1}{a^2} + \dfrac{2}{ab} + \dfrac{1}{b^2}}{\dfrac{1}{a^2} - \dfrac{1}{b^2}}$

**70.** $\dfrac{\dfrac{1}{x^2} - \dfrac{1}{y^2}}{\dfrac{1}{x^2} - \dfrac{2}{xy} + \dfrac{1}{y^2}}$

## Synthesis

*Simplify.*

**71.** $\dfrac{(x + h)^2 - x^2}{h}$

**72.** $\dfrac{\dfrac{1}{x + h} - \dfrac{1}{x}}{h}$

**73.** $\dfrac{(x + h)^3 - x^3}{h}$

**74.** $\dfrac{\dfrac{1}{(x + h)^2} - \dfrac{1}{x^2}}{h}$

**75.** $\left[ \dfrac{\dfrac{x + 1}{x - 1} + 1}{\dfrac{x + 1}{x - 1} - 1} \right]^5$

**76.** $1 + \dfrac{1}{1 + \dfrac{1}{1 + \dfrac{1}{1 + \dfrac{1}{x}}}}$

*Perform the indicated operations and, if possible, simplify.*

**77.** $\dfrac{n(n + 1)(n + 2)}{2 \cdot 3} + \dfrac{(n + 1)(n + 2)}{2}$

**78.** $\dfrac{n(n + 1)(n + 2)(n + 3)}{2 \cdot 3 \cdot 4} + \dfrac{(n + 1)(n + 2)(n + 3)}{2 \cdot 3}$

**79.** $\dfrac{x^2 - 9}{x^3 + 27} \cdot \dfrac{5x^2 - 15x + 45}{x^2 - 2x - 3} + \dfrac{x^2 + x}{4 + 2x}$

**80.** $\dfrac{x^2 + 2x - 3}{x^2 - x - 12} \div \dfrac{x^2 - 1}{x^2 - 16} - \dfrac{2x + 1}{x^2 + 2x + 1}$

## R.7 Radical Notation and Rational Exponents

- Simplify radical expressions.
- Rationalize denominators or numerators in rational expressions.
- Convert between exponential notation and radical notation.
- Simplify expressions with rational exponents.

A number $c$ is said to be a **square root** of $a$ if $c^2 = a$. Thus, 3 is a square root of 9, because $3^2 = 9$, and $-3$ is also a square root of 9, because $(-3)^2 = 9$. Similarly, 5 is a third root (called a **cube root**) of 125, because $5^3 = 125$. The number 125 has no other real-number cube root.

> ***n*th ROOT**
>
> A number $c$ is said to be an ***n*th root** of $a$ if $c^n = a$.

The symbol $\sqrt{a}$ denotes the nonnegative square root of $a$, and the symbol $\sqrt[3]{a}$ denotes the real-number cube root of $a$. The symbol $\sqrt[n]{a}$ denotes the $n$th root of $a$; that is, a number whose $n$th power is $a$. The symbol $\sqrt[n]{\phantom{x}}$ is called a **radical**, and the expression under the radical is called the **radicand**. The number $n$ (which is omitted when it is 2) is called the **index**. Examples of roots for $n = 3, 4$, and 2, respectively, are

$$\sqrt[3]{125}, \quad \sqrt[4]{16}, \quad \text{and} \quad \sqrt{3600}.$$

Any real number has only one real-number odd root. Any positive number has two square roots, one positive and one negative. Similarly, for any even index, a positive number has two real-number roots. The positive root is called the **principal root**. When an expression such as $\sqrt{4}$ or $\sqrt[6]{23}$ is used, it is understood to represent the principal (nonnegative) root. To denote a negative root, we use $-\sqrt{4}, -\sqrt[6]{23}$, and so on.

GCM **EXAMPLE 1** Simplify each of the following.

a) $\sqrt{36}$        b) $-\sqrt{36}$        c) $\sqrt[3]{-8}$

d) $\sqrt[5]{\dfrac{32}{243}}$        e) $\sqrt[4]{-16}$

***Solution***

a) $\sqrt{36} = 6$, because $6^2 = 36$.

b) $-\sqrt{36} = -6$, because $6^2 = 36$ and $-(\sqrt{36}) = -(6) = -6$.

c) $\sqrt[3]{-8} = -2$, because $(-2)^3 = -8$.

d) $\sqrt[5]{\dfrac{32}{243}} = \dfrac{2}{3}$, because $\left(\dfrac{2}{3}\right)^5 = \dfrac{2^5}{3^5} = \dfrac{32}{243}$.

e) $\sqrt[4]{-16}$ is not a real number, because we cannot find a real number that can be raised to the fourth power to get $-16$.    ▸ **Now Try Exercise 11.**

We can generalize Example 1(e) and say that when $a$ is negative and $n$ is even, $\sqrt[n]{a}$ is not a real number. For example, $\sqrt{-4}$ and $\sqrt[4]{-81}$ are not real numbers.

We can find $\sqrt{36}$ and $-\sqrt{36}$ in Example 1 using the square-root feature on the keypad of a graphing calculator, and we can use the cube-root feature to find $\sqrt[3]{-8}$. We can use the $x$th-root feature to find higher roots.

   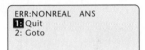

When we try to find $\sqrt[4]{-16}$ on a graphing calculator set in REAL mode, we get an error message indicating that the answer is nonreal.

## ■ Simplifying Radical Expressions

Consider the expression $\sqrt{(-3)^2}$. This is equivalent to $\sqrt{9}$, or 3. Similarly, $\sqrt{3^2} = \sqrt{9} = 3$. This illustrates the first of several properties of radicals, listed below.

---

### PROPERTIES OF RADICALS

Let $a$ and $b$ be any real numbers or expressions for which the given roots exist. For any natural numbers $m$ and $n$ ($n \neq 1$):

1. If $n$ is even, $\sqrt[n]{a^n} = |a|$.
2. If $n$ is odd, $\sqrt[n]{a^n} = a$.
3. $\sqrt[n]{a} \cdot \sqrt[n]{b} = \sqrt[n]{ab}$.
4. $\sqrt[n]{\dfrac{a}{b}} = \dfrac{\sqrt[n]{a}}{\sqrt[n]{b}}$   ($b \neq 0$).
5. $\sqrt[n]{a^m} = (\sqrt[n]{a})^m$.

---

**EXAMPLE 2**   Simplify each of the following.

a) $\sqrt{(-5)^2}$   b) $\sqrt[3]{(-5)^3}$   c) $\sqrt[4]{4} \cdot \sqrt[4]{5}$   d) $\sqrt{50}$

e) $\dfrac{\sqrt{72}}{\sqrt{6}}$   f) $\sqrt[3]{8^5}$   g) $\sqrt{216x^5y^3}$   h) $\sqrt{\dfrac{x^2}{16}}$

*Solution*

a) $\sqrt{(-5)^2} = |-5| = 5$   **Using Property 1**

b) $\sqrt[3]{(-5)^3} = -5$   **Using Property 2**

c) $\sqrt[4]{4} \cdot \sqrt[4]{5} = \sqrt[4]{4 \cdot 5} = \sqrt[4]{20}$   **Using Property 3**

**d)** $\sqrt{50} = \sqrt{25 \cdot 2} = \sqrt{25} \cdot \sqrt{2} = 5\sqrt{2}$    **Using Property 3**

**e)** $\dfrac{\sqrt{72}}{\sqrt{6}} = \sqrt{\dfrac{72}{6}} = \sqrt{12}$    **Using Property 4**

$\phantom{\dfrac{\sqrt{72}}{\sqrt{6}}} = \sqrt{4 \cdot 3} = \sqrt{4} \cdot \sqrt{3}$    **Using Property 3**

$\phantom{\dfrac{\sqrt{72}}{\sqrt{6}}} = 2\sqrt{3}$

**f)** $\sqrt[3]{8^5} = (\sqrt[3]{8})^5$    **Using Property 5**

$\phantom{\sqrt[3]{8^5}} = 2^5 = 32$

**g)** $\sqrt{216x^5y^3} = \sqrt{36 \cdot 6 \cdot x^4 \cdot x \cdot y^2 \cdot y}$

$\phantom{\sqrt{216x^5y^3}} = \sqrt{36x^4y^2}\sqrt{6xy}$    **Using Property 3**

$\phantom{\sqrt{216x^5y^3}} = |6x^2y|\sqrt{6xy}$    **Using Property 1**

$\phantom{\sqrt{216x^5y^3}} = 6x^2|y|\sqrt{6xy}$    $6x^2$ cannot be negative, so absolute-value bars are not needed for it.

**h)** $\sqrt{\dfrac{x^2}{16}} = \dfrac{\sqrt{x^2}}{\sqrt{16}}$    **Using Property 4**

$\phantom{\sqrt{\dfrac{x^2}{16}}} = \dfrac{|x|}{4}$    **Using Property 1**

> **Now Try Exercise 19.**

In many situations, radicands are never formed by raising negative quantities to even powers. In such cases, absolute-value notation is not required. For this reason, **we will henceforth assume that no radicands are formed by raising negative quantities to even powers** and, consequently, we will not use absolute-value notation when we simplify radical expressions. For example, we will write $\sqrt{x^2} = x$ and $\sqrt[4]{a^5b} = a\sqrt[4]{ab}$.

Radical expressions with the same index and the same radicand can be combined (added or subtracted) in much the same way that we combine like terms.

**EXAMPLE 3**   Perform the operations indicated.

**a)** $3\sqrt{8x^2} - 5\sqrt{2x^2}$ **b)** $(4\sqrt{3} + \sqrt{2})(\sqrt{3} - 5\sqrt{2})$

*Solution*

**a)** $3\sqrt{8x^2} - 5\sqrt{2x^2} = 3\sqrt{4x^2 \cdot 2} - 5\sqrt{x^2 \cdot 2}$

$\phantom{3\sqrt{8x^2} - 5\sqrt{2x^2}} = 3 \cdot 2x\sqrt{2} - 5x\sqrt{2}$

$\phantom{3\sqrt{8x^2} - 5\sqrt{2x^2}} = 6x\sqrt{2} - 5x\sqrt{2}$

$\phantom{3\sqrt{8x^2} - 5\sqrt{2x^2}} = (6x - 5x)\sqrt{2}$    **Using the distributive property**

$\phantom{3\sqrt{8x^2} - 5\sqrt{2x^2}} = x\sqrt{2}$

**b)** $(4\sqrt{3} + \sqrt{2})(\sqrt{3} - 5\sqrt{2}) = 4(\sqrt{3})^2 - 20\sqrt{6} + \sqrt{6} - 5(\sqrt{2})^2$

**Multiplying**

$\phantom{(4\sqrt{3} + \sqrt{2})(\sqrt{3} - 5\sqrt{2})} = 4 \cdot 3 + (-20 + 1)\sqrt{6} - 5 \cdot 2$

$\phantom{(4\sqrt{3} + \sqrt{2})(\sqrt{3} - 5\sqrt{2})} = 12 - 19\sqrt{6} - 10$

$\phantom{(4\sqrt{3} + \sqrt{2})(\sqrt{3} - 5\sqrt{2})} = 2 - 19\sqrt{6}$

> **Now Try Exercise 45.**

## ■ An Application

The Pythagorean theorem relates the lengths of the sides of a right triangle. The side opposite the triangle's right angle is the **hypotenuse**. The other sides are the **legs**.

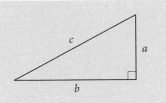

**THE PYTHAGOREAN THEOREM**

The sum of the squares of the lengths of the legs of a right triangle is equal to the square of the length of the hypotenuse:

$$a^2 + b^2 = c^2.$$

**EXAMPLE 4** *Building a Doll House.* Roy is building a doll house for his granddaughter. The doll house measures 26 in. across, and the slanted side of the roof measures 15 in. Find the height of the roof.

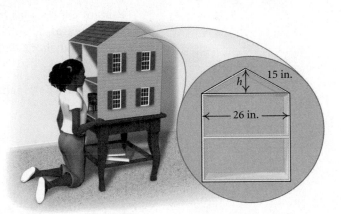

*Solution* We have a right triangle with hypotenuse 15 in. and one leg that measures 26/2, or 13 in. We use the Pythagorean theorem to find the length of the other leg:

$$c^2 = a^2 + b^2$$
$$15^2 = 13^2 + h^2 \qquad \text{Substituting 15 for } c, \text{ 13 for } a, \text{ and } h \text{ for } b$$
$$15^2 - 13^2 = h^2$$
$$225 - 169 = h^2$$
$$56 = h^2$$
$$\sqrt{56} = h$$
$$7.5 \approx h.$$

The height of the roof is about 7.5 in.

**Now Try Exercise 59.**

$\sqrt{15^2-13^2}$
　　　7.483314774

## ■ Rationalizing Denominators or Numerators

There are times when we need to remove the radicals in a denominator or a numerator. This is called **rationalizing the denominator** or **rationalizing the numerator**. It is done by multiplying by 1 in such a way as to obtain a perfect $n$th power.

**EXAMPLE 5**  Rationalize the denominator of each of the following.

a) $\sqrt{\dfrac{3}{2}}$
b) $\dfrac{\sqrt[3]{7}}{\sqrt[3]{9}}$

*Solution*

a) $\sqrt{\dfrac{3}{2}} = \sqrt{\dfrac{3}{2} \cdot \dfrac{2}{2}} = \sqrt{\dfrac{6}{4}} = \dfrac{\sqrt{6}}{\sqrt{4}} = \dfrac{\sqrt{6}}{2}$

b) $\dfrac{\sqrt[3]{7}}{\sqrt[3]{9}} = \dfrac{\sqrt[3]{7}}{\sqrt[3]{9}} \cdot \dfrac{\sqrt[3]{3}}{\sqrt[3]{3}} = \dfrac{\sqrt[3]{21}}{\sqrt[3]{27}} = \dfrac{\sqrt[3]{21}}{3}$

**Now Try Exercise 65.**

Pairs of expressions of the form $a\sqrt{b} + c\sqrt{d}$ and $a\sqrt{b} - c\sqrt{d}$ are called **conjugates**. The product of such a pair contains no radicals and can be used to rationalize a denominator or a numerator.

**EXAMPLE 6**  Rationalize the numerator: $\dfrac{\sqrt{x} - \sqrt{y}}{5}$.

*Solution*

$$\dfrac{\sqrt{x} - \sqrt{y}}{5} = \dfrac{\sqrt{x} - \sqrt{y}}{5} \cdot \dfrac{\sqrt{x} + \sqrt{y}}{\sqrt{x} + \sqrt{y}} \qquad \text{The conjugate of } \sqrt{x} - \sqrt{y} \text{ is } \sqrt{x} + \sqrt{y}.$$

$$= \dfrac{(\sqrt{x})^2 - (\sqrt{y})^2}{5\sqrt{x} + 5\sqrt{y}} \qquad (A + B)(A - B) = A^2 - B^2$$

$$= \dfrac{x - y}{5\sqrt{x} + 5\sqrt{y}}$$

**Now Try Exercise 85.**

## ■ Rational Exponents

We are motivated to define *rational exponents* so that the properties for integer exponents hold for them as well. For example, we must have

$$a^{1/2} \cdot a^{1/2} = a^{1/2+1/2} = a^1 = a.$$

Thus we are led to define $a^{1/2}$ to mean $\sqrt{a}$. Similarly, $a^{1/n}$ would mean $\sqrt[n]{a}$. Again, if the laws of exponents are to hold, we must have

$$(a^{1/n})^m = (a^m)^{1/n} = a^{m/n}.$$

Thus we are led to define $a^{m/n}$ to mean $(\sqrt[n]{a})^m$, or, equivalently, $\sqrt[n]{a^m}$.

---

**RATIONAL EXPONENTS**

For any real number $a$ and any natural numbers $m$ and $n$, $n \neq 1$, for which $\sqrt[n]{a}$ exists:

$$a^{1/n} = \sqrt[n]{a},$$
$$a^{m/n} = (\sqrt[n]{a})^m = \sqrt[n]{a^m}, \quad \text{and}$$
$$a^{-m/n} = \frac{1}{a^{m/n}}.$$

---

We can use the definition of rational exponents to convert between radical notation and exponential notation.

**EXAMPLE 7** Convert to radical notation and, if possible, simplify each of the following.

a) $7^{3/4}$                 b) $8^{-5/3}$

c) $m^{1/6}$                d) $(-32)^{2/5}$

**Solution**

a) $7^{3/4} = \sqrt[4]{7^3}$, or $(\sqrt[4]{7})^3$

b) $8^{-5/3} = \dfrac{1}{8^{5/3}} = \dfrac{1}{(\sqrt[3]{8})^5} = \dfrac{1}{2^5} = \dfrac{1}{32}$

c) $m^{1/6} = \sqrt[6]{m}$

d) $(-32)^{2/5} = \sqrt[5]{(-32)^2} = \sqrt[5]{1024} = 4$,   or

    $(-32)^{2/5} = (\sqrt[5]{-32})^2 = (-2)^2 = 4$

> **Now Try Exercise 87.**

**EXAMPLE 8** Convert each of the following to exponential notation.

a) $(\sqrt[4]{7xy})^5$                  b) $\sqrt[6]{x^3}$

**Solution**

a) $(\sqrt[4]{7xy})^5 = (7xy)^{5/4}$

b) $\sqrt[6]{x^3} = x^{3/6} = x^{1/2}$

> **Now Try Exercise 97.**

We can use the laws of exponents to simplify exponential expressions and radical expressions.

**GCM**   **EXAMPLE 9** Simplify and then, if appropriate, write radical notation for each of the following.

a) $x^{5/6} \cdot x^{2/3}$               b) $(x + 3)^{5/2}(x + 3)^{-1/2}$

c) $\sqrt[3]{\sqrt{7}}$

**Solution**

a) $x^{5/6} \cdot x^{2/3} = x^{5/6+2/3} = x^{9/6} = x^{3/2} = \sqrt{x^3} = \sqrt{x^2}\sqrt{x} = x\sqrt{x}$

b) $(x+3)^{5/2}(x+3)^{-1/2} = (x+3)^{5/2-1/2} = (x+3)^2$

c) $\sqrt[3]{\sqrt{7}} = \sqrt[3]{7^{1/2}} = (7^{1/2})^{1/3} = 7^{1/6} = \sqrt[6]{7}$

> **Now Try Exercise 107.**

We can add and subtract rational exponents on a graphing calculator. The FRAC feature from the MATH menu allows us to express the result as a fraction. The addition of the exponents in Example 9(a) is shown here.

```
5/6+2/3▶Frac
                    3
                    ─
                    2
```

**EXAMPLE 10**   Write an expression containing a single radical: $\sqrt{a}\,\sqrt[6]{b^5}$.

**Solution**   $\sqrt{a}\,\sqrt[6]{b^5} = a^{1/2}b^{5/6} = a^{3/6}b^{5/6} = (a^3b^5)^{1/6} = \sqrt[6]{a^3b^5}$

> **Now Try Exercise 117.**

# R.7   Exercise Set

*Simplify. Assume that variables can represent any real number.*

1. $\sqrt{(-21)^2}$

2. $\sqrt{(-7)^2}$

3. $\sqrt{9y^2}$

4. $\sqrt{64t^2}$

5. $\sqrt{(a-2)^2}$

6. $\sqrt{(2b+5)^2}$

7. $\sqrt[3]{-27x^3}$

8. $\sqrt[3]{-8y^3}$

9. $\sqrt[4]{81x^8}$

10. $\sqrt[4]{16z^{12}}$

11. $\sqrt[5]{32}$

12. $\sqrt[5]{-32}$

13. $\sqrt{180}$

14. $\sqrt{48}$

15. $\sqrt{72}$

16. $\sqrt{250}$

17. $\sqrt[3]{54}$

18. $\sqrt[3]{135}$

19. $\sqrt{128c^2d^4}$

20. $\sqrt{162c^4d^6}$

21. $\sqrt[4]{48x^6y^4}$

22. $\sqrt[4]{243m^5n^{10}}$

23. $\sqrt{x^2-4x+4}$

24. $\sqrt{x^2+16x+64}$

*Simplify. Assume that no radicands were formed by raising negative quantities to even powers.*

25. $\sqrt{15}\,\sqrt{35}$

26. $\sqrt{21}\,\sqrt{6}$

27. $\sqrt{8}\,\sqrt{10}$

28. $\sqrt{12}\,\sqrt{15}$

29. $\sqrt{2x^3y}\,\sqrt{12xy}$

30. $\sqrt{3y^4z}\,\sqrt{20z}$

31. $\sqrt[3]{3x^2y}\,\sqrt[3]{36x}$

32. $\sqrt[5]{8x^3y^4}\,\sqrt[5]{4x^4y}$

33. $\sqrt[3]{2(x+4)}\,\sqrt[3]{4(x+4)^4}$

34. $\sqrt[3]{4(x+1)^2}\,\sqrt[3]{18(x+1)^2}$

35. $\sqrt[8]{\dfrac{m^{16}n^{24}}{2^8}}$

36. $\sqrt[6]{\dfrac{m^{12}n^{24}}{64}}$

37. $\dfrac{\sqrt{40xy}}{\sqrt{8x}}$

38. $\dfrac{\sqrt[3]{40m}}{\sqrt[3]{5m}}$

39. $\dfrac{\sqrt[3]{3x^2}}{\sqrt[3]{24x^5}}$

40. $\dfrac{\sqrt{128a^2b^4}}{\sqrt{16ab}}$

41. $\sqrt[3]{\dfrac{64a^4}{27b^3}}$

42. $\sqrt{\dfrac{9x^7}{16y^8}}$

43. $\sqrt{\dfrac{7x^3}{36y^6}}$

44. $\sqrt[3]{\dfrac{2yz}{250z^4}}$

45. $5\sqrt{2}+3\sqrt{32}$

46. $7\sqrt{12}-2\sqrt{3}$

47. $6\sqrt{20}-4\sqrt{45}+\sqrt{80}$

**48.** $2\sqrt{32} + 3\sqrt{8} - 4\sqrt{18}$

**49.** $8\sqrt{2x^2} - 6\sqrt{20x} - 5\sqrt{8x^2}$

**50.** $2\sqrt[3]{8x^2} + 5\sqrt[3]{27x^2} - 3\sqrt{x^3}$

**51.** $(\sqrt{8} + 2\sqrt{5})(\sqrt{8} - 2\sqrt{5})$

**52.** $(\sqrt{3} - \sqrt{2})(\sqrt{3} + \sqrt{2})$

**53.** $(2\sqrt{3} + \sqrt{5})(\sqrt{3} - 3\sqrt{5})$

**54.** $(\sqrt{6} - 4\sqrt{7})(3\sqrt{6} + 2\sqrt{7})$

**55.** $(\sqrt{2} - 5)^2$

**56.** $(1 + \sqrt{3})^2$

**57.** $(\sqrt{5} - \sqrt{6})^2$

**58.** $(\sqrt{3} + \sqrt{2})^2$

**59.** *Surveying.* A surveyor places poles at points $A$, $B$, and $C$ in order to measure the distance across a pond. The distances $AC$ and $BC$ are measured as shown. Find the distance $AB$ across the pond.

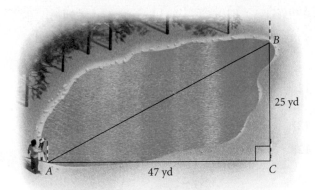

25 yd

47 yd

**60.** *Distance from Airport.* An airplane is flying at an altitude of 3700 ft. The slanted distance directly to the airport is 14,200 ft. How far horizontally is the airplane from the airport?

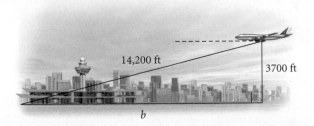

14,200 ft

3700 ft

$b$

**61.** An *equilateral triangle* is shown below.

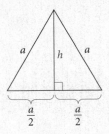

**a)** Find an expression for its height $h$ in terms of $a$.
**b)** Find an expression for its area $A$ in terms of $a$.

**62.** An isosceles right triangle has legs of length $s$. Find an expression for the length of the hypotenuse in terms of $s$.

**63.** The diagonal of a square has length $8\sqrt{2}$. Find the length of a side of the square.

**64.** The area of square $PQRS$ is 100 ft$^2$, and $A$, $B$, $C$, and $D$ are the midpoints of the sides. Find the area of square $ABCD$.

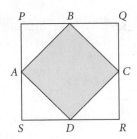

*Rationalize the denominator.*

**65.** $\sqrt{\dfrac{3}{7}}$

**66.** $\sqrt{\dfrac{2}{3}}$

**67.** $\dfrac{\sqrt[3]{7}}{\sqrt[3]{25}}$

**68.** $\dfrac{\sqrt[3]{5}}{\sqrt[3]{4}}$

**69.** $\sqrt[3]{\dfrac{16}{9}}$

**70.** $\sqrt[3]{\dfrac{3}{5}}$

**71.** $\dfrac{2}{\sqrt{3} - 1}$

**72.** $\dfrac{6}{3 + \sqrt{5}}$

**73.** $\dfrac{1 - \sqrt{2}}{2\sqrt{3} - \sqrt{6}}$

**74.** $\dfrac{\sqrt{5} + 4}{\sqrt{2} + 3\sqrt{7}}$

**75.** $\dfrac{6}{\sqrt{m} - \sqrt{n}}$

**76.** $\dfrac{3}{\sqrt{v} + \sqrt{w}}$

*Rationalize the numerator.*

**77.** $\dfrac{\sqrt{50}}{3}$

**78.** $\dfrac{\sqrt{12}}{5}$

**79.** $\sqrt[3]{\dfrac{2}{5}}$

**80.** $\sqrt[3]{\dfrac{7}{2}}$

**81.** $\dfrac{\sqrt{11}}{\sqrt{3}}$

**82.** $\dfrac{\sqrt{5}}{\sqrt{2}}$

**83.** $\dfrac{9 - \sqrt{5}}{3 - \sqrt{3}}$

**84.** $\dfrac{8 - \sqrt{6}}{5 - \sqrt{2}}$

**85.** $\dfrac{\sqrt{a} + \sqrt{b}}{3a}$

**86.** $\dfrac{\sqrt{p} - \sqrt{q}}{1 + \sqrt{q}}$

*Convert to radical notation and, if possible, simplify.*

**87.** $y^{5/6}$

**88.** $x^{2/3}$

**89.** $16^{3/4}$

**90.** $4^{7/2}$

**91.** $125^{-1/3}$

**92.** $32^{-4/5}$

**93.** $a^{5/4}b^{-3/4}$

**94.** $x^{2/5}y^{-1/5}$

**95.** $m^{5/3}n^{7/3}$

**96.** $p^{7/6}q^{11/6}$

*Convert to exponential notation.*

**97.** $\sqrt[5]{17^3}$

**98.** $\left(\sqrt[4]{13}\right)^5$

**99.** $\left(\sqrt[5]{12}\right)^4$

**100.** $\sqrt[3]{20^2}$

**101.** $\sqrt[3]{\sqrt{11}}$

**102.** $\sqrt[3]{\sqrt[4]{7}}$

**103.** $\sqrt{5}\,\sqrt[3]{5}$

**104.** $\sqrt[3]{2}\,\sqrt{2}$

**105.** $\sqrt[5]{32^2}$

**106.** $\sqrt[3]{64^2}$

*Simplify and then, if appropriate, write radical notation.*

**107.** $\left(2a^{3/2}\right)\left(4a^{1/2}\right)$

**108.** $\left(3a^{5/6}\right)\left(8a^{2/3}\right)$

**109.** $\left(\dfrac{x^6}{9b^{-4}}\right)^{1/2}$

**110.** $\left(\dfrac{x^{2/3}}{4y^{-2}}\right)^{1/2}$

**111.** $\dfrac{x^{2/3}y^{5/6}}{x^{-1/3}y^{1/2}}$

**112.** $\dfrac{a^{1/2}b^{5/8}}{a^{1/4}b^{3/8}}$

**113.** $\left(m^{1/2}n^{5/2}\right)^{2/3}$

**114.** $\left(x^{5/3}y^{1/3}z^{2/3}\right)^{3/5}$

**115.** $a^{3/4}\left(a^{2/3} + a^{4/3}\right)$

**116.** $m^{2/3}\left(m^{7/4} - m^{5/4}\right)$

*Write an expression containing a single radical and simplify.*

**117.** $\sqrt[3]{6}\sqrt{2}$

**118.** $\sqrt{2}\,\sqrt[4]{8}$

**119.** $\sqrt[4]{xy}\,\sqrt[3]{x^2y}$

**120.** $\sqrt[3]{ab^2}\,\sqrt{ab}$

**121.** $\sqrt[3]{a^4\sqrt{a^3}}$

**122.** $\sqrt{a^3\sqrt[3]{a^2}}$

**123.** $\dfrac{\sqrt{(a + x)^3}\,\sqrt[3]{(a + x)^2}}{\sqrt[4]{a + x}}$

**124.** $\dfrac{\sqrt[4]{(x + y)^2}\,\sqrt[3]{x + y}}{\sqrt{(x + y)^3}}$

## Synthesis

*Simplify.*

**125.** $\sqrt{1 + x^2} + \dfrac{1}{\sqrt{1 + x^2}}$

**126.** $\sqrt{1 - x^2} - \dfrac{x^2}{2\sqrt{1 - x^2}}$

**127.** $\left(\sqrt{a^{\sqrt{a}}}\right)^{\sqrt{a}}$

**128.** $\left(2a^3b^{5/4}c^{1/7}\right)^4 \div \left(54a^{-2}b^{2/3}c^{6/5}\right)^{-1/3}$

## STUDY GUIDE

### IMPORTANT PROPERTIES AND FORMULAS

#### Properties of the Real Numbers

| | |
|---|---|
| *Commutative:* | $a + b = b + a;$ |
| | $ab = ba$ |
| *Associative:* | $a + (b + c) = (a + b) + c;$ |
| | $a(bc) = (ab)c$ |
| *Additive Identity:* | $a + 0 = 0 + a = a$ |
| *Additive Inverse:* | $-a + a = a + (-a) = 0$ |
| *Multiplicative Identity:* | $a \cdot 1 = 1 \cdot a = a$ |
| *Multiplicative Inverse:* | $a \cdot \dfrac{1}{a} = \dfrac{1}{a} \cdot a = 1 \quad (a \neq 0)$ |
| *Distributive:* | $a(b + c) = ab + ac$ |

#### Absolute Value

For any real number $a$,

$$|a| = \begin{cases} a, & \text{if } a \geq 0, \\ -a, & \text{if } a < 0. \end{cases}$$

#### Properties of Exponents

For any real numbers $a$ and $b$ and any integers $m$ and $n$, assuming 0 is not raised to a nonpositive power:

| | |
|---|---|
| *The Product Rule:* | $a^m \cdot a^n = a^{m+n}$ |
| *The Quotient Rule:* | $\dfrac{a^m}{a^n} = a^{m-n} \quad (a \neq 0)$ |
| *The Power Rule:* | $(a^m)^n = a^{mn}$ |
| *Raising a Product to a Power:* | $(ab)^m = a^m b^m$ |
| *Raising a Quotient to a Power:* | |

$$\left(\frac{a}{b}\right)^m = \frac{a^m}{b^m} \quad (b \neq 0)$$

#### Compound Interest Formula

$$A = P\left(1 + \frac{r}{n}\right)^{nt}$$

#### Special Products of Binomials

$(A + B)^2 = A^2 + 2AB + B^2$

$(A - B)^2 = A^2 - 2AB + B^2$

$(A + B)(A - B) = A^2 - B^2$

#### Sum or Difference of Cubes

$A^3 + B^3 = (A + B)(A^2 - AB + B^2)$

$A^3 - B^3 = (A - B)(A^2 + AB + B^2)$

#### Equation-Solving Principles

*The Addition Principle:* If $a = b$ is true, then $a + c = b + c$ is true.

*The Multiplication Principle:* If $a = b$ is true, then $ac = bc$ is true.

*The Principle of Zero Products:* If $ab = 0$ is true, then $a = 0$ or $b = 0$, and if $a = 0$ or $b = 0$, then $ab = 0$.

*The Principle of Square Roots:* If $x^2 = k$, then $x = \sqrt{k}$ or $x = -\sqrt{k}$.

#### Properties of Radicals

Let $a$ and $b$ be any real numbers or expressions for which the given roots exist. For any natural numbers $m$ and $n$ ($n \neq 1$):

If $n$ is even, $\sqrt[n]{a^n} = |a|$.

If $n$ is odd, $\sqrt[n]{a^n} = a$.

$\sqrt[n]{a} \cdot \sqrt[n]{b} = \sqrt[n]{ab}$.

$\sqrt[n]{\dfrac{a}{b}} = \dfrac{\sqrt[n]{a}}{\sqrt[n]{b}} \quad (b \neq 0)$.

$\sqrt[n]{a^m} = (\sqrt[n]{a})^m$.

#### Pythagorean Theorem

$$a^2 + b^2 = c^2$$

#### Rational Exponents

For any real number $a$ and any natural numbers $m$ and $n$, $n \neq 1$, for which $\sqrt[n]{a}$ exists:

$$a^{1/n} = \sqrt[n]{a},$$

$$a^{m/n} = (\sqrt[n]{a})^m = \sqrt[n]{a^m}, \quad \text{and}$$

$$a^{-m/n} = \frac{1}{a^{m/n}}, \; a \neq 0.$$

## REVIEW EXERCISES

*Answers for all the review exercises appear in the answer section at the back of the book. If your answer is incorrect, restudy the section indicated in red next to the exercise or the direction line that precedes it.*

*Determine whether the statement is true or false.*

1. If $a < 0$, then $|a| = -a$. [R.1]

2. For any real number $a$, $a \neq 0$, and any integers $m$ and $n$, $a^m \cdot a^n = a^{mn}$. [R.2]

3. If $a = b$ is true, then $a + c = b + c$ is true. [R.5]

4. The domain of an algebraic expression is the set of all real numbers for which the expression is defined. [R.6]

*In Exercises 5–10, consider the numbers*

$$-7, \ 43, \ -\tfrac{4}{9}, \ \sqrt{17}, \ 0, \ 2.191191119\ldots, \ \sqrt[3]{64},$$
$$-\sqrt{2}, \ 4\tfrac{3}{4}, \ \tfrac{12}{7}, \ 102, \ \sqrt[5]{5}.$$

5. Which are rational numbers? [R.1]

6. Which are whole numbers? [R.1]

7. Which are integers? [R.1]

8. Which are real numbers? [R.1]

9. Which are natural numbers? [R.1]

10. Which are irrational numbers? [R.1]

11. Write interval notation for $\{x \mid -4 < x \leq 7\}$. [R.1]

*Simplify.* [R.1]

12. $|24|$

13. $\left|-\tfrac{7}{8}\right|$

14. Find the distance between $-5$ and $5$ on the number line. [R.1]

*Calculate.* [R.2]

15. $3 \cdot 2 - 4 \cdot 2^2 + 6(3 - 1)$

16. $\dfrac{3^4 - (6 - 7)^4}{2^3 - 2^4}$

*Convert to decimal notation.* [R.2]

17. $8.3 \times 10^{-5}$

18. $2.07 \times 10^7$

*Convert to scientific notation.* [R.2]

19. 405,000

20. 0.00000039

*Compute. Write the answer using scientific notation.* [R.2]

21. $(3.1 \times 10^5)(4.5 \times 10^{-3})$

22. $\dfrac{2.5 \times 10^{-8}}{3.2 \times 10^{13}}$

*Simplify.*

23. $(-3x^4 y^{-5})(4x^{-2}y)$ [R.2]

24. $\dfrac{48a^{-3}b^2 c^5}{6a^3 b^{-1} c^4}$ [R.2]

25. $\sqrt[4]{81}$ [R.7]

26. $\sqrt[5]{-32}$ [R.7]

27. $\dfrac{b - a^{-1}}{a - b^{-1}}$ [R.6]

28. $\dfrac{\dfrac{x^2}{y} + \dfrac{y^2}{x}}{y^2 - xy + x^2}$ [R.6]

29. $(\sqrt{3} - \sqrt{7})(\sqrt{3} + \sqrt{7})$ [R.7]

30. $(5 - \sqrt{2})^2$ [R.7]

31. $8\sqrt{5} + \dfrac{25}{\sqrt{5}}$ [R.7]

32. $(x + t)(x^2 - xt + t^2)$ [R.3]

33. $(5a + 4b)(2a - 3b)$ [R.3]

34. $(6x^2 y - 3xy^2 + 5xy - 3) - (-4x^2 y - 4xy^2 + 3xy + 8)$ [R.3]

*Factor.* [R.4]

35. $32x^4 - 40xy^3$

36. $y^3 + 3y^2 - 2y - 6$

37. $24x + 144 + x^2$

38. $9x^3 + 35x^2 - 4x$

39. $9x^2 - 30x + 25$

40. $8x^3 - 1$

41. $18x^2 - 3x + 6$

42. $4x^3 - 4x^2 - 9x + 9$

**43.** $6x^3 + 48$

**44.** $a^2b^2 - ab - 6$

**45.** $2x^2 + 5x - 3$

*Solve.* [R.5]

**46.** $2x - 7 = 7$

**47.** $5x - 7 = 3x - 9$

**48.** $8 - 3x = -7 + 2x$

**49.** $6(2x - 1) = 3 - (x + 10)$

**50.** $y^2 + 16y + 64 = 0$

**51.** $x^2 - x = 20$

**52.** $2x^2 + 11x - 6 = 0$

**53.** $x(x - 2) = 3$

**54.** $y^2 - 16 = 0$

**55.** $n^2 - 7 = 0$

**56.** $2a - 5b = 10$, for $b$

**57.** $xy = x + 3$, for $x$

**58.** Divide and simplify:

$$\frac{3x^2 - 12}{x^2 + 4x + 4} \div \frac{x - 2}{x + 2}. \quad [R.6]$$

**59.** Subtract and simplify:

$$\frac{x}{x^2 + 9x + 20} - \frac{4}{x^2 + 7x + 12}. \quad [R.6]$$

*Write an expression containing a single radical.* [R.7]

**60.** $\sqrt{y^5}\,\sqrt[3]{y^2}$

**61.** $\dfrac{\sqrt{(a + b)^3}\,\sqrt[3]{a + b}}{\sqrt[6]{(a + b)^7}}$

**62.** Convert to radical notation: $b^{7/5}$. [R.7]

**63.** Convert to exponential notation:

$$\sqrt[8]{\frac{m^{32}n^{16}}{3^8}}. \quad [R.7]$$

**64.** Rationalize the denominator:

$$\frac{4 - \sqrt{3}}{5 + \sqrt{3}}. \quad [R.7]$$

**65.** How long is a guy wire that reaches from the top of a 17-ft pole to a point on the ground 8 ft from the bottom of the pole? [R.7]

**66.** Calculate: $128 \div (-2)^3 \div (-2) \cdot 3$. [R.2]

  **A.** $\frac{8}{3}$   **B.** 24   **C.** 96   **D.** $\frac{512}{3}$

**67.** Factor completely: $9x^2 - 36y^2$. [R.4]

  **A.** $(3x + 6y)(3x - 6y)$
  **B.** $3(x + 2y)(x - 2y)$
  **C.** $9(x + 2y)(x - 2y)$
  **D.** $9(x - 2y)^2$

## Synthesis

*Mortgage Payments.* The formula

$$M = P\left[\frac{\frac{r}{12}\left(1 + \frac{r}{12}\right)^n}{\left(1 + \frac{r}{12}\right)^n - 1}\right]$$

*gives the monthly mortgage payment M on a home loan of P dollars at interest rate r, where n is the total number of payments (12 times the number of years). Use this formula in Exercises 68 and 69.* [R.2]

**68.** The cost of a house is \$98,000. The down payment is \$16,000, the interest rate is $6\frac{1}{2}$ %, and the loan period is 25 years. What is the monthly mortgage payment?

**69.** The cost of a house is \$124,000. The down payment is \$20,000, the interest rate is $5\frac{3}{4}$ %, and the loan period is 30 years. What is the monthly mortgage payment?

*Multiply. Assume that all exponents are integers.* [R.3]

**70.** $(t^a + t^{-a})^2$

**71.** $(a^n - b^n)^3$

*Factor.* [R.4]

**72.** $x^{2t} - 3x^t - 28$

**73.** $m^{6n} - m^{3n}$

1. Consider the numbers
$$6\tfrac{6}{7}, \ \sqrt{12}, \ 0, \ -\tfrac{13}{4}, \ \sqrt[3]{8}, \ -1.2, \ 29, \ -5.$$

   a) Which are whole numbers?
   b) Which are irrational numbers?
   c) Which are integers but not natural numbers?
   d) Which are rational numbers but not integers?

*Simplify.*

2. $|-17.6|$     3. $\left|\tfrac{15}{11}\right|$     4. $|0|$

5. Write interval notation for $\{x \mid -3 < x \le 6\}$. Then graph the interval.

6. Find the distance between $-9$ and $6$ on the number line.

7. Calculate: $32 \div 2^3 - 12 \div 4 \cdot 3$.

8. Convert to scientific notation: $4{,}509{,}000$.

9. Convert to decimal notation: $8.6 \times 10^{-5}$.

10. Compute and write scientific notation for the answer:
$$\frac{2.7 \times 10^4}{3.6 \times 10^{-3}}.$$

*Simplify.*

11. $x^{-8} \cdot x^5$

12. $(2y^2)^3 (3y^4)^2$

13. $(-3a^5 b^{-4})(5a^{-1} b^3)$

14. $(5xy^4 - 7xy^2 + 4x^2 - 3) - (-3xy^4 + 2xy^2 - 2y + 4)$

15. $(y - 2)(3y + 4)$

16. $(4x - 3)^2$

17. $\dfrac{\dfrac{x}{y} - \dfrac{y}{x}}{x + y}$     18. $\sqrt{45}$

19. $\sqrt[3]{56}$     20. $3\sqrt{75} + 2\sqrt{27}$

21. $\sqrt{18}\,\sqrt{10}$

22. $(2 + \sqrt{3})(5 - 2\sqrt{3})$

*Factor.*

23. $8x^2 - 18$

24. $y^2 - 3y - 18$

25. $2n^2 + 5n - 12$

26. $x^3 + 10x^2 + 25x$

27. $m^3 - 8$

*Solve.*

28. $7x - 4 = 24$

29. $3(y - 5) + 6 = 8 - (y + 2)$

30. $2x^2 + 5x + 3 = 0$

31. $z^2 - 11 = 0$

32. $8x + 3y = 24$, for $y$

33. Multiply and simplify:
$$\frac{x^2 + x - 6}{x^2 + 8x + 15} \cdot \frac{x^2 - 25}{x^2 - 4x + 4}.$$

34. Subtract and simplify: $\dfrac{x}{x^2 - 1} - \dfrac{3}{x^2 + 4x - 5}$.

35. Rationalize the denominator: $\dfrac{5}{7 - \sqrt{3}}$.

36. Convert to radical notation: $m^{3/8}$.

37. Convert to exponential notation: $\sqrt[6]{3^5}$.

38. How long is a guy wire that reaches from the top of a 12-ft pole to a point on the ground 5 ft from the bottom of the pole?

## Synthesis

39. Multiply: $(x - y - 1)^2$.

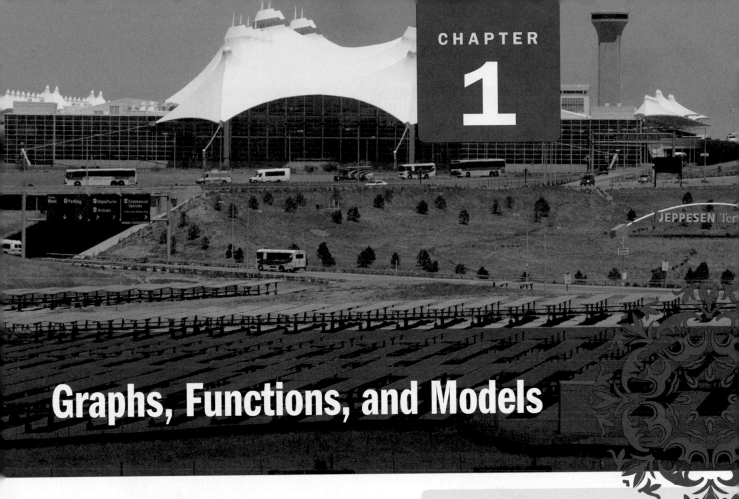

# Graphs, Functions, and Models

## APPLICATION

In December 2009, a solar energy farm was completed at the Denver International Airport. More than 9200 rectangular solar panels were installed (*Sources*: Woods Allee, Denver International Airport; www.solarpanelstore.com; *The Denver Post*). A solar panel, or photovoltaic panel, converts sunlight into electricity. The length of a panel is 13.6 in. less than twice the width, and the perimeter is 207.4 in. Find the length and the width.

This problem appears as Example 9 in Section 1.5.

## 1.1 Introduction to Graphing

- Plot points.
- Determine whether an ordered pair is a solution of an equation.
- Find the *x*- and *y*-intercepts of an equation of the form $Ax + By = C$.
- Graph equations.
- Find the distance between two points in the plane, and find the midpoint of a segment.
- Find an equation of a circle with a given center and radius, and given an equation of a circle in standard form, find the center and the radius.
- Graph equations of circles.

### Graphs

Graphs provide a means of displaying, interpreting, and analyzing data in a visual format. It is not uncommon to open a newspaper or a magazine and encounter graphs. Examples of line, circle, and bar graphs are shown below.

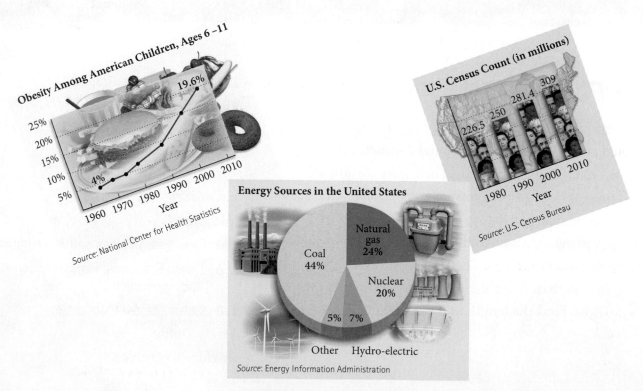

Many real-world situations can be modeled, or described mathematically, using equations in which two variables appear. We use a plane to graph a pair of numbers. To locate points on a plane, we use two perpendicular number lines, called **axes**, which intersect at $(0, 0)$. We call this point the

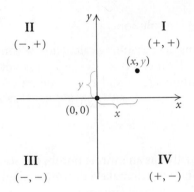

**origin**. The horizontal axis is called the *x*-**axis**, and the vertical axis is called the *y*-**axis**. (Other variables, such as *a* and *b*, can also be used.) The axes divide the plane into four regions, called **quadrants**, denoted by Roman numerals and numbered counterclockwise from the upper right. Arrows show the positive direction of each axis.

Each point $(x, y)$ in the plane is described by an **ordered pair**. The first number, *x*, indicates the point's horizontal location with respect to the *y*-axis, and the second number, *y*, indicates the point's vertical location with respect to the *x*-axis. We call *x* the **first coordinate, *x*-coordinate,** or **abscissa**. We call *y* the **second coordinate, *y*-coordinate,** or **ordinate**. Such a representation is called the **Cartesian coordinate system** in honor of the French mathematician and philosopher René Descartes (1596–1650).

In the first quadrant, both coordinates of a point are positive. In the second quadrant, the first coordinate is negative and the second is positive. In the third quadrant, both coordinates are negative, and in the fourth quadrant, the first coordinate is positive and the second is negative.

**EXAMPLE 1** Graph and label the points $(-3, 5)$, $(4, 3)$, $(3, 4)$, $(-4, -2)$, $(3, -4)$, $(0, 4)$, $(-3, 0)$, and $(0, 0)$.

**Solution** To graph or **plot** $(-3, 5)$, we note that the *x*-coordinate, $-3$, tells us to move from the origin 3 units to the left of the *y*-axis. Then we move 5 units up from the *x*-axis.* To graph the other points, we proceed in a similar manner. (See the graph at left.) Note that the point $(4, 3)$ is different from the point $(3, 4)$.

Now Try Exercise 3.

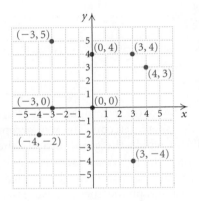

## ■ Solutions of Equations

Equations in two variables, like $2x + 3y = 18$, have solutions $(x, y)$ that are ordered pairs such that when the first coordinate is substituted for *x* and the second coordinate is substituted for *y*, the result is a true equation. The first coordinate in an ordered pair generally represents the variable that occurs first alphabetically.

**EXAMPLE 2** Determine whether each ordered pair is a solution of $2x + 3y = 18$.

**a)** $(-5, 7)$                               **b)** $(3, 4)$

**Solution** We substitute the ordered pair into the equation and determine whether the resulting equation is true.

**a)**
$$2x + 3y = 18$$
$$\frac{2(-5) + 3(7) \; ? \; 18}{\begin{array}{c|c} -10 + 21 & \\ 11 & 18 \end{array}}$$  We substitute $-5$ for *x* and 7 for *y* (alphabetical order).

                                     **FALSE**

The equation $11 = 18$ is false, so $(-5, 7)$ is not a solution.

---

*We first saw notation such as $(-3, 5)$ in Section R.1. There the notation represented an open interval. Here the notation represents an ordered pair. The context in which the notation appears usually makes the meaning clear.

**b)**

$$2x + 3y = 18$$

$$\begin{array}{c|c} 2(3) + 3(4) \;?\; 18 \\ 6 + 12 \\ 18 & 18 \quad \text{TRUE} \end{array}$$    We substitute 3 for *x* and 4 for *y*.

The equation $18 = 18$ is true, so $(3, 4)$ is a solution.

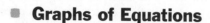

We can also perform these substitutions on a graphing calculator. When we substitute $-5$ for *x* and 7 for *y*, we get 11. Since $11 \neq 18$, $(-5, 7)$ is not a solution of the equation. When we substitute 3 for *x* and 4 for *y*, we get 18, so $(3, 4)$ is a solution.

**Now Try Exercise 11.**

### ■ Graphs of Equations

The equation considered in Example 2 actually has an infinite number of solutions. Since we cannot list all the solutions, we will make a drawing, called a **graph**, that represents them. On the following page are some suggestions for drawing graphs.

> ### TO GRAPH AN EQUATION
>
> To **graph an equation** is to make a drawing that represents the solutions of that equation.

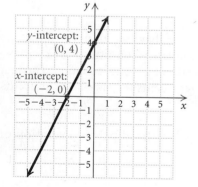

Graphs of equations of the type $Ax + By = C$ are straight lines. Many such equations can be graphed conveniently using intercepts. The **x-intercept** of the graph of an equation is the point at which the graph crosses the *x*-axis. The **y-intercept** is the point at which the graph crosses the *y*-axis. We know from geometry that only one line can be drawn through two given points. Thus, if we know the intercepts, we can graph the line. To ensure that a computational error has not been made, it is a good idea to calculate and plot a third point as a check.

> ### *x*-INTERCEPT AND *y*-INTERCEPT
>
> An **x-intercept** is a point $(a, 0)$. To find *a*, let $y = 0$ and solve for *x*.
> A **y-intercept** is a point $(0, b)$. To find *b*, let $x = 0$ and solve for *y*.

**EXAMPLE 3**    Graph: $2x + 3y = 18$.

*Solution*    The graph is a line. To find ordered pairs that are solutions of this equation, we can replace either *x* or *y* with any number and then solve for the other variable. In this case, it is convenient to find the intercepts of the graph. For instance, if $x$ is replaced with 0, then

$$2 \cdot 0 + 3y = 18$$
$$3y = 18$$
$$y = 6. \qquad \text{Dividing by 3 on both sides}$$

Thus, $(0, 6)$ is a solution. It is the *y-intercept* of the graph. If $y$ is replaced with 0, then

$$2x + 3 \cdot 0 = 18$$
$$2x = 18$$
$$x = 9. \qquad \text{Dividing by 2 on both sides}$$

Thus, $(9, 0)$ is a solution. It is the *x-intercept* of the graph. We find a third solution as a check. If $x$ is replaced with 3, then

$$2 \cdot 3 + 3y = 18$$
$$6 + 3y = 18$$
$$3y = 12 \qquad \text{Subtracting 6 on both sides}$$
$$y = 4. \qquad \text{Dividing by 3 on both sides}$$

Thus, $(3, 4)$ is a solution.

We list the solutions in a table and then plot the points. Note that the points appear to lie on a straight line.

| $x$ | $y$ | $(x, y)$ |
|-----|-----|----------|
| 0 | 6 | $(0, 6)$ |
| 9 | 0 | $(9, 0)$ |
| 3 | 4 | $(3, 4)$ |

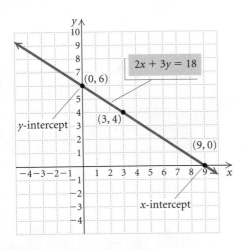

### Suggestions for Drawing Graphs

1. Calculate solutions and list the ordered pairs in a table.
2. Use graph paper.
3. Draw axes and label them with the variables.
4. Use arrows on the axes to indicate positive directions.
5. Scale the axes; that is, label the tick marks on the axes. Consider the ordered pairs found in part (1) above when choosing the scale.
6. Plot the ordered pairs, look for patterns, and complete the graph. Label the graph with the equation being graphed.

Were we to graph additional solutions of $2x + 3y = 18$, they would be on the same straight line. Thus, to complete the graph, we use a straightedge to draw a line, as shown in the figure. This line represents all solutions of the equation. Every point on the line represents a solution; every solution is represented by a point on the line. **Now Try Exercise 17.**

When graphing some equations, it is convenient to first solve for $y$ and then find ordered pairs. We can use the addition and multiplication principles to solve for $y$.

GCM

**EXAMPLE 4**    Graph: $3x - 5y = -10$.

*Solution*    We first solve for $y$:

$$3x - 5y = -10$$
$$-5y = -3x - 10 \qquad \text{Subtracting } 3x \text{ on both sides}$$
$$y = \tfrac{3}{5}x + 2. \qquad \text{Multiplying by } -\tfrac{1}{5} \text{ on both sides}$$

By choosing multiples of 5 for $x$, we can avoid fraction values when calculating $y$. For example, if we choose $-5$ for $x$, we get

$$y = \tfrac{3}{5}x + 2 = \tfrac{3}{5}(-5) + 2 = -3 + 2 = -1.$$

The following table lists a few more points. We plot the points and draw the graph.

| $x$ | $y$ | $(x, y)$ |
|-----|-----|----------|
| $-5$ | $-1$ | $(-5, -1)$ |
| $0$ | $2$ | $(0, 2)$ |
| $5$ | $5$ | $(5, 5)$ |

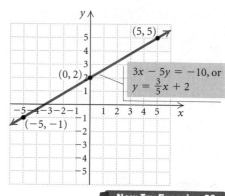

> **Now Try Exercise 29.**

---

| | |
|---|---|
| **EQUATION SOLVING** | |
| REVIEW SECTION **R.5.** | |

In the equation $y = \tfrac{3}{5}x + 2$ in Example 4, the value of $y$ *depends* on the value chosen for $x$, so $x$ is said to be the **independent variable** and $y$ the **dependent variable**.

We can graph an equation on a graphing calculator. Many calculators require an equation to be entered in the form $y =$ . In such a case, if the equation is not initially given in this form, it must be solved for $y$ before it is entered in the calculator. For the equation $3x - 5y = -10$ in Example 4, we enter $y = \tfrac{3}{5}x + 2$ on the equation-editor, or $y =$ , screen in the form $y = (3/5)x + 2$, as shown in the window at left.

Next, we determine the portion of the $xy$-plane that will appear on the calculator's screen. That portion of the plane is called the **viewing window**. The notation used in this text to denote a window setting consists of four numbers $[L, R, B, T]$, which represent the **L**eft and **R**ight endpoints of the $x$-axis and the **B**ottom and **T**op endpoints of the $y$-axis, respectively. The window with the settings $[-10, 10, -10, 10]$ is the **standard viewing window**. On some graphing calculators, the standard window can be selected quickly using the ZSTANDARD feature from the ZOOM menu.

Xmin and Xmax are used to set the left and right endpoints of the $x$-axis, respectively; Ymin and Ymax are used to set the bottom and top endpoints of the $y$-axis, respectively. The settings Xscl and Yscl give the scales for the axes. For example, Xscl $= 1$ and Yscl $= 1$ means that there is 1 unit between tick marks on each of the axes. In this text, scaling factors other than 1 will be listed by the window unless they are readily apparent.

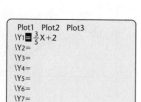

```
Plot1  Plot2  Plot3
\Y1 ▆ 3/5 X+2
\Y2=
\Y3=
\Y4=
\Y5=
\Y6=
\Y7=
```

```
WINDOW
Xmin = -10
Xmax = 10
Xscl = 1
Ymin = -10
Ymax = 10
Yscl = 1
Xres = 1
```

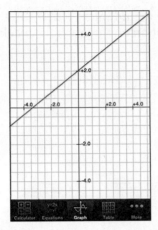

An equation such as $3x - 5y = -10$, or $y = \frac{3}{5}x + 2$, can be graphed using the Graphing Calculator, Appcylon LLC, app.

After entering the equation $y = (3/5)x + 2$ and choosing a viewing window, we can then draw the graph.

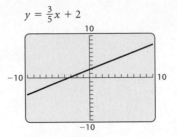

The advent of the iPhone, the iPod Touch, and the iPad has made available many inexpensive mathematical applications that can be purchased at the iTunes store. Most apps graph the functions encountered in this text. Throughout the text, we occasionally include technology windows from an app titled Graphing Calculator, Appcylon LLC.

**GCM**  **EXAMPLE 5**   Graph: $y = x^2 - 9x - 12$.

**Solution**   Note that since this equation is not of the form $Ax + By = C$, its graph is not a straight line. We make a table of values, plot enough points to obtain an idea of the shape of the curve, and connect them with a smooth curve. It is important to scale the axes to include most of the ordered pairs listed in the table. Here it is appropriate to use a larger scale on the $y$-axis than on the $x$-axis.

| $x$ | $y$ | $(x, y)$ |
|-----|-----|----------|
| $-3$ | $24$ | $(-3, 24)$ |
| $-1$ | $-2$ | $(-1, -2)$ |
| $0$ | $-12$ | $(0, -12)$ |
| $2$ | $-26$ | $(2, -26)$ |
| $4$ | $-32$ | $(4, -32)$ |
| $5$ | $-32$ | $(5, -32)$ |
| $10$ | $-2$ | $(10, -2)$ |
| $12$ | $24$ | $(12, 24)$ |

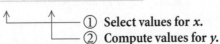

① **Select values for $x$.**
② **Compute values for $y$.**

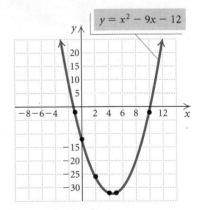

**Now Try Exercise 39.**

---

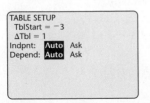

**FIGURE 1**

```
Plot1  Plot2  Plot3
\Y1◼X²−9X−12
\Y2=
\Y3=
\Y4=
\Y5=
\Y6=
\Y7=
```

```
TABLE SETUP
 TblStart = −3
 ΔTbl = 1
Indpnt:  Auto  Ask
Depend:  Auto  Ask
```

**FIGURE 2**

**GCM**

A graphing calculator can be used to create a table of ordered pairs that are solutions of an equation. For the equation in Example 5, $y = x^2 - 9x - 12$, we first enter the equation on the equation-editor screen. (See Fig. 1.) Then we set up a table in AUTO mode by designating a value for TBLSTART and a value for ΔTBL. The calculator will produce a table beginning with the value of TBLSTART and continuing by adding ΔTBL to supply succeeding $x$-values. For the equation $y = x^2 - 9x - 12$, we let TBLSTART $= -3$ and ΔTBL $= 1$. (See Fig. 2.)

| x | y1 |
|---|---|
| -3.000000 | 24.000000 |
| -2.000000 | 10.000000 |
| -1.000000 | -2.000000 |
| 0.000000 | -12.000000 |
| 1.000000 | -20.000000 |
| 2.000000 | -26.000000 |
| 3.000000 | -30.000000 |
| 4.000000 | -32.000000 |
| 5.000000 | -32.000000 |
| 6.000000 | -30.000000 |
| 7.000000 | -26.000000 |
| 8.000000 | -20.000000 |
| 9.000000 | -12.000000 |
| 10.000000 | -2.000000 |
| 11.000000 | 10.000000 |
| 12.000000 | 24.000000 |
| 13.000000 | 40.000000 |
| 14.000000 | 58.000000 |
| 15.000000 | 78.000000 |
| 16.000000 | 100.000000 |

The Graphing Calculator, Appcylon LLC, app has a table feature. Here we see a table of values for the equation in Example 5.

We can scroll up and down in the table to find values other than those shown in Fig. 3. We can also graph this equation on the graphing calculator, as shown in Fig. 4.

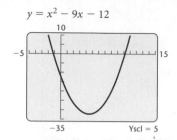

**FIGURE 3**

**FIGURE 4**

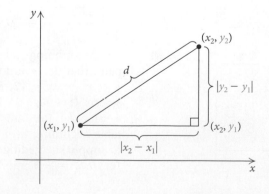

*The $5.25 billion expansion of the Panama Canal will soon double its capacity. A third canal lane is scheduled to open in August 2014. (Source: Panama Canal Authority)*

### ■ The Distance Formula

Suppose that an architect must determine the distance between two points, $A$ and $B$, on opposite sides of a lane of the Panama Canal. One way in which he or she might proceed is to measure two legs of a right triangle that is situated as shown below. The Pythagorean equation, $a^2 + b^2 = c^2$, where $c$ is the length of the hypotenuse and $a$ and $b$ are the lengths of the legs, can then be used to find the length of the hypotenuse, which is the distance from $A$ to $B$.

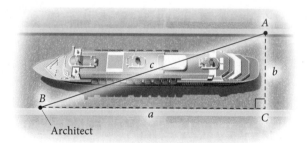

A similar strategy is used to find the distance between two points in a plane. For two points $(x_1, y_1)$ and $(x_2, y_2)$, we can draw a right triangle in which the legs have lengths $|x_2 - x_1|$ and $|y_2 - y_1|$.

Using the Pythagorean equation, we have

$$d^2 = |x_2 - x_1|^2 + |y_2 - y_1|^2.$$  Substituting $d$ for $c$, $|x_2 - x_1|$ for $a$, and $|y_2 - y_1|$ for $b$ in the Pythagorean equation, $c^2 = a^2 + b^2$

Because we are squaring, parentheses can replace the absolute-value symbols:

$$d^2 = (x_2 - x_1)^2 + (y_2 - y_1)^2.$$

Taking the principal square root, we obtain the distance formula.

---

**THE DISTANCE FORMULA**

The **distance** $d$ between any two points $(x_1, y_1)$ and $(x_2, y_2)$ is given by

$$d = \sqrt{(x_2 - x_1)^2 + (y_2 - y_1)^2}.$$

---

The subtraction of the $x$-coordinates can be done in any order, as can the subtraction of the $y$-coordinates. Although we derived the distance formula by considering two points not on a horizontal line or a vertical line, the distance formula holds for *any* two points.

**EXAMPLE 6**   Find the distance between each pair of points.

**a)** $(-2, 2)$ and $(3, -6)$           **b)** $(-1, -5)$ and $(-1, 2)$

*Solution*   We substitute into the distance formula.

**a)** $d = \sqrt{[3 - (-2)]^2 + (-6 - 2)^2}$
$= \sqrt{5^2 + (-8)^2} = \sqrt{25 + 64}$
$= \sqrt{89} \approx 9.4$

**b)** $d = \sqrt{[-1 - (-1)]^2 + (-5 - 2)^2}$
$= \sqrt{0^2 + (-7)^2} = \sqrt{0 + 49}$
$= \sqrt{49} = 7$

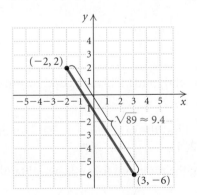

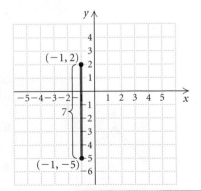

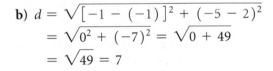

**Now Try Exercises 63 and 71.**

**EXAMPLE 7** The point $(-2, 5)$ is on a circle that has $(3, -1)$ as its center. Find the length of the radius of the circle.

*Solution* Since the length of the radius is the distance from the center to a point on the circle, we substitute into the distance formula:

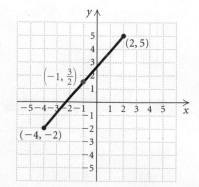

$$d = \sqrt{(x_2 - x_1)^2 + (y_2 - y_1)^2};$$

$$r = \sqrt{[3 - (-2)]^2 + (-1 - 5)^2}$$

Substituting $r$ for $d$, $(3, -1)$ for $(x_2, y_2)$, and $(-2, 5)$ for $(x_1, y_1)$. Either point can serve as $(x_1, y_1)$.

$$= \sqrt{5^2 + (-6)^2} = \sqrt{25 + 36}$$
$$= \sqrt{61} \approx 7.8.$$

Rounding to the nearest tenth

The radius of the circle is approximately 7.8.

Now Try Exercise 77.

## Midpoints of Segments

The distance formula can be used to develop a method of determining the *midpoint* of a segment when the endpoints are known. We state the formula and leave its proof to the exercises.

> **THE MIDPOINT FORMULA**
>
> If the endpoints of a segment are $(x_1, y_1)$ and $(x_2, y_2)$, then the coordinates of the **midpoint** of the segment are
>
> $$\left( \frac{x_1 + x_2}{2}, \frac{y_1 + y_2}{2} \right).$$
>
>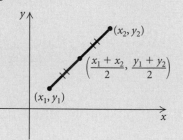

Note that we obtain the coordinates of the midpoint by averaging the coordinates of the endpoints. This is a good way to remember the midpoint formula.

**EXAMPLE 8** Find the midpoint of the segment whose endpoints are $(-4, -2)$ and $(2, 5)$.

*Solution* Using the midpoint formula, we obtain

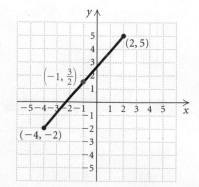

$$\left( \frac{-4 + 2}{2}, \frac{-2 + 5}{2} \right) = \left( \frac{-2}{2}, \frac{3}{2} \right) = \left( -1, \frac{3}{2} \right).$$

Now Try Exercise 83.

**EXAMPLE 9** The diameter of a circle connects the points $(2, -3)$ and $(6, 4)$ on the circle. Find the coordinates of the center of the circle.

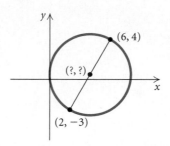

**Solution** Since the center of the circle is the midpoint of the diameter, we use the midpoint formula:

$$\left( \frac{2 + 6}{2}, \frac{-3 + 4}{2} \right), \quad \text{or} \quad \left( \frac{8}{2}, \frac{1}{2} \right), \quad \text{or} \quad \left( 4, \frac{1}{2} \right).$$

The coordinates of the center are $\left( 4, \frac{1}{2} \right)$. **Now Try Exercise 95.**

## ■ Circles

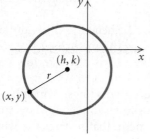

A **circle** is the set of all points in a plane that are a fixed distance $r$ from a *center* $(h, k)$. Thus if a point $(x, y)$ is to be $r$ units from the center, we must have

$$r = \sqrt{(x - h)^2 + (y - k)^2}. \qquad \underline{\text{Using the distance formula,}}$$
$$d = \sqrt{(x_2 - x_1)^2 + (y_2 - y_1)^2}$$

Squaring both sides gives an equation of a circle. The distance $r$ is the length of a *radius* of the circle.

> **THE EQUATION OF A CIRCLE**
>
> The standard form of the equation of a circle with center $(h, k)$ and radius $r$ is
>
> $$(x - h)^2 + (y - k)^2 = r^2.$$

### Interactive Figures

Icons throughout the text direct you to MyMathLab whenever an Interactive Figure is available. Math concepts come to life, letting you adjust variables and other settings to see how the visual representation is affected by the math.

**EXAMPLE 10** Find an equation of the circle having radius 5 and center $(3, -7)$.

**Solution** Using the standard form, we have

$$[x - 3]^2 + [y - (-7)]^2 = 5^2 \qquad \text{Substituting}$$
$$(x - 3)^2 + (y + 7)^2 = 25. \qquad \boxed{\text{Now Try Exercise 99.}}$$

**EXAMPLE 11** Graph the circle $(x + 5)^2 + (y - 2)^2 = 16$.

**Solution** We write the equation in standard form to determine the center and the radius:

$$[x - (-5)]^2 + [y - 2]^2 = 4^2.$$

The center is $(-5, 2)$ and the radius is 4. We locate the center and draw the circle using a compass.

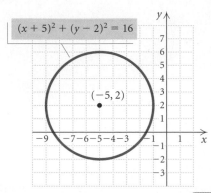

**Now Try Exercise 111.**

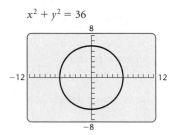

FIGURE 1

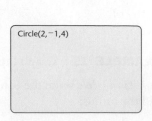

FIGURE 2

Circles can also be graphed using a graphing calculator. We show one method of doing so here. Another method is discussed in the conics chapter.

When we graph a circle, we select a viewing window in which the distance between units is visually the same on both axes. This procedure is called **squaring the viewing window**. We do this so that the graph will not be distorted. A graph of the circle $x^2 + y^2 = 36$ in a nonsquared window is shown in Fig. 1.

On many graphing calculators, the ratio of the height to the width of the viewing screen is $\frac{2}{3}$. When we choose a window in which Xscl = Yscl and the length of the $y$-axis is $\frac{2}{3}$ the length of the $x$-axis, the window will be squared. The windows with dimensions $[-6, 6, -4, 4]$, $[-9, 9, -6, 6]$, and $[-12, 12, -8, 8]$ are examples of squared windows. A graph of the circle $x^2 + y^2 = 36$ in a squared window is shown in Fig. 2. Many graphing calculators have an option on the ZOOM menu that squares the window automatically.

**GCM** **EXAMPLE 12** Graph the circle $(x - 2)^2 + (y + 1)^2 = 16$.

**Solution** The circle $(x - 2)^2 + (y + 1)^2 = 16$ has center $(2, -1)$ and radius 4, so the viewing window $[-9, 9, -6, 6]$ is a good choice for the graph.

To graph a circle, we select the CIRCLE feature from the DRAW menu and enter the coordinates of the center and the length of the radius. The graph of the circle $(x - 2)^2 + (y + 1)^2 = 16$ is shown here. For more on graphing circles with a graphing calculator, see the material on conic sections in a later chapter.

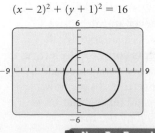

**Now Try Exercise 113.**

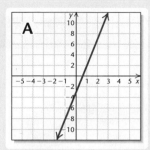

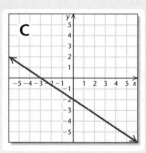

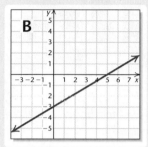

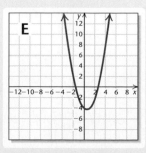

# Visualizing the Graph

Match the equation with its graph.

1. $y = -x^2 + 5x - 3$

2. $3x - 5y = 15$

3. $(x - 2)^2 + (y - 4)^2 = 36$

4. $y - 5x = -3$

5. $x^2 + y^2 = \dfrac{25}{4}$

6. $15y - 6x = 90$

7. $y = -\dfrac{2}{3}x - 2$

8. $(x + 3)^2 + (y - 1)^2 = 16$

9. $3x + 5y = 15$

10. $y = x^2 - x - 4$

**Answers on page A-3**

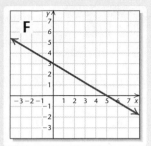

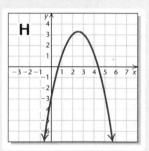

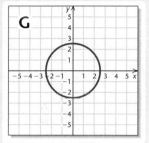

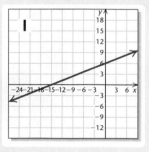

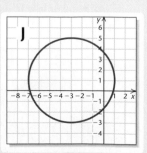

## 1.1 Exercise Set

*Use this graph for Exercises 1 and 2.*

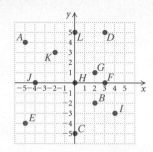

1. Find the coordinates of points $A$, $B$, $C$, $D$, $E$, and $F$.

2. Find the coordinates of points $G$, $H$, $I$, $J$, $K$, and $L$.

*Graph and label the given points by hand.*

3. $(4, 0)$, $(-3, -5)$, $(-1, 4)$, $(0, 2)$, $(2, -2)$

4. $(1, 4)$, $(-4, -2)$, $(-5, 0)$, $(2, -4)$, $(0, 3)$

5. $(-5, 1)$, $(5, 1)$, $(2, 3)$, $(2, -1)$, $(0, 1)$

6. $(0, -1)$, $(4, -3)$, $(-5, 2)$, $(2, 0)$, $(-1, -5)$

*Express the data pictured in the graph as ordered pairs, letting the first coordinate represent the year and the second coordinate the amount.*

7. **Percentage of U.S. Population That Is Foreign-Born**

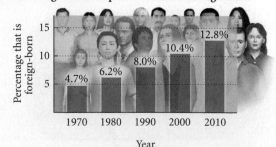

Sources: U.S. Census Bureau; U.S. Department of Commerce

8. **Sprint Cup Series: Tony Stewart in the Top 5**

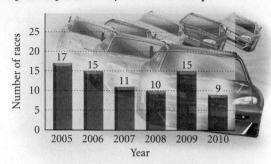

*Source: Racing-Reference.info*

*Use substitution to determine whether the given ordered pairs are solutions of the given equation.*

9. $(-1, -9)$, $(0, 2)$; $y = 7x - 2$

10. $\left(\frac{1}{2}, 8\right)$, $(-1, 6)$; $y = -4x + 10$

11. $\left(\frac{2}{3}, \frac{3}{4}\right)$, $\left(1, \frac{3}{2}\right)$; $6x - 4y = 1$

12. $(1.5, 2.6)$, $(-3, 0)$; $x^2 + y^2 = 9$

13. $\left(-\frac{1}{2}, -\frac{4}{5}\right)$, $\left(0, \frac{3}{5}\right)$; $2a + 5b = 3$

14. $\left(0, \frac{3}{2}\right)$, $\left(\frac{2}{3}, 1\right)$; $3m + 4n = 6$

15. $(-0.75, 2.75)$, $(2, -1)$; $x^2 - y^2 = 3$

16. $(2, -4)$, $(4, -5)$; $5x + 2y^2 = 70$

*Find the intercepts and then graph the line.*

17. $5x - 3y = -15$

18. $2x - 4y = 8$

19. $2x + y = 4$

20. $3x + y = 6$

21. $4y - 3x = 12$

22. $3y + 2x = -6$

*Graph the equation.*

23. $y = 3x + 5$

24. $y = -2x - 1$

25. $x - y = 3$

26. $x + y = 4$

27. $y = -\frac{3}{4}x + 3$

28. $3y - 2x = 3$

**29.** $5x - 2y = 8$

**30.** $y = 2 - \frac{4}{3}x$

**31.** $x - 4y = 5$

**32.** $6x - y = 4$

**33.** $2x + 5y = -10$

**34.** $4x - 3y = 12$

**35.** $y = -x^2$

**36.** $y = x^2$

**37.** $y = x^2 - 3$

**38.** $y = 4 - x^2$

**39.** $y = -x^2 + 2x + 3$

**40.** $y = x^2 + 2x - 1$

*In Exercises 41–44, use a graphing calculator to match the equation with one of the graphs (a)–(d) that follow.*

a)
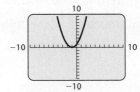

b)
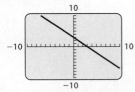

c)
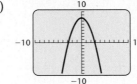

d)
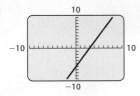

**41.** $y = 3 - x$

**42.** $2x - y = 6$

**43.** $y = x^2 + 2x + 1$

**44.** $y = 8 - x^2$

*Use a graphing calculator to graph the equation in the standard window.*

**45.** $y = 2x + 1$

**46.** $y = 3x - 4$

**47.** $4x + y = 7$

**48.** $5x + y = -8$

**49.** $y = \frac{1}{3}x + 2$

**50.** $y = \frac{3}{2}x - 4$

**51.** $2x + 3y = -5$

**52.** $3x + 4y = 1$

**53.** $y = x^2 + 6$

**54.** $y = x^2 - 8$

**55.** $y = 2 - x^2$

**56.** $y = 5 - x^2$

**57.** $y = x^2 + 4x - 2$

**58.** $y = x^2 - 5x + 3$

*Graph the equation in the standard window and in the given window. Determine which window better shows the shape of the graph and the x- and y-intercepts.*

**59.** $y = 3x^2 - 6$
$$[-4, 4, -4, 4]$$

**60.** $y = -2x + 24$
$$[-15, 15, -10, 30], \text{ with Xscl} = 3 \text{ and Yscl} = 5$$

**61.** $y = -\frac{1}{6}x^2 + \frac{1}{12}$
$$[-1, 1, -0.3, 0.3], \text{ with Xscl} = 0.1 \text{ and Yscl} = 0.1$$

**62.** $y = 6 - x^2$
$$[-3, 3, -3, 3]$$

*Find the distance between the pair of points. Give an exact answer and, where appropriate, an approximation to three decimal places.*

**63.** $(4, 6)$ and $(5, 9)$

**64.** $(-3, 7)$ and $(2, 11)$

**65.** $(-11, -8)$ and $(1, -13)$

**66.** $(-60, 5)$ and $(-20, 35)$

**67.** $(6, -1)$ and $(9, 5)$

**68.** $(-4, -7)$ and $(-1, 3)$

**69.** $\left(-8, \frac{7}{11}\right)$ and $\left(8, \frac{7}{11}\right)$

**70.** $\left(\frac{1}{2}, -\frac{4}{25}\right)$ and $\left(\frac{1}{2}, -\frac{13}{25}\right)$

**71.** $\left(-\frac{3}{5}, -4\right)$ and $\left(-\frac{3}{5}, \frac{2}{3}\right)$

**72.** $\left(-\frac{11}{3}, -\frac{1}{2}\right)$ and $\left(\frac{1}{3}, \frac{5}{2}\right)$

**73.** $(-4.2, 3)$ and $(2.1, -6.4)$

**74.** $(0.6, -1.5)$ and $(-8.1, -1.5)$

**75.** $(0, 0)$ and $(a, b)$

**76.** $(r, s)$ and $(-r, -s)$

**77.** The points $(-3, -1)$ and $(9, 4)$ are the endpoints of the diameter of a circle. Find the length of the radius of the circle.

**78.** The point $(0, 1)$ is on a circle that has center $(-3, 5)$. Find the length of the diameter of the circle.

*The converse of the Pythagorean theorem is also a true statement: If the sum of the squares of the lengths of two sides of a triangle is equal to the square of the length of the third side, then the triangle is a right triangle. Use the distance formula and the Pythagorean theorem to determine whether the set of points could be vertices of a right triangle.*

**79.** $(-4, 5)$, $(6, 1)$, and $(-8, -5)$

**80.** $(-3, 1)$, $(2, -1)$, and $(6, 9)$

**81.** $(-4, 3)$, $(0, 5)$, and $(3, -4)$

**82.** The points $(-3, 4)$, $(2, -1)$, $(5, 2)$, and $(0, 7)$ are vertices of a quadrilateral. Show that the quadrilateral is a rectangle. (*Hint:* Show that the opposite sides of the quadrilateral are the same length and that the two diagonals are the same length.)

*Find the midpoint of the segment having the given endpoints.*

**83.** $(4, -9)$ and $(-12, -3)$

**84.** $(7, -2)$ and $(9, 5)$

**85.** $\left(0, \frac{1}{2}\right)$ and $\left(-\frac{2}{5}, 0\right)$

**86.** $(0, 0)$ and $\left(-\frac{7}{13}, \frac{2}{7}\right)$

**87.** $(6.1, -3.8)$ and $(3.8, -6.1)$

**88.** $(-0.5, -2.7)$ and $(4.8, -0.3)$

**89.** $(-6, 5)$ and $(-6, 8)$

**90.** $(1, -2)$ and $(-1, 2)$

**91.** $\left(-\frac{1}{6}, -\frac{3}{5}\right)$ and $\left(-\frac{2}{3}, \frac{5}{4}\right)$

**92.** $\left(\frac{2}{9}, \frac{1}{3}\right)$ and $\left(-\frac{2}{5}, \frac{4}{5}\right)$

**93.** Graph the rectangle described in Exercise 82. Then determine the coordinates of the midpoint of each of the four sides. Are the midpoints vertices of a rectangle?

**94.** Graph the square with vertices $(-5, -1)$, $(7, -6)$, $(12, 6)$, and $(0, 11)$. Then determine the midpoint of each of the four sides. Are the midpoints vertices of a square?

**95.** The points $(\sqrt{7}, -4)$ and $(\sqrt{2}, 3)$ are endpoints of the diameter of a circle. Determine the center of the circle.

**96.** The points $(-3, \sqrt{5})$ and $(1, \sqrt{2})$ are endpoints of the diagonal of a square. Determine the center of the square.

*In Exercises 97 and 98, how would you change the window so that the circle is not distorted? Answers may vary.*

**97.** $(x + 3)^2 + (y - 2)^2 = 36$

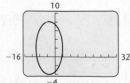

**98.** $(x - 4)^2 + (y + 5)^2 = 49$

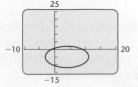

*Find an equation for a circle satisfying the given conditions.*

**99.** Center $(2, 3)$, radius of length $\frac{5}{3}$

**100.** Center $(4, 5)$, diameter of length 8.2

**101.** Center $(-1, 4)$, passes through $(3, 7)$

**102.** Center $(6, -5)$, passes through $(1, 7)$

**103.** The points $(7, 13)$ and $(-3, -11)$ are at the ends of a diameter.

**104.** The points $(-9, 4)$, $(-2, 5)$, $(-8, -3)$, and $(-1, -2)$ are vertices of an inscribed square.

**105.** Center $(-2, 3)$, tangent (touching at one point) to the $y$-axis

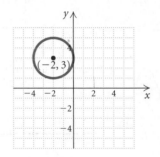

**106.** Center $(4, -5)$, tangent to the $x$-axis

*Find the center and the radius of the circle. Then graph the circle by hand. Check your graph with a graphing calculator.*

**107.** $x^2 + y^2 = 4$          **108.** $x^2 + y^2 = 81$

**109.** $x^2 + (y - 3)^2 = 16$

**110.** $(x + 2)^2 + y^2 = 100$

**111.** $(x - 1)^2 + (y - 5)^2 = 36$

**112.** $(x - 7)^2 + (y + 2)^2 = 25$

**113.** $(x + 4)^2 + (y + 5)^2 = 9$

**114.** $(x + 1)^2 + (y - 2)^2 = 64$

*Find the equation of the circle.*

**115.**

**116.**

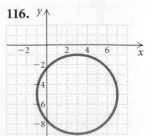

**117.**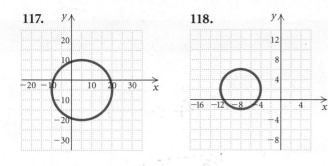

**118.**

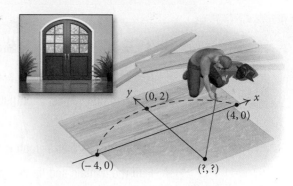

**a)** Using a coordinate system, locate the center of the circle.

**b)** What radius should Matt use to draw the arch?

## Synthesis

*To the student and the instructor: The Synthesis exercises found at the end of every exercise set challenge students to combine concepts or skills studied in that section or in preceding parts of the text.*

**119.** If the point $(p, q)$ is in the fourth quadrant, in which quadrant is the point $(q, -p)$?

*Find the distance between the pair of points and find the midpoint of the segment having the given points as endpoints.*

**120.** $\left( a, \dfrac{1}{a} \right)$ and $\left( a + h, \dfrac{1}{a + h} \right)$

**121.** $(a, \sqrt{a})$ and $(a + h, \sqrt{a + h})$

*Find an equation of a circle satisfying the given conditions.*

**122.** Center $(-5, 8)$ with a circumference of $10\pi$ units

**123.** Center $(2, -7)$ with an area of $36\pi$ square units

**124.** Find the point on the $x$-axis that is equidistant from the points $(-4, -3)$ and $(-1, 5)$.

**125.** Find the point on the $y$-axis that is equidistant from the points $(-2, 0)$ and $(4, 6)$.

**126.** Determine whether the points $(-1, -3)$, $(-4, -9)$, and $(2, 3)$ are collinear.

**127.** *An Arch of a Circle in Carpentry.* Matt is remodeling the front entrance to his home and needs to cut an arch for the top of an entranceway. The arch needs to be 8 ft wide and 2 ft high. To draw the arch, he will use a stretched string with chalk attached at an end as a compass.

**128.** Consider any right triangle with base $b$ and height $h$, situated as shown. Show that the midpoint of the hypotenuse $P$ is equidistant from the three vertices of the triangle.

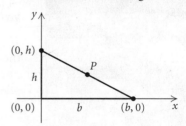

*Determine whether each of the following points lies on the **unit circle,** $x^2 + y^2 = 1$.*

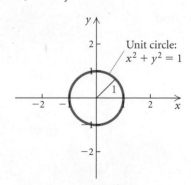

**129.** $\left( \dfrac{\sqrt{3}}{2}, -\dfrac{1}{2} \right)$

**130.** $(0, -1)$

**131.** $\left( -\dfrac{\sqrt{2}}{2}, \dfrac{\sqrt{2}}{2} \right)$

**132.** $\left( \dfrac{1}{2}, -\dfrac{\sqrt{3}}{2} \right)$

**133.** Prove the midpoint formula by showing that $\left( \dfrac{x_1 + x_2}{2}, \dfrac{y_1 + y_2}{2} \right)$ is equidistant from the points $(x_1, y_1)$ and $(x_2, y_2)$.

## 1.2

# Functions and Graphs

- Determine whether a correspondence or a relation is a function.
- Find function values, or outputs, using a formula or a graph.
- Graph functions.
- Determine whether a graph is that of a function.
- Find the domain and the range of a function.
- Solve applied problems using functions.

We now focus our attention on a concept that is fundamental to many areas of mathematics—the idea of a *function*.

### ■ Functions

We first consider an application.

*Used-Book Co-op.*    A community center operates a used-book co-op, and the proceeds are donated to summer youth programs. The total cost of a purchase is $2.50 per book plus a flat-rate surcharge of $3. If a customer selects 6 books, the total cost of the purchase is

$$\$2.50(6) + \$3, \quad \text{or} \quad \$18.$$

We can express this relationship with a set of ordered pairs, a graph, and an equation. A few ordered pairs are listed in the table below.

| $x$ | $y$ | Ordered Pairs: $(x, y)$ | Correspondence |
|-----|-----|-------------------------|----------------|
| 1 | 5.50 | (1, 5.50) | $1 \rightarrow 5.50$ |
| 2 | 8.00 | (2, 8) | $2 \rightarrow 8$ |
| 4 | 13.00 | (4, 13) | $4 \rightarrow 13$ |
| 7 | 20.50 | (7, 20.50) | $7 \rightarrow 20.50$ |
| 10 | 28.00 | (10, 28) | $10 \rightarrow 28$ |

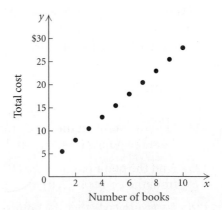

The ordered pairs express a relationship, or correspondence, between the first coordinate and the second coordinate. We can see this relationship in the graph as well. The equation that describes the correspondence is

$$y = 2.50x + 3, \quad \text{where } x \text{ is a natural number.}$$

This is an example of a *function*. In this case, the total cost of the purchase $y$ is a function of the number of books purchased; that is, $y$ is a function of $x$, where $x$ is the independent variable and $y$ is the dependent variable.

Let's consider some other correspondences before giving the definition of a function.

| First Set | Correspondence | Second Set |
|---|---|---|
| To each person | there corresponds | that person's DNA. |
| To each blue spruce sold | there corresponds | its price. |
| To each real number | there corresponds | the square of that number. |

In each correspondence, the first set is called the **domain** and the second set is called the **range**. For each member, or **element**, in the domain, there is *exactly one* member in the range to which it corresponds. That is, each person has exactly *one* DNA, each blue spruce has exactly *one* price, and each real number has exactly *one* square. Each correspondence is a *function*.

> ### FUNCTION
>
> A **function** is a correspondence between a first set, called the **domain**, and a second set, called the **range**, such that each member of the domain corresponds to *exactly one* member of the range.

It is important to note that not every correspondence between two sets is a function.

**EXAMPLE 1** Determine whether each of the following correspondences is a function.

a)
$$\begin{array}{ll} -6 \\ 6 \end{array} \searrow 36$$
$$\begin{array}{ll} -3 \\ 3 \end{array} \searrow 9$$
$$0 \longrightarrow 0$$

b)

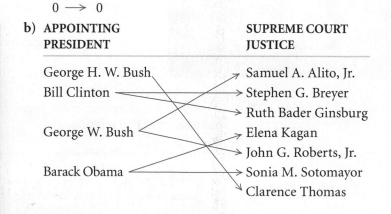

| APPOINTING PRESIDENT | SUPREME COURT JUSTICE |
|---|---|
| George H. W. Bush | Samuel A. Alito, Jr. |
| Bill Clinton | Stephen G. Breyer |
|  | Ruth Bader Ginsburg |
| George W. Bush | Elena Kagan |
|  | John G. Roberts, Jr. |
| Barack Obama | Sonia M. Sotomayor |
|  | Clarence Thomas |

*Solution*

a) This correspondence *is* a function because each member of the domain corresponds to exactly one member of the range. Note that the definition of a function allows more than one member of the domain to correspond to the same member of the range.

b) This correspondence *is not* a function because there is at least one member of the domain that is paired with more than one member of the range (Bill Clinton with Stephen G. Breyer and Ruth Bader Ginsburg; George W. Bush with Samuel A. Alito, Jr., and John G. Roberts, Jr.; Barack H. Obama with Elena Kagan and Sonia M. Sotomayor).

> **Now Try Exercises 5 and 7.**

**EXAMPLE 2**    Determine whether each of the following correspondences is a function.

| DOMAIN | CORRESPONDENCE | RANGE |
|---|---|---|
| a) Years in which a presidential election occurs | The person elected | A set of presidents |
| b) All automobiles produced in 2012 | Each automobile's VIN | A set of VIN numbers |
| c) The set of all drivers who won a NASCAR race in 2011 | The race won | The set of all NASCAR races in 2011 |
| d) The set of all NASCAR races in 2011 | The winner of the race | The set of all drivers who won a NASCAR race in 2011 |

*Solution*

a) This correspondence *is* a function because in each presidential election *exactly one* president is elected.

b) This correspondence *is* a function because each automobile has *exactly one* VIN number.

c) This correspondence *is not* a function because a winning driver could be paired with more than one race.

d) This correspondence *is* a function because each race has only one winning driver.

> **Now Try Exercises 11 and 13.**

When a correspondence between two sets is not a function, it may still be an example of a **relation**.

---

**RELATION**

A **relation** is a correspondence between a first set, called the **domain**, and a second set, called the **range**, such that each member of the domain corresponds to *at least one* member of the range.

All the correspondences in Examples 1 and 2 are relations, but, as we have seen, not all are functions. Relations are sometimes written as sets of ordered pairs (as we saw earlier in the example on used books) in which elements of the domain are the first coordinates of the ordered pairs and elements of the range are the second coordinates. For example, instead of writing $-3 \to 9$, as we did in Example 1(a), we could write the ordered pair $(-3, 9)$.

**EXAMPLE 3**   Determine whether each of the following relations is a function. Identify the domain and the range.

**a)** $\{(9, -5), (9, 5), (2, 4)\}$

**b)** $\{(-2, 5), (5, 7), (0, 1), (4, -2)\}$

**c)** $\{(-5, 3), (0, 3), (6, 3)\}$

*Solution*

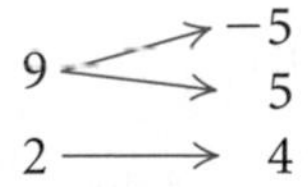

**a)** The relation *is not* a function because the ordered pairs $(9, -5)$ and $(9, 5)$ have the same first coordinate and different second coordinates. (See Fig. 1.)

The domain is the set of all first coordinates: $\{9, 2\}$.

The range is the set of all second coordinates: $\{-5, 5, 4\}$.

**b)** The relation *is* a function because *no* two ordered pairs have the same first coordinate and different second coordinates. (See Fig. 2.)

The domain is the set of all first coordinates: $\{-2, 5, 0, 4\}$.

The range is the set of all second coordinates: $\{5, 7, 1, -2\}$.

**c)** The relation *is* a function because *no* two ordered pairs have the same first coordinate and different second coordinates. (See Fig. 3.)

The domain is $\{-5, 0, 6\}$.

The range is $\{3\}$.

**Now Try Exercises 17 and 19.**

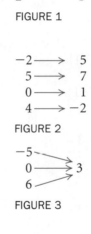

FIGURE 1

FIGURE 2

FIGURE 3

## ■ Notation for Functions

Functions used in mathematics are often given by equations. They generally require that certain calculations be performed in order to determine which member of the range is paired with each member of the domain. For example, in Section 1.1 we graphed the function $y = x^2 - 9x - 12$ by doing calculations like the following:

for $x = -2$, $y = (-2)^2 - 9(-2) - 12 = 10$,

for $x = 0$, $y = 0^2 - 9 \cdot 0 - 12 = -12$,   and

for $x = 1$, $y = 1^2 - 9 \cdot 1 - 12 = -20$.

A more concise notation is often used. For $y = x^2 - 9x - 12$, the **inputs** (members of the domain) are values of $x$ substituted into the equation. The **outputs** (members of the range) are the resulting values of $y$. If we call the function $f$, we can use $x$ to represent an arbitrary *input* and $f(x)$—read "$f$ of $x$," or "$f$ at $x$," or "the value of $f$ at $x$"—to represent the corresponding

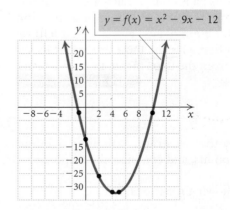

$y = f(x) = x^2 - 9x - 12$

*output.* In this notation, the function given by $y = x^2 - 9x - 12$ is written as $f(x) = x^2 - 9x - 12$, and the above calculations would be

$$f(-2) = (-2)^2 - 9(-2) - 12 = 10,$$
$$f(0) = 0^2 - 9 \cdot 0 - 12 = -12,$$
$$f(1) = 1^2 - 9 \cdot 1 - 12 = -20.$$

> Keep in mind that $f(x)$ *does not* mean $f \cdot x$.

Thus, instead of writing "when $x = -2$, the value of $y$ is $10$," we can simply write "$f(-2) = 10$," which can be read as "$f$ of $-2$ is $10$" or "for the input $-2$, the output of $f$ is $10$." The letters $g$ and $h$ are also often used to name functions.

**GCM**    **EXAMPLE 4**    A function $f$ is given by $f(x) = 2x^2 - x + 3$. Find each of the following.

**a)** $f(0)$                          **b)** $f(-7)$
**c)** $f(5a)$                         **d)** $f(a - 4)$

**Solution**    We can think of this formula as follows:

$$f(\blacksquare) = 2(\blacksquare)^2 - (\blacksquare) + 3.$$

Thus to find an output for a given input we think: "Whatever goes in the blank on the left goes in the blank(s) on the right." This gives us a "recipe" for finding outputs.

**a)** $f(0) = 2(0)^2 - 0 + 3 = 0 - 0 + 3 = 3$
**b)** $f(-7) = 2(-7)^2 - (-7) + 3 = 2 \cdot 49 + 7 + 3 = 108$
**c)** $f(5a) = 2(5a)^2 - 5a + 3 = 2 \cdot 25a^2 - 5a + 3 = 50a^2 - 5a + 3$
**d)** $f(a - 4) = 2(a - 4)^2 - (a - 4) + 3$
$$= 2(a^2 - 8a + 16) - a + 4 + 3$$
$$= 2a^2 - 16a + 32 - a + 4 + 3$$
$$= 2a^2 - 17a + 39$$

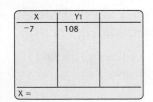

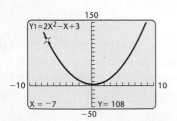

We can also find function values with a graphing calculator. Most calculators do not use function notation "$f(x) = \ldots$" to enter a function formula. Instead, we must enter the function using "$y = \ldots$." At left, we illustrate finding $f(-7)$ from part (b), first with the TABLE feature set in ASK mode and then with the VALUE feature from the CALC menu. In both screens, we see that $f(-7) = 108$.    **Now Try Exercise 21.**

## ▪ Graphs of Functions

We graph functions in the same way that we graph equations. We find ordered pairs $(x, y)$, or $(x, f(x))$, plot points, and complete the graph.

**EXAMPLE 5**    Graph each of the following functions.

**a)** $f(x) = x^2 - 5$          **b)** $f(x) = x^3 - x$          **c)** $f(x) = \sqrt{x + 4}$

**Solution**    We select values for $x$ and find the corresponding values of $f(x)$. Then we plot the points and connect them with a smooth curve.

**a)** $f(x) = x^2 - 5$

| $x$ | $f(x)$ | $(x, f(x))$ |
|-----|--------|-------------|
| $-3$ | $4$ | $(-3, 4)$ |
| $-2$ | $-1$ | $(-2, -1)$ |
| $-1$ | $-4$ | $(-1, -4)$ |
| $0$ | $-5$ | $(0, -5)$ |
| $1$ | $-4$ | $(1, -4)$ |
| $2$ | $-1$ | $(2, -1)$ |
| $3$ | $4$ | $(3, 4)$ |

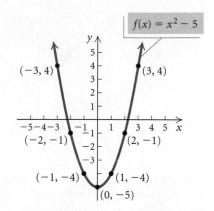

$y = x^3 - x$

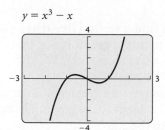

$y = \sqrt{x + 4}$

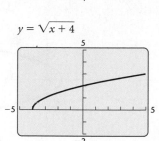

**b)** $f(x) = x^3 - x$

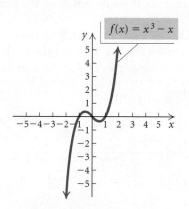

**c)** $f(x) = \sqrt{x + 4}$

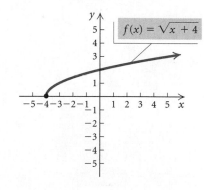

We can check the graphs with a graphing calculator. The checks for parts (b) and (c) are shown at left.    **Now Try Exercise 33.**

Function values can also be determined from a graph.

**EXAMPLE 6**   For the function $f(x) = x^2 - 6$, use the graph to find each of the following function values.

**a)** $f(-3)$                    **b)** $f(1)$

*Solution*

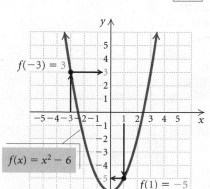

**a)** To find the function value $f(-3)$ from the graph at left, we locate the input $-3$ on the horizontal axis, move vertically to the graph of the function, and then move horizontally to find the output on the vertical axis. We see that $f(-3) = 3$.

**b)** To find the function value $f(1)$, we locate the input $1$ on the horizontal axis, move vertically to the graph, and then move horizontally to find the output on the vertical axis. We see that $f(1) = -5$.

**Now Try Exercise 37.**

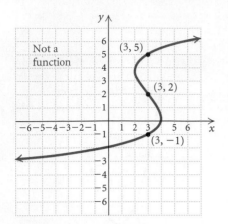

*Since 3 is paired with more than one member of the range, the graph does not represent a function.*

We know that when one member of the domain is paired with two or more different members of the range, the correspondence *is not* a function. Thus, when a graph contains two or more different points with the same first coordinate, the graph cannot represent a function. (See the graph at left. Note that 3 is paired with −1, 2, and 5.) Points sharing a common first coordinate are vertically above or below each other. This leads us to the *vertical-line test*.

---

### THE VERTICAL-LINE TEST

If it is possible for a vertical line to cross a graph more than once, then the graph *is not* the graph of a function.

---

To apply the vertical-line test, we try to find a vertical line that crosses the graph more than once. If we succeed, then the graph is not that of a function. If we do not, then the graph is that of a function.

**EXAMPLE 7** Which of graphs (a)–(f) (in red) are graphs of functions? In graph (f), the solid dot shows that $(-1, 1)$ belongs to the graph. The open circle shows that $(-1, -2)$ does *not* belong to the graph.

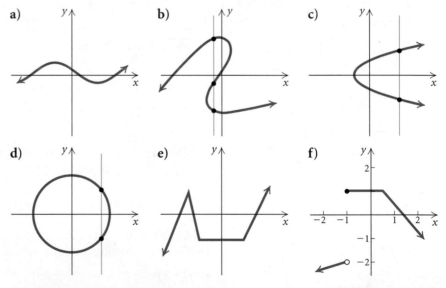

*Solution* Graphs (a), (e), and (f) are graphs of functions because we cannot find a vertical line that crosses any of them more than once. In (b), the vertical line drawn crosses the graph at three points, so graph (b) is not that of a function. Also, in (c) and (d), we can find a vertical line that crosses the graph more than once, so these are not graphs of functions.

**Now Try Exercises 45 and 49.**

## ▪ Finding Domains of Functions

When a function $f$ whose inputs and outputs are real numbers is given by a formula, the *domain* is understood to be the set of all inputs for which the expression is defined as a real number. When an input results in an expression that is not defined as a real number, we say that the function value *does not exist* and that the number being substituted *is not* in the domain of the function.

**DOMAINS OF EXPRESSIONS**

REVIEW SECTION **R.6.**

**EXAMPLE 8**   Find the indicated function values, if possible, and determine whether the given values are in the domain of the function.

**a)** $f(1)$ and $f(3)$, for $f(x) = \dfrac{1}{x - 3}$

**b)** $g(16)$ and $g(-7)$, for $g(x) = \sqrt{x} + 5$

*Solution*

**a)** $f(1) = \dfrac{1}{1 - 3} = \dfrac{1}{-2} = -\dfrac{1}{2}$

$y = 1/(x - 3)$

| X | Y₁ |
| --- | --- |
| 1 | −.5 |
| 3 | ERROR |

X =

Since $f(1)$ is defined, 1 is in the domain of $f$.

$$f(3) = \frac{1}{3 - 3} = \frac{1}{0}$$

Since division by 0 is not defined, $f(3)$ does not exist and the number 3 is not in the domain of $f$. In a table from a graphing calculator, this is indicated with an ERROR message.

**b)** $g(16) = \sqrt{16} + 5 = 4 + 5 = 9$

$y = \sqrt{x} + 5$

| X | Y₁ |
| --- | --- |
| 16 | 9 |
| −7 | ERROR |

X =

Since $g(16)$ is defined, 16 is in the domain of $g$.

$$g(-7) = \sqrt{-7} + 5$$

Since $\sqrt{-7}$ is not defined as a real number, $g(-7)$ does not exist and the number $-7$ is not in the domain of $g$. Note the ERROR message in the table at left.

As we see in Example 8, inputs that make a denominator 0 or that yield a negative radicand in an even root are not in the domain of a function.

**EXAMPLE 9**   Find the domain of each of the following functions.

**a)** $f(x) = \dfrac{1}{x - 7}$

**b)** $h(x) = \dfrac{3x^2 - x + 7}{x^2 + 2x - 3}$

**c)** $f(x) = x^3 + |x|$

**d)** $g(x) = \sqrt[3]{x - 1}$

*Solution*

**INTERVAL NOTATION**

REVIEW SECTION **R.1.**

**a)** Because $x - 7 = 0$ when $x = 7$, the only input that results in a denominator of 0 is 7. The domain is $\{x | x \neq 7\}$. We can also write the solution using interval notation and the symbol $\cup$ for the **union**, or inclusion, of both sets: $(-\infty, 7) \cup (7, \infty)$.

**b)** We can substitute any real number in the numerator, but we must avoid inputs that make the denominator 0. To find those inputs, we solve $x^2 + 2x - 3 = 0$, or $(x + 3)(x - 1) = 0$. Since $x^2 + 2x - 3$ is 0 for $-3$ and 1, the domain consists of the set of all real numbers except $-3$ and 1, or $\{x | x \neq -3 \text{ and } x \neq 1\}$, or $(-\infty, -3) \cup (-3, 1) \cup (1, \infty)$.

**c)** We can substitute any real number for $x$. The domain is the set of all real numbers, or $(-\infty, \infty)$.

**d)** Because the index is odd, the radicand, $x - 1$, can be any real number. Thus $x$ can be any real number. The domain is all real numbers, or $(-\infty, \infty)$.

> **Now Try Exercises 53, 59, and 65.**

## ■ Visualizing Domain and Range

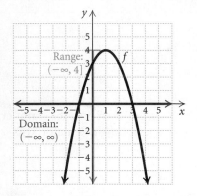

> Keep the following in mind regarding the *graph* of a function:
>
> **Domain** = the set of a function's inputs, found on the horizontal $x$-axis;
>
> **Range** = the set of a function's outputs, found on the vertical $y$-axis.

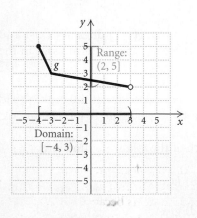

Consider the graph of function $f$, shown at left. To determine the domain of $f$, we look for the inputs on the $x$-axis that correspond to a point on the graph. We see that they include the entire set of real numbers, illustrated in red on the $x$-axis. Thus the domain is $(-\infty, \infty)$. To find the range, we look for the outputs on the $y$-axis that correspond to a point on the graph. We see that they include 4 and all real numbers less than 4, illustrated in blue on the $y$-axis. The bracket at 4 indicates that 4 is included in the interval. The range is $\{x | x \leq 4\}$, or $(-\infty, 4]$.

Let's now consider the graph of function $g$, shown at left. The solid dot shows that $(-4, 5)$ belongs to the graph. The open circle shows that $(3, 2)$ does *not* belong to the graph.

We see that the inputs of the function are $-4$ and all real numbers between $-4$ and 3, illustrated in red on the $x$-axis. The bracket at $-4$ indicates that $-4$ is included in the domain. The parenthesis at 3 indicates that 3 is not included in the domain. The domain is $\{x | -4 \leq x < 3\}$, or $[-4, 3)$. The outputs of the function are 5 and all real numbers between 2 and 5, illustrated in blue on the $y$-axis. The parenthesis at 2 indicates that 2 is not included in the range. The bracket at 5 indicates that 5 is included in the range. The range is $\{x | 2 < x \leq 5\}$, or $(2, 5]$.

**EXAMPLE 10**    Using the graph of the function, find the domain and the range of the function.

**a)** $f(x) = \frac{1}{2}x + 1$

**b)** $f(x) = \sqrt{x + 4}$

**c)** $f(x) = x^3 - x$

**d)** $f(x) = \dfrac{1}{x - 2}$

**e)** $f(x) = x^4 - 2x^2 - 3$

**f)** $f(x) = \sqrt{4 - (x - 3)^2}$

*Solution*

**a)**

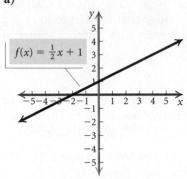

Domain = all real numbers, $(-\infty, \infty)$; range = all real numbers, $(-\infty, \infty)$

**b)**

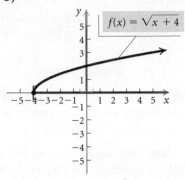

Domain = $[-4, \infty)$; range = $[0, \infty)$

**c)**

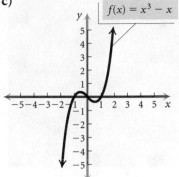

Domain = all real numbers, $(-\infty, \infty)$; range = all real numbers, $(-\infty, \infty)$

**d)**

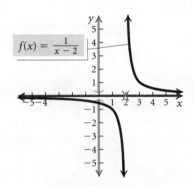

Since the graph does not touch or cross either the vertical line $x = 2$ or the $x$-axis $y = 0$, 2 is excluded from the domain and 0 is excluded from the range. Domain = $(-\infty, 2) \cup (2, \infty)$; range = $(-\infty, 0) \cup (0, \infty)$

**e)**

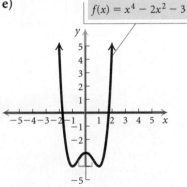

Domain = all real numbers, $(-\infty, \infty)$; range = $[-4, \infty)$

**f)**

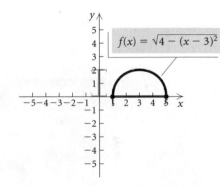

Domain = $[1, 5]$; range = $[0, 2]$

**Now Try Exercises 75 and 83.**

Always consider adding the reasoning of Example 9 to a graphical analysis. Think, "What can I input?" to find the domain. Think, "What do I get out?" to find the range. Thus, in Examples 10(c) and 10(e), it might not appear as though the domain is all real numbers because the graph rises steeply, but by examining the equation we see that we can indeed substitute any real number for $x$.

## Applications of Functions

**EXAMPLE 11**   *Linear Expansion of a Bridge.*   The linear expansion $L$ of the steel center span of a suspension bridge that is 1420 m long is a function of the change in temperature $t$, in degrees Celsius, from winter to summer and is given by

$$L(t) = 0.000013 \cdot 1420 \cdot t,$$

where 0.000013 is the coefficient of linear expansion for steel and $L$ is in meters. Find the linear expansion of the steel center span when the change in temperature from winter to summer is 30°, 42°, 50°, and 56° Celsius.

*Solution*   We use a graphing calculator with the TABLE feature set in ASK mode to compute the function values. We find that

| X | Y1 | |
|---|---|---|
| 30 | .5338 | |
| 42 | .77532 | |
| 50 | .923 | |
| 56 | 1.0338 | |
| X = | | |

$L(30) = 0.5538$ m,

$L(42) = 0.77532$ m,

$L(50) = 0.923$ m,   and

$L(56) = 1.0338$ m.

**Now Try Exercise 87.**

---

## CONNECTING THE CONCEPTS

**FUNCTION CONCEPTS**

Formula for $f$: $f(x) = 5 + 2x^2 - x^4$.

For every input, there is exactly one output.

$(1, 6)$ is on the graph.

For the input 1, the output is 6.

$f(1) = 6$

Domain: set of all inputs $= (-\infty, \infty)$

Range: set of all outputs $= (-\infty, 6]$

**GRAPH**

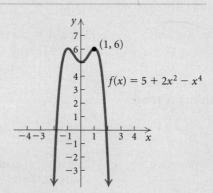

## 1.2 Exercise Set

*In Exercises 1–14, determine whether the correspondence is a function.*

**1.** $a \longrightarrow w$
$b \longrightarrow y$
$c \longrightarrow z$

**2.** $m \longrightarrow q$
$n \underset{o}{\overset{r}{\times}} s$

**3.** $-6 \longrightarrow 36$
$-2 \longrightarrow 4$
$2 \nearrow$

**4.** $-3 \longrightarrow 2$
$1 \searrow 4$
$5 \longrightarrow 6$
$9 \longrightarrow 8$

**5.** $m \longrightarrow A$
$n \searrow B$
$r \longrightarrow C$
$s \nearrow D$

**6.** $a \longrightarrow r$
$b \searrow s$
$c \nearrow t$
$d$

**7.**

| BOOK TITLE | AMERICAN NOVELIST |
|---|---|
| *The Color Purple*, 1982 (Pulitzer Prize 1983) | Ray Bradbury |
| *East of Eden*, 1952 | Pearl Buck |
| *Fahrenheit 451*, 1953 | |
| *The Good Earth*, 1931 (Pulitzer Prize 1932) | Ernest Hemingway |
| *For Whom the Bell Tolls*, 1940 | Harper Lee |
| *The Grapes of Wrath*, 1939 (Pulitzer Prize 1940) | John Steinbeck |
| *To Kill a Mockingbird*, 1960 (Pulitzer Prize 1961) | Alice Walker |
| *The Old Man and the Sea*, 1952 (Pulitzer Prize 1953) | Eudora Welty |
| *The Optimist's Daughter*, 1972 (Pulitzer Prize 1973) | |

**8.**

**ACTOR PORTRAYING JAMES BOND** — **MOVIE TITLE**

Sean Connery — *Goldfinger*, 1964
George Lazenby — *On Her Majesty's Secret Service*, 1969
*Diamonds Are Forever*, 1971
Roger Moore — *Moonraker*, 1979
Timothy Dalton — *For Your Eyes Only*, 1981
*The Living Daylights*, 1987
Pierce Brosnan — *GoldenEye*, 1995
*The World Is Not Enough*, 1999
Daniel Craig — *Quantum of Solace*, 2008

| | DOMAIN | CORRESPONDENCE | RANGE |
|---|---|---|---|
| **9.** | A set of cars in a parking lot | Each car's license number | A set of letters and numbers |
| **10.** | A set of people in a town | A doctor a person uses | A set of doctors |
| **11.** | The integers less than 9 | Five times the integer | A subset of integers |
| **12.** | A set of members of a rock band | An instrument each person plays | A set of instruments |
| **13.** | A set of students in a class | A student sitting in a neighboring seat | A set of students |
| **14.** | A set of bags of chips on a shelf | Each bag's weight | A set of weights |

*Determine whether the relation is a function. Identify the domain and the range.*

**15.** $\{(2, 10), (3, 15), (4, 20)\}$

**16.** $\{(3, 1), (5, 1), (7, 1)\}$

**17.** $\{(-7, 3), (-2, 1), (-2, 4), (0, 7)\}$

**18.** $\{(1, 3), (1, 5), (1, 7), (1, 9)\}$

**19.** $\{(-2, 1), (0, 1), (2, 1), (4, 1), (-3, 1)\}$

**20.** $\{(5, 0), (3, -1), (0, 0), (5, -1), (3, -2)\}$

**21.** Given that $g(x) = 3x^2 - 2x + 1$, find each of the following.
  **a)** $g(0)$  **b)** $g(-1)$
  **c)** $g(3)$  **d)** $g(-x)$
  **e)** $g(1 - t)$

**22.** Given that $f(x) = 5x^2 + 4x$, find each of the following.
  **a)** $f(0)$  **b)** $f(-1)$
  **c)** $f(3)$  **d)** $f(t)$
  **e)** $f(t - 1)$

**23.** Given that $g(x) = x^3$, find each of the following.

**a)** $g(2)$          **b)** $g(-2)$
**c)** $g(-x)$       **d)** $g(3y)$
**e)** $g(2 + h)$

**24.** Given that $f(x) = 2|x| + 3x$, find each of the following.

**a)** $f(1)$          **b)** $f(-2)$
**c)** $f(-x)$       **d)** $f(2y)$
**e)** $f(2 - h)$

**25.** Given that

$$g(x) = \frac{x - 4}{x + 3},$$

find each of the following.

**a)** $g(5)$          **b)** $g(4)$
**c)** $g(-3)$      **d)** $g(-16.25)$
**e)** $g(x + h)$

**26.** Given that

$$f(x) = \frac{x}{2 - x},$$

find each of the following.

**a)** $f(2)$          **b)** $f(1)$
**c)** $f(-16)$     **d)** $f(-x)$
**e)** $f\left(-\frac{2}{3}\right)$

**27.** Find $g(0)$, $g(-1)$, $g(5)$, and $g\left(\frac{1}{2}\right)$ for

$$g(x) = \frac{x}{\sqrt{1 - x^2}}.$$

**28.** Find $h(0)$, $h(2)$, and $h(-x)$ for

$$h(x) = x + \sqrt{x^2 - 1}.$$

*In Exercises 29 and 30, use a graphing calculator and the* TABLE *feature set in* ASK *mode.*

**29.** Given that

$$g(x) = 0.06x^3 - 5.2x^2 - 0.8x,$$

find $g(-2.1)$, $g(5.08)$, and $g(10.003)$. Round answers to the nearest tenth.

**30.** Given that

$$h(x) = 3x^4 - 10x^3 + 5x^2 - x + 6,$$

find $h(-11)$, $h(7)$, and $h(15)$.

*Graph the function.*

**31.** $f(x) = \frac{1}{2}x + 3$      **32.** $f(x) = \sqrt{x} - 1$

**33.** $f(x) = -x^2 + 4$      **34.** $f(x) = x^2 + 1$

**35.** $f(x) = \sqrt{x - 1}$      **36.** $f(x) = x - \frac{1}{2}x^3$

*In Exercises 37–42, a graph of a function is shown. Using the graph, find the indicated function values; that is, given the inputs, find the outputs.*

**37.** $h(1)$, $h(3)$, and $h(4)$

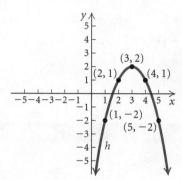

**38.** $t(-4)$, $t(0)$, and $t(3)$

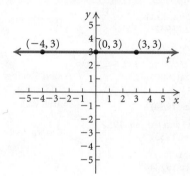

**39.** $s(-4)$, $s(-2)$, and $s(0)$

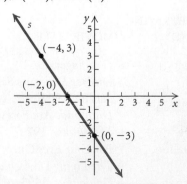

**40.** $g(-4)$, $g(-1)$, and $g(0)$

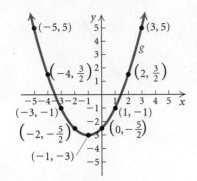

**41.** $f(-1)$, $f(0)$, and $f(1)$

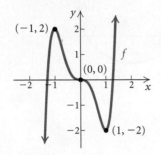

**42.** $g(-2)$, $g(0)$, and $g(2.4)$

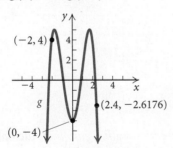

*In Exercises 43–50, determine whether the graph is that of a function. An open circle indicates that the point does not belong to the graph.*

**43.**

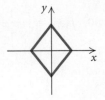

**44.**

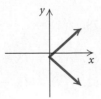

**45.**

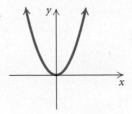

**46.**

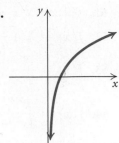

**47.**

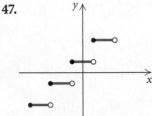

**48.**

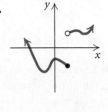

**49.**

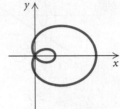

**50.**

*Find the domain of the function. Do not use a graphing calculator.*

**51.** $f(x) = 7x + 4$

**52.** $f(x) = |3x - 2|$

**53.** $f(x) = |6 - x|$

**54.** $f(x) = \dfrac{1}{x^4}$

**55.** $f(x) = 4 - \dfrac{2}{x}$

**56.** $f(x) = \dfrac{1}{5}x^2 - 5$

**57.** $f(x) = \dfrac{x + 5}{2 - x}$

**58.** $f(x) = \dfrac{8}{x + 4}$

**59.** $f(x) = \dfrac{1}{x^2 - 4x - 5}$

**60.** $f(x) = \dfrac{(x - 2)(x + 9)}{x^3}$

**61.** $f(x) = \sqrt[3]{x + 10} - 1$

**62.** $f(x) = \sqrt[3]{4 - x}$

**63.** $f(x) = \dfrac{8 - x}{x^2 - 7x}$

**64.** $f(x) = \dfrac{x^4 - 2x^3 + 7}{3x^2 - 10x - 8}$

**65.** $f(x) = \frac{1}{10}|x|$

**66.** $f(x) = x^2 - 2x$

*In Exercises 67–74, determine the domain and the range of the function.*

**67.**
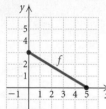

**68.**
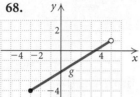

**69.**
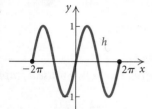

**70.**
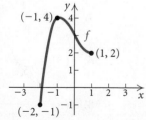

**71.**
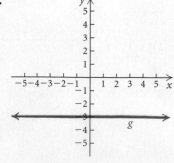

**72.**
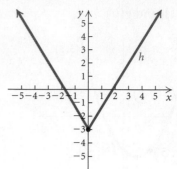

**73.**
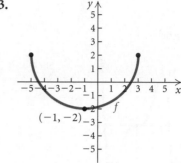

**74.**
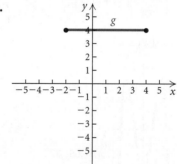

*Graph the function with a graphing calculator. Then visually estimate the domain and the range.*

**75.** $f(x) = |x|$

**76.** $f(x) = |x| - 2$

**77.** $f(x) = 3x - 2$

**78.** $f(x) = 5 - 3x$

**79.** $f(x) = \dfrac{1}{x - 3}$

**80.** $f(x) = \dfrac{1}{x + 1}$

**81.** $f(x) = (x - 1)^3 + 2$

**82.** $f(x) = (x - 2)^4 + 1$

**83.** $f(x) = \sqrt{7 - x}$

**84.** $f(x) = \sqrt{x + 8}$

**85.** $f(x) = -x^2 + 4x - 1$

**86.** $f(x) = 2x^2 - x^4 + 5$

**87.** *Decreasing Value of the Dollar.* In 2008, it took $21.57 to equal the value of $1 in 1913. In 1985, it took only $10.87 to equal the value of $1 in 1913. The amount it takes to equal the value of $1 in 1913 can be estimated by the linear function $V$ given by

$$V(x) = 0.4652x + 10.87,$$

where $x$ is the number of years since 1985. Thus, $V(10)$ gives the amount it took in 1995 to equal the value of $1 in 1913.

*Source:* U.S. Bureau of Labor Statistics

**a)** Use this function to predict the amount it will take in 2018 and in 2025 to equal the value of $1 in 1913.
**b)** When will it take approximately $40 to equal the value of $1 in 1913?

**88.** *Windmill Power.* Under certain conditions, the power $P$, in watts per hour, generated by a windmill with winds blowing $v$ miles per hour is given by

$$P(v) = 0.015v^3.$$

Find the power generated by 15-mph winds and by 35-mph winds.

**89.** *Boiling Point and Elevation.* The elevation $E$, in meters, above sea level at which the boiling point of water is $t$ degrees Celsius is given by the function

$$E(t) = 1000(100 - t) + 580(100 - t)^2.$$

At what elevation is the boiling point 99.5°? 100°?

## Skill Maintenance

*To the student and the instructor: The Skill Maintenance exercises review skills covered previously in the text. You can expect such exercises in every exercise set. They provide excellent review for a final examination. Answers to all skill maintenance exercises, along with section references, appear in the answer section at the back of the book.*

*Use substitution to determine whether the given ordered pairs are solutions of the given equation.*

**90.** $(-3, -2), (2, -3)$; $y^2 - x^2 = -5$

**91.** $(0, -7), (8, 11)$; $y = 0.5x + 7$

**92.** $\left(\frac{4}{5}, -2\right), \left(\frac{11}{5}, \frac{1}{10}\right)$; $15x - 10y = 32$

*Graph the equation.*

**93.** $y = (x - 1)^2$
**94.** $y = \frac{1}{3}x - 6$
**95.** $-2x - 5y = 10$
**96.** $(x - 3)^2 + y^2 = 4$

## Synthesis

*Find the domain of the function. Do not use a graphing calculator.*

**97.** $f(x) = \sqrt[4]{2x + 5} + 3$

**98.** $f(x) = \dfrac{\sqrt{x + 1}}{x}$

**99.** $f(x) = \dfrac{\sqrt{x+6}}{(x+2)(x-3)}$

**100.** $f(x) = \sqrt{x} - \sqrt{4-x}$

**101.** Give an example of two different functions that have the same domain and the same range, but have no pairs in common. Answers may vary.

**102.** Draw a graph of a function for which the domain is $[-4, 4]$ and the range is $[1, 2] \cup [3, 5]$. Answers may vary.

**103.** Suppose that for some function $g$, $g(x + 3) = 2x + 1$. Find $g(-1)$.

**104.** Suppose $f(x) = |x + 3| - |x - 4|$. Write $f(x)$ without using absolute-value notation if $x$ is in each of the following intervals.

a) $(-\infty, -3)$

b) $[-3, 4)$

c) $[4, \infty)$

---

# 1.3    Linear Functions, Slope, and Applications

- Determine the slope of a line given two points on the line.
- Solve applied problems involving slope, or average rate of change.
- Find the slope and the $y$-intercept of a line given the equation $y = mx + b$, or $f(x) = mx + b$.
- Graph a linear equation using the slope and the $y$-intercept.
- Solve applied problems involving linear functions.

In real-life situations, we often need to make decisions on the basis of limited information. When the given information is used to formulate an equation or an inequality that at least approximates the situation mathematically, we have created a **model**. One of the most frequently used mathematical models is *linear*. The graph of a linear model is a straight line.

## Linear Functions

Let's examine the connections among equations, functions, and graphs that are *straight lines*. First, examine the graphs of linear functions and nonlinear functions shown here. Note that the graphs of the two types of functions are quite different.

**Linear Functions**

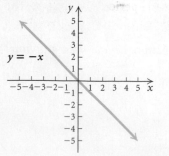

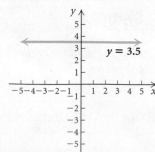

**Nonlinear Functions**

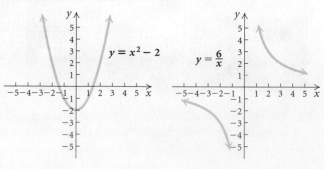

We begin with the definition of a linear function and related terminology that are illustrated with graphs below.

> ### LINEAR FUNCTION
>
> A function $f$ is a **linear function** if it can be written as
>
> $$f(x) = mx + b,$$
>
> where $m$ and $b$ are constants.
>
> If $m = 0$, the function is a **constant function** $f(x) = b$. If $m = 1$ and $b = 0$, the function is the **identity function** $f(x) = x$.

Linear function:
$y = mx + b$

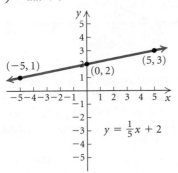

Identity function:
$y = 1 \cdot x + 0$, or $y = x$

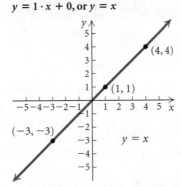

Constant function:
$y = 0 \cdot x + b$, or $y = b$ (Horizontal line)

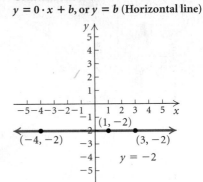

Vertical line: $x = a$
(*not* a function)

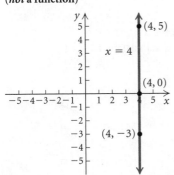

> ### HORIZONTAL LINES AND VERTICAL LINES
>
> **Horizontal lines** are given by equations of the type $y = b$ or $f(x) = b$. (They *are* functions.)
>
> **Vertical lines** are given by equations of the type $x = a$. (They *are not* functions.)

## ▪ The Linear Function $f(x) = mx + b$ and Slope

To attach meaning to the constant $m$ in the equation $f(x) = mx + b$, we first consider an application. Suppose Quality Foods is a wholesale supplier to restaurants that currently has stores in locations A and B in a large city. Their total operating costs for the same time period are given by the two functions shown in the tables and graphs that follow. The variable $x$

represents time, in months. The variable $y$ represents total costs, in thousands of dollars, over that period of time. Look for a pattern.

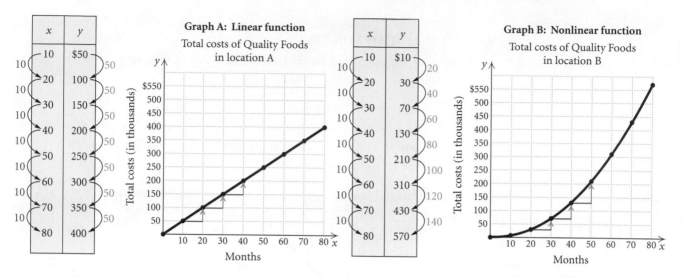

We see in graph A that *every* change of 10 months results in a $50 thousand change in total costs. But in graph B, changes of 10 months do *not* result in constant changes in total costs. This is a way to distinguish linear functions from nonlinear functions. The rate at which a linear function changes, or the steepness of its graph, is constant.

Mathematically, we define the steepness, or **slope**, of a line as the ratio of its vertical change (*rise*) to the corresponding horizontal change (*run*). Slope represents the **rate of change** of $y$ with respect to $x$.

**SLOPE**

The **slope $m$** of a line containing points $(x_1, y_1)$ and $(x_2, y_2)$ is given by

$$m = \frac{\text{rise}}{\text{run}}$$

$$= \frac{\text{the change in } y}{\text{the change in } x}$$

$$= \frac{y_2 - y_1}{x_2 - x_1} = \frac{y_1 - y_2}{x_1 - x_2}.$$

**EXAMPLE 1**  Graph the function $f(x) = -\frac{2}{3}x + 1$ and determine its slope.

**Solution**  Since the equation for $f$ is in the form $f(x) = mx + b$, we know that it is a linear function. We can graph it by connecting two points

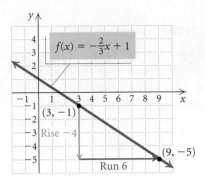

on the graph with a straight line. We calculate two ordered pairs, plot the points, graph the function, and determine the slope:

$$f(3) = -\frac{2}{3} \cdot 3 + 1 = -2 + 1 = -1;$$

$$f(9) = -\frac{2}{3} \cdot 9 + 1 = -6 + 1 = -5;$$

Pairs: $(3, -1)$, $(9, -5)$;

$$\text{Slope} = m = \frac{f(x_2) - f(x_1)}{x_2 - x_1} = \frac{y_2 - y_1}{x_2 - x_1}$$

$$= \frac{-5 - (-1)}{9 - 3} = \frac{-4}{6} = -\frac{2}{3}.$$

The slope is the same for any two points on a line. Thus, to check our work, note that $f(6) = -\frac{2}{3} \cdot 6 + 1 = -4 + 1 = -3$. Using the points $(6, -3)$ and $(3, -1)$, we have

$$m = \frac{-1 - (-3)}{3 - 6} = \frac{2}{-3} = -\frac{2}{3}.$$

We can also use the points in the opposite order when computing slope:

$$m = \frac{-3 - (-1)}{6 - 3} = \frac{-2}{3} = -\frac{2}{3}.$$

Note too that the slope of the line is the number $m$ in the equation for the function $f(x) = -\frac{2}{3}x + 1$.

**Now Try Exercise 7.**

The *slope* of the line given by $f(x) = mx + b$ is $m$.

If a line slants up from left to right, the change in $x$ and the change in $y$ have the same sign, so the line has a positive slope. The larger the slope, the steeper the line, as shown in Fig. 1. If a line slants down from left to right, the change in $x$ and the change in $y$ are of opposite signs, so the line has a negative slope. The larger the absolute value of the slope, the steeper the line, as shown in Fig. 2. Considering $y = mx$ when $m = 0$, we have $y = 0x$, or $y = 0$. Note that this horizontal line is the $x$-axis, as shown in Fig. 3.

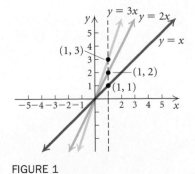

FIGURE 1

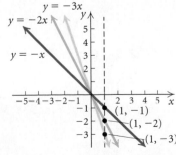

FIGURE 2

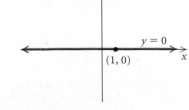

FIGURE 3

## HORIZONTAL LINES AND VERTICAL LINES

If a line is horizontal, the change in $y$ for any two points is 0 and the change in $x$ is nonzero. Thus a horizontal line has slope 0. (See Fig. 4.)

If a line is vertical, the change in $x$ for any two points is 0. Thus the slope is *not defined* because we cannot divide by 0. (See Fig. 5.)

**Horizontal lines**

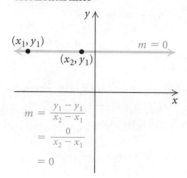

FIGURE 4

**Vertical lines**

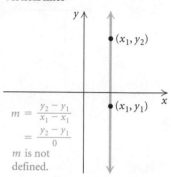

FIGURE 5

Note that zero slope and an undefined slope are two very different concepts.

**EXAMPLE 2**   Graph each linear equation and determine its slope.

**a)** $x = -2$                    **b)** $y = \frac{5}{2}$

*Solution*

**a)** Since $y$ is missing in $x = -2$, any value for $y$ will do.

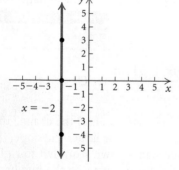

FIGURE 6

| $x$ | $y$ |
|-----|-----|
| $-2$ | 0 |
| $-2$ | 3 |
| $-2$ | $-4$ |

**Choose any number for $y$; $x$ must be $-2$.**

The graph is a *vertical line* 2 units to the left of the $y$-axis. (See Fig. 6.) The slope is not defined. The graph is *not* the graph of a function.

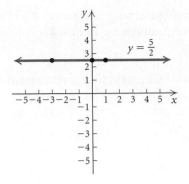

FIGURE 7

**b)** Since $x$ is missing in $y = \frac{5}{2}$, any value for $x$ will do.

| $x$ | $y$ |
|-----|-----|
| 0 | $\frac{5}{2}$ |
| $-3$ | $\frac{5}{2}$ |
| 1 | $\frac{5}{2}$ |

Choose any number for $x$; $y$ must be $\frac{5}{2}$.

The graph is a *horizontal line* $\frac{5}{2}$, or $2\frac{1}{2}$, units above the $x$-axis. (See Fig. 7.) The slope is 0. The graph is the graph of a constant function.

**Now Try Exercises 17 and 23.**

## ■ Applications of Slope

Slope has many real-world applications. Numbers like 2%, 4%, and 7% are often used to represent the **grade** of a road. Such a number is meant to tell how steep a road is on a hill or a mountain. For example, a 4% grade means that the road rises (or falls) 4 ft for every horizontal distance of 100 ft.

Road grade $= \frac{a}{b}$
(Expressed as a percent)

Whistler's bobsled/luge course for the 2010 Winter Olympics in Vancouver, British Columbia, was the fastest-ever Olympic course. The vertical drop for the course is 499 ft. The maximum slope of the track is 20% at Curve 2. (*Source*: Ron Judd, The *Seattle Times*)

The concept of grade is also used with a treadmill. During a treadmill test, a cardiologist might change the slope, or grade, of the treadmill to measure its effect on heart rate.

Another example occurs in hydrology. The strength or force of a river depends on how far the river falls vertically compared to how far it flows horizontally.

**EXAMPLE 3**    *Ramp for the Disabled.*    Construction laws regarding access ramps for the disabled state that every vertical rise of 1 ft requires a horizontal run of at least 12 ft. What is the grade, or slope, of such a ramp?

*Solution*    The grade, or slope, is given by $m = \frac{1}{12} \approx 0.083 = 8.3\%$.

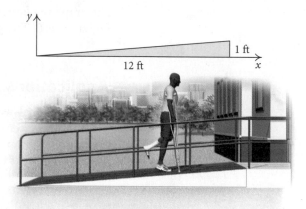

> ### AVERAGE RATE OF CHANGE
> Slope can also be considered as an **average rate of change**. To find the average rate of change between any two data points on a graph, we determine the slope of the line that passes through the two points.

**EXAMPLE 4**    *Adolescent Obesity.*    The percent of American adolescents ages 12 to 19 who are obese increased from about 6.5% in 1985 to 18% in 2008. The graph below illustrates this trend. Find the average rate of change from 1985 to 2008 in the percent of adolescents who are obese.

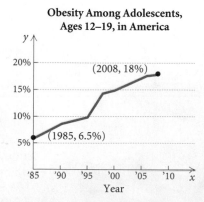

*Source*: National Center for Health Statistics, a division of the U.S. Centers for Disease Control and Prevention

***Solution*** We use the coordinates of two points on the graph. In this case, we use $(1985, 6.5\%)$ and $(2008, 18\%)$. Then we compute the slope, or average rate of change, as follows:

$$\text{Slope} = \text{Average rate of change} = \frac{\text{Change in } y}{\text{Change in } x}$$

$$= \frac{18\% - 6.5\%}{2008 - 1985} = \frac{11.5\%}{23} = 0.5\%.$$

The result tells us that each year from 1985 to 2008, the percent of adolescents who are obese increased an average of 0.5%. The average rate of change over this 23-year period was an increase of 0.5% per year.

**Now Try Exercise 41.**

**EXAMPLE 5** *Credit-Card Late Fees.* On February 22, 2010, the Credit Card Accountability, Responsibility and Disclosure Act of 2009 went into effect. The Credit Card Act requires that card issuers make it clear to their customers under what circumstances they will be charged a late fee and the limits on these fees put in place by the Federal Reserve. The amount of late fees paid by cardholders dropped from $901 million in January 2010 to $427 million in November 2010 (*Source*: Reuters News Service). Find the average rate of change in credit-card late fees per month from January 2010 to November 2010.

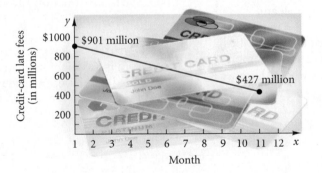

***Solution*** Using the points $(1, 901)$ and $(11, 427)$, where $x$ is the month and $y$ is the total amount of credit-card late fees per month, in millions of dollars, we compute the slope of the line containing these two points.

$$\text{Slope} = \text{Average rate of change} = \frac{\text{Change in } y}{\text{Change in } x}$$

$$= \frac{427 - 901}{11 - 1} = \frac{-474}{10} = -47.4.$$

The result tells us that each month from January 2010 to November 2010, the amount of credit-card late fees decreased on average $47.4 million. The average rate of change over the 10-month period was a decrease of $47.4 million per month.

**Now Try Exercise 47.**

## Slope–Intercept Equations of Lines

*y*-INTERCEPT

REVIEW SECTION **1.1.**

Compare the graphs of the equations

$$y = 3x \quad \text{and} \quad y = 3x - 2.$$

Note that the graph of $y = 3x - 2$ is a shift of the graph of $y = 3x$ down 2 units and that $y = 3x - 2$ has *y*-intercept $(0, -2)$. That is, the graph is parallel to $y = 3x$ and it crosses the *y*-axis at $(0, -2)$. The point $(0, -2)$ is the **y-intercept** of the graph.

### Exploring with Technology

We can use a graphing calculator to explore the effect of the constant *b* in linear equations of the type $f(x) = mx + b$. Begin with the graph of $y = x$. Now graph the lines $y = x + 3$ and $y = x - 4$ in the same viewing window. Try entering these equations as $y = x + \{0, 3, -4\}$ and compare the graphs. How do the last two lines differ from $y = x$? What do you think the line $y = x - 6$ will look like?

Clear the first set of equations and graph $y = -0.5x$, $y = -0.5x - 4$, and $y = -0.5x + 3$ in the same viewing window. Describe what happens to the graph of $y = -0.5x$ when a number *b* is added.

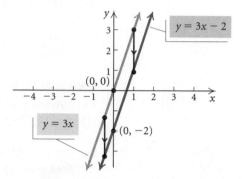

### THE SLOPE–INTERCEPT EQUATION

The linear function *f* given by

$$f(x) = mx + b$$

is written in slope–intercept form. The graph of an equation in this form is a straight line parallel to $f(x) = mx$. The constant *m* is called the slope, and the *y*-intercept is $(0, b)$.

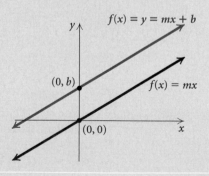

We can read the slope *m* and the *y*-intercept $(0, b)$ directly from the equation of a line written in slope–intercept form $y = mx + b$.

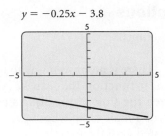

$y = -0.25x - 3.8$

**EXAMPLE 6** Find the slope and the $y$-intercept of the line with equation $y = -0.25x - 3.8$.

**Solution**

$$y = \underbrace{-0.25x}_{} \ \underbrace{- \ 3.8}_{}$$

$$\text{Slope} = -0.25; \quad y\text{-intercept} = (0, -3.8)$$

Now Try Exercise 49.

Any equation whose graph is a straight line is a **linear equation**. To find the slope and the $y$-intercept of the graph of a linear equation, we can solve for $y$, and then read the information from the equation.

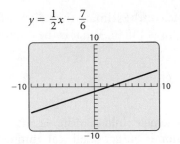

$y = \frac{1}{2}x - \frac{7}{6}$

**EXAMPLE 7** Find the slope and the $y$-intercept of the line with equation $3x - 6y - 7 = 0$.

**Solution** We solve for $y$:

$$3x - 6y - 7 = 0$$
$$-6y = -3x + 7 \qquad \text{Adding } -3x \text{ and 7 on both sides}$$
$$-\tfrac{1}{6}(-6y) = -\tfrac{1}{6}(-3x + 7) \qquad \text{Multiplying by } -\tfrac{1}{6} \text{ on both sides}$$
$$y = \tfrac{1}{2}x - \tfrac{7}{6}.$$

Thus the slope is $\frac{1}{2}$, and the $y$-intercept is $\left(0, -\frac{7}{6}\right)$.

Now Try Exercise 61.

## ■ Graphing $f(x) = mx + b$ Using $m$ and $b$

We can also graph a linear equation using its slope and $y$-intercept.

**EXAMPLE 8** Graph: $y = -\frac{2}{3}x + 4$.

**Solution** This equation is in slope–intercept form, $y = mx + b$. The $y$-intercept is $(0, 4)$. We plot this point. We can think of the slope $\left(m = -\frac{2}{3}\right)$ as $\frac{-2}{3}$.

$$m = \frac{\text{rise}}{\text{run}} = \frac{\text{change in } y}{\text{change in } x} = \frac{-2}{3} \quad \begin{matrix} \leftarrow \text{Move 2 units down.} \\ \leftarrow \text{Move 3 units to the right.} \end{matrix}$$

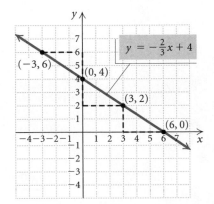

$y = -\frac{2}{3}x + 4$

Starting at the $y$-intercept and using the slope, we find another point by moving 2 units down and 3 units to the right. We get a new point $(3, 2)$. In a similar manner, we can move from $(3, 2)$ to find another point, $(6, 0)$.

We could also think of the slope $\left(m = -\frac{2}{3}\right)$ as $\frac{2}{-3}$. Then we can start at $(0, 4)$ and move 2 units up and 3 units to the left. We get to another point on the graph, $(-3, 6)$. We now plot the points and draw the line. Note that we need only the $y$-intercept and one other point in order to graph the line, but it's a good idea to find a third point as a check that the first two points are correct.

Now Try Exercise 63.

### ■ Applications of Linear Functions

We now consider an application of linear functions.

**EXAMPLE 9**    *Estimating Adult Height.*    There is no proven way to predict a child's adult height, but there is a linear function that can be used to estimate the adult height of a child, given the sum of the child's parents' heights. The adult height $M$, in inches, of a male child whose parents' combined height is $x$, in inches, can be estimated with the function

$$M(x) = 0.5x + 2.5.$$

The adult height $F$, in inches, of a female child whose parents' combined height is $x$, in inches, can be estimated with the function

$$F(x) = 0.5x - 2.5.$$

(*Source*: Jay L. Hoecker, M.D., MayoClinic.com) Estimate the height of a female child whose parents' combined height is 135 in. What is the domain of this function?

**Solution**    We substitute in the function:

$$F(135) = 0.5(135) - 2.5 = 65.$$

Thus we can estimate the adult height of the female child as 65 in., or 5 ft 5 in.

Theoretically, the domain of the function is the set of all real numbers. However, the context of the problem dictates a different domain. Thus the domain consists of all positive real numbers—that is, the interval $(0, \infty)$. A more realistic domain might be 100 in. to 170 in.—that is, the interval $[100, 170]$.

**Now Try Exercise 73.**

A
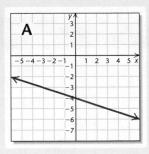

B
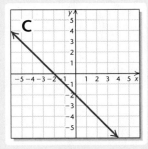

C
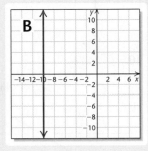

D
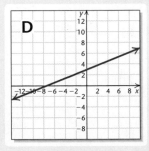

E
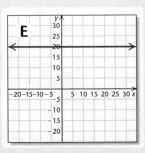

# Visualizing the Graph

Match the equation with its graph.

1. $y = 20$

2. $5y = 2x + 15$

3. $y = -\dfrac{1}{3}x - 4$

4. $x = \dfrac{5}{3}$

5. $y = -x - 2$

6. $y = 2x$

7. $y = -3$

8. $3y = -4x$

9. $x = -10$

10. $y = x + \dfrac{7}{2}$

Answers on page A-6

F
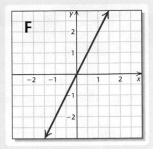

G
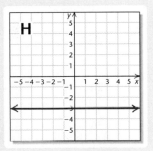

H
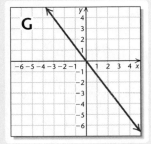

I
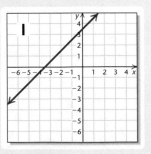

J
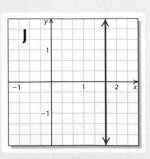

# 1.3 Exercise Set

*In Exercises 1–4, the table of data contains input–output values for a function. Answer the following questions for each table.*

**a)** Is the change in the inputs *x* the same?
**b)** Is the change in the outputs *y* the same?
**c)** Is the function linear?

**1.**

| x | y |
|----|----|
| −3 | 7 |
| −2 | 10 |
| −1 | 13 |
| 0 | 16 |
| 1 | 19 |
| 2 | 22 |
| 3 | 25 |

**2.**

| x | y |
|----|-------|
| 20 | 12.4 |
| 30 | 24.8 |
| 40 | 49.6 |
| 50 | 99.2 |
| 60 | 198.4 |
| 70 | 396.8 |
| 80 | 793.6 |

**3.**

| x | y |
|-----|------|
| 11 | 3.2 |
| 26 | 5.7 |
| 41 | 8.2 |
| 56 | 9.3 |
| 71 | 11.3 |
| 86 | 13.7 |
| 101 | 19.1 |

**4.**

| x | y |
|----|-----|
| 2 | −8 |
| 4 | −12 |
| 6 | −16 |
| 8 | −20 |
| 10 | −24 |
| 12 | −28 |
| 14 | −32 |

*Find the slope of the line containing the given points.*

**5.**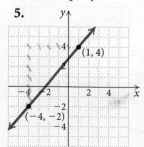

**6.**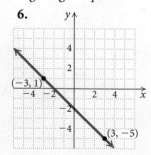

**7.**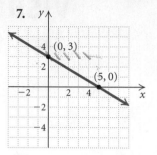

**8.**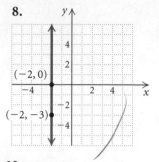

**9.**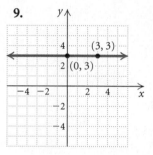

**10.**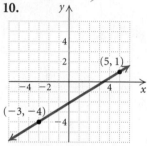

**11.** $(9, 4)$ and $(-1, 2)$

**12.** $(-3, 7)$ and $(5, -1)$

**13.** $(4, -9)$ and $(4, 6)$

**14.** $(-6, -1)$ and $(2, -13)$

**15.** $(0.7, -0.1)$ and $(-0.3, -0.4)$

**16.** $\left(-\frac{3}{4}, -\frac{1}{4}\right)$ and $\left(\frac{2}{7}, -\frac{5}{7}\right)$

**17.** $(2, -2)$ and $(4, -2)$

**18.** $(-9, 8)$ and $(7, -6)$

**19.** $\left(\frac{1}{2}, -\frac{3}{5}\right)$ and $\left(-\frac{1}{2}, \frac{3}{5}\right)$

**20.** $(-8.26, 4.04)$ and $(3.14, -2.16)$

**21.** $(16, -13)$ and $(-8, -5)$

**22.** $(\pi, -3)$ and $(\pi, 2)$

**23.** $(-10, -7)$ and $(-10, 7)$

**24.** $(\sqrt{2}, -4)$ and $(0.56, -4)$

**25.** $f(4) = 3$ and $f(-2) = 15$

26. $f(-4) = -5$ and $f(4) = 1$

27. $f\left(\frac{1}{5}\right) = \frac{1}{2}$ and $f(-1) = -\frac{11}{2}$

28. $f(8) = -1$ and $f\left(-\frac{2}{3}\right) = \frac{10}{3}$

29. $f(-6) = \frac{4}{5}$ and $f(0) = \frac{4}{5}$

30. $g\left(-\frac{9}{2}\right) = \frac{2}{9}$ and $g\left(\frac{2}{5}\right) = -\frac{5}{2}$

*Determine the slope, if it exists, of the graph of the given linear equation.*

31. $y = 1.3x - 5$

32. $y = -\frac{2}{5}x + 7$

33. $x = -2$

34. $f(x) = 4x - \frac{1}{4}$

35. $f(x) = -\frac{1}{2}x + 3$

36. $y = \frac{3}{4}$

37. $y = 9 - x$

38. $x = 8$

39. $y = 0.7$

40. $y = \frac{4}{5} - 2x$

41. *Fireworks Revenue.* The fireworks industry in the United States had a revenue of $945 million in 2009. This figure had increased from approximately $425 million in 1998. (*Source:* American Pyrotechnics Association) Find the average rate of change in the U.S. fireworks industry revenue from 1998 to 2009.

42. *Population Loss.* The population of Flint, Michigan, decreased from 124,943 in 2000 to 102,434 in 2010 (*Source:* U.S. Census Bureau). Find the average rate of change in the population of Flint, Michigan, over the 10-year period.

43. *Population Loss.* The population of St. Louis, Missouri, decreased from 348,194 in 2000 to 319,294 in 2010 (*Source:* U.S. Census Bureau). Find the average rate of change in the population of St. Louis, Missouri, over the 10-year period.

44. *Private Jets.* To cut costs, many corporations have been selling their private jets. The number of used

jets for sale worldwide has increased from 1022 in 1999 to 3014 in 2009 (*Source:* UBS Investment Research). Find the average rate of change in the number of used jets for sale from 1999 to 2009.

45. *Electric Bike Sales.* In 2009, electric bike sales in China totaled 21.0 million. Sales are estimated to rise to 25.0 million by 2012. (*Source: Electric Bike Worldwide Reports*) Find the average rate of change in sales of electric bikes in China from 2009 to 2012.

46. *Highway Deaths.* In 2010, the traffic fatality rate in the United States was the lowest since the federal government began keeping records in 1966. There were 43,510 highway deaths in 2005. This number had decreased to 32,788 in 2010. (*Source:* National Highway Traffic Safety Administration) Find the average rate of change in the number of highway deaths from 2005 to 2010.

47. *Consumption of Broccoli.* The U.S. annual per-capita consumption of broccoli was 3.1 lb in 1990. By 2008, this amount had risen to 5.5 lb. (*Source:* Economic Research Service, U.S. Department of Agriculture) Find the average rate of change in the consumption of broccoli per capita from 1990 to 2008.

**48.** *Account Overdraft Fees.* Bank revenue from overdraft fees for checking accounts, ATMs, and debit cards is increasing in the United States. In 2003, account overdraft revenue was $27.1 billion. This amount is estimated to increase to $38.0 billion by 2011. (*Source*: Moebs Services, R. K. Hammer Investment Bank) Find the average rate of change in account overdraft fees from 2003 to 2011.

*Find the slope and the y-intercept of the line with the given equation.*

**49.** $y = \frac{3}{5}x - 7$

**50.** $f(x) = -2x + 3$

**51.** $x = -\frac{2}{5}$

**52.** $y = \frac{4}{7}$

**53.** $f(x) = 5 - \frac{1}{2}x$

**54.** $y = 2 + \frac{3}{7}x$

**55.** $3x + 2y = 10$

**56.** $2x - 3y = 12$

**57.** $y = -6$

**58.** $x = 10$

**59.** $5y - 4x = 8$

**60.** $5x - 2y + 9 = 0$

**61.** $4y - x + 2 = 0$

**62.** $f(x) = 0.3 + x$

*Graph the equation using the slope and the y-intercept.*

**63.** $y = -\frac{1}{2}x - 3$

**64.** $y = \frac{3}{2}x + 1$

**65.** $f(x) = 3x - 1$

**66.** $f(x) = -2x + 5$

**67.** $3x - 4y = 20$

**68.** $2x + 3y = 15$

**69.** $x + 3y = 18$

**70.** $5y - 2x = -20$

**71.** *Whales and Pressure at Sea Depth.* Whales can withstand extreme atmospheric pressure changes because their bodies are flexible. Their rib cages and lungs can collapse safely under pressure. Sperm whales can hunt for squid at depths of 7000 ft or more. (*Sources*: National Ocean Service, National Oceanic and Atmospheric Administration) The function $P$, given by

$$P(d) = \frac{1}{33}d + 1,$$

gives the pressure, in atmospheres (atm), at a given depth $d$, in feet, under the sea.

**a)** Graph $P$.
**b)** Find $P(0)$, $P(33)$, $P(1000)$, $P(5000)$, and $P(7000)$.

**72.** *Stopping Distance on Glare Ice.* The stopping distance (at some fixed speed) of regular tires on glare ice is a function of the air temperature $F$, in degrees Fahrenheit. This function is estimated by

$$D(F) = 2F + 115,$$

where $D(F)$ is the stopping distance, in feet, when the air temperature is $F$, in degrees Fahrenheit.

**a)** Graph $D$.
**b)** Find $D(0°)$, $D(-20°)$, $D(10°)$, and $D(32°)$.
**c)** Explain why the domain should be restricted to $[-57.5°, 32°]$.

**73.** *Reaction Time.* Suppose that while driving a car, you suddenly see a deer standing in the road. Your brain registers the information and sends a signal to your foot to hit the brake. The car travels a distance $D$, in feet, during this time, where $D$ is a function of the speed $r$, in miles per hour, of the car when you see the deer. That reaction distance is a linear function given by

$$D(r) = \frac{11}{10}r + \frac{1}{2}.$$

**a)** Find the slope of this line and interpret its meaning in this application.
**b)** Graph $D$.

**c)** Find $D(5)$, $D(10)$, $D(20)$, $D(50)$, and $D(65)$.

**d)** What is the domain of this function? Explain.

**74.** *Straight-Line Depreciation.* A contractor buys a new truck for $23,000. The truck is purchased on January 1 and is expected to last 5 years, at the end of which time its *trade-in*, or *salvage, value* will be $4500. If the company figures the decline or depreciation in value to be the same each year, then the salvage value $V$, after $t$ years, is given by the linear function

$$V(t) = \$23{,}000 - \$3700t, \quad \text{for } 0 \le t \le 5.$$

**a)** Graph $V$.

**b)** Find $V(0)$, $V(1)$, $V(2)$, $V(3)$, and $V(5)$.

**c)** Find the domain and the range of this function.

**75.** *Total Cost.* Steven buys a phone for $89 and signs up for a Verizon nationwide plus Mexico single-line phone plan with 2000 monthly anytime minutes. The plan costs $114.99 per month (*Source:* verizonwireless.com). Write an equation that can be used to determine the total cost $C(t)$ of operating this Verizon phone plan for $t$ months. Then find the cost for 24 months, assuming that the number of minutes Steven uses does not exceed 2000 per month.

**76.** *Total Cost.* Superior Cable Television charges a $95 installation fee and $125 per month for the Star plan. Write an equation that can be used to determine the total cost $C(t)$ for $t$ months of the Star plan. Then find the total cost for 18 months of service.

*In Exercises 77 and 78, the term **fixed costs** refers to the start-up costs of operating a business. This includes machinery and building costs. The term **variable costs** refers to what it costs a business to produce or service one item.*

**77.** Max's Custom Lacrosse Stringing experienced fixed costs of $750 and variable costs of $15 for each lacrosse stick that was restrung. Write an equation that can be used to determine the total cost when $x$ sticks are restrung. Then

determine the total cost of restringing 32 lacrosse sticks.

**78.** Soosie's Cookie Company had fixed costs of $1250 and variable costs of $4.25 per dozen gourmet cookies that were baked and packaged for sale. Write an equation that can be used to determine the total cost when $x$ dozens of cookies are baked and sold. Then determine the total cost of baking and selling 85 dozen gourmet cookies.

## Skill Maintenance

*If $f(x) = x^2 - 3x$, find each of the following.*

**79.** $f\left(\frac{1}{2}\right)$

**80.** $f(5)$

**81.** $f(-5)$

**82.** $f(-a)$

**83.** $f(a + h)$

## Synthesis

**84.** *Grade of Treadmills.* A treadmill is 5 ft long and is set at an 8% grade. How high is the end of the treadmill?

*Find the slope of the line containing the given points.*

**85.** $(a, a^2)$ and $(a + h, (a + h)^2)$

**86.** $(r, s + t)$ and $(r, s)$

*Suppose that f is a linear function. Determine whether the statement is true or false.*

**87.** $f(c - d) = f(c) - f(d)$

**88.** $f(kx) = kf(x)$

*Let $f(x) = mx + b$. Find a formula for $f(x)$ given each of the following.*

**89.** $f(x + 2) = f(x) + 2$

**90.** $f(3x) = 3f(x)$

## Mid-Chapter Mixed Review

*Determine whether the statement is true or false.*

**1.** The $x$-intercept of the line that passes through $\left(-\frac{2}{3}, \frac{3}{2}\right)$ and the origin is $\left(-\frac{2}{3}, 0\right)$. [1.1]

**2.** All functions are relations, but not all relations are functions. [1.2]

**3.** The line parallel to the $y$-axis that passes through $(-5, 25)$ is $y = -5$. [1.3]

**4.** Find the intercepts of the graph of the line $-8x + 5y = -40$. [1.1]

*For each pair of points, find the distance between the points and the midpoint of the segment having the points as endpoints.* [1.1]

**5.** $(-8, -15)$ and $(3, 7)$

**6.** $\left(-\frac{3}{4}, \frac{1}{5}\right)$ and $\left(\frac{1}{4}, -\frac{4}{5}\right)$

**7.** Find an equation for a circle with center $(-5, 2)$ and radius of length 13. [1.1]

**8.** Find the center and the radius of the circle given by the equation $(x - 3)^2 + (y + 1)^2 = 4$. [1.1]

*Graph the equation.*

**9.** $3x - 6y = 6$ [1.1]

**10.** $y = -\frac{1}{2}x + 3$ [1.3]

**11.** $y = 2 - x^2$ [1.1]

**12.** $(x + 4)^2 + y^2 = 4$ [1.1]

**13.** Given that $f(x) = x - 2x^2$, find $f(-4), f(0)$, and $f(1)$. [1.2]

**14.** Given that $g(x) = \dfrac{x + 6}{x - 3}$, find $g(-6), g(0)$, and $g(3)$. [1.2]

*Find the domain of the function.* [1.2]

**15.** $g(x) = x + 9$

**16.** $f(x) = \dfrac{-5}{x + 5}$

**17.** $h(x) = \dfrac{1}{x^2 + 2x - 3}$

*Graph the function.* [1.2]

**18.** $f(x) = -2x$

**19.** $g(x) = x^2 - 1$

**20.** Determine the domain and the range of the function. [1.1]

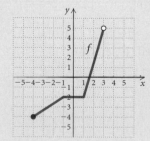

*Find the slope of the line containing the given points.* [1.3]

**21.** $(-2, 13)$ and $(-2, -5)$     **22.** $(10, -1)$ and $(-6, 3)$     **23.** $\left(\frac{5}{7}, \frac{1}{3}\right)$ and $\left(\frac{2}{7}, \frac{1}{3}\right)$

*Determine the slope, if it exists, and the y-intercept of the line with the given equation.* [1.3]

**24.** $f(x) = -\frac{1}{9}x + 12$            **25.** $y = -6$

**26.** $x = 2$                     **27.** $3x - 16y + 1 = 0$

## Collaborative Discussion and Writing

*To the student and the instructor: The Collaborative Discussion and Writing exercises are meant to be answered with one or more sentences. They can be discussed and answered collaboratively by the entire class or by small groups.*

**28.** Explain as you would to a fellow student how the numerical value of the slope of a line can be used to describe the slant and the steepness of that line. [1.3]

**29.** Discuss why the graph of a vertical line $x = a$ cannot represent a function. [1.3]

**30.** Explain in your own words the difference between the domain of a function and the range of a function. [1.2]

**31.** Explain how you could find the coordinates of a point $\frac{7}{8}$ of the way from point $A$ to point $B$. [1.1]

# Equations of Lines and Modeling

## 1.4

- Determine equations of lines.
- Given the equations of two lines, determine whether their graphs are parallel or perpendicular.
- Model a set of data with a linear function.
- Fit a regression line to a set of data. Then use the linear model to make predictions.

### ■ Slope–Intercept Equations of Lines

In Section 1.3, we developed the slope–intercept equation $y = mx + b$, or $f(x) = mx + b$. If we know the slope and the $y$-intercept of a line, we can find an equation of the line using the slope–intercept equation.

**EXAMPLE 1** A line has slope $-\frac{7}{9}$ and $y$-intercept $(0, 16)$. Find an equation of the line.

***Solution*** We use the slope–intercept equation and substitute $-\frac{7}{9}$ for $m$ and 16 for $b$:

$y = mx + b$

$y = -\frac{7}{9}x + 16$, or $f(x) = -\frac{7}{9}x + 16$.

> **Now Try Exercise 7.**

**EXAMPLE 2** A line has slope $-\frac{2}{3}$ and contains the point $(-3, 6)$. Find an equation of the line.

***Solution*** We use the slope–intercept equation, $y = mx + b$, and substitute $-\frac{2}{3}$ for $m$: $y = -\frac{2}{3}x + b$. Then, using the point $(-3, 6)$, we substitute $-3$ for $x$ and $6$ for $y$ in $y = -\frac{2}{3}x + b$. Finally, we solve for $b$.

$$y = mx + b$$
$$y = -\tfrac{2}{3}x + b \qquad \text{Substituting } -\tfrac{2}{3} \text{ for } m$$
$$6 = -\tfrac{2}{3}(-3) + b \qquad \text{Substituting } -3 \text{ for } x \text{ and } 6 \text{ for } y$$
$$6 = 2 + b$$
$$4 = b \qquad \text{Solving for } b. \text{ The } y\text{-intercept is } (0, b).$$

The equation of the line is $y = -\frac{2}{3}x + 4$, or $f(x) = -\frac{2}{3}x + 4$.

**Now Try Exercise 13.**

## ■ Point–Slope Equations of Lines

Another formula that can be used to determine an equation of a line is the *point–slope equation*. Suppose that we have a nonvertical line and that the coordinates of point $P_1$ on the line are $(x_1, y_1)$. We can think of $P_1$ as fixed and imagine another point $P$ on the line with coordinates $(x, y)$. Thus the slope is given by

$$\frac{y - y_1}{x - x_1} = m.$$

Multiplying by $x - x_1$ on both sides, we get the *point–slope equation* of the line:

$$(x - x_1) \cdot \frac{y - y_1}{x - x_1} = m \cdot (x - x_1)$$
$$y - y_1 = m(x - x_1).$$

### POINT–SLOPE EQUATION

The **point–slope equation** of the line with slope $m$ passing through $(x_1, y_1)$ is

$$y - y_1 = m(x - x_1).$$

If we know the slope of a line and the coordinates of one point on the line, we can find an equation of the line using either the point–slope equation,

$$y - y_1 = m(x - x_1),$$

or the slope–intercept equation,

$$y - mx + b.$$

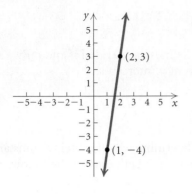

**EXAMPLE 3** Find an equation of the line containing the points $(2, 3)$ and $(1, -4)$.

**Solution** We first determine the slope:

$$m = \frac{-4 - 3}{1 - 2} = \frac{-7}{-1} = 7.$$

*Using the Point–Slope Equation:* We substitute 7 for $m$ and either of the points $(2, 3)$ or $(1, -4)$ for $(x_1, y_1)$ in the point–slope equation. In this case, we use $(2, 3)$.

$$y - y_1 = m(x - x_1) \qquad \textbf{Point–slope equation}$$
$$y - 3 = 7(x - 2) \qquad \textbf{Substituting}$$
$$y - 3 = 7x - 14$$
$$y = 7x - 11, \text{ or } f(x) = 7x - 11$$

*Using the Slope–Intercept Equation:* We substitute 7 for $m$ and either of the points $(2, 3)$ or $(1, -4)$ for $(x, y)$ in the slope–intercept equation and solve for $b$. Here we use $(1, -4)$.

$$y = mx + b \qquad \textbf{Slope–intercept equation}$$
$$-4 = 7 \cdot 1 + b \qquad \textbf{Substituting}$$
$$-4 = 7 + b$$
$$-11 = b \qquad \textbf{Solving for } b$$

We substitute 7 for $m$ and $-11$ for $b$ in $y = mx + b$ to get

$$y = 7x - 11, \text{ or } f(x) = 7x - 11.$$

**Now Try Exercise 19.**

## ▪ Parallel Lines

Can we determine whether the graphs of two linear equations are parallel without graphing them? Let's look at three pairs of equations and their graphs.

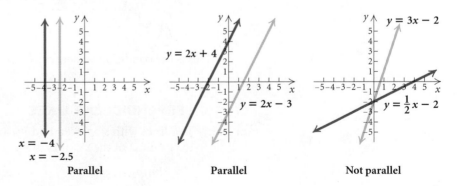

If two different lines, such as $x = -4$ and $x = -2.5$, are vertical, then they are parallel. Thus two equations such as $x = a_1$ and $x = a_2$, where $a_1 \neq a_2$, have graphs that are *parallel lines*. Two nonvertical lines, such as $y = 2x + 4$ and $y = 2x - 3$, or, in general, $y = mx + b_1$ and $y = mx + b_2$, where the slopes are the *same* and $b_1 \neq b_2$, also have graphs that are *parallel lines*.

> **PARALLEL LINES**
>
> Vertical lines are **parallel**. Nonvertical lines are **parallel** if and only if they have the same slope and different $y$-intercepts.

## ■ Perpendicular Lines

Can we examine a pair of equations to determine whether their graphs are perpendicular without graphing the equations? Let's look at the following pairs of equations and their graphs.

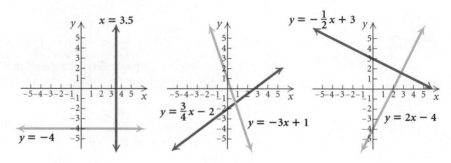

Perpendicular               Not perpendicular               Perpendicular

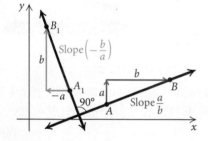

If one line is vertical and another is horizontal, they are perpendicular. For example, the lines $x = 5$ and $y = -3$ are perpendicular. Otherwise, how can we tell whether two lines are perpendicular? Consider a line $\overleftrightarrow{AB}$, as shown in the figure at left, with slope $a/b$. Then think of rotating the line $90°$ to get a line $\overleftrightarrow{A_1B_1}$ perpendicular to $\overleftrightarrow{AB}$. For the new line, the rise and the run are interchanged, but the run is now negative. Thus the slope of the new line is $-b/a$, which is the opposite of the reciprocal of the slope of the first line. Also note that when we multiply the slopes, we get

$$\frac{a}{b}\left(-\frac{b}{a}\right) = -1.$$

This is the condition under which lines will be perpendicular.

> **PERPENDICULAR LINES**
>
> Two lines with slopes $m_1$ and $m_2$ are **perpendicular** if and only if the product of their slopes is $-1$:
>
> $$m_1 m_2 = -1.$$
>
> Lines are also **perpendicular** if one is vertical ($x = a$) and the other is horizontal ($y = b$).

If a line has slope $m_1$, the slope $m_2$ of a line perpendicular to it is $-1/m_1$. The slope of one line is the *opposite of the reciprocal* of the other:

$$m_2 = -\frac{1}{m_1}, \quad \text{or} \quad m_1 = -\frac{1}{m_2}.$$

**EXAMPLE 4** Determine whether each of the following pairs of lines is parallel, perpendicular, or neither.

**a)** $y + 2 = 5x$, $5y + x = -15$
**b)** $2y + 4x = 8$, $5 + 2x = -y$
**c)** $2x + 1 = y$, $y + 3x = 4$

**Solution** We use the slopes of the lines to determine whether the lines are parallel or perpendicular.

**a)** We solve each equation for $y$:

$$y = 5x - 2, \qquad y = -\tfrac{1}{5}x - 3.$$

The slopes are $5$ and $-\tfrac{1}{5}$. Their product is $-1$, so the lines are perpendicular. (See Fig. 1.) Note in the graphs at left that the graphing calculator windows have been squared to avoid distortion. (Review squaring windows in Section 1.1.)

**b)** Solving each equation for $y$, we get

$$y = -2x + 4, \qquad y = -2x - 5.$$

We see that $m_1 = -2$ and $m_2 = -2$. Since the slopes are the same and the $y$-intercepts, $(0, 4)$ and $(0, -5)$, are different, the lines are parallel. (See Fig. 2.)

**c)** Rewriting the first equation and solving the second equation for $y$, we have

$$y = 2x + 1, \qquad y = -3x + 4.$$

We see that $m_1 = 2$ and $m_2 = -3$. Since the slopes are not the same and their product is not $-1$, it follows that the lines are neither parallel nor perpendicular. (See Fig. 3.) **Now Try Exercises 35 and 39.**

**EXAMPLE 5** Write equations of the lines **(a)** parallel and **(b)** perpendicular to the graph of the line $4y - x = 20$ and containing the point $(2, -3)$.

**Solution** We first solve $4y - x = 20$ for $y$ to get $y = \tfrac{1}{4}x + 5$. Thus the slope of the given line is $\tfrac{1}{4}$.

**a)** The line parallel to the given line will have slope $\tfrac{1}{4}$. We use either the slope–intercept equation or the point–slope equation for a line with slope $\tfrac{1}{4}$ and containing the point $(2, -3)$. Here we use the point–slope equation:

$$y - y_1 = m(x - x_1)$$
$$y - (-3) = \tfrac{1}{4}(x - 2)$$
$$y + 3 = \tfrac{1}{4}x - \tfrac{1}{2}$$
$$y = \tfrac{1}{4}x - \tfrac{7}{2}.$$

$y_1 = 5x - 2, \quad y_2 = -\frac{1}{5}x - 3$

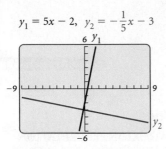

**FIGURE 1**

$y_1 = -2x + 4, \quad y_2 = -2x - 5$

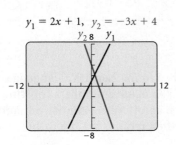

**FIGURE 2**

$y_1 = 2x + 1, \quad y_2 = -3x + 4$

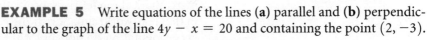

**FIGURE 3**

$y_1 = \frac{1}{4}x + 5, \quad y_2 = \frac{1}{4}x - \frac{7}{2},$

$y_3 = -4x + 5$

**b)** The slope of the perpendicular line is the opposite of the reciprocal of $\frac{1}{4}$, or $-4$. Again we use the point–slope equation to write an equation for a line with slope $-4$ and containing the point $(2, -3)$:

$$y - y_1 = m(x - x_1)$$
$$y - (-3) = -4(x - 2)$$
$$y + 3 = -4x + 8$$
$$y = -4x + 5.$$

**Now Try Exercise 43.**

---

### Summary of Terminology about Lines

| TERMINOLOGY | MATHEMATICAL INTERPRETATION |
|---|---|
| Slope | $m = \dfrac{y_2 - y_1}{x_2 - x_1}, \text{ or } \dfrac{y_1 - y_2}{x_1 - x_2}$ |
| Slope–intercept equation | $y = mx + b$ |
| Point–slope equation | $y - y_1 = m(x - x_1)$ |
| Horizontal line | $y = b$ |
| Vertical line | $x = a$ |
| Parallel lines | $m_1 = m_2, \ b_1 \neq b_2;$ <br> or $x = a_1, x = a_2, a_1 \neq a_2$ |
| Perpendicular lines | $m_1 m_2 = -1, \text{ or } m_2 = -\dfrac{1}{m_1};$ <br> or $x = a, y = b$ |

---

**Creating a Mathematical Model**

**1.** Recognize real-world problem.

**2.** Collect data.

**3.** Analyze data.

**4.** Construct model.

**5.** Test and refine model.

**6.** Explain and predict.

## Mathematical Models

When a real-world problem can be described in mathematical language, we have a **mathematical model**. For example, the natural numbers constitute a mathematical model for situations in which counting is essential. Situations in which algebra can be brought to bear often require the use of functions as models.

Mathematical models are abstracted from real-world situations. The mathematical model gives results that allow one to predict what will happen in that real-world situation. If the predictions are inaccurate or the results of experimentation do not conform to the model, the model must be changed or discarded.

Mathematical modeling can be an ongoing process. For example, finding a mathematical model that will provide an accurate prediction of population growth is not a simple task. Any population model that one might devise would need to be reshaped as further information is acquired.

## Curve Fitting

We will develop and use many kinds of mathematical models in this text. In this chapter, we have used *linear* functions as models. Other types of functions, such as quadratic, cubic, and exponential functions, can also model data. These functions are *nonlinear*. Modeling with quadratic and cubic functions is discussed in Chapter 4. Modeling with exponential functions is discussed in Chapter 5.

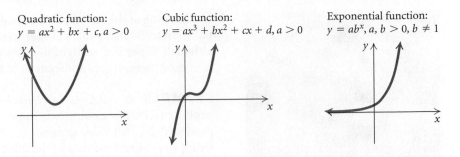

Quadratic function:
$y = ax^2 + bx + c, a > 0$

Cubic function:
$y = ax^3 + bx^2 + cx + d, a > 0$

Exponential function:
$y = ab^x, a, b > 0, b \neq 1$

In general, we try to find a function that fits, as well as possible, observations (data), theoretical reasoning, and common sense. We call this **curve fitting**; it is one aspect of mathematical modeling.

Let's look at some data and related graphs or **scatterplots** and determine whether a linear function seems to fit the set of data.

| Year, $x$ | Gross Domestic Product (GDP) (in trillions) | Scatterplot |
|---|---|---|
| 1980,  0 | $ 2.8 | **Gross Domestic Product** |
| 1985,  5 | 4.2 | It appears that the data points can be represented or modeled by a linear function. |
| 1990, 10 | 5.8 | |
| 1995, 15 | 7.4 | The graph is **linear**. |
| 2000, 20 | 10.0 | |
| 2005, 25 | 12.6 | |
| 2010, 30 | 14.7 | |

*Sources*: Bureau of Economic Analysis; U.S. Department of Commerce

| Year, $x$ | Installed Wind Power Capacity (in megawatts) | Scatterplot |
|---|---|---|
| 2001, 0 | 24,322 | It appears that the data points cannot be modeled accurately by a linear function. |
| 2002, 1 | 31,181 | |
| 2003, 2 | 39,295 | The graph is **nonlinear**. |
| 2004, 3 | 47,693 | |
| 2005, 4 | 59,024 | |
| 2006, 5 | 74,122 | |
| 2007, 6 | 93,930 | |
| 2008, 7 | 120,903 | |
| 2009, 8 | 159,213 | |
| 2010, 9 | 203,500 | |

*Source*: World Wind Energy Association

Looking at the scatterplots, we see that the data on gross domestic product seem to be rising in a manner to suggest that a *linear function* might fit, although a "perfect" straight line cannot be drawn through the data points. A linear function does not seem to fit the data on wind power capacity.

**EXAMPLE 6** *U.S. Gross Domestic Product.* The **gross domestic product** (GDP) of a country is the market value of final goods and services produced. Market value depends on the quantity of goods and services and their price. Model the data in the table on p. 113 on the U.S. Gross Domestic Product with a linear function. Then estimate the GDP in 2014.

*Solution* We can choose any two of the data points to determine an equation. Note that the first coordinate is the number of years since 1980 and the second coordinate is the corresponding GDP in trillions of dollars. Let's use $(5, 4.2)$ and $(25, 12.6)$.

We first determine the slope of the line:

$$m = \frac{12.6 - 4.2}{25 - 5} = \frac{8.4}{20} = 0.42.$$

Then we substitute 0.42 for $m$ and either of the points $(5, 4.2)$ or $(25, 12.6)$ for $(x_1, y_1)$ in the point–slope equation. In this case, we use $(5, 4.2)$. We get

$$y - y_1 = m(x - x_1) \qquad \text{Point–slope equation}$$
$$y - 4.2 = 0.42(x - 5), \qquad \text{Substituting}$$

which simplifies to

$$y = 0.42x + 2.1,$$

where $x$ is the number of years after 1980 and $y$ is in trillions of dollars.

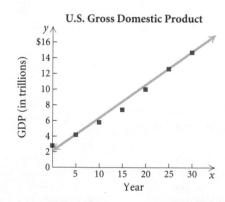

Next, we estimate the GDP in 2014 by substituting 34 $(2014 - 1980 = 34)$ for $x$ in the model:

$$y = 0.42x + 2.1 \qquad \text{Model}$$
$$= 0.42(34) + 2.1 \qquad \text{Substituting}$$
$$= 16.38.$$

We estimate that the gross domestic product will be $16.38 trillion in 2014.

**Now Try Exercise 61.**

In Example 6, if we were to use the data points $(10, 5.8)$ and $(30, 14.7)$, our model would be

$$y = 0.445x + 1.35,$$

and our estimate for the GDP in 2014 would be $16.48 trillion, about $0.10 trillion more than the estimate provided by the first model. This illustrates that a model and the estimates it produces are dependent on the data points used.

Models that consider all the data points, not just two, are generally better models. The linear model that best fits the data can be found using a graphing calculator and a procedure called **linear regression**.

## ◼ Linear Regression

Although discussion leading to a complete understanding of linear regression belongs in a statistics course, we present the procedure here because we can carry it out easily using technology. The graphing calculator gives us the powerful capability to find linear models and to make predictions using them.

Consider the data presented before Example 6 on the gross domestic product. We can fit a regression line of the form $y = mx + b$ to the data using the LINEAR REGRESSION feature on a graphing calculator.

**EXAMPLE 7**   *U.S. Gross Domestic Product.*   Fit a regression line to the data given in the table on gross domestic product on p. 113. Then use the function to estimate the GDP in 2014.

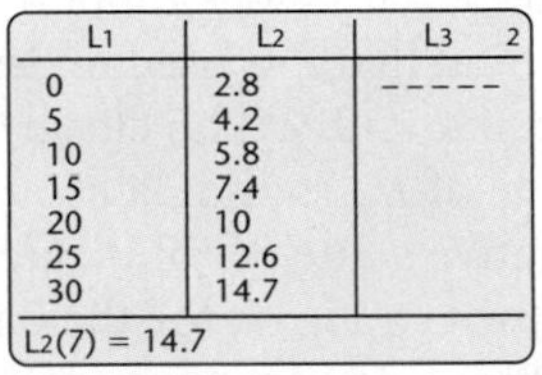

**Solution**   First, we enter the data in lists on the calculator. We enter the values of the independent variable $x$ in list L1 and the corresponding values of the dependent variable $y$ in L2. (See Fig. 1.) The graphing calculator can then create a scatterplot of the data, as shown in Fig. 2.

When we select the LINEAR REGRESSION feature from the STAT CALC menu, we find the linear equation that best models the data. It is

$$y = 0.405x + 2.139285714. \qquad \text{Regression line}$$

(See Figs. 3 and 4.) We can then graph the regression line on the same graph as the scatterplot, as shown in Fig. 5.

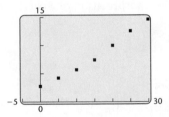

FIGURE 1

FIGURE 2

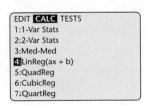

FIGURE 3

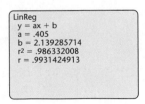

FIGURE 4

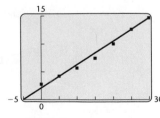

FIGURE 5

To estimate the gross domestic product in 2014, we substitute 34 for $x$ in the regression equation, which has been saved as Y1. Using this model, we see that the gross domestic product in 2014 is estimated to be about $15.91 trillion. (See Fig. 6.)

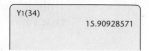

Y1(34)
           15.90928571

FIGURE 6

Note that $15.91 trillion is closer to the value $16.38 trillion found with data points $(5, 4.2)$ and $(25, 12.6)$ in Example 6 than to the value $16.48 trillion found with the data points $(10, 5.8)$ and $(30, 14.7)$ following Example 6.

**Now Try Exercises 69(a) and 69(b).**

## The Correlation Coefficient

On some graphing calculators with the DIAGNOSTIC feature turned on, a constant $r$ between $-1$ and $1$, called the **coefficient of linear correlation**, appears with the equation of the regression line. Though we cannot develop a formula for calculating $r$ in this text, keep in mind that it is used to describe the strength of the linear relationship between $x$ and $y$. The closer $|r|$ is to 1, the better the correlation. A positive value of $r$ also indicates that the regression line has a positive slope, and a negative value of $r$ indicates that the regression line has a negative slope. As shown in Fig. 4, for the data on gross domestic product just discussed, $r = 0.9931424913$, which indicates a very good linear correlation.

The following scatterplots summarize the interpretation of a correlation coefficient.

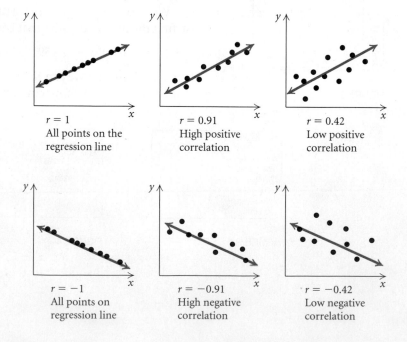

$r = 1$
All points on the regression line

$r = 0.91$
High positive correlation

$r = 0.42$
Low positive correlation

$r = -1$
All points on regression line

$r = -0.91$
High negative correlation

$r = -0.42$
Low negative correlation

## 1.4 Exercise Set

*Find the slope and the y-intercept of the graph of the linear equation. Then write the equation of the line in slope–intercept form.*

**1.**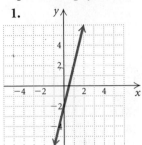

**2.**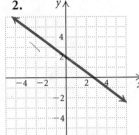

**3.**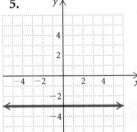

**4.**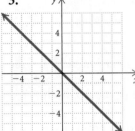

**5.**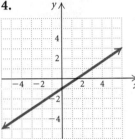

**6.**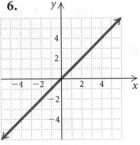

*Write a slope–intercept equation for a line with the given characteristics.*

**7.** $m = \frac{2}{9}$, y-intercept $(0, 4)$

**8.** $m = -\frac{3}{8}$, y-intercept $(0, 5)$

**9.** $m = -4$, y-intercept $(0, -7)$

**10.** $m = \frac{2}{7}$, y-intercept $(0, -6)$

**11.** $m = -4.2$, y-intercept $\left(0, \frac{3}{4}\right)$

**12.** $m = -4$, y-intercept $\left(0, -\frac{3}{2}\right)$

**13.** $m = \frac{2}{9}$, passes through $(3, 7)$

**14.** $m = -\frac{3}{8}$, passes through $(5, 6)$

**15.** $m = 0$, passes through $(-2, 8)$

**16.** $m = -2$, passes through $(-5, 1)$

**17.** $m = -\frac{3}{5}$, passes through $(-4, -1)$

**18.** $m = \frac{2}{3}$, passes through $(-4, -5)$

**19.** Passes through $(-1, 5)$ and $(2, -4)$

**20.** Passes through $\left(-3, \frac{1}{2}\right)$ and $\left(1, \frac{1}{2}\right)$

**21.** Passes through $(7, 0)$ and $(-1, 4)$

**22.** Passes through $(-3, 7)$ and $(-1, -5)$

**23.** Passes through $(0, -6)$ and $(3, -4)$

**24.** Passes through $(-5, 0)$ and $\left(0, \frac{4}{5}\right)$

**25.** Passes through $(-4, 7.3)$ and $(0, 7.3)$

**26.** Passes through $(-13, -5)$ and $(0, 0)$

*Write equations of the horizontal line and the vertical line that pass through the given point.*

**27.** $(0, -3)$

**28.** $\left(-\frac{1}{4}, 7\right)$

**29.** $\left(\frac{2}{11}, -1\right)$

**30.** $(0.03, 0)$

**31.** Find a linear function $h$ given $h(1) = 4$ and $h(-2) = 13$. Then find $h(2)$.

**32.** Find a linear function $g$ given $g\left(-\frac{1}{4}\right) = -6$ and $g(2) = 3$. Then find $g(-3)$.

**33.** Find a linear function $f$ given $f(5) = 1$ and $f(-5) = -3$. Then find $f(0)$.

**34.** Find a linear function $h$ given $h(-3) = 3$ and $h(0) = 2$. Then find $h(-6)$.

*Determine whether the pair of lines is parallel, perpendicular, or neither.*

**35.** $y = \frac{26}{3}x - 11$,
$y = -\frac{3}{26}x - 11$

**36.** $y = -3x + 1$,
$y = -\frac{1}{3}x + 1$

**37.** $y = \frac{2}{5}x - 4$,
$y = -\frac{2}{5}x + 4$

**38.** $y = \frac{3}{2}x - 8$,
$y = 8 + 1.5x$

**39.** $x + 2y = 5$,
$2x + 4y = 8$

**40.** $2x - 5y = -3$,
$2x + 5y = 4$

**41.** $y = 4x - 5$,
$4y = 8 - x$

**42.** $y = 7 - x$,
$y = x + 3$

*Write a slope–intercept equation for a line passing through the given point that is parallel to the given line. Then write a second equation for a line passing through the given point that is perpendicular to the given line.*

**43.** $(3, 5)$, $y = \frac{2}{7}x + 1$

**44.** $(-1, 6)$, $f(x) = 2x + 9$

**45.** $(-7, 0)$, $y = -0.3x + 4.3$

**46.** $(-4, -5)$, $2x + y = -4$

**47.** $(3, -2)$, $3x + 4y = 5$

**48.** $(8, -2)$, $y = 4.2(x - 3) + 1$

**49.** $(3, -3)$, $x = -1$

**50.** $(4, -5)$, $y = -1$

*In Exercises 51–56, determine whether the statement is true or false.*

**51.** The lines $x = -3$ and $y = 5$ are perpendicular.

**52.** The lines $y = 2x - 3$ and $y = -2x - 3$ are perpendicular.

**53.** The lines $y = \frac{2}{5}x + 4$ and $y = \frac{2}{5}x - 4$ are parallel.

**54.** The intersection of the lines $y = 2$ and $x = -\frac{3}{4}$ is $\left(-\frac{3}{4}, 2\right)$.

**55.** The lines $x = -1$ and $x = 1$ are perpendicular.

**56.** The lines $2x + 3y = 4$ and $3x - 2y = 4$ are perpendicular.

*In Exercises 57–60, determine whether a linear model might fit the data.*

**57.**

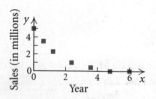

**58.**

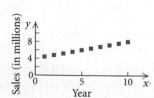

**59.**

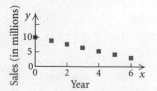

**60.**

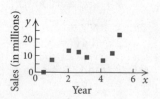

**61.** *Internet Use.* The table below illustrates the growth in worldwide Internet use.

| Year, $x$ | Number of Internet Users Worldwide, $y$ (in millions) |
|---|---|
| 2001, 0 | 495 |
| 2002, 1 | 677 |
| 2003, 2 | 785 |
| 2004, 3 | 914 |
| 2005, 4 | 1036 |
| 2006, 5 | 1159 |
| 2007, 6 | 1393 |
| 2008, 7 | 1611 |
| 2009, 8 | 1858 |
| 2010, 9 | 2084* |

*Estimate
*Sources:* International Telecommunication Union; ITU Statistics

**a)** Model the data with a linear function. Let the independent variable represent the number of years after 2001; that is, the data points are $(0, 495)$, $(3, 914)$, and so on. Answers may vary depending on the data points used.

**b)** Using the function found in part (a), estimate the number of Internet users worldwide in 2013 and in 2018.

**62.** *Cremations.* The table below illustrates the upward trend in America to choose cremation.

| Year, $x$ | Percentage of Deaths Followed by Cremation, $y$ |
|---|---|
| 2005, 0 | 32.3% |
| 2006, 1 | 33.6 |
| 2007, 2 | 34.3 |
| 2008, 3 | 35.8 |
| 2009, 4 | 36.9 |

*Source:* Cremation Association of North America

**a)** Model the data with a linear function. Let the independent variable represent the number of years after 2005. Answers may vary depending on the data points used.

**b)** Using the function found in part (a), estimate the percentage of deaths followed by cremation in 2013 and in 2016.

**63.** *Expenditures on Pets.* Data on total U.S. expenditures on pets, pet products, and related services are given in the table below. Model the data with a linear function. Then, using that function, estimate total U.S. expenditures on pets in 2005 and predict total expenditures in 2015. Answers may vary depending on the data points used.

| Year, x | Total U.S. Expenditures on Pets (in billions) |
|---|---|
| 1991, 0 | $19.6 |
| 1994, 3 | 24.9 |
| 1997, 6 | 32.5 |
| 2000, 9 | 39.7 |
| 2003, 12 | 46.8 |
| 2006, 15 | 56.9 |
| 2009, 18 | 67.1 |

*Sources*: Bureau of Economic Analysis; U.S. Department of Commerce

**64.** *Median Age.* Data on the median age of the U.S. population in selected years are listed in the table below. Model the data with a linear function, estimate the median age in 1998, and predict the median age in 2020. Answers may vary depending on the data points used.

| Year, x | Median Age, y |
|---|---|
| 1970, 0 | 28.0 |
| 1980, 10 | 30.0 |
| 1990, 20 | 32.8 |
| 2000, 30 | 35.3 |
| 2008, 38 | 36.8 |
| 2011, 41 | 36.9* |

*Estimated
*Sources*: U.S. Census Bureau; CIA *World Factbook*

**65.** *Nike Net Sales.* The net sales data in several years for Nike are given in the table at right. Model the data with a linear function, and predict the net sales in 2015. Answers may vary depending on the data points used.

| Year, x | Net Sales (in billions) |
|---|---|
| 2005, 0 | $13.7 |
| 2006, 1 | 15.0 |
| 2007, 2 | 16.3 |
| 2008, 3 | 18.6 |
| 2009, 4 | 19.2 |
| 2010, 5 | 19.0 |

*Source*: Nike

**66.** *Credit-Card Debt.* Data on average credit-card debt per U.S. household are given in the table below. Model the data with a linear function, and estimate the average debt in 2007 and in 2014. Answers may vary depending on the data points used.

| Year, x | Credit-Card Debt per Household, y |
|---|---|
| 1992, 0 | $ 3,803 |
| 1996, 4 | 6,912 |
| 2000, 8 | 8,308 |
| 2004, 12 | 9,577 |
| 2008, 16 | 10,691 |
| 2011, 19 | 14,750 |

*Sources*: CardTrak.com; www.creditcards.com

**67. a)** Use a graphing calculator to fit a regression line to the data in Exercise 61.
   **b)** Estimate the number of Internet users worldwide in 2013 and compare the value with the result found in Exercise 61.
   **c)** Find the correlation coefficient for the regression line and determine whether the line fits the data closely.

**68. a)** Use a graphing calculator to fit a regression line to the data in Exercise 62.
   **b)** Estimate the percentage of deaths followed by cremation in 2013 and compare the result with the estimate found with the model in Exercise 62.
   **c)** Find the correlation coefficient for the regression line and determine whether the line fits the data closely.

**69. a)** Use a graphing calculator to fit a regression line to the data in Exercise 63.

**b)** Estimate total U.S. expenditures on pets in 2015 and compare the value with the result found in Exercise 63.

**c)** Find the correlation coefficient for the regression line and determine whether the line fits the data closely.

**70. a)** Use a graphing calculator to fit a regression line to the data in Exercise 66.

**b)** Estimate average credit-card debt per U.S. household in 2014 and compare the result with the estimate found with the model in Exercise 66.

**c)** Find the correlation coefficient for the regression line and determine whether the line fits the data closely.

**71.** *Maximum Heart Rate.*   A person who is exercising should not exceed his or her maximum heart rate, which is determined on the basis of that person's sex, age, and resting heart rate. The table below relates resting heart rate and maximum heart rate for a 20-year-old man.

| Resting Heart Rate, $H$ (in beats per minute) | Maximum Heart Rate, $M$ (in beats per minute) |
|---|---|
| 50 | 166 |
| 60 | 168 |
| 70 | 170 |
| 80 | 172 |

*Source*: American Heart Association

**a)** Use a graphing calculator to model the data with a linear function.

**b)** Estimate the maximum heart rate if the resting heart rate is 40, 65, 76, and 84.

**c)** What is the correlation coefficient? How confident are you about using the regression line to estimate function values?

**72.** *Study Time versus Grades.*   A math instructor asked her students to keep track of how much time each spent studying a chapter on functions in her algebra–trigonometry course. She collected the information together with test scores from that chapter's test. The data are listed in the table below.

| Study Time, $x$ (in hours) | Test Grade, $y$ (in percent) |
|---|---|
| 23 | 81% |
| 15 | 85 |
| 17 | 80 |
| 9 | 75 |
| 21 | 86 |
| 13 | 80 |
| 16 | 85 |
| 11 | 93 |

**a)** Use a graphing calculator to model the data with a linear function.

**b)** Predict a student's score if he or she studies 24 hr, 6 hr, and 18 hr.

**c)** What is the correlation coefficient? How confident are you about using the regression line to predict function values?

## Skill Maintenance

*Find the slope of the line containing the given points.*

**73.** $(2, -8)$ and $(-5, -1)$    **74.** $(5, 7)$ and $(5, -7)$

*Find an equation for a circle satisfying the given conditions.*

**75.** Center $(-7, -1)$, radius of length $\frac{9}{5}$

**76.** Center $(0, 3)$, diameter of length 5

## Synthesis

**77.** Find $k$ so that the line containing the points $(-3, k)$ and $(4, 8)$ is parallel to the line containing the points $(5, 3)$ and $(1, -6)$.

**78.** *Road Grade.*   Using the figure below, find the road grade and an equation giving the height $y$ as a function of the horizontal distance $x$.

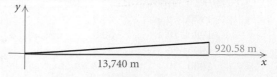

**79.** Find an equation of the line passing through the point $(4, 5)$ and perpendicular to the line passing through the points $(-1, 3)$ and $(2, 9)$.

# 1.5 Linear Equations, Functions, Zeros, and Applications

- Solve linear equations.
- Solve applied problems using linear models.
- Find zeros of linear functions.

An **equation** is a statement that two expressions are equal. To **solve** an equation in one variable is to find all the values of the variable that make the equation true. Each of these values is a **solution** of the equation. The set of all solutions of an equation is its **solution set**. Some examples of **equations in one variable** are

$$2x + 3 = 5, \quad 3(x - 1) = 4x + 5,$$

$$x^2 - 3x + 2 = 0, \quad \text{and} \quad \frac{x - 3}{x + 4} = 1.$$

## ▪ Linear Equations

The first two equations above are *linear equations* in one variable. We define such equations as follows.

> A **linear equation in one variable** is an equation that can be expressed in the form $mx + b = 0$, where $m$ and $b$ are real numbers and $m \neq 0$.

Equations that have the same solution set are **equivalent equations**. For example, $2x + 3 = 5$ and $x = 1$ are equivalent equations because 1 is the solution of each equation. On the other hand, $x^2 - 3x + 2 = 0$ and $x = 1$ are not equivalent equations because 1 and 2 are both solutions of $x^2 - 3x + 2 = 0$ but 2 is not a solution of $x = 1$.

To solve an equation, we find an equivalent equation in which the variable is isolated. The following principles allow us to solve linear equations.

> **EQUATION-SOLVING PRINCIPLES**
>
> For any real numbers $a$, $b$, and $c$:
>
> ***The Addition Principle***: If $a = b$ is true, then $a + c = b + c$ is true.
> ***The Multiplication Principle***: If $a = b$ is true, then $ac = bc$ is true.

*Note to the student and the instructor*: We assume that students come to a College Algebra course with some equation-solving skills from their study of Intermediate Algebra. Thus a portion of the material in this section might be considered by some to be review in nature. We present this material here in order to use linear functions, with which students are familiar, to lay the groundwork for zeros of higher-order polynomial functions and their connection to solutions of equations and *x*-intercepts of graphs.

GCM

**EXAMPLE 1** Solve: $\frac{3}{4}x - 1 = \frac{7}{5}$.

*Solution* When we have an equation that contains fractions, it is often convenient to multiply by the least common denominator (LCD) of the fractions on both sides of the equation in order to clear the equation of fractions. We have

$$\frac{3}{4}x - 1 = \frac{7}{5}$$  The LCD is $4 \cdot 5$, or 20.

$$20\left(\frac{3}{4}x - 1\right) = 20 \cdot \frac{7}{5}$$  Multiplying by the LCD on both sides to clear fractions

$$20 \cdot \frac{3}{4}x - 20 \cdot 1 = 28$$

$$15x - 20 = 28$$

$$15x - 20 + 20 = 28 + 20$$  Using the addition principle to add 20 on both sides

$$15x = 48$$

$$\frac{15x}{15} = \frac{48}{15}$$  Using the multiplication principle to multiply by $\frac{1}{15}$, or divide by 15, on both sides

$$x = \frac{48}{15}$$

$$x = \frac{16}{5}.$$  Simplifying. Note that $\frac{3}{4}x - 1 = \frac{7}{5}$ and $x = \frac{16}{5}$ are equivalent equations.

*Check*:
$$\frac{3}{4}x - 1 = \frac{7}{5}$$

$$\frac{3}{4} \cdot \frac{16}{5} - 1 \; ? \; \frac{7}{5}$$  Substituting $\frac{16}{5}$ for $x$

$$\frac{12}{5} - \frac{5}{5}$$

$$\frac{7}{5} \;\Big|\; \frac{7}{5} \quad \text{TRUE}$$

The solution is $\frac{16}{5}$.

We can use the INTERSECT feature on a graphing calculator to solve equations. We call this the **Intersect method**. To use the Intersect method to solve the equation in Example 1, for instance, we graph $y_1 = \frac{3}{4}x - 1$ and $y_2 = \frac{7}{5}$. The value of $x$ for which $y_1 = y_2$ is the solution of the equation $\frac{3}{4}x - 1 = \frac{7}{5}$. This value of $x$ is the first coordinate of the point of intersection of the graphs of $y_1$ and $y_2$. Using the INTERSECT feature, we find that the first coordinate of this point is 3.2. We can find fraction notation for the solution by using the ▶FRAC feature. The solution is 3.2, or $\frac{16}{5}$.

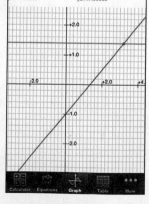

The Graphing Calculator Appcylon LLC app has an intersect feature. The intersection of $y = \frac{3}{4}x - 1$ and $y = \frac{7}{5}$ is shown in this window.

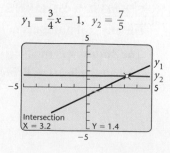

$y_1 = \frac{3}{4}x - 1, \; y_2 = \frac{7}{5}$

**EXAMPLE 2** Solve: $2(5 - 3x) = 8 - 3(x + 2)$.

## Algebraic Solution

We have

$$2(5 - 3x) = 8 - 3(x + 2)$$

| $10 - 6x = 8 - 3x - 6$ | Using the distributive property |
| $10 - 6x = 2 - 3x$ | Collecting like terms |
| $10 - 6x + 6x = 2 - 3x + 6x$ | Using the addition principle to add $6x$ on both sides |
| $10 = 2 + 3x$ | |
| $10 - 2 = 2 + 3x - 2$ | Using the addition principle to add $-2$, or subtract 2, on both sides |
| $8 = 3x$ | |
| $\dfrac{8}{3} = \dfrac{3x}{3}$ | Using the multiplication principle to multiply by $\frac{1}{3}$, or divide by 3, on both sides |
| $\dfrac{8}{3} = x.$ | |

*Check:*

$$2(5 - 3x) = 8 - 3(x + 2)$$

$$\begin{array}{c|c} 2\left(5 - 3 \cdot \frac{8}{3}\right) \;?\; 8 - 3\left(\frac{8}{3} + 2\right) & \text{Substituting } \frac{8}{3} \text{ for } x \\ 2(5 - 8) & 8 - 3\left(\frac{14}{3}\right) \\ 2(-3) & 8 - 14 \\ -6 & -6 \qquad \text{TRUE} \end{array}$$

The solution is $\dfrac{8}{3}$.

## Graphical Solution

We graph $y_1 = 2(5 - 3x)$ and $y_2 = 8 - 3(x + 2)$. The first coordinate of the point of intersection of the graphs is the value of $x$ for which $2(5 - 3x) = 8 - 3(x + 2)$ and is thus the solution of the equation.

$$y_1 = 2(5 - 3x), \quad y_2 = 8 - 3(x + 2)$$

$y_2 \; 5 \; y_1$

Intersection
X = 2.6666667    Y = -6

The solution is approximately 2.6666667.

We can find fraction notation for the exact solution by using the ▶FRAC feature. The solution is $\frac{8}{3}$.

X ▶ Frac                $\dfrac{8}{3}$

**Now Try Exercise 27.**

| X | Y1 | Y2 |
|---|----|----|
| 2.6667 | -6 | -6 |
| | | |
| X = | | |

We can use the TABLE feature on a graphing calculator, set in ASK mode, to check the solutions of equations. In Example 2, for instance, we let $y_1 = 2(5 - 3x)$ and $y_2 = 8 - 3(x + 2)$. When $\frac{8}{3}$ is entered for $x$, we see that $y_1 = y_2$, or $2(5 - 3x) = 8 - 3(x + 2)$. Thus, $\frac{8}{3}$ is the solution of the equation. (Note that the calculator converts $\frac{8}{3}$ to decimal notation in the table.)

## Special Cases

Some equations have *no* solution.

**EXAMPLE 3**  Solve: $-24x + 7 = 17 - 24x$.

*Solution*  We have

$$-24x + 7 = 17 - 24x$$
$$24x - 24x + 7 = 24x + 17 - 24x \qquad \text{Adding } 24x$$
$$7 = 17. \qquad \text{We get a false equation.}$$

No matter what number we substitute for $x$, we get a false equation. Thus the equation has *no* solution.  **Now Try Exercise 11.**

There are some equations for which *any* real number is a solution.

**EXAMPLE 4**  Solve: $3 - \frac{1}{3}x = -\frac{1}{3}x + 3$.

*Solution*  We have

$$3 - \frac{1}{3}x = -\frac{1}{3}x + 3$$
$$\frac{1}{3}x + 3 - \frac{1}{3}x = \frac{1}{3}x - \frac{1}{3}x + 3 \qquad \text{Adding } \frac{1}{3}x$$
$$3 = 3. \qquad \text{We get a true equation.}$$

Replacing $x$ with any real number gives a true equation. Thus *any* real number is a solution. This equation has *infinitely* many solutions. The solution set is the set of real numbers, $\{x \,|\, x \text{ is a real number}\}$, or $(-\infty, \infty)$.  **Now Try Exercise 3.**

## ▪ Applications Using Linear Models

Mathematical techniques can be used to answer questions arising from real-world situations. Linear equations and linear functions *model* many of these situations.

The following strategy is of great assistance in problem solving.

**FIVE STEPS FOR PROBLEM SOLVING**

1. **Familiarize** yourself with the problem situation. If the problem is presented in words, this means to read carefully. Some or all of the following can also be helpful.

   a) Make a drawing, if it makes sense to do so.
   b) Make a written list of the known facts and a list of what you wish to find out.
   c) Assign variables to represent unknown quantities.
   d) Organize the information in a chart or a table, if appropriate.
   e) Find further information. Look up a formula, consult a reference book or an expert in the field, or do research on the Internet.
   f) Guess or estimate the answer and check your guess or estimate.

   *(continued)*

2. **Translate** the problem situation to mathematical language or symbolism. For most of the problems you will encounter in algebra, this means to write one or more equations, but sometimes an inequality or some other mathematical symbolism may be appropriate.

3. **Carry out** some type of mathematical manipulation. Use your mathematical skills to find a possible solution. In algebra, this usually means to solve an equation, an inequality, or a system of equations or inequalities.

4. **Check** to see whether your possible solution actually fits the problem situation and is thus really a solution of the problem. Although you may have solved an equation, the solution(s) of the equation might not be solution(s) of the original problem.

5. **State** the answer clearly using a complete sentence.

**EXAMPLE 5** *Words in Languages.* There are about 232,000 words in the Japanese language. This is 19% more than the number of words in the Russian language. (*Source*: Global Language Monitor) How many words are in the Russian language?

*Solution*

1. **Familiarize.** Let's estimate that there are 200,000 words in the Russian language. Then the number of words in the Japanese language would be

$$200{,}000 + 19\% \cdot 200{,}000 = 1(200{,}000) + 0.19(200{,}000)$$
$$= 1.19(200{,}000) = 238{,}000.$$

Since we know that there are actually 232,000 words in the Japanese language, our estimate of 200,000 is too high. Nevertheless, the calculations performed indicate how we can translate the problem to an equation. We let $x =$ the number of words in the Russian language. Then $x + 19\%x$, or $1 \cdot x + 0.19x$, or $1.19x$, is the number of words in the Japanese language.

2. **Translate.** We translate to an equation:

$$\underbrace{\text{Number of words in Japanese language}}\; \underset{\downarrow}{\text{is}}\; \underset{\downarrow}{232{,}000.}$$
$$1.19x \qquad\qquad = \qquad 232{,}000$$

**3. Carry out.** We solve the equation, as follows:

$$1.19x = 232{,}000$$

$$x = \frac{232{,}000}{1.19} \quad \text{Dividing by 1.19 on both sides}$$

$$x \approx 195{,}000.$$

**4. Check.** 19% of 195,000 is 37,050, and 195,000 + 37,050 = 232,050. Since 232,050 ≈ 232,000, the answer checks. (Remember: We rounded the value of $x$.)

**5. State.** There are about 195,000 words in the Russian language.

Now Try Exercise 33.

**EXAMPLE 6** *Convenience Stores.* In 2011, there were over 146,000 convenience stores in the United States. The total number of convenience stores in Texas and California was 25,047. There were 3885 more convenience stores in Texas than in California. (*Source*: The Nielsen Company) Find the number of convenience stores in Texas and in California.

**Solution**

**1. Familiarize.** The number of convenience stores in Texas is described in terms of the number in California, so we let $x =$ the number of convenience stores in California. Then $x + 3885 =$ the number of convenience stores in Texas.

**2. Translate.** We translate to an equation:

| Number of convenience stores in California | plus | number of convenience stores in Texas | is | 25,047. |
|:---:|:---:|:---:|:---:|:---:|
| ↓ | | ↓ | ↓ | ↓ |
| $x$ | $+$ | $x + 3885$ | $=$ | 25,047 |

**3. Carry out.** We solve the equation, as follows:

$$x + x + 3885 = 25{,}047$$

$$2x + 3885 = 25{,}047 \quad \text{Collecting like terms}$$

$$2x = 21{,}162 \quad \text{Subtracting 3885 on both sides}$$

$$x = 10{,}581. \quad \text{Dividing by 2 on both sides}$$

If $x = 10{,}581$, then $x + 3885 = 10{,}581 + 3885 = 14{,}466$.

**4. Check.** If there were 14,466 convenience stores in Texas and 10,581 in California, then the total number of convenience stores in Texas and California was 14,466 + 10,581, or 25,047. Also, 14,466 is 3885 more than 10,581. The answer checks.

**5. State.** In 2011, there were 14,466 convenience stores in Texas and 10,581 convenience stores in California.

Now Try Exercise 35.

In some applications, we need to use a formula that describes the relationships among variables. When a situation involves distance, rate (also called speed or velocity), and time, for example, we use the following formula.

---

**THE MOTION FORMULA**

The distance $d$ traveled by an object moving at rate $r$ in time $t$ is given by

$$d = r \cdot t.$$

---

**EXAMPLE 7** *Airplane Speed.* Delta Airlines' fleet includes B737/800's, each with a cruising speed of 531 mph, and Saab 340B's, each with a cruising speed of 290 mph (*Source*: Delta Airlines). Suppose that a Saab 340B takes off and travels at its cruising speed. One hour later, a B737/800 takes off and follows the same route, traveling at its cruising speed. How long will it take the B737/800 to overtake the Saab 340B?

**Solution**

1. **Familiarize.** We make a drawing showing both the known and the unknown information. We let $t =$ the time, in hours, that the B737/800 travels before it overtakes the Saab 340B. Since the Saab 340B takes off 1 hr before the 737, it will travel for $t + 1$ hr before being overtaken. The planes will have traveled the same distance, $d$, when one overtakes the other.

We can also organize the information in a table, as follows.

$$d = r \cdot t$$

|           | Distance | Rate | Time    |
|-----------|----------|------|---------|
| **B737/800** | $d$      | 531  | $t$     |
| **Saab 340B** | $d$      | 290  | $t + 1$ |

$\longrightarrow d = 531t$

$\longrightarrow d = 290(t + 1)$

2. **Translate.** Using the formula $d = rt$ in each row of the table, we get two expressions for $d$:

$$d = 531t \quad \text{and} \quad d = 290(t + 1).$$

Since the distances are the same, we have the following equation:

$$531t = 290(t + 1).$$

3. **Carry out.** We solve the equation, as follows:

$$531t = 290(t + 1)$$

$531t = 290t + 290$     **Using the distributive property**

$241t = 290$     **Subtracting 290*t* on both sides**

$t \approx 1.2.$     **Dividing by 241 on both sides and rounding to the nearest tenth**

4. **Check.** If the B737/800 travels for about 1.2 hr, then the Saab 340B travels for about 1.2 + 1, or 2.2 hr. In 2.2 hr, the Saab 340B travels 290(2.2), or 638 mi, and in 1.2 hr, the B737/800 travels 531(1.2), or 637.2 mi. Since 637.2 mi $\approx$ 638 mi, the answer checks. (Remember: We rounded the value of *t*.)

5. **State.** About 1.2 hr after the B737/800 has taken off, it will overtake the Saab 340B.

**Now Try Exercise 57.**

For some applications, we need to use a formula to find the amount of interest earned by an investment or the amount of interest due on a loan.

---

**THE SIMPLE-INTEREST FORMULA**

The **simple interest** *I* on a principal of *P* dollars at interest rate *r* for *t* years is given by

$$I = Prt.$$

---

**EXAMPLE 8** *Student Loans.* Demarion's two student loans total $12,000. One loan is at 5% simple interest and the other is at 8% simple interest. After 1 year, Demarion owes $750 in interest. What is the amount of each loan?

*Solution*

1. **Familiarize.** We let $x =$ the amount borrowed at 5% interest. Then the remainder of the $12,000, or 12,000 $- x$, is borrowed at 8%. We organize the information in a table, keeping in mind the formula $I = Prt$.

| | Amount Borrowed | Interest Rate | Time | Amount of Interest |
|---|---|---|---|---|
| **5% Loan** | $x$ | 5%, or 0.05 | 1 year | $x(0.05)(1)$, or $0.05x$ |
| **8% Loan** | 12,000 $- x$ | 8%, or 0.08 | 1 year | $(12{,}000 - x)(0.08)(1)$, or $0.08(12{,}000 - x)$ |
| **Total** | 12,000 | | | 750 |

2. **Translate.** The total amount of interest on the two loans is $750. Thus we write the following equation:

Interest on 5% loan    plus    interest on 8% loan    is    $750.

$$0.05x + 0.08(12{,}000 - x) = 750$$

3. **Carry out.** We solve the equation, as follows:

$$0.05x + 0.08(12{,}000 - x) = 750$$

$$0.05x + 960 - 0.08x = 750 \qquad \text{Using the distributive property}$$

$$-0.03x + 960 = 750 \qquad \text{Collecting like terms}$$

$$-0.03x = -210 \qquad \text{Subtracting 960 on both sides}$$

$$x = 7000. \qquad \text{Dividing by } -0.03 \text{ on both sides}$$

If $x = 7000$, then $12{,}000 - x = 12{,}000 - 7000 = 5000$.

4. **Check.** The interest on $7000 at 5% for 1 year is $7000(0.05)(1)$, or $350. The interest on $5000 at 8% for 1 year is $5000(0.08)(1)$, or $400. Since $350 + $400 = $750, the answer checks.

5. **State.** Demarion borrowed $7000 at 5% interest and $5000 at 8% interest.

**Now Try Exercise 63.**

Sometimes we use formulas from geometry in solving applied problems. In the following example, we use the formula for the perimeter $P$ of a rectangle with length $l$ and width $w$: $P = 2l + 2w$.

**EXAMPLE 9** *Solar Panels.* In December 2009, a solar energy farm was completed at the Denver International Airport. More than 9200 rectangular solar panels were installed (*Sources:* Woods Allee, Denver International Airport; www.solarpanelstore.com; *The Denver Post*). A solar panel, or photovoltaic panel, converts sunlight into electricity. The length of a panel is 13.6 in. less than twice the width, and the perimeter is 207.4 in. Find the length and the width.

*Solution*

1. **Familiarize.** We first make a drawing. Since the length of the panel is described in terms of the width, we let $w =$ the width, in inches. Then $2w - 13.6 =$ the length, in inches.

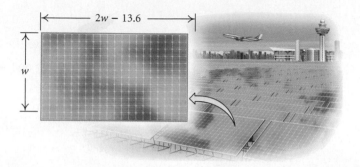

2. **Translate.** We use the formula for the perimeter of a rectangle:

$$P = 2l + 2w$$
$$207.4 = 2(2w - 13.6) + 2w. \quad \text{Substituting 207.4 for } P \text{ and } 2w - 13.6 \text{ for } l$$

3. **Carry out.** We solve the equation:

$$207.4 = 2(2w - 13.6) + 2w$$
$$207.4 = 4w - 27.2 + 2w \qquad \text{Using the distributive property}$$
$$207.4 = 6w - 27.2 \qquad \text{Collecting like terms}$$
$$234.6 = 6w \qquad \text{Adding 27.2 on both sides}$$
$$39.1 = w. \qquad \text{Dividing by 6 on both sides}$$

If $w = 39.1$, then $2w - 13.6 = 2(39.1) - 13.6 = 78.2 - 13.6 = 64.6$.

4. **Check.** The length, 64.6 in., is 13.6 in. less than twice the width, 39.1 in. Also

$$2 \cdot 64.6 \text{ in.} + 2 \cdot 39.1 \text{ in.} = 129.2 \text{ in.} + 78.2 \text{ in.} = 207.4 \text{ in.}$$

The answer checks.

5. **State.** The length of the solar panel is 64.6 in., and the width is 39.1 in.

**Now Try Exercise 53.**

**EXAMPLE 10** *Cab Fare.* Metro Taxi charges a $2.50 pickup fee and $2 per mile traveled. Grayson's cab fare from the airport to his hotel is $32.50. How many miles did he travel in the cab?

*Solution*

1. **Familiarize.** Let's guess that Grayson traveled 12 mi in the cab. Then his fare would be

$$\$2.50 + \$2 \cdot 12 = \$2.50 + \$24 = \$26.50.$$

We see that our guess is low, but the calculation shows us how to translate the problem to an equation. We let $m =$ the number of miles that Grayson traveled in the cab.

2. **Translate.** We translate to an equation:

| Pickup fee | plus | cost per mile | times | number of miles traveled | is | total charge. |
|---|---|---|---|---|---|---|
| 2.50 | + | 2 | · | m | = | 32.50 |

3. **Carry out.** We solve the equation:

$$2.50 + 2 \cdot m = 32.50$$
$$2m = 30 \qquad \text{Subtracting 2.50 on both sides}$$
$$m = 15. \qquad \text{Dividing by 2 on both sides}$$

4. **Check.** If Grayson travels 15 mi in the cab, the mileage charge is $2 · 15, or $30. Then, with the $2.50 pickup fee included, his total charge is $2.50 + $30, or $32.50. The answer checks.

5. **State.** Grayson traveled 15 mi in the cab. **Now Try Exercise 47.**

## ■ Zeros of Linear Functions

An input for which a function's output is 0 is called a **zero** of the function. We will restrict our attention in this section to zeros of linear functions. This allows us to become familiar with the concept of a zero, and it lays the groundwork for working with zeros of other types of functions in succeeding chapters.

> **ZEROS OF FUNCTIONS**
>
> An input $c$ of a function $f$ is called a **zero** of the function if the output for the function is 0 when the input is $c$. That is, $c$ is a zero of $f$ if $f(c) = 0$.

**LINEAR FUNCTIONS**

REVIEW SECTION **1.3.**

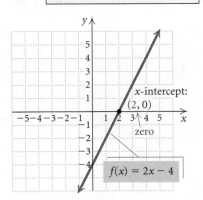

Recall that a linear function is given by $f(x) = mx + b$, where $m$ and $b$ are constants. For the linear function $f(x) = 2x - 4$, we have $f(2) = 2 \cdot 2 - 4 = 0$, so 2 is a **zero** of the function. In fact, 2 is the *only* zero of this function. In general, a **linear function $f(x) = mx + b$, with $m \neq 0$, has exactly one zero.**

The zero, 2, is the first coordinate of the point at which the graph crosses the $x$-axis. This point, $(2, 0)$, is the *x-intercept* of the graph. Thus when we find the zero of a linear function, we are also finding the first coordinate of the $x$-intercept of the graph of the function.

For every linear function $f(x) = mx + b$, there is an associated linear equation $mx + b = 0$. When we find the zero of a function $f(x) = mx + b$, we are also finding the solution of the equation $mx + b = 0$.

GCM    **EXAMPLE 11**    Find the zero of $f(x) = 5x - 9$.

## Algebraic Solution

We find the value of $x$ for which $f(x) = 0$:

$$5x - 9 = 0 \qquad \text{Setting } f(x) = 0$$
$$5x = 9 \qquad \text{Adding 9 on both sides}$$
$$x = \tfrac{9}{5}, \text{ or } 1.8. \qquad \text{Dividing by 5 on both sides}$$

Using a table, set in ASK mode, we can check the solution. We enter $y = 5x - 9$ on the equation-editor screen and then enter the value $x = \frac{9}{5}$, or 1.8, in the table.

| X   | Y1 |  |
|-----|----|--|
| 1.8 | 0  |  |
|     |    |  |
| X = |    |  |

We see that $y = 0$ when $x = 1.8$, so the number 1.8 checks. The zero is $\frac{9}{5}$, or 1.8. This means that $f\left(\frac{9}{5}\right) = 0$, or $f(1.8) = 0$. Note that the *zero* of the function $f(x) = 5x - 9$ is the *solution* of the equation $5x - 9 = 0$.

## Graphical Solution

The solution of $5x - 9 = 0$ is also the zero of $f(x) = 5x - 9$. Thus we can solve an equation by finding the zeros of the function associated with it. We call this the **zero method**.

We graph $y = 5x - 9$ in the standard window and use the ZERO feature from the CALC menu to find the zero of $f(x) = 5x - 9$. (See Fig. 1.) Note that the $x$-intercept must appear in the window when the ZERO feature is used.

We also can find the zero of this function with an app. (See Fig. 2.) Note that the app shown here uses the term "root" rather than "zero."

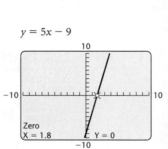

$y = 5x - 9$

FIGURE 1                    FIGURE 2

We can check algebraically by substituting 1.8 for $x$:

$$f(1.8) = 5(1.8) - 9 = 9 - 9 = 0.$$

The zero of $f(x) = 5x - 9$ is 1.8, or $\frac{9}{5}$.

**Now Try Exercise 73.**

## CONNECTING THE CONCEPTS

### The Intersect Method and the Zero Method

An equation such as $x - 1 = 2x - 6$ can be solved using the Intersect method by graphing $y_1 = x - 1$ and $y_2 = 2x - 6$ and using the INTERSECT feature to find the first coordinate of the point of intersection of the graphs.

The equation can also be solved using the Zero method by writing it with 0 on one side of the equals sign and then using the ZERO feature.

Solve: $x - 1 = 2x - 6$.

**The Intersect Method**
Graph $y_1 = x - 1$ and $y_2 = 2x - 6$.
Point of intersection: $(5, 4)$
Solution: 5

**The Zero Method**
First, add $-2x$ and 6 on both sides of the equation to get 0 on one side:

$$x - 1 = 2x - 6$$
$$x - 1 - 2x + 6 = 0.$$

Graph

$$y_3 = x - 1 - 2x + 6.$$

Zero: 5
Solution: 5

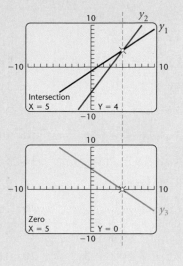

## CONNECTING THE CONCEPTS

### Zeros, Solutions, and Intercepts

The zero of a linear function $f(x) = mx + b$, with $m \neq 0$, is the solution of the linear equation $mx + b = 0$ and is the first coordinate of the $x$-intercept of the graph of $f(x) = mx + b$. To find the zero of $f(x) = mx + b$, we solve $f(x) = 0$, or $mx + b = 0$.

| FUNCTION | ZERO OF THE FUNCTION; SOLUTION OF THE EQUATION | ZERO OF THE FUNCTION; $x$-INTERCEPT OF THE GRAPH |
|---|---|---|
| **Linear Function** $f(x) = 2x - 4$, or $y = 2x - 4$ | To find the **zero** of $f(x)$, we solve $f(x) = 0$: $$2x - 4 = 0$$ $$2x = 4$$ $$x = 2.$$ The **solution** of $2x - 4 = 0$ is 2. This is the zero of the function $f(x) = 2x - 4$. That is, $f(2) = 0$. | The zero of $f(x)$ is the first coordinate of the **$x$-intercept** of the graph of $y = f(x)$. 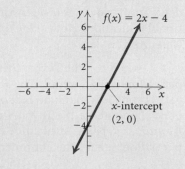 |

## 1.5 Exercise Set

*Solve.*

**1.** $4x + 5 = 21$

**2.** $2y - 1 = 3$

**3.** $23 - \frac{2}{5}x = -\frac{2}{5}x + 23$

**4.** $\frac{6}{5}y + 3 = \frac{3}{10}$

**5.** $4x + 3 = 0$

**6.** $3x - 16 = 0$

**7.** $3 - x = 12$

**8.** $4 - x = -5$

**9.** $3 - \frac{1}{4}x = \frac{3}{2}$

**10.** $10x - 3 = 8 + 10x$

**11.** $\frac{2}{11} - 4x = -4x + \frac{9}{11}$

**12.** $8 - \frac{2}{9}x = \frac{5}{6}$

**13.** $8 = 5x - 3$

**14.** $9 = 4x - 8$

**15.** $\frac{2}{5}y - 2 = \frac{1}{3}$

**16.** $-x + 1 = 1 - x$

**17.** $y + 1 = 2y - 7$

**18.** $5 - 4x = x - 13$

**19.** $2x + 7 = x + 3$

**20.** $5x - 4 = 2x + 5$

**21.** $3x - 5 = 2x + 1$

**22.** $4x + 3 = 2x - 7$

**23.** $4x - 5 = 7x - 2$

**24.** $5x + 1 = 9x - 7$

**25.** $5x - 2 + 3x = 2x + 6 - 4x$

**26.** $5x - 17 - 2x = 6x - 1 - x$

**27.** $7(3x + 6) = 11 - (x + 2)$

**28.** $4(5y + 3) = 3(2y - 5)$

**29.** $3(x + 1) = 5 - 2(3x + 4)$

**30.** $4(3x + 2) - 7 = 3(x - 2)$

**31.** $2(x - 4) = 3 - 5(2x + 1)$

**32.** $3(2x - 5) + 4 = 2(4x + 3)$

**33.** *Foreign Students in the United States.* In the 2009–2010 school year in the United States, there were 128,000 students from China. This number is 22% more than the number of students from India. (*Source*: Institute of International Education) How many foreign students were from India?

**34.** *Fuel Economy.* The Toyota Prius gets 44 miles per gallon (mpg) overall. The Hummer H2 gets 11 mpg less than one-half of the miles-per-gallon rate for the Toyota Prius. (*Source: Consumer Reports*, April 2010) Find the miles-per-gallon rate for the Hummer H2.

**35.** *Olive Oil.* It is estimated that 710,000 metric tons of olive oil were consumed in Italy in 2009–2010. This is 60,000 metric tons more than 2.5 times the amount consumed in the United States during the same time period. (*Source*: International Olive Council) Find the amount of olive oil consumed in the United States in 2009–2010.

**36.** *Salary Comparison.* The average salary of a landscape architect for the federal government is $80,830 per year. This is about 38.5% higher than the yearly salary of a private-sector landscape architect. (*Source*: U.S. Bureau of Labor Statistics) Find the salary of a private-sector landscape architect.

**37.** *Ocean Depth.* The average depth of the Pacific Ocean is 14,040 ft. This is 8890 ft less than the sum of the average depths of the Atlantic Ocean and the Indian Ocean. The average depth of the Indian Ocean is 272 ft less than four-fifths of the average depth of the Atlantic Ocean. (*Source: Time Almanac*, 2010) Find the average depth of the Indian Ocean.

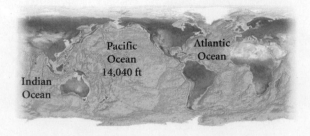

**38.** *Where the Textbook Dollar Goes.* Of each dollar spent on textbooks at college bookstores, 22.3 cents goes to the college store for profit, store operations, and personnel. On average, a college student at a four-year college spends $667 per year for textbooks. (*Source*: National Association of College Stores) How much of this expenditure goes to the college store?

**39.** *Vehicle Curb Weight.* The total curb weight of a Toyota Tundra truck, a Ford Mustang car, and a Smart For Two car is 11,150 lb. The weight of a Mustang is 5 lb less than the weight of two Smart For Two cars. The Tundra weighs 2135 lb more than the Mustang. (*Source*: *Consumer Reports*, April 2010) What is the curb weight of each vehicle?

Total: 11,150 lb

**40.** *Nutrition.* A slice of carrot cake from the popular restaurant The Cheesecake Factory contains 1560 calories. This is three-fourths of the average daily calorie requirement for many adults. (*Source*: The Center for Science in the Public Interest) Find the average daily calorie requirement for these adults.

**41.** *Television Viewers.* In a recent week, the television networks CBS, ABC, and NBC together averaged a total of 29.1 million viewers. CBS had 1.7 million more viewers than ABC, and NBC had 1.7 million fewer viewers than ABC. (*Source*: Nielsen Media Research) How many viewers did each network have?

**42.** *Nielsen Ratings.* Nielsen Media Research surveys TV-watching habits and provides a list of the 20 most-watched TV programs each week. Each rating point in the survey represents 1,102,000 households. One week "60 Minutes" had a rating of 11.0. How many households did this represent?

**43.** *Amount Borrowed.* Kendal borrowed money from her father at 5% simple interest to help pay her tuition at Wellington Community College. At the end of 1 year, she owed a total of

$1365 in principal and interest. How much did she borrow?

**44.** *Amount of an Investment.* Khalid makes an investment at 4% simple interest. At the end of 1 year, the total value of the investment is $1560. How much was originally invested?

**45.** *Sales Commission.* Ryan, a consumer electronics salesperson, earns a base salary of $1500 per month and a commission of 8% on the amount of sales he makes. One month Ryan received a paycheck for $2284. Find the amount of his sales for the month.

**46.** *Commission vs. Salary.* Juliet has a choice between receiving a monthly salary of $1800 from Furniture by Design or a base salary of $1600 and a 4% commission on the amount of furniture she sells during the month. For what amount of sales will the two choices be equal?

**47.** *Hourly Wage.* Soledad worked 48 hr one week and earned a $442 paycheck. She earns time and a half (1.5 times her regular hourly wage) for the number of hours she works in excess of 40. What is Soledad's regular hourly wage?

**48.** *Cab Fare.* City Cabs charges a $1.75 pickup fee and $1.50 per mile traveled. Diego's fare for a cross-town cab ride is $19.75. How far did he travel in the cab?

**49.** *Angle Measure.* In triangle *ABC*, angle *B* is five times as large as angle *A*. The measure of angle *C* is 2° less than that of angle *A*. Find the measures of the angles. (*Hint*: The sum of the angle measures is 180°.)

**50.** *Angle Measure.* In triangle *ABC*, angle *B* is twice as large as angle *A*. Angle *C* measures 20° more than angle *A*. Find the measures of the angles.

**51.** *Test-Plot Dimensions.* Morgan's Seeds has a rectangular test plot with a perimeter of 322 m. The length is 25 m more than the width. Find the dimensions of the plot.

**52.** *Garden Dimensions.* The children at Tiny Tots Day Care planted a rectangular vegetable garden with a perimeter of 39 m. The length is twice the width. Find the dimensions of the garden.

**53.** *Soccer-Field Dimensions.* The width of the soccer field recommended for players under the age of 12 is 35 yd less than the length. The perimeter of the field is 330 yd. (*Source*: U.S. Youth Soccer) Find the dimensions of the field.

**54.** *Poster Dimensions.* Marissa is designing a poster to promote the Talbot Street Art Fair. The width of the poster will be two-thirds of its height, and its perimeter will be 100 in. Find the dimensions of the poster.

**55.** *Water Weight.* Water accounts for 55% of a woman's weight (*Source*: ga.water.usgs.gov/edu). Lily weighs 135 lb. How much of her body weight is water?

**56.** *Water Weight.* Water accounts for 60% of a man's weight (*Source*: ga.water.usgs.gov/edu). Jake weighs 186 lb. How much of his body weight is water?

**57.** *Train Speeds.* A Central Railway freight train leaves a station and travels due north at a speed of 60 mph. One hour later, an Amtrak passenger train leaves the same station and travels due north on a parallel track at a speed of 80 mph. How long will it take the passenger train to overtake the freight train?

**58.** *Distance Traveled.* A private airplane leaves Midway Airport and flies due east at a speed of 180 km/h. Two hours later, a jet leaves Midway and flies due east at a speed of 900 km/h. How far from the airport will the jet overtake the private plane?

**59.** *Traveling Upstream.* A kayak moves at a rate of 12 mph in still water. If the river's current flows at a rate of 4 mph, how long does it take the boat to travel 36 mi upstream?

**60.** *Traveling Downstream.* Angelo's kayak travels 14 km/h in still water. If the river's current flows at a rate of 2 km/h, how long will it take him to travel 20 km downstream?

**61.** *Flying into a Headwind.* An airplane that travels 450 mph in still air encounters a 30-mph headwind. How long will it take the plane to travel 1050 mi into the wind?

**62.** *Flying with a Tailwind.* An airplane that can travel 375 mph in still air is flying with a 25-mph tailwind. How long will it take the plane to travel 700 mi with the wind?

**63.** *Investment Income.* Erica invested a total of $5000, part at 3% simple interest and part at 4% simple interest. At the end of 1 year, the investments had earned $176 interest. How much was invested at each rate?

**64.** *Student Loans.* Dimitri's two student loans total $9000. One loan is at 5% simple interest and the other is at 6% simple interest. At the end of 1 year, Dimitri owes $492 in interest. What is the amount of each loan?

**65.** *Networking Sites.* In February 2011, facebook.com had 134,078,221 unique visitors. This number of visitors was 51,604,956 less than the total number of visitors to YouTube.com and amazon.com. The number who visited YouTube.com was 42,826,225 more than the number who visited amazon.com. (*Source*: lists. compete.com) Find the number of visitors to YouTube.com and to amazon.com.

**66.** *Calcium Content of Foods.* Together, one 8-oz serving of plain nonfat yogurt and one 1-oz serving of Swiss cheese contain 676 mg of calcium. The yogurt contains 4 mg more than twice the calcium in the cheese. (*Source*: U.S. Department of Agriculture) Find the calcium content of each food.

**67.** *NFL Stadium Elevation.* The elevations of the 31 NFL stadiums range from 3 ft at MetLife Stadium in East Rutherford, New Jersey, to 5210 ft at Sports Authority Field at Mile High in Denver, Colorado. The elevation of Sports Authority Field at Mile High is 247 ft higher than seven times the elevation of Lucas Oil Stadium in Indianapolis, Indiana. What is the elevation of Lucas Oil Stadium?

**68.** *Public Libraries.* There is a total of 1525 public libraries in New York and Wisconsin. There are 151 more libraries in New York than twice the number in Wisconsin. (*Source*: Institute of Museums and Library Services) Find the number of public libraries in New York and in Wisconsin.

**69.** *Source of Drinking Water.* In the Dominican Republic, factory-bottled water is the primary source of drinking water for 67% of the urban population (*Source*: *National Geographic*, April 2010). In 2009, the population of the Dominican Republic was 9,650,054, of which 66.8% was urban. For how many in the urban population of the Dominican Republic was bottled water the primary source of drinking water?

**70.** *Volcanic Activity.* A volcano that is currently about one-half mile below the surface of the Pacific Ocean near the Big Island of Hawaii will eventually become a new Hawaiian island, Loihi. The volcano will break the surface of the ocean in about 50,000 years. (*Source*: U.S. Geological Survey) On average, how many inches does the volcano rise in a year?

*Find the zero of the linear function.*

**71.** $f(x) = x + 5$         **72.** $f(x) = 5x + 20$

**73.** $f(x) = -2x + 11$     **74.** $f(x) = 8 + x$

**75.** $f(x) = 16 - x$        **76.** $f(x) = -2x + 7$

**77.** $f(x) = x + 12$        **78.** $f(x) = 8x + 2$

**79.** $f(x) = -x + 6$        **80.** $f(x) = 4 + x$

**81.** $f(x) = 20 - x$        **82.** $f(x) = -3x + 13$

**83.** $f(x) = \frac{2}{5}x - 10$     **84.** $f(x) = 3x - 9$

**85.** $f(x) = -x + 15$       **86.** $f(x) = 4 - x$

*In Exercises 87–92, use the given graph to find each of the following:* **(a)** *the x-intercept and* **(b)** *the zero of the function.*

**87.**

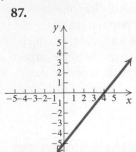

**88.**

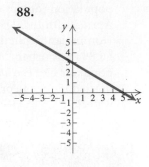

**89.**

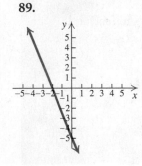

**90.**

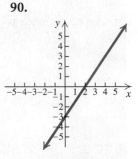

**91.**

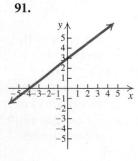

**92.**

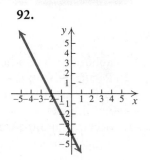

## Skill Maintenance

**93.** Write a slope–intercept equation for the line containing the point $(-1, 4)$ and parallel to the line $3x + 4y = 7$.

**94.** Write an equation of the line containing the points $(-5, 4)$ and $(3, -2)$.

**95.** Find the distance between $(2, 2)$ and $(-3, -10)$.

**96.** Find the midpoint of the segment with end points $\left(-\frac{1}{2}, \frac{2}{5}\right)$ and $\left(-\frac{3}{2}, \frac{3}{5}\right)$.

**97.** Given that $f(x) = \dfrac{x}{x - 3}$, find $f(-3)$, $f(0)$, and $f(3)$.

**98.** Find the slope and the $y$-intercept of the line with the equation $7x - y = \frac{1}{2}$.

## Synthesis

*State whether each of the following is a linear function.*

**99.** $f(x) = 7 - \frac{3}{2}x$

**100.** $f(x) = \dfrac{3}{2x} + 5$

**101.** $f(x) = x^2 + 1$

**102.** $f(x) = \frac{3}{4}x - (2.4)^2$

*Solve.*

**103.** $2x - \left\{x - \left[3x - (6x + 5)\right]\right\} = 4x - 1$

**104.** $14 - 2\left[3 + 5(x - 1)\right] = 3\left\{x - 4\left[1 + 6(2 - x)\right]\right\}$

**105.** *Packaging and Price.*   Dannon recently replaced its 8-oz cup of yogurt with a 6-oz cup and reduced the suggested retail price from 89 cents to 71 cents (*Source*: IRI). Was the price per ounce reduced by the same percent as the size of the cup? If not, find the price difference per ounce in terms of a percent.

**106.** *Bestsellers.*   One week 10 copies of the novel *The Last Song* by Nicholas Sparks were sold for every 7.9 copies of David Baldacci's *Deliver Us from Evil* that were sold (*Source*: USA Today Best-Selling Books). If a total of 10,919 copies of the two books were sold, how many copies of each were sold?

**107.** *Running vs. Walking.*   A 150-lb person who runs at 6 mph for 1 hr burns about 720 calories. The same person, walking at 4 mph for 90 min, burns about 480 calories. (*Source*: FitSmart, *USA Weekend*, July 19–21, 2002) Suppose a 150-lb person runs at 6 mph for 75 min. How far would the person have to walk at 4 mph in order to burn the same number of calories used running?

## 1.6 Solving Linear Inequalities

- Solve linear inequalities.
- Solve compound inequalities.
- Solve applied problems using inequalities.

An **inequality** is a sentence with $<$, $>$, $\leq$, or $\geq$ as its verb. An example is $3x - 5 < 6 - 2x$. To **solve** an inequality is to find all values of the variable that make the inequality true. Each of these values is a **solution** of the inequality, and the set of all such solutions is its **solution set.** Inequalities that have the same solution set are called **equivalent inequalities.**

### Linear Inequalities

The principles for solving inequalities are similar to those for solving equations.

> **PRINCIPLES FOR SOLVING INEQUALITIES**
>
> For any real numbers $a$, $b$, and $c$:
>
> *The Addition Principle for Inequalities:*
> If $a < b$ is true, then $a + c < b + c$ is true.
>
> *The Multiplication Principle for Inequalities:*
> **a)** If $a < b$ and $c > 0$ are true, then $ac < bc$ is true.
> **b)** If $a < b$ and $c < 0$ are true, then $ac > bc$ is true.
> (When both sides of an inequality are multiplied by a negative number, the inequality sign must be reversed.)
>
> Similar statements hold for $a \leq b$.

First-degree inequalities with one variable, like those in Example 1 below, are **linear inequalities.**

**EXAMPLE 1** Solve the inequality. Then graph the solution set.

**a)** $3x - 5 < 6 - 2x$  **b)** $13 - 7x \geq 10x - 4$

*Solution*

**a)** $3x - 5 < 6 - 2x$

$\qquad 5x - 5 < 6$ Using the addition principle for inequalities; adding $2x$

$\qquad\qquad 5x < 11$ Using the addition principle for inequalities; adding 5

$\qquad\qquad x < \frac{11}{5}$ Using the multiplication principle for inequalities; multiplying by $\frac{1}{5}$, or dividing by 5

**INTERVAL NOTATION**

REVIEW SECTION **R.1.**

Any number less than $\frac{11}{5}$ is a solution. The solution set is $\left\{ x \mid x < \frac{11}{5} \right\}$, or $\left( -\infty, \frac{11}{5} \right)$. The graph of the solution set is shown below.

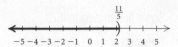

To check, we can graph $y_1 = 3x - 5$ and $y_2 = 6 - 2x$. The graph at left shows that for $x < 2.2$, or $x < \frac{11}{5}$, the graph of $y_1$ lies below the graph of $y_2$, or $y_1 < y_2$.

$y_1 = 3x - 5, \quad y_2 = 6 - 2x$

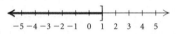

**b)** $13 - 7x \geq 10x - 4$

$\qquad 13 - 17x \geq -4$      Subtracting $10x$

$\qquad\qquad -17x \geq -17$      Subtracting $13$

$\qquad\qquad\qquad x \leq 1$      Dividing by $-17$ and reversing the inequality sign

The solution set is $\left\{ x \mid x \leq 1 \right\}$, or $(-\infty, 1]$. The graph of the solution set is shown below.

**Now Try Exercises 1 and 3.**

**EXAMPLE 2** Find the domain of the function.

**a)** $f(x) = \sqrt{x - 6}$          **b)** $h(x) = \dfrac{x}{\sqrt{3 - x}}$

*Solution*

**a)** The radicand, $x - 6$, must be greater than or equal to 0. We solve the inequality $x - 6 \geq 0$:

$\qquad x - 6 \geq 0$

$\qquad\qquad x \geq 6.$

The domain is $\left\{ x \mid x \geq 6 \right\}$, or $[6, \infty)$.

**b)** Any real number can be an input for $x$ in the numerator, but inputs for $x$ must be restricted in the denominator. We must have $3 - x \geq 0$ and $\sqrt{3 - x} \neq 0$. Thus, $3 - x > 0$. We solve for $x$:

$\qquad 3 - x > 0$

$\qquad\quad -x > -3$      Subtracting $3$

$\qquad\qquad x < 3.$      Multiplying by $-1$ and reversing the inequality sign

The domain is $\left\{ x \mid x < 3 \right\}$, or $(-\infty, 3)$.

**Now Try Exercises 17 and 21.**

## ■ Compound Inequalities

When two inequalities are joined by the word *and* or the word *or*, a **compound inequality** is formed. A compound inequality like

$$-3 < 2x + 5 \quad and \quad 2x + 5 \leq 7$$

is called a **conjunction,** because it uses the word *and.* The sentence $-3 < 2x + 5 \leq 7$ is an abbreviation for the preceding conjunction.

Compound inequalities can be solved using the addition and multiplication principles for inequalities.

**EXAMPLE 3** Solve $-3 < 2x + 5 \leq 7$. Then graph the solution set.

*Solution* We have

$$-3 < 2x + 5 \leq 7$$
$$-8 < 2x \leq 2 \qquad \text{Subtracting 5}$$
$$-4 < x \leq 1. \qquad \text{Dividing by 2}$$

The solution set is $\{x | -4 < x \leq 1\}$, or $(-4, 1]$. The graph of the solution set is shown below.

Now Try Exercise 23.

A compound inequality like $2x - 5 \leq -7$ *or* $2x - 5 > 1$ is called a **disjunction,** because it contains the word *or.* Unlike some conjunctions, it cannot be abbreviated; that is, it cannot be written without the word *or.*

**EXAMPLE 4** Solve: $2x - 5 \leq -7$ *or* $2x - 5 > 1$. Then graph the solution set.

*Solution* We have

$$2x - 5 \leq -7 \quad or \quad 2x - 5 > 1$$
$$2x \leq -2 \quad or \qquad 2x > 6 \qquad \text{Adding 5}$$
$$x \leq -1 \quad or \qquad x > 3. \qquad \text{Dividing by 2}$$

The solution set is $\{x | x \leq -1 \ or \ x > 3\}$. We can also write the solution set using interval notation and the symbol $\cup$ for the **union** or inclusion of both sets: $(-\infty, -1] \cup (3, \infty)$. The graph of the solution set is shown below.

$y_1 = 2x - 5, \ y_2 = -7, \ y_3 = 1$

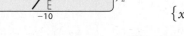

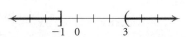

To check, we graph $y_1 = 2x - 5$, $y_2 = -7$, and $y_3 = 1$. Note that for $\{x | x \leq -1 \ or \ x > 3\}$, $y_1 \leq y_2 \ or \ y_1 > y_3$.

Now Try Exercise 35.

## ■ An Application

**EXAMPLE 5** *Income Plans.* For her house-painting job, Erica can be paid in one of two ways:

Plan A: $250 plus $10 per hour;

Plan B: $20 per hour.

Suppose that a job takes $n$ hours. For what values of $n$ is plan B better for Erica?

### Solution

1. **Familiarize.** Suppose that a job takes 20 hr. Then $n = 20$, and under plan A, Erica would earn $\$250 + \$10 \cdot 20$, or $\$250 + \$200$, or $\$450$. Her earnings under plan B would be $\$20 \cdot 20$, or $\$400$. This shows that plan A is better for Erica if a job takes 20 hr. If a job takes 30 hr, then $n = 30$, and under plan A, Erica would earn $\$250 + \$10 \cdot 30$, or $\$250 + \$300$, or $\$550$. Under plan B, she would earn $\$20 \cdot 30$, or $\$600$, so plan B is better in this case. To determine *all* values of $n$ for which plan B is better for Erica, we solve an inequality. Our work in this step helps us write the inequality.

2. **Translate.** We translate to an inequality:

$$\underbrace{\text{Income from plan B}}_{20n} \quad \underbrace{\text{is greater than}}_{>} \quad \underbrace{\text{income from plan A.}}_{250 + 10n}$$

3. **Carry out.** We solve the inequality:

$$20n > 250 + 10n$$
$$10n > 250 \qquad \text{Subtracting } 10n \text{ on both sides}$$
$$n > 25. \qquad \text{Dividing by 10 on both sides}$$

4. **Check.** For $n = 25$, the income from plan A is $\$250 + \$10 \cdot 25$, or $\$250 + \$250$, or $\$500$, and the income from plan B is $\$20 \cdot 25$, or $\$500$. This shows that for a job that takes 25 hr to complete, the income is the same under either plan. In the *Familiarize* step, we saw that plan B pays more for a 30-hr job. Since $30 > 25$, this provides a partial check of the result. We cannot check all values of $n$.

5. **State.** For values of $n$ greater than 25 hr, plan B is better for Erica.

**Now Try Exercise 45.**

## 1.6 Exercise Set

*Solve and graph the solution set.*

**1.** $4x - 3 > 2x + 7$

**2.** $8x + 1 \geq 5x - 5$

**3.** $x + 6 < 5x - 6$

**4.** $3 - x < 4x + 7$

**5.** $4 - 2x \leq 2x + 16$

**6.** $3x - 1 > 6x + 5$

**7.** $14 - 5y \leq 8y - 8$

**8.** $8x - 7 < 6x + 3$

**9.** $7x - 7 > 5x + 5$

**10.** $12 - 8y \geq 10y - 6$

**11.** $3x - 3 + 2x \geq 1 - 7x - 9$

**12.** $5y - 5 + y \leq 2 - 6y - 8$

**13.** $-\frac{3}{4}x \geq -\frac{5}{8} + \frac{2}{3}x$

**14.** $-\frac{5}{6}x \leq \frac{3}{4} + \frac{8}{3}x$

**15.** $4x(x - 2) < 2(2x - 1)(x - 3)$

**16.** $(x + 1)(x + 2) > x(x + 1)$

*Find the domain of the function.*

**17.** $h(x) = \sqrt{x - 7}$

**18.** $g(x) = \sqrt{x + 8}$

**19.** $f(x) = \sqrt{1 - 5x} + 2$

**20.** $f(x) = \sqrt{2x + 3} - 4$

**21.** $g(x) = \dfrac{5}{\sqrt{4 + x}}$

**22.** $h(x) = \dfrac{x}{\sqrt{8 - x}}$

*Solve and write interval notation for the solution set. Then graph the solution set.*

**23.** $-2 \leq x + 1 < 4$

**24.** $-3 < x + 2 \leq 5$

**25.** $5 \leq x - 3 \leq 7$

**26.** $-1 < x - 4 < 7$

**27.** $-3 \leq x + 4 \leq 3$

**28.** $-5 < x + 2 < 15$

**29.** $-2 < 2x + 1 < 5$ **30.** $-3 \leq 5x + 1 \leq 3$

**31.** $-4 \leq 6 - 2x < 4$ **32.** $-3 < 1 - 2x \leq 3$

**33.** $-5 < \frac{1}{2}(3x + 1) < 7$ **34.** $\frac{2}{3} \leq -\frac{4}{5}(x - 3) < 1$

**35.** $3x \leq -6 \text{ or } x - 1 > 0$

**36.** $2x < 8 \text{ or } x + 3 \geq 10$

**37.** $2x + 3 \leq -4 \text{ or } 2x + 3 \geq 4$

**38.** $3x - 1 < -5 \text{ or } 3x - 1 > 5$

**39.** $2x - 20 < -0.8 \text{ or } 2x - 20 > 0.8$

**40.** $5x + 11 \leq -4 \text{ or } 5x + 11 \geq 4$

**41.** $x + 14 \leq -\frac{1}{4} \text{ or } x + 14 \geq \frac{1}{4}$

**42.** $x - 9 < -\frac{1}{2} \text{ or } x - 9 > \frac{1}{2}$

**43.** *Information Technology.* The equation $y = 31.7x + 487$ estimates the amount that small and midsize businesses spend, in billions of dollars, on information technology, where $x$ is the number of years after 2007 (*Source*: IDC SMB Research). For what years will the spending be more than $775 billion?

**44.** *Televisions per Household.* The equation $y = 1.393x + 19.593$ estimates the percentage of U.S. households that have three or more televisions, where $x$ is the number of years after 1985 (*Source*: The Nielsen Company). For what years will the percentage of U.S. households with three or more televisions be at least 65%?

**45.** *Moving Costs.* Acme Movers charges $100 plus $30 per hour to move a household across town. Hank's Movers charges $55 per hour. For what lengths of time does it cost less to hire Hank's Movers?

**46.** *Investment Income.* Gina plans to invest $12,000, part at 4% simple interest and the rest at 6% simple interest. What is the most that she can invest at 4% and still be guaranteed at least $650 in interest per year?

**47.** *Investment Income.* Dillon plans to invest $7500, part at 4% simple interest and the rest at 5% simple interest. What is the most that he can invest at 4% and still be guaranteed at least $325 in interest per year?

**48.** *Investment Income.* A university invests $600,000 at simple interest, part at 6%, half that amount at 4.5%, and the rest at 3.5%. What is the most that the university can invest at 3.5% and be guaranteed $24,600 in interest per year?

**49.** *Investment Income.* A foundation invests $50,000 at simple interest, part at 7%, twice that amount at 4%, and the rest at 5.5%. What is the most that the foundation can invest at 4% and be guaranteed $2660 in interest per year?

**50.** *Income Plans.* Tori can be paid in one of two ways for selling insurance policies:

Plan A: A salary of $750 per month, plus a commission of 10% of sales;

Plan B: A salary of $1000 per month, plus a commission of 8% of sales in excess of $2000.

For what amount of monthly sales is plan A better than plan B if we can assume that sales are always more than $2000?

**51.** *Income Plans.* Achal can be paid in one of two ways for the furniture he sells:

Plan A: A salary of $900 per month, plus a commission of 10% of sales;

Plan B: A salary of $1200 per month, plus a commission of 15% of sales in excess of $8000.

For what amount of monthly sales is plan B better than plan A if we can assume that Achal's sales are always more than $8000?

**52.** *Income Plans.* Jeanette can be paid in one of two ways for painting a house:

Plan A: $200 plus $12 per hour;
Plan B: $20 per hour.

Suppose a job takes $n$ hours to complete. For what values of $n$ is plan A better for Jeanette?

## Skill Maintenance

### Vocabulary Reinforcement

*In each of Exercises 53–56, fill in the blank(s) with the correct term(s). Some of the given choices will not be used; others will be used more than once.*

| | |
|---|---|
| constant | domain |
| function | distance formula |
| any | exactly one |
| midpoint formula | identity |
| $y$-intercept | $x$-intercept |
| range | |

**53.** A(n) _____ is a correspondence between a first set, called the _____, and a second set, called the _____, such that each member of the _____ corresponds to _____ member of the _____.

**54.** The _____ is $\left( \dfrac{x_1 + x_2}{2}, \dfrac{y_1 + y_2}{2} \right)$.

**55.** A(n) _____ is a point $(a, 0)$.

**56.** A function $f$ is a linear function if it can be written as $f(x) = mx + b$, where $m$ and $b$ are constants. If $m = 0$, the function is a(n) _____ function $f(x) = b$. If $m = 1$ and $b = 0$, the function is the _____ function $f(x) = x$.

## Synthesis

*Solve.*

**57.** $2x \leq 5 - 7x < 7 + x$

**58.** $x \leq 3x - 2 \leq 2 - x$

**59.** $3y < 4 - 5y < 5 + 3y$

**60.** $y - 10 < 5y + 6 \leq y + 10$

# Chapter 1 Summary and Review

## STUDY GUIDE

| KEY TERMS AND CONCEPTS | EXAMPLES |
|---|---|

### SECTION 1.1: INTRODUCTION TO GRAPHING

**Graphing Equations**

To **graph** an equation is to make a drawing that represents the solutions of that equation. We can graph an equation by selecting values for one variable and finding the corresponding values for the other variable. We list the solutions (ordered pairs) in a table, plot the points, and draw the graph.

Graph: $y = 4 - x^2$.

| $x$ | $y = 4 - x^2$ | $(x, y)$ |
|---|---|---|
| 0 | 4 | $(0, 4)$ |
| $-1$ | 3 | $(-1, 3)$ |
| 1 | 3 | $(1, 3)$ |
| $-2$ | 0 | $(-2, 0)$ |
| 2 | 0 | $(2, 0)$ |

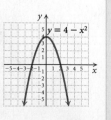

**Intercepts**

An *x*-intercept is a point $(a, 0)$.
To find $a$, let $y = 0$ and solve for $x$.

A *y*-intercept is a point $(0, b)$.
To find $b$, let $x = 0$ and solve for $y$.

We can graph a straight line by plotting the intercepts and drawing the line containing them.

Graph using intercepts: $2x - y = 4$.

Let $y = 0$:

$$2x - 0 = 4$$
$$2x = 4$$
$$x = 2.$$

The *x*-intercept is $(2, 0)$.

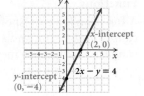

Let $x = 0$:

$$2 \cdot 0 - y = 4$$
$$-y = 4$$
$$y = -4.$$

The *y*-intercept is $(0, -4)$.

---

**Distance Formula**

The **distance** $d$ between any two points $(x_1, y_1)$ and $(x_2, y_2)$ is given by

$$d = \sqrt{(x_2 - x_1)^2 + (y_2 - y_1)^2}.$$

Find the distance between $(-5, 7)$ and $(2, -3)$.

$$\begin{aligned} d &= \sqrt{[2 - (-5)]^2 + (-3 - 7)^2} \\ &= \sqrt{7^2 + (-10)^2} \\ &= \sqrt{49 + 100} \\ &= \sqrt{149} \approx 12.2 \end{aligned}$$

---

**Midpoint Formula**

If the endpoints of a segment are $(x_1, y_1)$ and $(x_2, y_2)$, then the coordinates of the **midpoint** of the segment are

$$\left( \frac{x_1 + x_2}{2}, \frac{y_1 + y_2}{2} \right).$$

Find the midpoint of the segment whose endpoints are $(-10, 4)$ and $(3, 8)$.

$$\begin{aligned} \left( \frac{x_1 + x_2}{2}, \frac{y_1 + y_2}{2} \right) &= \left( \frac{-10 + 3}{2}, \frac{4 + 8}{2} \right) \\ &= \left( -\frac{7}{2}, 6 \right) \end{aligned}$$

---

**Circles**

The **standard form** of the equation of a circle with center $(h, k)$ and radius $r$ is

$$(x - h)^2 + (y - k)^2 = r^2.$$

Find an equation of a circle with center $(1, -6)$ and radius 8.

$$\begin{aligned} (x - h)^2 + (y - k)^2 &= r^2 \\ (x - 1)^2 + [y - (-6)]^2 &= 8^2 \\ (x - 1)^2 + (y + 6)^2 &= 64 \end{aligned}$$

Given the circle

$$(x + 9)^2 + (y - 2)^2 = 121,$$

determine the center and the radius.

Writing in standard form, we have

$$[x - (-9)]^2 + (y - 2)^2 = 11^2.$$

The center is $(-9, 2)$, and the radius is 11.

**SECTION 1.2: FUNCTIONS AND GRAPHS**

### Functions

A **function** is a correspondence between a first set, called the **domain**, and a second set, called the **range**, such that each member of the domain corresponds to *exactly one* member of the range.

Consider the function given by

$$g(x) = |x| - 1.$$
$$g(-3) = |-3| - 1$$
$$= 3 - 1$$
$$= 2$$

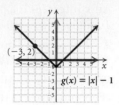

For the input $-3$, the output is $2$:

$$f(-3) = 2.$$

The point $(-3, 2)$ is on the graph.

Domain: Set of all inputs = all real numbers, or $(-\infty, \infty)$.

Range: Set of all outputs: $\{y \mid y \geq -1\}$, or $[-1, \infty)$.

### The Vertical-Line Test

If it is possible for a vertical line to cross a graph more than once, then the graph *is not* the graph of a function.

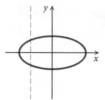

This *is not* the graph of a function because a vertical line can cross it more than once, as shown.

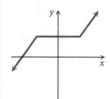

This *is* the graph of a function because no vertical line can cross it more than once.

### Domain

When a function $f$ whose inputs and outputs are real numbers is given by a formula, the **domain** is the set of all inputs for which the expression is defined as a real number.

Find the domain of the function given by

$$h(x) = \frac{(x - 1)}{(x + 5)(x - 10)}.$$

Division by 0 is not defined. Since $x + 5 = 0$ when $x = -5$ and $x - 10 = 0$ when $x = 10$, the domain of $h$ is

$$\{x \mid x \neq -5 \text{ and } x \neq 10\},$$
$$\text{or } (-\infty, -5) \cup (-5, 10) \cup (10, \infty).$$

## SECTION 1.3: LINEAR FUNCTIONS, SLOPE, AND APPLICATIONS

**Slope**

$$m = \frac{\text{rise}}{\text{run}} = \frac{y_2 - y_1}{x_2 - x_1} = \frac{y_1 - y_2}{x_1 - x_2}$$

Slope can also be considered as an **average rate of change**. To find the average rate of change between two data points on a graph, we determine the slope of the line that passes through the points.

The slope of the line containing the points $(3, -10)$ and $(-2, 6)$ is

$$m = \frac{y_2 - y_1}{x_2 - x_1} = \frac{6 - (-10)}{-2 - 3} = \frac{16}{-5} = -\frac{16}{5}.$$

In 2000, the population of Flint, Michigan, was 124,943. By 2008, the population had decreased to 112,900. Find the average rate of change in population from 2000 to 2008.

$$\text{Average rate of change} = m = \frac{112{,}900 - 124{,}943}{2008 - 2000}$$

$$= \frac{-12{,}043}{8} \approx -1505$$

The average rate of change in population over the 8-year period was a decrease of 1505 people per year.

---

**Slope–Intercept Form of an Equation**

$$f(x) = mx + b$$

The slope of the line is $m$.

The $y$-intercept of the line is $(0, b)$.

Determine the slope and the $y$-intercept of the line given by $5x - 7y = 14$.

We first find the slope–intercept form:

$$5x - 7y = 14$$
$$-7y = -5x + 14 \qquad \textbf{Adding } -5x$$
$$y = \tfrac{5}{7} x - 2. \qquad \textbf{Multiplying by } -\tfrac{1}{7}$$

The slope is $\tfrac{5}{7}$, and the $y$-intercept is $(0, -2)$.

---

To graph an equation written in slope–intercept form, we plot the $y$-intercept and use the slope to find another point. Then we draw the line.

Graph: $f(x) = -\tfrac{2}{3}x + 4$.

We plot the $y$-intercept, $(0, 4)$. Think of the slope as $\tfrac{-2}{3}$. From the $y$-intercept, we find another point by moving 2 units down and 3 units to the right to the point $(3, 2)$. We then draw the graph.

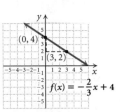

---

**Horizontal Lines**

The graph of $y = b$, or $f(x) = b$, is a horizontal line with $y$-intercept $(0, b)$. The slope of a horizontal line is 0.

**Vertical Lines**

The graph of $x = a$ is a vertical line with $x$-intercept $(a, 0)$. The slope of a vertical line is *not* defined.

Graph $y = -4$ and determine its slope.

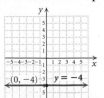

The slope is 0.

Graph $x = 3$ and determine its slope.

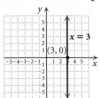

The slope is not defined.

## SECTION 1.4: EQUATIONS OF LINES AND MODELING

**Slope–Intercept Form of an Equation**

$$y = mx + b, \text{ or } f(x) = mx + b$$

The slope of the line is $m$.

The $y$-intercept of the line is $(0, b)$.

**Point–Slope Form of an Equation**

$$y - y_1 = m(x - x_1)$$

The slope of the line is $m$.

The line passes through $(x_1, y_1)$.

Write the slope–intercept equation for a line with slope $-\frac{2}{9}$ and $y$-intercept $(0, 4)$.

$$y = mx + b \qquad \text{Using the slope–intercept form}$$

$$y = -\frac{2}{9}x + 4 \qquad \text{Substituting } -\frac{2}{9} \text{ for } m \text{ and } 4 \text{ for } b$$

Write the slope–intercept equation for a line that passes through $(-5, 7)$ and $(3, -9)$.

We first determine the slope:

$$m = \frac{-9 - 7}{3 - (-5)} = \frac{-16}{8} = -2.$$

*Using the slope–intercept form:* We substitute $-2$ for $m$ and either $(-5, 7)$ or $(3, -9)$ for $(x, y)$ and solve for $b$:

$$y = mx + b$$
$$7 = -2 \cdot (-5) + b \qquad \text{Using } (-5, 7)$$
$$7 = 10 + b$$
$$-3 = b.$$

The slope–intercept equation is $y = -2x - 3$.

*Using the point–slope equation:* We substitute $-2$ for $m$ and either $(-5, 7)$ or $(3, -9)$ for $(x_1, y_1)$:

$$y - y_1 = m(x - x_1)$$
$$y - (-9) = -2(x - 3) \qquad \text{Using } (3, -9)$$
$$y + 9 = -2x + 6$$
$$y = -2x - 3.$$

The slope–intercept equation is $y = -2x - 3$.

**Parallel Lines**

Vertical lines are parallel. Nonvertical lines are **parallel** if and only if they have the same slope and different $y$-intercepts.

Write the slope–intercept equation for a line passing through $(-3, 1)$ that is parallel to the line $y = \frac{2}{3}x + 5$.

The slope of $y = \frac{2}{3}x + 5$ is $\frac{2}{3}$, so the slope of a line parallel to this line is also $\frac{2}{3}$. We use either the slope–intercept equation or the point–slope equation for a line with slope $\frac{2}{3}$ and containing the point $(-3, 1)$. Here we use the point–slope equation and substitute $\frac{2}{3}$ for $m$, $-3$ for $x_1$, and $1$ for $y_1$.

$$y - y_1 = m(x - x_1)$$
$$y - 1 = \frac{2}{3}\left[x - (-3)\right]$$
$$y - 1 = \frac{2}{3}x + 2$$
$$y = \frac{2}{3}x + 3 \qquad \text{Slope–intercept form}$$

**Perpendicular Lines**

Two lines are **perpendicular** if and only if the product of their slopes is $-1$ or if one line is vertical ($x = a$) and the other is horizontal ($y = b$).

Write the slope–intercept equation for a line that passes through $(-3, 1)$ and is perpendicular to the line $y = \frac{2}{3}x + 5$.

The slope of $y = \frac{2}{3}x + 5$ is $\frac{2}{3}$, so the slope of a line perpendicular to this line is the opposite of the reciprocal of $\frac{2}{3}$, or $-\frac{3}{2}$. Here we use the point–slope equation and substitute $-\frac{3}{2}$ for $m$, $-3$ for $x_1$, and 1 for $y_1$.

$$y - y_1 = m(x - x_1)$$
$$y - 1 = -\frac{3}{2}\left[x - (-3)\right]$$
$$y - 1 = -\frac{3}{2}(x + 3)$$
$$y - 1 = -\frac{3}{2}x - \frac{9}{2}$$
$$y = -\frac{3}{2}x - \frac{7}{2} \quad \text{Slope–intercept form}$$

## SECTION 1.5: LINEAR EQUATIONS, FUNCTIONS, ZEROS, AND APPLICATIONS

**Equation–Solving Principles**

*The Addition Principle:*
If $a = b$ is true, then $a + c = b + c$ is true.

*The Multiplication Principle:*
If $a = b$ is true, then $ac = bc$ is true.

Solve: $2(3x - 7) = 15 - (x + 1)$.

$$2(3x - 7) = 15 - (x + 1)$$
$$6x - 14 = 15 - x - 1 \quad \text{Using the distributive property}$$
$$6x - 14 = 14 - x \quad \text{Collecting like terms}$$
$$6x - 14 + x = 14 - x + x \quad \text{Adding } x \text{ on both sides}$$
$$7x - 14 = 14$$
$$7x - 14 + 14 = 14 + 14 \quad \text{Adding 14 on both sides}$$
$$7x = 28$$
$$\tfrac{1}{7} \cdot 7x = \tfrac{1}{7} \cdot 28 \quad \text{Multiplying by } \tfrac{1}{7} \text{ on both sides}$$
$$x = 4$$

*Check:* 
$$\frac{2(3x - 7) = 15 - (x + 1)}{2(3 \cdot 4 - 7) \;?\; 15 - (4 + 1)}$$
$$2(12 - 7) \;\Big|\; 15 - 5$$
$$2 \cdot 5 \;\Big|\; 10$$
$$10 \;\Big|\; 10 \qquad \text{TRUE}$$

The solution is 4.

**Special Cases**

Some equations have *no* solution.

Solve: $2 + 17x = 17x - 9$.

$$2 + 17x = 17x - 9$$
$$2 + 17x - 17x = 17x - 9 - 17x \quad \text{Subtracting } 17x \text{ on both sides}$$
$$2 = -9 \quad \text{False equation}$$

We get a false equation. Thus the equation has *no* solution.

There are some equations for which *any* real number is a solution.

Solve: $5 - \frac{1}{2}x = -\frac{1}{2}x + 5$.

$$5 - \frac{1}{2}x = -\frac{1}{2}x + 5$$

$$5 - \frac{1}{2}x + \frac{1}{2}x = -\frac{1}{2}x + 5 + \frac{1}{2}x \qquad \text{Adding } \tfrac{1}{2}x \text{ on both sides}$$

$$5 = 5 \qquad \text{True equation}$$

We get a true equation. Thus any real number is a solution. The solution set is

$$\{x \,|\, x \text{ is a real number}\}, \quad \text{or} \quad (-\infty, \infty).$$

### Zeros of Functions

An input $c$ of a function $f$ is called a zero of the function if the output for the function is 0 when the input is $c$. That is,

$c$ is a zero of $f$ if $f(c) = 0$.

A linear function $f(x) = mx + b$, with $m \neq 0$, has exactly one zero.

Find the zero of the linear function

$$f(x) = \tfrac{5}{8}x - 40.$$

We find the value of $x$ for which $f(x) = 0$:

$$\tfrac{5}{8}x - 40 = 0 \qquad \text{Setting } f(x) = 0$$

$$\tfrac{5}{8}x = 40 \qquad \text{Adding 40 on both sides}$$

$$\tfrac{8}{5} \cdot \tfrac{5}{8}x = \tfrac{8}{5} \cdot 40 \qquad \text{Multiplying by } \tfrac{8}{5} \text{ on both sides}$$

$$x = 64.$$

We can check by substituting 64 for $x$:

$$f(64) = \tfrac{5}{8} \cdot 64 - 40 = 40 - 40 = 0.$$

The zero of $f(x) = \tfrac{5}{8}x - 40$ is 64.

## SECTION 1.6: SOLVING LINEAR INEQUALITIES

### Principles for Solving Linear Inequalities

*The Addition Principle:*
If $a < b$ is true, then $a + c < b + c$ is true.

*The Multiplication Principle:*
If $a < b$ and $c > 0$ are true, then $ac < bc$ is true.
If $a < b$ and $c < 0$ are true, then $ac > bc$ is true.
Similar statements hold for $a \leq b$.

Solve $3x - 2 \leq 22 - 5x$ and graph the solution set.

$$3x - 2 \leq 22 - 5x$$

$$3x - 2 + 5x \leq 22 - 5x + 5x \qquad \text{Adding } 5x \text{ on both sides}$$

$$8x - 2 \leq 22$$

$$8x - 2 + 2 \leq 22 + 2 \qquad \text{Adding 2 on both sides}$$

$$8x \leq 24$$

$$\frac{8x}{8} \leq \frac{24}{8} \qquad \text{Dividing by 8 on both sides}$$

$$x \leq 3$$

The solution set is

$$\{x \,|\, x \leq 3\}, \quad \text{or} \quad (-\infty, 3].$$

The graph of the solution set is as follows.

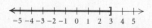

**Compound Inequalities**

When two inequalities are joined by the word *and* or the word *or*, a compound inequality is formed.

*A Conjunction:*

$$1 < 3x - 20 \text{ and } 3x - 20 \le 40, \text{ or}$$
$$1 < 3x - 20 \le 40$$

*A Disjunction:*

$$8x - 1 \le -17 \text{ or } 8x - 1 > 7$$

Solve: $1 < 3x - 20 \le 40$.

$$1 < 3x - 20 \le 40$$
$$21 < 3x \le 60 \qquad \text{Adding 20}$$
$$7 < x \le 20 \qquad \text{Dividing by 3}$$

The solution set is

$$\{x \mid 7 < x \le 20\}, \quad \text{or} \quad (7, 20].$$

Solve: $8x - 1 \le -17$ or $8x - 1 > 7$.

$$8x - 1 \le -17 \quad \text{or} \quad 8x - 1 > 7$$
$$8x \le -16 \quad \text{or} \qquad 8x > 8 \qquad \text{Adding 1}$$
$$x \le -2 \quad \text{or} \qquad x > 1 \qquad \text{Dividing by 8}$$

The solution set is

$$\{x \mid x \le -2 \text{ or } x > 1\}, \quad \text{or} \quad (-\infty, -2] \cup (1, \infty).$$

# REVIEW EXERCISES

*Answers to all of the review exercises appear in the answer section at the back of the book. If you get an incorrect answer, restudy the objective indicated in red next to the exercise or the direction line that precedes it.*

*Determine whether the statement is true or false.*

1. If the line $ax + y = c$ is perpendicular to the line $x - by = d$, then $\dfrac{a}{b} = 1$. [1.4]

2. The intersection of the lines $y = \frac{1}{2}$ and $x = -5$ is $\left(-5, \frac{1}{2}\right)$. [1.3]

3. The domain of the function $f(x) = \dfrac{\sqrt{3-x}}{x}$ does not contain $-3$ and $0$. [1.2]

4. The line parallel to the $x$-axis that passes through $\left(-\frac{1}{4}, 7\right)$ is $x = -\frac{1}{4}$. [1.3]

5. The zero of a linear function $f$ is the first coordinate of the $x$-intercept of the graph of $y = f(x)$. [1.5]

6. If $a < b$ is true and $c \ne 0$, then $ac < bc$ is true. [1.6]

*Use substitution to determine whether the given ordered pairs are solutions of the given equation.* [1.1]

7. $\left(3, \frac{24}{9}\right), (0, -9); \ 2x - 9y = -18$

8. $(0, 7), (7, 1); \ y = 7$

*Find the intercepts and then graph the line.* [1.1]

9. $2x - 3y = 6$

10. $10 - 5x = 2y$

*Graph the equation.* [1.1]

11. $y = -\frac{2}{3}x + 1$

12. $2x - 4y = 8$

13. $y = 2 - x^2$

14. Find the distance between $(3, 7)$ and $(-2, 4)$. [1.1]

15. Find the midpoint of the segment with endpoints $(3, 7)$ and $(-2, 4)$. [1.1]

16. Find the center and the radius of the circle with equation $(x + 1)^2 + (y - 3)^2 = 9$. Then graph the circle. [1.1]

*Find an equation for a circle satisfying the given conditions.* [1.1]

**17.** Center: $(0, -4)$, radius of length $\frac{3}{2}$

**18.** Center: $(-2, 6)$, radius of length $\sqrt{13}$

**19.** Diameter with endpoints $(-3, 5)$ and $(7, 3)$

*Determine whether the correspondence is a function.* [1.2]

**20.** $-6 \to 1$
$-1 \to 3$
$2 \nearrow 10$
$7 \to 12$

**21.** $h \to r$
$i \to s$
$j \nearrow t$
$k \nearrow$

*Determine whether the relation is a function. Identify the domain and the range.* [1.2]

**22.** $\{(3, 1), (5, 3), (7, 7), (3, 5)\}$

**23.** $\{(2, 7), (-2, -7), (7, -2), (0, 2), (1, -4)\}$

**24.** Given that $f(x) = x^2 - x - 3$, find each of the following. [1.2]

  **a)** $f(0)$       **b)** $f(-3)$

  **c)** $f(a - 1)$   **d)** $f(-x)$

**25.** Given that $f(x) = \dfrac{x - 7}{x + 5}$, find each of the following. [1.2]

  **a)** $f(7)$       **b)** $f(x + 1)$

  **c)** $f(-5)$     **d)** $f\left(-\frac{1}{2}\right)$

**26.** A graph of a function is shown below. Find $f(2)$, $f(-4)$, and $f(0)$. [1.2]

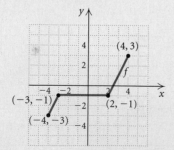

*Determine whether the graph is that of a function.* [1.2]

**27.**

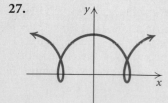

**28.**

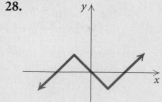

**29.**             **30.**

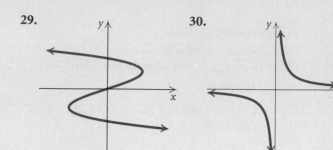

*Find the domain of the function.* [1.2]

**31.** $f(x) = 4 - 5x + x^2$

**32.** $f(x) = \dfrac{3}{x} + 2$

**33.** $f(x) = \dfrac{1}{x^2 - 6x + 5}$

**34.** $f(x) = \dfrac{-5x}{|16 - x^2|}$

*Graph the function. Then visually estimate the domain and the range.* [1.2]

**35.** $f(x) = \sqrt{16 - x^2}$

**36.** $g(x) = |x - 5|$

**37.** $f(x) = x^3 - 7$

**38.** $h(x) = x^4 + x^2$

*In Exercises 39 and 40, the table of data contains input–output values for a function. Answer the following questions.* [1.3]

  **a)** Is the change in the inputs, $x$, the same?

  **b)** Is the change in the outputs, $y$, the same?

  **c)** Is the function linear?

**39.**

| $x$ | $y$ |
|-----|-----|
| $-3$ | 8 |
| $-2$ | 11 |
| $-1$ | 14 |
| 0 | 17 |
| 1 | 20 |
| 2 | 22 |
| 3 | 26 |

**40.**

| $x$ | $y$ |
|-----|-----|
| 20 | 11.8 |
| 30 | 24.2 |
| 40 | 36.6 |
| 50 | 49.0 |
| 60 | 61.4 |
| 70 | 73.8 |
| 80 | 86.2 |

*Find the slope of the line containing the given points.* [1.3]

**41.** $(2, -11), (5, -6)$      **42.** $(5, 4), (-3, 4)$

**43.** $\left(\frac{1}{2}, 3\right), \left(\frac{1}{2}, 0\right)$

**44.** *Minimum Wage.* The minimum wage was $5.15 in 1990 and $7.25 in 2010 (*Source*: U.S. Department of Labor). Find the average rate of change in the minimum wage from 1990 to 2010. [1.3]

*Find the slope and the y-intercept of the line with the given equation.* [1.3]

**45.** $y = -\frac{7}{11}x - 6$  **46.** $-2x - y = 7$

**47.** Graph $y = -\frac{1}{4}x + 3$ using the slope and the y-intercept. [1.3]

**48.** *Total Cost.* Clear County Cable Television charges a $110 installation fee and $85 per month for basic service. Write an equation that can be used to determine the total cost $C(t)$ of $t$ months of basic cable television service. Find the total cost of 1 year of service. [1.3]

**49.** *Temperature and Depth of the Earth.* The function $T$ given by $T(d) = 10d + 20$ can be used to determine the temperature $T$, in degrees Celsius, at a depth $d$, in kilometers, inside the earth.
  **a)** Find $T(5)$, $T(20)$, and $T(1000)$. [1.3]
  **b)** The radius of the earth is about 5600 km. Use this fact to determine the domain of the function. [1.3]

*Write a slope–intercept equation for a line with the following characteristics.* [1.4]

**50.** $m = -\frac{2}{3}$, y-intercept $(0, -4)$

**51.** $m = 3$, passes through $(-2, -1)$

**52.** Passes through $(4, 1)$ and $(-2, -1)$

**53.** Write equations of the horizontal line and the vertical line that pass through $\left(-4, \frac{2}{5}\right)$. [1.4]

**54.** Find a linear function $h$ given $h(-2) = -9$ and $h(4) = 3$. Then find $h(0)$. [1.4]

*Determine whether the lines are parallel, perpendicular, or neither.* [1.4]

**55.** $3x - 2y = 8$,
$6x - 4y = 2$

**56.** $y - 2x = 4$,
$2y - 3x = -7$

**57.** $y = \frac{3}{2}x + 7$,
$y = -\frac{2}{3}x - 4$

*Given the point $(1, -1)$ and the line $2x + 3y = 4$:*

**58.** Find an equation of the line containing the given point and parallel to the given line. [1.4]

**59.** Find an equation of the line containing the given point and perpendicular to the given line. [1.4]

**60.** *Height of an Indian Elephant.* Sample data in the table below show the height at the shoulders $H$, in centimeters, of an adult Indian elephant (Elephas maximus indicus) that corresponds to the right forefoot circumference $c$, in centimeters.

| Right Forefoot Circumference, c (in centimeters) | Height at the Shoulders, H (in centimeters) |
|---|---|
| 92.3 | 182.1 |
| 98.7 | 194.8 |
| 105.0 | 207.4 |
| 114.2 | 225.7 |
| 120.7 | 238.6 |
| 125.0 | 247.2 |

*Source*: Forest Department of the Government of Kerala (India); Sreekumar, K. P., and G. Nirmalan, *Veterinary Research Communications*: Springer Netherlands, Volume 13, Number 1, January 1989.

**a)** Without using the regression feature on a graphing calculator, model the data with a linear function where the height $H$ is a function of the circumference $c$ of the right forefoot. Then using this function, estimate the height at the shoulders of an Indian elephant with a right forefoot circumference of 118 cm. Round the answer to the nearest hundredth. Answers may vary depending on the data points used.

**b)** Using a graphing calculator, fit a regression line to the data and use it to estimate the height at the shoulders of an Indian elephant with a right forefoot circumference of 118 cm. Round the answer to the nearest hundredth. What is the correlation coefficient for the regression line? How close a fit is the regression line?

*Solve.* [1.5]

**61.** $4y - 5 = 1$

**62.** $3x - 4 = 5x + 8$

**63.** $5(3x + 1) = 2(x - 4)$

**64.** $2(n - 3) = 3(n + 5)$

**65.** $\frac{3}{5}y - 2 = \frac{3}{8}$

**66.** $5 - 2x = -2x + 3$

**67.** $x - 13 = -13 + x$

**68.** *Salt Consumption.* Americans' salt consumption is increasing. The recommended daily intake of salt is 2300 milligrams per person. In 2006, American men consumed an average of 4300 milligrams of salt daily. This was a 54.7% increase over the average daily salt intake in 1974. (*Source*: National Health and Nutrition Examination Survey) What was the average daily salt intake for men in 1974? Round your answer to the nearest 10 milligrams. [1.5]

**69.** *Amount of Investment.* Candaleria makes an investment at 5.2% simple interest. At the end of 1 year, the total value of the investment is $2419.60. How much was originally invested? [1.5]

**70.** *Flying into a Headwind.* An airplane that can travel 550 mph in still air encounters a 20-mph headwind. How long will it take the plane to travel 1802 mi? [1.5]

*In Exercises 71–74, find the zero(s) of the function.* [1.5]

**71.** $f(x) = 6x - 18$

**72.** $f(x) = x - 4$

**73.** $f(x) = 2 - 10x$

**74.** $f(x) = 8 - 2x$

*Solve and write interval notation for the solution set. Then graph the solution set.* [1.6]

**75.** $2x - 5 < x + 7$

**76.** $3x + 1 \geq 5x + 9$

**77.** $-3 \leq 3x + 1 \leq 5$

**78.** $-2 < 5x - 4 \leq 6$

**79.** $2x < -1 \text{ or } x - 3 > 0$

**80.** $3x + 7 \leq 2 \text{ or } 2x + 3 \geq 5$

**81.** *Homeschooled Children in the United States.* The equation $y = 0.08x + 0.83$ estimates the number of homeschooled children in the United States, in millions, where $x$ is the number of years after 1999 (*Source*: Department of Education's National Center for Education Statistics). For what years will the number of homeschooled children exceed 2.0 million? [1.6]

**82.** *Temperature Conversion.* The formula $C = \frac{5}{9}(F - 32)$ can be used to convert Fahrenheit temperatures $F$ to Celsius temperatures $C$. For what Fahrenheit temperatures is the Celsius temperature lower than 45°C? [1.6]

**83.** The domain of the function
$$f(x) = \frac{x + 3}{8 - 4x}$$
is which of the following? [1.2]
  **A.** $(-3, 2)$
  **B.** $(-\infty, 2) \cup (2, \infty)$
  **C.** $(-\infty, -3) \cup (-3, 2) \cup (2, \infty)$
  **D.** $(-\infty, -3) \cup (-3, \infty)$

**84.** The center of the circle described by the equation $(x - 1)^2 + y^2 = 9$ is which of the following? [1.1]
  **A.** $(-1, 0)$
  **B.** $(1, 0)$
  **C.** $(0, -3)$
  **D.** $(-1, 3)$

**85.** The graph of $f(x) = -\frac{1}{2}x - 2$ is which of the following? [1.3]

**A.**

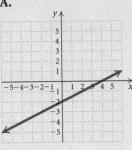

**B.**

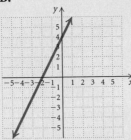

**C.**

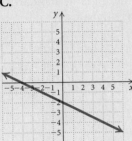

**D.**

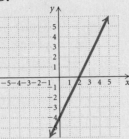

## Synthesis

**86.** Find the point on the $x$-axis that is equidistant from the points $(1, 3)$ and $(4, -3)$. [1.1]

*Find the domain.* [1.2]

**87.** $f(x) = \dfrac{\sqrt{1 - x}}{x - |x|}$

**88.** $f(x) = (x - 9x^{-1})^{-1}$

## Collaborative Discussion and Writing

**89.** Discuss why the graph of $f(x) = -\frac{3}{5}x + 4$ is steeper than the graph of $g(x) = \frac{1}{2}x - 6$. [1.3]

**90.** As the first step in solving

$$3x - 1 = 8,$$

Stella multiplies by $\frac{1}{3}$ on both sides. What advice would you give her about the procedure for solving equations? [1.5]

**91.** Is it possible for a disjunction to have no solution? Why or why not? [1.6]

**92.** Explain in your own words why a linear function $f(x) = mx + b$, with $m \neq 0$, has exactly one zero. [1.5]

**93.** Why can the conjunction $3 < x$ *and* $x < 4$ be written as $3 < x < 4$, but the disjunction $x < 3$ *or* $x > 4$ cannot be written $3 > x > 4$? [1.6]

**94.** Explain in your own words what a function is. [1.2]

## Chapter 1 Test

**1.** Determine whether the ordered pair $\left(\frac{1}{2}, \frac{9}{10}\right)$ is a solution of the equation $5y - 4 = x$.

**2.** Find the intercepts of $5x - 2y = -10$ and graph the line.

**3.** Find the distance between $(5, 8)$ and $(-1, 5)$.

**4.** Find the midpoint of the segment with endpoints $(-2, 6)$ and $(-4, 3)$.

**5.** Find the center and the radius of the circle
$$(x + 4)^2 + (y - 5)^2 = 36.$$

**6.** Find an equation of the circle with center $(-1, 2)$ and radius $\sqrt{5}$.

**7. a)** Determine whether the relation
$$\{(-4, 7), (3, 0), (1, 5), (0, 7)\}$$
is a function. Answer yes or no.
**b)** Find the domain of the relation.
**c)** Find the range of the relation.

**8.** Given that $f(x) = 2x^2 - x + 5$, find each of the following.
**a)** $f(-1)$
**b)** $f(a + 2)$

**9.** Given that $f(x) = \dfrac{1 - x}{x}$, find each of the following.

    **a)** $f(0)$             **b)** $f(1)$

**10.** Using the graph below, find $f(-3)$.

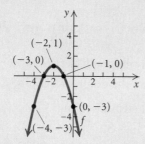

**11.** Determine whether each graph is that of a function. Answer yes or no.

  **a)**

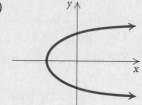

  **b)**

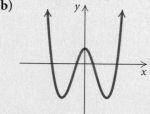

*Find the domain of the function.*

**12.** $f(x) = \dfrac{1}{x - 4}$

**13.** $g(x) = x^3 + 2$

**14.** $h(x) = \sqrt{25 - x^2}$

**15. a)** Graph: $f(x) = |x - 2| + 3$.
    **b)** Visually estimate the domain of $f(x)$.
    **c)** Visually estimate the range of $f(x)$.

*Find the slope of the line containing the given points.*

**16.** $\left(-2, \frac{2}{3}\right)$, $(-2, 5)$

**17.** $(4, -10)$, $(-8, 12)$

**18.** $(-5, 6)$, $\left(\frac{3}{4}, 6\right)$

**19.** *Declining Number Who Smoke.* Daily use of cigarettes by U.S. 12th graders is declining. In 1995, 21.6% of 12th graders smoked daily. This number decreased to 11.4% in 2008. (*Source: Monitoring the Future,* University of Michigan Institute for Social Research and National Institute on Drug Abuse) Find the average rate of change in the percent of 12th graders who smoke from 1995 to 2008.

**20.** Find the slope and the *y*-intercept of the line with equation $-3x + 2y = 5$.

**21.** *Total Cost.* Clear Signal charges $80 for a cell phone and $49.95 per month under its standard plan. Write an equation that can be used to determine the total cost $C(t)$ of operating a Clear Signal cell phone for $t$ months. Then find the total cost for 2 years.

**22.** Write an equation for the line with $m = -\frac{5}{8}$ and *y*-intercept $(0, -5)$.

**23.** Write an equation for the line that passes through $(-5, 4)$ and $(3, -2)$.

**24.** Write the equation of the vertical line that passes through $\left(-\frac{3}{8}, 11\right)$.

**25.** Determine whether the lines are parallel, perpendicular, or neither.

$$2x + 3y = -12,$$
$$2y - 3x = 8$$

**26.** Find an equation of the line containing the point $(-1, 3)$ and parallel to the line $x + 2y = -6$.

**27.** Find an equation of the line containing the point $(-1, 3)$ and perpendicular to the line $x + 2y = -6$.

**28.** *Miles per Car.* The data in the table below show a decrease in the average number of miles per passenger car from 2005 to 2008.

| Year, $x$ | Average Number of Miles per Passenger Car |
|---|---|
| 2005, 0 | 12,510 |
| 2006, 1 | 12,485 |
| 2007, 2 | 12,304 |
| 2008, 3 | 11,788 |

*Source:* Energy Information Administration
Monthly Energy Review, March 2010

a) Without using the regression feature on a graphing calculator, model the data with a linear function and using this function, estimate the average number of miles per passenger car in 2010 and in 2013. Answers may vary depending on the data points used.

b) Using a graphing calculator, fit a regression line to the data and use it to estimate, to the nearest mile, the average number of miles per passenger car in 2010 and in 2013. What is the correlation coefficient for the regression line?

*Solve.*

**29.** $6x + 7 = 1$

**30.** $2.5 - x = -x + 2.5$

**31.** $\frac{3}{2}y - 4 = \frac{5}{3}y + 6$

**32.** $2(4x + 1) = 8 - 3(x - 5)$

**33.** *Parking-Lot Dimensions.* The parking lot behind Kai's Kafé has a perimeter of 210 m. The width is three-fourths of the length. What are the dimensions of the parking lot?

**34.** *Pricing.* Jessie's Juice Bar prices its bottled juices by raising the wholesale price 50% and then adding 25¢. What is the wholesale price of a bottle of juice that sells for $2.95?

**35.** Find the zero(s) of the function
$$f(x) = 3x + 9.$$

*Solve and write interval notation for the solution set. Then graph the solution set.*

**36.** $5 - x \geq 4x + 20$

**37.** $-7 < 2x + 3 < 9$

**38.** $2x - 1 \leq 3 \text{ or } 5x + 6 \geq 26$

**39.** *Moving Costs.* Morgan Movers charges $90 plus $25 per hour to move households across town. McKinley Movers charges $40 per hour for cross-town moves. For what lengths of time does it cost less to hire Morgan Movers?

**40.** The graph of $g(x) = 1 - \frac{1}{2}x$ is which of the following ?

**A.**

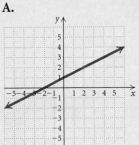

**B.**

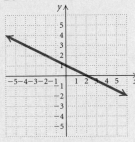

**C.**

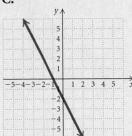

**D.**

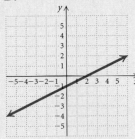

## Synthesis

**41.** Suppose that for some function $h$, $h(x + 2) = \frac{1}{2}x$. Find $h(-2)$.

# More on Functions

## APPLICATION

A wholesale garden center estimates that it will sell $N$ units of a basket of three potted amaryllises after spending $a$ dollars on advertising, where

$$N(a) = -a^2 + 300a + 6, \quad 0 \le a \le 300,$$

and $a$ is measured in thousands of dollars. Graph the function using a graphing calculator and then use the MAXIMUM feature to find the relative maximum. For what advertising expenditure will the greatest number of baskets be sold? How many baskets will be sold for that amount?

This problem appears as Exercise 27 in Section 2.1.

## 2.1 Increasing, Decreasing, and Piecewise Functions; Applications

- Graph functions, looking for intervals on which the function is increasing, decreasing, or constant, and estimate relative maxima and minima.

- Given an application, find a function that models the application. Find the domain of the function and function values, and then graph the function.

- Graph functions defined piecewise.

Because functions occur in so many real-world situations, it is important to be able to analyze them carefully.

### Increasing, Decreasing, and Constant Functions

On a given interval, if the graph of a function rises from left to right, it is said to be **increasing** on that interval. If the graph drops from left to right, it is said to be **decreasing**. If the function values stay the same from left to right, the function is said to be **constant**.

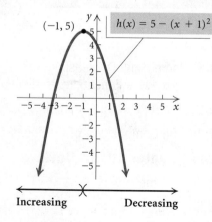

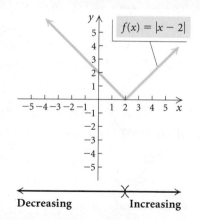

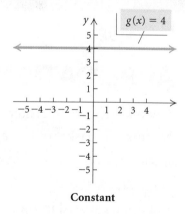

We are led to the following definitions.

> ### INCREASING, DECREASING, AND CONSTANT FUNCTIONS
>
> A function $f$ is said to be **increasing** on an *open* interval $I$, if for all $a$ and $b$ in that interval, $a < b$ implies $f(a) < f(b)$. (See Fig. 1 on the following page.)
>
> A function $f$ is said to be **decreasing** on an *open* interval $I$, if for all $a$ and $b$ in that interval, $a < b$ implies $f(a) > f(b)$. (See Fig. 2.)
>
> A function $f$ is said to be **constant** on an *open* interval $I$, if for all $a$ and $b$ in that interval, $f(a) = f(b)$. (See Fig. 3.)

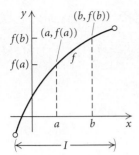

For $a < b$ in $I$, $f(a) < f(b)$;
$f$ is **increasing** on $I$.

**FIGURE 1**

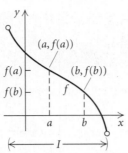

For $a < b$ in $I$, $f(a) > f(b)$;
$f$ is **decreasing** on $I$.

**FIGURE 2**

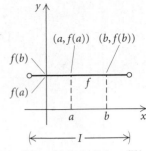

For all $a$ and $b$ in $I$, $f(a) = f(b)$;
$f$ is **constant** on $I$.

**FIGURE 3**

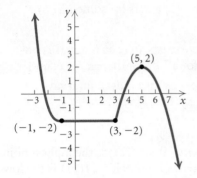

**EXAMPLE 1** Determine the intervals on which the function in the figure at left is (a) increasing; (b) decreasing; (c) constant.

*Solution* When expressing interval(s) on which a function is increasing, decreasing, or constant, we consider only values in the *domain* of the function. Since the domain of this function is $(-\infty, \infty)$, we consider all real values of $x$.

a) As $x$-values (that is, values in the domain) increase from $x = 3$ to $x = 5$, the $y$-values (that is, values in the range) increase from $-2$ to $2$. Thus the function is increasing on the interval $(3, 5)$.

b) As $x$-values increase from negative infinity to $-1$, $y$-values decrease; $y$-values also decrease as $x$-values increase from $5$ to positive infinity. Thus the function is decreasing on the intervals $(-\infty, -1)$ and $(5, \infty)$.

c) As $x$-values increase from $-1$ to $3$, $y$ remains $-2$. The function is constant on the interval $(-1, 3)$.

**Now Try Exercise 5.**

### Study Tips

Success can be planned. Combine goals and good study habits to create a plan for success that works for you. A list of study tips that your authors consider most helpful are included in the Guide to Success in the front of the text before Chapter R.

In calculus, the slope of a line tangent to the graph of a function at a particular point is used to determine whether the function is increasing, decreasing, or constant at that point. If the slope is positive, the function is increasing; if the slope is negative, the function is decreasing; if the slope is 0 over an interval, the function is constant. Since slope cannot be both positive and negative at the same point, a function cannot be both increasing and decreasing at a specific point. For this reason, increasing, decreasing, and constant intervals are expressed in *open interval* notation. In Example 1, if $[3, 5]$ had been used for the increasing interval and $[5, \infty)$ for a decreasing interval, the function would be both increasing and decreasing at $x = 5$. This is not possible.

### ■ Relative Maximum and Minimum Values

Consider the graph shown at the top of the following page. Note the "peaks" and "valleys" at the $x$-values $c_1$, $c_2$, and $c_3$. The function value $f(c_2)$ is called a **relative maximum** (plural, **maxima**). Each of the function values $f(c_1)$ and $f(c_3)$ is called a **relative minimum** (plural, **minima**).

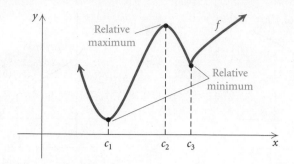

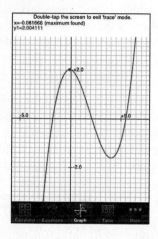

**RELATIVE MAXIMA AND MINIMA**

Suppose that $f$ is a function for which $f(c)$ exists for some $c$ in the domain of $f$. Then:

> $f(c)$ is a **relative maximum** if there exists an *open* interval $I$ containing $c$ such that $f(c) > f(x)$, for all $x$ in $I$ where $x \neq c$; and
>
> $f(c)$ is a **relative minimum** if there exists an *open* interval $I$ containing $c$ such that $f(c) < f(x)$, for all $x$ in $I$ where $x \neq c$.

Simply stated, $f(c)$ is a *relative maximum* if $(c, f(c))$ is the highest point in some *open* interval, and $f(c)$ is a *relative minimum* if $(c, f(c))$ is the lowest point in some *open* interval.

If you take a calculus course, you will learn a method for determining exact values of relative maxima and minima. In Section 3.3, we will find exact maximum and minimum values of quadratic functions algebraically. The MAXIMUM and MINIMUM features on a graphing calculator can be used to approximate relative maxima and minima.

**GCM**

**EXAMPLE 2**   Use a graphing calculator to determine any relative maxima and minima of the function $f(x) = 0.1x^3 - 0.6x^2 - 0.1x + 2$ and to determine intervals on which the function is increasing or decreasing.

*Solution*   We first graph the function, experimenting with the window dimensions as needed. The curvature is seen fairly well with window settings of $[-4, 6, -3, 3]$. Using the MAXIMUM and MINIMUM features, we determine the relative maximum value and the relative minimum value of the function.

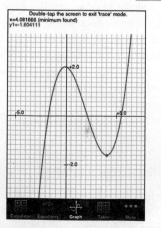

Apps for the iPhone and the iPod Touch also can be used to find the relative maximum values and the relative minimum values of the function in Example 2.

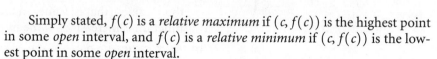

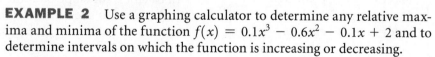

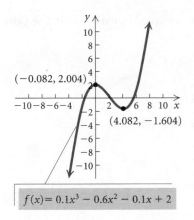

$$f(x) = 0.1x^3 - 0.6x^2 - 0.1x + 2$$

We see that the relative maximum value of the function is about 2.004. It occurs when $x \approx -0.082$. We also approximate the relative minimum, $-1.604$ at $x \approx 4.082$.

We note that the graph rises, or increases, from the left and stops increasing at the relative maximum. From this point, the graph decreases to the relative minimum and then begins to rise again. Thus the function is *increasing* on the intervals

$$(-\infty, -0.082) \quad \text{and} \quad (4.082, \infty)$$

and *decreasing* on the interval

$$(-0.082, 4.082).$$

**Now Try Exercise 23.**

## ▪ Applications of Functions

Many real-world situations can be modeled by functions.

**EXAMPLE 3** *Car Distance.* Elena and Thomas drive away from a restaurant at right angles to each other. Elena's speed is 35 mph and Thomas' is 40 mph.

**a)** Express the distance between the cars as a function of time, $d(t)$.

**b)** Find the domain of the function.

*Solution*

**a)** Suppose 1 hr goes by. At that time, Elena has traveled 35 mi and Thomas has traveled 40 mi. We can use the Pythagorean theorem to find the distance between them. This distance would be the length of the hypotenuse of a triangle with legs measuring 35 mi and 40 mi. After 2 hr, the triangle's legs would measure $2 \cdot 35$, or 70 mi, and $2 \cdot 40$, or 80 mi. Noting that the distances will always be changing, we make a drawing and let $t = $ the time, in hours, that Elena and Thomas have been driving since leaving the restaurant.

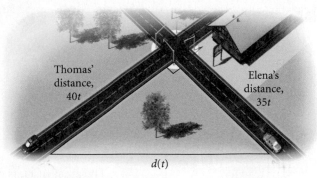

After $t$ hours, Elena has traveled $35t$ miles and Thomas $40t$ miles. We now use the Pythagorean theorem:

$$[d(t)]^2 = (35t)^2 + (40t)^2.$$

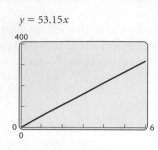

$y = 53.15x$

Because distance must be nonnegative, we need consider only the positive square root when solving for $d(t)$:

$$d(t) = \sqrt{(35t)^2 + (40t)^2}$$
$$= \sqrt{1225t^2 + 1600t^2}$$
$$= \sqrt{2825t^2}$$
$$\approx 53.15|t| \quad \text{Approximating the root to two decimal places}$$
$$\approx 53.15t. \quad \text{Since } t \geq 0, |t| = t.$$

Thus, $d(t) = 53.15t,\ t \geq 0$.

**b)** Since the time traveled, $t$, must be nonnegative, the domain is the set of nonnegative real numbers $[0, \infty)$.     **◀ Now Try Exercise 35.**

**EXAMPLE 4    *Storage Area.*** Jenna's restaurant supply store has 20 ft of dividers with which to set off a rectangular area for the storage of overstock. If a corner of the store is used for the storage area, the partition need only form two sides of a rectangle.

**a)** Express the floor area of the storage space as a function of the length of the partition.

**b)** Find the domain of the function.

**c)** Graph the function.

**d)** Find the dimensions that maximize the floor area.

***Solution***

**a)** Note that the dividers will form two sides of a rectangle. If, for example, 14 ft of dividers are used for the length of the rectangle, that would leave $20 - 14$, or 6 ft of dividers for the width. Thus if $x =$ the length, in feet, of the rectangle, then $20 - x =$ the width. We represent this information in a drawing, as shown below.

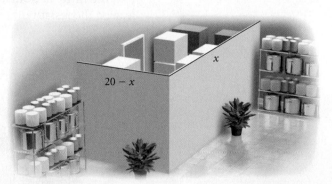

The area, $A(x)$, is given by

$$A(x) = x(20 - x) \quad \text{Area} = \text{length} \cdot \text{width.}$$
$$= 20x - x^2.$$

The function $A(x) = 20x - x^2$ can be used to express the rectangle's area as a function of the length of the partition.

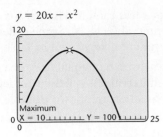

$y = 20x - x^2$

**b)** Because the rectangle's length and width must be positive and only 20 ft of dividers are available, we restrict the domain of $A$ to $\{x \mid 0 < x < 20\}$, that is, the interval $(0, 20)$.

**c)** The graph is shown at left.

**d)** We use the MAXIMUM feature as shown on the graph at left. The maximum value of the area function on the interval $(0, 20)$ is 100 when $x = 10$. Thus the dimensions that maximize the area are

$$\text{Length} = x = 10 \text{ ft} \quad \text{and}$$
$$\text{Width} = 20 - x = 20 - 10 = 10 \text{ ft}. \quad \boxed{\text{Now Try Exercise 41.}}$$

## ■ Functions Defined Piecewise

Sometimes functions are defined **piecewise** using different output formulas for different pieces, or parts, of the domain.

**EXAMPLE 5** For the function defined as

$$f(x) = \begin{cases} x + 1, & \text{for } x < -2, \\ 5, & \text{for } -2 \le x \le 3, \\ x^2, & \text{for } x > 3, \end{cases}$$

find $f(-5)$, $f(-3)$, $f(0)$, $f(3)$, $f(4)$, and $f(10)$.

**Solution** First, we determine which part of the domain contains the given input. Then we use the corresponding formula to find the output.

Since $-5 < -2$, we use the formula $f(x) = x + 1$:

$$f(-5) = -5 + 1 = -4.$$

Since $-3 < -2$, we use the formula $f(x) = x + 1$ again:

$$f(-3) = -3 + 1 = -2.$$

Since $-2 \le 0 \le 3$, we use the formula $f(x) = 5$:

$$f(0) = 5.$$

Since $-2 \le 3 \le 3$, we use the formula $f(x) = 5$ a second time:

$$f(3) = 5.$$

Since $4 > 3$, we use the formula $f(x) = x^2$:

$$f(4) = 4^2 = 16.$$

Since $10 > 3$, we once again use the formula $f(x) = x^2$:

$$f(10) = 10^2 = 100. \quad \boxed{\text{Now Try Exercise 47.}}$$

**EXAMPLE 6** Graph the function defined as

$$g(x) = \begin{cases} \frac{1}{3}x + 3, & \text{for } x < 3, \\ -x, & \text{for } x \ge 3. \end{cases}$$

**Solution** Since the function is defined in two pieces, or parts, we create the graph in two parts.

### Interactive Figures

Icons throughout the text direct you to MyMathLab whenever an Interactive Figure is available. Math concepts come to life, letting you adjust variables and other settings to see how the visual representation is affected by the math.

**TABLE 1**

| $x$ $(x < 3)$ | $g(x) = \frac{1}{3}x + 3$ |
|---|---|
| $-3$ | $2$ |
| $0$ | $3$ |
| $2$ | $3\frac{2}{3}$ |

**a)** We graph $g(x) = \frac{1}{3}x + 3$ *only* for inputs $x$ less than 3. That is, we use $g(x) = \frac{1}{3}x + 3$ only for $x$-values in the interval $(-\infty, 3)$. Some ordered pairs that are solutions of this piece of the function are shown in Table 1.

**b)** We graph $g(x) = -x$ *only* for inputs $x$ greater than or equal to 3. That is, we use $g(x) = -x$ only for $x$-values in the interval $[3, \infty)$. Some ordered pairs that are solutions of this piece of the function are shown in Table 2.

**TABLE 2**

| $x$ $(x \geq 3)$ | $g(x) = -x$ |
|---|---|
| $3$ | $-3$ |
| $4$ | $-4$ |
| $6$ | $-6$ |

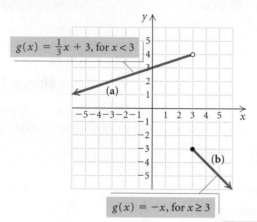

$g(x) = \frac{1}{3}x + 3$, for $x < 3$

(a)

(b)

$g(x) = -x$, for $x \geq 3$

Now Try Exercise 51.

**TABLE 3**

| $x$ $(x \leq 0)$ | $f(x) = 4$ |
|---|---|
| $-5$ | $4$ |
| $-2$ | $4$ |
| $0$ | $4$ |

GCM

**EXAMPLE 7** Graph the function defined as

$$f(x) = \begin{cases} 4, & \text{for } x \leq 0, \\ 4 - x^2, & \text{for } 0 < x \leq 2, \\ 2x - 6, & \text{for } x > 2. \end{cases}$$

*Solution* We create the graph in three pieces, or parts.

**a)** We graph $f(x) = 4$ *only* for inputs $x$ less than or equal to 0. That is, we use $f(x) = 4$ only for $x$-values in the interval $(-\infty, 0]$. Some ordered pairs that are solutions of this piece of the function are shown in Table 3.

**b)** We graph $f(x) = 4 - x^2$ *only* for inputs $x$ greater than 0 and less than or equal to 2. That is, we use $f(x) = 4 - x^2$ only for $x$-values in the interval $(0, 2]$. Some ordered pairs that are solutions of this piece of the function are shown in Table 4.

**TABLE 4**

| $x$ $(0 < x \leq 2)$ | $f(x) = 4 - x^2$ |
|---|---|
| $\frac{1}{2}$ | $3\frac{3}{4}$ |
| $1$ | $3$ |
| $2$ | $0$ |

**c)** We graph $f(x) = 2x - 6$ *only* for inputs $x$ greater than 2. That is, we use $f(x) = 2x - 6$ only for $x$-values in the interval $(2, \infty)$. Some ordered pairs that are solutions of this piece of the function are shown in Table 5.

**TABLE 5**

| $x$ $(x > 2)$ | $f(x) = 2x - 6$ |
|---|---|
| $2\frac{1}{2}$ | $-1$ |
| $3$ | $0$ |
| $5$ | $4$ |

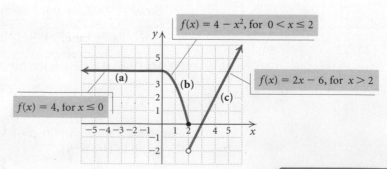

$f(x) = 4 - x^2$, for $0 < x \leq 2$

(a)

(b)

(c)

$f(x) = 2x - 6$, for $x > 2$

$f(x) = 4$, for $x \leq 0$

Now Try Exercise 55.

**EXAMPLE 8**   Graph the function defined as

$$f(x) = \begin{cases} \dfrac{x^2 - 4}{x + 2}, & \text{for } x \neq -2, \\ 3, & \text{for } x = -2. \end{cases}$$

**Solution**   When $x \neq -2$, the denominator of $(x^2 - 4)/(x + 2)$ is non-zero, so we can simplify:

$$\frac{x^2 - 4}{x + 2} = \frac{(x + 2)(x - 2)}{x + 2} = x - 2.$$

Thus,

$$f(x) = x - 2, \quad \text{for } x \neq -2.$$

The graph of this part of the function consists of a line with a "hole" at the point $(-2, -4)$, indicated by the open circle. The hole occurs because the piece of the function represented by $(x^2 - 4)/(x + 2)$ is not defined for $x = -2$. By the definition of the function, we see that $f(-2) = 3$, so we plot the point $(-2, 3)$ above the open circle.

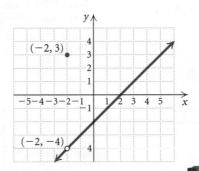

Now Try Exercise 59.

The hand-drawn graph of the piece of the function in Example 8 represented by $y = (x^2 - 4)/(x + 2)$, $x \neq 2$, can be checked using a graphing calculator. When $y = (x^2 - 4)/(x + 2)$ is graphed, the hole may or may not be visible, depending on the window dimensions chosen. When we use the ZDECIMAL feature from the ZOOM menu, note that the hole will appear, as shown in the graph at left. (See the *Graphing Calculator Manual* that accompanies this text for further details on selecting window dimensions.) If we examine a table of values for this function, we see that an ERROR message corresponds to the $x$-value $-2$. This indicates that $-2$ is *not* in the domain of the function $y = (x^2 - 4)/(x + 2)$. However, $-2$ *is* in the domain of the function $f$ in Example 8 because $f(-2)$ is defined to be 3.

A piecewise function with importance in calculus and computer programming is the **greatest integer function**, denoted $f(x) = [\![x]\!]$, or $\text{int}(x)$.

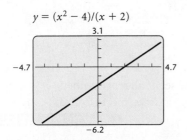

## GREATEST INTEGER FUNCTION

$f(x) = [\![x]\!] = $ the greatest integer *less than or equal to x.*

The greatest integer function pairs each input with the greatest integer *less than or equal to* that input. Thus, *x*-values 1, $1\frac{1}{2}$, and 1.8 are all paired with the *y*-value 1. Other pairings are shown below.

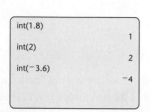

$$
\begin{array}{llll}
-4 & -1 & 0 & 2 \\
-3.6 \longrightarrow -4 & -\frac{7}{8} \longrightarrow -1 & \frac{1}{10} \longrightarrow 0 & 2.1 \longrightarrow 2 \\
-3\frac{1}{4} & -0.25 & 0.99 & 2\frac{3}{4}
\end{array}
$$

These values can be checked with a graphing calculator using the int( feature from the NUM submenu in the MATH menu.

**GCM** **EXAMPLE 9** Graph $f(x) = [\![x]\!]$ and determine its domain and range.

*Solution* The greatest integer function can also be defined as a piecewise function with an infinite number of statements. When plotting points by hand, it can be helpful to use the TABLE feature on a graphing calculator to find ordered pairs.

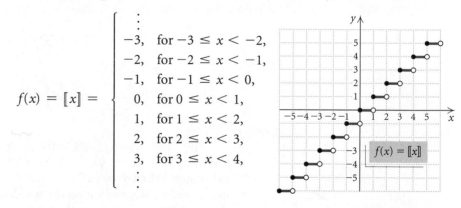

$$
f(x) = [\![x]\!] = \begin{cases}
\;\;\vdots \\
-3, & \text{for } -3 \le x < -2, \\
-2, & \text{for } -2 \le x < -1, \\
-1, & \text{for } -1 \le x < 0, \\
\;\;\;0, & \text{for } 0 \le x < 1, \\
\;\;\;1, & \text{for } 1 \le x < 2, \\
\;\;\;2, & \text{for } 2 \le x < 3, \\
\;\;\;3, & \text{for } 3 \le x < 4, \\
\;\;\vdots
\end{cases}
$$

We see that the domain of this function is the set of all real numbers, $(-\infty, \infty)$, and the range is the set of all integers, $\{\ldots, -3, -2, -1, 0, 1, 2, 3, \ldots\}$.

**Now Try Exercise 63.**

If we had used a calculator for Example 9, we would see the graph shown below.

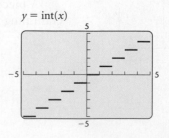

$y = \text{int}(x)$

*Determine the intervals on which the function is*
**(a)** *increasing,* **(b)** *decreasing, and* **(c)** *constant.*

**1.**

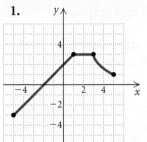

**2.**

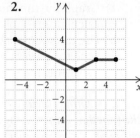

**3.**

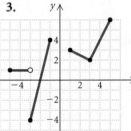

**4.**

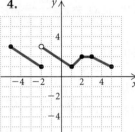

**5.**

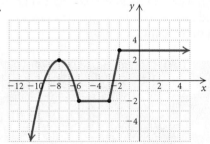

**6.**

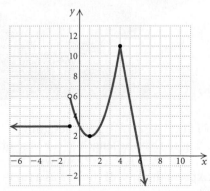

**7.–12.** Determine the domain and the range of each
of the functions graphed in Exercises 1–6.

*Using the graph, determine any relative maxima or
minima of the function and the intervals on which the
function is increasing or decreasing.*

**13.** $f(x) = -x^2 + 5x - 3$

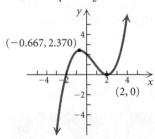

**14.** $f(x) = x^2 - 2x + 3$

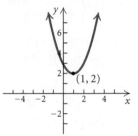

**15.** $f(x) = \frac{1}{4}x^3 - \frac{1}{2}x^2 - x + 2$

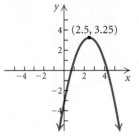

**16.** $f(x) = -0.09x^3 + 0.5x^2 - 0.1x + 1$

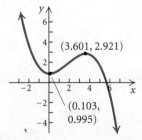

*Graph the function. Estimate the intervals on which the
function is increasing or decreasing and any relative
maxima or minima.*

**17.** $f(x) = x^2$

**18.** $f(x) = 4 - x^2$

**19.** $f(x) = 5 - |x|$

**20.** $f(x) = |x + 3| - 5$

**21.** $f(x) = x^2 - 6x + 10$

**22.** $f(x) = -x^2 - 8x - 9$

*Graph the function using the given viewing window. Find the intervals on which the function is increasing or decreasing and find any relative maxima or minima. Change the viewing window if it seems appropriate for further analysis.*

**23.** $f(x) = -x^3 + 6x^2 - 9x - 4$,
$[-3, 7, -20, 15]$

**24.** $f(x) = 0.2x^3 - 0.2x^2 - 5x - 4$,
$[-10, 10, -30, 20]$

**25.** $f(x) = 1.1x^4 - 5.3x^2 + 4.07$,
$[-4, 4, -4, 8]$

**26.** $f(x) = 1.2(x + 3)^4 + 10.3(x + 3)^2 + 9.78$,
$[-9, 3, -40, 100]$

**27.** *Advertising Effect.*    A wholesale garden center estimates that it will sell $N$ units of a basket of three potted amaryllises after spending $a$ dollars on advertising, where

$$N(a) = -a^2 + 300a + 6, \quad 0 \le a \le 300,$$

and $a$ is measured in thousands of dollars.

**a)** Graph the function using a graphing calculator.
**b)** Use the MAXIMUM feature to find the relative maximum.
**c)** For what advertising expenditure will the greatest number of baskets be sold? How many baskets will be sold for that amount?

**28.** *Temperature During an Illness.*    The temperature of a patient during an illness is given by the function

$$T(t) = -0.1t^2 + 1.2t + 98.6, \quad 0 \le t \le 12,$$

where $T$ is the temperature, in degrees Fahrenheit, at time $t$, in days, after the onset of the illness.

**a)** Graph the function using a graphing calculator.
**b)** Use the MAXIMUM feature to determine at what time the patient's temperature was the highest. What was the highest temperature?

*Use a graphing calculator to find the intervals on which the function is increasing or decreasing. Consider the entire set of real numbers if no domain is given.*

**29.** $f(x) = \dfrac{8x}{x^2 + 1}$

**30.** $f(x) = \dfrac{-4}{x^2 + 1}$

**31.** $f(x) = x\sqrt{4 - x^2}$, for $-2 \le x \le 2$

**32.** $f(x) = -0.8x\sqrt{9 - x^2}$, for $-3 \le x \le 3$

**33.** *Lumberyard.*    Rick's lumberyard has 480 yd of fencing with which to enclose a rectangular area. If the enclosed area is $x$ yards long, express its area as a function of its length.

**34.** *Triangular Flag.*    A seamstress is designing a triangular flag so that the length of the base of the triangle, in inches, is 7 less than twice the height $h$. Express the area of the flag as a function of the height.

**35.** *Rising Balloon.* A hot-air balloon rises straight up from the ground at a rate of 120 ft/min. The balloon is tracked from a rangefinder on the ground at point $P$, which is 400 ft from the release point $Q$ of the balloon. Let $d =$ the distance from the balloon to the rangefinder and $t =$ the time, in minutes, since the balloon was released. Express $d$ as a function of $t$.

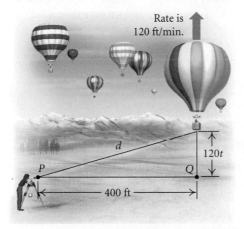

**36.** *Airplane Distance.* An airplane is flying at an altitude of 3700 ft. The slanted distance directly to the airport is $d$ feet. Express the horizontal distance $h$ as a function of $d$.

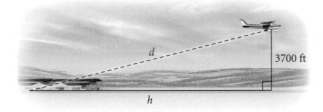

**37.** *Inscribed Rhombus.* A rhombus is inscribed in a rectangle that is $w$ meters wide with a perimeter of 40 m. Each vertex of the rhombus is a midpoint of a side of the rectangle. Express the area of the rhombus as a function of the rectangle's width.

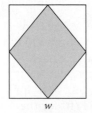

**38.** *Carpet Area.* A carpet installer uses 46 ft of linen tape to bind the edges of a

rectangular hall runner. If the runner is $w$ feet wide, express its area as a function of the width.

**39.** *Golf Distance Finder.* A device used in golf to estimate the distance $d$, in yards, to a hole measures the size $s$, in inches, that the 7-ft pin appears to be in a viewfinder. Express the distance $d$ as a function of $s$.

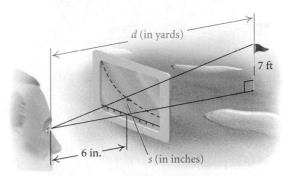

**40.** *Gas Tank Volume.* A gas tank has ends that are hemispheres of radius $r$ feet. The cylindrical midsection is 6 ft long. Express the volume of the tank as a function of $r$.

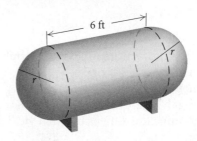

**41.** *Play Space.* A car dealership has 30 ft of dividers with which to enclose a rectangular play space in a corner of a customer lounge. The sides against the wall require no partition. Suppose the play space is $x$ feet long.

**a)** Express the area of the play space as a function of *x.*
**b)** Find the domain of the function.
**c)** Using the graph shown below, determine the dimensions that yield the maximum area.

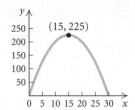

**42.** *Corral Design.* A rancher has 360 yd of fencing with which to enclose two adjacent rectangular corrals, one for horses and one for cattle. A river forms one side of the corrals. Suppose the width of each corral is *x* yards.

**a)** Express the total area of the two corrals as a function of *x.*
**b)** Find the domain of the function.
**c)** Using the graph of the function shown below, determine the dimensions that yield the maximum area.

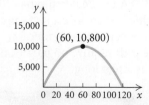

**43.** *Office File.* Designs Unlimited plans to produce a one-component vertical file by bending the long side of an 8-in. by 14-in. sheet of plastic along two lines to form a ⊔ shape.

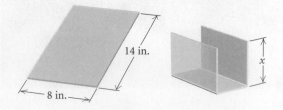

**a)** Express the volume of the file as a function of the height *x*, in inches, of the file.
**b)** Find the domain of the function.
**c)** Graph the function with a graphing calculator.
**d)** How tall should the file be in order to maximize the volume that the file can hold?

**44.** *Volume of a Box.* From a 12-cm by 12-cm piece of cardboard, square corners are cut out so that the sides can be folded up to make a box.

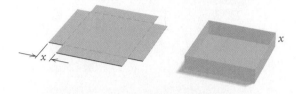

**a)** Express the volume of the box as a function of the side *x*, in centimeters, of a cut-out square.
**b)** Find the domain of the function.
**c)** Graph the function with a graphing calculator.
**d)** What dimensions yield the maximum volume?

**45.** *Area of an Inscribed Rectangle.* A rectangle that is *x* feet wide is inscribed in a circle of radius 8 ft.

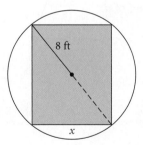

**a)** Express the area of the rectangle as a function of *x.*
**b)** Find the domain of the function.
**c)** Graph the function with a graphing calculator.
**d)** What dimensions maximize the area of the rectangle?

**46.** *Cost of Material.* A rectangular box with volume 320 ft³ is built with a square base and top. The cost is $1.50/ft² for the bottom, $2.50/ft² for the sides, and $1/ft² for the top. Let *x* = the length of the base, in feet.

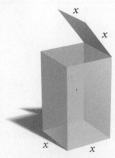

**a)** Express the cost of the box as a function of $x$.
**b)** Find the domain of the function.
**c)** Graph the function with a graphing calculator.
**d)** What dimensions minimize the cost of the box?

*For each piecewise function, find the specified function values.*

**47.** $g(x) = \begin{cases} x + 4, & \text{for } x \leq 1, \\ 8 - x, & \text{for } x > 1 \end{cases}$
$g(-4), g(0), g(1)$, and $g(3)$

**48.** $f(x) = \begin{cases} 3, & \text{for } x \leq -2, \\ \frac{1}{2}x + 6, & \text{for } x > -2 \end{cases}$
$f(-5), f(-2), f(0)$, and $f(2)$

**49.** $h(x) = \begin{cases} -3x - 18, & \text{for } x < -5, \\ 1, & \text{for } -5 \leq x < 1, \\ x + 2, & \text{for } x \geq 1 \end{cases}$
$h(-5), h(0), h(1)$, and $h(4)$

**50.** $f(x) = \begin{cases} -5x - 8, & \text{for } x < -2, \\ \frac{1}{2}x + 5, & \text{for } -2 \leq x \leq 4, \\ 10 - 2x, & \text{for } x > 4 \end{cases}$
$f(-4), f(-2), f(4)$, and $f(6)$

*Make a hand-drawn graph of each of the following. Check your results using a graphing calculator.*

**51.** $f(x) = \begin{cases} \frac{1}{2}x, & \text{for } x < 0, \\ x + 3, & \text{for } x \geq 0 \end{cases}$

**52.** $f(x) = \begin{cases} -\frac{1}{3}x + 2, & \text{for } x \leq 0, \\ x - 5, & \text{for } x > 0 \end{cases}$

**53.** $f(x) = \begin{cases} -\frac{3}{4}x + 2, & \text{for } x < 4, \\ -1, & \text{for } x \geq 4 \end{cases}$

**54.** $h(x) = \begin{cases} 2x - 1, & \text{for } x < 2, \\ 2 - x, & \text{for } x \geq 2 \end{cases}$

**55.** $f(x) = \begin{cases} x + 1, & \text{for } x \leq -3, \\ -1, & \text{for } -3 < x < 4, \\ \frac{1}{2}x, & \text{for } x \geq 4 \end{cases}$

**56.** $f(x) = \begin{cases} 4, & \text{for } x \leq -2, \\ x + 1, & \text{for } -2 < x < 3, \\ -x, & \text{for } x \geq 3 \end{cases}$

**57.** $g(x) = \begin{cases} \frac{1}{2}x - 1, & \text{for } x < 0, \\ 3, & \text{for } 0 \leq x \leq 1, \\ -2x, & \text{for } x > 1 \end{cases}$

**58.** $f(x) = \begin{cases} \dfrac{x^2 - 9}{x + 3}, & \text{for } x \neq -3, \\ 5, & \text{for } x = -3 \end{cases}$

**59.** $f(x) = \begin{cases} 2, & \text{for } x = 5, \\ \dfrac{x^2 - 25}{x - 5}, & \text{for } x \neq 5 \end{cases}$

**60.** $f(x) = \begin{cases} \dfrac{x^2 + 3x + 2}{x + 1}, & \text{for } x \neq -1, \\ 7, & \text{for } x = -1 \end{cases}$

**61.** $f(x) = [\![x]\!]$    **62.** $f(x) = 2[\![x]\!]$

**63.** $g(x) = 1 + [\![x]\!]$    **64.** $h(x) = \frac{1}{2}[\![x]\!] - 2$

**65.–70.** Find the domain and the range of each of the functions defined in Exercises 51–56.

*Determine the domain and the range of the piecewise function. Then write an equation for the function.*

**71.**
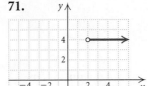

**72.**
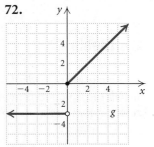

**73.**
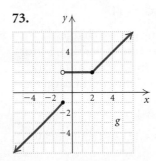

**74.**
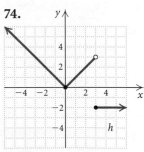

**75.**
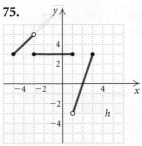

**76.**
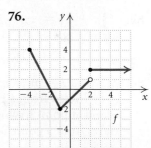

## Skill Maintenance

**77.** Given $f(x) = 5x^2 - 7$, find each of the following.

a) $f(-3)$
b) $f(3)$
c) $f(a)$
d) $f(-a)$

**78.** Given $f(x) = 4x^3 - 5x$, find each of the following.

a) $f(2)$
b) $f(-2)$
c) $f(a)$
d) $f(-a)$

**79.** Write an equation of the line perpendicular to the graph of the line $8x - y = 10$ and containing the point $(-1, 1)$.

**80.** Find the slope and the $y$-intercept of the line with equation $2x - 9y + 1 = 0$.

## Synthesis

*Using a graphing calculator, estimate the interval on which the function is increasing or decreasing and any relative maxima or minima.*

**81.** $f(x) = x^4 + 4x^3 - 36x^2 - 160x + 400$

**82.** $f(x) = 3.22x^5 - 5.208x^3 - 11$

**83.** *Parking Costs.* A parking garage charges $2 for up to (but not including) 1 hr of parking, $4 for up to 2 hr of parking, $6 for up to 3 hr of parking, and so on. Let $C(t) =$ the cost of parking for $t$ hours.

a) Graph the function.
b) Write an equation for $C(t)$ using the greatest integer notation $[\![t]\!]$.

**84.** If $[\![x + 2]\!] = -3$, what are the possible inputs for $x$?

**85.** If $([\![x]\!])^2 = 25$, what are the possible inputs for $x$?

**86.** *Minimizing Power Line Costs.* A power line is constructed from a power station at point A to an island at point I, which is 1 mi directly out in the water from a point B on the shore. Point B is 4 mi downshore from the power station at A. It costs $5000 per mile to lay the power line under water and $3000 per mile to lay the power line under ground. The line comes to the shore at point S downshore from A. Let $x =$ the distance from B to S.

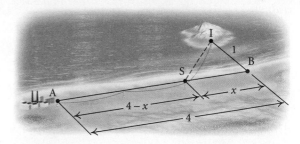

a) Express the cost $C$ of laying the line as a function of $x$.
b) At what distance $x$ from point B should the line come to shore in order to minimize cost?

**87.** *Volume of an Inscribed Cylinder.* A right circular cylinder of height $h$ and radius $r$ is inscribed in a right circular cone with a height of 10 ft and a base with radius 6 ft.

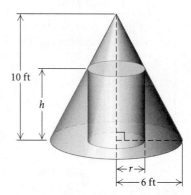

a) Express the height $h$ of the cylinder as a function of $r$.
b) Express the volume $V$ of the cylinder as a function of $r$.
c) Express the volume $V$ of the cylinder as a function of $h$.

## 2.2 The Algebra of Functions

- Find the sum, the difference, the product, and the quotient of two functions, and determine the domains of the resulting functions.
- Find the difference quotient for a function.

### The Algebra of Functions: Sums, Differences, Products, and Quotients

We now use addition, subtraction, multiplication, and division to combine functions and obtain new functions.

Consider the following two functions $f$ and $g$:

$$f(x) = x + 2 \quad \text{and} \quad g(x) = x^2 + 1.$$

Since $f(3) = 3 + 2 = 5$ and $g(3) = 3^2 + 1 = 10$, we have

$$f(3) + g(3) = 5 + 10 = 15,$$
$$f(3) - g(3) = 5 - 10 = -5,$$
$$f(3) \cdot g(3) = 5 \cdot 10 = 50,$$

and

$$\frac{f(3)}{g(3)} = \frac{5}{10} = \frac{1}{2}.$$

In fact, so long as $x$ is in the domain of *both* $f$ and $g$, we can easily compute $f(x) + g(x)$, $f(x) - g(x)$, $f(x) \cdot g(x)$, and, assuming $g(x) \neq 0$, $f(x)/g(x)$. We use the notation shown below.

---

**SUMS, DIFFERENCES, PRODUCTS, AND QUOTIENTS OF FUNCTIONS**

If $f$ and $g$ are functions and $x$ is in the domain of each function, then:

$$(f + g)(x) = f(x) + g(x),$$
$$(f - g)(x) = f(x) - g(x),$$
$$(fg)(x) = f(x) \cdot g(x),$$
$$(f/g)(x) = f(x)/g(x), \text{ provided } g(x) \neq 0.$$

---

**EXAMPLE 1**  Given that $f(x) = x + 1$ and $g(x) = \sqrt{x + 3}$, find each of the following.

**a)** $(f + g)(x)$       **b)** $(f + g)(6)$       **c)** $(f + g)(-4)$

*Solution*

**a)** $(f + g)(x) = f(x) + g(x)$
$$= x + 1 + \sqrt{x + 3} \qquad \text{This cannot be simplified.}$$

**b)** We can find $(f + g)(6)$ provided 6 is in the domain of *each* function. The domain of $f$ is all real numbers. The domain of $g$ is all real numbers $x$ for which $x + 3 \geq 0$, or $x \geq -3$. This is the interval $[-3, \infty)$. We see that 6 is in both domains, so we have

$$f(6) = 6 + 1 = 7, \qquad g(6) = \sqrt{6 + 3} = \sqrt{9} = 3,$$
$$(f + g)(6) = f(6) + g(6) = 7 + 3 = 10.$$

Another method is to use the formula found in part (a):

$$(f + g)(6) = 6 + 1 + \sqrt{6 + 3} = 7 + \sqrt{9} = 7 + 3 = 10.$$

We can check our work using a graphing calculator by entering

$$y_1 = x + 1, \qquad y_2 = \sqrt{x + 3}, \quad \text{and} \quad y_3 = y_1 + y_2$$

on the $y =$ screen. Then on the home screen, we find Y3(6).

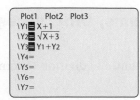

**c)** To find $(f + g)(-4)$, we must first determine whether $-4$ is in the domain of each function. We note that $-4$ is not in the domain of $g$, $[-3, \infty)$. That is, $\sqrt{-4 + 3}$ is not a real number. Thus, $(f + g)(-4)$ does not exist.

**Now Try Exercise 15.**

It is useful to view the concept of the sum of two functions graphically. In the graph below, we see the graphs of two functions $f$ and $g$ and their sum, $f + g$. Consider finding $(f + g)(4)$, or $f(4) + g(4)$. We can locate $g(4)$ on the graph of $g$ and measure it. Then we add that length on top of $f(4)$ on the graph of $f$. The sum gives us $(f + g)(4)$.

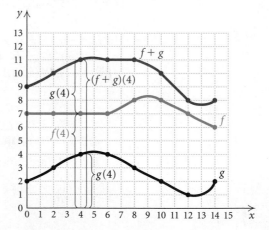

With this in mind, let's view Example 1 from a graphical perspective. Let's look at the graphs of

$$f(x) = x + 1, \qquad g(x) = \sqrt{x + 3}, \quad \text{and}$$
$$(f + g)(x) = x + 1 + \sqrt{x + 3}.$$

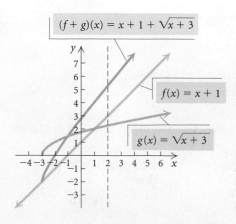

See the graph at left. Note that the domain of $f$ is the set of all real numbers. The domain of $g$ is $[-3, \infty)$. The domain of $f + g$ is the set of numbers in the intersection of the domains. This is the set of numbers in both domains.

Domain of $f$      $(-\infty, \infty)$

$-5\ -4\ -3\ -2\ -1\ \ 0\ \ 1\ \ 2\ \ 3\ \ 4\ \ 5$

Domain of $g$      $[-3, \infty)$

$-5\ -4\ -3\ -2\ -1\ \ 0\ \ 1\ \ 2\ \ 3\ \ 4\ \ 5$

Domain of $f + g$      $[-3, \infty)$

$-5\ -4\ -3\ -2\ -1\ \ 0\ \ 1\ \ 2\ \ 3\ \ 4\ \ 5$

Thus the domain of $f + g$ is $[-3, \infty)$.

We can confirm that the $y$-coordinates of the graph of $(f + g)(x)$ are the sums of the corresponding $y$-coordinates of the graphs of $f(x)$ and $g(x)$. Here we confirm it for $x = 2$.

$$f(x) = x + 1 \qquad\qquad g(x) = \sqrt{x + 3}$$
$$f(2) = 2 + 1 = 3; \qquad g(2) = \sqrt{2 + 3} = \sqrt{5};$$

$$(f + g)(x) = x + 1 + \sqrt{x + 3}$$
$$(f + g)(2) = 2 + 1 + \sqrt{2 + 3}$$
$$= 3 + \sqrt{5} = f(2) + g(2).$$

Let's also examine the domains of $f - g$, $fg$, and $f/g$ for the functions $f(x) = x + 1$ and $g(x) = \sqrt{x + 3}$ of Example 1. The domains of $f - g$ and $fg$ are the same as the domain of $f + g$, $[-3, \infty)$, because numbers in this interval are in the domains of *both* functions. For $f/g$, $g(x)$ cannot be 0. Since $\sqrt{x + 3} = 0$ when $x = -3$, we must exclude $-3$ so the domain of $f/g$ is $(-3, \infty)$.

## DOMAINS OF $f + g$, $f - g$, $fg$, AND $f/g$

If $f$ and $g$ are functions, then the domain of the functions $f + g$, $f - g$, and $fg$ is the intersection of the domain of $f$ and the domain of $g$. The domain of $f/g$ is also the intersection of the domains of $f$ and $g$ with the exclusion of any $x$-values for which $g(x) = 0$.

**EXAMPLE 2**   Given that $f(x) = x^2 - 4$ and $g(x) = x + 2$, find each of the following.

**a)** The domain of $f + g$, $f - g$, $fg$, and $f/g$
**b)** $(f + g)(x)$              **c)** $(f - g)(x)$
**d)** $(fg)(x)$                 **e)** $(f/g)(x)$
**f)** $(gg)(x)$

*Solution*

**a)** The domain of $f$ is the set of all real numbers. The domain of $g$ is also the set of all real numbers. The domains of $f + g$, $f - g$, and $fg$ are the set of numbers in the intersection of the domains—that is, the set of numbers in both domains, which is again the set of real numbers. For $f/g$, we must exclude $-2$, since $g(-2) = 0$. Thus the domain of $f/g$ is the set of real numbers excluding $-2$, or $(-\infty, -2) \cup (-2, \infty)$.

**b)** $(f + g)(x) = f(x) + g(x) = (x^2 - 4) + (x + 2) = x^2 + x - 2$

**c)** $(f - g)(x) = f(x) - g(x) = (x^2 - 4) - (x + 2) = x^2 - x - 6$

**d)** $(fg)(x) = f(x) \cdot g(x) = (x^2 - 4)(x + 2) = x^3 + 2x^2 - 4x - 8$

**e)** $(f/g)(x) = \dfrac{f(x)}{g(x)}$

$\qquad = \dfrac{x^2 - 4}{x + 2}$    Note that $g(x) = 0$ when $x = -2$, so $(f/g)(x)$ is *not defined* when $x = -2$.

$\qquad = \dfrac{(x + 2)(x - 2)}{x + 2}$    Factoring

$\qquad = x - 2$    Removing a factor of 1: $\dfrac{x + 2}{x + 2} = 1$

Thus, $(f/g)(x) = x - 2$ with the added stipulation that $x \neq -2$ since $-2$ is not in the domain of $(f/g)(x)$.

**f)** $(gg)(x) = g(x) \cdot g(x) = \left[g(x)\right]^2 = (x + 2)^2 = x^2 + 4x + 4$

> **Now Try Exercise 21.**

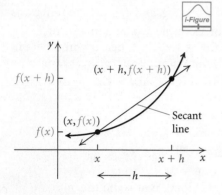

## ■ Difference Quotients

In Section 1.3, we learned that the slope of a line can be considered as an average *rate of change*. Here let's consider a nonlinear function $f$ and draw a line through two points $(x, f(x))$ and $(x + h, f(x + h))$ as shown at left.

The slope of the line, called a **secant line**, is

$$\frac{f(x + h) - f(x)}{x + h - x},$$

which simplifies to

$$\frac{f(x + h) - f(x)}{h}.$$    Difference quotient

This ratio is called the **difference quotient**, or the **average rate of change**. In calculus, it is important to be able to find and simplify difference quotients.

**EXAMPLE 3**   For the function $f$ given by $f(x) = 2x - 3$, find and simplify the difference quotient

$$\frac{f(x + h) - f(x)}{h}.$$

*Solution*

$$\frac{f(x + h) - f(x)}{h} = \frac{2(x + h) - 3 - (2x - 3)}{h}$$    Substituting

$$= \frac{2x + 2h - 3 - 2x + 3}{h}$$    Removing parentheses

$$= \frac{2h}{h}$$

$$= 2$$    Simplifying

> **Now Try Exercise 49.**

**EXAMPLE 4** For the function $f$ given by $f(x) = \dfrac{1}{x}$, find and simplify the difference quotient

$$\frac{f(x + h) - f(x)}{h}.$$

*Solution*

$$\frac{f(x + h) - f(x)}{h} = \frac{\dfrac{1}{x + h} - \dfrac{1}{x}}{h} \qquad \text{Substituting}$$

$$= \frac{\dfrac{1}{x + h} \cdot \dfrac{x}{x} - \dfrac{1}{x} \cdot \dfrac{x + h}{x + h}}{h} \qquad \begin{array}{l}\text{The LCD of } \dfrac{1}{x + h} \\[6pt] \text{and } \dfrac{1}{x} \text{ is } x(x + h).\end{array}$$

$$= \frac{\dfrac{x}{x(x + h)} - \dfrac{x + h}{x(x + h)}}{h}$$

$$= \frac{\dfrac{x - (x + h)}{x(x + h)}}{h} \qquad \text{Subtracting in the numerator}$$

$$= \frac{\dfrac{x - x - h}{x(x + h)}}{h} \qquad \text{Removing parentheses}$$

$$= \frac{\dfrac{-h}{x(x + h)}}{h} \qquad \text{Simplifying the numerator}$$

$$= \frac{-h}{x(x + h)} \cdot \frac{1}{h} \qquad \begin{array}{l}\text{Multiplying by the} \\ \text{reciprocal of the divisor}\end{array}$$

$$= \frac{-h \cdot 1}{x \cdot (x + h) \cdot h}$$

$$= \frac{-1 \cdot \cancel{h}}{x \cdot (x + h) \cdot \cancel{h}} \qquad \text{Simplifying}$$

$$= \frac{-1}{x(x + h)}, \text{ or } -\frac{1}{x(x + h)}$$

**Now Try Exercise 55.**

**EXAMPLE 5** For the function $f$ given by $f(x) = 2x^2 - x - 3$, find and simplify the difference quotient

$$\frac{f(x + h) - f(x)}{h}.$$

***Solution*** We first find $f(x + h)$:

$$f(x + h) = 2(x + h)^2 - (x + h) - 3 \quad \text{Substituting } x + h \text{ for } x \\ \text{in } f(x) = 2x^2 - x - 3$$
$$= 2[x^2 + 2xh + h^2] - (x + h) - 3$$
$$= 2x^2 + 4xh + 2h^2 - x - h - 3.$$

Then

$$\frac{f(x + h) - f(x)}{h} = \frac{[2x^2 + 4xh + 2h^2 - x - h - 3] - [2x^2 - x - 3]}{h}$$

$$= \frac{2x^2 + 4xh + 2h^2 - x - h - 3 - 2x^2 + x + 3}{h}$$

$$= \frac{4xh + 2h^2 - h}{h}$$

$$= \frac{h(4x + 2h - 1)}{h \cdot 1} = \frac{h}{h} \cdot \frac{4x + 2h - 1}{1} = 4x + 2h - 1.$$

Now Try Exercise 63.

## 2.2 Exercise Set

*Given that* $f(x) = x^2 - 3$ *and* $g(x) = 2x + 1$, *find each of the following, if it exists.*

**1.** $(f + g)(5)$

**2.** $(fg)(0)$

**3.** $(f - g)(-1)$

**4.** $(fg)(2)$

**5.** $(f/g)\left(-\frac{1}{2}\right)$

**6.** $(f - g)(0)$

**7.** $(fg)\left(-\frac{1}{2}\right)$

**8.** $(f/g)(-\sqrt{3})$

**9.** $(g - f)(-1)$

**10.** $(g/f)\left(-\frac{1}{2}\right)$

*Given that* $h(x) = x + 4$ *and* $g(x) = \sqrt{x - 1}$, *find each of the following, if it exists.*

**11.** $(h - g)(-4)$

**12.** $(gh)(10)$

**13.** $(g/h)(1)$

**14.** $(h/g)(1)$

**15.** $(g + h)(1)$

**16.** $(hg)(3)$

*For each pair of functions in Exercises 17–34:*

**a)** *Find the domain of* $f, g, f + g, f - g, fg, ff, f/g,$ *and* $g/f$.

**b)** *Find* $(f + g)(x), (f - g)(x), (fg)(x), (ff)(x),$ $(f/g)(x),$ *and* $(g/f)(x)$.

**17.** $f(x) = 2x + 3, \ g(x) = 3 - 5x$

**18.** $f(x) = -x + 1, \ g(x) = 4x - 2$

**19.** $f(x) = x - 3, \ g(x) = \sqrt{x + 4}$

**20.** $f(x) = x + 2, \ g(x) = \sqrt{x - 1}$

**21.** $f(x) = 2x - 1, \ g(x) = -2x^2$

**22.** $f(x) = x^2 - 1, \ g(x) = 2x + 5$

**23.** $f(x) = \sqrt{x - 3}, \ g(x) = \sqrt{x + 3}$

**24.** $f(x) = \sqrt{x}, \ g(x) = \sqrt{2 - x}$

**25.** $f(x) = x + 1, \ g(x) = |x|$

**26.** $f(x) = 4|x|, \ g(x) = 1 - x$

**27.** $f(x) = x^3, \ g(x) = 2x^2 + 5x - 3$

**28.** $f(x) = x^2 - 4, \ g(x) = x^3$

**29.** $f(x) = \dfrac{4}{x + 1}, \ g(x) = \dfrac{1}{6 - x}$

**30.** $f(x) = 2x^2, \ g(x) = \dfrac{2}{x - 5}$

**31.** $f(x) = \dfrac{1}{x}, \ g(x) = x - 3$

**32.** $f(x) = \sqrt{x + 6}, \ g(x) = \dfrac{1}{x}$

**33.** $f(x) = \dfrac{3}{x - 2}, \ g(x) = \sqrt{x - 1}$

**34.** $f(x) = \dfrac{2}{4 - x}, \ g(x) = \dfrac{5}{x - 1}$

*In Exercises 35–40, consider the functions F and G as shown in the graph below.*

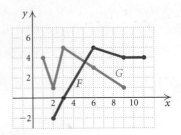

**35.** Find the domain of $F$, the domain of $G$, and the domain of $F + G$.

**36.** Find the domain of $F - G$, $FG$, and $F/G$.

**37.** Find the domain of $G/F$.

**38.** Graph $F + G$.

**39.** Graph $G - F$.

**40.** Graph $F - G$.

*In Exercises 41–46, consider the functions F and G as shown in the graph below.*

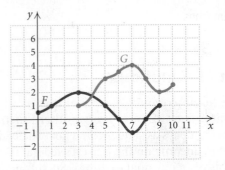

**41.** Find the domain of $F$, the domain of $G$, and the domain of $F + G$.

**42.** Find the domain of $F - G$, $FG$, and $F/G$.

**43.** Find the domain of $G/F$.

**44.** Graph $F + G$.

**45.** Graph $G - F$.

**46.** Graph $F - G$.

**47.** *Total Cost, Revenue, and Profit.* In economics, functions that involve revenue, cost, and profit are used. For example, suppose that $R(x)$ and $C(x)$ denote the total revenue and the total cost, respectively, of producing a new kind of tool for King Hardware Wholesalers. Then the difference

$$P(x) = R(x) - C(x)$$

represents the total profit for producing $x$ tools. Given

$$R(x) = 60x - 0.4x^2 \quad \text{and} \quad C(x) = 3x + 13,$$

find each of the following.

a) $P(x)$

b) $R(100)$, $C(100)$, and $P(100)$

c) Using a graphing calculator, graph the three functions in the viewing window $[0, 160, 0, 3000]$.

**48.** *Total Cost, Revenue, and Profit.* Given that

$$R(x) = 200x - x^2 \quad \text{and} \quad C(x) = 5000 + 8x$$

for a new weather radio produced by Clear Communication, find each of the following. (See Exercise 47.)

a) $P(x)$

b) $R(175)$, $C(175)$, and $P(175)$

c) Using a graphing calculator, graph the three functions in the viewing window $[0, 200, 0, 10{,}000]$.

*For each function f, construct and simplify the difference quotient*

$$\frac{f(x + h) - f(x)}{h}.$$

**49.** $f(x) = 3x - 5$

**50.** $f(x) = 4x - 1$

**51.** $f(x) = 6x + 2$

**52.** $f(x) = 5x + 3$

**53.** $f(x) = \frac{1}{3}x + 1$

**54.** $f(x) = -\frac{1}{2}x + 7$

**55.** $f(x) = \dfrac{1}{3x}$

**56.** $f(x) = \dfrac{1}{2x}$

**57.** $f(x) = -\dfrac{1}{4x}$

**58.** $f(x) = -\dfrac{1}{x}$

**59.** $f(x) = x^2 + 1$

**60.** $f(x) = x^2 - 3$

**61.** $f(x) = 4 - x^2$

**62.** $f(x) = 2 - x^2$

**63.** $f(x) = 3x^2 - 2x + 1$

**64.** $f(x) = 5x^2 + 4x$

**65.** $f(x) = 4 + 5|x|$

**66.** $f(x) = 2|x| + 3x$

**67.** $f(x) = x^3$

**68.** $f(x) = x^3 - 2x$

**69.** $f(x) = \dfrac{x - 4}{x + 3}$

**70.** $f(x) = \dfrac{x}{2 - x}$

## Skill Maintenance

*Graph the equation.*

**71.** $y = 3x - 1$

**72.** $2x + y = 4$

**73.** $x - 3y = 3$

**74.** $y = x^2 + 1$

## Synthesis

**75.** Write equations for two functions $f$ and $g$ such that the domain of $f - g$ is
$$\{x \mid x \neq -7 \text{ and } x \neq 3\}.$$

**76.** For functions $h$ and $f$, find the domain of $h + f$, $h - f$, $hf$, and $h/f$ if
$$h = \{(-4, 13), (-1, 7), (0, 5), (\tfrac{5}{2}, 0), (3, -5)\}, \quad \text{and}$$
$$f = \{(-4, -7), (-2, -5), (0, -3), (3, 0), (5, 2), (9, 6)\}.$$

**77.** Find the domain of $(h/g)(x)$ given that
$$h(x) = \frac{5x}{3x - 7} \quad \text{and} \quad g(x) = \frac{x^4 - 1}{5x - 15}.$$

---

## 2.3 The Composition of Functions

- Find the composition of two functions and the domain of the composition.
- Decompose a function as a composition of two functions.

---

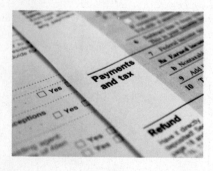

### The Composition of Functions

In real-world situations, it is not uncommon for the output of a function to depend on some input that is itself an output of another function. For instance, the amount that a person pays as state income tax usually depends on the amount of adjusted gross income on the person's federal tax return, which, in turn, depends on his or her annual earnings. Such functions are called **composite functions**.

To see how composite functions work, suppose a chemistry student needs a formula to convert Fahrenheit temperatures to Kelvin units. The formula

$$c(t) = \tfrac{5}{9}(t - 32)$$

gives the Celsius temperature $c(t)$ that corresponds to the Fahrenheit temperature $t$. The formula

$$k(c(t)) = c(t) + 273$$

gives the Kelvin temperature $k(c(t))$ that corresponds to the Celsius temperature $c(t)$. Thus, 50° Fahrenheit corresponds to

$$c(50) = \tfrac{5}{9}(50 - 32) = \tfrac{5}{9}(18) = 10° \text{ Celsius,}$$

and 10° Celsius corresponds to

$$k(c(50)) = k(10) = 10 + 273 = 283 \text{ Kelvin units,}$$

which is usually written 283K. We see that 50° Fahrenheit is the same as 283K. This two-step procedure can be used to convert any Fahrenheit temperature to Kelvin units.

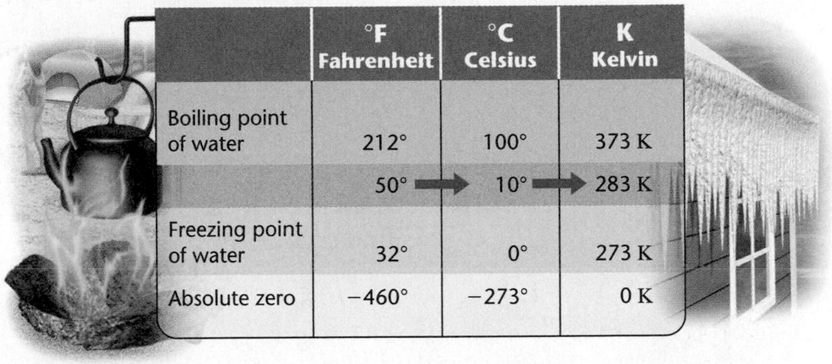

| | °F<br>Fahrenheit | °C<br>Celsius | K<br>Kelvin |
|---|---|---|---|
| Boiling point of water | 212° | 100° | 373 K |
| | 50° ➡ | 10° ➡ | 283 K |
| Freezing point of water | 32° | 0° | 273 K |
| Absolute zero | −460° | −273° | 0 K |

$y_1 = \frac{5}{9}(x - 32), \quad y_2 = y_1 + 273$

| X | Y₁ | Y₂ |
|---|---|---|
| 50 | 10 | 283 |
| 59 | 15 | 288 |
| 68 | 20 | 293 |
| 77 | 25 | 298 |
| 86 | 30 | 303 |
| 95 | 35 | 308 |
| 104 | 40 | 313 |
| X = 50 | | |

In the table shown at left, we use a graphing calculator to convert Fahrenheit temperatures $x$ to Celsius temperatures $y_1$, using $y_1 = \frac{5}{9}(x - 32)$. We also convert Celsius temperatures to Kelvin units $y_2$, using $y_2 = y_1 + 273$.

A student making numerous conversions might look for a formula that converts directly from Fahrenheit to Kelvin. Such a formula can be found by substitution:

$$
\begin{aligned}
k(c(t)) &= c(t) + 273 \\
&= \frac{5}{9}(t - 32) + 273 \qquad \text{Substituting } \frac{5}{9}(t - 32) \text{ for } c(t) \\
&= \frac{5}{9}t - \frac{160}{9} + 273 \\
&= \frac{5}{9}t - \frac{160}{9} + \frac{2457}{9} \\
&= \frac{5t + 2297}{9}. \qquad \text{Simplifying}
\end{aligned}
$$

We can show on a graphing calculator that the same values that appear in the table for $y_2$ will appear when $y_2$ is entered as

$$y_2 = \frac{5x + 2297}{9}.$$

Since the formula found above expresses the Kelvin temperature as a new function $K$ of the Fahrenheit temperature $t$, we can write

$$K(t) = \frac{5t + 2297}{9},$$

where $K(t)$ is the Kelvin temperature corresponding to the Fahrenheit temperature, $t$. Here we have $K(t) = k(c(t))$. The new function $K$ is called the **composition** of $k$ and $c$ and can be denoted $k \circ c$ (read "$k$ composed with $c$," "the composition of $k$ and $c$," or "$k$ circle $c$").

> ### COMPOSITION OF FUNCTIONS
>
> The **composite function** $f \circ g$, the **composition** of $f$ and $g$, is defined as
>
> $$(f \circ g)(x) = f(g(x)),$$
>
> where $x$ is in the domain of $g$ and $g(x)$ is in the domain of $f$.

**GCM**  **EXAMPLE 1**  Given that $f(x) = 2x - 5$ and $g(x) = x^2 - 3x + 8$, find each of the following.

**a)** $(f \circ g)(x)$ and $(g \circ f)(x)$    **b)** $(f \circ g)(7)$ and $(g \circ f)(7)$
**c)** $(g \circ g)(1)$                          **d)** $(f \circ f)(x)$

***Solution***  Consider each function separately:

$$f(x) = 2x - 5 \qquad \text{This function multiplies each input by 2 and then subtracts 5.}$$

and

$$g(x) = x^2 - 3x + 8. \qquad \text{This function squares an input, subtracts three times the input from the result, and then adds 8.}$$

**a)** To find $(f \circ g)(x)$, we substitute $g(x)$ for $x$ in the equation for $f(x)$:

$$(f \circ g)(x) = f(g(x)) = f(x^2 - 3x + 8) \qquad \begin{array}{l} x^2 - 3x + 8 \text{ is the} \\ \text{input for } f. \end{array}$$

$$= 2(x^2 - 3x + 8) - 5 \qquad \begin{array}{l} f \text{ multiplies the} \\ \text{input by 2 and then} \\ \text{subtracts 5.} \end{array}$$

$$= 2x^2 - 6x + 16 - 5$$

$$= 2x^2 - 6x + 11.$$

To find $(g \circ f)(x)$, we substitute $f(x)$ for $x$ in the equation for $g(x)$:

$$(g \circ f)(x) = g(f(x)) = g(2x - 5) \qquad \begin{array}{l} 2x - 5 \text{ is the} \\ \text{input for } g. \end{array}$$

$$= (2x - 5)^2 - 3(2x - 5) + 8 \qquad \begin{array}{l} g \text{ squares the} \\ \text{input, subtracts} \\ \text{three times the} \\ \text{input, and then} \\ \text{adds 8.} \end{array}$$

$$= 4x^2 - 20x + 25 - 6x + 15 + 8$$

$$= 4x^2 - 26x + 48.$$

**b)** To find $(f \circ g)(7)$, we first find $g(7)$. Then we use $g(7)$ as an input for $f$:

$$(f \circ g)(7) = f(g(7)) = f(7^2 - 3 \cdot 7 + 8)$$

$$= f(36) = 2 \cdot 36 - 5$$

$$= 67.$$

To find $(g \circ f)(7)$, we first find $f(7)$. Then we use $f(7)$ as an input for $g$:

$$(g \circ f)(7) = g(f(7)) = g(2 \cdot 7 - 5)$$
$$= g(9) = 9^2 - 3 \cdot 9 + 8$$
$$= 62.$$

We could also find $(f \circ g)(7)$ and $(g \circ f)(7)$ by substituting 7 for $x$ in the equations that we found in part (a):

$$(f \circ g)(x) = 2x^2 - 6x + 11$$
$$(f \circ g)(7) = 2 \cdot 7^2 - 6 \cdot 7 + 11 = 67;$$

$$(g \circ f)(x) = 4x^2 - 26x + 48$$
$$(g \circ f)(7) = 4 \cdot 7^2 - 26 \cdot 7 + 48 = 62.$$

$y_1 = 2x - 5, \ y_2 = x^2 - 3x + 8$

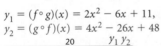

We can check our work using a graphing calculator. On the equation-editor screen, we enter $f(x)$ as $y_1 = 2x - 5$ and $g(x)$ as $y_2 = x^2 - 3x + 8$. Then, on the home screen, we find $(f \circ g)(7)$ and $(g \circ f)(7)$ using the function notations Y₁(Y₂(7)) and Y₂(Y₁(7)), respectively.

**c)** $(g \circ g)(1) = g(g(1)) = g(1^2 - 3 \cdot 1 + 8)$
$$= g(1 - 3 + 8) = g(6)$$
$$= 6^2 - 3 \cdot 6 + 8$$
$$= 36 - 18 + 8 = 26$$

**d)** $(f \circ f)(x) = f(f(x)) = f(2x - 5)$
$$= 2(2x - 5) - 5$$
$$= 4x - 10 - 5 = 4x - 15$$

**Now Try Exercises 1 and 15.**

$y_1 = (f \circ g)(x) = 2x^2 - 6x + 11,$
$y_2 = (g \circ f)(x) = 4x^2 - 26x + 48$

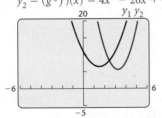

Example 1 illustrates that, as a rule, $(f \circ g)(x) \neq (g \circ f)(x)$. We can see this graphically, as shown in the graphs at left.

**EXAMPLE 2**   Given that $f(x) = \sqrt{x}$ and $g(x) = x - 3$:

**a)** Find $f \circ g$ and $g \circ f$.

**b)** Find the domain of $f \circ g$ and the domain of $g \circ f$.

*Solution*

**a)** $(f \circ g)(x) = f(g(x)) = f(x - 3) = \sqrt{x - 3}$
  $(g \circ f)(x) = g(f(x)) = g(\sqrt{x}) = \sqrt{x} - 3$

**b)** Since $f(x)$ is not defined for negative radicands, the domain of $f(x)$ is $\{x | x \geq 0\}$, or $[0, \infty)$. Any real number can be an input for $g(x)$, so the domain of $g(x)$ is $(-\infty, \infty)$.

Since the inputs of $f \circ g$ are outputs of $g$, the domain of $f \circ g$ consists of the values of $x$ in the domain of $g$, $(-\infty, \infty)$, for which $g(x)$ is nonnegative. (Recall that the inputs of $f(x)$ must be nonnegative.) Thus we have

$$g(x) \geq 0$$
$$x - 3 \geq 0 \qquad \text{Substituting } x - 3 \text{ for } g(x)$$
$$x \geq 3.$$

We see that the domain of $f \circ g$ is $\{x | x \geq 3\}$, or $[3, \infty)$.

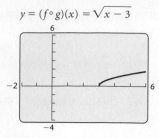

$y = (f \circ g)(x) = \sqrt{x} - 3$

**FIGURE 1**

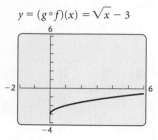

$y = (g \circ f)(x) = \sqrt{x} - 3$

**FIGURE 2**

We can also find the domain of $f \circ g$ by examining the composite function itself, $(f \circ g)(x) = \sqrt{x - 3}$, keeping in mind any restrictions on the domain of $g$. Since any real number can be an input for $g$, the only restriction on $f \circ g$ is that the radicand must be nonnegative. We have

$$x - 3 \geq 0$$
$$x \geq 3.$$

Again, we see that the domain of $f \circ g$ is $\{x | x \geq 3\}$, or $[3, \infty)$. The graph in Fig. 1 confirms this.

The inputs of $g \circ f$ are outputs of $f$, so the domain of $g \circ f$ consists of the values of $x$ in the domain of $f$, $[0, \infty)$, for which $g(x)$ is defined. Since $g$ can accept *any* real number as an input, any output from $f$ is acceptable, so the entire domain of $f$ is the domain of $g \circ f$. That is, the domain of $g \circ f$ is $\{x | x \geq 0\}$, or $[0, \infty)$.

We can also examine the composite function itself to find its domain, keeping in mind any restrictions on the domain of $f$. First, recall that the domain of $f$ is $\{x | x \geq 0\}$, or $[0, \infty)$. Then consider $(g \circ f)(x) = \sqrt{x} - 3$. The radicand cannot be negative, so we have $x \geq 0$. As above, we see that the domain of $g \circ f$ is the domain of $f$, $\{x | x \geq 0\}$, or $[0, \infty)$. The graph in Fig. 2 confirms this.

> ◼ **Now Try Exercise 27.**

**EXAMPLE 3**    Given that $f(x) = \dfrac{1}{x - 2}$ and $g(x) = \dfrac{5}{x}$, find $f \circ g$ and $g \circ f$ and the domain of each.

*Solution*    We have

$$(f \circ g)(x) = f(g(x)) = f\left(\frac{5}{x}\right) = \frac{1}{\dfrac{5}{x} - 2} = \frac{1}{\dfrac{5 - 2x}{x}} = \frac{x}{5 - 2x};$$

$$(g \circ f)(x) = g(f(x)) = g\left(\frac{1}{x - 2}\right) = \frac{5}{\dfrac{1}{x - 2}} = 5(x - 2).$$

Values of $x$ that make the denominator 0 are not in the domains of these functions. Since $x - 2 = 0$ when $x = 2$, the domain of $f$ is $\{x | x \neq 2\}$. The denominator of $g$ is $x$, so the domain of $g$ is $\{x | x \neq 0\}$.

The inputs of $f \circ g$ are outputs of $g$, so the domain of $f \circ g$ consists of the values of $x$ in the domain of $g$ for which $g(x) \neq 2$. (Recall that 2 cannot be an input of $f$.) Since the domain of $g$ is $\{x | x \neq 0\}$, 0 is not in the domain of $f \circ g$. In addition, we must find the value(s) of $x$ for which $g(x) = 2$. We have

$$g(x) = 2$$
$$\frac{5}{x} = 2 \qquad \text{Substituting } \frac{5}{x} \text{ for } g(x)$$
$$5 = 2x$$
$$\frac{5}{2} = x.$$

This tells us that $\frac{5}{2}$ is also *not* in the domain of $f \circ g$. Then the domain of $f \circ g$ is $\{x | x \neq 0 \text{ and } x \neq \frac{5}{2}\}$, or $(-\infty, 0) \cup \left(0, \frac{5}{2}\right) \cup \left(\frac{5}{2}, \infty\right)$.

We can also examine the composite function $f \circ g$ to find its domain, keeping in mind any restrictions on the domain of $g$. First, recall that 0 is not in the domain of $g$, so it cannot be in the domain of $(f \circ g)(x) = x/(5 - 2x)$. We must also exclude the value(s) of $x$ for which the denominator of $f \circ g$ is 0. We have

$$5 - 2x = 0$$
$$5 = 2x$$
$$\tfrac{5}{2} = x.$$

Again, we see that $\tfrac{5}{2}$ is also not in the domain, so the domain of $f \circ g$ is $\{x \,|\, x \neq 0 \text{ and } x \neq \tfrac{5}{2}\}$, or $(-\infty, 0) \cup (0, \tfrac{5}{2}) \cup (\tfrac{5}{2}, \infty)$.

Since the inputs of $g \circ f$ are outputs of $f$, the domain of $g \circ f$ consists of the values of $x$ in the domain of $f$ for which $f(x) \neq 0$. (Recall that 0 cannot be an input of $g$.) The domain of $f$ is $\{x \,|\, x \neq 2\}$, so 2 is not in the domain of $g \circ f$. Next, we determine whether there are values of $x$ for which $f(x) = 0$:

$$f(x) = 0$$

$$\frac{1}{x - 2} = 0 \qquad\qquad \text{Substituting } \frac{1}{x - 2} \text{ for } f(x)$$

$$(x - 2) \cdot \frac{1}{x - 2} = (x - 2) \cdot 0 \qquad \text{Multiplying by } x - 2$$

$$1 = 0. \qquad\qquad\qquad \text{False equation}$$

We see that there are no values of $x$ for which $f(x) = 0$, so there are no additional restrictions on the domain of $g \circ f$. Thus the domain of $g \circ f$ is $\{x \,|\, x \neq 2\}$, or $(-\infty, 2) \cup (2, \infty)$.

We can also examine $g \circ f$ to find its domain. First, recall that 2 is not in the domain of $f$, so it cannot be in the domain of $(g \circ f)(x) = 5(x - 2)$. Since $5(x - 2)$ is defined for all real numbers, there are no additional restrictions on the domain of $g \circ f$. The domain is $\{x \,|\, x \neq 2\}$, or $(-\infty, 2) \cup (2, \infty)$.

**Now Try Exercise 23.**

## ■ Decomposing a Function as a Composition

In calculus, one often needs to recognize how a function can be expressed as the composition of two functions. In this way, we are "decomposing" the function.

**EXAMPLE 4** If $h(x) = (2x - 3)^5$, find $f(x)$ and $g(x)$ such that $h(x) = (f \circ g)(x)$.

*Solution* The function $h(x)$ raises $(2x - 3)$ to the 5th power. Two functions that can be used for the composition are

$$f(x) = x^5 \quad \text{and} \quad g(x) = 2x - 3.$$

We can check by forming the composition:

$$h(x) = (f \circ g)(x) = f(g(x)) = f(2x - 3) = (2x - 3)^5.$$

This is the most "obvious" solution. There can be other less obvious solutions. For example, if

$$f(x) = (x + 7)^5 \quad \text{and} \quad g(x) = 2x - 10,$$

then

$$h(x) = (f \circ g)(x) = f(g(x))$$
$$= f(2x - 10)$$
$$= [2x - 10 + 7]^5 = (2x - 3)^5.$$

**Now Try Exercise 39.**

**EXAMPLE 5** If $h(x) = \dfrac{1}{(x + 3)^3}$, find $f(x)$ and $g(x)$ such that $h(x) = (f \circ g)(x)$.

*Solution* Two functions that can be used are

$$f(x) = \frac{1}{x^3} \quad \text{and} \quad g(x) = x + 3.$$

We check by forming the composition:

$$h(x) = (f \circ g)(x) = f(g(x)) = f(x + 3) = \frac{1}{(x + 3)^3}.$$

There are other functions that can be used as well. For example, if

$$f(x) = \frac{1}{x} \quad \text{and} \quad g(x) = (x + 3)^3,$$

then

$$h(x) = (f \circ g)(x) = f(g(x)) = f((x + 3)^3) = \frac{1}{(x + 3)^3}.$$

**Now Try Exercise 41.**

## 2.3 Exercise Set

*Given that $f(x) = 3x + 1$, $g(x) = x^2 - 2x - 6$, and $h(x) = x^3$, find each of the following.*

**1.** $(f \circ g)(-1)$

**2.** $(g \circ f)(-2)$

**3.** $(h \circ f)(1)$

**4.** $(g \circ h)\left(\frac{1}{2}\right)$

**5.** $(g \circ f)(5)$

**6.** $(f \circ g)\left(\frac{1}{3}\right)$

**7.** $(f \circ h)(-3)$

**8.** $(h \circ g)(3)$

**9.** $(g \circ g)(-2)$

**10.** $(g \circ g)(3)$

**11.** $(h \circ h)(2)$

**12.** $(h \circ h)(-1)$

**13.** $(f \circ f)(-4)$

**14.** $(f \circ f)(1)$

**15.** $(h \circ h)(x)$

**16.** $(f \circ f)(x)$

*Find $(f \circ g)(x)$ and $(g \circ f)(x)$ and the domain of each.*

**17.** $f(x) = x + 3$, $g(x) = x - 3$

**18.** $f(x) = \frac{4}{5}x$, $g(x) = \frac{5}{4}x$

**19.** $f(x) = x + 1$, $g(x) = 3x^2 - 2x - 1$

**20.** $f(x) = 3x - 2$, $g(x) = x^2 + 5$

**21.** $f(x) = x^2 - 3$, $g(x) = 4x - 3$

**22.** $f(x) = 4x^2 - x + 10$, $g(x) = 2x - 7$

**23.** $f(x) = \dfrac{4}{1 - 5x}$, $g(x) = \dfrac{1}{x}$

**24.** $f(x) = \dfrac{6}{x}$, $g(x) = \dfrac{1}{2x + 1}$

**25.** $f(x) = 3x - 7$, $g(x) = \dfrac{x + 7}{3}$

**26.** $f(x) = \frac{2}{3}x - \frac{4}{5}$, $g(x) = 1.5x + 1.2$

**27.** $f(x) = 2x + 1$, $g(x) = \sqrt{x}$

**28.** $f(x) = \sqrt{x}$, $g(x) = 2 - 3x$

**29.** $f(x) = 20$, $g(x) = 0.05$

**30.** $f(x) = x^4$, $g(x) = \sqrt[4]{x}$

**31.** $f(x) = \sqrt{x + 5}$, $g(x) = x^2 - 5$

**32.** $f(x) = x^5 - 2$, $g(x) = \sqrt[5]{x + 2}$

**33.** $f(x) = x^2 + 2$, $g(x) = \sqrt{3 - x}$

**34.** $f(x) = 1 - x^2$, $g(x) = \sqrt{x^2 - 25}$

**35.** $f(x) = \dfrac{1 - x}{x}$, $g(x) = \dfrac{1}{1 + x}$

**36.** $f(x) = \dfrac{1}{x - 2}$, $g(x) = \dfrac{x + 2}{x}$

**37.** $f(x) = x^3 - 5x^2 + 3x + 7$, $g(x) = x + 1$

**38.** $f(x) = x - 1$, $g(x) = x^3 + 2x^2 - 3x - 9$

*Find $f(x)$ and $g(x)$ such that $h(x) = (f \circ g)(x)$.*
*Answers may vary.*

**39.** $h(x) = (4 + 3x)^5$

**40.** $h(x) = \sqrt[3]{x^2 - 8}$

**41.** $h(x) = \dfrac{1}{(x - 2)^4}$

**42.** $h(x) = \sqrt{\dfrac{1}{3x + 7}}$

**43.** $h(x) = \dfrac{x^3 - 1}{x^3 + 1}$

**44.** $h(x) = |9x^2 - 4|$

**45.** $h(x) = \left(\dfrac{2 + x^3}{2 - x^3}\right)^6$

**46.** $h(x) = (\sqrt{x} - 3)^4$

**47.** $h(x) = \sqrt{\dfrac{x - 5}{x + 2}}$

**48.** $h(x) = \sqrt{1 + \sqrt{1 + x}}$

**49.** $h(x) = (x + 2)^3 - 5(x + 2)^2 + 3(x + 2) - 1$

**50.** $h(x) = 2(x - 1)^{5/3} + 5(x - 1)^{2/3}$

**51.** *Ripple Spread.* A stone is thrown into a pond, creating a circular ripple that spreads over the pond in such a way that the radius is increasing at a rate of 3 ft/sec.

**a)** Find a function $r(t)$ for the radius in terms of $t$.

**b)** Find a function $A(r)$ for the area of the ripple in terms of the radius $r$.

**c)** Find $(A \circ r)(t)$. Explain the meaning of this function.

**52.** The surface area $S$ of a right circular cylinder is given by the formula $S = 2\pi rh + 2\pi r^2$. If the height is twice the radius, find each of the following.

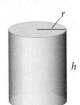

**a)** A function $S(r)$ for the surface area as a function of $r$

**b)** A function $S(h)$ for the surface area as a function of $h$

**53.** A manufacturer of tools, selling rechargeable drills to a chain of home improvement stores, charges $2 more per drill than its manufacturing cost $m$. The chain then sells each drill for 150% of the price that it paid the manufacturer. Find a function $P(m)$ for the price at the home improvement stores.

**54.** *Blouse Sizes.* A blouse that is size $x$ in Japan is size $s(x)$ in the United States, where $s(x) = x - 3$. A blouse that is size $x$ in the United States is size $t(x)$ in Australia, where $t(x) = x + 4$. (*Source:* www.onlineconversion.com) Find a function that will convert Japanese blouse sizes to Australian blouse sizes.

## Skill Maintenance

*Consider the following linear equations. Without graphing them, answer the questions below.*

a) $y = x$                       b) $y = -5x + 4$
c) $y = \frac{2}{3}x + 1$        d) $y = -0.1x + 6$
e) $y = 3x - 5$                  f) $y = -x - 1$
g) $2x - 3y = 6$                 h) $6x + 3y = 9$

**55.** Which, if any, have $y$-intercept $(0, 1)$?

**56.** Which, if any, have the same $y$-intercept?

**57.** Which slope down from left to right?

**58.** Which has the steepest slope?

**59.** Which pass(es) through the origin?

**60.** Which, if any, have the same slope?

**61.** Which, if any, are parallel?

**62.** Which, if any, are perpendicular?

## Synthesis

**63.** Let $p(a)$ represent the number of pounds of grass seed required to seed a lawn with area $a$. Let $c(s)$ represent the cost of $s$ pounds of grass seed. Which composition makes sense: $(c \circ p)(a)$ or $(p \circ c)(s)$? What does it represent?

**64.** Write equations of two functions $f$ and $g$ such that $f \circ g = g \circ f = x$. (In Section 5.1, we will study inverse functions. If $f \circ g = g \circ f = x$, functions $f$ and $g$ are *inverses* of each other.)

# Mid-Chapter Mixed Review

*Determine whether the statement is true or false.*

**1.** $f(c)$ is a relative maximum if $(c, f(c))$ is the highest point in some open interval containing $c$. [2.1]

**2.** If $f$ and $g$ are functions, then the domain of the functions $f + g$, $f - g$, $fg$, and $f/g$ is the intersection of the domain of $f$ and the domain of $g$. [2.2]

**3.** In general, $(f \circ g)(x) \neq (g \circ f)(x)$. [2.3]

**4.** Determine the intervals on which the function is **(a)** increasing; **(b)** decreasing; **(c)** constant. [2.1]

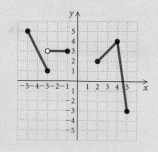

**5.** Using the graph, determine any relative maxima or minima of the function and the intervals on which the function is increasing or decreasing. [2.1]

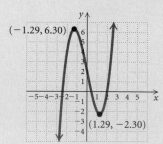

**6.** Determine the domain and the range of the function graphed in Exercise 4. [2.1]

**7.** *Pendant Design.*  Ellen is designing a triangular pendant so that the length of the base is 2 less than the height $h$. Express the area of the pendant as a function of the height. [2.1]

**8.** For the function defined as

$$f(x) = \begin{cases} x - 5, & \text{for } x \leq -3, \\ 2x + 3, & \text{for } -3 < x \leq 0, \\ \frac{1}{2}x, & \text{for } x > 0, \end{cases}$$

find $f(-5), f(-3), f(-1)$, and $f(6)$. [2.1]

**9.** Graph the function defined as

$$g(x) = \begin{cases} x + 2, & \text{for } x < -4, \\ -x, & \text{for } x \geq -4. \end{cases} \quad [2.1]$$

*Given that $f(x) = 3x - 1$ and $g(x) = x^2 + 4$, find each of the following, if it exists.* [2.2]

**10.** $(f + g)(-1)$

**11.** $(fg)(0)$

**12.** $(g - f)(3)$

**13.** $(g/f)\left(\frac{1}{3}\right)$

*For each pair of functions in Exercises 14 and 15:*

**a)** *Find the domains of $f, g, f + g, f - g, fg, ff, f/g$, and $g/f$.*

**b)** *Find $(f + g)(x), (f - g)(x), (fg)(x), (ff)(x), (f/g)(x)$, and $(g/f)(x)$.* [2.2]

**14.** $f(x) = 2x + 5;\ g(x) = -x - 4$

**15.** $f(x) = x - 1;\ g(x) = \sqrt{x + 2}$

*For each function $f$, construct and simplify the difference quotient*

$$\frac{f(x + h) - f(x)}{h}. \quad [2.2]$$

**16.** $f(x) = 4x - 3$

**17.** $f(x) = 6 - x^2$

*Given that $f(x) = 5x - 4, g(x) = x^3 + 1$, and $h(x) = x^2 - 2x + 3$, find each of the following.* [2.3]

**18.** $(f \circ g)(1)$

**19.** $(g \circ h)(2)$

**20.** $(f \circ f)(0)$

**21.** $(h \circ f)(-1)$

*Find $(f \circ g)(x)$ and $(g \circ f)(x)$ and the domain of each.* [2.3]

**22.** $f(x) = \frac{1}{2}x, g(x) = 6x + 4$

**23.** $f(x) = 3x + 2, g(x) = \sqrt{x}$

## Collaborative Discussion and Writing

**24.** If $g(x) = b$, where $b$ is a positive constant, describe how the graphs of $y = h(x)$ and $y = (h - g)(x)$ will differ. [2.2]

**25.** If the domain of a function $f$ is the set of real numbers and the domain of a function $g$ is also the set of real numbers, under what circumstances do $(f + g)(x)$ and $(f/g)(x)$ have different domains? [2.2]

**26.** If $f$ and $g$ are linear functions, what can you say about the domain of $f \circ g$ and the domain of $g \circ f$? [2.3]

**27.** Nora determines the domain of $f \circ g$ by examining only the formula for $(f \circ g)(x)$. Is her approach valid? Why or why not? [2.3]

# 2.4

## Symmetry

- Determine whether a graph is symmetric with respect to the *x*-axis, the *y*-axis, and the origin.

- Determine whether a function is even, odd, or neither even nor odd.

### Symmetry

Symmetry occurs often in nature and in art. For example, when viewed from the front, the bodies of most animals are at least approximately symmetric. This means that each eye is the same distance from the center of the bridge of the nose, each shoulder is the same distance from the center of the chest, and so on. Architects have used symmetry for thousands of years to enhance the beauty of buildings.

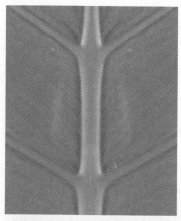

A knowledge of symmetry in mathematics helps us graph and analyze equations and functions.

Consider the points $(4, 2)$ and $(4, -2)$ that appear on the graph of $x = y^2$. (See Fig. 1 on the following page.) Points like these have the same *x*-value but opposite *y*-values and are **reflections** of each other across the *x*-axis. If, for any point $(x, y)$ on a graph, the point $(x, -y)$ is also on the graph, then the graph is said to be **symmetric with respect to the *x*-axis**. If we fold the graph on the *x*-axis, the parts above and below the *x*-axis will coincide.

Consider the points $(3, 4)$ and $(-3, 4)$ that appear on the graph of $y = x^2 - 5$. (See Fig. 2.) Points like these have the same $y$-value but opposite $x$-values and are **reflections** of each other across the $y$-axis. If, for any point $(x, y)$ on a graph, the point $(-x, y)$ is also on the graph, then the graph is said to be **symmetric with respect to the $y$-axis**. If we fold the graph on the $y$-axis, the parts to the left and right of the $y$-axis will coincide.

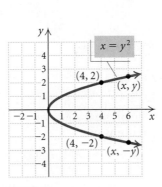

FIGURE 1

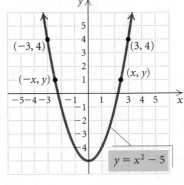

FIGURE 2

Consider the points $(-3, \sqrt{7})$ and $(3, -\sqrt{7})$ that appear on the graph of $x^2 = y^2 + 2$. (See Fig. 3.) Note that if we take the opposites of the coordinates of one pair, we get the other pair. If, for any point $(x, y)$ on a graph, the point $(-x, -y)$ is also on the graph, then the graph is said to be **symmetric with respect to the origin**. Visually, if we rotate the graph $180°$ about the origin, the resulting figure coincides with the original.

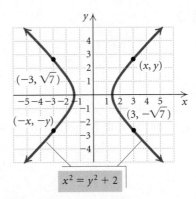

FIGURE 3

## ALGEBRAIC TESTS OF SYMMETRY

*x-axis:* If replacing $y$ with $-y$ produces an equivalent equation, then the graph is *symmetric with respect to the x-axis.*

*y-axis:* If replacing $x$ with $-x$ produces an equivalent equation, then the graph is *symmetric with respect to the y-axis.*

*Origin:* If replacing $x$ with $-x$ and $y$ with $-y$ produces an equivalent equation, then the graph is *symmetric with respect to the origin.*

**EXAMPLE 1**  Test $y = x^2 + 2$ for symmetry with respect to the $x$-axis, the $y$-axis, and the origin.

## Algebraic Solution

*x-Axis:*
We replace $y$ with $-y$:

$$y = x^2 + 2$$
$$-y = x^2 + 2$$
$$y = -x^2 - 2. \quad \text{Multiplying by } -1 \text{ on both sides}$$

The resulting equation *is not* equivalent to the original equation, so the graph *is not* symmetric with respect to the $x$-axis.

*y-Axis:*
We replace $x$ with $-x$:

$$y = x^2 + 2$$
$$y = (-x)^2 + 2$$
$$y = x^2 + 2. \quad \text{Simplifying}$$

The resulting equation *is* equivalent to the original equation, so the graph *is* symmetric with respect to the $y$-axis.

*Origin:*
We replace $x$ with $-x$ and $y$ with $-y$:

$$y = x^2 + 2$$
$$-y = (-x)^2 + 2$$
$$-y = x^2 + 2 \quad \text{Simplifying}$$
$$y = -x^2 - 2.$$

The resulting equation *is not* equivalent to the original equation, so the graph *is not* symmetric with respect to the origin.

## Graphical Solution

We use a graphing calculator to graph the equation.

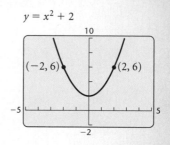

$y = x^2 + 2$

Note that if the graph were folded on the $x$-axis, the parts above and below the $x$-axis would not co-incide so the graph *is not* symmetric with respect to the $x$-axis. If it were folded on the $y$-axis, the parts to the left and right of the $y$-axis would coincide so the graph *is* symmetric with respect to the $y$-axis. If we rotated it 180° around the origin, the resulting graph would not coincide with the original graph so the graph *is not* symmetric with respect to the origin.

**Now Try Exercise 11.**

The algebraic method is often easier to apply than the graphical method, especially with equations that we may not be able to graph easily. It is also often more precise.

**EXAMPLE 2**  Test $x^2 + y^4 = 5$ for symmetry with respect to the $x$-axis, the $y$-axis, and the origin.

## Algebraic Solution

**$x$-Axis:**
We replace $y$ with $-y$:

$$x^2 + y^4 = 5$$
$$x^2 + (-y)^4 = 5$$
$$x^2 + y^4 = 5.$$

The resulting equation *is* equivalent to the original equation. Thus the graph *is* symmetric with respect to the $x$-axis.

**$y$-Axis:**
We replace $x$ with $-x$:

$$x^2 + y^4 = 5$$
$$(-x)^2 + y^4 = 5$$
$$x^2 + y^4 = 5.$$

The resulting equation *is* equivalent to the original equation, so the graph *is* symmetric with respect to the $y$-axis.

**Origin:**
We replace $x$ with $-x$ and $y$ with $-y$:

$$x^2 + y^4 = 5$$
$$(-x)^2 + (-y)^4 = 5$$
$$x^2 + y^4 = 5.$$

The resulting equation *is* equivalent to the original equation, so the graph *is* symmetric with respect to the origin.

## Graphical Solution

To graph $x^2 + y^4 = 5$ using a graphing calculator, we first solve the equation for $y$:

$$y = \pm\sqrt[4]{5 - x^2}.$$

Then on the Y= screen we enter the equations

$$y_1 = \sqrt[4]{5 - x^2} \quad \text{and}$$
$$y_2 = -\sqrt[4]{5 - x^2}.$$

From the graph of the equation, we see symmetry with respect to both axes and with respect to the origin.

**Now Try Exercise 21.**

## Even Functions and Odd Functions

Now we relate symmetry to graphs of functions.

---

**Algebraic Procedure for Determining Even and Odd Functions**

Given the function $f(x)$:

1. Find $f(-x)$ and simplify. If $f(x) = f(-x)$, then $f$ is even.
2. Find $-f(x)$, simplify, and compare with $f(-x)$ from step (1). If $f(-x) = -f(x)$, then $f$ is odd.

Except for the function $f(x) = 0$, a function cannot be *both* even and odd. Thus if $f(x) \neq 0$ and we see in step (1) that $f(x) = f(-x)$ (that is, $f$ is even), we need not continue.

---

### EVEN FUNCTIONS AND ODD FUNCTIONS

If the graph of a function $f$ is symmetric with respect to the $y$-axis, we say that it is an **even function**. That is, for each $x$ in the domain of $f$, $f(x) = f(-x)$.

If the graph of a function $f$ is symmetric with respect to the origin, we say that it is an **odd function**. That is, for each $x$ in the domain of $f$, $f(-x) = -f(x)$.

An algebraic procedure for determining even functions and odd functions is shown at left. Below we show an even function and an odd function. Many functions are neither even nor odd.

**EXAMPLE 3**   Determine whether each of the following functions is even, odd, or neither.

**a)** $f(x) = 5x^7 - 6x^3 - 2x$      **b)** $h(x) = 5x^6 - 3x^2 - 7$

---

### Algebraic Solution

**a)** $f(x) = 5x^7 - 6x^3 - 2x$

1. $f(-x) = 5(-x)^7 - 6(-x)^3 - 2(-x)$
   $= 5(-x^7) - 6(-x^3) + 2x$
   $(-x)^7 = (-1 \cdot x)^7 = (-1)^7 x^7 = -x^7;$
   $(-x)^3 = (-1 \cdot x)^3 = (-1)^3 x^3 = -x^3$
   $= -5x^7 + 6x^3 + 2x$

   We see that $f(x) \neq f(-x)$. Thus, $f$ is *not* even.

2. $-f(x) = -(5x^7 - 6x^3 - 2x)$
   $= -5x^7 + 6x^3 + 2x$

   We see that $f(-x) = -f(x)$. Thus, $f$ is odd.

**b)** $h(x) = 5x^6 - 3x^2 - 7$

1. $h(-x) = 5(-x)^6 - 3(-x)^2 - 7$
   $= 5x^6 - 3x^2 - 7$

   We see that $h(x) = h(-x)$. Thus the function is even.

---

### Graphical Solution

**a)** We see that the graph appears to be symmetric with respect to the origin. The function is odd.

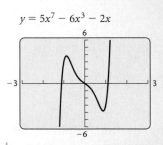

$y = 5x^7 - 6x^3 - 2x$

**b)** We see that the graph appears to be symmetric with respect to the $y$ axis. The function is even.

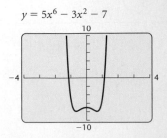

$y = 5x^6 - 3x^2 - 7$

---

**Now Try Exercises 39 and 41.**

# 2.4 Exercise Set

*Determine visually whether the graph is symmetric with respect to the x-axis, the y-axis, and the origin.*

**1.**

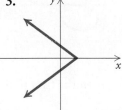

**2.**

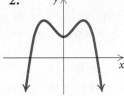

**3.**

**4.**

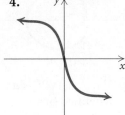

**5.**

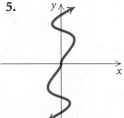

**6.**

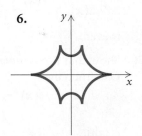

*First, graph the equation and determine visually whether it is symmetric with respect to the x-axis, the y-axis, and the origin. Then verify your assertion algebraically.*

**7.** $y = |x| - 2$      **8.** $y = |x + 5|$

**9.** $5y = 4x + 5$      **10.** $2x - 5 = 3y$

**11.** $5y = 2x^2 - 3$      **12.** $x^2 + 4 = 3y$

**13.** $y = \dfrac{1}{x}$      **14.** $y = -\dfrac{4}{x}$

*Test algebraically whether the graph is symmetric with respect to the x-axis, the y-axis, and the origin. Then check your work graphically, if possible, using a graphing calculator.*

**15.** $5x - 5y = 0$      **16.** $6x + 7y = 0$

**17.** $3x^2 - 2y^2 = 3$      **18.** $5y = 7x^2 - 2x$

**19.** $y = |2x|$      **20.** $y^3 = 2x^2$

**21.** $2x^4 + 3 = y^2$      **22.** $2y^2 = 5x^2 + 12$

**23.** $3y^3 = 4x^3 + 2$      **24.** $3x = |y|$

**25.** $xy = 12$      **26.** $xy - x^2 = 3$

*Find the point that is symmetric to the given point with respect to the x-axis, the y-axis, and the origin.*

**27.** $(-5, 6)$      **28.** $\left(\frac{7}{2}, 0\right)$

**29.** $(-10, -7)$      **30.** $\left(1, \frac{3}{8}\right)$

**31.** $(0, -4)$      **32.** $(8, -3)$

*Determine visually whether the function is even, odd, or neither even nor odd.*

**33.**

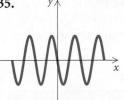

**34.**

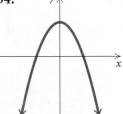

**35.**

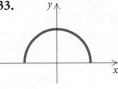

**36.**

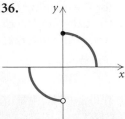

**37.**

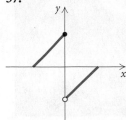

**38.**

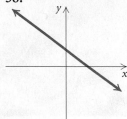

*Determine algebraically whether the function is even, odd, or neither even nor odd. Then check your work graphically, where possible, using a graphing calculator.*

**39.** $f(x) = -3x^3 + 2x$

**40.** $f(x) = 7x^3 + 4x - 2$

**41.** $f(x) = 5x^2 + 2x^4 - 1$

**42.** $f(x) = x + \dfrac{1}{x}$

**43.** $f(x) = x^{17}$

**44.** $f(x) = \sqrt[3]{x}$

**45.** $f(x) = x - |x|$

**46.** $f(x) = \dfrac{1}{x^2}$

**47.** $f(x) = 8$

**48.** $f(x) = \sqrt{x^2 + 1}$

## Skill Maintenance

**49.** Graph: $f(x) = \begin{cases} x - 2, & \text{for } x \le -1, \\ 3, & \text{for } -1 < x \le 2, \\ x, & \text{for } x > 2. \end{cases}$

**50.** *Olympics Ticket Prices.*   Together, the price of a ticket for the opening ceremonies of the 2010 Winter Olympics in Vancouver, British Columbia, Canada, and a ticket to the closing ceremonies cost $1875 (in Canadian dollars). The ticket for the opening ceremonies cost $325 more than the ticket for the closing ceremonies. (*Source:* www.vancouver2010.com) Find the price of each ticket.

## Synthesis

*Determine whether the function is even, odd, or neither even nor odd.*

**51.** $f(x) = x\sqrt{10 - x^2}$

**52.** $f(x) = \dfrac{x^2 + 1}{x^3 - 1}$

*Determine whether the graph is symmetric with respect to the x-axis, the y-axis, and the origin.*

**53.** $y^2 + 4xy^2 - y^4 = x^4 - 4x^3 + 3x^2 + 2x^2y^2$

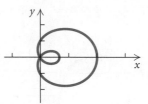

**54.** $(x^2 + y^2)^2 = 2xy$

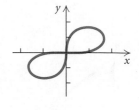

**55.** Show that if $f$ is *any* function, then the function $E$ defined by
$$E(x) = \frac{f(x) + f(-x)}{2}$$
is even.

**56.** Show that if $f$ is *any* function, then the function $O$ defined by
$$O(x) = \frac{f(x) - f(-x)}{2}$$
is odd.

**57.** Consider the functions $E$ and $O$ of Exercises 55 and 56.

**a)** Show that $f(x) = E(x) + O(x)$. This means that every function can be expressed as the sum of an even function and an odd function.

**b)** Let $f(x) = 4x^3 - 11x^2 + \sqrt{x} - 10$. Express $f$ as a sum of an even function and an odd function.

*State whether each of the following is true or false.*

**58.** The product of two odd functions is odd.

**59.** The sum of two even functions is even.

**60.** The product of an even function and an odd function is odd.

## 2.5 Transformations

- Given the graph of a function, graph its transformation under translations, reflections, stretchings, and shrinkings.

### Transformations of Functions

The graphs of some basic functions are shown below. Others can be seen on the inside back cover.

Identity function:
$y = x$

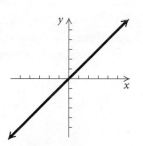

Squaring function:
$y = x^2$

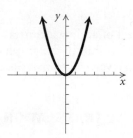

Square root function:
$y = \sqrt{x}$

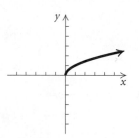

Cubing function:
$y = x^3$

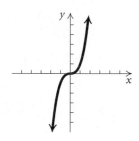

Cube root function:
$y = \sqrt[3]{x}$

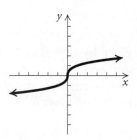

Reciprocal function:
$y = \frac{1}{x}$

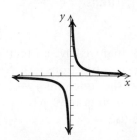

Absolute-value function:
$y = |x|$

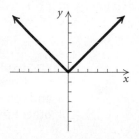

These functions can be considered building blocks for many other functions. We can create graphs of new functions by shifting them horizontally or vertically, stretching or shrinking them, and reflecting them across an axis. We now consider these **transformations**.

### Vertical Translations and Horizontal Translations

Suppose that we have a function given by $y = f(x)$. Let's explore the graphs of the new functions $y = f(x) + b$ and $y = f(x) - b$, for $b > 0$.

Consider the functions $y = \frac{1}{5}x^4$, $y = \frac{1}{5}x^4 + 5$, and $y = \frac{1}{5}x^4 - 3$ and compare their graphs. What pattern do you see? Test it with some other functions.

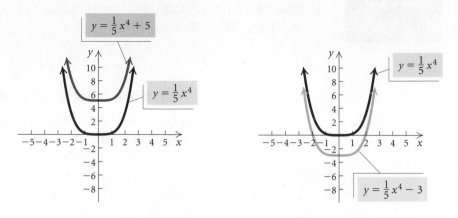

The effect of adding a constant to or subtracting a constant from $f(x)$ in $y = f(x)$ is a shift of the graph of $f(x)$ up or down, respectively. Such a shift is called a **vertical translation**.

> ### VERTICAL TRANSLATION
> For $b > 0$:
>
>   the graph of $y = f(x) + b$ is the graph of $y = f(x)$ shifted *up b* units;
>
>   the graph of $y = f(x) - b$ is the graph of $y = f(x)$ shifted *down b* units.

Suppose that we have a function given by $y = f(x)$. Let's explore the graphs of the new functions $y = f(x - d)$ and $y = f(x + d)$, for $d > 0$.

Consider the functions $y = \frac{1}{5}x^4$, $y = \frac{1}{5}(x - 3)^4$, and $y = \frac{1}{5}(x + 7)^4$ and compare their graphs. What pattern do you observe? Test it with some other functions.

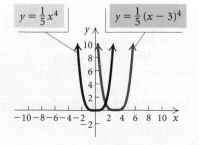
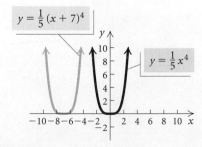

The effect of subtracting a constant from the $x$-value or adding a constant to the $x$-value in $y = f(x)$ is a shift of the graph of $f(x)$ to the right or to the left, respectively. Such a shift is called a **horizontal translation**.

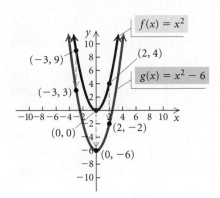

FIGURE 1

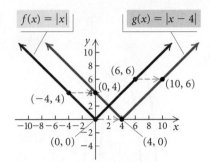

FIGURE 2

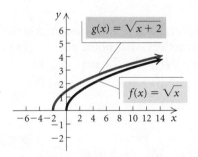

FIGURE 3

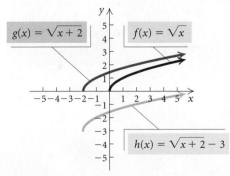

FIGURE 4

## HORIZONTAL TRANSLATION

For $d > 0$:

the graph of $y = f(x - d)$ is the graph of $y = f(x)$ shifted *right* $d$ units;

the graph of $y = f(x + d)$ is the graph of $y = f(x)$ shifted *left* $d$ units.

**EXAMPLE 1** Graph each of the following. Before doing so, describe how each graph can be obtained from one of the basic graphs shown on the preceding pages.

**a)** $g(x) = x^2 - 6$      **b)** $g(x) = |x - 4|$

**c)** $g(x) = \sqrt{x + 2}$      **d)** $h(x) = \sqrt{x + 2} - 3$

*Solution*

**a)** To graph $g(x) = x^2 - 6$, first think of the graph of $f(x) = x^2$. Since $g(x) = f(x) - 6$, the graph of $g(x) = x^2 - 6$ is the graph of $f(x) = x^2$ shifted, or translated, *down* 6 units. (See Fig. 1.)

Let's compare some points on the graphs of $f$ and $g$.

Points on $f$:    $(-3, 9)$,    $(0, 0)$,    $(2, 4)$

Corresponding points on $g$:    $(-3, 3)$,    $(0, -6)$,    $(2, -2)$

We note that the $y$-coordinate of a point on the graph of $g$ is 6 less than the corresponding $y$-coordinate on the graph of $f$.

**b)** To graph $g(x) = |x - 4|$, first think of the graph of $f(x) = |x|$. Since $g(x) = f(x - 4)$, the graph of $g(x) = |x - 4|$ is the graph of $f(x) = |x|$ shifted *right* 4 units. (See Fig. 2.)

Let's again compare points on the two graphs.

Points on $f$    Corresponding points on $g$

$(-4, 4) \longrightarrow (0, 4)$

$(0, 0) \longrightarrow (4, 0)$

$(6, 6) \longrightarrow (10, 6)$

Observing points on $f$ and $g$, we see that the $x$-coordinate of a point on the graph of $g$ is 4 more than the $x$-coordinate of the corresponding point on $f$.

**c)** To graph $g(x) = \sqrt{x + 2}$, first think of the graph of $f(x) = \sqrt{x}$. Since $g(x) = f(x + 2)$, the graph of $g(x) = \sqrt{x + 2}$ is the graph of $f(x) = \sqrt{x}$ shifted *left* 2 units. (See Fig. 3.)

**d)** To graph $h(x) = \sqrt{x + 2} - 3$, first think of the graph of $f(x) = \sqrt{x}$. In part (c), we found that the graph of $g(x) = \sqrt{x + 2}$ is the graph of $f(x) = \sqrt{x}$ shifted *left* 2 units. Since $h(x) = g(x) - 3$, we shift the graph of $g(x) = \sqrt{x + 2}$ *down* 3 units. Together, the graph of $f(x) = \sqrt{x}$ is shifted *left* 2 units and *down* 3 units. (See Fig. 4.)

**Now Try Exercises 1 and 59.**

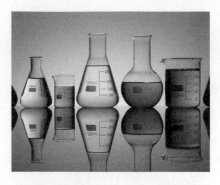

### ◼ Reflections

Suppose that we have a function given by $y = f(x)$. Let's explore the graphs of the new functions $y = -f(x)$ and $y = f(-x)$.

Compare the functions $y = f(x)$ and $y = -f(x)$ by observing the graphs of $y = \frac{1}{5}x^4$ and $y = -\frac{1}{5}x^4$ shown on the left below. What do you see? Test your observation with some other functions $y_1$ and $y_2$, where $y_2 = -y_1$.

Compare the functions $y = f(x)$ and $y = f(-x)$ by observing the graphs of $y = 2x^3 - x^4 + 5$ and $y = 2(-x)^3 - (-x)^4 + 5$ shown on the right below. What do you see? Test your observation with some other functions in which $x$ is replaced with $-x$.

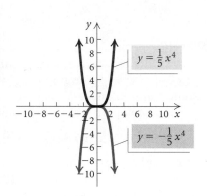

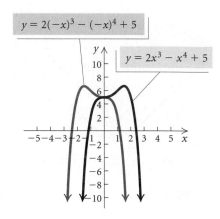

Given the graph of $y = f(x)$, we can reflect each point *across the x-axis* to obtain the graph of $y = -f(x)$. We can reflect each point of $y = f(x)$ *across the y-axis* to obtain the graph of $y = f(-x)$. The new graphs are called **reflections** of $y = f(x)$.

---

**REFLECTIONS**

The graph of $y = -f(x)$ is the **reflection** of the graph of $y = f(x)$ across the $x$-axis.

The graph of $y = f(-x)$ is the **reflection** of the graph of $y = f(x)$ across the $y$-axis.

If a point $(x, y)$ is on the graph of $y = f(x)$, then $(x, -y)$ is on the graph of $y = -f(x)$, and $(-x, y)$ is on the graph of $y = f(-x)$.

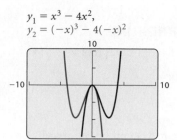

$$y_1 = x^3 - 4x^2,$$
$$y_2 = (-x)^3 - 4(-x)^2$$

FIGURE 1

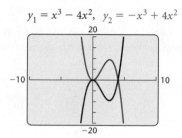

$$y_1 = x^3 - 4x^2, \quad y_2 = -x^3 + 4x^2$$

FIGURE 2

**EXAMPLE 2**   Graph each of the following. Before doing so, describe how each graph can be obtained from the graph of $f(x) = x^3 - 4x^2$.

**a)** $g(x) = (-x)^3 - 4(-x)^2$        **b)** $h(x) = 4x^2 - x^3$

*Solution*

**a)** We first note that

$$f(-x) = (-x)^3 - 4(-x)^2 = g(x).$$

Thus the graph of *g* is a *reflection* of the graph of *f* across the *y*-axis. (See Fig. 1.) If $(x, y)$ is on the graph of *f*, then $(-x, y)$ is on the graph of *g*. For example, $(2, -8)$ is on *f* and $(-2, -8)$ is on *g*.

**b)** We first note that

$$-f(x) = -(x^3 - 4x^2)$$
$$= -x^3 + 4x^2$$
$$= h(x).$$

Thus the graph of *h* is a reflection of the graph of *f* across the *x*-axis. (See Fig. 2.) If $(x, y)$ is on the graph of *f*, then $(x, -y)$ is on the graph of *h*. For example, $(2, -8)$ is on *f* and $(2, 8)$ is on *h*.

## Vertical and Horizontal Stretchings and Shrinkings

Suppose that we have a function given by $y = f(x)$. Let's explore the graphs of the new functions $y = af(x)$ and $y = f(cx)$.

Consider the functions $y = f(x) = x^3 - x, y = \frac{1}{10}(x^3 - x) = \frac{1}{10}f(x)$, $y = 2(x^3 - x) = 2f(x)$, and $y = -2(x^3 - x) = -2f(x)$ and compare their graphs. What pattern do you observe? Test it with some other functions.

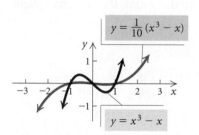

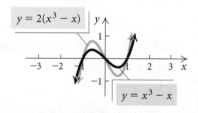

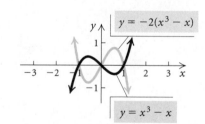

Consider any function *f* given by $y = f(x)$. Multiplying $f(x)$ by any constant *a*, where $|a| > 1$, to obtain $g(x) = af(x)$ will *stretch* the graph vertically away from the *x*-axis. If $0 < |a| < 1$, then the graph will be flattened or *shrunk* vertically toward the *x*-axis. If $a < 0$, the graph is also reflected across the *x*-axis.

## VERTICAL STRETCHING AND SHRINKING

The graph of $y = af(x)$ can be obtained from the graph of $y = f(x)$ by

stretching vertically for $|a| > 1$, or
shrinking vertically for $0 < |a| < 1$.

For $a < 0$, the graph is also reflected across the $x$-axis.
(The $y$-coordinates of the graph of $y = af(x)$ can be obtained by multiplying the $y$-coordinates of $y = f(x)$ by $a$.)

Consider the functions $y = f(x) = x^3 - x$, $y = (2x)^3 - (2x) = f(2x)$, $y = \left(\frac{1}{2}x\right)^3 - \left(\frac{1}{2}x\right) = f\left(\frac{1}{2}x\right)$, and $y = \left(-\frac{1}{2}x\right)^3 - \left(-\frac{1}{2}x\right) = f\left(-\frac{1}{2}x\right)$ and compare their graphs. What pattern do you observe? Test it with some other functions.

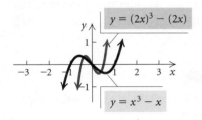 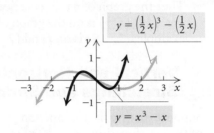 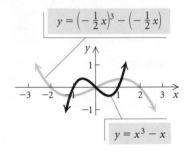

The constant $c$ in the equation $g(x) = f(cx)$ will *shrink* the graph of $y = f(x)$ horizontally toward the $y$-axis if $|c| > 1$. If $0 < |c| < 1$, the graph will be *stretched* horizontally away from the $y$-axis. If $c < 0$, the graph is also reflected across the $y$-axis.

## HORIZONTAL STRETCHING AND SHRINKING

The graph of $y = f(cx)$ can be obtained from the graph of $y = f(x)$ by

shrinking horizontally for $|c| > 1$, or
stretching horizontally for $0 < |c| < 1$.

For $c < 0$, the graph is also reflected across the $y$-axis.
(The $x$-coordinates of the graph of $y = f(cx)$ can be obtained by dividing the $x$-coordinates of the graph of $y = f(x)$ by $c$.)

It is instructive to use these concepts to create transformations of a given graph.

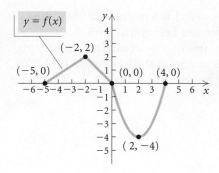

**EXAMPLE 3** Shown at left is a graph of $y = f(x)$ for some function $f$. No formula for $f$ is given. Graph each of the following.

**a)** $g(x) = 2f(x)$        **b)** $h(x) = \frac{1}{2}f(x)$

**c)** $r(x) = f(2x)$        **d)** $s(x) = f\left(\frac{1}{2}x\right)$

**e)** $t(x) = f\left(-\frac{1}{2}x\right)$

*Solution*

**a)** Since $|2| > 1$, the graph of $g(x) = 2f(x)$ is a vertical stretching of the graph of $y = f(x)$ by a factor of 2. We can consider the key points $(-5, 0)$, $(-2, 2)$, $(0, 0)$, $(2, -4)$, and $(4, 0)$ on the graph of $y = f(x)$. The transformation multiplies each $y$-coordinate by 2 to obtain the key points $(-5, 0)$, $(-2, 4)$, $(0, 0)$, $(2, -8)$, and $(4, 0)$ on the graph of $g(x) = 2f(x)$. The graph is shown below.

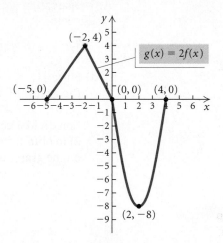

**b)** Since $\left|\frac{1}{2}\right| < 1$, the graph of $h(x) = \frac{1}{2}f(x)$ is a vertical shrinking of the graph of $y = f(x)$ by a factor of $\frac{1}{2}$. We again consider the key points $(-5, 0)$, $(-2, 2)$, $(0, 0)$, $(2, -4)$, and $(4, 0)$ on the graph of $y = f(x)$. The transformation multiplies each $y$-coordinate by $\frac{1}{2}$ to obtain the key points $(-5, 0)$, $(-2, 1)$, $(0, 0)$, $(2, -2)$, and $(4, 0)$ on the graph of $h(x) = \frac{1}{2}f(x)$. The graph is shown below.

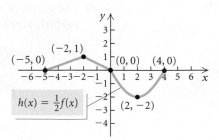

c) Since $|2| > 1$, the graph of $r(x) = f(2x)$ is a horizontal shrinking of the graph of $y = f(x)$. We consider the key points $(-5, 0)$, $(-2, 2)$, $(0, 0)$, $(2, -4)$, and $(4, 0)$ on the graph of $y = f(x)$. The transformation divides each $x$-coordinate by 2 to obtain the key points $(-2.5, 0)$, $(-1, 2)$, $(0, 0)$, $(1, -4)$, and $(2, 0)$ on the graph of $r(x) = f(2x)$. The graph is shown below.

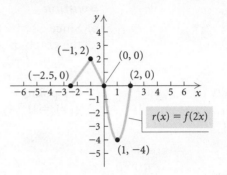

d) Since $\left|\frac{1}{2}\right| < 1$, the graph of $s(x) = f\left(\frac{1}{2}x\right)$ is a horizontal stretching of the graph of $y = f(x)$. We consider the key points $(-5, 0)$, $(-2, 2)$, $(0, 0)$, $(2, -4)$, and $(4, 0)$ on the graph of $y = f(x)$. The transformation divides each $x$-coordinate by $\frac{1}{2}$ (which is the same as multiplying by 2) to obtain the key points $(-10, 0)$, $(-4, 2)$, $(0, 0)$, $(4, -4)$, and $(8, 0)$ on the graph of $s(x) = f\left(\frac{1}{2}x\right)$. The graph is shown below.

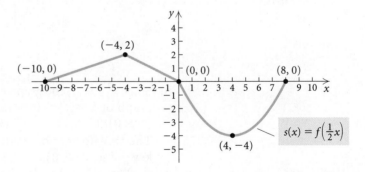

e) The graph of $t(x) = f\left(-\frac{1}{2}x\right)$ can be obtained by reflecting the graph in part (d) across the $y$-axis.

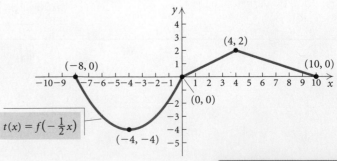

**Now Try Exercises 59 and 61.**

**EXAMPLE 4** Use the graph of $y = f(x)$ shown at left to graph
$y = -2f(x - 3) + 1$.

*Solution*

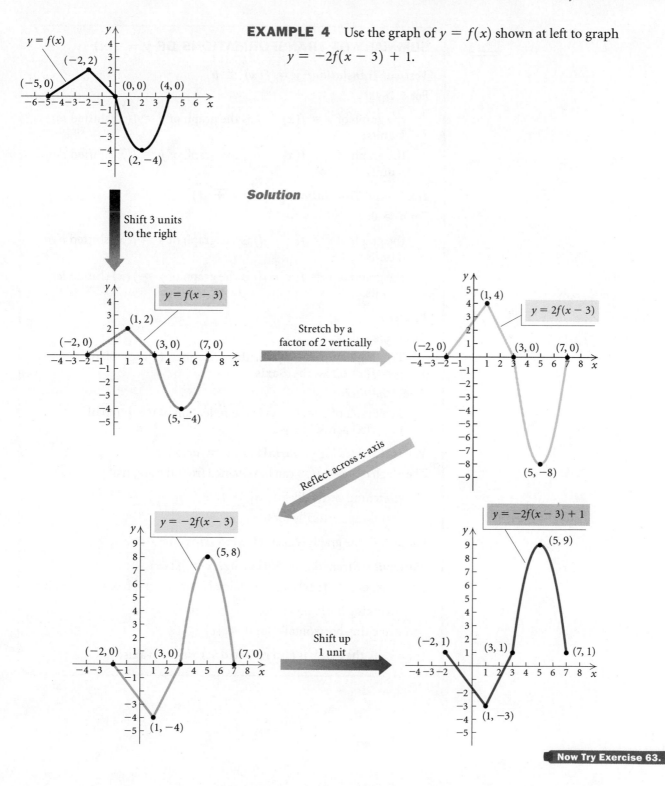

Now Try Exercise 63.

## SUMMARY OF TRANSFORMATIONS OF $y = f(x)$

***Vertical Translation:*** $y = f(x) \pm b$

For $b > 0$:

the graph of $y = f(x) + b$ is the graph of $y = f(x)$ shifted *up* $b$ units;

the graph of $y = f(x) - b$ is the graph of $y = f(x)$ shifted *down* $b$ units.

***Horizontal Translation:*** $y = f(x \mp d)$

For $d > 0$:

the graph of $y = f(x - d)$ is the graph of $y = f(x)$ shifted *right* $d$ units;

the graph of $y = f(x + d)$ is the graph of $y = f(x)$ shifted *left* $d$ units.

***Reflections***

*Across the x-axis*:

The graph of $y = -f(x)$ is the reflection of the graph of $y = f(x)$ across the $x$-axis.

*Across the y-axis*:

The graph of $y = f(-x)$ is the reflection of the graph of $y = f(x)$ across the $y$-axis.

***Vertical Stretching or Shrinking:*** $y = af(x)$

The graph of $y = af(x)$ can be obtained from the graph of $y = f(x)$ by

stretching vertically for $|a| > 1$, or

shrinking vertically for $0 < |a| < 1$.

For $a < 0$, the graph is also reflected across the $x$-axis.

***Horizontal Stretching or Shrinking:*** $y = f(cx)$

The graph of $y = f(cx)$ can be obtained from the graph of $y = f(x)$ by

shrinking horizontally for $|c| > 1$, or

stretching horizontally for $0 < |c| < 1$.

For $c < 0$, the graph is also reflected across the $y$-axis.

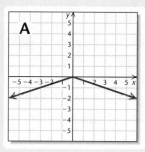

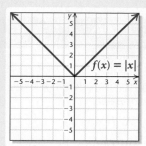

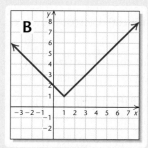

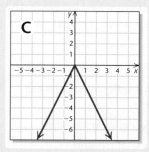

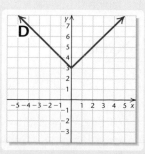

# Visualizing the Graph

Match the function with its graph. Use transformation graphing techniques to obtain the graph of $g$ from the basic function $f(x) = |x|$ shown at top left.

**1.** $g(x) = -2|x|$

**2.** $g(x) = |x - 1| + 1$

**3.** $g(x) = -\left|\dfrac{1}{3}x\right|$

**4.** $g(x) = |2x|$

**5.** $g(x) = |x + 2|$

**6.** $g(x) = |x| + 3$

**7.** $g(x) = -\dfrac{1}{2}|x - 4|$

**8.** $g(x) = \dfrac{1}{2}|x| - 3$

**9.** $g(x) = -|x| - 2$

**Answers on page A-14**

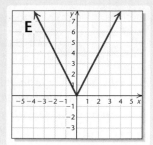

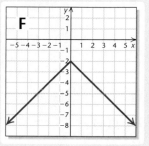

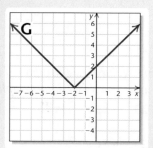

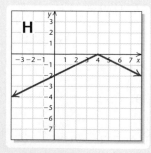

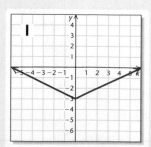

## 2.5  Exercise Set

*Describe how the graph of the function can be obtained from one of the basic graphs on p. 199. Then graph the function by hand or with a graphing calculator.*

**1.** $f(x) = (x - 3)^2$

**2.** $g(x) = x^2 + \frac{1}{2}$

**3.** $g(x) = x - 3$

**4.** $g(x) = -x - 2$

**5.** $h(x) = -\sqrt{x}$

**6.** $g(x) = \sqrt{x - 1}$

**7.** $h(x) = \frac{1}{x} + 4$

**8.** $g(x) = \frac{1}{x - 2}$

**9.** $h(x) = -3x + 3$

**10.** $f(x) = 2x + 1$

**11.** $h(x) = \frac{1}{2}|x| - 2$

**12.** $g(x) = -|x| + 2$

**13.** $g(x) = -(x - 2)^3$

**14.** $f(x) = (x + 1)^3$

**15.** $g(x) = (x + 1)^2 - 1$

**16.** $h(x) = -x^2 - 4$

**17.** $g(x) = \frac{1}{3}x^3 + 2$

**18.** $h(x) = (-x)^3$

**19.** $f(x) = \sqrt{x + 2}$

**20.** $f(x) = -\frac{1}{2}\sqrt{x - 1}$

**21.** $f(x) = \sqrt[3]{x} - 2$

**22.** $h(x) = \sqrt[3]{x + 1}$

*Describe how the graph of the function can be obtained from one of the basic graphs on p. 199.*

**23.** $g(x) = |3x|$

**24.** $f(x) = \frac{1}{2}\sqrt[3]{x}$

**25.** $h(x) = \frac{2}{x}$

**26.** $f(x) = |x - 3| - 4$

**27.** $f(x) = 3\sqrt{x} - 5$

**28.** $f(x) = 5 - \frac{1}{x}$

**29.** $g(x) = \left|\frac{1}{3}x\right| - 4$

**30.** $f(x) = \frac{2}{3}x^3 - 4$

**31.** $f(x) = -\frac{1}{4}(x - 5)^2$

**32.** $f(x) = (-x)^3 - 5$

**33.** $f(x) = \frac{1}{x + 3} + 2$

**34.** $g(x) = \sqrt{-x} + 5$

**35.** $h(x) = -(x - 3)^2 + 5$

**36.** $f(x) = 3(x + 4)^2 - 3$

*The point $(-12, 4)$ is on the graph of $y = f(x)$. Find the corresponding point on the graph of $y = g(x)$.*

**37.** $g(x) = \frac{1}{2}f(x)$

**38.** $g(x) = f(x - 2)$

**39.** $g(x) = f(-x)$

**40.** $g(x) = f(4x)$

**41.** $g(x) = f(x) - 2$

**42.** $g(x) = f\left(\frac{1}{2}x\right)$

**43.** $g(x) = 4f(x)$

**44.** $g(x) = -f(x)$

*Given that $f(x) = x^2 + 3$, match the function g with a transformation of f from one of A–D.*

**45.** $g(x) = x^2 + 4$  **A.** $f(x - 2)$

**46.** $g(x) = 9x^2 + 3$  **B.** $f(x) + 1$

**47.** $g(x) = (x - 2)^2 + 3$  **C.** $2f(x)$

**48.** $g(x) = 2x^2 + 6$  **D.** $f(3x)$

*Write an equation for a function that has a graph with the given characteristics.*

**49.** The shape of $y = x^2$, but reflected across the $x$-axis and shifted right 8 units

**50.** The shape of $y = \sqrt{x}$, but shifted left 6 units and down 5 units

**51.** The shape of $y = |x|$, but shifted left 7 units and up 2 units

**52.** The shape of $y = x^3$, but reflected across the $x$-axis and shifted right 5 units

**53.** The shape of $y = 1/x$, but shrunk horizontally by a factor of 2 and shifted down 3 units

**54.** The shape of $y = x^2$, but shifted right 6 units and up 2 units

**55.** The shape of $y = x^2$, but reflected across the $x$-axis and shifted right 3 units and up 4 units

**56.** The shape of $y = |x|$, but stretched horizontally by a factor of 2 and shifted down 5 units

**57.** The shape of $y = \sqrt{x}$, but reflected across the $y$-axis and shifted left 2 units and down 1 unit

**58.** The shape of $y = 1/x$, but reflected across the $x$-axis and shifted up 1 unit

*A graph of $y = f(x)$ follows. No formula for f is given. In Exercises 59–66, graph the given equation.*

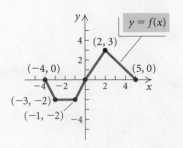

**59.** $g(x) = -2f(x)$  **60.** $g(x) = \frac{1}{2}f(x)$

**61.** $g(x) = f\left(-\frac{1}{2}x\right)$  **62.** $g(x) = f(2x)$

**63.** $g(x) = -\frac{1}{2}f(x - 1) + 3$

**64.** $g(x) = -3f(x + 1) - 4$

**65.** $g(x) = f(-x)$

**66.** $g(x) = -f(x)$

*A graph of* $y = g(x)$ *follows. No formula for g is given. In Exercises 67–70, graph the given equation.*

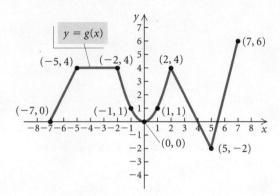

**67.** $h(x) = -g(x + 2) + 1$

**68.** $h(x) = \frac{1}{2}g(-x)$  **69.** $h(x) = g(2x)$

**70.** $h(x) = 2g(x - 1) - 3$

*The graph of the function f is shown in figure (a). In Exercises 71–78, match the function g with one of the graphs (a)–(h), which follow. Some graphs may be used more than once and some may not be used at all.*

**a)**

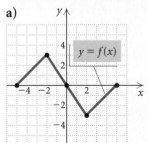

**b)**

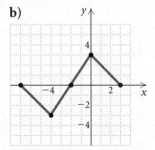

**c)**

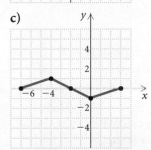

**d)**

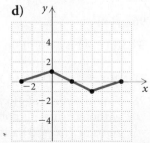

**e)**

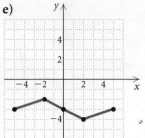

**f)**

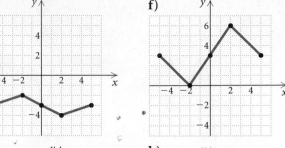

**g)**

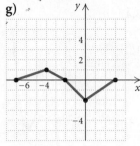

**h)**

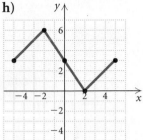

**71.** $g(x) = f(-x) + 3$

**72.** $g(x) = f(x) + 3$

**73.** $g(x) = -f(x) + 3$

**74.** $g(x) = -f(-x)$

**75.** $g(x) = \frac{1}{3}f(x - 2)$

**76.** $g(x) = \frac{1}{3}f(x) - 3$

**77.** $g(x) = \frac{1}{3}f(x + 2)$

**78.** $g(x) = -f(x + 2)$

*For each pair of functions, determine if* $g(x) = f(-x)$.

**79.** $f(x) = 2x^4 - 35x^3 + 3x - 5$,
$g(x) = 2x^4 + 35x^3 - 3x - 5$

**80.** $f(x) = \frac{1}{4}x^4 + \frac{1}{5}x^3 - 81x^2 - 17$,
$g(x) = \frac{1}{4}x^4 + \frac{1}{5}x^3 + 81x^2 - 17$

*A graph of the function* $f(x) = x^3 - 3x^2$ *is shown below. Exercises 81–84 show graphs of functions transformed from this one. Find a formula for each function.*

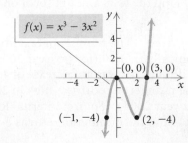

**81.**

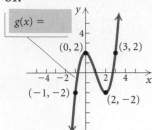

**82.**

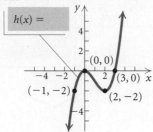

**83.**

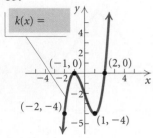

**84.**

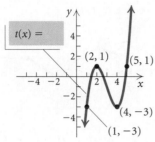

## Skill Maintenance

*Determine algebraically whether the graph is symmetric with respect to the x-axis, the y-axis, and the origin.*

**85.** $y = 3x^4 - 3$

**86.** $y^2 = x$

**87.** $2x - 5y = 0$

**88.** *Video Game Sales.*    Sales of the video game Wii Fit totaled 3.5 million games in the first eleven months of 2009. This was 1 million less than three times the number of Madden NFL 10 games sold during the same period. (*Source*: The PPB Group) Find the number of Madden games sold.

**89.** *Gift Cards.*    It is estimated that about $5 billion in gift cards given as Christmas gifts in 2009 went unspent. This is about 6% of the total amount spent on gift cards. (*Source*: Tower Group) Find the total amount spent on gift cards.

**90.** *e-filing Taxes.*    The number of tax returns filed electronically in 2010 (for tax year 2009) was 98.3 million. This was an increase of 43.9% over the number of returns e-filed in 2005 (for tax year 2004). (*Source*: Internal Revenue Service) Find the number of returns e-filed in 2005.

## Synthesis

*Use the graph of the function f shown below in Exercises 91 and 92.*

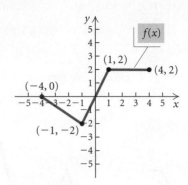

**91.** Graph: $y = |f(x)|$.

**92.** Graph: $y = f(|x|)$.

*Use the graph of the function g shown below in Exercises 93 and 94.*

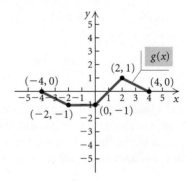

**93.** Graph: $y = g(|x|)$.

**94.** Graph: $y = |g(x)|$.

*Graph each of the following using a graphing calculator. Before doing so, describe how the graph can be obtained from a more basic graph. Give the domain and the range of the function.*

**95.** $f(x) = [\![ x - \frac{1}{2} ]\!]$

**96.** $f(x) = |\sqrt{x} - 1|$

**97.** If $(3, 4)$ is a point on the graph of $y = f(x)$, what point do you know is on the graph of $y = 2f(x)$? of $y = 2 + f(x)$? of $y = f(2x)$?

**98.** Find the zeros of $f(x) = 3x^5 - 20x^3$. Then, without using a graphing calculator, state the zeros of $f(x - 3)$ and $f(x + 8)$.

## 2.6 Variation and Applications

- Find equations of direct variation, inverse variation, and combined variation given values of the variables.
- Solve applied problems involving variation.

We now extend our study of formulas and functions by considering applications involving variation.

### Direct Variation

Suppose a veterinary assistant earns $10 per hour. In 1 hr, $10 is earned; in 2 hr, $20 is earned; in 3 hr, $30 is earned; and so on. This gives rise to a set of ordered pairs:

$$(1, 10), \quad (2, 20), \quad (3, 30), \quad (4, 40), \quad \text{and so on.}$$

**Veterinary Assistant's Earnings**

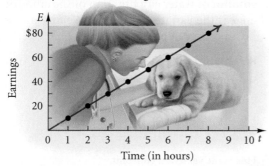

Note that the ratio of the second coordinate to the first coordinate is the same number for each pair:

$$\frac{10}{1} = 10, \qquad \frac{20}{2} = 10, \qquad \frac{30}{3} = 10, \qquad \frac{40}{4} = 10, \quad \text{and so on.}$$

Whenever a situation produces pairs of numbers in which the *ratio is constant*, we say that there is **direct variation**. Here the amount earned $E$ varies directly as the time worked $t$:

$$\frac{E}{t} = 10 \text{ (a constant)}, \quad \text{or} \quad E = 10t,$$

or, using function notation, $E(t) = 10t$. This equation is an equation of **direct variation**. The coefficient, 10, is called the **variation constant**. In this case, it is the rate of change of earnings with respect to time.

The graph of $y = kx$, $k > 0$, always goes through the origin and rises from left to right. Note that as $x$ increases, $y$ increases; that is, the function is increasing on the interval $(0, \infty)$. The constant $k$ is also the slope of the line.

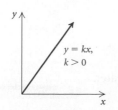

### DIRECT VARIATION

If a situation gives rise to a linear function $f(x) = kx$, or $y = kx$, where $k$ is a positive constant, we say that we have **direct variation**, or that **$y$ varies directly as $x$**, or that **$y$ is directly proportional to $x$**. The number $k$ is called the **variation constant**, or the **constant of proportionality**.

**EXAMPLE 1**　Find the variation constant and an equation of variation in which $y$ varies directly as $x$, and $y = 32$ when $x = 2$.

***Solution***　We know that $(2, 32)$ is a solution of $y = kx$. Thus,

$$y = kx$$
$$32 = k \cdot 2 \qquad \textbf{Substituting}$$
$$\frac{32}{2} = k \qquad \textbf{Solving for } k$$
$$16 = k. \qquad \textbf{Simplifying}$$

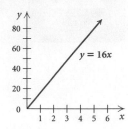

The variation constant, 16, is the rate of change of $y$ with respect to $x$. The equation of variation is $y = 16x$.

> **Now Try Exercise 1.**

**EXAMPLE 2**　*Water from Melting Snow.*　The number of centimeters of water $W$ produced from melting snow varies directly as $S$, the number of centimeters of snow. Meteorologists have found that under certain conditions 150 cm of snow will melt to 16.8 cm of water. To how many centimeters of water will 200 cm of snow melt under the same conditions?

***Solution***　We can express the amount of water as a function of the amount of snow. Thus, $W(S) = kS$, where $k$ is the variation constant. We first find $k$ using the given data and then find an equation of variation:

$$W(S) = kS \qquad \textbf{\textit{W} varies directly as \textit{S}.}$$
$$W(150) = k \cdot 150 \qquad \textbf{Substituting 150 for \textit{S}}$$
$$16.8 = k \cdot 150 \qquad \textbf{Replacing \textit{W}(150) with 16.8}$$
$$\frac{16.8}{150} = k \qquad \textbf{Solving for \textit{k}}$$
$$0.112 = k. \qquad \textbf{This is the variation constant.}$$

The equation of variation is $W(S) = 0.112S$.

Next, we use the equation to find how many centimeters of water will result from melting 200 cm of snow:

$$W(S) = 0.112S$$
$$W(200) = 0.112(200) \qquad \textbf{Substituting}$$
$$W = 22.4.$$

Thus, 200 cm of snow will melt to 22.4 cm of water.

> **Now Try Exercise 17.**

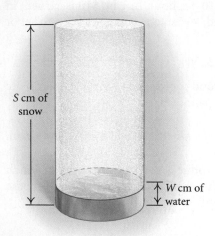

*S* cm of snow

*W* cm of water

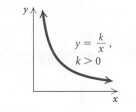

Time (in hours) / Speed (in miles per hour)

(5, 4)
(10, 2)
(20, 1)
$\left(40, \frac{1}{2}\right)$

## ■ Inverse Variation

Suppose a bus is traveling a distance of 20 mi. At a speed of 5 mph, the trip will take 4 hr; at 10 mph, it will take 2 hr; at 20 mph, it will take 1 hr; at 40 mph, it will take $\frac{1}{2}$ hr; and so on. We plot this information on a graph, using speed as the first coordinate and time as the second coordinate to determine a set of ordered pairs:

$$(5, 4), \quad (10, 2), \quad (20, 1), \quad \left(40, \tfrac{1}{2}\right), \quad \text{and so on.}$$

Note that the products of the coordinates are all the same number:

$$5 \cdot 4 = 20, \quad 10 \cdot 2 = 20, \quad 20 \cdot 1 = 20, \quad 40 \cdot \tfrac{1}{2} = 20, \quad \text{and so on.}$$

Whenever a situation produces pairs of numbers in which the *product is constant*, we say that there is **inverse variation**. Here the time varies inversely as the speed, or rate:

$$rt = 20 \, (\text{a constant}), \quad \text{or} \quad t = \frac{20}{r},$$

or, using function notation, $t(r) = 20/r$. This equation is an equation of **inverse variation**. The coefficient, 20, is called the **variation constant**. Note that as the first number increases, the second number decreases.

The graph of $y = k/x$, $k > 0$, is like the one shown below. Note that as $x$ increases, $y$ decreases; that is, the function is decreasing on the interval $(0, \infty)$.

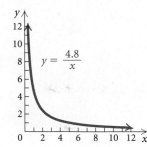

$y = \dfrac{k}{x}$, $k > 0$

### INVERSE VARIATION

If a situation gives rise to a function $f(x) = k/x$, or $y = k/x$, where $k$ is a positive constant, we say that we have **inverse variation**, or that $y$ **varies inversely as** $x$, or that $y$ **is inversely proportional to** $x$. The number $k$ is called the **variation constant**, or the **constant of proportionality**.

**EXAMPLE 3** Find the variation constant and an equation of variation in which $y$ varies inversely as $x$, and $y = 16$ when $x = 0.3$.

*Solution* We know that $(0.3, 16)$ is a solution of $y = k/x$. We substitute:

$$y = \frac{k}{x}$$

$$16 = \frac{k}{0.3} \qquad \text{Substituting}$$

$$(0.3)16 = k \qquad \text{Solving for } k$$

$$4.8 = k.$$

$y = \dfrac{4.8}{x}$

The variation constant is 4.8. The equation of variation is $y = 4.8/x$.

**Now Try Exercise 3.**

There are many real-world problems that translate to an equation of inverse variation.

**EXAMPLE 4**   *Filling a Swimming Pool.*   The time $t$ required to fill a swimming pool varies inversely as the rate of flow $r$ of water into the pool. A tank truck can fill a pool in 90 min at a rate of 1500 L/min. How long would it take to fill the pool at a rate of 1800 L/min?

**Solution**   We can express the amount of time required as a function of the rate of flow. Thus we have $t(r) = k/r$. We first find $k$ using the given information and then find an equation of variation:

$$t(r) = \frac{k}{r} \qquad \text{\textit{t} varies inversely as \textit{r}.}$$

$$t(1500) = \frac{k}{1500} \qquad \textbf{Substituting 1500 for } \textit{r}$$

$$90 = \frac{k}{1500} \qquad \textbf{Replacing } t(1500) \textbf{ with 90}$$

$$90 \cdot 1500 = k \qquad \textbf{Solving for } k$$

$$135{,}000 = k. \qquad \textbf{This is the variation constant.}$$

The equation of variation is

$$t(r) = \frac{135{,}000}{r}.$$

Next, we use the equation to find the time that it would take to fill the pool at a rate of 1800 L/min:

$$t(r) = \frac{135{,}000}{r}$$

$$t(1800) = \frac{135{,}000}{1800} \qquad \textbf{Substituting}$$

$$t = 75.$$

Thus it would take 75 min to fill the pool at a rate of 1800 L/min.

**Now Try Exercise 15.**

## ◾ Combined Variation

We now look at other kinds of variation.

---

$y$ varies **directly as the *n*th power of *x*** if there is some positive constant $k$ such that

$$y = kx^n.$$

$y$ varies **inversely as the *n*th power of *x*** if there is some positive constant $k$ such that

$$y = \frac{k}{x^n}.$$

$y$ varies **jointly as *x* and *z*** if there is some positive constant $k$ such that

$$y = kxz.$$

There are other types of combined variation as well. Consider the formula for the volume of a right circular cylinder, $V = \pi r^2 h$, in which $V$, $r$, and $h$ are variables and $\pi$ is a constant. We say that $V$ varies jointly as $h$ and the square of $r$. In this formula, $\pi$ is the variation constant.

**EXAMPLE 5**   Find an equation of variation in which $y$ varies directly as the square of $x$, and $y = 12$ when $x = 2$.

*Solution*   We write an equation of variation and find $k$:

$$y = kx^2$$
$$12 = k \cdot 2^2 \qquad \text{Substituting}$$
$$12 = k \cdot 4$$
$$3 = k.$$

Thus, $y = 3x^2$.

> **Now Try Exercise 27.**

**EXAMPLE 6**   Find an equation of variation in which $y$ varies jointly as $x$ and $z$, and $y = 42$ when $x = 2$ and $z = 3$.

*Solution*   We have

$$y = kxz$$
$$42 = k \cdot 2 \cdot 3 \qquad \text{Substituting}$$
$$42 = k \cdot 6$$
$$7 = k.$$

Thus, $y = 7xz$.

> **Now Try Exercise 29.**

**EXAMPLE 7**   Find an equation of variation in which $y$ varies jointly as $x$ and $z$ and inversely as the square of $w$, and $y = 105$ when $x = 3$, $z = 20$, and $w = 2$.

*Solution*   We have

$$y = k \cdot \frac{xz}{w^2}$$
$$105 = k \cdot \frac{3 \cdot 20}{2^2} \qquad \text{Substituting}$$
$$105 = k \cdot 15$$
$$7 = k.$$

Thus, $y = 7\dfrac{xz}{w^2}$, or $y = \dfrac{7xz}{w^2}$.

> **Now Try Exercise 33.**

Many applied problems can be modeled using equations of combined variation.

**EXAMPLE 8**   *Volume of a Tree.*   The volume of wood $V$ in a tree varies jointly as the height $h$ and the square of the girth $g$. (Girth is distance around.) If the volume of a redwood tree is 216 m³ when the height is 30 m and the girth

is 1.5 m, what is the height of a tree whose volume is 960 m³ and whose girth is 2 m?

**Solution**    We first find $k$ using the first set of data. Then we solve for $h$ using the second set of data.

$$V = khg^2$$
$$216 = k \cdot 30 \cdot 1.5^2$$
$$216 = k \cdot 30 \cdot 2.25$$
$$216 = k \cdot 67.5$$
$$3.2 = k$$

Then the equation of variation is $V = 3.2hg^2$. We substitute the second set of data into the equation:

$$960 = 3.2 \cdot h \cdot 2^2$$
$$960 = 3.2 \cdot h \cdot 4$$
$$960 = 12.8 \cdot h$$
$$75 = h.$$

The height of the tree is 75 m.

**Now Try Exercise 35.**

---

## 2.6    Exercise Set

*Find the variation constant and an equation of variation for the given situation.*

1. $y$ varies directly as $x$, and $y = 54$ when $x = 12$

2. $y$ varies directly as $x$, and $y = 0.1$ when $x = 0.2$

3. $y$ varies inversely as $x$, and $y = 3$ when $x = 12$

4. $y$ varies inversely as $x$, and $y = 12$ when $x = 5$

5. $y$ varies directly as $x$, and $y = 1$ when $x = \frac{1}{4}$

6. $y$ varies inversely as $x$, and $y = 0.1$ when $x = 0.5$

7. $y$ varies inversely as $x$, and $y = 32$ when $x = \frac{1}{8}$

8. $y$ varies directly as $x$, and $y = 3$ when $x = 33$

9. $y$ varies directly as $x$, and $y = \frac{3}{4}$ when $x = 2$

10. $y$ varies inversely as $x$, and $y = \frac{1}{5}$ when $x = 35$

11. $y$ varies inversely as $x$, and $y = 1.8$ when $x = 0.3$

12. $y$ varies directly as $x$, and $y = 0.9$ when $x = 0.4$

13. *Sales Tax.*    The amount of sales tax paid on a product is directly proportional to its purchase price. In Indiana, the sales tax on a Sony Reader that sells for $260 is $17.50. What is the sales tax on an e-book that sells for $21?

14. *Child's Allowance.*    The Gemmers decide to give their children a weekly allowance that is directly proportional to each child's age. Their 6-year-old daughter receives an allowance of $4.50. What is their 11-year-old son's allowance?

15. *Beam Weight.*    The weight $W$ that a horizontal beam can support varies inversely as the length $L$ of the beam. Suppose an 8-m beam can support

1200 kg. How many kilograms can a 14-m beam support?

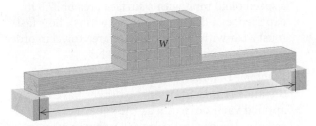

**16.** *Rate of Travel.* The time $t$ required to drive a fixed distance varies inversely as the speed $r$. It takes 5 hr at a speed of 80 km/h to drive a fixed distance. How long will it take to drive the same distance at a speed of 70 km/h?

**17.** *Fat Intake.* The maximum number of grams of fat that should be in a diet varies directly as a person's weight. A person weighing 120 lb should have no more than 60 g of fat per day. What is the maximum daily fat intake for a person weighing 180 lb?

**18.** *U.S. House of Representatives.* The number of representatives $N$ that each state has varies directly as the number of people $P$ living in the state. If New York, with 19,254,630 residents, has 29 representatives, how many representatives does Colorado, with a population of 4,665,177, have?

**19.** *Work Rate.* The time $T$ required to do a job varies inversely as the number of people $P$ working. It takes 5 hr for 7 bricklayers to build a park wall. How long will it take 10 bricklayers to complete the job?

**20.** *Pumping Rate.* The time $t$ required to empty a tank varies inversely as the rate $r$ of pumping. If a pump can empty a tank in 45 min at the rate of 600 kL/min, how long will it take the pump to empty the same tank at the rate of 1000 kL/min?

**21.** *Hooke's Law.* Hooke's law states that the distance $d$ that a spring will stretch varies directly as the mass $m$ of an object hanging from the spring. If a 3-kg mass stretches a spring 40 cm, how far will a 5-kg mass stretch the spring?

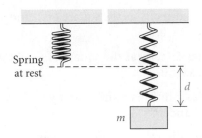

**22.** *Relative Aperture.* The relative aperture, or f-stop, of a 23.5-mm diameter lens is directly proportional to the focal length $F$ of the lens. If a 150-mm focal length has an f-stop of 6.3, find the f-stop of a 23.5-mm diameter lens with a focal length of 80 mm.

**23.** *Musical Pitch.* The pitch $P$ of a musical tone varies inversely as its wavelength $W$. One tone has a pitch of 330 vibrations per second and a wavelength of 3.2 ft. Find the wavelength of another tone that has a pitch of 550 vibrations per second.

**24.** *Weight on Mars.* The weight $M$ of an object on Mars varies directly as its weight $E$ on Earth. A person who weighs 95 lb on Earth weighs 38 lb on Mars. How much would a 100-lb person weigh on Mars?

*Find an equation of variation for the given situation.*

**25.** $y$ varies inversely as the square of $x$, and $y = 0.15$ when $x = 0.1$

**26.** $y$ varies inversely as the square of $x$, and $y = 6$ when $x = 3$

**27.** $y$ varies directly as the square of $x$, and $y = 0.15$ when $x = 0.1$

**28.** $y$ varies directly as the square of $x$, and $y = 6$ when $x = 3$

**29.** $y$ varies jointly as $x$ and $z$, and $y = 56$ when $x = 7$ and $z = 8$

**30.** $y$ varies directly as $x$ and inversely as $z$, and $y = 4$ when $x = 12$ and $z = 15$

**31.** $y$ varies jointly as $x$ and the square of $z$, and $y = 105$ when $x = 14$ and $z = 5$

**32.** $y$ varies jointly as $x$ and $z$ and inversely as $w$, and $y = \frac{3}{2}$ when $x = 2$, $z = 3$, and $w = 4$

**33.** $y$ varies jointly as $x$ and $z$ and inversely as the product of $w$ and $p$, and $y = \frac{3}{28}$ when $x = 3$, $z = 10$, $w = 7$, and $p = 8$

**34.** $y$ varies jointly as $x$ and $z$ and inversely as the square of $w$, and $y = \frac{12}{5}$ when $x = 16$, $z = 3$, and $w = 5$

**35.** *Intensity of Light.* The intensity $I$ of light from a light bulb varies inversely as the square of the distance $d$ from the bulb. Suppose that $I$ is 90 W/m$^2$ (watts per square meter) when the distance is 5 m. How much *farther* would it be to a point where the intensity is 40 W/m$^2$?

**36.** *Atmospheric Drag.* Wind resistance, or atmospheric drag, tends to slow down moving objects. Atmospheric drag varies jointly as an object's surface area $A$ and velocity $v$. If a car traveling at a speed of 40 mph with a surface area of 37.8 ft$^2$ experiences a drag of 222 N (Newtons), how fast must a car with 51 ft$^2$ of surface area travel in order to experience a drag force of 430 N?

**37.** *Stopping Distance of a Car.* The stopping distance $d$ of a car after the brakes have been applied varies directly as the square of the speed $r$. If a car traveling 60 mph can stop in 200 ft, how fast can a car travel and still stop in 72 ft?

**38.** *Weight of an Astronaut.* The weight $W$ of an object varies inversely as the square of the distance $d$ from the center of the earth. At sea level (3978 mi from the center of the earth), an astronaut weighs 220 lb. Find his weight when he is 200 mi above the surface of the earth.

**39.** *Earned-Run Average.* A pitcher's earned-run average $E$ varies directly as the number $R$ of earned runs allowed and inversely as the number $I$ of innings pitched. In 2010, Bronson Arroyo of the Cincinnati Reds had an earned-run average of 3.89. He gave up 93 earned runs in 215.2 innings. How many earned runs would he have given up had he pitched 238 innings with the same average? Round to the nearest whole number.

**40.** *Boyle's Law.* The volume $V$ of a given mass of a gas varies directly as the temperature $T$ and inversely as the pressure $P$. If $V = 231$ cm$^3$ when $T = 42°$ and $P = 20$ kg/cm$^2$, what is the volume when $T = 30°$ and $P = 15$ kg/cm$^2$?

## Skill Maintenance

### Vocabulary Review

*In each of Exercises 41–45, fill in the blank with the correct term. Some of the given choices will not be used.*

| | |
|---|---|
| even function | relative maximum |
| odd function | relative minimum |
| constant function | solution |
| composite function | zero |
| direct variation | perpendicular |
| inverse variation | parallel |

**41.** Nonvertical lines are _____ if and only if they have the same slope and different $y$-intercepts.

**42.** An input $c$ of a function $f$ is a(n) _____ of the function if $f(c) = 0$.

**43.** For a function $f$ for which $f(c)$ exists, $f(c)$ is a(n) _____ if $f(c)$ is the lowest point in some open interval.

**44.** If the graph of a function is symmetric with respect to the origin, then $f$ is a(n) _____.

**45.** An equation $y = k/x$ is an equation of _____.

## Synthesis

**46.** In each of the following equations, state whether $y$ varies directly as $x$, inversely as $x$, or neither directly nor inversely as $x$.

**a)** $7xy = 14$
**b)** $x - 2y = 12$
**c)** $-2x + 3y = 0$
**d)** $x = \frac{3}{4}y$
**e)** $\dfrac{x}{y} = 2$

**47.** *Volume and Cost.* An 18-oz jar of peanut butter in the shape of a right circular cylinder is 5 in. high and 3 in. in diameter and sells for $2.89. In the same store, a 28-oz jar of the same brand is $5\frac{1}{2}$ in. high and $3\frac{1}{4}$ in. in diameter. If the cost is directly proportional to volume, what should the price of the larger jar be? If the cost is directly proportional to weight, what should the price of the larger jar be?

**48.** Describe in words the variation given by the equation

$$Q = \frac{kp^2}{q^3}.$$

**49.** *Area of a Circle.* The area of a circle varies directly as the square of the length of a diameter. What is the variation constant?

# Chapter 2 Summary and Review

## STUDY GUIDE

KEY TERMS AND CONCEPTS          EXAMPLES

### SECTION 2.1: INCREASING, DECREASING, AND PIECEWISE FUNCTIONS; APPLICATIONS

**Increasing, Decreasing, and Constant Functions**

A function $f$ is said to be **increasing** on an *open* interval $I$, if for all $a$ and $b$ in that interval, $a < b$ implies $f(a) < f(b)$.

A function $f$ is said to be **decreasing** on an *open* interval $I$, if for all $a$ and $b$ in that interval, $a < b$ implies $f(a) > f(b)$.

A function $f$ is said to be **constant** on an *open* interval $I$, if for all $a$ and $b$ in that interval, $f(a) = f(b)$.

Determine the intervals on which the function is (a) increasing; (b) decreasing; (c) constant.

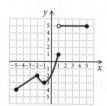

a) As $x$-values increase from $-5$ to $-2$, $y$-values increase from $-4$ to $-2$; $y$-values also increase as $x$-values increase from $-1$ to $1$. Thus the function is increasing on the intervals $(-5, -2)$ and $(-1, 1)$.

b) As $x$-values increase from $-2$ to $-1$, $y$-values decrease from $-2$ to $-3$, so the function is decreasing on the interval $(-2, -1)$.

c) As $x$-values increase from $1$ to $5$, $y$ remains $5$, so the function is constant on the interval $(1, 5)$.

---

**Relative Maxima and Minima**

Suppose that $f$ is a function for which $f(c)$ exists for some $c$ in the domain of $f$. Then:

$f(c)$ is a **relative maximum** if there exists an *open* interval $I$ containing $c$ such that $f(c) > f(x)$ for all $x$ in $I$, where $x \neq c$; and

$f(c)$ is a **relative minimum** if there exists an *open* interval $I$ containing $c$ such that $f(c) < f(x)$ for all $x$ in $I$, where $x \neq c$.

Determine any relative maxima and minima of the function.

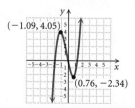

We see from the graph that the function has one relative maximum, $4.05$. It occurs when $x = -1.09$. We also see that there is one relative minimum, $-2.34$. It occurs when $x = 0.76$.

---

Some applied problems can be modeled by functions.

See Examples 3 and 4 on pp. 163 and 164.

---

To graph a function that is defined **piecewise**, we graph the function in parts as defined by its output formulas.

Graph the function defined as

$$f(x) = \begin{cases} 2x - 3, & \text{for } x < 1, \\ x + 1, & \text{for } x \geq 1. \end{cases}$$

*(continued)*

We create the graph in two parts. First, we graph $f(x) = 2x - 3$ for inputs $x$ less than 1. Then we graph $f(x) = x + 1$ for inputs $x$ greater than or equal to 1.

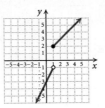

**Greatest Integer Function**

$f(x) = [x] =$ the greatest integer less than or equal to $x$.

The graph of the greatest integer function is shown below. Each input is paired with the greatest integer less than or equal to that input.

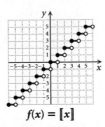

$$f(x) = [x]$$

## SECTION 2.2: THE ALGEBRA OF FUNCTIONS

**Sums, Differences, Products, and Quotients of Functions**

If $f$ and $g$ are functions and $x$ is in the domain of each function, then:

$(f + g)(x) = f(x) + g(x),$
$(f - g)(x) = f(x) - g(x),$
$(fg)(x) = f(x) \cdot g(x),$
$(f/g)(x) = f(x)/g(x),$ provided $g(x) \neq 0.$

Given that $f(x) = x - 4$ and $g(x) = \sqrt{x + 5}$, find each of the following.

**a)** $(f + g)(x)$  **b)** $(f - g)(x)$
**c)** $(fg)(x)$  **d)** $(f/g)(x)$

**a)** $(f + g)(x) = f(x) + g(x) = x - 4 + \sqrt{x + 5}$
**b)** $(f - g)(x) = f(x) - g(x) = x - 4 - \sqrt{x + 5}$
**c)** $(fg)(x) = f(x) \cdot g(x) = (x - 4)\sqrt{x + 5}$

**d)** $(f/g)(x) = f(x)/g(x) = \dfrac{x - 4}{\sqrt{x + 5}}$

**Domains of $f + g$, $f - g$, $fg$, and $f/g$**

If $f$ and $g$ are functions, then the domain of the functions $f + g$, $f - g$, and $fg$ is the intersection of the domain of $f$ and the domain of $g$. The domain of $f/g$ is also the intersection of the domain of $f$ and the domain of $g$, with the exclusion of any $x$-values for which $g(x) = 0$.

For the functions $f$ and $g$ above, find the domains of $f + g$, $f - g$, $fg$, and $f/g$.

The domain of $f(x) = x - 4$ is the set of all real numbers.

The domain of $g(x) = \sqrt{x + 5}$ is the set of all real numbers for which $x + 5 \geq 0$, or $x \geq -5$, or $[-5, \infty)$. Then the domain of $f + g$, $f - g$, and $fg$ is the set of numbers in the intersection of these domains, or $[-5, \infty)$.

Since $g(-5) = 0$, we must exclude $-5$ from the domain of $f/g$. Thus the domain of $f/g$ is $[-5, \infty)$ excluding $-5$, or $(-5, \infty)$.

The **difference quotient** for a function $f(x)$ is the ratio

$$\frac{f(x + h) - f(x)}{h}.$$

For the function $f(x) = x^2 - 4$, construct and simplify the difference quotient.

$$\frac{f(x + h) - f(x)}{h} = \frac{[(x + h)^2 - 4] - (x^2 - 4)}{h}$$

$$= \frac{x^2 + 2xh + h^2 - 4 - x^2 + 4}{h}$$

$$= \frac{2xh + h^2}{h} = \frac{h(2x + h)}{h}$$

$$= 2x + h$$

## SECTION 2.3: THE COMPOSITION OF FUNCTIONS

The **composition of functions**, $f \circ g$, is defined as

$$(f \circ g)(x) = f(g(x)),$$

where $x$ is in the domain of $g$ and $g(x)$ is in the domain of $f$.

Given that $f(x) = 2x - 1$ and $g(x) = \sqrt{x}$, find each of the following.

**a)** $(f \circ g)(4)$          **b)** $(g \circ g)(625)$

**c)** $(f \circ g)(x)$          **d)** $(g \circ f)(x)$

**e)** The domain of $f \circ g$ and the domain of $g \circ f$

**a)** $(f \circ g)(4) = f(g(4)) = f(\sqrt{4}) = f(2) = 2 \cdot 2 - 1 = 4 - 1 = 3$

**b)** $(g \circ g)(625) = g(g(625)) = g(\sqrt{625}) = g(25) = \sqrt{25} = 5$

**c)** $(f \circ g)(x) = f(g(x)) = f(\sqrt{x}) = 2\sqrt{x} - 1$

**d)** $(g \circ f)(x) = g(f(x)) = g(2x - 1) = \sqrt{2x - 1}$

**e)** The domain and the range of $f(x)$ are both $(-\infty, \infty)$, and the domain and the range of $g(x)$ are both $[0, \infty)$. Since the inputs of $f \circ g$ are outputs of $g$ and since $f$ can accept any real number as an input, the domain of $f \circ g$ consists of all real numbers that are outputs of $g$, or $[0, \infty)$.

The inputs of $g \circ f$ consist of all real numbers $f(x)$ that are in the domain of $g$. Thus we must have $f(x) = 2x - 1 \geq 0$, or

$x \geq \frac{1}{2}$, so the domain of $g \circ f$ is $\left[\frac{1}{2}, \infty\right)$.

When we **decompose** a function, we write it as the composition of two functions.

If $h(x) = \sqrt{3x + 7}$, find $f(x)$ and $g(x)$ such that $h(x) = (f \circ g)(x)$.

This function finds the square root of $3x + 7$, so one decomposition is $f(x) = \sqrt{x}$ and $g(x) = 3x + 7$.

There are other correct answers, but this one is probably the most obvious.

## SECTION 2.4: SYMMETRY

**Algebraic Tests of Symmetry**

*x-axis*: If replacing $y$ with $-y$ produces an equivalent equation, then the graph is *symmetric with respect to the x-axis.*

*y-axis*: If replacing $x$ with $-x$ produces an equivalent equation, then the graph is *symmetric with respect to the y-axis.*

*Origin*: If replacing $x$ with $-x$ and $y$ with $-y$ produces an equivalent equation, then the graph is *symmetric with respect to the origin.*

Test $y = 2x^3$ for symmetry with respect to the $x$-axis, the $y$-axis, and the origin.

*x-axis*: We replace $y$ with $-y$:

$$-y = 2x^3$$
$$y = -2x^3. \qquad \textbf{Multiplying by } -1$$

The resulting equation *is not* equivalent to the original equation, so the graph *is not* symmetric with respect to the $x$-axis.

*y-axis*: We replace $x$ with $-x$:

$$y = 2(-x)^3$$
$$y = -2x^3.$$

The resulting equation *is not* equivalent to the original equation, so the graph *is not* symmetric with respect to the $y$-axis.

*Origin*: We replace $x$ with $-x$ and $y$ with $-y$:

$$-y = 2(-x)^3$$
$$-y = -2x^3$$
$$y = 2x^3.$$

The resulting equation *is* equivalent to the original equation, so the graph *is* symmetric with respect to the origin.

**Even Functions and Odd Functions**

If the graph of a function is symmetric with respect to the $y$-axis, we say that it is an **even function**. That is, for each $x$ in the domain of $f$, $f(x) = f(-x)$.

If the graph of a function is symmetric with respect to the origin, we say that it is an **odd function**. That is, for each $x$ in the domain of $f$, $f(-x) = -f(x)$.

Determine whether each function is even, odd, or neither.

**a)** $g(x) = 2x^2 - 4$        **b)** $h(x) = x^5 - 3x^3 - x$

**a)** We first find $g(-x)$ and simplify:

$$g(-x) = 2(-x)^2 - 4 = 2x^2 - 4.$$

$g(x) = g(-x)$, so $g$ is even. Since a function other than $f(x) = 0$ cannot be *both* even and odd and $g$ is even, we need not test to see if it is an odd function.

**b)** We first find $h(-x)$ and simplify:

$$h(-x) = (-x)^5 - 3(-x)^3 - (-x)$$
$$= -x^5 + 3x^3 + x.$$

$h(x) \neq h(-x)$, so $h$ *is not* even.

Next, we find $-h(x)$ and simplify:

$$-h(x) = -(x^5 - 3x^3 - x) = -x^5 + 3x^3 + x.$$

$h(-x) = -h(x)$, so $h$ *is* odd.

## SECTION 2.5: TRANSFORMATIONS

**Vertical Translation**

For $b > 0$:

the graph of $y = f(x) + b$ is the graph of $y = f(x)$ shifted *up* $b$ units;

the graph of $y = f(x) - b$ is the graph of $y = f(x)$ shifted *down* $b$ units.

**Horizontal Translation**

For $d > 0$:

the graph of $y = f(x - d)$ is the graph of $y = f(x)$ shifted *right* $d$ units;

the graph of $y = f(x + d)$ is the graph of $y = f(x)$ shifted *left* $d$ units.

Graph $g(x) = (x - 2)^2 + 1$. Before doing so, describe how the graph can be obtained from the graph of $f(x) = x^2$.

First, we note that the graph of $h(x) = (x - 2)^2$ is the graph of $f(x) = x^2$ shifted right 2 units. Then the graph of $g(x) = (x - 2)^2 + 1$ is the graph of $h(x) = (x - 2)^2$ shifted up 1 unit. Thus the graph of $g$ is obtained by shifting the graph of $f(x) = x^2$ right 2 units and up 1 unit.

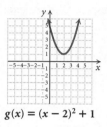

$$g(x) = (x - 2)^2 + 1$$

**Reflections**

The graph of $y = -f(x)$ is the **reflection** of $y = f(x)$ across the $x$-axis.

The graph of $y = f(-x)$ is the **reflection** of $y = f(x)$ across the $y$-axis.

If a point $(x, y)$ is on the graph of $y = f(x)$, then $(x, -y)$ is on the graph of $y = -f(x)$, and $(-x, y)$ is on the graph of $y = f(-x)$.

Graph each of the following. Before doing so, describe how each graph can be obtained from the graph of $f(x) = x^2 - x$.

**a)** $g(x) = x - x^2$       **b)** $h(x) = (-x)^2 - (-x)$

**a)** Note that

$$\begin{aligned}
-f(x) &= -(x^2 - x) \\
&= -x^2 + x \\
&= x - x^2 \\
&= g(x).
\end{aligned}$$

Thus the graph is a reflection of the graph of $f(x) = x^2 - x$ across the $x$-axis.

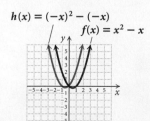

$f(x) = x^2 - x$

$g(x) = x - x^2$

**b)** Note that

$$f(-x) = (-x)^2 - (-x) = h(x).$$

Thus the graph is a reflection of the graph of $f(x) = x^2 - x$ across the $y$-axis.

$$h(x) = (-x)^2 - (-x)$$
$$f(x) = x^2 - x$$

## Vertical Stretching and Shrinking

The graph of $y = af(x)$ can be obtained from the graph of $y = f(x)$ by:

> stretching vertically for $|a| > 1$, or
> shrinking vertically for $0 < |a| < 1$.

For $a < 0$, the graph is also reflected across the $x$-axis.

(The $y$-coordinates of the graph of $y = af(x)$ can be obtained by multiplying the $y$-coordinates of $y = f(x)$ by $a$.)

## Horizontal Stretching and Shrinking

The graph of $y = f(cx)$ can be obtained from the graph of $y = f(x)$ by:

> shrinking horizontally for $|c| > 1$, or
> stretching horizontally for $0 < |c| < 1$.

For $c < 0$, the graph is also reflected across the $y$-axis.

(The $x$-coordinates of the graph of $y = f(cx)$ can be obtained by dividing the $x$-coordinates of $y = f(x)$ by $c$.)

A graph of $y = g(x)$ is shown below. Use this graph to graph each of the given equations.

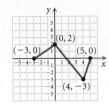

a) $f(x) = g(2x)$      b) $f(x) = -2g(x)$

c) $f(x) = \frac{1}{2}g(x)$      d) $f(x) = g\left(\frac{1}{2}x\right)$

a) Since $|2| > 1$, the graph of $f(x) = g(2x)$ is a horizontal shrinking of the graph of $y = g(x)$. The transformation divides each $x$-coordinate of $g$ by 2.

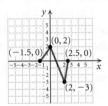

b) Since $|-2| > 1$, the graph of $f(x) = -2g(x)$ is a vertical stretching of the graph of $y = g(x)$. The transformation multiplies each $y$-coordinate of $g$ by 2. Since $-2 < 0$, the graph is also reflected across the $x$-axis.

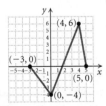

c) Since $\left|\frac{1}{2}\right| < 1$, the graph of $f(x) = \frac{1}{2}g(x)$ is a vertical shrinking of the graph of $y = g(x)$. The transformation multiplies each $y$-coordinate of $g$ by $\frac{1}{2}$.

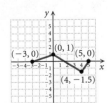

*(continued)*

**d)** Since $\left|\frac{1}{2}\right| < 1$, the graph of $f(x) = g\left(\frac{1}{2}x\right)$ is a horizontal stretching of the graph of $y = g(x)$. The transformation divides each $x$-coordinate of $g$ by $\frac{1}{2}$ (which is the same as multiplying by 2).

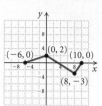

---

## SECTION 2.6: VARIATION AND APPLICATIONS

### Direct Variation

If a situation gives rise to a linear function $f(x) = kx$, or $y = kx$, where $k$ is a positive constant, we say that we have **direct variation**, or that $y$ **varies directly as** $x$, or that $y$ **is directly proportional to** $x$. The number $k$ is called the **variation constant**, or the **constant of proportionality**.

Find an equation of variation in which $y$ varies directly as $x$, and $y = 24$ when $x = 8$. Then find the value of $y$ when $x = 5$.

$$y = kx \qquad y \text{ varies directly as } x.$$
$$24 = k \cdot 8 \qquad \text{Substituting}$$
$$3 = k \qquad \text{Variation constant}$$

The equation of variation is $y = 3x$. Now we use the equation to find the value of $y$ when $x = 5$:

$$y = 3x$$
$$y = 3 \cdot 5 \qquad \text{Substituting}$$
$$y = 15.$$

When $x = 5$, the value of $y$ is 15.

### Inverse Variation

If a situation gives rise to a linear function $f(x) = k/x$, or $y = k/x$, where $k$ is a positive constant, we say that we have **inverse variation**, or that $y$ **varies inversely as** $x$, or that $y$ **is inversely proportional to** $x$. The number $k$ is called the **variation constant**, or the **constant of proportionality**.

Find an equation of variation in which $y$ varies inversely as $x$, and $y = 5$ when $x = 0.1$. Then find the value of $y$ when $x = 10$.

$$y = \frac{k}{x} \qquad y \text{ varies inversely as } x.$$
$$5 = \frac{k}{0.1} \qquad \text{Substituting}$$
$$0.5 = k \qquad \text{Variation constant}$$

The equation of variation is $y = \dfrac{0.5}{x}$. Now we use the equation to find the value of $y$ when $x = 10$:

$$y = \frac{0.5}{x}$$
$$y = \frac{0.5}{10} \qquad \text{Substituting}$$
$$y = 0.05.$$

When $x = 10$, the value of $y$ is 0.05.

**Combined Variation**

$y$ varies **directly as the $n$th power of $x$** if there is some positive constant $k$ such that

$$y = kx^n.$$

$y$ varies **inversely as the $n$th power of $x$** if there is some positive constant $k$ such that

$$y = \frac{k}{x^n}.$$

$y$ varies **jointly as $x$ and $z$** if there is some positive constant $k$ such that

$$y = kxz.$$

Find an equation of variation in which $y$ varies jointly as $w$ and the square of $x$ and inversely as $z$, and $y = 8$ when $w = 3$, $x = 2$, and $z = 6$.

$$y = k \cdot \frac{wx^2}{z}$$

$$8 = k \cdot \frac{3 \cdot 2^2}{6} \qquad \text{Substituting}$$

$$8 = k \cdot \frac{3 \cdot 4}{6}$$

$$8 = 2k$$

$$4 = k \qquad \text{Variation constant}$$

The equation of variation is $y = 4\dfrac{wx^2}{z}$, or $y = \dfrac{4wx^2}{z}$.

# REVIEW EXERCISES

*Determine whether the statement is true or false.*

1. The greatest integer function pairs each input with the greatest integer less than or equal to that input. [2.1]

2. In general, for functions $f$ and $g$, the domain of $f \circ g$ = the domain of $g \circ f$. [2.3]

3. The graph of $y = (x - 2)^2$ is the graph of $y = x^2$ shifted right 2 units. [2.5]

4. The graph of $y = -x^2$ is the reflection of the graph of $y = x^2$ across the $x$-axis. [2.5]

*Determine the intervals on which the function is (a) increasing, (b) decreasing, and (c) constant.* [2.1]

5.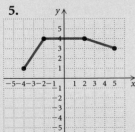

6.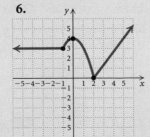

*Graph the function. Estimate the intervals on which the function is increasing or decreasing, and estimate any relative maxima or minima.* [2.1]

7. $f(x) = x^2 - 1$

8. $f(x) = 2 - |x|$

*Use a graphing calculator to find the intervals on which the function is increasing or decreasing, and find any relative maxima or minima.* [2.1]

9. $f(x) = x^2 - 4x + 3$

10. $f(x) = -x^2 + x + 6$

11. $f(x) = x^3 - 4x$

12. $f(x) = 2x - 0.5x^3$

13. *Tablecloth Area.* A seamstress uses 20 ft of lace to trim the edges of a rectangular tablecloth. If the tablecloth is $l$ feet long, express its area as a function of the length. [2.1]

14. *Inscribed Rectangle.* A rectangle is inscribed in a semicircle of radius 2, as shown. The variable $x = $ half the length of the rectangle. Express the area of the rectangle as a function of $x$. [2.1]

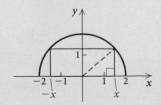

**15.** *Dog Pen.* Mamie has 66 ft of fencing with which to enclose a rectangular dog pen. The side of her garage forms one side of the pen. Suppose the side of the pen parallel to the garage is $x$ feet long.

**a)** Express the area of the dog pen as a function of $x$. [2.1]

**b)** Find the domain of the function. [2.1]

**c)** Graph the function using a graphing calculator. [2.1]

**d)** Determine the dimensions that yield the maximum area. [2.1]

**16.** *Minimizing Surface Area.* A container firm is designing an open-top rectangular box, with a square base, that will hold 108 in³. Let $x =$ the length of a side of the base.

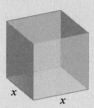

**a)** Express the surface area as a function of $x$. [2.1]

**b)** Find the domain of the function. [2.1]

**c)** Using the graph below, determine the dimensions that will minimize the surface area of the box. [2.1]

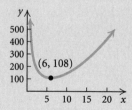

*Graph each of the following.* [2.1]

**17.** $f(x) = \begin{cases} -x, & \text{for } x \le -4, \\ \frac{1}{2}x + 1, & \text{for } x > -4 \end{cases}$

**18.** $f(x) = \begin{cases} x^3, & \text{for } x < -2, \\ |x|, & \text{for } -2 \le x \le 2, \\ \sqrt{x - 1}, & \text{for } x > 2 \end{cases}$

**19.** $f(x) = \begin{cases} \dfrac{x^2 - 1}{x + 1}, & \text{for } x \ne -1, \\ 3, & \text{for } x = -1 \end{cases}$

**20.** $f(x) = [\![x]\!]$           **21.** $f(x) = [\![x - 3]\!]$

**22.** For the function in Exercise 18, find $f(-1)$, $f(5)$, $f(-2)$, and $f(-3)$. [2.1]

**23.** For the function in Exercise 19, find $f(-2)$, $f(-1)$, $f(0)$, and $f(4)$. [2.1]

*Given that $f(x) = \sqrt{x - 2}$ and $g(x) = x^2 - 1$, find each of the following, if it exists.* [2.2]

**24.** $(f - g)(6)$

**25.** $(fg)(2)$

**26.** $(f + g)(-1)$

*For each pair of functions in Exercises 27 and 28:*

**a)** *Find the domain of $f$, $g$, $f + g$, $f - g$, $fg$, and $f/g$.* [2.2]

**b)** *Find $(f + g)(x)$, $(f - g)(x)$, $(fg)(x)$, and $(f/g)(x)$.* [2.2]

**27.** $f(x) = \dfrac{4}{x^2}$; $g(x) = 3 - 2x$

**28.** $f(x) = 3x^2 + 4x$; $g(x) = 2x - 1$

**29.** Given the total-revenue and total-cost functions $R(x) = 120x - 0.5x^2$ and $C(x) = 15x + 6$, find the total-profit function $P(x)$. [2.2]

*For each function $f$, construct and simplify the difference quotient.* [2.2]

**30.** $f(x) = 2x + 7$

**31.** $f(x) = 3 - x^2$

**32.** $f(x) = \dfrac{4}{x}$

*Given that $f(x) = 2x - 1$, $g(x) = x^2 + 4$, and $h(x) = 3 - x^3$, find each of the following.* [2.3]

**33.** $(f \circ g)(1)$           **34.** $(g \circ f)(1)$

**35.** $(h \circ f)(-2)$           **36.** $(g \circ h)(3)$

**37.** $(f \circ h)(-1)$           **38.** $(h \circ g)(2)$

**39.** $(f \circ f)(x)$           **40.** $(h \circ h)(x)$

*In Exercises 41 and 42, for the pair of functions:*

**a)** *Find $(f \circ g)(x)$ and $(g \circ f)(x)$.* [2.3]

**b)** *Find the domain of $f \circ g$ and the domain of $g \circ f$.* [2.3]

**41.** $f(x) = \dfrac{4}{x^2}$; $g(x) = 3 - 2x$

**42.** $f(x) = 3x^2 + 4x$; $g(x) = 2x - 1$

*Find $f(x)$ and $g(x)$ such that $h(x) = (f \circ g)(x)$.* [2.3]

**43.** $h(x) = \sqrt{5x + 2}$

**44.** $h(x) = 4(5x - 1)^2 + 9$

*Graph the given equation and determine visually whether it is symmetric with respect to the x-axis, the y-axis, and the origin. Then verify your assertion algebraically.* [2.4]

**45.** $x^2 + y^2 = 4$        **46.** $y^2 = x^2 + 3$

**47.** $x + y = 3$          **48.** $y = x^2$

**49.** $y = x^3$            **50.** $y = x^4 - x^2$

*Determine visually whether the function is even, odd, or neither even nor odd.* [2.4]

**51.**

**52.**

**53.**
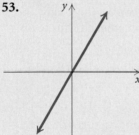

**54.**
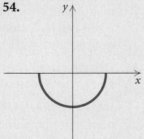

*Determine whether the function is even, odd, or neither even nor odd.* [2.4]

**55.** $f(x) = 9 - x^2$         **56.** $f(x) = x^3 - 2x + 4$

**57.** $f(x) = x^7 - x^5$       **58.** $f(x) = |x|$

**59.** $f(x) = \sqrt{16 - x^2}$     **60.** $f(x) = \dfrac{10x}{x^2 + 1}$

*Write an equation for a function that has a graph with the given characteristics.* [2.5]

**61.** The shape of $y = x^2$, but shifted left 3 units

**62.** The shape of $y = \sqrt{x}$, but reflected across the x-axis and shifted right 3 units and up 4 units

**63.** The shape of $y = |x|$, but stretched vertically by a factor of 2 and shifted right 3 units

*A graph of $y = f(x)$ is shown below. No formula for f is given. Graph each of the following.* [2.5]

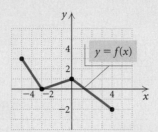

**64.** $y = f(x - 1)$          **65.** $y = f(2x)$

**66.** $y = -2f(x)$            **67.** $y = 3 + f(x)$

*Find an equation of variation for the given situation.* [2.6]

**68.** $y$ varies directly as $x$, and $y = 100$ when $x = 25$.

**69.** $y$ varies directly as $x$, and $y = 6$ when $x = 9$.

**70.** $y$ varies inversely as $x$, and $y = 100$ when $x = 25$.

**71.** $y$ varies inversely as $x$, and $y = 6$ when $x = 9$.

**72.** $y$ varies inversely as the square of $x$, and $y = 12$ when $x = 2$.

**73.** $y$ varies jointly as $x$ and the square of $z$ and inversely as $w$, and $y = 2$ when $x = 16$, $w = 0.2$, and $z = \frac{1}{2}$.

**74.** *Pumping Time.* The time $t$ required to empty a tank varies inversely as the rate $r$ of pumping. If a pump can empty a tank in 35 min at the rate of 800 kL/min, how long will it take the pump to empty the same tank at the rate of 1400 kL/min? [2.6]

**75.** *Test Score.* The score $N$ on a test varies directly as the number of correct responses $a$. Ellen answers 29 questions correctly and earns a score of 87. What would Ellen's score have been if she had answered 25 questions correctly? [2.6]

**76.** *Power of Electric Current.* The power $P$ expended by heat in an electric circuit of fixed resistance varies directly as the square of the current $C$ in the circuit. A circuit expends 180 watts when a current of 6 amperes is flowing. What is the amount of heat expended when the current is 10 amperes? [2.6]

**77.** For $f(x) = x + 1$ and $g(x) = \sqrt{x}$, the domain of $(g \circ f)(x)$ is which of the following? [2.3]

    **A.** $[-1, \infty)$      **B.** $[-1, 0)$

    **C.** $[0, \infty)$      **D.** $(-\infty, \infty)$

**78.** For $b > 0$, the graph of $y = f(x) + b$ is the graph of $y = f(x)$ shifted in which of the following ways? [2.5]

    **A.** Right $b$ units      **B.** Left $b$ units

    **C.** Up $b$ units      **D.** Down $b$ units

**79.** The graph of the function $f$ is shown below.

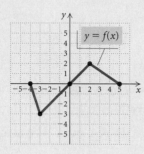

The graph of $g(x) = -\frac{1}{2}f(x) + 1$ is which of the following? [2.5]

**A.**

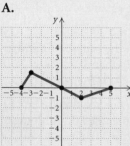

**B.**

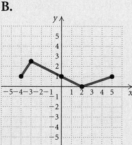

**C.**

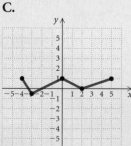

**D.**

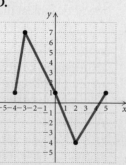

## Synthesis

**80.** Prove that the sum of two odd functions is odd. [2.2], [2.4]

**81.** Describe how the graph of $y = -f(-x)$ is obtained from the graph of $y = f(x)$. [2.5]

## Collaborative Discussion and Writing

**82.** Given that $f(x) = 4x^3 - 2x + 7$, find each of the following. Then discuss how each expression differs from the other. [1.2], [2.5]

    **a)** $f(x) + 2$

    **b)** $f(x + 2)$

    **c)** $f(x) + f(2)$

**83.** Given the graph of $y = f(x)$, explain and contrast the effect of the constant $c$ on the graphs of $y = f(cx)$ and $y = cf(x)$. [2.5]

**84.** Consider the constant function $f(x) = 0$. Determine whether the graph of this function is symmetric with respect to the $x$-axis, the $y$-axis, and/or the origin. Determine whether this function is even or odd. [2.4]

**85.** Describe conditions under which you would know whether a polynomial function

$$f(x) = a_n x^n + a_{n-1} x^{n-1} + \cdots + a_2 x^2 + a_1 x + a_0$$

is even or odd without using an algebraic procedure. Explain. [2.4]

**86.** If $y$ varies directly as $x^2$, explain why doubling $x$ would not cause $y$ to be doubled as well. [2.6]

**87.** If $y$ varies directly as $x$ and $x$ varies inversely as $z$, how does $y$ vary with regard to $z$? Why? [2.6]

1. Determine the intervals on which the function is (a) increasing, (b) decreasing, and (c) constant.

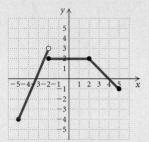

2. Graph the function $f(x) = 2 - x^2$. Estimate the intervals on which the function is increasing or decreasing, and estimate any relative maxima or minima.

3. Use a graphing calculator to find the intervals on which the function $f(x) = x^3 + 4x^2$ is increasing or decreasing, and find any relative maxima or minima.

4. *Triangular Pennant.* A softball team is designing a triangular pennant such that the height is 6 in. less than four times the length of the base $b$. Express the area of the pennant as a function of $b$.

5. Graph:
$$f(x) = \begin{cases} x^2, & \text{for } x < -1, \\ |x|, & \text{for } -1 \le x \le 1, \\ \sqrt{x - 1}, & \text{for } x > 1. \end{cases}$$

6. For the function in Exercise 5, find $f\left(-\frac{7}{8}\right)$, $f(5)$, and $f(-4)$.

*Given that $f(x) = x^2 - 4x + 3$ and $g(x) = \sqrt{3 - x}$, find each of the following, if it exists.*

7. $(f + g)(-6)$

8. $(f - g)(-1)$

9. $(fg)(2)$

10. $(f/g)(1)$

*For $f(x) = x^2$ and $g(x) = \sqrt{x - 3}$, find each of the following.*

11. The domain of $f$

12. The domain of $g$

13. The domain of $f + g$

14. The domain of $f - g$

15. The domain of $fg$

16. The domain of $f/g$

17. $(f + g)(x)$

18. $(f - g)(x)$

19. $(fg)(x)$

20. $(f/g)(x)$

*For each function, construct and simplify the difference quotient.*

21. $f(x) = \frac{1}{2}x + 4$

22. $f(x) = 2x^2 - x + 3$

*Given that $f(x) = x^2 - 1$, $g(x) = 4x + 3$, and $h(x) = 3x^2 + 2x + 4$, find each of the following.*

23. $(g \circ h)(2)$

24. $(f \circ g)(-1)$

25. $(h \circ f)(1)$

26. $(g \circ g)(x)$

*For $f(x) = \sqrt{x - 5}$ and $g(x) = x^2 + 1$:*

27. Find $(f \circ g)(x)$ and $(g \circ f)(x)$.

28. Find the domain of $(f \circ g)(x)$ and the domain of $(g \circ f)(x)$.

29. Find $f(x)$ and $g(x)$ such that
$$h(x) = (f \circ g)(x) = (2x - 7)^4.$$

30. Determine whether the graph of $y = x^4 - 2x^2$ is symmetric with respect to the $x$-axis, the $y$-axis, and the origin.

31. Determine whether the function
$$f(x) = \frac{2x}{x^2 + 1}$$
is even, odd, or neither even nor odd. Show your work.

32. Write an equation for a function that has the shape of $y = x^2$, but shifted right 2 units and down 1 unit.

33. Write an equation for a function that has the shape of $y = x^2$, but reflected across the $x$-axis and shifted left 2 units and up 3 units.

**34.** The graph of a function $y = f(x)$ is shown below. No formula for $f$ is given. Graph $y = -\frac{1}{2}f(x)$.

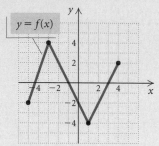

**35.** Find an equation of variation in which $y$ varies inversely as $x$, and $y = 5$ when $x = 6$.

**36.** Find an equation of variation in which $y$ varies directly as $x$, and $y = 60$ when $x = 12$.

**37.** Find an equation of variation where $y$ varies jointly as $x$ and the square of $z$ and inversely as $w$, and $y = 100$ when $x = 0.1$, $z = 10$, and $w = 5$.

**38.** The stopping distance $d$ of a car after the brakes have been applied varies directly as the square of the speed $r$. If a car traveling 60 mph can stop in 200 ft, how long will it take a car traveling 30 mph to stop?

**39.** The graph of the function $f$ is shown below.

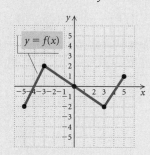

The graph of $g(x) = 2f(x) - 1$ is which of the following?

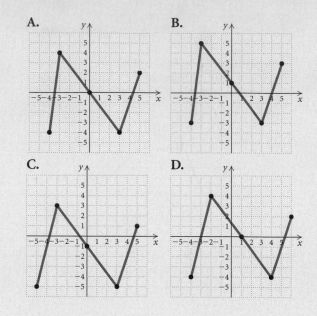

## Synthesis

**40.** If $(-3, 1)$ is a point on the graph of $y = f(x)$, what point do you know is on the graph of $y = f(3x)$?

# Quadratic Functions and Equations; Inequalities

## APPLICATION

The number of TV channels that the average U.S. home receives has been soaring in recent years. The function $t(x) = 0.16x^2 + 0.46x + 21.36$ can be used to estimate this number, where $x$ is the number of years after 1985 (*Source: Nielsen Media Research, National People Meter Sample*). Using this function, estimate in what year the average U.S. household received 130 channels.

**This problem appears as Exercise 109 in Section 3.2.**

# 3.1 The Complex Numbers

- Perform computations involving complex numbers.

Some functions have zeros that are not real numbers. In order to find the zeros of such functions, we must consider the **complex-number system**.

## The Complex-Number System

We know that the square root of a negative number is not a real number. For example, $\sqrt{-1}$ is not a real number because there is no real number $x$ such that $x^2 = -1$. This means that certain equations, like $x^2 = -1$, or $x^2 + 1 = 0$, do not have real-number solutions, and certain functions, like $f(x) = x^2 + 1$, do not have real-number zeros. Consider the graph of $f(x) = x^2 + 1$.

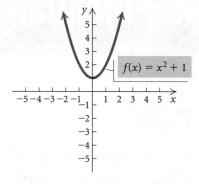

RADICAL EXPRESSIONS

REVIEW SECTION **R.7.**

We see that the graph does not cross the $x$-axis and thus has no $x$-intercepts. This illustrates that the function $f(x) = x^2 + 1$ has no real-number zeros. Thus there are no real-number solutions of the corresponding equation $x^2 + 1 = 0$.

We can define a nonreal number that is a solution of the equation $x^2 + 1 = 0$.

### THE NUMBER $i$

The number $i$ is defined such that

$$i = \sqrt{-1} \quad \text{and} \quad i^2 = -1.$$

To express roots of negative numbers in terms of $i$, we can use the fact that

$$\sqrt{-p} = \sqrt{-1 \cdot p} = \sqrt{-1} \cdot \sqrt{p} = i\sqrt{p}$$

when $p$ is a positive real number.

**EXAMPLE 1**   Express each number in terms of $i$.

**a)** $\sqrt{-7}$        **b)** $\sqrt{-16}$        **c)** $-\sqrt{-13}$

**d)** $-\sqrt{-64}$        **e)** $\sqrt{-48}$

*Solution*

**a)** $\sqrt{-7} = \sqrt{-1 \cdot 7} = \sqrt{-1} \cdot \sqrt{7}$

$\qquad\qquad = i\sqrt{7}, \text{ or } \sqrt{7}i \longleftarrow$

$i$ is *not* under the radical.

**b)** $\sqrt{-16} = \sqrt{-1 \cdot 16} = \sqrt{-1} \cdot \sqrt{16}$

$\qquad\qquad = i \cdot 4 = 4i$

**c)** $-\sqrt{-13} = -\sqrt{-1 \cdot 13} = -\sqrt{-1} \cdot \sqrt{13}$

$\qquad\qquad = -i\sqrt{13}, \text{ or } -\sqrt{13}i \longleftarrow$

**d)** $-\sqrt{-64} = -\sqrt{-1 \cdot 64} = -\sqrt{-1} \cdot \sqrt{64}$

$\qquad\qquad = -i \cdot 8 = -8i$

**e)** $\sqrt{-48} = \sqrt{-1 \cdot 48} = \sqrt{-1} \cdot \sqrt{48}$

$\qquad\qquad = i\sqrt{16 \cdot 3}$

$\qquad\qquad = i \cdot 4\sqrt{3}$

$\qquad\qquad = 4i\sqrt{3}, \text{ or } 4\sqrt{3}i \longleftarrow$

**Now Try Exercises 1, 7, and 9.**

The complex numbers are formed by adding real numbers and multiples of $i$.

> **COMPLEX NUMBERS**
>
> A **complex number** is a number of the form $a + bi$, where $a$ and $b$ are real numbers. The number $a$ is said to be the **real part** of $a + bi$, and the number $b$ is said to be the **imaginary part** of $a + bi$.*

Note that either $a$ or $b$ or both can be 0. When $b = 0$, $a + bi = a + 0i = a$, so every real number is a complex number. A complex number like $3 + 4i$ or $17i$, in which $b \neq 0$, is called an **imaginary number**. A complex number like $17i$ or $-4i$, in which $a = 0$ and $b \neq 0$, is sometimes called a **pure imaginary number**. The relationships among various types of complex numbers are shown in the figure on the following page.

---

*Sometimes $bi$ is considered to be the imaginary part.

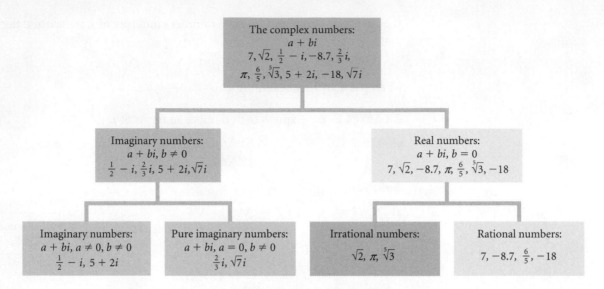

### Addition and Subtraction

The complex numbers obey the commutative, associative, and distributive laws. Thus we can add and subtract them as we do binomials. We collect the real parts and the imaginary parts of complex numbers just as we collect like terms in binomials.

GCM

**EXAMPLE 2**   Add or subtract and simplify each of the following.

**a)** $(8 + 6i) + (3 + 2i)$ **b)** $(4 + 5i) - (6 - 3i)$

*Solution*

**a)** $(8 + 6i) + (3 + 2i) = (8 + 3) + (6i + 2i)$

   Collecting the real parts and the imaginary parts

   $= 11 + (6 + 2)i = 11 + 8i$

**b)** $(4 + 5i) - (6 - 3i) = (4 - 6) + [5i - (-3i)]$

   Note that 6 and $-3i$ are both being subtracted.

   $= -2 + 8i$     ◼ **Now Try Exercises 11 and 21.**

When set in $a + bi$ mode, most graphing calculators can perform operations on complex numbers. The operations in Example 2 are shown in the window at left. Some calculators will express a complex number in the form $(a, b)$ rather than $a + bi$.

```
(8+6i)+(3+2i)
                    11+8i
(4+5i)−(6−3i)
                    −2+8i
```

### Multiplication

When $\sqrt{a}$ and $\sqrt{b}$ are real numbers, $\sqrt{a} \cdot \sqrt{b} = \sqrt{ab}$, but this is not true when $\sqrt{a}$ and $\sqrt{b}$ are not real numbers. Thus,

$$\sqrt{-2} \cdot \sqrt{-5} = \sqrt{-1} \cdot \sqrt{2} \cdot \sqrt{-1} \cdot \sqrt{5}$$
$$= i\sqrt{2} \cdot i\sqrt{5}$$
$$= i^2\sqrt{10} = -1\sqrt{10} = -\sqrt{10} \quad \text{is correct!}$$

But

$$\sqrt{-2} \cdot \sqrt{-5} = \sqrt{(-2)(-5)} = \sqrt{10} \quad \text{is wrong!}$$

Keeping this and the fact that $i^2 = -1$ in mind, we multiply with imaginary numbers in much the same way that we do with real numbers.

**GCM**   **EXAMPLE 3**   Multiply and simplify each of the following.

**a)** $\sqrt{-16} \cdot \sqrt{-25}$     **b)** $(1 + 2i)(1 + 3i)$     **c)** $(3 - 7i)^2$

*Solution*

**a)** $\sqrt{-16} \cdot \sqrt{-25} = \sqrt{-1} \cdot \sqrt{16} \cdot \sqrt{-1} \cdot \sqrt{25}$

$$= i \cdot 4 \cdot i \cdot 5$$
$$= i^2 \cdot 20$$
$$= -1 \cdot 20 \qquad i^2 = -1$$
$$= -20$$

**b)** $(1 + 2i)(1 + 3i) = 1 + 3i + 2i + 6i^2$     Multiplying each term of one number by every term of the other (FOIL)

$$= 1 + 3i + 2i - 6 \qquad i^2 = -1$$
$$= -5 + 5i \qquad \text{Collecting like terms}$$

**c)** $(3 - 7i)^2 = 3^2 - 2 \cdot 3 \cdot 7i + (7i)^2$     Recall that $(A - B)^2 = A^2 - 2AB + B^2$.

$$= 9 - 42i + 49i^2$$
$$= 9 - 42i - 49 \qquad i^2 = -1$$
$$= -40 - 42i$$

**Now Try Exercises 31 and 39.**

We can multiply complex numbers on a graphing calculator set in $a + bi$ mode. The products found in Example 3 are shown below.

Recall that $-1$ raised to an *even* power is 1, and $-1$ raised to an *odd* power is $-1$. Simplifying powers of $i$ can then be done by using the fact that $i^2 = -1$ and expressing the given power of $i$ in terms of $i^2$. Consider the following:

$$i = \sqrt{-1},$$
$$i^2 = -1,$$
$$i^3 = i^2 \cdot i = (-1)i = -i,$$
$$i^4 = (i^2)^2 = (-1)^2 = 1,$$
$$i^5 = i^4 \cdot i = (i^2)^2 \cdot i = (-1)^2 \cdot i = 1 \cdot i = i,$$
$$i^6 = (i^2)^3 = (-1)^3 = -1,$$
$$i^7 = i^6 \cdot i = (i^2)^3 \cdot i = (-1)^3 \cdot i = -1 \cdot i = -i,$$
$$i^8 = (i^2)^4 = (-1)^4 = 1.$$

Note that the powers of $i$ cycle through the values $i$, $-1$, $-i$, and 1.

**EXAMPLE 4** Simplify each of the following.

**a)** $i^{37}$                                         **b)** $i^{58}$

**c)** $i^{75}$                                         **d)** $i^{80}$

*Solution*

**a)** $i^{37} = i^{36} \cdot i = (i^2)^{18} \cdot i = (-1)^{18} \cdot i = 1 \cdot i = i$

**b)** $i^{58} = (i^2)^{29} = (-1)^{29} = -1$

**c)** $i^{75} = i^{74} \cdot i = (i^2)^{37} \cdot i = (-1)^{37} \cdot i = -1 \cdot i = -i$

**d)** $i^{80} = (i^2)^{40} = (-1)^{40} = 1$      **Now Try Exercises 79 and 83.**

These powers of $i$ can also be simplified in terms of $i^4$ rather than $i^2$. Consider $i^{37}$ in Example 4(a), for instance. When we divide 37 by 4, we get 9 with a remainder of 1. Then $37 = 4 \cdot 9 + 1$, so

$$i^{37} = (i^4)^9 \cdot i = 1^9 \cdot i = 1 \cdot i = i.$$

The other examples shown above can be done in a similar manner.

## ■ Conjugates and Division

*Conjugates* of complex numbers are defined as follows.

---

### CONJUGATE OF A COMPLEX NUMBER

The **conjugate** of a complex number $a + bi$ is $a - bi$. The numbers $a + bi$ and $a - bi$ are **complex conjugates**.

---

Each of the following pairs of numbers are complex conjugates:

$$-3 + 7i \text{ and } -3 - 7i; \quad 14 - 5i \text{ and } 14 + 5i; \quad \text{and} \quad 8i \text{ and } -8i.$$

*The product of a complex number and its conjugate is a real number.*

**EXAMPLE 5** Multiply each of the following.

**a)** $(5 + 7i)(5 - 7i)$                   **b)** $(8i)(-8i)$

*Solution*

**a)** $(5 + 7i)(5 - 7i) = 5^2 - (7i)^2$     Using $(A + B)(A - B) = A^2 - B^2$

$$\begin{aligned} &= 25 - 49i^2 \\ &= 25 - 49(-1) \\ &= 25 + 49 \\ &= 74 \end{aligned}$$

**b)** $(8i)(-8i) = -64i^2$

$$\begin{aligned} &= -64(-1) \\ &= 64 \end{aligned}$$      **Now Try Exercise 49.**

Conjugates are used when we divide complex numbers.

---

(5+7i)(5−7i)

                         74

(8i)(−8i)

                         64

GCM    **EXAMPLE 6**    Divide $2 - 5i$ by $1 - 6i$.

**Solution**    We write fraction notation and then multiply by 1, using the conjugate of the denominator to form the symbol for 1.

$$\frac{2 - 5i}{1 - 6i} = \frac{2 - 5i}{1 - 6i} \cdot \frac{1 + 6i}{1 + 6i} \qquad \text{Note that } 1 + 6i \text{ is the conjugate of the divisor, } 1 - 6i.$$

$$= \frac{(2 - 5i)(1 + 6i)}{(1 - 6i)(1 + 6i)}$$

$$= \frac{2 + 12i - 5i - 30i^2}{1 - 36i^2}$$

$$= \frac{2 + 7i + 30}{1 + 36} \qquad i^2 = -1$$

$$= \frac{32 + 7i}{37}$$

$$= \frac{32}{37} + \frac{7}{37}i. \qquad \text{Writing the quotient in the form } a + bi$$

**Now Try Exercise 69.**

$(2-5i)/(1-6i) \blacktriangleright \text{Frac}$
$\qquad \frac{32}{37} + \frac{7}{37}i$

With a graphing calculator set in $a + bi$ mode, we can divide complex numbers and express the real parts and the imaginary parts in fraction form, just as we did in Example 6.

# 3.1    Exercise Set

*Express the number in terms of i.*

**1.** $\sqrt{-3}$

**2.** $\sqrt{-21}$

**3.** $\sqrt{-25}$

**4.** $\sqrt{-100}$

**5.** $-\sqrt{-33}$

**6.** $-\sqrt{-59}$

**7.** $-\sqrt{-81}$

**8.** $-\sqrt{-9}$

**9.** $\sqrt{-98}$

**10.** $\sqrt{-28}$

*Simplify. Write answers in the form $a + bi$, where a and b are real numbers.*

**11.** $(-5 + 3i) + (7 + 8i)$

**12.** $(-6 - 5i) + (9 + 2i)$

**13.** $(4 - 9i) + (1 - 3i)$

**14.** $(7 - 2i) + (4 - 5i)$

**15.** $(12 + 3i) + (-8 + 5i)$

**16.** $(-11 + 4i) + (6 + 8i)$

**17.** $(-1 - i) + (-3 - i)$

**18.** $(-5 - i) + (6 + 2i)$

**19.** $(3 + \sqrt{-16}) + (2 + \sqrt{-25})$

**20.** $(7 - \sqrt{-36}) + (2 + \sqrt{-9})$

**21.** $(10 + 7i) - (5 + 3i)$

**22.** $(-3 - 4i) - (8 - i)$

**23.** $(13 + 9i) - (8 + 2i)$

**24.** $(-7 + 12i) - (3 - 6i)$

**25.** $(6 - 4i) - (-5 + i)$

**26.** $(8 - 3i) - (9 - i)$

**27.** $(-5 + 2i) - (-4 - 3i)$

**28.** $(-6 + 7i) - (-5 - 2i)$

**29.** $(4 - 9i) - (2 + 3i)$

**30.** $(10 - 4i) - (8 + 2i)$

**31.** $\sqrt{-4} \cdot \sqrt{-36}$      **32.** $\sqrt{-49} \cdot \sqrt{-9}$

**33.** $\sqrt{-81} \cdot \sqrt{-25}$      **34.** $\sqrt{-16} \cdot \sqrt{-100}$

**35.** $7i(2 - 5i)$      **36.** $3i(6 + 4i)$

**37.** $-2i(-8 + 3i)$      **38.** $-6i(-5 + i)$

**39.** $(1 + 3i)(1 - 4i)$      **40.** $(1 - 2i)(1 + 3i)$

**41.** $(2 + 3i)(2 + 5i)$      **42.** $(3 - 5i)(8 - 2i)$

**43.** $(-4 + i)(3 - 2i)$      **44.** $(5 - 2i)(-1 + i)$

**45.** $(8 - 3i)(-2 - 5i)$

**46.** $(7 - 4i)(-3 - 3i)$

**47.** $(3 + \sqrt{-16})(2 + \sqrt{-25})$

**48.** $(7 - \sqrt{-16})(2 + \sqrt{-9})$

**49.** $(5 - 4i)(5 + 4i)$      **50.** $(5 + 9i)(5 - 9i)$

**51.** $(3 + 2i)(3 - 2i)$      **52.** $(8 + i)(8 - i)$

**53.** $(7 - 5i)(7 + 5i)$      **54.** $(6 - 8i)(6 + 8i)$

**55.** $(4 + 2i)^2$      **56.** $(5 - 4i)^2$

**57.** $(-2 + 7i)^2$      **58.** $(-3 + 2i)^2$

**59.** $(1 - 3i)^2$      **60.** $(2 - 5i)^2$

**61.** $(-1 - i)^2$      **62.** $(-4 - 2i)^2$

**63.** $(3 + 4i)^2$      **64.** $(6 + 5i)^2$

**65.** $\dfrac{3}{5 - 11i}$      **66.** $\dfrac{i}{2 + i}$

**67.** $\dfrac{5}{2 + 3i}$      **68.** $\dfrac{-3}{4 - 5i}$

**69.** $\dfrac{4 + i}{-3 - 2i}$      **70.** $\dfrac{5 - i}{-7 + 2i}$

**71.** $\dfrac{5 - 3i}{4 + 3i}$      **72.** $\dfrac{6 + 5i}{3 - 4i}$

**73.** $\dfrac{2 + \sqrt{3}i}{5 - 4i}$      **74.** $\dfrac{\sqrt{5} + 3i}{1 - i}$

**75.** $\dfrac{1 + i}{(1 - i)^2}$      **76.** $\dfrac{1 - i}{(1 + i)^2}$

**77.** $\dfrac{4 - 2i}{1 + i} + \dfrac{2 - 5i}{1 + i}$      **78.** $\dfrac{3 + 2i}{1 - i} + \dfrac{6 + 2i}{1 - i}$

*Simplify.*

**79.** $i^{11}$    **80.** $i^7$    **81.** $i^{35}$

**82.** $i^{24}$    **83.** $i^{64}$    **84.** $i^{42}$

**85.** $(-i)^{71}$    **86.** $(-i)^6$    **87.** $(5i)^4$

**88.** $(2i)^5$

## Skill Maintenance

**89.** Write a slope–intercept equation for the line containing the point $(3, -5)$ and perpendicular to the line $3x - 6y = 7$.

*Given that $f(x) = x^2 + 4$ and $g(x) = 3x + 5$, find each of the following.*

**90.** The domain of $f - g$

**91.** The domain of $f/g$

**92.** $(f - g)(x)$

**93.** $(f/g)(2)$

**94.** For the function $f(x) = x^2 - 3x + 4$, construct and simplify the difference quotient

$$\frac{f(x + h) - f(x)}{h}.$$

## Synthesis

*Determine whether the statement is true or false.*

**95.** The sum of two numbers that are complex conjugates of each other is always a real number.

**96.** The conjugate of a sum is the sum of the conjugates of the individual complex numbers.

**97.** The conjugate of a product is the product of the conjugates of the individual complex numbers.

*Let $z = a + bi$ and $\bar{z} = a - bi$.*

**98.** Find a general expression for $1/z$.

**99.** Find a general expression for $z\bar{z}$.

**100.** Solve $z + 6\bar{z} = 7$ for $z$.

**101.** Multiply and simplify:

$$[x - (3 + 4i)][x - (3 - 4i)].$$

# Quadratic Equations, Functions, Zeros, and Models

- Find zeros of quadratic functions and solve quadratic equations by using the principle of zero products, by using the principle of square roots, by completing the square, and by using the quadratic formula.
- Solve equations that are reducible to quadratic.
- Solve applied problems using quadratic equations.

## ◼ Quadratic Equations and Quadratic Functions

In this section, we will explore the relationship between the solutions of quadratic equations and the zeros of quadratic functions. We define quadratic equations and quadratic functions as follows.

### QUADRATIC EQUATIONS

A **quadratic equation** is an equation that can be written in the form

$$ax^2 + bx + c = 0, \quad a \neq 0,$$

where $a$, $b$, and $c$ are real numbers.

### QUADRATIC FUNCTIONS

A **quadratic function** $f$ is a function that can be written in the form

$$f(x) = ax^2 + bx + c, \quad a \neq 0,$$

where $a$, $b$, and $c$ are real numbers.

A quadratic equation written in the form $ax^2 + bx + c = 0$ is said to be in **standard form**.

The *zeros* of a quadratic function $f(x) = ax^2 + bx + c$ are the *solutions* of the associated quadratic equation $ax^2 + bx + c = 0$. (These solutions are sometimes called *roots* of the equation.) Quadratic functions can have real-number zeros or imaginary-number zeros and quadratic equations can have real-number solutions or imaginary-number solutions. If the zeros or solutions are real numbers, they are also the first coordinates of the $x$-intercepts of the graph of the quadratic function.

The following principles allow us to solve many quadratic equations.

| ZEROS OF A FUNCTION |
| --- |
| REVIEW SECTION **1.5**. |

### EQUATION-SOLVING PRINCIPLES

*The Principle of Zero Products*: If $ab = 0$ is true, then $a = 0$ or $b = 0$, and if $a = 0$ or $b = 0$, then $ab = 0$.

*The Principle of Square Roots*: If $x^2 = k$, then $x = \sqrt{k}$ or $x = -\sqrt{k}$.

> FACTORING TRINOMIALS
>
> REVIEW SECTION **R.4.**

**EXAMPLE 1** Solve: $2x^2 - x = 3$.

## Algebraic Solution

We factor and use the principle of zero products:

$$2x^2 - x = 3$$
$$2x^2 - x - 3 = 0$$
$$(x + 1)(2x - 3) = 0$$
$$x + 1 = 0 \quad or \quad 2x - 3 = 0$$
$$x = -1 \quad or \quad 2x = 3$$
$$x = -1 \quad or \quad x = \tfrac{3}{2}.$$

*Check:*

For $x = -1$:

$$\frac{2x^2 - x = 3}{2(-1)^2 - (-1) \;?\; 3}$$
$$2 \cdot 1 + 1$$
$$2 + 1$$
$$3 \;\big|\; 3 \quad \textbf{TRUE}$$

For $x = \tfrac{3}{2}$:

$$\frac{2x^2 - x = 3}{2\left(\tfrac{3}{2}\right)^2 - \tfrac{3}{2} \;?\; 3}$$
$$2 \cdot \tfrac{9}{4} - \tfrac{3}{2}$$
$$\tfrac{9}{2} - \tfrac{3}{2}$$
$$\tfrac{6}{2}$$
$$3 \;\big|\; 3 \quad \textbf{TRUE}$$

The solutions are $-1$ and $\tfrac{3}{2}$.

## Graphical Solution

The solutions of the equation $2x^2 - x = 3$, or the equivalent equation $2x^2 - x - 3 = 0$, are the zeros of the function $f(x) = 2x^2 - x - 3$. They are also the first coordinates of the $x$-intercepts of the graph of $f(x) = 2x^2 - x - 3$.

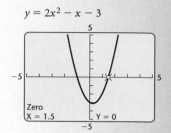

The solutions are $-1$ and $1.5$, or $-1$ and $\tfrac{3}{2}$.

We also can find the zeros, also called roots, of the function with the Graphing Calculator, Appcylon LCC, app. We again see that the solutions of the equation $2x^2 - x = 3$ are $-1$ and $1.5$.

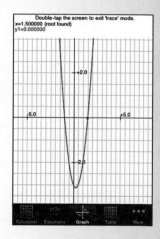

> **Now Try Exercise 3.**

**EXAMPLE 2** Solve: $2x^2 - 10 = 0$.

*Solution* We have

$$2x^2 - 10 = 0$$
$$2x^2 = 10 \qquad \text{Adding 10 on both sides}$$
$$x^2 = 5 \qquad \text{Dividing by 2 on both sides}$$
$$x = \sqrt{5} \quad or \quad x = -\sqrt{5}. \qquad \text{Using the principle of square roots}$$

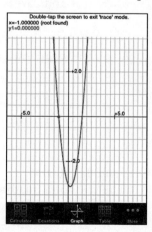

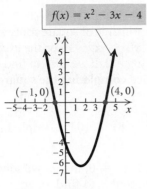

$f(x) = x^2 - 3x - 4$

Two real-number zeros
Two x-intercepts

**FIGURE 1**

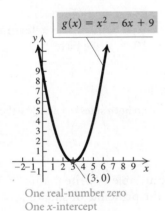

$g(x) = x^2 - 6x + 9$

One real-number zero
One x-intercept

**FIGURE 2**

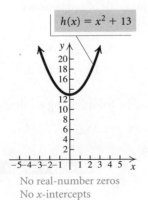

$h(x) = x^2 + 13$

No real-number zeros
No x-intercepts

**FIGURE 3**

*Check*:

$$2x^2 - 10 = 0$$

$$\frac{}{2(\pm\sqrt{5})^2 - 10 \overset{?}{=} 0} \qquad \text{We can check both solutions at once.}$$

$$2 \cdot 5 - 10$$

$$10 - 10$$

$$0 \mid 0 \quad \text{TRUE}$$

The solutions are $\sqrt{5}$ and $-\sqrt{5}$, or $\pm\sqrt{5}$.

> **Now Try Exercise 7.**

We have seen that some quadratic equations can be solved by factoring and using the principle of zero products. For example, consider the equation $x^2 - 3x - 4 = 0$:

$$x^2 - 3x - 4 = 0$$
$$(x + 1)(x - 4) = 0 \qquad \text{Factoring}$$
$$x + 1 = 0 \quad or \quad x - 4 = 0 \qquad \text{Using the principle of zero products}$$
$$x = -1 \quad or \qquad x = 4.$$

The equation $x^2 - 3x - 4 = 0$ has *two real-number* solutions, $-1$ and $4$. These are the zeros of the associated quadratic function $f(x) = x^2 - 3x - 4$ and the first coordinates of the x-intercepts of the graph of this function. (See Fig. 1.)

Next, consider the equation $x^2 - 6x + 9 = 0$. Again, we factor and use the principle of zero products:

$$x^2 - 6x + 9 = 0$$
$$(x - 3)(x - 3) = 0 \qquad \text{Factoring}$$
$$x - 3 = 0 \quad or \quad x - 3 = 0 \qquad \text{Using the principle of zero products}$$
$$x = 3 \quad or \qquad x = 3.$$

The equation $x^2 - 6x + 9 = 0$ has *one real-number* solution, $3$. It is the zero of the quadratic function $g(x) = x^2 - 6x + 9$ and the first coordinate of the x-intercept of the graph of this function. (See Fig. 2.)

The principle of square roots can be used to solve quadratic equations like $x^2 + 13 = 0$:

$$x^2 + 13 = 0$$
$$x^2 = -13$$
$$x = \pm\sqrt{-13} \qquad \text{Using the principle of square roots}$$
$$x = \pm\sqrt{13}i. \qquad \sqrt{-13} = \sqrt{-1} \cdot \sqrt{13} = i \cdot \sqrt{13} = \sqrt{13}i$$

The equation has *two imaginary-number* solutions, $-\sqrt{13}i$ and $\sqrt{13}i$. These are the zeros of the associated quadratic function $h(x) = x^2 + 13$. Since the zeros are not real numbers, the graph of the function has no x-intercepts. (See Fig. 3.)

### ■ Completing the Square

Neither the principle of zero products nor the principle of square roots would yield the *exact* zeros of a function like $f(x) = x^2 - 6x - 10$ or the *exact* solutions of the associated equation $x^2 - 6x - 10 = 0$. If we wish to find exact zeros or solutions, we can use a procedure called **completing the square** and then use the principle of square roots.

**EXAMPLE 3** Find the zeros of $f(x) = x^2 - 6x - 10$ by completing the square.

**Solution** We find the values of $x$ for which $f(x) = 0$; that is, we solve the associated equation $x^2 - 6x - 10 = 0$. Our goal is to find an equivalent equation of the form $x^2 + bx + c = d$ in which $x^2 + bx + c$ is a perfect square. Since

$$x^2 + bx + \left(\frac{b}{2}\right)^2 = \left(x + \frac{b}{2}\right)^2,$$

the number $c$ is found by taking half the coefficient of the $x$-term and squaring it. Thus for the equation $x^2 - 6x - 10 = 0$, we have

$$x^2 - 6x - 10 = 0$$
$$x^2 - 6x \phantom{+ 9} = 10 \qquad \text{Adding 10}$$
$$x^2 - 6x + 9 = 10 + 9 \qquad \text{Adding 9 on both sides to complete the}$$
$$\phantom{x^2 - 6x + 9 = 10 + 9 \qquad} \text{square: } \left(\frac{b}{2}\right)^2 = \left(\frac{-6}{2}\right)^2 = (-3)^2 = 9$$
$$x^2 - 6x + 9 = 19.$$

Because $x^2 - 6x + 9$ is a perfect square, we are able to write it as $(x - 3)^2$, the square of a binomial. We can then use the principle of square roots to finish the solution:

$$(x - 3)^2 = 19 \qquad \text{Factoring}$$
$$x - 3 = \pm\sqrt{19} \qquad \text{Using the principle of square roots}$$
$$x = 3 \pm \sqrt{19}. \qquad \text{Adding 3}$$

Therefore, the solutions of the equation are $3 + \sqrt{19}$ and $3 - \sqrt{19}$, or simply $3 \pm \sqrt{19}$. The zeros of $f(x) = x^2 - 6x - 10$ are also $3 + \sqrt{19}$ and $3 - \sqrt{19}$, or $3 \pm \sqrt{19}$.

Decimal approximations for $3 \pm \sqrt{19}$ can be found using a calculator:

$$3 + \sqrt{19} \approx 7.359 \quad \text{and} \quad 3 - \sqrt{19} \approx -1.359.$$

The zeros are approximately 7.359 and $-1.359$.

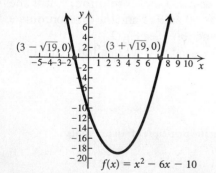

$f(x) = x^2 - 6x - 10$

**Now Try Exercise 31.**

Approximations for the zeros of the quadratic function $f(x) = x^2 - 6x - 10$ in Example 3 can be found using the Zero method.

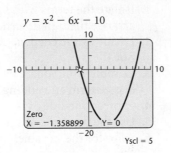

$y = x^2 - 6x - 10$

Zero
X = −1.358899    Y= 0

Yscl = 5

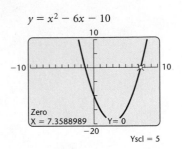

$y = x^2 - 6x - 10$

Zero
X = 7.3588989    Y= 0

Yscl = 5

Before we can complete the square, the coefficient of the $x^2$-term must be 1. When it is not, we divide by the $x^2$-coefficient on both sides of the equation.

**EXAMPLE 4**  Solve: $2x^2 - 1 = 3x$.

*Solution*  We have

$$2x^2 - 1 = 3x$$

$$2x^2 - 3x - 1 = 0$$  Subtracting $3x$. We are unable to factor the result.

$$2x^2 - 3x = 1$$  Adding 1

$$x^2 - \frac{3}{2}x = \frac{1}{2}$$  Dividing by 2 to make the $x^2$-coefficient 1

$$x^2 - \frac{3}{2}x + \frac{9}{16} = \frac{1}{2} + \frac{9}{16}$$  Completing the square: $\frac{1}{2}\left(-\frac{3}{2}\right) = -\frac{3}{4}$ and $\left(-\frac{3}{4}\right)^2 = \frac{9}{16}$; adding $\frac{9}{16}$

$$\left(x - \frac{3}{4}\right)^2 = \frac{17}{16}$$  Factoring and simplifying

$$x - \frac{3}{4} = \pm\frac{\sqrt{17}}{4}$$  Using the principle of square roots and the quotient rule for radicals

$$x = \frac{3}{4} \pm \frac{\sqrt{17}}{4}$$  Adding $\frac{3}{4}$

$$x = \frac{3 \pm \sqrt{17}}{4}.$$

The solutions are

$$\frac{3 + \sqrt{17}}{4} \quad \text{and} \quad \frac{3 - \sqrt{17}}{4}, \quad \text{or} \quad \frac{3 \pm \sqrt{17}}{4}.$$

**Now Try Exercise 35.**

> To solve a quadratic equation by completing the square:
>
> 1. Isolate the terms with variables on one side of the equation and arrange them in descending order.
> 2. Divide by the coefficient of the squared term if that coefficient is not 1.
> 3. Complete the square by finding half the coefficient of the first-degree term and adding its square on both sides of the equation.
> 4. Express one side of the equation as the square of a binomial.
> 5. Use the principle of square roots.
> 6. Solve for the variable.

## Using the Quadratic Formula

Because completing the square works for *any* quadratic equation, it can be used to solve the general quadratic equation $ax^2 + bx + c = 0$ for $x$. The result will be a formula that can be used to solve any quadratic equation quickly.

Consider any quadratic equation in standard form:

$$ax^2 + bx + c = 0, \quad a \neq 0.$$

For now, we assume that $a > 0$ and solve by completing the square. As the steps are carried out, compare them with those of Example 4.

$$\begin{array}{ll} ax^2 + bx + c = 0 & \text{Standard form} \\[4pt] ax^2 + bx = -c & \text{Adding } -c \\[4pt] x^2 + \dfrac{b}{a}x = -\dfrac{c}{a} & \text{Dividing by } a \end{array}$$

Half of $\dfrac{b}{a}$ is $\dfrac{b}{2a}$, and $\left(\dfrac{b}{2a}\right)^2 = \dfrac{b^2}{4a^2}$. Thus we add $\dfrac{b^2}{4a^2}$:

$$x^2 + \frac{b}{a}x + \frac{b^2}{4a^2} = -\frac{c}{a} + \frac{b^2}{4a^2} \qquad \text{Adding } \frac{b^2}{4a^2} \text{ to complete the square}$$

$$\left(x + \frac{b}{2a}\right)^2 = -\frac{4ac}{4a^2} + \frac{b^2}{4a^2} \qquad \begin{array}{l}\text{Factoring on the left; finding a} \\ \text{common denominator on the} \\ \text{right: } -\frac{c}{a} = -\frac{c}{a} \cdot \frac{4a}{4a} = -\frac{4ac}{4a^2}\end{array}$$

$$\left(x + \frac{b}{2a}\right)^2 = \frac{b^2 - 4ac}{4a^2}$$

$$x + \frac{b}{2a} = \pm \frac{\sqrt{b^2 - 4ac}}{2a} \qquad \begin{array}{l}\text{Using the principle of square roots} \\ \text{and the quotient rule for radicals.} \\ \text{Since } a > 0, \sqrt{4a^2} = 2a.\end{array}$$

$$x = -\frac{b}{2a} \pm \frac{\sqrt{b^2 - 4ac}}{2a} \qquad \text{Adding } -\frac{b}{2a}$$

$$x = \frac{-b \pm \sqrt{b^2 - 4ac}}{2a}.$$

It can also be shown that this result holds if $a < 0$.

---

**THE QUADRATIC FORMULA**

The solutions of $ax^2 + bx + c = 0$, $a \neq 0$, are given by

$$x = \frac{-b \pm \sqrt{b^2 - 4ac}}{2a}.$$

---

**EXAMPLE 5**   Solve $3x^2 + 2x = 7$. Find exact solutions and approximate solutions rounded to three decimal places.

## Algebraic Solution

After writing the equation in standard form, we are unable to factor, so we identify $a$, $b$, and $c$ in order to use the quadratic formula:

$$3x^2 + 2x = 7$$
$$3x^2 + 2x - 7 = 0;$$
$$a = 3, \quad b = 2, \quad c = -7.$$

We then use the quadratic formula:

$$x = \frac{-b \pm \sqrt{b^2 - 4ac}}{2a}$$

$$= \frac{-2 \pm \sqrt{2^2 - 4(3)(-7)}}{2(3)} \quad \text{Substituting}$$

$$= \frac{-2 \pm \sqrt{4 + 84}}{6} = \frac{-2 \pm \sqrt{88}}{6}$$

$$= \frac{-2 \pm \sqrt{4 \cdot 22}}{6} = \frac{-2 \pm 2\sqrt{22}}{6} = \frac{2(-1 \pm \sqrt{22})}{2 \cdot 3}$$

$$= \frac{2}{2} \cdot \frac{-1 \pm \sqrt{22}}{3} = \frac{-1 \pm \sqrt{22}}{3}.$$

The exact solutions are

$$\frac{-1 - \sqrt{22}}{3} \quad \text{and} \quad \frac{-1 + \sqrt{22}}{3}.$$

Using a calculator, we approximate the solutions to be $-1.897$ and $1.230$.

## Graphical Solution

Using the Intersect method, we graph $y_1 = 3x^2 + 2x$ and $y_2 = 7$ and use the INTERSECT feature to find the coordinates of the points of intersection. The first coordinates of these points are the solutions of the equation $y_1 = y_2$, or $3x^2 + 2x = 7$.

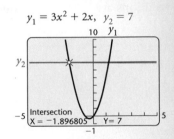

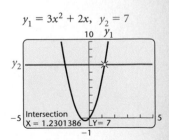

The solutions of $3x^2 + 2x = 7$ are approximately $-1.897$ and $1.230$. We could also write the equation in standard form, $3x^2 + 2x - 7 = 0$, and use the Zero method.

**Now Try Exercise 41.**

Not all quadratic equations can be solved graphically.

**EXAMPLE 6**   Solve: $x^2 + 5x + 8 = 0$.

### Algebraic Solution

To find the solutions, we use the quadratic formula. For $x^2 + 5x + 8 = 0$, we have

$$a = 1, \quad b = 5, \quad c = 8;$$

$$x = \frac{-b \pm \sqrt{b^2 - 4ac}}{2a}$$

$$= \frac{-5 \pm \sqrt{5^2 - 4(1)(8)}}{2 \cdot 1} \qquad \text{Substituting}$$

$$= \frac{-5 \pm \sqrt{25 - 32}}{2} \qquad \text{Simplifying}$$

$$= \frac{-5 \pm \sqrt{-7}}{2}$$

$$= \frac{-5 \pm \sqrt{7}i}{2}.$$

The solutions are $-\dfrac{5}{2} - \dfrac{\sqrt{7}}{2}i$ and $-\dfrac{5}{2} + \dfrac{\sqrt{7}}{2}i$.

### Graphical Solution

The graph of the function $f(x) = x^2 + 5x + 8$ shows no $x$-intercepts.

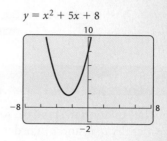

$y = x^2 + 5x + 8$

Thus the function has no real-number zeros and there are no real-number solutions of the associated equation $x^2 + 5x + 8 = 0$. This is a quadratic equation that cannot be solved graphically.

**Now Try Exercise 47.**

## ■ The Discriminant

From the quadratic formula, we know that the solutions $x_1$ and $x_2$ of a quadratic equation are given by

$$x_1 = \frac{-b + \sqrt{b^2 - 4ac}}{2a} \quad \text{and} \quad x_2 = \frac{-b - \sqrt{b^2 - 4ac}}{2a}.$$

The expression $b^2 - 4ac$ shows the nature of the solutions. This expression is called the **discriminant**. If it is 0, then it makes no difference whether we choose the plus sign or the minus sign in the formula. That is, $x_1 = -\dfrac{b}{2a} = x_2$, so there is just one solution. In this case, we sometimes say that there is one repeated real solution. If the discriminant is positive, there will be two different real solutions. If it is negative, we will be taking the square root of a negative number; hence there will be two imaginary-number solutions, and they will be complex conjugates.

> **DISCRIMINANT**
>
> For $ax^2 + bx + c = 0$, where $a$, $b$, and $c$ are real numbers:
>
> $b^2 - 4ac = 0 \rightarrow$ One real-number solution;
>
> $b^2 - 4ac > 0 \rightarrow$ Two different real-number solutions;
>
> $b^2 - 4ac < 0 \rightarrow$ Two different imaginary-number solutions, complex conjugates.

In Example 5, the discriminant, 88, is positive, indicating that there are two different real-number solutions. The negative discriminant, $-7$, in Example 6 indicates that there are two different imaginary-number solutions.

### ■ Equations Reducible to Quadratic

Some equations can be treated as quadratic, provided that we make a suitable substitution. For example, consider the following:

$$x^4 - 5x^2 + 4 = 0$$
$$(x^2)^2 - 5x^2 + 4 = 0 \qquad x^4 = (x^2)^2$$
$$u^2 - 5u + 4 = 0. \qquad \text{Substituting } u \text{ for } x^2$$

The equation $u^2 - 5u + 4 = 0$ can be solved for $u$ by factoring or using the quadratic formula. Then we can reverse the substitution, replacing $u$ with $x^2$, and solve for $x$. Equations like the one above are said to be **reducible to quadratic**, or **quadratic in form**.

**EXAMPLE 7** Solve: $x^4 - 5x^2 + 4 = 0$.

**Algebraic Solution**

We let $u = x^2$ and substitute:

$$u^2 - 5u + 4 = 0 \qquad \text{Substituting } u \text{ for } x^2$$
$$(u - 1)(u - 4) = 0 \qquad \text{Factoring}$$
$$u - 1 = 0 \quad or \quad u - 4 = 0 \qquad \text{Using the principle of zero products}$$
$$u = 1 \quad or \qquad u = 4.$$

Don't stop here! We must solve for the original variable. We substitute $x^2$ for $u$ and solve for $x$:

$$x^2 = 1 \quad or \quad x^2 = 4$$
$$x = \pm 1 \quad or \quad x = \pm 2. \qquad \text{Using the principle of square roots}$$

The solutions are $-1$, $1$, $-2$, and $2$.

**Graphical Solution**

Using the Zero method, we graph the function $y = x^4 - 5x^2 + 4$ and use the ZERO feature to find the zeros.

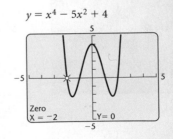

$$y = x^4 - 5x^2 + 4$$

The leftmost zero is $-2$. Using the ZERO feature three more times, we find that the other zeros are $-1$, $1$, and $2$. Thus the solutions of $x^4 - 5x^2 + 4 = 0$ are $-2$, $-1$, $1$, and $2$.

**Now Try Exercise 91.**

**EXAMPLE 8**  Solve: $t^{2/3} - 2t^{1/3} - 3 = 0$.

***Solution***  We let $u = t^{1/3}$ and substitute:

$$t^{2/3} - 2t^{1/3} - 3 = 0$$
$$(t^{1/3})^2 - 2t^{1/3} - 3 = 0$$
$$u^2 - 2u - 3 = 0 \qquad \text{Substituting } u \text{ for } t^{1/3}$$
$$(u + 1)(u - 3) = 0 \qquad \text{Factoring}$$
$$u + 1 = 0 \quad or \quad u - 3 = 0 \qquad \text{Using the principle of}$$
$$\qquad\qquad\qquad\qquad\qquad\qquad\qquad \text{zero products}$$
$$u = -1 \quad or \qquad u = 3.$$

Now we must solve for the original variable, $t$. We substitute $t^{1/3}$ for $u$ and solve for $t$:

$$t^{1/3} = -1 \quad or \quad t^{1/3} = 3$$
$$(t^{1/3})^3 = (-1)^3 \quad or \quad (t^{1/3})^3 = 3^3 \qquad \text{Cubing on both sides}$$
$$t = -1 \quad or \qquad t = 27.$$

The solutions are $-1$ and $27$.

**Now Try Exercise 97.**

## ■ Applications

Some applied problems can be translated to quadratic equations.

**EXAMPLE 9**  *Time of a Free Fall.*  The Burj Khalifa tower (also known as the Burj Dubai or the Dubai Tower) in the United Arab Emirates is 2717 ft tall. How long would it take an object dropped from the top to reach the ground?

***Solution***

1. **Familiarize.** The formula $s = 16t^2$ is used to approximate the distance $s$, in feet, that an object falls freely from rest in $t$ seconds. In this case, the distance is 2717 ft.

2. **Translate.** We substitute 2717 for $s$ in the formula:

$$2717 = 16t^2.$$

**3. Carry out.**

| **Algebraic Solution** | **Graphical Solution** |
|---|---|
| We use the principle of square roots: | Using the Intersect method (see Fig. 1), we replace $t$ with $x$, graph $y_1 = 2717$ and $y_2 = 16x^2$, and find the first coordinates of the points of intersection. Time cannot be negative in this application, so we need find only the point of intersection with a positive first coordinate. Since $y_1 = 2717$, we must choose a viewing window with Ymax greater than this value. Trial and error shows that a good choice is $[-20, 20, 0, 3000]$, with Xscl = 5 and Yscl = 500. |

$$2717 = 16t^2$$

$$\frac{2717}{16} = t^2 \qquad \text{Dividing by 16}$$

$$\sqrt{\frac{2717}{16}} = t \qquad \begin{array}{l}\text{Taking the positive}\\\text{square root. Time}\\\text{cannot be negative}\\\text{in this application.}\end{array}$$

$$13.03 \approx t.$$

We see that $x \approx 13.03$. Finding the solution with the Graphing Calculator, Appcylon LLC, app is shown in Fig. 2.

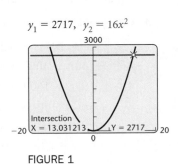

$y_1 = 2717, \quad y_2 = 16x^2$

**FIGURE 1**

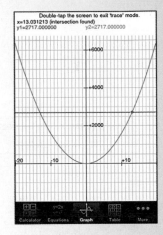

**FIGURE 2**

**4. Check.** In 13.03 sec, a dropped object would travel a distance of $16(13.03)^2$, or about 2717 ft. The answer checks.

**5. State.** It would take about 13.03 sec for an object dropped from the top of the Burj Khalifa to reach the ground.

**Now Try Exercise 111.**

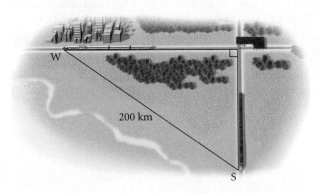

200 km

**EXAMPLE 10** *Train Speeds.* Two trains leave a station at the same time. One train travels due west, and the other travels due south. The train traveling west travels 20 km/h faster than the train traveling south. After 2 hr, the trains are 200 km apart. Find the speed of each train.

*Solution*

1. **Familiarize.** We let $r =$ the speed of the train traveling south, in kilometers per hour. Then $r + 20 =$ the speed of the train traveling west, in kilometers per hour. We use the motion formula $d = rt$, where $d$ is the distance, $r$ is the rate (or speed), and $t$ is the time. After 2 hr, the train traveling south has traveled $2r$ kilometers, and the train traveling west has traveled $2(r + 20)$ kilometers. We add these distances to the drawing.

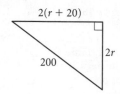

2. **Translate.** We use the Pythagorean theorem, $a^2 + b^2 = c^2$, where $a$ and $b$ are the lengths of the legs of a right triangle and $c$ is the length of the hypotenuse:

$$[2(r + 20)]^2 + (2r)^2 = 200^2.$$

3. **Carry out.** We solve the equation:

$$\begin{aligned}
[2(r + 20)]^2 + (2r)^2 &= 200^2 \\
4(r^2 + 40r + 400) + 4r^2 &= 40{,}000 \\
4r^2 + 160r + 1600 + 4r^2 &= 40{,}000 \\
8r^2 + 160r + 1600 &= 40{,}000 \qquad \text{Collecting like terms}\\
8r^2 + 160r - 38{,}400 &= 0 \qquad \text{Subtracting 40,000}\\
r^2 + 20r - 4800 &= 0 \qquad \text{Dividing by 8}\\
(r + 80)(r - 60) &= 0 \qquad \text{Factoring}\\
r + 80 = 0 \quad or \quad r - 60 &= 0 \qquad \text{Principle of zero products}\\
r = -80 \quad or \qquad\quad r &= 60.
\end{aligned}$$

We also can solve this equation with a graphing calculator using the Intersect method, as shown in the window at left.

4. **Check.** Since speed cannot be negative, we need check only 60. If the speed of the train traveling south is 60 km/h, then the speed of the train traveling west is $60 + 20$, or 80 km/h. In 2 hr, the train heading south travels $60 \cdot 2$, or 120 km, and the train heading west travels $80 \cdot 2$, or 160 km. Then they are $\sqrt{120^2 + 160^2}$, or $\sqrt{40{,}000}$, or 200 km apart. The answer checks.

5. **State.** The speed of the train heading south is 60 km/h, and the speed of the train heading west is 80 km/h.    **Now Try Exercise 113.**

---

| THE PYTHAGOREAN THEOREM |
| :---: |
| REVIEW SECTION **R.7.** |

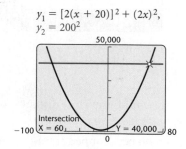

$y_1 = [2(x + 20)]^2 + (2x)^2,$
$y_2 = 200^2$

## CONNECTING THE CONCEPTS

### Zeros, Solutions, and Intercepts

The zeros of a function $y = f(x)$ are also the solutions of the equation $f(x) = 0$, and the real-number zeros are the first coordinates of the x-intercepts of the graph of the function.

| FUNCTION | ZEROS OF THE FUNCTION; SOLUTIONS OF THE EQUATION | X-INTERCEPTS OF THE GRAPH |
|---|---|---|
| **Linear Function** $f(x) = 2x - 4$, or $y = 2x - 4$ | To find the **zero** of $f(x)$, we solve $f(x) = 0$: $$2x - 4 = 0$$ $$2x = 4$$ $$x = 2.$$ The **solution** of the equation $2x - 4 = 0$ is $2$. This is the zero of the function $f(x) = 2x - 4$; that is, $f(2) = 0$. | The zero of $f(x)$ is the first coordinate of the **x-intercept** of the graph of $y = f(x)$. 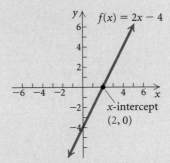 |
| **Quadratic Function** $g(x) = x^2 - 3x - 4$, or $y = x^2 - 3x - 4$ | To find the **zeros** of $g(x)$, we solve $g(x) = 0$: $$x^2 - 3x - 4 = 0$$ $$(x + 1)(x - 4) = 0$$ $$x + 1 = 0 \quad or \quad x - 4 = 0$$ $$x = -1 \quad or \quad x = 4.$$ The **solutions** of the equation $x^2 - 3x - 4 = 0$ are $-1$ and $4$. They are the zeros of the function $g(x) = x^2 - 3x - 4$; that is, $g(-1) = 0$ and $g(4) = 0$. | The real-number zeros of $g(x)$ are the first coordinates of the **x-intercepts** of the graph of $y = g(x)$. 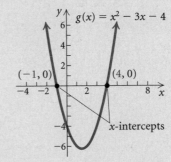 |

## 3.2 Exercise Set

*Solve.*

**1.** $(2x - 3)(3x - 2) = 0$

**2.** $(5x - 2)(2x + 3) = 0$

**3.** $x^2 - 8x - 20 = 0$

**4.** $x^2 + 6x + 8 = 0$

**5.** $3x^2 + x - 2 = 0$

**6.** $10x^2 - 16x + 6 = 0$

**7.** $4x^2 - 12 = 0$

**8.** $6x^2 = 36$

**9.** $3x^2 = 21$

**10.** $2x^2 - 20 = 0$

**11.** $5x^2 + 10 = 0$

**12.** $4x^2 + 12 = 0$

**13.** $x^2 + 16 = 0$

**14.** $x^2 + 25 = 0$

**15.** $2x^2 = 6x$

**16.** $18x + 9x^2 = 0$

**17.** $3y^3 - 5y^2 - 2y = 0$

**18.** $3t^3 + 2t = 5t^2$

**19.** $7x^3 + x^2 - 7x - 1 = 0$
(*Hint*: Factor by grouping.)

**20.** $3x^3 + x^2 - 12x - 4 = 0$
(*Hint*: Factor by grouping.)

In Exercises 21–28, use the given graph to find (**a**) the
*x*-intercepts and (**b**) the zeros of the function.

**21.**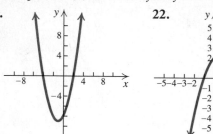

**22.**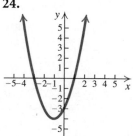

**23.**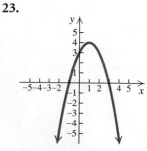

**24.**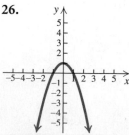

**25.**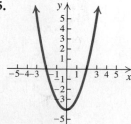

**26.**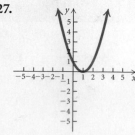

**27.**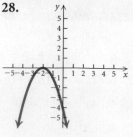

**28.**

*Solve by completing the square to obtain exact solutions.*

**29.** $x^2 + 6x = 7$

**30.** $x^2 + 8x = -15$

**31.** $x^2 = 8x - 9$

**32.** $x^2 = 22 + 10x$

**33.** $x^2 + 8x + 25 = 0$

**34.** $x^2 + 6x + 13 = 0$

**35.** $3x^2 + 5x - 2 = 0$

**36.** $2x^2 - 5x - 3 = 0$

*Use the quadratic formula to find exact solutions.*

**37.** $x^2 - 2x = 15$

**38.** $x^2 + 4x = 5$

**39.** $5m^2 + 3m = 2$

**40.** $2y^2 - 3y - 2 = 0$

**41.** $3x^2 + 6 = 10x$

**42.** $3t^2 + 8t + 3 = 0$

**43.** $x^2 + x + 2 = 0$

**44.** $x^2 + 1 = x$

**45.** $5t^2 - 8t = 3$

**46.** $5x^2 + 2 = x$

**47.** $3x^2 + 4 = 5x$

**48.** $2t^2 - 5t = 1$

**49.** $x^2 - 8x + 5 = 0$

**50.** $x^2 - 6x + 3 = 0$

**51.** $3x^2 + x = 5$

**52.** $5x^2 + 3x = 1$

**53.** $2x^2 + 1 = 5x$

**54.** $4x^2 + 3 = x$

**55.** $5x^2 + 2x = -2$

**56.** $3x^2 + 3x = -4$

*For each of the following, find the discriminant, $b^2 - 4ac$,
and then determine whether one real-number solution,
two different real-number solutions, or two different
imaginary-number solutions exist.*

**57.** $4x^2 = 8x + 5$

**58.** $4x^2 - 12x + 9 = 0$

**59.** $x^2 + 3x + 4 = 0$

**60.** $x^2 - 2x + 4 = 0$

**61.** $5t^2 - 7t = 0$

**62.** $5t^2 - 4t = 11$

*Solve graphically. Round solutions to three decimal
places, where appropriate.*

**63.** $x^2 - 8x + 12 = 0$

**64.** $5x^2 + 42x + 16 = 0$

**65.** $7x^2 - 43x + 6 = 0$

**66.** $10x^2 - 23x + 12 = 0$

**67.** $6x + 1 = 4x^2$

**68.** $3x^2 + 5x = 3$

**69.** $2x^2 - 4 = 5x$

**70.** $4x^2 - 2 = 3x$

*Find the zeros of the function algebraically. Give exact
answers.*

**71.** $f(x) = x^2 + 6x + 5$

**72.** $f(x) = x^2 - x - 2$

**73.** $f(x) = x^2 - 3x - 3$

**74.** $f(x) = 3x^2 + 8x + 2$

**75.** $f(x) = x^2 - 5x + 1$

**76.** $f(x) = x^2 - 3x - 7$

**77.** $f(x) = x^2 + 2x - 5$

**78.** $f(x) = x^2 - x - 4$

**79.** $f(x) = 2x^2 - x + 4$

**80.** $f(x) = 2x^2 + 3x + 2$

**81.** $f(x) = 3x^2 - x - 1$

**82.** $f(x) = 3x^2 + 5x + 1$

**83.** $f(x) = 5x^2 - 2x - 1$

**84.** $f(x) = 4x^2 - 4x - 5$

**85.** $f(x) = 4x^2 + 3x - 3$

**86.** $f(x) = x^2 + 6x - 3$

*Use a graphing calculator to find the zeros of the function. Round to three decimal places.*

**87.** $f(x) = 3x^2 + 2x - 4$

**88.** $f(x) = 9x^2 - 8x - 7$

**89.** $f(x) = 5.02x^2 - 4.19x - 2.057$

**90.** $f(x) = 1.21x^2 - 2.34x - 5.63$

*Solve.*

**91.** $x^4 - 3x^2 + 2 = 0$

**92.** $x^4 + 3 = 4x^2$

**93.** $x^4 + 3x^2 = 10$

**94.** $x^4 - 8x^2 = 9$

**95.** $y^4 + 4y^2 - 5 = 0$

**96.** $y^4 - 15y^2 - 16 = 0$

**97.** $x - 3\sqrt{x} - 4 = 0$
(*Hint*: Let $u = \sqrt{x}$.)

**98.** $2x - 9\sqrt{x} + 4 = 0$

**99.** $m^{2/3} - 2m^{1/3} - 8 = 0$
(*Hint*: Let $u = m^{1/3}$.)

**100.** $t^{2/3} + t^{1/3} - 6 = 0$

**101.** $x^{1/2} - 3x^{1/4} + 2 = 0$

**102.** $x^{1/2} - 4x^{1/4} = -3$

**103.** $(2x - 3)^2 - 5(2x - 3) + 6 = 0$
(*Hint*: Let $u = 2x - 3$.)

**104.** $(3x + 2)^2 + 7(3x + 2) - 8 = 0$

**105.** $(2t^2 + t)^2 - 4(2t^2 + t) + 3 = 0$

**106.** $12 = (m^2 - 5m)^2 + (m^2 - 5m)$

*Multigenerational Households.* *After declining between 1940 and 1980, the number of multigenerational American households has been increasing since 1980. The function $h(x) = 0.012x^2 - 0.583x + 35.727$ can be used to estimate the number of multigenerational households in the United States, in millions, x years after 1940 (Source: Pew Research Center). Use this function for Exercises 107 and 108.*

**107.** In what year were there 40 million multigenerational households?

**108.** In what year were there 55 million multigenerational households?

*TV Channels.* *The number of TV channels that the average U.S. home receives has been soaring in recent years. The function $t(x) = 0.16x^2 + 0.46x + 21.36$ can be used to estimate this number, where x is the number of years after 1985 (Source: Nielsen Media Research, National People Meter Sample). Use this function for Exercises 109 and 110.*

**109.** In what year did the average U.S. household receive 130 channels?

**110.** In what year did the average U.S. household receive 50 channels?

*Time of a Free Fall.    The formula $s = 16t^2$ is used to approximate the distance s, in feet, that an object falls freely from rest in t seconds. Use this formula for Exercises 111 and 112.*

**111.** The Taipei 101 Tower, also known as the Taipei Financial Center, in Taipei, Taiwan, is 1670 ft tall. How long would it take an object dropped from the top to reach the ground?

**112.** At 630 ft, the Gateway Arch in St. Louis is the tallest man-made monument in the United States. How long would it take an object dropped from the top to reach the ground?

**113.** The length of a rectangular poster is 1 ft more than the width, and a diagonal of the poster is 5 ft. Find the length and the width.

**114.** One leg of a right triangle is 7 cm less than the length of the other leg. The length of the hypotenuse is 13 cm. Find the lengths of the legs.

**115.** One number is 5 greater than another. The product of the numbers is 36. Find the numbers.

**116.** One number is 6 less than another. The product of the numbers is 72. Find the numbers.

**117.** *Box Construction.*    An open box is made from a 10-cm by 20-cm piece of tin by cutting a square from each corner and folding up the edges. The area of the resulting base is 96 cm². What is the length of the sides of the squares?

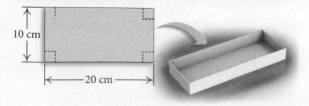

**118.** *Petting Zoo Dimensions.*    At the Glen Island Zoo, 170 m of fencing was used to enclose a petting area of 1750 m². Find the dimensions of the petting area.

**119.** *Dimensions of a Rug.*    Find the dimensions of a Persian rug whose perimeter is 28 ft and whose area is 48 ft².

**120.** *Picture Frame Dimensions.*    The frame on a picture is 8 in. by 10 in. outside and is of uniform width. What is the width of the frame if 48 in² of the picture shows?

8 in.

10 in.

*State whether the function is linear or quadratic.*

**121.** $f(x) = 4 - 5x$

**122.** $f(x) = 4 - 5x^2$

**123.** $f(x) = 7x^2$

**124.** $f(x) = 23x + 6$

**125.** $f(x) = 1.2x - (3.6)^2$

**126.** $f(x) = 2 - x - x^2$

## Skill Maintenance

*Spending on Antipsychotic Drugs.    The amount of spending on antipsychotic drugs, used to treat schizophrenia and other conditions, recently edged out cholesterol medications at the top of U.S. sales charts. The function $a(x) = 1.24x + 9.24$ can be used to estimate the amount of spending per year on antipsychotic drugs in the United States, in billions of dollars, x years after 2004 (Source: IMS Health). Use this function for Exercises 127 and 128.*

**127.** Estimate the amount spent on antipsychotic drugs in the United States in 2010.

**128.** When will the amount of spending on antipsychotic drugs reach $24 billion?

*Determine whether the graph is symmetric with respect to the x-axis, the y-axis, and the origin.*

**129.** $3x^2 + 4y^2 = 5$        **130.** $y^3 = 6x^2$

*Determine whether the function is even, odd, or neither even nor odd.*

**131.** $f(x) = 2x^3 - x$

**132.** $f(x) = 4x^2 + 2x - 3$

## Synthesis

*For each equation in Exercises 133–136, under the given condition:* **(a)** *Find k and* **(b)** *find a second solution.*

**133.** $kx^2 - 17x + 33 = 0$; one solution is 3

**134.** $kx^2 - 2x + k = 0$; one solution is $-3$

**135.** $x^2 - kx + 2 = 0$; one solution is $1 + i$

**136.** $x^2 - (6 + 3i)x + k = 0$; one solution is 3

*Solve.*

**137.** $(x - 2)^3 = x^3 - 2$

**138.** $(x + 1)^3 = (x - 1)^3 + 26$

**139.** $(6x^3 + 7x^2 - 3x)(x^2 - 7) = 0$

**140.** $\left(x - \frac{1}{5}\right)\left(x^2 - \frac{1}{4}\right) + \left(x - \frac{1}{5}\right)\left(x^2 + \frac{1}{8}\right) = 0$

**141.** $x^2 + x - \sqrt{2} = 0$

**142.** $x^2 + \sqrt{5}x - \sqrt{3} = 0$

**143.** $2t^2 + (t - 4)^2 = 5t(t - 4) + 24$

**144.** $9t(t + 2) - 3t(t - 2) = 2(t + 4)(t + 6)$

**145.** $\sqrt{x - 3} - \sqrt[4]{x - 3} = 2$

**146.** $x^2 + 3x + 1 - \sqrt{x^2 + 3x + 1} = 8$

**147.** $\left(y + \frac{2}{y}\right)^2 + 3y + \frac{6}{y} = 4$

**148.** Solve $\frac{1}{2}at^2 + v_0 t + x_0 = 0$ for $t$.

---

## 3.3

# Analyzing Graphs of Quadratic Functions

- Find the vertex, the axis of symmetry, and the maximum or minimum value of a quadratic function using the method of completing the square.

- Graph quadratic functions.

- Solve applied problems involving maximum and minimum function values.

---

### Graphing Quadratic Functions of the Type $f(x) = a(x - h)^2 + k$

The graph of a quadratic function is called a **parabola**. The graph of every parabola evolves from the graph of the squaring function $f(x) = x^2$ using transformations.

### Exploring with Technology

Think of transformations and look for patterns. Consider the following functions:

$$y_1 = x^2, \quad y_2 = -0.4x^2,$$
$$y_3 = -0.4(x - 2)^2, \quad y_4 = -0.4(x - 2)^2 + 3.$$

Graph $y_1$ and $y_2$. How do you get from the graph of $y_1$ to $y_2$?

Graph $y_2$ and $y_3$. How do you get from the graph of $y_2$ to $y_3$?

Graph $y_3$ and $y_4$. How do you get from the graph of $y_3$ to $y_4$?

Consider the following functions:

$$y_1 = x^2, \quad y_2 = 2x^2,$$
$$y_3 = 2(x + 3)^2, \quad y_4 = 2(x + 3)^2 - 5.$$

Graph $y_1$ and $y_2$. How do you get from the graph of $y_1$ to $y_2$?

Graph $y_2$ and $y_3$. How do you get from the graph of $y_2$ to $y_3$?

Graph $y_3$ and $y_4$. How do you get from the graph of $y_3$ to $y_4$?

> **TRANSFORMATIONS**
>
> REVIEW SECTION **2.5.**

We get the graph of $f(x) = a(x - h)^2 + k$ from the graph of $f(x) = x^2$ as follows:

$$f(x) = x^2$$
$$\downarrow$$
$$f(x) = ax^2 \qquad \text{Vertical stretching or shrinking with a reflection across the } x\text{-axis if } a < 0$$
$$\downarrow$$
$$f(x) = a(x - h)^2 \qquad \text{Horizontal translation}$$
$$\downarrow$$
$$f(x) = a(x - h)^2 + k. \qquad \text{Vertical translation}$$

Consider the following graphs of the form $f(x) = a(x - h)^2 + k$. The point $(h, k)$ at which the graph turns is called the **vertex**. The maximum or minimum value of $f(x)$ occurs at the vertex. Each graph has a line $x = h$ that is called the **axis of symmetry**.

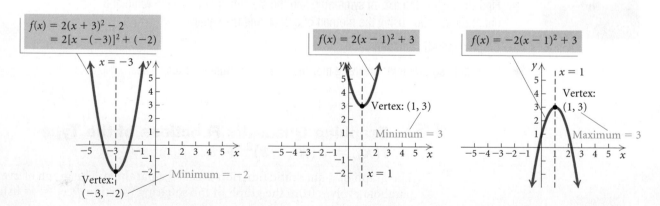

## CONNECTING THE CONCEPTS

### Graphing Quadratic Functions

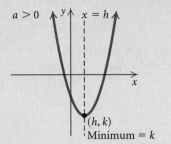

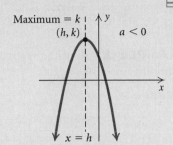

The graph of the function $f(x) = a(x - h)^2 + k$ is a parabola that

- opens up if $a > 0$ and down if $a < 0$;
- has $(h, k)$ as the vertex;
- has $x = h$ as the axis of symmetry;
- has $k$ as a minimum value (output) if $a > 0$;
- has $k$ as a maximum value if $a < 0$.

As we saw in Section 2.5, the constant $a$ serves to stretch or shrink the graph vertically. As a parabola is stretched vertically, it becomes narrower, and as it is shrunk vertically, it becomes wider. That is, as $|a|$ increases, the graph becomes narrower, and as $|a|$ gets close to 0, the graph becomes wider.

If the equation is in the form $f(x) = a(x - h)^2 + k$, we can learn a great deal about the graph without actually graphing the function.

| Function | $f(x) = 3\left(x - \frac{1}{4}\right)^2 - 2$ <br> $\quad = 3\left(x - \frac{1}{4}\right)^2 + (-2)$ | $g(x) = -3(x + 5)^2 + 7$ <br> $\quad = -3\left[x - (-5)\right]^2 + 7$ |
|---|---|---|
| **Vertex** | $\left(\frac{1}{4}, -2\right)$ | $(-5, 7)$ |
| **Axis of Symmetry** | $x = \frac{1}{4}$ | $x = -5$ |
| **Maximum** | None ($3 > 0$, so the graph opens up.) | 7 ($-3 < 0$, so the graph opens down.) |
| **Minimum** | $-2$ ($3 > 0$, so the graph opens up.) | None ($-3 < 0$, so the graph opens down.) |

Note that the vertex $(h, k)$ is used to find the maximum or minimum value of the function. The maximum or minimum value is the number $k$, *not* the ordered pair $(h, k)$.

### ■ Graphing Quadratic Functions of the Type $f(x) = ax^2 + bx + c, a \neq 0$

We now use a modification of the method of completing the square as an aid in graphing and analyzing quadratic functions of the form $f(x) = ax^2 + bx + c, a \neq 0$.

**EXAMPLE 1** Find the vertex, the axis of symmetry, and the maximum or minimum value of $f(x) = x^2 + 10x + 23$. Then graph the function.

*Solution* To express

$$f(x) = x^2 + 10x + 23$$

in the form

$$f(x) = a(x - h)^2 + k,$$

we complete the square on the terms involving $x$. To do so, we take half the coefficient of $x$ and square it, obtaining $(10/2)^2$, or 25. We now add and subtract that number on the *right side*:

$$f(x) = x^2 + 10x + 23 = x^2 + 10x + 25 - 25 + 23.$$

Since $25 - 25 = 0$, the new expression for the function is equivalent to the original expression. Note that this process differs from the one we used to complete the square in order to solve a quadratic equation, where we added the same number on both sides of the equation to obtain an equivalent equation. Instead, when we complete the square to write a function in the form $f(x) = a(x - h)^2 + k$, we add and subtract the same number on one side. The entire process is shown below:

$$
\begin{aligned}
f(x) &= x^2 + 10x + 23 & &\text{Note that 25 completes the square for } x^2 + 10x. \\
&= x^2 + 10x + 25 - 25 + 23 & &\text{Adding } 25 - 25, \text{ or 0, to the right side} \\
&= (x^2 + 10x + 25) - 25 + 23 & &\text{Regrouping} \\
&= (x + 5)^2 - 2 & &\text{Factoring and simplifying} \\
&= [x - (-5)]^2 + (-2). & &\text{Writing in the form } f(x) = a(x - h)^2 + k
\end{aligned}
$$

Keeping in mind that this function will have a minimum value since $a > 0$ ($a = 1$), from this form of the function we know the following:

Vertex: $(-5, -2)$;

Axis of symmetry: $x = -5$;

Minimum value of the function: $-2$.

To graph the function by hand, we first plot the vertex, $(-5, -2)$, and find several points on either side of $x = -5$. Then we plot these points and connect them with a smooth curve. We see that the points $(-4, -1)$ and $(-3, 2)$ are reflections of the points $(-6, -1)$ and $(-7, 2)$, respectively, across the axis of symmetry, $x = -5$.

| $x$ | $f(x)$ |
|-----|--------|
| $-5$ | $-2$ | ← Vertex |
| $-6$ | $-1$ |
| $-4$ | $-1$ |
| $-7$ | $2$ |
| $-3$ | $2$ |

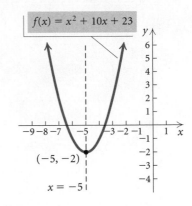

The graph of $f(x) = x^2 + 10x + 23$, or

$$f(x) = [x - (-5)]^2 + (-2),$$

shown above, is a shift of the graph of $y = x^2$ left 5 units and down 2 units.

**Now Try Exercise 3.**

Keep in mind that the axis of symmetry is not part of the graph; it is a characteristic of the graph. If you fold the graph on its axis of symmetry, the two halves of the graph will coincide.

**EXAMPLE 2** Find the vertex, the axis of symmetry, and the maximum or minimum value of $g(x) = x^2/2 - 4x + 8$. Then graph the function.

***Solution*** We complete the square in order to write the function in the form $g(x) = a(x - h)^2 + k$. First, we factor $\frac{1}{2}$ out of the first two terms. This makes the coefficient of $x^2$ within the parentheses 1:

$$g(x) = \frac{x^2}{2} - 4x + 8$$

$$= \frac{1}{2}(x^2 - 8x) + 8. \qquad \text{Factoring } \tfrac{1}{2} \text{ out of the first two terms:} \\ x^2/2 - 4x = \tfrac{1}{2} \cdot x^2 - \tfrac{1}{2} \cdot 8x$$

Next, we complete the square inside the parentheses: Half of $-8$ is $-4$, and $(-4)^2 = 16$. We add and subtract 16 inside the parentheses:

$$g(x) = \tfrac{1}{2}(x^2 - 8x + 16 - 16) + 8$$
$$= \tfrac{1}{2}(x^2 - 8x + 16) - \tfrac{1}{2} \cdot 16 + 8 \qquad \begin{array}{l}\textbf{Using the distributive}\\ \textbf{law to remove } -16 \textbf{ from}\\ \textbf{within the parentheses}\end{array}$$

$$= \tfrac{1}{2}(x^2 - 8x + 16) - 8 + 8$$
$$= \tfrac{1}{2}(x - 4)^2 + 0, \text{ or } \tfrac{1}{2}(x - 4)^2. \qquad \textbf{Factoring and simplifying}$$

We know the following:

Vertex: $(4, 0)$;

Axis of symmetry: $x = 4$;

Minimum value of the function: $0$.

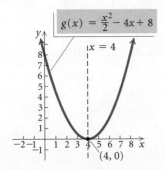

Finally, we plot the vertex and several points on either side of it and draw the graph of the function. The graph of $g$ is a vertical shrinking of the graph of $y = x^2$ along with a shift 4 units to the right.

**Now Try Exercise 9.**

**EXAMPLE 3** Find the vertex, the axis of symmetry, and the maximum or minimum value of $f(x) = -2x^2 + 10x - \frac{23}{2}$. Then graph the function.

***Solution*** We have

$$f(x) = -2x^2 + 10x - \frac{23}{2}$$
$$= -2(x^2 - 5x) - \frac{23}{2} \qquad \text{Factoring } -2 \text{ out of the first two terms}$$
$$= -2\left(x^2 - 5x + \frac{25}{4} - \frac{25}{4}\right) - \frac{23}{2} \qquad \text{Completing the square inside the parentheses}$$
$$= -2\left(x^2 - 5x + \frac{25}{4}\right) - 2\left(-\frac{25}{4}\right) - \frac{23}{2} \qquad \text{Using the distributive law to remove } -\frac{25}{4} \text{ from within the parentheses}$$
$$= -2\left(x^2 - 5x + \frac{25}{4}\right) + \frac{25}{2} - \frac{23}{2}$$
$$= -2\left(x - \frac{5}{2}\right)^2 + 1.$$

This form of the function yields the following:

Vertex: $\left(\frac{5}{2}, 1\right)$;

Axis of symmetry: $x = \frac{5}{2}$;

Maximum value of the function: 1.

The graph is found by shifting the graph of $f(x) = x^2$ right $\frac{5}{2}$ units, reflecting it across the $x$-axis, stretching it vertically, and shifting it up 1 unit.

**Now Try Exercise 13.**

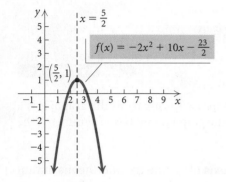

In many situations, we want to use a formula to find the coordinates of the vertex directly from the equation $f(x) = ax^2 + bx + c$. One way to develop such a formula is to first note that the $x$-coordinate of the vertex is centered between the $x$-intercepts, or zeros, of the function. By averaging the two solutions of $ax^2 + bx + c = 0$, we find a formula for the $x$-coordinate of the vertex:

$$x\text{-coordinate of vertex} = \frac{\dfrac{-b - \sqrt{b^2 - 4ac}}{2a} + \dfrac{-b + \sqrt{b^2 - 4ac}}{2a}}{2}$$

$$= \frac{\dfrac{-2b}{2a}}{2} = \frac{-\dfrac{b}{a}}{2}$$

$$= -\frac{b}{a} \cdot \frac{1}{2} = -\frac{b}{2a}.$$

We use this value of $x$ to find the $y$-coordinate of the vertex, $f\left(-\dfrac{b}{2a}\right)$.

---

**THE VERTEX OF A PARABOLA**

The **vertex** of the graph of $f(x) = ax^2 + bx + c$ is

$$\left(-\frac{b}{2a}, f\left(-\frac{b}{2a}\right)\right).$$

We calculate the     We substitute to
$x$-coordinate.       find the $y$-coordinate.

---

**EXAMPLE 4** For the function $f(x) = -x^2 + 14x - 47$:

**a)** Find the vertex.

**b)** Determine whether there is a maximum or minimum value and find that value.

**c)** Find the range.

**d)** On what intervals is the function increasing? decreasing?

*Solution* There is no need to graph the function.

**a)** The $x$-coordinate of the vertex is

$$-\frac{b}{2a} = -\frac{14}{2(-1)} = -\frac{14}{-2} = 7.$$

Since

$$f(7) = -7^2 + 14 \cdot 7 - 47 = -49 + 98 - 47 = 2,$$

the vertex is $(7, 2)$.

**b)** Since $a$ is negative $(a = -1)$, the graph opens down so the second coordinate of the vertex, 2, is the maximum value of the function.

**c)** The range is $(-\infty, 2]$.

**d)** Since the graph opens down, function values increase as we approach the vertex from the left and decrease as we move away from the vertex on the right. Thus the function is increasing on the interval $(-\infty, 7)$ and decreasing on $(7, \infty)$.    **Now Try Exercise 31.**

We can use a graphing calculator to do Example 4. Once we have graphed $y = -x^2 + 14x - 47$, we see that the graph opens down and thus has a maximum value. We can use the MAXIMUM feature to find the coordinates of the vertex. Using these coordinates, we can then find the maximum value and the range of the function along with the intervals on which the function is increasing or decreasing.

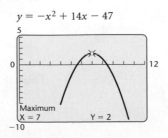

$$y = -x^2 + 14x - 47$$

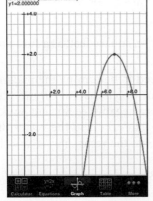

The Graphing Calculator, Appcylon LLC, app can be used to find the maximum value of the function in Example 4.

### ■ Applications

Many real-world situations involve finding the maximum or minimum value of a quadratic function.

**EXAMPLE 5** *Maximizing Area.* A landscaper has enough bricks to enclose a rectangular flower box below the picture window of the Jacobsens' house with 12 ft of brick wall. If the house forms one side of the rectangle, what is the maximum area that the landscaper can enclose? What dimensions of the flower box will yield this area?

*Solution* We will use the five-step problem-solving strategy.

**PROBLEM-SOLVING STRATEGY**

REVIEW SECTION **1.5.**

1. **Familiarize.** We first make a drawing of the situation, using $w$ to represent the width of the flower box, in feet. Then $(12 - 2w)$ feet of brick is available for the length. Suppose the flower box were 1 ft wide. Then its length would be $12 - 2 \cdot 1 = 10$ ft, and its area would be $(10 \text{ ft})(1 \text{ ft}) = 10 \text{ ft}^2$. If the flower box were 2 ft wide, its length would be $12 - 2 \cdot 2 = 8$ ft, and its area would be $(8 \text{ ft})(2 \text{ ft}) = 16 \text{ ft}^2$. This is larger than the first area we found, but we do not know if it is the maximum possible area. To find the maximum area, we will find a function that represents the area and then determine its maximum value.

2. **Translate.** We write a function for the area of the flower box. We have

$$A(w) = (12 - 2w)w \qquad A = lw; l = 12 - 2w$$
$$= -2w^2 + 12w,$$

where $A(w)$ is the area of the flower box, in square feet, as a function of the width $w$.

3. **Carry out.** To solve this problem, we need to determine the maximum value of $A(w)$ and find the dimensions for which that maximum occurs. Since $A$ is a quadratic function and $w^2$ has a negative coefficient, we know that the function has a maximum value that occurs at the vertex of the graph of the function. The first coordinate of the vertex, $(w, A(w))$, is

$$w = -\frac{b}{2a} = -\frac{12}{2(-2)} = -\frac{12}{-4} = 3.$$

Thus, if $w = 3$ ft, then the length $l = 12 - 2 \cdot 3 = 6$ ft, and the area is $(6 \text{ ft})(3 \text{ ft}) = 18 \text{ ft}^2$.

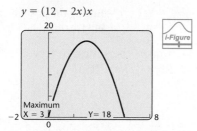

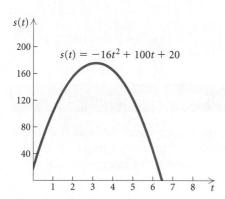

4. **Check.** As a partial check, we note that $18 \text{ ft}^2 > 16 \text{ ft}^2$, which is the larger area we found in a guess in the *Familiarize* step. As a more complete check, assuming that the function $A(w)$ is correct, we could examine a table of values for $A(w) = (12 - 2w)w$ and/or examine its graph.

5. **State.** The maximum possible area is $18 \text{ ft}^2$ when the flower box is 3 ft wide and 6 ft long. **Now Try Exercise 45.**

**EXAMPLE 6** *Height of a Rocket.* A model rocket is launched with an initial velocity of 100 ft/sec from the top of a hill that is 20 ft high. Its height, in feet, $t$ seconds after it has been launched is given by the function $s(t) = -16t^2 + 100t + 20$. Determine the time at which the rocket reaches its maximum height and find the maximum height.

*Solution*

1., 2. **Familiarize** and **Translate.** We are given the function in the statement of the problem: $s(t) = -16t^2 + 100t + 20$.

3. **Carry out.** We need to find the maximum value of the function and the value of $t$ for which it occurs. Since $s(t)$ is a quadratic function and $t^2$ has a negative coefficient, we know that the maximum value of the function occurs at the vertex of the graph of the function. The first coordinate of the vertex gives the time $t$ at which the rocket reaches its maximum height. It is

$$t = -\frac{b}{2a} = -\frac{100}{2(-16)} = -\frac{100}{-32} = 3.125.$$

The second coordinate of the vertex gives the maximum height of the rocket. We substitute in the function to find it:

$$s(3.125) = -16(3.125)^2 + 100(3.125) + 20 = 176.25.$$

4. **Check.** As a check, we can complete the square to write the function in the form $s(t) = a(t - h)^2 + k$ and determine the coordinates of the vertex from this form of the function. We get

$$s(t) = -16(t - 3.125)^2 + 176.25.$$

This confirms that the vertex is $(3.125, 176.25)$, so the answer checks.

5. **State.** The rocket reaches a maximum height of 176.25 ft 3.125 sec after it has been launched. **Now Try Exercise 41.**

**EXAMPLE 7** *Finding the Depth of a Well.* Two seconds after a chlorine tablet has been dropped into a well, a splash is heard. The speed of sound is 1100 ft/sec. How far is the top of the well from the water?

*Solution*

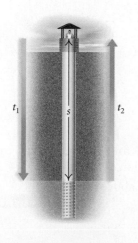

1. **Familiarize.** We first make a drawing and label it with known and unknown information. We let $s =$ the depth of the well, in feet, $t_1 =$ the time, in seconds, that it takes for the tablet to hit the water, and $t_2 =$ the time, in seconds, that it takes for the sound to reach the top of the well. This gives us the equation

$$t_1 + t_2 = 2. \qquad (1)$$

2. **Translate.** Can we find any relationship between the two times and the distance $s$? Often in problem solving you may need to look up related formulas in a physics book, another mathematics book, or on the Internet. We find that the formula $s = 16t^2$ gives the distance, in feet, that a dropped object falls in $t$ seconds. The time $t_1$ that it takes the tablet to hit the water can be found as follows:

$$s = 16t_1^2, \quad \text{or} \quad \frac{s}{16} = t_1^2, \quad \text{so} \quad t_1 = \frac{\sqrt{s}}{4}. \qquad \text{Taking the positive square root} \quad (2)$$

To find an expression for $t_2$, the time it takes the sound to travel to the top of the well, recall that *Distance = Rate · Time*. Thus,

$$s = 1100t_2, \quad \text{or} \quad t_2 = \frac{s}{1100}. \qquad (3)$$

We now have expressions for $t_1$ and $t_2$, both in terms of $s$. Substituting into equation (1), we obtain

$$t_1 + t_2 = 2, \quad \text{or} \quad \frac{\sqrt{s}}{4} + \frac{s}{1100} = 2. \qquad (4)$$

3. **Carry out.**

### Algebraic Solution

We solve equation (4) for $s$. Multiplying by 1100, we get

$$275\sqrt{s} + s = 2200, \quad \text{or} \quad s + 275\sqrt{s} - 2200 = 0.$$

This equation is reducible to quadratic with $u = \sqrt{s}$. Substituting, we get

$$u^2 + 275u - 2200 = 0.$$

Using the quadratic formula, we can solve for $u$:

$$u = \frac{-b \pm \sqrt{b^2 - 4ac}}{2a}$$

$$= \frac{-275 + \sqrt{275^2 - 4 \cdot 1 \cdot (-2200)}}{2 \cdot 1}$$

We want only the positive solution.

$$= \frac{-275 + \sqrt{84{,}425}}{2} \approx 7.78.$$

Since $u \approx 7.78$, we have

$$\sqrt{s} = 7.78$$

$$s \approx 60.5. \qquad \text{Squaring both sides}$$

### Graphical Solution

We use the Intersect method. It will probably require some trial and error to determine an appropriate window.

$$y_1 = \frac{\sqrt{x}}{4} + \frac{x}{1100}, \quad y_2 = 2$$

Intersection
X = 60.526879    Y = 2

4. **Check.** To check, we can substitute 60.5 for $s$ in equation (4) and see that $t_1 + t_2 \approx 2$. We leave the computation to the student.

5. **State.** The top of the well is about 60.5 ft above the water.

**Now Try Exercise 55.**

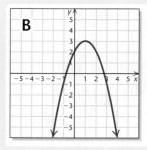

A

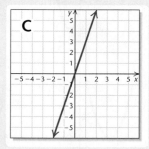

B

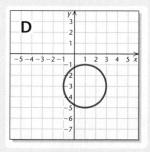

C

D

E

# Visualizing the Graph

Match the equation with its graph.

**1.** $y = 3x$

**2.** $y = -(x - 1)^2 + 3$

**3.** $(x + 2)^2 + (y - 2)^2 = 9$

**4.** $y = 3$

**5.** $2x - 3y = 6$

**6.** $(x - 1)^2 + (y + 3)^2 = 4$

**7.** $y = -2x + 1$

**8.** $y = 2x^2 - x - 4$

**9.** $x = -2$

**10.** $y = -3x^2 + 6x - 2$

**Answers on page A-19**

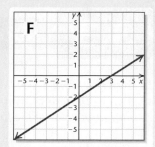

F

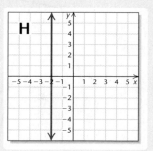

G

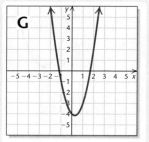

H

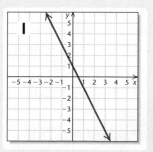

I

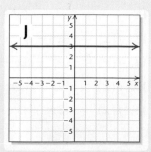

J

## 3.3 Exercise Set

In Exercises 1 and 2, use the given graph to find (**a**) the vertex; (**b**) the axis of symmetry; and (**c**) the maximum or minimum value of the function.

**1.**

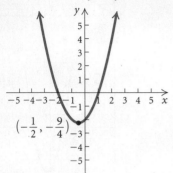

$\left(-\frac{1}{2}, -\frac{9}{4}\right)$

**2.**

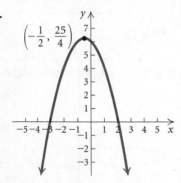

$\left(-\frac{1}{2}, \frac{25}{4}\right)$

In Exercises 3–16, (**a**) find the vertex; (**b**) find the axis of symmetry; (**c**) determine whether there is a maximum or minimum value and find that value; and (**d**) graph the function.

**3.** $f(x) = x^2 - 8x + 12$  **4.** $g(x) = x^2 + 7x - 8$

**5.** $f(x) = x^2 - 7x + 12$  **6.** $g(x) = x^2 - 5x + 6$

**7.** $f(x) = x^2 + 4x + 5$  **8.** $f(x) = x^2 + 2x + 6$

**9.** $g(x) = \dfrac{x^2}{2} + 4x + 6$  **10.** $g(x) = \dfrac{x^2}{3} - 2x + 1$

**11.** $g(x) = 2x^2 + 6x + 8$

**12.** $f(x) = 2x^2 - 10x + 14$

**13.** $f(x) = -x^2 - 6x + 3$

**14.** $f(x) = -x^2 - 8x + 5$

**15.** $g(x) = -2x^2 + 2x + 1$

**16.** $f(x) = -3x^2 - 3x + 1$

In Exercises 17–24, match the equation with one of the graphs (a)–(h), which follow.

**a)**

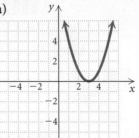

**b)**

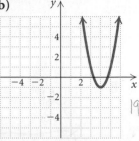

19

**c)**

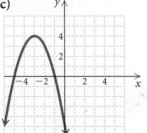

**d)**

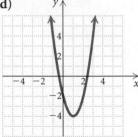

**e)**

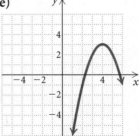

**f)**

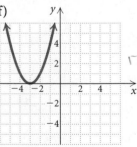

17

**g)**

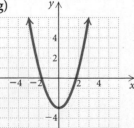

**h)**

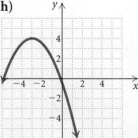

**17.** $y = (x + 3)^2$  **18.** $y = -(x - 4)^2 + 3$

**19.** $y = 2(x - 4)^2 - 1$  **20.** $y = x^2 - 3$

**21.** $y = -\frac{1}{2}(x + 3)^2 + 4$  **22.** $y = (x - 3)^2$

**23.** $y = -(x + 3)^2 + 4$  **24.** $y = 2(x - 1)^2 - 4$

*Determine whether the statement is true or false.*

**25.** The function $f(x) = -3x^2 + 2x + 5$ has a maximum value.

**26.** The vertex of the graph of $f(x) = ax^2 + bx + c$ is $-\dfrac{b}{2a}$.

**27.** The graph of $h(x) = (x + 2)^2$ can be obtained by translating the graph of $h(x) = x^2$ right 2 units.

**28.** The vertex of the graph of the function $g(x) = 2(x - 4)^2 - 1$ is $(-4, -1)$.

**29.** The axis of symmetry of the function $f(x) = -(x + 2)^2 - 4$ is $x = -2$.

**30.** The minimum value of the function $f(x) = 3(x - 1)^2 + 5$ is 5.

*In Exercises 31–40:*

**a)** *Find the vertex.*
**b)** *Determine whether there is a maximum or minimum value and find that value.*
**c)** *Find the range.*
**d)** *Find the intervals on which the function is increasing and the intervals on which the function is decreasing.*

**31.** $f(x) = x^2 - 6x + 5$

**32.** $f(x) = x^2 + 4x - 5$

**33.** $f(x) = 2x^2 + 4x - 16$

**34.** $f(x) = \frac{1}{2}x^2 - 3x + \frac{5}{2}$

**35.** $f(x) = -\frac{1}{2}x^2 + 5x - 8$

**36.** $f(x) = -2x^2 - 24x - 64$

**37.** $f(x) = 3x^2 + 6x + 5$

**38.** $f(x) = -3x^2 + 24x - 49$

**39.** $g(x) = -4x^2 - 12x + 9$

**40.** $g(x) = 2x^2 - 6x + 5$

**41.** *Height of a Ball.*   A ball is thrown directly upward from a height of 6 ft with an initial velocity of 20 ft/sec. The function $s(t) = -16t^2 + 20t + 6$ gives the height of the ball, in feet, $t$ seconds after it has been thrown. Determine the time at which the ball reaches its maximum height and find the maximum height.

**42.** *Height of a Projectile.*   A stone is thrown directly upward from a height of 30 ft with an initial velocity of 60 ft/sec. The height of the stone, in feet, $t$ seconds after it has been thrown is given by the function $s(t) = -16t^2 + 60t + 30$. Determine the time at which the stone reaches its maximum height and find the maximum height.

**43.** *Height of a Rocket.*   A model rocket is launched with an initial velocity of 120 ft/sec from a height of 80 ft. The height of the rocket, in feet, $t$ seconds after it has been launched is given by the function $s(t) = -16t^2 + 120t + 80$. Determine the time at which the rocket reaches its maximum height and find the maximum height.

**44.** *Height of a Rocket.*   A model rocket is launched with an initial velocity of 150 ft/sec from a height of 40 ft. The function $s(t) = -16t^2 + 150t + 40$ gives the height of the rocket, in feet, $t$ seconds after it has been launched. Determine the time at which the rocket reaches its maximum height and find the maximum height.

**45.** *Maximizing Volume.*   Mendoza Manufacturing plans to produce a one-compartment vertical file by bending the long side of a 10-in. by 18-in. sheet of plastic along two lines to form a ⊔-shape. How tall should the file be in order to maximize the volume that it can hold?

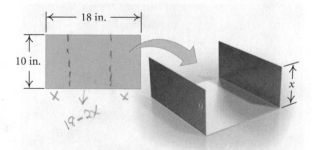

**46.** *Maximizing Area.*   A fourth-grade class decides to enclose a rectangular garden, using the side of the school as one side of the rectangle. What is the maximum area that the class can enclose using 32 ft of fence? What should the dimensions of the garden be in order to yield this area?

**47.** *Maximizing Area.* The sum of the base and the height of a triangle is 20 cm. Find the dimensions for which the area is a maximum.

**48.** *Maximizing Area.* The sum of the base and the height of a parallelogram is 69 cm. Find the dimensions for which the area is a maximum.

**49.** *Minimizing Cost.* Classic Furniture Concepts has determined that when $x$ hundred wooden chairs are built, the average cost per chair is given by

$$C(x) = 0.1x^2 - 0.7x + 1.625,$$

where $C(x)$ is in hundreds of dollars. How many chairs should be built in order to minimize the average cost per chair?

*Maximizing Profit.* *In business, profit is the difference between revenue and cost; that is,*

$$Total\ profit = Total\ revenue - Total\ cost,$$
$$P(x) = R(x) - C(x),$$

where $x$ is the number of units sold. Find the maximum profit and the number of units that must be sold in order to yield the maximum profit for each of the following.

**50.** $R(x) = 5x,\ C(x) = 0.001x^2 + 1.2x + 60$

**51.** $R(x) = 50x - 0.5x^2,\ C(x) = 10x + 3$

**52.** $R(x) = 20x - 0.1x^2,\ C(x) = 4x + 2$

**53.** *Maximizing Area.* A rancher needs to enclose two adjacent rectangular corrals, one for cattle and one for sheep. If a river forms one side of the corrals and 240 yd of fencing is available, what is the largest total area that can be enclosed?

**54.** *Norman Window.* A Norman window is a rectangle with a semicircle on top. Sky Blue Windows is designing a Norman window that will require 24 ft of trim on the outer edges. What dimensions will allow the maximum amount of light to enter a house?

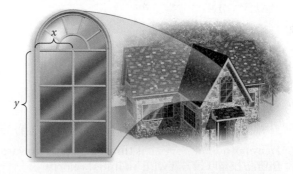

**55.** *Finding the Height of an Elevator Shaft.* Jenelle drops a screwdriver from the top of an elevator shaft. Exactly 5 sec later, she hears the sound of the screwdriver hitting the bottom of the shaft. How tall is the elevator shaft? (*Hint:* See Example 7.)

**56.** *Finding the Height of a Cliff.* A water balloon is dropped from a cliff. Exactly 3 sec later, the sound of the balloon hitting the ground reaches the top of the cliff. How high is the cliff? (*Hint:* See Example 7.)

## Skill Maintenance

*For each function f, construct and simplify the difference quotient*

$$\frac{f(x + h) - f(x)}{h}.$$

**57.** $f(x) = 3x - 7$

**58.** $f(x) = 2x^2 - x + 4$

*A graph of y = f(x) follows. No formula is given for f. Make a hand-drawn graph of each of the following.*

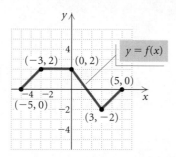

**59.** $g(x) = -2f(x)$

**60.** $g(x) = f(2x)$

## Synthesis

**61.** Find *c* such that

$$f(x) = -0.2x^2 - 3x + c$$

has a maximum value of $-225$.

**62.** Find *b* such that

$$f(x) = -4x^2 + bx + 3$$

has a maximum value of 50.

**63.** Graph: $f(x) = (|x| - 5)^2 - 3$.

**64.** Find a quadratic function with vertex $(4, -5)$ and containing the point $(-3, 1)$.

**65.** *Minimizing Area.* A 24-in. piece of string is cut into two pieces. One piece is used to form a circle while the other is used to form a square. How should the string be cut so that the sum of the areas is a minimum?

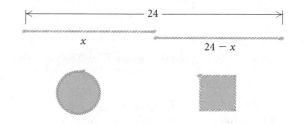

---

## Mid-Chapter Mixed Review

*Determine whether the statement is true or false.*

**1.** The product of a complex number and its conjugate is a real number. [3.1]

**2.** Every quadratic equation has at least one *x*-intercept. [3.2]

**3.** If a quadratic equation has two different real-number solutions, then its discriminant is positive. [3.2]

**4.** The vertex of the graph of the function $f(x) = 3(x + 4)^2 + 5$ is $(4, 5)$. [3.3]

*Express the number in terms of i.* [3.1]

**5.** $\sqrt{-36}$

**6.** $\sqrt{-5}$

**7.** $-\sqrt{-16}$

**8.** $\sqrt{-32}$

*Simplify. Write answers in the form a + bi, where a and b are real numbers.* [3.1]

**9.** $(3 - 2i) + (-4 + 3i)$

**10.** $(-5 + i) - (2 - 4i)$

**11.** $(2 + 3i)(4 - 5i)$

**12.** $\dfrac{3 + i}{-2 + 5i}$

*Simplify.* [3.1]

**13.** $i^{13}$

**14.** $i^{44}$

**15.** $(-i)^5$

**16.** $(2i)^6$

*Solve.* [3.2]

**17.** $x^2 + 3x - 4 = 0$

**18.** $2x^2 + 6 = -7x$

**19.** $4x^2 = 24$

**20.** $x^2 + 100 = 0$

**21.** Find the zeros of $f(x) = 4x^2 - 8x - 3$ by completing the square. Show your work. [3.2]

*In Exercises 22–24,* **(a)** *find the discriminant $b^2 - 4ac$, and then determine whether one real-number solution, two different real-number solutions, or two different imaginary-number solutions exist; and* **(b)** *solve the equation, finding exact solutions and approximate solutions rounded to three decimal places, where appropriate.* [3.2]

**22.** $x^2 - 3x - 5 = 0$

**23.** $4x^2 - 12x + 9 = 0$

**24.** $3x^2 + 2x = -1$

*Solve.* [3.2]

**25.** $x^4 + 5x^2 - 6 = 0$

**26.** $2x - 5\sqrt{x} + 2 = 0$

**27.** One number is 2 more than another. The product of the numbers is 35. Find the numbers. [3.2]

*In Exercises 28 and 29:*

**a)** *Find the vertex.* [3.3]
**b)** *Find the axis of symmetry.* [3.3]
**c)** *Determine whether there is a maximum or minimum value, and find that value.* [3.3]
**d)** *Find the range.* [3.3]
**e)** *Find the intervals on which the function is increasing and the intervals on which the function is decreasing.* [3.3]
**f)** *Graph the function.* [3.3]

**28.** $f(x) = x^2 - 6x + 7$

**29.** $f(x) = -2x^2 - 4x - 5$

**30.** The sum of the base and the height of a triangle is 16 in. Find the dimensions for which the area is a maximum. [3.3]

## Collaborative Discussion and Writing

**31.** Is the sum of two imaginary numbers always an imaginary number? Explain your answer. [3.1]

**32.** The graph of a quadratic function can have 0, 1, or 2 $x$-intercepts. How can you predict the number of $x$-intercepts without drawing the graph or (completely) solving an equation? [3.2]

**33.** Discuss two ways in which we used completing the square in this chapter. [3.2], [3.3]

**34.** Suppose that the graph of $f(x) = ax^2 + bx + c$ has $x$-intercepts $(x_1, 0)$ and $(x_2, 0)$. What are the $x$-intercepts of $g(x) = -ax^2 - bx - c$? Explain. [3.3]

## 3.4 Solving Rational Equations and Radical Equations

- Solve rational equations.
- Solve radical equations.

### ■ Rational Equations

Equations containing rational expressions are called **rational equations**. Solving such equations involves multiplying on both sides by the least common denominator (LCD) to *clear the equation of fractions*.

**EXAMPLE 1** Solve: $\dfrac{x-8}{3} + \dfrac{x-3}{2} = 0$.

### Algebraic Solution

We have

$$\frac{x-8}{3} + \frac{x-3}{2} = 0 \qquad \text{The LCD is } 3 \cdot 2, \text{ or } 6.$$

$$6\left(\frac{x-8}{3} + \frac{x-3}{2}\right) = 6 \cdot 0 \qquad \begin{array}{l}\text{Multiplying by the LCD on both} \\ \text{sides to clear fractions}\end{array}$$

$$6 \cdot \frac{x-8}{3} + 6 \cdot \frac{x-3}{2} = 0$$

$$2(x-8) + 3(x-3) = 0$$

$$2x - 16 + 3x - 9 = 0$$

$$5x - 25 = 0$$

$$5x = 25$$

$$x = 5.$$

The possible solution is 5. We check using a table in ASK mode.

$$y = \frac{x-8}{3} + \frac{x-3}{2}$$

| X | Y₁ | |
|---|---|---|
| 5 | 0 | |
| | | |
| | | |
| X = | | |

Since the value of $\dfrac{x-8}{3} + \dfrac{x-3}{2}$ is 0 when $x = 5$, the number 5 is the solution.

### Graphical Solution

We use the Zero method. The solution of the equation

$$\frac{x-8}{3} + \frac{x-3}{2} = 0$$

is the zero of the function

$$f(x) = \frac{x-8}{3} + \frac{x-3}{2}.$$

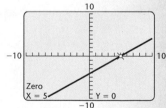

$$y = \frac{x-8}{3} + \frac{x-3}{2}$$

The zero of the function is 5. Thus the solution of the equation is 5.

**Now Try Exercise 3.**

> **CAUTION!** Clearing fractions is a valid procedure when solving rational equations but not when adding, subtracting, multiplying, or dividing rational expressions. A rational expression may have operation signs but it will have no equals sign. A rational equation *always* has an equals sign. For example, $\dfrac{x-8}{3} + \dfrac{x-3}{2}$ is a rational expression but $\dfrac{x-8}{3} + \dfrac{x-3}{2} = 0$ is a rational equation.
>
> To *simplify* the rational *expression* $\dfrac{x-8}{3} + \dfrac{x-3}{2}$, we first find the LCD and write each fraction with that denominator. The final result is usually a rational expression.
>
> To *solve* the rational *equation* $\dfrac{x-8}{3} + \dfrac{x-3}{2} = 0$, we first multiply on both sides by the LCD to clear fractions. The final result is one or more numbers. As we will see in Example 2, these numbers must be checked in the original equation.

When we use the multiplication principle to multiply (or divide) on both sides of an equation by an expression with a variable, we might not obtain an equivalent equation. We must check the possible solutions obtained in this manner by substituting them in the original equation. The next example illustrates this.

**EXAMPLE 2** Solve: $\dfrac{x^2}{x-3} = \dfrac{9}{x-3}$.

**Solution** The LCD is $x - 3$.

$$(x-3) \cdot \frac{x^2}{x-3} = (x-3) \cdot \frac{9}{x-3}$$

$$x^2 = 9$$

$$x = -3 \quad or \quad x = 3 \qquad \text{Using the principle of square roots}$$

The possible solutions are $-3$ and $3$. We check.

*Check:*  For $-3$:

$$\frac{x^2}{x-3} = \frac{9}{x-3}$$

$$\frac{(-3)^2}{-3-3} \;\overset{?}{\bigm|}\; \frac{9}{-3-3}$$

$$\frac{9}{-6} \;\bigm|\; \frac{9}{-6} \qquad \text{TRUE}$$

For $3$:

$$\frac{x^2}{x-3} = \frac{9}{x-3}$$

$$\frac{3^2}{3-3} \;\overset{?}{\bigm|}\; \frac{9}{3-3}$$

$$\frac{9}{0} \;\bigm|\; \frac{9}{0} \qquad \text{NOT DEFINED}$$

The number $-3$ checks, so it is a solution. Since division by 0 is not defined, 3 is not a solution. Note that 3 is not in the domain of either $x^2/(x-3)$ or $9/(x-3)$.

We can also use a table on a graphing calculator to check the possible solutions.

$$y_1 = \frac{x^2}{x-3}, \quad y_2 = \frac{9}{x-3}$$

| X | Y₁ | Y₂ |
|---|---|---|
| −3 | −1.5 | −1.5 |
| 3 | ERROR | ERROR |

X =

When $x = -3$, we see that $y_1 = -1.5 = y_2$, so $-3$ is a solution. When $x = 3$, we get ERROR messages. This indicates that 3 is not in the domain of $y_1$ or $y_2$ and thus is not a solution.

> **Now Try Exercise 9.**

**EXAMPLE 3**  Solve: $\dfrac{2}{3x+6} + \dfrac{1}{x^2-4} = \dfrac{4}{x-2}$.

*Solution*  We first factor the denominators in order to determine the LCD:

$$\frac{2}{3(x+2)} + \frac{1}{(x+2)(x-2)} = \frac{4}{x-2} \qquad \begin{array}{l} \text{The LCD is} \\ 3(x+2)(x-2). \end{array}$$

$$3(x+2)(x-2)\left(\frac{2}{3(x+2)} + \frac{1}{(x+2)(x-2)}\right) = 3(x+2)(x-2) \cdot \frac{4}{x-2}$$

**Multiplying by the LCD to clear fractions**

$$2(x-2) + 3 = 3 \cdot 4(x+2)$$
$$2x - 4 + 3 = 12x + 24$$
$$2x - 1 = 12x + 24$$
$$-10x = 25$$
$$x = -\tfrac{5}{2}.$$

The possible solution is $-\tfrac{5}{2}$. We check this on a graphing calculator.

$$y_1 = \frac{2}{3x+6} + \frac{1}{x^2-4}, \quad y_2 = \frac{4}{x-2}$$

| X | Y₁ | Y₂ |
|---|---|---|
| −2.5 | −.8889 | −.8889 |

X =

We see that $y_1 = y_2$ when $x = -\tfrac{5}{2}$, or $-2.5$, so $-\tfrac{5}{2}$ is the solution.

> **Now Try Exercise 21.**

## ▪ Radical Equations

A **radical equation** is an equation in which variables appear in one or more radicands. For example,

$$\sqrt{2x - 5} - \sqrt{x - 3} = 1$$

is a radical equation. The following principle is used to solve such equations.

---

**THE PRINCIPLE OF POWERS**

For any positive integer $n$:

If $a = b$ is true, then $a^n = b^n$ is true.

---

**EXAMPLE 4**   Solve: $\sqrt{3x + 1} = 4$.

### Algebraic Solution

We use the principle of powers and square both sides:

$$\sqrt{3x + 1} = 4$$
$$\left(\sqrt{3x + 1}\right)^2 = 4^2$$
$$3x + 1 = 16$$
$$3x = 15$$
$$x = 5.$$

*Check:*

$$\begin{array}{c|c} \sqrt{3x + 1} = 4 \\ \hline \sqrt{3 \cdot 5 + 1} \;?\; 4 \\ \sqrt{15 + 1} \\ \sqrt{16} \\ 4 \;\bigm|\; 4 \quad \text{TRUE} \end{array}$$

The solution is 5.

### Graphical Solution

We graph $y_1 = \sqrt{3x + 1}$ and $y_2 = 4$ and then use the INTERSECT feature. We see that the solution is 5. The check shown in the table below confirms that the solution is 5.

$$y_1 = \sqrt{3x + 1}, \;\; y_2 = 4$$

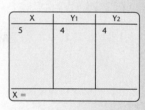

| X | Y₁ | Y₂ |
|---|----|----|
| 5 | 4  | 4  |
| X = |  |  |

We can also find the solution using the Graphing Calculator, Appcylon LLC, app.

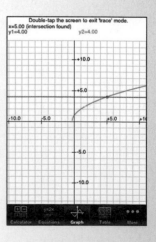

**Now Try Exercise 31.**

In Example 4, the radical was isolated on one side of the equation. If this had not been the case, our first step would have been to isolate the radical. We do so in the next example.

**EXAMPLE 5**  Solve: $5 + \sqrt{x + 7} = x$.

## Algebraic Solution

We first isolate the radical and then use the principle of powers:

$$5 + \sqrt{x + 7} = x$$

$$\sqrt{x + 7} = x - 5 \qquad \text{Subtracting 5 on both sides to isolate the radical}$$

$$\left(\sqrt{x + 7}\right)^2 = (x - 5)^2 \qquad \text{Using the principle of powers; squaring both sides}$$

$$x + 7 = x^2 - 10x + 25$$

$$0 = x^2 - 11x + 18 \qquad \text{Subtracting } x \text{ and 7}$$

$$0 = (x - 9)(x - 2) \qquad \text{Factoring}$$

$$x - 9 = 0 \quad or \quad x - 2 = 0$$

$$x = 9 \quad or \qquad x = 2.$$

The possible solutions are 9 and 2.

*Check:*

For 9:
$$\begin{array}{c|c} 5 + \sqrt{x + 7} = x \\ \hline 5 + \sqrt{9 + 7} \;?\; 9 \\ 5 + \sqrt{16} \\ 5 + 4 \\ 9 \;\big|\; 9 \quad \text{TRUE} \end{array}$$

For 2:
$$\begin{array}{c|c} 5 + \sqrt{x + 7} = x \\ \hline 5 + \sqrt{2 + 7} \;?\; 2 \\ 5 + \sqrt{9} \\ 5 + 3 \\ 8 \;\big|\; 2 \quad \text{FALSE} \end{array}$$

Since 9 checks but 2 does not, the only solution is 9.

## Graphical Solution

We graph $y_1 = 5 + \sqrt{x + 7}$ and $y_2 = x$. Using the INTERSECT feature, we see that the solution is 9.

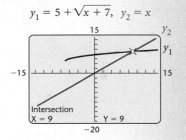

$y_1 = 5 + \sqrt{x + 7}, \; y_2 = x$

We can also use the ZERO feature to get this result. To do so, we first write the equivalent equation $5 + \sqrt{x + 7} - x = 0$. The zero of the function $f(x) = 5 + \sqrt{x + 7} - x$ is 9, so the solution of the original equation is 9.

$y = 5 + \sqrt{x + 7} - x$

Note that the graphs show that the equation has only one solution.

**Now Try Exercise 55.**

When we raise both sides of an equation to an even power, the resulting equation can have solutions that the original equation does not. This is because the converse of the principle of powers is not necessarily true. That is, if $a^n = b^n$ is true, we do not know that $a = b$ is true. For example, $(-2)^2 = 2^2$, but $-2 \neq 2$. Thus, as we saw in Example 5, it is necessary to check the possible solutions in the original equation when the principle of powers is used to raise both sides of an equation to an even power.

When a radical equation has two radical terms on one side, we isolate one of them and then use the principle of powers. If, after doing so, a radical term remains, we repeat these steps.

**EXAMPLE 6**   Solve: $\sqrt{x-3} + \sqrt{x+5} = 4$.

*Solution*   We have

$$\sqrt{x-3} = 4 - \sqrt{x+5} \qquad \text{Isolating one radical}$$

$$\left(\sqrt{x-3}\right)^2 = \left(4 - \sqrt{x+5}\right)^2 \qquad \text{Using the principle of powers; squaring both sides}$$

$$x - 3 = 16 - 8\sqrt{x+5} + (x+5)$$

$$x - 3 = 21 - 8\sqrt{x+5} + x \qquad \text{Collecting like terms}$$

$$-24 = -8\sqrt{x+5} \qquad \text{Isolating the remaining radical; subtracting } x \text{ and } 21 \text{ on both sides}$$

$$3 = \sqrt{x+5} \qquad \text{Dividing by } -8 \text{ on both sides}$$

$$3^2 = \left(\sqrt{x+5}\right)^2 \qquad \text{Using the principle of powers; squaring both sides}$$

$$9 = x + 5$$

$$4 = x. \qquad \text{Subtracting 5 on both sides}$$

We check the possible solution, 4, on a graphing calculator.

$$y_1 = \sqrt{x-3} + \sqrt{x+5}, \quad y_2 = 4$$

| X | Y1 | Y2 |
|---|----|----|
| 4 | 4  | 4  |
|   |    |    |

X =

Since $y_1 = y_2$ when $x = 4$, the number 4 checks. It is the solution.

**Now Try Exercise 65.**

# 3.4 Exercise Set

*Solve.*

**1.** $\dfrac{1}{4} + \dfrac{1}{5} = \dfrac{1}{t}$

**2.** $\dfrac{1}{3} - \dfrac{5}{6} = \dfrac{1}{x}$

**3.** $\dfrac{x+2}{4} - \dfrac{x-1}{5} = 15$

**4.** $\dfrac{t+1}{3} - \dfrac{t-1}{2} = 1$

**5.** $\dfrac{1}{2} + \dfrac{2}{x} = \dfrac{1}{3} + \dfrac{3}{x}$

**6.** $\dfrac{1}{t} + \dfrac{1}{2t} + \dfrac{1}{3t} = 5$

**7.** $\dfrac{5}{3x+2} = \dfrac{3}{2x}$

**8.** $\dfrac{2}{x-1} = \dfrac{3}{x+2}$

**9.** $\dfrac{y^2}{y+4} = \dfrac{16}{y+4}$

**10.** $\dfrac{49}{w-7} = \dfrac{w^2}{w-7}$

**11.** $x + \dfrac{6}{x} = 5$

**12.** $x - \dfrac{12}{x} = 1$

**13.** $\dfrac{6}{y+3} + \dfrac{2}{y} = \dfrac{5y-3}{y^2-9}$

**14.** $\dfrac{3}{m+2} + \dfrac{2}{m} = \dfrac{4m-4}{m^2-4}$

**15.** $\dfrac{2x}{x-1} = \dfrac{5}{x-3}$

**16.** $\dfrac{2x}{x+7} = \dfrac{5}{x+1}$

**17.** $\dfrac{2}{x+5} + \dfrac{1}{x-5} = \dfrac{16}{x^2-25}$

**18.** $\dfrac{2}{x^2-9} + \dfrac{5}{x-3} = \dfrac{3}{x+3}$

**19.** $\dfrac{3x}{x+2} + \dfrac{6}{x} = \dfrac{12}{x^2+2x}$

**20.** $\dfrac{3y+5}{y^2+5y} + \dfrac{y+4}{y+5} = \dfrac{y+1}{y}$

**21.** $\dfrac{1}{5x+20} - \dfrac{1}{x^2-16} = \dfrac{3}{x-4}$

**22.** $\dfrac{1}{4x+12} - \dfrac{1}{x^2-9} = \dfrac{5}{x-3}$

**23.** $\dfrac{2}{5x+5} - \dfrac{3}{x^2-1} = \dfrac{4}{x-1}$

**24.** $\dfrac{1}{3x+6} - \dfrac{1}{x^2-4} = \dfrac{3}{x-2}$

**25.** $\dfrac{8}{x^2-2x+4} = \dfrac{x}{x+2} + \dfrac{24}{x^3+8}$

**26.** $\dfrac{18}{x^2-3x+9} - \dfrac{x}{x+3} = \dfrac{81}{x^3+27}$

**27.** $\dfrac{x}{x-4} - \dfrac{4}{x+4} = \dfrac{32}{x^2-16}$

**28.** $\dfrac{x}{x-1} - \dfrac{1}{x+1} = \dfrac{2}{x^2-1}$

**29.** $\dfrac{1}{x-6} - \dfrac{1}{x} = \dfrac{6}{x^2-6x}$

**30.** $\dfrac{1}{x-15} - \dfrac{1}{x} = \dfrac{15}{x^2-15x}$

**31.** $\sqrt{3x-4} = 1$

**32.** $\sqrt{4x+1} = 3$

**33.** $\sqrt{2x-5} = 2$

**34.** $\sqrt{3x+2} = 6$

**35.** $\sqrt{7-x} = 2$

**36.** $\sqrt{5-x} = 1$

**37.** $\sqrt{1-2x} = 3$

**38.** $\sqrt{2-7x} = 2$

**39.** $\sqrt[3]{5x-2} = -3$

**40.** $\sqrt[3]{2x+1} = -5$

**41.** $\sqrt[4]{x^2-1} = 1$

**42.** $\sqrt[5]{3x+4} = 2$

**43.** $\sqrt{y-1} + 4 = 0$

**44.** $\sqrt{m+1} - 5 = 8$

**45.** $\sqrt{b+3} - 2 = 1$

**46.** $\sqrt{x-4} + 1 = 5$

**47.** $\sqrt{z+2} + 3 = 4$

**48.** $\sqrt{y-5} - 2 = 3$

**49.** $\sqrt{2x+1} - 3 = 3$

**50.** $\sqrt{3x-1} + 2 = 7$

**51.** $\sqrt{2-x} - 4 = 6$

**52.** $\sqrt{5-x} + 2 = 8$

**53.** $\sqrt[3]{6x+9} + 8 = 5$

**54.** $\sqrt[5]{2x-3} - 1 = 1$

**55.** $\sqrt{x+4} + 2 = x$

**56.** $\sqrt{x+1} + 1 = x$

**57.** $\sqrt{x-3} + 5 = x$

**58.** $\sqrt{x+3} - 1 = x$

**59.** $\sqrt{x+7} = x + 1$

**60.** $\sqrt{6x+7} = x + 2$

**61.** $\sqrt{3x+3} = x + 1$

**62.** $\sqrt{2x+5} = x - 5$

**63.** $\sqrt{5x+1} = x - 1$

**64.** $\sqrt{7x+4} = x + 2$

**65.** $\sqrt{x-3} + \sqrt{x+2} = 5$

**66.** $\sqrt{x} - \sqrt{x - 5} = 1$

**67.** $\sqrt{3x - 5} + \sqrt{2x + 3} + 1 = 0$

**68.** $\sqrt{2m - 3} = \sqrt{m + 7} - 2$

**69.** $\sqrt{x} - \sqrt{3x - 3} = 1$

**70.** $\sqrt{2x + 1} - \sqrt{x} = 1$

**71.** $\sqrt{2y - 5} - \sqrt{y - 3} = 1$

**72.** $\sqrt{4p + 5} + \sqrt{p + 5} = 3$

**73.** $\sqrt{y + 4} - \sqrt{y - 1} = 1$

**74.** $\sqrt{y + 7} + \sqrt{y + 16} = 9$

**75.** $\sqrt{x + 5} + \sqrt{x + 2} = 3$

**76.** $\sqrt{6x + 6} = 5 + \sqrt{21 - 4x}$

**77.** $x^{1/3} = -2$      **78.** $t^{1/5} = 2$

**79.** $t^{1/4} = 3$      **80.** $m^{1/2} = -7$

*Solve.*

**81.** $\dfrac{P_1 V_1}{T_1} = \dfrac{P_2 V_2}{T_2}$, for $T_1$
(A chemistry formula for gases)

**82.** $\dfrac{1}{F} = \dfrac{1}{m} + \dfrac{1}{p}$, for $F$
(A formula from optics)

**83.** $W = \sqrt{\dfrac{1}{LC}}$, for $C$
(An electricity formula)

**84.** $s = \sqrt{\dfrac{A}{6}}$, for $A$
(A geometry formula)

**85.** $\dfrac{1}{R} = \dfrac{1}{R_1} + \dfrac{1}{R_2}$, for $R_2$
(A formula for resistance)

**86.** $\dfrac{1}{t} = \dfrac{1}{a} + \dfrac{1}{b}$, for $t$
(A formula for work rate)

**87.** $I = \sqrt{\dfrac{A}{P}} - 1$, for $P$
(A compound-interest formula)

**88.** $T = 2\pi \sqrt{\dfrac{1}{g}}$, for $g$
(A pendulum formula)

**89.** $\dfrac{1}{F} = \dfrac{1}{m} + \dfrac{1}{p}$, for $p$
(A formula from optics)

**90.** $\dfrac{V^2}{R^2} = \dfrac{2g}{R + h}$, for $h$
(A formula for escape velocity)

## Skill Maintenance

*Find the zero of the function.*

**91.** $f(x) = 15 - 2x$

**92.** $f(x) = -3x + 9$

**93.** *Deadly Distractions.* Drivers who were distracted by such things as text-messaging, talking on a cell phone, conversing with passengers, and eating were involved in 5870 highway fatalities in 2008. This was an increase of about 18% over the number of distracted-driving fatalities in 2004 and is attributed largely to the increased number of drivers who texted in 2008. (*Source*: NHTSA's National Center for Statistics and Analysis) How many highway fatalities involved distracted driving in 2004?

**94.** *Big Sites.* Together, the Mall of America in Minnesota and the Disneyland theme park in California occupy 181 acres of land. The Mall of America occupies 11 acres more than Disneyland. (*Sources*: Mall of America; Disneyland) How much land does each occupy?

## Synthesis

*Solve.*

**95.** $(x - 3)^{2/3} = 2$

**96.** $\dfrac{x + 3}{x + 2} - \dfrac{x + 4}{x + 3} = \dfrac{x + 5}{x + 4} - \dfrac{x + 6}{x + 5}$

**97.** $\sqrt{x + 5} + 1 = \dfrac{6}{\sqrt{x + 5}}$

**98.** $\sqrt{15 + \sqrt{2x + 80}} = 5$

**99.** $x^{2/3} = x$

# 3.5

# Solving Equations and Inequalities with Absolute Value

- Solve equations with absolute value.
- Solve inequalities with absolute value.

ABSOLUTE VALUE

REVIEW SECTION **R.1.**

## ■ Equations with Absolute Value

Recall that the absolute value of a number is its distance from 0 on the number line. We use this concept to solve equations with absolute value.

> For $a > 0$ and an algebraic expression $X$:
>
> $$|X| = a \text{ is equivalent to } X = -a \text{ or } X = a.$$

**EXAMPLE 1** Solve: $|x| = 5$.

| **Algebraic Solution** | **Graphical Solution** |
|---|---|

### Algebraic Solution

We have

$$|x| = 5$$
$$x = -5 \quad \text{or} \quad x = 5.$$
**Writing an equivalent statement**

The solutions are $-5$ and $5$.

To check, note that $-5$ and $5$ are both 5 units from 0 on the number line.

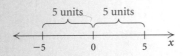

### Graphical Solution

Using the Intersect method, we graph $y_1 = |x|$ and $y_2 = 5$ and find the first coordinates of the points of intersection.

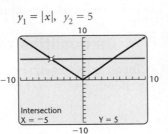

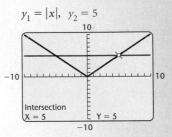

The solutions are $-5$ and $5$.

We could also have used the Zero method to get this result, graphing $y = |x| - 5$ and using the ZERO feature twice.

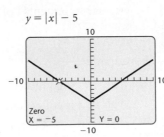

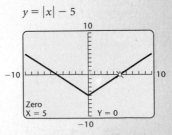

The zeros of $f(x) = |x| - 5$ are $-5$ and $5$, so the solutions of the original equation are $-5$ and $5$.

**Now Try Exercise 1.**

**EXAMPLE 2** Solve: $|x - 3| - 1 = 4$.

***Solution*** First, we add 1 on both sides to get an expression of the form $|X| = a$:

$$|x - 3| - 1 = 4$$
$$|x - 3| = 5$$
$$x - 3 = -5 \quad or \quad x - 3 = 5 \qquad |X| = a \text{ is equivalent to } X = -a \text{ or } X = a.$$
$$x = -2 \quad or \qquad x = 8. \qquad \text{Adding 3}$$

*Check:*

For $-2$:

$$\begin{array}{c|c} |x - 3| - 1 = 4 \\ \hline |-2 - 3| - 1 \overset{?}{\phantom{=}} 4 \\ |-5| - 1 \\ 5 - 1 \\ 4 & 4 \quad \text{TRUE} \end{array}$$

For 8:

$$\begin{array}{c|c} |x - 3| - 1 = 4 \\ \hline |8 - 3| - 1 \overset{?}{\phantom{=}} 4 \\ |5| - 1 \\ 5 - 1 \\ 4 & 4 \quad \text{TRUE} \end{array}$$

The solutions are $-2$ and 8.

> **Now Try Exercise 21.**

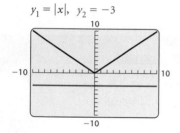

$y_1 = |x|, \ y_2 = -3$

When $a = 0$, $|X| = a$ is equivalent to $X = 0$. Note that for $a < 0$, $|X| = a$ has no solution, because the absolute value of an expression is never negative. We can use a graph to illustrate the last statement for a specific value of $a$. For example, if we let $a = -3$ and graph $y_1 = |x|$ and $y_2 = -3$, we see that the graphs do not intersect, as shown at left. Thus the equation $|x| = -3$ has no solution. The solution set is the **empty set**, denoted $\varnothing$.

## ▪ Inequalities with Absolute Value

Inequalities sometimes contain absolute-value notation. The following properties are used to solve them.

---

For $a > 0$ and an algebraic expression $X$:

$|X| < a$    is equivalent to    $-a < X < a$.

$|X| > a$    is equivalent to    $X < -a \ or \ X > a$.

Similar statements hold for $|X| \leq a$ and $|X| \geq a$.

---

For example,

$|x| < 3$ is equivalent to $-3 < x < 3$;

$|y| \geq 1$ is equivalent to $y \leq -1 \ or \ y \geq 1$; and

$|2x + 3| \leq 4$ is equivalent to $-4 \leq 2x + 3 \leq 4$.

---

> **INTERVAL NOTATION**
>
> REVIEW SECTION **R.1.**

**EXAMPLE 3** Solve and graph the solution set: $|3x + 2| < 5$.

*Solution* We have

$$|3x + 2| < 5$$

| | |
|---|---|
| $-5 < 3x + 2 < 5$ | Writing an equivalent inequality |
| $-7 < 3x < 3$ | Subtracting 2 |
| $-\frac{7}{3} < x < 1.$ | Dividing by 3 |

The solution set is $\left\{ x \mid -\frac{7}{3} < x < 1 \right\}$, or $\left( -\frac{7}{3}, 1 \right)$. The graph of the solution set is shown below.

**Now Try Exercise 45.**

**EXAMPLE 4** Solve and graph the solution set: $|5 - 2x| \geq 1$.

*Solution* We have

$$|5 - 2x| \geq 1$$

| | |
|---|---|
| $5 - 2x \leq -1$ *or* $5 - 2x \geq 1$ | Writing an equivalent inequality |
| $-2x \leq -6$ *or* $-2x \geq -4$ | Subtracting 5 |
| $x \geq 3$ *or* $x \leq 2.$ | Dividing by $-2$ and reversing the inequality signs |

The solution set is $\{ x \mid x \leq 2 \text{ or } x \geq 3 \}$, or $(-\infty, 2] \cup [3, \infty)$. The graph of the solution set is shown below.

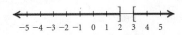

**Now Try Exercise 47.**

---

## 3.5  Exercise Set

*Solve.*

**1.** $|x| = 7$

**2.** $|x| = 4.5$

**3.** $|x| = 0$

**4.** $|x| = \frac{3}{2}$

**5.** $|x| = \frac{5}{6}$

**6.** $|x| = -\frac{3}{5}$

**7.** $|x| = -10.7$

**8.** $|x| = 12$

**9.** $|3x| = 1$

**10.** $|5x| = 4$

**11.** $|8x| = 24$

**12.** $|6x| = 0$

**13.** $|x - 1| = 4$

**14.** $|x - 7| = 5$

**15.** $|x + 2| = 6$

**16.** $|x + 5| = 1$

**17.** $|3x + 2| = 1$

**18.** $|7x - 4| = 8$

**19.** $\left| \frac{1}{2}x - 5 \right| = 17$

**20.** $\left| \frac{1}{3}x - 4 \right| = 13$

**21.** $|x - 1| + 3 = 6$

**22.** $|x + 2| - 5 = 9$

**23.** $|x + 3| - 2 = 8$

**24.** $|x - 4| + 3 = 9$

**25.** $|3x + 1| - 4 = -1$

**26.** $|2x - 1| - 5 = -3$

**27.** $|4x - 3| + 1 = 7$

**28.** $|5x + 4| + 2 = 5$

**29.** $12 - |x + 6| = 5$          **30.** $9 - |x - 2| = 7$

**31.** $7 - |2x - 1| = 6$          **32.** $5 - |4x + 3| = 2$

*Solve and write interval notation for the solution set.*
*Then graph the solution set.*

**33.** $|x| < 7$                    **34.** $|x| \leq 4.5$

**35.** $|x| \leq 2$                 **36.** $|x| < 3$

**37.** $|x| \geq 4.5$               **38.** $|x| > 7$

**39.** $|x| > 3$                    **40.** $|x| \geq 2$

**41.** $|3x| < 1$                   **42.** $|5x| \leq 4$

**43.** $|2x| \geq 6$                **44.** $|4x| > 20$

**45.** $|x + 8| < 9$                **46.** $|x + 6| \leq 10$

**47.** $|x + 8| \geq 9$             **48.** $|x + 6| > 10$

**49.** $|x - \frac{1}{4}| < \frac{1}{2}$          **50.** $|x - 0.5| \leq 0.2$

**51.** $|2x + 3| \leq 9$            **52.** $|3x + 4| < 13$

**53.** $|x - 5| > 0.1$              **54.** $|x - 7| \geq 0.4$

**55.** $|6 - 4x| \geq 8$            **56.** $|5 - 2x| > 10$

**57.** $|x + \frac{2}{3}| \leq \frac{5}{3}$       **58.** $|x + \frac{3}{4}| < \frac{1}{4}$

**59.** $\left|\dfrac{2x + 1}{3}\right| > 5$        **60.** $\left|\dfrac{2x - 1}{3}\right| \geq \dfrac{5}{6}$

**61.** $|2x - 4| < -5$              **62.** $|3x + 5| < 0$

**63.** $|7 - x| \geq -4$            **64.** $|2x + 1| > -\frac{1}{2}$

## Skill Maintenance

### Vocabulary Reinforcement

*In each of Exercises 65–72, fill in the blank with the*
*correct term. Some of the given choices will not be used.*

distance formula
midpoint formula
function
relation
*x*-intercept
*y*-intercept
perpendicular
parallel
horizontal lines
vertical lines

symmetric with respect
  to the *x*-axis
symmetric with respect
  to the *y*-axis
symmetric with respect
  to the origin
increasing
decreasing
constant

**65.** A(n) _____ is a point $(0, b)$.

**66.** The _____ is
$d = \sqrt{(x_2 - x_1)^2 + (y_2 - y_1)^2}$.

**67.** A(n) _____ is a correspondence such
that each member of the domain corresponds to
at least one member of the range.

**68.** A(n) _____ is a correspondence such
that each member of the domain corresponds to
exactly one member of the range.

**69.** _____ are given by equations of the
type $y = b$, or $f(x) = b$.

**70.** Nonvertical lines are _____ if and
only if they have the same slope and different
*y*-intercepts.

**71.** A function $f$ is said to be _____ on an
open interval $I$ if, for all $a$ and $b$ in that interval,
$a < b$ implies $f(a) > f(b)$.

**72.** For an equation $y = f(x)$, if replacing $x$ with $-x$
produces an equivalent equation, then the graph
is _____.

## Synthesis

*Solve.*

**73.** $|3x - 1| > 5x - 2$

**74.** $|x + 2| \leq |x - 5|$

**75.** $|p - 4| + |p + 4| < 8$

**76.** $|x| + |x + 1| < 10$

**77.** $|x - 3| + |2x + 5| > 6$

# Chapter 3 Summary and Review

## STUDY GUIDE

| KEY TERMS AND CONCEPTS | EXAMPLES |
|---|---|

### SECTION 3.1: THE COMPLEX NUMBERS

| | |
|---|---|
| The number $i$ is defined such that $i = \sqrt{-1}$ and $i^2 = -1$. | Express each number in terms of $i$.<br><br>$\sqrt{-5} = \sqrt{-1 \cdot 5} = \sqrt{-1} \cdot \sqrt{5} = i\sqrt{5}$, or $\sqrt{5}i$;<br>$-\sqrt{-36} = -\sqrt{-1 \cdot 36} = -\sqrt{-1} \cdot \sqrt{36}$<br>$\qquad\qquad\qquad = -i \cdot 6 = -6i$ |
| A **complex number** is a number of the form $a + bi$, where $a$ and $b$ are real numbers. The number $a$ is said to be the **real part** of $a + bi$, and the number $b$ is said to be the **imaginary part** of $a + bi$.<br><br>To **add** or **subtract complex numbers**, we add or subtract the real parts, and we add or subtract the imaginary parts. | Add or subtract.<br><br>$(-3 + 4i) + (5 - 8i) = (-3 + 5) + (4i - 8i)$<br>$\qquad\qquad\qquad\qquad\quad = 2 - 4i$;<br>$(6 - 7i) - (10 + 3i) = (6 - 10) + (-7i - 3i)$<br>$\qquad\qquad\qquad\qquad\quad\; = -4 - 10i$ |
| When we **multiply complex numbers**, we must keep in mind the fact that $i^2 = -1$.<br>Note that $\sqrt{a} \cdot \sqrt{b} \neq \sqrt{ab}$ when $\sqrt{a}$ and $\sqrt{b}$ are not real numbers. | Multiply.<br><br>$\sqrt{-4} \cdot \sqrt{-100} = \sqrt{-1} \cdot \sqrt{4} \cdot \sqrt{-1} \cdot \sqrt{100}$<br>$\qquad\qquad\qquad = i \cdot 2 \cdot i \cdot 10$<br>$\qquad\qquad\qquad = i^2 \cdot 20$<br>$\qquad\qquad\qquad = -1 \cdot 20 \qquad i^2 = -1$<br>$\qquad\qquad\qquad = -20$;<br><br>$(2 - 5i)(3 + i) = 6 + 2i - 15i - 5i^2$<br>$\qquad\qquad\qquad = 6 - 13i - 5(-1)$<br>$\qquad\qquad\qquad = 6 - 13i + 5$<br>$\qquad\qquad\qquad = 11 - 13i$ |
| The **conjugate of a complex number** $a + bi$ is $a - bi$. The numbers $a + bi$ and $a - bi$ are **complex conjugates**.<br><br>Conjugates are used when we **divide complex numbers**. | Divide.<br><br>$\dfrac{5 - 2i}{3 + i} = \dfrac{5 - 2i}{3 + i} \cdot \dfrac{3 - i}{3 - i}$ $\quad$ $3 - i$ is the conjugate of the divisor, $3 + i$.<br><br>$\qquad = \dfrac{15 - 5i - 6i + 2i^2}{9 - i^2}$<br><br>$\qquad = \dfrac{15 - 11i - 2}{9 + 1}$ $\quad i^2 = -1$<br><br>$\qquad = \dfrac{13 - 11i}{10} = \dfrac{13}{10} - \dfrac{11}{10}i$ |

## SECTION 3.2: QUADRATIC EQUATIONS, FUNCTIONS, ZEROS, AND MODELS

A **quadratic equation** is an equation that can be written in the form

$$ax^2 + bx + c = 0, \quad a \neq 0,$$

where $a$, $b$, and $c$ are real numbers.

A **quadratic function** $f$ is a function that can be written in the form

$$f(x) = ax^2 + bx + c, \quad a \neq 0,$$

where $a$, $b$, and $c$ are real numbers.

The **zeros** of a quadratic function $f(x) = ax^2 + bx + c$ are the *solutions* of the associated quadratic equation $ax^2 + bx + c = 0$.

$3x^2 - 2x + 4 = 0$ and $5 - 4x = x^2$ are examples of quadratic equations. The equation $3x^2 - 2x + 4 = 0$ is written in **standard form**.

The functions $f(x) = 2x^2 + x + 1$ and $f(x) = 5x^2 - 4$ are examples of quadratic functions.

---

**The Principle of Zero Products**

If $ab = 0$ is true, then $a = 0$ *or* $b = 0$, and if $a = 0$ *or* $b = 0$, then $ab = 0$.

Solve: $3x^2 - 4 = 11x$.

$$3x^2 - 4 = 11x$$

$$3x^2 - 11x - 4 = 0 \qquad \text{Subtracting } 11x \text{ on both sides to get 0 on one side of the equation}$$

$$(3x + 1)(x - 4) = 0 \qquad \text{Factoring}$$

$$3x + 1 = 0 \quad or \quad x - 4 = 0 \qquad \text{Using the principle of zero products}$$

$$3x = -1 \quad or \qquad x = 4$$

$$x = -\tfrac{1}{3} \quad or \qquad x = 4$$

---

**The Principle of Square Roots**

If $x^2 = k$, then $x = \sqrt{k}$ or $x = -\sqrt{k}$.

Solve: $3x^2 - 18 = 0$.

$$3x^2 - 18 = 0$$

$$3x^2 = 18 \qquad \text{Adding 18 on both sides}$$

$$x^2 = 6 \qquad \text{Dividing by 3 on both sides}$$

$$x = \sqrt{6} \quad or \quad x = -\sqrt{6} \qquad \text{Using the principle of square roots}$$

To solve a quadratic equation by **completing the square:**

1. Isolate the terms with variables on one side of the equation and arrange them in descending order.
2. Divide by the coefficient of the squared term if that coefficient is not 1.
3. Complete the square by taking half the coefficient of the first-degree term and adding its square on both sides of the equation.
4. Express one side of the equation as the square of a binomial.
5. Use the principle of square roots.
6. Solve for the variable.

Solve: $2x^2 - 3 = 6x$.

$$2x^2 - 3 = 6x$$

$$2x^2 - 6x - 3 = 0 \qquad \text{Subtracting } 6x$$

$$2x^2 - 6x = 3 \qquad \text{Adding 3}$$

$$x^2 - 3x = \frac{3}{2} \qquad \begin{array}{l}\text{Dividing by 2 to make}\\ \text{the } x^2\text{-coefficient 1}\end{array}$$

$$x^2 - 3x + \frac{9}{4} = \frac{3}{2} + \frac{9}{4} \qquad \begin{array}{l}\text{Completing the square:}\\ \frac{1}{2}(-3) = -\frac{3}{2} \text{ and}\\ \left(-\frac{3}{2}\right)^2 = \frac{9}{4}; \text{ adding } \frac{9}{4}\end{array}$$

$$\left(x - \frac{3}{2}\right)^2 = \frac{15}{4} \qquad \text{Factoring and simplifying}$$

$$x - \frac{3}{2} = \pm\frac{\sqrt{15}}{2} \qquad \begin{array}{l}\text{Using the principle of}\\ \text{square roots and}\\ \text{the quotient rule for}\\ \text{radicals}\end{array}$$

$$x = \frac{3}{2} \pm \frac{\sqrt{15}}{2}$$

$$= \frac{3 \pm \sqrt{15}}{2}$$

---

The solutions of $ax^2 + bx + c = 0$, $a \neq 0$, can be found using the **quadratic formula:**

$$x = \frac{-b \pm \sqrt{b^2 - 4ac}}{2a}.$$

Solve: $x^2 - 6 = 3x$.

$$x^2 - 6 = 3x$$

$$x^2 - 3x - 6 = 0 \qquad \text{Standard form}$$

$$a = 1, b = -3, c = -6$$

$$x = \frac{-b \pm \sqrt{b^2 - 4ac}}{2a}$$

$$= \frac{-(-3) \pm \sqrt{(-3)^2 - 4(1)(-6)}}{2 \cdot 1}$$

$$= \frac{3 \pm \sqrt{9 + 24}}{2}$$

$$= \frac{3 \pm \sqrt{33}}{2} \qquad \text{Exact solutions}$$

Using a calculator, we approximate the solutions to be 4.372 and $-1.372$.

**Discriminant**

For $ax^2 + bx + c = 0$, where $a$, $b$, and $c$ are real numbers:

$b^2 - 4ac = 0 \rightarrow$ One real-number solution;

$b^2 - 4ac > 0 \rightarrow$ Two different real-number solutions;

$b^2 - 4ac < 0 \rightarrow$ Two different imaginary-number solutions, complex conjugates.

For the equation above, $x^2 - 6 = 3x$, we see that $b^2 - 4ac$ is 33. Since 33 is positive, there are two different real-number solutions.

For $2x^2 - x + 4 = 0$, with $a = 2$, $b = -1$, and $c = 4$, the discriminant, $(-1)^2 - 4 \cdot 2 \cdot 4 = 1 - 32 = -31$, is negative, so there are two different imaginary-number (or nonreal) solutions.

For $x^2 - 6x + 9 = 0$, with $a = 1$, $b = -6$, and $c = 9$, the discriminant, $(-6)^2 - 4 \cdot 1 \cdot 9 = 36 - 36 = 0$, is 0 so there is one real-number solution.

Equations **reducible to quadratic**, or **quadratic in form**, can be treated as quadratic equations if a suitable substitution is made.

Solve: $x^4 - x^2 - 12 = 0$.

$$x^4 - x^2 - 12 = 0 \qquad \text{Let } u = x^2. \text{ Then } u^2 = (x^2)^2 = x^4.$$

$$u^2 - u - 12 = 0 \qquad \text{Substituting}$$

$$(u - 4)(u + 3) = 0$$

$$u - 4 = 0 \quad or \quad u + 3 = 0$$

$$u = 4 \quad or \quad u = -3 \qquad \text{Solving for } u$$

$$x^2 = 4 \quad or \quad x^2 = -3$$

$$x = \pm 2 \quad or \quad x = \pm\sqrt{3}i \qquad \text{Solving for } x$$

## SECTION 3.3: ANALYZING GRAPHS OF QUADRATIC FUNCTIONS

**Graphing Quadratic Functions**

The graph of the function $f(x) = a(x - h)^2 + k$ is a parabola that:

- opens up if $a > 0$ and down if $a < 0$;
- has $(h, k)$ as the vertex;
- has $x = h$ as the axis of symmetry;
- has $k$ as a minimum value (output) if $a > 0$;
- has $k$ as a maximum value if $a < 0$.

We can use a modification of the technique of completing the square as an aid in analyzing and graphing quadratic functions.

Find the vertex, the axis of symmetry, and the maximum or minimum value of $f(x) = 2x^2 + 12x + 12$.

$$f(x) = 2x^2 + 12x + 12$$

$$= 2(x^2 + 6x) + 12 \qquad \text{Note that 9 completes the square for } x^2 + 6x.$$

$$= 2(x^2 + 6x + 9 - 9) + 12 \qquad \text{Adding } 9 - 9, \text{ or } 0, \text{ inside the parentheses}$$

$$= 2(x^2 + 6x + 9) - 2 \cdot 9 + 12$$

$$\qquad \text{Using the distributive law to remove } -9 \text{ from within the parentheses}$$

$$= 2(x + 3)^2 - 6$$

$$= 2[x - (-3)]^2 + (-6)$$

The function is now written in the form $f(x) = a(x - h)^2 + k$ with $a = 2$, $h = -3$, and $k = -6$. Because $a > 0$, we know the graph opens up and thus the function has a minimum value. We also know the following:

Vertex $(h, k)$: $(-3, -6)$;

Axis of symmetry $x = h$: $x = -3$;

Minimum value of the function $k$: $-6$.

To graph the function, we first plot the vertex and then find several points on either side of it. We plot these points and connect them with a smooth curve.

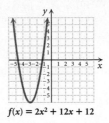

$$f(x) = 2x^2 + 12x + 12$$

**The Vertex of a Parabola**

The **vertex** of the graph of $f(x) = ax^2 + bx + c$ is

$$\left( -\frac{b}{2a}, f\left( -\frac{b}{2a} \right) \right).$$

We calculate the $x$-coordinate.   We substitute to find the $y$-coordinate.

Find the vertex of the function

$$f(x) = -3x^2 + 6x + 1.$$

$$-\frac{b}{2a} = -\frac{6}{2(-3)} = 1$$

$$f(1) = -3 \cdot 1^2 + 6 \cdot 1 + 1 = 4$$

The vertex is $(1, 4)$.

Some applied problems can be solved by finding the maximum or minimum value of a quadratic function.

See Examples 5–7 on pp. 266–268.

### SECTION 3.4: SOLVING RATIONAL EQUATIONS AND RADICAL EQUATIONS

A **rational equation** is an equation containing one or more rational expressions. When we solve a rational equation, we usually first multiply by the least common denominator (LCD) of all the denominators to clear the fractions.

CAUTION! When we multiply by an expression containing a variable, we might not obtain an equation equivalent to the original equation, so we *must* check the possible solutions obtained by substituting them in the original equation.

Solve: $\dfrac{5}{x + 2} - \dfrac{4}{x^2 - 4} = \dfrac{x - 3}{x - 2}.$

$$\frac{5}{x + 2} - \frac{4}{(x + 2)(x - 2)} = \frac{x - 3}{x - 2} \qquad \text{The LCD is } (x + 2)(x - 2).$$

$$(x + 2)(x - 2)\left( \frac{5}{x + 2} - \frac{4}{(x + 2)(x - 2)} \right)$$

$$= (x + 2)(x - 2) \cdot \frac{x - 3}{x - 2}$$

$$5(x - 2) - 4 = (x + 2)(x - 3)$$

$$5x - 10 - 4 = x^2 - x - 6$$

$$5x - 14 = x^2 - x - 6$$

$$0 = x^2 - 6x + 8$$

$$0 = (x - 2)(x - 4)$$

$$x - 2 = 0 \quad or \quad x - 4 = 0$$

$$x = 2 \quad or \qquad x = 4$$

The number 2 does not check, but 4 does. The solution is 4.

A **radical equation** is an equation that contains one or more radicals. We use the **principle of powers** to solve radical equations.

For any positive integer $n$:

If $a = b$ is true, then $a^n = b^n$ is true.

CAUTION! If $a^n = b^n$ is true, it is not necessarily true that $a = b$, so we *must* check the possible solutions obtained by substituting them in the original equation.

Solve: $\sqrt{x + 2} + \sqrt{x - 1} = 3$.

$$\sqrt{x + 2} = 3 - \sqrt{x - 1} \qquad \text{Isolating one radical}$$
$$(\sqrt{x + 2})^2 = (3 - \sqrt{x - 1})^2$$
$$x + 2 = 9 - 6\sqrt{x - 1} + (x - 1)$$
$$x + 2 = 8 - 6\sqrt{x - 1} + x$$
$$-6 = -6\sqrt{x - 1} \qquad \text{Isolating the remaining radical}$$
$$1 = \sqrt{x - 1} \qquad \text{Dividing by } -6$$
$$1^2 = (\sqrt{x - 1})^2$$
$$1 = x - 1$$
$$2 = x$$

The number 2 checks. It is the solution.

## SECTION 3.5: SOLVING EQUATIONS AND INEQUALITIES WITH ABSOLUTE VALUE

We use the following property to **solve equations with absolute value**.

For $a > 0$ and an algebraic expression $X$:

$|X| = a$   is equivalent to   $X = -a$  or  $X = a$.

Solve: $|x + 1| = 4$.

$$|x + 1| = 4$$
$$x + 1 = -4 \quad or \quad x + 1 = 4$$
$$x = -5 \quad or \qquad x = 3$$

Both numbers check.

The following properties are used to **solve inequalities with absolute value**.

For $a > 0$ and an algebraic expression $X$:

$|X| < a$   is equivalent to   $-a < X < a$.
$|X| > a$   is equivalent to   $X < -a \ or \ X > a$.

Similar statements hold for

$|X| \leq a$   and   $|X| \geq a$.

Solve: $|x - 2| < 3$.

$$|x - 2| < 3$$
$$-3 < x - 2 < 3$$
$$-1 < x < 5 \qquad \text{Adding 2}$$

The solution set is $\{x | -1 < x < 5\}$, or $(-1, 5)$.

$$|3x| \geq 6$$
$$3x \leq -6 \quad or \quad 3x \geq 6$$
$$x \leq -2 \quad or \quad x \geq 2 \qquad \text{Dividing by 3}$$

The solution set is $\{x | x \leq -2 \ or \ x \geq 2\}$, or $(-\infty, -2] \cup [2, \infty)$.

## REVIEW EXERCISES

*Determine whether the statement is true or false.*

1. We can use the quadratic formula to solve any quadratic equation.  [3.2]

2. The function $f(x) = -3(x + 4)^2 - 1$ has a maximum value.  [3.3]

3. For any positive integer $n$, if $a^n = b^n$ is true, then $a = b$ is true.  [3.4]

4. An equation with absolute value cannot have two negative-number solutions.  [3.5]

*Solve.*  [3.2]

5. $(2y + 5)(3y - 1) = 0$

6. $x^2 + 4x - 5 = 0$

7. $3x^2 + 2x = 8$

8. $5x^2 = 15$

9. $x^2 + 10 = 0$

*Find the zero(s) of the function.*  [3.2]

10. $f(x) = x^2 - 2x + 1$

11. $f(x) = x^2 + 2x - 15$

12. $f(x) = 2x^2 - x - 5$

13. $f(x) = 3x^2 + 2x + 3$

*Solve.*

14. $\dfrac{5}{2x + 3} + \dfrac{1}{x - 6} = 0$  [3.4]

15. $\dfrac{3}{8x + 1} + \dfrac{8}{2x + 5} = 1$  [3.4]

16. $\sqrt{5x + 1} - 1 = \sqrt{3x}$  [3.4]

17. $\sqrt{x - 1} - \sqrt{x - 4} = 1$  [3.4]

18. $|x - 4| = 3$  [3.5]

19. $|2y + 7| = 9$  [3.5]

*Solve and write interval notation for the solution set. Then graph the solution set.*  [3.5]

20. $|5x| \geq 15$

21. $|3x + 4| < 10$

22. $|6x - 1| < 5$

23. $|x + 4| \geq 2$

24. Solve $\dfrac{1}{M} + \dfrac{1}{N} = \dfrac{1}{P}$ for $P$.  [3.4]

*Express in terms of i.*  [3.1]

25. $-\sqrt{-40}$

26. $\sqrt{-12} \cdot \sqrt{-20}$

27. $\dfrac{\sqrt{-49}}{-\sqrt{-64}}$

*Simplify each of the following. Write the answer in the form $a + bi$, where a and b are real numbers.*  [3.1]

28. $(6 + 2i) + (-4 - 3i)$

29. $(3 - 5i) - (2 - i)$

30. $(6 + 2i)(-4 - 3i)$

31. $\dfrac{2 - 3i}{1 - 3i}$

32. $i^{23}$

*Solve by completing the square to obtain exact solutions. Show your work.*  [3.2]

33. $x^2 - 3x = 18$

34. $3x^2 - 12x - 6 = 0$

*Solve. Give exact solutions.*  [3.2]

35. $3x^2 + 10x = 8$

36. $r^2 - 2r + 10 = 0$

37. $x^2 = 10 + 3x$

38. $x = 2\sqrt{x - 1}$

39. $y^4 - 3y^2 + 1 = 0$

40. $(x^2 - 1)^2 - (x^2 - 1) - 2 = 0$

41. $(p - 3)(3p + 2)(p + 2) = 0$

42. $x^3 + 5x^2 - 4x - 20 = 0$

*In Exercises 43 and 44, complete the square to:*

a) *find the vertex;*
b) *find the axis of symmetry;*
c) *determine whether there is a maximum or minimum value and find that value;*
d) *find the range; and*
e) *graph the function.*  [3.3]

43. $f(x) = -4x^2 + 3x - 1$

44. $f(x) = 5x^2 - 10x + 3$

*In Exercises 45–48, match the equation with one of the figures (a)–(d), which follow.* [3.3]

a)
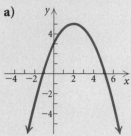

b)
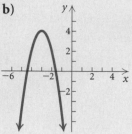

c)
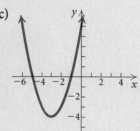

d)
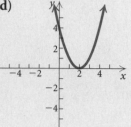

**45.** $y = (x - 2)^2$

**46.** $y = (x + 3)^2 - 4$

**47.** $y = -2(x + 3)^2 + 4$

**48.** $y = -\frac{1}{2}(x - 2)^2 + 5$

**49.** *Legs of a Right Triangle.*    The hypotenuse of a right triangle is 50 ft. One leg is 10 ft longer than the other. What are the lengths of the legs? [3.2]

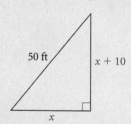

**50.** *Bicycling Speed.*    Logan and Cassidy leave a campsite, Logan biking due north and Cassidy biking due east. Logan bikes 7 km/h slower than Cassidy. After 4 hr, they are 68 km apart. Find the speed of each bicyclist. [3.2]

**51.** *Sidewalk Width.*    A 60-ft by 80-ft parking lot is torn up to install a sidewalk of uniform width around its perimeter. The new area of the parking lot is two-thirds of the old area. How wide is the sidewalk? [3.2]

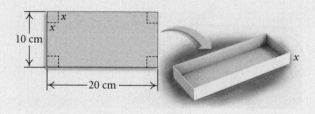

**52.** *Maximizing Volume.*    The Berniers have 24 ft of flexible fencing with which to build a rectangular "toy corral." If the fencing is 2 ft high, what dimensions should the corral have in order to maximize its volume? [3.3]

**53.** *Dimensions of a Box.*    An open box is made from a 10-cm by 20-cm piece of aluminum by cutting a square from each corner and folding up the edges. The area of the resulting base is 90 cm$^2$. What is the length of the sides of the squares? [3.2]

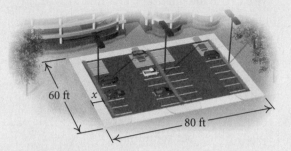

**54.** Find the zeros of $f(x) = 2x^2 - 5x + 1$. [3.2]

**A.** $\dfrac{5 \pm \sqrt{17}}{2}$        **B.** $\dfrac{5 \pm \sqrt{17}}{4}$

**C.** $\dfrac{5 \pm \sqrt{33}}{4}$        **D.** $\dfrac{-5 \pm \sqrt{17}}{4}$

**55.** Solve: $\sqrt{4x + 1} + \sqrt{2x} = 1$. [3.4]

    **A.** There are two solutions.
    **B.** There is only one solution. It is less than 1.
    **C.** There is only one solution. It is greater than 1.
    **D.** There is no solution.

**56.** The graph of $f(x) = (x - 2)^2 - 3$ is which of the following? [3.3]

**A.**
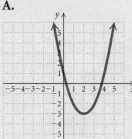

**B.**
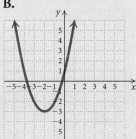

**C.**
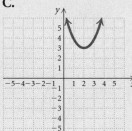

**D.**
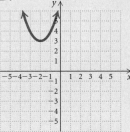

---

## Synthesis

*Solve.*

**57.** $\sqrt{\sqrt{\sqrt{x}}} = 2$ [3.4]

**58.** $(t - 4)^{4/5} = 3$ [3.4]

**59.** $(x - 1)^{2/3} = 4$ [3.4]

**60.** $(2y - 2)^2 + y - 1 = 5$ [3.2]

**61.** $\sqrt{x + 2} + \sqrt[4]{x + 2} - 2 = 0$ [3.2]

**62.** At the beginning of the year, $3500 was deposited in a savings account. One year later, $4000 was deposited in another account. The interest rate was the same for both accounts. At the end of the second year, there was a total of $8518.35 in the accounts. What was the annual interest rate? [3.2]

**63.** Find $b$ such that $f(x) = -3x^2 + bx - 1$ has a maximum value of 2. [3.3]

---

## Collaborative Discussion and Writing

**64.** Is the product of two imaginary numbers always an imaginary number? Explain your answer. [3.1]

**65.** Is it possible for a quadratic function to have one real zero and one imaginary zero? Why or why not? [3.2]

**66.** If the graphs of
$$f(x) = a_1(x - h_1)^2 + k_1$$
and
$$g(x) = a_2(x - h_2)^2 + k_2$$
have the same shape, what, if anything, can you conclude about the $a$'s, the $h$'s, and the $k$'s? Explain your answer. [3.3]

**67.** Explain why it is necessary to check the possible solutions of a rational equation. [3.4]

**68.** Explain why it is necessary to check the possible solutions when the principle of powers is used to solve an equation. [3.4]

**69.** Explain why $|x| < p$ has no solution for $p \le 0$. [3.5]

**70.** Explain why all real numbers are solutions of $|x| > p$, for $p < 0$. [3.5]

---

## Chapter 3 Test

*Solve. Find exact solutions.*

**1.** $(2x - 1)(x + 5) = 0$      **2.** $6x^2 - 36 = 0$

**3.** $x^2 + 4 = 0$      **4.** $x^2 - 2x - 3 = 0$

**5.** $x^2 - 5x + 3 = 0$      **6.** $2t^2 - 3t + 4 = 0$

**7.** $x + 5\sqrt{x} - 36 = 0$

**8.** $\dfrac{3}{3x + 4} + \dfrac{2}{x - 1} = 2$

**9.** $\sqrt{x + 4} - 2 = 1$

**10.** $\sqrt{x + 4} - \sqrt{x - 4} = 2$

**11.** $|x + 4| = 7$

**12.** $|4y - 3| = 5$

*Solve and write interval notation for the solution set. Then graph the solution set.*

**13.** $|x + 3| \le 4$         **14.** $|2x - 1| < 5$

**15.** $|x + 5| > 2$         **16.** $|3x - 5| \ge 7$

**17.** Solve $\dfrac{1}{A} + \dfrac{1}{B} = \dfrac{1}{C}$ for $B$.

**18.** Solve $R = \sqrt{3np}$ for $n$.

**19.** Solve $x^2 + 4x = 1$ by completing the square. Find the exact solutions. Show your work.

**20.** The tallest structure in the United States, at 2063 ft, is the KTHI-TV tower in North Dakota (*Source: The Cambridge Fact Finder*). How long would it take an object falling freely from the top to reach the ground? (Use the formula $s = 16t^2$.)

*Express in terms of i.*

**21.** $\sqrt{-43}$

**22.** $-\sqrt{-25}$

*Simplify.*

**23.** $(5 - 2i) - (2 + 3i)$

**24.** $(3 + 4i)(2 - i)$

**25.** $\dfrac{1 - i}{6 + 2i}$

**26.** $i^{33}$

*Find the zeros of each function.*

**27.** $f(x) = 4x^2 - 11x - 3$

**28.** $f(x) = 2x^2 - x - 7$

**29.** For the graph of the function
$$f(x) = -x^2 + 2x + 8:$$
   a) Find the vertex.
   b) Find the axis of symmetry.
   c) State whether there is a maximum or minimum value and find that value.
   d) Find the range.
   e) Graph the function.

**30.** *Maximizing Area.*   A homeowner wants to fence a rectangular play yard using 80 ft of fencing. The side of the house will be used as one side of the rectangle. Find the dimensions for which the area is a maximum.

**31.** The graph of $f(x) = x^2 - 2x - 1$ is which of the following?

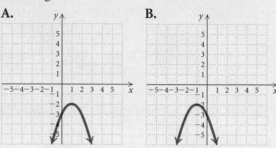

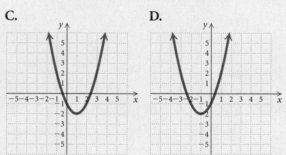

---

## Synthesis

**32.** Find $a$ such that $f(x) = ax^2 - 4x + 3$ has a maximum value of 12.

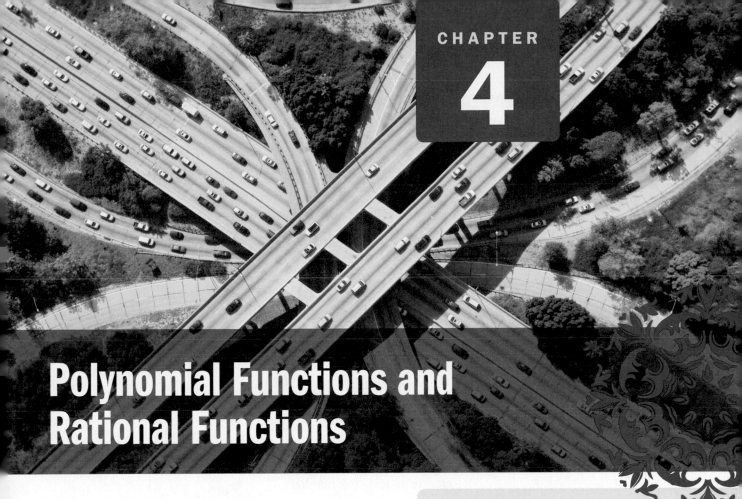

# Polynomial Functions and Rational Functions

## APPLICATION

The goal of the U.S. automakers is to meet the CAFE (corporate average fuel economy) standard of 35.5 mpg (miles per gallon) by 2016. Adjusted industry fuel economy levels reached in various years from 2000–2010 and future CAFE standards are illustrated in the graph on p. 308. (*Source*: U.S. Environmental Protection Agency). Model the data with a quadratic function, a cubic function, and a quartic function. Then using $R^2$, the coefficient of determination, decide which function is the better fit; and with that function, estimate the fuel efficiency in 2005, in 2011, in 2017, and in 2021.

**This problem appears as Example 9 in Section 4.1.**

## 4.1

# Polynomial Functions and Modeling

- Determine the behavior of the graph of a polynomial function using the leading-term test.
- Factor polynomial functions and find their zeros and their multiplicities.
- Use a graphing calculator to graph a polynomial function and find its real-number zeros, its relative maximum and minimum values, and its domain and range.
- Solve applied problems using polynomial models; fit linear, quadratic, power, cubic, and quartic polynomial functions to data.

There are many different kinds of functions. The constant, linear, and quadratic functions that we studied in Chapters 1 and 3 are part of a larger group of functions called *polynomial functions*.

---

### POLYNOMIAL FUNCTION

A **polynomial function** $P$ is given by

$$P(x) = a_n x^n + a_{n-1} x^{n-1} + a_{n-2} x^{n-2} + \cdots + a_1 x + a_0,$$

where the coefficients $a_n, a_{n-1}, \ldots, a_1, a_0$ are real numbers and the exponents are whole numbers.

---

The first nonzero coefficient, $a_n$, is called the **leading coefficient**. The term $a_n x^n$ is called the **leading term**. The **degree** of the polynomial function is $n$. Some examples of polynomial functions follow.

| POLYNOMIAL FUNCTION | EXAMPLE | DEGREE | LEADING TERM | LEADING COEFFICIENT |
|---|---|---|---|---|
| Constant | $f(x) = 3$ $\quad(f(x) = 3 = 3x^0)$ | 0 | 3 | 3 |
| Linear | $f(x) = \frac{2}{3}x + 5$ $\left(f(x) = \frac{2}{3}x + 5 = \frac{2}{3}x^1 + 5\right)$ | 1 | $\frac{2}{3}x$ | $\frac{2}{3}$ |
| Quadratic | $f(x) = 4x^2 - x + 3$ | 2 | $4x^2$ | 4 |
| Cubic | $f(x) = x^3 + 2x^2 + x - 5$ | 3 | $x^3$ | 1 |
| Quartic | $f(x) = -x^4 - 1.1x^3 + 0.3x^2 - 2.8x - 1.7$ | 4 | $-x^4$ | $-1$ |

The function $f(x) = 0$ can be described in many ways:

$$f(x) = 0 = 0x^2 = 0x^{15} = 0x^{48},$$

and so on. For this reason, we say that the constant function $f(x) = 0$ has no degree.

Functions such as

$$f(x) = \frac{2}{x} + 5, \text{ or } 2x^{-1} + 5, \quad \text{and} \quad g(x) = \sqrt{x} - 6, \text{ or } x^{1/2} - 6,$$

are *not* polynomial functions because the exponents $-1$ and $\frac{1}{2}$ are *not* whole numbers.

From our study of functions in Chapters 1–3, we know how to find or at least estimate many characteristics of a polynomial function. Let's consider two examples for review.

## Quadratic Function

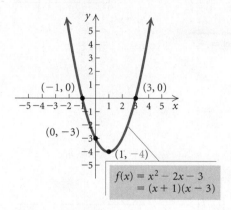

Function: $f(x) = x^2 - 2x - 3$
$\qquad\qquad\;\; = (x + 1)(x - 3)$

Zeros: $-1, 3$

$x$-intercepts: $(-1, 0), (3, 0)$

$y$-intercept: $(0, -3)$

Minimum: $-4$ at $x = 1$

Maximum: None

Domain: All real numbers, $(-\infty, \infty)$

Range: $[-4, \infty)$

## Cubic Function

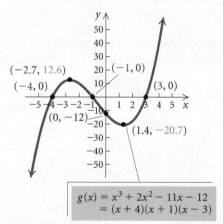

Function: $g(x) = x^3 + 2x^2 - 11x - 12$
$\qquad\qquad\;\; = (x + 4)(x + 1)(x - 3)$

Zeros: $-4, -1, 3$

$x$-intercepts: $(-4, 0), (-1, 0), (3, 0)$

$y$-intercept: $(0, -12)$

Relative minimum: $-20.7$ at $x = 1.4$

Relative maximum: $12.6$ at $x = -2.7$

Domain: All real numbers, $(-\infty, \infty)$

Range: All real numbers, $(-\infty, \infty)$

All graphs of polynomial functions have some characteristics in common. Compare the following graphs. How do the graphs of polynomial functions differ from the graphs of nonpolynomial functions? Describe some characteristics of the graphs of polynomial functions that you observe.

## Polynomial Functions

$f(x) = x^2 + 3x + 1$

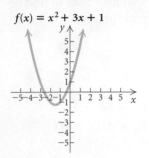

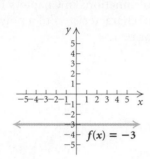

$f(x) = -3$

$f(x) = 2x^3 + x^2 + x - 1$

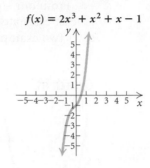

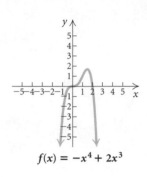

$f(x) = -x^4 + 2x^3$

## Nonpolynomial Functions

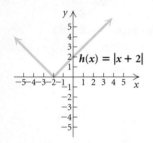

$h(x) = |x + 2|$

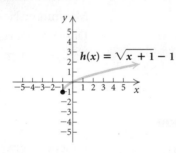

$h(x) = \sqrt{x + 1} - 1$

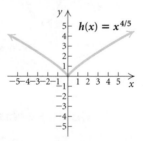

$h(x) = x^{4/5}$

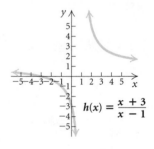

$h(x) = \dfrac{x + 3}{x - 1}$

You probably noted that the graph of a polynomial function is *continuous*; that is, it has no holes or breaks. It is also smooth; there are no sharp corners.

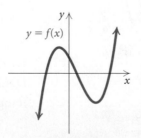

A continuous function

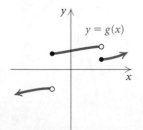

A discontinuous function

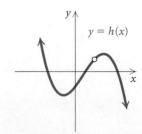

A discontinuous function

The *domain* of a polynomial function is the set of all real numbers, $(-\infty, \infty)$.

## The Leading-Term Test

The behavior of the graph of a polynomial function as $x$ becomes very large $(x \rightarrow \infty)$ or very small $(x \rightarrow -\infty)$ is referred to as the end behavior of the graph. The leading term of a polynomial function determines its end behavior.

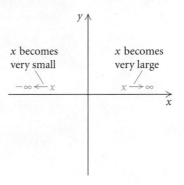

$x$ becomes very small

$x$ becomes very large

$-\infty \leftarrow x$

$x \xrightarrow{} \infty$

Using the graphs shown below, let's see if we can discover some general patterns by comparing the end behavior of even-degree and odd-degree functions. We also observe the effect of positive and negative leading coefficients.

### Even Degree

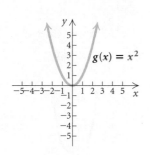

$g(x) = x^2$

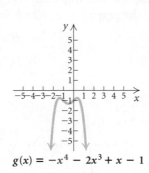

$g(x) = -x^4 - 2x^3 + x - 1$

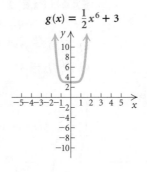

$g(x) = \frac{1}{2}x^6 + 3$

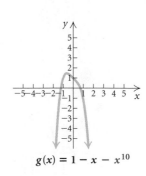

$g(x) = 1 - x - x^{10}$

### Odd Degree

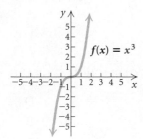

$f(x) = x^3$

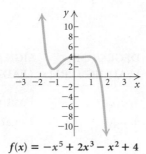

$f(x) = -x^5 + 2x^3 - x^2 + 4$

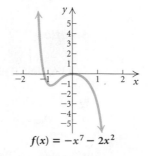

$f(x) = -x^7 - 2x^2$

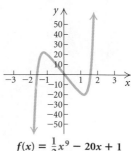

$f(x) = \frac{1}{2}x^9 - 20x + 1$

We can summarize our observations as follows.

### THE LEADING-TERM TEST

If $a_n x^n$ is the leading term of a polynomial function, then the behavior of the graph as $x \to \infty$ or as $x \to -\infty$ can be described in one of the four following ways.

| $n$ | $a_n > 0$ | $a_n < 0$ |
|---|---|---|
| Even | | |
| Odd | | |

The ᴧᴧᴧᴧ portion of the graph is not determined by this test.

**EXAMPLE 1** Using the leading-term test, match each of the following functions with one of the graphs A–D, which follow.

**a)** $f(x) = 3x^4 - 2x^3 + 3$      **b)** $f(x) = -5x^3 - x^2 + 4x + 2$

**c)** $f(x) = x^5 + \frac{1}{4}x + 1$      **d)** $f(x) = -x^6 + x^5 - 4x^3$

**A.** 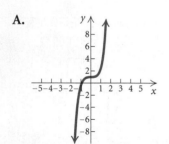      **B.** 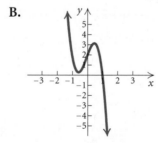      **C.** 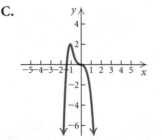      **D.**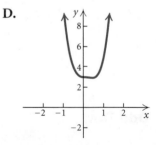

*Solution*

| | LEADING TERM | DEGREE OF LEADING TERM | SIGN OF LEADING COEFFICIENT | GRAPH |
|---|---|---|---|---|
| **a)** | $3x^4$ | 4, even | Positive | D |
| **b)** | $-5x^3$ | 3, odd | Negative | B |
| **c)** | $x^5$ | 5, odd | Positive | A |
| **d)** | $-x^6$ | 6, even | Negative | C |

**Now Try Exercise 19.**

### ■ Finding Zeros of Factored Polynomial Functions

Let's review the meaning of the real zeros of a function and their connection to the $x$-intercepts of the function's graph.

## CONNECTING THE CONCEPTS

### Zeros, Solutions, and Intercepts

| FUNCTION | ZEROS OF THE FUNCTION; SOLUTIONS OF THE EQUATION | ZEROS OF THE FUNCTION; X-INTERCEPTS OF THE GRAPH |
|---|---|---|

**Quadratic Polynomial**

$g(x) = x^2 - 2x - 8$
$\quad = (x + 2)(x - 4),$

or

$\quad y = (x + 2)(x - 4)$

To find the **zeros** of $g(x)$, we solve $g(x) = 0$:

$$x^2 - 2x - 8 = 0$$
$$(x + 2)(x - 4) = 0$$
$$x + 2 = 0 \quad or \quad x - 4 = 0$$
$$x = -2 \quad or \quad x = 4.$$

The **solutions** of $x^2 - 2x - 8 = 0$ are $-2$ and $4$. They are the zeros of the function $g(x)$; that is,

$$g(-2) = 0 \quad and \quad g(4) = 0.$$

The real-number zeros of $g(x)$ are the $x$-coordinates of the **$x$-intercepts** of the graph of $y = g(x)$.

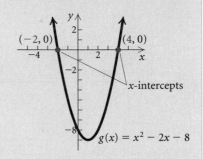

**Cubic Polynomial**

$h(x)$
$\quad = x^3 + 2x^2 - 5x - 6$
$\quad = (x + 3)(x + 1)(x - 2),$

or

$y = (x + 3)(x + 1)(x - 2)$

To find the **zeros** of $h(x)$, we solve $h(x) = 0$:

$$x^3 + 2x^2 - 5x - 6 = 0$$
$$(x + 3)(x + 1)(x - 2) = 0$$
$$x + 3 = 0 \quad or \quad x + 1 = 0 \quad or \quad x - 2 = 0$$
$$x = -3 \ or \quad x = -1 \ or \quad x = 2.$$

The **solutions** of $x^3 + 2x^2 - 5x - 6 = 0$ are $-3, -1,$ and $2$. They are the zeros of the function $h(x)$; that is,

$$h(-3) = 0,$$
$$h(-1) = 0, \quad and$$
$$h(2) = 0.$$

The real-number zeros of $h(x)$ are the $x$-coordinates of the **$x$-intercepts** of the graph of $y = h(x)$.

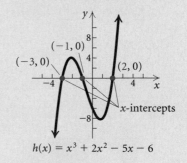

The connection between the real-number zeros of a function and the $x$-intercepts of the graph of the function is easily seen in the preceding examples. If $c$ is a real zero of a function (that is, if $f(c) = 0$), then $(c, 0)$ is an $x$-intercept of the graph of the function.

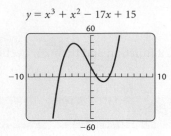

$y = x^3 + x^2 - 17x + 15$

**EXAMPLE 2** Consider $P(x) = x^3 + x^2 - 17x + 15$. Determine whether each of the numbers 2 and $-5$ is a zero of $P(x)$.

**Solution** We first evaluate $P(2)$:

$$P(2) = (2)^3 + (2)^2 - 17(2) + 15 = -7.$$ Substituting 2 into the polynomial

Since $P(2) \neq 0$, we know that 2 is *not* a zero of the polynomial function. We then evaluate $P(-5)$:

$$P(-5) = (-5)^3 + (-5)^2 - 17(-5) + 15 = 0.$$ Substituting $-5$ into the polynomial

Since $P(-5) = 0$, we know that $-5$ is a zero of $P(x)$.

Now Try Exercise 23.

Let's take a closer look at the polynomial function

$$h(x) = x^3 + 2x^2 - 5x - 6$$

(see Connecting the Concepts on p. 303). The factors of $h(x)$ are

$$x + 3, \qquad x + 1, \quad \text{and} \quad x - 2,$$

and the zeros are

$$-3, \qquad -1, \quad \text{and} \quad 2.$$

We note that when the polynomial is expressed as a product of linear factors, each factor determines a zero of the function. Thus if we know the linear factors of a polynomial function $f(x)$, we can easily find the zeros of $f(x)$ by solving the equation $f(x) = 0$ using the principle of zero products.

---

PRINCIPLE OF ZERO PRODUCTS

REVIEW SECTION **3.2.**

---

$y = 5(x - 2)^3 (x + 1)$

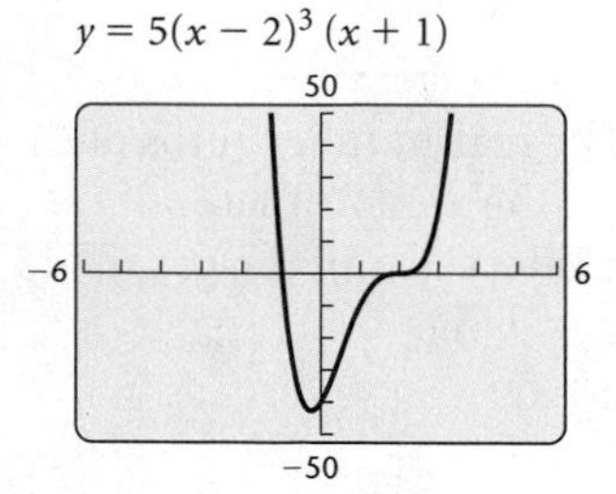

FIGURE 1

**EXAMPLE 3** Find the zeros of

$$\begin{aligned} f(x) &= 5(x - 2)(x - 2)(x - 2)(x + 1) \\ &= 5(x - 2)^3(x + 1). \end{aligned}$$

**Solution** To solve the equation $f(x) = 0$, we use the principle of zero products, solving $x - 2 = 0$ and $x + 1 = 0$. The zeros of $f(x)$ are 2 and $-1$. (See Fig. 1.)

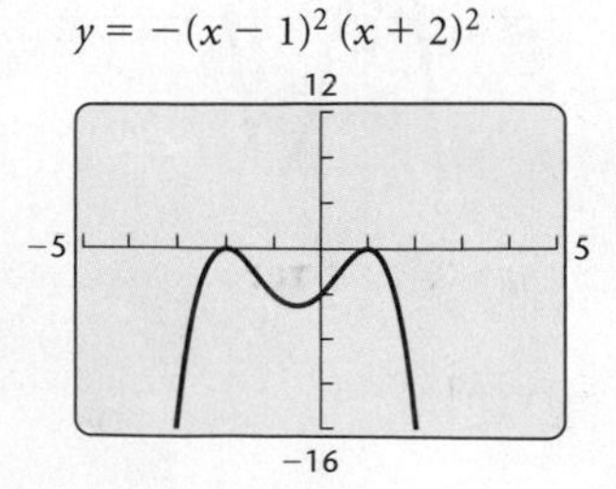

$y = -(x - 1)^2 (x + 2)^2$

FIGURE 2

**EXAMPLE 4** Find the zeros of

$$\begin{aligned} g(x) &= -(x - 1)(x - 1)(x + 2)(x + 2) \\ &= -(x - 1)^2(x + 2)^2. \end{aligned}$$

**Solution** To solve the equation $g(x) = 0$, we use the principle of zero products, solving $x - 1 = 0$ and $x + 2 = 0$. The zeros of $g(x)$ are 1 and $-2$. (See Fig. 2.)

Let's consider the occurrences of the zeros in the functions in Examples 3 and 4 and their relationship to the graphs of those functions. In Example 3, the factor $x - 2$ occurs three times. In a case like this, we say that the zero we obtain from this factor, 2, has a **multiplicity** of 3. The factor $x + 1$ occurs one time. The zero we obtain from this factor, $-1$, has a *multiplicity* of 1.

In Example 4, the factors $x - 1$ and $x + 2$ each occur two times. Thus both zeros, 1 and $-2$, have a *multiplicity* of 2.

Note, in Example 3, that the zeros have odd multiplicities and the graph crosses the $x$-axis at both $-1$ and 2. But in Example 4, the zeros have even multiplicities and the graph is tangent to (touches but does not cross) the $x$-axis at $-2$ and 1. This leads us to the following generalization.

---

**EVEN AND ODD MULTIPLICITY**

If $(x - c)^k$, $k \geq 1$, is a factor of a polynomial function $P(x)$ and $(x - c)^{k+1}$ is not a factor and:

- $k$ is odd, then the graph crosses the $x$-axis at $(c, 0)$;
- $k$ is even, then the graph is tangent to the $x$-axis at $(c, 0)$.

---

Some polynomials can be factored by grouping. Then we use the principle of zero products to find their zeros.

**EXAMPLE 5**  Find the zeros of

$$f(x) = x^3 - 2x^2 - 9x + 18.$$

**Solution**  We factor by grouping, as follows:

$$\begin{aligned}
f(x) &= x^3 - 2x^2 - 9x + 18 \\
&= x^2(x - 2) - 9(x - 2) && \text{Grouping } x^3 \text{ with } -2x^2 \text{ and } -9x \\
& && \text{with 18 and factoring each group} \\
&= (x - 2)(x^2 - 9) && \text{Factoring out } x - 2 \\
&= (x - 2)(x + 3)(x - 3). && \text{Factoring } x^2 - 9
\end{aligned}$$

Then, by the principle of zero products, the solutions of the equation $f(x) = 0$ are 2, $-3$, and 3. These are the zeros of $f(x)$.

**Now Try Exercise 39.**

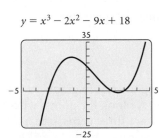

$y = x^3 - 2x^2 - 9x + 18$

Other factoring techniques can also be used.

**EXAMPLE 6**  Find the zeros of

$$f(x) = x^4 + 4x^2 - 45.$$

**Solution**  We factor as follows:

$$f(x) = x^4 + 4x^2 - 45 = (x^2 - 5)(x^2 + 9).$$

We now solve the equation $f(x) = 0$ to determine the zeros. We use the principle of zero products:

$$\begin{aligned}
(x^2 - 5)(x^2 + 9) &= 0 \\
x^2 - 5 = 0 \quad &or \quad x^2 + 9 = 0 \\
x^2 = 5 \quad &or \quad x^2 = -9 \\
x = \pm\sqrt{5} \quad &or \quad x = \pm\sqrt{-9} = \pm 3i.
\end{aligned}$$

The solutions are $\pm\sqrt{5}$ and $\pm 3i$. These are the zeros of $f(x)$.

**Now Try Exercise 37.**

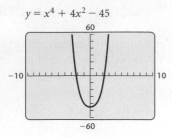

$$y = x^4 + 4x^2 - 45$$

Only the real-number zeros of a function correspond to the *x*-intercepts of its graph. For instance, the real-number zeros of the function in Example 6, $-\sqrt{5}$ and $\sqrt{5}$, can be seen on the graph of the function at left, but the non-real zeros, $-3i$ and $3i$, cannot.

> Every polynomial function of degree *n*, with $n \geq 1$, has at least one zero and at most *n* zeros.

This is often stated as follows: "Every polynomial function of degree *n*, with $n \geq 1$, has *exactly n* zeros." This statement is compatible with the preceding statement, if one takes multiplicities into account.

## Finding Real Zeros on a Calculator

Finding exact values of the real zeros of a function can be difficult. We can find approximations using a graphing calculator.

**EXAMPLE 7** Find the real zeros of the function *f* given by

$$f(x) = 0.1x^3 - 0.6x^2 - 0.1x + 2.$$

Approximate the zeros to three decimal places.

*Solution* We use a graphing calculator, trying to create a graph that clearly shows the curvature and the intercepts. Then we look for points where the graph crosses the *x*-axis. It appears that there are three zeros, one near $-2$, one near 2, and one near 6. We know that there are no more than 3 because the degree of the polynomial is 3. We use the ZERO feature to find them.

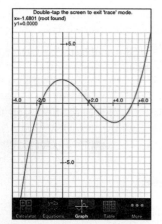

The zeros of a function can be found with a graphing calculator app. One of the three zeros of the function in Example 7 is shown here.

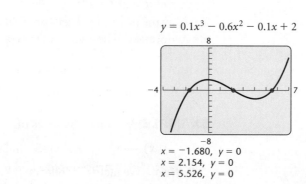

$$y = 0.1x^3 - 0.6x^2 - 0.1x + 2$$

$x = -1.680, \; y = 0$
$x = 2.154, \; y = 0$
$x = 5.526, \; y = 0$

The zeros are approximately $-1.680$, $2.154$, and $5.526$.

**Now Try Exercise 43.**

## Polynomial Models

Polynomial functions have many uses as models in science, engineering, and business. The simplest use of polynomial functions in applied problems occurs when we merely evaluate a polynomial function. In such cases, a model has already been developed.

**EXAMPLE 8** *Ibuprofen in the Bloodstream.* The polynomial function

$$M(t) = 0.5t^4 + 3.45t^3 - 96.65t^2 + 347.7t$$

can be used to estimate the number of milligrams of the pain relief medication ibuprofen in the bloodstream $t$ hours after 400 mg of the medication has been taken.

**a)** Find the number of milligrams in the bloodstream at $t = 0, 0.5, 1, 1.5$, and so on, up to 6 hr. Round the function values to the nearest tenth.

**b)** Find the domain, the relative maximum and where it occurs, and the range.

**Solution**

**a)** We can evaluate the function with the TABLE feature of a graphing calculator set in AUTO mode. We start at 0 and use a step-value of 0.5.

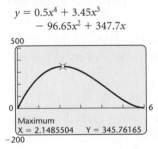

$$M(0) = 0, \qquad M(3.5) = 255.9,$$
$$M(0.5) = 150.2, \qquad M(4) = 193.2,$$
$$M(1) = 255, \qquad M(4.5) = 126.9,$$
$$M(1.5) = 318.3, \qquad M(5) = 66,$$
$$M(2) = 344.4, \qquad M(5.5) = 20.2,$$
$$M(2.5) = 338.6, \qquad M(6) = 0.$$
$$M(3) = 306.9,$$

**b)** Recall that the domain of a polynomial function, unless restricted by a statement of the function, is $(-\infty, \infty)$. The implications of this application restrict the domain of the function. If we assume that a patient had not taken any of the medication before, it seems reasonable that $M(0) = 0$; that is, at time 0, there is 0 mg of the medication in the bloodstream. After the medication has been taken, $M(t)$ will be positive for a period of time and eventually decrease back to 0 when $t = 6$ and not increase again (unless another dose is taken). Thus the restricted domain is $[0, 6]$.

To determine the range, we find the relative maximum value of the function using the MAXIMUM feature. The maximum is about 345.8 mg. It occurs approximately 2.15 hr, or 2 hr 9 min, after the initial dose has been taken. The range is about $[0, 345.8]$.

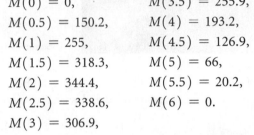

**Now Try Exercise 63.**

In Chapter 1, we used regression to model data with linear functions. We now expand that procedure to include quadratic, cubic, and quartic models.

**EXAMPLE 9** *Increasing Vehicle Fuel Efficiency.*   The goal of the U.S. automakers is to meet the CAFE (corporate average fuel economy) standard of 35.5 mpg (miles per gallon) by 2016. Adjusted industry fuel economy levels reached in various years from 2000–2010 and future CAFE standards are illustrated in the graph below (*Source*: U.S. Environmental Protection Agency).

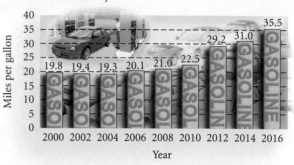

**Vehicle Fuel Efficiency**

*Source*: U.S. Environmental Protection Agency

Looking at the graph above, we note that the data, in miles per gallon, could be modeled with a quadratic function, a cubic function, or a quartic function.

**a)** Model the data with a quadratic function, a cubic function, and a quartic function. Let the first coordinate of each data point be the number of years after 2000; that is, enter the data as $(0, 19.8)$, $(2, 19.4)$, $(4, 19.3)$, and so on. Then using $R^2$, the **coefficient of determination**, decide which function is the best fit.

**b)** Graph the function with the scatterplot of the data.

**c)** Use the answer to part (a) to estimate the number of miles per gallon in 2005, in 2011, in 2017, and in 2021.

### Solution

**a)** Using the REGRESSION feature with DIAGNOSTIC turned on, we get the following.

```
     QuadReg
y=ax²+bx+c
a=.0998376623
b=−.5990692641
c=19.94060606
R²=.9764318496
```

```
       CubicReg
y=ax³+bx²+cx+d
a=−.0012310606
b=.1293831169
c=−.7773268398
d=20.10606061
R²=.9769001262
```

```
       QuarticReg
y=ax⁴+bx³+...+e
a=−.0010252768
b=.0315777972
c=−.1969478438
d=.2444347319
↓e=19.71235431
```

```
       QuarticReg
y=ax⁴+bx³+...+e
↑b=.0315777972
c=−.1969478438
d=.2444347319
e=19.71235431
R²=.9822620259
```

Since the $R^2$-value for the quartic function, 0.9822620259, is closer to 1 than the $R^2$-values for the quadratic function and the cubic function, the quartic function is the best fit:

$$f(x) = -0.0010252768x^4 + 0.0315777972x^3$$
$$- 0.1969478438x^2 + 0.2444347319x$$
$$+ 19.71235431.$$

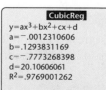

**b)** The scatterplot and the graph are shown at left.

c) We evaluate the function found in part (a).

| X | Y1 | |
|---|---|---|
| 5 | 19.317 | |
| 11 | 25.589 | |
| 17 | 36.459 | |
| 21 | 31.037 | |
| X = | | |

With this function, we can estimate the fuel efficiency at 19.3 mpg in 2005, 25.6 mpg in 2011, and 36.5 mpg in 2017. Looking at the bar graph shown on the preceding page, we see that these estimates appear to be fairly accurate.

If we use the function to estimate the number of miles per gallon in 2021, we get about 31.0 mpg. This estimate is not realistic since it is unreasonable to expect the number of miles per gallon to decrease in the future. The quartic model has a higher value for $R^2$ than the quadratic function or the cubic function over the domain of the data, but this number does not reflect the degree of accuracy for extended values. It is always important when using regression to evaluate predictions with common sense and knowledge of current trends.

**Now Try Exercise 77.**

## 4.1 Exercise Set

*Determine the leading term, the leading coefficient, and the degree of the polynomial. Then classify the polynomial function as constant, linear, quadratic, cubic, or quartic.*

**1.** $g(x) = \frac{1}{2}x^3 - 10x + 8$

**2.** $f(x) = 15x^2 - 10 + 0.11x^4 - 7x^3$

**3.** $h(x) = 0.9x - 0.13$

**4.** $f(x) = -6$

**5.** $g(x) = 305x^4 + 4021$

**6.** $h(x) = 2.4x^3 + 5x^2 - x + \frac{7}{8}$

**7.** $h(x) = -5x^2 + 7x^3 + x^4$

**8.** $f(x) = 2 - x^2$

**9.** $g(x) = 4x^3 - \frac{1}{2}x^2 + 8$

**10.** $f(x) = 12 + x$

*In Exercises 11–18, select one of the four sketches (a)–(d), which follow, to describe the end behavior of the graph of the function.*

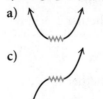

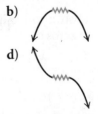

**11.** $f(x) = -3x^3 - x + 4$

**12.** $f(x) = \frac{1}{4}x^4 + \frac{1}{2}x^3 - 6x^2 + x - 5$

**13.** $f(x) = -x^6 + \frac{3}{4}x^4$

**14.** $f(x) = \frac{2}{5}x^5 - 2x^4 + x^3 - \frac{1}{2}x + 3$

**15.** $f(x) = -3.5x^4 + x^6 + 0.1x^7$

**16.** $f(x) = -x^3 + x^5 - 0.5x^6$

**17.** $f(x) = 10 + \frac{1}{10}x^4 - \frac{2}{5}x^3$

**18.** $f(x) = 2x + x^3 - x^5$

*In Exercises 19–22, use the leading-term test to match the function with one of the graphs (a)–(d), which follow.*

**a)**

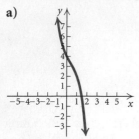

**b)**

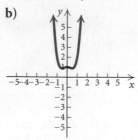

**c)**

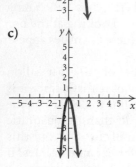

**d)**

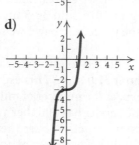

**19.** $f(x) = -x^6 + 2x^5 - 7x^2$

**20.** $f(x) = 2x^4 - x^2 + 1$

**21.** $f(x) = x^5 + \frac{1}{10}x - 3$

**22.** $f(x) = -x^3 + x^2 - 2x + 4$

**23.** Use substitution to determine whether 4, 5, and $-2$ are zeros of
$$f(x) = x^3 - 9x^2 + 14x + 24.$$

**24.** Use substitution to determine whether 2, 3, and $-1$ are zeros of
$$f(x) = 2x^3 - 3x^2 + x + 6.$$

**25.** Use substitution to determine whether 2, 3, and $-1$ are zeros of
$$g(x) = x^4 - 6x^3 + 8x^2 + 6x - 9.$$

**26.** Use substitution to determine whether 1, $-2$, and 3 are zeros of
$$g(x) = x^4 - x^3 - 3x^2 + 5x - 2.$$

*Find the zeros of the polynomial function and state the multiplicity of each.*

**27.** $f(x) = (x + 3)^2(x - 1)$

**28.** $f(x) = (x + 5)^3(x - 4)(x + 1)^2$

**29.** $f(x) = -2(x - 4)(x - 4)(x - 4)(x + 6)$

**30.** $f(x) = \left(x + \frac{1}{2}\right)(x + 7)(x + 7)(x + 5)$

**31.** $f(x) = (x^2 - 9)^3$

**32.** $f(x) = (x^2 - 4)^2$

**33.** $f(x) = x^3(x - 1)^2(x + 4)$

**34.** $f(x) = x^2(x + 3)^2(x - 4)(x + 1)^4$

**35.** $f(x) = -8(x - 3)^2(x + 4)^3x^4$

**36.** $f(x) = (x^2 - 5x + 6)^2$

**37.** $f(x) = x^4 - 4x^2 + 3$

**38.** $f(x) = x^4 - 10x^2 + 9$

**39.** $f(x) = x^3 + 3x^2 - x - 3$

**40.** $f(x) = x^3 - x^2 - 2x + 2$

**41.** $f(x) = 2x^3 - x^2 - 8x + 4$

**42.** $f(x) = 3x^3 + x^2 - 48x - 16$

*Using a graphing calculator, find the real zeros of the function. Approximate the zeros to three decimal places.*

**43.** $f(x) = x^3 - 3x - 1$

**44.** $f(x) = x^3 + 3x^2 - 9x - 13$

**45.** $f(x) = x^4 - 2x^2$

**46.** $f(x) = x^4 - 2x^3 - 5.6$

**47.** $f(x) = x^3 - x$

**48.** $f(x) = 2x^3 - x^2 - 14x - 10$

**49.** $f(x) = x^8 + 8x^7 - 28x^6 - 56x^5 + 70x^4$
$\qquad + 56x^3 - 28x^2 - 8x + 1$

**50.** $f(x) = x^6 - 10x^5 + 13x^3 - 4x^2 - 5$

*Using a graphing calculator, estimate the real zeros, the relative maxima and minima, and the range of the polynomial function.*

**51.** $g(x) = x^3 - 1.2x + 1$

**52.** $h(x) = -\frac{1}{2}x^4 + 3x^3 - 5x^2 + 3x + 6$

**53.** $f(x) = x^6 - 3.8$

**54.** $h(x) = 2x^3 - x^4 + 20$

**55.** $f(x) = x^2 + 10x - x^5$

**56.** $f(x) = 2x^4 - 5.6x^2 + 10$

*Determine whether the statement is true or false.*

**57.** If $P(x) = (x - 3)^4(x + 1)^3$, then the graph of the polynomial function $y = P(x)$ crosses the $x$-axis at $(3, 0)$.

**58.** If $P(x) = (x + 2)^2\left(x - \frac{1}{4}\right)^5$, then the graph of the polynomial function $y = P(x)$ crosses the $x$-axis at $\left(\frac{1}{4}, 0\right)$.

**59.** If $P(x) = (x - 2)^3(x + 5)^6$, then the graph of $y = P(x)$ is tangent to the $x$-axis at $(-5, 0)$.

**60.** If $P(x) = (x + 4)^2(x - 1)^2$, then the graph of $y = P(x)$ is tangent to the $x$-axis at $(4, 0)$.

**61.** *Twin Births.* As a result of a greater number of births to older women and the increased use of fertility drugs, the number of twin births in the United States increased approximately 42% from 1990 to 2005 (*Source*: National Center for Health Statistics, U.S. Department of Health and Human Services). The quartic function

$$f(x) = -0.056316x^4 - 19.500154x^3$$
$$+ 584.892054x^2 - 1518.5717x$$
$$+ 94,299.1990,$$

where $x$ is the number of years since 1990, can be used to estimate the number of twin births from 1990 to 2006. Estimate the number of twin births in 1995 and in 2005.

**62.** *Railroad Miles.* The greatest combined length of U.S.-owned operating railroad track existed in 1916, when industrial activity increased during World War I. The total length has decreased ever since. The data over the years 1900 to 2008 are modeled by the quartic function

$$f(x) = -0.004091x^4 + 1.275179x^3$$
$$- 142.589291x^2 + 5069.1067x$$
$$+ 197,909.1675,$$

where $x$ is the number of years since 1900 and $f(x)$ is in miles (*Source*: Association of American Railroads). Find the number of miles of operating railroad track in the United States in 1916, in 1960, and in 1985, and estimate the number in 2010. (*Note*: The lengths exclude yard tracks, sidings, and parallel tracks.)

**63.** *Dog Years.* A dog's life span is typically much shorter than that of a human. The cubic function

$$d(x) = 0.010255x^3 - 0.340119x^2$$
$$+ 7.397499x + 6.618361,$$

where $x$ is the dog's age, in years, approximates the equivalent human age in years. Estimate the equivalent human age for dogs that are 3, 12, and 16 years old.

**64.** *Threshold Weight.* In a study performed by Alvin Shemesh, it was found that the **threshold weight $W$**, defined as the weight above which the risk of death rises dramatically, is given by

$$W(h) = \left(\frac{h}{12.3}\right)^3,$$

where $W$ is in pounds and $h$ is a person's height, in inches. Find the threshold weight of a person who is 5 ft 7 in. tall.

**65.** *Projectile Motion.* A stone thrown downward with an initial velocity of 34.3 m/sec will travel a distance of $s$ meters, where

$$s(t) = 4.9t^2 + 34.3t$$

and $t$ is in seconds. If a stone is thrown downward at 34.3 m/sec from a height of 294 m, how long will it take the stone to hit the ground?

**66.** *Games in a Sports League.* If there are $x$ teams in a sports league and all the teams play each other twice, a total of $N(x)$ games are played, where

$$N(x) = x^2 - x.$$

A softball league has 9 teams, each of which plays the others twice. If the league pays $110 per game for the field and the umpires, how much will it cost to play the entire schedule?

**67.** *Median Home Prices.* The median price for an existing home in the United States peaked at $221,900 in 2006 (*Source*: National Association of REALTORS®). The quartic function

$$h(x) = 56.8328x^4 - 1554.7494x^3$$
$$+ 10{,}451.8211x^2 - 5655.7692x$$
$$+ 140{,}589.1608,$$

where $x$ is the number of years since 2000, can be used to estimate the median existing-home price from 2000 to 2009. Estimate the median existing-home price in 2002, in 2005, in 2008, and in 2009.

**68.** *Circulation of Daily Newspapers.* In 1985, the circulation of daily newspapers reached its highest level (*Source*: Newspaper Association of America). The quartic function

$$f(x) = -0.006093x^4 + 0.849362x^3$$
$$- 51.892087x^2 + 1627.3581x$$
$$+ 41{,}334.7289,$$

where $x$ is the number of years since 1940, can be used to estimate the circulation of daily newspapers, in thousands, from 1940 to 2008. Using this function, estimate the circulation of daily newspapers in 1945, in 1985, and in 2008.

**69.** *Interest Compounded Annually.* When $P$ dollars is invested at interest rate $i$, compounded annually, for $t$ years, the investment grows to $A$ dollars, where

$$A = P(1 + i)^t.$$

Trevor's parents deposit $8000 in a savings account when Trevor is 16 years old. The principal plus interest is to be used for a truck when Trevor is 18 years old. Find the interest rate $i$ if the $8000 grows to $9039.75 in 2 years.

**70.** *Interest Compounded Annually.* When $P$ dollars is invested at interest rate $i$, compounded annually, for $t$ years, the investment grows to $A$ dollars, where

$$A = P(1 + i)^t.$$

When Sara enters the 11th grade, her grandparents deposit $10,000 in a college savings account. Find the interest rate $i$ if the $10,000 grows to $11,193.64 in 2 years.

*For the scatterplots and graphs in Exercises 71–76, determine which, if any, of the following functions might be used as a model for the data.*

a) *Linear,* $f(x) = mx + b$
b) *Quadratic,* $f(x) = ax^2 + bx + c, a > 0$
c) *Quadratic,* $f(x) = ax^2 + bx + c, a < 0$
d) *Polynomial, not linear or quadratic*

**71.**

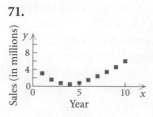

**72.**

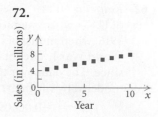

**73.**

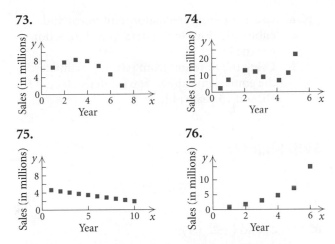

**74.**

**75.**

**76.**

**77.** *Foreign Adoptions.* The number of foreign adoptions in the United States has declined in recent years, as shown in the table below.

| Year, $x$ | Number of U.S. Foreign Adoptions from Top 15 Countries, $y$ |
|---|---|
| 2000, 0 | 18,120 |
| 2001, 1 | 19,087 |
| 2002, 2 | 20,100 |
| 2003, 3 | 21,320 |
| 2004, 4 | 22,911 |
| 2005, 5 | 22,710 |
| 2006, 6 | 20,705 |
| 2007, 7 | 19,741 |
| 2008, 8 | 17,229 |
| 2009, 9 | 12,782 |

*Sources*: Office of Immigration Statistics; Department of Homeland Security

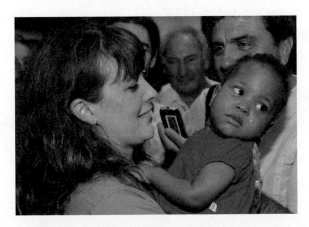

**a)** Use a graphing calculator to fit quadratic, cubic, and quartic functions to the data. Let $x$ represent

the number of years since 2000. Using $R^2$-values, determine which function is the best fit.

**b)** Using the function found in part (a), estimate the number of U.S. foreign adoptions in 2010.

**78.** *U.S. Farm Acreage.* As the number of farms has decreased in the United States, the average size of the remaining farms has grown larger, as shown in the table below.

| Year | Average Acreage per Farm |
|---|---|
| 1900 | 147 |
| 1910 | 139 |
| 1920 | 149 |
| 1930 | 157 |
| 1940 | 174 |
| 1950 | 213 |
| 1960 | 297 |
| 1970 | 374 |
| 1980 | 426 |
| 1990 | 460 |
| 1995 | 438 |
| 2000 | 436 |
| 2003 | 441 |
| 2005 | 445 |
| 2009 | 418 |

*Sources: Statistical History of the United States* (1970); National Agricultural Statistics Service; U.S. Department of Agriculture

**a)** Use a graphing calculator to fit quadratic, cubic, and quartic functions to the data. Let $x$ represent the number of years since 1900. Using $R^2$-values, determine which function is the best fit.

**b)** Using the function found in part (a), estimate the average acreage in 1955, in 1998, and in 2011.

**79.** *Classified Ad Revenue.*    The table below lists the newspaper revenue from classified ads for selected years from 1975 to 2009.

| Year, x | Newspaper Revenue from Classified Ads, y (in billions of dollars) |
|---|---|
| 1975,  0 | $2.159 |
| 1980,  5 | 4.222 |
| 1985, 10 | 8.375 |
| 1990, 15 | 11.506 |
| 1995, 20 | 13.742 |
| 2000, 25 | 19.608 |
| 2005, 30 | 17.312 |
| 2006, 31 | 16.986 |
| 2007, 32 | 14.186 |
| 2008, 33 | 9.975 |
| 2009, 34 | 6.179 |

*Source*: *Editor & Publisher International Yearbook*, 2010

a) Use a graphing calculator to fit cubic and quartic functions to the data. Let $x$ represent the number of years since 1975. Using $R^2$-values, determine which function is the best fit.

b) Using the function found in part (a), estimate the newspaper revenue from classified ads in 1988, in 2002, and in 2010.

**80.** *Dog Years.*    A dog's life span is typically much shorter than that of a human. Age equivalents for dogs and humans are listed in the table below.

| Age of Dog, x (in years) | Human Age, h(x) (in years) |
|---|---|
| 0.25 | 5 |
| 0.5 | 10 |
| 1 | 15 |
| 2 | 24 |
| 4 | 32 |
| 6 | 40 |
| 8 | 48 |
| 10 | 56 |
| 14 | 72 |
| 18 | 91 |
| 21 | 106 |

*Source*: Based on "How to Determine a Dog's Age in Human Years," by Melissa Maroff, eHow Contributor, www.ehow.com

a) Use a graphing calculator to fit linear and cubic functions to the data. Which function has the better fit?

b) Using the function from part (a), estimate the equivalent human age for dogs that are 5, 10, and 15 years old.

## Skill Maintenance

*Find the distance between the pair of points.*

**81.** $(3, -5)$ and $(0, -1)$

**82.** $(4, 2)$ and $(-2, -4)$

**83.** Find the center and the radius of the circle
$$(x - 3)^2 + (y + 5)^2 = 49.$$

**84.** The diameter of a circle connects the points $(-6, 5)$ and $(-2, 1)$ on the circle. Find the coordinates of the center of the circle and the length of the radius.

*Solve.*

**85.** $2y - 3 \geq 1 - y + 5$

**86.** $(x - 2)(x + 5) > x(x - 3)$

**87.** $|x + 6| \geq 7$

**88.** $\left|x + \frac{1}{4}\right| \leq \frac{2}{3}$

## Synthesis

*Determine the degree and the leading term of the polynomial function.*

**89.** $f(x) = (x^5 - 1)^2(x^2 + 2)^3$

**90.** $f(x) = (10 - 3x^5)^2(5 - x^4)^3(x + 4)$

## Graphing Polynomial Functions

### 4.2

- Graph polynomial functions.
- Use the intermediate value theorem to determine whether a function has a real zero between two given real numbers.

### ■ Graphing Polynomial Functions

In addition to using the leading-term test and finding the zeros of the function, it is helpful to consider the following facts when graphing a polynomial function.

> If $P(x)$ is a polynomial function of degree $n$, then the graph of the function has:
>
> - at most $n$ real zeros, and thus at most $n$ $x$-intercepts;
> - at most $n - 1$ turning points.
>
> (Turning points on a graph, also called relative maxima and minima, occur when the function changes from decreasing to increasing or from increasing to decreasing.)

**EXAMPLE 1** Graph the polynomial function $h(x) = -2x^4 + 3x^3$.

*Solution*

1. First, we use the leading-term test to determine the end behavior of the graph. The leading term is $-2x^4$. The degree, 4, is even, and the coefficient, $-2$, is negative. Thus the end behavior of the graph as $x \rightarrow \infty$ and as $x \rightarrow -\infty$ can be sketched as follows.

2. The zeros of the function are the first coordinates of the $x$-intercepts of the graph. To find the zeros, we solve $h(x) = 0$ by factoring and using the principle of zero products.

$$-2x^4 + 3x^3 = 0$$
$$-x^3(2x - 3) = 0 \qquad \text{Factoring}$$
$$-x^3 = 0 \quad or \quad 2x - 3 = 0 \qquad \text{Using the principle of zero products}$$
$$x = 0 \quad or \qquad x = \tfrac{3}{2}.$$

The zeros of the function are 0 and $\frac{3}{2}$. Note that the multiplicity of 0 is 3 and the multiplicity of $\frac{3}{2}$ is 1. The $x$-intercepts are $(0, 0)$ and $\left(\frac{3}{2}, 0\right)$.

**3.** The zeros divide the *x*-axis into three intervals:

$$(-\infty, 0), \quad \left(0, \tfrac{3}{2}\right), \quad \text{and} \quad \left(\tfrac{3}{2}, \infty\right).$$

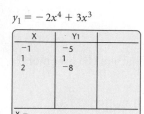

The sign of $h(x)$ is the same for all values of $x$ in an interval. That is, $h(x)$ is positive for all *x*-values in an interval or $h(x)$ is negative for all *x*-values in an interval. To determine which, we choose a test value for $x$ from each interval and find $h(x)$.

$y_1 = -2x^4 + 3x^3$

| X | Y1 |
|---|---|
| −1 | −5 |
| 1 | 1 |
| 2 | −8 |

X =

| Interval | $(-\infty, 0)$ | $\left(0, \tfrac{3}{2}\right)$ | $\left(\tfrac{3}{2}, \infty\right)$ |
|---|---|---|---|
| **Test Value, *x*** | −1 | 1 | 2 |
| **Function Value, $h(x)$** | −5 | 1 | −8 |
| **Sign of $h(x)$** | − | + | − |
| **Location of Points on Graph** | Below *x*-axis | Above *x*-axis | Below *x*-axis |

This test-point procedure also gives us three points to plot. In this case, we have $(-1, -5), (1, 1)$, and $(2, -8)$.

**4.** To determine the *y*-intercept, we find $h(0)$:

$$h(x) = -2x^4 + 3x^3$$
$$h(0) = -2 \cdot 0^4 + 3 \cdot 0^3 = 0.$$

The *y*-intercept is $(0, 0)$.

**5.** A few additional points are helpful when completing the graph.

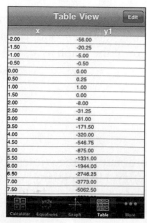

| Table View | Edit |
|---|---|
| x | y1 |
| -2.00 | -56.00 |
| -1.50 | -20.25 |
| -1.00 | -5.00 |
| -0.50 | -0.50 |
| 0.00 | 0.00 |
| 0.50 | 0.25 |
| 1.00 | 1.00 |
| 1.50 | 0.00 |
| 2.00 | -8.00 |
| 2.50 | -31.25 |
| 3.00 | -81.00 |
| 3.50 | -171.50 |
| 4.00 | -320.00 |
| 4.50 | -546.75 |
| 5.00 | -875.00 |
| 5.50 | -1331.00 |
| 6.00 | -1944.00 |
| 6.50 | -2746.25 |
| 7.00 | -3773.00 |
| 7.50 | -5062.50 |

A table of ordered pairs for $h(x) = -2x^4 + 3x^3$ can be created with a graphing calculator app.

| *x* | $h(x)$ |
|---|---|
| −1.5 | −20.25 |
| −0.5 | −0.5 |
| 0.5 | 0.25 |
| 2.5 | −31.25 |

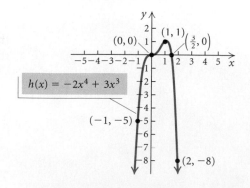

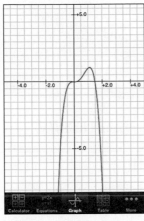

We can check the graph of $h(x) =$ $-2x^4 + 3x^3$ in Example 1 with a graphing calculator app.

**6.** The degree of $h$ is 4. The graph of $h$ can have at most 4 $x$-intercepts and at most 3 turning points. In fact, it has 2 $x$-intercepts and 1 turning point. The zeros, 0 and $\frac{3}{2}$, each have odd multiplicities: 3 for 0 and 1 for $\frac{3}{2}$. Since the multiplicities are odd, the graph crosses the $x$-axis at 0 and $\frac{3}{2}$. The end behavior of the graph is what we described in step (1). As $x \to \infty$ and also as $x \to -\infty$, $h(x) \to -\infty$. The graph appears to be correct.

**Now Try Exercise 23.**

The following is a procedure for graphing polynomial functions.

To graph a polynomial function:

1. Use the leading-term test to determine the end behavior.
2. Find the zeros of the function by solving $f(x) = 0$. Any real zeros are the first coordinates of the $x$-intercepts.
3. Use the $x$-intercepts (zeros) to divide the $x$-axis into intervals, and choose a test point in each interval to determine the sign of all function values in that interval.
4. Find $f(0)$. This gives the $y$-intercept of the function.
5. If necessary, find additional function values to determine the general shape of the graph and then draw the graph.
6. As a partial check, use the facts that the graph has at most $n$ $x$-intercepts and at most $n - 1$ turning points. Multiplicity of zeros can also be considered in order to check where the graph crosses or is tangent to the $x$-axis. We can also check the graph with a graphing calculator.

**EXAMPLE 2** Graph the polynomial function

$$f(x) = 2x^3 + x^2 - 8x - 4.$$

*Solution*

1. The leading term is $2x^3$. The degree, 3, is odd, and the coefficient, 2, is positive. Thus the end behavior of the graph will appear as follows.

2. To find the zeros, we solve $f(x) = 0$. Here we can use factoring by grouping.

$$2x^3 + x^2 - 8x - 4 = 0$$
$$x^2(2x + 1) - 4(2x + 1) = 0 \qquad \text{Factoring by grouping}$$
$$(2x + 1)(x^2 - 4) = 0$$
$$(2x + 1)(x + 2)(x - 2) = 0 \qquad \text{Factoring a difference of squares}$$

The zeros are $-\frac{1}{2}, -2$, and $2$. Each is of multiplicity 1. The $x$-intercepts are $(-2, 0)$, $\left(-\frac{1}{2}, 0\right)$, and $(2, 0)$.

**3.** The zeros divide the $x$-axis into four intervals:

$$(-\infty, -2), \quad \left(-2, -\tfrac{1}{2}\right), \quad \left(-\tfrac{1}{2}, 2\right), \quad \text{and} \quad (2, \infty).$$

We choose a test value for $x$ from each interval and find $f(x)$.

| Interval | $(-\infty, -2)$ | $\left(-2, -\tfrac{1}{2}\right)$ | $\left(-\tfrac{1}{2}, 2\right)$ | $(2, \infty)$ |
|---|---|---|---|---|
| **Test Value, $x$** | $-3$ | $-1$ | $1$ | $3$ |
| **Function Value, $f(x)$** | $-25$ | $3$ | $-9$ | $35$ |
| **Sign of $f(x)$** | $-$ | $+$ | $-$ | $+$ |
| **Location of Points on Graph** | Below $x$-axis | Above $x$-axis | Below $x$-axis | Above $x$-axis |

$y_1 = 2x^3 + x^2 - 8x - 4$

| X | Y1 |
|---|---|
| -3 | -25 |
| -1 | 3 |
| 1 | -9 |
| 3 | 35 |

X =

The test values and corresponding function values also give us four points on the graph: $(-3, -25), (-1, 3), (1, -9),$ and $(3, 35)$.

**4.** To determine the $y$-intercept, we find $f(0)$:

$$f(x) = 2x^3 + x^2 - 8x - 4$$
$$f(0) = 2 \cdot 0^3 + 0^2 - 8 \cdot 0 - 4 = -4.$$

The $y$-intercept is $(0, -4)$.

**5.** We find a few additional points and complete the graph.

| $x$ | $f(x)$ |
|---|---|
| $-2.5$ | $-9$ |
| $-1.5$ | $3.5$ |
| $0.5$ | $-7.5$ |
| $1.5$ | $-7$ |

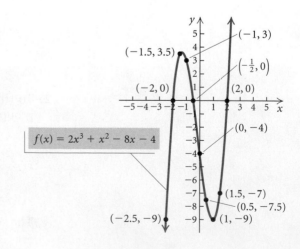

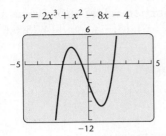

$y = 2x^3 + x^2 - 8x - 4$

6. The degree of $f$ is 3. The graph of $f$ can have at most 3 $x$-intercepts and at most 2 turning points. It has 3 $x$-intercepts and 2 turning points. Each zero has a multiplicity of 1. Thus the graph crosses the $x$-axis at $-2$, $-\frac{1}{2}$, and 2. The graph has the end behavior described in step (1). As $x \to -\infty$, $h(x) \to -\infty$, and as $x \to \infty$, $h(x) \to \infty$. The graph appears to be correct.

**Now Try Exercise 33.**

Some polynomials are difficult to factor. In the next example, the polynomial is given in factored form. In Sections 4.3 and 4.4, we will learn methods that facilitate determining factors of such polynomials.

**EXAMPLE 3**   Graph the polynomial function

$$g(x) = x^4 - 7x^3 + 12x^2 + 4x - 16 = (x + 1)(x - 2)^2(x - 4).$$

**Solution**

1. The leading term is $x^4$. The degree, 4, is even, and the coefficient, 1, is positive. The sketch below shows the end behavior.

2. To find the zeros, we solve $g(x) = 0$:

$$(x + 1)(x - 2)^2(x - 4) = 0.$$

The zeros are $-1$, 2, and 4; 2 is of multiplicity 2; the others are of multiplicity 1. The $x$-intercepts are $(-1, 0)$, $(2, 0)$, and $(4, 0)$.

$y_1 = x^4 - 7x^3 + 12x^2 + 4x - 16$

| X | Y1 | |
|---|---|---|
| −1.25 | 13.863 | |
| 1 | −6 | |
| 3 | −4 | |
| 4.25 | 6.6445 | |
| | | |
| X = | | |

3. The zeros divide the $x$-axis into four intervals:

$$(-\infty, -1), \quad (-1, 2), \quad (2, 4), \quad \text{and} \quad (4, \infty).$$

We choose a test value for $x$ from each interval and find $g(x)$.

| Interval | $(-\infty, -1)$ | $(-1, 2)$ | $(2, 4)$ | $(4, \infty)$ |
|---|---|---|---|---|
| Test Value, $x$ | −1.25 | 1 | 3 | 4.25 |
| Function Value, $g(x)$ | $\approx 13.9$ | −6 | −4 | $\approx 6.6$ |
| Sign of $g(x)$ | + | − | − | + |
| Location of Points on Graph | Above $x$-axis | Below $x$-axis | Below $x$-axis | Above $x$-axis |

The test values and corresponding function values also give us four points on the graph: $(-1.25, 13.9)$, $(1, -6)$, $(3, -4)$, and $(4.25, 6.6)$.

4. To determine the $y$-intercept, we find $g(0)$:

$$g(x) = x^4 - 7x^3 + 12x^2 + 4x - 16$$
$$g(0) = 0^4 - 7 \cdot 0^3 + 12 \cdot 0^2 + 4 \cdot 0 - 16 = -16.$$

The $y$-intercept is $(0, -16)$.

5. We find a few additional points and draw the graph.

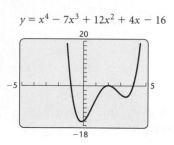

$y = x^4 - 7x^3 + 12x^2 + 4x - 16$

| $x$ | $g(x)$ |
|------|--------|
| $-0.5$ | $-14.1$ |
| $0.5$ | $-11.8$ |
| $1.5$ | $-1.6$ |
| $2.5$ | $-1.3$ |
| $3.5$ | $-5.1$ |

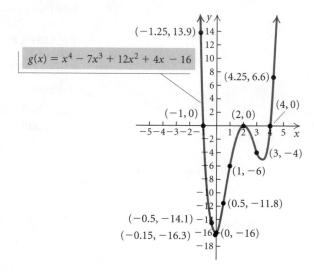

$g(x) = x^4 - 7x^3 + 12x^2 + 4x - 16$

6. The degree of $g$ is 4. The graph of $g$ can have at most 4 $x$-intercepts and at most 3 turning points. It has 3 $x$-intercepts and 3 turning points. One of the zeros, 2, has a multiplicity of 2, so the graph is tangent to the $x$-axis at 2. The other zeros, $-1$ and 4, each have a multiplicity of 1 so the graph crosses the $x$-axis at $-1$ and 4. The graph has the end behavior described in step (1). As $x \to \infty$ and as $x \to -\infty$, $g(x) \to \infty$. The graph appears to be correct.

**Now Try Exercise 19.**

## The Intermediate Value Theorem

Polynomial functions are continuous, hence their graphs are unbroken. The domain of a polynomial function, unless restricted by the statement of the function, is $(-\infty, \infty)$. Suppose two polynomial function values $P(a)$ and $P(b)$ have opposite signs. Since $P$ is continuous, its graph must be a curve from $(a, P(a))$ to $(b, P(b))$ without a break. Then it follows that the curve must cross the $x$-axis at some point $c$ between $a$ and $b$; that is, the function has a zero at $c$ between $a$ and $b$.

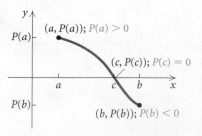

> ### THE INTERMEDIATE VALUE THEOREM
>
> For any polynomial function $P(x)$ with real coefficients, suppose that for $a \neq b$, $P(a)$ and $P(b)$ are of opposite signs. Then the function has a real zero between $a$ and $b$.
>
> The intermediate value theorem *cannot* be used to determine whether there is a real zero between $a$ and $b$ when $P(a)$ and $P(b)$ have the *same* sign.

**EXAMPLE 4**  Using the intermediate value theorem, determine, if possible, whether the function has a real zero between $a$ and $b$.

**a)** $f(x) = x^3 + x^2 - 6x$; $a = -4, b = -2$
**b)** $f(x) = x^3 + x^2 - 6x$; $a = -1, b = 3$
**c)** $g(x) = \frac{1}{3}x^4 - x^3$; $a = -\frac{1}{2}, b = \frac{1}{2}$
**d)** $g(x) = \frac{1}{3}x^4 - x^3$; $a = 1, b = 2$

**Solution**  We find $f(a)$ and $f(b)$ or $g(a)$ and $g(b)$ and determine whether they differ in sign. The graphs of $f(x)$ and $g(x)$ at left provide a visual check of the conclusions.

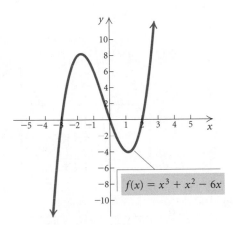

**a)** $f(-4) = (-4)^3 + (-4)^2 - 6(-4) = -24$,
  $f(-2) = (-2)^3 + (-2)^2 - 6(-2) = 8$

Note that $f(-4)$ is negative and $f(-2)$ is positive. By the intermediate value theorem, since $f(-4)$ and $f(-2)$ have opposite signs, then $f(x)$ has a zero between $-4$ and $-2$. The graph at left confirms this.

**b)** $f(-1) = (-1)^3 + (-1)^2 - 6(-1) = 6$,
  $f(3) = 3^3 + 3^2 - 6(3) = 18$

Both $f(-1)$ and $f(3)$ are positive. Thus the intermediate value theorem *does not allow* us to determine whether there is a real zero between $-1$ and 3. Note that the graph of $f(x)$ shows that there are two zeros between $-1$ and 3.

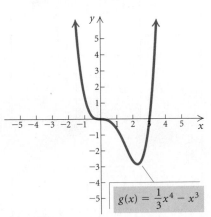

**c)** $g\left(-\frac{1}{2}\right) = \frac{1}{3}\left(-\frac{1}{2}\right)^4 - \left(-\frac{1}{2}\right)^3 = \frac{7}{48}$,
  $g\left(\frac{1}{2}\right) = \frac{1}{3}\left(\frac{1}{2}\right)^4 - \left(\frac{1}{2}\right)^3 = -\frac{5}{48}$

Since $g\left(-\frac{1}{2}\right)$ and $g\left(\frac{1}{2}\right)$ have opposite signs, $g(x)$ has a zero between $-\frac{1}{2}$ and $\frac{1}{2}$. The graph at left confirms this.

**d)** $g(1) = \frac{1}{3}(1)^4 - 1^3 = -\frac{2}{3}$,
  $g(2) = \frac{1}{3}(2)^4 - 2^3 = -\frac{8}{3}$

Both $g(1)$ and $g(2)$ are negative. Thus the intermediate value theorem *does not allow* us to determine whether there is a real zero between 1 and 2. Note that the graph of $g(x)$ at left shows that there are no zeros between 1 and 2.

**Now Try Exercises 39 and 43.**

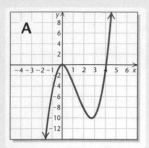

A

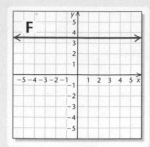

F

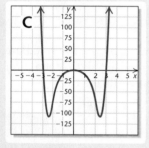

B

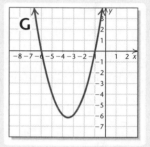

G

# Visualizing the Graph

Match the function with its graph.

**1.** $f(x) = -x^4 - x + 5$

**2.** $f(x) = -3x^2 + 6x - 3$

**3.** $f(x) = x^4 - 4x^3 + 3x^2 + 4x - 4$

**4.** $f(x) = -\dfrac{2}{5}x + 4$

**5.** $f(x) = x^3 - 4x^2$

**6.** $f(x) = x^6 - 9x^4$

**7.** $f(x) = x^5 - 3x^3 + 2$

**8.** $f(x) = -x^3 - x - 1$

**9.** $f(x) = x^2 + 7x + 6$

**10.** $f(x) = \dfrac{7}{2}$

**Answers on page A-23**

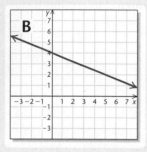

C

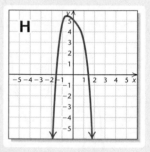

H

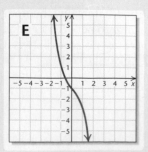

D

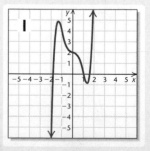

I

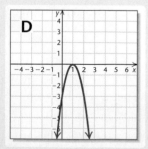

E

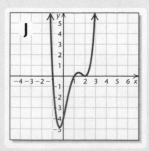

J

# 4.2 Exercise Set

*For each function in Exercises 1–6, state:*

a) *the maximum number of real zeros that the function can have;*

b) *the maximum number of x-intercepts that the graph of the function can have; and*

c) *the maximum number of turning points that the graph of the function can have.*

1. $f(x) = x^5 - x^2 + 6$

2. $f(x) = -x^2 + x^4 - x^6 + 3$

3. $f(x) = x^{10} - 2x^5 + 4x - 2$

4. $f(x) = \frac{1}{4}x^3 + 2x^2$

5. $f(x) = -x - x^3$

6. $f(x) = -3x^4 + 2x^3 - x - 4$

*In Exercises 7–12, use the leading-term test and your knowledge of y-intercepts to match the function with one of the graphs (a)–(f), which follow.*

a)
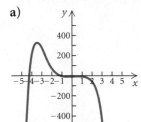

b)
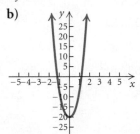

c)
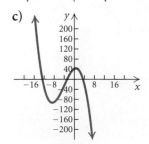

d)
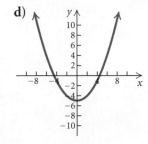

e)
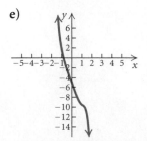

f)
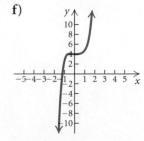

7. $f(x) = \frac{1}{4}x^2 - 5$

8. $f(x) = -0.5x^6 - x^5 + 4x^4 - 5x^3 - 7x^2 + x - 3$

9. $f(x) = x^5 - x^4 + x^2 + 4$

10. $f(x) = -\frac{1}{3}x^3 - 4x^2 + 6x + 42$

11. $f(x) = x^4 - 2x^3 + 12x^2 + x - 20$

12. $f(x) = -0.3x^7 + 0.11x^6 - 0.25x^5 + x^4 + x^3 - 6x - 5$

*Graph the polynomial function. Follow the steps outlined in the procedure on p. 317.*

13. $f(x) = -x^3 - 2x^2$

14. $g(x) = x^4 - 4x^3 + 3x^2$

15. $h(x) = x^2 + 2x - 3$

16. $f(x) = x^2 - 5x + 4$

17. $h(x) = x^5 - 4x^3$

18. $f(x) = x^3 - x$

19. $h(x) = x(x - 4)(x + 1)(x - 2)$

20. $f(x) = x(x - 1)(x + 3)(x + 5)$

21. $g(x) = -\frac{1}{4}x^3 - \frac{3}{4}x^2$

22. $f(x) = \frac{1}{2}x^3 + \frac{5}{2}x^2$

23. $g(x) = -x^4 - 2x^3$

24. $h(x) = x^3 - 3x^2$

25. $f(x) = -\frac{1}{2}(x - 2)(x + 1)^2(x - 1)$

26. $g(x) = (x - 2)^3(x + 3)$

27. $g(x) = -x(x - 1)^2(x + 4)^2$

28. $h(x) = -x(x - 3)(x - 3)(x + 2)$

29. $f(x) = (x - 2)^2(x + 1)^4$

30. $g(x) = x^4 - 9x^2$

31. $g(x) = -(x - 1)^4$

32. $h(x) = (x + 2)^3$

33. $h(x) = x^3 + 3x^2 - x - 3$

34. $g(x) = -x^3 + 2x^2 + 4x - 8$

**35.** $f(x) = 6x^3 - 8x^2 - 54x + 72$

**36.** $h(x) = x^5 - 5x^3 + 4x$

*Graph the piecewise function.*

**37.** $g(x) = \begin{cases} -x + 3, & \text{for } x \le -2, \\ 4, & \text{for } -2 < x < 1, \\ \frac{1}{2}x^3, & \text{for } x \ge 1 \end{cases}$

**38.** $h(x) = \begin{cases} -x^2, & \text{for } x < -2, \\ x + 1, & \text{for } -2 \le x < 0, \\ x^3 - 1, & \text{for } x \ge 0 \end{cases}$

*Using the intermediate value theorem, determine, if possible, whether the function f has a real zero between a and b.*

**39.** $f(x) = x^3 + 3x^2 - 9x - 13$; $a = -5, b = -4$

**40.** $f(x) = x^3 + 3x^2 - 9x - 13$; $a = 1, b = 2$

**41.** $f(x) = 3x^2 - 2x - 11$; $a = -3, b = -2$

**42.** $f(x) = 3x^2 - 2x - 11$; $a = 2, b = 3$

**43.** $f(x) = x^4 - 2x^2 - 6$; $a = 2, b = 3$

**44.** $f(x) = 2x^5 - 7x + 1$; $a = 1, b = 2$

**45.** $f(x) = x^3 - 5x^2 + 4$; $a = 4, b = 5$

**46.** $f(x) = x^4 - 3x^2 + x - 1$; $a = -3, b = -2$

## Skill Maintenance

c)
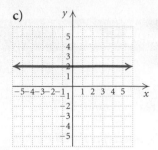

d)
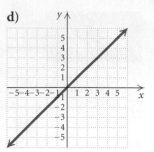

e)
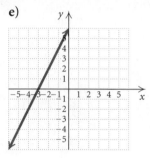

f)
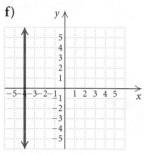

**47.** $y = x$

**48.** $x = -4$

**49.** $y - 2x = 6$

**50.** $3x + 2y = -6$

**51.** $y = 1 - x$

**52.** $y = 2$

*Solve.*

**53.** $2x - \frac{1}{2} = 4 - 3x$

**54.** $x^3 - x^2 - 12x = 0$

**55.** $6x^2 - 23x - 55 = 0$

**56.** $\frac{3}{4}x + 10 = \frac{1}{5} + 2x$

*Match the equation with one of the graphs (a)–(f), which follow.*

a)
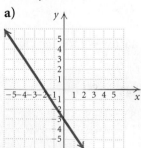

b)
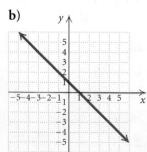

# 4.3 Polynomial Division; The Remainder Theorem and the Factor Theorem

- Perform long division with polynomials and determine whether one polynomial is a factor of another.
- Use synthetic division to divide a polynomial by $x - c$.
- Use the remainder theorem to find a function value $f(c)$.
- Use the factor theorem to determine whether $x - c$ is a factor of $f(x)$.

In general, finding exact zeros of many polynomial functions is neither easy nor straightforward. In this section and the one that follows, we develop concepts that help us find exact zeros of certain polynomial functions with degree 3 or greater. Consider the polynomial

$$h(x) = x^3 + 2x^2 - 5x - 6 = (x + 3)(x + 1)(x - 2).$$

The factors are

$$x + 3, \quad x + 1, \quad \text{and} \quad x - 2,$$

and the zeros are

$$-3, \quad -1, \quad \text{and} \quad 2.$$

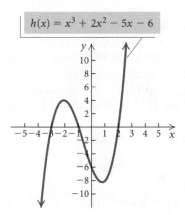

$h(x) = x^3 + 2x^2 - 5x - 6$

When a polynomial is expressed in factored form, each factor determines a zero of the function. Thus if we know the factors of a polynomial, we can easily find the zeros. The "reverse" is also true: If we know the zeros of a polynomial function, we can find the factors of the polynomial.

## Division and Factors

When we divide one polynomial by another, we obtain a quotient and a remainder. If the remainder is 0, then the divisor is a factor of the dividend.

**EXAMPLE 1** Divide to determine whether $x + 1$ and $x - 3$ are factors of

$$x^3 + 2x^2 - 5x - 6.$$

**Solution** We divide $x^3 + 2x^2 - 5x - 6$ by $x + 1$.

$$
\begin{array}{r}
\overbrace{x^2 + x - 6}^{\text{Quotient}} \\
x + 1 \overline{)\; x^3 + 2x^2 - 5x - 6} \quad \longleftarrow \text{Dividend} \\
\underline{x^3 + x^2} \phantom{- 5x - 6} \\
x^2 - 5x \phantom{- 6} \\
\underline{x^2 + x} \phantom{- 6} \\
-6x - 6 \\
\underline{-6x - 6} \\
0 \quad \longleftarrow \text{Remainder}
\end{array}
$$

Divisor $\longrightarrow$ $x + 1$

Since the remainder is 0, we know that $x + 1$ is a factor of $x^3 + 2x^2 - 5x - 6$. In fact, we know that

$$x^3 + 2x^2 - 5x - 6 = (x + 1)(x^2 + x - 6).$$

We divide $x^3 + 2x^2 - 5x - 6$ by $x - 3$.

$$
\begin{array}{r}
x^2 + 5x + 10 \\
x - 3 \overline{\smash{)}\ x^3 + 2x^2 - 5x - 6} \\
\underline{x^3 - 3x^2} \\
5x^2 - 5x \\
\underline{5x^2 - 15x} \\
10x - 6 \\
\underline{10x - 30} \\
24 \quad \longleftarrow \text{Remainder}
\end{array}
$$

Since the remainder is not 0, we know that $x - 3$ is *not* a factor of $x^3 + 2x^2 - 5x - 6$.

**Now Try Exercise 3.**

When we divide a polynomial $P(x)$ by a divisor $d(x)$, a polynomial $Q(x)$ is the quotient and a polynomial $R(x)$ is the remainder. The quotient $Q(x)$ must have degree less than that of the dividend $P(x)$. The remainder $R(x)$ must either be 0 or have degree less than that of the divisor $d(x)$.

As in arithmetic, to check a division, we multiply the quotient by the divisor and add the remainder, to see if we get the dividend. Thus these polynomials are related as follows:

$$P(x) = d(x) \cdot Q(x) + R(x)$$

$$\text{Dividend} \quad \text{Divisor} \quad \text{Quotient} \quad \text{Remainder}$$

For instance, if $P(x) = x^3 + 2x^2 - 5x - 6$ and $d(x) = x - 3$, as in Example 1, then $Q(x) = x^2 + 5x + 10$ and $R(x) = 24$, and

$$
\begin{aligned}
P(x) &= d(x) \cdot Q(x) + R(x) \\
x^3 + 2x^2 - 5x - 6 &= (x - 3) \cdot (x^2 + 5x + 10) + 24 \\
&= x^3 + 5x^2 + 10x - 3x^2 - 15x - 30 + 24 \\
&= x^3 + 2x^2 - 5x - 6.
\end{aligned}
$$

## ■ The Remainder Theorem and Synthetic Division

Consider the function

$$h(x) = x^3 + 2x^2 - 5x - 6.$$

When we divided $h(x)$ by $x + 1$ and $x - 3$ in Example 1, the remainders were 0 and 24, respectively. Let's now find the function values $h(-1)$ and $h(3)$:

$$
\begin{aligned}
h(-1) &= (-1)^3 + 2(-1)^2 - 5(-1) - 6 = 0; \\
h(3) &= (3)^3 + 2(3)^2 - 5(3) - 6 = 24.
\end{aligned}
$$

Note that the function values are the same as the remainders. This suggests the following theorem.

> ### THE REMAINDER THEOREM
> If a number $c$ is substituted for $x$ in the polynomial $f(x)$, then the result $f(c)$ is the remainder that would be obtained by dividing $f(x)$ by $x - c$. That is, if $f(x) = (x - c) \cdot Q(x) + R$, then $f(c) = R$.

**Proof (Optional).** The equation $f(x) = d(x) \cdot Q(x) + R(x)$, where $d(x) = x - c$, is the basis of this proof. If we divide $f(x)$ by $x - c$, we obtain a quotient $Q(x)$ and a remainder $R(x)$ related as follows:

$$f(x) = (x - c) \cdot Q(x) + R(x).$$

The remainder $R(x)$ must either be 0 or have degree less than $x - c$. Thus, $R(x)$ must be a constant. Let's call this constant $R$. The equation above is true for any replacement of $x$, so we replace $x$ with $c$. We get

$$\begin{aligned} f(c) &= (c - c) \cdot Q(c) + R \\ &= 0 \cdot Q(c) + R \\ &= R. \end{aligned}$$

Thus the function value $f(c)$ is the remainder obtained when we divide $f(x)$ by $x - c$. ∎

The remainder theorem motivates us to find a rapid way of dividing by $x - c$ in order to find function values. To streamline division, we can arrange the work so that duplicate and unnecessary writing is avoided. Consider the following:

$$(4x^3 - 3x^2 + x + 7) \div (x - 2).$$

**A.**
$$
\begin{array}{r}
4x^2 + 5x + 11 \\
x - 2\overline{)4x^3 - 3x^2 + \phantom{1}x + 7} \\
\underline{4x^3 - 8x^2} \phantom{++++++} \\
5x^2 + \phantom{1}x \phantom{++} \\
\underline{5x^2 - 10x} \phantom{++} \\
11x + \phantom{1}7 \\
\underline{11x - 22} \\
29
\end{array}
$$

**B.**
$$
\begin{array}{r}
4 \quad 5 \quad 11 \\
1 - 2\overline{)4 - 3 + \phantom{1}1 + 7} \\
\underline{4 - 8} \phantom{+++++} \\
5 + \phantom{1}1 \phantom{++} \\
\underline{5 - 10} \phantom{++} \\
11 + \phantom{1}7 \\
\underline{11 - 22} \\
29
\end{array}
$$

The division in (B) is the same as that in (A), but we wrote only the coefficients. The red numerals are duplicated, so we look for an arrangement in which they are not duplicated. In place of the divisor in the form $x - c$, we can simply use $c$ and then add rather than subtract. When the procedure is "collapsed," we have the algorithm known as **synthetic division**.

**C.** *Synthetic Division*

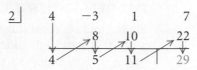

The divisor is $x - 2$; thus we use 2 in synthetic division.

We "bring down" the 4. Then we multiply it by the 2 to get 8 and add to get 5. We then multiply 5 by 2 to get 10, add, and so on. The last number, 29, is the remainder. The others, 4, 5, and 11, are the coefficients of the quotient, $4x^2 + 5x + 11$. (Note that the degree of the quotient is 1 less than the degree of the dividend when the degree of the divisor is 1.)

When using synthetic division, we write a 0 for a missing term in the dividend.

**EXAMPLE 2** Use synthetic division to find the quotient and the remainder:

$$(2x^3 + 7x^2 - 5) \div (x + 3).$$

***Solution*** First, we note that $x + 3 = x - (-3)$.

$$
\begin{array}{r|rrrr}
-3 & 2 & 7 & 0 & -5 \\
   &   & -6 & -3 & 9 \\
\hline
   & 2 & 1 & -3 & 4
\end{array}
$$

*Note:* We must write a 0 for the missing $x$-term.

The quotient is $2x^2 + x - 3$. The remainder is 4.

▶ Now Try Exercise 11.

We can now use synthetic division to find polynomial function values.

**EXAMPLE 3** Given that $f(x) = 2x^5 - 3x^4 + x^3 - 2x^2 + x - 8$, find $f(10)$.

***Solution*** By the remainder theorem, $f(10)$ is the remainder when $f(x)$ is divided by $x - 10$. We use synthetic division to find that remainder.

$$
\begin{array}{r|rrrrrr}
10 & 2 & -3 & 1 & -2 & 1 & -8 \\
   &   & 20 & 170 & 1710 & 17,080 & 170,810 \\
\hline
   & 2 & 17 & 171 & 1708 & 17,081 & 170,802
\end{array}
$$

Thus, $f(10) = 170,802$.

▶ Now Try Exercise 25.

Compare the computations in Example 3 with those in a direct substitution:

$$
\begin{aligned}
f(10) &= 2(10)^5 - 3(10)^4 + (10)^3 - 2(10)^2 + 10 - 8 \\
&= 2 \cdot 100{,}000 - 3 \cdot 10{,}000 + 1000 - 2 \cdot 100 + 10 - 8 \\
&= 200{,}000 - 30{,}000 + 1000 - 200 + 10 - 8 \\
&= 170{,}802.
\end{aligned}
$$

The computations in synthetic division are less complicated than those involved in substituting. The easiest way to find $f(10)$ is to use one of the methods for evaluating a function on a graphing calculator. In the figure at left, we show the result when we enter $y_1 = 2x^5 - 3x^4 + x^3 - 2x^2 + x - 8$ and then use function notation on the home screen.

| Y₁(10) | |
|---|---|
| | 170802 |

**EXAMPLE 4** Determine whether 5 is a zero of $g(x)$, where

$$g(x) = x^4 - 26x^2 + 25.$$

**Solution** We use synthetic division and the remainder theorem to find $g(5)$.

$$
\begin{array}{r|rrrrr}
5 & 1 & 0 & -26 & 0 & 25 \\
  &   & 5 & 25 & -5 & -25 \\
\hline
  & 1 & 5 & -1 & -5 & \;\;0
\end{array}
$$

Writing 0's for missing terms:
$x^4 + 0x^3 - 26x^2 + 0x + 25$

Since $g(5) = 0$, the number 5 is a zero of $g(x)$.

> **Now Try Exercise 31.**

**EXAMPLE 5** Determine whether $i$ is a zero of $f(x)$, where

$$f(x) = x^3 - 3x^2 + x - 3.$$

**Solution** We use synthetic division and the remainder theorem to find $f(i)$.

$$
\begin{array}{r|rrrr}
i & 1 & -3 & 1 & -3 \\
  &   & i & -3i-1 & 3 \\
\hline
  & 1 & -3+i & -3i & \;\;0
\end{array}
$$

$i(-3 + i) = -3i + i^2 = -3i - 1$
$i(-3i) = -3i^2 = 3$

Since $f(i) = 0$, the number $i$ is a zero of $f(x)$.

> **Now Try Exercise 35.**

## ▪ Finding Factors of Polynomials

We now consider a useful result that follows from the remainder theorem.

> **THE FACTOR THEOREM**
>
> For a polynomial $f(x)$, if $f(c) = 0$, then $x - c$ is a factor of $f(x)$.

**Proof (Optional).** If we divide $f(x)$ by $x - c$, we obtain a quotient and a remainder, related as follows:

$$f(x) = (x - c) \cdot Q(x) + f(c).$$

Then if $f(c) = 0$, we have

$$f(x) = (x - c) \cdot Q(x),$$

so $x - c$ is a factor of $f(x)$. ▪

The factor theorem is very useful in factoring polynomials and hence in solving polynomial equations and finding zeros of polynomial functions. If we know a zero of a polynomial function, we know a factor.

**EXAMPLE 6** Let $f(x) = x^3 - 3x^2 - 6x + 8$. Factor $f(x)$ and solve the equation $f(x) = 0$.

**Solution** We look for linear factors of the form $x - c$. Let's try $x + 1$, or $x - (-1)$. (In the next section, we will learn a method for choosing the numbers to try for *c*.) We use synthetic division to determine whether $f(-1) = 0$.

$$\begin{array}{r|rrrr} -1 & 1 & -3 & -6 & 8 \\ & & -1 & 4 & 2 \\ \hline & 1 & -4 & -2 & | \quad 10 \end{array}$$

Since $f(-1) \neq 0$, we know that $x + 1$ *is not a factor* of $f(x)$. We now try $x - 1$.

$$\begin{array}{r|rrrr} 1 & 1 & -3 & -6 & 8 \\ & & 1 & -2 & -8 \\ \hline & 1 & -2 & -8 & | \quad 0 \end{array}$$

Since $f(1) = 0$, we know that $x - 1$ *is one factor* of $f(x)$ and the quotient, $x^2 - 2x - 8$, is another. Thus,

$$f(x) = (x - 1)(x^2 - 2x - 8).$$

The trinomial $x^2 - 2x - 8$ is easily factored in this case, so we have

$$f(x) = (x - 1)(x - 4)(x + 2).$$

We now solve the equation $f(x) = 0$. To do so, we use the principle of zero products:

$$(x - 1)(x - 4)(x + 2) = 0$$
$$x - 1 = 0 \quad or \quad x - 4 = 0 \quad or \quad x + 2 = 0$$
$$x = 1 \quad or \qquad x = 4 \quad or \qquad x = -2.$$

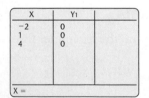

The solutions of the equation $x^3 - 3x^2 - 6x + 8 = 0$ are $-2, 1$, and $4$. They are also the zeros of the function $f(x) = x^3 - 3x^2 - 6x + 8$. We can use a table set in ASK mode to check the solutions. (See the table at left.)

> **Now Try Exercise 41.**

## CONNECTING THE CONCEPTS

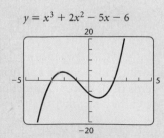

$y = x^3 + 2x^2 - 5x - 6$

Consider the function

$$f(x) = (x - 2)(x + 3)(x + 1), \quad or \quad f(x) = x^3 + 2x^2 - 5x - 6,$$

and its graph.

We can make the following statements:

- $-3$ is a zero of *f*.
- $f(-3) = 0$.
- $-3$ is a solution of $f(x) = 0$.
- $(-3, 0)$ is an *x*-intercept of the graph of *f*.
- $0$ is the remainder when $f(x)$ is divided by $x + 3$, or $x - (-3)$.
- $x - (-3)$, or $x + 3$, is a factor of *f*.

Similar statements are also true for $-1$ and $2$.

# 4.3  Exercise Set

1. For the function
$$f(x) = x^4 - 6x^3 + x^2 + 24x - 20,$$
use long division to determine whether each of the following is a factor of $f(x)$.
   **a)** $x + 1$   **b)** $x - 2$   **c)** $x + 5$

2. For the function
$$h(x) = x^3 - x^2 - 17x - 15,$$
use long division to determine whether each of the following is a factor of $h(x)$.
   **a)** $x + 5$   **b)** $x + 1$   **c)** $x + 3$

3. For the function
$$g(x) = x^3 - 2x^2 - 11x + 12,$$
use long division to determine whether each of the following is a factor of $g(x)$.
   **a)** $x - 4$   **b)** $x - 3$   **c)** $x - 1$

4. For the function
$$f(x) = x^4 + 8x^3 + 5x^2 - 38x + 24,$$
use long division to determine whether each of the following is a factor of $f(x)$.
   **a)** $x + 6$   **b)** $x + 1$   **c)** $x - 4$

*In each of the following, a polynomial $P(x)$ and a divisor $d(x)$ are given. Use long division to find the quotient $Q(x)$ and the remainder $R(x)$ when $P(x)$ is divided by $d(x)$. Express $P(x)$ in the form $d(x) \cdot Q(x) + R(x)$.*

5. $P(x) = x^3 - 8,$
$d(x) = x + 2$

6. $P(x) = 2x^3 - 3x^2 + x - 1,$
$d(x) = x - 3$

7. $P(x) = x^3 + 6x^2 - 25x + 18,$
$d(x) = x + 9$

8. $P(x) = x^3 - 9x^2 + 15x + 25,$
$d(x) = x - 5$

9. $P(x) = x^4 - 2x^2 + 3,$
$d(x) = x + 2$

10. $P(x) = x^4 + 6x^3,$
$d(x) = x - 1$

*Use synthetic division to find the quotient and the remainder.*

11. $(2x^4 + 7x^3 + x - 12) \div (x + 3)$

12. $(x^3 - 7x^2 + 13x + 3) \div (x - 2)$

13. $(x^3 - 2x^2 - 8) \div (x + 2)$

14. $(x^3 - 3x + 10) \div (x - 2)$

15. $(3x^3 - x^2 + 4x - 10) \div (x + 1)$

16. $(4x^4 - 2x + 5) \div (x + 3)$

17. $(x^5 + x^3 - x) \div (x - 3)$

18. $(x^7 - x^6 + x^5 - x^4 + 2) \div (x + 1)$

19. $(x^4 - 1) \div (x - 1)$

20. $(x^5 + 32) \div (x + 2)$

21. $\left(2x^4 + 3x^2 - 1\right) \div \left(x - \frac{1}{2}\right)$

22. $\left(3x^4 - 2x^2 + 2\right) \div \left(x - \frac{1}{4}\right)$

*Use synthetic division to find the function values. Then check your work using a graphing calculator.*

23. $f(x) = x^3 - 6x^2 + 11x - 6$; find $f(1)$, $f(-2)$, and $f(3)$.

24. $f(x) = x^3 + 7x^2 - 12x - 3$; find $f(-3)$, $f(-2)$, and $f(1)$.

25. $f(x) = x^4 - 3x^3 + 2x + 8$; find $f(-1)$, $f(4)$, and $f(-5)$.

26. $f(x) = 2x^4 + x^2 - 10x + 1$; find $f(-10)$, $f(2)$, and $f(3)$.

27. $f(x) = 2x^5 - 3x^4 + 2x^3 - x + 8$; find $f(20)$ and $f(-3)$.

28. $f(x) = x^5 - 10x^4 + 20x^3 - 5x - 100$; find $f(-10)$ and $f(5)$.

29. $f(x) = x^4 - 16$; find $f(2)$, $f(-2)$, $f(3)$, and $f(1 - \sqrt{2}\,)$.

30. $f(x) = x^5 + 32$; find $f(2)$, $f(-2)$, $f(3)$, and $f(2 + 3i)$.

*Using synthetic division, determine whether the numbers are zeros of the polynomial function.*

**31.** $-3, 2;\ f(x) = 3x^3 + 5x^2 - 6x + 18$

**32.** $-4, 2;\ f(x) = 3x^3 + 11x^2 - 2x + 8$

**33.** $-3, 1;\ h(x) = x^4 + 4x^3 + 2x^2 - 4x - 3$

**34.** $2, -1;\ g(x) = x^4 - 6x^3 + x^2 + 24x - 20$

**35.** $i, -2i;\ g(x) = x^3 - 4x^2 + 4x - 16$

**36.** $\frac{1}{3}, 2;\ h(x) = x^3 - x^2 - \frac{1}{9}x + \frac{1}{9}$

**37.** $-3, \frac{1}{2};\ f(x) = x^3 - \frac{7}{2}x^2 + x - \frac{3}{2}$

**38.** $i, -i, -2;\ f(x) = x^3 + 2x^2 + x + 2$

*Factor the polynomial function $f(x)$. Then solve the equation $f(x) = 0$.*

**39.** $f(x) = x^3 + 4x^2 + x - 6$

**40.** $f(x) = x^3 + 5x^2 - 2x - 24$

**41.** $f(x) = x^3 - 6x^2 + 3x + 10$

**42.** $f(x) = x^3 + 2x^2 - 13x + 10$

**43.** $f(x) = x^3 - x^2 - 14x + 24$

**44.** $f(x) = x^3 - 3x^2 - 10x + 24$

**45.** $f(x) = x^4 - 7x^3 + 9x^2 + 27x - 54$

**46.** $f(x) = x^4 - 4x^3 - 7x^2 + 34x - 24$

**47.** $f(x) = x^4 - x^3 - 19x^2 + 49x - 30$

**48.** $f(x) = x^4 + 11x^3 + 41x^2 + 61x + 30$

*Sketch the graph of the polynomial function. Follow the procedure outlined on p. 317. Use synthetic division and the remainder theorem to find the zeros.*

**49.** $f(x) = x^4 - x^3 - 7x^2 + x + 6$

**50.** $f(x) = x^4 + x^3 - 3x^2 - 5x - 2$

**51.** $f(x) = x^3 - 7x + 6$

**52.** $f(x) = x^3 - 12x + 16$

**53.** $f(x) = -x^3 + 3x^2 + 6x - 8$

**54.** $f(x) = -x^4 + 2x^3 + 3x^2 - 4x - 4$

## Skill Maintenance

*Solve. Find exact solutions.*

**55.** $2x^2 + 12 = 5x$

**56.** $7x^2 + 4x = 3$

*In Exercises 57–59, consider the function*
$$g(x) = x^2 + 5x - 14.$$

**57.** What are the inputs if the output is $-14$?

**58.** What is the output if the input is 3?

**59.** Given an output of $-20$, find the corresponding inputs.

**60.** *Movie Ticket Prices.*    The average price of a movie ticket has increased linearly over the years, rising from \$5.39 in 2000 to \$7.89 in 2010 (*Source*: National Association of Theatre Owners). Using these two data points, find a linear function, $f(x) = mx + b$, that models the data. Let $x$ represent the number of years since 2000. Then use this function to estimate the average price of a movie ticket in 2005 and in 2015.

**61.** The sum of the base and the height of a triangle is 30 in. Find the dimensions for which the area is a maximum.

## Synthesis

*In Exercises 62 and 63, a graph of a polynomial function is given. On the basis of the graph:*

a)  *Find as many factors of the polynomial as you can.*

b)  *Construct a polynomial function with the zeros shown in the graph.*

c)  *Can you find any other polynomial functions with the given zeros?*

d)  *Can you find more than one polynomial function with the given zeros and the same graph?*

**62.**

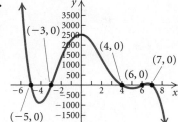

**63.**

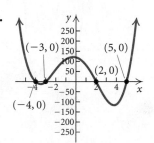

**64.** For what values of $k$ will the remainder be the same when $x^2 + kx + 4$ is divided by $x - 1$ and by $x + 1$?

**65.** Find $k$ such that $x + 2$ is a factor of $x^3 - kx^2 + 3x + 7k$.

**66.** *Beam Deflection.* A beam rests at two points $A$ and $B$ and has a concentrated load applied to its center. Let $y =$ the deflection, in feet, of the beam at a distance of $x$ feet from $A$. Under certain conditions, this deflection is given by

$$y = \tfrac{1}{13}x^3 - \tfrac{1}{14}x.$$

Find the zeros of the polynomial in the interval $[0, 2]$.

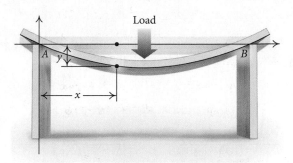

*Solve.*

**67.** $\dfrac{2x^2}{x^2 - 1} + \dfrac{4}{x + 3} = \dfrac{12x - 4}{x^3 + 3x^2 - x - 3}$

**68.** $\dfrac{6x^2}{x^2 + 11} + \dfrac{60}{x^3 - 7x^2 + 11x - 77} = \dfrac{1}{x - 7}$

**69.** Find a 15th-degree polynomial for which $x - 1$ is a factor. Answers may vary.

*Use synthetic division to divide.*

**70.** $(x^4 - y^4) \div (x - y)$

**71.** $(x^3 + 3ix^2 - 4ix - 2) \div (x + i)$

**72.** $(x^2 - 4x - 2) \div [x - (3 + 2i)]$

**73.** $(x^2 - 3x + 7) \div (x - i)$

## Mid-Chapter Mixed Review

*Determine whether the statement is true or false.*

**1.** The $y$-intercept of the graph of the function $P(x) = 5 - 2x^3$ is $(5, 0)$. [4.2]

**2.** The degree of the polynomial $x - \tfrac{1}{2}x^4 - 3x^6 + x^5$ is 6. [4.1]

**3.** If $f(x) = (x + 7)(x - 8)$, then $f(8) = 0$. [4.3]

**4.** If $f(12) = 0$, then $x + 12$ is a factor of $f(x)$. [4.3]

*Find the zeros of the polynomial function and state the multiplicity of each.* [4.1]

**5.** $f(x) = (x^2 - 10x + 25)^3$

**6.** $h(x) = 2x^3 + x^2 - 50x - 25$

**7.** $g(x) = x^4 - 3x^2 + 2$

**8.** $f(x) = -6(x - 3)^2(x + 4)$

*In Exercises 9–12, match the function with one of the graphs (a)–(d), which follow.* [4.2]

**a)**

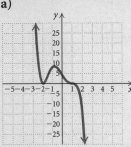

**b)**

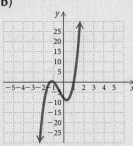

**c)**

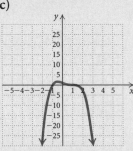

**d)**

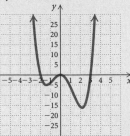

**9.** $f(x) = x^4 - x^3 - 6x^2$

**10.** $f(x) = -(x - 1)^3(x + 2)^2$

**11.** $f(x) = 6x^3 + 8x^2 - 6x - 8$

**12.** $f(x) = -(x - 1)^3(x + 1)$

*Using the intermediate value theorem, determine, if possible, whether the function has at least one real zero between a and b.* [4.2]

**13.** $f(x) = x^3 - 2x^2 + 3; \ a = -2, b = 0$

**14.** $f(x) = x^3 - 2x^2 + 3; \ a = -\frac{1}{2}, b = 1$

**15.** For the polynomial $P(x) = x^4 - 6x^3 + x - 2$ and the divisor $d(x) = x - 1$, use long division to find the quotient $Q(x)$ and the remainder $R(x)$ when $P(x)$ is divided by $d(x)$. Express $P(x)$ in the form $d(x) \cdot Q(x) + R(x)$. [4.3]

*Use synthetic division to find the quotient and the remainder.* [4.3]

**16.** $(3x^4 - x^3 + 2x^2 - 6x + 6) \div (x - 2)$

**17.** $(x^5 - 5) \div (x + 1)$

*Use synthetic division to find the function values.* [4.3]

**18.** $g(x) = x^3 - 9x^2 + 4x - 10;$ find $g(-5)$

**19.** $f(x) = 20x^2 - 40x;$ find $f\left(\frac{1}{2}\right)$

**20.** $f(x) = 5x^4 + x^3 - x;$ find $f(-\sqrt{2})$

*Using synthetic division, determine whether the numbers are zeros of the polynomial function.* [4.3]

**21.** $-3i, 3; \ f(x) = x^3 - 4x^2 + 9x - 36$

**22.** $-1, 5; \ f(x) = x^6 - 35x^4 + 259x^2 - 225$

*Factor the polynomial function $f(x)$. Then solve the equation $f(x) = 0$.* [4.3]

**23.** $h(x) = x^3 - 2x^2 - 55x + 56$

**24.** $g(x) = x^4 - 2x^3 - 13x^2 + 14x + 24$

## Collaborative Discussion and Writing

**25.** How is the range of a polynomial function related to the degree of the polynomial? [4.1]

**26.** Is it possible for the graph of a polynomial function to have no $y$-intercept? no $x$-intercepts? Explain your answer. [4.2]

**27.** Explain why values of a function must be all positive or all negative between consecutive zeros. [4.2]

**28.** In synthetic division, why is the degree of the quotient 1 less than that of the dividend? [4.3]

## Theorems about Zeros of Polynomial Functions

**4.4**

- Find a polynomial with specified zeros.

- For a polynomial function with integer coefficients, find the rational zeros and the other zeros, if possible.

- Use Descartes' rule of signs to find information about the number of real zeros of a polynomial function with real coefficients.

We will now allow the coefficients of a polynomial to be complex numbers. In certain cases, we will restrict the coefficients to be real numbers, rational numbers, or integers, as shown in the following examples.

| Polynomial | Type of Coefficient |
|---|---|
| $5x^3 - 3x^2 + (2 + 4i)x + i$ | Complex |
| $5x^3 - 3x^2 + \sqrt{2}x - \pi$ | Real |
| $5x^3 - 3x^2 + \frac{2}{3}x - \frac{7}{4}$ | Rational |
| $5x^3 - 3x^2 + 8x - 11$ | Integer |

### ■ The Fundamental Theorem of Algebra

A linear, or first-degree, polynomial function $f(x) = mx + b$ (where $m \neq 0$) has just one zero, $-b/m$. It can be shown that any quadratic polynomial function $f(x) = ax^2 + bx + c$ with complex numbers for coefficients has at least one, and at most two, complex zeros. The following theorem is a generalization. No proof is given in this text.

> **THE FUNDAMENTAL THEOREM OF ALGEBRA**
>
> Every polynomial function of degree $n$, with $n \geq 1$, has at least one zero in the set of complex numbers.

Note that although the fundamental theorem of algebra guarantees that a zero exists, it does not tell how to find it. Recall that the zeros of a polynomial function $f(x)$ are the solutions of the polynomial equation $f(x) = 0$. We now develop some concepts that can help in finding zeros. First, we consider one of the results of the fundamental theorem of algebra.

> Every polynomial function $f$ of degree $n$, with $n \geq 1$, can be factored into $n$ linear factors (not necessarily unique); that is,
>
> $$f(x) = a_n(x - c_1)(x - c_2) \cdots (x - c_n).$$

### Finding Polynomials with Given Zeros

Given several numbers, we can find a polynomial function with those numbers as its zeros.

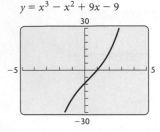

$y = x^3 - x^2 + 9x - 9$

**EXAMPLE 1**     Find a polynomial function of degree 3, having the zeros 1, $3i$, and $-3i$.

**Solution**     Such a function has factors $x - 1$, $x - 3i$, and $x + 3i$, so we have

$$f(x) = a_n(x - 1)(x - 3i)(x + 3i).$$

The number $a_n$ can be any nonzero number. The simplest polynomial function will be obtained if we let it be 1. If we then multiply the factors, we obtain

$$\begin{aligned}
f(x) &= (x - 1)(x^2 - 9i^2) &&\text{Multiplying } (x - 3i)(x + 3i) \\
&= (x - 1)(x^2 + 9) &&-9i^2 = -9(-1) = 9 \\
&= x^3 - x^2 + 9x - 9.
\end{aligned}$$

**Now Try Exercise 3.**

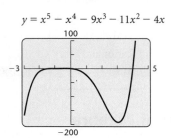

$y = x^5 - x^4 - 9x^3 - 11x^2 - 4x$

**EXAMPLE 2**     Find a polynomial function of degree 5 with $-1$ as a zero of multiplicity 3, 4 as a zero of multiplicity 1, and 0 as a zero of multiplicity 1.

**Solution**     Proceeding as in Example 1, letting $a_n = 1$, we obtain

$$\begin{aligned}
f(x) &= \left[x - (-1)\right]^3(x - 4)(x - 0) \\
&= (x + 1)^3(x - 4)x \\
&= (x^3 + 3x^2 + 3x + 1)(x^2 - 4x) \\
&= x^5 - x^4 - 9x^3 - 11x^2 - 4x.
\end{aligned}$$

**Now Try Exercise 13.**

### Zeros of Polynomial Functions with Real Coefficients

Consider the quadratic equation $x^2 - 2x + 2 = 0$, with real coefficients. Its solutions are $1 + i$ and $1 - i$. Note that they are complex conjugates. This generalizes to any polynomial equation with real coefficients.

---

**NONREAL ZEROS: $a + bi$ AND $a - bi$, $b \neq 0$**

If a complex number $a + bi$, $b \neq 0$, is a zero of a polynomial function $f(x)$ with *real* coefficients, then its conjugate, $a - bi$, is also a zero. For example, if $2 + 7i$ is a zero of a polynomial function $f(x)$, with real coefficients, then its conjugate, $2 - 7i$, is also a zero. (Nonreal zeros occur in conjugate pairs.)

---

In order for the preceding to be true, it is essential that the coefficients be *real* numbers.

## ▪ Rational Coefficients

When a polynomial function has rational numbers for coefficients, certain irrational zeros also occur in pairs, as described in the following theorem.

> ### IRRATIONAL ZEROS: $a + c\sqrt{b}$ AND $a - c\sqrt{b}$, $b$ IS NOT A PERFECT SQUARE
>
> If $a + c\sqrt{b}$, where $a$, $b$, and $c$ are rational and $b$ is not a perfect square, is a zero of a polynomial function $f(x)$ with *rational* coefficients, then its conjugate, $a - c\sqrt{b}$, is also a zero. For example, if $-3 + 5\sqrt{2}$ is a zero of a polynomial function $f(x)$ with rational coefficients, then its conjugate, $-3 - 5\sqrt{2}$, is also a zero. (Irrational zeros occur in conjugate pairs.)

**EXAMPLE 3**  Suppose that a polynomial function of degree 6 with rational coefficients has

$$-2 + 5i, \quad -2i, \quad \text{and} \quad 1 - \sqrt{3}$$

as three of its zeros. Find the other zeros.

**Solution**  Since the coefficients are rational, and thus real, the other zeros are the conjugates of the given zeros,

$$-2 - 5i, \quad 2i, \quad \text{and} \quad 1 + \sqrt{3}.$$

There are no other zeros because a polynomial function of degree 6 can have at most 6 zeros.  `Now Try Exercise 19.`

**EXAMPLE 4**  Find a polynomial function of lowest degree with rational coefficients that has $-\sqrt{3}$ and $1 + i$ as two of its zeros.

**Solution**  The function must also have the zeros $\sqrt{3}$ and $1 - i$. Because we want to find the polynomial function of lowest degree with the given zeros, we will not include additional zeros; that is, we will write a polynomial function of degree 4. Thus if we let $a_n = 1$, the polynomial function is

$$
\begin{aligned}
f(x) &= \left[x - (-\sqrt{3})\right]\left[x - \sqrt{3}\right]\left[x - (1 + i)\right]\left[x - (1 - i)\right] \\
&= (x + \sqrt{3})(x - \sqrt{3})\left[(x - 1) - i\right]\left[(x - 1) + i\right] \\
&= (x^2 - 3)\left[(x - 1)^2 - i^2\right] \\
&= (x^2 - 3)\left[x^2 - 2x + 1 + 1\right] \\
&= (x^2 - 3)(x^2 - 2x + 2) \\
&= x^4 - 2x^3 - x^2 + 6x - 6.
\end{aligned}
$$

`Now Try Exercise 39.`

$y = x^4 - 2x^3 - x^2 + 6x - 6$

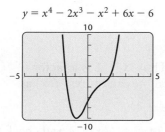

## ▪ Integer Coefficients and the Rational Zeros Theorem

It is not always easy to find the zeros of a polynomial function. However, if a polynomial function has integer coefficients, there is a procedure that will yield all the rational zeros.

> **THE RATIONAL ZEROS THEOREM**
> Let
>
> $$P(x) = a_n x^n + a_{n-1} x^{n-1} + \cdots + a_1 x + a_0,$$
>
> where all the coefficients are integers. Consider a rational number denoted by $p/q$, where $p$ and $q$ are relatively prime (having no common factor besides $-1$ and $1$). If $p/q$ is a zero of $P(x)$, then $p$ is a factor of $a_0$ and $q$ is a factor of $a_n$.

**EXAMPLE 5**    Given $f(x) = 3x^4 - 11x^3 + 10x - 4$:

**a)** Find the rational zeros and then the other zeros; that is, solve $f(x) = 0$.

**b)** Factor $f(x)$ into linear factors.

*Solution*

**a)** Because the degree of $f(x)$ is 4, there are at most 4 distinct zeros. The rational zeros theorem says that if a rational number $p/q$ is a zero of $f(x)$, then $p$ must be a factor of $-4$ and $q$ must be a factor of 3. Thus the possibilities for $p/q$ are

$$\frac{\text{Possibilities for } p}{\text{Possibilities for } q}: \quad \frac{\pm 1, \pm 2, \pm 4}{\pm 1, \pm 3};$$

Possibilities for $p/q$:    $1, -1, 2, -2, 4, -4, \frac{1}{3}, -\frac{1}{3}, \frac{2}{3}, -\frac{2}{3}, \frac{4}{3}, -\frac{4}{3}$.

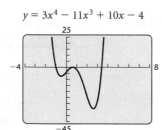

$y = 3x^4 - 11x^3 + 10x - 4$

We could use the TABLE feature or some other method to find function values. However, if we use synthetic division, the quotient polynomial becomes a beneficial by-product if a zero is found. Rather than use synthetic division to check *each* of these possibilities, we graph the function and inspect the graph for zeros that appear to be near any of the possible rational zeros. (See the graph at left.)

From the graph, we see that of the possibilities in the list, only the numbers $-1$, $\frac{1}{3}$, and $\frac{2}{3}$ might be rational zeros.

We try $-1$.

$$
\begin{array}{r|rrrrr}
-1 & 3 & -11 & 0 & 10 & -4 \\
   &   & -3 & 14 & -14 & 4 \\
\hline
   & 3 & -14 & 14 & -4 & \;\;0
\end{array}
$$

We have $f(-1) = 0$, so $-1$ is a zero. Thus, $x + 1$ is a factor of $f(x)$. Using the results of the synthetic division, we can express $f(x)$ as

$$f(x) = (x + 1)(3x^3 - 14x^2 + 14x - 4).$$

We now consider the factor $3x^3 - 14x^2 + 14x - 4$ and check the other possible zeros. We try $\frac{1}{3}$.

$$
\begin{array}{r|rrrr}
1/3 & 3 & -14 & 14 & -4 \\
    &   & 1 & -\frac{13}{3} & \frac{29}{9} \\
\hline
    & 3 & -13 & \frac{29}{3} & -\frac{7}{9}
\end{array}
$$

Since $f\!\left(\frac{1}{3}\right) \neq 0$, we know that $\frac{1}{3}$ is not a zero.

Let's now try $\frac{2}{3}$.

$$\begin{array}{r|rrrr} 2/3 & 3 & -14 & 14 & -4 \\ & & 2 & -8 & 4 \\ \hline & 3 & -12 & 6 & \,|\,\ 0 \end{array}$$

Since the remainder is 0, we know that $x - \frac{2}{3}$ is a factor of $3x^3 - 14x^2 + 14x - 4$ and is also a factor of $f(x)$. Thus, $\frac{2}{3}$ is a zero of $f(x)$.

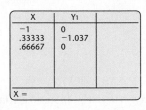

We can check the zeros with the TABLE feature. (See the window at left.) Note that the graphing calculator converts $\frac{1}{3}$ and $\frac{2}{3}$ to decimal notation. Since $f(-1) = 0$ and $f\left(\frac{2}{3}\right) = 0$, $-1$ and $\frac{2}{3}$ are zeros. Since $f\left(\frac{1}{3}\right) \neq 0$, $\frac{1}{3}$ is not a zero.

Using the results of the synthetic division, we can factor further:

$$f(x) = (x + 1)\left(x - \tfrac{2}{3}\right)(3x^2 - 12x + 6) \quad \begin{array}{l}\text{\small Using the results of the}\\ \text{\small last synthetic division}\end{array}$$

$$= (x + 1)\left(x - \tfrac{2}{3}\right) \cdot 3 \cdot (x^2 - 4x + 2). \quad \text{\small Removing a factor of 3}$$

The quadratic formula can be used to find the values of $x$ for which $x^2 - 4x + 2 = 0$. Those values are also zeros of $f(x)$:

$$x = \frac{-b \pm \sqrt{b^2 - 4ac}}{2a}$$

$$= \frac{-(-4) \pm \sqrt{(-4)^2 - 4 \cdot 1 \cdot 2}}{2 \cdot 1} \quad a = 1, b = -4, \text{ and } c = 2$$

$$= \frac{4 \pm \sqrt{8}}{2} = \frac{4 \pm 2\sqrt{2}}{2} = \frac{2(2 \pm \sqrt{2})}{2}$$

$$= 2 \pm \sqrt{2}.$$

The rational zeros are $-1$ and $\frac{2}{3}$. The other zeros are $2 \pm \sqrt{2}$.

**b)** The complete factorization of $f(x)$ is

$$f(x) = 3(x + 1)\left(x - \tfrac{2}{3}\right)\left[x - (2 - \sqrt{2})\right]\left[x - (2 + \sqrt{2})\right], \text{ or}$$
$$f(x) = (x + 1)(3x - 2)\left[x - (2 - \sqrt{2})\right]\left[x - (2 + \sqrt{2})\right].$$

Replacing $3\left(x - \tfrac{2}{3}\right)$ with $(3x - 2)$

> **Now Try Exercise 55.**

**EXAMPLE 6**   Given $f(x) = 2x^5 - x^4 - 4x^3 + 2x^2 - 30x + 15$:

**a)** Find the rational zeros and then the other zeros; that is, solve $f(x) = 0$.

**b)** Factor $f(x)$ into linear factors.

*Solution*

**a)** Because the degree of $f(x)$ is 5, there are at most 5 distinct zeros. According to the rational zeros theorem, any rational zero of $f$ must be of the form $p/q$, where $p$ is a factor of 15 and $q$ is a factor of 2. The possibilities are

$$\frac{\text{\textit{Possibilities for p}}}{\text{\textit{Possibilities for q}}}: \quad \frac{\pm 1, \pm 3, \pm 5, \pm 15}{\pm 1, \pm 2};$$

*Possibilities for* $p/q$:   $1, -1, 3, -3, 5, -5, 15, -15, \frac{1}{2}, -\frac{1}{2}, \frac{3}{2}, -\frac{3}{2},$
$\frac{5}{2}, -\frac{5}{2}, \frac{15}{2}, -\frac{15}{2}.$

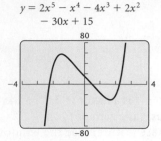

$y = 2x^5 - x^4 - 4x^3 + 2x^2 - 30x + 15$

| X | Y1 | |
|---|---|---|
| -2.5 | -69.38 | |
| .5 | 0 | |
| 2.5 | 46.25 | |

X =

Rather than use synthetic division to check each of these possibilities, we graph $y = 2x^5 - x^4 - 4x^3 + 2x^2 - 30x + 15$. (See the graph at left.) We can then inspect the graph for zeros that appear to be near any of the possible rational zeros.

From the graph, we see that of the possibilities in the list, only the numbers $-\frac{5}{2}, \frac{1}{2}$, and $\frac{5}{2}$ might be rational zeros. By synthetic division or using the TABLE feature of a graphing calculator (see the table at left), we see that only $\frac{1}{2}$ is actually a rational zero.

$$\begin{array}{r|rrrrrr} 1/2 & 2 & -1 & -4 & 2 & -30 & 15 \\ & & 1 & 0 & -2 & 0 & -15 \\ \hline & 2 & 0 & -4 & 0 & -30 & 0 \end{array}$$

This means that $x - \frac{1}{2}$ is a factor of $f(x)$. We write the factorization and try to factor further:

$$\begin{aligned} f(x) &= \left(x - \tfrac{1}{2}\right)(2x^4 - 4x^2 - 30) \\ &= \left(x - \tfrac{1}{2}\right) \cdot 2 \cdot (x^4 - 2x^2 - 15) &&\text{Factoring out the 2} \\ &= \left(x - \tfrac{1}{2}\right) \cdot 2 \cdot (x^2 - 5)(x^2 + 3). &&\text{Factoring the trinomial} \end{aligned}$$

We now solve the equation $f(x) = 0$ to determine the zeros. We use the principle of zero products:

$$\left(x - \tfrac{1}{2}\right) \cdot 2 \cdot (x^2 - 5)(x^2 + 3) = 0$$

$$\begin{aligned} x - \tfrac{1}{2} = 0 \quad &\textit{or} \quad x^2 - 5 = 0 \quad &\textit{or} \quad x^2 + 3 = 0 \\ x = \tfrac{1}{2} \quad &\textit{or} \quad x^2 = 5 \quad &\textit{or} \quad x^2 = -3 \\ x = \tfrac{1}{2} \quad &\textit{or} \quad x = \pm\sqrt{5} \quad &\textit{or} \quad x = \pm\sqrt{3}i. \end{aligned}$$

There is only one rational zero, $\frac{1}{2}$. The other zeros are $\pm\sqrt{5}$ and $\pm\sqrt{3}i$.

**b)** The factorization into linear factors is

$$f(x) = 2\left(x - \tfrac{1}{2}\right)(x + \sqrt{5})(x - \sqrt{5})(x + \sqrt{3}i)(x - \sqrt{3}i), \quad \text{or}$$
$$f(x) = (2x - 1)(x + \sqrt{5})(x - \sqrt{5})(x + \sqrt{3}i)(x - \sqrt{3}i).$$

Replacing $2\left(x - \tfrac{1}{2}\right)$ with $(2x - 1)$

**Now Try Exercise 63.**

## ■ Descartes' Rule of Signs

The development of a rule that helps determine the number of positive real zeros and the number of negative real zeros of a polynomial function is credited to the French mathematician René Descartes. To use the rule, we must have the polynomial arranged in descending order or ascending order, with no zero terms written in and the constant term not 0. Then we determine the number of *variations of sign*, that is, the number of times, in reading through the polynomial, that successive coefficients are of different signs.

**EXAMPLE 7** Determine the number of variations of sign in the polynomial function $P(x) = 2x^5 - 3x^2 + x + 4$.

*Solution* We have

$$P(x) = \underbrace{2x^5 - 3x^2 + x}_{} + 4$$

From positive to
negative; a variation

From negative to positive;
a variation

Both positive; no variation

The number of variations of sign is 2.

Note the following:

$$P(-x) = 2(-x)^5 - 3(-x)^2 + (-x) + 4$$
$$= -2x^5 - 3x^2 - x + 4.$$

We see that the number of variations of sign in $P(-x)$ is 1. It occurs as we go from $-x$ to 4.

We now state Descartes' rule, without proof.

---

### DESCARTES' RULE OF SIGNS

Let $P(x)$, written in descending order or ascending order, be a polynomial function with real coefficients and a nonzero constant term. The number of positive real zeros of $P(x)$ is either:

1. The same as the number of variations of sign in $P(x)$, or
2. Less than the number of variations of sign in $P(x)$ by a positive even integer.

The number of negative real zeros of $P(x)$ is either:

3. The same as the number of variations of sign in $P(-x)$, or
4. Less than the number of variations of sign in $P(-x)$ by a positive even integer.

A zero of multiplicity $m$ must be counted $m$ times.

---

In each of Examples 8–10, what does Descartes' rule of signs tell you about the number of positive real zeros and the number of negative real zeros?

### EXAMPLE 8   $P(x) = 2x^5 - 5x^2 - 3x + 6$

*Solution*   The number of variations of sign in $P(x)$ is 2. Therefore, the number of positive real zeros is either 2 or less than 2 by 2, 4, 6, and so on. Thus the number of positive real zeros is either 2 or 0, since a negative number of zeros has no meaning.

$$P(-x) = -2x^5 - 5x^2 + 3x + 6$$

The number of variations of sign in $P(-x)$ is 1. Thus there is exactly 1 negative real zero. Since nonreal, complex conjugates occur in pairs, we also know the possible ways in which nonreal zeros might occur. The table shown at left summarizes all the possibilities for real zeros and nonreal zeros of $P(x)$.

| Total Number of Zeros | 5 | |
|---|---|---|
| Positive Real | 2 | 0 |
| Negative Real | 1 | 1 |
| Nonreal | 2 | 4 |

**Now Try Exercise 93.**

| Total Number of Zeros | 4 | | |
|---|---|---|---|
| Positive Real | 4 | 2 | 0 |
| Negative Real | 0 | 0 | 0 |
| Nonreal | 0 | 2 | 4 |

**EXAMPLE 9** $P(x) = 5x^4 - 3x^3 + 7x^2 - 12x + 4$

*Solution* There are 4 variations of sign. Thus the number of positive real zeros is either

$$4 \quad \text{or} \quad 4 - 2 \quad \text{or} \quad 4 - 4;$$

that is, the number of positive real zeros is 4, 2, or 0.

$$P(-x) = 5x^4 + 3x^3 + 7x^2 + 12x + 4$$

There are 0 changes in sign, so there are no negative real zeros. The table at left summarizes all the possibilities for real zeros and nonreal zeros of $P(x)$.

**Now Try Exercise 81.**

**EXAMPLE 10** $P(x) = 6x^6 - 2x^2 - 5x$

*Solution* As written, the polynomial does not satisfy the conditions of Descartes' rule of signs because the constant term is 0. But because $x$ is a factor of every term, we know that the polynomial has 0 as a zero. We can then factor as follows:

$$P(x) = x(6x^5 - 2x - 5).$$

| Total Number of Zeros | 6 | |
|---|---|---|
| 0 as a Zero | 1 | 1 |
| Positive Real | 1 | 1 |
| Negative Real | 2 | 0 |
| Nonreal | 2 | 4 |

Now we analyze $Q(x) = 6x^5 - 2x - 5$ and $Q(-x) = -6x^5 + 2x - 5$. The number of variations of sign in $Q(x)$ is 1. Therefore, there is exactly 1 positive real zero. The number of variations of sign in $Q(-x)$ is 2. Thus the number of negative real zeros is 2 or 0. The same results apply to $P(x)$. Since nonreal, complex conjugates occur in pairs, we know the possible ways in which nonreal zeros might occur. The table at left summarizes all the possibilities for real zeros and nonreal zeros of $P(x)$.

**Now Try Exercise 95.**

# 4.4 Exercise Set

*Find a polynomial function of degree 3 with the given numbers as zeros.*

1. $-2, 3, 5$

2. $-1, 0, 4$

3. $-3, 2i, -2i$

4. $2, i, -i$

5. $\sqrt{2}, -\sqrt{2}, 3$

6. $-5, \sqrt{3}, -\sqrt{3}$

7. $1 - \sqrt{3}, 1 + \sqrt{3}, -2$

8. $-4, 1 - \sqrt{5}, 1 + \sqrt{5}$

9. $1 + 6i, 1 - 6i, -4$

10. $1 + 4i, 1 - 4i, -1$

11. $-\frac{1}{3}, 0, 2$

12. $-3, 0, \frac{1}{2}$

13. Find a polynomial function of degree 5 with $-1$ as a zero of multiplicity 3, 0 as a zero of multiplicity 1, and 1 as a zero of multiplicity 1.

14. Find a polynomial function of degree 4 with $-2$ as a zero of multiplicity 1, 3 as a zero of multiplicity 2, and $-1$ as a zero of multiplicity 1.

**15.** Find a polynomial function of degree 4 with $-1$ as a zero of multiplicity 3 and 0 as a zero of multiplicity 1.

**16.** Find a polynomial function of degree 5 with $-\frac{1}{2}$ as a zero of multiplicity 2, 0 as a zero of multiplicity 1, and 1 as a zero of multiplicity 2.

*Suppose that a polynomial function of degree 4 with rational coefficients has the given numbers as zeros. Find the other zero(s).*

**17.** $-1, \sqrt{3}, \frac{11}{3}$

**18.** $-\sqrt{2}, -1, \frac{4}{5}$

**19.** $-i, 2 - \sqrt{5}$

**20.** $i, -3 + \sqrt{3}$

**21.** $3i, 0, -5$

**22.** $3, 0, -2i$

**23.** $-4 - 3i, 2 - \sqrt{3}$

**24.** $6 - 5i, -1 + \sqrt{7}$

*Suppose that a polynomial function of degree 5 with rational coefficients has the given numbers as zeros. Find the other zero(s).*

**25.** $-\frac{1}{2}, \sqrt{5}, -4i$

**26.** $\frac{3}{4}, -\sqrt{3}, 2i$

**27.** $-5, 0, 2 - i, 4$

**28.** $-2, 3, 4, 1 - i$

**29.** $6, -3 + 4i, 4 - \sqrt{5}$

**30.** $-3 - 3i, 2 + \sqrt{13}, 6$

**31.** $-\frac{3}{4}, \frac{3}{4}, 0, 4 - i$

**32.** $-0.6, 0, 0.6, -3 + \sqrt{2}$

*Find a polynomial function of lowest degree with rational coefficients that has the given numbers as some of its zeros.*

**33.** $1 + i, 2$

**34.** $2 - i, -1$

**35.** $4i$

**36.** $-5i$

**37.** $-4i, 5$

**38.** $3, -i$

**39.** $1 - i, -\sqrt{5}$

**40.** $2 - \sqrt{3}, 1 + i$

**41.** $\sqrt{5}, -3i$

**42.** $-\sqrt{2}, 4i$

*Given that the polynomial function has the given zero, find the other zeros.*

**43.** $f(x) = x^3 + 5x^2 - 2x - 10; -5$

**44.** $f(x) = x^3 - x^2 + x - 1; 1$

**45.** $f(x) = x^4 - 5x^3 + 7x^2 - 5x + 6; -i$

**46.** $f(x) = x^4 - 16; 2i$

**47.** $f(x) = x^3 - 6x^2 + 13x - 20; 4$

**48.** $f(x) = x^3 - 8; 2$

*List all possible rational zeros of the function.*

**49.** $f(x) = x^5 - 3x^2 + 1$

**50.** $f(x) = x^7 + 37x^5 - 6x^2 + 12$

**51.** $f(x) = 2x^4 - 3x^3 - x + 8$

**52.** $f(x) = 3x^3 - x^2 + 6x - 9$

**53.** $f(x) = 15x^6 + 47x^2 + 2$

**54.** $f(x) = 10x^{25} + 3x^{17} - 35x + 6$

*For each polynomial function:*

**a)** *Find the rational zeros and then the other zeros; that is, solve $f(x) = 0$.*

**b)** *Factor $f(x)$ into linear factors.*

**55.** $f(x) = x^3 + 3x^2 - 2x - 6$

**56.** $f(x) = x^3 - x^2 - 3x + 3$

**57.** $f(x) = 3x^3 - x^2 - 15x + 5$

**58.** $f(x) = 4x^3 - 4x^2 - 3x + 3$

**59.** $f(x) = x^3 - 3x + 2$

**60.** $f(x) = x^3 - 2x + 4$

**61.** $f(x) = 2x^3 + 3x^2 + 18x + 27$

**62.** $f(x) = 2x^3 + 7x^2 + 2x - 8$

**63.** $f(x) = 5x^4 - 4x^3 + 19x^2 - 16x - 4$

**64.** $f(x) = 3x^4 - 4x^3 + x^2 + 6x - 2$

**65.** $f(x) = x^4 - 3x^3 - 20x^2 - 24x - 8$

**66.** $f(x) = x^4 + 5x^3 - 27x^2 + 31x - 10$

**67.** $f(x) = x^3 - 4x^2 + 2x + 4$

**68.** $f(x) = x^3 - 8x^2 + 17x - 4$

**69.** $f(x) = x^3 + 8$

**70.** $f(x) = x^3 - 8$

**71.** $f(x) = \frac{1}{3}x^3 - \frac{1}{2}x^2 - \frac{1}{6}x + \frac{1}{6}$

**72.** $f(x) = \frac{2}{3}x^3 - \frac{1}{2}x^2 + \frac{2}{3}x - \frac{1}{2}$

*Find only the rational zeros of the function.*

**73.** $f(x) = x^4 + 2x^3 - 5x^2 - 4x + 6$

**74.** $f(x) = x^4 - 3x^3 - 9x^2 - 3x - 10$

**75.** $f(x) = x^3 - x^2 - 4x + 3$

**76.** $f(x) = 2x^3 + 3x^2 + 2x + 3$

**77.** $f(x) = x^4 + 2x^3 + 2x^2 - 4x - 8$

**78.** $f(x) = x^4 + 6x^3 + 17x^2 + 36x + 66$

**79.** $f(x) = x^5 - 5x^4 + 5x^3 + 15x^2 - 36x + 20$

**80.** $f(x) = x^5 - 3x^4 - 3x^3 + 9x^2 - 4x + 12$

*What does Descartes' rule of signs tell you about the number of positive real zeros and the number of negative real zeros of the function?*

**81.** $f(x) = 3x^5 - 2x^2 + x - 1$

**82.** $g(x) = 5x^6 - 3x^3 + x^2 - x$

**83.** $h(x) = 6x^7 + 2x^2 + 5x + 4$

**84.** $P(x) = -3x^5 - 7x^3 - 4x - 5$

**85.** $F(p) = 3p^{18} + 2p^4 - 5p^2 + p + 3$

**86.** $H(t) = 5t^{12} - 7t^4 + 3t^2 + t + 1$

**87.** $C(x) = 7x^6 + 3x^4 - x - 10$

**88.** $g(z) = -z^{10} + 8z^7 + z^3 + 6z - 1$

**89.** $h(t) = -4t^5 - t^3 + 2t^2 + 1$

**90.** $P(x) = x^6 + 2x^4 - 9x^3 - 4$

**91.** $f(y) = y^4 + 13y^3 - y + 5$

**92.** $Q(x) = x^4 - 2x^2 + 12x - 8$

**93.** $r(x) = x^4 - 6x^2 + 20x - 24$

**94.** $f(x) = x^5 - 2x^3 - 8x$

**95.** $R(x) = 3x^5 - 5x^3 - 4x$

**96.** $f(x) = x^4 - 9x^2 - 6x + 4$

*Sketch the graph of the polynomial function. Follow the procedure outlined on p. 317. Use the rational zeros theorem when finding the zeros.*

**97.** $f(x) = 4x^3 + x^2 - 8x - 2$

**98.** $f(x) = 3x^3 - 4x^2 - 5x + 2$

**99.** $f(x) = 2x^4 - 3x^3 - 2x^2 + 3x$

**100.** $f(x) = 4x^4 - 37x^2 + 9$

## Skill Maintenance

*For Exercises 101 and 102, complete the square to:*

**a)** *find the vertex;*

**b)** *find the axis of symmetry; and*

**c)** *determine whether there is a maximum or minimum function value and find that value.*

**101.** $f(x) = x^2 - 8x + 10$

**102.** $f(x) = 3x^2 - 6x - 1$

*Find the zeros of the function.*

**103.** $f(x) = -\frac{4}{5}x + 8$

**104.** $g(x) = x^2 - 8x - 33$

*Determine the leading term, the leading coefficient, and the degree of the polynomial. Then describe the end behavior of the function's graph and classify the polynomial function as constant, linear, quadratic, cubic, or quartic.*

**105.** $g(x) = -x^3 - 2x^2$

**106.** $f(x) = -x^2 - 3x + 6$

**107.** $f(x) = -\frac{4}{9}$

**108.** $h(x) = x - 2$

**109.** $g(x) = x^4 - 2x^3 + x^2 - x + 2$

**110.** $h(x) = x^3 + \frac{1}{2}x^2 - 4x - 3$

## Synthesis

**111.** Consider $f(x) = 2x^3 - 5x^2 - 4x + 3$. Find the solutions of each equation.

**a)** $f(x) = 0$      **b)** $f(x - 1) = 0$

**c)** $f(x + 2) = 0$      **d)** $f(2x) = 0$

**112.** Use the rational zeros theorem and the equation $x^4 - 12 = 0$ to show that $\sqrt[4]{12}$ is irrational.

*Find the rational zeros of the function.*

**113.** $P(x) = 2x^5 - 33x^4 - 84x^3 + 2203x^2 - 3348x - 10{,}080$

**114.** $P(x) = x^6 - 6x^5 - 72x^4 - 81x^2 + 486x + 5832$

## 4.5 Rational Functions

- For a rational function, find the domain and graph the function, identifying all of the asymptotes.

- Solve applied problems involving rational functions.

Now we turn our attention to functions that represent the quotient of two polynomials. Whereas the sum, difference, or product of two polynomials is a polynomial, in general the quotient of two polynomials is *not* itself a polynomial.

A *rational number* can be expressed as the quotient of two integers, $p/q$, where $q \neq 0$. A *rational function* is formed by the quotient of two polynomials, $p(x)/q(x)$, where $q(x) \neq 0$. Here are some examples of rational functions and their graphs.

$$f(x) = \frac{1}{x}$$

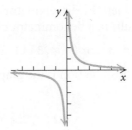

$$f(x) = \frac{1}{x^2}$$

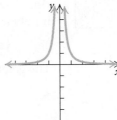

$$f(x) = \frac{x - 3}{x^2 + x - 2}$$

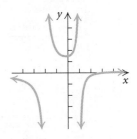

$$f(x) = \frac{2x + 5}{2x - 6}$$

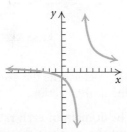

$$f(x) = \frac{x^2 + 2x - 3}{x^2 - x - 2}$$

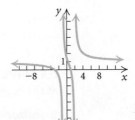

$$f(x) = \frac{-x^2}{x + 1}$$

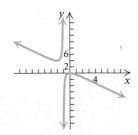

> ### RATIONAL FUNCTION
>
> A **rational function** is a function $f$ that is a quotient of two polynomials. That is,
>
> $$f(x) = \frac{p(x)}{q(x)},$$
>
> where $p(x)$ and $q(x)$ are polynomials and where $q(x)$ is not the zero polynomial. The domain of $f$ consists of all inputs $x$ for which $q(x) \neq 0$.

## The Domain of a Rational Function

GCM    **EXAMPLE 1**    Consider

$$f(x) = \frac{1}{x - 3}.$$

Find the domain and graph $f$.

> DOMAINS OF FUNCTIONS
>
> REVIEW SECTION **1.2.**

*Solution*    When the denominator $x - 3$ is 0, we have $x = 3$, so the only input that results in a denominator of 0 is 3. Thus the domain is

$$\{x \mid x \neq 3\}, \text{ or } (-\infty, 3) \cup (3, \infty).$$

The graph of this function is the graph of $y = 1/x$ translated right 3 units.

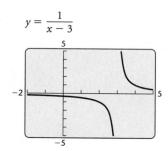

$$y = \frac{1}{x - 3}$$

**EXAMPLE 2**    Determine the domain of each of the functions illustrated at the beginning of this section.

*Solution*    The domain of each rational function will be the set of all real numbers except those values that make the denominator 0. To determine those exceptions, we set the denominator equal to 0 and solve for $x$.

| FUNCTION | DOMAIN |
|---|---|
| $f(x) = \dfrac{1}{x}$ | $\{x \mid x \neq 0\}$, or $(-\infty, 0) \cup (0, \infty)$ |
| $f(x) = \dfrac{1}{x^2}$ | $\{x \mid x \neq 0\}$, or $(-\infty, 0) \cup (0, \infty)$ |
| $f(x) = \dfrac{x-3}{x^2+x-2} = \dfrac{x-3}{(x+2)(x-1)}$ | $\{x \mid x \neq -2 \text{ and } x \neq 1\}$, or $(-\infty, -2) \cup (-2, 1) \cup (1, \infty)$ |
| $f(x) = \dfrac{2x+5}{2x-6} = \dfrac{2x+5}{2(x-3)}$ | $\{x \mid x \neq 3\}$, or $(-\infty, 3) \cup (3, \infty)$ |
| $f(x) = \dfrac{x^2+2x-3}{x^2-x-2} = \dfrac{x^2+2x-3}{(x+1)(x-2)}$ | $\{x \mid x \neq -1 \text{ and } x \neq 2\}$, or $(-\infty, -1) \cup (-1, 2) \cup (2, \infty)$ |
| $f(x) = \dfrac{-x^2}{x+1}$ | $\{x \mid x \neq -1\}$, or $(-\infty, -1) \cup (-1, \infty)$ |

As a partial check of the domains, we can observe the discontinuities (breaks) in the graphs of these functions. (See p. 345.)

## Asymptotes

### Vertical Asymptotes

Look at the graph of $f(x) = 1/(x - 3)$, shown at left. (Also see Example 1.) Let's explore what happens as $x$-values get closer and closer to 3 from the left. We then explore what happens as $x$-values get closer and closer to 3 from the right.

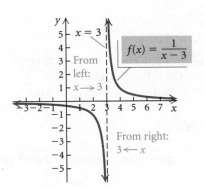

Vertical asymptote: $x = 3$

From the left:

| $x$ | 2 | $2\frac{1}{2}$ | $2\frac{99}{100}$ | $2\frac{9999}{10,000}$ | $2\frac{999,999}{1,000,000}$ | $\longrightarrow 3$ |
|---|---|---|---|---|---|---|
| $f(x)$ | $-1$ | $-2$ | $-100$ | $-10,000$ | $-1,000,000$ | $\longrightarrow -\infty$ |

From the right:

| $x$ | 4 | $3\frac{1}{2}$ | $3\frac{1}{100}$ | $3\frac{1}{10,000}$ | $3\frac{1}{1,000,000}$ | $\longrightarrow 3$ |
|---|---|---|---|---|---|---|
| $f(x)$ | 1 | 2 | 100 | 10,000 | 1,000,000 | $\longrightarrow \infty$ |

We see that as $x$-values get closer and closer to 3 from the left, the function values ($y$-values) decrease without bound (that is, they approach negative infinity, $-\infty$). Similarly, as the $x$-values approach 3 from the right, the function values increase without bound (that is, they approach positive infinity, $\infty$). We write this as

$$f(x) \to -\infty \text{ as } x \to 3^- \quad \text{and} \quad f(x) \to \infty \text{ as } x \to 3^+.$$

We read "$f(x) \to -\infty$ as $x \to 3^-$" as "$f(x)$ decreases without bound as $x$ approaches 3 from the left." We read "$f(x) \to \infty$ as $x \to 3^+$" as "$f(x)$ increases without bound as $x$ approaches 3 from the right." The notation

$x \longrightarrow 3$ means that $x$ gets as close to 3 as possible without being equal to 3. The vertical line $x = 3$ is said to be a *vertical asymptote* for this curve.

In general, the line $x = a$ is a **vertical asymptote** for the graph of $f$ if any of the following is true:

$$f(x) \rightarrow \infty \text{ as } x \rightarrow a^-, \quad \text{or} \quad f(x) \rightarrow -\infty \text{ as } x \rightarrow a^-, \quad \text{or}$$
$$f(x) \rightarrow \infty \text{ as } x \rightarrow a^+, \quad \text{or} \quad f(x) \rightarrow -\infty \text{ as } x \rightarrow a^+.$$

The following figures show the four ways in which a vertical asymptote can occur.

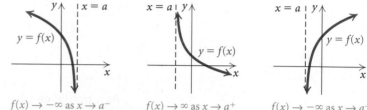

$f(x) \rightarrow \infty \text{ as } x \rightarrow a^-$    $f(x) \rightarrow -\infty \text{ as } x \rightarrow a^-$    $f(x) \rightarrow \infty \text{ as } x \rightarrow a^+$    $f(x) \rightarrow -\infty \text{ as } x \rightarrow a^+$

The vertical asymptotes of a rational function $f(x) = p(x)/q(x)$ are found by determining the zeros of $q(x)$ that are not also zeros of $p(x)$. If $p(x)$ and $q(x)$ are polynomials with no common factors other than constants, we need determine only the zeros of the denominator $q(x)$.

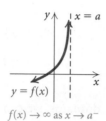

## DETERMINING VERTICAL ASYMPTOTES

For a rational function $f(x) = p(x)/q(x)$, where $p(x)$ and $q(x)$ are polynomials with no common factors other than constants, if $a$ is a zero of the denominator, then the line $x = a$ is a vertical asymptote for the graph of the function.

**EXAMPLE 3**    Determine the vertical asymptotes for the graph of each of the following functions.

**a)** $f(x) = \dfrac{2x - 11}{x^2 + 2x - 8}$    **b)** $h(x) = \dfrac{x^2 - 4x}{x^3 - x}$

**c)** $g(x) = \dfrac{x - 2}{x^3 - 5x}$

***Solution***

**a)**  First, we factor the denominator:

$$f(x) = \frac{2x - 11}{x^2 + 2x - 8} = \frac{2x - 11}{(x + 4)(x - 2)}.$$

The numerator and the denominator have no common factors. The zeros of the denominator are $-4$ and $2$. Thus the vertical asymptotes for the graph of $f(x)$ are the lines $x = -4$ and $x = 2$. (See Fig. 1.)

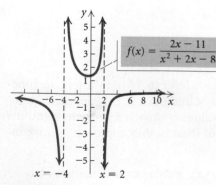

FIGURE 1

**b)** We factor the numerator and the denominator:

$$h(x) = \frac{x^2 - 4x}{x^3 - x} = \frac{x(x - 4)}{x(x^2 - 1)} = \frac{x(x - 4)}{x(x + 1)(x - 1)}.$$

The domain of the function is $\{x \,|\, x \neq -1 \text{ and } x \neq 0 \text{ and } x \neq 1\}$, or $(-\infty, -1) \cup (-1, 0) \cup (0, 1) \cup (1, \infty)$. Note that the numerator and the denominator share a common factor, $x$. The vertical asymptotes of $h(x)$ are found by determining the zeros of the denominator, $x(x + 1)(x - 1)$, that are *not* also zeros of the numerator, $x(x - 4)$. The zeros of $x(x + 1)(x - 1)$ are $0$, $-1$, and $1$. The zeros of $x(x - 4)$ are $0$ and $4$. Thus, although the denominator has three zeros, the graph of $h(x)$ has only two vertical asymptotes, $x = -1$ and $x = 1$. (See Fig. 2.)

The rational expression $[x(x - 4)] / [x(x + 1)(x - 1)]$ can be simplified. Thus,

$$h(x) = \frac{x(x - 4)}{x(x + 1)(x - 1)} = \frac{x - 4}{(x + 1)(x - 1)},$$

where $x \neq 0$, $x \neq -1$, and $x \neq 1$. The graph of $h(x)$ is the graph of

$$h(x) = \frac{x - 4}{(x + 1)(x - 1)}$$

with the point where $x = 0$ missing. To determine the $y$-coordinate of the "hole," we substitute $0$ for $x$:

$$h(0) = \frac{0 - 4}{(0 + 1)(0 - 1)} = \frac{-4}{1 \cdot (-1)} = 4.$$

Thus the "hole" is located at $(0, 4)$.

**c)** We factor the denominator:

$$g(x) = \frac{x - 2}{x^3 - 5x} = \frac{x - 2}{x(x^2 - 5)}.$$

The numerator and the denominator have no common factors. We find the zeros of the denominator, $x(x^2 - 5)$. Solving $x(x^2 - 5) = 0$, we get

$$x = 0 \quad or \quad x^2 - 5 = 0$$
$$x = 0 \quad or \quad x^2 = 5$$
$$x = 0 \quad or \quad x = \pm\sqrt{5}.$$

The zeros of the denominator are $0$, $\sqrt{5}$, and $-\sqrt{5}$. Thus the vertical asymptotes are the lines $x = 0$, $x = \sqrt{5}$, and $x = -\sqrt{5}$. (See Fig. 3.)

**Now Try Exercises 15 and 19.**

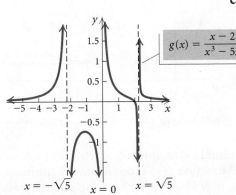

$x = -\sqrt{5}$   $x = 0$   $x = \sqrt{5}$

FIGURE 3

## Horizontal Asymptotes

Looking again at the graph of $f(x) = 1/(x - 3)$ (also see Example 1), let's explore what happens to $f(x) = 1/(x - 3)$ as $x$ increases without bound (approaches positive infinity, $\infty$) and as $x$ decreases without bound (approaches negative infinity, $-\infty$).

---

(0, 4)

$h(x) = \dfrac{x^2 - 4x}{x^3 - x}$

FIGURE 2

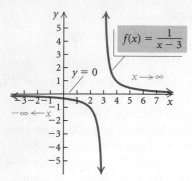

**Horizontal asymptote: $y = 0$**

*x* increases without bound:

| $x$ | 100 | 5000 | 1,000,000 | $\longrightarrow \infty$ |
|------|------|------|-----------|------|
| $f(x)$ | $\approx 0.0103$ | $\approx 0.0002$ | $\approx 0.000001$ | $\longrightarrow 0$ |

*x* decreases without bound:

| $x$ | $-300$ | $-8000$ | $-1,000,000$ | $\longrightarrow -\infty$ |
|------|--------|---------|--------------|------|
| $f(x)$ | $\approx -0.0033$ | $\approx -0.0001$ | $\approx -0.000001$ | $\longrightarrow 0$ |

We see that

$$\frac{1}{x-3} \to 0 \text{ as } x \to \infty \quad \text{and} \quad \frac{1}{x-3} \to 0 \text{ as } x \to -\infty.$$

Since $y = 0$ is the equation of the *x*-axis, we say that the curve approaches the *x*-axis asymptotically and that the *x*-axis is a *horizontal asymptote* for the curve.

In general, the line $y = b$ is a **horizontal asymptote** for the graph of *f* if either or both of the following are true:

$$f(x) \to b \text{ as } x \to \infty \quad \text{or} \quad f(x) \to b \text{ as } x \to -\infty.$$

The following figures illustrate four ways in which horizontal asymptotes can occur. In each case, the curve gets close to the line $y = b$ either as $x \to \infty$ or as $x \to -\infty$. Keep in mind that the symbols $\infty$ and $-\infty$ convey the idea of increasing without bound and decreasing without bound, respectively.

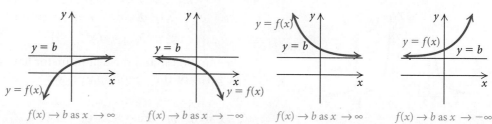

$f(x) \to b \text{ as } x \to \infty$  $f(x) \to b \text{ as } x \to -\infty$  $f(x) \to b \text{ as } x \to \infty$  $f(x) \to b \text{ as } x \to -\infty$

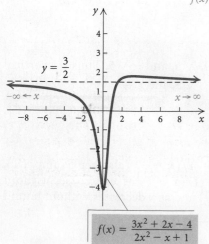

$$f(x) = \frac{3x^2 + 2x - 4}{2x^2 - x + 1}$$

How can we determine a horizontal asymptote? As *x* gets very large or very small, the value of a polynomial function $p(x)$ is dominated by the function's leading term. Because of this, if $p(x)$ and $q(x)$ have the *same* degree, the value of $p(x)/q(x)$ as $x \to \infty$ or as $x \to -\infty$ is dominated by the ratio of the numerator's leading coefficient to the denominator's leading coefficient.

For $f(x) = (3x^2 + 2x - 4)/(2x^2 - x + 1)$, we see that the numerator, $3x^2 + 2x - 4$, is dominated by $3x^2$ and the denominator, $2x^2 - x + 1$, is dominated by $2x^2$, so $f(x)$ approaches $3x^2/2x^2$, or $3/2$ as *x* gets very large or very small:

$$\frac{3x^2 + 2x - 4}{2x^2 - x + 1} \to \frac{3}{2}, \text{ or } 1.5, \text{ as } x \to \infty, \quad \text{and}$$

$$\frac{3x^2 + 2x - 4}{2x^2 - x + 1} \to \frac{3}{2}, \text{ or } 1.5, \text{ as } x \to -\infty.$$

We say that the curve approaches the horizontal line $y = \frac{3}{2}$ asymptotically and that $y = \frac{3}{2}$ is a *horizontal asymptote* for the curve.

It follows that when the numerator and the denominator of a rational function have the same degree, the line $y = a/b$ is the horizontal asymptote, where $a$ and $b$ are the leading coefficients of the numerator and the denominator, respectively.

**EXAMPLE 4**   Find the horizontal asymptote: $f(x) = \dfrac{-7x^4 - 10x^2 + 1}{11x^4 + x - 2}$.

**Solution**   The numerator and the denominator have the same degree. The ratio of the leading coefficients is $-\frac{7}{11}$, so the line $y = -\frac{7}{11}$, or $-0.\overline{63}$, is the horizontal asymptote.     [ **Now Try Exercise 21.** ]

| X | Y₁ |
|---|---|
| 100000 | −.6364 |
| −80000 | −.6364 |

X =

As a partial check of Example 4, we could use a graphing calculator to evaluate the function for a very large value of $x$ and a very small value of $x$. (See the window at left.) It is useful in calculus to multiply by 1. Here we use $(1/x^4)/(1/x^4)$:

$$f(x) = \frac{-7x^4 - 10x^2 + 1}{11x^4 + x - 2} \cdot \frac{\dfrac{1}{x^4}}{\dfrac{1}{x^4}} = \frac{\dfrac{-7x^4}{x^4} - \dfrac{10x^2}{x^4} + \dfrac{1}{x^4}}{\dfrac{11x^4}{x^4} + \dfrac{x}{x^4} - \dfrac{2}{x^4}}$$

$$= \frac{-7 - \dfrac{10}{x^2} + \dfrac{1}{x^4}}{11 + \dfrac{1}{x^3} - \dfrac{2}{x^4}}.$$

As $|x|$ becomes very large, each expression whose denominator is a power of $x$ tends toward 0. Specifically, as $x \to \infty$ or as $x \to -\infty$, we have

$$f(x) \to \frac{-7 - 0 + 0}{11 + 0 - 0}, \quad \text{or} \quad f(x) \to -\frac{7}{11}.$$

The horizontal asymptote is $y = -\frac{7}{11}$, or $-0.\overline{63}$.

We now investigate the occurrence of a horizontal asymptote when the degree of the numerator is less than the degree of the denominator.

**EXAMPLE 5**   Find the horizontal asymptote: $f(x) = \dfrac{2x + 3}{x^3 - 2x^2 + 4}$.

**Solution**   We let $p(x) = 2x + 3$, $q(x) = x^3 - 2x^2 + 4$, and $f(x) = p(x)/q(x)$. Note that as $x \to \infty$, the value of $q(x)$ grows much faster than the value of $p(x)$. Because of this, the ratio $p(x)/q(x)$ shrinks toward 0. As $x \to -\infty$, the ratio $p(x)/q(x)$ behaves in a similar manner. The horizontal asymptote is $y = 0$, the $x$-axis. This is the case for all rational functions for which the degree of the numerator is less than the degree of the denominator. Note in Example 1 that $y = 0$, the $x$-axis, is the horizontal asymptote of $f(x) = 1/(x - 3)$.     [ **Now Try Exercise 23.** ]

The following statements describe the two ways in which a horizontal asymptote occurs.

---

### DETERMINING A HORIZONTAL ASYMPTOTE

- When the numerator and the denominator of a rational function have the same degree, the line $y = a/b$ is the horizontal asymptote, where $a$ and $b$ are the leading coefficients of the numerator and the denominator, respectively.
- When the degree of the numerator of a rational function is less than the degree of the denominator, the $x$-axis, or $y = 0$, is the horizontal asymptote.
- When the degree of the numerator of a rational function is greater than the degree of the denominator, there is no horizontal asymptote.

---

**EXAMPLE 6** Graph

$$g(x) = \frac{2x^2 + 1}{x^2}.$$

Include and label all asymptotes.

***Solution*** Since 0 is the zero of the denominator and not of the numerator, the $y$-axis, $x = 0$, is the vertical asymptote. Note also that the degree of the numerator is the same as the degree of the denominator. Thus, $y = 2/1$, or 2, is the horizontal asymptote.

To draw the graph, we first draw the asymptotes with dashed lines. Then we compute and plot some ordered pairs and draw the two branches of the curve. We can check the graph with a graphing calculator.

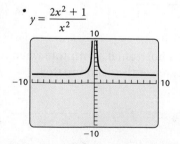

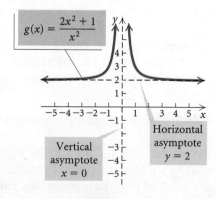

| $x$ | $g(x)$ |
|------|--------|
| $-2$ | 2.25 |
| $-1.5$ | $2.\overline{4}$ |
| $-1$ | 3 |
| $-0.5$ | 6 |
| 0.5 | 6 |
| 1 | 3 |
| 1.5 | $2.\overline{4}$ |
| 2 | 2.25 |

**Now Try Exercise 41.**

## Oblique Asymptotes

Sometimes a line that is neither horizontal nor vertical is an asymptote. Such a line is called an **oblique asymptote**, or a **slant asymptote**.

**EXAMPLE 7**   Find all the asymptotes of

$$f(x) = \frac{2x^2 - 3x - 1}{x - 2}.$$

**Solution**   The line $x = 2$ is the vertical asymptote because 2 is the zero of the denominator and is not a zero of the numerator. There is no horizontal asymptote because the degree of the numerator is greater than the degree of the denominator. When the degree of the numerator is 1 greater than the degree of the denominator, we divide to find an equivalent expression:

$$\frac{2x^2 - 3x - 1}{x - 2} = (2x + 1) + \frac{1}{x - 2}. \qquad \begin{array}{r} 2x + 1 \\ x - 2\overline{)2x^2 - 3x - 1} \\ \underline{2x^2 - 4x} \\ x - 1 \\ \underline{x - 2} \\ 1 \end{array}$$

We see that when $x \to \infty$ or $x \to -\infty$, $1/(x - 2) \to 0$ and the value of $f(x) \to 2x + 1$. This means that as $|x|$ becomes very large, the graph of $f(x)$ gets very close to the graph of $y = 2x + 1$. Thus the line $y = 2x + 1$ is the oblique asymptote.

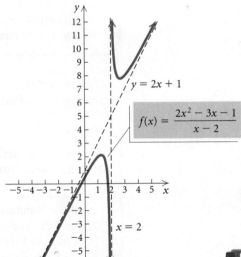

$$y = 2x + 1$$

$$f(x) = \frac{2x^2 - 3x - 1}{x - 2}$$

$$x = 2$$

**Now Try Exercise 59.**

## Occurrence of Lines as Asymptotes of Rational Functions

For a rational function $f(x) = p(x)/q(x)$, where $p(x)$ and $q(x)$ have no common factors other than constants:

**Vertical asymptotes** occur at any $x$-values that make the denominator 0.

**The $x$-axis is the horizontal asymptote** when the degree of the numerator is less than the degree of the denominator.

**A horizontal asymptote other than the $x$-axis** occurs when the numerator and the denominator have the same degree.

**An oblique asymptote** occurs when the degree of the numerator is 1 greater than the degree of the denominator.

There can be only one horizontal asymptote or one oblique asymptote and never both.

An asymptote is *not* part of the graph of the function.

The following statements are also true.

## Crossing an Asymptote

- The graph of a rational function *never crosses* a vertical asymptote.
- The graph of a rational function *might cross* a horizontal asymptote but does not necessarily do so.

Shown below is an outline of a procedure that we can follow to create accurate graphs of rational functions.

To graph a rational function $f(x) = p(x)/q(x)$, where $p(x)$ and $q(x)$ have no common factor other than constants:

1. Find any real zeros of the denominator. Determine the domain of the function and sketch any vertical asymptotes.
2. Find the horizontal asymptote or the oblique asymptote, if there is one, and sketch it.
3. Find any zeros of the function. The zeros are found by determining the zeros of the numerator. These are the first coordinates of the $x$-intercepts of the graph.
4. Find $f(0)$. This gives the $y$-intercept, $(0, f(0))$, of the function.
5. Find other function values to determine the general shape. Then draw the graph.

**EXAMPLE 8** Graph: $f(x) = \dfrac{2x + 3}{3x^2 + 7x - 6}$.

*Solution*

1. We find the zeros of the denominator by solving $3x^2 + 7x - 6 = 0$. Since

$$3x^2 + 7x - 6 = (3x - 2)(x + 3),$$

the zeros are $\frac{2}{3}$ and $-3$. Thus the domain excludes $\frac{2}{3}$ and $-3$ and is

$$(-\infty, -3) \cup \left(-3, \tfrac{2}{3}\right) \cup \left(\tfrac{2}{3}, \infty\right).$$

Since neither zero of the denominator is a zero of the numerator, the graph has vertical asymptotes $x = -3$ and $x = \frac{2}{3}$. We sketch these as dashed lines.

2. Because the degree of the numerator is less than the degree of the denominator, the $x$-axis, $y = 0$, is the horizontal asymptote.

3. To find the zeros of the numerator, we solve $2x + 3 = 0$ and get $x = -\frac{3}{2}$. Thus, $-\frac{3}{2}$ is the zero of the function, and the pair $\left(-\frac{3}{2}, 0\right)$ is the $x$-intercept.

4. We find $f(0)$:

$$f(0) = \dfrac{2 \cdot 0 + 3}{3 \cdot 0^2 + 7 \cdot 0 - 6}$$

$$= \dfrac{3}{-6} = -\dfrac{1}{2}.$$

The point $\left(0, -\frac{1}{2}\right)$ is the $y$-intercept.

5. We find other function values to determine the general shape. We choose values in each interval of the domain, as shown in the table below, and then draw the graph. Note that the graph of this function crosses its horizontal asymptote at $x = -\frac{3}{2}$.

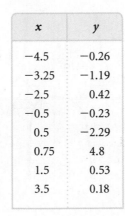

We can check the graph of the function in Example 8 with a graphing calculator app. This app connects points on either side of each zero, $-3$ and $\frac{2}{3}$, resulting in two lines that "appear" to be the vertical lines $x = -3$ and $x = \frac{2}{3}$. These lines are not part of the graph.

| $x$ | $y$ |
|------|-------|
| $-4.5$ | $-0.26$ |
| $-3.25$ | $-1.19$ |
| $-2.5$ | $0.42$ |
| $-0.5$ | $-0.23$ |
| $0.5$ | $-2.29$ |
| $0.75$ | $4.8$ |
| $1.5$ | $0.53$ |
| $3.5$ | $0.18$ |

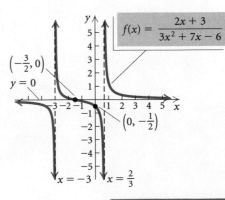

**Now Try Exercise 65.**

**EXAMPLE 9** Graph: $g(x) = \dfrac{x^2 - 1}{x^2 + x - 6}$.

*Solution*

1. We find the zeros of the denominator by solving $x^2 + x - 6 = 0$. Since

   $$x^2 + x - 6 = (x + 3)(x - 2),$$

   the zeros are $-3$ and $2$. Thus the domain excludes the $x$-values $-3$ and $2$ and is

   $$(-\infty, -3) \cup (-3, 2) \cup (2, \infty).$$

   Since neither zero of the denominator is a zero of the numerator, the graph has vertical asymptotes $x = -3$ and $x = 2$. We sketch these as dashed lines.

2. The numerator and the denominator have the same degree, so the horizontal asymptote is determined by the ratio of the leading coefficients: $1/1$, or $1$. Thus, $y = 1$ is the horizontal asymptote. We sketch it with a dashed line.

3. To find the zeros of the numerator, we solve $x^2 - 1 = 0$. The solutions are $-1$ and $1$. Thus, $-1$ and $1$ are the zeros of the function and the pairs $(-1, 0)$ and $(1, 0)$ are the $x$-intercepts.

4. We find $g(0)$:

   $$g(0) = \frac{0^2 - 1}{0^2 + 0 - 6} = \frac{-1}{-6} = \frac{1}{6}.$$

   Thus, $\left(0, \frac{1}{6}\right)$ is the $y$-intercept.

5. We find other function values to determine the general shape and then draw the graph.

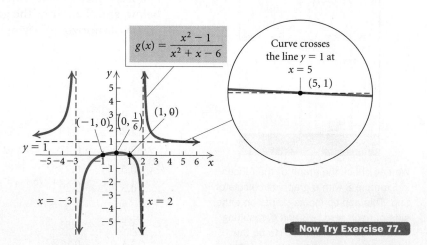

Now Try Exercise 77.

The magnified portion of the graph in Example 9 above shows another situation in which a graph can cross its horizontal asymptote. The point where $g(x)$ crosses $y = 1$ can be found by setting $g(x) = 1$ and solving for $x$:

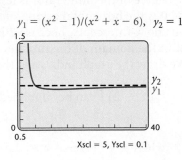

$y_1 = (x^2 - 1)/(x^2 + x - 6),\ \ y_2 = 1$

Xscl = 5, Yscl = 0.1

$$\frac{x^2 - 1}{x^2 + x - 6} = 1$$

$$x^2 - 1 = x^2 + x - 6$$

$$-1 = x - 6 \qquad \text{Subtracting } x^2$$

$$5 = x. \qquad \text{Adding } 6$$

The point of intersection is $(5, 1)$. Let's observe the behavior of the curve after it crosses the horizontal asymptote at $x = 5$. (See the graph at left.) It continues to decrease for a short interval and then begins to increase, getting closer and closer to $y = 1$ as $x \to \infty$.

Graphs of rational functions can also cross an oblique asymptote. The graph of

$$f(x) = \frac{2x^3}{x^2 + 1}$$

shown below crosses its oblique asymptote $y = 2x$. **Remember: Graphs can cross horizontal asymptotes or oblique asymptotes, but they cannot cross vertical asymptotes.**

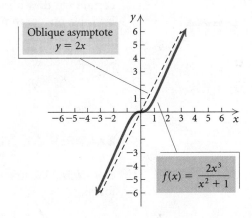

Oblique asymptote
$y = 2x$

$f(x) = \dfrac{2x^3}{x^2 + 1}$

Let's now graph a rational function $f(x) = p(x)/q(x)$, where $p(x)$ and $q(x)$ have a common factor, $x - c$. The graph of such a function has a "hole" in it. We first saw this situation in Example 3(b), where the common factor was $x$.

**GCM**

**EXAMPLE 10**  Graph: $g(x) = \dfrac{x - 2}{x^2 - x - 2}$.

*Solution*  We first express the denominator in factored form:

$$g(x) = \frac{x - 2}{x^2 - x - 2} = \frac{x - 2}{(x + 1)(x - 2)}.$$

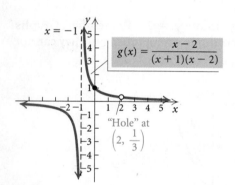

| X | Y1 | |
|---|---|---|
| −3 | −.5 | |
| −2 | −1 | |
| −1 | ERROR | |
| 0 | 1 | |
| 1 | .5 | |
| 2 | ERROR | |
| 3 | .25 | |
| X = | | |

The domain of the function is $\{x \mid x \neq -1 \ and \ x \neq 2\}$, or $(-\infty, -1) \cup (-1, 2) \cup (2, \infty)$. Note that both the numerator and the denominator have the common factor $x - 2$. The zeros of the denominator are $-1$ and $2$, and the zero of the numerator is $2$. Since $-1$ is the only zero of the denominator that is *not* a zero of the numerator, the graph of the function has $x = -1$ as its only vertical asymptote. The degree of the numerator is less than the degree of the denominator, so $y = 0$ is the horizontal asymptote. There are no zeros of the function and thus no $x$-intercepts, because 2 is the only zero of the numerator and 2 is not in the domain of the function. Since $g(0) = 1$, $(0, 1)$ is the $y$-intercept. We draw the graph indicating the "hole" when $x = 2$ with an open circle.

The rational expression $(x - 2)/[(x + 1)(x - 2)]$ can be simplified. Thus,

$$g(x) = \frac{x - 2}{(x + 1)(x - 2)} = \frac{1}{x + 1}, \quad \text{where } x \neq -1 \text{ and } x \neq 2.$$

The graph of $g(x)$ is the graph of $y = 1/(x + 1)$ with the point where $x = 2$ missing. To determine the coordinates of the "hole," we substitute 2 for $x$ in $g(x) = 1/(x + 1)$:

$$g(2) = \frac{1}{2 + 1} = \frac{1}{3}.$$

Thus the hole is located at $\left(2, \frac{1}{3}\right)$. With certain window dimensions, the hole is visible on a graphing calculator, as shown at right.

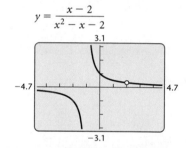

$$y = \frac{x - 2}{x^2 - x - 2}$$

**Now Try Exercise 49.**

**EXAMPLE 11** Graph: $f(x) = \dfrac{-2x^2 - x + 15}{x^2 - x - 12}$.

**Solution** We first express the numerator and the denominator in factored form:

$$f(x) = \frac{-2x^2 - x + 15}{x^2 - x - 12} = \frac{-(2x^2 + x - 15)}{x^2 - x - 12} = \frac{-(2x - 5)(x + 3)}{(x - 4)(x + 3)}.$$

The domain of the function is $\{x \mid x \neq -3 \ and \ x \neq 4\}$, or $(-\infty, -3) \cup (-3, 4) \cup (4, \infty)$. The numerator and the denominator have the common factor $x + 3$. The zeros of the denominator are $-3$ and $4$, and the zeros of the numerator are $-3$ and $\frac{5}{2}$. Since 4 is the only zero of the denominator that is *not* a zero of the numerator, the graph of the function has $x = 4$ as its *only* vertical asymptote.

The degrees of the numerator and the denominator are the same, so the line $y = \frac{-2}{1} = -2$ is the horizontal asymptote. The zeros of the numerator are $\frac{5}{2}$ and $-3$. Because $-3$ is not in the domain of the function, the only $x$-intercept is $\left(\frac{5}{2}, 0\right)$. Since $f(0) = \frac{15}{-12} = -\frac{5}{4}$, then $\left(0, -\frac{5}{4}\right)$ is the $y$-intercept.

The rational function

$$\frac{-(2x - 5)(x + 3)}{(x - 4)(x + 3)}$$

can be simplified. Thus,

$$f(x) = \frac{-(2x - 5)(x + 3)}{(x - 4)(x + 3)} = \frac{-(2x - 5)}{x - 4}, \quad \text{where } x \neq -3 \text{ and } x \neq 4.$$

The graph of $f(x)$ is the graph of $y = -(2x - 5)/(x - 4)$ with the point where $x = -3$ missing. To determine the coordinates of the hole, we substitute $-3$ for $x$ in $f(x) = -(2x - 5)/(x - 4)$:

$$f(-3) = \frac{-[2(-3) - 5]}{-3 - 4}$$

$$= \frac{-[-11]}{-7} = \frac{11}{-7} = -\frac{11}{7}.$$

Thus the hole is located at $\left(-3, -\frac{11}{7}\right)$. We draw the graph indicating the hole when $x = -3$ with an open circle.

From this table of ordered pairs for the function in Example 11, we see that there is a vertical asymptote $x = 4$ and a hole in the graph when $x = -3$.

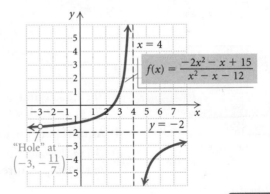

**Now Try Exercise 67.**

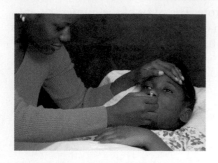

## ■ Applications

**EXAMPLE 12** *Temperature During an Illness.* A person's temperature $T$, in degrees Fahrenheit, during an illness is given by the function

$$T(t) = \frac{4t}{t^2 + 1} + 98.6,$$

where time $t$ is given in hours since the onset of the illness.

**a)** Graph the function on the interval $[0, 48]$.

**b)** Find the temperature at $t = 0, 1, 2, 5, 12,$ and $24$..

**c)** Find the horizontal asymptote of the graph of $T(t)$. Complete:

$$T(t) \to \boxed{\phantom{xxx}} \text{ as } t \to \infty.$$

**d)** Give the meaning of the answer to part (c) in terms of the application.

**e)** Find the maximum temperature during the illness.

*Solution*

**a)** The graph is shown at left.

**b)** We have

$$T(0) = 98.6, \qquad T(1) = 100.6, \qquad T(2) = 100.2,$$
$$T(5) \approx 99.369, \qquad T(12) \approx 98.931, \quad \text{and} \quad T(24) \approx 98.766.$$

**c)** Since

$$T(t) = \frac{4t}{t^2 + 1} + 98.6,$$
$$= \frac{98.6t^2 + 4t + 98.6}{t^2 + 1},$$

the horizontal asymptote is $y = 98.6/1$, or $98.6$. Then it follows that $T(t) \to 98.6$ as $t \to \infty$.

**d)** As time goes on, the temperature returns to "normal," which is $98.6°$F.

**e)** Using the MAXIMUM feature on a graphing calculator, we find the maximum temperature to be $100.6°$F at $t = 1$ hr.

**Now Try Exercise 83.**

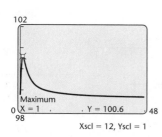

$$T(x) = \frac{4x}{x^2 + 1} + 98.6$$

Xscl = 12, Yscl = 1

Maximum  X = 1  Y = 100.6

Xscl = 12, Yscl = 1

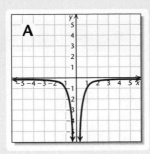

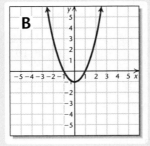

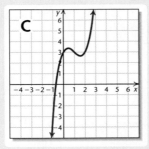

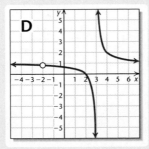

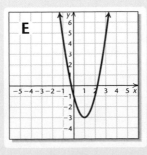

# Visualizing the Graph

Match the function with its graph.

1. $f(x) = -\dfrac{1}{x^2}$

2. $f(x) = x^3 - 3x^2 + 2x + 3$

3. $f(x) = \dfrac{x^2 - 4}{x^2 - x - 6}$

4. $f(x) = -x^2 + 4x - 1$

5. $f(x) = \dfrac{x - 3}{x^2 + x - 6}$

6. $f(x) = \dfrac{3}{4}x + 2$

7. $f(x) = x^2 - 1$

8. $f(x) = x^4 - 2x^2 - 5$

9. $f(x) = \dfrac{8x - 4}{3x + 6}$

10. $f(x) = 2x^2 - 4x - 1$

**Answers on page A-25**

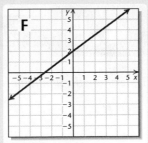

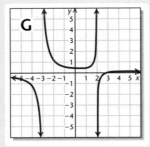

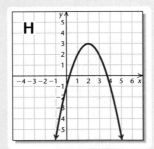

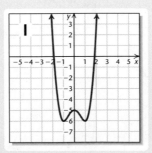

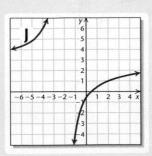

## 4.5 Exercise Set

*Determine the domain of the function.*

**1.** $f(x) = \dfrac{x^2}{2 - x}$

**2.** $f(x) = \dfrac{1}{x^3}$

**3.** $f(x) = \dfrac{x + 1}{x^2 - 6x + 5}$

**4.** $f(x) = \dfrac{(x + 4)^2}{4x - 3}$

**5.** $f(x) = \dfrac{3x - 4}{3x + 15}$

**6.** $f(x) = \dfrac{x^2 + 3x - 10}{x^2 + 2x}$

*In Exercises 7–12, use your knowledge of asymptotes and intercepts to match the equation with one of the graphs (a)–(f), which follow. List all asymptotes. Check your work using a graphing calculator.*

a)
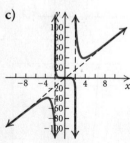

b)
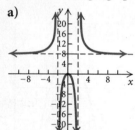

c)
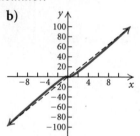

d)
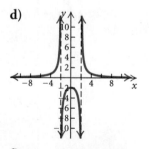

e)
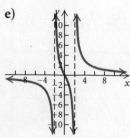

f)
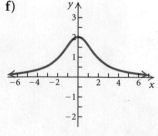

**7.** $f(x) = \dfrac{8}{x^2 - 4}$

**8.** $f(x) = \dfrac{8}{x^2 + 4}$

**9.** $f(x) = \dfrac{8x}{x^2 - 4}$

**10.** $f(x) = \dfrac{8x^2}{x^2 - 4}$

**11.** $f(x) = \dfrac{8x^3}{x^2 - 4}$

**12.** $f(x) = \dfrac{8x^3}{x^2 + 4}$

*Determine the vertical asymptotes of the graph of the function.*

**13.** $g(x) = \dfrac{1}{x^2}$

**14.** $f(x) = \dfrac{4x}{x^2 + 10x}$

**15.** $h(x) = \dfrac{x + 7}{2 - x}$

**16.** $g(x) = \dfrac{x^4 + 2}{x}$

**17.** $f(x) = \dfrac{3 - x}{(x - 4)(x + 6)}$

**18.** $h(x) = \dfrac{x^2 - 4}{x(x + 5)(x - 2)}$

**19.** $g(x) = \dfrac{x^3}{2x^3 - x^2 - 3x}$

**20.** $f(x) = \dfrac{x + 5}{x^2 + 4x - 32}$

*Determine the horizontal asymptote of the graph of the function.*

**21.** $f(x) = \dfrac{3x^2 + 5}{4x^2 - 3}$

**22.** $g(x) = \dfrac{x + 6}{x^3 + 2x^2}$

**23.** $h(x) = \dfrac{x^2 - 4}{2x^4 + 3}$

**24.** $f(x) = \dfrac{x^5}{x^5 + x}$

**25.** $g(x) = \dfrac{x^3 - 2x^2 + x - 1}{x^2 - 16}$

**26.** $h(x) = \dfrac{8x^4 + x - 2}{2x^4 - 10}$

*Determine the oblique asymptote of the graph of the function.*

**27.** $g(x) = \dfrac{x^2 + 4x - 1}{x + 3}$

**28.** $f(x) = \dfrac{x^2 - 6x}{x - 5}$

**29.** $h(x) = \dfrac{x^4 - 2}{x^3 + 1}$

**30.** $g(x) = \dfrac{12x^3 - x}{6x^2 + 4}$

**31.** $f(x) = \dfrac{x^3 - x^2 + x - 4}{x^2 + 2x - 1}$

**32.** $h(x) = \dfrac{5x^3 - x^2 + x - 1}{x^2 - x + 2}$

*Make a hand-drawn graph for each of Exercises 33–78. Be sure to label all the asymptotes. List the domain and the x-intercepts and the y-intercepts. Check your work using a graphing calculator.*

**33.** $f(x) = \dfrac{1}{x}$

**34.** $g(x) = \dfrac{1}{x^2}$

**35.** $h(x) = -\dfrac{4}{x^2}$

**36.** $f(x) = -\dfrac{6}{x}$

**37.** $g(x) = \dfrac{x^2 - 4x + 3}{x + 1}$

**38.** $h(x) = \dfrac{2x^2 - x - 3}{x - 1}$

**39.** $f(x) = \dfrac{-2}{x - 5}$

**40.** $f(x) = \dfrac{1}{x - 5}$

**41.** $f(x) = \dfrac{2x + 1}{x}$

**42.** $f(x) = \dfrac{3x - 1}{x}$

**43.** $f(x) = \dfrac{x + 3}{x^2 - 9}$

**44.** $f(x) = \dfrac{x - 1}{x^2 - 1}$

**45.** $f(x) = \dfrac{x}{x^2 + 3x}$

**46.** $f(x) = \dfrac{3x}{3x - x^2}$

**47.** $f(x) = \dfrac{1}{(x - 2)^2}$

**48.** $f(x) = \dfrac{-2}{(x - 3)^2}$

**49.** $f(x) = \dfrac{x^2 + 2x - 3}{x^2 + 4x + 3}$

**50.** $f(x) = \dfrac{x^2 - x - 2}{x^2 - 5x - 6}$

**51.** $f(x) = \dfrac{1}{x^2 + 3}$

**52.** $f(x) = \dfrac{-1}{x^2 + 2}$

**53.** $f(x) = \dfrac{x^2 - 4}{x - 2}$

**54.** $f(x) = \dfrac{x^2 - 9}{x + 3}$

**55.** $f(x) = \dfrac{x - 1}{x + 2}$

**56.** $f(x) = \dfrac{x - 2}{x + 1}$

**57.** $f(x) = \dfrac{x^2 + 3x}{2x^3 - 5x^2 - 3x}$

**58.** $f(x) = \dfrac{3x}{x^2 + 5x + 4}$

**59.** $f(x) = \dfrac{x^2 - 9}{x + 1}$

**60.** $f(x) = \dfrac{x^3 - 4x}{x^2 - x}$

**61.** $f(x) = \dfrac{x^2 + x - 2}{2x^2 + 1}$

**62.** $f(x) = \dfrac{x^2 - 2x - 3}{3x^2 + 2}$

**63.** $g(x) = \dfrac{3x^2 - x - 2}{x - 1}$

**64.** $f(x) = \dfrac{2x^2 - 5x - 3}{2x + 1}$

**65.** $f(x) = \dfrac{x - 1}{x^2 - 2x - 3}$

**66.** $f(x) = \dfrac{x + 2}{x^2 + 2x - 15}$

**67.** $f(x) = \dfrac{3x^2 + 11x - 4}{x^2 + 2x - 8}$

**68.** $f(x) = \dfrac{2x^2 - 3x - 9}{x^2 - 2x - 3}$

**69.** $f(x) = \dfrac{x - 3}{(x + 1)^3}$

**70.** $f(x) = \dfrac{x + 2}{(x - 1)^3}$

**71.** $f(x) = \dfrac{x^3 + 1}{x}$

**72.** $f(x) = \dfrac{x^3 - 1}{x}$

**73.** $f(x) = \dfrac{x^3 + 2x^2 - 15x}{x^2 - 5x - 14}$

**74.** $f(x) = \dfrac{x^3 + 2x^2 - 3x}{x^2 - 25}$

**75.** $f(x) = \dfrac{5x^4}{x^4 + 1}$

**76.** $f(x) = \dfrac{x + 1}{x^2 + x - 6}$

**77.** $f(x) = \dfrac{x^2}{x^2 - x - 2}$

**78.** $f(x) = \dfrac{x^2 - x - 2}{x + 2}$

*Find a rational function that satisfies the given conditions. Answers may vary, but try to give the simplest answer possible.*

**79.** Vertical asymptotes $x = -4$, $x = 5$

**80.** Vertical asymptotes $x = -4$, $x = 5$; $x$-intercept $(-2, 0)$

**81.** Vertical asymptotes $x = -4$, $x = 5$; horizontal asymptote $y = \frac{3}{2}$; $x$-intercept $(-2, 0)$

**82.** Oblique asymptote $y = x - 1$

**83.** *Medical Dosage.* The function

$$N(t) = \frac{0.8t + 1000}{5t + 4}, \quad t \geq 15,$$

gives the body concentration $N(t)$, in parts per million, of a certain dosage of medication after time $t$, in hours.

**a)** Graph the function on the interval $[15, \infty)$ and complete the following:

$$N(t) \to \boxed{\phantom{x}} \text{ as } t \to \infty.$$

**b)** Explain the meaning of the answer to part (a) in terms of the application.

**84.** *Average Cost.* The average cost per disc, in dollars, for a company to produce $x$ discs on exercising is given by the function

$$A(x) = \frac{2x + 100}{x}, \quad x > 0.$$

**a)** Graph the function on the interval $(0, \infty)$ and complete the following:

$$A(x) \to \boxed{\phantom{x}} \text{ as } x \to \infty.$$

**b)** Explain the meaning of the answer to part (a) in terms of the application.

**85.** *Population Growth.* The population $P$, in thousands, of a senior community is given by

$$P(t) = \frac{500t}{2t^2 + 9},$$

where $t$ is the time, in months.

**a)** Graph the function on the interval $[0, \infty)$.

**b)** Find the population at $t = 0, 1, 3,$ and 8 months.

**c)** Find the horizontal asymptote of the graph and complete the following:

$$P(t) \to \boxed{\phantom{x}} \text{ as } t \to \infty.$$

**d)** Explain the meaning of the answer to part (c) in terms of the application.

**e)** Find the maximum population and the value of $t$ that will yield it.

**86.** *Minimizing Surface Area.* The Hold-It Container Co. is designing an open-top rectangular box, with a square base, that will hold 108 cubic centimeters.

**a)** Express the surface area $S$ as a function of the length $x$ of a side of the base.

**b)** Use a graphing calculator to graph the function on the interval $(0, \infty)$.

**c)** Estimate the minimum surface area and the value of $x$ that will yield it.

## Skill Maintenance

### Vocabulary Reinforcement

*In each of Exercises 87–95, fill in the blank with the correct term. Some of the given choices will not be used. Others will be used more than once.*

| | |
|---|---|
| $x$-intercept | vertical lines |
| $y$-intercept | point–slope |
| odd function | equation |
| even function | slope–intercept |
| domain | equation |
| range | difference |
| slope | quotient |
| distance formula | $f(x) = f(-x)$ |
| midpoint formula | $f(-x) = -f(x)$ |
| horizontal lines | |

**87.** A function is a correspondence between a first set, called the _____, and a second set, called the _____, such that each member of the _____ corresponds to exactly one member of the _____.

**88.** The _____ of a line containing $(x_1, y_1)$ and $(x_2, y_2)$ is given by $(y_2 - y_1)/(x_2 - x_1)$.

**89.** The _____ of the line with slope $m$ and $y$-intercept $(0, b)$ is $y = mx + b$.

**90.** The _____ of the line with slope $m$ passing through $(x_1, y_1)$ is $y - y_1 = m(x - x_1)$.

**91.** A(n) _____ is a point $(a, 0)$.

**92.** For each $x$ in the domain of an odd function $f$, _____.

**93.** _____ are given by equations of the type $x = a$.

**94.** The _____ is $\left( \dfrac{x_1 + x_2}{2}, \dfrac{y_1 + y_2}{2} \right)$.

**95.** A(n) _____ is a point $(0, b)$.

**Synthesis**

**96.** Graph

$$y_1 = \frac{x^3 + 4}{x} \quad \text{and} \quad y_2 = x^2$$

using the same viewing window. Explain how the parabola $y_2 = x^2$ can be thought of as a nonlinear asymptote for $y_1$.

*Find the nonlinear asymptote of the function.*

**97.** $f(x) = \dfrac{x^5 + 2x^3 + 4x^2}{x^2 + 2}$

**98.** $f(x) = \dfrac{x^4 + 3x^2}{x^2 + 1}$

*Graph the function.*

**99.** $f(x) = \dfrac{2x^3 + x^2 - 8x - 4}{x^3 + x^2 - 9x - 9}$

**100.** $f(x) = \dfrac{x^3 + 4x^2 + x - 6}{x^2 - x - 2}$

---

## 4.6 Polynomial Inequalities and Rational Inequalities

- Solve polynomial inequalities.
- Solve rational inequalities.

We will use a combination of algebraic and graphical methods to solve polynomial inequalities and rational inequalities.

### Polynomial Inequalities

Just as a quadratic equation can be written in the form $ax^2 + bx + c = 0$, a **quadratic inequality** can be written in the form $ax^2 + bx + c \ \square \ 0$, where $\square$ is $<, >, \leq$, or $\geq$. Here are some examples of quadratic inequalities:

$$x^2 - 4x - 5 < 0 \quad \text{and} \quad -\tfrac{1}{2}x^2 + 4x - 7 \geq 0.$$

When the inequality symbol in a polynomial inequality is replaced with an equals sign, a **related equation** is formed. Polynomial inequalities can be solved once the related equation has been solved.

**EXAMPLE 1** Solve: $x^2 - 4x - 5 > 0$.

**Solution** We are asked to find all $x$-values for which $x^2 - 4x - 5 > 0$. To locate these values, we graph $f(x) = x^2 - 4x - 5$. Then we note that whenever the graph passes through an $x$-intercept, the function changes sign. Thus to solve $x^2 - 4x - 5 > 0$, we first solve the *related equation* $x^2 - 4x - 5 = 0$ to find all zeros of the function:

$$x^2 - 4x - 5 = 0$$
$$(x + 1)(x - 5) = 0.$$

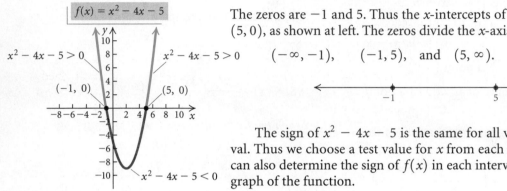

The zeros are $-1$ and $5$. Thus the $x$-intercepts of the graph are $(-1, 0)$ and $(5, 0)$, as shown at left. The zeros divide the $x$-axis into three intervals:

$$(-\infty, -1), \quad (-1, 5), \quad \text{and} \quad (5, \infty).$$

The sign of $x^2 - 4x - 5$ is the same for all values of $x$ in a given interval. Thus we choose a test value for $x$ from each interval and find $f(x)$. We can also determine the sign of $f(x)$ in each interval by simply looking at the graph of the function.

| Interval | $(-\infty, -1)$ | $(-1, 5)$ | $(5, \infty)$ |
|---|---|---|---|
| **Test Value** | $f(-2) = 7$ | $f(0) = -5$ | $f(7) = 16$ |
| **Sign of $f(x)$** | Positive | Negative | Positive |

Since we are solving $x^2 - 4x - 5 > 0$, the solution set consists of only two of the three intervals, those in which the sign of $f(x)$ is positive. Since the inequality sign is $>$, we do not include the endpoints of the intervals in the solution set. The solution set is

$$(-\infty, -1) \cup (5, \infty), \quad \text{or} \quad \{x \,|\, x < -1 \text{ or } x > 5\}.$$

**Now Try Exercise 27.**

**EXAMPLE 2** Solve: $x^2 + 3x - 5 \le x + 3$.

**Solution** By subtracting $x + 3$ on both sides, we form an equivalent inequality:

$$x^2 + 3x - 5 - x - 3 \le 0$$
$$x^2 + 2x - 8 \le 0.$$

We need to find all $x$-values for which $x^2 + 2x - 8 \le 0$. To visualize these values, we first graph $f(x) = x^2 + 2x - 8$ and then determine the zeros of the function.

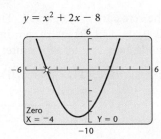

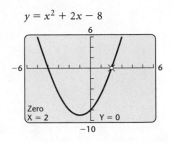

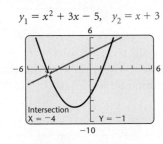

Using the ZERO feature, we see that the zeros are $-4$ and 2.

The intervals to be considered are $(-\infty, -4)$, $(-4, 2)$, and $(2, \infty)$. Using test values for $f(x)$, we determine the sign of $f(x)$ in each interval. (See the table at left.)

Function values are negative in the interval $(-4, 2)$. We can also note on the graph where the function values are negative. Since the inequality symbol is $\leq$, we include the endpoints of the interval in the solution set. The solution set of $x^2 + 3x - 5 \leq x + 3$ is $[-4, 2]$, or $\{x \,|\, -4 \leq x \leq 2\}$.

An alternative approach to solving the inequality

$$x^2 + 3x - 5 \leq x + 3$$

is to graph both sides: $y_1 = x^2 + 3x - 5$ and $y_2 = x + 3$ and determine where the graph of $y_1$ is below the graph of $y_2$. Using the INTERSECT feature, we determine the points of intersection, $(-4, -1)$ and $(2, 5)$.

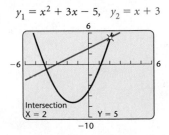

The graph of $y_1$ is below the graph of $y_2$ over the interval $(-4, 2)$. Since the inequality symbol is $\leq$, we include the endpoints of the interval in the solution. Thus the solution set of $x^2 + 3x - 5 \leq x + 3$ is $[-4, 2]$.

> **Now Try Exercise 29.**

Quadratic inequalities are one type of **polynomial inequality**. Other examples of polynomial inequalities are

$$-2x^4 + x^2 - 3 < 7, \quad \tfrac{2}{3}x + 4 \geq 0, \quad \text{and} \quad 4x^3 - 2x^2 > 5x + 7.$$

**EXAMPLE 3**   Solve: $x^3 - x > 0$.

***Solution***   We are asked to find all $x$-values for which $x^3 - x > 0$. To locate these values, we graph $f(x) = x^3 - x$. Then we note that whenever the function changes sign, its graph passes through an $x$-intercept. Thus to solve $x^3 - x > 0$, we first solve the related equation $x^3 - x = 0$ to find all zeros of the function:

$$x^3 - x = 0$$
$$x(x^2 - 1) = 0$$
$$x(x + 1)(x - 1) = 0.$$

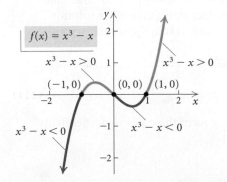

The zeros are $-1$, $0$, and $1$. Thus the $x$-intercepts of the graph are $(-1, 0)$, $(0, 0)$, and $(1, 0)$, as shown in the figure on the preceding page. The zeros divide the $x$-axis into four intervals:

$$(-\infty, -1), \qquad (-1, 0), \qquad (0, 1), \quad \text{and} \quad (1, \infty).$$

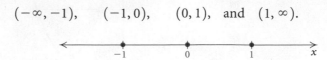

The sign of $x^3 - x$ is the same for all values of $x$ in a given interval. Thus we choose a test value for $x$ from each interval and find $f(x)$. We can use the TABLE feature set in ASK mode to determine the sign of $f(x)$ in each interval. (See the table at left.) We can also determine the sign of $f(x)$ in each interval by simply looking at the graph of the function.

| X | Y1 |
|----|-------|
| -2 | -6 |
| -.5 | .375 |
| .5 | -.375 |
| 2 | 6 |

X =

| Interval | $(-\infty, -1)$ | $(-1, 0)$ | $(0, 1)$ | $(1, \infty)$ |
|----------|-----------------|-----------|----------|---------------|
| **Test Value** | $f(-2) = -6$ | $f(-0.5) = 0.375$ | $f(0.5) = -0.375$ | $f(2) = 6$ |
| **Sign of $f(x)$** | Negative | Positive | Negative | Positive |

Since we are solving $x^3 - x > 0$, the solution set consists of only two of the four intervals, those in which the sign of $f(x)$ is *positive*. We see that the solution set is $(-1, 0) \cup (1, \infty)$, or $\{x | -1 < x < 0 \ or \ x > 1\}$.

**Now Try Exercise 39.**

To solve a polynomial inequality:

1. Find an equivalent inequality with $P(x)$ on one side and 0 on the other.
2. Change the inequality symbol to an equals sign and solve the related equation; that is, solve $P(x) = 0$.
3. Use the solutions to divide the $x$-axis into intervals. Then select a test value from each interval and determine the sign of the polynomial on the interval.
4. Determine the intervals for which the inequality is satisfied and write interval notation or set-builder notation for the solution set. Include the endpoints of the intervals in the solution set if the inequality symbol is $\leq$ or $\geq$.

**EXAMPLE 4**   Solve: $3x^4 + 10x \leq 11x^3 + 4$.

***Solution***   By subtracting $11x^3 + 4$, we form the equivalent inequality

$$3x^4 - 11x^3 + 10x - 4 \leq 0.$$

## Algebraic Solution

To solve the related equation

$$3x^4 - 11x^3 + 10x - 4 = 0,$$

we need to use the theorems of Section 4.4. We solved this equation in Example 5 in Section 4.4. The solutions are

$$-1, \quad 2 - \sqrt{2}, \quad \tfrac{2}{3}, \quad \text{and} \quad 2 + \sqrt{2},$$

or approximately

$$-1, \quad 0.586, \quad 0.667, \quad \text{and} \quad 3.414.$$

These numbers divide the *x*-axis into five intervals: $(-\infty, -1)$, $(-1, 2 - \sqrt{2})$, $\left(2 - \sqrt{2}, \tfrac{2}{3}\right)$, $\left(\tfrac{2}{3}, 2 + \sqrt{2}\right)$, and $(2 + \sqrt{2}, \infty)$.

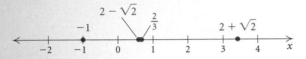

We then let $f(x) = 3x^4 - 11x^3 + 10x - 4$ and, using test values for *x*, determine the sign of $f(x)$ in each interval.

| X | Y1 | |
|----|-------|--|
| −2 | 112 | |
| 0 | −4 | |
| .6 | .0128 | |
| 1 | −2 | |
| 4 | 100 | |

X =

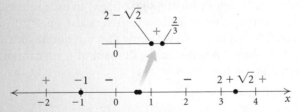

Function values are negative in the intervals $(-1, 2 - \sqrt{2})$ and $\left(\tfrac{2}{3}, 2 + \sqrt{2}\right)$. Since the inequality sign is $\leq$, we include the endpoints of the intervals in the solution set. The solution set is

$$\left[-1, 2 - \sqrt{2}\right] \cup \left[\tfrac{2}{3}, 2 + \sqrt{2}\right], \quad \text{or}$$

$$\left\{x \mid -1 \leq x \leq 2 - \sqrt{2} \text{ or } \tfrac{2}{3} \leq x \leq 2 + \sqrt{2}\right\}.$$

## Graphical Solution

We graph $y = 3x^4 - 11x^3 + 10x - 4$ using a viewing window that reveals the curvature of the graph.

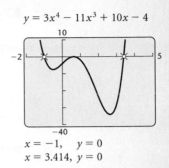

$x = -1, \quad y = 0$
$x = 3.414, \quad y = 0$

Using the ZERO feature, we see that two of the zeros are $-1$ and approximately 3.414 $(2 + \sqrt{2} \approx 3.414)$. However, this window leaves us uncertain about the number of zeros of the function in the interval $[0, 1]$. The window below shows another view of the zeros in the interval $[0, 1]$. Those zeros are about 0.586 and 0.667 $\left(2 - \sqrt{2} \approx 0.586; \tfrac{2}{3} \approx 0.667\right)$.

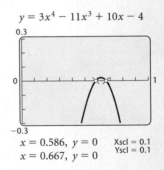

$x = 0.586, \quad y = 0$   Xscl = 0.1
$x = 0.667, \quad y = 0$   Yscl = 0.1

The intervals to be considered are $(-\infty, -1)$, $(-1, 0.586)$, $(0.586, 0.667)$, $(0.667, 3.414)$, and $(3.414, \infty)$. We note on the graph where the function is negative. Then including appropriate endpoints, we find that the solution set is approximately

$$[-1, 0.586] \cup [0.667, 3.414], \quad \text{or}$$

$$\{x \mid -1 \leq x \leq 0.586 \text{ or } 0.667 \leq x \leq 3.414\}.$$

**Now Try Exercise 45.**

## ◼ Rational Inequalities

Some inequalities involve rational expressions and functions. These are called **rational inequalities.** To solve rational inequalities, we need to make some adjustments to the preceding method.

**EXAMPLE 5** Solve: $\dfrac{3x}{x+6} < 0$.

**Solution** We look for all values of $x$ for which the related function

$$f(x) = \frac{3x}{x+6}$$

is not defined or is 0. These are called **critical values.**

The denominator tells us that $f(x)$ is not defined when $x = -6$. Next, we solve $f(x) = 0$:

$$\frac{3x}{x+6} = 0$$

$$(x+6) \cdot \frac{3x}{x+6} = (x+6) \cdot 0$$

$$3x = 0$$

$$x = 0.$$

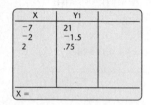

The critical values are $-6$ and $0$. These values divide the $x$-axis into three intervals:

$$(-\infty, -6), \quad (-6, 0), \quad \text{and} \quad (0, \infty).$$

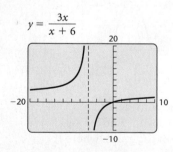

$$y = \frac{3x}{x+6}$$

We then use a test value to determine the sign of $f(x)$ in each interval.

Function values are negative in only the interval $(-6, 0)$. Since $f(0) = 0$ and the inequality symbol is $<$, we know that 0 is not included in the solution set. Note that since $-6$ is not in the domain of $f$, $-6$ cannot be part of the solution set. The solution set is

$$(-6, 0), \quad \text{or} \quad \{x \mid -6 < x < 0\}.$$

The graph of $f(x)$ shows where $f(x)$ is positive and where it is negative.

**Now Try Exercise 57.**

**EXAMPLE 6** Solve: $\dfrac{x-3}{x+4} \geq \dfrac{x+2}{x-5}$.

**Solution** We first subtract $(x+2)/(x-5)$ on both sides in order to find an equivalent inequality with 0 on one side:

$$\frac{x-3}{x+4} - \frac{x+2}{x-5} \geq 0.$$

## Algebraic Solution

We look for all values of $x$ for which the related function

$$f(x) = \frac{x-3}{x+4} - \frac{x+2}{x-5}$$

is not defined or is 0. These are the critical values.

A look at the denominators shows that $f(x)$ is not defined for $x = -4$ and $x = 5$. Next, we solve $f(x) = 0$:

$$\frac{x-3}{x+4} - \frac{x+2}{x-5} = 0$$

$$(x+4)(x-5)\left(\frac{x-3}{x+4} - \frac{x+2}{x-5}\right) = (x+4)(x-5)\cdot 0$$

$$(x-5)(x-3) - (x+4)(x+2) = 0$$

$$x^2 - 8x + 15 - (x^2 + 6x + 8) = 0$$

$$-14x + 7 = 0$$

$$x = \tfrac{1}{2}.$$

The critical values are $-4, \tfrac{1}{2}$, and 5. These values divide the $x$-axis into four intervals:

$$(-\infty, -4), \quad \left(-4, \tfrac{1}{2}\right), \quad \left(\tfrac{1}{2}, 5\right), \quad \text{and} \quad (5, \infty).$$

We then use a test value to determine the sign of $f(x)$ in each interval.

| X | Y1 |
|---|---|
| -5 | 7.7 |
| -2 | -2.5 |
| 3 | 2.5 |
| 6 | -7.7 |

X =

Function values are positive in the intervals $(-\infty, -4)$ and $\left(\tfrac{1}{2}, 5\right)$. Since $f\left(\tfrac{1}{2}\right) = 0$ and the inequality symbol is $\geq$, we know that $\tfrac{1}{2}$ must be in the solution set. Note that since neither $-4$ nor 5 is in the domain of $f$, they cannot be part of the solution set.

The solution set is $(-\infty, -4) \cup \left[\tfrac{1}{2}, 5\right)$.

## Graphical Solution

We graph

$$y = \frac{x-3}{x+4} - \frac{x+2}{x-5}$$

in the standard window, which shows the curvature of the function.

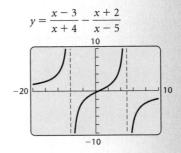

$$y = \frac{x-3}{x+4} - \frac{x+2}{x-5}$$

Using the ZERO feature, we find that 0.5 is a zero.

We then look for values where the function is not defined. By examining the denominators $x + 4$ and $x - 5$, we see that $f(x)$ is not defined for $x = -4$ and $x = 5$.

The critical values are $-4$, 0.5, and 5.

The graph shows where $y$ is positive and where it is negative. Note that $-4$ and 5 cannot be in the solution set since $y$ is not defined for these values. We do include 0.5, however, since the inequality symbol is $\geq$ and $f(0.5) = 0$. This solution set is

$$(-\infty, -4) \cup [0.5, 5).$$

**Now Try Exercise 61.**

The following is a method for solving rational inequalities.

> **To solve a rational inequality:**
>
> 1. Find an equivalent inequality with 0 on one side.
> 2. Change the inequality symbol to an equals sign and solve the related equation.
> 3. Find values of the variable for which the related rational function is not defined.
> 4. The numbers found in steps (2) and (3) are called critical values. Use the critical values to divide the $x$-axis into intervals. Then determine the function's sign in each interval using an $x$-value from the interval or the graph of the equation.
> 5. Select the intervals for which the inequality is satisfied and write interval notation or set-builder notation for the solution set. If the inequality symbol is $\leq$ or $\geq$, then the solutions to step (2) should be included in the solution set. The $x$-values found in step (3) are never included in the solution set.

It works well to use a combination of algebraic and graphical methods to solve polynomial inequalities and rational inequalities. The algebraic methods give exact numbers for the critical values, and the graphical methods usually allow us to see easily what intervals satisfy the inequality.

## 4.6    Exercise Set

*For the function $f(x) = x^2 + 2x - 15$, solve each of the following.*

**1.** $f(x) = 0$

**2.** $f(x) < 0$

**3.** $f(x) \leq 0$

**4.** $f(x) > 0$

**5.** $f(x) \geq 0$

*For the function*

$$g(x) = \frac{x - 2}{x + 4},$$

*solve each of the following.*

**6.** $g(x) = 0$

**7.** $g(x) > 0$

**8.** $g(x) \leq 0$

**9.** $g(x) \geq 0$

**10.** $g(x) < 0$

*For the function*

$$h(x) = \frac{7x}{(x - 1)(x + 5)},$$

*solve each of the following.*

**11.** $h(x) = 0$

**12.** $h(x) \leq 0$

**13.** $h(x) \geq 0$

**14.** $h(x) > 0$

**15.** $h(x) < 0$

*For the function $g(x) = x^5 - 9x^3$, solve each of the following.*

**16.** $g(x) = 0$

**17.** $g(x) < 0$

**18.** $g(x) \leq 0$

**19.** $g(x) > 0$

**20.** $g(x) \geq 0$

*In Exercises 21–24, a related function is graphed. Solve the given inequality.*

**21.** $x^3 + 6x^2 < x + 30$

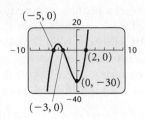

**22.** $x^4 - 27x^2 - 14x + 120 \geq 0$

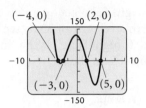

**23.** $\dfrac{8x}{x^2 - 4} \geq 0$

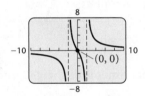

**24.** $\dfrac{8}{x^2 - 4} < 0$

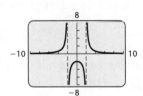

*Solve.*

**25.** $(x - 1)(x + 4) < 0$

**26.** $(x + 3)(x - 5) < 0$

**27.** $x^2 + x - 2 > 0$

**28.** $x^2 - x - 6 > 0$

**29.** $x^2 - x - 5 \geq x - 2$

**30.** $x^2 + 4x + 7 \geq 5x + 9$

**31.** $x^2 > 25$

**32.** $x^2 \leq 1$

**33.** $4 - x^2 \leq 0$

**34.** $11 - x^2 \geq 0$

**35.** $6x - 9 - x^2 < 0$

**36.** $x^2 + 2x + 1 \leq 0$

**37.** $x^2 + 12 < 4x$

**38.** $x^2 - 8 > 6x$

**39.** $4x^3 - 7x^2 \leq 15x$

**40.** $2x^3 - x^2 < 5x$

**41.** $x^3 + 3x^2 - x - 3 \geq 0$

**42.** $x^3 + x^2 - 4x - 4 \geq 0$

**43.** $x^3 - 2x^2 < 5x - 6$

**44.** $x^3 + x \leq 6 - 4x^2$

**45.** $x^5 + x^2 \geq 2x^3 + 2$

**46.** $x^5 + 24 > 3x^3 + 8x^2$

**47.** $2x^3 + 6 \leq 5x^2 + x$

**48.** $2x^3 + x^2 < 10 + 11x$

**49.** $x^3 + 5x^2 - 25x \leq 125$

**50.** $x^3 - 9x + 27 \geq 3x^2$

**51.** $0.1x^3 - 0.6x^2 - 0.1x + 2 < 0$

**52.** $19.2x^3 + 12.8x^2 + 144 \geq 172.8x + 3.2x^4$

*List the critical values of the related function. Then solve the inequality.*

**53.** $\dfrac{1}{x + 4} > 0$

**54.** $\dfrac{1}{x - 3} \leq 0$

**55.** $\dfrac{-4}{2x + 5} < 0$

**56.** $\dfrac{-2}{5 - x} \geq 0$

**57.** $\dfrac{2x}{x - 4} \geq 0$

**58.** $\dfrac{5x}{x + 1} < 0$

**59.** $\dfrac{x - 4}{x + 3} - \dfrac{x + 2}{x - 1} \leq 0$

**60.** $\dfrac{x + 1}{x - 2} - \dfrac{x - 3}{x - 1} < 0$

**61.** $\dfrac{x + 6}{x - 2} > \dfrac{x - 8}{x - 5}$

**62.** $\dfrac{x - 7}{x + 2} \geq \dfrac{x - 9}{x + 3}$

**63.** $\dfrac{x + 1}{x - 2} \geq 3$

**64.** $\dfrac{x}{x - 5} < 2$

**65.** $x - 2 > \dfrac{1}{x}$

**66.** $4 \geq \dfrac{4}{x} + x$

**67.** $\dfrac{2}{x^2 - 4x + 3} \le \dfrac{5}{x^2 - 9}$

**68.** $\dfrac{3}{x^2 - 4} \le \dfrac{5}{x^2 + 7x + 10}$

**69.** $\dfrac{3}{x^2 + 1} \ge \dfrac{6}{5x^2 + 2}$    **70.** $\dfrac{4}{x^2 - 9} < \dfrac{3}{x^2 - 25}$

**71.** $\dfrac{5}{x^2 + 3x} < \dfrac{3}{2x + 1}$    **72.** $\dfrac{2}{x^2 + 3} > \dfrac{3}{5 + 4x^2}$

**73.** $\dfrac{5x}{7x - 2} > \dfrac{x}{x + 1}$    **74.** $\dfrac{x^2 - x - 2}{x^2 + 5x + 6} < 0$

**75.** $\dfrac{x}{x^2 + 4x - 5} + \dfrac{3}{x^2 - 25} \le \dfrac{2x}{x^2 - 6x + 5}$

**76.** $\dfrac{2x}{x^2 - 9} + \dfrac{x}{x^2 + x - 12} \ge \dfrac{3x}{x^2 + 7x + 12}$

**77.** *Temperature During an Illness.*   A person's temperature $T$, in degrees Fahrenheit, during an illness is given by the function

$$T(t) = \dfrac{4t}{t^2 + 1} + 98.6,$$

where $t$ is the time since the onset of the illness, in hours. Find the interval on which the temperature was over 100°F. (See Example 12 in Section 4.5.)

**78.** *Population Growth.*   The population $P$, in thousands, of a new senior community is given by

$$P(t) = \dfrac{500t}{2t^2 + 9},$$

where $t$ is the time, in months. Find the interval on which the population was 40,000 or greater. (See Exercise 85 in Exercise Set 4.5.)

**79.** *Total Profit.*   Flexl, Inc., determines that its total profit is given by the function

$$P(x) = -3x^2 + 630x - 6000.$$

**a)** Flexl makes a profit for those nonnegative values of $x$ for which $P(x) > 0$. Find the values of $x$ for which Flexl makes a profit.

**b)** Flexl loses money for those nonnegative values of $x$ for which $P(x) < 0$. Find the values of $x$ for which Flexl loses money.

**80.** *Height of a Thrown Object.*   The function

$$S(t) = -16t^2 + 32t + 1920$$

gives the height $S$, in feet, of an object thrown upward with a velocity of 32 ft/sec from a cliff that is 1920 ft high. Here $t$ is the time, in seconds, that the object is in the air.

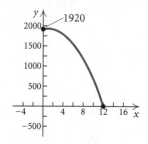

**a)** For what times is the height greater than 1920 ft?

**b)** For what times is the height less than 640 ft?

**81.** *Number of Diagonals.*   A polygon with $n$ sides has $D$ diagonals, where $D$ is given by the function

$$D(n) = \dfrac{n(n - 3)}{2}.$$

Find the number of sides $n$ if

$$27 \le D \le 230.$$

**82.** *Number of Handshakes.*   If there are $n$ people in a room, the number $N$ of possible handshakes by all the people in the room is given by the function

$$N(n) = \dfrac{n(n - 1)}{2}.$$

For what number $n$ of people is

$$66 \le N \le 300?$$

## Skill Maintenance

*Find an equation for a circle satisfying the given conditions.*

**83.** Center: $(-2, 4)$; radius of length 3

**84.** Center: $(0, -3)$; diameter of length $\frac{7}{2}$

*In Exercises 85 and 86:*

a) *Find the vertex.*
b) *Determine whether there is a maximum or minimum value and find that value.*
c) *Find the range.*

**85.** $h(x) = -2x^2 + 3x - 8$

**86.** $g(x) = x^2 - 10x + 2$

## Synthesis

*Solve.*

**87.** $|x^2 - 5| = 5 - x^2$

**88.** $x^4 - 6x^2 + 5 > 0$

**89.** $2|x|^2 - |x| + 2 \le 5$

**90.** $(7 - x)^{-2} < 0$

**91.** $\left|1 + \dfrac{1}{x}\right| < 3$

**92.** $\left|2 - \dfrac{1}{x}\right| \le 2 + \left|\dfrac{1}{x}\right|$

**93.** Write a quadratic inequality for which the solution set is $(-4, 3)$.

**94.** Write a polynomial inequality for which the solution set is $[-4, 3] \cup [7, \infty)$.

*Find the domain of the function.*

**95.** $f(x) = \sqrt{\dfrac{72}{x^2 - 4x - 21}}$

**96.** $f(x) = \sqrt{x^2 - 4x - 21}$

# Chapter 4 Summary and Review

## STUDY GUIDE

| KEY TERMS AND CONCEPTS | EXAMPLES |
|---|---|

### SECTION 4.1: POLYNOMIAL FUNCTIONS AND MODELING

**Polynomial Function**

A polynomial function is given by

$$P(x) = a_n x^n + a_{n-1}x^{n-1} + a_{n-2}x^{n-2} + \cdots + a_1 x + a_0,$$

where the coefficients $a_n, a_{n-1}, \ldots, a_1, a_0$ are real numbers and the exponents are whole numbers.

The first nonzero coefficient, $a_n$, is called the **leading coefficient**. The term $a_n x^n$ is called the **leading term**. The **degree** of the polynomial function is $n$.

Consider the polynomial function

$$P(x) = \tfrac{1}{3}x^2 + x - 4x^5 + 2.$$

Leading term: $-4x^5$
Leading coefficient: $-4$
Degree of polynomial: 5

Classifying polynomial functions by degree:

| Type | Degree |
|------|--------|
| Constant | 0 |
| Linear | 1 |
| Quadratic | 2 |
| Cubic | 3 |
| Quartic | 4 |

Classify the following polynomial functions:

| Function | Type |
|----------|------|
| $f(x) = -2$ | Constant |
| $f(x) = 0.6x - 11$ | Linear |
| $f(x) = 5x^2 + x - 4$ | Quadratic |
| $f(x) = 5x^3 - x + 10$ | Cubic |
| $f(x) = -x^4 + 8x^3 + x$ | Quartic |

**The Leading-Term Test**

If $a_n x^n$ is the leading term of a polynomial function, then the behavior of the graph as $x \rightarrow \infty$ and as $x \rightarrow -\infty$ can be described in one of the four following ways.

**a)** If $n$ is even, and $a_n > 0$:

**b)** If $n$ is even, and $a_n < 0$:

**c)** If $n$ is odd, and $a_n > 0$:

**d)** If $n$ is odd, and $a_n < 0$:

Using the leading-term test, describe the end behavior of the graph of each function by selecting one of (a)–(d) shown at left.

$$h(x) = -2x^6 + x^4 - 3x^2 + x$$

The leading term, $a_n x^n$, is $-2x^6$. Since 6 is even and $-2 < 0$, the shape is shown in (b).

$$g(x) = 4x^3 - 8x + 1$$

The leading term, $a_n x^n$, is $4x^3$. Since 3 is odd and $4 > 0$, the shape is shown in (c).

**Zeros of Functions**

If $c$ is a real zero of a function $f(x)$ (that is, if $f(c) = 0$), then $x - c$ is a factor of $f(x)$ and $(c, 0)$ is an $x$-intercept of the graph of the function.

If we know the linear factors of a polynomial function $f(x)$, we can find the zeros of $f(x)$ by solving the equation $f(x) = 0$ using the principle of zero products.

Every function of degree $n$, with $n \geq 1$, has at least one zero and at most $n$ zeros.

To find the zeros of

$$f(x) = -2(x - 3)(x + 8)^2,$$

we solve $-2(x - 3)(x + 8)^2 = 0$ using the principle of zero products:

$x - 3 = 0$   or   $x + 8 = 0$   or   $x + 8 = 0$
$x = 3$   or   $x = -8$   or   $x = -8.$

The zeros of $f(x)$ are 3 and $-8$.

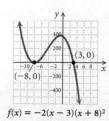

$f(x) = -2(x - 3)(x + 8)^2$

To find the zeros of

$$h(x) = x^4 - 12x^2 - 64,$$

we solve $h(x) = 0$:

$$x^4 - 12x^2 - 64 = 0$$
$$(x^2 - 16)(x^2 + 4) = 0$$
$$(x + 4)(x - 4)(x^2 + 4) = 0$$

$x + 4 = 0$   *or*   $x - 4 = 0$   *or*   $x^2 + 4 = 0$

$x = -4$  *or*      $x = 4$  *or*      $x^2 = -4$

$x = -4$  *or*      $x = 4$  *or*      $x = \pm\sqrt{-4}$

$x = -4$  *or*      $x = 4$  *or*      $x = \pm 2i.$

The zeros of $h(x)$ are $-4, 4$, and $\pm 2i$.

---

**Even and Odd Multiplicity**

If $(x - c)^k, k \geq 1$, is a factor of a polynomial function $P(x)$ and $(x - c)^{k+1}$ is not a factor and:

- $k$ is odd, then the graph crosses the $x$-axis at $(c, 0)$;

- $k$ is even, then the graph is tangent to the $x$-axis at $(c, 0)$.

For $f(x) = -2(x - 3)(x + 8)^2$ graphed above, note that for the factor $x - 3$, or $(x - 3)^1$, the exponent 1 is odd and the graph crosses the $x$-axis at $(3, 0)$. For the factor $(x + 8)^2$, the exponent 2 is even and the graph is tangent to the $x$-axis at $(-8, 0)$.

---

## SECTION 4.2: GRAPHING POLYNOMIAL FUNCTIONS

If $P(x)$ is a polynomial function of degree $n$, the graph of the function has:

- at most $n$ real zeros, and thus at most $n$ $x$-intercepts, and

- at most $n - 1$ turning points.

**To Graph a Polynomial Function**

1. Use the leading-term test to determine the end behavior.

2. Find the zeros of the function by solving $f(x) = 0$. Any real zeros are the first coordinates of the $x$-intercepts.

Graph:

$$h(x) = x^4 - 12x^2 - 16x = x(x - 4)(x + 2)^2.$$

1. The leading term is $x^4$. Since 4 is even and $1 > 0$, the end behavior of the graph can be sketched as follows.

2. We solve $x(x - 4)(x + 2)^2 = 0$. The solutions are $0, 4$, and $-2$. The zeros of $h(x)$ are $0, 4$, and $-2$. The $x$-intercepts are $(0, 0), (4, 0)$, and $(-2, 0)$. The multiplicity of 0 and 4 is 1, so the graph will cross the $x$-axis at 0 and 4. The multiplicity of $-2$ is 2, so the graph is tangent to the $x$-axis at $-2$.

*(continued)*

3. Use the *x*-intercepts (zeros) to divide the *x*-axis into intervals and choose a test point in each interval to determine the sign of all function values in that interval. For all *x*-values in an interval, $f(x)$ is either always positive for all values or always negative for all values.

4. Find $f(0)$. This gives the *y*-intercept of the function.

5. If necessary, find additional function values to determine the general shape of the graph and then draw the graph.

3. The zeros divide the *x*-axis into four intervals.

| Interval | $(-\infty, -2)$ | $(-2, 0)$ | $(0, 4)$ | $(4, \infty)$ |
|---|---|---|---|---|
| Test Value | $-3$ | $-1$ | $1$ | $5$ |
| Function Value, $h(x)$ | $21$ | $5$ | $-27$ | $245$ |
| Sign of $h(x)$ | $+$ | $+$ | $-$ | $+$ |
| Location of Points on Graph | Above *x*-axis | Above *x*-axis | Below *x*-axis | Above *x*-axis |

Four points on the graph are $(-3, 21)$, $(-1, 5)$, $(1, -27)$, and $(5, 245)$.

4. We find $h(0)$:
$$h(0) = 0(0 - 4)(0 + 2)^2 = 0.$$

The *y*-intercept is $(0, 0)$.

5. We find additional points and draw the graph.

| $x$ | $h(x)$ |
|---|---|
| $-2.5$ | $4.1$ |
| $-1.5$ | $2.1$ |
| $-0.5$ | $5.1$ |
| $0.5$ | $-10.9$ |
| $2$ | $-64$ |
| $3$ | $-75$ |

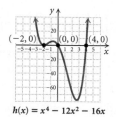

$h(x) = x^4 - 12x^2 - 16x$

### The Intermediate Value Theorem

For any polynomial function $P(x)$ with real coefficients, suppose that for $a \neq b$, $P(a)$ and $P(b)$ are of opposite signs. Then the function has at least one real zero between $a$ and $b$.

The intermediate value theorem *cannot* be used to determine whether there is a real zero between $a$ and $b$ when $P(a)$ and $P(b)$ have the *same* sign.

Use the intermediate value theorem to determine, if possible, whether the function has a real zero between $a$ and $b$.

$$f(x) = 2x^3 - 5x^2 + x - 2, \quad a = 2, b = 3;$$
$$f(2) = -4; \quad f(3) = 10$$

Since $f(2)$ and $f(3)$ have opposite signs, $f(x)$ has at least one zero between 2 and 3.

$$f(x) = 2x^3 - 5x^2 + x - 2, \quad a = -2, b = -1;$$
$$f(-2) = -40; \quad f(-1) = -10$$

Both $f(-2)$ and $f(-1)$ are negative. Thus the intermediate value theorem does not allow us to determine whether there is a real zero between $-2$ and $-1$.

## SECTION 4.3: POLYNOMIAL DIVISION; THE REMAINDER THEOREM AND THE FACTOR THEOREM

**Polynomial Division**

$$P(x) = d(x) \cdot Q(x) + R(x)$$

Dividend  Divisor  Quotient  Remainder

When we divide a polynomial $P(x)$ by a divisor $d(x)$, a polynomial $Q(x)$ is the quotient and a polynomial $R(x)$ is the remainder. The quotient $Q(x)$ must have degree less than that of the dividend $P(x)$. The remainder $R(x)$ must either be 0 or have degree less than that of the divisor $d(x)$. If $R(x) = 0$, then the divisor $d(x)$ is a factor of the dividend.

Given $P(x) = x^4 - 6x^3 + 9x^2 + 4x - 12$ and $d(x) = x + 2$, use long division to find the quotient and the remainder when $P(x)$ is divided by $d(x)$. Express $P(x)$ in the form $d(x) \cdot Q(x) + R(x)$.

$$
\begin{array}{r}
x^3 - 8x^2 + 25x - 46 \\
x + 2 \overline{)\, x^4 - 6x^3 + 9x^2 + 4x - 12} \\
\underline{x^4 + 2x^3} \phantom{aaaaaaaaaaaaaaaa} \\
-8x^3 + 9x^2 \phantom{aaaaaaaaa} \\
\underline{-8x^3 - 16x^2} \phantom{aaaaaaaa} \\
25x^2 + 4x \phantom{aaaa} \\
\underline{25x^2 + 50x} \phantom{aaaa} \\
-46x - 12 \\
\underline{-46x - 92} \\
80
\end{array}
$$

$Q(x) = x^3 - 8x^2 + 25x - 46$ and $R(x) = 80$. Thus, $P(x) = (x + 2)(x^3 - 8x^2 + 25x - 46) + 80$. Since $R(x) \neq 0$, $x + 2$ is not a factor of $P(x)$.

**The Remainder Theorem**

If a number $c$ is substituted for $x$ in the polynomial $f(x)$, then the result, $f(c)$, is the remainder that would be obtained by dividing $f(x)$ by $x - c$. That is, if $f(x) = (x - c) \cdot Q(x) + R$, then $f(c) = R$.

The long-division process can be streamlined with synthetic division. Synthetic division also can be used to find polynomial function values.

We repeat the division shown above using synthetic division. Note that the divisor $x + 2 = x - (-2)$.

$$
\begin{array}{r|rrrrr}
-2 & 1 & -6 & 9 & 4 & -12 \\
   &   & -2 & 16 & -50 & 92 \\
\hline
   & 1 & -8 & 25 & -46 & \,|\, 80
\end{array}
$$

Again, note that $Q(x) = x^3 - 8x^2 + 25x - 46$ and $R(x) = 80$. Since $R(x) \neq 0$, $x - (-2)$, or $x + 2$, is not a factor of $P(x)$.

We now divide $P(x)$ by $x - 3$.

$$
\begin{array}{r|rrrrr}
3 & 1 & -6 & 9 & 4 & -12 \\
  &   & 3 & -9 & 0 & 12 \\
\hline
  & 1 & -3 & 0 & 4 & \,|\, 0
\end{array}
$$

$Q(x) = x^3 - 3x^2 + 4$ and $R(x) = 0$. Since $R(x) = 0$, $x - 3$ is a factor of $P(x)$.

For $f(x) = 2x^5 - x^3 - 3x^2 - 4x + 15$, find $f(-2)$.

$$
\begin{array}{r|rrrrrr}
-2 & 2 & 0 & -1 & -3 & -4 & 15 \\
   &   & -4 & 8 & -14 & 34 & -60 \\
\hline
   & 2 & -4 & 7 & -17 & 30 & \,|\, -45
\end{array}
$$

Thus, $f(-2) = -45$.

**The Factor Theorem**

For a polynomial $f(x)$, if $f(c) = 0$, then $x - c$ is a factor of $f(x)$.

Let $g(x) = x^4 + 8x^3 + 6x^2 - 40x + 25$. Factor $g(x)$ and solve $g(x) = 0$.

Use synthetic division to look for factors of the form $x - c$. Let's try $x + 5$.

$$
\begin{array}{r|rrrrr}
-5 & 1 & 8 & 6 & -40 & 25 \\
   &   & -5 & -15 & 45 & -25 \\
\hline
   & 1 & 3 & -9 & 5 & 0
\end{array}
$$

Since $g(-5) = 0$, the number $-5$ is a zero of $g(x)$ and $x - (-5)$, or $x + 5$, is a factor of $g(x)$. This gives us

$$g(x) = (x + 5)(x^3 + 3x^2 - 9x + 5).$$

Let's try $x + 5$ again with the factor $x^3 + 3x^2 - 9x + 5$.

$$
\begin{array}{r|rrrr}
-5 & 1 & 3 & -9 & 5 \\
   &   & -5 & 10 & -5 \\
\hline
   & 1 & -2 & 1 & 0
\end{array}
$$

Now we have

$$g(x) = (x + 5)^2(x^2 - 2x + 1).$$

The trinomial $x^2 - 2x + 1$ easily factors, so

$$g(x) = (x + 5)^2(x - 1)^2.$$

Solve $g(x) = 0$. The solutions of $(x + 5)^2(x - 1)^2 = 0$ are $-5$ and $1$. They are also the zeros of $g(x)$.

## SECTION 4.4: THEOREMS ABOUT ZEROS OF POLYNOMIAL FUNCTIONS

**The Fundamental Theorem of Algebra**

Every polynomial function of degree $n$, $n \geq 1$, with complex coefficients has at least one zero in the system of complex numbers.

Every polynomial function $f$ of degree $n$, with $n \geq 1$, can be factored into $n$ linear factors (not necessarily unique); that is,

$$f(x) = a_n(x - c_1)(x - c_2) \cdots (x - c_n).$$

**Nonreal Zeros**

$a + bi$ and $a - bi$, $b \neq 0$

If a complex number $a + bi$, $b \neq 0$, is a zero of a polynomial function $f(x)$ with *real* coefficients, then its conjugate, $a - bi$, is also a zero. (Nonreal zeros occur in conjugate pairs.)

Find a polynomial function of degree 5 with $-4$ and $2$ as zeros of multiplicity 1, and $-1$ as a zero of multiplicity 3.

$$
\begin{aligned}
f(x) &= [x - (-4)][x - 2][x - (-1)]^3 \\
     &= (x + 4)(x - 2)(x + 1)^3 \\
     &= x^5 + 5x^4 + x^3 - 17x^2 - 22x - 8
\end{aligned}
$$

Find a polynomial function with rational coefficients of lowest degree with $1 - i$ and $\sqrt{7}$ as two of its zeros.

If $1 - i$ is a zero, then $1 + i$ is also a zero. If $\sqrt{7}$ is a zero, then $-\sqrt{7}$ is also a zero.

$$
\begin{aligned}
f(x) &= [x - (1 - i)][x - (1 + i)] \cdot \\
     &\quad [x - (-\sqrt{7})][x - \sqrt{7}] \\
     &= [(x - 1) + i][(x - 1) - i] \cdot \\
     &\quad (x + \sqrt{7})(x - \sqrt{7})
\end{aligned}
$$

*(continued)*

**Irrational Zeros**

$a + c\sqrt{b}$, and $a - c\sqrt{b}$,

**b not a perfect square**

If $a + c\sqrt{b}$, where $a$, $b$, and $c$ are rational and $b$ is not a perfect square, is a zero of a polynomial function $f(x)$ with *rational* coefficients, then its conjugate, $a - c\sqrt{b}$, is also a zero. (Irrational zeros occur in conjugate pairs.)

$$= [(x-1)^2 - i^2](x^2 - 7)$$
$$= (x^2 - 2x + 1 + 1)(x^2 - 7)$$
$$= (x^2 - 2x + 2)(x^2 - 7)$$
$$= x^4 - 2x^3 - 5x^2 + 14x - 14$$

**The Rational Zeros Theorem**

Consider the polynomial function

$$P(x) = a_n x^n + a_{n-1} x^{n-1} + a_{n-2} x^{n-2}$$
$$+ \cdots + a_1 x + a_0,$$

where all the coefficients are integers and $n \geq 1$. Also, consider a rational number $p/q$, where $p$ and $q$ have no common factor other than $-1$ and $1$. If $p/q$ is a zero of $P(x)$, then $p$ is a factor of $a_0$ and $q$ is a factor of $a_n$.

For $f(x) = 2x^4 - 9x^3 - 16x^2 - 9x - 18$, solve $f(x) = 0$ and factor $f(x)$ into linear factors.

There are at most 4 distinct zeros. Any rational zeros of $f$ must be of the form $p/q$, where $p$ is a factor of $-18$ and $q$ is a factor of 2.

$\dfrac{\text{Possibilities for } p}{\text{Possibilities for } q}$:  $\dfrac{\pm 1, \pm 2, \pm 3, \pm 6, \pm 9, \pm 18}{\pm 1, \pm 2}$

Possibilities for $p/q$:  $1, -1, 2, -2, 3, -3, 6, -6, 9,$
$-9, 18, -18, \frac{1}{2}, -\frac{1}{2}, \frac{3}{2}, -\frac{3}{2}, \frac{9}{2}, -\frac{9}{2}$

Use synthetic division to check the possibilities. We leave it to the student to verify that $\pm 1$, $\pm 2$, and $\pm 3$ are not zeros. Let's try 6.

$$
\begin{array}{r|rrrrr}
6 & 2 & -9 & -16 & -9 & -18 \\
  &   & 12 & 18 & 12 & 18 \\
\hline
  & 2 & 3 & 2 & 3 & 0
\end{array}
$$

Since $f(6) = 0$, 6 is a zero and $x - 6$ is a factor of $f(x)$. Now express $f(x)$ as

$$f(x) = (x - 6)(2x^3 + 3x^2 + 2x + 3).$$

Consider the factor $2x^3 + 3x^2 + 2x + 3$ and check the other possibilities. Let's try $-\frac{3}{2}$.

$$
\begin{array}{r|rrrr}
-\frac{3}{2} & 2 & 3 & 2 & 3 \\
            &   & -3 & 0 & -3 \\
\hline
            & 2 & 0 & 2 & 0
\end{array}
$$

Since $f\left(-\frac{3}{2}\right) = 0$, $-\frac{3}{2}$ is also a zero and $x + \frac{3}{2}$ is a factor of $f(x)$. We express $f(x)$ as

$$f(x) = (x - 6)\left(x + \tfrac{3}{2}\right)(2x^2 + 2)$$
$$= 2(x - 6)\left(x + \tfrac{3}{2}\right)(x^2 + 1).$$

Now we solve the equation $f(x) = 0$ to determine the zeros. We see that the only rational zeros are 6 and $-\frac{3}{2}$. The other zeros are $\pm i$.

The factorization into linear factors is

$$f(x) = 2(x - 6)\left(x + \tfrac{3}{2}\right)(x - i)(x + i), \quad \text{or}$$
$$(x - 6)(2x + 3)(x - i)(x + i).$$

**Descartes' Rule of Signs**

Let $P(x)$, written in descending or ascending order, be a polynomial function with real coefficients and a nonzero constant term. The number of positive real zeros of $P(x)$ is either:

1. The same as the number of variations of sign in $P(x)$, or

2. Less than the number of variations of sign in $P(x)$ by a positive even integer.

The number of negative real zeros of $P(x)$ is either:

3. The same as the number of variations of sign in $P(-x)$, or

4. Less than the number of variations of sign in $P(-x)$ by a positive even integer.

A zero of multiplicity $m$ must be counted $m$ times.

Determine the number of positive real zeros and the number of negative real zeros of

$$P(x) = 4x^5 - x^4 - 2x^3 + 8x - 10.$$

There are 3 variations of sign in $P(x)$. Thus the number of positive real zeros is 3 or 1.

$$P(-x) = -4x^5 - x^4 + 2x^3 - 8x - 10$$

There are 2 variations of sign in $P(-x)$. Thus the number of negative real zeros is 2 or 0.

## SECTION 4.5: RATIONAL FUNCTIONS

**Rational Function**

$$f(x) = \frac{p(x)}{q(x)},$$

where $p(x)$ and $q(x)$ are polynomials and $q(x)$ is not the zero polynomial, is a rational function. The domain of $f(x)$ consists of all $x$ for which $q(x) \neq 0$.

Determine the domain of each function.

| FUNCTION | DOMAIN |
|---|---|
| $f(x) = \dfrac{1}{x^5}$ | $(-\infty, 0) \cup (0, \infty)$ |
| $f(x) = \dfrac{x + 6}{x^2 + 2x - 8}$ | |
| $\quad = \dfrac{x + 6}{(x - 2)(x + 4)}$ | $(-\infty, -4) \cup (-4, 2) \cup$ $(2, \infty)$ |

**Vertical Asymptotes**

For a rational function $f(x) = p(x)/q(x)$, where $p(x)$ and $q(x)$ are polynomials with *no common factors* other than constants, if $a$ is a zero of the denominator, then the line $x = a$ is a vertical asymptote for the graph of the function.

**Horizontal Asymptotes**

When the numerator and the denominator have the same degree, the line $y = a/b$ is the horizontal asymptote, where $a$ and $b$ are the leading coefficients of the numerator and the denominator, respectively.

When the degree of the numerator is less than the degree of the denominator, the $x$-axis, or $y = 0$, is the horizontal asymptote.

When the degree of the numerator is greater than the degree of the denominator, there is *no* horizontal asymptote.

Determine the vertical, horizontal, and oblique asymptotes of the graph of the function.

| FUNCTION | ASYMPTOTES |
|---|---|
| $f(x) = \dfrac{x^2 - 2}{x - 1}$ | Vertical: $x = 1$ <br> Horizontal: None <br> Oblique: $y = x + 1$ |
| $f(x) = \dfrac{3x - 4}{x^2 + 6x - 7}$ | Vertical: $x = -7$; $x = 1$ <br> Horizontal: $y = 0$ <br> Oblique: None |
| $f(x) = \dfrac{2x^2 + 9x - 5}{3x^2 + 13x + 12}$ | Vertical: $x = -\frac{4}{3}$; $x = -3$ <br> Horizontal: $y = \frac{2}{3}$ <br> Oblique: None |

**Oblique Asymptotes**

When the degree of the numerator is 1 greater than the degree of the denominator, there is an oblique asymptote.

---

**To Graph a Rational Function**

$f(x) = p(x)/q(x)$, where $p(x)$ and $q(x)$ have no common factor other than constants:

1. Find any real zeros of the denominator. Determine the domain of the function and sketch any vertical asymptotes.

2. Find the horizontal asymptote or the oblique asymptote, if there is one, and sketch it.

3. Find any zeros of the function. The zeros are found by determining the zeros of the numerator. These are the first coordinates of the $x$-intercepts of the graph.

4. Find $f(0)$. This gives the $y$-intercept, $(0, f(0))$, of the function.

5. Find other function values to determine the general shape. Then draw the graph.

**Crossing an Asymptote**

The graph of a rational function never crosses a vertical asymptote.

The graph of a rational function might cross a horizontal asymptote but does not necessarily do so.

---

Graph: $g(x) = \dfrac{x^2 - 4}{x^2 + 4x - 5}$.

- *Domain*: The zeros of the denominator are $-5$ and $1$. The domain is $(-\infty, -5) \cup (-5, 1) \cup (1, \infty)$.
- *Vertical asymptotes*: Since neither zero of the denominator is a zero of the numerator, the graph has vertical asymptotes at $x = -5$ and $x = 1$.
- *Horizontal asymptote*: The degree of the numerator is the same as the degree of the denominator, so the horizontal asymptote is determined by the ratio of the leading coefficients: $1/1$, or $1$. The horizontal asymptote is $y = 1$.
- *Oblique asymptote*: None
- *Zeros of g*: Solving $g(x) = 0$ gives us $-2$ and $2$, so the zeros are $-2$ and $2$.
- *x-intercepts*: $(-2, 0)$ and $(2, 0)$
- *y-intercept*: $\left(0, \frac{4}{5}\right)$ because $g(0) = \frac{4}{5}$.

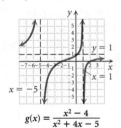

$$g(x) = \frac{x^2 - 4}{x^2 + 4x - 5}$$

---

Graph: $f(x) = \dfrac{x + 3}{x^2 - x - 12}$.

- *Domain*: The zeros of the denominator are $-3$ and $4$. The domain is $(-\infty, -3) \cup (-3, 4) \cup (4, \infty)$.
- *Vertical asymptote*: Since $4$ is the only zero of the denominator that is not a zero of the numerator, the only vertical asymptote is $x = 4$.
- *Horizontal asymptote*: Because the degree of the numerator is less than the degree of the denominator, the $x$-axis, $y = 0$, is the horizontal asymptote.
- *Oblique asymptote*: None
- *Zeros of f*: The equation $f(x) = 0$ has no solutions, so there are no zeros of $f$.
- *x-intercepts*: None
- *y-intercept*: $\left(0, -\frac{1}{4}\right)$ because $f(0) = -\frac{1}{4}$.

*"Hole" in the graph*:

$$f(x) = \frac{x+3}{(x+3)(x-4)} = \frac{1}{x-4}, \quad \text{where}$$

$x \neq -3$ and $x \neq 4$.

To determine the coordinates of the hole, we substitute $-3$ for $x$ in $f(x) = 1/(x-4)$:

$$f(-3) = \frac{1}{-3-4} = -\frac{1}{7}.$$

The hole is located at $\left(-3, -\frac{1}{7}\right)$.

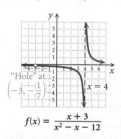

$$f(x) = \frac{x+3}{x^2-x-12}$$

---

## SECTION 4.6: POLYNOMIAL INEQUALITIES AND RATIONAL INEQUALITIES

**To Solve a Polynomial Inequality**

1. Find an equivalent inequality with 0 on one side.

2. Change the inequality symbol to an equals sign and solve the related equation.

3. Use the solutions to divide the $x$-axis into intervals. Then select a test value from each interval and determine the polynomial's sign on the interval.

4. Determine the intervals for which the inequality is satisfied and write interval notation or set-builder notation for the solution set. Include the endpoints of the intervals in the solution set if the inequality symbol is $\leq$ or $\geq$.

Solve: $x^3 - 3x^2 \leq 6x - 8$.

*Equivalent inequality*: $x^3 - 3x^2 - 6x + 8 \leq 0$.

First, we solve the related equation:

$$x^3 - 3x^2 - 6x + 8 = 0.$$

The solutions are $-2$, $1$, and $4$. The numbers divide the $x$-axis into 4 intervals. Next, we let $f(x) = x^3 - 3x^2 - 6x + 8$ and, using test values for $f(x)$, we determine the sign of $f(x)$ in each interval.

| INTERVAL | TEST VALUE | SIGN OF $f(x)$ |
|---|---|---|
| $(-\infty, -2)$ | $f(-3) = -28$ | $-$ |
| $(-2, 1)$ | $f(0) = 8$ | $+$ |
| $(1, 4)$ | $f(2) = -8$ | $-$ |
| $(4, \infty)$ | $f(6) = 80$ | $+$ |

Test values are negative in the intervals $(-\infty, -2)$ and $(1, 4)$. Since the inequality symbol is $\leq$, we include the endpoints of the intervals in the solution set. The solution set is

$$(-\infty, -2] \cup [1, 4].$$

**To Solve a Rational Inequality**

1. Find an equivalent inequality with 0 on one side.

2. Change the inequality symbol to an equals sign and solve the related equation.

3. Find values of the variable for which the related rational function is not defined.

4. The numbers found in steps (2) and (3) are called *critical values*. Use the critical values to divide the $x$-axis into intervals. Then determine the function's sign in each interval using an $x$-value from the interval or the graph of the equation.

5. Select the intervals for which the inequality is satisfied and write interval notation or set-builder notation for the solution set. If the inequality symbol is $\leq$ or $\geq$, then the solutions to step (2) should be included in the solution set. The $x$-values found in step (3) are never included in the solution set.

Solve: $\dfrac{x-1}{x+5} > \dfrac{x+3}{x-2}$.

*Equivalent inequality*: $\dfrac{x-1}{x+5} - \dfrac{x+3}{x-2} > 0$

*Related function*: $f(x) = \dfrac{x-1}{x+5} - \dfrac{x+3}{x-2}$

The function is not defined for $x = -5$ and $x = 2$. Solving $f(x) = 0$, we get $x = -\frac{13}{11}$. The critical values are $-5$, $-\frac{13}{11}$, and $2$. These divide the $x$-axis into four intervals.

| INTERVAL | TEST VALUE | SIGN OF $f(x)$ |
|---|---|---|
| $(-\infty, -5)$ | $f(-6) = 6.625$ | $+$ |
| $\left(-5, -\frac{13}{11}\right)$ | $f(-2) = -0.75$ | $-$ |
| $\left(-\frac{13}{11}, 2\right)$ | $f(0) = 1.3$ | $+$ |
| $(2, \infty)$ | $f(3) = -5.75$ | $-$ |

Test values are positive in the intervals $(-\infty, -5)$ and $\left(-\frac{13}{11}, 2\right)$. Since $f\left(-\frac{13}{11}\right) = 0$ and $-5$ and $2$ are not in the domain of $f$, $-5$, $-\frac{13}{11}$, and $2$ cannot be part of the solution set. The solution set is

$$(-\infty, -5) \cup \left(-\tfrac{13}{11}, 2\right).$$

## REVIEW EXERCISES

*Determine whether the statement is true or false.*

1. If $f(x) = (x + a)(x + b)(x - c)$, then $f(-b) = 0$. [4.3]

2. The graph of a rational function never crosses a vertical asymptote. [4.5]

3. For the function $g(x) = x^4 - 8x^2 - 9$, the only possible rational zeros are $1, -1, 3$, and $-3$. [4.4]

4. The graph of $P(x) = x^6 - x^8$ has at most 6 $x$-intercepts. [4.2]

5. The domain of the function

$$f(x) = \frac{x - 4}{(x + 2)(x - 3)}$$

is $(-\infty, -2) \cup (3, \infty)$. [4.5]

*Use a graphing calculator to graph the polynomial function. Then estimate the function's* **(a)** *zeros,* **(b)** *relative maxima,* **(c)** *relative minima, and* **(d)** *domain and range.* [4.1]

6. $f(x) = -2x^2 - 3x + 6$

7. $f(x) = x^3 + 3x^2 - 2x - 6$

8. $f(x) = x^4 - 3x^3 + 2x^2$

*Determine the leading term, the leading coefficient, and the degree of the polynomial. Then classify the polynomial function as constant, linear, quadratic, cubic, or quartic.* [4.1]

9. $f(x) = 7x^2 - 5 + 0.45x^4 - 3x^3$

10. $h(x) = -25$

**11.** $g(x) = 6 - 0.5x$

**12.** $f(x) = \frac{1}{3}x^3 - 2x + 3$

*Use the leading-term test to describe the end behavior of the graph of the function.* [4.1]

**13.** $f(x) = -\frac{1}{2}x^4 + 3x^2 + x - 6$

**14.** $f(x) = x^5 + 2x^3 - x^2 + 5x + 4$

*Find the zeros of the polynomial function and state the multiplicity of each.* [4.1]

**15.** $g(x) = \left(x - \frac{2}{3}\right)(x + 2)^3(x - 5)^2$

**16.** $f(x) = x^4 - 26x^2 + 25$

**17.** $h(x) = x^3 + 4x^2 - 9x - 36$

**18.** *Interest Compounded Annually.* When $P$ dollars is invested at interest rate $i$, compounded annually, for $t$ years, the investment grows to $A$ dollars, where

$$A = P(1 + i)^t.$$

a) Find the interest rate $i$ if $6250 grows to $6760 in 2 years. [4.1]

b) Find the interest rate $i$ if $1,000,000 grows to $1,215,506.25 in 4 years. [4.1]

**19.** *Cholesterol Level and the Risk of Heart Attack.* The table below lists data concerning the relationship of cholesterol level in men to the risk of a heart attack.

| Cholesterol Level | Number of Men per 100,000 Who Suffer a Heart Attack |
|---|---|
| 100 | 30 |
| 200 | 65 |
| 250 | 100 |
| 275 | 130 |

*Source: Nutrition Action Newsletter*

a) Use regression on a graphing calculator to fit linear, quadratic, and cubic functions to the data. [4.1]

b) It is also known that 180 of 100,000 men with a cholesterol level of 300 have a heart attack. Which function in part (a) would best make this prediction? [4.1]

c) Use the answer to part (b) to predict the heart attack rate for men with cholesterol levels of 350 and of 400. [4.1]

*Sketch the graph of the polynomial function.*

**20.** $f(x) = -x^4 + 2x^3$ [4.2]

**21.** $g(x) = (x - 1)^3(x + 2)^2$ [4.2]

**22.** $h(x) = x^3 + 3x^2 - x - 3$ [4.2]

**23.** $f(x) = x^4 - 5x^3 + 6x^2 + 4x - 8$ [4.2], [4.3], [4.4]

**24.** $g(x) = 2x^3 + 7x^2 - 14x + 5$ [4.2], [4.4]

*Using the intermediate value theorem, determine, if possible, whether the function $f$ has a zero between $a$ and $b$.* [4.2]

**25.** $f(x) = 4x^2 - 5x - 3;\ a = 1,\ b = 2$

**26.** $f(x) = x^3 - 4x^2 + \frac{1}{2}x + 2;\ a = -1,\ b = 1$

*In each of the following, a polynomial $P(x)$ and a divisor $d(x)$ are given. Use long division to find the quotient $Q(x)$ and the remainder $R(x)$ when $P(x)$ is divided by $d(x)$. Express $P(x)$ in the form $d(x) \cdot Q(x) + R(x)$.* [4.3]

**27.** $P(x) = 6x^3 - 2x^2 + 4x - 1$,
$d(x) = x - 3$

**28.** $P(x) = x^4 - 2x^3 + x + 5$,
$d(x) = x + 1$

*Use synthetic division to find the quotient and the remainder.* [4.3]

**29.** $(x^3 + 2x^2 - 13x + 10) \div (x - 5)$

**30.** $(x^4 + 3x^3 + 3x^2 + 3x + 2) \div (x + 2)$

**31.** $(x^5 - 2x) \div (x + 1)$

*Use synthetic division to find the indicated function value.* [4.3]

**32.** $f(x) = x^3 + 2x^2 - 13x + 10;\ f(-2)$

**33.** $f(x) = x^4 - 16;\ f(-2)$

**34.** $f(x) = x^5 - 4x^4 + x^3 - x^2 + 2x - 100$;
$f(-10)$

*Using synthetic division, determine whether the given numbers are zeros of the polynomial function.* [4.3]

**35.** $-i, -5;\ f(x) = x^3 - 5x^2 + x - 5$

**36.** $-1, -2;\ f(x) = x^4 - 4x^3 - 3x^2 + 14x - 8$

**37.** $\frac{1}{3}, 1;\ f(x) = x^3 - \frac{4}{3}x^2 - \frac{5}{3}x + \frac{2}{3}$

**38.** $2, -\sqrt{3};\ f(x) = x^4 - 5x^2 + 6$

*Factor the polynomial $f(x)$. Then solve the equation $f(x) = 0$.* [4.3], [4.4]

**39.** $f(x) = x^3 + 2x^2 - 7x + 4$

**40.** $f(x) = x^3 + 4x^2 - 3x - 18$

**41.** $f(x) = x^4 - 4x^3 - 21x^2 + 100x - 100$

**42.** $f(x) = x^4 - 3x^2 + 2$

*Find a polynomial function of degree 3 with the given numbers as zeros.* [4.4]

**43.** $-4, -1, 2$

**44.** $-3, 1 - i, 1 + i$

**45.** $\frac{1}{2}, 1 - \sqrt{2}, 1 + \sqrt{2}$

**46.** Find a polynomial function of degree 4 with $-5$ as a zero of multiplicity 3 and $\frac{1}{2}$ as a zero of multiplicity 1. [4.4]

**47.** Find a polynomial function of degree 5 with $-3$ as a zero of multiplicity 2, 2 as a zero of multiplicity 1, and 0 as a zero of multiplicity 2. [4.4]

*Suppose that a polynomial function of degree 5 with rational coefficients has the given zeros. Find the other zero(s).* [4.4]

**48.** $-\frac{2}{3}, \sqrt{5}, 4 + i$

**49.** $0, 1 + \sqrt{3}, -\sqrt{3}$

**50.** $-\sqrt{2}, \frac{1}{2}, 1, 2$

*Find a polynomial function of lowest degree with rational coefficients and the following as some of its zeros.* [4.4]

**51.** $\sqrt{11}$

**52.** $-i, 6$

**53.** $-1, 4, 1 + i$

**54.** $\sqrt{5}, -2i$

**55.** $\frac{1}{3}, 0, -3$

*List all possible rational zeros.* [4.4]

**56.** $h(x) = 4x^5 - 2x^3 + 6x - 12$

**57.** $g(x) = 3x^4 - x^3 + 5x^2 - x + 1$

**58.** $f(x) = x^3 - 2x^2 + x - 24$

*For each polynomial function:*

**a)** *Find the rational zeros and then the other zeros; that is, solve $f(x) = 0$.* [4.4]

**b)** *Factor $f(x)$ into linear factors.* [4.4]

**59.** $f(x) = 3x^5 + 2x^4 - 25x^3 - 28x^2 + 12x$

**60.** $f(x) = x^3 - 2x^2 - 3x + 6$

**61.** $f(x) = x^4 - 6x^3 + 9x^2 + 6x - 10$

**62.** $f(x) = x^3 + 3x^2 - 11x - 5$

**63.** $f(x) = 3x^3 - 8x^2 + 7x - 2$

**64.** $f(x) = x^5 - 8x^4 + 20x^3 - 8x^2 - 32x + 32$

**65.** $f(x) = x^6 + x^5 - 28x^4 - 16x^3 + 192x^2$

**66.** $f(x) = 2x^5 - 13x^4 + 32x^3 - 38x^2 + 22x - 5$

*What does Descartes' rule of signs tell you about the number of positive real zeros and the number of negative real zeros of each of the following polynomial functions?* [4.4]

**67.** $f(x) = 2x^6 - 7x^3 + x^2 - x$

**68.** $h(x) = -x^8 + 6x^5 - x^3 + 2x - 2$

**69.** $g(x) = 5x^5 - 4x^2 + x - 1$

*Graph the function. Be sure to label all the asymptotes. List the domain and the x- and y-intercepts.* [4.5]

**70.** $f(x) = \dfrac{x^2 - 5}{x + 2}$

**71.** $f(x) = \dfrac{5}{(x - 2)^2}$

**72.** $f(x) = \dfrac{x^2 + x - 6}{x^2 - x - 20}$

**73.** $f(x) = \dfrac{x - 2}{x^2 - 2x - 15}$

*In Exercises 74 and 75, find a rational function that satisfies the given conditions. Answers may vary, but try to give the simplest answer possible.* [4.5]

**74.** Vertical asymptotes $x = -2, x = 3$

**75.** Vertical asymptotes $x = -2, x = 3$; horizontal asymptote $y = 4$; x-intercept $(-3, 0)$

**76.** *Medical Dosage.* The function

$$N(t) = \frac{0.7t + 2000}{8t + 9}, \quad t \geq 5,$$

gives the body concentration $N(t)$, in parts per million, of a certain dosage of medication after time $t$, in hours.

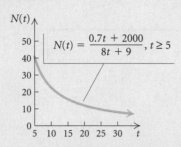

**a)** Find the horizontal asymptote of the graph and complete the following:

$$N(t) \rightarrow \boxed{\phantom{00}} \text{ as } t \rightarrow \infty. \ [4.5]$$

**b)** Explain the meaning of the answer to part (a) in terms of the application. [4.5]

*Solve.* [4.6]

**77.** $x^2 - 9 < 0$

**78.** $2x^2 > 3x + 2$

**79.** $(1 - x)(x + 4)(x - 2) \leq 0$

**80.** $\dfrac{x - 2}{x + 3} < 4$

**81.** *Height of a Rocket.* The function

$$S(t) = -16t^2 + 80t + 224$$

gives the height $S$, in feet, of a model rocket launched with a velocity of 80 ft/sec from a hill that is 224 ft high, where $t$ is the time, in seconds.

**a)** Determine when the rocket reaches the ground. [4.1]

**b)** On what interval is the height greater than 320 ft? [4.1], [4.6]

**82.** *Population Growth.* The population $P$, in thousands, of Novi is given by

$$P(t) = \frac{8000t}{4t^2 + 10},$$

where $t$ is the time, in months. Find the interval on which the population was 400,000 or greater. [4.6]

**83.** Determine the domain of the function

$$g(x) = \frac{x^2 + 2x - 3}{x^2 - 5x + 6}. \ [4.5]$$

**A.** $(-\infty, 2) \cup (2, 3) \cup (3, \infty)$
**B.** $(-\infty, -3) \cup (-3, 1) \cup (1, \infty)$
**C.** $(-\infty, 2) \cup (3, \infty)$
**D.** $(-\infty, -3) \cup (1, \infty)$

**84.** Determine the vertical asymptotes of the function

$$f(x) = \frac{x - 4}{(x + 1)(x - 2)(x + 4)}. \ [4.5]$$

**A.** $x = 1, x = -2$, and $x = 4$
**B.** $x = -1, x = 2, x = -4$, and $x = 4$
**C.** $x = -1, x = 2$, and $x = -4$
**D.** $x = 4$

**85.** The graph of $f(x) = -\frac{1}{2}x^4 + x^3 + 1$ is which of the following? [4.2]

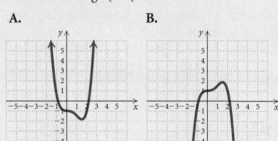

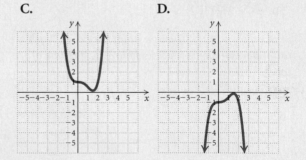

**Synthesis**

*Solve.*

**86.** $x^2 \geq 5 - 2x$ [4.6]    **87.** $\left|1 - \dfrac{1}{x^2}\right| < 3$ [4.6]

**88.** $x^4 - 2x^3 + 3x^2 - 2x + 2 = 0$ [4.4], [4.5]

**89.** $(x - 2)^{-3} < 0$ [4.6]

**90.** Express $x^3 - 1$ as a product of linear factors. [4.4]

**91.** Find $k$ such that $x + 3$ is a factor of $x^3 + kx^2 + kx - 15$. [4.3]

**92.** When $x^2 - 4x + 3k$ is divided by $x + 5$, the remainder is 33. Find the value of $k$. [4.3]

*Find the domain of the function.* [4.6]

**93.** $f(x) = \sqrt{x^2 + 3x - 10}$

**94.** $f(x) = \sqrt{x^2 - 3.1x + 2.2} + 1.75$

**95.** $f(x) = \dfrac{1}{\sqrt{5 - |7x + 2|}}$

## Collaborative Discussion and Writing

**96.** Explain the difference between a polynomial function and a rational function. [4.1], [4.5]

**97.** Is it possible for a third-degree polynomial with rational coefficients to have no real zeros? Why or why not? [4.4]

**98.** Explain and contrast the three types of asymptotes considered for rational functions. [4.5]

**99.** If $P(x)$ is an even function, and by Descartes' rule of signs, $P(x)$ has one positive real zero, how many negative real zeros does $P(x)$ have? Explain. [4.4]

**100.** Explain why the graph of a rational function cannot have both a horizontal asymptote and an oblique asymptote. [4.5]

**101.** Under what circumstances would a quadratic inequality have a solution set that is a closed interval? [4.6]

## Chapter 4 Test

*Determine the leading term, the leading coefficient, and the degree of the polynomial. Then classify the polynomial as constant, linear, quadratic, cubic, or quartic.*

**1.** $f(x) = 2x^3 + 6x^2 - x^4 + 11$

**2.** $h(x) = -4.7x + 29$

**3.** Find the zeros of the polynomial function and state the multiplicity of each:
$$f(x) = x(3x - 5)(x - 3)^2(x + 1)^3.$$

**4.** *Foreign-Born Population.* In 1970, only 4.7% of the U.S. population was foreign-born, while in 2007, 12.6% of the population was foreign-born (*Sources*: Annual Social and Economic Supplements, Current Population Surveys, U.S. Census Bureau, U.S. Department of Commerce). The quartic function

$$f(x) = -0.0000007623221x^4 + 0.00021189064x^3$$
$$- 0.016314058x^2 + 0.2440779643x$$
$$+ 13.59260684,$$

where $x$ is the number of years since 1900, can be used to estimate the percent of the U.S. population for years 1900 to 2007 that was foreign-born. Using this function, estimate the percent of the population that was foreign-born in 1930, in 1990, and in 2000.

*Sketch the graph of the polynomial function.*

**5.** $f(x) = x^3 - 5x^2 + 2x + 8$

**6.** $f(x) = -2x^4 + x^3 + 11x^2 - 4x - 12$

*Using the intermediate value theorem, determine, if possible, whether the function has a zero between a and b.*

**7.** $f(x) = -5x^2 + 3$; $a = 0$, $b = 2$

**8.** $g(x) = 2x^3 + 6x^2 - 3$; $a = -2$, $b = -1$

9. Use long division to find the quotient $Q(x)$ and the remainder $R(x)$ when $P(x)$ is divided by $d(x)$. Express $P(x)$ in the form $d(x) \cdot Q(x) + R(x)$. Show your work.

$$P(x) = x^4 + 3x^3 + 2x - 5,$$
$$d(x) = x - 1$$

10. Use synthetic division to find the quotient and the remainder. Show your work.

$$(3x^3 - 12x + 7) \div (x - 5)$$

11. Use synthetic division to find $P(-3)$ for $P(x) = 2x^3 - 6x^2 + x - 4$. Show your work.

12. Use synthetic division to determine whether $-2$ is a zero of $f(x) = x^3 + 4x^2 + x - 6$. Answer yes or no. Show your work.

13. Find a polynomial of degree 4 with $-3$ as a zero of multiplicity 2 and 0 and 6 as zeros of multiplicity 1.

14. Suppose that a polynomial function of degree 5 with rational coefficients has 1, $\sqrt{3}$, and $2 - i$ as zeros. Find the other zeros.

*Find a polynomial function of lowest degree with rational coefficients and the following as some of its zeros.*

15. $-10, 3i$

16. $0, -\sqrt{3}, 1 - i$

*List all possible rational zeros.*

17. $f(x) = 2x^3 + x^2 - 2x + 12$

18. $h(x) = 10x^4 - x^3 + 2x - 5$

*For each polynomial function:*

a) *Find the rational zeros and then the other zeros; that is, solve $f(x) = 0$.*

b) *Factor $f(x)$ into linear factors.*

19. $f(x) = x^3 + x^2 - 5x - 5$

20. $f(x) = 2x^4 - 11x^3 + 16x^2 - x - 6$

21. $f(x) = x^3 + 4x^2 + 4x + 16$

22. $f(x) = 3x^4 - 11x^3 + 15x^2 - 9x + 2$

23. What does Descartes' rule of signs tell you about the number of positive real zeros and the number of negative real zeros of the following function?

$$g(x) = -x^8 + 2x^6 - 4x^3 - 1$$

*Graph the function. Be sure to label all the asymptotes. List the domain and the x- and y-intercepts.*

24. $f(x) = \dfrac{2}{(x - 3)^2}$

25. $f(x) = \dfrac{x + 3}{x^2 - 3x - 4}$

26. Find a rational function that has vertical asymptotes $x = -1$ and $x = 2$ and x-intercept $(-4, 0)$.

*Solve.*

27. $2x^2 > 5x + 3$

28. $\dfrac{x + 1}{x - 4} \le 3$

29. The function $S(t) = -16t^2 + 64t + 192$ gives the height $S$, in feet, of a model rocket launched with a velocity of 64 ft/sec from a hill that is 192 ft high.

a) Determine how long it will take the rocket to reach the ground.

b) Find the interval on which the height of the rocket is greater than 240 ft.

30. The graph of $f(x) = x^3 - x^2 - 2$ is which of the following?

**A.** **B.**

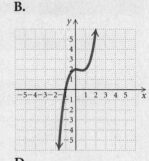

**C.** **D.**

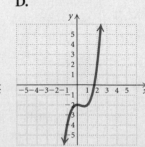

## Synthesis

31. Find the domain of $f(x) = \sqrt{x^2 + x - 12}$.

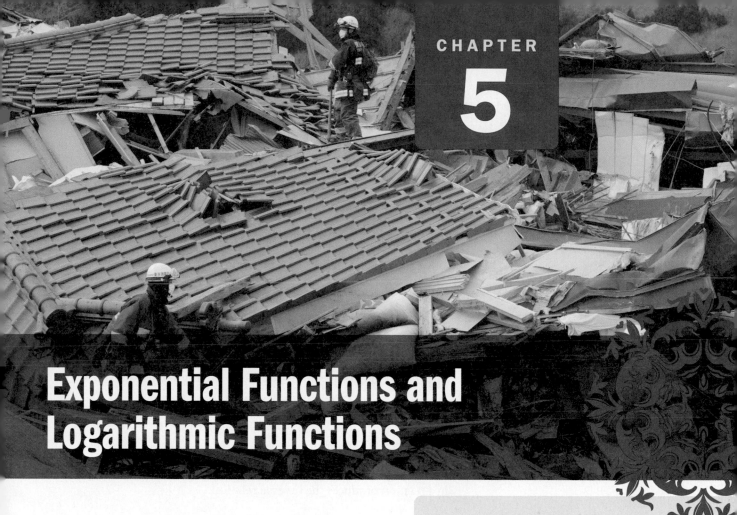

# Exponential Functions and Logarithmic Functions

## APPLICATION

The undersea Tohoku earthquake and tsunami, near the northeast coast of Honshu, Japan, on March 11, 2011, had an intensity of $10^{9.0} \cdot I_0$ (*Source*: earthquake.usgs.gov). They caused extensive loss of life and severe structural damage to buildings, railways, and roads. What was the magnitude on the Richter scale?

This problem appears as Example 13 in Section 5.3.

# 5.1

## Inverse Functions

- Determine whether a function is one-to-one, and if it is, find a formula for its inverse.
- Simplify expressions of the type $(f \circ f^{-1})(x)$ and $(f^{-1} \circ f)(x)$.

### Inverses

When we go from an output of a function back to its input or inputs, we get an inverse relation. When that relation is a function, we have an inverse function.

Consider the relation $h$ given as follows:

$$h = \{(-8, 5), (4, -2), (-7, 1), (3.8, 6.2)\}.$$

Suppose we *interchange* the first and second coordinates. The relation we obtain is called the **inverse** of the relation $h$ and is given as follows:

Inverse of $h = \{(5, -8), (-2, 4), (1, -7), (6.2, 3.8)\}.$

| RELATIONS |
|---|
| REVIEW SECTION **1.2.** |

### INVERSE RELATION

Interchanging the first and second coordinates of each ordered pair in a relation produces the **inverse relation**.

**EXAMPLE 1** Consider the relation $g$ given by

$$g = \{(2, 4), (-1, 3), (-2, 0)\}.$$

Graph the relation in blue. Find the inverse and graph it in red.

*Solution* The relation $g$ is shown in blue in the figure at left. The inverse of the relation is

$$\{(4, 2), (3, -1), (0, -2)\}$$

and is shown in red. The pairs in the inverse are reflections of the pairs in $g$ across the line $y = x$.

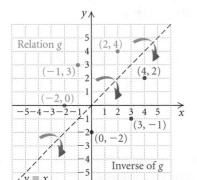

**Now Try Exercise 1.**

### INVERSE RELATION

If a relation is defined by an equation, interchanging the variables produces an equation of the **inverse relation**.

**EXAMPLE 2** Find an equation for the inverse of the relation

$$y = x^2 - 5x.$$

**Solution** We interchange $x$ and $y$ and obtain an equation of the inverse:

$$x = y^2 - 5y.$$

**Now Try Exercise 9.**

If a relation is given by an equation, then the solutions of the inverse can be found from those of the original equation by interchanging the first and second coordinates of each ordered pair. Thus the graphs of a relation and its inverse are always reflections of each other across the line $y = x$. This is illustrated with the equations of Example 2 in the tables and graph below. We will explore inverses and their graphs later in this section.

| $x = y^2 - 5y$ | $y$ |
|---|---|
| 6 | −1 |
| 0 | 0 |
| −6 | 2 |
| −4 | 4 |

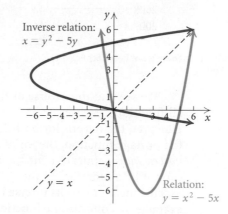

| $x$ | $y = x^2 - 5x$ |
|---|---|
| −1 | 6 |
| 0 | 0 |
| 2 | −6 |
| 4 | −4 |

## ■ Inverses and One-to-One Functions

Let's consider the following two functions.

| Year (domain) | First-Class Postage Cost, in cents (range) |
|---|---|
| 1996 | 32 |
| 1998 | 32 |
| 2000 | 33 |
| 2002 | 37 |
| 2004 | 37 |
| 2006 | 39 |
| 2007 | 41 |
| 2008 | 42 |
| 2009 | 44 |

| Number (domain) | Cube (range) |
|---|---|
| −3 | −27 |
| −2 | −8 |
| −1 | −1 |
| 0 | 0 |
| 1 | 1 |
| 2 | 8 |
| 3 | 27 |

*Source*: U.S. Postal Service

Suppose we reverse the arrows. Are these inverse relations functions?

| Year (range) | First-Class Postage Cost, in cents (domain) |
|---|---|
| 1996 ← | |
| 1998 ← | 32 |
| 2000 ← | 33 |
| 2002 ← | |
| 2004 ← | 37 |
| 2006 ← | 39 |
| 2007 ← | 41 |
| 2008 ← | 42 |
| 2009 ← | 44 |

| Number (range) | Cube (domain) |
|---|---|
| −3 ← | −27 |
| −2 ← | −8 |
| −1 ← | −1 |
| 0 ← | 0 |
| 1 ← | 1 |
| 2 ← | 8 |
| 3 ← | 27 |

*Source*: U.S. Postal Service

We see that the inverse of the postage function is not a function. Like all functions, each input in the postage function has exactly one output. However, the output for both 1996 and 1998 is 32. Thus in the inverse of the postage function, the input 32 has *two* outputs, 1996 and 1998. When two or more inputs of a function have the same output, the inverse relation cannot be a function. In the cubing function, each output corresponds to exactly one input, so its inverse is also a function. The cubing function is an example of a **one-to-one function**.

> ### ONE-TO-ONE FUNCTIONS
>
> A function $f$ is **one-to-one** if different inputs have different outputs—that is,
>
> $$\text{if} \quad a \neq b, \quad \text{then} \quad f(a) \neq f(b).$$
>
> Or, a function $f$ is **one-to-one** if when the outputs are the same, the inputs are the same—that is,
>
> $$\text{if} \quad f(a) = f(b), \quad \text{then} \quad a = b.$$

If the inverse of a function $f$ is also a function, it is named $f^{-1}$ (read "$f$-inverse").

**The $-1$ in $f^{-1}$ is *not* an exponent!**

Do *not* misinterpret the $-1$ in $f^{-1}$ as a negative exponent: $f^{-1}$ does *not* mean the reciprocal of $f$ and $f^{-1}(x)$ is *not* equal to $\dfrac{1}{f(x)}$.

## ONE-TO-ONE FUNCTIONS AND INVERSES

- If a function $f$ is one-to-one, then its inverse $f^{-1}$ is a function.
- The domain of a one-to-one function $f$ is the range of the inverse $f^{-1}$.
- The range of a one-to-one function $f$ is the domain of the inverse $f^{-1}$.

$$\begin{matrix} D_f & \nearrow & D_{f^{-1}} \\ R_f & \searrow & R_{f^{-1}} \end{matrix}$$

- A function that is increasing over its entire domain or is decreasing over its entire domain is a one-to-one function.

**EXAMPLE 3**   Given the function $f$ described by $f(x) = 2x - 3$, prove that $f$ is one-to-one (that is, it has an inverse that is a function).

***Solution***   To show that $f$ is one-to-one, we show that if $f(a) = f(b)$, then $a = b$. Assume that $f(a) = f(b)$ for $a$ and $b$ in the domain of $f$. Since $f(a) = 2a - 3$ and $f(b) = 2b - 3$, we have

$$2a - 3 = 2b - 3$$
$$2a = 2b \qquad \text{Adding 3}$$
$$a = b. \qquad \text{Dividing by 2}$$

Thus, if $f(a) = f(b)$, then $a = b$. This shows that $f$ is one-to-one.

**Now Try Exercise 17.**

**EXAMPLE 4**   Given the function $g$ described by $g(x) = x^2$, prove that $g$ is not one-to-one.

***Solution***   We can prove that $g$ is not one-to-one by finding two numbers $a$ and $b$ for which $a \neq b$ and $g(a) = g(b)$. Two such numbers are $-3$ and $3$, because $-3 \neq 3$ and $g(-3) = g(3) = 9$. Thus $g$ is not one-to-one.

**Now Try Exercise 21.**

The graphs below show a function, in blue, and its inverse, in red. To determine whether the inverse is a function, we can apply the vertical-line test to its graph. By reflecting each such vertical line back across the line $y = x$, we obtain an equivalent **horizontal-line test** for the original function.

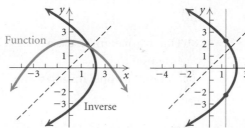

The vertical-line test shows that the inverse is not a function.

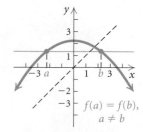

The horizontal-line test shows that the function is not one-to-one.

> ## HORIZONTAL-LINE TEST
>
> If it is possible for a horizontal line to intersect the graph of a function more than once, then the function is *not* one-to-one and its inverse is *not* a function.

**EXAMPLE 5** From the graph shown, determine whether each function is one-to-one and thus has an inverse that is a function.

**a)**

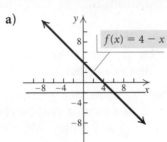

**b)**

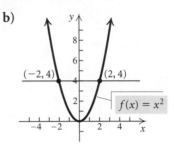

**c)**

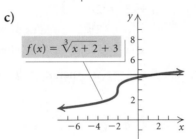

**d)**

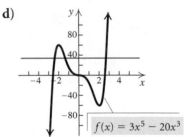

**Solution** For each function, we apply the horizontal-line test.

| RESULT | REASON |
| --- | --- |
| **a)** One-to-one; inverse is a function | No horizontal line intersects the graph more than once. |
| **b)** Not one-to-one; inverse is not a function | There are many horizontal lines that intersect the graph more than once. Note that where the line $y = 4$ intersects the graph, the first coordinates are $-2$ and 2. Although these are different inputs, they have the same output, 4. |
| **c)** One-to-one; inverse is a function | No horizontal line intersects the graph more than once. |
| **d)** Not one-to-one; inverse is not a function | There are many horizontal lines that intersect the graph more than once. |

**Now Try Exercises 25 and 27.**

### ■ Finding Formulas for Inverses

Suppose that a function is described by a formula. If it has an inverse that is a function, we proceed as follows to find a formula for $f^{-1}$.

---

**Obtaining a Formula for an Inverse**

If a function $f$ is one-to-one, a formula for its inverse can generally be found as follows:

1. Replace $f(x)$ with $y$.
2. Interchange $x$ and $y$.
3. Solve for $y$.
4. Replace $y$ with $f^{-1}(x)$.

---

**EXAMPLE 6**   Determine whether the function $f(x) = 2x - 3$ is one-to-one, and if it is, find a formula for $f^{-1}(x)$.

*Solution*   The graph of $f$ is shown at left. It passes the horizontal-line test. Thus it is one-to-one and its inverse is a function. We also proved that $f$ is one-to-one in Example 3. We find a formula for $f^{-1}(x)$.

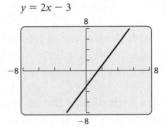

$y = 2x - 3$

1. Replace $f(x)$ with $y$:       $y = 2x - 3$
2. Interchange $x$ and $y$:       $x = 2y - 3$
3. Solve for $y$:            $x + 3 = 2y$

$$\frac{x + 3}{2} = y$$

4. Replace $y$ with $f^{-1}(x)$:   $f^{-1}(x) = \dfrac{x + 3}{2}.$

**Now Try Exercise 57.**

Consider

$$f(x) = 2x - 3 \quad \text{and} \quad f^{-1}(x) = \frac{x + 3}{2}$$

from Example 6. For the input 5, we have

$$f(5) = 2 \cdot 5 - 3 = 10 - 3 = 7.$$

The output is 7. Now we use 7 for the input in the inverse:

$$f^{-1}(7) = \frac{7 + 3}{2} = \frac{10}{2} = 5.$$

The function $f$ takes the number 5 to 7. The inverse function $f^{-1}$ takes the number 7 back to 5.

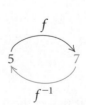

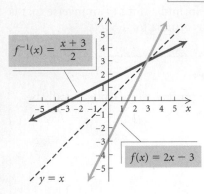

$$f^{-1}(x) = \frac{x+3}{2}$$

$$f(x) = 2x - 3$$

$$y = x$$

**EXAMPLE 7**   Graph

$$f(x) = 2x - 3 \quad \text{and} \quad f^{-1}(x) = \frac{x + 3}{2}$$

using the same set of axes. Then compare the two graphs.

***Solution***   The graphs of $f$ and $f^{-1}$ are shown at left. The solutions of the inverse function can be found from those of the original function by interchanging the first and second coordinates of each ordered pair.

| $x$ | $f(x) = 2x - 3$ | |
|---|---|---|
| $-1$ | $-5$ | |
| $0$ | $-3$ | ← $y$-intercept |
| $2$ | $1$ | |
| $3$ | $3$ | |

| $x$ | $f^{-1}(x) = \dfrac{x+3}{2}$ | |
|---|---|---|
| $-5$ | $-1$ | |
| $-3$ | $0$ | ← $x$-intercept |
| $1$ | $2$ | |
| $3$ | $3$ | |

On some graphing calculators, we can graph the inverse of a function after graphing the function itself by accessing a drawing feature. Consult your user's manual or the *Graphing Calculator Manual* that accompanies this text for the procedure.

When we interchange $x$ and $y$ in finding a formula for the inverse of $f(x) = 2x - 3$, we are in effect reflecting the graph of that function across the line $y = x$. For example, when the coordinates of the $y$-intercept of the graph of $f$, $(0, -3)$, are reversed, we get the $x$-intercept of the graph of $f^{-1}$, $(-3, 0)$, If we were to graph $f(x) = 2x - 3$ in wet ink and fold along the line $y = x$, the graph of $f^{-1}(x) = (x + 3)/2$ would be formed by the ink transferred from $f$.

*i-Figure*

> The graph of $f^{-1}$ is a reflection of the graph of $f$ across the line $y = x$.

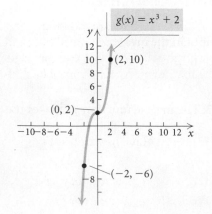

$$g(x) = x^3 + 2$$

$(2, 10)$

$(0, 2)$

$(-2, -6)$

**EXAMPLE 8**   Consider $g(x) = x^3 + 2$.

**a)** Determine whether the function is one-to-one.

**b)** If it is one-to-one, find a formula for its inverse.

**c)** Graph the function and its inverse.

***Solution***

**a)** The graph of $g(x) = x^3 + 2$ is shown at left. It passes the horizontal-line test and thus has an inverse that is a function. We also know that $g(x)$ is one-to-one because it is an increasing function over its entire domain.

**b)** We follow the procedure for finding an inverse.

**1.** Replace $g(x)$ with $y$: $\qquad y = x^3 + 2$

**2.** Interchange $x$ and $y$: $\qquad x = y^3 + 2$

**3.** Solve for $y$: $\qquad x - 2 = y^3$

$$\sqrt[3]{x - 2} = y$$

**4.** Replace $y$ with $g^{-1}(x)$: $\quad g^{-1}(x) = \sqrt[3]{x - 2}$.

We can test a point as a partial check:

$$g(x) = x^3 + 2$$
$$g(3) = 3^3 + 2 = 27 + 2 = 29.$$

Will $g^{-1}(29) = 3$? We have

$$g^{-1}(x) = \sqrt[3]{x - 2}$$
$$g^{-1}(29) = \sqrt[3]{29 - 2} = \sqrt[3]{27} = 3.$$

Since $g(3) = 29$ and $g^{-1}(29) = 3$, we can be reasonably certain that the formula for $g^{-1}(x)$ is correct.

**c)** To find the graph, we reflect the graph of $g(x) = x^3 + 2$ across the line $y = x$. This can be done by plotting points or by using a graphing calculator.

| $x$ | $g(x)$ |
|-----|--------|
| $-2$ | $-6$ |
| $-1$ | $1$ |
| $0$ | $2$ |
| $1$ | $3$ |
| $2$ | $10$ |

| $x$ | $g^{-1}(x)$ |
|-----|-------------|
| $-6$ | $-2$ |
| $1$ | $-1$ |
| $2$ | $0$ |
| $3$ | $1$ |
| $10$ | $2$ |

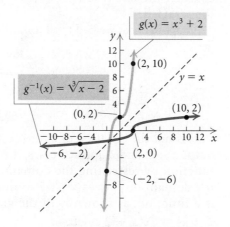

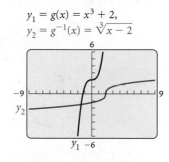

Now Try Exercise 79.

## Inverse Functions and Composition

Suppose that we were to use some input $a$ for a one-to-one function $f$ and find its output, $f(a)$. The function $f^{-1}$ would then take that output back to $a$. Similarly, if we began with an input $b$ for the function $f^{-1}$ and found its output, $f^{-1}(b)$, the original function $f$ would then take that output back to $b$. This is summarized as follows.

If a function $f$ is one-to-one, then $f^{-1}$ is the unique function such that each of the following holds:

$(f^{-1} \circ f)(x) = f^{-1}(f(x)) = x$, for each $x$ in the domain of $f$, and
$(f \circ f^{-1})(x) = f(f^{-1}(x)) = x$, for each $x$ in the domain of $f^{-1}$.

**EXAMPLE 9** Given that $f(x) = 5x + 8$, use composition of functions to show that

$$f^{-1}(x) = \frac{x - 8}{5}.$$

*Solution* We find $(f^{-1} \circ f)(x)$ and $(f \circ f^{-1})(x)$ and check to see that each is $x$:

$$(f^{-1} \circ f)(x) = f^{-1}(f(x))$$

$$= f^{-1}(5x + 8) = \frac{(5x + 8) - 8}{5} = \frac{5x}{5} = x;$$

$$(f \circ f^{-1})(x) = f(f^{-1}(x))$$

$$= f\left(\frac{x - 8}{5}\right) = 5\left(\frac{x - 8}{5}\right) + 8 = x - 8 + 8 = x.$$

> **Now Try Exercise 87.**

## ■ Restricting a Domain

In the case in which the inverse of a function is not a function, the domain of the function can be restricted to allow the inverse to be a function. Let's consider the function $f(x) = x^2 - 2$. It is not one-to-one. The graph is shown at left.

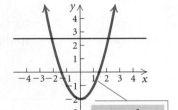
$f(x) = x^2 - 2$

$f^{-1}(x) = \sqrt{x + 2}$

$f(x) = x^2 - 2$,
for inputs $x \geq 0$

Suppose that we had tried to find a formula for the inverse as follows:

$$y = x^2 - 2 \qquad \text{Replacing } f(x) \text{ with } y$$
$$x = y^2 - 2 \qquad \text{Interchanging } x \text{ and } y$$
$$x + 2 = y^2$$
$$\pm\sqrt{x + 2} = y. \qquad \text{Solving for } y$$

This is not the equation of a function. An input of, say, 2 would yield two outputs, $-2$ and 2. In such cases, it is convenient to consider "part" of the function by restricting the domain of $f(x)$. For example, if we restrict the domain of $f(x) = x^2 - 2$ to nonnegative numbers, then its inverse is a function, as shown with the graphs of $f(x) = x^2 - 2$, $x \geq 0$, and $f^{-1}(x) = \sqrt{x + 2}$ at left.

## 5.1 Exercise Set

*Find the inverse of the relation.*

1. $\{(7, 8), (-2, 8), (3, -4), (8, -8)\}$

2. $\{(0, 1), (5, 6), (-2, -4)\}$

3. $\{(-1, -1), (-3, 4)\}$

4. $\{(-1, 3), (2, 5), (-3, 5), (2, 0)\}$

*Find an equation of the inverse relation.*

5. $y = 4x - 5$

6. $2x^2 + 5y^2 = 4$

7. $x^3 y = -5$

8. $y = 3x^2 - 5x + 9$

**9.** $x = y^2 - 2y$

**10.** $x = \frac{1}{2}y + 4$

*Graph the equation by substituting and plotting points. Then reflect the graph across the line $y = x$ to obtain the graph of its inverse.*

**11.** $x = y^2 - 3$       **12.** $y = x^2 + 1$

**13.** $y = 3x - 2$       **14.** $x = -y + 4$

**15.** $y = |x|$       **16.** $x + 2 = |y|$

*Given the function $f$, prove that $f$ is one-to-one using the definition of a one-to-one function on p. 394.*

**17.** $f(x) = \frac{1}{3}x - 6$       **18.** $f(x) = 4 - 2x$

**19.** $f(x) = x^3 + \frac{1}{2}$       **20.** $f(x) = \sqrt[3]{x}$

*Given the function $g$, prove that $g$ is not one-to-one using the definition of a one-to-one function on p. 394.*

**21.** $g(x) = 1 - x^2$

**22.** $g(x) = 3x^2 + 1$

**23.** $g(x) = x^4 - x^2$

**24.** $g(x) = \dfrac{1}{x^6}$

*Using the horizontal-line test, determine whether the function is one-to-one.*

**25.** $f(x) = 2.7^x$       **26.** $f(x) = 2^{-x}$

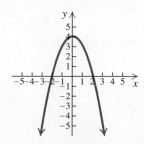 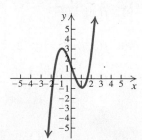

**27.** $f(x) = 4 - x^2$       **28.** $f(x) = x^3 - 3x + 1$

**29.** $f(x) = \dfrac{8}{x^2 - 4}$       **30.** $f(x) = \sqrt{\dfrac{10}{4 + x}}$

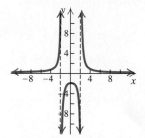 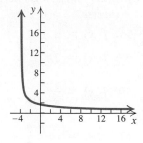

**31.** $f(x) = \sqrt[3]{x + 2} - 2$       **32.** $f(x) = \dfrac{8}{x}$

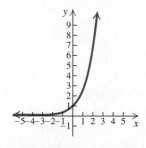 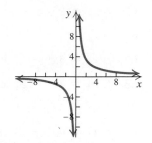

*Graph the function and determine whether the function is one-to-one using the horizontal-line test.*

**33.** $f(x) = 5x - 8$       **34.** $f(x) = 3 + 4x$

**35.** $f(x) = 1 - x^2$       **36.** $f(x) = |x| - 2$

**37.** $f(x) = |x + 2|$       **38.** $f(x) = -0.8$

**39.** $f(x) = -\dfrac{4}{x}$       **40.** $f(x) = \dfrac{2}{x + 3}$

**41.** $f(x) = \dfrac{2}{3}$       **42.** $f(x) = \frac{1}{2}x^2 + 3$

**43.** $f(x) = \sqrt{25 - x^2}$       **44.** $f(x) = -x^3 + 2$

*Graph the function and its inverse using a graphing calculator. Use an inverse drawing feature, if available. Find the domain and the range of $f$ and of $f^{-1}$.*

**45.** $f(x) = 0.8x + 1.7$       **46.** $f(x) = 2.7 - 1.08x$

**47.** $f(x) = \frac{1}{2}x - 4$       **48.** $f(x) = x^3 - 1$

**49.** $f(x) = \sqrt{x - 3}$       **50.** $f(x) = -\dfrac{2}{x}$

**51.** $f(x) = x^2 - 4, \ x \geq 0$

**52.** $f(x) = 3 - x^2, \ x \geq 0$

**53.** $f(x) = (3x - 9)^3$

**54.** $f(x) = \sqrt[3]{\dfrac{x - 3.2}{1.4}}$

*In Exercises 55–70, for each function:*

**a)** *Determine whether it is one-to-one.*

**b)** *If the function is one-to-one, find a formula for the inverse.*

**55.** $f(x) = x + 4$

**56.** $f(x) = 7 - x$

**57.** $f(x) = 2x - 1$

**58.** $f(x) = 5x + 8$

**59.** $f(x) = \dfrac{4}{x + 7}$

**60.** $f(x) = -\dfrac{3}{x}$

**61.** $f(x) = \dfrac{x + 4}{x - 3}$

**62.** $f(x) = \dfrac{5x - 3}{2x + 1}$

**63.** $f(x) = x^3 - 1$

**64.** $f(x) = (x + 5)^3$

**65.** $f(x) = x\sqrt{4 - x^2}$

**66.** $f(x) = 2x^2 - x - 1$

**67.** $f(x) = 5x^2 - 2,\ x \geq 0$

**68.** $f(x) = 4x^2 + 3,\ x \geq 0$

**69.** $f(x) = \sqrt{x + 1}$

**70.** $f(x) = \sqrt[3]{x - 8}$

*Find the inverse by thinking about the operations of the function and then reversing, or undoing, them. Check your work algebraically.*

| FUNCTION | INVERSE |
|---|---|
| **71.** $f(x) = 3x$ | $f^{-1}(x) = $ ▨ |
| **72.** $f(x) = \frac{1}{4}x + 7$ | $f^{-1}(x) = $ ▨ |
| **73.** $f(x) = -x$ | $f^{-1}(x) = $ ▨ |
| **74.** $f(x) = \sqrt[3]{x} - 5$ | $f^{-1}(x) = $ ▨ |
| **75.** $f(x) = \sqrt[3]{x - 5}$ | $f^{-1}(x) = $ ▨ |
| **76.** $f(x) = x^{-1}$ | $f^{-1}(x) = $ ▨ |

*Each graph in Exercises 77–82 is the graph of a one-to-one function f. Sketch the graph of the inverse function $f^{-1}$.*

**77.**
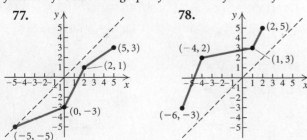

**78.**

**79.**
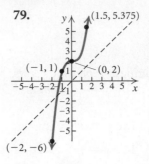

**80.**
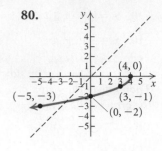

**81.**
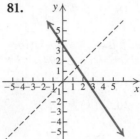

**82.**
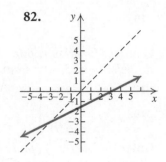

*For the function f, use composition of functions to show that $f^{-1}$ is as given.*

**83.** $f(x) = \frac{7}{8}x,\ f^{-1}(x) = \frac{8}{7}x$

**84.** $f(x) = \dfrac{x + 5}{4},\ f^{-1}(x) = 4x - 5$

**85.** $f(x) = \dfrac{1 - x}{x},\ f^{-1}(x) = \dfrac{1}{x + 1}$

**86.** $f(x) = \sqrt[3]{x + 4},\ f^{-1}(x) = x^3 - 4$

**87.** $f(x) = \dfrac{2}{5}x + 1,\ f^{-1}(x) = \dfrac{5x - 5}{2}$

**88.** $f(x) = \dfrac{x + 6}{3x - 4},\ f^{-1}(x) = \dfrac{4x + 6}{3x - 1}$

*Find the inverse of the given one-to-one function f. Give the domain and the range of f and of $f^{-1}$, and then graph both f and $f^{-1}$ on the same set of axes.*

**89.** $f(x) = 5x - 3$

**90.** $f(x) = 2 - x$

**91.** $f(x) = \dfrac{2}{x}$

**92.** $f(x) = -\dfrac{3}{x + 1}$

**93.** $f(x) = \frac{1}{3}x^3 - 2$

**94.** $f(x) = \sqrt[3]{x} - 1$

**95.** $f(x) = \dfrac{x + 1}{x - 3}$

**96.** $f(x) = \dfrac{x - 1}{x + 2}$

**97.** Find $f(f^{-1}(5))$ and $f^{-1}(f(a))$:

$$f(x) = x^3 - 4.$$

**98.** Find $f^{-1}(f(p))$ and $f(f^{-1}(1253))$:

$$f(x) = \sqrt[5]{\frac{2x - 7}{3x + 4}}.$$

**99.** *Swimming Lessons.* A city swimming league determines that the cost per person of a group swim lesson is given by the formula

$$C(x) = \frac{60 + 2x}{x},$$

where $x$ is the number of people in the group and $C(x)$ is in dollars. Find $C^{-1}(x)$ and explain what it represents.

**100.** *Spending on Pets.* The total amount of spending per year, in billions of dollars, on pets in the United States $x$ years after 2000 is given by the function

$$P(x) = 2.1782x + 25.3$$

(*Source:* Animal Pet Products Manufacturing Association).

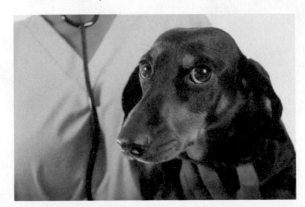

**a)** Determine the total amount of spending per year on pets in 2005 and in 2010.

**b)** Find $P^{-1}(x)$ and explain what it represents.

**101.** *Converting Temperatures.* The following formula can be used to convert Fahrenheit temperatures $x$ to Celsius temperatures $T(x)$:

$$T(x) = \tfrac{5}{9}(x - 32).$$

**a)** Find $T(-13°)$ and $T(86°)$.

**b)** Find $T^{-1}(x)$ and explain what it represents.

**102.** *Women's Shoe Sizes.* A function that will convert women's shoe sizes in the United States to those in Australia is

$$s(x) = \frac{2x - 3}{2}$$

(*Source:* OnlineConversion.com).

**a)** Determine the women's shoe sizes in Australia that correspond to sizes 5, $7\tfrac{1}{2}$, and 8 in the United States.

**b)** Find a formula for the inverse of the function.

**c)** Use the inverse function to determine the women's shoe sizes in the United States that correspond to sizes 3, $5\tfrac{1}{2}$, and 7 in Australia.

## Skill Maintenance

*Consider the following quadratic functions. Without graphing them, answer the questions below.*

**a)** $f(x) = 2x^2$
**b)** $f(x) = -x^2$
**c)** $f(x) = \tfrac{1}{4}x^2$
**d)** $f(x) = -5x^2 + 3$
**e)** $f(x) = \tfrac{2}{3}(x - 1)^2 - 3$
**f)** $f(x) = -2(x + 3)^2 + 1$
**g)** $f(x) = (x - 3)^2 + 1$
**h)** $f(x) = -4(x + 1)^2 - 3$

**103.** Which functions have a maximum value?

**104.** Which graphs open up?

**105.** Consider (a) and (c). Which graph is narrower?

**106.** Consider (d) and (e). Which graph is narrower?

**107.** Which graph has vertex $(-3, 1)$?

**108.** For which is the line of symmetry $x = 0$?

## Synthesis

*Using only a graphing calculator, determine whether the functions are inverses of each other.*

**109.** $f(x) = \sqrt[3]{\dfrac{x - 3.2}{1.4}}$, $g(x) = 1.4x^3 + 3.2$

**110.** $f(x) = \dfrac{2x - 5}{4x + 7}$, $g(x) = \dfrac{7x - 4}{5x + 2}$

**111.** The function $f(x) = x^2 - 3$ is not one-to-one. Restrict the domain of $f$ so that its inverse is a function. Find the inverse and state the restriction on the domain of the inverse.

**112.** Consider the function $f$ given by
$$f(x) = \begin{cases} x^3 + 2, & x \le -1, \\ x^2, & -1 < x < 1, \\ x + 1, & x \ge 1. \end{cases}$$
Does $f$ have an inverse that is a function? Why or why not?

**113.** Find three examples of functions that are their own inverses; that is, $f = f^{-1}$.

**114.** Given the function $f(x) = ax + b$, $a \ne 0$, find the values of $a$ and $b$ for which $f^{-1}(x) = f(x)$.

---

## 5.2 Exponential Functions and Graphs

- Graph exponential equations and exponential functions.
- Solve applied problems involving exponential functions and their graphs.

We now turn our attention to the study of a set of functions that are very rich in application. Consider the following graphs. Each one illustrates an *exponential function*. In this section, we consider such functions and some important applications.

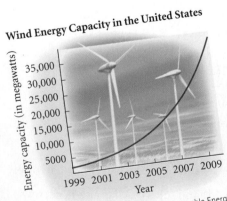

**Wind Energy Capacity in the United States**

*Source: Office of Energy Efficiency and Renewable Energy*

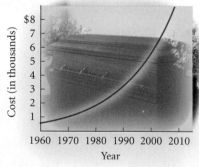

**Cost of a Funeral for an Adult**

*Source: National Funeral Association Directors*

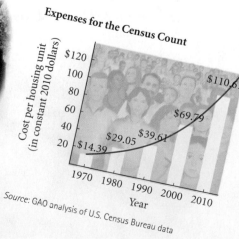

**Expenses for the Census Count**

*Source: GAO analysis of U.S. Census Bureau data*

**EXPONENTS**

REVIEW SECTIONS **R.2** AND **R.7.**

## ■ Graphing Exponential Functions

We now define exponential functions. We assume that $a^x$ has meaning for any real number $x$ and any positive real number $a$ and that the laws of exponents still hold, though we will not prove them here.

> ### EXPONENTIAL FUNCTION
>
> The function $f(x) = a^x$, where $x$ is a real number, $a > 0$ and $a \neq 1$, is called the **exponential function, base $a$.**

We require the **base** to be positive in order to avoid the imaginary numbers that would occur by taking even roots of negative numbers: An example is $(-1)^{1/2}$, the square root of $-1$, which is not a real number. The restriction $a \neq 1$ is made to exclude the constant function $f(x) = 1^x = 1$, which does not have an inverse that is a function because it is not one-to-one.

The following are examples of exponential functions:

$$f(x) = 2^x, \qquad f(x) = \left(\tfrac{1}{2}\right)^x, \qquad f(x) = 3.57^{(x-2)}.$$

Note that, in contrast to functions like $f(x) = x^5$ and $f(x) = x^{1/2}$ in which the variable is the base of an exponential expression, the variable in an exponential function is *in the exponent.*

Let's now consider graphs of exponential functions.

**EXAMPLE 1** Graph the exponential function

$$y = f(x) = 2^x.$$

*Solution* We compute some function values and list the results in a table.

$$f(0) = 2^0 = 1; \qquad f(-1) = 2^{-1} = \frac{1}{2^1} = \frac{1}{2};$$
$$f(1) = 2^1 = 2;$$
$$f(2) = 2^2 = 4; \qquad f(-2) = 2^{-2} = \frac{1}{2^2} = \frac{1}{4};$$
$$f(3) = 2^3 = 8;$$
$$f(-3) = 2^{-3} = \frac{1}{2^3} = \frac{1}{8}.$$

| $x$ | $y = f(x) = 2^x$ | $(x, y)$ |
|-----|------------------|----------|
| 0 | 1 | $(0, 1)$ |
| 1 | 2 | $(1, 2)$ |
| 2 | 4 | $(2, 4)$ |
| 3 | 8 | $(3, 8)$ |
| $-1$ | $\frac{1}{2}$ | $\left(-1, \frac{1}{2}\right)$ |
| $-2$ | $\frac{1}{4}$ | $\left(-2, \frac{1}{4}\right)$ |
| $-3$ | $\frac{1}{8}$ | $\left(-3, \frac{1}{8}\right)$ |

Next, we plot these points and connect them with a smooth curve. Be sure to plot enough points to determine how steeply the curve rises.

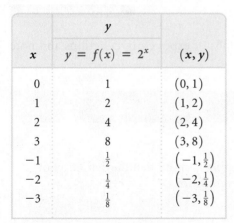

The curve comes very close to the x-axis, but does not touch or cross it.

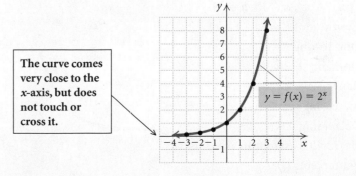

$y = f(x) = 2^x$

HORIZONTAL ASYMPTOTES

REVIEW SECTION **4.5.**

Note that as $x$ increases, the function values increase without bound. As $x$ decreases, the function values decrease, getting close to 0. That is, as $x \to -\infty$, $y \to 0$. Thus the $x$-axis, or the line $y = 0$, is a horizontal asymptote. As the $x$-inputs decrease, the curve gets closer and closer to this line, but does not cross it.     **Now Try Exercise 11.**

| Points of $f(x) = 2^x$ | Points of $f(x) = \left(\frac{1}{2}\right)^x = 2^{-x}$ |
|---|---|
| $(0, 1)$ | $(0, 1)$ |
| $(1, 2)$ | $(-1, 2)$ |
| $(2, 4)$ | $(-2, 4)$ |
| $(3, 8)$ | $(-3, 8)$ |
| $\left(-1, \frac{1}{2}\right)$ | $\left(1, \frac{1}{2}\right)$ |
| $\left(-2, \frac{1}{4}\right)$ | $\left(2, \frac{1}{4}\right)$ |
| $\left(-3, \frac{1}{8}\right)$ | $\left(3, \frac{1}{8}\right)$ |

**EXAMPLE 2**   Graph the exponential function $y = f(x) = \left(\frac{1}{2}\right)^x$.

*Solution*   Before we plot points and draw the curve, note that

$$y = f(x) = \left(\tfrac{1}{2}\right)^x = (2^{-1})^x = 2^{-x}.$$

This tells us that this graph is a reflection of the graph of $y = 2^x$ across the $y$-axis. For example, if $(3, 8)$ is a point of the graph of $f(x) = 2^x$, then $(-3, 8)$ is a point of the graph of $f(x) = 2^{-x}$. Selected points are listed in the table at left.

Next, we plot points and connect them with a smooth curve.

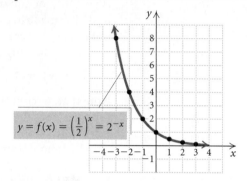

$$y = f(x) = \left(\tfrac{1}{2}\right)^x = 2^{-x}$$

**Exploring with Technology**

Check the graphs of $y = f(x) = 2^x$ and $y = f(x) = \left(\frac{1}{2}\right)^x$ using a graphing calculator. Use the TRACE and TABLE features to confirm that the graphs never cross the $x$-axis.

Note that as $x$ increases, the function values decrease, getting close to 0. The $x$-axis, $y = 0$, is the horizontal asymptote. As $x$ decreases, the function values increase without bound.     **Now Try Exercise 15.**

Take a look at the following graphs of exponential functions and note the patterns in them.

## CONNECTING THE CONCEPTS

### Properties of Exponential Functions

Let's list and compare some characteristics of exponential functions, keeping in mind that the definition of an exponential function, $f(x) = a^x$, requires that $a$ be positive and different from 1.

$f(x) = a^x, a > 0, a \neq 1$

Continuous

One-to-one

Domain: $(-\infty, \infty)$

Range: $(0, \infty)$

Increasing if $a > 1$

Decreasing if $0 < a < 1$

Horizontal asymptote is $x$-axis

$y$-intercept: $(0, 1)$

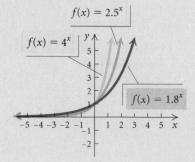

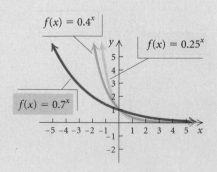

<table>
<tr><td>TRANSFORMATIONS OF<br>FUNCTIONS</td></tr>
<tr><td>REVIEW SECTION 2.4.</td></tr>
</table>

To graph other types of exponential functions, keep in mind the ideas of translation, stretching, and reflection. All these concepts allow us to visualize the graph before drawing it.

**EXAMPLE 3** Graph each of the following by hand. Before doing so, describe how each graph can be obtained from the graph of $f(x) = 2^x$. Then check your graph with a graphing calculator.

**a)** $f(x) = 2^{x-2}$  **b)** $f(x) = 2^x - 4$  **c)** $f(x) = 5 - 0.5^x$

*Solution*

**a)** The graph of $f(x) = 2^{x-2}$ is the graph of $y = 2^x$ shifted *right* 2 units.

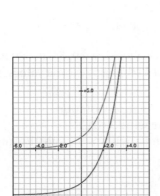

$y_1 = 2^x$, $y_2 = 2^{x-2}$

| $x$ | $f(x)$ |
|-----|--------|
| $-1$ | $\frac{1}{8}$ |
| $0$ | $\frac{1}{4}$ |
| $1$ | $\frac{1}{2}$ |
| $2$ | $1$ |
| $3$ | $2$ |
| $4$ | $4$ |
| $5$ | $8$ |

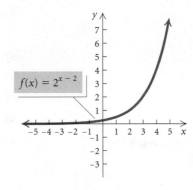

**b)** The graph of $f(x) = 2^x - 4$ is the graph of $y = 2^x$ shifted *down* 4 units.

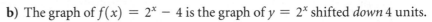

| $x$ | $f(x)$ |
|-----|--------|
| $-2$ | $-3\frac{3}{4}$ |
| $-1$ | $-3\frac{1}{2}$ |
| $0$ | $-3$ |
| $1$ | $-2$ |
| $2$ | $0$ |
| $3$ | $4$ |

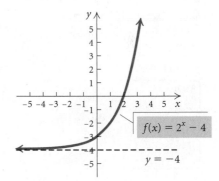

**c)** The graph of $f(x) = 5 - 0.5^x = 5 - \left(\frac{1}{2}\right)^x = 5 - 2^{-x}$ is a reflection of the graph of $y = 2^x$ across the $y$-axis, followed by a reflection across the $x$-axis and then a shift *up* 5 units.

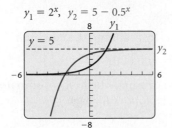

$y_1 = 2^x$, $y_2 = 5 - 0.5^x$

| $x$ | $f(x)$ |
|-----|--------|
| $-3$ | $-3$ |
| $-2$ | $1$ |
| $-1$ | $3$ |
| $0$ | $4$ |
| $1$ | $4\frac{1}{2}$ |
| $2$ | $4\frac{3}{4}$ |

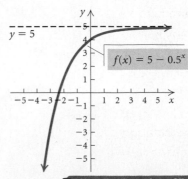

**Now Try Exercises 27 and 31.**

## ■ Applications

Graphing calculators are especially helpful when we are working with exponential functions. They not only facilitate computations but they also allow us to visualize the functions. One of the most frequent applications of exponential functions occurs with compound interest.

**EXAMPLE 4**   *Compound Interest.*   The amount of money $A$ to which a principal $P$ will grow after $t$ years at interest rate $r$ (in decimal form), compounded $n$ times per year, is given by the function

$$A(t) = P\left(1 + \frac{r}{n}\right)^{nt}.$$

Suppose that $100,000 is invested at 6.5% interest, compounded semiannually.

**a)** Find a function for the amount to which the investment grows after $t$ years.

**b)** Graph the function.

**c)** Find the amount of money in the account at $t = 0$, 4, 8, and 10 years.

**d)** When will the amount of money in the account reach $400,000?

*Solution*

**a)** Since $P = \$100,000$, $r = 6.5\% = 0.065$, and $n = 2$, we can substitute these values and write the following function:

$$A(t) = 100,000\left(1 + \frac{0.065}{2}\right)^{2\cdot t} = \$100,000(1.0325)^{2t}.$$

**b)** For the graph shown at left, we use the viewing window $[0, 30, 0, 500,000]$ because of the large numbers and the fact that negative time values have no meaning in this application.

$y = 100,000(1.0325)^{2x}$

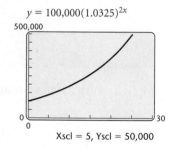

$y = 100,000(1.0325)^{2x}$

| Y₁(0) | |
| --- | --- |
| | 100000 |
| Y₁(4) | |
| | 129157.7535 |
| Y₁(8) | |
| | 166817.253 |

**c)** We can compute function values using function notation on the home screen of a graphing calculator. (See the window at left.) We can also calculate the values directly on a graphing calculator by substituting in the expression for $A(t)$:

$$A(0) = 100,000(1.0325)^{2\cdot 0} = \$100,000;$$
$$A(4) = 100,000(1.0325)^{2\cdot 4} \approx \$129,157.75;$$
$$A(8) = 100,000(1.0325)^{2\cdot 8} \approx \$166,817.25;$$
$$A(10) = 100,000(1.0325)^{2\cdot 10} \approx \$189,583.79.$$

**d)** To find the amount of time it takes for the account to grow to $400,000, we set

$$100,000(1.0325)^{2t} = 400,000$$

and solve for $t$. One way we can do this is by graphing the equations

$$y_1 = 100,000(1.0325)^{2x} \quad \text{and} \quad y_2 = 400,000.$$

Then we can use the Intersect method to estimate the first coordinate of the point of intersection. (See Fig. 1 on the following page.)

We can also use the Zero method to estimate the zero of the function $y = 100{,}000(1.0325)^{2x} - 400{,}000$. (See Fig. 2 below.)

Regardless of the method we use, we see that the account grows to $400,000 after about 21.67 years, or about 21 years, 8 months, and 2 days.

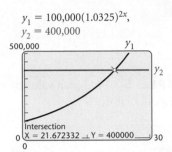

INTERSECT METHOD

REVIEW SECTION **1.5.**

ZERO METHOD

REVIEW SECTION **1.5.**

FIGURE 1

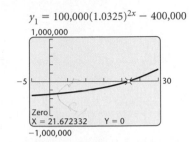

FIGURE 2

**Now Try Exercise 51.**

## ■ The Number *e*

We now consider a very special number in mathematics. In 1741, Leonhard Euler named this number *e*. Though you may not have encountered it before, you will see here and in future mathematics courses that it has many important applications. To explain this number, we use the compound interest formula $A = P(1 + r/n)^{nt}$ discussed in Example 4. Suppose that $1 is invested at 100% interest for 1 year. Since $P = 1$, $r = 100\% = 1$, and $t = 1$, the formula above becomes a function $A$ defined in terms of the number of compounding periods $n$:

$$A = P\left(1 + \frac{r}{n}\right)^{nt} = 1\left(1 + \frac{1}{n}\right)^{n \cdot 1} = \left(1 + \frac{1}{n}\right)^{n}.$$

Let's visualize this function with its graph shown at left and explore the values of $A(n)$ as $n \to \infty$. Consider the graph for larger and larger values of $n$. Does this function have a horizontal asymptote?

Let's find some function values using a calculator.

$$A(n) = \left(1 + \frac{1}{n}\right)^{n}$$

| *n*, Number of Compounding Periods | $A(n) = \left(1 + \dfrac{1}{n}\right)^{n}$ |
|---|---|
| 1 (compounded annually) | $2.00 |
| 2 (compounded semiannually) | 2.25 |
| 3 | 2.3704 |
| 4 (compounded quarterly) | 2.4414 |
| 5 | 2.4883 |
| 100 | 2.7048 |
| 365 (compounded daily) | 2.7146 |
| 8760 (compounded hourly) | 2.7181 |

It appears from these values that the graph does have a horizontal asymptote, $y \approx 2.7$. As the values of $n$ get larger and larger, the function values get closer and closer to the number Euler named $e$. Its decimal representation does not terminate or repeat; it is irrational.

$$e = 2.7182818284 \ldots$$

GCM **EXAMPLE 5** Find each value of $e^x$, to four decimal places, using the ⟨e^x⟩ key on a calculator.

a) $e^3$     b) $e^{-0.23}$     c) $e^0$

*Solution*

| FUNCTION VALUE | READOUT | ROUNDED |
|---|---|---|
| a) $e^3$ | e^3      20.08553692 | 20.0855 |
| b) $e^{-0.23}$ | e^-.23      .7945336025 | 0.7945 |
| c) $e^0$ | e^0      1 | 1 |

**Now Try Exercises 1 and 3.**

### ■ Graphs of Exponential Functions, Base *e*

We demonstrate ways in which to graph exponential functions.

**EXAMPLE 6** Graph $f(x) = e^x$ and $g(x) = e^{-x}$.

*Solution* We can compute points for each equation using a table or the ⟨e^x⟩ key on a calculator. Then we plot these points and draw the graphs of the functions.

$y_1 = e^x, \ y_2 = e^{-x}$

| X | Y1 | Y2 |
|---|---|---|
| −3 | .04979 | 20.086 |
| −2 | .13534 | 7.3891 |
| −1 | .36788 | 2.7183 |
| 0 | 1 | 1 |
| 1 | 2.7183 | .36788 |
| 2 | 7.3891 | .13534 |
| 3 | 20.086 | .04979 |

X =

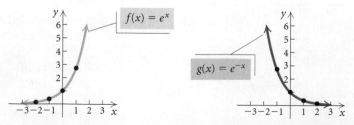

Note that the graph of $g$ is the reflection of the graph of $f$ across the $y$-axis.

**Now Try Exercise 23.**

**EXAMPLE 7**   Graph each of the following by hand. Before doing so, describe how each graph can be obtained from the graph of $y = e^x$.

**a)** $f(x) = e^{x+3}$       **b)** $f(x) = e^{-0.5x}$       **c)** $f(x) = 1 - e^{-2x}$

*Solution*

**a)** The graph of $f(x) = e^{x+3}$ is a translation of the graph of $y = e^x$ left 3 units.

$y_1 = e^x, \ y_2 = e^{x+3}$

| $x$ | $f(x)$ |
|-----|--------|
| $-7$ | 0.018 |
| $-5$ | 0.135 |
| $-3$ | 1 |
| $-1$ | 7.389 |
| $0$ | 20.086 |

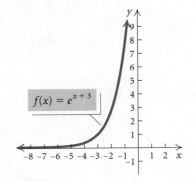

**b)** We note that the graph of $f(x) = e^{-0.5x}$ is a horizontal stretching of the graph of $y = e^x$ followed by a reflection across the $y$-axis.

$y_1 = e^x, \ y_2 = e^{-0.5x}$

| $x$ | $f(x)$ |
|-----|--------|
| $-2$ | 2.718 |
| $-1$ | 1.649 |
| $0$ | 1 |
| $1$ | 0.607 |
| $2$ | 0.368 |

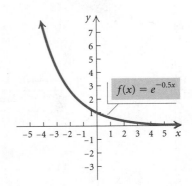

**c)** The graph of $f(x) = 1 - e^{-2x}$ is a horizontal shrinking of the graph of $y = e^x$, followed by a reflection across the $y$-axis, then across the $x$-axis, followed by a translation up 1 unit.

| $x$ | $f(x)$ |
|-----|--------|
| $-1$ | $-6.389$ |
| $0$ | 0 |
| $1$ | 0.865 |
| $2$ | 0.982 |
| $3$ | 0.998 |

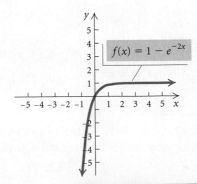

**Now Try Exercises 41 and 47.**

## 5.2 Exercise Set

*Find each of the following, to four decimal places, using a calculator.*

**1.** $e^4$

**2.** $e^{10}$

**3.** $e^{-2.458}$

**4.** $\left(\dfrac{1}{e^3}\right)^2$

*In Exercises 5–10, match the function with one of the graphs (a)–(f), which follow.*

**a)**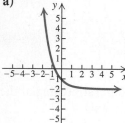

**b)**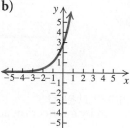

**c)**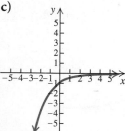

**d)**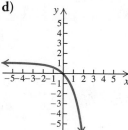

**e)**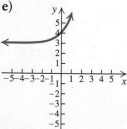

**f)**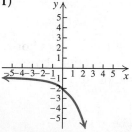

**5.** $f(x) = -2^x - 1$

**6.** $f(x) = -\left(\frac{1}{2}\right)^x$

**7.** $f(x) = e^x + 3$

**8.** $f(x) = e^{x+1}$

**9.** $f(x) = 3^{-x} - 2$

**10.** $f(x) = 1 - e^x$

*Graph the function by substituting and plotting points. Then check your work using a graphing calculator.*

**11.** $f(x) = 3^x$

**12.** $f(x) = 5^x$

**13.** $f(x) = 6^x$

**14.** $f(x) = 3^{-x}$

**15.** $f(x) = \left(\frac{1}{4}\right)^x$

**16.** $f(x) = \left(\frac{2}{3}\right)^x$

**17.** $y = -2^x$

**18.** $y = 3 - 3^x$

**19.** $f(x) = -0.25^x + 4$

**20.** $f(x) = 0.6^x - 3$

**21.** $f(x) = 1 + e^{-x}$

**22.** $f(x) = 2 - e^{-x}$

**23.** $y = \frac{1}{4}e^x$

**24.** $y = 2e^{-x}$

**25.** $f(x) = 1 - e^{-x}$

**26.** $f(x) = e^x - 2$

*Sketch the graph of the function and check the graph with a graphing calculator. Describe how each graph can be obtained from the graph of a basic exponential function.*

**27.** $f(x) = 2^{x+1}$

**28.** $f(x) = 2^{x-1}$

**29.** $f(x) = 2^x - 3$

**30.** $f(x) = 2^x + 1$

**31.** $f(x) = 2^{1-x} + 2$

**32.** $f(x) = 5 - 2^{-x}$

**33.** $f(x) = 4 - 3^{-x}$

**34.** $f(x) = 2^{x-1} - 3$

**35.** $f(x) = \left(\frac{3}{2}\right)^{x-1}$

**36.** $f(x) = 3^{4-x}$

**37.** $f(x) = 2^{x+3} - 5$

**38.** $f(x) = -3^{x-2}$

**39.** $f(x) = 3 \cdot 2^{x-1} + 1$

**40.** $f(x) = 2 \cdot 3^{x+1} - 2$

**41.** $f(x) = e^{2x}$

**42.** $f(x) = e^{-0.2x}$

**43.** $f(x) = \frac{1}{2}(1 - e^x)$

**44.** $f(x) = 3(1 + e^x) - 2$

**45.** $y = e^{-x+1}$

**46.** $y = e^{2x} + 1$

**47.** $f(x) = 2(1 - e^{-x})$

**48.** $f(x) = 1 - e^{-0.01x}$

*Graph the piecewise function.*

**49.** $f(x) = \begin{cases} e^{-x} - 4, & \text{for } x < -2, \\ x + 3, & \text{for } -2 \le x < 1, \\ x^2, & \text{for } x \ge 1 \end{cases}$

**50.** $g(x) = \begin{cases} 4, & \text{for } x \le -3, \\ x^2 - 6, & \text{for } -3 < x < 0, \\ e^x, & \text{for } x \ge 0 \end{cases}$

**51.** *Compound Interest.* Suppose that $82,000 is invested at $4\frac{1}{2}\%$ interest, compounded quarterly.

**a)** Find the function for the amount to which the investment grows after $t$ years.

**b)** Graph the function.

**c)** Find the amount of money in the account at $t = 0, 2, 5,$ and $10$ years.

**d)** When will the amount of money in the account reach $100,000?

**52.** *Compound Interest.* Suppose that $750 is invested at 7% interest, compounded semiannually.

**a)** Find the function for the amount to which the investment grows after *t* years.
**b)** Graph the function.
**c)** Find the amount of money in the account at $t = 1, 6, 10, 15,$ and 25 years.
**d)** When will the amount of money in the account reach $3000?

**53.** *Interest on a CD.* On Will's sixth birthday, his grandparents present him with a $3000 certificate of deposit (CD) that earns 5% interest, compounded quarterly. If the CD matures on his sixteenth birthday, what amount will be available then?

**54.** *Interest in a College Trust Fund.* Following the birth of her child, Sophia deposits $10,000 in a college trust fund where interest is 3.9%, compounded semiannually.

**a)** Find a function for the amount in the account after *t* years.
**b)** Find the amount of money in the account at $t = 0, 4, 8, 10,$ and 18 years.

In Exercises 55–62, use the compound-interest formula to find the account balance A with the given conditions:

$P$ = principal,
$r$ = interest rate,
$n$ = number of compounding periods per year,
$t$ = time, in years,
$A$ = account balance.

| | $P$ | $r$ | Compounded | $n$ | $t$ | $A$ |
|---|---|---|---|---|---|---|
| **55.** | $3,000 | 4% | Semiannually | | 2 | |
| **56.** | $12,500 | 3% | Quarterly | | 3 | |
| **57.** | $120,000 | 2.5% | Annually | | 10 | |
| **58.** | $120,000 | 2.5% | Quarterly | | 10 | |
| **59.** | $53,500 | $5\frac{1}{2}$% | Quarterly | | $6\frac{1}{2}$ | |
| **60.** | $6,250 | $6\frac{3}{4}$% | Semiannually | | $4\frac{1}{2}$ | |
| **61.** | $17,400 | 8.1% | Daily | | 5 | |
| **62.** | $900 | 7.3% | Daily | | $7\frac{1}{4}$ | |

**63.** *Overweight Troops.* The number of U.S. troops diagnosed as overweight has more than doubled since 1998 (*Source*: U.S. Department of Defense). The exponential function

$$W(x) = 23{,}672.16(1.112)^x,$$

where *x* is the number of years since 1998, can be used to estimate the number of overweight active troops. Find the number of overweight troops in 2000, in 2008, and in 2013.

**64.** *Foreign Students from India.* The number of students from India enrolled in U.S. colleges and universities has grown exponentially from about 9000 in 1980 to about 95,000 in 2008 (*Source*: Institute of International Education, New York). The exponential function

$$I(t) = 9000(1.0878)^t,$$

where *t* is the number of years since 1980, can be used to estimate the number of students from India enrolled in U.S. colleges and universities. Find the number of students from India enrolled in 1985, in 1999, in 2006, and in 2012.

**65.** *IRS Tax Code.* The number of pages in the IRS Tax Code has increased from 400 pages in 1913 to 70,320 pages in 2009 (*Source*: IRS). The increase can be modeled by the exponential function

$$T(x) = 400(1.055)^x,$$

where *x* is the number of years since 1913. Use this function to estimate the number of pages in the tax code for 1950, for 1990, and for 2000. Round to the nearest one.

**66.** *Medicare Premiums.* The monthly Medicare Part B health-care premium for most beneficiaries ages 65 and older has increased significantly since 1975. The monthly premium has increased from about $7 in 1975 to $110.50 in 2011 (*Source*: Centers for Medicare and Medicaid Services). The following exponential function models the premium increases:

$$M(x) = 7(1.080)^x,$$

where $x$ is the number of years since 1975. Estimate the monthly Medicare Part B premium in 1985, in 1992, and in 2002. Round to the nearest dollar.

**67.** *Taxi Medallions.*    The business of taxi medallions, required licenses fastened to the hood of all New York City cabs, has proven itself to be a valuable investment. The average price for a corporate-license taxi medallion in New York has risen from about $200,000 in 2001 to $753,000 in 2009 (*Sources*: New York City Taxi & Limousine Commission; Andrew Murstein, president of Medallion Financial). The exponential function

$$M(x) = 200,000(1.1802)^x,$$

where $x$ is the number of years since 2001, models the data for the average price of a medallion in recent years. Estimate the price of a taxi medallion in 2003, in 2007, and in 2013. Round to the nearest dollar.

**68.** *Cost of Census.*    The cost per household for taking the U.S. census has increased exponentially since 1970 (*Source*: GAO analysis of U.S. Census Bureau data). The following function models the increase in cost:

$$C(x) = 15.5202(1.0508)^x,$$

where $x$ is the number of years since 1970. Estimate the cost per household in 1990 and in 2010.

**69.** *Recycling Plastic Bags.*    It is estimated that 90 billion plastic carry-out bags are produced annually in the United States. In recent years, the number of tons of plastic bags that are recycled has grown exponentially from approximately 80,000 tons in 1996 to 380,000 tons in 2007. (*Source*: Environmental Protection Agency) The following exponential function models the amount recycled, in tons:

$$R(x) = 80,000(1.1522)^x,$$

where $x$ is the number of years since 1996. Find the total amount recycled, in tons, in 1999, in 2007, and in 2012. Then use the function to estimate the total number of tons that will be recycled in 2012.

**70.** *Master's Degrees Earned by Women.*    The function

$$D(t) = 43.1224(1.0475)^t$$

gives the number of master's degrees, in thousands, conferred on women in the United States $t$ years after 1960 (*Sources*: National Center for Educational Statistics; Digest of Education Statistics). Find the number of master's degrees earned by women in 1984, in 2002, and in 2010. Then estimate the number of master's degrees that will be earned by women in 2015.

**71.** *Tennis Participation.*    In recent years, U.S. tennis participation has increased. The function

$$T(x) = 23.7624(1.0752)^x,$$

where $T(x)$ is in millions and $x$ is the number of years since 2006, models the number of people who had played tennis at least once in the year (*Source*: USTA/Coyne Public Relations). Estimate the number who had played tennis at least once in a year for 2007. Then use this function to project participation in tennis in 2014.

**72.** *Charitable Giving.* Over the last four decades, charitable giving in the United States has grown exponentially from approximately $20.7 billion in 1969 to approximately $303.8 billion in 2009 (*Sources*: Giving USA Foundation; Volunteering in America by the Corporation for National & Community Service). The function

$$G(x) = 20.7(1.06946)^x,$$

where $x$ is the number of years since 1969, can be used to estimate charitable giving, in billions of dollars. Find the amount of charitable giving in 1978, in 1985, and in 2000.

**73.** *Salvage Value.* A landscape company purchased a backhoe for $56,395. The value of the backhoe each year is 90% of the value of the preceding year. After $t$ years, its value, in dollars, is given by the exponential function

$$V(t) = 56,395(0.9)^t.$$

**a)** Graph the function.
**b)** Find the value of the backhoe after 0, 1, 3, 6, and 10 years. Round to the nearest dollar.

**c)** The company decides to replace the backhoe when its value has declined to $30,000. After how long will the machine be replaced?

**74.** *Salvage Value.* A restaurant purchased a 72-in. range with six burners for $6982. The value of the range each year is 85% of the value of the preceding year. After $t$ years, its value, in dollars, is given by the exponential function

$$V(t) = 6982(0.85)^t.$$

**a)** Graph the function.
**b)** Find the value of the range after 0, 1, 2, 5, and 8 years.
**c)** The restaurant decides to replace the range when its value has declined to $1000. After how long will the range be replaced?

**75.** *Advertising.* A company begins an Internet advertising campaign to market a new phone. The percentage of the target market that buys a product is generally a function of the length of the advertising campaign. The estimated percentage is given by

$$f(t) = 100(1 - e^{-0.04t}),$$

where $t$ is the number of days of the campaign.

**a)** Graph the function.
**b)** Find $f(25)$, the percentage of the target market that has bought the phone after a 25-day advertising campaign.
**c)** After how long will 90% of the target market have bought the phone?

**76.** *Growth of a Stock.* The value of a stock is given by the function

$$V(t) = 58(1 - e^{-1.1t}) + 20,$$

where $V$ is the value of the stock after time $t$, in months.

**a)** Graph the function.
**b)** Find $V(1)$, $V(2)$, $V(4)$, $V(6)$, and $V(12)$.
**c)** After how long will the value of the stock be $75?

*In Exercises 77–90, use a graphing calculator to match the equation with one of the figures (a)–(n), which follow.*

**a)**

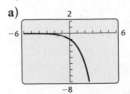

**b)**

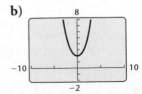

**c)**

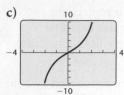

**d)**

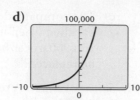

**e)**

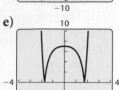

**f)**

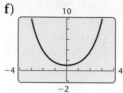

**g)**

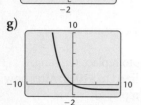

**h)**

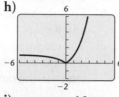

**i)**

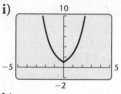

**j)**

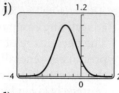

**k)**

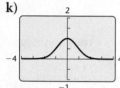

**l)**

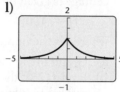

**m)**

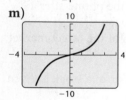

**n)**

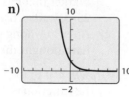

**77.** $y = 3^x - 3^{-x}$

**78.** $y = 3^{-(x+1)^2}$

**79.** $f(x) = -2.3^x$

**80.** $f(x) = 30{,}000(1.4)^x$

**81.** $y = 2^{-|x|}$

**82.** $y = 2^{-(x-1)}$

**83.** $f(x) = (0.58)^x - 1$

**84.** $y = 2^x + 2^{-x}$

**85.** $g(x) = e^{|x|}$

**86.** $f(x) = |2^x - 1|$

**87.** $y = 2^{-x^2}$

**88.** $y = |2^{x^2} - 8|$

**89.** $g(x) = \dfrac{e^x - e^{-x}}{2}$

**90.** $f(x) = \dfrac{e^x + e^{-x}}{2}$

*Use a graphing calculator to find the point(s) of intersection of the graphs of each of the following pairs of equations.*

**91.** $y = |1 - 3^x|, \ y = 4 + 3^{-x}$

**92.** $y = 4^x + 4^{-x}, \ y = 8 - 2x - x^2$

**93.** $y = 2e^x - 3, \ y = \dfrac{e^x}{x}$

**94.** $y = \dfrac{1}{e^x + 1}, \ y = 0.3x + \dfrac{7}{9}$

*Solve graphically.*

**95.** $5.3^x - 4.2^x = 1073$

**96.** $e^x = x^3$

**97.** $2^x > 1$

**98.** $3^x \le 1$

**99.** $2^x + 3^x = x^2 + x^3$

**100.** $31{,}245e^{-3x} = 523{,}467$

## Skill Maintenance

*Simplify.*

**101.** $(1 - 4i)(7 + 6i)$

**102.** $\dfrac{2 - i}{3 + i}$

*Find the x-intercepts and the zeros of the function.*

**103.** $f(x) = 2x^2 - 13x - 7$

**104.** $h(x) = x^3 - 3x^2 + 3x - 1$

**105.** $h(x) = x^4 - x^2$

**106.** $g(x) = x^3 + x^2 - 12x$

*Solve.*

**107.** $x^3 + 6x^2 - 16x = 0$

**108.** $3x^2 - 6 = 5x$

## Synthesis

**109.** Which is larger, $7^\pi$ or $\pi^7$? $70^{80}$ or $80^{70}$?

*In Exercises 110 and 111:*

**a)** *Graph using a graphing calculator.*
**b)** *Approximate the zeros.*
**c)** *Approximate the relative maximum and minimum values. If your graphing calculator has a* MAX–MIN *feature, use it.*

**110.** $f(x) = x^2 e^{-x}$

**111.** $f(x) = e^{-x^2}$

**112.** Graph $f(x) = x^{1/(x-1)}$ for $x > 0$. Use a graphing calculator and the TABLE feature to identify the horizontal asymptote.

**113.** For the function $f$, construct and simplify the difference quotient. (See Section 2.2 for review.)

$$f(x) = 2e^x - 3$$

## Logarithmic Functions and Graphs

**5.3**

- Find common logarithms and natural logarithms with and without a calculator.
- Convert between exponential equations and logarithmic equations.
- Change logarithm bases.
- Graph logarithmic functions.
- Solve applied problems involving logarithmic functions.

We now consider *logarithmic*, or *logarithm*, *functions*. These functions are inverses of exponential functions and have many applications.

### Logarithmic Functions

We have noted that every exponential function (with $a > 0$ and $a \neq 1$) is one-to-one. Thus such a function has an inverse that is a function. In this section, we will name these inverse functions logarithmic functions and use them in applications. We can draw the graph of the inverse of an exponential function by interchanging $x$ and $y$.

**EXAMPLE 1** Graph: $x = 2^y$.

**Solution** Note that $x$ is alone on one side of the equation. We can find ordered pairs that are solutions by choosing values for $y$ and then computing the corresponding $x$-values.

For $y = 0, x = 2^0 = 1.$
For $y = 1, x = 2^1 = 2.$
For $y = 2, x = 2^2 = 4.$
For $y = 3, x = 2^3 = 8.$
For $y = -1, x = 2^{-1} = \dfrac{1}{2^1} = \dfrac{1}{2}.$
For $y = -2, x = 2^{-2} = \dfrac{1}{2^2} = \dfrac{1}{4}.$
For $y = -3, x = 2^{-3} = \dfrac{1}{2^3} = \dfrac{1}{8}.$

| $x$ | | |
| --- | --- | --- |
| $x = 2^y$ | $y$ | $(x, y)$ |
| 1 | 0 | $(1, 0)$ |
| 2 | 1 | $(2, 1)$ |
| 4 | 2 | $(4, 2)$ |
| 8 | 3 | $(8, 3)$ |
| $\dfrac{1}{2}$ | $-1$ | $\left(\dfrac{1}{2}, -1\right)$ |
| $\dfrac{1}{4}$ | $-2$ | $\left(\dfrac{1}{4}, -2\right)$ |
| $\dfrac{1}{8}$ | $-3$ | $\left(\dfrac{1}{8}, -3\right)$ |

↑ (1) Choose values for $y$.
↑ (2) Compute values for $x$.

We plot the points and connect them with a smooth curve. Note that the curve does not touch or cross the $y$-axis. The $y$-axis is a vertical asymptote.

Note too that this curve looks just like the graph of $y = 2^x$, except that it is reflected across the line $y = x$, as we would expect for an inverse. The inverse of $y = 2^x$ is $x = 2^y$.

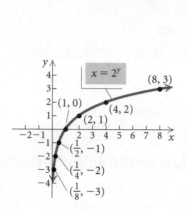

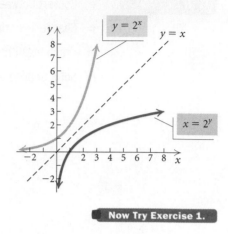

Now Try Exercise 1.

To find a formula for $f^{-1}$ when $f(x) = 2^x$, we try to use the method of Section 5.1:

1. Replace $f(x)$ with $y$:       $y = 2^x$
2. Interchange $x$ and $y$:       $x = 2^y$
3. Solve for $y$:       $y =$ the power to which we raise 2 to get $x$
4. Replace $y$ with $f^{-1}(x)$:       $f^{-1}(x) =$ the power to which we raise 2 to get $x$.

Mathematicians have defined a new symbol to replace the words "the power to which we raise 2 to get $x$." That symbol is "$\log_2 x$," read "the logarithm, base 2, of $x$."

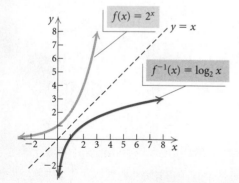

---

### Logarithmic Function, Base 2

"$\log_2 x$," read "the logarithm, base 2, of $x$," means "the power to which we raise 2 to get $x$."

---

Thus if $f(x) = 2^x$, then $f^{-1}(x) = \log_2 x$. For example,

$$f^{-1}(8) = \log_2 8 = 3,$$

because

*3 is the power to which we raise 2 to get 8.*

Similarly, $\log_2 13$ is the power to which we raise 2 to get 13. As yet, we have no simpler way to say this other than

"$\log_2 13$ is the power to which we raise 2 to get 13."

Later, however, we will learn how to approximate this expression using a calculator.

For any exponential function $f(x) = a^x$, its inverse is called a **logarithmic function, base $a$**. The graph of the inverse can be obtained by reflecting the graph of $y = a^x$ across the line $y = x$, to obtain $x = a^y$. Then $x = a^y$ is equivalent to $y = \log_a x$. We read $\log_a x$ as "the logarithm, base $a$, of $x$."

**The inverse of $f(x) = a^x$ is given by $f^{-1}(x) = \log_a x$.**

---

### LOGARITHMIC FUNCTION, BASE $a$

We define $y = \log_a x$ as that number $y$ such that $x = a^y$, where $x > 0$ and $a$ is a positive constant other than 1.

---

Let's look at the graphs of $f(x) = a^x$ and $f^{-1}(x) = \log_a x$ for $a > 1$ and $0 < a < 1$.

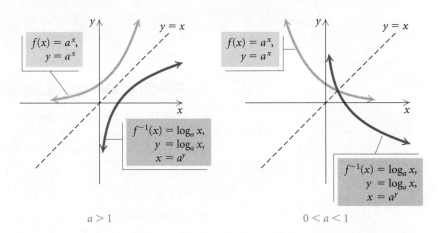

Note that the graphs of $f(x)$ and $f^{-1}(x)$ are reflections of each other across the line $y = x$.

## CONNECTING THE CONCEPTS

### Comparing Exponential Functions and Logarithmic Functions

In the following table, we compare exponential functions and logarithmic functions with bases $a$ greater than 1. Similar statements could be made for $a$, where $0 < a < 1$. It is helpful to visualize the differences by carefully observing the graphs.

**EXPONENTIAL FUNCTION**

$y = a^x$
$f(x) = a^x$
$a > 1$
Continuous
One-to-one
Domain: All real
    numbers, $(-\infty, \infty)$
Range: All positive
    real numbers, $(0, \infty)$
Increasing
Horizontal asymptote is $x$-axis:
    $(a^x \to 0$ as $x \to -\infty)$
$y$-intercept: $(0, 1)$
There is no $x$-intercept.

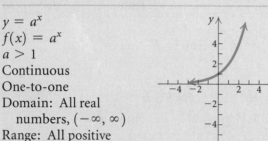

**LOGARITHMIC FUNCTION**

$x = a^y$
$f^{-1}(x) = \log_a x$
$a > 1$
Continuous
One-to-one
Domain: All positive
    real numbers, $(0, \infty)$
Range: All real
    numbers, $(-\infty, \infty)$
Increasing
Vertical asymptote is $y$-axis:
    $(\log_a x \to -\infty$ as $x \to 0^+)$
$x$-intercept: $(1, 0)$
There is no $y$-intercept.

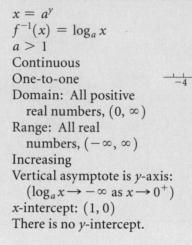

### ■ Finding Certain Logarithms

Let's use the definition of logarithms to find some logarithmic values.

**EXAMPLE 2**    Find each of the following logarithms.

**a)** $\log_{10} 10{,}000$      **b)** $\log_{10} 0.01$      **c)** $\log_5 125$

**d)** $\log_9 3$      **e)** $\log_6 1$      **f)** $\log_8 8$

*Solution*

**a)** The exponent to which we raise 10 to obtain 10,000 is 4; thus, $\log_{10} 10{,}000 = 4$.

**b)** We have $0.01 = \dfrac{1}{100} = \dfrac{1}{10^2} = 10^{-2}$. The exponent to which we raise 10 to get 0.01 is $-2$, so $\log_{10} 0.01 = -2$.

**c)** $125 = 5^3$. The exponent to which we raise 5 to get 125 is 3, so $\log_5 125 = 3$.

**d)** $3 = \sqrt{9} = 9^{1/2}$. The exponent to which we raise 9 to get 3 is $\frac{1}{2}$, so $\log_9 3 = \frac{1}{2}$.

**e)** $1 = 6^0$. The exponent to which we raise 6 to get 1 is 0, so $\log_6 1 = 0$.

**f)** $8 = 8^1$. The exponent to which we raise 8 to get 8 is 1, so $\log_8 8 = 1$.

**Now Try Exercises 9 and 15.**

Examples 2(e) and 2(f) illustrate two important properties of logarithms. The property $\log_a 1 = 0$ follows from the fact that $a^0 = 1$. Thus, $\log_5 1 = 0$, $\log_{10} 1 = 0$, and so on. The property $\log_a a = 1$ follows from the fact that $a^1 = a$. Thus, $\log_5 5 = 1$, $\log_{10} 10 = 1$, and so on.

$$\log_a 1 = 0 \quad \text{and} \quad \log_a a = 1, \quad \text{for any logarithmic base } a.$$

## Converting Between Exponential Equations and Logarithmic Equations

It is helpful in dealing with logarithmic functions to remember that a logarithm of a number is an *exponent*. It is the exponent $y$ in $x = a^y$. You might think to yourself, "the logarithm, base $a$, of a number $x$ is the power to which $a$ must be raised to get $x$."

We are led to the following. (The symbol $\longleftrightarrow$ means that the two statements are equivalent; that is, when one is true, the other is true. The words "if and only if" can be used in place of $\longleftrightarrow$.)

$$\log_a x = y \longleftrightarrow x = a^y \quad \text{A logarithm is an exponent!}$$

**EXAMPLE 3** Convert each of the following to a logarithmic equation.

**a)** $16 = 2^x$        **b)** $10^{-3} = 0.001$        **c)** $e^t = 70$

*Solution*

                           The exponent is the logarithm.

**a)** $16 = 2^x \qquad \log_2 16 = x$

                           The base remains the same.

**b)** $10^{-3} = 0.001 \rightarrow \log_{10} 0.001 = -3$

**c)** $e^t = 70 \rightarrow \log_e 70 = t$

> **Now Try Exercise 37.**

**EXAMPLE 4** Convert each of the following to an exponential equation.

**a)** $\log_2 32 = 5$        **b)** $\log_a Q = 8$        **c)** $x = \log_t M$

*Solution*

                           The logarithm is the exponent.

**a)** $\log_2 32 = 5 \qquad 2^5 = 32$

                           The base remains the same.

**b)** $\log_a Q = 8 \rightarrow a^8 = Q$

**c)** $x = \log_t M \rightarrow t^x = M$

> **Now Try Exercise 45.**

### ■ Finding Logarithms on a Calculator

Before calculators became so widely available, base-10 logarithms, or **common logarithms**, were used extensively to simplify complicated calculations. In fact, that is why logarithms were invented. The abbreviation **log**, with no base written, is used to represent common logarithms, or base-10 logarithms. Thus,

$$\log x \quad \text{means} \quad \log_{10} x.$$

For example, $\log 29$ means $\log_{10} 29$. Let's compare $\log 29$ with $\log 10$ and $\log 100$:

$$\left.\begin{array}{l} \log 10 = \log_{10} 10 = 1 \\ \log 29 = \,? \\ \log 100 = \log_{10} 100 = 2 \end{array}\right\} \quad \begin{array}{l} \text{Since 29 is between 10 and 100, it} \\ \text{seems reasonable that } \log 29 \text{ is} \\ \text{between 1 and 2.} \end{array}$$

On a calculator, the key for common logarithms is generally marked **LOG**. Using that key, we find that

$$\log 29 \approx 1.462397998 \approx 1.4624$$

rounded to four decimal places. Since $1 < 1.4624 < 2$, our answer seems reasonable. This also tells us that $10^{1.4624} \approx 29$.

GCM **EXAMPLE 5** Find each of the following common logarithms on a calculator. If you are using a graphing calculator, set the calculator in REAL mode. Round to four decimal places.

**a)** $\log 645{,}778$      **b)** $\log 0.0000239$      **c)** $\log(-3)$

*Solution*

| FUNCTION VALUE | READOUT | ROUNDED |
|---|---|---|
| **a)** $\log 645{,}778$ | log(645778)<br>　　　　5.810083246 | 5.8101 |
| **b)** $\log 0.0000239$ | log(0.0000239)<br>　　　−4.621602099 | −4.6216 |
| **c)** $\log(-3)$ | ERR:NONREAL ANS<br>**1:**Quit<br>2:GoTo | * Does not exist |

log(645778)<br>　　　　5.810083246<br>10^Ans<br>　　　　645778

A check for part (a) is shown at left. Since 5.810083246 is the power to which we raise 10 to get 645,778, we can check part (a) by finding $10^{5.810083246}$. We can check part (b) in a similar manner. In part (c), $\log(-3)$ does not exist as a real number because there is no real-number power to which we can raise 10 to get $-3$. The number 10 raised to any real-number power is positive. The common logarithm of a negative number does not exist as a real number. Recall that logarithmic functions are inverses of exponential functions, and since the range of an exponential function is $(0, \infty)$, the domain of $f(x) = \log_a x$ is $(0, \infty)$.

**Now Try Exercises 57 and 61.**

*If the graphing calculator is set in $a + bi$ mode, the readout is $.4771212547 + 1.364376354i$.

## ◾ **Natural Logarithms**

Logarithms, base $e$, are called **natural logarithms**. The abbreviation "ln" is generally used for natural logarithms. Thus,

> $\ln x$   means   $\log_e x$.

For example, $\ln 53$ means $\log_e 53$. On a calculator, the key for natural logarithms is generally marked **LN**. Using that key, we find that

$$\ln 53 \approx 3.970291914$$
$$\approx 3.9703$$

rounded to four decimal places. This also tells us that $e^{3.9703} \approx 53$.

**GCM**

**EXAMPLE 6**   Find each of the following natural logarithms on a calculator. If you are using a graphing calculator, set the calculator in REAL mode. Round to four decimal places.

**a)** $\ln 645{,}778$      **b)** $\ln 0.0000239$      **c)** $\ln(-5)$

**d)** $\ln e$      **e)** $\ln 1$

*Solution*

| FUNCTION VALUE | READOUT | ROUNDED |
|---|---|---|
| **a)** $\ln 645{,}778$ | ln(645778)<br>13.37821107 | 13.3782 |
| **b)** $\ln 0.0000239$ | ln(0.0000239)<br>−10.6416321 | −10.6416 |
| **c)** $\ln(-5)$ | ERR:NONREAL ANS<br>**1:**Quit<br>2:GoTo | * Does not exist |
| **d)** $\ln e$ | ln(e)<br>1 | 1 |
| **e)** $\ln 1$ | ln(1)<br>0 | 0 |

ln(0.0000239)
          −10.6416321
$e^{\text{Ans}}$
          2.39ᴇ−5

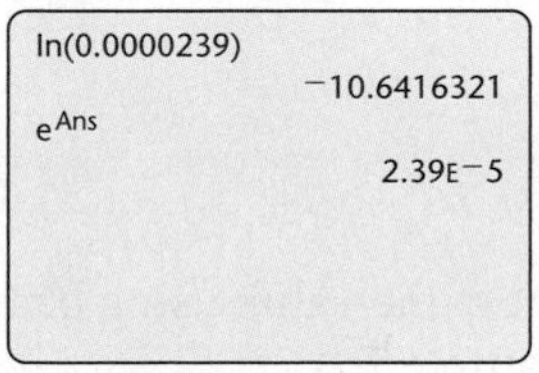

Since 13.37821107 is the power to which we raise $e$ to get 645,778, we can check part (a) by finding $e^{13.37821107}$. We can check parts (b), (d), and (e) in a similar manner. A check for part (b) is shown at left. In parts (d) and (e), note that $\ln e = \log_e e = 1$ and $\ln 1 = \log_e 1 = 0$.

**Now Try Exercises 65 and 67.**

> $\ln 1 = 0$   and   $\ln e = 1$, for the logarithmic base $e$.

---

*If the graphing calculator is set in $a + bi$ mode, the readout is $1.609437912 + 3.141592654i$.

With some calculators, it is possible to find a logarithm with any logarithmic base using the logBASE operation from the MATH MATH menu. The computation of $\log_4 15$ is shown in the window below.

log₄(15)
                1.953445298

## ■ Changing Logarithmic Bases

Most calculators give the values of both common logarithms and natural logarithms. To find a logarithm with a base other than 10 or $e$, we can use the following conversion formula.

---

### THE CHANGE-OF-BASE FORMULA

For any logarithmic bases $a$ and $b$, and any positive number $M$,

$$\log_b M = \frac{\log_a M}{\log_a b}.$$

---

We will prove this result in the next section.

**GCM**    **EXAMPLE 7**    Find $\log_5 8$ using common logarithms.

***Solution***    First, we let $a = 10$, $b = 5$, and $M = 8$. Then we substitute into the change-of-base formula:

$$\log_5 8 = \frac{\log_{10} 8}{\log_{10} 5} \qquad \text{Substituting}$$

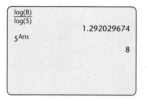

log(8)
log(5)
                1.292029674
5^Ans
                        8

$$\approx 1.2920. \qquad \text{Using a calculator}$$

Since $\log_5 8$ is the power to which we raise 5 to get 8, we would expect this power to be greater than 1 ($5^1 = 5$) and less than 2 ($5^2 = 25$), so the result is reasonable. The check is shown in the window at left.

**Now Try Exercise 69.**

We can also use base $e$ for a conversion.

**EXAMPLE 8**    Find $\log_5 8$ using natural logarithms.

***Solution***    Substituting $e$ for $a$, 5 for $b$, and 8 for $M$, we have

$$\log_5 8 = \frac{\log_e 8}{\log_e 5}$$

$$= \frac{\ln 8}{\ln 5} \approx 1.2920.$$

ln(8)
ln(5)
                1.292029674
5^Ans
                        8

The check is shown at left. Note that we get the same value using base $e$ for the conversion that we did using base 10 in Example 7.

**Now Try Exercise 75.**

## Graphs of Logarithmic Functions

We demonstrate several ways to graph logarithmic functions.

**EXAMPLE 9**  Graph: $y = f(x) = \log_5 x$.

***Solution***  The equation $y = \log_5 x$ is equivalent to $x = 5^y$. We can find ordered pairs that are solutions by choosing values for $y$ and computing the corresponding $x$-values. We then plot points, remembering that $x$ is still the first coordinate.

To use a graphing calculator to graph $y = \log_5 x$ in Example 9, we first change the base or use the logBASE operation. Here we change from base 5 to base $e$:

$$y = \log_5 x = \frac{\ln x}{\ln 5}.$$

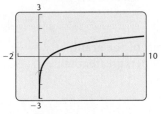

Some graphing calculators have a feature that graphs inverses automatically. If we begin with $y_1 = 5^x$, the graphs of both $y_1$ and its inverse $y_2 = \log_5 x$ will be drawn.

$y_1 = 5^x$,  $y_2 = \log_5 x$

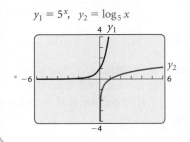

For $y = 0$, $x = 5^0 = 1$.
For $y = 1$, $x = 5^1 = 5$.
For $y = 2$, $x = 5^2 = 25$.
For $y = 3$, $x = 5^3 = 125$.
For $y = -1$, $x = 5^{-1} = \frac{1}{5}$.
For $y = -2$, $x = 5^{-2} = \frac{1}{25}$.

| $x$, or $5^y$ | $y$ |
|---|---|
| 1 | 0 |
| 5 | 1 |
| 25 | 2 |
| 125 | 3 |
| $\frac{1}{5}$ | $-1$ |
| $\frac{1}{25}$ | $-2$ |

(1) Select $y$.
(2) Compute $x$.

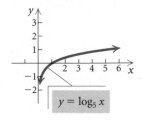

Two other methods for graphing this function are shown at left.

**Now Try Exercise 5.**

**EXAMPLE 10**  Graph: $g(x) = \ln x$.

***Solution***  To graph $y = g(x) = \ln x$, we select values for $x$ and use the **LN** key on a calculator to find the corresponding values of $\ln x$. We then plot points and draw the curve. We can graph the function $y = \ln x$ directly using a graphing calculator.

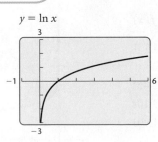

| $x$ | $g(x)$<br>$g(x) = \ln x$ |
|---|---|
| 0.5 | $-0.7$ |
| 1 | 0 |
| 2 | 0.7 |
| 3 | 1.1 |
| 4 | 1.4 |
| 5 | 1.6 |

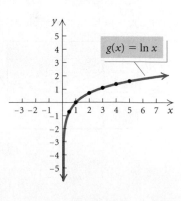

We could also write $g(x) = \ln x$, or $y = \ln x$, as $x = e^y$, select values for $y$, and use a calculator to find the corresponding values of $x$.

**Now Try Exercise 7.**

Recall that the graph of $f(x) = \log_a x$, for any base $a$, has the $x$-intercept $(1, 0)$. The domain is the set of positive real numbers, and the range is the set of all real numbers. The $y$-axis is the vertical asymptote.

**EXAMPLE 11** Graph each of the following by hand. Before doing so, describe how each graph can be obtained from the graph of $y = \ln x$. Give the domain and the vertical asymptote of each function. Then check your graph with a graphing calculator.

**a)** $f(x) = \ln (x + 3)$
**b)** $f(x) = 3 - \frac{1}{2} \ln x$
**c)** $f(x) = |\ln (x - 1)|$

**Solution**

**a)** The graph of $f(x) = \ln (x + 3)$ is a shift of the graph of $y = \ln x$ left 3 units. The domain is the set of all real numbers greater than $-3$, $(-3, \infty)$. The line $x = -3$ is the vertical asymptote.

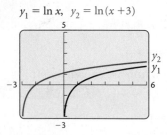

$y_1 = \ln x, \ y_2 = \ln(x + 3)$

| $x$ | $f(x)$ |
|------|---------|
| $-2.9$ | $-2.303$ |
| $-2$ | $0$ |
| $0$ | $1.099$ |
| $2$ | $1.609$ |
| $4$ | $1.946$ |

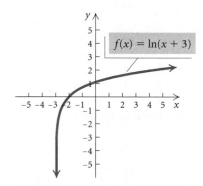

**b)** The graph of $f(x) = 3 - \frac{1}{2} \ln x$ is a vertical shrinking of the graph of $y = \ln x$, followed by a reflection across the $x$-axis, and then a translation up 3 units. The domain is the set of all positive real numbers, $(0, \infty)$. The $y$-axis is the vertical asymptote.

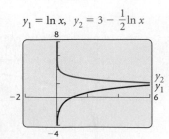

$y_1 = \ln x, \ y_2 = 3 - \frac{1}{2} \ln x$

| $x$ | $f(x)$ |
|------|---------|
| $0.1$ | $4.151$ |
| $1$ | $3$ |
| $3$ | $2.451$ |
| $6$ | $2.104$ |
| $9$ | $1.901$ |

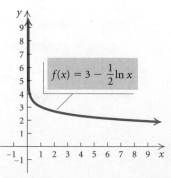

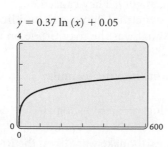

**c)** The graph of $f(x) = |\ln(x - 1)|$ is a translation of the graph of $y = \ln x$ right 1 unit. Then the absolute value has the effect of reflecting negative outputs across the $x$-axis. The domain is the set of all real numbers greater than 1, $(1, \infty)$. The line $x = 1$ is the vertical asymptote.

| $x$ | $f(x)$ |
|-----|--------|
| 1.1 | 2.303  |
| 2   | 0      |
| 4   | 1.099  |
| 6   | 1.609  |
| 8   | 1.946  |

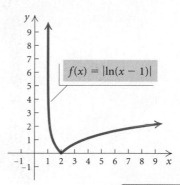

$f(x) = |\ln(x - 1)|$

**Now Try Exercise 89.**

## ■ Applications

**EXAMPLE 12**   *Walking Speed.*   In a study by psychologists Bornstein and Bornstein, it was found that the average walking speed $w$, in feet per second, of a person living in a city of population $P$, in thousands, is given by the function

$$w(P) = 0.37 \ln P + 0.05$$

(*Source*: *International Journal of Psychology*).

**a)** The population of Savannah, Georgia, is 136,286. Find the average walking speed of people living in Savannah.

**b)** The population of Philadelphia, Pennsylvania, is 1,526,006. Find the average walking speed of people living in Philadelphia.

**c)** Graph the function.

**d)** A sociologist computes the average walking speed in a city to be approximately 2.0 ft/sec. Use this information to estimate the population of the city.

*Solution*

**a)** Since $P$ is in thousands and $136{,}286 = 136.286$ thousand, we substitute 136.286 for $P$:

$$w(136.286) = 0.37 \ln 136.286 + 0.05 \approx 1.9.$$

The average walking speed of people living in Savannah is about 1.9 ft/sec.

**b)** We substitute 1526.006 for $P$:

$$w(1526.006) = 0.37 \ln 1526.006 + 0.05 \approx 2.7.$$

The average walking speed of people living in Philadelphia is about 2.7 ft/sec.

**c)** We graph with a viewing window of $[0, 600, 0, 4]$ because inputs are very large and outputs are very small by comparison.

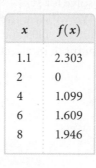

$y = 0.37 \ln(x) + 0.05$

**d)** To find the population for which the average walking speed is 2.0 ft/sec, we substitute 2.0 for $w(P)$,

$$2.0 = 0.37 \ln P + 0.05,$$

and solve for $P$.

We will use the Intersect method. We graph the equations $y_1 = 0.37 \ln x + 0.05$ and $y_2 = 2$ and use the INTERSECT feature to approximate the point of intersection.

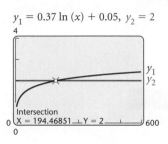

$y_1 = 0.37 \ln (x) + 0.05, \; y_2 = 2$

Intersection
X = 194.46851    Y = 2

We see that in a city in which the average walking speed is 2.0 ft/sec, the population is about 194.5 thousand, or 194,500.

**Now Try Exercise 95(d).**

**EXAMPLE 13** *Earthquake Magnitude.* Measured on the Richter scale, the magnitude $R$ of an earthquake of intensity $I$ is defined as

$$R = \log \frac{I}{I_0},$$

where $I_0$ is a minimum intensity used for comparison. We can think of $I_0$ as a threshold intensity that is the weakest earthquake that can be recorded on a seismograph. If one earthquake is 10 times as intense as another, its magnitude on the Richter scale is 1 greater than that of the other. If one earthquake is 100 times as intense as another, its magnitude on the Richter scale is 2 higher, and so on. Thus an earthquake whose magnitude is 7 on the Richter scale is 10 times as intense as an earthquake whose magnitude is 6. Earthquake intensities can be interpreted as multiples of the minimum intensity $I_0$.

The undersea Tohoku earthquake and tsunami, near the northeast coast of Honshu, Japan, on March 11, 2011, had an intensity of $10^{9.0} \cdot I_0$ (*Source*: earthquake.usgs.gov). They caused extensive loss of life and severe structural damage to buildings, railways, and roads. What was the magnitude on the Richter scale?

**Solution** We substitute into the formula:

$$R = \log \frac{I}{I_0} = \log \frac{10^{9.0} \cdot I_0}{I_0} = \log 10^{9.0} = 9.0.$$

The magnitude of the earthquake was 9.0 on the Richter scale.

**Now Try Exercise 97(a).**

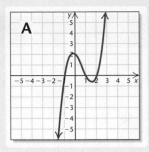

A

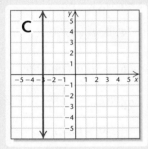

B

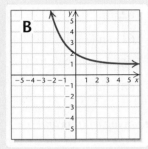

C

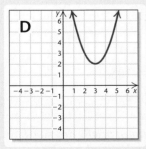

D

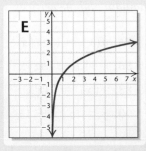

E

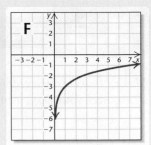

F

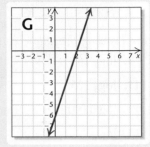

G

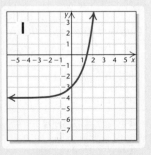

H

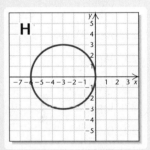

I

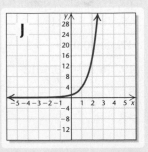

J

# Visualizing the Graph

Match the equation or function with its graph.

**1.** $f(x) = 4^x$

**2.** $f(x) = \ln x - 3$

**3.** $(x + 3)^2 + y^2 = 9$

**4.** $f(x) = 2^{-x} + 1$

**5.** $f(x) = \log_2 x$

**6.** $f(x) = x^3 - 2x^2 - x + 2$

**7.** $x = -3$

**8.** $f(x) = e^x - 4$

**9.** $f(x) = (x - 3)^2 + 2$

**10.** $3x = 6 + y$

Answers on page A-34

## 5.3 Exercise Set

*Make a hand-drawn graph of each of the following. Then check your work using a graphing calculator.*

**1.** $x = 3^y$

**2.** $x = 4^y$

**3.** $x = \left(\frac{1}{2}\right)^y$

**4.** $x = \left(\frac{4}{3}\right)^y$

**5.** $y = \log_3 x$

**6.** $y = \log_4 x$

**7.** $f(x) = \log x$

**8.** $f(x) = \ln x$

*Find each of the following. Do not use a calculator.*

**9.** $\log_2 16$

**10.** $\log_3 9$

**11.** $\log_5 125$

**12.** $\log_2 64$

**13.** $\log 0.001$

**14.** $\log 100$

**15.** $\log_2 \frac{1}{4}$

**16.** $\log_8 2$

**17.** $\ln 1$

**18.** $\ln e$

**19.** $\log 10$

**20.** $\log 1$

**21.** $\log_5 5^4$

**22.** $\log \sqrt{10}$

**23.** $\log_3 \sqrt[4]{3}$

**24.** $\log 10^{8/5}$

**25.** $\log 10^{-7}$

**26.** $\log_5 1$

**27.** $\log_{49} 7$

**28.** $\log_3 3^{-2}$

**29.** $\ln e^{3/4}$

**30.** $\log_2 \sqrt{2}$

**31.** $\log_4 1$

**32.** $\ln e^{-5}$

**33.** $\ln \sqrt{e}$

**34.** $\log_{64} 4$

*Convert to a logarithmic equation.*

**35.** $10^3 = 1000$

**36.** $5^{-3} = \frac{1}{125}$

**37.** $8^{1/3} = 2$

**38.** $10^{0.3010} = 2$

**39.** $e^3 = t$

**40.** $Q^t = x$

**41.** $e^2 = 7.3891$

**42.** $e^{-1} = 0.3679$

**43.** $p^k = 3$

**44.** $e^{-t} = 4000$

*Convert to an exponential equation.*

**45.** $\log_5 5 = 1$

**46.** $t = \log_4 7$

**47.** $\log 0.01 = -2$

**48.** $\log 7 = 0.845$

**49.** $\ln 30 = 3.4012$

**50.** $\ln 0.38 = -0.9676$

**51.** $\log_a M = -x$

**52.** $\log_t Q = k$

**53.** $\log_a T^3 = x$

**54.** $\ln W^5 = t$

*Find each of the following using a calculator. Round to four decimal places.*

**55.** $\log 3$

**56.** $\log 8$

**57.** $\log 532$

**58.** $\log 93{,}100$

**59.** $\log 0.57$

**60.** $\log 0.082$

**61.** $\log (-2)$

**62.** $\ln 50$

**63.** $\ln 2$

**64.** $\ln (-4)$

**65.** $\ln 809.3$

**66.** $\ln 0.00037$

**67.** $\ln (-1.32)$

**68.** $\ln 0$

*Find the logarithm using common logarithms and the change-of-base formula.*

**69.** $\log_4 100$

**70.** $\log_3 20$

**71.** $\log_{100} 0.3$

**72.** $\log_\pi 100$

**73.** $\log_{200} 50$

**74.** $\log_{5.3} 1700$

*Find the logarithm using natural logarithms and the change-of-base formula.*

**75.** $\log_3 12$

**76.** $\log_4 25$

**77.** $\log_{100} 15$

**78.** $\log_9 100$

*Graph the function and its inverse using the same set of axes. Use any method.*

**79.** $f(x) = 3^x,\ f^{-1}(x) = \log_3 x$

**80.** $f(x) = \log_4 x,\ f^{-1}(x) = 4^x$

**81.** $f(x) = \log x,\ f^{-1}(x) = 10^x$

**82.** $f(x) = e^x,\ f^{-1}(x) = \ln x$

*For each of the following functions, briefly describe how the graph can be obtained from the graph of a basic logarithmic function. Then graph the function using a graphing calculator. Give the domain and the vertical asymptote of each function.*

**83.** $f(x) = \log_2 (x + 3)$

**84.** $f(x) = \log_3 (x - 2)$

**85.** $y = \log_3 x - 1$

**86.** $y = 3 + \log_2 x$

**87.** $f(x) = 4 \ln x$

**88.** $f(x) = \frac{1}{2} \ln x$

**89.** $y = 2 - \ln x$

**90.** $y = \ln (x + 1)$

**91.** $f(x) = \frac{1}{2} \log (x - 1) - 2$

**92.** $f(x) = 5 - 2 \log (x + 1)$

*Graph the piecewise function.*

**93.** $g(x) = \begin{cases} 5, & \text{for } x \leq 0, \\ \log x + 1, & \text{for } x > 0 \end{cases}$

**94.** $f(x) = \begin{cases} 1 - x, & \text{for } x \leq -1, \\ \ln (x + 1), & \text{for } x > -1 \end{cases}$

**95.** *Walking Speed.* Refer to Example 12. Various cities and their populations are given below. Find the average walking speed in each city.

a) Seattle, Washington: 608,660
b) Los Angeles, California: 3,792,621
c) Virginia Beach, Virginia: 437,994
d) Houston, Texas: 2,099,451
e) Memphis, Tennessee: 646,889
f) Tampa, Florida: 335,709
g) Indianapolis, Indiana: 820,445
h) Anchorage, Alaska: 291,826

**96.** *Forgetting.* Students in a computer science class took a final exam and then took equivalent forms of the exam at monthly intervals thereafter. The average score $S(t)$, as a percent, after $t$ months was found to be given by the function

$$S(t) = 78 - 15 \log (t + 1), \quad t \geq 0.$$

a) What was the average score when the students initially took the test, $t = 0$?
b) What was the average score after 4 months? after 24 months?
c) Graph the function.
d) After what time $t$ was the average score 50%?

**97.** *Earthquake Magnitude.* Refer to Example 13. Various locations of earthquakes and their intensities are given below. What was the magnitude on the Richter scale?

a) San Francisco, California, 1906: $10^{7.7} \cdot I_0$
b) Chile, 1960: $10^{9.5} \cdot I_0$
c) Iran, 2003: $10^{6.6} \cdot I_0$
d) Turkey, 1999: $10^{7.4} \cdot I_0$
e) Peru, 2007: $10^{8.0} \cdot I_0$
f) China, 2008: $10^{7.9} \cdot I_0$
g) Indonesia, 2004: $10^{9.1} \cdot I_0$
h) Sumatra, 2004: $10^{9.3} \cdot I_0$

**98.** *pH of Substances in Chemistry.* In chemistry, the pH of a substance is defined as

$$\text{pH} = -\log[\text{H}^+],$$

where $\text{H}^+$ is the hydrogen ion concentration, in moles per liter. Find the pH of each substance.

| SUBSTANCE | HYDROGEN ION CONCENTRATION |
|---|---|
| a) Pineapple juice | $1.6 \times 10^{-4}$ |
| b) Hair conditioner | 0.0013 |
| c) Mouthwash | $6.3 \times 10^{-7}$ |
| d) Eggs | $1.6 \times 10^{-8}$ |
| e) Tomatoes | $6.3 \times 10^{-5}$ |

**99.** Find the hydrogen ion concentration of each substance, given the pH. (See Exercise 98.) Express the answer in scientific notation.

| SUBSTANCE | pH |
|---|---|
| a) Tap water | 7 |
| b) Rainwater | 5.4 |
| c) Orange juice | 3.2 |
| d) Wine | 4.8 |

**100.** *Advertising.*   A model for advertising response is given by the function

$$N(a) = 1000 + 200 \ln a, \quad a \geq 1,$$

where $N(a)$ is the number of units sold when $a$ is the amount spent on advertising, in thousands of dollars.

a) How many units were sold after spending $1000 ($a = 1$) on advertising?
b) How many units were sold after spending $5000?
c) Graph the function.
d) How much would have to be spent in order to sell 2000 units?

**101.** *Loudness of Sound.*   The **loudness L**, in bels (after Alexander Graham Bell), of a sound of intensity $I$ is defined to be

$$L = \log \frac{I}{I_0},$$

where $I_0$ is the minimum intensity detectable by the human ear (such as the tick of a watch at 20 ft under quiet conditions). If a sound is 10 times as intense as another, its loudness is 1 bel greater than that of the other. If a sound is 100 times as intense as another, its loudness is 2 bels greater, and so on. The bel is a large unit, so a subunit, the **decibel**, is generally used. For $L$, in decibels, the formula is

$$L = 10 \log \frac{I}{I_0}.$$

Find the loudness, in decibels, of each sound with the given intensity.

| SOUND | INTENSITY |
|---|---|
| a) Jet engine at 100 ft | $10^{14} \cdot I_0$ |
| b) Loud rock concert | $10^{11.5} \cdot I_0$ |
| c) Train whistle at 500 ft | $10^{9} \cdot I_0$ |
| d) Normal conversation | $10^{6.5} \cdot I_0$ |
| e) Trombone | $10^{10} \cdot I_0$ |
| f) Loudest sound possible | $10^{19.4} \cdot I_0$ |

## Skill Maintenance

*Find the slope and the y-intercept of the line.*

**102.** $3x - 10y = 14$

**103.** $y = 6$

**104.** $x = -4$

*Use synthetic division to find the function values.*

**105.** $g(x) = x^3 - 6x^2 + 3x + 10$; find $g(-5)$

**106.** $f(x) = x^4 - 2x^3 + x - 6$; find $f(-1)$

*Find a polynomial function of degree 3 with the given numbers as zeros. Answers may vary.*

**107.** $\sqrt{7}, -\sqrt{7}, 0$

**108.** $4i, -4i, 1$

## Synthesis

*Simplify.*

**109.** $\dfrac{\log_5 8}{\log_5 2}$

**110.** $\dfrac{\log_3 64}{\log_3 16}$

*Find the domain of the function.*

**111.** $f(x) = \log_5 x^3$

**112.** $f(x) = \log_4 x^2$

**113.** $f(x) = \ln |x|$

**114.** $f(x) = \log (3x - 4)$

*Solve.*

**115.** $\log_2 (2x + 5) < 0$

**116.** $\log_2 (x - 3) \geq 4$

*In Exercises 117–120, match the equation with one of the figures (a)–(d), which follow.*

a)
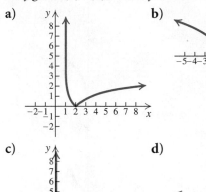

b)
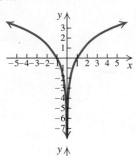

c)
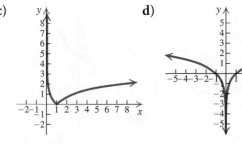

d)
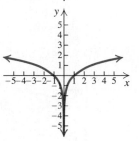

**117.** $f(x) = \ln|x|$

**118.** $f(x) = |\ln x|$

**119.** $f(x) = \ln x^2$

**120.** $g(x) = |\ln(x - 1)|$

*For Exercises 121–124:*

a) *Graph the function.*

b) *Estimate the zeros.*

c) *Estimate the relative maximum values and the relative minimum values.*

**121.** $f(x) = x \ln x$

**122.** $f(x) = x^2 \ln x$

**123.** $f(x) = \dfrac{\ln x}{x^2}$

**124.** $f(x) = e^{-x} \ln x$

**125.** Using a graphing calculator, find the point(s) of intersection of the graphs of the following.

$$y = 4 \ln x, \qquad y = \frac{4}{e^x + 1}$$

## Mid-Chapter Mixed Review

*Determine whether the statement is true or false.*

**1.** The domain of all logarithmic functions is $[1, \infty)$. [5.3]

**2.** The range of a one-to-one function $f$ is the domain of its inverse $f^{-1}$. [5.1]

**3.** The $y$-intercept of $f(x) = e^{-x}$ is $(0, -1)$. [5.2]

*For each function, determine whether it is one-to-one, and if the function is one-to-one, find a formula for its inverse.* [5.1]

**4.** $f(x) = -\dfrac{2}{x}$

**5.** $f(x) = 3 + x^2$

**6.** $f(x) = \dfrac{5}{x - 2}$

**7.** Given the function $f(x) = \sqrt{x - 5}$, use composition of functions to show that $f^{-1}(x) = x^2 + 5$. [5.1]

**8.** Given the one-to-one function $f(x) = x^3 + 2$, find the inverse and give the domain and the range of $f$ and of $f^{-1}$. [5.1]

*Match the function with one of the graphs (a)–(h), which follow.* [5.2], [5.3]

a)
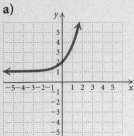

b)
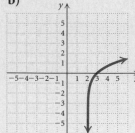

c)
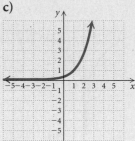

d)
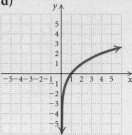

e)
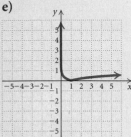

f)

g)

h)
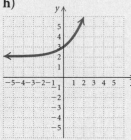

**9.** $y = \log_2 x$

**10.** $f(x) = 2^x + 2$

**11.** $f(x) = e^{x-1}$

**12.** $f(x) = \ln x - 2$

**13.** $f(x) = \ln(x - 2)$

**14.** $y = 2^{-x}$

**15.** $f(x) = |\log x|$

**16.** $f(x) = e^x + 1$

**17.** Suppose that \$3200 is invested at $4\frac{1}{2}\%$ interest, compounded quarterly. Find the amount of money in the account in 6 years. [5.2]

*Find each of the following without a calculator.* [5.3]

**18.** $\log_4 1$

**19.** $\ln e^{-4/5}$

**20.** $\log 0.01$

**21.** $\ln e^2$

**22.** $\ln 1$

**23.** $\log_2 \dfrac{1}{16}$

**24.** $\log 1$

**25.** $\log_3 27$

**26.** $\log \sqrt[4]{10}$

**27.** $\ln e$

**28.** Convert $e^{-6} = 0.0025$ to a logarithmic equation. [5.3]

**29.** Convert $\log T = r$ to an exponential equation. [5.3]

*Find the logarithm using the change-of-base formula.* [5.3]

**30.** $\log_3 20$

**31.** $\log_\pi 10$

## Collaborative Discussion and Writing

**32.** Explain why an even function $f$ does not have an inverse $f^{-1}$ that is a function. [5.1]

**33.** Suppose that \$10,000 is invested for 8 years at 6.4% interest, compounded annually. In what year will the most interest be earned? Why? [5.2]

**34.** Describe the differences between the graphs of $f(x) = x^3$ and $g(x) = 3^x$. [5.2]

**35.** If $\log b < 0$, what can you say about $b$? [5.3]

## 5.4

# Properties of Logarithmic Functions

- Convert from logarithms of products, powers, and quotients to expressions in terms of individual logarithms, and conversely.
- Simplify expressions of the type $\log_a a^x$ and $a^{\log_a x}$.

We now establish some properties of logarithmic functions. These properties are based on the corresponding rules for exponents.

## ■ Logarithms of Products

The first property of logarithms corresponds to the product rule for exponents: $a^m \cdot a^n = a^{m+n}$.

> **THE PRODUCT RULE**
>
> For any positive numbers $M$ and $N$ and any logarithmic base $a$,
>
> $$\log_a MN = \log_a M + \log_a N.$$
>
> (The logarithm of a product is the sum of the logarithms of the factors.)

**EXAMPLE 1** Express as a sum of logarithms: $\log_3 (9 \cdot 27)$.

*Solution* We have

$$\log_3 (9 \cdot 27) = \log_3 9 + \log_3 27. \qquad \text{Using the product rule}$$

As a check, note that

$$\log_3 (9 \cdot 27) = \log_3 243 = 5 \qquad 3^5 = 243$$
$$\text{and} \quad \log_3 9 + \log_3 27 = 2 + 3 = 5. \qquad 3^2 = 9; 3^3 = 27$$

> **Now Try Exercise 1.**

**EXAMPLE 2** Express as a single logarithm: $\log_2 p^3 + \log_2 q$.

*Solution* We have

$$\log_2 p^3 + \log_2 q = \log_2 (p^3 q).$$

> **Now Try Exercise 35.**

**A Proof of the Product Rule:** Let $\log_a M = x$ and $\log_a N = y$. Converting to exponential equations, we have $a^x = M$ and $a^y = N$. Then

$$MN = a^x \cdot a^y = a^{x+y}.$$

Converting back to a logarithmic equation, we get

$$\log_a MN = x + y.$$

Remembering what $x$ and $y$ represent, we know it follows that

$$\log_a MN = \log_a M + \log_a N.$$

## ◼ Logarithms of Powers

The second property of logarithms corresponds to the power rule for exponents: $(a^m)^n = a^{mn}$.

> ### THE POWER RULE
>
> For any positive number $M$, any logarithmic base $a$, and any real number $p$,
>
> $$\log_a M^p = p \log_a M.$$
>
> (The logarithm of a power of $M$ is the exponent times the logarithm of $M$.)

**EXAMPLE 3**   Express each of the following as a product.

**a)** $\log_a 11^{-3}$          **b)** $\log_a \sqrt[4]{7}$          **c)** $\ln x^6$

*Solution*

| RATIONAL EXPONENTS |
|---|
| REVIEW SECTION **R.7.** |

**a)** $\log_a 11^{-3} = -3 \log_a 11$      Using the power rule

**b)** $\log_a \sqrt[4]{7} = \log_a 7^{1/4}$      Writing exponential notation

            $= \frac{1}{4} \log_a 7$      Using the power rule

**c)** $\ln x^6 = 6 \ln x$      Using the power rule

> **Now Try Exercises 13 and 15.**

**A Proof of the Power Rule:**   Let $x = \log_a M$. The equivalent exponential equation is $a^x = M$. Raising both sides to the power $p$, we obtain

$$(a^x)^p = M^p, \quad \text{or} \quad a^{xp} = M^p.$$

Converting back to a logarithmic equation, we get

$$\log_a M^p = xp.$$

But $x = \log_a M$, so substituting gives us

$$\log_a M^p = (\log_a M)p = p \log_a M. \qquad ◼$$

## ◼ Logarithms of Quotients

The third property of logarithms corresponds to the quotient rule for exponents: $a^m/a^n = a^{m-n}$.

> ### THE QUOTIENT RULE
>
> For any positive numbers $M$ and $N$, and any logarithmic base $a$,
>
> $$\log_a \frac{M}{N} = \log_a M - \log_a N.$$
>
> (The logarithm of a quotient is the logarithm of the numerator minus the logarithm of the denominator.)

**EXAMPLE 4** Express as a difference of logarithms: $\log_t \dfrac{8}{w}$.

*Solution* We have

$$\log_t \frac{8}{w} = \log_t 8 - \log_t w. \qquad \text{Using the quotient rule}$$

**Now Try Exercise 17.**

**EXAMPLE 5** Express as a single logarithm: $\log_b 64 - \log_b 16$.

*Solution* We have

$$\log_b 64 - \log_b 16 = \log_b \frac{64}{16} = \log_b 4.$$

**Now Try Exercise 37.**

**A Proof of the Quotient Rule:** The proof follows from both the product rule and the power rule:

$$\log_a \frac{M}{N} = \log_a MN^{-1}$$

$$= \log_a M + \log_a N^{-1} \qquad \text{Using the product rule}$$

$$= \log_a M + (-1)\log_a N \qquad \text{Using the power rule}$$

$$= \log_a M - \log_a N.$$

**Common Errors**

| | |
|---|---|
| $\log_a MN \neq (\log_a M)(\log_a N)$ | The logarithm of a product is *not* the product of the logarithms. |
| $\log_a (M + N) \neq \log_a M + \log_a N$ | The logarithm of a sum is *not* the sum of the logarithms. |
| $\log_a \dfrac{M}{N} \neq \dfrac{\log_a M}{\log_a N}$ | The logarithm of a quotient is *not* the quotient of the logarithms. |
| $(\log_a M)^p \neq p \log_a M$ | The power of a logarithm is *not* the exponent times the logarithm. |

## ■ Applying the Properties

**EXAMPLE 6** Express each of the following in terms of sums and differences of logarithms.

a) $\log_a \dfrac{x^2 y^5}{z^4}$      b) $\log_a \sqrt[3]{\dfrac{a^2 b}{c^5}}$      c) $\log_b \dfrac{a y^5}{m^3 n^4}$

*Solution*

a) $\log_a \dfrac{x^2 y^5}{z^4} = \log_a (x^2 y^5) - \log_a z^4 \qquad \text{Using the quotient rule}$

$$= \log_a x^2 + \log_a y^5 - \log_a z^4 \qquad \text{Using the product rule}$$

$$= 2\log_a x + 5\log_a y - 4\log_a z \qquad \text{Using the power rule}$$

**b)** $\log_a \sqrt[3]{\dfrac{a^2b}{c^5}} = \log_a \left(\dfrac{a^2b}{c^5}\right)^{1/3}$     **Writing exponential notation**

$= \dfrac{1}{3} \log_a \dfrac{a^2b}{c^5}$     **Using the power rule**

$= \dfrac{1}{3} \left(\log_a a^2b - \log_a c^5\right)$     **Using the quotient rule. The parentheses are necessary.**

$= \dfrac{1}{3} \left(2 \log_a a + \log_a b - 5 \log_a c\right)$     **Using the product rule and the power rule**

$= \dfrac{1}{3} \left(2 + \log_a b - 5 \log_a c\right)$     $\log_a a = 1$

$= \dfrac{2}{3} + \dfrac{1}{3} \log_a b - \dfrac{5}{3} \log_a c$     **Multiplying to remove parentheses**

**c)** $\log_b \dfrac{ay^5}{m^3n^4} = \log_b ay^5 - \log_b m^3n^4$     **Using the quotient rule**

$= \left(\log_b a + \log_b y^5\right) - \left(\log_b m^3 + \log_b n^4\right)$     **Using the product rule**

$= \log_b a + \log_b y^5 - \log_b m^3 - \log_b n^4$     **Removing parentheses**

$= \log_b a + 5 \log_b y - 3 \log_b m - 4 \log_b n$     **Using the power rule**

> **Now Try Exercises 25 and 31.**

**EXAMPLE 7** Express as a single logarithm:

$$5 \log_b x - \log_b y + \tfrac{1}{4} \log_b z.$$

*Solution* We have

$5 \log_b x - \log_b y + \tfrac{1}{4} \log_b z = \log_b x^5 - \log_b y + \log_b z^{1/4}$

    **Using the power rule**

$= \log_b \dfrac{x^5}{y} + \log_b z^{1/4}$     **Using the quotient rule**

$= \log_b \dfrac{x^5 z^{1/4}}{y}$, or $\log_b \dfrac{x^5 \sqrt[4]{z}}{y}$.

    **Using the product rule**

> **Now Try Exercise 41.**

**EXAMPLE 8** Express as a single logarithm:

$$\ln (3x + 1) - \ln (3x^2 - 5x - 2).$$

*Solution* We have

$\ln (3x + 1) - \ln (3x^2 - 5x - 2) = \ln \dfrac{3x + 1}{3x^2 - 5x - 2}$     **Using the quotient rule**

$= \ln \dfrac{3x + 1}{(3x + 1)(x - 2)}$     **Factoring**

$= \ln \dfrac{1}{x - 2}.$     **Simplifying**

> **Now Try Exercise 45.**

**EXAMPLE 9** Given that $\log_a 2 \approx 0.301$ and $\log_a 3 \approx 0.477$, find each of the following, if possible.

**a)** $\log_a 6$          **b)** $\log_a \frac{2}{3}$          **c)** $\log_a 81$

**d)** $\log_a \frac{1}{4}$          **e)** $\log_a 5$          **f)** $\dfrac{\log_a 3}{\log_a 2}$

*Solution*

**a)** $\log_a 6 = \log_a (2 \cdot 3) = \log_a 2 + \log_a 3$      Using the product rule

$$\approx 0.301 + 0.477$$

$$\approx 0.778$$

**b)** $\log_a \frac{2}{3} = \log_a 2 - \log_a 3$      Using the quotient rule

$$\approx 0.301 - 0.477 \approx -0.176$$

**c)** $\log_a 81 = \log_a 3^4 = 4 \log_a 3$      Using the power rule

$$\approx 4(0.477) \approx 1.908$$

**d)** $\log_a \frac{1}{4} = \log_a 1 - \log_a 4$      Using the quotient rule

$$= 0 - \log_a 2^2 \qquad \log_a 1 = 0$$

$$= -2 \log_a 2 \qquad \text{Using the power rule}$$

$$\approx -2(0.301) \approx -0.602$$

**e)** $\log_a 5$ *cannot* be found using these properties and the given information.

$$\log_a 5 \neq \log_a 2 + \log_a 3 \qquad \log_a 2 + \log_a 3 = \log_a (2 \cdot 3) = \log_a 6$$

**f)** $\dfrac{\log_a 3}{\log_a 2} \approx \dfrac{0.477}{0.301} \approx 1.585$      We simply divide, not using any of the properties.

> **Now Try Exercises 53 and 55.**

### ■ Simplifying Expressions of the Type $\log_a a^x$ and $a^{\log_a x}$

We have two final properties of logarithms to consider. The first follows from the product rule: Since $\log_a a^x = x \log_a a = x \cdot 1 = x$, we have $\log_a a^x = x$. This property also follows from the definition of a logarithm: $x$ is the power to which we raise $a$ in order to get $a^x$.

---

**THE LOGARITHM OF A BASE TO A POWER**

For any base $a$ and any real number $x$,

$$\log_a a^x = x.$$

(The logarithm, base $a$, of $a$ to a power is the power.)

**EXAMPLE 10** Simplify each of the following.

**a)** $\log_a a^8$          **b)** $\ln e^{-t}$          **c)** $\log 10^{3k}$

*Solution*

**a)** $\log_a a^8 = 8$     8 is the power to which we raise $a$ in order to get $a^8$.

**b)** $\ln e^{-t} = \log_e e^{-t} = -t$     $\ln e^x = x$

**c)** $\log 10^{3k} = \log_{10} 10^{3k} = 3k$

**Now Try Exercises 65 and 73.**

Let $M = \log_a x$. Then $a^M = x$. Substituting $\log_a x$ for $M$, we obtain $a^{\log_a x} = x$. This also follows from the definition of a logarithm: $\log_a x$ is the power to which $a$ is raised in order to get $x$.

---

### A BASE TO A LOGARITHMIC POWER

For any base $a$ and any positive real number $x$,

$$a^{\log_a x} = x.$$

(The number $a$ raised to the power $\log_a x$ is $x$.)

---

**EXAMPLE 11** Simplify each of the following.

**a)** $4^{\log_4 k}$          **b)** $e^{\ln 5}$          **c)** $10^{\log 7t}$

*Solution*

**a)** $4^{\log_4 k} = k$

**b)** $e^{\ln 5} = e^{\log_e 5} = 5$

**c)** $10^{\log 7t} = 10^{\log_{10} 7t} = 7t$

**Now Try Exercises 69 and 71.**

**CHANGE-OF-BASE FORMULA**

REVIEW SECTION **5.3**.

**A Proof of the Change-of-Base Formula:** We close this section by proving the change-of-base formula and summarizing the properties of logarithms considered thus far in this chapter. In Section 5.3, we used the change-of-base formula,

$$\log_b M = \frac{\log_a M}{\log_a b},$$

to make base conversions in order to find logarithmic values using a calculator. Let $x = \log_b M$. Then

| | |
|---|---|
| $b^x = M$ | Definition of logarithm |
| $\log_a b^x = \log_a M$ | Taking the logarithm on both sides |
| $x \log_a b = \log_a M$ | Using the power rule |
| $x = \dfrac{\log_a M}{\log_a b},$ | Dividing by $\log_a b$ |

so

$$x = \log_b M = \frac{\log_a M}{\log_a b}.$$

Following is a summary of the properties of logarithms.

---

**Summary of the Properties of Logarithms**

| | |
|---|---|
| *The Product Rule*: | $\log_a MN = \log_a M + \log_a N$ |
| *The Power Rule*: | $\log_a M^p = p \log_a M$ |
| *The Quotient Rule*: | $\log_a \dfrac{M}{N} = \log_a M - \log_a N$ |
| *The Change-of-Base Formula*: | $\log_b M = \dfrac{\log_a M}{\log_a b}$ |
| *Other Properties*: | $\log_a a = 1, \quad \log_a 1 = 0,$ |
| | $\log_a a^x = x, \quad a^{\log_a x} = x$ |

---

# 5.4 Exercise Set

*Express as a sum of logarithms.*

**1.** $\log_3 (81 \cdot 27)$

**2.** $\log_2 (8 \cdot 64)$

**3.** $\log_5 (5 \cdot 125)$

**4.** $\log_4 (64 \cdot 4)$

**5.** $\log_t 8Y$

**6.** $\log 0.2x$

**7.** $\ln xy$

**8.** $\ln ab$

*Express as a product.*

**9.** $\log_b t^3$

**10.** $\log_a x^4$

**11.** $\log y^8$

**12.** $\ln y^5$

**13.** $\log_c K^{-6}$

**14.** $\log_b Q^{-8}$

**15.** $\ln \sqrt[3]{4}$

**16.** $\ln \sqrt{a}$

*Express as a difference of logarithms.*

**17.** $\log_t \dfrac{M}{8}$

**18.** $\log_a \dfrac{76}{13}$

**19.** $\log \dfrac{x}{y}$

**20.** $\ln \dfrac{a}{b}$

**21.** $\ln \dfrac{r}{s}$

**22.** $\log_b \dfrac{3}{w}$

*Express in terms of sums and differences of logarithms.*

**23.** $\log_a 6xy^5 z^4$

**24.** $\log_a x^3 y^2 z$

**25.** $\log_b \dfrac{p^2 q^5}{m^4 b^9}$

**26.** $\log_b \dfrac{x^2 y}{b^3}$

**27.** $\ln \dfrac{2}{3x^3 y}$

**28.** $\log \dfrac{5a}{4b^2}$

**29.** $\log \sqrt{r^3 t}$

**30.** $\ln \sqrt[3]{5x^5}$

**31.** $\log_a \sqrt{\dfrac{x^6}{p^5 q^8}}$

**32.** $\log_c \sqrt[3]{\dfrac{y^3 z^2}{x^4}}$

**33.** $\log_a \sqrt[4]{\dfrac{m^8 n^{12}}{a^3 b^5}}$

**34.** $\log_a \sqrt{\dfrac{a^6 b^8}{a^2 b^5}}$

*Express as a single logarithm and, if possible, simplify.*

**35.** $\log_a 75 + \log_a 2$

**36.** $\log 0.01 + \log 1000$

**37.** $\log 10{,}000 - \log 100$

**38.** $\ln 54 - \ln 6$

**39.** $\frac{1}{2} \log n + 3 \log m$

**40.** $\frac{1}{2} \log a - \log 2$

**41.** $\frac{1}{2} \log_a x + 4 \log_a y - 3 \log_a x$

**42.** $\frac{2}{5} \log_a x - \frac{1}{3} \log_a y$

**43.** $\ln x^2 - 2 \ln \sqrt{x}$

**44.** $\ln 2x + 3(\ln x - \ln y)$

**45.** $\ln (x^2 - 4) - \ln (x + 2)$

**46.** $\log (x^3 - 8) - \log (x - 2)$

**47.** $\log(x^2 - 5x - 14) - \log(x^2 - 4)$

**48.** $\log_a \dfrac{a}{\sqrt{x}} - \log_a \sqrt{ax}$

**49.** $\ln x - 3\left[\ln(x - 5) + \ln(x + 5)\right]$

**50.** $\frac{2}{3}\left[\ln(x^2 - 9) - \ln(x + 3)\right] + \ln(x + y)$

**51.** $\frac{3}{2}\ln 4x^6 - \frac{4}{5}\ln 2y^{10}$

**52.** $120(\ln \sqrt[5]{x^3} + \ln \sqrt[3]{y^2} - \ln \sqrt[4]{16z^5})$

*Given that $\log_a 2 \approx 0.301$, $\log_a 7 \approx 0.845$, and $\log_a 11 \approx 1.041$, find each of the following, if possible. Round the answer to the nearest thousandth.*

**53.** $\log_a \frac{2}{11}$

**54.** $\log_a 14$

**55.** $\log_a 98$

**56.** $\log_a \frac{1}{7}$

**57.** $\dfrac{\log_a 2}{\log_a 7}$

**58.** $\log_a 9$

*Given that $\log_b 2 \approx 0.693$, $\log_b 3 \approx 1.099$, and $\log_b 5 \approx 1.609$, find each of the following, if possible. Round the answer to the nearest thousandth.*

**59.** $\log_b 125$

**60.** $\log_b \frac{5}{3}$

**61.** $\log_b \frac{1}{6}$

**62.** $\log_b 30$

**63.** $\log_b \dfrac{3}{b}$

**64.** $\log_b 15b$

*Simplify.*

**65.** $\log_p p^3$

**66.** $\log_t t^{2713}$

**67.** $\log_e e^{|x-4|}$

**68.** $\log_q q^{\sqrt{3}}$

**69.** $3^{\log_3 4x}$

**70.** $5^{\log_5 (4x-3)}$

**71.** $10^{\log w}$

**72.** $e^{\ln x^3}$

**73.** $\ln e^{8t}$

**74.** $\log 10^{-k}$

**75.** $\log_b \sqrt{b}$

**76.** $\log_b \sqrt{b^3}$

## Skill Maintenance

*In each of Exercises 77–86, classify the function as linear, quadratic, cubic, quartic, rational, exponential, or logarithmic.*

**77.** $f(x) = 5 - x^2 + x^4$

**78.** $f(x) = 2^x$

**79.** $f(x) = -\frac{3}{4}$

**80.** $f(x) = 4^x - 8$

**81.** $f(x) = -\dfrac{3}{x}$

**82.** $f(x) = \log x + 6$

**83.** $f(x) = -\frac{1}{3}x^3 - 4x^2 + 6x + 42$

**84.** $f(x) = \dfrac{x^2 - 1}{x^2 + x - 6}$

**85.** $f(x) = \frac{1}{2}x + 3$

**86.** $f(x) = 2x^2 - 6x + 3$

## Synthesis

*Solve for x.*

**87.** $5^{\log_5 8} = 2x$

**88.** $\ln e^{3x-5} = -8$

*Express as a single logarithm and, if possible, simplify.*

**89.** $\log_a(x^2 + xy + y^2) + \log_a(x - y)$

**90.** $\log_a(a^{10} - b^{10}) - \log_a(a + b)$

*Express as a sum or a difference of logarithms.*

**91.** $\log_a \dfrac{x - y}{\sqrt{x^2 - y^2}}$

**92.** $\log_a \sqrt{9 - x^2}$

**93.** Given that $\log_a x = 2$, $\log_a y = 3$, and $\log_a z = 4$, find

$$\log_a \dfrac{\sqrt[4]{y^2 z^5}}{\sqrt[4]{x^3 z^{-2}}}.$$

*Determine whether each of the following is true or false. Assume that $a$, $x$, $M$, and $N$ are positive.*

**94.** $\log_a M + \log_a N = \log_a(M + N)$

**95.** $\log_a M - \log_a N = \log_a \dfrac{M}{N}$

**96.** $\dfrac{\log_a M}{\log_a N} = \log_a M - \log_a N$

**97.** $\dfrac{\log_a M}{x} = \log_a M^{1/x}$

**98.** $\log_a x^3 = 3\log_a x$

**99.** $\log_a 8x = \log_a x + \log_a 8$

**100.** $\log_N (MN)^x = x\log_N M + x$

*Suppose that* $\log_a x = 2$. *Find each of the following.*

**101.** $\log_a\left(\dfrac{1}{x}\right)$      **102.** $\log_{1/a} x$

**103.** Simplify:

$\log_{10} 11 \cdot \log_{11} 12 \cdot \log_{12} 13 \cdots \log_{998} 999 \cdot \log_{999} 1000.$

*Write each of the following without using logarithms.*

**104.** $\log_a x + \log_a y - mz = 0$

**105.** $\ln a - \ln b + xy = 0$

*Prove each of the following for any base a and any positive number x.*

**106.** $\log_a\left(\dfrac{1}{x}\right) = -\log_a x = \log_{1/a} x$

**107.** $\log_a\left(\dfrac{x + \sqrt{x^2 - 5}}{5}\right) = -\log_a\left(x - \sqrt{x^2 - 5}\right)$

---

## 5.5 Solving Exponential Equations and Logarithmic Equations

- Solve exponential equations.
- Solve logarithmic equations.

---

### Solving Exponential Equations

Equations with variables in the exponents, such as

$$3^x = 20 \quad \text{and} \quad 2^{5x} = 64,$$

are called **exponential equations**.

Sometimes, as is the case with the equation $2^{5x} = 64$, we can write each side as a power of the same number:

$$2^{5x} = 2^6.$$

We can then set the exponents equal and solve:

$$5x = 6$$
$$x = \tfrac{6}{5}, \text{ or } 1.2.$$

We use the following property to solve exponential equations.

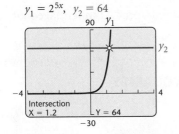

$y_1 = 2^{5x}, \quad y_2 = 64$

Intersection
X = 1.2    Y = 64

> **BASE–EXPONENT PROPERTY**
> For any $a > 0, a \neq 1,$
> $$a^x = a^y \longleftrightarrow x = y.$$

> **ONE-TO-ONE FUNCTIONS**
>
> REVIEW SECTION **5.1.**

This property follows from the fact that for any $a > 0, a \neq 1$, $f(x) = a^x$ is a one-to-one function. If $a^x = a^y$, then $f(x) = f(y)$. Then since $f$ is one-to-one, it follows that $x = y$. Conversely, if $x = y$, it follows that $a^x = a^y$, since we are raising $a$ to the same power in each case.

**EXAMPLE 1** Solve: $2^{3x-7} = 32$.

## Algebraic Solution

Note that $32 = 2^5$. Thus we can write each side as a power of the same number:

$$2^{3x-7} = 2^5.$$

Since the bases are the same number, 2, we can use the base–exponent property and set the exponents equal:

$$3x - 7 = 5$$
$$3x = 12$$
$$x = 4.$$

*Check:*

$$\begin{array}{c|c} 2^{3x-7} = 32 \\ \hline 2^{3(4)-7} \ ? \ 32 \\ 2^{12-7} \\ 2^5 \\ 32 & 32 \quad \text{TRUE} \end{array}$$

The solution is 4.

## Graphical Solution

We will use the Intersect method. We graph

$$y_1 = 2^{3x-7} \quad \text{and} \quad y_2 = 32$$

to find the coordinates of the point of intersection. The first coordinate of this point is the solution of the equation $y_1 = y_2$, or $2^{3x-7} = 32$.

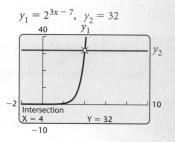

$$y_1 = 2^{3x-7}, \ y_2 = 32$$

The solution is 4.

We could also write the equation in the form $2^{3x-7} - 32 = 0$ and use the Zero method.

**Now Try Exercise 7.**

Another property that is used when solving some exponential equations and logarithmic equations is as follows.

> ### PROPERTY OF LOGARITHMIC EQUALITY
> For any $M > 0$, $N > 0$, $a > 0$, and $a \neq 1$,
> $$\log_a M = \log_a N \longleftrightarrow M = N.$$

This property follows from the fact that for any $a > 0$, $a \neq 1$, $f(x) = \log_a x$ is a one-to-one function. If $\log_a x = \log_a y$, then $f(x) = f(y)$. Then since $f$ is one-to-one, it follows that $x = y$. Conversely, if $x = y$, it follows that $\log_a x = \log_a y$, since we are taking the logarithm of the same number in each case.

When it does not seem possible to write each side as a power of the same base, we can use the property of logarithmic equality and take the logarithm with any base on each side and then use the power rule for logarithms.

**EXAMPLE 2** Solve: $3^x = 20$.

## Algebraic Solution

We have

$$3^x = 20$$

$$\log 3^x = \log 20 \qquad \text{Taking the common logarithm on both sides}$$

$$x \log 3 = \log 20 \qquad \text{Using the power rule}$$

$$x = \frac{\log 20}{\log 3}. \qquad \text{Dividing by log 3}$$

This is an exact answer. We cannot simplify further, but we can approximate using a calculator:

$$x = \frac{\log 20}{\log 3} \approx 2.7268.$$

We can check this by finding $3^{2.7268}$:

$$3^{2.7268} \approx 20.$$

The solution is about 2.7268.

## Graphical Solution

We will use the Intersect method. We graph

$$y_1 = 3^x \quad \text{and} \quad y_2 = 20$$

to find the $x$-coordinate of the point of intersection. That $x$-coordinate is the value of $x$ for which $3^x = 20$ and is thus the solution of the equation.

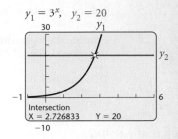

$y_1 = 3^x, \quad y_2 = 20$

Intersection
X = 2.726833   Y = 20

The solution is approximately 2.7268.

We could also write the equation in the form $3^x - 20 = 0$ and use the Zero method.

**Now Try Exercise 11.**

In Example 2, we took the common logarithm on both sides of the equation. Any base will give the same result. Let's try base 3. We have

$$3^x = 20$$

$$\log_3 3^x = \log_3 20$$

$$x = \log_3 20 \qquad \log_a a^x = x$$

$$x = \frac{\log 20}{\log 3} \qquad \text{Using the change-of-base formula}$$

$$x \approx 2.7268.$$

Note that we changed the base to do the final calculation. We also could find $\log_3 20$ directly using the logBASE operation.

log₃(20)
                    2.726833028

**EXAMPLE 3**    Solve: $100e^{0.08t} = 2500$.

### Algebraic Solution

It will make our work easier if we take the natural logarithm when working with equations that have $e$ as a base.

We have

$$100e^{0.08t} = 2500$$

$$e^{0.08t} = 25 \qquad \text{Dividing by 100}$$

$$\ln e^{0.08t} = \ln 25 \qquad \begin{array}{l}\text{Taking the natural logarithm}\\\text{on both sides}\end{array}$$

$$0.08t = \ln 25 \qquad \begin{array}{l}\text{Finding the logarithm of a base}\\\text{to a power: } \log_a a^x = x\end{array}$$

$$t = \frac{\ln 25}{0.08} \qquad \text{Dividing by 0.08}$$

$$\approx 40.2.$$

The solution is about 40.2.

### Graphical Solution

Using the Intersect method, we graph the equations

$$y_1 = 100e^{0.08x} \quad \text{and} \quad y_2 = 2500$$

and determine the point of intersection. The first coordinate of the point of intersection is the solution of the equation $100e^{0.08x} = 2500$.

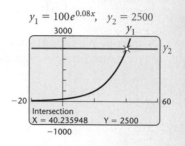

$y_1 = 100e^{0.08x}, \quad y_2 = 2500$

Intersection
X = 40.235948    Y = 2500

The solution is approximately 40.2.

**Now Try Exercise 19.**

---

**EXAMPLE 4**    Solve: $4^{x+3} = 3^{-x}$.

### Algebraic Solution

We have

$$4^{x+3} = 3^{-x}$$

$$\log 4^{x+3} = \log 3^{-x} \qquad \begin{array}{l}\text{Taking the common}\\\text{logarithm on both}\\\text{sides}\end{array}$$

$$(x + 3)\log 4 = -x\log 3 \qquad \begin{array}{l}\text{Using the power}\\\text{rule}\end{array}$$

$$x\log 4 + 3\log 4 = -x\log 3 \qquad \begin{array}{l}\text{Removing}\\\text{parentheses}\end{array}$$

$$x\log 4 + x\log 3 = -3\log 4 \qquad \begin{array}{l}\text{Adding } x\log 3 \text{ and}\\\text{subtracting } 3\log 4\end{array}$$

$$x(\log 4 + \log 3) = -3\log 4 \qquad \text{Factoring on the left}$$

$$x = \frac{-3\log 4}{\log 4 + \log 3} \qquad \begin{array}{l}\text{Dividing by}\\\log 4 + \log 3\end{array}$$

$$\approx -1.6737.$$

### Graphical Solution

We will use the Intersect method. We graph

$$y_1 = 4^{x+3} \quad \text{and} \quad y_2 = 3^{-x}$$

to find the $x$-coordinate of the point of intersection. That $x$-coordinate is the value of $x$ for which $4^{x+3} = 3^{-x}$ and is the solution of the equation.

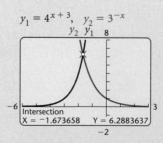

$y_1 = 4^{x+3}, \quad y_2 = 3^{-x}$

Intersection
X = -1.673658    Y = 6.2883637

The solution is approximately $-1.6737$.

We could also write the equation in the form $4^{x+3} - 3^{-x} = 0$ and use the Zero method.

**Now Try Exercise 21.**

| EQUATIONS REDUCIBLE TO QUADRATIC |
|---|
| REVIEW SECTION **3.2.** |

**EXAMPLE 5** Solve: $e^x + e^{-x} - 6 = 0$.

## Algebraic Solution

In this case, we have more than one term with $x$ in the exponent:

$$e^x + e^{-x} - 6 = 0$$

$$e^x + \frac{1}{e^x} - 6 = 0 \qquad \text{Rewriting } e^{-x} \text{ with a positive exponent}$$

$$e^{2x} + 1 - 6e^x = 0. \qquad \text{Multiplying by } e^x \text{ on both sides}$$

This equation is reducible to quadratic with $u = e^x$:

$$u^2 - 6u + 1 = 0.$$

We use the quadratic formula with $a = 1$, $b = -6$, and $c = 1$:

$$u = \frac{-b \pm \sqrt{b^2 - 4ac}}{2a}$$

$$u = \frac{-(-6) \pm \sqrt{(-6)^2 - 4 \cdot 1 \cdot 1}}{2 \cdot 1}$$

$$u = \frac{6 \pm \sqrt{32}}{2} = \frac{6 \pm 4\sqrt{2}}{2}$$

$$u = 3 \pm 2\sqrt{2}$$

$$e^x = 3 \pm 2\sqrt{2}. \qquad \text{Replacing } u \text{ with } e^x$$

We now take the natural logarithm on both sides:

$$\ln e^x = \ln (3 \pm 2\sqrt{2})$$

$$x = \ln (3 + 2\sqrt{2}). \qquad \text{Using } \ln e^x = x$$

Approximating each of the solutions, we obtain 1.76 and $-1.76$.

## Graphical Solution

Using the Zero method, we begin by graphing the function

$$y = e^x + e^{-x} - 6.$$

Then we find the zeros of the function.

$y = e^x + e^{-x} - 6$

```
                    10

  -6 |_|_|_|_×|_|_|_|_| 6

     Zero
     X = -1.762747  Y = 0
                   -6
```

The leftmost zero is about $-1.76$. Using the ZERO feature one more time, we find that the other zero is about 1.76.

The solutions are about $-1.76$ and 1.76.

**Now Try Exercise 25.**

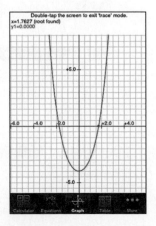

It is possible that when encountering an equation like the one in Example 5, you might not recognize that it could be solved in the algebraic manner shown. This points out the value of the graphical solution. The window shown at left is from an app on an iPhone.

### ■ Solving Logarithmic Equations

Equations containing variables in logarithmic expressions, such as $\log_2 x = 4$ and $\log x + \log (x + 3) = 1$, are called **logarithmic equations**. To solve logarithmic equations algebraically, we first try to obtain a single logarithmic expression on one side and then write an equivalent exponential equation.

**EXAMPLE 6** Solve: $\log_3 x = -2$.

---

**Algebraic Solution**

We have

$\log_3 x = -2$

$\quad 3^{-2} = x$     **Converting to an exponential equation**

$\quad \dfrac{1}{3^2} = x$

$\quad \dfrac{1}{9} = x.$

*Check:*

$$\begin{array}{c|c} \log_3 x = -2 \\ \hline \log_3 \frac{1}{9} \;?\; -2 \\ \log_3 3^{-2} \quad\bigg| \\ \hspace{2em} -2 \;\bigg|\; -2 \quad \textbf{TRUE} \end{array}$$

The solution is $\frac{1}{9}$.

---

**Graphical Solution**

We use the change-of-base formula and graph the equations

$$y_1 = \log_3 x = \frac{\ln x}{\ln 3} \quad \text{and} \quad y_2 = -2.$$

We could also graph $y_1 = \log_3 x$ directly with the logBASE operation. Then we use the Intersect method with a graphing calculator or with an app on an iPhone.

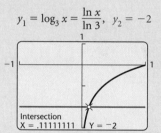

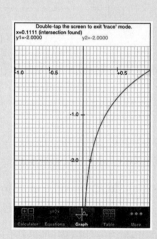

If the solution is a rational number, we can usually find fraction notation for the exact solution by using the FRAC feature from the MATH submenu of the MATH menu.

The solution is $\frac{1}{9}$.

---

**Now Try Exercise 33.**

**EXAMPLE 7** Solve: $\log x + \log(x + 3) = 1$.

## Algebraic Solution

In this case, we have common logarithms. Writing the base of 10 will help us understand the problem:

$$\log_{10} x + \log_{10}(x + 3) = 1$$
$$\log_{10}[x(x + 3)] = 1 \qquad \text{Using the product rule to obtain a single logarithm}$$
$$x(x + 3) = 10^1 \qquad \text{Writing an equivalent exponential equation}$$
$$x^2 + 3x = 10$$
$$x^2 + 3x - 10 = 0$$
$$(x - 2)(x + 5) = 0 \qquad \text{Factoring}$$
$$x - 2 = 0 \quad or \quad x + 5 = 0$$
$$x = 2 \quad or \qquad x = -5.$$

*Check:*   For 2:

$$\dfrac{\log x + \log(x + 3) = 1}{\log 2 + \log(2 + 3) \;\overset{?}{\vert}\; 1}$$
$$\log 2 + \log 5$$
$$\log(2 \cdot 5)$$
$$\log 10$$
$$1 \;\vert\; 1 \quad \textbf{TRUE}$$

For −5:

$$\dfrac{\log x + \log(x + 3) = 1}{\log(-5) + \log(-5 + 3) \;\overset{?}{}\; 1} \quad \textbf{FALSE}$$

The number −5 is not a solution because negative numbers do not have real-number logarithms. The solution is 2.

## Graphical Solution

We can graph the equations

$$y_1 = \log x + \log(x + 3)$$

and

$$y_2 = 1$$

and use the Intersect method. The first coordinate of the point of intersection is the solution of the equation.

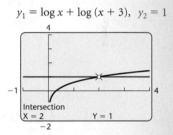

$y_1 = \log x + \log(x + 3), \quad y_2 = 1$

We could also graph the function

$$y = \log x + \log(x + 3) - 1$$

and use the Zero method. The zero of the function is the solution of the equation.

$y_1 = \log x + \log(x + 3) - 1$

The solution of the equation is 2. From the graph, we can easily see that there is only one solution.

**Now Try Exercise 41.**

**EXAMPLE 8** Solve: $\log_3(2x - 1) - \log_3(x - 4) = 2$.

## Algebraic Solution

We have

$$\log_3(2x - 1) - \log_3(x - 4) = 2$$

$$\log_3 \frac{2x - 1}{x - 4} = 2 \qquad \text{Using the quotient rule}$$

$$\frac{2x - 1}{x - 4} = 3^2 \qquad \text{Writing an equivalent exponential equation}$$

$$\frac{2x - 1}{x - 4} = 9$$

$$(x - 4) \cdot \frac{2x - 1}{x - 4} = 9(x - 4) \qquad \begin{array}{l}\text{Multiplying} \\ \text{by the LCD,} \\ x - 4\end{array}$$

$$2x - 1 = 9x - 36$$

$$35 = 7x$$

$$5 = x.$$

*Check:*

$$\log_3(2x - 1) - \log_3(x - 4) = 2$$

$$\begin{array}{c|c} \log_3(2 \cdot 5 - 1) - \log_3(5 - 4) & \overset{?}{=} 2 \\ \log_3 9 - \log_3 1 & \\ 2 - 0 & \\ 2 & 2 \quad \text{TRUE} \end{array}$$

The solution is 5.

## Graphical Solution

Here we use the Intersect method to find the solution of the equation. We use the change-of-base formula and graph the equations

$$y_1 = \frac{\ln(2x - 1)}{\ln 3} - \frac{\ln(x - 4)}{\ln 3}$$

and

$y_2 = 2$. We could also graph $y_1$ directly with the logBASE operation.

$$y_1 = \frac{\ln(2x - 1)}{\ln 3} - \frac{\ln(x - 4)}{\ln 3}, \quad y_2 = 2$$

The solution is 5.

**Now Try Exercise 45.**

**EXAMPLE 9**   Solve: $\ln (4x + 6) - \ln (x + 5) = \ln x$.

## Algebraic Solution

We have

$$\ln (4x + 6) - \ln (x + 5) = \ln x$$

$$\ln \frac{4x + 6}{x + 5} = \ln x \qquad \text{Using the quotient rule}$$

$$\frac{4x + 6}{x + 5} = x \qquad \begin{array}{l}\text{Using the property}\\ \text{of logarithmic}\\ \text{equality}\end{array}$$

$$(x + 5) \cdot \frac{4x + 6}{x + 5} = x(x + 5) \qquad \text{Multiplying by } x + 5$$

$$4x + 6 = x^2 + 5x$$

$$0 = x^2 + x - 6$$

$$0 = (x + 3)(x - 2) \qquad \text{Factoring}$$

$$x + 3 = 0 \quad or \quad x - 2 = 0$$

$$x = -3 \quad or \qquad x = 2.$$

The number $-3$ is not a solution because $\ln \left[ 4(-3) + 6 \right]$, or $\ln (-6)$, on the left side of the equation and $\ln (-3)$ on the right side of the equation are not real numbers. The value 2 checks and is the solution.

## Graphical Solution

The solution of the equation

$$\ln (4x + 6) - \ln (x + 5) = \ln x$$

is the zero of the function

$$f(x) = \ln (4x + 6) - \ln (x + 5) - \ln x.$$

The solution is also the first coordinate of the $x$-intercept of the graph of the function. Here we use the Zero method.

$$y_1 = \ln (4x + 6) - \ln (x + 5) - \ln x$$

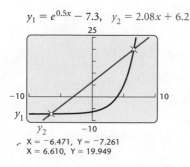

Zero
X = 2          Y = 0

The solution of the equation is 2. From the graph, we can easily see that there is only one solution.

**Now Try Exercise 43.**

Sometimes we encounter equations for which an algebraic solution seems difficult or impossible.

**EXAMPLE 10**   Solve: $e^{0.5x} - 7.3 = 2.08x + 6.2$.

*Graphical Solution*   We graph the equations

$$y_1 = e^{0.5x} - 7.3 \quad \text{and} \quad y_2 = 2.08x + 6.2$$

on a graphing calculator and use the Intersect method. (See the window at left.)

We can also consider the equation

$$y = e^{0.5x} - 7.3 - 2.08x - 6.2, \quad \text{or} \quad y = e^{0.5x} - 2.08x - 13.5,$$

and use the Zero method. The approximate solutions are $-6.471$ and $6.610$.

**Now Try Exercise 67.**

$y_1 = e^{0.5x} - 7.3, \quad y_2 = 2.08x + 6.2$

X = −6.471, Y = −7.261
X = 6.610, Y = 19.949

## 5.5 Exercise Set

*Solve the exponential equation algebraically. Then check using a graphing calculator.*

**1.** $3^x = 81$

**2.** $2^x = 32$

**3.** $2^{2x} = 8$

**4.** $3^{7x} = 27$

**5.** $2^x = 33$

**6.** $2^x = 40$

**7.** $5^{4x-7} = 125$

**8.** $4^{3x-5} = 16$

**9.** $27 = 3^{5x} \cdot 9^{x^2}$

**10.** $3^{x^2+4x} = \frac{1}{27}$

**11.** $84^x = 70$

**12.** $28^x = 10^{-3x}$

**13.** $10^{-x} = 5^{2x}$

**14.** $15^x = 30$

**15.** $e^{-c} = 5^{2c}$

**16.** $e^{4t} = 200$

**17.** $e^t = 1000$

**18.** $e^{-t} = 0.04$

**19.** $e^{-0.03t} = 0.08$

**20.** $1000e^{0.09t} = 5000$

**21.** $3^x = 2^{x-1}$

**22.** $5^{x+2} = 4^{1-x}$

**23.** $(3.9)^x = 48$

**24.** $250 - (1.87)^x = 0$

**25.** $e^x + e^{-x} = 5$

**26.** $e^x - 6e^{-x} = 1$

**27.** $3^{2x-1} = 5^x$

**28.** $2^{x+1} = 5^{2x}$

**29.** $2e^x = 5 - e^{-x}$

**30.** $e^x + e^{-x} = 4$

*Solve the logarithmic equation algebraically. Then check using a graphing calculator.*

**31.** $\log_5 x = 4$

**32.** $\log_2 x = -3$

**33.** $\log x = -4$

**34.** $\log x = 1$

**35.** $\ln x = 1$

**36.** $\ln x = -2$

**37.** $\log_{64} \frac{1}{4} = x$

**38.** $\log_{125} \frac{1}{25} = x$

**39.** $\log_2 (10 + 3x) = 5$

**40.** $\log_5 (8 - 7x) = 3$

**41.** $\log x + \log (x - 9) = 1$

**42.** $\log_2 (x + 1) + \log_2 (x - 1) = 3$

**43.** $\log_2 (x + 20) - \log_2 (x + 2) = \log_2 x$

**44.** $\log (x + 5) - \log (x - 3) = \log 2$

**45.** $\log_8 (x + 1) - \log_8 x = 2$

**46.** $\log x - \log (x + 3) = -1$

**47.** $\log x + \log (x + 4) = \log 12$

**48.** $\log_3 (x + 14) - \log_3 (x + 6) = \log_3 x$

**49.** $\log (x + 8) - \log (x + 1) = \log 6$

**50.** $\ln x - \ln (x - 4) = \ln 3$

**51.** $\log_4 (x + 3) + \log_4 (x - 3) = 2$

**52.** $\ln (x + 1) - \ln x = \ln 4$

**53.** $\log (2x + 1) - \log (x - 2) = 1$

**54.** $\log_5 (x + 4) + \log_5 (x - 4) = 2$

**55.** $\ln (x + 8) + \ln (x - 1) = 2 \ln x$

**56.** $\log_3 x + \log_3 (x + 1) = \log_3 2 + \log_3 (x + 3)$

*Solve.*

**57.** $\log_6 x = 1 - \log_6 (x - 5)$

**58.** $2^{x^2-9x} = \frac{1}{256}$

**59.** $9^{x-1} = 100(3^x)$

**60.** $2 \ln x - \ln 5 = \ln (x + 10)$

**61.** $e^x - 2 = -e^{-x}$

**62.** $2 \log 50 = 3 \log 25 + \log (x - 2)$

*Use a graphing calculator to find the approximate solutions of the equation.*

**63.** $2^x - 5 = 3x + 1$

**64.** $0.082e^{0.05x} = 0.034$

**65.** $xe^{3x} - 1 = 3$

**66.** $4x - 3^x = -6$

**67.** $5e^{5x} + 10 = 3x + 40$

**68.** $4 \ln (x + 3.4) = 2.5$

**69.** $\log_8 x + \log_8 (x + 2) = 2$

**70.** $\ln x^2 = -x^2$

**71.** $\log_5 (x + 7) - \log_5 (2x - 3) = 1$

**72.** $\log_3 x + 7 = 4 - \log_5 x$

*Approximate the point(s) of intersection of the pair of equations.*

**73.** $2.3x + 3.8y = 12.4$, $y = 1.1 \ln (x - 2.05)$

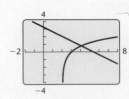

**74.** $y = \ln 3x,\ y = 3x - 8$

**75.** $y = 2.3 \ln (x + 10.7),\ y = 10e^{-0.007x^2}$

**76.** $y = 2.3 \ln (x + 10.7),\ y = 10e^{-0.07x^2}$

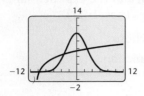

## Skill Maintenance

*In Exercises 77–80:*

**a)** *Find the vertex.*
**b)** *Find the axis of symmetry.*
**c)** *Determine whether there is a maximum or minimum value and find that value.*

**77.** $g(x) = x^2 - 6$

**78.** $f(x) = -x^2 + 6x - 8$

**79.** $G(x) = -2x^2 - 4x - 7$

**80.** $H(x) = 3x^2 - 12x + 16$

## Synthesis

*Solve using any method.*

**81.** $\dfrac{e^x + e^{-x}}{e^x - e^{-x}} = 3$

**82.** $\ln (\ln x) = 2$

**83.** $\sqrt{\ln x} = \ln \sqrt{x}$

**84.** $\ln \sqrt[4]{x} = \sqrt{\ln x}$

**85.** $(\log_3 x)^2 - \log_3 x^2 = 3$

**86.** $\log_3 (\log_4 x) = 0$

**87.** $\ln x^2 = (\ln x)^2$

**88.** $x \left( \ln \tfrac{1}{6} \right) = \ln 6$

**89.** $5^{2x} - 3 \cdot 5^x + 2 = 0$

**90.** $x^{\log x} = \dfrac{x^3}{100}$

**91.** $\ln x^{\ln x} = 4$

**92.** $|2^{x^2} - 8| = 3$

**93.** $\dfrac{\sqrt{(e^{2x} \cdot e^{-5x})^{-4}}}{e^x \div e^{-x}} = e^7$

**94.** Given that $a = (\log_{125} 5)^{\log_5 125}$, find the value of $\log_3 a$.

**95.** Given that $a = \log_8 225$ and $b = \log_2 15$, express $a$ as a function of $b$.

**96.** Given that $f(x) = e^x - e^{-x}$, find $f^{-1}(x)$ if it exists.

---

## Applications and Models: Growth and Decay; Compound Interest

# 5.6

- Solve applied problems involving exponential growth and decay.
- Solve applied problems involving compound interest.
- Find models involving exponential functions and logarithmic functions.

Exponential functions and logarithmic functions with base *e* are rich in applications to many fields such as business, science, psychology, and sociology.

### ● Population Growth

The function

$$P(t) = P_0 e^{kt}, \quad k > 0,$$

is a model of many kinds of population growth, whether it be a population of people, bacteria, cell phones, or money. In this function, $P_0$ is the population at

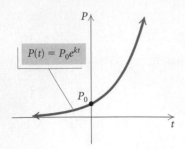

time 0, $P$ is the population after time $t$, and $k$ is called the **exponential growth rate**. The graph of such an equation is shown at left.

**EXAMPLE 1**    *Population Growth of Mexico.*    In 2011, the population of Mexico was about 113.7 million, and the exponential growth rate was 1.1% per year (*Source*: CIA *World Factbook*, 2011).

**a)** Find the exponential growth function.

**b)** Graph the exponential growth function.

**c)** Estimate the population in 2015.

**d)** After how long will the population be double what it was in 2011?

### Solution

**a)** At $t = 0$ (2011), the population was 113.7 million, and the exponential growth rate was 1.1% per year. We substitute 113.7 for $P_0$ and 1.1%, or 0.011, for $k$ to obtain the exponential growth function

$$P(t) = 113.7e^{0.011t},$$

where $t$ is the number of years after 2011 and $P(t)$ is in millions.

**b)** Using a graphing calculator, we obtain the graph of the exponential growth function, shown at left.

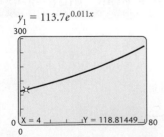

**c)** In 2015, $t = 4$; that is, 4 years have passed since 2011. To find the population in 2015, we substitute 4 for $t$:

$$P(4) = 113.7e^{0.011(4)} = 113.7e^{0.044} \approx 118.8.$$

We can also use the VALUE feature from the CALC menu on a graphing calculator to find $P(4)$. (See the window at left.) The population will be about 118.8 million, or 118,800,000, in 2015.

**d)** We are looking for the time $T$ for which $P(T) = 2 \cdot 113.7$, or 227.4. The number $T$ is called the **doubling time**. To find $T$, we solve the equation

$$227.4 = 113.7e^{0.011T}$$

using both an algebraic method and a graphical method.

## Algebraic Solution

We have

$$227.4 = 113.7e^{0.011T}$$  **Substituting 227.4 for $P(T)$**

$$2 = e^{0.011T}$$  **Dividing by 113.7**

$$\ln 2 = \ln e^{0.011T}$$  **Taking the natural logarithm on both sides**

$$\ln 2 = 0.011T$$  **$\ln e^x = x$**

$$\frac{\ln 2}{0.011} = T$$  **Dividing by 0.011**

$$63.0 \approx T.$$

The population of Mexico will be double what it was in 2011 about 63.0 years after 2011.

## Graphical Solution

Using the Intersect method, we graph the equations

$$y_1 = 113.7e^{0.011x} \quad \text{and} \quad y_2 = 227.4$$

and find the first coordinate of their point of intersection.

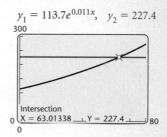

$y_1 = 113.7e^{0.011x}, \quad y_2 = 227.4$

Intersection
X = 63.01338    Y = 227.4

The solution is about 63.0, so the population of Mexico will be double what it was in 2011 about 63.0 years after 2011.

**Now Try Exercise 1.**

## ▪ Interest Compounded Continuously

When interest is paid on interest, we call it **compound interest.** Suppose that an amount $P_0$ is invested in a savings account at interest rate $k$ **compounded continuously.** The amount $P(t)$ in the account after $t$ years is given by the exponential function

$$P(t) = P_0 e^{kt}.$$

**EXAMPLE 2** *Interest Compounded Continuously.* Suppose that $2000 is invested at interest rate $k$, compounded continuously, and grows to $2504.65 in 5 years.

**a)** What is the interest rate?
**b)** Find the exponential growth function.
**c)** What will the balance be after 10 years?
**d)** After how long will the $2000 have doubled?

*Solution*

**a)** At $t = 0$, $P(0) = P_0 = \$2000$. Thus the exponential growth function is of the form

$$P(t) = 2000e^{kt}.$$

We know that $P(5) = \$2504.65$. We substitute and solve for $k$:

$$2504.65 = 2000e^{k(5)}$$  **Substituting 2504.65 for $P(t)$ and 5 for $t$**

$$2504.65 = 2000e^{5k}$$

$$\frac{2504.65}{2000} = e^{5k}.$$  **Dividing by 2000**

Then

$$\ln \frac{2504.65}{2000} = \ln e^{5k} \qquad \textbf{Taking the natural logarithm}$$

$$\ln \frac{2504.65}{2000} = 5k \qquad \textbf{Using } \ln e^x = x$$

$$\frac{\ln \dfrac{2504.65}{2000}}{5} = k \qquad \textbf{Dividing by 5}$$

$$0.045 \approx k.$$

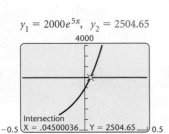

$y_1 = 2000e^{5x}, \quad y_2 = 2504.65$

The interest rate is about 0.045, or 4.5%.

We can also find $k$ by graphing the equations

$$y_1 = 2000e^{5x} \quad \text{and} \quad y_2 = 2504.65$$

and using the Intersect feature to approximate the first coordinate of the point of intersection. The interest rate is about 0.045, or 4.5%.

**b)** Substituting 0.045 for $k$ in the function $P(t) = 2000e^{kt}$, we see that the exponential growth function is

$$P(t) = 2000e^{0.045t}.$$

**c)** The balance after 10 years is

$$P(10) = 2000e^{0.045(10)} = 2000e^{0.45} \approx \$3136.62.$$

**d)** To find the doubling time $T$, we set $P(T) = 2 \cdot P_0 = 2 \cdot \$2000 = \$4000$ and solve for $T$. We solve

$$4000 = 2000e^{0.045T}$$

using both an algebraic method and a graphical method.

## Algebraic Solution

We have

$$4000 = 2000e^{0.045T}$$

$$2 = e^{0.045T} \qquad \textbf{Dividing by 2000}$$

$$\ln 2 = \ln e^{0.045T} \qquad \textbf{Taking the natural logarithm}$$

$$\ln 2 = 0.045T \qquad \textbf{ln } e^x = x$$

$$\frac{\ln 2}{0.045} = T \qquad \textbf{Dividing by 0.045}$$

$$15.4 \approx T.$$

Thus the original investment of $2000 will double in about 15.4 years.

## Graphical Solution

We use the Zero method. We graph the equation

$$y = 2000e^{0.045x} - 4000$$

and find the zero of the function. The zero of the function is the solution of the equation.

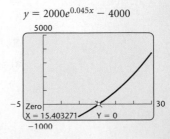

$y = 2000e^{0.045x} - 4000$

The solution is about 15.4, so the original investment of $2000 will double in about 15.4 years.

Now Try Exercise 7.

We can find a general expression relating the growth rate $k$ and the doubling time $T$ by solving the following equation:

$$2P_0 = P_0 e^{kT} \qquad \text{Substituting } 2P_0 \text{ for } P \text{ and } T \text{ for } t$$
$$2 = e^{kT} \qquad \text{Dividing by } P_0$$
$$\ln 2 = \ln e^{kT} \qquad \text{Taking the natural logarithm}$$
$$\ln 2 = kT \qquad \text{Using } \ln e^x = x$$
$$\frac{\ln 2}{k} = T.$$

---

### GROWTH RATE AND DOUBLING TIME

The **growth rate** $k$ and the **doubling time** $T$ are related by

$$kT = \ln 2, \quad \text{or} \quad k = \frac{\ln 2}{T}, \quad \text{or} \quad T = \frac{\ln 2}{k}.$$

---

Note that the relationship between $k$ and $T$ does not depend on $P_0$.

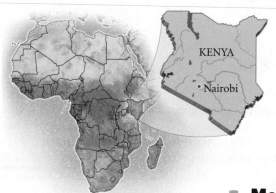

**EXAMPLE 3** *Population Growth.* The population of Kenya is now doubling every 28.2 years (*Source: CIA World Factbook, 2011*). What is the exponential growth rate?

*Solution* We have

$$k = \frac{\ln 2}{T} = \frac{\ln 2}{28.2} \approx 0.0246 \approx 2.46\%.$$

The growth rate of the population of Kenya is about 2.46% per year.

**Now Try Exercise 3(e).**

## ■ Models of Limited Growth

The model $P(t) = P_0 e^{kt}$, $k > 0$, has many applications involving unlimited population growth. However, in some populations, there can be factors that prevent a population from exceeding some limiting value—perhaps a limitation on food, living space, or other natural resources. One model of such growth is

$$P(t) = \frac{a}{1 + be^{-kt}}.$$

This is called a **logistic function**. This function increases toward a *limiting value* $a$ as $t \to \infty$. Thus, $y = a$ is a horizontal asymptote of the graph of $P(t)$.

**EXAMPLE 4** *Limited Population Growth in a Lake.* A lake is stocked with 400 fish of a new variety. The size of the lake, the availability of food, and the number of other fish restrict the growth of that type of fish in the lake to a limiting value of 2500. The population gets closer and closer to this limiting value, but never reaches it. The population of fish in the lake after time $t$, in months, is given by the logistic function

$$P(t) = \frac{2500}{1 + 5.25e^{-0.32t}}.$$

**a)** Graph the function.

**b)** Find the population after 0, 1, 5, 10, 15, and 20 months.

**Solution**

**a)** We use a graphing calculator to graph the function. The graph is shown below. Note that this function increases toward a limiting value of 2500. The graph has $y = 2500$ as a horizontal asymptote.

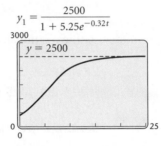

| X | Y₁ |
|---|---|
| 0 | 400 |
| 1 | 519.5 |
| 5 | 1213.6 |
| 10 | 2059.3 |
| 15 | 2396.5 |
| 20 | 2478.4 |

X =

**b)** We can use the TABLE feature on a graphing calculator set in ASK mode to find the function values. (See the window on the right above.) Thus the population will be about 400 after 0 months, 520 after 1 month, 1214 after 5 months, 2059 after 10 months, 2397 after 15 months, and 2478 after 20 months.

> **Now Try Exercise 17.**

Another model of limited growth is provided by the function

$$P(t) = L(1 - e^{-kt}), \quad k > 0,$$

which is shown graphed at right. This function also increases toward a limiting value $L$, as $t \to \infty$, so $y = L$ is the horizontal asymptote of the graph of $P(t)$.

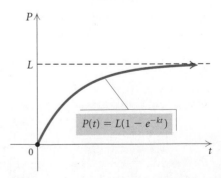

## ◼ Exponential Decay

The function

$$P(t) = P_0 e^{-kt}, \quad k > 0,$$

is an effective model of the decline, or decay, of a population. An example is the decay of a radioactive substance. In this case, $P_0$ is the amount of the substance at time $t = 0$, and $P(t)$ is the amount of the substance left after time $t$, where $k$ is a positive constant that depends on the situation. The constant $k$ is called the **decay rate**.

How can scientists determine that an animal bone has lost 30% of its carbon-14? The assumption is that the percentage of carbon-14 in the atmosphere is the same as that in living plants and animals. When a plant or an animal dies, the amount of carbon-14 that it contains decays exponentially. A scientist can burn an animal bone and use a Geiger counter to determine the percentage of the smoke that is carbon-14. The amount by which this varies from the percentage in the atmosphere tells how much carbon-14 has been lost.

The process of carbon-14 dating was developed by the American chemist Willard E. Libby in 1952. It is known that the radioactivity in a living plant is 16 disintegrations per gram per minute. Since the half-life of carbon-14 is 5750 years, an object with an activity of 8 disintegrations per gram per minute is 5750 years old, one with an activity of 4 disintegrations per gram per minute is 11,500 years old, and so on. Carbon-14 dating can be used to measure the age of objects up to 40,000 years old. Beyond such an age, it is too difficult to measure the radioactivity and some other method would have to be used.

Carbon-14 dating was used to find the age of the Dead Sea Scrolls. It was also used to refute the authenticity of the Shroud of Turin, presumed to have covered the body of Christ.

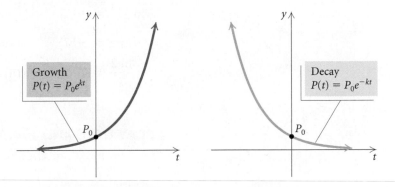

The **half-life** of bismuth (Bi-210) is 5 days. This means that half of an amount of radioactive bismuth will cease to be radioactive in 5 days. The effect of half-life $T$ for nonnegative inputs is shown in the graph below. The exponential function gets close to 0, but never reaches 0, as $t$ gets very large. Thus, according to an exponential decay model, a radioactive substance never completely decays.

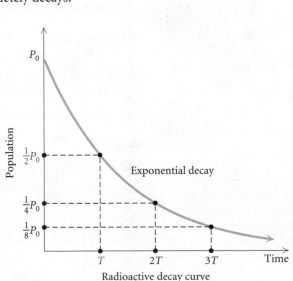

Radioactive decay curve

We can find a general expression relating the decay rate $k$ and the half-life time $T$ by solving the following equation:

$$\frac{1}{2}P_0 = P_0 e^{-kT} \qquad \text{Substituting } \frac{1}{2}P_0 \text{ for } P \text{ and } T \text{ for } t$$

$$\frac{1}{2} = e^{-kT} \qquad \text{Dividing by } P_0$$

$$\ln \frac{1}{2} = \ln e^{-kT} \qquad \text{Taking the natural logarithm on both sides}$$

$$\ln 2^{-1} = -kT \qquad \frac{1}{2} = 2^{-1}; \ln e^x = x$$

$$-\ln 2 = -kT \qquad \text{Using the power rule}$$

$$\frac{\ln 2}{k} = T. \qquad \text{Dividing by } -k$$

---

### DECAY RATE AND HALF-LIFE

The decay rate $k$ and the half-life $T$ are related by

$$kT = \ln 2, \quad \text{or} \quad k = \frac{\ln 2}{T}, \quad \text{or} \quad T = \frac{\ln 2}{k}.$$

---

Note that the relationship between decay rate and half-life is the same as that between growth rate and doubling time.

**EXAMPLE 5**    *Carbon Dating.*    The radioactive element carbon-14 has a half-life of 5750 years. The percentage of carbon-14 present in the remains of organic matter can be used to determine the age of that organic matter. Archaeologists discovered that the linen wrapping from one of the Dead Sea Scrolls had lost 22.3% of its carbon-14 at the time it was found. How old was the linen wrapping?

*Solution*    We first find $k$ when the half-life $T$ is 5750 years:

$$k = \frac{\ln 2}{T}$$

$$= \frac{\ln 2}{5750} \qquad \text{Substituting 5750 for } T$$

$$= 0.00012.$$

Now we have the function

$$P(t) = P_0 e^{-0.00012t}.$$

(This function can be used for any subsequent carbon-dating problem.) If the linen wrapping has lost 22.3% of its carbon-14 from an initial amount $P_0$, then $77.7\% P_0$ is the amount present. To find the age $t$ of the wrapping, we solve the equation for $t$:

In 1947, a Bedouin youth looking for a stray goat climbed into a cave at Kirbet Qumran on the shores of the Dead Sea near Jericho and came upon earthenware jars containing an incalculable treasure of ancient manuscripts. Shown here are fragments of those Dead Sea Scrolls, a portion of some 600 or so texts found so far and which concern the Jewish books of the Bible. Officials date them before 70 A.D., making them the oldest Biblical manuscripts by 1000 years.

$$77.7\% P_0 = P_0 e^{-0.00012t}$$     **Substituting 77.7% $P_0$ for $P$**

$$0.777 = e^{-0.00012t}$$     **Dividing by $P_0$ and writing 77.7% as 0.777**

$$\ln 0.777 = \ln e^{-0.00012t}$$     **Taking the natural logarithm on both sides**

$$\ln 0.777 = -0.00012t$$     **$\ln e^x = x$**

$$\frac{\ln 0.777}{-0.00012} = t$$     **Dividing by $-0.00012$**

$$2103 \approx t.$$

Thus the linen wrapping on the Dead Sea Scrolls was about 2103 years old when it was found.

**Now Try Exercise 9.**

## ■ Exponential and Logarithmic Curve Fitting

We have added several new functions that can be considered when we fit curves to data. Let's review some of them.

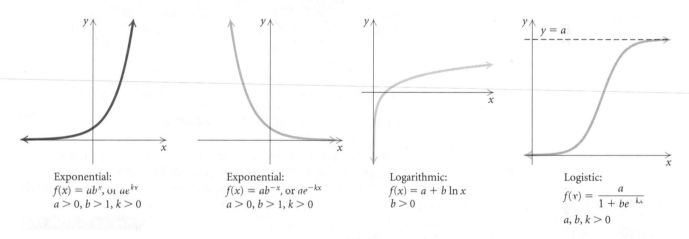

Exponential:
$f(x) = ab^x$, or $ae^{kr}$
$a > 0, b > 1, k > 0$

Exponential:
$f(x) = ab^{-x}$, or $ae^{-kx}$
$a > 0, b > 1, k > 0$

Logarithmic:
$f(x) = a + b \ln x$
$b > 0$

Logistic:
$f(x) = \dfrac{a}{1 + be^{-kx}}$
$a, b, k > 0$

Now, when we analyze a set of data for curve fitting, these models can be considered as well as polynomial functions (such as linear, quadratic, cubic, and quartic functions) and rational functions.

GCM

**EXAMPLE 6** *Spending on Video Rentals.* The number of video stores is declining rapidly. Consumer spending on video rentals from subscription services, including Netflix, kiosks, and cable and satellite video-on-demand services has increased greatly in recent years, as shown in the table below.

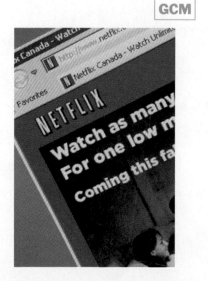

| YEAR, $x$ | U.S. Consumer Spending (in billions) on Video Rentals through Subscription Services |
|-----------|-----------------------------------------------------------------------------------|
| 2005, 0 | $2.2 |
| 2006, 1 | 2.6 |
| 2007, 2 | 3.1 |
| 2008, 3 | 3.8 |
| 2009, 4 | 4.5 |
| 2010, 5 | 5.6 |
| 2011, 6 | 6.6* |

*Estimate
Source: *Screen Digest*

a) Use a graphing calculator to fit an exponential function to the data.

b) Graph the function with the scatterplot of the data.

c) Estimate consumer spending on video rentals through subscription services in 2014.

### Solution

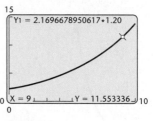

FIGURE 1

a) We will fit an equation of the type $y = a \cdot b^x$ to the data, where $x$ is the number of years since 2005. Entering the data into the calculator and carrying out the regression procedure, we find that the equation in

$$y = 2.169667895(1.20420817)^x.$$

(See Fig. 1.) The correlation coefficient, $r$, is very close to 1. This gives us an indication that the exponential function fits the data well.

b) The scatterplot and the graph are shown in Fig. 2 on the left below.

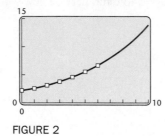

FIGURE 2                    FIGURE 3

c) Using the VALUE feature in the CALC menu (see Fig. 3 on the right above), we evaluate the function found in part (a) for $x = 9$ $(2014 - 2005 = 9)$, and estimate consumer spending on video rentals from subscription services in 2014 to be about $11.6 billion.

**Now Try Exercise 29.**

On some graphing calculators, there may be a REGRESSION feature that yields an exponential function, base $e$. If not, and you wish to find such a function, a conversion can be done using the following.

---

### CONVERTING FROM BASE *b* TO BASE *e*

$$b^x = e^{x(\ln b)}$$

---

Then, for the equation in Example 6, we have

$$y = 2.169667895(1.20420817)^x$$
$$= 2.169667895e^{x(\ln 1.20420817)}$$
$$= 2.169667895e^{0.1858222306x}. \qquad \ln 1.20420817 \approx 0.1858222306$$

We can prove this conversion formula using properties of logarithms, as follows:

$$e^{x(\ln b)} = e^{\ln b^x} = b^x.$$

# 5.6 Exercise Set

**1.** *Population Growth of Houston.*    The Houston–Sugar Land–Baytown metropolitan area is the fifth largest metropolitan area in the United States. In 2010, the population of this area was 5.9 million, and the exponential growth rate was 1.76% per year.

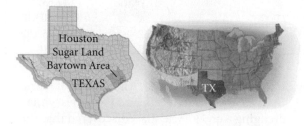

Houston
Sugar Land
Baytown Area
TEXAS
TX

**a)** Find the exponential growth function.
**b)** Estimate the population of the Houston–Sugar Land–Baytown metropolitan area in 2016.
**c)** When will the population of this metropolitan area be 7 million?
**d)** Find the doubling time.

**2.** *Population Growth of Rabbits.*    Under ideal conditions, a population of rabbits has an exponential growth rate of 11.7% per day. Consider an initial population of 100 rabbits.

**a)** Find the exponential growth function.
**b)** Graph the function.
**c)** What will the population be after 7 days? after 2 weeks?
**d)** Find the doubling time.

**3.** *Population Growth.*    Complete the following table.

| Country | Growth Rate, $k$ | Doubling Time, $T$ |
|---|---|---|
| **a)** United States | | 70.7 years |
| **b)** Venezuela | | 45.9 years |
| **c)** Ethiopia | 3.21% | |
| **d)** Australia | 1.20% | |
| **e)** Denmark | | 248 years |
| **f)** Laos | 2.32% | |
| **g)** India | 1.40% | |
| **h)** China | | 105.0 years |
| **i)** Tanzania | | 34.0 years |
| **j)** Haiti | 0.79% | |

**4.** *U.S. Exports.*    United States exports increased from $12 billion in 1950 to $1.550 trillion, or $1550 billion, in 2009 (*Sources*: www.dlc.org; U.S. Department of Commerce).

Assuming that the exponential growth model applies:

**a)** Find the value of $k$ and write the function.
**b)** Estimate the value of exports, in billions, in 1972, in 2001, and in 2015. Round to the nearest tenth of a billion.

**5.** *Population Growth of Haiti.*    The population of Haiti has a growth rate of 0.787% per year.

In 2010, the population was 10,032,619. The land area of Haiti is 32,961,561,600 square yards. (*Source*: *Statistical Abstract of the United States*) Assuming this growth rate continues and is exponential, after how long will there be one person for every 1000 square yards of land?

6. *Picasso Painting.* In May 2010, a 1932 painting, *Nude, Green Leaves, and Bust,* by Pablo Picasso, sold at a New York City art auction for $106.5 million to an anonymous buyer. This is a record price for any work of art sold at auction. The painting had belonged to the estate of Sydney and Francis Brody, who bought it for $17,000 in 1952 from a New York art dealer, who had acquired it from Picasso in 1936. (*Sources*: Associated Press, "Picasso Painting Fetches World Record $106.5M at NYC Auction," by Ula Ilnytzky, May 4, 2010; Online Associated Newspapers, May 6, 2010)

Assuming that the value $A_0$ of the painting has grown exponentially:

a) Find the value of $k$, and determine the exponential growth function, assuming $A_0 = 17,000$ and $t$ is the number of years since 1952.
b) Estimate the value of the painting in 2020.
c) What is the doubling time for the value of the painting?
d) After how long will the value of the painting be $240 million, assuming there is no change in the growth rate?

7. *Interest Compounded Continuously.* Suppose that $10,000 is invested at an interest rate of 5.4% per year, compounded continuously.

a) Find the exponential function that describes the amount in the account after time $t$, in years.
b) What is the balance after 1 year? 2 years? 5 years? 10 years?
c) What is the doubling time?

8. *Interest Compounded Continuously.* Complete the following table.

| Initial Investment at $t = 0, P_0$ | Interest Rate, $k$ | Doubling Time, $T$ | Amount After 5 Years |
|---|---|---|---|
| a) $35,000 | 3.2% | | |
| b) $5000 | | | $7,130.90 |
| c) | 5.6% | | $9,923.47 |
| d) | | 11 years | $17,539.32 |
| e) $109,000 | | | $136,503.18 |
| f) | | 46.2 years | $19,552.82 |

9. *Carbon Dating.* In 1970, Amos Flora of Flora, Indiana, discovered teeth and jawbones while dredging a creek. Scientists discovered that the bones were from a mastodon and that they had lost 77.2% of their carbon-14. How old were the bones at the time that they were discovered? (*Sources*: "Farm Yields Bones Thousands of Years Old," by Dan McFeely, *Indianapolis Star*, October 20, 2008; Field Museum of Chicago, Bill Turnbull, anthropologist)

10. *Tomb in the Valley of the Kings.* In February 2006, in the Valley of the Kings in Egypt, a team of archaeologists uncovered the first tomb since King Tut's tomb was found in 1922. The tomb contained five wooden sarcophagi that contained mummies. The archaeologists believe that the mummies are from the 18th Dynasty, about 3300 to 3500 years ago. Determine the

amount of carbon-14 that the mummies have lost.

**11.** *Radioactive Decay.* Complete the following table.

| Radioactive Substance | Decay Rate, $k$ | Half-Life $T$ |
|---|---|---|
| **a)** Polonium (Po-218) | | 3.1 min |
| **b)** Lead (Pb-210) | | 22.3 years |
| **c)** Iodine (I-125) | 1.15% per day | |
| **d)** Krypton (Kr-85) | 6.5% per year | |
| **e)** Strontium (Sr-90) | | 29.1 years |
| **f)** Uranium (U-232) | | 70.0 years |
| **g)** Plutonium (Pu-239) | | 24,100 years |

**12.** *Number of Farms.* The number $N$ of farms in the United States has declined continually since 1950. In 1950, there were 5.6 million farms, and in 2008 that number had decreased to 2.2 million (*Sources*: U.S. Department of Agriculture; National Agricultural Statistics Service).

Assuming that the number of farms decreased according to the exponential decay model:

**a)** Find the value of $k$, and write an exponential function that describes the number of farms, in millions, after time $t$, in years, where $t$ is the number of years since 1950.

**b)** Estimate the number of farms in 2011 and in 2015.

**c)** At this decay rate, when will only 1,000,000 farms remain?

**13.** *Solar Power.* Photovoltaic solar power has become more affordable in recent years. Solar-power plants project that the average cost per watt of electricity generated by solar panels will decrease exponentially as shown in the graph below.

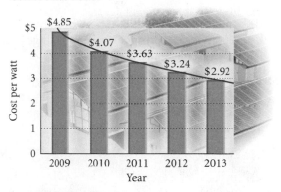

**Cost of Solar Power**

Source: *USA TODAY*, April 12, 2010

Assuming that the cost per watt of electricity generated by solar panels will decrease according to the exponential decay model:

**a)** Using the data for 2009 and 2013, find the value of $k$, and write an exponential function that describes the cost per watt of electricity after time $t$, in years, where $t$ is the number of years since 2009.

**b)** Estimate the cost per watt of electricity in 2015 and in 2018.

**c)** At this decay rate, in which year will the average cost per watt be $1.85?

**14.** *Lamborghini 350 GT.* The market value of the 1964–1965 Lamborghini 350 GT has had a recent upswing. The car's value increased from $66,000 in 1999 to $220,000 in 2009

(*Source*: "1964–1965 Lamborghini 350 GT," by David LaChance, *Hemmings Motor News*, July, 2010, p. 28).

Assuming that the value $V_0$ of the car has grown exponentially:

**a)** Find the value of $k$, and determine the exponential growth function, assuming $V_0 = 66,000$ and $t$ is the number of years since 1999.

**b)** Estimate the value of the car in 2011.

**c)** What is the doubling time for the value of the car?

**d)** After how long will the value of the car be $300,000, assuming there is no change in the growth rate?

**15.** *Cable TV Networks.*   The number of cable TV networks has grown exponentially from 28 in 1980 to 565 in 2010 (*Source*: National Cable & Telecommunications Association).

**a)** Using the data for 1980 and 2010, find the value of $k$, and write an exponential function that describes the number of cable TV networks after time $t$, in years, where $t$ is the number of years since 1980.

**b)** Estimate the number of cable TV networks in 2013 and in 2017.

**c)** At this growth rate, in which year will the number of cable networks be 1500?

**16.** *T206 Wagner Baseball Card.*   In 1909, the Pittsburgh Pirates shortstop Honus Wagner forced the American Tobacco Company to withdraw his baseball card that was packaged with cigarettes. Fewer than 60 of the Wagner cards still exist. In 1971, a Wagner card sold for $1000; and in September 2007, a card in near-mint condition was purchased for a record $2.8 million (*Source*: *USA Today*, 9/6/07; Kathy Willens/AP).

Assuming that the value $W_0$ of the baseball card has grown exponentially:

**a)** Find the value of $k$, and determine the exponential growth function, assuming $W_0 = 1000$ and $t$ is the number of years since 1971.

**b)** Estimate the value of the Wagner card in 2011.

**c)** What is the doubling time for the value of the card?

**d)** After how long will the value of the Wagner card be $10.8 million, assuming there is no change in the growth rate?

**17.** *Spread of an Epidemic.*   In a town whose population is 3500, a disease creates an epidemic. The number of people $N$ infected $t$ days after the disease has begun is given by the function

$$N(t) = \frac{3500}{1 + 19.9e^{-0.6t}}.$$

**a)** Graph the function.

**b)** How many are initially infected with the disease $(t = 0)$?

**c)** Find the number infected after 2 days, 5 days, 8 days, 12 days, and 16 days.

**d)** Using this model, can you say whether all 3500 people will ever be infected? Explain.

18. *Limited Population Growth in a Lake.* A lake is stocked with 640 fish of a new variety. The size of the lake, the availability of food, and the number of other fish restrict the growth of that type of fish in the lake to a limiting value of 3040. The population of fish in the lake after time *t*, in months, is given by the function

$$P(t) = \frac{3040}{1 + 3.75e^{-0.32t}}.$$

a) Graph the function.
b) Find the population after 0, 1, 5, 10, 15, and 20 months.

*Newton's Law of Cooling.* Suppose that a body with temperature $T_1$ is placed in surroundings with temperature $T_0$ different from that of $T_1$. The body will either cool or warm to temperature $T(t)$ after time t, in minutes, where

$$T(t) = T_0 + (T_1 - T_0)e^{-kt}.$$

*Use this law in Exercises 19–22.*

19. A cup of coffee with temperature 105°F is placed in a freezer with temperature 0°F. After 5 min, the temperature of the coffee is 70°F. What will its temperature be after 10 min?

20. A dish of lasagna baked at 375°F is taken out of the oven at 11:15 A.M. into a kitchen that is 72°F. After 3 min, the temperature of the lasagna is 365°F. What will the temperature of the lasagna be at 11:30 A.M.?

21. A chilled gelatin salad that has a temperature of 43°F is taken from the refrigerator and placed on the dining room table in a room that is 68°F. After 12 min, the temperature of the salad is 55°F. What will the temperature of the salad be after 20 min?

22. *When Was the Murder Committed?* The police discover the body of a murder victim. Critical to solving the crime is determining when the murder was committed. The coroner arrives at the murder scene at 12:00 P.M. She immediately takes the temperature of the body and finds it to be 94.6°F. She then takes the temperature 1 hr later and finds it to be 93.4°F. The temperature of the room is 70°F. When was the murder committed?

*In Exercises 23–28, determine which, if any, of these functions might be used as a model for the data in the scatterplot.*

a) Quadratic, $f(x) = ax^2 + bx + c$
b) Polynomial, not quadratic
c) Exponential, $f(x) = ab^x$, or $P_0e^{kx}$, $k > 0$
d) Exponential, $f(x) = ab^{-x}$, or $P_0e^{-kx}$, $k > 0$
e) Logarithmic, $f(x) = a + b\ln x$
f) Logistic, $f(x) = \dfrac{a}{1 + be^{-kx}}$

23. 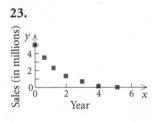  24.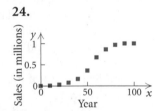

25. 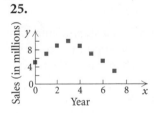  26.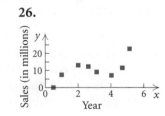

27. 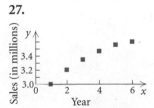  28.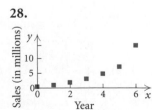

29. *Percent of Americans Ages 85 and Older.* In 1900, 0.2% of the U.S. population, or 122,000 people, were ages 85 and older. This number grew to 5,751,000 in 2010. The table on the following page lists data regarding the percentage of the U.S. population ages 85 and older in selected years from 1900 to 2010.

| Year, x | Percent of U.S. Population Ages 85 and Older, y |
|---|---|
| 1900, 0 | 0.2% |
| 1910, 10 | 0.2 |
| 1920, 20 | 0.2 |
| 1930, 30 | 0.2 |
| 1940, 40 | 0.3 |
| 1950, 50 | 0.4 |
| 1960, 60 | 0.5 |
| 1970, 70 | 0.7 |
| 1980, 80 | 1.0 |
| 1990, 90 | 1.2 |
| 1995, 95 | 1.4 |
| 2000, 100 | 1.5 |
| 2005, 105 | 1.7 |
| 2010, 110 | 1.9 |

*Sources*: U.S. Census Bureau; U.S. Department of Commerce

a) Use a graphing calculator to fit an exponential function to the data, where *x* is the number of years after 1900. Determine whether the function is a good fit.
b) Graph the function found in part (a) with a scatterplot of the data.
c) Estimate the percentage of the U.S. population ages 85 and older in 2007, in 2015, and in 2020.

**30.** *Forgetting.* In an economics class, students were given a final exam at the end of the course. Then they were retested with an equivalent test at subsequent time intervals. Their scores after time *x*, in months, are listed in the table below.

GCM

| Time, x (in months) | Score, y |
|---|---|
| 1 | 84.9% |
| 2 | 84.6 |
| 3 | 84.4 |
| 4 | 84.2 |
| 5 | 84.1 |
| 6 | 83.9 |

a) Use a graphing calculator to fit a logarithmic function $y = a + b \ln x$ to the data.
b) Use the function to predict test scores after 8, 10, 24, and 36 months.
c) After how long will the test scores fall below 82%?

**31.** *Unoccupied Homes in a Census.* After door-to-door canvassing, census workers determine the number of unoccupied homes. The table below lists the number of homes, in millions, that were unoccupied in four census years.

| Year, x | Number of Unoccupied Homes in Census (in millions) |
|---|---|
| 1980, 0 | 5.8 |
| 1990, 10 | 7.3 |
| 2000, 20 | 9.9 |
| 2010, 30 | 14.3 |

*Source*: U.S. Census Bureau

a) Use a graphing calculator to fit an exponential function to the data, where *x* is the number of years after 1980.
b) Graph the function found in part (a).
c) Use the function found in part (a) to project the number of unoccupied homes in the 2020 Census.

**32.** *Older Americans in Debt.* The debt among older Americans has been exponentially increasing in recent years. The table below lists, for selected years, the average debt per household, where the primary head is age 55 or older.

| Year, x | Average Debt per Household Where Primary Head Is Age 55 or Older |
|---|---|
| 2000, 0 | $34,000 |
| 2002, 2 | 38,000 |
| 2004, 4 | 42,000 |
| 2006, 6 | 59,000 |
| 2008, 8 | 66,000 |

*Source*: MacroMonitor, by Strategic Business Insights

a) Use a graphing calculator to fit the data with an exponential function, where *x* is the number of years since 2000.
b) Graph the function found in part (a).
c) Use the function in part (a) to estimate the debt in 2014.
d) In what year is average debt among older Americans expected to exceed $140,000?

**33.** *Farmland Prices.* The average price of an average-quality acre of farmland has increased over

$2000 in the last ten years. The table below lists the price of an acre of average-quality farmland in selected years.

| Year, $x$ | Price of an Acre of Average-Quality Farmland |
|---|---|
| 2000, 0 | $2150 |
| 2004, 4 | 2692 |
| 2005, 5 | 2945 |
| 2006, 6 | 3162 |
| 2007, 7 | 3688 |
| 2008, 8 | 4240 |
| 2009, 9 | 4188 |
| 2010, 10 | 4270 |

*Source of data: The Indianapolis Star*; Purdue University Department of Agricultural Economics

a) Use a graphing calculator to model the data with an exponential function, where $x$ is the number of years since 2000.

b) Use the function found in part (a) to estimate the price per acre of average-quality farmland in 2012 and in 2015.

c) In what year will the price per acre of average-quality farmland exceed $7300?

**34.** *Effect of Advertising.* A company introduced a new software product on a trial run in a city. They advertised the product on television and found the following data regarding the percent $P$ of people who bought the product after $x$ ads were run.

| Number of Ads, $x$ | Percentage Who Bought, $P$ |
|---|---|
| 0 | 0.2% |
| 10 | 0.7 |
| 20 | 2.7 |
| 30 | 9.2 |
| 40 | 27.0 |
| 50 | 57.6 |
| 60 | 83.3 |
| 70 | 94.8 |
| 80 | 98.5 |
| 90 | 99.6 |

a) Use a graphing calculator to fit a logistic function
$$P(x) = \frac{a}{1 + be^{-kx}}$$ to the data.

b) What percent of people bought the product when 55 ads were run? 100 ads?

c) Find the horizontal asymptote for the graph. Interpret the asymptote in terms of the advertising situation.

## Skill Maintenance

### Vocabulary Reinforcement

*In Exercises 35–40, choose the correct name of the principle or rule from the given choices.*

principle of zero products
multiplication principle for equations
product rule
addition principle for inequalities
power rule
multiplication principle for inequalities
principle of square roots
quotient rule

**35.** For any real numbers $a$, $b$, and $c$: If $a < b$ and $c > 0$ are true, then $ac < bc$ is true. If $a < b$ and $c < 0$ are true, then $ac > bc$ is true.

_____

**36.** For any positive numbers $M$ and $N$ and any logarithm base $a$, $\log_a MN = \log_a M + \log_a N$.

_____

**37.** If $ab = 0$ is true, then $a = 0$ or $b = 0$, and if $a = 0$ or $b = 0$, then $ab = 0$.

_____

**38.** If $x^2 = k$, then $x = \sqrt{k}$ or $x = -\sqrt{k}$.

_____

**39.** For any positive number $M$, any logarithm base $a$, and any real number $p$, $\log_a M^p = p \log_a M$.

_____

**40.** For any real numbers $a$, $b$, and $c$: If $a = b$ is true, then $ac = bc$ is true.

_____

## Synthesis

**41.** *Present Value.* Following the birth of a child, a parent wants to make an initial investment $P_0$ that will grow to $50,000 for the child's education at age 18. Interest is compounded continuously at 7%. What should the initial investment be? Such an amount is called the **present value** of $50,000 due 18 years from now.

42. *Supply and Demand.* The supply function and the demand function for the sale of a certain type of DVD player are given by

$$S(p) = 150e^{0.004p} \quad \text{and} \quad D(p) = 480e^{-0.003p},$$

where $S(p)$ is the number of DVD players that the company is willing to sell at price $p$ and $D(p)$ is the quantity that the public is willing to buy at price $p$. Find $p$ such that $D(p) = S(p)$. This is called the **equilibrium price**.

43. *The Beer–Lambert Law.* A beam of light enters a medium such as water or smog with initial intensity $I_0$. Its intensity decreases depending on the thickness (or concentration) of the medium.

The intensity $I$ at a depth (or concentration) of $x$ units is given by

$$I = I_0 e^{-\mu x}.$$

The constant $\mu$ (the Greek letter "mu") is called the **coefficient of absorption**, and it varies with the medium. For sea water, $\mu = 1.4$.

a) What percentage of light intensity $I_0$ remains in sea water at a depth of 1 m? 3 m? 5 m? 50 m?

b) Plant life cannot exist below 10 m. What percentage of $I_0$ remains at 10 m?

44. Given that $y = ax^b$, take the natural logarithm on both sides. Let $Y = \ln y$ and $X = \ln x$. Consider $Y$ as a function of $X$. What kind of function is $Y$?

45. Given that $y = ae^x$, take the natural logarithm on both sides. Let $Y = \ln y$. Consider $Y$ as a function of $x$. What kind of function is $Y$?

# Chapter 5 Summary and Review

## STUDY GUIDE

| KEY TERMS AND CONCEPTS | EXAMPLES |
|---|---|
| **SECTION 5.1: INVERSE FUNCTIONS** | |

**Inverse Relation**

If a relation is defined by an equation, interchanging the variables produces an equation of the inverse relation.

Given $y = -5x + 7$, find an equation of the inverse relation.

$$y = -5x + 7 \qquad \text{Relation}$$
$$x = -5y + 7 \qquad \text{Inverse relation}$$

**One-to-One Functions**

A function $f$ is one-to-one if different inputs have different outputs—that is,

$$\text{if } a \neq b, \quad \text{then} \quad f(a) \neq f(b).$$

Or a function $f$ is one-to-one if when the outputs are the same, the inputs are the same—that is,

$$\text{if } f(a) = f(b), \quad \text{then} \quad a = b.$$

Prove that $f(x) = 16 - 3x$ is one-to-one.

Show that if $f(a) = f(b)$, then $a = b$. Assume $f(a) = f(b)$. Since $f(a) = 16 - 3a$ and $f(b) = 16 - 3b$, we have

$$16 - 3a = 16 - 3b$$
$$-3a = -3b$$
$$a = b.$$

Thus, if $f(a) = f(b)$, then $a = b$ and $f$ is one-to-one.

### Horizontal-Line Test

If it is possible for a horizontal line to intersect the graph of a function more than once, then the function is *not* one-to-one and its inverse is *not* a function.

### One-to-One Functions and Inverses

- If a function $f$ is one-to-one, then its inverse $f^{-1}$ is a function.
- The domain of a one-to-one function $f$ is the range of the inverse $f^{-1}$.
- The range of a one-to-one function $f$ is the domain of the inverse $f^{-1}$.
- A function that is increasing over its entire domain or is decreasing over its entire domain is a one-to-one function.

Using its graph, determine whether each function is one-to-one.

**a)** 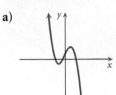   **b)**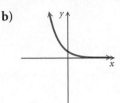

**a)** There are many horizontal lines that intersect the graph more than once. Thus the function is *not* one-to-one and its inverse is *not* a function.

**b)** No horizontal line intersects the graph more than once. Thus the function is one-to-one and its inverse is a function.

---

### Obtaining a Formula for an Inverse

If a function $f$ is one-to-one, a formula for its inverse can generally be found as follows:

1. Replace $f(x)$ with $y$.
2. Interchange $x$ and $y$.
3. Solve for $y$.
4. Replace $y$ with $f^{-1}(x)$.

The $-1$ in $f^{-1}$ is *not* an exponent. The graph of $f^{-1}$ is a reflection of the graph of $f$ across the line $y = x$.

Given the one-to-one function $f(x) = 2 - x^3$, find a formula for its inverse. Then graph the function and its inverse on the same set of axes.

$$f(x) = 2 - x^3$$
$$\downarrow$$

**1.**  $\quad y = 2 - x^3$
$$\downarrow \qquad\qquad \downarrow$$

**2.**  $\quad x = 2 - y^3$

**3.** Solve for $y$:

$$y^3 = 2 - x \qquad \text{Adding } y^3 \text{ and subtracting } x$$
$$y = \sqrt[3]{2 - x}.$$

**4.** $f^{-1}(x) = \sqrt[3]{2 - x}$

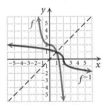

---

If a function $f$ is one-to-one, then $f^{-1}$ is the unique function such that each of the following holds:

$$(f^{-1} \circ f)(x) = f^{-1}(f(x)) = x,$$
for each $x$ in the domain of $f$, and
$$(f \circ f^{-1})(x) = f(f^{-1}(x)) = x,$$
for each $x$ in the domain of $f^{-1}$.

Given $f(x) = \dfrac{3 + x}{x}$, use composition of functions to show that $f^{-1}(x) = \dfrac{3}{x - 1}$.

$$(f^{-1} \circ f)(x) = f^{-1}(f(x))$$
$$= f^{-1}\left(\frac{3 + x}{x}\right) = \frac{3}{\dfrac{3 + x}{x} - 1}$$
$$= \frac{3}{\dfrac{3 + x - x}{x}} = \frac{3}{\dfrac{3}{x}} = 3 \cdot \frac{x}{3} = x;$$

*(continued)*

$$(f \circ f^{-1})(x) = f(f^{-1}(x)) = f\left(\frac{3}{x-1}\right)$$

$$= \frac{3 + \dfrac{3}{x-1}}{\dfrac{3}{x-1}} = \frac{3(x-1)+3}{x-1} \cdot \frac{x-1}{3}$$

$$= \frac{3x - 3 + 3}{3} = \frac{3x}{3} = x$$

## SECTION 5.2: EXPONENTIAL FUNCTIONS AND GRAPHS

**Exponential Function**

$$y = a^x, \text{ or } f(x) = a^x, \quad a > 0, a \neq 1$$

Continuous
One-to-one
Domain: $(-\infty, \infty)$
Range: $(0, \infty)$
Increasing if $a > 1$
Decreasing if $0 < a < 1$
Horizontal asymptote is $x$-axis
$y$-intercept: $(0, 1)$

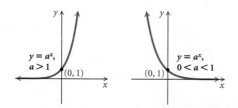

Graph: $f(x) = 2^x$, $g(x) = 2^{-x}$, $h(x) = 2^{x-1}$, and $t(x) = 2^x - 1$.

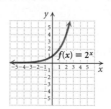

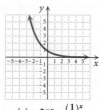

$$g(x) = 2^{-x} = \left(\frac{1}{2}\right)^x$$

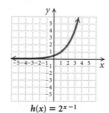

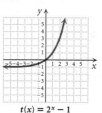

$h(x) = 2^{x-1}$

$t(x) = 2^x - 1$

**Compound Interest**

The amount of money $A$ to which a principal $P$ will grow after $t$ years at interest rate $r$ (in decimal form), compounded $n$ times per year, is given by the formula

$$A = P\left(1 + \frac{r}{n}\right)^{nt}.$$

Suppose that \$5000 is invested at 3.5% interest, compounded quarterly. Find the money in the account after 3 years.

$$A = P\left(1 + \frac{r}{n}\right)^{nt} = 5000\left(1 + \frac{0.035}{4}\right)^{4 \cdot 3}$$

$$\approx \$5551.02$$

**The Number $e$**

$$e = 2.7182818284\ldots$$

Find each of the following, to four decimal places, using a calculator.

$$e^{-3} \approx 0.0498;$$
$$e^{4.5} \approx 90.0171$$

Graph: $f(x) = e^x$ and $g(x) = e^{-x+2} - 4$.

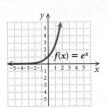

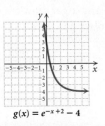

$$g(x) = e^{-x+2} - 4$$

## SECTION 5.3: LOGARITHMIC FUNCTIONS AND GRAPHS

**Logarithmic Function**

$$y = \log_a x, \quad x > 0, a > 0, a \neq 1$$

Continuous
One-to-one
Domain: $(0, \infty)$
Range: $(-\infty, \infty)$
Increasing if $a > 1$
Vertical asymptote is $y$-axis
$x$-intercept: $(1, 0)$

The inverse of an exponential function $f(x) = a^x$ is given by $f^{-1}(x) = \log_a x$.

Graph: $f(x) = \log_2 x$ and $g(x) = \ln(x - 1) + 2$.

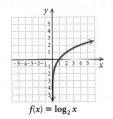

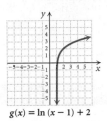

$$f(x) = \log_2 x \qquad g(x) = \ln(x - 1) + 2$$

A logarithm is an exponent:

$$\log_a x = y \longleftrightarrow x = a^y.$$

Convert each logarithmic equation to an exponential equation.

$$\log_4 \tfrac{1}{16} = -2 \longleftrightarrow 4^{-2} = \tfrac{1}{16};$$
$$\ln R = 3 \longleftrightarrow e^3 = R$$

Convert each exponential equation to a logarithmic equation.

$$e^{-5} = 0.0067 \longleftrightarrow \ln 0.0067 = -5;$$
$$7^2 = 49 \longleftrightarrow \log_7 49 = 2$$

$\log x$ means $\log_{10} x$    **Common logarithms**
$\ln x$ means $\log_e x$    **Natural logarithms**

For any logarithm base $a$,

$$\log_a 1 = 0 \quad \text{and} \quad \log_a a = 1.$$

For the logarithm base $e$,

$$\ln 1 = 0 \quad \text{and} \quad \ln e = 1.$$

Find each of the following without using a calculator.

$$\log 100 = 2; \qquad \log 10^{-5} = -5;$$
$$\ln 1 = 0; \qquad \log_9 9 = 1;$$
$$\ln \sqrt[3]{e} = \tfrac{1}{3}; \qquad \log_2 64 = 6;$$
$$\log_8 1 = 0; \qquad \ln e = 1$$

Find each of the following using a calculator and rounding to four decimal places.

$$\ln 223 \approx 5.4072; \qquad \log \tfrac{2}{9} \approx -0.6532;$$
$$\log(-8) \quad \text{Does not exist}; \qquad \ln 0.06 \approx -2.8134$$

**The Change-of–Base Formula**

For any logarithmic bases $a$ and $b$, and any positive number $M$,

$$\log_b M = \frac{\log_a M}{\log_a b}.$$

Find $\log_3 11$ using common logarithms:

$$\log_3 11 = \frac{\log 11}{\log 3} \approx 2.1827.$$

Find $\log_3 11$ using natural logarithms:

$$\log_3 11 = \frac{\ln 11}{\ln 3} \approx 2.1827.$$

**Earthquake Magnitude**

The magnitude $R$, measured on the Richter scale, of an earthquake of intensity $I$ is defined as

$$R = \log \frac{I}{I_0},$$

where $I_0$ is a minimum intensity used for comparison.

What is the magnitude on the Richter scale of an earthquake of intensity $10^{6.8} \cdot I_0$?

$$R = \log \frac{I}{I_0} = \log \frac{10^{6.8} \cdot I_0}{I_0} = \log 10^{6.8} = 6.8$$

## SECTION 5.4: PROPERTIES OF LOGARITHMIC FUNCTIONS

**The Product Rule**

For any positive numbers $M$ and $N$, and any logarithmic base $a$,

$$\log_a MN = \log_a M + \log_a N.$$

**The Power Rule**

For any positive number $M$, any logarithmic base $a$, and any real number $p$,

$$\log_a M^p = p \log_a M.$$

**The Quotient Rule**

For any positive numbers $M$ and $N$, and any logarithmic base $a$,

$$\log_a \frac{M}{N} = \log_a M - \log_a N.$$

Express $\log_c \sqrt{\dfrac{c^2 r}{b^3}}$ in terms of sums and differences of logarithms.

$$\begin{aligned}
\log_c \sqrt{\frac{c^2 r}{b^3}} &= \log_c \left( \frac{c^2 r}{b^3} \right)^{1/2} \\
&= \tfrac{1}{2} \log_c \left( \frac{c^2 r}{b^3} \right) \\
&= \tfrac{1}{2} (\log_c c^2 r - \log_c b^3) \\
&= \tfrac{1}{2} (\log_c c^2 + \log_c r - 3 \log_c b) \\
&= \tfrac{1}{2} (2 + \log_c r - 3 \log_c b) \\
&= 1 + \tfrac{1}{2} \log_c r - \tfrac{3}{2} \log_c b
\end{aligned}$$

Express $\ln (3x^2 + 5x - 2) - \ln (x + 2)$ as a single logarithm.

$$\begin{aligned}
&\ln (3x^2 + 5x - 2) - \ln (x + 2) \\
&= \ln \frac{3x^2 + 5x - 2}{x + 2} = \ln \frac{(3x - 1)(x + 2)}{x + 2} \\
&= \ln (3x - 1)
\end{aligned}$$

Given $\log_a 7 \approx 0.8451$ and $\log_a 5 \approx 0.6990$, find $\log_a \frac{1}{7}$ and $\log_a 35$.

$$\log_a \tfrac{1}{7} = \log_a 1 - \log_a 7 \approx 0 \quad 0.8451 \approx -0.8451;$$

$$\log_a 35 = \log_a (7 \cdot 5) = \log_a 7 + \log_a 5$$
$$\approx 0.8451 + 0.6990 \approx 1.5441$$

For any base $a$ and any real number $x$,

$$\log_a a^x = x.$$

For any base $a$ and any positive real number $x$,

$$a^{\log_a x} = x.$$

Simplify each of the following.

$$8^{\log_8 k} = k; \qquad \log 10^{43} = 43;$$
$$\log_a a^4 = 4; \qquad e^{\ln 2} = 2$$

## SECTION 5.5: SOLVING EXPONENTIAL EQUATIONS AND LOGARITHMIC EQUATIONS

**The Base–Exponent Property**
For any $a > 0$, $a \neq 1$,

$$a^x = a^y \longleftrightarrow x = y.$$

Solve: $3^{2x-3} = 81$.

$$3^{2x-3} = 81$$
$$3^{2x-3} = 3^4 \qquad 81 = 3^4$$
$$2x - 3 = 4$$
$$2x = 7$$
$$x = \tfrac{7}{2}$$

The solution is $\tfrac{7}{2}$.

**The Property of Logarithmic Equality**
For any $M > 0$, $N > 0$, $a > 0$, and $a \neq 1$,

$$\log_a M = \log_a N \longleftrightarrow M = N.$$

Solve: $6^{x-2} = 2^{-3x}$.

$$6^{x-2} = 2^{-3x}$$
$$\log 6^{x-2} = \log 2^{-3x}$$
$$(x - 2)\log 6 = -3x \log 2$$
$$x \log 6 - 2 \log 6 = -3x \log 2$$
$$x \log 6 + 3x \log 2 = 2 \log 6$$
$$x(\log 6 + 3 \log 2) = 2 \log 6$$
$$x = \frac{2 \log 6}{\log 6 + 3 \log 2}$$
$$x \approx 0.9257$$

Solve: $\log_3 (x - 2) + \log_3 x = 1$.

$$\log_3 (x - 2) + \log_3 x = 1$$
$$\log_3 [(x - 2)x] = 1$$
$$(x - 2)x = 3^1$$
$$x^2 - 2x - 3 = 0$$
$$(x - 3)(x + 1) = 0$$
$$x - 3 = 0 \quad or \quad x + 1 = 0$$
$$x = 3 \quad or \qquad x = -1$$

The number $-1$ is not a solution because negative numbers do not have real-number logarithms. The value 3 checks and is the solution.

Solve: $\ln(x + 10) - \ln(x + 4) = \ln x$.

$$\ln(x + 10) - \ln(x + 4) = \ln x$$

$$\ln\frac{x + 10}{x + 4} = \ln x$$

$$\frac{x + 10}{x + 4} = x$$

$$x + 10 = x(x + 4)$$

$$x + 10 = x^2 + 4x$$

$$0 = x^2 + 3x - 10$$

$$0 = (x + 5)(x - 2)$$

$$x + 5 = 0 \quad or \quad x - 2 = 0$$

$$x = -5 \quad or \quad x = 2$$

The number $-5$ is not a solution because $\ln(-5 + 4)$, or $\ln(-1)$, on the left and $\ln(-5)$ on the right are not real numbers. The value 2 checks and is the solution.

## SECTION 5.6: APPLICATIONS AND MODELS: GROWTH AND DECAY; COMPOUND INTEREST

**Exponential Growth Model**

$$P(t) = P_0 e^{kt}, \quad k > 0$$

**Doubling Time**

$$kT = \ln 2, \quad or \quad k = \frac{\ln 2}{T}, \quad or \quad T = \frac{\ln 2}{k}$$

In July 2010, the population of the United States was 310.2 million, and the exponential growth rate was 0.97% per year (*Sources*: CIA World Factbook; *The New York Times Almanac* 2010). After how long will the population be double what it was in 2010? Estimate the population in 2015.

With a population growth rate of 0.97%, or 0.0097, the doubling time $T$ is

$$T = \frac{\ln 2}{k} = \frac{\ln 2}{0.0097} \approx 71.$$

The population of the United States will be double what it was in 2010 about 71 years after 2010.

The exponential growth function is

$$P(t) = 310.2 e^{0.0097t}$$

where $t$ is the number of years after 2010 and $P(t)$ is in millions. Since in 2015, $t = 5$, we substitute 5 for $t$:

$$P(5) = 310.2 e^{0.0097 \cdot 5} = 310.2 e^{0.0485} \approx 325.6.$$

The population will be about 325.6 million, or 325,600,000, in 2015.

**Interest Compounded Continuously**

$$P(t) = P_0 e^{kt}, \quad k > 0$$

Suppose that $20,000 is invested at interest rate $k$, compounded continuously, and grows to $23,236.68 in 3 years. What is the interest rate? What will the balance be in 8 years?

The exponential growth function is of the form $P(t) = 20{,}000e^{kt}$. Given that $P(3) = \$23{,}236.68$, substituting 3 for $t$ and 23,236.68 for $P(t)$ gives

$$23{,}236.68 = 20{,}000e^{k(3)}$$

to get $k \approx 0.05$, or 5%.

Next, we substitute 0.05 for $k$ and 8 for $t$ and determine $P(8)$:

$$P(8) = 20{,}000e^{0.05(8)} = 20{,}000e^{0.4} \approx \$29{,}836.49.$$

**Exponential Decay Model**

$$P(t) = P_0 e^{-kt}, \quad k > 0$$

**Half-Life**

$$kT = \ln 2, \quad \text{or} \quad k = \frac{\ln 2}{T}, \quad \text{or} \quad T = \frac{\ln 2}{k}$$

Archaeologists discovered an animal bone that had lost 65.2% of its carbon-14 at the time it was found. How old was the bone?

The decay rate for carbon-14 is 0.012%, or 0.00012. If the bone has lost 65.2% of its carbon-14 from an initial amount $P_0$, then 34.8% $P_0$ is the amount present. We substitute 34.8% $P_0$ for $P(t)$ and solve:

$$34.8\% \ P_0 = P_0 e^{-0.00012t}$$
$$0.348 = e^{-0.00012t}$$
$$\ln 0.348 = -0.00012t$$
$$\frac{\ln 0.348}{-0.00012} = t$$
$$8796 \approx t.$$

The bone was about 8796 years old when it was found.

# REVIEW EXERCISES

*Determine whether the statement is true or false.*

1. The domain of a one-to-one function $f$ is the range of the inverse $f^{-1}$. [5.1]

2. The $x$-intercept of $f(x) = \log x$ is $(0, 1)$. [5.3]

3. The graph of $f^{-1}$ is a reflection of the graph of $f$ across $y = 0$. [5.1]

4. If it is not possible for a horizontal line to intersect the graph of a function more than once, then the function is one-to-one and its inverse is a function. [5.1]

5. The range of all exponential functions is $[0, \infty)$. [5.2]

6. The horizontal asymptote of $y = 2^x$ is $y = 0$. [5.2]

7. Find the inverse of the relation
$$\{(1.3, -2.7), (8, -3), (-5, 3), (6, -3), (7, -5)\}.$$
[5.1]

8. Find an equation of the inverse relation. [5.1]
   a) $y = -2x + 3$
   b) $y = 3x^2 + 2x - 1$
   c) $0.8x^3 - 5.4y^2 = 3x$

*Graph the function and determine whether the function is one-to-one using the horizontal-line test.* [5.1]

9. $f(x) = -|x| + 3$

10. $f(x) = x^2 + 1$

11. $f(x) = 2x - \frac{3}{4}$

12. $f(x) = -\dfrac{6}{x + 1}$

*In Exercises 13–18, given the function:*

**a)** *Sketch the graph and determine whether the function is one-to-one.* [5.1], [5.3]

**b)** *If it is one-to-one, find a formula for the inverse.* [5.1], [5.3]

**13.** $f(x) = 2 - 3x$

**14.** $f(x) = \dfrac{x + 2}{x - 1}$

**15.** $f(x) = \sqrt{x - 6}$

**16.** $f(x) = x^3 - 8$

**17.** $f(x) = 3x^2 + 2x - 1$    **18.** $f(x) = e^x$

*For the function f, use composition of functions to show that $f^{-1}$ is as given.* [5.1]

**19.** $f(x) = 6x - 5$, $f^{-1}(x) = \dfrac{x + 5}{6}$

**20.** $f(x) = \dfrac{x + 1}{x}$, $f^{-1}(x) = \dfrac{1}{x - 1}$

*Find the inverse of the given one-to-one function f. Give the domain and the range of f and of $f^{-1}$ and then graph both f and $f^{-1}$ on the same set of axes.* [5.1]

**21.** $f(x) = 2 - 5x$

**22.** $f(x) = \dfrac{x - 3}{x + 2}$

**23.** Find $f(f^{-1}(657))$:
$$f(x) = \dfrac{4x^5 - 16x^{37}}{119x}, \quad x > 1. \; [5.1]$$

**24.** Find $f(f^{-1}(a))$: $f(x) = \sqrt[3]{3x - 4}$. [5.1]

*Graph the function.*

**25.** $f(x) = \left(\frac{1}{3}\right)^x$ [5.2]

**26.** $f(x) = 1 + e^x$ [5.2]

**27.** $f(x) = -e^{-x}$ [5.2]

**28.** $f(x) = \log_2 x$ [5.3]

**29.** $f(x) = \frac{1}{2}\ln x$ [5.3]

**30.** $f(x) = \log x - 2$ [5.3]

*In Exercises 31–36, match the equation with one of the figures (a)–(f), which follow*

**a)**
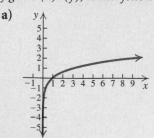

**b)**
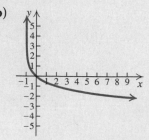

**c)**
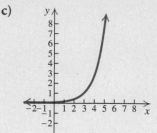

**d)**
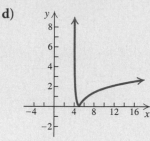

**e)**
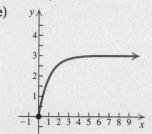

**f)**
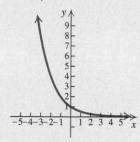

**31.** $f(x) = e^{x-3}$ [5.2]

**32.** $f(x) = \log_3 x$ [5.3]

**33.** $y = -\log_3(x + 1)$ [5.3]

**34.** $y = \left(\frac{1}{2}\right)^x$ [5.2]

**35.** $f(x) = 3(1 - e^{-x})$, $x \geq 0$ [5.2]

**36.** $f(x) = |\ln(x - 4)|$ [5.3]

*Find each of the following. Do not use a calculator.* [5.3]

**37.** $\log_5 125$

**38.** $\log 100{,}000$

**39.** $\ln e$

**40.** $\ln 1$

**41.** $\log 10^{1/4}$

**42.** $\log_3 \sqrt{3}$

**43.** $\log 1$

**44.** $\log 10$

**45.** $\log_2 \sqrt[3]{2}$

**46.** $\log 0.01$

*Convert to an exponential equation.* [5.3]

**47.** $\log_4 x = 2$

**48.** $\log_a Q = k$

*Convert to a logarithmic equation.* [5.3]

**49.** $4^{-3} = \frac{1}{64}$

**50.** $e^x = 80$

*Find each of the following using a calculator. Round to four decimal places.* [5.3]

**51.** $\log 11$

**52.** $\log 0.234$

**53.** $\ln 3$

**54.** $\ln 0.027$

**55.** $\log(-3)$

**56.** $\ln 0$

*Find the logarithm using the change-of-base formula.* [5.3]

**57.** $\log_5 24$

**58.** $\log_8 3$

*Express as a single logarithm and, if possible, simplify.* [5.4]

**59.** $3 \log_b x - 4 \log_b y + \frac{1}{2} \log_b z$

**60.** $\ln (x^3 - 8) - \ln (x^2 + 2x + 4) + \ln (x + 2)$

*Express in terms of sums and differences of logarithms.* [5.4]

**61.** $\ln \sqrt[4]{wr^2}$

**62.** $\log \sqrt[3]{\dfrac{M^2}{N}}$

*Given that* $\log_a 2 = 0.301$, $\log_a 5 = 0.699$, *and* $\log_a 6 = 0.778$, *find each of the following.* [5.4]

**63.** $\log_a 3$

**64.** $\log_a 50$

**65.** $\log_a \frac{1}{5}$

**66.** $\log_a \sqrt[3]{5}$

*Simplify.* [5.4]

**67.** $\ln e^{-5k}$

**68.** $\log_5 5^{-6t}$

*Solve.* [5.5]

**69.** $\log_4 x = 2$

**70.** $3^{1-x} = 9^{2x}$

**71.** $e^x = 80$

**72.** $4^{2x-1} - 3 = 61$

**73.** $\log_{16} 4 = x$

**74.** $\log_x 125 = 3$

**75.** $\log_2 x + \log_2 (x - 2) = 3$

**76.** $\log (x^2 - 1) - \log (x - 1) = 1$

**77.** $\log x^2 = \log x$

**78.** $e^{-x} = 0.02$

**79.** *Saving for College.* Following the birth of triplets, the grandparents deposit $30,000 in a college trust fund that earns 4.2% interest, compounded quarterly.

   **a)** Find a function for the amount in the account after *t* years. [5.2]

   **b)** Find the amount in the account at $t = 0, 6,$ 12, and 18 years. [5.2]

**80.** *Personal Breathalyzer Sales.* In recent years, the sales of personal breathalyzers have been increasing exponentially. The total amount of sales, in millions of dollars, is estimated by the function

$$B(t) = 27.9(1.3299)^t,$$

where *t* is the number of years since 2005 (*Source*: WinterGreen Research). Estimate the sales in 2008 and in 2014. [5.2]

**81.** How long will it take an investment to double if it is invested at 8.6%, compounded continuously? [5.6]

**82.** The population of a metropolitan area consisting of 8 counties doubled in 26 years. What was the exponential growth rate? [5.6]

**83.** How old is a skeleton that has lost 27% of its carbon-14? [5.6]

**84.** The hydrogen ion concentration of milk is $2.3 \times 10^{-6}$. What is the pH? (See Exercise 98 in Exercise Set 5.3.) [5.3]

**85.** *Earthquake Magnitude.* The earthquake in Kashgar, China, on February 25, 2003, had an intensity of $10^{6.3} \cdot I_0$ (*Source*: U.S. Geological Survey). What is the magnitude on the Richter scale? [5.3]

**86.** What is the loudness, in decibels, of a sound whose intensity is $1000 I_0$? (See Exercise 101 in Exercise Set 5.3.) [5.3]

**87.** *Walking Speed.* The average walking speed *w*, in feet per second, of a person living in a city of population *P*, in thousands, is given by the function

$$w(P) = 0.37 \ln P + 0.05.$$

   **a)** The population of Wichita, Kansas, is 353,823. Find the average walking speed. [5.3]

   **b)** A city's population has an average walking speed of 3.4 ft/sec. Find the population. [5.6]

**88.** *Social Security Distributions.* Cash Social Security distributions were $35 million, or $0.035 billion, in 1940. This amount has increased exponentially to $683 billion in 2009. (*Source*: Social Security Administration) Assuming the exponential growth model applies:

   **a)** Find the exponential growth rate *k*. [5.6]

   **b)** Find the exponential growth function. [5.6]

   **c)** Estimate the total cash distributions in 1965, in 1995, and in 2015. [5.6]

   **d)** In what year will the cash benefits reach $2 trillion? [5.6]

**89.** *Population of Papua New Guinea.* The population of Papua New Guinea was 6.188 million in 2011, and the exponential growth rate was 1.985% per year (*Source: CIA World Factbook*)

a) Find the exponential growth function. [5.6]
b) Estimate the population in 2013 and in 2015. [5.6]
c) When will the population be 10 million? [5.6]
d) What is the doubling time? [5.6]

**90.** *Price of a First-Class Stamp.* The cost of mailing a letter increased 40¢ from 1958 to 2011. The table below lists the cost of a first-class stamp in selected years.

| Year, x | First-Class Postage Cost C (in cents) |
|---------|----------------------------------------|
| 1958, 0 | 4¢ |
| 1968, 10 | 6 |
| 1978, 20 | 15 |
| 1988, 30 | 25 |
| 1995, 37 | 32 |
| 2002, 44 | 37 |
| 2006, 48 | 39 |
| 2008, 50 | 42 |
| 2011, 53 | 44 |

*Source*: U.S. Postal Service

a) Use a graphing calculator to fit an exponential function to the data, where *x* is the number of years after 1958. [5.6]
b) Graph the function with a scatterplot of the data. [5.6]
c) Estimate the cost of a first-class stamp in 1970, in 2000, and in 2015.

**91.** Using only a graphing calculator, determine whether the following functions are inverses of each other:

$$f(x) = \frac{4 + 3x}{x - 2}, \qquad g(x) = \frac{x + 4}{x - 3}. \quad [5.1]$$

**92. a)** Use a graphing calculator to graph $f(x) = 5e^{-x} \ln x$ in the viewing window $[-1, 10, -5, 5]$. [5.2], [5.3]
**b)** Estimate the relative maximum and minimum values of the function. [5.2], [5.3]

**93.** Which of the following is the horizontal asymptote of the graph of $f(x) = e^{x-3} + 2$? [5.2]
A. $y = -2$     B. $y = -3$
C. $y = 3$      D. $y = 2$

**94.** Which of the following is the domain of the logarithmic function $f(x) = \log(2x - 3)$? [5.3]
A. $\left(\frac{3}{2}, \infty\right)$     B. $\left(-\infty, \frac{3}{2}\right)$
C. $(3, \infty)$     D. $(-\infty, \infty)$

**95.** The graph of $f(x) = 2^{x-2}$ is which of the following? [5.2]

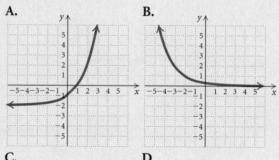

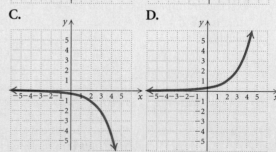

**96.** The graph of $f(x) = \log_2 x$ is which of the following? [5.3]

**A.**

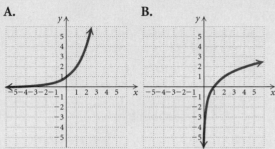

**B.**

**C.**

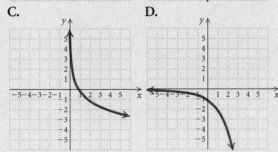

**D.**

**100.** Find the domain: $f(x) = \log_3(\ln x)$. [5.3]

## Collaborative Discussion and Writing

**101.** Explain how the graph of $f(x) = \ln x$ can be used to obtain the graph of $g(x) = e^{x-2}$. [5.3]

**102.** *Atmospheric Pressure.*   Atmospheric pressure $P$ at an altitude $a$ is given by

$$P = P_0 e^{-0.00005a},$$

where $P_0$ is the pressure at sea level, approximately $14.7 \text{ lb}/\text{in}^2$ (pounds per square inch). Explain how a barometer, or some device for measuring atmospheric pressure, can be used to find the height of a skyscraper. [5.6]

**103.** Explain the errors, if any, in the following: [5.4]
$$\log_a ab^3 = (\log_a a)(\log_a b^3) = 3 \log_a b.$$

**104.** Describe the difference between $f^{-1}(x)$ and $[f(x)]^{-1}$. [5.1]

## Synthesis

*Solve.* [5.5]

**97.** $|\log_4 x| = 3$

**98.** $\log x = \ln x$

**99.** $5^{\sqrt{x}} = 625$

## Chapter 5 Test

**1.** Find the inverse of the relation
$$\{(-2, 5), (4, 3), (0, -1), (-6, -3)\}.$$

*Determine whether the function is one-to-one. Answer yes or no.*

**2.**

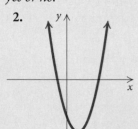

**3.**

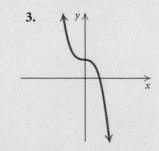

*In Exercises 4–7, given the function:*

**a)** *Sketch the graph and determine whether the function is one-to-one.*

**b)** *If it is one-to-one, find a formula for the inverse.*

**4.** $f(x) = x^3 + 1$

**5.** $f(x) = 1 - x$

**6.** $f(x) = \dfrac{x}{2 - x}$

**7.** $f(x) = x^2 + x - 3$

**8.** Use composition of functions to show that $f^{-1}$ is as given:

$$f(x) = -4x + 3, \qquad f^{-1}(x) = \dfrac{3 - x}{4}.$$

**9.** Find the inverse of the one-to-one function

$$f(x) = \frac{1}{x - 4}.$$

Give the domain and the range of $f$ and of $f^{-1}$ and then graph both $f$ and $f^{-1}$ on the same set of axes.

*Graph the function.*

**10.** $f(x) = 4^{-x}$

**11.** $f(x) = \log x$

**12.** $f(x) = e^x - 3$

**13.** $f(x) = \ln(x + 2)$

*Find each of the following. Do not use a calculator.*

**14.** $\log 0.00001$

**15.** $\ln e$

**16.** $\ln 1$

**17.** $\log_4 \sqrt[5]{4}$

**18.** Convert to an exponential equation: $\ln x = 4$.

**19.** Convert to a logarithmic equation: $3^x = 5.4$.

*Find each of the following using a calculator. Round to four decimal places.*

**20.** $\ln 16$

**21.** $\log 0.293$

**22.** Find $\log_6 10$ using the change-of-base formula.

**23.** Express as a single logarithm:

$$2\log_a x - \log_a y + \tfrac{1}{2}\log_a z.$$

**24.** Express $\ln \sqrt[5]{x^2 y}$ in terms of sums and differences of logarithms.

**25.** Given that $\log_a 3 = 1.585$ and $\log_a 15 = 3.907$, find $\log_a 5$.

**26.** Simplify: $\ln e^{-4t}$.

*Solve.*

**27.** $\log_{25} 5 = x$

**28.** $\log_3 x + \log_3 (x + 8) = 2$

**29.** $3^{4-x} = 27^x$

**30.** $e^x = 65$

**31.** *Earthquake Magnitude.* The earthquake in Bam, in southeast Iran, on December 26, 2003, had an intensity of $10^{6.6} \cdot I_0$ (*Source*: U.S. Geological Survey). What was its magnitude on the Richter scale?

**32.** *Growth Rate.* A country's population doubled in 45 years. What was the exponential growth rate?

**33.** *Compound Interest.* Suppose $1000 is invested at interest rate $k$, compounded continuously, and grows to $1144.54 in 3 years.

   **a)** Find the interest rate.

   **b)** Find the exponential growth function.

   **c)** Find the balance after 8 years.

   **d)** Find the doubling time.

**34.** The graph of $f(x) = 2^{x-1} + 1$ is which of the following?

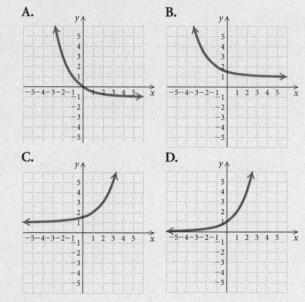

**A.**

**B.**

**C.**

**D.**

## Synthesis

**35.** Solve: $4^{\sqrt[3]{x}} = 8$.

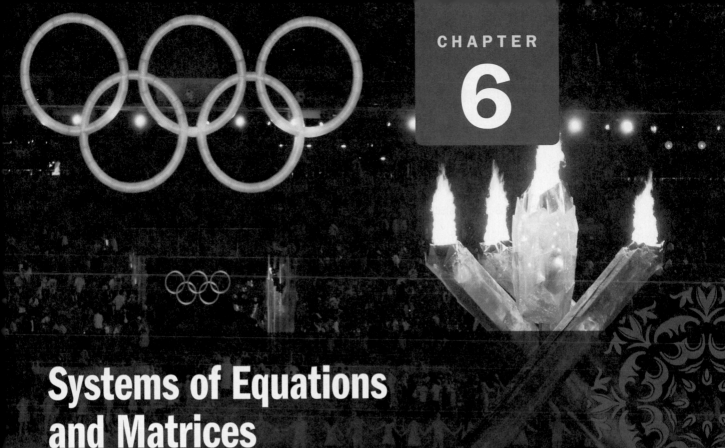

# Systems of Equations and Matrices

## APPLICATION

The Winter Olympics have been held a total of 21 times on the continents of North America, Europe, and Asia. The number of European sites is 5 more than the total number of sites in North America and Asia. There are 4 more sites in North America than in Asia. (*Source*: *USA Today* research) Find the number of Winter Olympic sites on each continent.

**This problem appears as Exercise 17 in Section 6.2.**

## 6.1 Systems of Equations in Two Variables

- Solve a system of two linear equations in two variables by graphing.
- Solve a system of two linear equations in two variables using the substitution method and the elimination method.
- Use systems of two linear equations to solve applied problems.

A **system of equations** is composed of two or more equations considered simultaneously. For example,

$$x - y = 5,$$
$$2x + y = 1$$

is a **system of two linear equations in two variables**. The solution set of this system consists of all ordered pairs that make *both* equations true. The ordered pair $(2, -3)$ is a solution of the system of equations above. We can verify this by substituting 2 for $x$ and $-3$ for $y$ in *each* equation.

| $x - y = 5$ | $2x + y = 1$ |
|---|---|
| $2 - (-3)\ ?\ 5$ | $2 \cdot 2 + (-3)\ ?\ 1$ |
| $2 + 3$ | $4 - 3$ |
| $5\ \vert\ 5$   TRUE | $1\ \vert\ 1$   TRUE |

### ■ Solving Systems of Equations Graphically

Recall that the graph of a linear equation is a line that contains all the ordered pairs in the solution set of the equation. When we graph a system of linear equations, each point at which the graphs intersect is a solution of *both* equations and therefore a **solution of the system of equations**.

**EXAMPLE 1**   Solve the following system of equations graphically:

$$x - y = 5,$$
$$2x + y = 1.$$

**Solution**   We graph the equations on the same set of axes, as shown below.

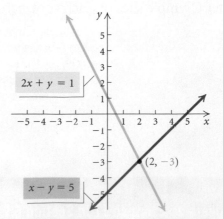

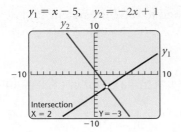

$y_1 = x - 5, \quad y_2 = -2x + 1$

We see that the graphs intersect at a single point, $(2, -3)$, so $(2, -3)$ is the solution of the system of equations. To check this solution, we substitute 2 for $x$ and $-3$ for $y$ in both equations as we did above.

To use a graphing calculator to solve this system of equations, it might be necessary to write each equation in "$Y = \cdots$" form. If so, we would graph $y_1 = x - 5$ and $y_2 = -2x + 1$ and then use the INTERSECT feature. We see in the window at left that the solution is $(2, -3)$.

**Now Try Exercise 7.**

The graphs of most of the systems of linear equations that we use to model applications intersect at a single point, like the system above. However, it is possible that the graphs will have no points in common or infinitely many points in common. Each of these possibilities is illustrated below.

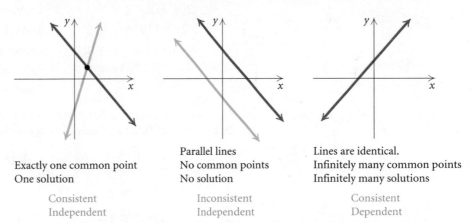

Exactly one common point
One solution

Consistent
Independent

Parallel lines
No common points
No solution

Inconsistent
Independent

Lines are identical.
Infinitely many common points
Infinitely many solutions

Consistent
Dependent

If a system of equations has at least one solution, it is **consistent**. If the system has no solutions, it is **inconsistent**. In addition, for a system of two linear equations in two variables, if one equation can be obtained by multiplying on both sides of the other equation by a constant, the equations are **dependent**. Otherwise, they are **independent**. A system of two dependent linear equations in two variables has an infinite number of solutions.

## ▪ The Substitution Method

Solving a system of equations graphically is not always accurate when the solutions are not integers. A solution like $\left(\frac{43}{27}, -\frac{19}{27}\right)$, for instance, will be difficult to determine from a hand-drawn graph.

Algebraic methods for solving systems of equations, when used correctly, always give accurate results. One such technique is the **substitution method**. It is used most often when a variable is alone on one side of an equation or when it is easy to solve for a variable. To apply the substitution method, we begin by using one of the equations to express one variable in terms of the other. Then we substitute that expression in the other equation of the system.

**EXAMPLE 2**   Use the substitution method to solve the system

$$x - y = 5, \qquad (1)$$
$$2x + y = 1. \qquad (2)$$

***Solution***   First, we solve equation (1) for $x$. (We could have solved for $y$ instead.) We have

$$x - y = 5, \qquad\qquad (1)$$
$$x = y + 5. \qquad \text{Solving for } x$$

Then we substitute $y + 5$ for $x$ in equation (2). This gives an equation in one variable, which we know how to solve:

$$2x + y = 1 \qquad\qquad (2)$$
$$2(y + 5) + y = 1 \qquad\quad \text{The parentheses are necessary.}$$
$$2y + 10 + y = 1 \qquad\quad \text{Removing parentheses}$$
$$3y + 10 = 1 \qquad\quad \text{Collecting like terms on the left}$$
$$3y = -9 \qquad\quad \text{Subtracting 10 on both sides}$$
$$y = -3. \qquad\quad \text{Dividing by 3 on both sides}$$

Now we substitute $-3$ for $y$ in either of the original equations (this is called **back-substitution**) and solve for $x$. We choose equation (1):

$$x - y = 5, \qquad\qquad (1)$$
$$x - (-3) = 5 \qquad\quad \text{Substituting } -3 \text{ for } y$$
$$x + 3 = 5$$
$$x = 2. \qquad\quad \text{Subtracting 3 on both sides}$$

We have previously checked the pair $(2, -3)$ in both equations. The solution of the system of equations is $(2, -3)$.   **Now Try Exercise 17.**

## ■ The Elimination Method

Another algebraic technique for solving systems of equations is the **elimination method**. With this method, we eliminate a variable by adding two equations. If the coefficients of a particular variable are opposites, we can eliminate that variable simply by adding the original equations. For example, if the $x$-coefficient is $-3$ in one equation and is $3$ in the other equation, then the sum of the $x$-terms will be $0$ and thus the variable $x$ will be eliminated when we add the equations.

**EXAMPLE 3**   Use the elimination method to solve the system

$$2x + y = 2, \qquad (1)$$
$$x - y = 7. \qquad (2)$$

## Algebraic Solution

Since the $y$-coefficients, 1 and $-1$, are opposites, we can eliminate $y$ by adding the equations:

$$\begin{aligned} 2x + y &= 2 \quad \textbf{(1)} \\ \underline{x - y = 7} \quad &\textbf{(2)} \\ 3x \quad &= 9 \quad \text{Adding} \\ x &= 3. \end{aligned}$$

We then back-substitute 3 for $x$ in either equation and solve for $y$. We choose equation (1):

$$\begin{aligned} 2x + y &= 2 \qquad \textbf{(1)} \\ 2 \cdot 3 + y &= 2 \qquad \text{Substituting 3 for } x \\ 6 + y &= 2 \\ y &= -4. \end{aligned}$$

We check the solution by substituting the pair $(3, -4)$ in both equations.

$$\begin{array}{c|c}
2x + y = 2 & x - y = 7 \\
\hline
2 \cdot 3 + (-4) \;?\; 2 & 3 - (-4) \;?\; 7 \\
6 - 4 \phantom{\;?\; 2} & 3 + 4 \phantom{\;?\; 7} \\
2 \mid 2 \;\text{ TRUE} & 7 \mid 7 \;\text{ TRUE}
\end{array}$$

The solution is $(3, -4)$. Since there is exactly one solution, the system of equations is consistent, and the equations are independent.

## Graphical Solution

We solve each equation for $y$, getting

$$y = -2x + 2 \quad \text{and} \quad y = x - 7.$$

Next, we graph these equations and find the point of intersection of the graphs using the INTERSECT feature.

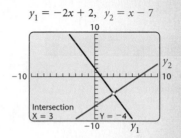

$y_1 = -2x + 2, \;\; y_2 = x - 7$

The graphs intersect at the point $(3, -4)$, so the solution of the system of equations is $(3, -4)$.

**Now Try Exercise 31.**

Before we add, it might be necessary to multiply one or both equations by suitable constants in order to find two equations in which the coefficients of a variable are opposites.

**EXAMPLE 4** Use the elimination method to solve the system

$$\begin{aligned} 4x + 3y &= 11, \quad \textbf{(1)} \\ -5x + 2y &= 15. \quad \textbf{(2)} \end{aligned}$$

**Solution** We can obtain $x$-coefficients that are opposites by multiplying the first equation by 5 and the second equation by 4:

$$\begin{aligned} 20x + 15y &= 55 \qquad \text{Multiplying equation (1) by 5} \\ \underline{-20x + 8y = 60} \qquad &\text{Multiplying equation (2) by 4} \\ 23y &= 115 \qquad \text{Adding} \\ y &= 5. \end{aligned}$$

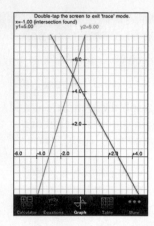

We can check the solution to the system in Example 4 with an app.

We then back-substitute 5 for $y$ in either equation (1) or (2) and solve for $x$. We choose equation (1):

$$4x + 3y = 11 \qquad \textbf{(1)}$$
$$4x + 3 \cdot 5 = 11 \qquad \text{Substituting 5 for } y$$
$$4x + 15 = 11$$
$$4x = -4$$
$$x = -1.$$

We can check the pair $(-1, 5)$ by substituting in both equations or by graphing, as shown at left. The solution is $(-1, 5)$. The system of equations is consistent and the equations are independent.

**Now Try Exercise 33.**

In Example 4, the two systems

$$\begin{array}{cc} 4x + 3y = 11, \\ -5x + 2y = 15 \end{array} \quad \text{and} \quad \begin{array}{cc} 20x + 15y = 55, \\ -20x + 8y = 60 \end{array}$$

are **equivalent** because they have exactly the same solutions. When we use the elimination method, we often multiply one or both equations by constants to find equivalent equations that allow us to eliminate a variable by adding.

**EXAMPLE 5**    Solve each of the following systems using the elimination method.

**a)**    $x - 3y = 1,$     **(1)**        **b)** $2x + 3y = 6,$     **(1)**
$\quad\quad -2x + 6y = 5$     **(2)**        $\quad\quad 4x + 6y = 12$     **(2)**

*Solution*

**a)**  We multiply equation (1) by 2 and add:

$$\begin{array}{ll} 2x - 6y = 2 & \text{Multiplying equation (1) by 2} \\ \underline{-2x + 6y = 5} & \textbf{(2)} \\ \quad\quad\quad 0 = 7. & \text{Adding} \end{array}$$

There are no values of $x$ and $y$ for which $0 = 7$ is true, so the system has *no solution*. The solution set is $\varnothing$. The system of equations is inconsistent and the equations are independent. The graphs of the equations are parallel lines, as shown in Fig. 1.

**b)**  We multiply equation (1) by $-2$ and add:

$$\begin{array}{ll} -4x - 6y = -12 & \text{Multiplying equation (1) by } -2 \\ \underline{\quad 4x + 6y = 12} & \textbf{(2)} \\ \quad\quad\quad\quad 0 = 0. & \text{Adding} \end{array}$$

We obtain the equation $0 = 0$, which is true for all values of $x$ and $y$. This tells us that the equations are dependent, so there are *infinitely many solutions*. That is, any solution of one equation of the system is also a solution of the other. The system of equations is consistent. The graphs of the equations are identical, as shown in Fig. 2.

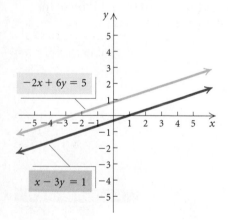

FIGURE 1

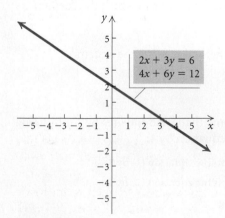

FIGURE 2

Solving either equation for $y$, we have $y = -\frac{2}{3}x + 2$, so we can write the solutions of the system as ordered pairs $(x, y)$, where $y$ is expressed as $-\frac{2}{3}x + 2$. Thus the solutions can be written in the form $\left(x, -\frac{2}{3}x + 2\right)$. Any real value that we choose for $x$ then gives us a value for $y$ and thus an ordered pair in the solution set. For example,

if $x = -3$, then $-\frac{2}{3}x + 2 = -\frac{2}{3}(-3) + 2 = 4$,

if $x = 0$, then $-\frac{2}{3}x + 2 = -\frac{2}{3} \cdot 0 + 2 = 2$, and

if $x = 6$, then $-\frac{2}{3}x + 2 = -\frac{2}{3} \cdot 6 + 2 = -2$.

Thus some of the solutions are $(-3, 4)$, $(0, 2)$, and $(6, -2)$.

Similarly, solving either equation for $x$, we have $x = -\frac{3}{2}y + 3$, so the solutions $(x, y)$ can also be written, expressing $x$ as $-\frac{3}{2}y + 3$, in the form $\left(-\frac{3}{2}y + 3, y\right)$.

Since the two forms of the solutions are equivalent, they yield the same solution set, as illustrated in the table at left. Note, for example, that when $y = 4$, we have the solution $(-3, 4)$; when $y = 2$, we have $(0, 2)$; and when $y = -2$, we have $(6, -2)$.

**Now Try Exercises 35 and 37.**

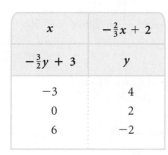

| $x$ | $-\frac{2}{3}x + 2$ |
|---|---|
| $-\frac{3}{2}y + 3$ | $y$ |
| $-3$ | $4$ |
| $0$ | $2$ |
| $6$ | $-2$ |

## ■ Applications

Frequently the most challenging and time-consuming step in the problem-solving process is translating a situation to mathematical language. However, in many cases, this task is made easier if we translate to more than one equation in more than one variable.

**EXAMPLE 6** *Snack Mixtures.* At Max's Munchies, caramel corn worth $2.50 per pound is mixed with honey roasted mixed nuts worth $7.50 per pound in order to get 20 lb of a mixture worth $4.50 per pound. How much of each snack is used?

$2.50 per lb

$4.50 per lb

$7.50 per lb

*Solution*  We use the five-step problem-solving process.

1. **Familiarize.** Let's begin by making a guess. Suppose 16 lb of caramel corn and 4 lb of nuts are used. Then the total weight of the mixture would be 16 lb + 4 lb, or 20 lb, the desired weight. The total values of these amounts of ingredients are found by multiplying the price per pound by the number of pounds used:

   Caramel corn: $2.50(16) = $40

   Nuts:             $7.50(4) = $30

   Total value:                $70.

   The desired value of the mixture is $4.50 per pound, so the value of 20 lb would be $4.50(20), or $90. Thus we see that our guess, which led to a total of $70, is incorrect. Nevertheless, these calculations will help us to translate.

2. **Translate.** We organize the information in a table. We let $x =$ the number of pounds of caramel corn in the mixture and $y =$ the number of pounds of nuts.

|  | Caramel Corn | Nuts | Mixture | |
|---|---|---|---|---|
| Price per Pound | $2.50 | $7.50 | $4.50 | |
| Number of Pounds | $x$ | $y$ | 20 | $\rightarrow x + y = 20$ |
| Value of Mixture | $2.50x$ | $7.50y$ | $4.50(20)$, or 90 | $\rightarrow 2.50x + 7.50y = 90$ |

From the second row of the table, we get one equation:

$$x + y = 20.$$

The last row of the table yields a second equation:

$$2.50x + 7.50y = 90, \quad \text{or} \quad 2.5x + 7.5y = 90.$$

We can multiply by 10 on both sides of the second equation to clear the decimals. This gives us the following system of equations:

$$
\begin{aligned}
x + y &= 20, &\textbf{(1)}\\
25x + 75y &= 900. &\textbf{(2)}
\end{aligned}
$$

**3. Carry out.** We carry out the solution as follows.

## Algebraic Solution

Using the elimination method, we multiply equation (1) by $-25$ and add it to equation (2):

$$
\begin{aligned}
-25x - 25y &= -500\\
25x + 75y &= 900\\
\hline
50y &= 400\\
y &= 8.
\end{aligned}
$$

Then we back-substitute to find $x$:

$$
\begin{aligned}
x + y &= 20 &\textbf{(1)}\\
x + 8 &= 20 &\text{Substituting 8 for } y\\
x &= 12.
\end{aligned}
$$

## Graphical Solution

We solve each equation for $y$, getting

$$y = 20 - x \quad \text{and} \quad y = \frac{900 - 25x}{75}.$$

Next, we graph these equations and find the point of intersection of the graphs.

$$y_1 = 20 - x, \quad y_2 = \frac{900 - 25x}{75}$$

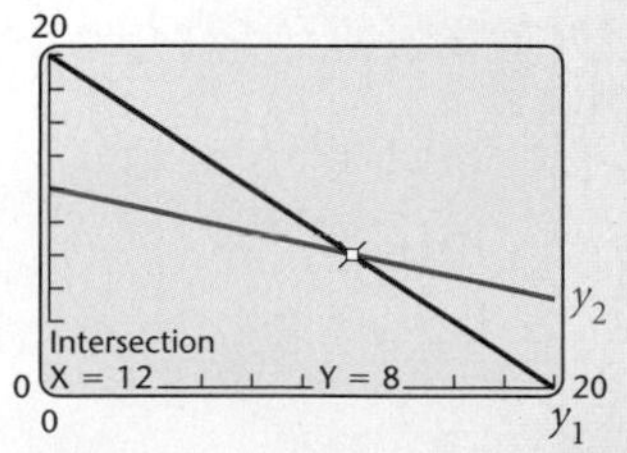

The graphs intersect at the point $(12, 8)$, so the possible solution is $(12, 8)$.

4. **Check.** If 12 lb of caramel corn and 8 lb of nuts are used, the mixture weighs $12 + 8$, or 20 lb. The value of the mixture is $2.50(12) + 7.50(8)$, or $30 + 60$, or $90. Since the possible solution yields the desired weight and value of the mixture, our result checks.

5. **State.** The mixture should consist of 12 lb of caramel corn and 8 lb of honey roasted mixed nuts.

**Now Try Exercise 63.**

**EXAMPLE 7** *Airplane Travel.* An airplane flies the 3000-mi distance from Los Angeles to New York, with a tailwind, in 5 hr. The return trip, against the wind, takes 6 hr. Find the speed of the airplane and the speed of the wind.

*Solution*

1. **Familiarize.** We first make a drawing, letting $p = $ the speed of the plane, in miles per hour, and $w = $ the speed of the wind, also in miles per hour. When the plane is traveling with a tailwind, the wind increases the speed of the plane, so the speed with the tailwind is $p + w$. On the other hand, the headwind slows the plane down, so the speed with the headwind is $p - w$.

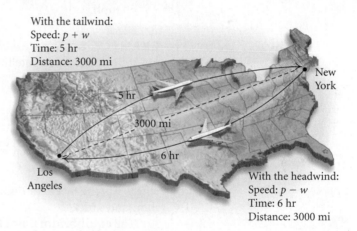

With the tailwind:
Speed: $p + w$
Time: 5 hr
Distance: 3000 mi

5 hr

3000 mi

6 hr

New York

Los Angeles

With the headwind:
Speed: $p - w$
Time: 6 hr
Distance: 3000 mi

2. **Translate.** We organize the information in a table. Using the formula *Distance* = *Rate* (or *Speed*) · *Time*, we find that each row of the table yields an equation.

|  | Distance | Rate | Time |  |
|---|---|---|---|---|
| **With Tailwind** | 3000 | $p + w$ | 5 | $\rightarrow 3000 = (p + w)5$ |
| **With Headwind** | 3000 | $p - w$ | 6 | $\rightarrow 3000 = (p - w)6$ |

We now have a system of equations:

$$3000 = (p + w)5, \qquad 600 = p + w, \quad (1) \qquad \text{Dividing by 5}$$
$$3000 = (p - w)6, \quad \text{or} \quad 500 = p - w. \quad (2) \qquad \text{Dividing by 6}$$

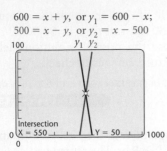

$600 = x + y$, or $y_1 = 600 - x$;
$500 = x - y$, or $y_2 = x - 500$

**3. Carry out.** We use the elimination method:

$$600 = p + w \quad \textbf{(1)}$$
$$\underline{500 = p - w} \quad \textbf{(2)}$$
$$1100 = 2p \qquad \text{Adding}$$
$$550 = p. \qquad \text{Dividing by 2 on both sides}$$

Now we substitute in one of the equations to find $w$:

$$600 = p + w \qquad \textbf{(1)}$$
$$600 = 550 + w \qquad \text{Substituting 550 for } p$$
$$50 = w. \qquad \text{Subtracting 550 on both sides}$$

**4. Check.** If $p = 550$ and $w = 50$, then the speed of the plane with the tailwind is $550 + 50$, or 600 mph, and the speed with the headwind is $550 - 50$, or 500 mph. At 600 mph, the time it takes to travel 3000 mi is $3000/600$, or 5 hr. At 500 mph, the time it takes to travel 3000 mi is $3000/500$, or 6 hr. The times check, so the answer is correct.

**5. State.** The speed of the plane is 550 mph, and the speed of the wind is 50 mph.

> **Now Try Exercise 67.**

**EXAMPLE 8** *Supply and Demand.* Suppose that the price and the supply of the Star Station satellite radio are related by the equation

$$y = 90 + 30x,$$

where $y$ is the price, in dollars, at which the seller is willing to supply $x$ thousand units. Also suppose that the price and the demand for the same model of satellite radio are related by the equation

$$y = 200 - 25x,$$

where $y$ is the price, in dollars, at which the consumer is willing to buy $x$ thousand units.

The **equilibrium point** for this radio is the pair $(x, y)$ that is a solution of both equations. The **equilibrium price** is the price at which the amount of the product that the seller is willing to supply is the same as the amount demanded by the consumer. Find the equilibrium point for this radio.

*Solution*

**1., 2. Familiarize** and **Translate.** We are given a system of equations in the statement of the problem, so no further translation is necessary.

$$y = 90 + 30x, \qquad \textbf{(1)}$$
$$y = 200 - 25x. \qquad \textbf{(2)}$$

We substitute some values for $x$ in each equation to get an idea of the corresponding prices. When $x = 1$,

$$y = 90 + 30 \cdot 1 = 120, \qquad \text{Substituting in equation (1)}$$
$$y = 200 - 25 \cdot 1 = 175. \qquad \text{Substituting in equation (2)}$$

This indicates that the price when 1 thousand units are supplied is lower than the price when 1 thousand units are demanded.

When $x = 4$,

$y = 90 + 30 \cdot 4 = 210,$     **Substituting in equation (1)**

$y = 200 - 25 \cdot 4 = 100.$     **Substituting in equation (2)**

In this case, the price related to supply is higher than the price related to demand. It would appear that the $x$-value we are looking for is between 1 and 4.

**3. Carry out.** We use the substitution method:

$$y = 90 + 30x \qquad \text{Equation (1)}$$

$$200 - 25x = 90 + 30x \qquad \text{Substituting } 200 - 25x \text{ for } y \text{ in equation (1)}$$

$$110 = 55x \qquad \text{Adding } 25x \text{ and subtracting } 90 \text{ on both sides}$$

$$2 = x. \qquad \text{Dividing by } 55 \text{ on both sides}$$

We now back-substitute 2 for $x$ in either equation and find $y$:

$$y = 200 - 25x \qquad \text{(2)}$$
$$= 200 - 25 \cdot 2 \qquad \text{Substituting 2 for } x$$
$$= 200 - 50$$
$$= 150.$$

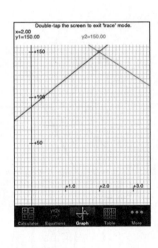

We can visualize the solution as the coordinates of the point of intersection of the graphs of the equations $y = 90 + 30x$ and $y = 200 - 25x$, as shown with an app at left.

**4. Check.** We can check by substituting 2 for $x$ and 150 for $y$ in both equations. Also note that 2 is between 1 and 4 as expected from the *Familiarize* and *Translate* steps.

**5. State.** The equilibrium point is $(2, \$150)$. That is, the equilibrium quantity is 2 thousand units and the equilibrium price is $150.

**Now Try Exercise 71.**

**A**
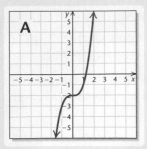

**B**
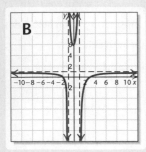

**C**
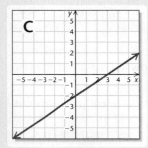

**D**
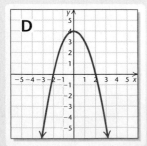

**E**
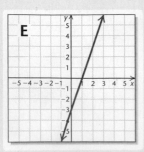

# Visualizing the Graph

Match the equation or system of equations with its graph.

**1.** $2x - 3y = 6$

**2.** $f(x) = x^2 - 2x - 3$

**3.** $f(x) = -x^2 + 4$

**4.** $(x - 2)^2 + (y + 3)^2 = 9$

**5.** $f(x) = x^3 - 2$

**6.** $f(x) = -(x - 1)^2(x + 1)^2$

**7.** $f(x) = \dfrac{x - 1}{x^2 - 4}$

**8.** $f(x) = \dfrac{x^2 - x - 6}{x^2 - 1}$

**9.** $\begin{aligned} x - y &= -1, \\ 2x - y &= 2 \end{aligned}$

**10.** $\begin{aligned} 3x - y &= 3, \\ 2y &= 6x - 6 \end{aligned}$

Answers on page A-38

**F**
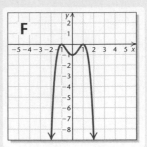

**G**
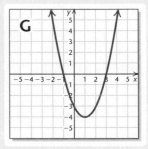

**H**
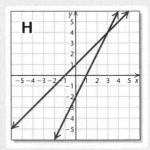

**I**
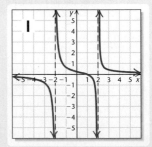

**J**
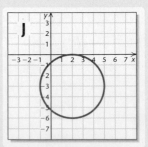

## 6.1 Exercise Set

*In Exercises 1–6, match the system of equations with one of the graphs (a)–(f), which follow.*

**a)**

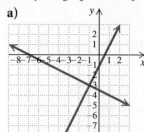

**b)**

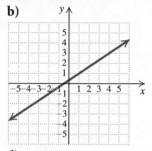

**c)**

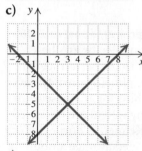

**d)**

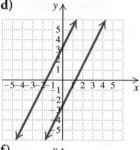

**e)**

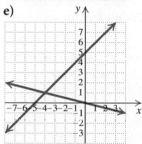

**f)**

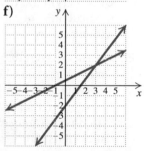

**1.** $x + y = -2,$
$\quad y = x - 8$

**2.** $x - y = -5,$
$\quad x = -4y$

**3.** $x - 2y = -1,$
$\quad 4x - 3y = 6$

**4.** $2x - y = 1,$
$\quad x + 2y = -7$

**5.** $2x - 3y = -1,$
$\quad -4x + 6y = 2$

**6.** $4x - 2y = 5,$
$\quad 6x - 3y = -10$

*Solve graphically.*

**7.** $x + y = 2,$
$\quad 3x + y = 0$

**8.** $x + y = 1,$
$\quad 3x + y = 7$

**9.** $x + 2y = 1,$
$\quad x + 4y = 3$

**10.** $3x + 4y = 5,$
$\quad x - 2y = 5$

**11.** $y + 1 = 2x,$
$\quad y - 1 = 2x$

**12.** $2x - y = 1,$
$\quad 3y = 6x - 3$

**13.** $x - y = -6,$
$\quad y = -2x$

**14.** $2x + y = 5,$
$\quad x = -3y$

**15.** $2y = x - 1,$
$\quad 3x = 6y + 3$

**16.** $y = 3x + 2,$
$\quad 3x - y = -3$

*Solve using the substitution method. Use a graphing calculator to check your answer.*

**17.** $x + y = 9,$
$\quad 2x - 3y = -2$

**18.** $3x - y = 5,$
$\quad x + y = \frac{1}{2}$

**19.** $x - 2y = 7,$
$\quad x = y + 4$

**20.** $x + 4y = 6,$
$\quad x = -3y + 3$

**21.** $y = 2x - 6,$
$\quad 5x - 3y = 16$

**22.** $3x + 5y = 2,$
$\quad 2x - y = -3$

**23.** $x + y = 3,$
$\quad y = 4 - x$

**24.** $x - 2y = 3,$
$\quad 2x = 4y + 6$

**25.** $x - 5y = 4,$
$\quad y = 7 - 2x$

**26.** $5x + 3y = -1,$
$\quad x + y = 1$

**27.** $x + 2y = 2,$
$\quad 4x + 4y = 5$

**28.** $2x - y = 2,$
$\quad 4x + y = 3$

**29.** $3x - y = 5,$
$\quad 3y = 9x - 15$

**30.** $2x - y = 7,$
$\quad y = 2x - 5$

*Solve using the elimination method. Also determine whether each system is consistent or inconsistent and whether the equations are dependent or independent. Use a graphing calculator to check your answer.*

**31.** $x + 2y = 7,$
$\quad x - 2y = -5$

**32.** $3x + 4y = -2,$
$\quad -3x - 5y = 1$

**33.** $x - 3y = 2,$
$\quad 6x + 5y = -34$

**34.** $x + 3y = 0,$
$\quad 20x - 15y = 75$

**35.** $3x - 12y = 6,$
$\quad 2x - 8y = 4$

**36.** $2x + 6y = 7,$
$\quad 3x + 9y = 10$

**37.** $4x - 2y = 3,$
$\quad 2x - y = 4$

**38.** $6x + 9y = 12,$
$\quad 4x + 6y = 8$

**39.** $2x = 5 - 3y,$
$\quad 4x = 11 - 7y$

**40.** $7(x - y) = 14,$
$\quad 2x = y + 5$

**41.** $0.3x - 0.2y = -0.9,$
$\quad 0.2x - 0.3y = -0.6$
(*Hint*: Since each coefficient has one decimal place, first multiply each equation by 10 to clear the decimals.)

**42.** $0.2x - 0.3y = 0.3,$
$0.4x + 0.6y = -0.2$
(*Hint*: Since each coefficient has one decimal place, first multiply each equation by 10 to clear the decimals.)

**43.** $\frac{1}{5}x + \frac{1}{2}y = 6,$
$\frac{3}{5}x - \frac{1}{2}y = 2$
(*Hint*: First multiply by the least common denominator to clear fractions.)

**44.** $\frac{2}{3}x + \frac{3}{5}y = -17,$
$\frac{1}{2}x - \frac{1}{3}y = -1$
(*Hint*: First multiply by the least common denominator to clear fractions.)

*In Exercises 45–50, determine whether the statement is true or false.*

**45.** If the graph of a system of equations is a pair of parallel lines, the system of equations is inconsistent.

**46.** If we obtain the equation $0 = 0$ when using the elimination method to solve a system of equations, the system has no solution.

**47.** If a system of two linear equations in two variables is consistent, it has exactly one solution.

**48.** If a system of two linear equations in two variables is dependent, it has infinitely many solutions.

**49.** It is possible for a system of two linear equations in two variables to be consistent and dependent.

**50.** It is possible for a system of two linear equations in two variables to be inconsistent and dependent.

**51.** *Joint-Replacement Surgery.*   It is estimated that in 2030, there will be a total of 4.072 million knee replacement and hip replacement surgeries, with knee replacements outnumbering hip replacements by 2.928 million (*Source: Indianapolis Star* research). Find the estimated number of each type of joint replacement.

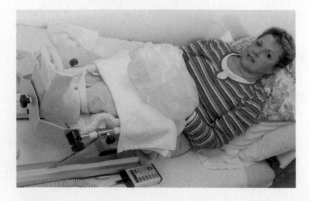

**52.** *Street Names.*   The two most common names for streets in the United States are Second Street and Main Street, with 15,684 streets bearing one of these names. There are 260 more streets named Second Street than Main Street. (*Source*: 2006 Tele Atlas digital map data) How many streets bear each name?

**53.** *Consumer Complaints.*   In 2009, the Better Business Bureau received the most complaints from consumers about cell-phone providers and cable/satellite TV providers. These two industries accounted for a total of 70,093 complaints, with cable/satellite TV providers receiving 4861 fewer complaints than cell-phone providers (*Source*: Better Business Bureau). How many complaints did each industry receive?

**54.** *Health-Care Costs.*   It is projected that in 2018, the average total of medical bills for an obese adult will be $2460 higher than the average for an adult with a healthy weight. Together, the average amount of medical bills for an obese adult and an adult with a healthy weight are projected to total $14,170. (*Source*: Kenneth Thorpe, Emory University) Find the projected average amount of medical bills for an obese adult and for an adult with a healthy weight.

**55.** *Amish.*   About 250,000 Amish live in the United States. Of that number, 73,950 live in Wisconsin and Ohio. The number of Amish in Ohio is 12,510 more than three times the number in Wisconsin. (*Source:* Young Center for Anabaptist & Pietist Studies, Elizabethtown College) How many Amish live in each state?

**56.** *Added Sugar in Soft Drinks.*   Together, a 20-fl oz Coca-Cola and a 20-fl oz Pepsi contain 34 tsp of added sugar. The Coca-Cola contains 1 tsp less of added sugar than the Pepsi. (*Source*: *Nutrition Action HealthLetter*, January/February, 2010) Find the amount of added sugar in each type of soft drink.

57. *Calories in Pie.* Using the calorie count of one-eighth of a 9-in. pie, we find that the number of calories in a piece of pecan pie is 221 less than twice the number of calories in a piece of lemon meringue pie. If one eats a piece of each, a total of 865 calories is consumed. (*Source*: *Good Housekeeping*, Good Health, p. 41, November 2007) How many calories are there in each piece of pie?

58. *Refugees in the United States.* In 2010, the number of refugees from Africa and Near East/South Asia was 49,087. The number from Near East/South Asia was 4133 less than three times the number of refugees from Africa. (*Source*: U.S. Refugee Admissions Report) How many refugees were from Near East/South Asia and how many from Africa?

59. *Mail-Order Business.* A mail-order garden-equipment business shipped 120 packages one day. Customers were charged $3.50 for each standard-delivery package and $7.50 for each express-delivery package. Total shipping charges for the day were $596. How many of each kind of package were shipped?

60. *Concert Ticket Prices.* One evening 1500 concert tickets were sold for the Fairmont Summer Jazz Festival. Tickets cost $25 for a covered pavilion seat and $15 for a lawn seat. Total receipts were $28,500. How many of each type of ticket were sold?

61. *Investment.* Quinn inherited $15,000 and invested it in two municipal bonds that pay 4% and 5% simple interest. The annual interest is $690. Find the amount invested at each rate.

62. *Tee Shirt Sales.* Turi's Tee Shirt Shack sold 36 shirts one day. Each short-sleeved tee shirt costs $12, and each long-sleeved tee shirt costs $18. Total receipts for the day were $522. How many of each kind of shirt were sold?

63. *Coffee Mixtures.* The owner of The Daily Grind coffee shop mixes French roast coffee worth $9.00 per pound with Kenyan coffee worth $7.50 per pound in order to get 10 lb of a mixture worth $8.40 per pound. How much of each type of coffee was used?

64. *Commissions.* Bentley's Office Solutions offers its sales representatives a choice between being paid a commission of 8% of sales or being paid a monthly salary of $1500 plus a commission of 1% of sales. For what monthly sales do the two plans pay the same amount?

65. *Nutrition.* A one-cup serving of spaghetti with meatballs contains 260 Cal (calories) and 32 g of carbohydrates. A one-cup serving of chopped iceberg lettuce contains 5 Cal and 1 g of carbohydrates. (*Source*: *Home and Garden Bulletin No. 72*, U.S. Government Printing Office,

Washington, D.C. 20402) How many servings of each would be required in order to obtain 400 Cal and 50 g of carbohydrates?

66. *Nutrition.* One serving of tomato soup contains 100 Cal and 18 g of carbohydrates. One slice of whole wheat bread contains 70 Cal and 13 g of carbohydrates. (*Source*: *Home and Garden Bulletin No. 72*, U.S. Government Printing Office, Washington, D.C. 20402) How many servings of each would be required in order to obtain 230 Cal and 42 g of carbohydrates?

67. *Motion.* A Leisure Time Cruises riverboat travels 46 km downstream in 2 hr. It travels 51 km upstream in 3 hr. Find the speed of the boat and the speed of the stream.

68. *Motion.* A DC10 airplane travels 3000 km with a tailwind in 3 hr. It travels 3000 km with a headwind in 4 hr. Find the speed of the plane and the speed of the wind.

69. *Motion.* Two private airplanes travel toward each other from cities that are 780 km apart at speeds of 190 km/h and 200 km/h. They leave at the same time. In how many hours will they meet?

70. *Motion.* Christopher's boat travels 45 mi downstream in 3 hr. The return trip upstream takes 5 hr. Find the speed of the boat in still water and the speed of the current.

71. *Supply and Demand.* The supply and demand for a video-game system are related to price by the equations

$$y = 70 + 2x,$$
$$y = 175 - 5x,$$

respectively, where $y$ is the price, in dollars, and $x$ is the number of units, in thousands. Find the equilibrium point for this product.

72. *Supply and Demand.* The supply and demand for a particular model of treadmill are related to price by the equations

$$y = 240 + 40x,$$
$$y = 500 - 25x,$$

respectively, where $y$ is the price, in dollars, and $x$ is the number of units, in thousands. Find the equilibrium point for this product.

*The point at which a company's costs equal its revenues is the* **break-even point**. *In Exercises 73–76, C represents the production cost, in dollars, of x units of a product and R represents the revenue, in dollars, from the sale of x units. Find the number of units that must be produced and sold in order to break even. That is, find the value of x for which C = R.*

73. $C = 14x + 350,$
    $R = 16.5x$

74. $C = 8.5x + 75,$
    $R = 10x$

75. $C = 15x + 12,000,$
    $R = 18x - 6000$

76. $C = 3x + 400,$
    $R = 7x - 600$

77. *Red Meat and Poultry Consumption.* The amount of red meat consumed in the United States has remained fairly constant in recent years while the amount of poultry consumed has increased during those years, as shown by the data in the table below.

| Year | U.S. Red Meat Consumption (in billions of pounds) | U.S. Poultry Consumption (in billions of pounds) |
|------|------|------|
| 1995 | 43.967 | 25.944 |
| 2000 | 46.560 | 30.508 |
| 2005 | 47.385 | 34.947 |
| 2009 | 47.227 | 34.116 |

*Source*: U.S. Department of Agriculture, Economic Research Service, *Food Consumption, Prices, and Expenditures*

**a)** Find linear regression functions $r(x)$ and $p(x)$ that represent red meat consumption and poultry consumption, respectively, in billions of pounds $x$ years after 1995.

**b)** Use the functions found in part (a) to estimate when poultry consumption will equal red meat consumption.

**78.** *Video Rentals.* The table below lists the revenue, in billions of dollars, from video rentals in video stores and with subscriptions in recent years.

| Year, $x$ | Video Rentals in Video Stores, $v(x)$ (in billions) | Video Rentals with Subscriptions, $s(x)$ (in billions) |
|---|---|---|
| 2005 | $6.1 | $1.4 |
| 2006 | 5.8 | 1.7 |
| 2007 | 4.9 | 2.0 |
| 2008 | 4.3 | 2.1 |
| 2009 | 3.3 | 2.2 |

*Source: Screen Digest*

**a)** Find linear functions $v(x)$ and $s(x)$ that represent the revenues from video rentals in video stores and with subscriptions $x$ years since 2005.

**b)** Use the functions found in part (a) to estimate when revenues from video rentals in video stores and with subscription services will be equal.

## Skill Maintenance

**79.** *Decline in Air Travel.* The number of people flying to, from, and within the United States in 2009 was at its lowest level since 2004. In 2009, the number of air travelers totaled 769.6 million, a decrease of 8.2% from 2004 (*Source:* Bureau of Transportation Statistics, Department of Transportation). Find the number of air travelers in 2004.

**80.** *Internet Retail Sales.* Online retail sales totaled $133.6 billion in 2008. This was about three times the total in 2002. (*Source:* U.S. Census Bureau) Find the online sales total in 2002.

*Consider the function*
$$f(x) = x^2 - 4x + 3$$
*in Exercises 81–84.*

**81.** What are the inputs if the output is 15?

**82.** Given an output of 8, find the corresponding inputs.

**83.** What is the output if the input is $-9$?

**84.** Find the zeros of the function.

## Synthesis

**85.** *Motion.* Nancy jogs and walks to campus each day. She averages 4 km/h walking and 8 km/h jogging. The distance from home to the campus is 6 km, and she makes the trip in 1 hr. How far does she jog on each trip?

**86.** *e-Commerce.* Shirts.com advertises a limited-time sale, offering 1 turtleneck for $15 and 2 turtlenecks for $25. A total of 1250 turtlenecks are sold and $16,750 is taken in. How many customers ordered 2 turtlenecks?

**87.** *Motion.* A train leaves Union Station for Central Station, 216 km away, at 9 A.M. One hour later, a train leaves Central Station for Union Station. They meet at noon. If the second train had started at 9 A.M. and the first train at 10:30 A.M., they would still have met at noon. Find the speed of each train.

**88.** *Antifreeze Mixtures.* An automobile radiator contains 16 L of antifreeze and water. This mixture is 30% antifreeze. How much of this mixture should be drained and replaced with pure antifreeze so that the final mixture will be 50% antifreeze?

**89.** Two solutions of the equation $Ax + By = 1$ are $(3, -1)$ and $(-4, -2)$. Find $A$ and $B$.

**90.** *Ticket Line.* You are in line at a ticket window. There are 2 more people ahead of you in line than there are behind you. In the entire line, there are three times as many people as there are behind you. How many people are ahead of you?

**91.** *Gas Mileage.* The Honda Civic Hybrid vehicle, powered by gasoline–electric technology, gets 49 miles per gallon (mpg) in city driving and 51 mpg in highway driving (*Source:* Honda.com). The car is driven 447 mi on 9 gal of gasoline. How many miles were driven in the city and how many were driven on the highway?

## 6.2

# Systems of Equations in Three Variables

- Solve systems of linear equations in three variables.
- Use systems of three equations to solve applied problems.
- Model a situation using a quadratic function.

A **linear equation in three variables** is an equation equivalent to one of the form $Ax + By + Cz = D$, where $A$, $B$, $C$, and $D$ are real numbers and none of $A$, $B$, and $C$ is 0. A **solution of a system of three equations in three variables** is an ordered triple that makes all three equations true. For example, the triple $(2, -1, 0)$ is a solution of the system of equations

$$4x + 2y + 5z = 6,$$
$$2x - y + z = 5,$$
$$3x + 2y - z = 4.$$

We can verify this by substituting 2 for $x$, $-1$ for $y$, and 0 for $z$ in each equation.

### ■ Solving Systems of Equations in Three Variables

We will solve systems of equations in three variables using an algebraic method called **Gaussian elimination**, named for the German mathematician Karl Friedrich Gauss (1777–1855). Our goal is to transform the original system to an equivalent system (one with the same solution set) of the form

$$Ax + By + Cz = D,$$
$$Ey + Fz = G,$$
$$Hz = K.$$

Then we solve the third equation for $z$ and back-substitute to find $y$ and then $x$.

Each of the following operations can be used to transform the original system to an equivalent system in the desired form.

---

1. Interchange any two equations.
2. Multiply on both sides of one of the equations by a nonzero constant.
3. Add a nonzero multiple of one equation to another equation.

---

**EXAMPLE 1** Solve the following system:

$$x - 2y + 3z = 11, \quad \textbf{(1)}$$
$$4x + 2y - 3z = 4, \quad \textbf{(2)}$$
$$3x + 3y - z = 4. \quad \textbf{(3)}$$

***Solution*** First, we choose one of the variables to eliminate using two different pairs of equations. Let's eliminate $x$ from equations (2) and (3). We multiply equation (1) by $-4$ and add it to equation (2). We also multiply equation (1) by $-3$ and add it to equation (3).

$$
\begin{array}{ll}
-4x + 8y - 12z = -44 & \text{Multiplying (1) by } -4 \\
\underline{4x + 2y - 3z = 4} & \textbf{(2)} \\
10y - 15z = -40; & \textbf{(4)}
\end{array}
$$

$$
\begin{array}{ll}
-3x + 6y - 9z = -33 & \text{Multiplying (1) by } -3 \\
\underline{3x + 3y - z = 4} & \textbf{(3)} \\
9y - 10z = -29. & \textbf{(5)}
\end{array}
$$

Now we have

$$
\begin{array}{ll}
x - 2y + 3z = 11, & \textbf{(1)} \\
10y - 15z = -40, & \textbf{(4)} \\
9y - 10z = -29. & \textbf{(5)}
\end{array}
$$

Next, we multiply equation (5) by 10 to make the $y$-coefficient a multiple of the $y$-coefficient in the equation above it:

$$
\begin{array}{ll}
x - 2y + 3z = 11, & \textbf{(1)} \\
10y - 15z = -40, & \textbf{(4)} \\
90y - 100z = -290. & \textbf{(6)}
\end{array}
$$

Next, we multiply equation (4) by $-9$ and add it to equation (6):

$$
\begin{array}{ll}
-90y + 135z = 360 & \text{Multiplying (4) by } -9 \\
\underline{90y - 100z = -290} & \textbf{(6)} \\
35z = 70. & \textbf{(7)}
\end{array}
$$

We now have the system of equations

$$
\begin{array}{ll}
x - 2y + 3z = 11, & \textbf{(1)} \\
10y - 15z = -40, & \textbf{(4)} \\
35z = 70. & \textbf{(7)}
\end{array}
$$

Now we solve equation (7) for $z$:

$$
\begin{aligned}
35z &= 70 \\
z &= 2.
\end{aligned}
$$

Then we back-substitute 2 for $z$ in equation (4) and solve for $y$:

$$
\begin{aligned}
10y - 15 \cdot 2 &= -40 \\
10y - 30 &= -40 \\
10y &= -10 \\
y &= -1.
\end{aligned}
$$

Finally, we back-substitute $-1$ for $y$ and 2 for $z$ in equation (1) and solve for $x$:

$$x - 2(-1) + 3 \cdot 2 = 11$$
$$x + 2 + 6 = 11$$
$$x = 3.$$

We can check the triple $(3, -1, 2)$ in each of the three original equations. Since it makes all three equations true, the solution is $(3, -1, 2)$.

**Now Try Exercise 1.**

**EXAMPLE 2**   Solve the following system:

$$x + \phantom{2}y + \phantom{2}z = 7, \qquad \textbf{(1)}$$
$$3x - 2y + \phantom{2}z = 3, \qquad \textbf{(2)}$$
$$x + 6y + 3z = 25. \qquad \textbf{(3)}$$

***Solution***   We multiply equation (1) by $-3$ and add it to equation (2). We also multiply equation (1) by $-1$ and add it to equation (3).

$$x + y + \phantom{2}z = 7, \qquad \textbf{(1)}$$
$$-5y - 2z = -18, \qquad \textbf{(4)}$$
$$5y + 2z = 18 \qquad \textbf{(5)}$$

Next, we add equation (4) to equation (5):

$$x + y + \phantom{2}z = 7, \qquad \textbf{(1)}$$
$$-5y - 2z = -18, \qquad \textbf{(4)}$$
$$0 = 0.$$

The equation $0 = 0$ tells us that equations (1), (2), and (3) are dependent. This means that the original system of three equations is equivalent to a system of two equations. One way to see this is to observe that four times equation (1) minus equation (2) is equation (3). Thus removing equation (3) from the system does not affect the solution of the system. We can say that the original system is equivalent to

$$x + \phantom{2}y + z = 7, \qquad \textbf{(1)}$$
$$3x - 2y + z = 3. \qquad \textbf{(2)}$$

In this particular case, the original system has infinitely many solutions. (In some cases, a system containing dependent equations is inconsistent.) To find an expression for these solutions, we first solve equation (4) for either $y$ or $z$. We choose to solve for $y$:

$$-5y - 2z = -18 \qquad \textbf{(4)}$$
$$-5y = 2z - 18$$
$$y = -\tfrac{2}{5}z + \tfrac{18}{5}.$$

Then we back-substitute in equation (1) to find an expression for $x$ in terms of $z$:

$$x - \tfrac{2}{5}z + \tfrac{18}{5} + z = 7 \qquad \text{Substituting } -\tfrac{2}{5}z + \tfrac{18}{5} \text{ for } y$$
$$x + \tfrac{3}{5}z + \tfrac{18}{5} = 7$$
$$x + \tfrac{3}{5}z = \tfrac{17}{5}$$
$$x = -\tfrac{3}{5}z + \tfrac{17}{5}.$$

The solutions of the system of equations are ordered triples of the form $\left(-\tfrac{3}{5}z + \tfrac{17}{5}, -\tfrac{2}{5}z + \tfrac{18}{5}, z\right)$, where $z$ can be any real number. Any real number that we use for $z$ then gives us values for $x$ and $y$ and thus an ordered triple in the solution set. For example, if we choose $z = 0$, we have the solution $\left(\tfrac{17}{5}, \tfrac{18}{5}, 0\right)$. If we choose $z = -1$, we have $(4, 4, -1)$.

**Now Try Exercise 9.**

If we get a false equation, such as $0 = -5$, at some stage of the elimination process, we conclude that the original system is *inconsistent*; that is, it has no solutions.

Although systems of three linear equations in three variables do not lend themselves well to graphical solutions, it is of interest to picture some possible solutions. The graph of a linear equation in three variables is a plane. Thus the solution set of such a system is the intersection of three planes. Some possibilities are shown below.

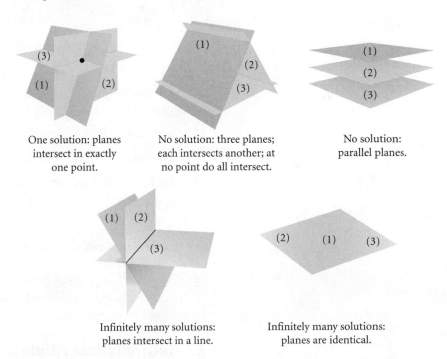

One solution: planes intersect in exactly one point.

No solution: three planes; each intersects another; at no point do all intersect.

No solution: parallel planes.

Infinitely many solutions: planes intersect in a line.

Infinitely many solutions: planes are identical.

## ■ Applications

Systems of equations in three or more variables allow us to solve many problems in fields such as business, the social and natural sciences, and engineering.

**EXAMPLE 3** *Investment.* Moira inherited $15,000 and invested part of it in a money market account, part in municipal bonds, and part in a mutual fund. After 1 year, she received a total of $730 in simple interest from the three investments. The money market account paid 4% annually, the bonds paid 5% annually, and the mutual fund paid 6% annually. There was $2000 more invested in the mutual fund than in bonds. Find the amount that Moira invested in each category.

*Solution*

1. **Familiarize.** We let $x$, $y$, and $z$ represent the amounts invested in the money market account, the bonds, and the mutual fund, respectively. Then the amounts of income produced annually by each investment are given by 4%$x$, 5%$y$, and 6%$z$, or $0.04x$, $0.05y$, and $0.06z$.

2. **Translate.** The fact that a total of $15,000 is invested gives us one equation:

   $$x + y + z = 15,000.$$

   Since the total interest is $730, we have a second equation:

   $$0.04x + 0.05y + 0.06z = 730.$$

   Another statement in the problem gives us a third equation.

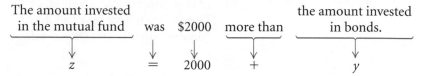

The amount invested in the mutual fund    was    $2000    more than    the amount invested in bonds.
$z$ $=$ $2000$ $+$ $y$

   We now have a system of three equations:

   $$\begin{aligned} x + y + z &= 15,000, & x + y + z &= 15,000, \\ 0.04x + 0.05y + 0.06z &= 730, & \text{or} \quad 4x + 5y + 6z &= 73,000, \\ z &= 2000 + y; & -y + z &= 2000. \end{aligned}$$

3. **Carry out.** Solving the system of equations, we get

   $$(7000, 3000, 5000).$$

4. **Check.** The sum of the numbers is 15,000. The income produced is

   $$0.04(7000) + 0.05(3000) + 0.06(5000) = 280 + 150 + 300, \quad \text{or} \quad \$730.$$

   Also the amount invested in the mutual fund, $5000, is $2000 more than the amount invested in bonds, $3000. Our solution checks in the original problem.

5. **State.** Moira invested $7000 in a money market account, $3000 in municipal bonds, and $5000 in a mutual fund.

   **Now Try Exercise 29.**

## ▪ Mathematical Models and Applications

Recall that when we model a situation using a linear function $f(x) = mx + b$, we need to know two data points in order to determine $m$ and $b$. For a quadratic model, $f(x) = ax^2 + bx + c$, we need three data points in order to determine $a$, $b$, and $c$.

**EXAMPLE 4** *Snow-Sports Sales.* The table below lists sales of winter sports equipment, apparel, and accessories in November of three recent years. Use the data to find a quadratic function that gives the snow-sports sales as a function of the number of years after 2006. Then use the function to estimate snow-sports sales in 2009.

| YEAR, $x$ | SNOW-SPORTS SALES (in millions) |
|---|---|
| 2006, 0 | $296 |
| 2007, 1 | 423 |
| 2008, 2 | 402 |

*Source:* SIA Retail Audit

**Solution** We let $x =$ the number of years after 2006 and $s(x) =$ snow-sports sales. Then $x = 0$ corresponds to 2006, $x = 1$ corresponds to 2007, and $x = 2$ corresponds to 2008. We use the three data points $(0, 296)$, $(1, 423)$, and $(2, 402)$ to find $a$, $b$, and $c$ in the function $s(x) = ax^2 + bx + c$.

First, we substitute:

$$s(x) = ax^2 + bx + c$$

For $(0, 296)$:  $296 = a \cdot 0^2 + b \cdot 0 + c,$

For $(1, 423)$:  $423 = a \cdot 1^2 + b \cdot 1 + c,$

For $(2, 402)$:  $402 = a \cdot 2^2 + b \cdot 2 + c.$

We now have a system of equations in the variables $a$, $b$, and $c$:

$$c = 296,$$
$$a + b + c = 423,$$
$$4a + 2b + c = 402.$$

Solving this system, we get $(-74, 201, 296)$. Thus,

$$s(x) = -74x^2 + 201x + 296.$$

To estimate snow-sports sales in 2009, we find $s(3)$, since 2009 is 3 years after 2006:

$$s(3) = -74 \cdot 3^2 + 201 \cdot 3 + 296$$
$$= \$233 \text{ million.}$$

**Now Try Exercise 33.**

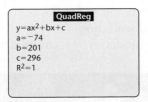

```
        QuadReg
y=ax²+bx+c
a=−74
b=201
c=296
R²=1
```

```
Y₁(3)
                    233
```

The function in Example 4 can also be found using the QUADRATIC REGRESSION feature on a graphing calculator. Note that the method of Example 4 works when we have exactly three data points, whereas the QUADRATIC REGRESSION feature on a graphing calculator can be used for *three or more* points.

## 6.2   Exercise Set

*Solve the system of equations.*

**1.** $x + y + z = 2,$
$6x - 4y + 5z = 31,$
$5x + 2y + 2z = 13$

**2.** $x + 6y + 3z = 4,$
$2x + y + 2z = 3,$
$3x - 2y + z = 0$

**3.** $x - y + 2z = -3,$
$x + 2y + 3z = 4,$
$2x + y + z = -3$

**4.** $x + y + z = 6,$
$2x - y - z = -3,$
$x - 2y + 3z = 6$

**5.** $x + 2y - z = 5,$
$2x - 4y + z = 0,$
$3x + 2y + 2z = 3$

**6.** $2x + 3y - z = 1,$
$x + 2y + 5z = 4,$
$3x - y - 8z = -7$

**7.** $x + 2y - z = -8,$
$2x - y + z = 4,$
$8x + y + z = 2$

**8.** $x + 2y - z = 4,$
$4x - 3y + z = 8,$
$5x - y = 12$

**9.** $2x + y - 3z = 1,$
$x - 4y + z = 6,$
$4x - 7y - z = 13$

**10.** $x + 3y + 4z = 1,$
$3x + 4y + 5z = 3,$
$x + 8y + 11z = 2$

**11.** $4a + 9b = 8,$
$8a + 6c = -1,$
$6b + 6c = -1$

**12.** $3p + 2r = 11,$
$q - 7r = 4,$
$p - 6q = 1$

**13.** $2x + z = 1,$
$3y - 2z = 6,$
$x - 2y = -9$

**14.** $3x + 4z = -11,$
$x - 2y = 5,$
$4y - z = -10$

**15.** $w + x + y + z = 2,$
$w + 2x + 2y + 4z = 1,$
$-w + x - y - z = -6,$
$-w + 3x + y - z = -2$

**16.** $w + x - y + z = 0,$
$-w + 2x + 2y + z = 5,$
$-w + 3x + y - z = -4,$
$-2w + x + y - 3z = -7$

**17.** *Winter Olympic Sites.* The Winter Olympics have been held a total of 21 times on the continents of North America, Europe, and Asia. The number of European sites is 5 more than the total number of sites in North America and Asia. There are 4 more sites in North America than in Asia.

(*Source: USA Today* research) Find the number of Winter Olympic sites on each continent.

**18.** *Restaurant Meals.* The total number of restaurant-purchased meals that the average person will eat in a restaurant, in a car, or at home in a year is 170. The total number of these meals eaten in a car or at home exceeds the number eaten in a restaurant by 14. Twenty more restaurant-purchased meals will be eaten in a restaurant than at home. (*Source:* The NPD Group) Find the number of restaurant-purchased meals eaten in a restaurant, the number eaten in a car, and the number eaten at home.

**19.** *Top Apple Growers.* The top three apple growers in the world—China, the United States, and Turkey—grew a total of about 74 billion lb of apples in a recent year. China produced 44 billion lb more than the combined production of the United States and Turkey. The United States produced twice as many pounds of apples as Turkey. (*Source:* U.S. Apple Association) Find the number of pounds of apples produced by each country.

**20.** *Adopting Abroad.* The three foreign countries from which the largest number of children were adopted in 2009 were China, Ethiopia, and Russia. A total of 6864 children were adopted from these countries. The number of children adopted from China was 862 fewer than the total number adopted from Ethiopia and Russia. Twice the number adopted from Russia is 171 more than the number adopted from China. (*Source*: U.S. Department of State) Find the number of children adopted from each country.

**21.** *Jolts of Caffeine.* One 8-oz serving each of brewed coffee, Red Bull energy drink, and Mountain Dew soda contains a total of 197 mg of caffeine. One serving of brewed coffee has 6 mg more caffeine than two servings of Mountain Dew. One serving of Red Bull contains 37 mg less caffeine than one serving each of brewed coffee and Mountain Dew. (*Source*: Australian Institute of Sport) Find the amount of caffeine in one serving of each beverage.

**22.** *Mother's Day Spending.* The top three Mother's Day gifts are flowers, jewelry, and gift certificates. The total of the average amounts spent on these gifts is $53.42. The average amount spent on jewelry is $4.40 more than the average amount spent on gift certificates. Together, the average amounts spent on flowers and gift certificates is $15.58 more than the average amount spent on jewelry. (*Source*: BIGresearch) What is the average amount spent on each type of gift?

**23.** *Favorite Pets.* Americans own a total of about 355 million fish, cats, and dogs as pets. The number of fish owned is 11 million more than the total number of cats and dogs owned, and 16 million more cats are owned than dogs. (*Source*: American Pet Products Manufacturers Association) How many of each type of pet do Americans own?

**24.** *Mail-Order Business.* Natural Fibers Clothing charges $4 for shipping orders of $25 or less, $8 for orders from $25.01 to $75, and $10 for orders over $75. One week, shipping charges for 600 orders totaled $4280. Eighty more orders for $25 or less were shipped than orders for more than $75. Find the number of orders shipped at each rate.

**25.** *Biggest Weekend at the Box Office.* Recently, the movies *The Dark Knight*, *Spider-Man 3*, and *The Twilight Saga: New Moon* held the record for having the three highest-grossing weekends at the box office, with a total of $452 million. Together, *Spider-Man 3* and *New Moon* earned $136 million more than *The Dark Knight*. *New Moon* earned $15 million less than *The Dark Knight*. (*Source*: the-numbers.com) Find the amount earned by each movie.

**26.** *Spring Cleaning.* In a group of 100 adults, 70 say they are most likely to do spring housecleaning in March, April, or May. Of these 70, the number who clean in April is 14 more than the total number who clean in March and May. The total number who clean in April and May is 2 more than three times the number who clean in March. (*Source*: Zoomerang online survey) Find the number who clean in each month.

**27.** *Nutrition.* A hospital dietician must plan a lunch menu that provides 485 Cal, 41.5 g of carbohydrates, and 35 mg of calcium. A 3-oz serving of broiled ground beef contains 245 Cal, 0 g of carbohydrates, and 9 mg of calcium. One baked potato contains 145 Cal, 34 g of carbohydrates, and 8 mg of calcium. A one-cup serving of strawberries contains 45 Cal, 10 g of carbohydrates, and 21 mg of calcium. (*Source*: *Home and Garden Bulletin No. 72*, U.S. Government Printing Office, Washington, D.C. 20402) How many servings of each are required to provide the desired nutritional values?

**28.** *Nutrition.* A diabetic patient wishes to prepare a meal consisting of roasted chicken breast, mashed potatoes, and peas. A 3-oz serving of roasted skinless chicken breast contains 140 Cal, 27 g of protein, and 64 mg of sodium. A one-cup serving of mashed potatoes contains 160 Cal, 4 g of protein, and 636 mg of sodium, and a one-cup serving of peas contains 125 Cal, 8 g of protein, and 139 mg of sodium. (*Source*: *Home and Garden Bulletin*

*No. 72*, U.S. Government Printing Office, Washington, D.C. 20402) How many servings of each should be used if the meal is to contain 415 Cal, 50.5 g of protein, and 553 mg of sodium?

**29.** *Investment.* Walter earns a year-end bonus of $5000 and puts it in 3 one-year investments that pay $243 in simple interest. Part is invested at 3%, part at 4%, and part at 6%. There is $1500 more invested at 6% than at 3%. Find the amount invested at each rate.

**30.** *Investment.* Natsumi receives $126 per year in simple interest from three investments. Part is invested at 2%, part at 3%, and part at 4%. There is $500 more invested at 3% than at 2%. The amount invested at 4% is three times the amount invested at 3%. Find the amount invested at each rate.

**31.** *Price Increases.* In New York City, orange juice, a raisin bagel, and a cup of coffee from Katie's Koffee Kart cost a total of $8.15. Katie posts a notice announcing that, effective the following week, the price of orange juice will increase 25% and the price of bagels will increase 20%. After the increase, the same purchase will cost a total of $9.30, and the raisin bagel will cost 30¢ more than coffee. Find the price of each item before the increase.

**32.** *Cost of Snack Food.* Martin and Eva pool their loose change to buy snacks on their coffee break. One day, they spent $6.75 on 1 carton of milk, 2 donuts, and 1 cup of coffee. The next day, they spent $8.50 on 3 donuts and 2 cups of coffee. The third day, they bought 1 carton of milk, 1 donut, and 2 cups of coffee and spent $7.25. On the fourth day, they have a total of $6.45 left. Is this enough to buy 2 cartons of milk and 2 donuts?

**33.** *Job Loss.* The table below lists the percent of American workers who responded that they were likely to be laid off from their jobs in the coming year, represented in terms of the number of years since 1990.

| Year, $x$ | Percent Who Say Lay-off Likely |
|---|---|
| 1990, 0 | 16 |
| 1997, 7 | 9 |
| 2010, 20 | 21 |

*Source*: Gallup Poll

**a)** Fit a quadratic function $f(x) = ax^2 + bx + c$ to the data, where $x$ is the number of years since 1990.
**b)** Use the function to estimate the percent of workers who responded that they were likely to be laid off in the coming year in 2003.

**34.** *Student Loans.* The table below lists the volume of nonfederal student loans, in billions of dollars, represented in terms of the number of years since 2004.

| Year, $x$ | Volume of Student Loans (in billions) |
|---|---|
| 2004, 0 | $15.1 |
| 2006, 2 | 20.5 |
| 2008, 4 | 11.0 |

*Source*: "Trends in Student Aid," The College Board

**a)** Fit a quadratic function $f(x) = ax^2 + bx + c$ to the data, where $x$ is the number of years since 2004.
**b)** Use the function to estimate the volume of nonfederal student loans in 2007.

**35.** *Farm Acreage.* The table below lists the average size of U.S. farms, represented in terms of the number of years since 1997.

| Year, $x$ | Average Size of U.S. Farms (in number of acres) |
|---|---|
| 1997, 0 | 431 |
| 2002, 5 | 441 |
| 2009, 12 | 418 |

*Sources*: 2007 Census of Agriculture; National Agricultural Statistics Service

**a)** Fit a quadratic function $f(x) = ax^2 + bx + c$ to the data, where $x$ is the number of years since 1997.

**b)** Use the function to estimate the average size of U.S. farms in 2000 and in 2010.

**36.** *Book Shipments.* The table below lists the number of books shipped by publishers, in billions, represented in terms of the number of years since 2007.

| Year, $x$ | Books Shipped (in billions) |
|---|---|
| 2007, 0 | 3.127 |
| 2008, 1 | 3.079 |
| 2009, 2 | 3.101 |

*Source*: Book Industry Study Group, Inc., www.bisg.org

**a)** Fit a quadratic function $f(x) = ax^2 + bx + c$ to the data, where $x$ is the number of years since 2007.

**b)** Use the function to estimate the number of books shipped in 2012.

**37.** *Unemployment Rate.* The table below lists the U.S. unemployment rate in October for selected years.

| Year | U.S. Unemployment Rate in October |
|---|---|
| 2002 | 5.7% |
| 2004 | 5.5 |
| 2005 | 5.0 |
| 2008 | 6.6 |
| 2010 | 9.7 |

*Source*: U.S. Bureau of Labor Statistics

**a)** Use a graphing calculator to fit a quadratic function $f(x) = ax^2 + bx + c$ to the data, where $x$ is the number of years since 2002.

**b)** Use the function found in part (a) to estimate the unemployment rate in 2003, in 2007, and in 2009.

**38.** *Box-Office Revenue.* The table below lists the box-office revenue for January through March, in millions of dollars, for years 2008–2011.

| Year | Box-Office Revenue for January–March (in millions) |
|---|---|
| 2008 | $2219.40 |
| 2009 | 2430.30 |
| 2010 | 2646.50 |
| 2011 | 2095.50 |

*Source*: Box Office Mojo, www.boxofficemojo.com

**a)** Use a graphing calculator to fit a quadratic function $f(x) = ax^2 + bx + c$ to the data, where $x$ is the number of years since 2008.

**b)** Use the function found in part (a) to estimate the box-office revenue for January through March in 2012.

## Skill Maintenance

### Vocabulary Reinforcement

*In each of Exercises 39–46, fill in the blank with the correct term. Some of the given choices will not be used.*

| | |
|---|---|
| Descartes' rule of signs | a vertical asymptote |
| the leading-term test | an oblique |
| the intermediate | asymptote |
| value theorem | direct variation |
| the fundamental | inverse variation |
| theorem of algebra | a horizontal line |
| a polynomial function | a vertical line |
| a rational function | parallel |
| a one-to-one function | perpendicular |
| a constant function | |
| a horizontal asymptote | |

**39.** Two lines with slopes $m_1$ and $m_2$ are _____ if and only if the product of their slopes is $-1$.

**40.** We can use _____ to determine the behavior of the graph of a polynomial function as $x \to \infty$ or as $x \to -\infty$.

**41.** If it is possible for _____ to cross a graph more than once, then the graph is not the graph of a function.

**42.** A function is _____ if different inputs have different outputs.

**43.** _____ is a function that is a quotient of two polynomials.

**44.** If a situation gives rise to a function $f(x) = k/x$, or $y = k/x$, where $k$ is a positive constant, we say that we have _____.

**45.** _____ of a rational function $p(x)/q(x)$, where $p(x)$ and $q(x)$ have no common factors other than constants, occurs at an $x$-value that makes the denominator 0.

**46.** When the numerator and the denominator of a rational function have the same degree, the graph of the function has _____.

## Synthesis

**47.** Let $u$ represent $1/x$, $v$ represent $1/y$, and $w$ represent $1/z$. Solve first for $u$, $v$, and $w$. Then solve the following system of equations:

$$\frac{2}{x} + \frac{2}{y} - \frac{3}{z} = 3,$$

$$\frac{1}{x} - \frac{2}{y} - \frac{3}{z} = 9,$$

$$\frac{7}{x} - \frac{2}{y} + \frac{9}{z} = -39.$$

**48.** *Transcontinental Railroad.* Use the following facts to find the year in which the first U.S. transcontinental railroad was completed: The sum of the digits in the year is 24. The units digit is 1 more than the hundreds digit. Both the tens and the units digits are multiples of three.

*In Exercises 49 and 50, three solutions of an equation are given. Use a system of three equations in three variables to find the constants and write the equation.*

**49.** $Ax + By + Cz = 12;$
$\left(1, \frac{3}{4}, 3\right)$, $\left(\frac{4}{3}, 1, 2\right)$, and $(2, 1, 1)$

**50.** $y = B - Mx - Nz;$
$(1, 1, 2)$, $(3, 2, -6)$, and $\left(\frac{3}{2}, 1, 1\right)$

*In Exercises 51 and 52, four solutions of the equation $y = ax^3 + bx^2 + cx + d$ are given. Use a system of four equations in four variables to find the constants $a$, $b$, $c$, and $d$ and write the equation.*

**51.** $(-2, 59)$, $(-1, 13)$, $(1, -1)$, and $(2, -17)$

**52.** $(-2, -39)$, $(-1, -12)$, $(1, -6)$, and $(3, 16)$

---

## 6.3

# Matrices and Systems of Equations

- Solve systems of equations using matrices.

### ■ Matrices and Row-Equivalent Operations

In this section, we consider additional techniques for solving systems of equations. You have probably observed that when we solve a system of equations, we perform computations with the coefficients and the constants and continually rewrite the variables. We can streamline the solution process by omitting the variables until a solution is found. For example, the system

$$2x - 3y = 7,$$
$$x + 4y = -2$$

can be written more simply as

$$\left[\begin{array}{rr|r} 2 & -3 & 7 \\ 1 & 4 & -2 \end{array}\right].$$

The vertical line replaces the equals signs.

A rectangular array of numbers like the one above is called a **matrix** (pl., **matrices**). The matrix above is called an **augmented matrix** for the given system of equations, because it contains not only the coefficients but also the constant terms. The matrix

$$\left[\begin{array}{rr} 2 & -3 \\ 1 & 4 \end{array}\right]$$

is called the **coefficient matrix** of the system.

The **rows** of a matrix are horizontal, and the **columns** are vertical. The augmented matrix above has 2 rows and 3 columns, and the coefficient matrix has 2 rows and 2 columns. A matrix with $m$ rows and $n$ columns is said to be of **order** $m \times n$. Thus the order of the augmented matrix above is $2 \times 3$, and the order of the coefficient matrix is $2 \times 2$. When $m = n$, a matrix is said to be **square**. The coefficient matrix above is a square matrix. The numbers 2 and 4 lie on the **main diagonal** of the coefficient matrix. The numbers in a matrix are called **entries** or **elements**.

## ■ Gaussian Elimination with Matrices

In Section 6.2, we described a series of operations that can be used to transform a system of equations to an equivalent system. Each of these operations corresponds to one that can be used to produce *row-equivalent matrices*.

---

**ROW-EQUIVALENT OPERATIONS**

1. Interchange any two rows.
2. Multiply each entry in a row by the same nonzero constant.
3. Add a nonzero multiple of one row to another row.

---

We can use these operations on the augmented matrix of a system of equations to solve the system.

GCM

**EXAMPLE 1** Solve the following system:

$$\begin{aligned} 2x - \phantom{2}y + \phantom{1}4z &= -3, \\ x - 2y - 10z &= -6, \\ 3x \phantom{- 2y} + \phantom{1}4z &= 7. \end{aligned}$$

*Solution* First, we write the augmented matrix, writing 0 for the missing $y$-term in the last equation:

$$\left[\begin{array}{rrr|r} 2 & -1 & 4 & -3 \\ 1 & -2 & -10 & -6 \\ 3 & 0 & 4 & 7 \end{array}\right].$$

$$\begin{bmatrix} 1 & a & b & | & c \\ 0 & 1 & d & | & e \\ 0 & 0 & 1 & | & f \end{bmatrix}$$

Our goal is to find a row-equivalent matrix of the form shown at left. The variables can then be reinserted to form equations from which we can complete the solution. This is done by working from the bottom equation to the top and using back-substitution.

The first step is to multiply and/or interchange rows so that each number in the first column below the first number is a multiple of that number. In this case, we interchange the first and second rows to obtain a 1 in the upper left-hand corner.

$$\begin{bmatrix} 1 & -2 & -10 & | & -6 \\ 2 & -1 & 4 & | & -3 \\ 3 & 0 & 4 & | & 7 \end{bmatrix} \qquad \begin{array}{l} \text{New row 1} = \text{row 2} \\ \text{New row 2} = \text{row 1} \end{array}$$

Next, we multiply the first row by $-2$ and add it to the second row. We also multiply the first row by $-3$ and add it to the third row.

$$\begin{bmatrix} 1 & -2 & -10 & | & -6 \\ 0 & 3 & 24 & | & 9 \\ 0 & 6 & 34 & | & 25 \end{bmatrix} \qquad \begin{array}{l} \text{Row 1 is unchanged.} \\ \text{New row 2} = -2(\text{row 1}) + \text{row 2} \\ \text{New row 3} = -3(\text{row 1}) + \text{row 3} \end{array}$$

Now we multiply the second row by $\frac{1}{3}$ to get a 1 in the second row, second column.

$$\begin{bmatrix} 1 & -2 & -10 & | & -6 \\ 0 & 1 & 8 & | & 3 \\ 0 & 6 & 34 & | & 25 \end{bmatrix} \qquad \text{New row 2} = \tfrac{1}{3}(\text{row 2})$$

Then we multiply the second row by $-6$ and add it to the third row.

$$\begin{bmatrix} 1 & -2 & -10 & | & -6 \\ 0 & 1 & 8 & | & 3 \\ 0 & 0 & -14 & | & 7 \end{bmatrix} \qquad \text{New row 3} = -6(\text{row 2}) + \text{row 3}$$

Finally, we multiply the third row by $-\frac{1}{14}$ to get a 1 in the third row, third column.

$$\begin{bmatrix} 1 & -2 & -10 & | & -6 \\ 0 & 1 & 8 & | & 3 \\ 0 & 0 & 1 & | & -\frac{1}{2} \end{bmatrix} \qquad \text{New row 3} = -\tfrac{1}{14}(\text{row 3})$$

Now we can write the system of equations that corresponds to the last matrix above:

$$\begin{array}{rcll} x - 2y - 10z &=& -6, & \textbf{(1)} \\ y + 8z &=& 3, & \textbf{(2)} \\ z &=& -\tfrac{1}{2}. & \textbf{(3)} \end{array}$$

We back-substitute $-\frac{1}{2}$ for $z$ in equation (2) and solve for $y$:

$$\begin{aligned} y + 8\left(-\tfrac{1}{2}\right) &= 3 \\ y - 4 &= 3 \\ y &= 7. \end{aligned}$$

Next, we back-substitute 7 for $y$ and $-\frac{1}{2}$ for $z$ in equation (1) and solve for $x$:

$$x - 2 \cdot 7 - 10\left(-\frac{1}{2}\right) = -6$$
$$x - 14 + 5 = -6$$
$$x - 9 = -6$$
$$x = 3.$$

The triple $\left(3, 7, -\frac{1}{2}\right)$ checks in the original system of equations, so it is the solution.

**Now Try Exercise 27.**

```
rowSwap([A],1,2)→[B]
  [1 −2 −10 −6]
  [2 −1   4 −3]
  [3  0   4  7]
```

Row-equivalent operations can be performed on a graphing calculator. For example, to interchange the first and second rows of the augmented matrix, as we did in the first step in Example 1, we enter the matrix as matrix **A** and select "rowSwap" from the MATRIX MATH menu. Some graphing calculators will not automatically store the matrix produced using a row-equivalent operation, so when several operations are to be performed in succession, it is helpful to store the result of each operation as it is produced. In the window at left, we see both the matrix produced by the rowSwap operation and the indication that this matrix is stored as matrix **B**.

The procedure followed in Example 1 is called **Gaussian elimination with matrices**. The last matrix in Example 1 is in **row-echelon form**. To be in this form, a matrix must have the following properties.

**ROW-ECHELON FORM**

1. If a row does not consist entirely of 0's, then the first nonzero element in the row is a 1 (called a **leading 1**).
2. For any two successive nonzero rows, the leading 1 in the lower row is farther to the right than the leading 1 in the higher row.
3. All the rows consisting entirely of 0's are at the bottom of the matrix.

If a fourth property is also satisfied, a matrix is said to be in **reduced row-echelon form**:

4. Each column that contains a leading 1 has 0's everywhere else.

**EXAMPLE 2**   Which of the following matrices are in row-echelon form? Which, if any, are in reduced row-echelon form?

a) $\begin{bmatrix} 1 & -3 & 5 & | & -2 \\ 0 & 1 & -4 & | & 3 \\ 0 & 0 & 1 & | & 10 \end{bmatrix}$

b) $\begin{bmatrix} 0 & -1 & | & 2 \\ 0 & 1 & | & 5 \end{bmatrix}$

c) $\begin{bmatrix} 1 & -2 & -6 & 4 & | & 7 \\ 0 & 3 & 5 & -8 & | & -1 \\ 0 & 0 & 1 & 9 & | & 2 \end{bmatrix}$

d) $\begin{bmatrix} 1 & 0 & 0 & | & -2.4 \\ 0 & 1 & 0 & | & 0.8 \\ 0 & 0 & 1 & | & 5.6 \end{bmatrix}$

e) $\begin{bmatrix} 1 & 0 & 0 & 0 & | & \frac{2}{3} \\ 0 & 1 & 0 & 0 & | & -\frac{1}{4} \\ 0 & 0 & 1 & 0 & | & \frac{6}{7} \\ 0 & 0 & 0 & 0 & | & 0 \end{bmatrix}$

f) $\begin{bmatrix} 1 & -4 & 2 & | & 5 \\ 0 & 0 & 0 & | & 0 \\ 0 & 1 & -3 & | & -8 \end{bmatrix}$

***Solution*** The matrices in (a), (d), and (e) satisfy the row-echelon criteria and, thus, are in row-echelon form. In (b) and (c), the first nonzero elements of the first and second rows, respectively, are not 1. In (f), the row consisting entirely of 0's is not at the bottom of the matrix. Thus the matrices in (b), (c), and (f) are not in row-echelon form. In (d) and (e), not only are the row-echelon criteria met but each column that contains a leading 1 also has 0's elsewhere, so these matrices are in reduced row-echelon form.

## Gauss–Jordan Elimination

We have seen that with Gaussian elimination we perform row-equivalent operations on a matrix to obtain a row-equivalent matrix in row-echelon form. When we continue to apply these operations until we have a matrix in *reduced* row-echelon form, we are using **Gauss–Jordan elimination**. This method is named for Karl Friedrich Gauss and Wilhelm Jordan (1842–1899).

**EXAMPLE 3** Use Gauss–Jordan elimination to solve the system of equations in Example 1.

***Solution*** Using Gaussian elimination in Example 1, we obtained the matrix

$$\begin{bmatrix} 1 & -2 & -10 & | & -6 \\ 0 & 1 & 8 & | & 3 \\ 0 & 0 & 1 & | & -\frac{1}{2} \end{bmatrix}.$$

We continue to perform row-equivalent operations until we have a matrix in reduced row-echelon form. We multiply the third row by 10 and add it to the first row. We also multiply the third row by $-8$ and add it to the second row.

$$\begin{bmatrix} 1 & -2 & 0 & | & -11 \\ 0 & 1 & 0 & | & 7 \\ 0 & 0 & 1 & | & -\frac{1}{2} \end{bmatrix}$$ New row 1 $= 10(\text{row } 3) + \text{row } 1$
New row 2 $= -8(\text{row } 3) + \text{row } 2$

Next, we multiply the second row by 2 and add it to the first row.

$$\begin{bmatrix} 1 & 0 & 0 & | & 3 \\ 0 & 1 & 0 & | & 7 \\ 0 & 0 & 1 & | & -\frac{1}{2} \end{bmatrix}$$ New row 1 $= 2(\text{row } 2) + \text{row } 1$

Writing the system of equations that corresponds to this matrix, we have

$$\begin{aligned} x \phantom{+y+z} &= 3, \\ y \phantom{+z} &= 7, \\ z &= -\tfrac{1}{2}. \end{aligned}$$

We can actually read the solution, $\left(3, 7, -\frac{1}{2}\right)$, directly from the last column of the reduced row-echelon matrix.

We can also use a graphing calculator to solve this system of equations. After the augmented matrix is entered, reduced row-echelon form can be found directly using the "rref" operation from the MATRIX MATH menu.

**Now Try Exercise 27.**

rref([A]) ▶ Frac
$$\begin{bmatrix} 1 & 0 & 0 & 3 \\ 0 & 1 & 0 & 7 \\ 0 & 0 & 1 & -\frac{1}{2} \end{bmatrix}$$

**EXAMPLE 4** Solve the following system:

$$3x - 4y - z = 6,$$
$$2x - y + z = -1,$$
$$4x - 7y - 3z = 13.$$

*Solution* We write the augmented matrix and use Gauss–Jordan elimination.

$$\begin{bmatrix} 3 & -4 & -1 & | & 6 \\ 2 & -1 & 1 & | & -1 \\ 4 & -7 & -3 & | & 13 \end{bmatrix}$$

We begin by multiplying the second and third rows by 3 so that each number in the first column below the first number, 3, is a multiple of that number.

$$\begin{bmatrix} 3 & -4 & -1 & | & 6 \\ 6 & -3 & 3 & | & -3 \\ 12 & -21 & -9 & | & 39 \end{bmatrix} \quad \begin{array}{l} \text{New row 2} = 3(\text{row 2}) \\ \text{New row 3} = 3(\text{row 3}) \end{array}$$

Next, we multiply the first row by $-2$ and add it to the second row. We also multiply the first row by $-4$ and add it to the third row.

$$\begin{bmatrix} 3 & -4 & -1 & | & 6 \\ 0 & 5 & 5 & | & -15 \\ 0 & -5 & -5 & | & 15 \end{bmatrix} \quad \begin{array}{l} \text{New row 2} = -2(\text{row 1}) + \text{row 2} \\ \text{New row 3} = -4(\text{row 1}) + \text{row 3} \end{array}$$

Now we add the second row to the third row.

$$\begin{bmatrix} 3 & -4 & -1 & | & 6 \\ 0 & 5 & 5 & | & -15 \\ 0 & 0 & 0 & | & 0 \end{bmatrix} \quad \text{New row 3} = \text{row 2} + \text{row 3}$$

We can stop at this stage because we have a row consisting entirely of 0's. The last row of the matrix corresponds to the equation $0 = 0$, which is true for all values of $x$, $y$, and $z$. Consequently, the equations are dependent and the system is equivalent to

$$3x - 4y - z = 6,$$
$$5y + 5z = -15.$$

This particular system has infinitely many solutions. (A system containing dependent equations could be inconsistent.)

Solving the second equation for $y$ gives us

$$y = -z - 3.$$

Substituting $-z - 3$ for $y$ in the first equation and solving for $x$, we get

$$3x - 4(-z - 3) - z = 6$$
$$x = -z - 2.$$

Then the solutions of this system are of the form

$$(-z - 2, -z - 3, z),$$

where $z$ can be any real number.

**Now Try Exercise 33.**

Similarly, if we obtain a row whose only nonzero entry occurs in the last column, we have an inconsistent system of equations. For example, in the matrix

$$\begin{bmatrix} 1 & 0 & 3 & | & -2 \\ 0 & 1 & 5 & | & 4 \\ 0 & 0 & 0 & | & 6 \end{bmatrix},$$

the last row corresponds to the false equation $0 = 6$, so we know the original system of equations has no solution.

## 6.3    Exercise Set

*Determine the order of the matrix.*

**1.** $\begin{bmatrix} 1 & -6 \\ -3 & 2 \\ 0 & 5 \end{bmatrix}$    **2.** $\begin{bmatrix} 7 \\ -5 \\ -1 \\ 3 \end{bmatrix}$

**3.** $\begin{bmatrix} 2 & -4 & 0 & 9 \end{bmatrix}$    **4.** $\begin{bmatrix} -8 \end{bmatrix}$

**5.** $\begin{bmatrix} 1 & -5 & -8 \\ 6 & 4 & -2 \\ -3 & 0 & 7 \end{bmatrix}$    **6.** $\begin{bmatrix} 13 & 2 & -6 & 4 \\ -1 & 18 & 5 & -12 \end{bmatrix}$

*Write the augmented matrix for the system of equations.*

**7.** $2x - y = 7,$
$\quad x + 4y = -5$

**8.** $3x + 2y = 8,$
$\quad 2x - 3y = 15$

**9.** $x - 2y + 3z = 12,$
$\quad 2x \qquad - 4z = 8,$
$\qquad 3y + z = 7$

**10.** $\quad x + y - z = 7,$
$\qquad\quad 3y + 2z = 1,$
$\quad -2x - 5y \qquad = 6$

*Write the system of equations that corresponds to the augmented matrix.*

**11.** $\begin{bmatrix} 3 & -5 & | & 1 \\ 1 & 4 & | & -2 \end{bmatrix}$

**12.** $\begin{bmatrix} 1 & 2 & | & -6 \\ 4 & 1 & | & -3 \end{bmatrix}$

**13.** $\begin{bmatrix} 2 & 1 & -4 & | & 12 \\ 3 & 0 & 5 & | & -1 \\ 1 & -1 & 1 & | & 2 \end{bmatrix}$

**14.** $\begin{bmatrix} -1 & -2 & 3 & | & 6 \\ 0 & 4 & 1 & | & 2 \\ 2 & -1 & 0 & | & 9 \end{bmatrix}$

*Solve the system of equations using Gaussian elimination or Gauss–Jordan elimination. Use a graphing calculator to check your answer.*

**15.** $4x + 2y = 11,$
$\quad 3x - y = 2$

**16.** $2x + y = 1,$
$\quad 3x + 2y = -2$

**17.** $5x - 2y = -3,$
$\quad 2x + 5y = -24$

**18.** $2x + y = 1,$
$\quad 3x - 6y = 4$

**19.** $\quad 3x + 4y = 7,$
$\quad -5x + 2y = 10$

**20.** $5x - 3y = -2,$
$\quad 4x + 2y = 5$

**21.** $3x + 2y = 6,$
$\quad 2x - 3y = -9$

**22.** $\quad x - 4y = 9,$
$\quad 2x + 5y = 5$

**23.** $\quad x - 3y = 8,$
$\quad -2x + 6y = 3$

**24.** $4x - 8y = 12,$
$\quad -x + 2y = -3$

**25.** $-2x + 6y = 4,$
$\quad 3x - 9y = -6$

**26.** $\quad 6x + 2y = -10,$
$\quad -3x - y = 6$

**27.** $x + 2y - 3z = 9,$
$\quad 2x - y + 2z = -8,$
$\quad 3x - y - 4z = 3$

**28.** $x - y + 2z = 0,$
$\quad x - 2y + 3z = -1,$
$\quad 2x - 2y + z = -3$

**29.** $4x - y - 3z = 1,$
$\quad 8x + y - z = 5,$
$\quad 2x + y + 2z = 5$

**30.** $3x + 2y + 2z = 3,$
$\quad x + 2y - z = 5,$
$\quad 2x - 4y + z = 0$

**31.** $x - 2y + 3z = -4,$
$3x + y - z = 0,$
$2x + 3y - 5z = 1$

**32.** $2x - 3y + 2z = 2,$
$x + 4y - z = 9,$
$-3x + y - 5z = 5$

**33.** $2x - 4y - 3z = 3,$
$x + 3y + z = -1,$
$5x + y - 2z = 2$

**34.** $x + y - 3z = 4,$
$4x + 5y + z = 1,$
$2x + 3y + 7z = -7$

**35.** $p + q + r = 1,$
$p + 2q + 3r = 4,$
$4p + 5q + 6r = 7$

**36.** $m + n + t = 9,$
$m - n - t = -15,$
$3m + n + t = 2$

**37.** $a + b - c = 7,$
$a - b + c = 5,$
$3a + b - c = -1$

**38.** $a - b + c = 3,$
$2a + b - 3c = 5,$
$4a + b - c = 11$

**39.** $-2w + 2x + 2y - 2z = -10,$
$w + x + y + z = -5,$
$3w + x - y + 4z = -2,$
$w + 3x - 2y + 2z = -6$

**40.** $-w + 2x - 3y + z = -8,$
$-w + x + y - z = -4,$
$w + x + y + z = 22,$
$-w + x - y - z = -14$

*Use Gaussian elimination or Gauss–Jordan elimination in Exercises 41–44.*

**41.** *Stamp Purchase.* Otto spent $22.35 on 44¢ and 17¢ stamps. He bought a total of 60 stamps. How many of each type did he buy?

**42.** *Time of Return.* The Coopers pay their babysitter $8 per hour before 11 P.M. and $9.50 per hour after 11 P.M. One evening they went out for 6 hr and paid the sitter $50.25. What time did they come home?

**43.** *Advertising Expense.* eAuction.com spent a total of $11 million on advertising in fiscal years 2010, 2011, and 2012. The amount spent in 2012 was three times the amount spent in 2010. The amount spent in 2011 was $3 million less than the amount spent in 2012. How much was spent on advertising each year?

**44.** *Borrowing.* Serrano Manufacturing borrowed $30,000 to buy a new piece of equipment. Part of the money was borrowed at 8%, part at 10%, and part at 12%. The annual interest was $3040, and the total amount borrowed at 8% and at 10% was twice the amount borrowed at 12%. How much was borrowed at each rate?

## Skill Maintenance

*Classify the function as linear, quadratic, cubic, quartic, rational, exponential, or logarithmic.*

**45.** $f(x) = 3^{x-1}$

**46.** $f(x) = 3x - 1$

**47.** $f(x) = \dfrac{3x - 1}{x^2 + 4}$

**48.** $f(x) = -\frac{3}{4}x^4 + \frac{9}{2}x^3 + 2x^2 - 4$

**49.** $f(x) = \ln(3x - 1)$

**50.** $f(x) = \frac{3}{4}x^3 - x$

**51.** $f(x) = 3$

**52.** $f(x) = 2 - x - x^2$

## Synthesis

*In Exercises 53 and 54, three solutions of the equation $y = ax^2 + bx + c$ are given. Use a system of three equations in three variables and Gaussian elimination or Gauss–Jordan elimination to find the constants a, b, and c and write the equation.*

**53.** $(-3, 12), (-1, -7),$ and $(1, -2)$

**54.** $(-1, 0), (1, -3),$ and $(3, -22)$

**55.** Find two different row-echelon forms of
$$\begin{bmatrix} 1 & 5 \\ 3 & 2 \end{bmatrix}.$$

**56.** Consider the system of equations
$$\begin{aligned} x - y + 3z &= -8, \\ 2x + 3y - z &= 5, \\ 3x + 2y + 2kz &= -3k. \end{aligned}$$
For what value(s) of $k$, if any, will the system have

a) no solution?
b) exactly one solution?
c) infinitely many solutions?

*Solve using matrices.*

**57.** $y = x + z$,
$3y + 5z = 4$,
$x + 4 = y + 3z$

**58.** $x + y = 2z$,
$2x - 5z = 4$,
$x - z = y + 8$

**59.** $x - 4y + 2z = 7$,
$3x + y + 3z = -5$

**60.** $x - y - 3z = 3$,
$-x + 3y + z = -7$

**61.** $4x + 5y = 3$,
$-2x + y = 9$,
$3x - 2y = -15$

**62.** $2x - 3y = -1$,
$-x + 2y = -2$,
$3x - 5y = 1$

## 6.4 Matrix Operations

- Add, subtract, and multiply matrices when possible.
- Write a matrix equation equivalent to a system of equations.

In Section 6.3, we used matrices to solve systems of equations. Matrices are useful in many other types of applications as well. In this section, we study matrices and some of their properties.

A capital letter is generally used to name a matrix, and lower-case letters with double subscripts generally denote its entries. For example, $a_{47}$, read "$a$ sub four seven," indicates the entry in the fourth row and the seventh column. A general term is represented by $a_{ij}$. The notation $a_{ij}$ indicates the entry in row $i$ and column $j$. In general, we can write a matrix as

$$\mathbf{A} = \begin{bmatrix} a_{ij} \end{bmatrix} = \begin{bmatrix} a_{11} & a_{12} & a_{13} & \cdots & a_{1n} \\ a_{21} & a_{22} & a_{23} & \cdots & a_{2n} \\ a_{31} & a_{32} & a_{33} & \cdots & a_{3n} \\ \vdots & \vdots & \vdots & & \vdots \\ a_{m1} & a_{m2} & a_{m3} & \cdots & a_{mn} \end{bmatrix}.$$

The matrix above has $m$ rows and $n$ columns; that is, its order is $m \times n$.

Two matrices are **equal** if they have the same order and corresponding entries are equal.

### ▪ Matrix Addition and Subtraction

To add or subtract matrices, we add or subtract their corresponding entries. For this to be possible, the matrices must have the same order.

**ADDITION AND SUBTRACTION OF MATRICES**

Given two $m \times n$ matrices $\mathbf{A} = \begin{bmatrix} a_{ij} \end{bmatrix}$ and $\mathbf{B} = \begin{bmatrix} b_{ij} \end{bmatrix}$, their sum is

$$\mathbf{A} + \mathbf{B} = \begin{bmatrix} a_{ij} + b_{ij} \end{bmatrix}$$

and their difference is

$$\mathbf{A} - \mathbf{B} = \begin{bmatrix} a_{ij} - b_{ij} \end{bmatrix}.$$

Addition of matrices is both commutative and associative.

GCM  **EXAMPLE 1** Find $\mathbf{A} + \mathbf{B}$ for each of the following.

a) $\mathbf{A} = \begin{bmatrix} -5 & 0 \\ 4 & \frac{1}{2} \end{bmatrix}$, $\mathbf{B} = \begin{bmatrix} 6 & -3 \\ 2 & 3 \end{bmatrix}$

b) $\mathbf{A} = \begin{bmatrix} 1 & 3 \\ -1 & 5 \\ 6 & 0 \end{bmatrix}$, $\mathbf{B} = \begin{bmatrix} -1 & -2 \\ 1 & -2 \\ -3 & 1 \end{bmatrix}$

***Solution*** We have a pair of $2 \times 2$ matrices in part (a) and a pair of $3 \times 2$ matrices in part (b). Since the matrices within each pair of matrices have the same order, we can add the corresponding entries.

a) $\mathbf{A} + \mathbf{B} = \begin{bmatrix} -5 & 0 \\ 4 & \frac{1}{2} \end{bmatrix} + \begin{bmatrix} 6 & -3 \\ 2 & 3 \end{bmatrix}$

$= \begin{bmatrix} -5 + 6 & 0 + (-3) \\ 4 + 2 & \frac{1}{2} + 3 \end{bmatrix} = \begin{bmatrix} 1 & -3 \\ 6 & 3\frac{1}{2} \end{bmatrix}$

```
[A]+[B]
         [1  -3]
         [6  3.5]
```

We can also enter $\mathbf{A}$ and $\mathbf{B}$ in a graphing calculator and then find $\mathbf{A} + \mathbf{B}$.

b) $\mathbf{A} + \mathbf{B} = \begin{bmatrix} 1 & 3 \\ -1 & 5 \\ 6 & 0 \end{bmatrix} + \begin{bmatrix} -1 & -2 \\ 1 & -2 \\ -3 & 1 \end{bmatrix}$

$= \begin{bmatrix} 1 + (-1) & 3 + (-2) \\ -1 + 1 & 5 + (-2) \\ 6 + (-3) & 0 + 1 \end{bmatrix} = \begin{bmatrix} 0 & 1 \\ 0 & 3 \\ 3 & 1 \end{bmatrix}$

This sum can also be found on a graphing calculator after $\mathbf{A}$ and $\mathbf{B}$ have been entered.

**Now Try Exercise 5.**

```
[A]+[B]
         [0  1]
         [0  3]
         [3  1]
```

GCM  **EXAMPLE 2** Find $\mathbf{C} - \mathbf{D}$ for each of the following.

a) $\mathbf{C} = \begin{bmatrix} 1 & 2 \\ -2 & 0 \\ -3 & -1 \end{bmatrix}$, $\mathbf{D} = \begin{bmatrix} 1 & -1 \\ 1 & 3 \\ 2 & 3 \end{bmatrix}$

b) $\mathbf{C} = \begin{bmatrix} 5 & -6 \\ -3 & 4 \end{bmatrix}$, $\mathbf{D} = \begin{bmatrix} -4 \\ 1 \end{bmatrix}$

*Solution*

**a)** Since the order of each matrix is $3 \times 2$, we can subtract corresponding entries:

$$\mathbf{C} - \mathbf{D} = \begin{bmatrix} 1 & 2 \\ -2 & 0 \\ -3 & -1 \end{bmatrix} - \begin{bmatrix} 1 & -1 \\ 1 & 3 \\ 2 & 3 \end{bmatrix}$$

$$= \begin{bmatrix} 1-1 & 2-(-1) \\ -2-1 & 0-3 \\ -3-2 & -1-3 \end{bmatrix} = \begin{bmatrix} 0 & 3 \\ -3 & -3 \\ -5 & -4 \end{bmatrix}.$$

[C]−[D]
$$\begin{bmatrix} 0 & 3 \\ -3 & -3 \\ -5 & -4 \end{bmatrix}$$

This subtraction can also be done using a graphing calculator.

**b)** $\mathbf{C}$ is a $2 \times 2$ matrix and $\mathbf{D}$ is a $2 \times 1$ matrix. Since the matrices do not have the same order, we cannot subtract.  ▶ **Now Try Exercise 13.**

ERR:DIM MISMATCH
**1:** Quit
2: Goto

The **opposite**, or **additive inverse**, of a matrix is obtained by replacing each entry with its opposite.

**EXAMPLE 3**  Find $-\mathbf{A}$ and $\mathbf{A} + (-\mathbf{A})$ for

$$\mathbf{A} = \begin{bmatrix} 1 & 0 & 2 \\ 3 & -1 & 5 \end{bmatrix}.$$

*Solution*  To find $-\mathbf{A}$, we replace each entry of $\mathbf{A}$ with its opposite.

$$-\mathbf{A} = \begin{bmatrix} -1 & 0 & -2 \\ -3 & 1 & -5 \end{bmatrix}.$$

[A]+(−[A])
$$\begin{bmatrix} 0 & 0 & 0 \\ 0 & 0 & 0 \end{bmatrix}$$

$$\mathbf{A} + (-\mathbf{A}) = \begin{bmatrix} 1 & 0 & 2 \\ 3 & -1 & 5 \end{bmatrix} + \begin{bmatrix} -1 & 0 & -2 \\ -3 & 1 & -5 \end{bmatrix}$$

$$= \begin{bmatrix} 0 & 0 & 0 \\ 0 & 0 & 0 \end{bmatrix}$$

A matrix having 0's for all its entries is called a **zero matrix**. When a zero matrix is added to a second matrix of the same order, the second matrix is unchanged. Thus a zero matrix is an **additive identity**. For example,

$$\begin{bmatrix} 2 & 3 & -4 \\ 0 & 6 & 5 \end{bmatrix} + \begin{bmatrix} 0 & 0 & 0 \\ 0 & 0 & 0 \end{bmatrix} = \begin{bmatrix} 2 & 3 & -4 \\ 0 & 6 & 5 \end{bmatrix}.$$

The matrix

$$\begin{bmatrix} 0 & 0 & 0 \\ 0 & 0 & 0 \end{bmatrix}$$

is the additive identity for any $2 \times 3$ matrix.

## ■ Scalar Multiplication

When we find the product of a number and a matrix, we obtain a **scalar product**.

> **SCALAR PRODUCT**
>
> The **scalar product** of a number $k$ and a matrix $\mathbf{A}$ is the matrix denoted $k\mathbf{A}$, obtained by multiplying each entry of $\mathbf{A}$ by the number $k$. The number $k$ is called a **scalar**.

GCM    **EXAMPLE 4**   Find $3\mathbf{A}$ and $(-1)\mathbf{A}$ for

$$\mathbf{A} = \begin{bmatrix} -3 & 0 \\ 4 & 5 \end{bmatrix}.$$

*Solution*   We have

$$3\mathbf{A} = 3\begin{bmatrix} -3 & 0 \\ 4 & 5 \end{bmatrix} = \begin{bmatrix} 3(-3) & 3 \cdot 0 \\ 3 \cdot 4 & 3 \cdot 5 \end{bmatrix} = \begin{bmatrix} -9 & 0 \\ 12 & 15 \end{bmatrix},$$

$$(-1)\mathbf{A} = -1\begin{bmatrix} -3 & 0 \\ 4 & 5 \end{bmatrix} = \begin{bmatrix} -1(-3) & -1 \cdot 0 \\ -1 \cdot 4 & -1 \cdot 5 \end{bmatrix} = \begin{bmatrix} 3 & 0 \\ -4 & -5 \end{bmatrix}.$$

These scalar products can also be found on a graphing calculator after $\mathbf{A}$ has been entered.    **Now Try Exercise 9.**

```
3[A]
              [ -9   0]
              [ 12  15]
(-1)[A]
              [  3   0]
              [ -4  -5]
```

The properties of matrix addition and scalar multiplication are similar to the properties of addition and multiplication of real numbers.

> **PROPERTIES OF MATRIX ADDITION AND SCALAR MULTIPLICATION**
>
> For any $m \times n$ matrices $\mathbf{A}$, $\mathbf{B}$, and $\mathbf{C}$ and any scalars $k$ and $l$:
>
> $\mathbf{A} + \mathbf{B} = \mathbf{B} + \mathbf{A}.$      Commutative property of addition
>
> $\mathbf{A} + (\mathbf{B} + \mathbf{C}) = (\mathbf{A} + \mathbf{B}) + \mathbf{C}.$      Associative property of addition
>
> $(kl)\mathbf{A} = k(l\mathbf{A}).$      Associative property of scalar multiplication
>
> $k(\mathbf{A} + \mathbf{B}) = k\mathbf{A} + k\mathbf{B}.$      Distributive property
>
> $(k + l)\mathbf{A} = k\mathbf{A} + l\mathbf{A}.$      Distributive property
>
> There exists a unique matrix $\mathbf{0}$ such that:
>
> $\mathbf{A} + \mathbf{0} = \mathbf{0} + \mathbf{A} = \mathbf{A}.$      Additive identity property
>
> There exists a unique matrix $-\mathbf{A}$ such that:
>
> $\mathbf{A} + (-\mathbf{A}) = -\mathbf{A} + \mathbf{A} = \mathbf{0}.$      Additive inverse property

**EXAMPLE 5**    *Production.*    Waterworks, Inc., manufactures three types of kayaks in its two plants. The table below lists the number of each style produced at each plant in April.

|  | Whitewater Kayak | Ocean Kayak | Crossover Kayak |
|---|---|---|---|
| **Madison Plant** | 150 | 120 | 100 |
| **Greensburg Plant** | 180 | 90 | 130 |

**a)** Write a $2 \times 3$ matrix **A** that represents the information in the table.

**b)** The manufacturer increased production by 20% in May. Find a matrix **M** that represents the increased production figures.

**c)** Find the matrix **A** + **M** and tell what it represents.

*Solution*

**a)** Write the entries in the table in a $2 \times 3$ matrix **A**.

$$\mathbf{A} = \begin{bmatrix} 150 & 120 & 100 \\ 180 & 90 & 130 \end{bmatrix}$$

**b)** The production in May will be represented by **A** + 20%**A**, or **A** + 0.2**A**, or 1.2**A**. Thus,

$$\mathbf{M} = (1.2)\begin{bmatrix} 150 & 120 & 100 \\ 180 & 90 & 130 \end{bmatrix} = \begin{bmatrix} 180 & 144 & 120 \\ 216 & 108 & 156 \end{bmatrix}.$$

**c)** $\mathbf{A} + \mathbf{M} = \begin{bmatrix} 150 & 120 & 100 \\ 180 & 90 & 130 \end{bmatrix} + \begin{bmatrix} 180 & 144 & 120 \\ 216 & 108 & 156 \end{bmatrix}$

$= \begin{bmatrix} 330 & 264 & 220 \\ 396 & 198 & 286 \end{bmatrix}$

The matrix **A** + **M** represents the total production of each of the three types of kayaks at each plant in April and in May.    **Now Try Exercise 29.**

### ■ Products of Matrices

Matrix multiplication is defined in such a way that it can be used in solving systems of equations and in many applications.

> **MATRIX MULTIPLICATION**
>
> For an $m \times n$ matrix $\mathbf{A} = \begin{bmatrix} a_{ij} \end{bmatrix}$ and an $n \times p$ matrix $\mathbf{B} = \begin{bmatrix} b_{ij} \end{bmatrix}$, the **product AB** $= \begin{bmatrix} c_{ij} \end{bmatrix}$ is an $m \times p$ matrix, where
>
> $$c_{ij} = a_{i1} \cdot b_{1j} + a_{i2} \cdot b_{2j} + a_{i3} \cdot b_{3j} + \cdots + a_{in} \cdot b_{nj}.$$

In other words, the entry $c_{ij}$ in **AB** is obtained by multiplying the entries in row $i$ of **A** by the corresponding entries in column $j$ of **B** and adding the results.

> Note that we can multiply two matrices only when the number of columns in the first matrix is equal to the number of rows in the second matrix.

GCM **EXAMPLE 6** For

$$\mathbf{A} = \begin{bmatrix} 3 & 1 & -1 \\ 2 & 0 & 3 \end{bmatrix}, \quad \mathbf{B} = \begin{bmatrix} 1 & 6 \\ 3 & -5 \\ -2 & 4 \end{bmatrix}, \quad \text{and} \quad \mathbf{C} = \begin{bmatrix} 4 & -6 \\ 1 & 2 \end{bmatrix},$$

find each of the following.

a) **AB**      b) **BA**

c) **BC**      d) **AC**

*Solution*

**a)** **A** is a $2 \times 3$ matrix and **B** is a $3 \times 2$ matrix, so **AB** will be a $2 \times 2$ matrix.

$$\mathbf{AB} = \begin{bmatrix} 3 & 1 & -1 \\ 2 & 0 & 3 \end{bmatrix}\begin{bmatrix} 1 & 6 \\ 3 & -5 \\ -2 & 4 \end{bmatrix}$$

$$= \begin{bmatrix} 3 \cdot 1 + 1 \cdot 3 + (-1)(-2) & 3 \cdot 6 + 1(-5) + (-1)(4) \\ 2 \cdot 1 + 0 \cdot 3 + 3(-2) & 2 \cdot 6 + 0(-5) + 3 \cdot 4 \end{bmatrix} = \begin{bmatrix} 8 & 9 \\ -4 & 24 \end{bmatrix}$$

**b)** **B** is a $3 \times 2$ matrix and **A** is a $2 \times 3$ matrix, so **BA** will be a $3 \times 3$ matrix.

$$\mathbf{BA} = \begin{bmatrix} 1 & 6 \\ 3 & -5 \\ -2 & 4 \end{bmatrix}\begin{bmatrix} 3 & 1 & -1 \\ 2 & 0 & 3 \end{bmatrix}$$

$$= \begin{bmatrix} 1 \cdot 3 + 6 \cdot 2 & 1 \cdot 1 + 6 \cdot 0 & 1(-1) + 6 \cdot 3 \\ 3 \cdot 3 + (-5)(2) & 3 \cdot 1 + (-5)(0) & 3(-1) + (-5)(3) \\ -2 \cdot 3 + 4 \cdot 2 & -2 \cdot 1 + 4 \cdot 0 & -2(-1) + 4 \cdot 3 \end{bmatrix} = \begin{bmatrix} 15 & 1 & 17 \\ -1 & 3 & -18 \\ 2 & -2 & 14 \end{bmatrix}$$

> Note in parts (a) and (b) that **AB** $\neq$ **BA**. Multiplication of matrices is generally not commutative.

[A][B]
$$\begin{bmatrix} 8 & 9 \\ -4 & 24 \end{bmatrix}$$
[B][A]
$$\begin{bmatrix} 15 & 1 & 17 \\ -1 & 3 & -18 \\ 2 & -2 & 14 \end{bmatrix}$$

Matrix multiplication can be performed on a graphing calculator. The products in parts (a) and (b) are shown at left.

**c)** **B** is a 3 $\times$ 2 matrix and **C** is a 2 $\times$ 2 matrix, so **BC** will be a 3 $\times$ 2 matrix.

$$\mathbf{BC} = \begin{bmatrix} 1 & 6 \\ 3 & -5 \\ -2 & 4 \end{bmatrix} \begin{bmatrix} 4 & -6 \\ 1 & 2 \end{bmatrix}$$

$$= \begin{bmatrix} 1 \cdot 4 + 6 \cdot 1 & 1(-6) + 6 \cdot 2 \\ 3 \cdot 4 + (-5)(1) & 3(-6) + (-5)(2) \\ -2 \cdot 4 + 4 \cdot 1 & -2(-6) + 4 \cdot 2 \end{bmatrix} = \begin{bmatrix} 10 & 6 \\ 7 & -28 \\ -4 & 20 \end{bmatrix}$$

**d)** The product **AC** is not defined because the number of columns of **A**, 3, is not equal to the number of rows of **C**, 2.

When the product **AC** is entered on a graphing calculator, an ERROR message is returned, indicating that the dimensions of the matrices are mismatched.

```
[A] [C]
```

```
ERR:DIM MISMATCH
1:Quit
2: Goto
```

> **Now Try Exercises 23 and 25.**

**EXAMPLE 7** *Bakery Profit.* Two of the items sold at Sweet Treats Bakery are gluten-free bagels and gluten-free doughnuts. The table below lists the number of dozens of each product that are sold at the bakery's three stores one week.

| | Main Street Store | Avon Road Store | Dalton Avenue Store |
|---|---|---|---|
| **Bagels (in dozens)** | 25 | 30 | 20 |
| **Doughnuts (in dozens)** | 40 | 35 | 15 |

The bakery's profit on one dozen bagels is $5, and its profit on one dozen doughnuts is $6. Use matrices to find the total profit on these items at each store for the given week.

*Solution* We can write the table showing the sales of the products as a 2 $\times$ 3 matrix:

$$\mathbf{S} = \begin{bmatrix} 25 & 30 & 20 \\ 40 & 35 & 15 \end{bmatrix}.$$

The profit per dozen for each product can also be written as a matrix:

$$\mathbf{P} = \begin{bmatrix} 5 & 6 \end{bmatrix}.$$

Then the total profit at each store is given by the matrix product **PS**:

$$\mathbf{PS} = \begin{bmatrix} 5 & 6 \end{bmatrix} \begin{bmatrix} 25 & 30 & 20 \\ 40 & 35 & 15 \end{bmatrix}$$

$$= \begin{bmatrix} 5 \cdot 25 + 6 \cdot 40 & 5 \cdot 30 + 6 \cdot 35 & 5 \cdot 20 + 6 \cdot 15 \end{bmatrix}$$

$$= \begin{bmatrix} 365 & 360 & 190 \end{bmatrix}.$$

The total profit on gluten-free bagels and gluten-free doughnuts for the given week was $365 at the Main Street store, $360 at the Avon Road store, and $190 at the Dalton Avenue store.

> **Now Try Exercise 33.**

A matrix that consists of a single row, like **P** in Example 7, is called a **row matrix**. Similarly, a matrix that consists of a single column, like

$$\begin{bmatrix} 8 \\ -3 \\ 5 \end{bmatrix},$$

is called a **column matrix**.

We have already seen that matrix multiplication is generally not commutative. Nevertheless, matrix multiplication does have some properties that are similar to those for multiplication of real numbers.

---

**PROPERTIES OF MATRIX MULTIPLICATION**

For matrices **A**, **B**, and **C**, assuming that the indicated operations are possible:

| | |
|---|---|
| $\mathbf{A}(\mathbf{BC}) = (\mathbf{AB})\mathbf{C}.$ | Associative property of multiplication |
| $\mathbf{A}(\mathbf{B} + \mathbf{C}) = \mathbf{AB} + \mathbf{AC}.$ | Distributive property |
| $(\mathbf{B} + \mathbf{C})\mathbf{A} = \mathbf{BA} + \mathbf{CA}.$ | Distributive property |

---

### ■ Matrix Equations

We can write a matrix equation equivalent to a system of equations.

**EXAMPLE 8** Write a matrix equation equivalent to the following system of equations:

$$\begin{aligned} 4x + 2y - \phantom{2}z &= 3, \\ 9x \phantom{{}+ 2y} + \phantom{2}z &= 5, \\ 4x + 5y - 2z &= 1. \end{aligned}$$

*Solution* We write the coefficients on the left in a matrix. We then write the product of that matrix and the column matrix containing the variables and set the result equal to the column matrix containing the constants on the right:

$$\begin{bmatrix} 4 & 2 & -1 \\ 9 & 0 & 1 \\ 4 & 5 & -2 \end{bmatrix} \begin{bmatrix} x \\ y \\ z \end{bmatrix} = \begin{bmatrix} 3 \\ 5 \\ 1 \end{bmatrix}.$$

If we let

$$\mathbf{A} = \begin{bmatrix} 4 & 2 & -1 \\ 9 & 0 & 1 \\ 4 & 5 & -2 \end{bmatrix}, \quad \mathbf{X} = \begin{bmatrix} x \\ y \\ z \end{bmatrix}, \quad \text{and} \quad \mathbf{B} = \begin{bmatrix} 3 \\ 5 \\ 1 \end{bmatrix},$$

we can write this matrix equation as $\mathbf{AX} = \mathbf{B}$.

**Now Try Exercise 39.**

In the next section, we will solve systems of equations using a matrix equation like the one in Example 8.

## 6.4  Exercise Set

*Find x and y.*

**1.** $\begin{bmatrix} 5 & x \end{bmatrix} = \begin{bmatrix} y & -3 \end{bmatrix}$

**2.** $\begin{bmatrix} 6x \\ 25 \end{bmatrix} = \begin{bmatrix} -9 \\ 5y \end{bmatrix}$

**3.** $\begin{bmatrix} 3 & 2x \\ y & -8 \end{bmatrix} = \begin{bmatrix} 3 & -2 \\ 1 & -8 \end{bmatrix}$

**4.** $\begin{bmatrix} x-1 & 4 \\ y+3 & -7 \end{bmatrix} = \begin{bmatrix} 0 & 4 \\ -2 & -7 \end{bmatrix}$

*For Exercises 5–20, let*

$$A = \begin{bmatrix} 1 & 2 \\ 4 & 3 \end{bmatrix}, \qquad B = \begin{bmatrix} -3 & 5 \\ 2 & -1 \end{bmatrix},$$

$$C = \begin{bmatrix} 1 & -1 \\ -1 & 1 \end{bmatrix}, \qquad D = \begin{bmatrix} 1 & 1 \\ 1 & 1 \end{bmatrix},$$

$$E = \begin{bmatrix} 1 & 3 \\ 2 & 6 \end{bmatrix}, \qquad F = \begin{bmatrix} 3 & 3 \\ -1 & -1 \end{bmatrix},$$

$$0 = \begin{bmatrix} 0 & 0 \\ 0 & 0 \end{bmatrix}, \qquad I = \begin{bmatrix} 1 & 0 \\ 0 & 1 \end{bmatrix}.$$

*Find each of the following.*

**5.** $A + B$

**6.** $B + A$

**7.** $E + 0$

**8.** $2A$

**9.** $3F$

**10.** $(-1)D$

**11.** $3F + 2A$

**12.** $A - B$

**13.** $B - A$

**14.** $AB$

**15.** $BA$

**16.** $0F$

**17.** $CD$

**18.** $EF$

**19.** $AI$

**20.** $IA$

*Find the product, if possible.*

**21.** $\begin{bmatrix} -1 & 0 & 7 \\ 3 & -5 & 2 \end{bmatrix} \begin{bmatrix} 6 \\ -4 \\ 1 \end{bmatrix}$

**22.** $\begin{bmatrix} 6 & -1 & 2 \end{bmatrix} \begin{bmatrix} 1 & 4 \\ -2 & 0 \\ 5 & -3 \end{bmatrix}$

**23.** $\begin{bmatrix} -2 & 4 \\ 5 & 1 \\ -1 & -3 \end{bmatrix} \begin{bmatrix} 3 & -6 \\ -1 & 4 \end{bmatrix}$

**24.** $\begin{bmatrix} 2 & -1 & 0 \\ 0 & 5 & 4 \end{bmatrix} \begin{bmatrix} -3 & 1 & 0 \\ 0 & 2 & -1 \\ 5 & 0 & 4 \end{bmatrix}$

**25.** $\begin{bmatrix} 1 \\ -5 \\ 3 \end{bmatrix} \begin{bmatrix} -6 & 5 & 8 \\ 0 & 4 & -1 \end{bmatrix}$

**26.** $\begin{bmatrix} 2 & 0 & 0 \\ 0 & -1 & 0 \\ 0 & 0 & 3 \end{bmatrix} \begin{bmatrix} 0 & -4 & 3 \\ 2 & 1 & 0 \\ -1 & 0 & 6 \end{bmatrix}$

**27.** $\begin{bmatrix} 1 & -4 & 3 \\ 0 & 8 & 0 \\ -2 & -1 & 5 \end{bmatrix} \begin{bmatrix} 3 & 0 & 0 \\ 0 & -4 & 0 \\ 0 & 0 & 1 \end{bmatrix}$

**28.** $\begin{bmatrix} 4 \\ -5 \end{bmatrix} \begin{bmatrix} 2 & 0 \\ 6 & -7 \\ 0 & -3 \end{bmatrix}$

**29.** *Produce.*  The produce manager at Dugan's Market orders 40 lb of tomatoes, 20 lb of zucchini, and 30 lb of onions from a local farmer one week.

**a)** Write a $1 \times 3$ matrix **A** that represents the amount of each item ordered.

**b)** The following week the produce manager increases her order by 10%. Find a matrix **B** that represents this order.

**c)** Find $A + B$ and tell what the entries represent.

**30.** *Budget.*  For the month of June, Madelyn budgets $400 for food, $160 for clothes, and $60 for entertainment.

**a)** Write a $1 \times 3$ matrix **B** that represents the amounts budgeted for these items.

**b)** After receiving a raise, Madelyn increases the amount budgeted for each item in July by 5%. Find a matrix **R** that represents the new amounts.

**c)** Find **B** + **R** and tell what the entries represent.

**31.** *Nutrition.* A 3-oz serving of roasted, skinless chicken breast contains 140 Cal, 27 g of protein, 3 g of fat, 13 mg of calcium, and 64 mg of sodium. One-half cup of potato salad contains 180 Cal, 4 g of protein, 11 g of fat, 24 mg of calcium, and 662 mg of sodium. One broccoli spear contains 50 Cal, 5 g of protein, 1 g of fat, 82 mg of calcium, and 20 mg of sodium. (*Source: Home and Garden Bulletin No. 72*, U.S. Government Printing Office, Washington, D.C. 20402)

**a)** Write $1 \times 5$ matrices **C**, **P**, and **B** that represent the nutritional values of each food.

**b)** Find **C** + 2**P** + 3**B** and tell what the entries represent.

**32.** *Nutrition.* One slice of cheese pizza contains 290 Cal, 15 g of protein, 9 g of fat, and 39 g of carbohydrates. One-half cup of gelatin dessert contains 70 Cal, 2 g of protein, 0 g of fat, and 17 g of carbohydrates. One cup of whole milk contains 150 Cal, 8 g of protein, 8 g of fat, and 11 g of carbohydrates. (*Source: Home and Garden Bulletin No. 72*, U.S. Government Printing Office, Washington, D.C. 20402)

**a)** Write $1 \times 4$ matrices **P**, **G**, and **M** that represent the nutritional values of each food.

**b)** Find 3**P** + 2**G** + 2**M** and tell what the entries represent.

**33.** *Food Service Management.* The food service manager at a large hospital is concerned about maintaining reasonable food costs. The table below lists the cost per serving, in dollars, for items on four menus.

| Menu | Meat | Potato | Vegetable | Salad | Dessert |
|------|------|--------|-----------|-------|---------|
| 1 | 1.50 | 0.15 | 0.26 | 0.23 | 0.64 |
| 2 | 1.55 | 0.14 | 0.24 | 0.21 | 0.75 |
| 3 | 1.62 | 0.22 | 0.31 | 0.28 | 0.53 |
| 4 | 1.70 | 0.20 | 0.29 | 0.33 | 0.68 |

On a particular day, a dietician orders 65 meals from menu 1, 48 from menu 2, 93 from menu 3, and 57 from menu 4.

**a)** Write the information in the table as a $4 \times 5$ matrix **M**.

**b)** Write a row matrix **N** that represents the number of each menu ordered.

**c)** Find the product **NM**.

**d)** State what the entries of **NM** represent.

**34.** *Food Service Management.* A college food service manager uses a table like the one below to list the number of units of ingredients, by weight, required for various menu items.

|  | White Cake | Bread | Coffee Cake | Sugar Cookies |
|--------|-----------|-------|-------------|---------------|
| **Flour** | 1 | 2.5 | 0.75 | 0.5 |
| **Milk** | 0 | 0.5 | 0.25 | 0 |
| **Eggs** | 0.75 | 0.25 | 0.5 | 0.5 |
| **Butter** | 0.5 | 0 | 0.5 | 1 |

The cost per unit of each ingredient is 25 cents for flour, 34 cents for milk, 54 cents for eggs, and 83 cents for butter.

**a)** Write the information in the table as a $4 \times 4$ matrix **M**.

**b)** Write a row matrix **C** that represents the cost per unit of each ingredient.

**c)** Find the product **CM**.

**d)** State what the entries of **CM** represent.

**35.** *Profit.* A manufacturer produces exterior plywood, interior plywood, and fiberboard, which are shipped to two distributors. The table below shows the number of units of each type of product that are shipped to each warehouse.

|  | Distributor 1 | Distributor 2 |
|------|---------------|---------------|
| **Exterior Plywood** | 900 | 500 |
| **Interior Plywood** | 450 | 1000 |
| **Fiberboard** | 600 | 700 |

The profits from each unit of exterior plywood, interior plywood, and fiberboard are $5, $8, and $4, respectively.

**a)** Write the information in the table as a 3 × 2 matrix **M**.
**b)** Write a row matrix **P** that represents the profit, per unit, of each type of product.
**c)** Find the product **PM**.
**d)** State what the entries of **PM** represent.

**36.** *Production Cost.*   Karin supplies two small campus coffee shops with homemade chocolate chip cookies, oatmeal cookies, and peanut butter cookies. The table below shows the number of each type of cookie, in dozens, that Karin sold in one week.

|  | Mugsy's Coffee Shop | The Coffee Club |
|---|---|---|
| **Chocolate Chip** | 8 | 15 |
| **Oatmeal** | 6 | 10 |
| **Peanut Butter** | 4 | 3 |

Karin spends $4 for the ingredients for one dozen chocolate chip cookies, $2.50 for the ingredients for one dozen oatmeal cookies, and $3 for the ingredients for one dozen peanut butter cookies.

**a)** Write the information in the table as a 3 × 2 matrix **S**.
**b)** Write a row matrix **C** that represents the cost, per dozen, of the ingredients for each type of cookie.

**c)** Find the product **CS**.
**d)** State what the entries of **CS** represent.

**37.** *Production Cost.*   In Exercise 35, suppose that the manufacturer's production costs for each unit of exterior plywood, interior plywood, and fiberboard are $20, $25, and $15, respectively.

**a)** Write a row matrix **C** that represents this information.
**b)** Use the matrices **M** and **C** to find the total production cost for the products shipped to each distributor.

**38.** *Profit.*   In Exercise 36, suppose that Karin's profits on one dozen chocolate chip, oatmeal, and peanut butter cookies are $6, $4.50, and $5.20, respectively.

**a)** Write a row matrix **P** that represents this information.
**b)** Use the matrices **S** and **P** to find Karin's total profit from each coffee shop.

*Write a matrix equation equivalent to the system of equations.*

**39.** $2x - 3y = 7,$
$\quad x + 5y = -6$

**40.** $-x + \ y = 3,$
$\quad 5x - 4y = 16$

**41.** $x + \ y - 2z = 6,$
$\quad 3x - \ y + \ z = 7,$
$\quad 2x + 5y - 3z = 8$

**42.** $3x - \ y + \ z = 1,$
$\quad x + 2y - \ z = 3,$
$\quad 4x + 3y - 2z = 11$

**43.** $3x - 2y + 4z = 17,$
$\quad 2x + \ y - 5z = 13$

**44.** $3x + 2y + 5z = 9,$
$\quad 4x - 3y + 2z = 10$

**45.** $-4w + \ x - \ y + 2z = 12,$
$\quad w + 2x - \ y - \ z = 0,$
$\quad -w + \ x + 4y - 3z = 1,$
$\quad 2w + 3x + 5y - 7z = 9$

**46.** $12w + 2x + \ 4y - \ 5z = 2,$
$\quad -w + 4x - \ y + 12z = 5,$
$\quad 2w - \ x + \ 4y \qquad\ = 13,$
$\quad 2x + 10y + \ z = 5$

## Skill Maintenance

*In Exercises 47–50:*

a) *Find the vertex.*
b) *Find the axis of symmetry.*
c) *Determine whether there is a maximum or minimum value and find that value.*
d) *Graph the function.*

**47.** $f(x) = x^2 - x - 6$

**48.** $f(x) = 2x^2 - 5x - 3$

**49.** $f(x) = -x^2 - 3x + 2$

**50.** $f(x) = -3x^2 + 4x + 4$

## Synthesis

*For Exercises 51–54, let*

$$\mathbf{A} = \begin{bmatrix} -1 & 0 \\ 2 & 1 \end{bmatrix} \quad and \quad \mathbf{B} = \begin{bmatrix} 1 & -1 \\ 0 & 2 \end{bmatrix}.$$

**51.** Show that
$$(\mathbf{A} + \mathbf{B})(\mathbf{A} - \mathbf{B}) \neq \mathbf{A}^2 - \mathbf{B}^2,$$
where
$$\mathbf{A}^2 = \mathbf{AA} \quad and \quad \mathbf{B}^2 = \mathbf{BB}.$$

**52.** Show that
$$(\mathbf{A} + \mathbf{B})(\mathbf{A} + \mathbf{B}) \neq \mathbf{A}^2 + 2\mathbf{AB} + \mathbf{B}^2.$$

**53.** Show that
$$(\mathbf{A} + \mathbf{B})(\mathbf{A} - \mathbf{B}) = \mathbf{A}^2 + \mathbf{BA} - \mathbf{AB} - \mathbf{B}^2.$$

**54.** Show that
$$(\mathbf{A} + \mathbf{B})(\mathbf{A} + \mathbf{B}) = \mathbf{A}^2 + \mathbf{BA} + \mathbf{AB} + \mathbf{B}^2.$$

## Mid-Chapter Mixed Review

*Determine whether the statement is true or false.*

**1.** For a system of two linear equations in two variables, if the graphs of the equations are parallel lines, then the system of equations has infinitely many solutions. [6.1]

**2.** One of the properties of a matrix written in row-echelon form is that all the rows consisting entirely of 0's are at the bottom of the matrix. [6.3]

**3.** We can multiply two matrices only when the number of columns in the first matrix is equal to the number of rows in the second matrix. [6.4]

**4.** Addition of matrices is not commutative. [6.4]

*Solve.* [6.1], [6.2]

**5.** $2x + y = -4,$
  $x = y - 5$

**6.** $x + y = 4,$
  $y = 2 - x$

**7.** $2x - 3y = 8,$
  $3x + 2y = -1$

**8.** $x - 3y = 1,$
  $6y = 2x - 2$

**9.** $x + 2y + 3z = 4,$
  $x - 2y + z = 2,$
  $2x - 6y + 4z = 7$

**10.** *e-Commerce.* computerwarehouse.com charges $3 for shipping orders up to 10 lb, $5 for orders from 10 lb up to 15 lb, and $7.50 for orders of 15 lb or more. One day shipping charges for 150 orders totaled $680. The number of orders under 10 lb was three times the number of orders weighing 15 lb or more. Find the number of packages shipped at each rate. [6.2]

*Solve the system of equations using Gaussian elimination or Gauss–Jordan elimination.* [6.3]

**11.** $2x + y = 5,$
$3x + 2y = 6$

**12.** $3x + 2y - 3z = -2,$
$2x + 3y + 2z = -2,$
$x + 4y + 4z = 1$

*For Exercises 13–20, let*

$$A = \begin{bmatrix} 3 & -1 \\ 5 & 4 \end{bmatrix}, \quad B = \begin{bmatrix} -2 & 6 \\ 1 & -3 \end{bmatrix}, \quad C = \begin{bmatrix} -4 & 1 & -1 \\ 2 & 3 & -2 \end{bmatrix}, \quad and \quad D = \begin{bmatrix} -2 & 3 & 0 \\ 1 & -1 & 2 \\ -3 & 4 & 1 \end{bmatrix}.$$

*Find each of the following.* [6.4]

**13.** $A + B$

**14.** $B - A$

**15.** $4D$

**16.** $2A + 3B$

**17.** $AB$

**18.** $BA$

**19.** $BC$

**20.** $DC$

**21.** Write a matrix equation equivalent to the following system of equations: [6.4]

$2x - y + 3z = 7,$
$x + 2y - z = 3,$
$3x - 4y + 2z = 5.$

## Collaborative Discussion and Writing

**22.** Explain in your own words when using the elimination method for solving a system of equations is preferable to using the substitution method. [6.1]

**23.** Given two linear equations in three variables,

$Ax + By + Cz = D$ and $Ex + Fy + Gz = H,$

explain how you could find a third equation such that the system contains dependent equations. [6.2]

**24.** Explain in your own words why the augmented matrix below represents a system of dependent equations. [6.3]

$$\begin{bmatrix} 1 & -3 & 2 & | & -5 \\ 0 & 1 & -4 & | & 8 \\ 0 & 0 & 0 & | & 0 \end{bmatrix}$$

**25.** Is it true that if $AB = 0$, for matrices $A$ and $B$, then $A = 0$ or $B = 0$? Why or why not? [6.4]

## 6.5

# Inverses of Matrices

- Find the inverse of a square matrix, if it exists.
- Use inverses of matrices to solve systems of equations.

In this section, we continue our study of matrix algebra, finding the **multiplicative inverse**, or simply **inverse**, of a square matrix, if it exists. Then we use such inverses to solve systems of equations.

### The Identity Matrix

Recall that, for real numbers, $a \cdot 1 = 1 \cdot a = a$; 1 is the multiplicative identity. A multiplicative identity matrix is very similar to the number 1.

> **IDENTITY MATRIX**
>
> For any positive integer $n$, the $n \times n$ **identity matrix** is an $n \times n$ matrix with 1's on the main diagonal and 0's elsewhere and is denoted by
>
> $$\mathbf{I} = \begin{bmatrix} 1 & 0 & 0 & \cdots & 0 \\ 0 & 1 & 0 & \cdots & 0 \\ 0 & 0 & 1 & \cdots & 0 \\ \vdots & \vdots & \vdots & & \vdots \\ 0 & 0 & 0 & \cdots & 1 \end{bmatrix}.$$
>
> Then $\mathbf{AI} = \mathbf{IA} = \mathbf{A}$, for any $n \times n$ matrix $\mathbf{A}$.

**EXAMPLE 1** For

$$\mathbf{A} = \begin{bmatrix} 4 & -7 \\ -3 & 2 \end{bmatrix} \quad \text{and} \quad \mathbf{I} = \begin{bmatrix} 1 & 0 \\ 0 & 1 \end{bmatrix},$$

find each of the following.

**a)** $\mathbf{AI}$                                             **b)** $\mathbf{IA}$

*Solution*

**a)** $\mathbf{AI} = \begin{bmatrix} 4 & -7 \\ -3 & 2 \end{bmatrix} \begin{bmatrix} 1 & 0 \\ 0 & 1 \end{bmatrix}$

$\qquad = \begin{bmatrix} 4 \cdot 1 - 7 \cdot 0 & 4 \cdot 0 - 7 \cdot 1 \\ -3 \cdot 1 + 2 \cdot 0 & -3 \cdot 0 + 2 \cdot 1 \end{bmatrix} = \begin{bmatrix} 4 & -7 \\ -3 & 2 \end{bmatrix} = \mathbf{A}$

**b)** $\mathbf{IA} = \begin{bmatrix} 1 & 0 \\ 0 & 1 \end{bmatrix} \begin{bmatrix} 4 & -7 \\ -3 & 2 \end{bmatrix}$

$\qquad = \begin{bmatrix} 1 \cdot 4 + 0(-3) & 1(-7) + 0 \cdot 2 \\ 0 \cdot 4 + 1(-3) & 0(-7) + 1 \cdot 2 \end{bmatrix} = \begin{bmatrix} 4 & -7 \\ -3 & 2 \end{bmatrix} = \mathbf{A}$

These products can also be found using a graphing calculator after **A** and **I** have been entered.

## The Inverse of a Matrix

Recall that for every nonzero real number $a$, there is a multiplicative inverse $1/a$, or $a^{-1}$, such that $a \cdot a^{-1} = a^{-1} \cdot a = 1$. The multiplicative inverse of a matrix behaves in a similar manner.

> ### INVERSE OF A MATRIX
>
> For an $n \times n$ matrix **A**, if there is a matrix $\mathbf{A}^{-1}$ for which $\mathbf{A}^{-1} \cdot \mathbf{A} = \mathbf{I} = \mathbf{A} \cdot \mathbf{A}^{-1}$, then $\mathbf{A}^{-1}$ is the **inverse** of **A**.

We read $\mathbf{A}^{-1}$ as "**A** inverse." Note that not every matrix has an inverse.

**EXAMPLE 2** Verify that

$$\mathbf{B} = \begin{bmatrix} 4 & -3 \\ 3 & -2 \end{bmatrix} \text{ is the inverse of } \mathbf{A} = \begin{bmatrix} -2 & 3 \\ -3 & 4 \end{bmatrix}.$$

*Solution* We show that $\mathbf{BA} = \mathbf{I} = \mathbf{AB}$.

$$\mathbf{BA} = \begin{bmatrix} 4 & -3 \\ 3 & -2 \end{bmatrix}\begin{bmatrix} -2 & 3 \\ -3 & 4 \end{bmatrix} = \begin{bmatrix} 1 & 0 \\ 0 & 1 \end{bmatrix},$$

$$\mathbf{AB} = \begin{bmatrix} -2 & 3 \\ -3 & 4 \end{bmatrix}\begin{bmatrix} 4 & -3 \\ 3 & -2 \end{bmatrix} = \begin{bmatrix} 1 & 0 \\ 0 & 1 \end{bmatrix}$$

**Now Try Exercise 1.**

We can find the inverse of a square matrix, if it exists, by using row-equivalent operations as in the Gauss–Jordan elimination method. For example, consider the matrix

$$\mathbf{A} = \begin{bmatrix} -2 & 3 \\ -3 & 4 \end{bmatrix}.$$

To find its inverse, we first form an **augmented matrix** consisting of **A** on the left side and the $2 \times 2$ identity matrix on the right side:

$$\begin{bmatrix} -2 & 3 & | & 1 & 0 \\ -3 & 4 & | & 0 & 1 \end{bmatrix}.$$

$\uparrow$ $\uparrow$

The $2 \times 2$ matrix A    The $2 \times 2$ identity matrix

Then we attempt to transform the augmented matrix to one of the form

$$\left[\begin{array}{cc|cc} 1 & 0 & a & b \\ 0 & 1 & c & d \end{array}\right].$$

The $2 \times 2$ identity matrix      The matrix $A^{-1}$

If we can do this, the matrix on the right, $\begin{bmatrix} a & b \\ c & d \end{bmatrix}$, is $A^{-1}$.

**GCM** **EXAMPLE 3** Find $A^{-1}$, where

$$A = \begin{bmatrix} -2 & 3 \\ -3 & 4 \end{bmatrix}.$$

*Solution* First, we write the augmented matrix. Then we transform it to the desired form.

$$\left[\begin{array}{cc|cc} -2 & 3 & 1 & 0 \\ -3 & 4 & 0 & 1 \end{array}\right]$$

$$\left[\begin{array}{cc|cc} 1 & -\frac{3}{2} & -\frac{1}{2} & 0 \\ -3 & 4 & 0 & 1 \end{array}\right] \qquad \text{New row 1} = -\frac{1}{2}(\text{row 1})$$

$$\left[\begin{array}{cc|cc} 1 & -\frac{3}{2} & -\frac{1}{2} & 0 \\ 0 & -\frac{1}{2} & -\frac{3}{2} & 1 \end{array}\right] \qquad \text{New row 2} = 3(\text{row 1}) + \text{row 2}$$

$$\left[\begin{array}{cc|cc} 1 & -\frac{3}{2} & -\frac{1}{2} & 0 \\ 0 & 1 & 3 & -2 \end{array}\right] \qquad \text{New row 2} = -2(\text{row 2})$$

$$\left[\begin{array}{cc|cc} 1 & 0 & 4 & -3 \\ 0 & 1 & 3 & -2 \end{array}\right] \qquad \text{New row 1} = \frac{3}{2}(\text{row 2}) + \text{row 1}$$

Thus,

$$A^{-1} = \begin{bmatrix} 4 & -3 \\ 3 & -2 \end{bmatrix},$$

which we verified in Example 2.

The $\boxed{x^{-1}}$ key on a graphing calculator can also be used to find the inverse of a matrix.

▸ **Now Try Exercise 5.**

$[A]^{-1}$

$\begin{bmatrix} 4 & -3 \\ 3 & -2 \end{bmatrix}$

**EXAMPLE 4** Find $A^{-1}$, where

$$A = \begin{bmatrix} 1 & 2 & -1 \\ 3 & 5 & 3 \\ 2 & 4 & 3 \end{bmatrix}.$$

*Solution* First, we write the augmented matrix. Then we transform it to the desired form.

$$\left[\begin{array}{rrr|rrr} 1 & 2 & -1 & 1 & 0 & 0 \\ 3 & 5 & 3 & 0 & 1 & 0 \\ 2 & 4 & 3 & 0 & 0 & 1 \end{array}\right]$$

$$\left[\begin{array}{rrr|rrr} 1 & 2 & -1 & 1 & 0 & 0 \\ 0 & -1 & 6 & -3 & 1 & 0 \\ 0 & 0 & 5 & -2 & 0 & 1 \end{array}\right]$$ New row 2 = $-3$(row 1) + row 2
New row 3 = $-2$(row 1) + row 3

$$\left[\begin{array}{rrr|rrr} 1 & 2 & -1 & 1 & 0 & 0 \\ 0 & -1 & 6 & -3 & 1 & 0 \\ 0 & 0 & 1 & -\frac{2}{5} & 0 & \frac{1}{5} \end{array}\right]$$ New row 3 = $\frac{1}{5}$(row 3)

$$\left[\begin{array}{rrr|rrr} 1 & 2 & 0 & \frac{3}{5} & 0 & \frac{1}{5} \\ 0 & -1 & 0 & -\frac{3}{5} & 1 & -\frac{6}{5} \\ 0 & 0 & 1 & -\frac{2}{5} & 0 & \frac{1}{5} \end{array}\right]$$ New row 1 = row 3 + row 1
New row 2 = $-6$(row 3) + row 2

$$\left[\begin{array}{rrr|rrr} 1 & 0 & 0 & -\frac{3}{5} & 2 & -\frac{11}{5} \\ 0 & -1 & 0 & -\frac{3}{5} & 1 & -\frac{6}{5} \\ 0 & 0 & 1 & -\frac{2}{5} & 0 & \frac{1}{5} \end{array}\right]$$ New row 1 = $2$(row 2) + row 1

$$\left[\begin{array}{rrr|rrr} 1 & 0 & 0 & -\frac{3}{5} & 2 & -\frac{11}{5} \\ 0 & 1 & 0 & \frac{3}{5} & -1 & \frac{6}{5} \\ 0 & 0 & 1 & -\frac{2}{5} & 0 & \frac{1}{5} \end{array}\right]$$ New row 2 = $-1$(row 2)

Thus,

$$\mathbf{A}^{-1} = \left[\begin{array}{rrr} -\frac{3}{5} & 2 & -\frac{11}{5} \\ \frac{3}{5} & -1 & \frac{6}{5} \\ -\frac{2}{5} & 0 & \frac{1}{5} \end{array}\right].$$

**Now Try Exercise 9.**

If a matrix has an inverse, we say that it is **invertible**, or **nonsingular**. When we cannot obtain the identity matrix on the left using the Gauss–Jordan method, then no inverse exists. This occurs when we obtain a row consisting entirely of 0's in either of the two matrices in the augmented matrix. In this case, we say that **A** is a **singular matrix**.

When we try to find the inverse of a noninvertible, or singular, matrix using a graphing calculator, the calculator returns an error message similar to ERR: SINGULAR MATRIX.

### ■ Solving Systems of Equations

**MATRIX EQUATIONS**

REVIEW SECTION **6.4.**

We can write a system of $n$ linear equations in $n$ variables as a matrix equation $\mathbf{AX} = \mathbf{B}$. If **A** has an inverse, then the system of equations has a unique solution that can be found by solving for **X**, as follows:

$$\mathbf{AX} = \mathbf{B}$$
$$\mathbf{A}^{-1}(\mathbf{AX}) = \mathbf{A}^{-1}\mathbf{B} \qquad \text{Multiplying by } \mathbf{A}^{-1} \text{ on the left on both sides}$$
$$(\mathbf{A}^{-1}\mathbf{A})\mathbf{X} = \mathbf{A}^{-1}\mathbf{B} \qquad \text{Using the associative property of matrix multiplication}$$
$$\mathbf{IX} = \mathbf{A}^{-1}\mathbf{B} \qquad \mathbf{A}^{-1}\mathbf{A} = \mathbf{I}$$
$$\mathbf{X} = \mathbf{A}^{-1}\mathbf{B}. \qquad \mathbf{IX} = \mathbf{X}$$

> **MATRIX SOLUTIONS OF SYSTEMS OF EQUATIONS**
>
> For a system of $n$ linear equations in $n$ variables, $\mathbf{AX} = \mathbf{B}$, if $\mathbf{A}$ is an invertible matrix, then the unique solution of the system is given by
>
> $$\mathbf{X} = \mathbf{A}^{-1}\mathbf{B}.$$

> Since matrix multiplication is not commutative in general, care must be taken to multiply *on the left* by $\mathbf{A}^{-1}$.

GCM

**EXAMPLE 5** Use an inverse matrix to solve the following system of equations:

$$-2x + 3y = 4,$$
$$-3x + 4y = 5.$$

*Solution* We write an equivalent matrix equation, $\mathbf{AX} = \mathbf{B}$:

$$\begin{bmatrix} -2 & 3 \\ -3 & 4 \end{bmatrix} \cdot \begin{bmatrix} x \\ y \end{bmatrix} = \begin{bmatrix} 4 \\ 5 \end{bmatrix}$$
$$\quad\ \mathbf{A} \qquad\quad \cdot\ \ \mathbf{X}\ \ =\ \ \mathbf{B}$$

In Example 3, we found that

$$\mathbf{A}^{-1} = \begin{bmatrix} 4 & -3 \\ 3 & -2 \end{bmatrix}.$$

We also verified this in Example 2. Now we have

$$\mathbf{X} = \mathbf{A}^{-1}\mathbf{B}$$
$$\begin{bmatrix} x \\ y \end{bmatrix} = \begin{bmatrix} 4 & -3 \\ 3 & -2 \end{bmatrix} \begin{bmatrix} 4 \\ 5 \end{bmatrix} = \begin{bmatrix} 1 \\ 2 \end{bmatrix}.$$

The solution of the system of equations is $(1, 2)$.

To use a graphing calculator to solve this system of equations, we enter $\mathbf{A}$ and $\mathbf{B}$ and then enter the notation $\mathbf{A}^{-1}\mathbf{B}$ on the home screen.

Now Try Exercises 25 and 35.

```
[A]⁻¹[B]
                    [1]
                    [2]
```

# 6.5  Exercise Set

*Determine whether* $\mathbf{B}$ *is the inverse of* $\mathbf{A}$.

**1.** $\mathbf{A} = \begin{bmatrix} 1 & -3 \\ -2 & 7 \end{bmatrix}$, $\mathbf{B} = \begin{bmatrix} 7 & 3 \\ 2 & 1 \end{bmatrix}$

**2.** $\mathbf{A} = \begin{bmatrix} 3 & 2 \\ 4 & 3 \end{bmatrix}$, $\mathbf{B} = \begin{bmatrix} 3 & -2 \\ -4 & 3 \end{bmatrix}$

**3.** $\mathbf{A} = \begin{bmatrix} -1 & -1 & 6 \\ 1 & 0 & -2 \\ 1 & 0 & -3 \end{bmatrix}$, $\mathbf{B} = \begin{bmatrix} 2 & 3 & 2 \\ 3 & 3 & 4 \\ 1 & 1 & 1 \end{bmatrix}$

**4.** $\mathbf{A} = \begin{bmatrix} -2 & 0 & -3 \\ 5 & 1 & 7 \\ -3 & 0 & 4 \end{bmatrix}$, $\mathbf{B} = \begin{bmatrix} 4 & 0 & -3 \\ 1 & 1 & 1 \\ -3 & 0 & 2 \end{bmatrix}$

*Use the Gauss–Jordan method to find $\mathbf{A}^{-1}$, if it exists. Check your answers by using a graphing calculator to find $\mathbf{A}^{-1}\mathbf{A}$ and $\mathbf{A}\mathbf{A}^{-1}$.*

**5.** $\mathbf{A} = \begin{bmatrix} 3 & 2 \\ 5 & 3 \end{bmatrix}$

**6.** $\mathbf{A} = \begin{bmatrix} 3 & 5 \\ 1 & 2 \end{bmatrix}$

**7.** $\mathbf{A} = \begin{bmatrix} 6 & 9 \\ 4 & 6 \end{bmatrix}$

**8.** $\mathbf{A} = \begin{bmatrix} -4 & -6 \\ 2 & 3 \end{bmatrix}$

**GCM**

**9.** $\mathbf{A} = \begin{bmatrix} 3 & 1 & 0 \\ 1 & 1 & 1 \\ 1 & -1 & 2 \end{bmatrix}$

**10.** $\mathbf{A} = \begin{bmatrix} 1 & 0 & 1 \\ 2 & 1 & 0 \\ 1 & -1 & 1 \end{bmatrix}$

**11.** $\mathbf{A} = \begin{bmatrix} 1 & -4 & 8 \\ 1 & -3 & 2 \\ 2 & -7 & 10 \end{bmatrix}$

**12.** $\mathbf{A} = \begin{bmatrix} -2 & 5 & 3 \\ 4 & -1 & 3 \\ 7 & -2 & 5 \end{bmatrix}$

*Use a graphing calculator to find $\mathbf{A}^{-1}$, if it exists.*

**13.** $\mathbf{A} = \begin{bmatrix} 4 & -3 \\ 1 & -2 \end{bmatrix}$

**14.** $\mathbf{A} = \begin{bmatrix} 0 & -1 \\ 1 & 0 \end{bmatrix}$

**15.** $\mathbf{A} = \begin{bmatrix} 2 & 3 & 2 \\ 3 & 3 & 4 \\ -1 & -1 & -1 \end{bmatrix}$

**16.** $\mathbf{A} = \begin{bmatrix} 1 & 2 & 3 \\ 2 & -1 & -2 \\ -1 & 3 & 3 \end{bmatrix}$

**17.** $\mathbf{A} = \begin{bmatrix} 1 & 2 & -1 \\ -2 & 0 & 1 \\ 1 & -1 & 0 \end{bmatrix}$

**18.** $\mathbf{A} = \begin{bmatrix} 7 & -1 & -9 \\ 2 & 0 & -4 \\ -4 & 0 & 6 \end{bmatrix}$

**19.** $\mathbf{A} = \begin{bmatrix} 1 & 3 & -1 \\ 0 & 2 & -1 \\ 1 & 1 & 0 \end{bmatrix}$

**20.** $\mathbf{A} = \begin{bmatrix} -1 & 0 & -1 \\ -1 & 1 & 0 \\ 0 & 1 & 1 \end{bmatrix}$

**21.** $\mathbf{A} = \begin{bmatrix} 1 & 2 & 3 & 4 \\ 0 & 1 & 3 & -5 \\ 0 & 0 & 1 & -2 \\ 0 & 0 & 0 & -1 \end{bmatrix}$

**22.** $\mathbf{A} = \begin{bmatrix} -2 & -3 & 4 & 1 \\ 0 & 1 & 1 & 0 \\ 0 & 4 & -6 & 1 \\ -2 & -2 & 5 & 1 \end{bmatrix}$

**23.** $\mathbf{A} = \begin{bmatrix} 1 & -14 & 7 & 38 \\ -1 & 2 & 1 & -2 \\ 1 & 2 & -1 & -6 \\ 1 & -2 & 3 & 6 \end{bmatrix}$

**24.** $\mathbf{A} = \begin{bmatrix} 10 & 20 & -30 & 15 \\ 3 & -7 & 14 & -8 \\ -7 & -2 & -1 & 2 \\ 4 & 4 & -3 & 1 \end{bmatrix}$

*In Exercises 25–28, a system of equations is given, together with the inverse of the coefficient matrix. Use the inverse of the coefficient matrix to solve the system of equations.*

**25.** $\begin{aligned} 11x + 3y &= -4, \\ 7x + 2y &= 5; \end{aligned}$ $\mathbf{A}^{-1} = \begin{bmatrix} 2 & -3 \\ -7 & 11 \end{bmatrix}$

**26.** $\begin{aligned} 8x + 5y &= -6, \\ 5x + 3y &= 2; \end{aligned}$ $\mathbf{A}^{-1} = \begin{bmatrix} -3 & 5 \\ 5 & -8 \end{bmatrix}$

**27.** $\begin{aligned} 3x + y &= 2, \\ 2x - y + 2z &= -5, \\ x + y + z &= 5; \end{aligned}$ $\mathbf{A}^{-1} = \frac{1}{9} \begin{bmatrix} 3 & 1 & -2 \\ 0 & -3 & 6 \\ -3 & 2 & 5 \end{bmatrix}$

**28.** $\begin{aligned} y - z &= -4, \\ 4x + y &= -3, \\ 3x - y + 3z &= 1; \end{aligned}$ $\mathbf{A}^{-1} = \frac{1}{5} \begin{bmatrix} -3 & 2 & -1 \\ 12 & -3 & 4 \\ 7 & -3 & 4 \end{bmatrix}$

*Solve the system of equations using the inverse of the coefficient matrix of the equivalent matrix equation.*

**29.** $\begin{aligned} 4x + 3y &= 2, \\ x - 2y &= 6 \end{aligned}$

**30.** $\begin{aligned} 2x - 3y &= 7, \\ 4x + y &= -7 \end{aligned}$

**31.** $\begin{aligned} 5x + y &= 2, \\ 3x - 2y &= -4 \end{aligned}$

**32.** $\begin{aligned} x - 6y &= 5, \\ -x + 4y &= -5 \end{aligned}$

**33.** $\begin{aligned} x + z &= 1, \\ 2x + y &= 3, \\ x - y + z &= 4 \end{aligned}$

**34.** $\begin{aligned} x + 2y + 3z &= -1, \\ 2x - 3y + 4z &= 2, \\ -3x + 5y - 6z &= 4 \end{aligned}$

**35.** $\begin{aligned} 2x + 3y + 4z &= 2, \\ x - 4y + 3z &= 2, \\ 5x + y + z &= -4 \end{aligned}$

**36.** $\begin{aligned} x + y \hspace{1.5em} &= 2, \\ 3x \hspace{1.5em} + 2z &= 5, \\ 2x + 3y - 3z &= 9 \end{aligned}$

**37.** $\begin{aligned} 2w - 3x + 4y - 5z &= 0, \\ 3w - 2x + 7y - 3z &= 2, \\ w + x - y + z &= 1, \\ -w - 3x - 6y + 4z &= 6 \end{aligned}$

**38.** $\begin{aligned} 5w - 4x + 3y - 2z &= -6, \\ w + 4x - 2y + 3z &= -5, \\ 2w - 3x + 6y - 9z &= 14, \\ 3w - 5x + 2y - 4z &= -3 \end{aligned}$

**39.** *Sales.* Kayla sold a total of 145 Italian sausages and hot dogs from her curbside pushcart and collected $242.05. She sold 45 more hot dogs than sausages. How many of each did she sell?

**40.** *Price of School Supplies.* Rubio bought 4 lab record books and 3 highlighters for $17.83. Marcus bought 3 lab record books and 2 highlighters for $13.05. Find the price of each item.

**41.** *Cost of Materials.* Green-Up Landscaping bought 4 tons of topsoil, 3 tons of mulch, and 6 tons of pea gravel for $2825. The next week the firm bought 5 tons of topsoil, 2 tons of mulch, and 5 tons of pea gravel for $2663. Pea gravel costs $17 less per ton than topsoil. Find the price per ton for each item.

**42.** *Investment.* Donna receives $230 per year in simple interest from three investments totaling $8500. Part is invested at 2.2%, part at 2.65%, and the rest at 3.05%. There is $1500 more invested at 3.05% than at 2.2%. Find the amount invested at each rate.

## Skill Maintenance

*Use synthetic division to find the function values.*

**43.** $f(x) = x^3 - 6x^2 + 4x - 8$; find $f(-2)$

**44.** $f(x) = 2x^4 - x^3 + 5x^2 + 6x - 4$; find $f(3)$

*Solve.*

**45.** $2x^2 + x = 7$

**46.** $\dfrac{1}{x + 1} - \dfrac{6}{x - 1} = 1$

**47.** $\sqrt{2x + 1} - 1 = \sqrt{2x - 4}$

**48.** $x - \sqrt{x - 6} = 0$

*Factor the polynomial $f(x)$.*

**49.** $f(x) = x^3 - 3x^2 - 6x + 8$

**50.** $f(x) = x^4 + 2x^3 - 16x^2 - 2x + 15$

## Synthesis

*State the conditions under which $\mathbf{A}^{-1}$ exists. Then find a formula for $\mathbf{A}^{-1}$.*

**51.** $\mathbf{A} = \begin{bmatrix} x \end{bmatrix}$

**52.** $\mathbf{A} = \begin{bmatrix} x & 0 \\ 0 & y \end{bmatrix}$

**53.** $\mathbf{A} = \begin{bmatrix} 0 & 0 & x \\ 0 & y & 0 \\ z & 0 & 0 \end{bmatrix}$

**54.** $\mathbf{A} = \begin{bmatrix} x & 1 & 1 & 1 \\ 0 & y & 0 & 0 \\ 0 & 0 & z & 0 \\ 0 & 0 & 0 & w \end{bmatrix}$

## Determinants and Cramer's Rule

### 6.6

- Evaluate determinants of square matrices.
- Use Cramer's rule to solve systems of equations.

---

### ■ Determinants of Square Matrices

With every square matrix, we associate a number called its *determinant*.

---

**DETERMINANT OF A 2 × 2 MATRIX**

The **determinant** of the matrix $\begin{bmatrix} a & c \\ b & d \end{bmatrix}$ is denoted $\begin{vmatrix} a & c \\ b & d \end{vmatrix}$ and is defined as

$$\begin{vmatrix} a & c \\ b & d \end{vmatrix} = ad - bc.$$

---

**EXAMPLE 1**   Evaluate: $\begin{vmatrix} \sqrt{2} & -3 \\ -4 & -\sqrt{2} \end{vmatrix}$.

*Solution*

$$\begin{vmatrix} \sqrt{2} & -3 \\ -4 & -\sqrt{2} \end{vmatrix}$$   The arrows indicate the products involved.

$$= \sqrt{2}(-\sqrt{2}) - (-4)(-3)$$
$$= -2 - 12$$
$$= -14$$

> **Now Try Exercise 1.**

We now consider a way to evaluate determinants of square matrices of order 3 × 3 or higher.

### ■ Evaluating Determinants Using Cofactors

Often we first find minors and cofactors of matrices in order to evaluate determinants.

---

**MINOR**

For a square matrix $\mathbf{A} = \begin{bmatrix} a_{ij} \end{bmatrix}$, the **minor** $M_{ij}$ of an entry $a_{ij}$ is the determinant of the matrix formed by deleting the $i$th row and the $j$th column of $\mathbf{A}$.

---

**EXAMPLE 2**   For the matrix

$$\mathbf{A} = \begin{bmatrix} a_{ij} \end{bmatrix} = \begin{bmatrix} -8 & 0 & 6 \\ 4 & -6 & 7 \\ -1 & -3 & 5 \end{bmatrix},$$

find each of the following.

**a)** $M_{11}$

**b)** $M_{23}$

*Solution*

**a)** For $M_{11}$, we delete the first row and the first column and find the determinant of the $2 \times 2$ matrix formed by the remaining entries.

$$\begin{bmatrix} -8 & 0 & 6 \\ 4 & -6 & 7 \\ -1 & -3 & 5 \end{bmatrix} \qquad \begin{aligned} M_{11} &= \begin{vmatrix} -6 & 7 \\ -3 & 5 \end{vmatrix} \\ &= (-6) \cdot 5 - (-3) \cdot 7 \\ &= -30 - (-21) \\ &= -30 + 21 \\ &= -9 \end{aligned}$$

**b)** For $M_{23}$, we delete the second row and the third column and find the determinant of the $2 \times 2$ matrix formed by the remaining entries.

$$\begin{bmatrix} -8 & 0 & 6 \\ 4 & -6 & 7 \\ -1 & -3 & 5 \end{bmatrix} \qquad \begin{aligned} M_{23} &= \begin{vmatrix} -8 & 0 \\ -1 & -3 \end{vmatrix} \\ &= -8(-3) - (-1)0 \\ &= 24 \end{aligned}$$

**Now Try Exercise 9.**

---

**COFACTOR**

For a square matrix $\mathbf{A} = \begin{bmatrix} a_{ij} \end{bmatrix}$, the **cofactor** $A_{ij}$ of an entry $a_{ij}$ is given by

$$A_{ij} = (-1)^{i+j} M_{ij},$$

where $M_{ij}$ is the minor of $a_{ij}$.

---

**EXAMPLE 3**   For the matrix given in Example 2, find each of the following.

**a)** $A_{11}$

**b)** $A_{23}$

*Solution*

**a)** In Example 2, we found that $M_{11} = -9$. Then

$$A_{11} = (-1)^{1+1} M_{11} = (1)(-9) = -9.$$

**b)** In Example 2, we found that $M_{23} = 24$. Then

$$A_{23} = (-1)^{2+3} M_{24} = (-1)(24) = -24.$$

**Now Try Exercise 11.**

> Note that minors and cofactors are *numbers*. They are *not matrices*.

Consider the matrix **A** given by

$$\mathbf{A} = \begin{bmatrix} a_{11} & a_{12} & a_{13} \\ a_{21} & a_{22} & a_{23} \\ a_{31} & a_{32} & a_{33} \end{bmatrix}.$$

The determinant of the matrix, denoted $|\mathbf{A}|$, can be found by multiplying each element of the first column by its cofactor and adding:

$$|\mathbf{A}| = a_{11}A_{11} + a_{21}A_{21} + a_{31}A_{31}.$$

Because

$$A_{11} = (-1)^{1+1}M_{11} = M_{11},$$
$$A_{21} = (-1)^{2+1}M_{21} = -M_{21},$$

and $\quad A_{31} = (-1)^{3+1}M_{31} = M_{31},$

we can write

$$|\mathbf{A}| = a_{11} \cdot \begin{vmatrix} a_{22} & a_{23} \\ a_{32} & a_{33} \end{vmatrix} - a_{21} \cdot \begin{vmatrix} a_{12} & a_{13} \\ a_{32} & a_{33} \end{vmatrix} + a_{31} \cdot \begin{vmatrix} a_{12} & a_{13} \\ a_{22} & a_{23} \end{vmatrix}.$$

It can be shown that we can determine $|\mathbf{A}|$ by choosing *any* row or column, multiplying each element in that row or column by its cofactor, and adding. This is called *expanding* across a row or down a column. We just expanded down the first column. We now define the determinant of a square matrix of any order.

---

### DETERMINANT OF ANY SQUARE MATRIX

For any square matrix **A** of order $n \times n \ (n > 1)$, we define the **determinant** of **A**, denoted $|\mathbf{A}|$, as follows. Choose any row or column. Multiply each element in that row or column by its cofactor and add the results. The determinant of a $1 \times 1$ matrix is simply the element of the matrix. The value of a determinant will be the same no matter which row or column is chosen.

---

**EXAMPLE 4** Evaluate $|\mathbf{A}|$ by expanding across the third row:

$$\mathbf{A} = \begin{bmatrix} -8 & 0 & 6 \\ 4 & -6 & 7 \\ -1 & -3 & 5 \end{bmatrix}.$$

*Solution*   We have

$$|\mathbf{A}| = (-1)A_{31} + (-3)A_{32} + 5A_{33}$$

$$= (-1)(-1)^{3+1} \cdot \begin{vmatrix} 0 & 6 \\ -6 & 7 \end{vmatrix} + (-3)(-1)^{3+2} \cdot \begin{vmatrix} -8 & 6 \\ 4 & 7 \end{vmatrix}$$

$$+ 5(-1)^{3+3} \cdot \begin{vmatrix} -8 & 0 \\ 4 & -6 \end{vmatrix}$$

$$= (-1) \cdot 1 \cdot \left[ 0 \cdot 7 - (-6)6 \right] + (-3)(-1)\left[ -8 \cdot 7 - 4 \cdot 6 \right]$$

$$+ 5 \cdot 1 \cdot \left[ -8(-6) - 4 \cdot 0 \right]$$

$$= -\left[ 36 \right] + 3\left[ -80 \right] + 5\left[ 48 \right]$$

$$= -36 - 240 + 240 = -36.$$

The value of this determinant is $-36$ no matter which row or column we expand on.    **Now Try Exercise 13.**

Determinants can also be evaluated using a graphing calculator.

**GCM**   **EXAMPLE 5**   Use a graphing calculator to evaluate $|\mathbf{A}|$:

$$\mathbf{A} = \begin{bmatrix} 1 & 6 & -1 \\ -3 & -5 & 3 \\ 0 & 4 & 2 \end{bmatrix}.$$

```
det ([A])
                26
```

*Solution*   First, we enter $\mathbf{A}$. Then we select the determinant operation, det, from the MATRIX MATH menu and enter the name of the matrix, $\mathbf{A}$. The calculator will return the value of the determinant of the matrix, 26.

**Now Try Exercise 17.**

## ▪ Cramer's Rule

Determinants can be used to solve systems of linear equations. Consider a system of two linear equations:

$$a_1 x + b_1 y = c_1,$$
$$a_2 x + b_2 y = c_2.$$

Solving this system using the elimination method, we obtain

$$x = \frac{c_1 b_2 - c_2 b_1}{a_1 b_2 - a_2 b_1} \quad \text{and} \quad y = \frac{a_1 c_2 - a_2 c_1}{a_1 b_2 - a_2 b_1}.$$

The numerators and the denominators of these expressions can be written as determinants:

$$x = \frac{\begin{vmatrix} c_1 & b_1 \\ c_2 & b_2 \end{vmatrix}}{\begin{vmatrix} a_1 & b_1 \\ a_2 & b_2 \end{vmatrix}} \quad \text{and} \quad y = \frac{\begin{vmatrix} a_1 & c_1 \\ a_2 & c_2 \end{vmatrix}}{\begin{vmatrix} a_1 & b_1 \\ a_2 & b_2 \end{vmatrix}}.$$

If we let

$$D = \begin{vmatrix} a_1 & b_1 \\ a_2 & b_2 \end{vmatrix}, \qquad D_x = \begin{vmatrix} c_1 & b_1 \\ c_2 & b_2 \end{vmatrix}, \quad \text{and} \quad D_y = \begin{vmatrix} a_1 & c_1 \\ a_2 & c_2 \end{vmatrix},$$

we have

$$x = \frac{D_x}{D} \quad \text{and} \quad y = \frac{D_y}{D}.$$

This procedure for solving systems of equations is known as *Cramer's rule.*

---

### CRAMER'S RULE FOR 2 × 2 SYSTEMS

The solution of the system of equations

$$a_1 x + b_1 y = c_1,$$
$$a_2 x + b_2 y = c_2$$

is given by

$$x = \frac{D_x}{D}, \qquad y = \frac{D_y}{D},$$

where

$$D = \begin{vmatrix} a_1 & b_1 \\ a_2 & b_2 \end{vmatrix}, \qquad D_x = \begin{vmatrix} c_1 & b_1 \\ c_2 & b_2 \end{vmatrix},$$

$$D_y = \begin{vmatrix} a_1 & c_1 \\ a_2 & c_2 \end{vmatrix}, \quad \text{and} \quad D \neq 0.$$

---

Note that the denominator $D$ contains the coefficients of $x$ and $y$, in the same position as in the original equations. For $x$, the numerator is obtained by replacing the $x$-coefficients in $D$ (the $a$'s) with the $c$'s. For $y$, the numerator is obtained by replacing the $y$-coefficients in $D$ (the $b$'s) with the $c$'s.

GCM    **EXAMPLE 6**    Solve using Cramer's rule:

$$2x + 5y = 7,$$
$$5x - 2y = -3.$$

## Algebraic Solution

We have

$$x = \frac{\begin{vmatrix} 7 & 5 \\ -3 & -2 \end{vmatrix}}{\begin{vmatrix} 2 & 5 \\ 5 & -2 \end{vmatrix}} = \frac{7(-2) - (-3)5}{2(-2) - 5 \cdot 5}$$

$$= \frac{-14 + 15}{-4 - 25}$$

$$= \frac{1}{-29} = -\frac{1}{29};$$

$$y = \frac{\begin{vmatrix} 2 & 7 \\ 5 & 3 \end{vmatrix}}{\begin{vmatrix} 2 & 5 \\ 5 & -2 \end{vmatrix}} = \frac{2(-3) - 5 \cdot 7}{-29}$$

$$= \frac{-6 - 35}{-29}$$

$$= \frac{-41}{-29} = \frac{41}{29}.$$

The solution is $\left(-\frac{1}{29}, \frac{41}{29}\right)$.

## Graphical Solution

To use Cramer's rule to solve this system of equations on a graphing calculator, we first enter the matrices corresponding to $D$, $D_x$, and $D_y$. We enter

$$\mathbf{A} = \begin{bmatrix} 2 & 5 \\ 5 & -2 \end{bmatrix}, \qquad \mathbf{B} = \begin{bmatrix} 7 & 5 \\ -3 & -2 \end{bmatrix},$$

and $\quad \mathbf{C} = \begin{bmatrix} 2 & 7 \\ 5 & -3 \end{bmatrix}.$

Then

$$x = \frac{\det(\mathbf{B})}{\det(\mathbf{A})} \quad \text{and} \quad y = \frac{\det(\mathbf{C})}{\det(\mathbf{A})}.$$

```
det ([B])/det([A]) ▶ Frac
                              -1/29
det ([C])/det([A]) ▶ Frac
                              41/29
```

The solution is $\left(-\frac{1}{29}, \frac{41}{29}\right)$.

**Now Try Exercise 31.**

Cramer's rule works only when a system of equations has a unique solution. This occurs when $D \neq 0$. If $D = 0$ and $D_x$ and $D_y$ are also 0, then the equations are dependent. If $D = 0$ and $D_x$ and/or $D_y$ is not 0, then the system is inconsistent.

Cramer's rule can be extended to a system of $n$ linear equations in $n$ variables. We consider a $3 \times 3$ system.

## CRAMER'S RULE FOR 3 × 3 SYSTEMS

The solution of the system of equations

$$a_1x + b_1y + c_1z = d_1,$$
$$a_2x + b_2y + c_2z = d_2,$$
$$a_3x + b_3y + c_3z = d_3$$

is given by

$$x = \frac{D_x}{D}, \qquad y = \frac{D_y}{D}, \qquad z = \frac{D_z}{D},$$

where

$$D = \begin{vmatrix} a_1 & b_1 & c_1 \\ a_2 & b_2 & c_2 \\ a_3 & b_3 & c_3 \end{vmatrix}, \qquad D_x = \begin{vmatrix} d_1 & b_1 & c_1 \\ d_2 & b_2 & c_2 \\ d_3 & b_3 & c_3 \end{vmatrix},$$

$$D_y = \begin{vmatrix} a_1 & d_1 & c_1 \\ a_2 & d_2 & c_2 \\ a_3 & d_3 & c_3 \end{vmatrix}, \qquad D_z = \begin{vmatrix} a_1 & b_1 & d_1 \\ a_2 & b_2 & d_2 \\ a_3 & b_3 & d_3 \end{vmatrix}, \quad \text{and} \quad D \neq 0.$$

Note that the determinant $D_x$ is obtained from $D$ by replacing the $x$-coefficients with $d_1$, $d_2$, and $d_3$. $D_y$ and $D_z$ are obtained in a similar manner. As with a system of two equations, Cramer's rule cannot be used if $D = 0$. If $D = 0$ and $D_x$, $D_y$, and $D_z$ are 0, the equations are dependent. If $D = 0$ and at least one of $D_x$, $D_y$, or $D_z$ is not 0, then the system is inconsistent.

**EXAMPLE 7**   Solve using Cramer's rule:

$$x - 3y + 7z = 13,$$
$$x + y + z = 1,$$
$$x - 2y + 3z = 4.$$

*Solution*   We have

$$D = \begin{vmatrix} 1 & -3 & 7 \\ 1 & 1 & 1 \\ 1 & -2 & 3 \end{vmatrix} = -10, \qquad D_x = \begin{vmatrix} 13 & -3 & 7 \\ 1 & 1 & 1 \\ 4 & -2 & 3 \end{vmatrix} = 20,$$

$$D_y = \begin{vmatrix} 1 & 13 & 7 \\ 1 & 1 & 1 \\ 1 & 4 & 3 \end{vmatrix} = -6, \qquad D_z = \begin{vmatrix} 1 & -3 & 13 \\ 1 & 1 & 1 \\ 1 & -2 & 4 \end{vmatrix} = -24.$$

Then

$$x = \frac{D_x}{D} = \frac{20}{-10} = -2, \qquad y = \frac{D_y}{D} = \frac{-6}{-10} = \frac{3}{5}, \qquad z = \frac{D_z}{D} = \frac{-24}{-10} = \frac{12}{5}.$$

The solution is $\left(-2, \frac{3}{5}, \frac{12}{5}\right)$.

In practice, it is not necessary to evaluate $D_z$. When we have found values for $x$ and $y$, we can substitute them into one of the equations to find $z$.

**Now Try Exercise 39.**

# 6.6 Exercise Set

*Evaluate the determinant.*

**1.** $\begin{vmatrix} 5 & 3 \\ -2 & -4 \end{vmatrix}$
    **2.** $\begin{vmatrix} -8 & 6 \\ -1 & 2 \end{vmatrix}$

**3.** $\begin{vmatrix} 4 & -7 \\ -2 & 3 \end{vmatrix}$
    **4.** $\begin{vmatrix} -9 & -6 \\ 5 & 4 \end{vmatrix}$

**5.** $\begin{vmatrix} 2 & -\sqrt{5} \\ -\sqrt{5} & 3 \end{vmatrix}$
    **6.** $\begin{vmatrix} \sqrt{5} & -3 \\ 4 & 2 \end{vmatrix}$

**7.** $\begin{vmatrix} x & 4 \\ x & x^2 \end{vmatrix}$
    **8.** $\begin{vmatrix} y^2 & -2 \\ y & 3 \end{vmatrix}$

*Use the following matrix for Exercises 9–17:*

$$\mathbf{A} = \begin{bmatrix} 7 & -4 & -6 \\ 2 & 0 & -3 \\ 1 & 2 & -5 \end{bmatrix}.$$

**9.** Find $M_{11}$, $M_{32}$, and $M_{22}$.

**10.** Find $M_{13}$, $M_{31}$, and $M_{23}$.

**11.** Find $A_{11}$, $A_{32}$, and $A_{22}$.

**12.** Find $A_{13}$, $A_{31}$, and $A_{23}$.

**13.** Evaluate $|\mathbf{A}|$ by expanding across the second row.

**14.** Evaluate $|\mathbf{A}|$ by expanding down the second column.

**15.** Evaluate $|\mathbf{A}|$ by expanding down the third column.

**16.** Evaluate $|\mathbf{A}|$ by expanding across the first row.

**17.** Use a graphing calculator to evaluate $|\mathbf{A}|$.

*Use the following matrix for Exercises 18–24:*

$$\mathbf{A} = \begin{bmatrix} 1 & 0 & 0 & -2 \\ 4 & 1 & 0 & 0 \\ 5 & 6 & 7 & 8 \\ -2 & -3 & -1 & 0 \end{bmatrix}.$$

**18.** Find $M_{12}$ and $M_{44}$.

**19.** Find $M_{41}$ and $M_{33}$.

**20.** Find $A_{22}$ and $A_{34}$.

**21.** Find $A_{24}$ and $A_{43}$.

**22.** Evaluate $|\mathbf{A}|$ by expanding down the third column.

**23.** Evaluate $|\mathbf{A}|$ by expanding across the first row.

**24.** Use a graphing calculator to evaluate $|\mathbf{A}|$.

*Evaluate the determinant.*

**25.** $\begin{vmatrix} 3 & 1 & 2 \\ -2 & 3 & 1 \\ 3 & 4 & -6 \end{vmatrix}$
    **26.** $\begin{vmatrix} 3 & -2 & 1 \\ 2 & 4 & 3 \\ -1 & 5 & 1 \end{vmatrix}$

**27.** $\begin{vmatrix} x & 0 & -1 \\ 2 & x & x^2 \\ -3 & x & 1 \end{vmatrix}$
    **28.** $\begin{vmatrix} x & 1 & -1 \\ x^2 & x & x \\ 0 & x & 1 \end{vmatrix}$

*Solve using Cramer's rule.*

**29.** $-2x + 4y = 3,$
     $3x - 7y = 1$
    **30.** $5x - 4y = -3,$
     $7x + 2y = 6$

**31.** $2x - y = 5,$
     $x - 2y = 1$
    **32.** $3x + 4y = -2,$
     $5x - 7y = 1$

**33.** $2x + 9y = -2,$
     $4x - 3y = 3$
    **34.** $2x + 3y = -1,$
     $3x + 6y = -0.5$

**35.** $2x + 5y = 7,$
     $3x - 2y = 1$
    **36.** $3x + 2y = 7,$
     $2x + 3y = -2$

**37.** $3x + 2y - z = 4,$
     $3x - 2y + z = 5,$
     $4x - 5y - z = -1$

**38.** $3x - y + 2z = 1,$
     $x - y + 2z = 3,$
     $-2x + 3y + z = 1$

**39.** $3x + 5y - z = -2,$
     $x - 4y + 2z = 13,$
     $2x + 4y + 3z = 1$

**40.** $3x + 2y + 2z = 1,$
     $5x - y - 6z = 3,$
     $2x + 3y + 3z = 4$

**41.** $x - 3y - 7z = 6,$
     $2x + 3y + z = 9,$
     $4x + y = 7$

**42.** $x - 2y - 3z = 4,$
     $3x - 2z = 8,$
     $2x + y + 4z = 13$

**43.**
$$6y + 6z = -1,$$
$$8x \quad\quad + 6z = -1,$$
$$4x + 9y \quad\quad = 8$$

**44.** $3x + 5y \quad\quad = 2,$
$\quad\; 2x \quad\quad - 3z = 7,$
$\quad\quad\quad 4y + 2z = -1$

## Skill Maintenance

*Determine whether the function is one-to-one, and if it is, find a formula for $f^{-1}(x)$.*

**45.** $f(x) = 3x + 2$

**46.** $f(x) = x^2 - 4$

**47.** $f(x) = |x| + 3$

**48.** $f(x) = \sqrt[3]{x} + 1$

*Simplify. Write answers in the form $a + bi$, where $a$ and $b$ are real numbers.*

**49.** $(3 - 4i) - (-2 - i)$  **50.** $(5 + 2i) + (1 - 4i)$

**51.** $(1 - 2i)(6 + 2i)$  **52.** $\dfrac{3 + i}{4 - 3i}$

## Synthesis

*Solve.*

**53.** $\begin{vmatrix} y & 2 \\ 3 & y \end{vmatrix} = y$

**54.** $\begin{vmatrix} x & -3 \\ -1 & x \end{vmatrix} \geq 0$

**55.** $\begin{vmatrix} 2 & x & 1 \\ 1 & 2 & -1 \\ 3 & 4 & -2 \end{vmatrix} = -6$

**56.** $\begin{vmatrix} m + 2 & -3 \\ m + 5 & -4 \end{vmatrix} = 3m - 5$

*Rewrite the expression using a determinant. Answers may vary.*

**57.** $a^2 + b^2$  **58.** $\frac{1}{2}h(a + b)$

**59.** $2\pi r^2 + 2\pi rh$  **60.** $x^2 y^2 - Q^2$

---

## 6.7 Systems of Inequalities and Linear Programming

- Graph linear inequalities.
- Graph systems of linear inequalities.
- Solve linear programming problems.

A graph of an inequality is a drawing that represents its solutions. We have already seen that an inequality in one variable can be graphed on the number line. An inequality in two variables can be graphed on a coordinate plane.

### ▪ Graphs of Linear Inequalities

A statement like $5x - 4y < 20$ is a linear inequality in two variables.

> **LINEAR INEQUALITY IN TWO VARIABLES**
>
> A **linear inequality in two variables** is an inequality that can be written in the form
>
> $$Ax + By < C,$$
>
> where $A$, $B$, and $C$ are real numbers and $A$ and $B$ are not both zero. The symbol $<$ may be replaced with $\leq$, $>$, or $\geq$.

A solution of a linear inequality in two variables is an ordered pair $(x, y)$ for which the inequality is true. For example, $(1, 3)$ is a solution of $5x - 4y < 20$ because $5 \cdot 1 - 4 \cdot 3 < 20$, or $-7 < 20$, is true. On the other hand, $(2, -6)$ is not a solution of $5x - 4y < 20$ because $5 \cdot 2 - 4 \cdot (-6) \not< 20$, or $34 \not< 20$.

The **solution set** of an inequality is the set of all ordered pairs that make it true. The **graph of an inequality** represents its solution set.

**GCM**

**EXAMPLE 1**   Graph: $y < x + 3$.

***Solution***   We begin by graphing the **related equation** $y = x + 3$. We use a dashed line because the inequality symbol is $<$. This indicates that the line itself is not in the solution set of the inequality.

Note that the line divides the coordinate plane into two regions called **half-planes**. One of these half-planes satisfies the inequality. Either *all* points in a half-plane are in the solution set of the inequality or *none* is.

To determine which half-plane satisfies the inequality, we try a test point in either region. The point $(0, 0)$ is usually a convenient choice so long as it does not lie on the line.

$$y < x + 3$$

$$\overline{0 \; \overset{?}{\vert} \; 0 + 3}$$

$$0 \; \vert \; 3 \qquad \textbf{TRUE} \qquad 0 < 3 \text{ is true.}$$

Since $(0, 0)$ satisfies the inequality, so do all points in the half-plane that contains $(0, 0)$. We shade this region to show the solution set of the inequality.

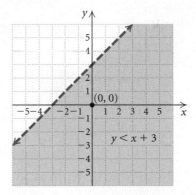

There are several ways to graph this inequality on a graphing calculator. One method is to first enter the related equation, $y = x + 3$, as shown at left. Then, after using a test point to determine which half-plane to shade as described above, we select the "shade below" graph style. Note that we must keep in mind that the line $y = x + 3$ is not included in the solution set. (Even if DOT mode or the dot graph style is selected, the line appears to be solid rather than dashed.)

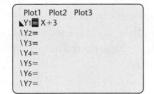

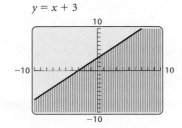

$y = x + 3$

We could also use the Shade option from the DRAW menu to graph this inequality. This method is described in the *Graphing Calculator Manual* that accompanies the text.

Some calculators have a pre-loaded application that can be used to graph an inequality. This application, Inequalz, is found on the APPS menu. The keystrokes for using Inequalz are found in the *Graphing Calculator Manual* that accompanies the text. Note that when this application is used, the inequality $y < x + 3$ is entered directly and the graph of the related equation appears as a dashed line.

$$y < x + 3$$

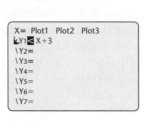
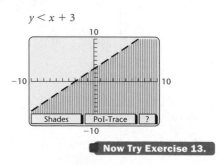

**Now Try Exercise 13.**

In general, we use the following procedure to graph linear inequalities in two variables by hand.

---

To graph a linear inequality in two variables:

1. Replace the inequality symbol with an equals sign and graph this related equation. If the inequality symbol is $<$ or $>$, draw the line dashed. If the inequality symbol is $\leq$ or $\geq$, draw the line solid.
2. The graph consists of a half-plane on one side of the line and, if the line is solid, the line as well. To determine which half-plane to shade, test a point not on the line in the original inequality. If that point is a solution, shade the half-plane containing that point. If not, shade the opposite half-plane.

---

**EXAMPLE 2** Graph: $3x + 4y \geq 12$.

*Solution*

1. First, we graph the related equation $3x + 4y = 12$. We use a solid line because the inequality symbol is $\geq$. This indicates that the line is included in the solution set.
2. To determine which half-plane to shade, we test a point in either region. We choose $(0, 0)$.

$$3x + 4y \geq 12$$
$$\overline{3 \cdot 0 + 4 \cdot 0 \overset{?}{\phantom{|}} 12}$$
$$0 \,\bigl|\, 12 \quad \text{FALSE} \qquad 0 \geq 12 \text{ is false.}$$

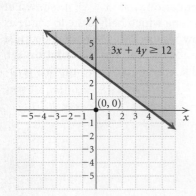

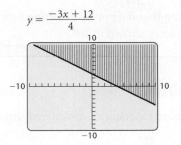

$$y = \frac{-3x + 12}{4}$$

Because $(0, 0)$ is *not* a solution, all the points in the half-plane that does *not* contain $(0, 0)$ are solutions. We shade that region, as shown in the figure on the preceding page.

To graph this inequality on a graphing calculator using either the "shade above" graph style or the Shade option from the DRAW menu, we must first solve the related equation for $y$ and enter $y = \frac{-3x + 12}{4}$.

Similarly, to use the Inequalz application, we solve the inequality for $y$ and enter $y \geq \frac{-3x + 12}{4}$.

**Now Try Exercise 17.**

GCM

**EXAMPLE 3**  Graph: $x > -3$ on a plane.

**Solution**

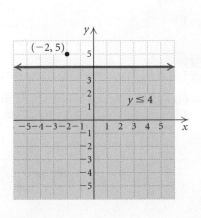

1. First, we graph the related equation $x = -3$. We use a dashed line because the inequality symbol is $>$. This indicates that the line is not included in the solution set.

2. The inequality tells us that all points $(x, y)$ for which $x > -3$ are solutions. These are the points to the right of the line. We can also use a test point to determine the solutions. We choose $(5, 1)$.

$$\frac{x > -3}{5 \ ? \ -3} \quad \text{TRUE} \qquad 5 > -3 \text{ is true.}$$

Because $(5, 1)$ is a solution, we shade the region containing that point— that is, the region to the right of the dashed line.

We can also graph this inequality on a graphing calculator that has the Inequalz application on the APPS menu.

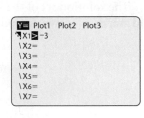    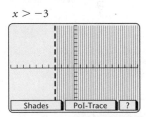

**Now Try Exercise 23.**

**EXAMPLE 4**  Graph $y \leq 4$ on a plane.

**Solution**

1. First, we graph the related equation $y = 4$. We use a solid line because the inequality symbol is $\leq$.

2. The inequality tells us that all points $(x, y)$ for which $y \leq 4$ are solutions of the inequality. These are the points on or below the line. We can also use a test point to determine the solutions. We choose $(-2, 5)$.

$$\frac{y \leq 4}{5 \ ? \ 4} \quad \text{FALSE} \qquad 5 \leq 4 \text{ is false.}$$

Because $(-2, 5)$ is not a solution, we shade the half-plane that does not contain that point.

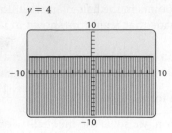

$y = 4$

We can graph the inequality $y \leq 4$ using Inequalz or by graphing $y = 4$ and then using the "shade below" graph style, as shown at left.

> **Now Try Exercise 25.**

### ■ Systems of Linear Inequalities

A system of inequalities consists of two or more inequalities considered simultaneously. For example,

$$x + y \leq 4,$$
$$x - y \geq 2$$

is a system of *two linear inequalities in two variables*.

A solution of a system of inequalities is an ordered pair that is a solution of each inequality in the system. To graph a system of linear inequalities, we graph each inequality and determine the region that is common to *all* the solution sets.

GCM    **EXAMPLE 5**    Graph the solution set of the system

$$x + y \leq 4,$$
$$x - y \geq 2.$$

**Solution**    We graph $x + y \leq 4$ by first graphing the equation $x + y = 4$ using a solid line. Next, we choose $(0, 0)$ as a test point and find that it is a solution of $x + y \leq 4$, so we shade the half-plane containing $(0, 0)$ using red. Next, we graph $x - y = 2$ using a solid line. We find that $(0, 0)$ is not a solution of $x - y \geq 2$, so we shade the half-plane that does not contain $(0, 0)$ using green. The arrows near the ends of each line help to indicate the half-plane that contains each solution set.

The solution set of the system of equations is the region shaded both red and green, or brown, including parts of the lines $x + y = 4$ and $x - y = 2$.

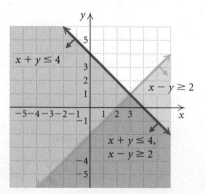

$y_1 = 4 - x, \quad y_2 = x - 2$

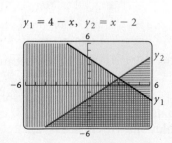

**FIGURE 1**

We can use different shading patterns on a graphing calculator to graph this system of inequalities. The solution set is the region shaded using both patterns. See Fig. 1 at left.

$y_1 \le 4 - x, \quad y_2 \le x - 2$

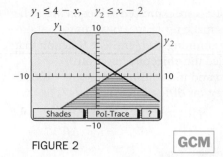

FIGURE 2 GCM

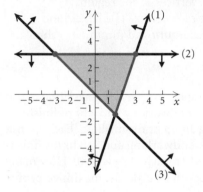

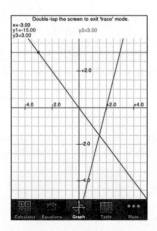

An app can also be used to find the vertices. Graph $y_1 = 3x - 6$, $y_2 = 3$, and $y_3 = -x$ and use the Intersect feature. The vertex, $(-3, 3)$, is illustrated here.

We can also use the Inequalz application to graph this system of inequalities. If we choose the Ineq Intersection option from the Shades menu, only the solution set is shaded, as shown in Fig. 2.

A system of inequalities may have a graph that consists of a polygon and its interior. As we will see later in this section, in many applications we will need to know the vertices of such a polygon.

**EXAMPLE 6** Graph the following system of inequalities and find the coordinates of any vertices formed:

$$3x - y \le 6, \quad \textbf{(1)}$$
$$y - 3 \le 0, \quad \textbf{(2)}$$
$$x + y \ge 0. \quad \textbf{(3)}$$

***Solution*** We graph the related equations $3x - y = 6$, $y - 3 = 0$, and $x + y = 0$ using solid lines. The half-plane containing the solution set for each inequality is indicated by the arrows near the ends of each line. We shade the region common to all three solution sets.

To find the vertices, we solve three systems of equations. The system of equations from inequalities (1) and (2) is

$$3x - y = 6,$$
$$y - 3 = 0.$$

Solving, we obtain the vertex $(3, 3)$.

The system of equations from inequalities (1) and (3) is

$$3x - y = 6,$$
$$x + y = 0.$$

Solving, we obtain the vertex $\left(\frac{3}{2}, -\frac{3}{2}\right)$.

The system of equations from inequalities (2) and (3) is

$$y - 3 = 0,$$
$$x + y = 0.$$

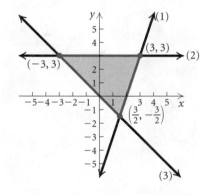

Solving, we obtain the vertex $(-3, 3)$.

If a system of inequalities is graphed on a graphing calculator using a shading option, the coordinates of the vertices can be found using the INTERSECT feature. If the Inequalz application from the APPS menu is used to graph a system of inequalities, the PoI-Trace feature can be used to find the coordinates of the vertices. This method is discussed in the *Graphing Calculator Manual* that accompanies the text.

**Now Try Exercise 55.**

## ■ Applications: Linear Programming

In many applications, we want to find a maximum value or a minimum value. In business, for example, we might want to maximize profit and minimize cost. **Linear programming** can tell us how to do this.

In our study of linear programming, we will consider linear functions of two variables that are to be maximized or minimized subject to several conditions, or **constraints**. These constraints are expressed as inequalities. The solution set of the system of inequalities made up of the constraints contains all the **feasible solutions** of a linear programming problem. The function that we want to maximize or minimize is called the **objective function**.

It can be shown that the maximum and minimum values of the objective function occur at a vertex of the region of feasible solutions. Thus we have the following procedure.

---

### LINEAR PROGRAMMING PROCEDURE

To find the maximum or minimum value of a linear objective function subject to a set of constraints:

1. Graph the region of feasible solutions.
2. Determine the coordinates of the vertices of the region.
3. Evaluate the objective function at each vertex. The largest and smallest of those values are the maximum and minimum values of the function, respectively.

---

**EXAMPLE 7**    *Maximizing Profit.*    Aspen Carpentry makes bookcases and desks. Each bookcase requires 5 hr of woodworking and 4 hr of finishing. Each desk requires 10 hr of woodworking and 3 hr of finishing. Each month the shop has 600 hr of labor available for woodworking and 240 hr for finishing. The profit on each bookcase is $75 and on each desk is $140. How many of each product should be made each month in order to maximize profit? Assume that all that are made are sold.

**Solution**    We let $x$ = the number of bookcases to be produced and $y$ = the number of desks. Then the profit $P$ is given by the function

$$P = 75x + 140y. \qquad \text{To emphasize that } P \text{ is a function of two variables,}$$
$$\text{we sometimes write } P(x, y) = 75x + 140y.$$

We know that $x$ bookcases require $5x$ hr of woodworking and $y$ desks require $10y$ hr of woodworking. Since there is no more than 600 hr of labor available for woodworking, we have one constraint:

$$5x + 10y \leq 600.$$

Similarly, the bookcases and desks require $4x$ hr and $3y$ hr of finishing, respectively. There is no more than 240 hr of labor available for finishing, so we have a second constraint:

$$4x + 3y \leq 240.$$

We also know that $x \geq 0$ and $y \geq 0$ because the carpentry shop cannot make a negative number of either product.

Thus we want to maximize the objective function $P = 75x + 140y$ subject to the constraints

$$5x + 10y \leq 600,$$
$$4x + 3y \leq 240,$$
$$x \geq 0,$$
$$y \geq 0.$$

We graph the system of inequalities and determine the vertices. Then we evaluate the objective function *P* at each vertex.

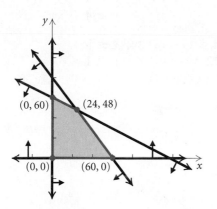

| Vertices $(x, y)$ | Profit $P = 75x + 140y$ | |
|---|---|---|
| $(0, 0)$ | $P = 75 \cdot 0 + 140 \cdot 0 = 0$ | |
| $(60, 0)$ | $P = 75 \cdot 60 + 140 \cdot 0 = 4500$ | |
| $(24, 48)$ | $P = 75 \cdot 24 + 140 \cdot 48 = 8520$ | ← Maximum |
| $(0, 60)$ | $P = 75 \cdot 0 + 140 \cdot 60 = 8400$ | |

The carpentry shop will make a maximum profit of $8520 when 24 bookcases and 48 desks are produced and sold.

We can create a table in which an objective function is evaluated at each vertex of a system of inequalities if the system has been graphed using the Inequalz application in the APPS menu. Refer to the *Graphing Calculator Manual* that accompanies the text for details.

**Now Try Exercise 65.**

---

## 6.7 Exercise Set

*In Exercises 1–8, match the inequality with one of the graphs (a)–(h), which follow.*

**a)**
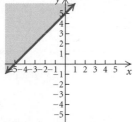

**b)**
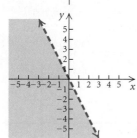

**c)**

**d)**
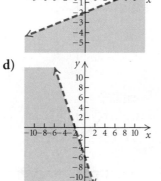

**e)**

**f)**
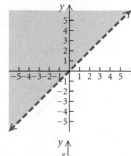

**g)**

**h)**
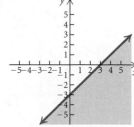

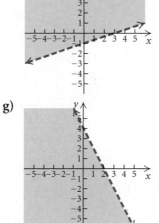

**1.** $y > x$

**2.** $y < -2x$

**3.** $y \leq x - 3$

**4.** $y \geq x + 5$

**5.** $2x + y < 4$

**6.** $3x + y < -6$

**7.** $2x - 5y > 10$

**8.** $3x - 9y < 9$

*Graph.*

**9.** $y > 2x$

**10.** $2y < x$

**11.** $y + x \geq 0$

**12.** $y - x < 0$

**13.** $y > x - 3$

**14.** $y \leq x + 4$

**15.** $x + y < 4$

**16.** $x - y \geq 5$

**17.** $3x - 2y \leq 6$

**18.** $2x - 5y < 10$

**19.** $3y + 2x \geq 6$

**20.** $2y + x \leq 4$

**21.** $3x - 2 \leq 5x + y$

**22.** $2x - 6y \geq 8 + 2y$

**23.** $x < -4$

**24.** $y > -3$

**25.** $y \geq 5$

**26.** $x \leq 5$

**27.** $-4 < y < -1$
(*Hint*: Think of this as $-4 < y$ and $y < -1$.)

**28.** $-3 \leq x \leq 3$
(*Hint*: Think of this as $-3 \leq x$ and $x \leq 3$.)

**29.** $y \geq |x|$

**30.** $y \leq |x + 2|$

*In Exercises 31–36, match the system of inequalities with one of the graphs (a)–(f), which follow.*

**a)**

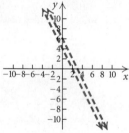

**b)**

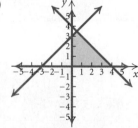

**c)**

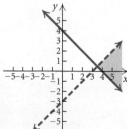

**d)**

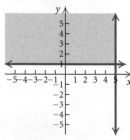

**e)**

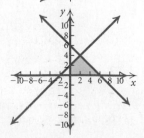

**f)**

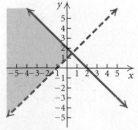

**31.** $y > x + 1$,
$y \leq 2 - x$

**32.** $y < x - 3$,
$y \geq 4 - x$

**33.** $2x + y < 4$,
$4x + 2y > 12$

**34.** $x \leq 5$,
$y \geq 1$

**35.** $x + y \leq 4$,
$x - y \geq -3$,
$x \geq 0$,
$y \geq 0$

**36.** $x - y \geq -2$,
$x + y \leq 6$,
$x \geq 0$,
$y \geq 0$

*Find a system of inequalities with the given graph. Answers may vary.*

**37.**

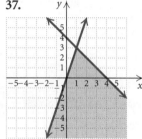

**38.**

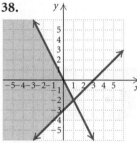

**39.**

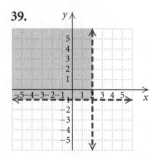

**40.**

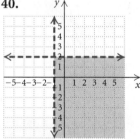

**41.**

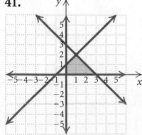

**42.**

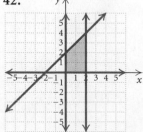

*Graph the system of inequalities. Then find the coordinates of the vertices.*

**43.** $y \leq x,$
$y \geq 3 - x$

**GCM**

**44.** $y \leq x,$
$y \geq 5 - x$

**45.** $y \geq x,$
$y \leq 4 - x$

**46.** $y \geq x,$
$y \leq 2 - x$

**47.** $y \geq -3,$
$x \geq 1$

**48.** $y \leq -2,$
$x \geq 2$

**49.** $x \leq 3,$
$y \geq 2 - 3x$

**50.** $x \geq -2,$
$y \leq 3 - 2x$

**51.** $x + y \leq 1,$
$x - y \leq 2$

**52.** $y + 3x \geq 0,$
$y + 3x \leq 2$

**53.** $2y - x \leq 2,$
$y + 3x \geq -1$

**54.** $\quad\quad y \leq 2x + 1,$
$\quad\quad y \geq -2x + 1,$
$x - 2 \leq 0$

**55.** $\quad x - y \leq 2,$
$x + 2y \geq 8,$
$y - 4 \leq 0$

**56.** $x + 2y \leq 12,$
$2x + y \leq 12,$
$x \geq 0,$
$y \geq 0$

**57.** $4y - 3x \geq -12,$
$4y + 3x \geq -36,$
$\quad\quad y \leq 0,$
$\quad\quad x \leq 0$

**58.** $8x + 5y \leq 40,$
$\quad x + 2y \leq 8,$
$\quad\quad x \geq 0,$
$\quad\quad y \geq 0$

**59.** $3x + 4y \geq 12,$
$5x + 6y \leq 30,$
$\quad 1 \leq x \leq 3$

**60.** $y - x \geq 1,$
$y - x \leq 3,$
$2 \leq x \leq 5$

*Find the maximum value and the minimum value of the function and the values of x and y for which they occur.*

**61.** $P = 17x - 3y + 60,$ subject to
$\quad\quad 6x + 8y \leq 48,$
$\quad\quad\quad 0 \leq y \leq 4,$
$\quad\quad\quad 0 \leq x \leq 7.$

**62.** $Q = 28x - 4y + 72,$ subject to
$\quad\quad 5x + 4y \geq 20,$
$\quad\quad\quad 0 \leq y \leq 4,$
$\quad\quad\quad 0 \leq x \leq 3.$

**63.** $F = 5x + 36y,$ subject to
$\quad\quad 5x + 3y \leq 34,$
$\quad\quad 3x + 5y \leq 30,$
$\quad\quad\quad x \geq 0,$
$\quad\quad\quad y \geq 0.$

**64.** $G = 16x + 14y,$ subject to
$\quad\quad 3x + 2y \leq 12,$
$\quad\quad 7x + 5y \leq 29,$
$\quad\quad\quad x \geq 0,$
$\quad\quad\quad y \geq 0.$

**65.** *Maximizing Income.* Golden Harvest Foods makes jumbo biscuits and regular biscuits. The oven can cook at most 200 biscuits per day. Each jumbo biscuit requires 2 oz of flour, each regular biscuit requires 1 oz of flour, and there is 300 oz of flour available. The income from each jumbo biscuit is $0.80 and from each regular biscuit is $0.50. How many of each size biscuit should be made in order to maximize income? What is the maximum income?

**66.** *Maximizing Mileage.* Omar owns a pickup truck and a moped. He can afford 12 gal of gasoline to be split between the truck and the moped. Omar's truck gets 20 mpg and, with the fuel currently in the tank, can hold at most an additional 10 gal of gas. His moped gets 100 mpg and can hold at most 3 gal of gas. How many gallons of gasoline should each vehicle use if Omar wants to travel as far as possible on the 12 gal of gas? What is the maximum number of miles that he can travel?

20 mpg

100 mpg

**67.** *Maximizing Profit.* Norris Mill can convert logs into lumber and plywood. In a given week, the mill can turn out 400 units of production, of which 100 units of lumber and 150 units of plywood are required by regular customers. The profit is $20 per unit of lumber and $30 per unit of plywood. How many units of each should the mill produce in order to maximize the profit?

**68.** *Maximizing Profit.*     Sunnydale Farm includes 240 acres of cropland. The farm owner wishes to plant this acreage in corn and oats. The profit per acre in corn production is $40 and in oats is $30. A total of 320 hr of labor is available. Each acre of corn requires 2 hr of labor, whereas each acre of oats requires 1 hr of labor. How should the land be divided between corn and oats in order to yield the maximum profit? What is the maximum profit?

**69.** *Minimizing Cost.*     An animal feed to be mixed from soybean meal and oats must contain at least 120 lb of protein, 24 lb of fat, and 10 lb of mineral ash. Each 100-lb sack of soybean meal costs $15 and contains 50 lb of protein, 8 lb of fat, and 5 lb of mineral ash. Each 100-lb sack of oats costs $5 and contains 15 lb of protein, 5 lb of fat, and 1 lb of mineral ash. How many sacks of each should be used to satisfy the minimum requirements at minimum cost?

**70.** *Minimizing Cost.*     Suppose that in the preceding problem the oats were replaced by alfalfa, which costs $8 per 100 lb and contains 20 lb of protein, 6 lb of fat, and 8 lb of mineral ash. How much of each is now required in order to minimize the cost?

**71.** *Maximizing Income.*     Clayton is planning to invest up to $40,000 in corporate and municipal bonds. The least he is allowed to invest in corporate bonds is $6000, and he does not want to invest more than $22,000 in corporate bonds. He also does not want to invest more than $30,000 in municipal bonds. The interest is 8% on corporate bonds and $7\frac{1}{2}$% on municipal bonds. This is simple interest for one year. How much should he invest in each type of bond in order to maximize his income? What is the maximum income?

**72.** *Maximizing Income.*     Margaret is planning to invest up to $22,000 in certificates of deposit at City Bank and People's Bank. She wants to invest at least $2000 but no more than $14,000 at City Bank. People's Bank does not insure more than a $15,000 investment, so she will invest no more than that in People's Bank. The interest is 6% at City Bank and $6\frac{1}{2}$% at People's Bank. This is

simple interest for one year. How much should she invest in each bank in order to maximize her income? What is the maximum income?

**73.** *Minimizing Transportation Cost.*     An airline with two types of airplanes, $P_1$ and $P_2$, has contracted with a tour group to provide transportation for a minimum of 2000 first-class, 1500 tourist-class, and 2400 economy-class passengers. For a certain trip, airplane $P_1$ costs $12 thousand to operate and can accommodate 40 first-class, 40 tourist-class, and 120 economy-class passengers, whereas airplane $P_2$ costs $10 thousand to operate and can accommodate 80 first-class, 30 tourist-class, and 40 economy-class passengers. How many of each type of airplane should be used in order to minimize the operating cost?

**74.** *Minimizing Transportation Cost.*     Suppose that in the preceding problem a new airplane $P_3$ becomes available, having an operating cost for the same trip of $15 thousand and accommodating 40 first-class, 40 tourist-class, and 80 economy-class passengers. If airplane $P_1$ were replaced by airplane $P_3$, how many of $P_2$ and $P_3$ should be used in order to minimize the operating cost?

**75.** *Maximizing Profit.*     It takes Just Sew 2 hr of cutting and 4 hr of sewing to make a knit suit. It takes 4 hr of cutting and 2 hr of sewing to make a worsted suit. At most 20 hr per day are available for cutting and at most 16 hr per day are available for sewing. The profit is $34 on a knit suit and $31 on a worsted suit. How many of each kind of suit should be made each day in order to maximize profit? What is the maximum profit?

**76.** *Maximizing Profit.* Cambridge Metal Works manufactures two sizes of gears. The smaller gear requires 4 hr of machining and 1 hr of polishing and yields a profit of $25. The larger gear requires 1 hr of machining and 1 hr of polishing and yields a profit of $10. The firm has available at most 24 hr per day for machining and 9 hr per day for polishing. How many of each type of gear should be produced each day in order to maximize profit? What is the maximum profit?

**77.** *Minimizing Nutrition Cost.* Suppose that it takes 12 units of carbohydrates and 6 units of protein to satisfy Jacob's minimum weekly requirements. A particular type of meat contains 2 units of carbohydrates and 2 units of protein per pound. A particular cheese contains 3 units of carbohydrates and 1 unit of protein per pound. The meat costs $3.50 per pound and the cheese costs $4.60 per pound. How many pounds of each are needed in order to minimize the cost and still meet the minimum requirements?

**78.** *Minimizing Salary Cost.* The Spring Hill school board is analyzing education costs for Hill Top School. It wants to hire teachers and teacher's aides to make up a faculty that satisfies its needs at minimum cost. The average annual salary for a teacher is $46,000 and for a teacher's aide is $21,000. The school building can accommodate a faculty of no more than 50 but needs at least 20 faculty members to function properly. The school must have at least 12 aides, but the number of teachers must be at least twice the number of aides in order to accommodate the expectations of the community. How many teachers and teacher's aides should be hired in order to minimize salary costs?

**79.** *Maximizing Animal Support in a Forest.* A certain area of forest is populated by two species of animal, which scientists refer to as A and B for simplicity. The forest supplies two kinds of food, referred to as $F_1$ and $F_2$. For one year, each member of species A requires 1 unit of $F_1$ and 0.5 unit of $F_2$. Each member of species B requires 0.2 unit of $F_1$ and 1 unit of $F_2$. The forest can normally supply at most 600 units of $F_1$ and 525 units of $F_2$ per year. What is the maximum total number of these animals that the forest can support?

**80.** *Maximizing Animal Support in a Forest.* Refer to Exercise 79. If there is a wet spring, then supplies of food increase to 1080 units of $F_1$ and 810 units of $F_2$. In this case, what is the maximum total number of these animals that the forest can support?

## Skill Maintenance

*Solve.*

**81.** $-5 \leq x + 2 < 4$

**82.** $|x - 3| \geq 2$

**83.** $x^2 - 2x \leq 3$

**84.** $\dfrac{x - 1}{x + 2} > 4$

## Synthesis

*Graph the system of inequalities.*

**85.** $y \geq x^2 - 2,$
$y \leq 2 - x^2$

**86.** $y < x + 1,$
$y \geq x^2$

*Graph the inequality.*

**87.** $|x + y| \leq 1$

**88.** $|x| + |y| \leq 1$

**89.** $|x| > |y|$

**90.** $|x - y| > 0$

**91.** *Allocation of Resources.* Dawson Handcrafted Furniture produces chairs and sofas. Each chair requires 20 ft of wood, 1 lb of foam rubber, and 2 yd$^2$ of fabric. Each sofa requires 100 ft of wood, 50 lb of foam rubber, and 20 yd$^2$ of fabric. The manufacturer has in stock 1900 ft of wood, 500 lb of foam rubber, and 240 yd$^2$ of fabric. The chairs can be sold for $120 each and the sofas for $575 each. How many of each should be produced in order to maximize income? What is the maximum income?

## 6.8

# Partial Fractions

■ Decompose rational expressions into partial fractions.

There are situations in calculus in which it is useful to write a rational expression as a sum of two or more simpler rational expressions. In the equation

$$\frac{4x - 13}{2x^2 + x - 6} = \frac{3}{x + 2} + \frac{-2}{2x - 3},$$

each fraction on the right side is called a **partial fraction**. The expression on the right side is the **partial fraction decomposition** of the rational expression on the left side. In this section, we learn how such decompositions are created.

## ■ Partial Fraction Decompositions

The procedure for finding the partial fraction decomposition of a rational expression involves factoring its denominator into linear factors and quadratic factors.

---

### PROCEDURE FOR DECOMPOSING A RATIONAL EXPRESSION INTO PARTIAL FRACTIONS

Consider any rational expression $P(x)/Q(x)$ such that $P(x)$ and $Q(x)$ have no common factor other than 1 or $-1$.

1. If the degree of $P(x)$ is greater than or equal to the degree of $Q(x)$, divide to express $P(x)/Q(x)$ as a quotient $+$ remainder$/Q(x)$ and follow steps (2)–(5) to decompose the resulting rational expression.

2. If the degree of $P(x)$ is less than the degree of $Q(x)$, factor $Q(x)$ into linear factors of the form $(px + q)^n$ and/or quadratic factors of the form $(ax^2 + bx + c)^m$. Any quadratic factor $ax^2 + bx + c$ must be *irreducible*, meaning that it cannot be factored into linear factors with rational coefficients.

3. Assign to each linear factor $(px + q)^n$ the sum of $n$ partial fractions:

$$\frac{A_1}{px + q} + \frac{A_2}{(px + q)^2} + \cdots + \frac{A_n}{(px + q)^n}.$$

4. Assign to each quadratic factor $(ax^2 + bx + c)^m$ the sum of $m$ partial fractions:

$$\frac{B_1x + C_1}{ax^2 + bx + c} + \frac{B_2x + C_2}{(ax^2 + bx + c)^2} + \cdots + \frac{B_mx + C_m}{(ax^2 + bx + c)^m}.$$

5. Apply algebraic methods, as illustrated in the following examples, to find the constants in the numerators of the partial fractions.

**EXAMPLE 1** Decompose into partial fractions:

$$\frac{4x - 13}{2x^2 + x - 6}.$$

**Solution** The degree of the numerator is less than the degree of the denominator. We begin by factoring the denominator: $(x + 2)(2x - 3)$. We find constants $A$ and $B$ such that

$$\frac{4x - 13}{(x + 2)(2x - 3)} = \frac{A}{x + 2} + \frac{B}{2x - 3}.$$

To determine $A$ and $B$, we add the expressions on the right:

$$\frac{4x - 13}{(x + 2)(2x - 3)} = \frac{A(2x - 3) + B(x + 2)}{(x + 2)(2x - 3)}.$$

Next, we equate the numerators:

$$4x - 13 = A(2x - 3) + B(x + 2).$$

Since the last equation containing $A$ and $B$ is true for all $x$, we can substitute any value of $x$ and still have a true equation. If we choose $x = \frac{3}{2}$, then $2x - 3 = 0$ and $A$ will be eliminated when we make the substitution. This gives us

$$4\left(\tfrac{3}{2}\right) - 13 = A\left(2 \cdot \tfrac{3}{2} - 3\right) + B\left(\tfrac{3}{2} + 2\right)$$
$$-7 = 0 + \tfrac{7}{2}B.$$

Solving, we obtain $B = -2$.

In order to have $x + 2 = 0$, we let $x = -2$. Then $B$ will be eliminated when we make the substitution. This gives us

$$4(-2) - 13 = A\left[2(-2) - 3\right] + B(-2 + 2)$$
$$-21 = -7A + 0.$$

Solving, we obtain $A = 3$.

The decomposition is as follows:

$$\frac{4x - 13}{2x^2 + x - 6} = \frac{3}{x + 2} + \frac{-2}{2x - 3}, \quad \text{or} \quad \frac{3}{x + 2} - \frac{2}{2x - 3}.$$

To check, we can add to see if we get the expression on the left. We can also use the TABLE feature on a graphing calculator, comparing values of

$$y_1 = \frac{4x - 13}{2x^2 + x - 6} \quad \text{and} \quad y_2 = \frac{3}{x + 2} - \frac{2}{2x - 3}$$

for the same values of $x$. Since $y_1 = y_2$ for the given values of $x$ as we scroll through the table, the decomposition appears to be correct.

**Now Try Exercise 3.**

| X | Y₁ | Y₂ |
|---|------|------|
| −1 | 3.4 | 3.4 |
| 0 | 2.1667 | 2.1667 |
| 1 | 3 | 3 |
| 2 | −1.25 | −1.25 |
| 3 | −.0667 | −.0667 |
| 4 | .1 | .1 |
| 5 | .14286 | .14286 |

X = −1

**EXAMPLE 2**    Decompose into partial fractions:

$$\frac{7x^2 - 29x + 24}{(2x - 1)(x - 2)^2}.$$

**Solution**    The degree of the numerator is 2 and the degree of the denominator is 3, so the degree of the numerator is less than the degree of the denominator. The denominator is given in factored form. The decomposition has the following form:

$$\frac{7x^2 - 29x + 24}{(2x - 1)(x - 2)^2} = \frac{A}{2x - 1} + \frac{B}{x - 2} + \frac{C}{(x - 2)^2}.$$

As in Example 1, we add the expressions on the right:

$$\frac{7x^2 - 29x + 24}{(2x - 1)(x - 2)^2} = \frac{A(x - 2)^2 + B(2x - 1)(x - 2) + C(2x - 1)}{(2x - 1)(x - 2)^2}.$$

Then we equate the numerators. This gives us

$$7x^2 - 29x + 24 = A(x - 2)^2 + B(2x - 1)(x - 2) + C(2x - 1).$$

Since the equation containing $A$, $B$, and $C$ is true for all $x$, we can substitute any value of $x$ and still have a true equation. In order to have $2x - 1 = 0$, we let $x = \frac{1}{2}$. This gives us

$$7\left(\tfrac{1}{2}\right)^2 - 29 \cdot \tfrac{1}{2} + 24 = A\left(\tfrac{1}{2} - 2\right)^2 + 0 + 0$$
$$\tfrac{45}{4} = \tfrac{9}{4}A.$$

Solving, we obtain $A = 5$.

In order to have $x - 2 = 0$, we let $x = 2$. Substituting gives us

$$7(2)^2 - 29(2) + 24 = 0 + 0 + C(2 \cdot 2 - 1)$$
$$-6 = 3C.$$

Solving, we obtain $C = -2$.

To find $B$, we choose any value for $x$ except $\frac{1}{2}$ or 2 and replace $A$ with 5 and $C$ with $-2$. We let $x = 1$:

$$7 \cdot 1^2 - 29 \cdot 1 + 24 = 5(1 - 2)^2 + B(2 \cdot 1 - 1)(1 - 2)$$
$$+ (-2)(2 \cdot 1 - 1)$$
$$2 = 5 - B - 2$$
$$B = 1.$$

The decomposition is as follows:

$$\frac{7x^2 - 29x + 24}{(2x - 1)(x - 2)^2} = \frac{5}{2x - 1} + \frac{1}{x - 2} - \frac{2}{(x - 2)^2}.$$

We can check the result using a table of values. We let

| X | Y₁ | Y₂ |
|---|-----|-----|
| -5 | -.6382 | -.6382 |
| -4 | -.7778 | -.7778 |
| -3 | -.9943 | -.9943 |
| -2 | -1.375 | -1.375 |
| -1 | -2.222 | -2.222 |
| 0 | -6 | -6 |
| 1 | 2 | 2 |

X = -5

$$y_1 = \frac{7x^2 - 29x + 24}{(2x - 1)(x - 2)^2} \quad \text{and} \quad y_2 = \frac{5}{2x - 1} + \frac{1}{x - 2} - \frac{2}{(x - 2)^2}.$$

Since $y_1 = y_2$ for given values of $x$ as we scroll through the table, the decomposition appears to be correct.

**Now Try Exercise 7.**

**EXAMPLE 3** Decompose into partial fractions:

$$\frac{6x^3 + 5x^2 - 7}{3x^2 - 2x - 1}.$$

**Solution** The degree of the numerator is greater than that of the denominator. Thus we divide and find an equivalent expression:

$$
\begin{array}{r}
2x + 3 \\
3x^2 - 2x - 1{\overline{\smash{\big)}\,6x^3 + 5x^2 \phantom{+2x} - 7}} \\
\underline{6x^3 - 4x^2 - 2x} \\
9x^2 + 2x - 7 \\
\underline{9x^2 - 6x - 3} \\
8x - 4.
\end{array}
$$

The original expression is thus equivalent to

$$2x + 3 + \frac{8x - 4}{3x^2 - 2x - 1}.$$

We decompose the fraction to get

$$\frac{8x - 4}{(3x + 1)(x - 1)} = \frac{5}{3x + 1} + \frac{1}{x - 1}.$$

The final result is

$$2x + 3 + \frac{5}{3x + 1} + \frac{1}{x - 1}.$$

**Now Try Exercise 17.**

Systems of equations can be used to decompose rational expressions. Let's reconsider Example 2.

**EXAMPLE 4** Decompose into partial fractions:

$$\frac{7x^2 - 29x + 24}{(2x - 1)(x - 2)^2}.$$

**Solution** The decomposition has the following form:

$$\frac{A}{2x - 1} + \frac{B}{x - 2} + \frac{C}{(x - 2)^2}.$$

We first add as in Example 2:

$$
\frac{7x^2 - 29x + 24}{(2x - 1)(x - 2)^2} = \frac{A}{2x - 1} + \frac{B}{x - 2} + \frac{C}{(x - 2)^2}
$$

$$
= \frac{A(x - 2)^2 + B(2x - 1)(x - 2) + C(2x - 1)}{(2x - 1)(x - 2)^2}.
$$

Then we equate numerators:

$$
7x^2 - 29x + 24
$$
$$
= A(x - 2)^2 + B(2x - 1)(x - 2) + C(2x - 1)
$$
$$
= A(x^2 - 4x + 4) + B(2x^2 - 5x + 2) + C(2x - 1)
$$
$$
= Ax^2 - 4Ax + 4A + 2Bx^2 - 5Bx + 2B + 2Cx - C,
$$

or, combining like terms,

$$7x^2 - 29x + 24$$
$$= (A + 2B)x^2 + (-4A - 5B + 2C)x + (4A + 2B - C).$$

Next, we equate corresponding coefficients:

$7 = A + 2B,$      **The coefficients of the $x^2$-terms must be the same.**

$-29 = -4A - 5B + 2C,$      **The coefficients of the $x$-terms must be the same.**

$24 = 4A + 2B - C.$      **The constant terms must be the same.**

> **SYSTEMS OF EQUATIONS IN THREE VARIABLES**
>
> REVIEW SECTION **6.2** OR **6.5.**

We now have a system of three equations. You should confirm that the solution of the system is

$$A = 5, \quad B = 1, \quad \text{and} \quad C = -2.$$

The decomposition is as follows:

$$\frac{7x^2 - 29x + 24}{(2x - 1)(x - 2)^2} = \frac{5}{2x - 1} + \frac{1}{x - 2} - \frac{2}{(x - 2)^2}.$$

> **Now Try Exercise 15.**

**EXAMPLE 5**    Decompose into partial fractions:

$$\frac{11x^2 - 8x - 7}{(2x^2 - 1)(x - 3)}.$$

*Solution*    The decomposition has the following form:

$$\frac{11x^2 - 8x - 7}{(2x^2 - 1)(x - 3)} = \frac{Ax + B}{2x^2 - 1} + \frac{C}{x - 3}.$$

Adding and equating numerators, we get

$$11x^2 - 8x - 7 = (Ax + B)(x - 3) + C(2x^2 - 1)$$
$$= Ax^2 - 3Ax + Bx - 3B + 2Cx^2 - C,$$

or      $11x^2 - 8x - 7 = (A + 2C)x^2 + (-3A + B)x + (-3B - C).$

We then equate corresponding coefficients:

$11 = A + 2C,$      **The coefficients of the $x^2$-terms**

$-8 = -3A + B,$      **The coefficients of the $x$-terms**

$-7 = -3B - C.$      **The constant terms**

We solve this system of three equations and obtain

$$A = 3, \quad B = 1, \quad \text{and} \quad C = 4.$$

The decomposition is as follows:

$$\frac{11x^2 - 8x - 7}{(2x^2 - 1)(x - 3)} = \frac{3x + 1}{2x^2 - 1} + \frac{4}{x - 3}.$$

> **Now Try Exercise 13.**

# 6.8 Exercise Set

*Decompose into partial fractions. Check your answers using a graphing calculator.*

1. $\dfrac{x + 7}{(x - 3)(x + 2)}$

2. $\dfrac{2x}{(x + 1)(x - 1)}$

3. $\dfrac{7x - 1}{6x^2 - 5x + 1}$

4. $\dfrac{13x + 46}{12x^2 - 11x - 15}$

5. $\dfrac{3x^2 - 11x - 26}{(x^2 - 4)(x + 1)}$

6. $\dfrac{5x^2 + 9x - 56}{(x - 4)(x - 2)(x + 1)}$

7. $\dfrac{9}{(x + 2)^2(x - 1)}$

8. $\dfrac{x^2 - x - 4}{(x - 2)^3}$

9. $\dfrac{2x^2 + 3x + 1}{(x^2 - 1)(2x - 1)}$

10. $\dfrac{x^2 - 10x + 13}{(x^2 - 5x + 6)(x - 1)}$

11. $\dfrac{x^4 - 3x^3 - 3x^2 + 10}{(x + 1)^2(x - 3)}$

12. $\dfrac{10x^3 - 15x^2 - 35x}{x^2 - x - 6}$

13. $\dfrac{-x^2 + 2x - 13}{(x^2 + 2)(x - 1)}$

14. $\dfrac{26x^2 + 208x}{(x^2 + 1)(x + 5)}$

15. $\dfrac{6 + 26x - x^2}{(2x - 1)(x + 2)^2}$

16. $\dfrac{5x^3 + 6x^2 + 5x}{(x^2 - 1)(x + 1)^3}$

17. $\dfrac{6x^3 + 5x^2 + 6x - 2}{2x^2 + x - 1}$

18. $\dfrac{2x^3 + 3x^2 - 11x - 10}{x^2 + 2x - 3}$

19. $\dfrac{2x^2 - 11x + 5}{(x - 3)(x^2 + 2x - 5)}$

20. $\dfrac{3x^2 - 3x - 8}{(x - 5)(x^2 + x - 4)}$

21. $\dfrac{-4x^2 - 2x + 10}{(3x + 5)(x + 1)^2}$

22. $\dfrac{26x^2 - 36x + 22}{(x - 4)(2x - 1)^2}$

23. $\dfrac{36x + 1}{12x^2 - 7x - 10}$

24. $\dfrac{-17x + 61}{6x^2 + 39x - 21}$

25. $\dfrac{-4x^2 - 9x + 8}{(3x^2 + 1)(x - 2)}$

26. $\dfrac{11x^2 - 39x + 16}{(x^2 + 4)(x - 8)}$

## Skill Maintenance

*Find the zeros of the polynomial function.*

27. $f(x) = x^3 + x^2 + 9x + 9$

28. $f(x) = x^3 - 3x^2 + x - 3$

29. $f(x) = x^3 + x^2 - 3x - 2$

30. $f(x) = x^4 - x^3 - 5x^2 - x - 6$

31. $f(x) = x^3 + 5x^2 + 5x - 3$

## Synthesis

*Decompose into partial fractions.*

32. $\dfrac{9x^3 - 24x^2 + 48x}{(x - 2)^4(x + 1)}$
    [*Hint*: Let the expression equal
    $$\frac{A}{x + 1} + \frac{P(x)}{(x - 2)^4}$$
    and find $P(x)$].

33. $\dfrac{x}{x^4 - a^4}$

34. $\dfrac{1}{e^{-x} + 3 + 2e^x}$

35. $\dfrac{1 + \ln x^2}{(\ln x + 2)(\ln x - 3)^2}$

# Chapter 6 Summary and Review

## STUDY GUIDE

| KEY TERMS AND CONCEPTS | EXAMPLES |
| --- | --- |

### SECTION 6.1: SYSTEMS OF EQUATIONS IN TWO VARIABLES

A **system of two linear equations in two variables** is composed of two linear equations that are considered simultaneously.

The **solutions** of the system of equations are all ordered pairs that make *both* equations true.

A system of equations is **consistent** if it has at least one solution. A system of equations that has no solution is **inconsistent**.

The equations are **dependent** if one equation can be obtained by multiplying by a constant on both sides of the other equation. Otherwise, the equations are **independent**.

Systems of two equations in two variables can be solved graphically.

Solve: $x + y = 2$,
$\quad\quad\; y = x - 4.$

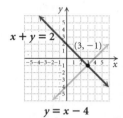

The solution is the point of intersection, $(3, -1)$. The system is consistent. The equations are independent.

Solve: $x + y = 2$,
$\quad\quad x + y = -2.$

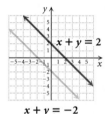

The graphs do not intersect, so there is no solution. The system is inconsistent. The equations are independent.

Solve: $\;x +\; y = 2$,
$\quad\quad 3x + 3y = 6.$

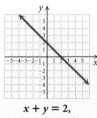

The graphs are the same. There are infinitely many common points, so there are infinitely many solutions. The solutions are of the form $(x, 2 - x)$ or $(2 - y, y)$. The system is consistent. The equations are dependent.

| | |
|---|---|
| Systems of two equations in two variables can be solved using substitution. | Solve: $x = y - 5,$ <br> $\quad\quad 2x + 3y = 5.$ |

*Substitute and solve for y:*

$2(y - 5) + 3y = 5$
$2y - 10 + 3y = 5$
$5y - 10 = 5$
$5y = 15$
$y = 3.$

*Then back-substitute and solve for x:*

$x = y - 5$
$x = 3 - 5$
$x = -2.$

The solution is $(-2, 3)$.

| | |
|---|---|
| Systems of two equations in two variables can be solved using elimination. | Solve: $3x + \;\;y = -1,$ <br> $\quad\quad\; x - 3y = 8.$ |

*Eliminate y and solve for x:*

$9x + 3y = -3$
$\underline{\;\;x - 3y = 8\;\;}$
$10x \quad\quad\; = 5$
$x = \frac{1}{2}.$

*Then back-substitute and solve for y:*

$3x + y = -1$
$3 \cdot \frac{1}{2} + y = -1$
$\frac{3}{2} + y = -1$
$y = -\frac{5}{2}.$

The solution is $\left(\frac{1}{2}, -\frac{5}{2}\right)$.

### SECTION 6.2: SYSTEMS OF EQUATIONS IN THREE VARIABLES

A **solution** of a system of equations in three variables is an ordered triple that makes *all* three equations true.

We can use **Gaussian elimination** to solve a system of three equations in three variables by using the operations listed on p. 500 to transform the original system to one of the form

$$Ax + By + Cz = D,$$
$$Ey + Fz = G,$$
$$Hz = K.$$

Then we solve the third equation for $z$ and back-substitute to find $y$ and $x$.

As we see in Example 1 on p. 500, Gaussian elimination can be used to transform the system of equations

$$x - 2y + 3z = 11,$$
$$4x + 2y - 3z = 4,$$
$$3x + 3y - \;z = 4$$

to the equivalent form

$$x - 2y + \;3z = 11,$$
$$10y - 15z = -40,$$
$$35z = 70.$$

Solve for $z$: $35z = 70$
$z = 2.$

*Back-substitute to find y and x:*

$10y - 15 \cdot 2 = -40$
$10y - 30 = -40$
$10y = -10$
$y = -1.$

$x - 2(-1) + 3 \cdot 2 = 11$
$x + 2 + 6 = 11$
$x + 8 = 11$
$x = 3.$

The solution is $(3, -1, 2)$.

We can use a system of three equations to model a situation with a quadratic function.

Find a quadratic function that fits the data points $(0, -5)$, $(1, -4)$, and $(2, 1)$.

Substitute in the function $f(x) = ax^2 + bx + c$:

For $(0, -5)$:   $-5 = a \cdot 0^2 + b \cdot 0 + c$,
For $(1, -4)$:   $-4 = a \cdot 1^2 + b \cdot 1 + c$,
For $(2, 1)$:    $1 = a \cdot 2^2 + b \cdot 2 + c$.

We have a system of equations:

$$c = -5,$$
$$a + b + c = -4,$$
$$4a + 2b + c = 1.$$

Solving this system of equations gives $(2, -1, -5)$. Thus,

$$f(x) = 2x^2 - x - 5.$$

## SECTION 6.3: MATRICES AND SYSTEMS OF EQUATIONS

A **matrix** (pl., **matrices**) is a rectangular array of numbers called **entries**, or **elements** of the matrix.

$$\begin{array}{c} \text{Row 1} \rightarrow \\ \text{Row 2} \rightarrow \end{array} \begin{bmatrix} 3 & -2 & 5 \\ -1 & 4 & -3 \end{bmatrix}$$

Column 1 Column 2 Column 3

This matrix has 2 rows and 3 columns. Its **order** is $2 \times 3$.

We can apply the **row-equivalent operations** on p. 511 to use **Gaussian elimination** with matrices to solve systems of equations.

Solve:   $x - 2y = 8,$
     $2x + y = 1.$

Write the augmented matrix and transform it to **row-echelon form** or **reduced row-echelon form**:

$$\begin{bmatrix} 1 & -2 & | & 8 \\ 2 & 1 & | & 1 \end{bmatrix} \rightarrow \begin{bmatrix} 1 & -2 & | & 8 \\ 0 & 1 & | & -3 \end{bmatrix} \rightarrow \begin{bmatrix} 1 & 0 & | & 2 \\ 0 & 1 & | & -3 \end{bmatrix}$$

Row-echelon form    Reduced row-echelon form

Then we have $x = 2$, $y = -3$. The solution is $(2, -3)$.

## SECTION 6.4: MATRIX OPERATIONS

Matrices of the same order can be added or subtracted by adding or subtracting their corresponding entries.

$$\begin{bmatrix} 3 & -4 \\ -1 & 2 \end{bmatrix} + \begin{bmatrix} -5 & -1 \\ 3 & 0 \end{bmatrix}$$

$$= \begin{bmatrix} 3 + (-5) & -4 + (-1) \\ -1 + 3 & 2 + 0 \end{bmatrix} = \begin{bmatrix} -2 & -5 \\ 2 & 2 \end{bmatrix};$$

$$\begin{bmatrix} 3 & -4 \\ -1 & 2 \end{bmatrix} - \begin{bmatrix} -5 & -1 \\ 3 & 0 \end{bmatrix}$$

$$= \begin{bmatrix} 3 - (-5) & -4 - (-1) \\ -1 - 3 & 2 - 0 \end{bmatrix} = \begin{bmatrix} 8 & -3 \\ -4 & 2 \end{bmatrix}$$

The **scalar product** of a number $k$ and a matrix $\mathbf{A}$ is the matrix $k\mathbf{A}$ obtained by multiplying each entry of $\mathbf{A}$ by $k$. The number $k$ is called a **scalar**.

The properties of matrix addition and scalar multiplication are given on p. 521.

For $\mathbf{A} = \begin{bmatrix} 2 & 3 & -1 \\ -4 & -2 & 5 \end{bmatrix}$, find $2\mathbf{A}$.

$$2\mathbf{A} = 2\begin{bmatrix} 2 & 3 & -1 \\ -4 & -2 & 5 \end{bmatrix}$$

$$= \begin{bmatrix} 2\cdot 2 & 2\cdot 3 & 2\cdot(-1) \\ 2\cdot(-4) & 2\cdot(-2) & 2\cdot 5 \end{bmatrix}$$

$$= \begin{bmatrix} 4 & 6 & -2 \\ -8 & -4 & 10 \end{bmatrix}$$

For an $m \times n$ matrix $\mathbf{A} = \begin{bmatrix} a_{ij} \end{bmatrix}$ and an $n \times p$ matrix $\mathbf{B} = \begin{bmatrix} b_{ij} \end{bmatrix}$, the **product** $\mathbf{AB} = \begin{bmatrix} c_{ij} \end{bmatrix}$ is an $m \times p$ matrix, where

$$c_{ij} = a_{i1} \cdot b_{1j} + a_{i2} \cdot b_{2j} + a_{i3} \cdot b_{3j} + \cdots + a_{in} \cdot b_{nj}.$$

The properties of matrix multiplication are given on p. 525.

For $\mathbf{A} = \begin{bmatrix} 4 & -1 & 3 \\ 0 & -2 & 1 \end{bmatrix}$ and $\mathbf{B} = \begin{bmatrix} -3 & 1 \\ 3 & 4 \\ 2 & -1 \end{bmatrix}$, find $\mathbf{AB}$.

$$\mathbf{AB} = \begin{bmatrix} 4 & -1 & 3 \\ 0 & -2 & 1 \end{bmatrix}\begin{bmatrix} -3 & 1 \\ 3 & 4 \\ 2 & -1 \end{bmatrix}$$

$$= \begin{bmatrix} 4\cdot(-3)+(-1)\cdot 3+3\cdot 2 & 4\cdot 1+(-1)\cdot 4+3\cdot(-1) \\ 0\cdot(-3)+(-2)\cdot 3+1\cdot 2 & 0\cdot 1+(-2)\cdot 4+1\cdot(-1) \end{bmatrix}$$

$$= \begin{bmatrix} -9 & -3 \\ -4 & -9 \end{bmatrix}$$

We can write a matrix equation equivalent to a system of equations.

Write a matrix equation equivalent to the system of equations:

$$2x - 3y = 6,$$
$$x - 4y = 1.$$

This system of equations can be written as

$$\begin{bmatrix} 2 & -3 \\ 1 & -4 \end{bmatrix}\begin{bmatrix} x \\ y \end{bmatrix} = \begin{bmatrix} 6 \\ 1 \end{bmatrix}.$$

## SECTION 6.5: INVERSES OF MATRICES

The $n \times n$ **identity matrix I** is an $n \times n$ matrix with 1's on the main diagonal and 0's elsewhere.

For any $n \times n$ matrix $\mathbf{A}$,

$$\mathbf{AI} = \mathbf{IA} = \mathbf{A}.$$

For an $n \times n$ matrix $\mathbf{A}$, if there is a matrix $\mathbf{A}^{-1}$ for which $\mathbf{A}^{-1} \cdot \mathbf{A} = \mathbf{I} = \mathbf{A} \cdot \mathbf{A}^{-1}$, then $\mathbf{A}^{-1}$ is the **inverse** of $\mathbf{A}$.

The inverse of an $n \times n$ matrix $\mathbf{A}$ can be found by first writing an augmented matrix consisting of $\mathbf{A}$ on the left side and the $n \times n$ identity matrix on the right side. Then row-equivalent operations are used to transform the augmented matrix to a matrix with the $n \times n$ identity matrix on the left side and the inverse on the right side.

See Examples 3 and 4 on pp. 533 and 534.

For a system of $n$ linear equations in $n$ variables, $\mathbf{AX} = \mathbf{B}$, if $\mathbf{A}$ has an inverse, then the solution of the system of equations is given by

$$\mathbf{X} = \mathbf{A}^{-1}\mathbf{B}.$$

Since matrix multiplication is not commutative, in general, $\mathbf{B}$ *must* be multiplied *on the left* by $\mathbf{A}^{-1}$.

Use an inverse matrix to solve the following system of equations:

$$\begin{aligned} x - y &= 1, \\ x - 2y &= -1. \end{aligned}$$

First, we write an equivalent matrix equation:

$$\underset{\mathbf{A}}{\begin{bmatrix} 1 & -1 \\ 1 & -2 \end{bmatrix}} \cdot \underset{\mathbf{X}}{\begin{bmatrix} x \\ y \end{bmatrix}} = \underset{\mathbf{B}}{\begin{bmatrix} 1 \\ -1 \end{bmatrix}}.$$

Then we find $\mathbf{A}^{-1}$ and multiply *on the left* by $\mathbf{A}^{-1}$:

$$\mathbf{X} = \mathbf{A}^{-1} \cdot \mathbf{B}$$

$$\begin{bmatrix} x \\ y \end{bmatrix} = \begin{bmatrix} 2 & -1 \\ 1 & -1 \end{bmatrix}\begin{bmatrix} 1 \\ -1 \end{bmatrix} = \begin{bmatrix} 3 \\ 2 \end{bmatrix}.$$

The solution is $(3, 2)$.

## SECTION 6.6: DETERMINANTS AND CRAMER'S RULE

**Determinant of a 2 × 2 Matrix**

The determinant of the matrix $\begin{bmatrix} a & c \\ b & d \end{bmatrix}$ is denoted by $\begin{vmatrix} a & c \\ b & d \end{vmatrix}$ and is defined as

$$\begin{vmatrix} a & c \\ b & d \end{vmatrix} = ad - bc.$$

Evaluate: $\begin{vmatrix} 3 & -4 \\ 2 & 1 \end{vmatrix}$.

$$\begin{vmatrix} 3 & -4 \\ 2 & 1 \end{vmatrix} = 3 \cdot 1 - 2(-4) = 3 + 8 = 11$$

---

The **determinant** of any **square matrix** can be found by *expanding across a row* or *down a column*. See p. 540.

See Example 4 on p. 540.

---

We can use determinants to solve systems of linear equations.

Cramer's rule for a 2 × 2 system is given on p. 542. Cramer's rule for a 3 × 3 system is given on p. 544.

Solve: $2x - 3y = 2$,
$\qquad 6x + 6y = 1$.

$$x = \frac{\begin{vmatrix} 2 & -3 \\ 1 & 6 \end{vmatrix}}{\begin{vmatrix} 2 & -3 \\ 6 & 6 \end{vmatrix}} = \frac{15}{30} = \frac{1}{2},$$

$$y = \frac{\begin{vmatrix} 2 & 2 \\ 6 & 1 \end{vmatrix}}{\begin{vmatrix} 2 & -3 \\ 6 & 6 \end{vmatrix}} = \frac{-10}{30} = -\frac{1}{3}.$$

The solution is $\left(\frac{1}{2}, -\frac{1}{3}\right)$.

## SECTION 6.7: SYSTEMS OF INEQUALITIES AND LINEAR PROGRAMMING

To graph a linear inequality in two variables:

1. Graph the related equation. Draw a dashed line if the inequality symbol is $<$ or $>$; draw a solid line if the inequality symbol is $\leq$ or $\geq$.

2. Use a test point to determine which half-plane to shade.

Graph: $x + y > 2$.

1. Graph $x + y = 2$ using a dashed line.

2. Test a point not on the line. We use $(0, 0)$.

$$\begin{array}{c} x + y > 2 \\ \hline 0 + 0 \; ? \; 2 \\ 0 \; \Big| \quad \textbf{FALSE} \end{array}$$

Since $0 > 2$ is false, we shade the half-plane that does not contain $(0, 0)$.

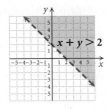

To graph a system of inequalities, graph each inequality and determine the region that is common to all the solution sets.

Graph the solution set of the system

$$x - y \leq 3,$$
$$2x + y \geq 4.$$

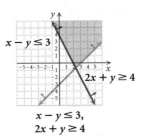

$$x - y \leq 3,$$
$$2x + y \geq 4$$

For a system of linear inequalities, the maximum or minimum value of an **objective function** over a region of **feasible solutions** is the maximum or minimum value of the function at a vertex of that region.

Maximize $G = 8x - 5y$ subject to

$$x + y \leq 3,$$
$$x \geq 0,$$
$$y \geq 1.$$

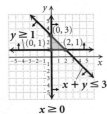

| Vertex | $G = 8x - 5y$ | |
|--------|---------------|--|
| $(0, 1)$ | $G = 8 \cdot 0 - 5 \cdot 1 = -5$ | |
| $(0, 3)$ | $G = 8 \cdot 0 - 5 \cdot 3 = -15$ | |
| $(2, 1)$ | $G = 8 \cdot 2 - 5 \cdot 1 = 11$ | $\longleftarrow$ Maximum |

## SECTION 6.8: PARTIAL FRACTIONS

The procedure for **decomposing a rational expression into partial fractions** is given on p. 558.

See Examples 1–5 on pp. 559–562.

## REVIEW EXERCISES

*Determine whether the statement is true or false.*

1. A system of equations with exactly one solution is consistent and has independent equations.  [6.1]

2. A system of two linear equations in two variables can have exactly two solutions.  [6.1]

3. For any $m \times n$ matrices **A** and **B**,
   $$\mathbf{A} + \mathbf{B} = \mathbf{B} + \mathbf{A}. \ [6.4]$$

4. In general, matrix multiplication is commutative.  [6.4]

*In Exercises 5–12, match the equations or inequalities with one of the graphs (a)–(h), which follow.*

a)
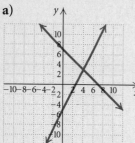

b)
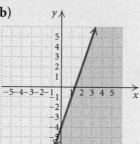

c)
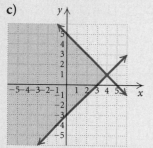

d)
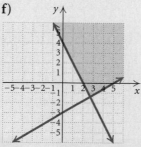

e)
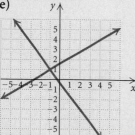

f)

g)
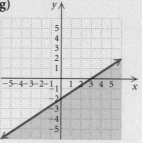

h)
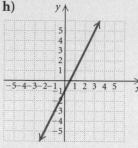

5. $x + y = 7,$
   $2x - y = 5$  [6.1]

6. $3x - 5y = -8,$
   $4x + 3y = -1$  [6.1]

7. $y = 2x - 1,$
   $4x - 2y = 2$  [6.1]

8. $6x - 3y = 5,$
   $y = 2x + 3$  [6.1]

9. $y \le 3x - 4$  [6.7]

10. $2x - 3y \ge 6$  [6.7]

11. $x - y \le 3,$
    $x + y \le 5$  [6.7]

12. $2x + y \ge 4,$
    $3x - 5y \le 15$  [6.7]

*Solve.*

13. $5x - 3y = -4,$
    $3x - y = -4$  [6.1]

14. $2x + 3y = 2,$
    $5x - y = -29$  [6.1]

15. $x + 5y = 12,$
    $5x + 25y = 12$  [6.1]

16. $x + y = -2,$
    $-3x - 3y = 6$
    [6.1]

17. $x + 5y - 3z = 4,$
    $3x - 2y + 4z = 3,$
    $2x + 3y - z = 5$  [6.2]

18. $2x - 4y + 3z = -3,$
    $-5x + 2y - z = 7,$
    $3x + 2y - 2z = 4$  [6.2]

19. $x - y = 5,$
    $y - z = 6,$
    $z - w = 7,$
    $x + w = 8$  [6.2]

20. Classify each of the systems in Exercises 13–19 as consistent or inconsistent.  [6.1], [6.2]

21. Classify each of the systems in Exercises 13–19 as having dependent or independent equations.
    [6.1], [6.2]

*Solve the system of equations using Gaussian elimination or Gauss–Jordan elimination.* [6.3]

**22.** $x + 2y = 5,$
$2x - 5y = -8$

**23.** $3x + 4y + 2z = 3,$
$5x - 2y - 13z = 3,$
$4x + 3y - 3z = 6$

**24.** $3x + 5y + z = 0,$
$2x - 4y - 3z = 0,$
$x + 3y + z = 0$

**25.** $w + x + y + z = -2,$
$-3w - 2x + 3y + 2z = 10,$
$2w + 3x + 2y - z = -12,$
$2w + 4x - y + z = 1$

**26.** *Coins.* The value of 75 coins, consisting of only nickels and dimes, is $5.95. How many of each kind are there? [6.1]

**27.** *Investment.* The Wong family invested $5000, part at 3% and the remainder at 3.5%. The annual income from both investments is $167. What is the amount invested at each rate? [6.1]

**28.** *Nutrition.* A dietician must plan a breakfast menu that provides 460 Cal, 9 g of fat, and 55 mg of calcium. One plain bagel contains 200 Cal, 2 g of fat, and 29 mg of calcium. A one-tablespoon serving of cream cheese contains 100 Cal, 10 g of fat, and 24 mg of calcium. One banana contains 105 Cal, 1 g of fat, and 7 g of calcium. (*Source: Home and Garden Bulletin No. 72*, U.S. Government Printing Office, Washington D.C. 20402) How many servings of each are required to provide the desired nutritional values? [6.2]

**29.** *Test Scores.* A student has a total of 226 points on three tests. The sum of the scores on the first and second tests exceeds the score on the third test by 62. The first score exceeds the second by 6. Find the three scores. [6.2]

**30.** *Trademarks.* The table below lists the number of trademarks renewed, in thousands, in the United States, represented in terms of the number of years since 2006.

| Year, x | Number of Trademarks Renewed (in thousands) |
|---------|----------------------------------------------|
| 2006, 0 | 40 |
| 2007, 1 | 48 |
| 2008, 2 | 40 |

*Source:* U.S. Patent and Trademark Office

**a)** Use a system of equations to fit a quadratic function $f(x) = ax^2 + bx + c$ to the data. [6.2]

**b)** Use the function to estimate the number of trademarks renewed in 2009. [6.2]

*For Exercises 31–38, let*

$$A = \begin{bmatrix} 1 & -1 & 0 \\ 2 & 3 & -2 \\ -2 & 0 & 1 \end{bmatrix}, \quad B = \begin{bmatrix} -1 & 0 & 6 \\ 1 & -2 & 0 \\ 0 & 1 & -3 \end{bmatrix},$$

*and*

$$C = \begin{bmatrix} -2 & 0 \\ 1 & 3 \end{bmatrix}.$$

*Find each of the following, if possible.* [6.4]

**31.** $A + B$

**32.** $-3A$

**33.** $-A$

**34.** $AB$

**35.** $B + C$

**36.** $A - B$

**37.** $BA$

**38.** $A + 3B$

**39.** *Food Service Management.* The table below lists the cost per serving, in dollars, for items on four menus that are served at an elder-care facility.

| Menu | Meat | Potato | Vegetable | Salad | Dessert |
|------|------|--------|-----------|-------|---------|
| 1 | 0.98 | 0.23 | 0.30 | 0.28 | 0.45 |
| 2 | 1.03 | 0.19 | 0.27 | 0.34 | 0.41 |
| 3 | 1.01 | 0.21 | 0.35 | 0.31 | 0.39 |
| 4 | 0.99 | 0.25 | 0.29 | 0.33 | 0.42 |

On a particular day, a dietician orders 32 meals from menu 1, 19 from menu 2, 43 from menu 3, and 38 from menu 4.

**a)** Write the information in the table as a 4 × 5 matrix **M**. [6.4]

**b)** Write a row matrix **N** that represents the number of each menu ordered. [6.4]

**c)** Find the product **NM**. [6.4]

**d)** State what the entries of **NM** represent. [6.4]

*Find* $A^{-1}$, *if it exists.* [6.5]

**40.** $A = \begin{bmatrix} -2 & 0 \\ 1 & 3 \end{bmatrix}$

**41.** $A = \begin{bmatrix} 0 & 0 & 3 \\ 0 & -2 & 0 \\ 4 & 0 & 0 \end{bmatrix}$

**42.** $A = \begin{bmatrix} 1 & 0 & 0 & 0 \\ 0 & 4 & -5 & 0 \\ 0 & 2 & 2 & 0 \\ 0 & 0 & 0 & 1 \end{bmatrix}$

**43.** Write a matrix equation equivalent to this system of equations:

$$3x - 2y + 4z = 13,$$
$$x + 5y - 3z = 7,$$
$$2x - 3y + 7z = -8. \quad [6.4]$$

*Solve the system of equations using the inverse of the co-efficient matrix of the equivalent matrix equation.* [6.5]

**44.** $2x + 3y = 5,$
$\quad 3x + 5y = 11$

**45.** $5x - y + 2z = 17,$
$\quad 3x + 2y - 3z = -16,$
$\quad 4x - 3y - z = 5$

**46.** $\quad w - x - y + z = -1,$
$\quad 2w + 3x - 2y - z = 2,$
$\quad -w + 5x + 4y - 2z = 3,$
$\quad 3w - 2x + 5y + 3z = 4$

*Evaluate the determinant.* [6.6]

**47.** $\begin{vmatrix} 1 & -2 \\ 3 & 4 \end{vmatrix}$

**48.** $\begin{vmatrix} \sqrt{3} & -5 \\ -3 & -\sqrt{3} \end{vmatrix}$

**49.** $\begin{vmatrix} 2 & -1 & 1 \\ 1 & 2 & -1 \\ 3 & 4 & -3 \end{vmatrix}$

**50.** $\begin{vmatrix} 1 & -1 & 2 \\ -1 & 2 & 0 \\ -1 & 3 & 1 \end{vmatrix}$

*Solve using Cramer's rule.* [6.6]

**51.** $5x - 2y = 19,$
$\quad 7x + 3y = 15$

**52.** $\quad x + y = 4,$
$\quad 4x + 3y = 11$

**53.** $3x - 2y + z = 5,$
$\quad 4x - 5y - z = -1,$
$\quad 3x + 2y - z = 4$

**54.** $2x - y - z = 2,$
$\quad 3x + 2y + 2z = 10,$
$\quad x - 5y - 3z = -2$

*Graph.* [6.7]

**55.** $y \le 3x + 6$

**56.** $4x - 3y \ge 12$

**57.** Graph this system of inequalities and find the coordinates of any vertices formed. [6.7]

$$2x + y \ge 9,$$
$$4x + 3y \ge 23,$$
$$x + 3y \ge 8,$$
$$x \ge 0,$$
$$y \ge 0$$

**58.** Find the maximum value and the minimum value of $T = 6x + 10y$ subject to

$$x + y \le 10,$$
$$5x + 10y \ge 50,$$
$$x \ge 2,$$
$$y \ge 0. \quad [6.7]$$

**59.** *Maximizing a Test Score.* Aiden is taking a test that contains questions in group A worth 7 points each and questions in group B worth 12 points each. The total number of questions answered must be at least 8. If Aiden knows that group A questions take 8 min each and group B questions take 10 min each and the maximum time for the test is 80 min, how many questions from each group must he answer correctly in order to maximize his score? What is the maximum score? [6.7]

*Decompose into partial fractions.* [6.8]

**60.** $\dfrac{5}{(x + 2)^2(x + 1)}$

**61.** $\dfrac{-8x + 23}{2x^2 + 5x - 12}$

**62.** Solve: $2x + y = 7,$
$\qquad\quad x - 2y = 6. \quad [6.1]$

**A.** $x$ and $y$ are both positive numbers.
**B.** $x$ and $y$ are both negative numbers.
**C.** $x$ is positive and $y$ is negative.
**D.** $x$ is negative and $y$ is positive.

**63.** Which of the following is *not* a row-equivalent operation on a matrix? [6.3]

**A.** Interchange any two columns.
**B.** Interchange any two rows.
**C.** Add two rows.
**D.** Multiply each entry in a row by $-3$.

**64.** The graph of the given system of inequalities is which of the following? [6.7]

$$x + y \leq 3,$$
$$x - y \leq 4$$

**A.**

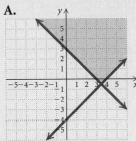

**B.**

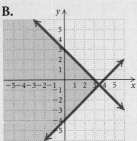

**C.**

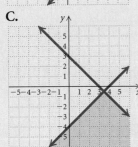

**D.**

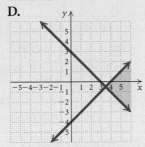

## Synthesis

**65.** One year, Don invested a total of $40,000, part at 4%, part at 5%, and the rest at $5\frac{1}{2}$%. The total amount of interest received on the investments was $1990. The interest received on the $5\frac{1}{2}$% investment was $590 more than the interest received on the 4% investment. How much was invested at each rate? [6.2]

*Solve.*

**66.** $\dfrac{2}{3x} + \dfrac{4}{5y} = 8,$

$\dfrac{5}{4x} - \dfrac{3}{2y} = -6$ [6.1]

**67.** $\dfrac{3}{x} - \dfrac{4}{y} + \dfrac{1}{z} = -2,$

$\dfrac{5}{x} + \dfrac{1}{y} - \dfrac{2}{z} = 1,$

$\dfrac{7}{x} + \dfrac{3}{y} + \dfrac{2}{z} = 19$

[6.2]

*Graph.* [6.7]

**68.** $|x| - |y| \leq 1$

**69.** $|xy| > 1$

## Collaborative Discussion and Writing

**70.** Cassidy solves the equation $2x + 5 = 3x - 7$ by finding the point of intersection of the graphs of $y_1 = 2x + 5$ and $y_2 = 3x - 7$. She finds the same point when she solves the system of equations

$$y = 2x + 5,$$
$$y = 3x - 7.$$

Explain the difference between the solution of the equation and the solution of the system of equations. [6.1]

**71.** For square matrices **A** and **B**, is it true, in general, that $(\mathbf{AB})^2 = \mathbf{A}^2\mathbf{B}^2$? Explain. [6.4]

**72.** Given the system of equations

$$a_1x + b_1y = c_1,$$
$$a_2x + b_2y = c_2,$$

explain why the equations are dependent or the system is inconsistent when

$$\begin{vmatrix} a_1 & b_1 \\ a_2 & b_2 \end{vmatrix} = 0. \ [6.6]$$

**73.** If the lines $a_1x + b_1y = c_1$ and $a_2x + b_2y = c_2$ are parallel, what can you say about the values of

$$\begin{vmatrix} a_1 & b_1 \\ a_2 & b_2 \end{vmatrix}, \quad \begin{vmatrix} c_1 & b_1 \\ c_2 & b_2 \end{vmatrix}, \quad \text{and} \quad \begin{vmatrix} a_1 & c_1 \\ a_2 & c_2 \end{vmatrix}? \ [6.6]$$

**74.** Describe how the graph of a linear inequality differs from the graph of a linear equation. [6.7]

**75.** What would you say to a classmate who tells you that the partial fraction decomposition of

$$\frac{3x^2 - 8x + 9}{(x + 3)(x^2 - 5x + 6)}$$

is

$$\frac{2}{x + 3} + \frac{x - 1}{x^2 - 5x + 6}?$$

Explain. [6.8]

*Solve. Use any method. Also determine whether the system is consistent or inconsistent and whether the equations are dependent or independent.*

**1.** $3x + 2y = 1,$
$2x - y = -11$

**2.** $2x - y = 3,$
$2y = 4x - 6$

**3.** $x - y = 4,$
$3y = 3x - 8$

**4.** $2x - 3y = 8,$
$5x - 2y = 9$

*Solve.*

**5.** $4x + 2y + z = 4,$
$3x - y + 5z = 4,$
$5x + 3y - 3z = -2$

**6.** *Ticket Sales.*   One evening 750 tickets were sold for Shortridge Community College's spring musical. Tickets cost $3 for students and $5 for nonstudents. Total receipts were $3066. How many of each type of ticket were sold?

**7.** Latonna, Cole, and Sam can process 352 online orders per day. Latonna and Cole together can process 224 orders per day while Latonna and Sam together can process 248 orders per day. How many orders can each of them process alone?

*For Exercises 8–13, let*

$$A = \begin{bmatrix} 1 & -1 & 3 \\ -2 & 5 & 2 \end{bmatrix}, \qquad B = \begin{bmatrix} -5 & 1 \\ -2 & 4 \end{bmatrix},$$

*and*

$$C = \begin{bmatrix} 3 & -4 \\ -1 & 0 \end{bmatrix}.$$

*Find each of the following, if possible.*

**8.** $B + C$

**9.** $A - C$

**10.** $CB$

**11.** $AB$

**12.** $2A$

**13.** $C^{-1}$

**14.** *Food Service Management.*   The table below lists the cost per serving, in dollars, for items on three lunch menus served at a senior citizens' center.

| Menu | Main Dish | Side Dish | Dessert |
|------|-----------|-----------|---------|
| 1 | 0.95 | 0.40 | 0.39 |
| 2 | 1.10 | 0.35 | 0.41 |
| 3 | 1.05 | 0.39 | 0.36 |

On a particular day, 26 Menu 1 meals, 18 Menu 2 meals, and 23 Menu 3 meals are served.

**a)** Write the information in the table as a 3 × 3 matrix **M**.

**b)** Write a row matrix **N** that represents the number of each menu served.

**c)** Find the product **NM**.

**d)** State what the entries of **NM** represent.

**15.** Write a matrix equation equivalent to the system of equations

$$3x - 4y + 2z = -8,$$
$$2x + 3y + z = 7,$$
$$x - 5y - 3z = 3.$$

**16.** Solve the system of equations using the inverse of the coefficient matrix of the equivalent matrix equation.

$$3x + 2y + 6z = 2,$$
$$x + y + 2z = 1,$$
$$2x + 2y + 5z = 3$$

*Evaluate the determinant.*

**17.** $\begin{vmatrix} 3 & -5 \\ 8 & 7 \end{vmatrix}$

**18.** $\begin{vmatrix} 2 & -1 & 4 \\ -3 & 1 & -2 \\ 5 & 3 & -1 \end{vmatrix}$

**19.** Solve using Cramer's rule. Show your work.

$$5x + 2y = -1,$$
$$7x + 6y = 1$$

**20.** Graph: $3x + 4y \leq -12$.

**21.** Find the maximum value and the minimum value of $Q = 2x + 3y$ subject to

$$x + y \leq 6,$$
$$2x - 3y \geq -3,$$
$$x \geq 1,$$
$$y \geq 0.$$

**22.** *Maximizing Profit.* Gracilyn's Gourmet Sweets prepares coffeecakes and cheesecakes for restaurants. In a given week, at most 100 cakes can be prepared, of which 25 coffeecakes and 15 cheesecakes are required by regular customers. The profit from each coffeecake is $8, and the profit from each cheesecake is $17. How many of each type of cake should be prepared in order to maximize the profit? What is the maximum profit?

**23.** Decompose into partial fractions:

$$\frac{3x - 11}{x^2 + 2x - 3}.$$

**24.** The graph of the given system of inequalities is which of the following?

$$x + 2y \geq 4,$$
$$x - y \leq 2$$

**A.**

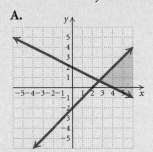

**B.**

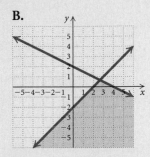

**C.**

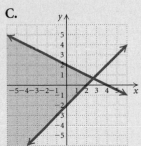

**D.**

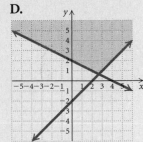

## Synthesis

**25.** Three solutions of the equation $Ax - By = Cz - 8$ are $(2, -2, 2)$, $(-3, -1, 1)$, and $(4, 2, 9)$. Find $A$, $B$, and $C$.

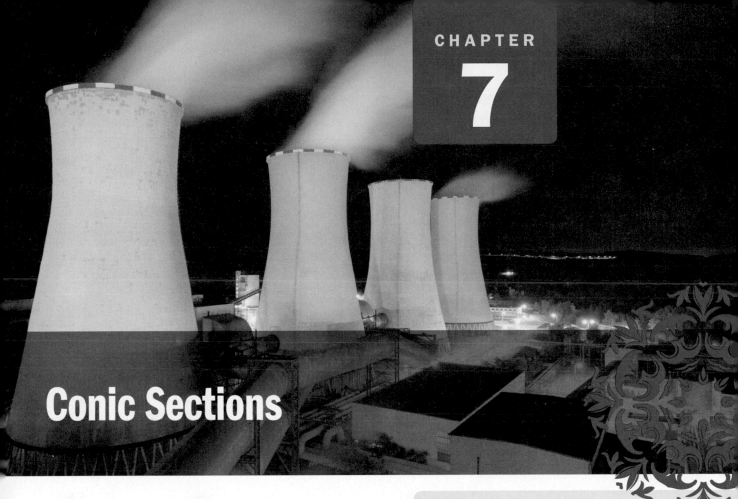

# Conic Sections

## APPLICATION

A cross section of a nuclear cooling tower is a hyperbola with equation

$$\frac{x^2}{90^2} - \frac{y^2}{130^2} = 1.$$

The tower is 450 ft tall, and the distance from the top of the tower to the center of the hyperbola is half the distance from the base of the tower to the center of the hyperbola. Find the diameter of the top and the base of the tower.

This problem appears as Exercise 39 in Section 7.3.

# The Parabola

## 7.1

- Given an equation of a parabola, complete the square, if necessary, and then find the vertex, the focus, and the directrix and graph the parabola.

A **conic section** is formed when a right circular cone with two parts, called *nappes*, is intersected by a plane. One of four types of curves can be formed: a parabola, a circle, an ellipse, or a hyperbola.

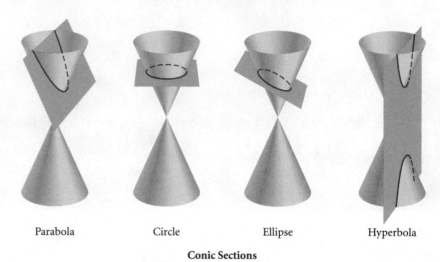

Parabola          Circle          Ellipse          Hyperbola

**Conic Sections**

Conic sections can be defined algebraically using second-degree equations of the form $Ax^2 + Bxy + Cy^2 + Dx + Ey + F = 0$. In addition, they can be defined geometrically as a set of points that satisfy certain conditions.

## ■ Parabolas

In Section 3.3, we saw that the graph of the quadratic function $f(x) = ax^2 + bx + c$, $a \neq 0$, is a parabola. A parabola can be defined geometrically.

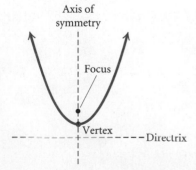

> ### PARABOLA
>
> A **parabola** is the set of all points in a plane equidistant from a fixed line (the **directrix**) and a fixed point not on the line (the **focus**).

The line that is perpendicular to the directrix and contains the focus is the **axis of symmetry**. The **vertex** is the midpoint of the segment between the focus and the directrix. (See the figure at left.)

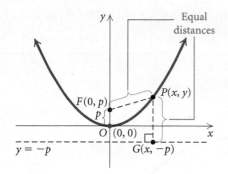

FIGURE 1

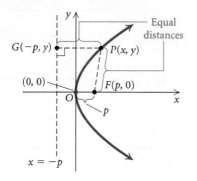

FIGURE 2

Let's derive the standard equation of a parabola with vertex $(0, 0)$ and directrix $y = -p$, where $p > 0$. We place the coordinate axes as shown in Fig. 1. The $y$-axis is the axis of symmetry and contains the focus $F$. The distance from the focus to the vertex is the same as the distance from the vertex to the directrix. Thus the coordinates of $F$ are $(0, p)$.

Let $P(x, y)$ be any point on the parabola and consider $\overline{PG}$ perpendicular to the line $y = -p$. The coordinates of $G$ are $(x, -p)$. By the definition of a parabola,

$$PF = PG. \qquad \text{The distance from } P \text{ to the focus is the same as the distance from } P \text{ to the directrix.}$$

Then using the distance formula, we have

$$\sqrt{(x - 0)^2 + (y - p)^2} = \sqrt{(x - x)^2 + [y - (-p)]^2}$$
$$x^2 + y^2 - 2py + p^2 = y^2 + 2py + p^2 \qquad \text{Squaring both sides and squaring the binomials}$$
$$x^2 = 4py.$$

We have shown that if $P(x, y)$ is on the parabola shown in Fig. 1, then its coordinates satisfy this equation. The converse is also true, but we will not prove it here.

Note that if $p > 0$, as above, the graph opens up. If $p < 0$, the graph opens down.

The equation of a parabola with vertex $(0, 0)$ and directrix $x = -p$ is derived in a similar manner. Such a parabola opens either to the right $(p > 0)$, as shown in Fig. 2, or to the left $(p < 0)$.

---

### STANDARD EQUATION OF A PARABOLA WITH VERTEX AT THE ORIGIN

The standard equation of a parabola with vertex $(0, 0)$ and directrix $y = -p$ is

$$x^2 = 4py.$$

The focus is $(0, p)$ and the $y$-axis is the axis of symmetry.

The standard equation of a parabola with vertex $(0, 0)$ and directrix $x = -p$ is

$$y^2 = 4px.$$

The focus is $(p, 0)$ and the $x$-axis is the axis of symmetry.

---

**EXAMPLE 1** Find the focus, the vertex, and the directrix of the parabola $y = -\frac{1}{12}x^2$. Then graph the parabola.

***Solution*** We write $y = -\frac{1}{12}x^2$ in the form $x^2 = 4py$:

$$-\tfrac{1}{12}x^2 = y \qquad \text{Given equation}$$
$$x^2 = -12y \qquad \text{Multiplying by } -12 \text{ on both sides}$$
$$x^2 = 4(-3)y. \qquad \text{Standard form}$$

Since the equation can be written in the form $x^2 = 4py$, we know that the vertex is $(0, 0)$.

We have $p = -3$, so the focus is $(0, p)$, or $(0, -3)$. The directrix is $y = -p = -(-3) = 3$.

| $x$ | $y$ |
|-----|-----|
| 0 | 0 |
| $\pm 1$ | $-\frac{1}{12}$ |
| $\pm 2$ | $-\frac{1}{3}$ |
| $\pm 3$ | $-\frac{3}{4}$ |
| $\pm 4$ | $-\frac{4}{3}$ |

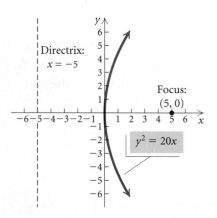

Directrix: $y = 3$

$y = -\frac{1}{12}x^2$

Focus: $(0, -3)$

**Now Try Exercise 7.**

**EXAMPLE 2** Find an equation of the parabola with vertex $(0, 0)$ and focus $(5, 0)$. Then graph the parabola.

**Solution** The focus is on the $x$-axis so the line of symmetry is the $x$-axis. The equation is of the type

$$y^2 = 4px.$$

Since the focus $(5, 0)$ is 5 units to the right of the vertex, $p = 5$ and the equation is

$$y^2 = 4(5)x, \quad \text{or} \quad y^2 = 20x.$$

| $x$ | $y^2$ | $y$ | $(x, y)$ |
|-----|-------|-----|----------|
| 0 | 0 | 0 | $(0, 0)$ |
| 1 | 20 | $\pm \sqrt{20}$ | $(1, 4.47)$ |
| | | | $(1, -4.47)$ |
| 2 | 40 | $\pm \sqrt{40}$ | $(2, 6.32)$ |
| | | | $(2, -6.32)$ |
| 3 | 60 | $\pm \sqrt{60}$ | $(3, 7.75)$ |
| | | | $(3, -7.75)$ |

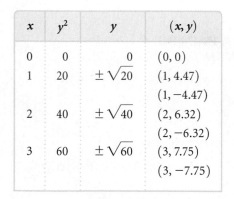

Directrix: $x = -5$

Focus: $(5, 0)$

$y^2 = 20x$

We can also use a graphing calculator to graph parabolas. It might be necessary to solve the equation for $y$ before entering it in the calculator:

$$y^2 = 20x$$
$$y = \pm \sqrt{20x}.$$

We now graph

$$y_1 = \sqrt{20x} \quad \text{and} \quad y_2 = -\sqrt{20x}, \quad \text{or} \quad y_1 = \sqrt{20x} \quad \text{and} \quad y_2 = -y_1,$$

in a squared viewing window.

On some graphing calculators, the Conics application from the APPS menu can be used to graph parabolas. This method will be discussed in Example 4.

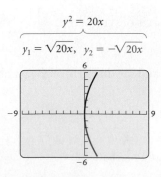

$y^2 = 20x$

$y_1 = \sqrt{20x}, \quad y_2 = -\sqrt{20x}$

**Now Try Exercise 15.**

### ■ Finding Standard Form by Completing the Square

If a parabola with vertex at the origin is translated horizontally $|h|$ units and vertically $|k|$ units, it has an equation as follows.

**STANDARD EQUATION OF A PARABOLA WITH VERTEX ($h$, $k$) AND VERTICAL AXIS OF SYMMETRY**

The standard equation of a parabola with vertex $(h, k)$ and vertical axis of symmetry is

$$(x - h)^2 = 4p(y - k),$$

where the vertex is $(h, k)$, the focus is $(h, k + p)$, and the directrix is $y = k - p$.

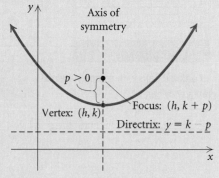

(When $p < 0$, the parabola opens down.)

**STANDARD EQUATION OF A PARABOLA WITH VERTEX ($h$, $k$) AND HORIZONTAL AXIS OF SYMMETRY**

The standard equation of a parabola with vertex $(h, k)$ and horizontal axis of symmetry is

$$(y - k)^2 = 4p(x - h),$$

where the vertex is $(h, k)$, the focus is $(h + p, k)$, and the directrix is $x = h - p$.

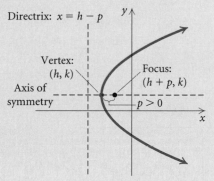

(When $p < 0$, the parabola opens to the left.)

---

**COMPLETING THE SQUARE**

REVIEW SECTION **3.2.**

We can complete the square on equations of the form

$$y = ax^2 + bx + c \quad \text{or} \quad x = ay^2 + by + c$$

in order to write them in standard form.

**EXAMPLE 3**    For the parabola

$$x^2 + 6x + 4y + 5 = 0,$$

find the vertex, the focus, and the directrix. Then draw the graph.

| Table View | Edit |
|---|---|
| x | y1 |
| -8.00 | -5.25 |
| -7.00 | -3.00 |
| -6.00 | -1.25 |
| -5.00 | 0.00 |
| -4.00 | 0.75 |
| -3.00 | 1.00 |
| -2.00 | 0.75 |
| -1.00 | 0.00 |
| 0.00 | -1.25 |
| 1.00 | -3.00 |
| 2.00 | -5.25 |
| 3.00 | -8.00 |
| 4.00 | -11.25 |
| 5.00 | -15.00 |
| 6.00 | -19.25 |
| 7.00 | -24.00 |
| 8.00 | -29.25 |
| 9.00 | -35.00 |
| 10.00 | -41.25 |
| 11.00 | -48.00 |

Calculator  Equations  Graph  Table  More

A table of ordered pairs with an app is helpful when graphing by hand.

**Solution** We first complete the square:

$$x^2 + 6x + 4y + 5 = 0$$
$$x^2 + 6x \qquad = -4y - 5 \qquad \text{Subtracting } 4y \text{ and } 5 \text{ on both sides}$$
$$x^2 + 6x + 9 = -4y - 5 + 9 \qquad \text{Adding 9 on both sides to complete the square on the left side}$$
$$x^2 + 6x + 9 = -4y + 4$$
$$(x + 3)^2 = -4(y - 1) \qquad \text{Factoring}$$
$$[x - (-3)]^2 = 4(-1)(y - 1). \qquad \text{Writing standard form:} \ (x - h)^2 = 4p(y - k)$$

We see that $h = -3$, $k = 1$, and $p = -1$, so we have the following:

Vertex $(h, k)$: $\qquad (-3, 1)$;
Focus $(h, k + p)$: $\qquad (-3, 1 + (-1))$, or $(-3, 0)$;
Directrix $y = k - p$: $\quad y = 1 - (-1)$, or $y = 2$.

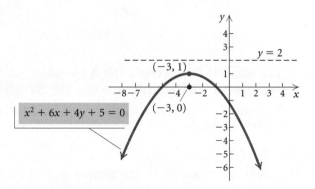

$y = \frac{1}{4}(-x^2 - 6x - 5)$

We can check the graph on a graphing calculator using a squared viewing window, as shown at left. It might be necessary to solve for $y$ first:

$$x^2 + 6x + 4y + 5 = 0$$
$$4y = -x^2 - 6x - 5$$
$$y = \frac{1}{4}(-x^2 - 6x - 5).$$

The hand-drawn graph appears to be correct.

**Now Try Exercise 25.**

**GCM**

**EXAMPLE 4** For the parabola

$$y^2 - 2y - 8x - 31 = 0,$$

find the vertex, the focus, and the directrix. Then draw the graph.

**Solution** We first complete the square:

$$y^2 - 2y - 8x - 31 = 0$$
$$y^2 - 2y \qquad = 8x + 31 \qquad \text{Adding } 8x \text{ and } 31 \text{ on both sides}$$
$$y^2 - 2y + 1 = 8x + 31 + 1. \qquad \text{Adding 1 on both sides to complete the square on the left side}$$

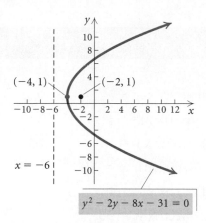

$$y^2 - 2y - 8x - 31 = 0$$

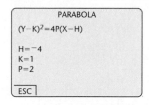

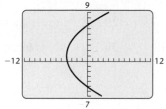

Then

$$y^2 - 2y + 1 = 8x + 32$$
$$(y - 1)^2 = 8(x + 4) \qquad \text{Factoring}$$
$$(y - 1)^2 = 4(2)[x - (-4)]. \qquad \text{Writing standard form:}$$
$$(y - k)^2 = 4p(x - h)$$

We see that $h = -4$, $k = 1$, and $p = 2$, so we have the following:

Vertex $(h, k)$:  $(-4, 1)$;

Focus $(h + p, k)$:  $(-4 + 2, 1)$, or $(-2, 1)$;

Directrix $x = h - p$:  $x = -4 - 2$, or $x = -6$.

When the equation of a parabola is written in standard form, we can use the Conics PARABOLA APP to graph it, as shown at left.

We can also draw the graph or check the hand-drawn graph on a graphing calculator by first solving the original equation for $y$ using the quadratic formula. For details, see the *Graphing Calculator Manual* that accompanies this text.

**Now Try Exercise 31.**

## ■ Applications

Parabolas have many applications. For example, cross sections of car headlights, flashlights, and searchlights are parabolas. The bulb is located at the focus and light from that point is reflected outward parallel to the axis of symmetry. Satellite dishes and field microphones used at sporting events often have parabolic cross sections. Incoming radio waves or sound waves parallel to the axis are reflected into the focus.

Similarly, in solar cooking, a parabolic mirror is mounted on a rack with a cooking pot hung in the focal area. Incoming sun rays parallel to the axis are reflected into the focus, producing a temperature high enough for cooking.

## 7.1 Exercise Set

*In Exercises 1–6, match the equation with one of the graphs (a)–(f), which follow.*

a)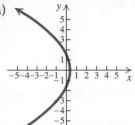

b)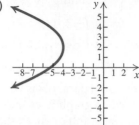

c)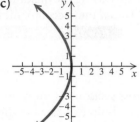

d)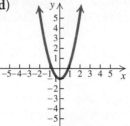

e)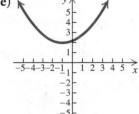

f)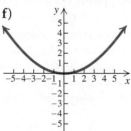

**1.** $x^2 = 8y$

**2.** $y^2 = -10x$

**3.** $(y - 2)^2 = -3(x + 4)$

**4.** $(x + 1)^2 = 5(y - 2)$

**5.** $13x^2 - 8y - 9 = 0$

**6.** $41x + 6y^2 = 12$

*Find the vertex, the focus, and the directrix. Then draw the graph.*

**7.** $x^2 = 20y$

**8.** $x^2 = 16y$

**9.** $y^2 = -6x$

**10.** $y^2 = -2x$

**11.** $x^2 - 4y = 0$

**12.** $y^2 + 4x = 0$

**13.** $x = 2y^2$

**14.** $y = \frac{1}{2}x^2$

*Find an equation of a parabola satisfying the given conditions.*

**15.** Vertex $(0, 0)$, focus $(-3, 0)$

**16.** Vertex $(0, 0)$, focus $(0, 10)$

**17.** Focus $(7, 0)$, directrix $x = -7$

**18.** Focus $\left(0, \frac{1}{4}\right)$, directrix $y = -\frac{1}{4}$

**19.** Focus $(0, -\pi)$, directrix $y = \pi$

**20.** Focus $(-\sqrt{2}, 0)$, directrix $x = \sqrt{2}$

**21.** Focus $(3, 2)$, directrix $x = -4$

**22.** Focus $(-2, 3)$, directrix $y = -3$

*Find the vertex, the focus, and the directrix. Then draw the graph.*

**23.** $(x + 2)^2 = -6(y - 1)$

**24.** $(y - 3)^2 = -20(x + 2)$

**25.** $x^2 + 2x + 2y + 7 = 0$

**26.** $y^2 + 6y - x + 16 = 0$

**27.** $x^2 - y - 2 = 0$       **28.** $x^2 - 4x - 2y = 0$

**29.** $y = x^2 + 4x + 3$       **30.** $y = x^2 + 6x + 10$

**31.** $y^2 - y - x + 6 = 0$

**32.** $y^2 + y - x - 4 = 0$

**33.** *Satellite Dish.*   An engineer designs a satellite dish with a parabolic cross section. The dish is 15 ft wide at the opening, and the focus is placed 4 ft from the vertex.

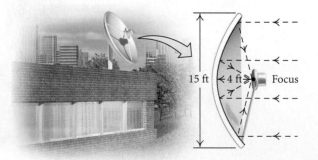

15 ft   4 ft   Focus

a) Position a coordinate system with the origin at the vertex and the $x$-axis on the parabola's axis of symmetry and find an equation of the parabola.

b) Find the depth of the satellite dish at the vertex.

**34.** *Flashlight Mirror.* A heavy-duty flashlight mirror has a parabolic cross section with diameter 6 in. and depth 1 in.

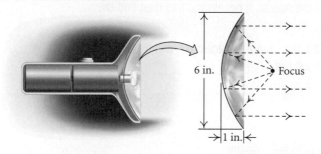

6 in.

Focus

1 in.

a) Position a coordinate system with the origin at the vertex and the *x*-axis on the parabola's axis of symmetry and find an equation of the parabola.

b) How far from the vertex should the bulb be positioned if it is to be placed at the focus?

**35.** *Ultrasound Receiver.* Information Unlimited designed and sells the Ultrasonic Receiver, which detects sounds unable to be heard by the human ear. The HT90P can detect mechanical and electrical sounds such as leaking gases, air, corona, and motor friction noises. It can also be used to hear bats, insects, and even beading water. The receiver has a parabolic cross section and is 2.625 in. deep. The focus is 3.287 in. from the vertex. (*Source*: Information Unlimited, Amherst, NH, Robert Iannini, President) Find the diameter of the outside edge of the receiver.

**36.** *Spotlight.* A spotlight has a parabolic cross section that is 4 ft wide at the opening and 1.5 ft deep at the vertex. How far from the vertex is the focus?

## Skill Maintenance

*Consider the following linear equations. Without graphing them, answer the questions below.*

**a)** $y = 2x$        **b)** $y = \frac{1}{3}x + 5$

**c)** $y = -3x - 2$      **d)** $y = -0.9x + 7$

**e)** $y = -5x + 3$      **f)** $y = x + 4$

**g)** $8x - 4y = 7$       **h)** $3x + 6y = 2$

**37.** Which has/have *x*-intercept $\left(\frac{2}{3}, 0\right)$?

**38.** Which has/have *y*-intercept $(0, 7)$?

**39.** Which slant up from left to right?

**40.** Which has/have the least steep slant?

**41.** Which has/have slope $\frac{1}{3}$?

**42.** Which, if any, contain the point $(3, 7)$?

**43.** Which, if any, are parallel?

**44.** Which, if any, are perpendicular?

## Synthesis

**45.** Find an equation of the parabola with a vertical axis of symmetry and vertex $(-1, 2)$ and containing the point $(-3, 1)$.

**46.** Find an equation of a parabola with a horizontal axis of symmetry and vertex $(-2, 1)$ and containing the point $(-3, 5)$.

*Use a graphing calculator to find the vertex, the focus, and the directrix of each of the following.*

**47.** $4.5x^2 - 7.8x + 9.7y = 0$

**48.** $134.1y^2 + 43.4x - 316.6y - 122.4 = 0$

**49.** *Suspension Bridge.* The cables of a 200-ft portion of the roadbed of a suspension bridge are positioned as shown below. Vertical cables are to be spaced every 20 ft along this portion of the roadbed. Calculate the lengths of these vertical cables.

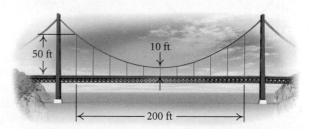

50 ft

10 ft

200 ft

# The Circle and the Ellipse

- Given an equation of a circle, complete the square, if necessary, and then find the center and the radius and graph the circle.
- Given an equation of an ellipse, complete the square, if necessary, and then find the center, the vertices, and the foci and graph the ellipse.

## Circles

| CIRCLES |
|---|
| REVIEW SECTION **1.1.** |

We can define a circle geometrically.

> ### CIRCLE
> A **circle** is the set of all points in a plane that are at a fixed distance from a fixed point (the **center**) in the plane.

Circles were introduced in Section 1.1. Recall the standard equation of a circle with center $(h, k)$ and radius $r$.

> ### STANDARD EQUATION OF A CIRCLE
> The standard equation of a circle with center $(h, k)$ and radius $r$ is
> $$(x - h)^2 + (y - k)^2 = r^2.$$

**GCM**  **EXAMPLE 1**   For the circle
$$x^2 + y^2 - 16x + 14y + 32 = 0,$$
find the center and the radius. Then graph the circle.

*Solution*   First, we complete the square twice:
$$x^2 + y^2 - 16x + 14y + 32 = 0$$
$$x^2 - 16x \qquad + y^2 + 14y \qquad = -32$$
$$x^2 - 16x + 64 + y^2 + 14y + 49 = -32 + 64 + 49$$

$$\left[\tfrac{1}{2}(-16)\right]^2 = (-8)^2 = 64 \text{ and } \left(\tfrac{1}{2} \cdot 14\right)^2 = 7^2 = 49;$$
adding 64 and 49 on both sides to complete the square twice on the left side

$$(x - 8)^2 + (y + 7)^2 = 81$$
$$(x - 8)^2 + [y - (-7)]^2 = 9^2. \qquad \textbf{Writing standard form}$$

The center is $(8, -7)$ and the radius is 9. We graph the circle as shown at left.

To use a graphing calculator to graph the circle, it might be necessary to solve for $y$ first. The original equation can be solved using the quadratic formula, or the standard form of the equation can be solved using the principle of square roots. See the *Graphing Calculator Manual* that accompanies this text for detailed instructions.

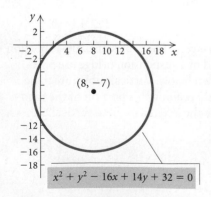

$x^2 + y^2 - 16x + 14y + 32 = 0$

When we use the Conics CIRCLE APP on a graphing calculator to graph a circle, it is not necessary to write the equation in standard form or to solve it for $y$ first. We enter the coefficients of $x^2$, $y^2$, $x$, and $y$ and also the constant term when the equation is written in the form $ax^2 + ay^2 + bx + cy + d = 0$.

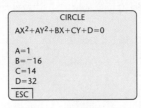
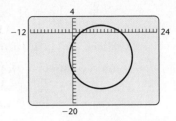

Some graphing calculators have a DRAW feature that provides a quick way to graph a circle when the center and the radius are known. This feature is described on p. 68.

**Now Try Exercise 7.**

## Ellipses

We have studied two conic sections, the parabola and the circle. Now we turn our attention to a third, the *ellipse*.

> ### ELLIPSE
>
> An **ellipse** is the set of all points in a plane, the sum of whose distances from two fixed points (the **foci**) is constant. The **center** of an ellipse is the midpoint of the segment between the foci.

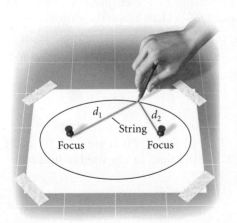

We can draw an ellipse by first placing two thumbtacks in a piece of cardboard, as shown at left. These are the foci (singular, *focus*). We then attach a piece of string to the tacks. Its length is the constant sum of the distances $d_1 + d_2$ from the foci to any point on the ellipse. Next, we trace a curve with a pencil held tight against the string. The figure traced is an ellipse.

Let's first consider the ellipse shown below with center at the origin. The points $F_1$ and $F_2$ are the foci. The segment $\overline{A'A}$ is the **major axis**, and the points $A'$ and $A$ are the **vertices**. The segment $\overline{B'B}$ is the **minor axis**, and the points $B'$ and $B$ are the **y-intercepts**. Note that the major axis of an ellipse is longer than the minor axis.

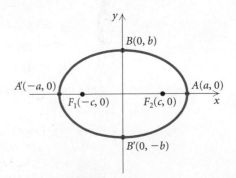

## STANDARD EQUATION OF AN ELLIPSE WITH CENTER AT THE ORIGIN

### *Major Axis Horizontal*

$$\frac{x^2}{a^2} + \frac{y^2}{b^2} = 1, \quad a > b > 0$$

Vertices: $(-a, 0)$, $(a, 0)$

$y$-intercepts: $(0, -b)$, $(0, b)$

Foci: $(-c, 0)$, $(c, 0)$, where $c^2 = a^2 - b^2$

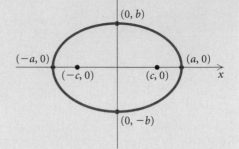

### *Major Axis Vertical*

$$\frac{x^2}{b^2} + \frac{y^2}{a^2} = 1, \quad a > b > 0$$

Vertices: $(0, -a)$, $(0, a)$

$x$-intercepts: $(-b, 0)$, $(b, 0)$

Foci: $(0, -c)$, $(0, c)$, where $c^2 = a^2 - b^2$

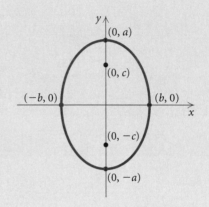

GCM    **EXAMPLE 2**    Find the standard equation of the ellipse with vertices $(-5, 0)$ and $(5, 0)$ and foci $(-3, 0)$ and $(3, 0)$. Then graph the ellipse.

*Solution*    Since the foci are on the $x$-axis and the origin is the midpoint of the segment between them, the major axis is horizontal and $(0, 0)$ is the center of the ellipse. Thus the equation is of the form

$$\frac{x^2}{a^2} + \frac{y^2}{b^2} = 1.$$

Since the vertices are $(-5, 0)$ and $(5, 0)$ and the foci are $(-3, 0)$ and $(3, 0)$, we know that $a = 5$ and $c = 3$. These values can be used to find $b^2$:

$$c^2 = a^2 - b^2$$
$$3^2 = 5^2 - b^2$$
$$9 = 25 - b^2$$
$$b^2 = 16.$$

Thus the equation of the ellipse is

$$\frac{x^2}{25} + \frac{y^2}{16} = 1, \quad \text{or} \quad \frac{x^2}{5^2} + \frac{y^2}{4^2} = 1.$$

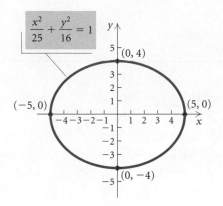

To graph the ellipse, we plot the vertices $(-5, 0)$ and $(5, 0)$. Since $b^2 = 16$, we know that $b = 4$ and the $y$-intercepts are $(0, -4)$ and $(0, 4)$. We plot these points as well and connect the four points that we have plotted with a smooth curve.

When the equation of an ellipse is written in standard form, we can use the Conics ELLIPSE APP on a graphing calculator to graph it. Note that the center is $(0, 0)$, so $H = 0$ and $K = 0$.

$$\frac{x^2}{5^2} + \frac{y^2}{4^2} = 1$$

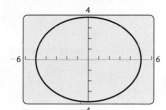

We can also draw the graph or check the hand-drawn graph on a graphing calculator by first solving for $y$. For detailed instruction, see the *Graphing Calculator Manual* that accompanies this text.

**Now Try Exercise 31.**

**EXAMPLE 3**   For the ellipse

$$9x^2 + 4y^2 = 36,$$

find the vertices and the foci. Then draw the graph.

*Solution*   We first find standard form:

$$9x^2 + 4y^2 = 36$$

$$\frac{9x^2}{36} + \frac{4y^2}{36} = \frac{36}{36}$$   **Dividing by 36 on both sides to get 1 on the right side**

$$\frac{x^2}{4} + \frac{y^2}{9} = 1$$

$$\frac{x^2}{2^2} + \frac{y^2}{3^2} = 1.$$   **Writing standard form**

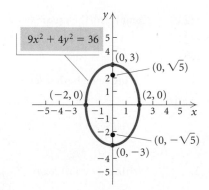

Thus, $a = 3$ and $b = 2$. The major axis is vertical, so the vertices are $(0, -3)$ and $(0, 3)$. Since we know that $c^2 = a^2 - b^2$, we have $c^2 = 3^2 - 2^2 = 9 - 4 = 5$, so $c = \sqrt{5}$ and the foci are $(0, -\sqrt{5})$ and $(0, \sqrt{5})$.

To graph the ellipse, we plot the vertices. Note also that since $b = 2$, the $x$-intercepts are $(-2, 0)$ and $(2, 0)$. We plot these points as well and connect the four points we have plotted with a smooth curve.

**Now Try Exercise 25.**

If the center of an ellipse is not at the origin but at some point $(h, k)$, then we can think of an ellipse with center at the origin being translated horizontally $|h|$ units and vertically $|k|$ units.

## STANDARD EQUATION OF AN ELLIPSE WITH CENTER AT $(h, k)$

### Major Axis Horizontal

$$\frac{(x - h)^2}{a^2} + \frac{(y - k)^2}{b^2} = 1, \quad a > b > 0$$

Vertices: $(h - a, k), (h + a, k)$

Length of minor axis: $2b$

Foci: $(h - c, k), (h + c, k)$, where $c^2 = a^2 - b^2$

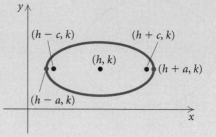

### Major Axis Vertical

$$\frac{(x - h)^2}{b^2} + \frac{(y - k)^2}{a^2} = 1, \quad a > b > 0$$

Vertices: $(h, k - a), (h, k + a)$

Length of minor axis: $2b$

Foci: $(h, k - c), (h, k + c)$, where $c^2 = a^2 - b^2$

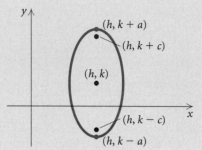

GCM **EXAMPLE 4**   For the ellipse

$$4x^2 + y^2 + 24x - 2y + 21 = 0,$$

find the center, the vertices, and the foci. Then draw the graph.

*Solution*   First, we complete the square twice to get standard form:

$$4x^2 + y^2 + 24x - 2y + 21 = 0$$

$$4(x^2 + 6x \quad\;\;) + (y^2 - 2y \quad\;\;) = -21$$

$$4(x^2 + 6x + 9 - 9) + (y^2 - 2y + 1 - 1) = -21 \qquad \text{Completing the square twice}$$

$$4(x^2 + 6x + 9) + 4(-9) + (y^2 - 2y + 1) + (-1) = -21$$

$$4(x + 3)^2 - 36 + (y - 1)^2 - 1 = -21$$

$$4(x + 3)^2 + (y - 1)^2 = 16 \qquad \text{Adding 37 on both sides}$$

$$\tfrac{1}{16}\big[4(x + 3)^2 + (y - 1)^2\big] = \tfrac{1}{16} \cdot 16$$

$$\frac{(x + 3)^2}{4} + \frac{(y - 1)^2}{16} = 1$$

$$\frac{[x - (-3)]^2}{2^2} + \frac{(y - 1)^2}{4^2} = 1. \qquad \text{Writing standard form}$$

The center is $(-3, 1)$. Note that $a = 4$ and $b = 2$. The major axis is vertical, so the vertices are 4 units above and below the center:

$$(-3, 1 + 4) \text{ and } (-3, 1 - 4), \quad \text{or} \quad (-3, 5) \text{ and } (-3, -3).$$

We know that $c^2 = a^2 - b^2$, so $c^2 = 4^2 - 2^2 = 16 - 4 = 12$ and $c = \sqrt{12}$, or $2\sqrt{3}$. Then the foci are $2\sqrt{3}$ units above and below the center:

$$(-3, 1 + 2\sqrt{3}) \quad \text{and} \quad (-3, 1 - 2\sqrt{3}).$$

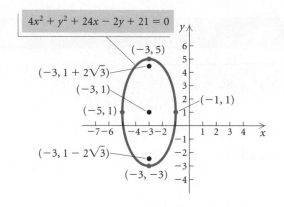

$$4x^2 + y^2 + 24x - 2y + 21 = 0$$

To graph the ellipse, we plot the vertices. Note also that since $b = 2$, two other points on the graph are the endpoints of the minor axis, 2 units right and left of the center:

$$(-3 + 2, 1) \quad \text{and} \quad (-3 - 2, 1),$$

or

$$(-1, 1) \quad \text{and} \quad (-5, 1).$$

We plot these points as well and connect the four points with a smooth curve, as shown at left.

When the equation of an ellipse is written in standard form, we can use the Conics ELLIPSE APP to graph it.

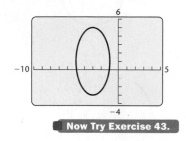

**Now Try Exercise 43.**

## ◾ Applications

An exciting medical application of an ellipse is a device called a *lithotripter*. One type of this device uses electromagnetic technology to generate a shock wave to pulverize kidney stones. The wave originates at one focus of an ellipse and is reflected to the kidney stone, which is positioned at the other focus. Recovery time following the use of this technique is much shorter than with conventional surgery, and the mortality rate is far lower.

Ellipses have many other applications. Planets travel around the sun in elliptical orbits with the sun at one focus, for example, and satellites travel around the earth in elliptical orbits as well.

A room with an ellipsoidal ceiling is known as a *whispering gallery*. In such a room, a word whispered at one focus can be clearly heard at the other. Whispering galleries are found in the rotunda of the Capitol Building in Washington, D.C., and in St. Paul's Cathedral in London.

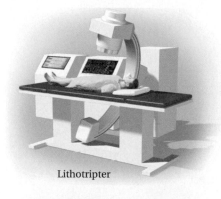

Lithotripter

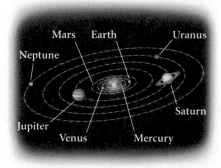

## 7.2 Exercise Set

*In Exercises 1–6, match the equation with one of the graphs (a)–(f), which follow.*

a)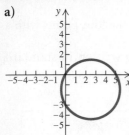

b)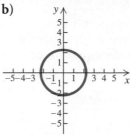

c)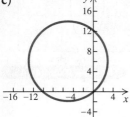

d)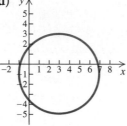

e)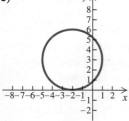

f)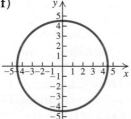

**1.** $x^2 + y^2 = 5$

**2.** $y^2 = 20 - x^2$

**3.** $x^2 + y^2 - 6x + 2y = 6$

**4.** $x^2 + y^2 + 10x - 12y = 3$

**5.** $x^2 + y^2 - 5x + 3y = 0$

**6.** $x^2 + 4x - 2 = 6y - y^2 - 6$

*Find the center and the radius of the circle with the given equation. Then draw the graph.*

**7.** $x^2 + y^2 - 14x + 4y = 11$

**8.** $x^2 + y^2 + 2x - 6y = -6$

**9.** $x^2 + y^2 + 6x - 2y = 6$

**10.** $x^2 + y^2 - 4x + 2y = 4$

**11.** $x^2 + y^2 + 4x - 6y - 12 = 0$

**12.** $x^2 + y^2 - 8x - 2y - 19 = 0$

**13.** $x^2 + y^2 - 6x - 8y + 16 = 0$

**14.** $x^2 + y^2 - 2x + 6y + 1 = 0$

**15.** $x^2 + y^2 + 6x - 10y = 0$

**16.** $x^2 + y^2 - 7x - 2y = 0$

**17.** $x^2 + y^2 - 9x = 7 - 4y$

**18.** $y^2 - 6y - 1 = 8x - x^2 + 3$

*In Exercises 19–22, match the equation with one of the graphs (a)–(d), which follow.*

a)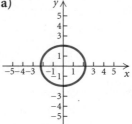

b)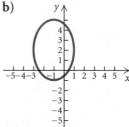

c)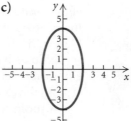

d)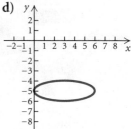

**19.** $16x^2 + 4y^2 = 64$

**20.** $4x^2 + 5y^2 = 20$

**21.** $x^2 + 9y^2 - 6x + 90y = -225$

**22.** $9x^2 + 4y^2 + 18x - 16y = 11$

*Find the vertices and the foci of the ellipse with the given equation. Then draw the graph.*

**23.** $\dfrac{x^2}{4} + \dfrac{y^2}{1} = 1$

**24.** $\dfrac{x^2}{25} + \dfrac{y^2}{36} = 1$

**25.** $16x^2 + 9y^2 = 144$

**26.** $9x^2 + 4y^2 = 36$

**27.** $2x^2 + 3y^2 = 6$

**28.** $5x^2 + 7y^2 = 35$

**29.** $4x^2 + 9y^2 = 1$

**30.** $25x^2 + 16y^2 = 1$

*Find an equation of an ellipse satisfying the given conditions.*

**31.** Vertices: $(-7, 0)$ and $(7, 0)$;
foci: $(-3, 0)$ and $(3, 0)$

**32.** Vertices: $(0, -6)$ and $(0, 6)$;
foci: $(0, -4)$ and $(0, 4)$

**33.** Vertices: $(0, -8)$ and $(0, 8)$;
length of minor axis: 10

**34.** Vertices: $(-5, 0)$ and $(5, 0)$;
length of minor axis: 6

**35.** Foci: $(-2, 0)$ and $(2, 0)$;
length of major axis: 6

**36.** Foci: $(0, -3)$ and $(0, 3)$;
length of major axis: 10

*Find the center, the vertices, and the foci of the ellipse. Then draw the graph.*

**37.** $\dfrac{(x - 1)^2}{9} + \dfrac{(y - 2)^2}{4} = 1$

**38.** $\dfrac{(x - 1)^2}{1} + \dfrac{(y - 2)^2}{4} = 1$

**39.** $\dfrac{(x + 3)^2}{25} + \dfrac{(y - 5)^2}{36} = 1$

**40.** $\dfrac{(x - 2)^2}{16} + \dfrac{(y + 3)^2}{25} = 1$

**41.** $3(x + 2)^2 + 4(y - 1)^2 = 192$

**42.** $4(x - 5)^2 + 3(y - 4)^2 = 48$

**43.** $4x^2 + 9y^2 - 16x + 18y - 11 = 0$

**44.** $x^2 + 2y^2 - 10x + 8y + 29 = 0$

**45.** $4x^2 + y^2 - 8x - 2y + 1 = 0$

**46.** $9x^2 + 4y^2 + 54x - 8y + 49 = 0$

*The **eccentricity** of an ellipse is defined as $e = c/a$. For an ellipse, $0 < c < a$, so $0 < e < 1$. When $e$ is close to 0, an ellipse appears to be nearly circular. When $e$ is close to 1, an ellipse is very flat.*

**47.** Observe the shapes of the ellipses in Examples 2 and 4. Which ellipse has the smaller eccentricity? Confirm your answer by computing the eccentricity of each ellipse.

**48.** Which ellipse has the smaller eccentricity? (Assume that the coordinate systems have the same scale.)

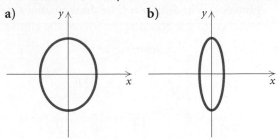

**a)**   **b)**

**49.** Find an equation of an ellipse with vertices $(0, -4)$ and $(0, 4)$ and $e = \frac{1}{4}$.

**50.** Find an equation of an ellipse with vertices $(-3, 0)$ and $(3, 0)$ and $e = \frac{7}{10}$.

**51.** *Bridge Supports.* The bridge support shown in the figure below is the top half of an ellipse. Assuming that a coordinate system is superimposed on the drawing in such a way that point $Q$, the center of the ellipse, is at the origin, find an equation of the ellipse.

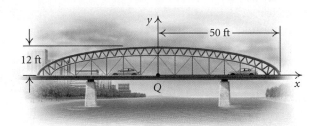

**52.** *The Ellipse.* In Washington, D.C., there is a large grassy area south of the White House known as the Ellipse. It is actually an ellipse with major axis of length 1048 ft and minor axis of length 898 ft. Assuming that a coordinate system is superimposed on the area in such a way that the center is at the origin and the major and minor axes are on the $x$- and $y$-axes of the coordinate system, respectively, find an equation of the ellipse.

**53.** *The Earth's Orbit.* The maximum distance of the earth from the sun is $9.3 \times 10^7$ mi. The minimum distance is $9.1 \times 10^7$ mi. The sun is at one focus of the elliptical orbit. Find the distance from the sun to the other focus.

**54.** *Carpentry.*    A carpenter is cutting a 3-ft by 4-ft elliptical sign from a 3-ft by 4-ft piece of plywood. The ellipse will be drawn using a string attached to the board at the foci of the ellipse.

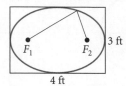

   **a)** How far from the ends of the board should the string be attached?
   **b)** How long should the string be?

## Skill Maintenance

### Vocabulary Reinforcement

*In each of Exercises 55–62, fill in the blank with the correct term. Some of the given choices will not be used.*

   piecewise function
   linear equation
   factor
   remainder
   solution
   zero
   x-intercept
   y-intercept
   parabola
   circle
   ellipse
   midpoint
   distance
   one real-number solution
   two different real-number solutions
   two different imaginary-number solutions

**55.** The _____ between two points $(x_1, y_1)$ and $(x_2, y_2)$ is given by $\left( \dfrac{x_1 + x_2}{2}, \dfrac{y_1 + y_2}{2} \right)$.

**56.** An input $c$ of a function $f$ is a(n) _____ of the function if $f(c) = 0$.

**57.** A(n) _____ of the graph of an equation is a point $(0, b)$.

**58.** For a quadratic equation $ax^2 + bx + c = 0$, if $b^2 - 4ac > 0$, the equation has _____.

**59.** Given a polynomial $f(x)$, then $f(c)$ is the _____ that would be obtained by dividing $f(x)$ by $x - c$.

**60.** A(n) _____ is the set of all points in a plane the sum of whose distances from two fixed points is constant.

**61.** A(n) _____ is the set of all points in a plane equidistant from a fixed line and a fixed point not on the line.

**62.** A(n) _____ is the set of all points in a plane that are at a fixed distance from a fixed point in the plane.

## Synthesis

*Find an equation of an ellipse satisfying the given conditions.*

**63.** Vertices: $(3, -4)$, $(3, 6)$;
endpoints of minor axis: $(1, 1)$, $(5, 1)$

**64.** Vertices: $(-1, -1)$, $(-1, 5)$;
endpoints of minor axis: $(-3, 2)$, $(1, 2)$

**65.** Vertices: $(-3, 0)$ and $(3, 0)$;
passing through $\left( 2, \frac{22}{3} \right)$

**66.** Center: $(-2, 3)$; major axis vertical;
length of major axis: 4;
length of minor axis: 1

*Use a graphing calculator to find the center and the vertices of each of the following.*

**67.** $4x^2 + 9y^2 - 16.025x + 18.0927y - 11.346 = 0$

**68.** $9x^2 + 4y^2 + 54.063x - 8.016y + 49.872 = 0$

**69.** *Bridge Arch.*    A bridge with a semielliptical arch spans a river as shown here. What is the clearance 6 ft from the riverbank?

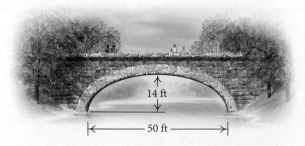

*Determine whether the statement is true or false.*

**1.** The graph of $(x + 3)^2 = 8(y - 2)$ is a parabola with vertex $(-3, 2)$.  [7.1]

**2.** A parabola must open up or down.  [7.1]

**3.** The graph of $(x - 4)^2 + (y + 1)^2 = 9$ is a circle with radius 9.  [7.2]

**4.** The major axis of the ellipse $\dfrac{x^2}{4} + \dfrac{y^2}{16} = 1$ is vertical.  [7.2]

*In Exercises 5–12, match the equation with one of the graphs (a)–(h), which follow.*  [7.1], [7.2]

a)
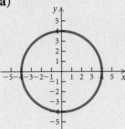

b)
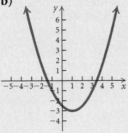

c)
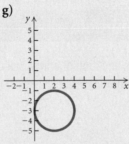

d)
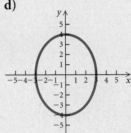

e)
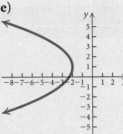

f)
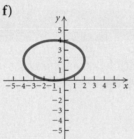

g)
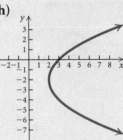

h)

**5.** $x^2 = -4y$

**6.** $(y + 2)^2 = 4(x - 2)$

**7.** $16x^2 + 9y^2 = 144$

**8.** $x^2 + y^2 = 16$

**9.** $(x - 1)^2 = 2(y + 3)$

**10.** $4(x + 1)^2 + 9(y - 2)^2 = 36$

**11.** $(x - 2)^2 + (y + 3)^2 = 4$

**12.** $y^2 - 2y + 3x + 7 = 0$

*Find the vertex, the focus, and the directrix of the parabola. Then draw the graph.*  [7.1]

**13.** $y^2 = 12x$

**14.** $x^2 - 6x - 4y = -17$

*Find the equation of a parabola satisfying the given conditions.*  [7.1]

**15.** Focus: $(0, 3)$; directrix: $y = 1$

**16.** Focus: $(-4, 6)$; directrix: $x = 2$

*Find the center and the radius of the circle. Then draw the graph.*  [7.2]

**17.** $x^2 + y^2 + 4x - 8y = 5$

**18.** $x^2 + y^2 - 6x + 2y - 6 = 0$

*Find the vertices and the foci of the ellipse. Then draw the graph.*  [7.2]

**19.** $\dfrac{x^2}{1} + \dfrac{y^2}{9} = 1$

**20.** $2x^2 + 3y^2 = 12$

**21.** $\dfrac{(x - 2)^2}{16} + \dfrac{(y + 1)^2}{4} = 1$

**22.** $25x^2 + 4y^2 - 50x + 8y = 71$

*Write an equation of the ellipse satisfying the given conditions.* [7.2]

**23.** Vertices: $(-5, 0)$, $(5, 0)$; foci: $(-2, 0)$, $(2, 0)$

**24.** Vertices: $(0, -3)$, $(0, 3)$; length of minor axis: 4

**25.** Foci: $(-3, 0)$, $(3, 0)$; length of major axis: 8

## Collaborative Discussion and Writing

**26.** Is a parabola always the graph of a function? Why or why not? [7.1]

**27.** Explain how the distance formula is used to find the standard equation of a parabola. [7.1]

**28.** Explain why function notation is not used in Section 7.2. [7.2]

**29.** Is the center of an ellipse part of the graph of the ellipse? Why or why not? [7.2]

---

# 7.3 The Hyperbola

■ Given an equation of a hyperbola, complete the square, if necessary, and then find the center, the vertices, and the foci and graph the hyperbola.

---

The last type of conic section that we will study is the *hyperbola*.

### HYPERBOLA

A **hyperbola** is the set of all points in a plane for which the absolute value of the difference of the distances from two fixed points (the **foci**) is constant. The midpoint of the segment between the foci is the **center** of the hyperbola.

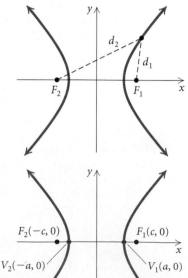

### ■ Standard Equations of Hyperbolas

We first consider the equation of a hyperbola with center at the origin. In the figure at left, $F_1$ and $F_2$ are the foci. The segment $\overline{V_2 V_1}$ is the **transverse axis,** and the points $V_2$ and $V_1$ are the **vertices**.

## STANDARD EQUATION OF A HYPERBOLA WITH CENTER AT THE ORIGIN

*Transverse Axis Horizontal*

$$\frac{x^2}{a^2} - \frac{y^2}{b^2} = 1$$

Vertices: $(-a, 0), (a, 0)$

Foci: $(-c, 0), (c, 0)$,
  where $c^2 = a^2 + b^2$

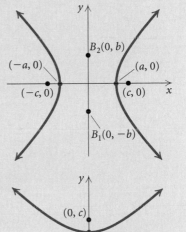

*Transverse Axis Vertical*

$$\frac{y^2}{a^2} - \frac{x^2}{b^2} = 1$$

Vertices: $(0, -a), (0, a)$

Foci: $(0, -c), (0, c)$,
  where $c^2 = a^2 + b^2$

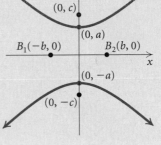

The segment $\overline{B_1 B_2}$ is the **conjugate axis** of the hyperbola.

To graph a hyperbola with a horizontal transverse axis, it is helpful to begin by graphing the lines $y = -(b/a)x$ and $y = (b/a)x$. These are the **asymptotes** of the hyperbola. For a hyperbola with a vertical transverse axis, the asymptotes are $y = -(a/b)x$ and $y = (a/b)x$. As $|x|$ gets larger and larger, the graph of the hyperbola gets closer and closer to the asymptotes.

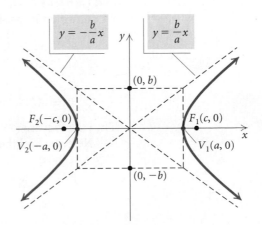

**EXAMPLE 1** Find an equation of the hyperbola with vertices $(0, -4)$ and $(0, 4)$ and foci $(0, -6)$ and $(0, 6)$.

*Solution* We know that $a = 4$ and $c = 6$. We find $b^2$:

$$c^2 = a^2 + b^2$$
$$6^2 = 4^2 + b^2$$
$$36 = 16 + b^2$$
$$20 = b^2.$$

Since the vertices and the foci are on the $y$-axis, we know that the transverse axis is vertical. We can now write the equation of the hyperbola:

$$\frac{y^2}{a^2} - \frac{x^2}{b^2} = 1$$
$$\frac{y^2}{16} - \frac{x^2}{20} = 1.$$

> **Now Try Exercise 7.**

GCM **EXAMPLE 2** For the hyperbola given by

$$9x^2 - 16y^2 = 144,$$

find the vertices, the foci, and the asymptotes. Then graph the hyperbola.

*Solution* First, we find standard form:

$$9x^2 - 16y^2 = 144$$

$$\frac{1}{144}(9x^2 - 16y^2) = \frac{1}{144} \cdot 144 \qquad \text{Multiplying by } \tfrac{1}{144} \text{ to get 1 on the right side}$$

$$\frac{x^2}{16} - \frac{y^2}{9} = 1$$

$$\frac{x^2}{4^2} - \frac{y^2}{3^2} = 1. \qquad \text{Writing standard form}$$

The hyperbola has a horizontal transverse axis, so the vertices are $(-a, 0)$ and $(a, 0)$, or $(-4, 0)$ and $(4, 0)$. From the standard form of the equation, we know that $a^2 = 4^2$, or 16, and $b^2 = 3^2$, or 9. We find the foci:

$$c^2 = a^2 + b^2$$
$$c^2 = 16 + 9$$
$$c^2 = 25$$
$$c = 5.$$

Thus the foci are $(-5, 0)$ and $(5, 0)$.

Next, we find the asymptotes:

$$y = -\frac{b}{a}x = -\frac{3}{4}x \quad \text{and} \quad y = \frac{b}{a}x = \frac{3}{4}x.$$

To draw the graph, we sketch the asymptotes first. This is easily done by drawing the rectangle with horizontal sides passing through $(0, 3)$ and $(0, -3)$ and vertical sides through $(4, 0)$ and $(-4, 0)$. Then we draw and extend the diagonals of this rectangle. The two extended diagonals are the asymptotes of the hyperbola. Next, we plot the vertices and draw the branches of the hyperbola outward from the vertices toward the asymptotes.

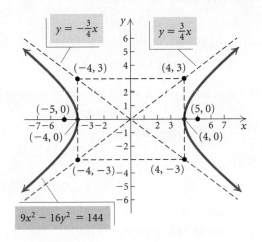

To graph this hyperbola on a graphing calculator, it might be necessary to solve for $y$ first and then graph the top and bottom halves of the hyperbola in the same squared viewing window. See the *Graphing Calculator Manual* that accompanies this text for detailed instructions.

On some graphing calculators, the Conics HYPERBOLA APP can be used to graph hyperbolas, as shown below. Note that the center is $(0, 0)$, so $H = 0$ and $K = 0$.

$$\frac{x^2}{4^2} - \frac{y^2}{3^2} = 1 \qquad\qquad \frac{x^2}{4^2} - \frac{y^2}{3^2} = 1$$

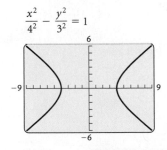

**Now Try Exercise 17.**

If a hyperbola with center at the origin is translated horizontally $|h|$ units and vertically $|k|$ units, the center is at the point $(h, k)$.

## STANDARD EQUATION OF A HYPERBOLA WITH CENTER AT $(h, k)$

**Transverse Axis Horizontal**

$$\frac{(x - h)^2}{a^2} - \frac{(y - k)^2}{b^2} = 1$$

Vertices: $(h - a, k), (h + a, k)$

Asymptotes: $y - k = \dfrac{b}{a}(x - h),$

$$y - k = -\frac{b}{a}(x - h)$$

Foci: $(h - c, k), (h + c, k)$, where $c^2 = a^2 + b^2$

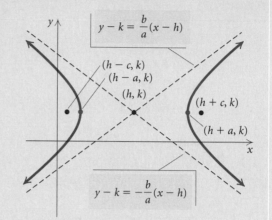

**Transverse Axis Vertical**

$$\frac{(y - k)^2}{a^2} - \frac{(x - h)^2}{b^2} = 1$$

Vertices: $(h, k - a), (h, k + a)$

Asymptotes: $y - k = \dfrac{a}{b}(x - h),$

$$y - k = -\frac{a}{b}(x - h)$$

Foci: $(h, k - c), (h, k + c)$, where $c^2 = a^2 + b^2$

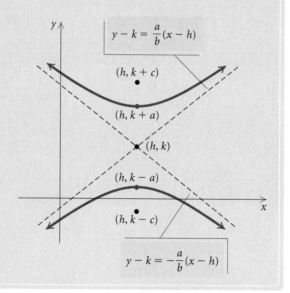

**GCM**  **EXAMPLE 3**  For the hyperbola given by

$$4y^2 - x^2 + 24y + 4x + 28 = 0,$$

find the center, the vertices, the foci, and the asymptotes. Then draw the graph.

**Solution**  First, we complete the square to get standard form:

$$4y^2 - x^2 + 24y + 4x + 28 = 0$$
$$4(y^2 + 6y \qquad) - (x^2 - 4x \qquad) = -28$$
$$4(y^2 + 6y + 9 - 9) - (x^2 - 4x + 4 - 4) = -28$$
$$4(y^2 + 6y + 9) + 4(-9) - (x^2 - 4x + 4) - (-4) = -28$$
$$4(y^2 + 6y + 9) - 36 - (x^2 - 4x + 4) + 4 = -28.$$

Then

$$4(y^2 + 6y + 9) - (x^2 - 4x + 4) = -28 + 36 - 4$$
$$4(y + 3)^2 - (x - 2)^2 = 4$$
$$\frac{(y + 3)^2}{1} - \frac{(x - 2)^2}{4} = 1 \qquad \text{Dividing by 4}$$
$$\frac{[y - (-3)]^2}{1^2} - \frac{(x - 2)^2}{2^2} = 1. \qquad \text{Standard form}$$

The center is $(2, -3)$. Note that $a = 1$ and $b = 2$. The transverse axis is vertical, so the vertices are 1 unit below and above the center:

$$(2, -3 - 1) \text{ and } (2, -3 + 1), \quad \text{or} \quad (2, -4) \text{ and } (2, -2).$$

We know that $c^2 = a^2 + b^2$, so $c^2 = 1^2 + 2^2 = 1 + 4 = 5$ and $c = \sqrt{5}$. Thus the foci are $\sqrt{5}$ units below and above the center:

$$(2, -3 - \sqrt{5}) \quad \text{and} \quad (2, -3 + \sqrt{5}).$$

The asymptotes are

$$y - (-3) = \tfrac{1}{2}(x - 2) \quad \text{and} \quad y - (-3) = -\tfrac{1}{2}(x - 2),$$

or

$$y + 3 = \tfrac{1}{2}(x - 2) \quad \text{and} \quad y + 3 = -\tfrac{1}{2}(x - 2).$$

We sketch the asymptotes, plot the vertices, and draw the graph.

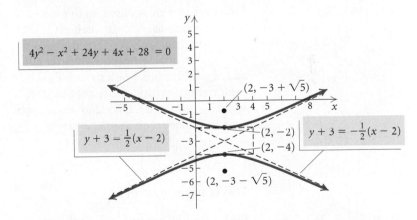

When the equation of a hyperbola is written in standard form, we can use the Conics HYPERBOLA APP to graph it, as shown in the windows below.

$$\frac{[y - (-3)]^2}{1^2} - \frac{(x - 2)^2}{2^2} = 1$$

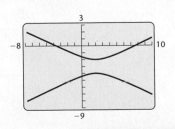

**Now Try Exercise 29.**

## CONNECTING THE CONCEPTS

### Classifying Equations of Conic Sections

| EQUATION | TYPE OF CONIC SECTION | GRAPH |
|---|---|---|
| $x - 4 + 4y = y^2$ | Only one variable is squared, so this cannot be a circle, an ellipse, or a hyperbola. Find an equivalent equation: $$x = (y - 2)^2.$$ This is an equation of a parabola. | 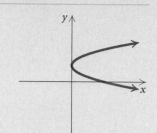 |
| $3x^2 + 3y^2 = 75$ | Both variables are squared, so this cannot be a parabola. The squared terms are added, so this cannot be a hyperbola. Divide by 3 on both sides to find an equivalent equation: $$x^2 + y^2 = 25.$$ This is an equation of a circle. | 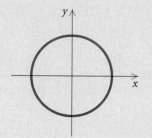 |
| $y^2 = 16 - 4x^2$ | Both variables are squared, so this cannot be a parabola. Add $4x^2$ on both sides to find an equivalent equation: $4x^2 + y^2 = 16$. The squared terms are added, so this cannot be a hyperbola. The coefficients of $x^2$ and $y^2$ are not the same, so this is not a circle. Divide by 16 on both sides to find an equivalent equation: $$\frac{x^2}{4} + \frac{y^2}{16} = 1.$$ This is an equation of an ellipse. | |
| $x^2 = 4y^2 + 36$ | Both variables are squared, so this cannot be a parabola. Subtract $4y^2$ on both sides to find an equivalent equation: $x^2 - 4y^2 = 36$. The squared terms are not added, so this cannot be a circle or an ellipse. Divide by 36 on both sides to find an equivalent equation: $$\frac{x^2}{36} - \frac{y^2}{9} = 1.$$ This is an equation of a hyperbola. | 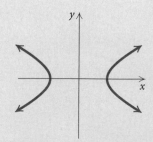 |

## ■ Applications

Some comets travel in hyperbolic paths with the sun at one focus. Such comets pass by the sun only one time, unlike those with elliptical orbits, which reappear at intervals. We also see hyperbolas in architecture, such as in a cross section of a planetarium, an amphitheater, or a cooling tower for a steam or nuclear power plant.

Another application of hyperbolas is in the long-range navigation system LORAN. This system uses transmitting stations in three locations to send out simultaneous signals to a ship or an aircraft. The difference in the arrival times of the signals from one pair of transmitters is recorded on the ship or aircraft. This difference is also recorded for signals from another pair of transmitters. For each pair, a computation is performed to determine the difference in the distances from each member of the pair to the ship or aircraft. If each pair of differences is kept constant, two hyperbolas can be drawn. Each has one of the pairs of transmitters as foci, and the ship or aircraft lies on the intersection of two of their branches.

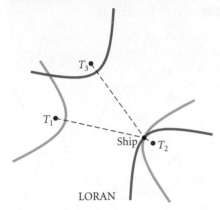

LORAN

---

## 7.3 Exercise Set

*In Exercises 1–6, match the equation with one of the graphs (a)–(f), which follow.*

**a)**
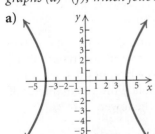

**b)**
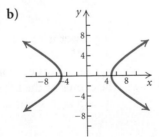

**c)**
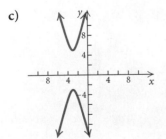

**d)**
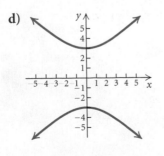

**e)**
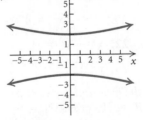

**f)**
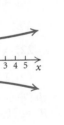

**1.** $\dfrac{x^2}{25} - \dfrac{y^2}{9} = 1$

**2.** $\dfrac{y^2}{4} - \dfrac{x^2}{36} = 1$

**3.** $\dfrac{(y-1)^2}{16} - \dfrac{(x+3)^2}{1} = 1$

**4.** $\dfrac{(x+4)^2}{100} - \dfrac{(y-2)^2}{81} = 1$

**5.** $25x^2 - 16y^2 = 400$

**6.** $y^2 - x^2 = 9$

*Find an equation of a hyperbola satisfying the given conditions.*

**7.** Vertices: $(0, 3)$ and $(0, -3)$;
 foci: $(0, 5)$ and $(0, -5)$

**8.** Vertices: $(1, 0)$ and $(-1, 0)$;
 foci: $(2, 0)$ and $(-2, 0)$

**9.** Asymptotes: $y = \frac{3}{2}x$, $y = -\frac{3}{2}x$;
 one vertex: $(2, 0)$

**10.** Asymptotes: $y = \frac{5}{4}x$, $y = -\frac{5}{4}x$;
 one vertex: $(0, 3)$

*Find the center, the vertices, the foci, and the asymptotes. Then draw the graph.*

**11.** $\dfrac{x^2}{4} - \dfrac{y^2}{4} = 1$

**12.** $\dfrac{x^2}{1} - \dfrac{y^2}{9} = 1$

**13.** $\dfrac{(x - 2)^2}{9} - \dfrac{(y + 5)^2}{1} = 1$

**14.** $\dfrac{(x - 5)^2}{16} - \dfrac{(y + 2)^2}{9} = 1$

**15.** $\dfrac{(y + 3)^2}{4} - \dfrac{(x + 1)^2}{16} = 1$

**16.** $\dfrac{(y + 4)^2}{25} - \dfrac{(x + 2)^2}{16} = 1$

**17.** $x^2 - 4y^2 = 4$   **18.** $4x^2 - y^2 = 16$

**19.** $9y^2 - x^2 = 81$   **20.** $y^2 - 4x^2 = 4$

**21.** $x^2 - y^2 = 2$   **22.** $x^2 - y^2 = 3$

**23.** $y^2 - x^2 = \frac{1}{4}$   **24.** $y^2 - x^2 = \frac{1}{9}$

*Find the center, the vertices, the foci, and the asymptotes of the hyperbola. Then draw the graph.*

**25.** $x^2 - y^2 - 2x - 4y - 4 = 0$

**26.** $4x^2 - y^2 + 8x - 4y - 4 = 0$

**27.** $36x^2 - y^2 - 24x + 6y - 41 = 0$

**28.** $9x^2 - 4y^2 + 54x + 8y + 41 = 0$

**29.** $9y^2 - 4x^2 - 18y + 24x - 63 = 0$

**30.** $x^2 - 25y^2 + 6x - 50y = 41$

**31.** $x^2 - y^2 - 2x - 4y = 4$

**32.** $9y^2 - 4x^2 - 54y - 8x + 41 = 0$

**33.** $y^2 - x^2 - 6x - 8y - 29 = 0$

**34.** $x^2 - y^2 = 8x - 2y - 13$

*The **eccentricity** of a hyperbola is defined as $e = c/a$. For a hyperbola, $c > a > 0$, so $e > 1$. When e is close to 1, a hyperbola appears to be very narrow. As the eccentricity increases, the hyperbola becomes "wider."*

**35.** Observe the shapes of the hyperbolas in Examples 2 and 3. Which hyperbola has the larger eccentricity? Confirm your answer by computing the eccentricity of each hyperbola.

**36.** Which hyperbola has the larger eccentricity? (Assume that the coordinate systems have the same scale.)

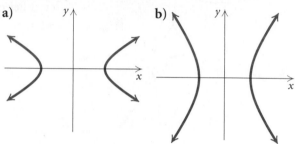

a)   b)

**37.** Find an equation of a hyperbola with vertices $(3, 7)$ and $(-3, 7)$ and $e = \frac{5}{3}$.

**38.** Find an equation of a hyperbola with vertices $(-1, 3)$ and $(-1, 7)$ and $e = 4$.

**39.** *Nuclear Cooling Tower.* A cross section of a nuclear cooling tower is a hyperbola with equation

$$\frac{x^2}{90^2} - \frac{y^2}{130^2} = 1.$$

The tower is 450 ft tall, and the distance from the top of the tower to the center of the hyperbola is half the distance from the base of the tower to the center of the hyperbola. Find the diameter of the top and the base of the tower.

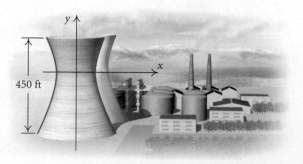

450 ft

**40.** *Hyperbolic Mirror.* Certain telescopes contain both a parabolic mirror and a hyperbolic mirror. In the telescope shown in the figure, the parabola and the hyperbola share focus $F_1$, which is 14 m above the vertex of the parabola. The hyperbola's second focus $F_2$ is 2 m above the parabola's vertex. The vertex of the hyperbolic mirror is 1 m below $F_1$. Position a coordinate system with the origin at the center of the hyperbola and with the foci on the $y$-axis. Then find the equation of the hyperbola.

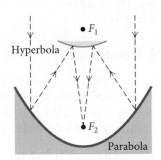

## Skill Maintenance

*In Exercises 41–44, given the function:*

**a)** *Determine whether it is one-to-one.*
**b)** *If it is one-to-one, find a formula for the inverse.*

**41.** $f(x) = 2x - 3$     **42.** $f(x) = x^3 + 2$

**43.** $f(x) = \dfrac{5}{x - 1}$     **44.** $f(x) = \sqrt{x + 4}$

*Solve.*

**45.** $x + y = 5,$
      $x - y = 7$

**46.** $3x - 2y = 5,$
      $5x + 2y = 3$

**47.** $2x - 3y = 7,$
      $3x + 5y = 1$

**48.** $3x + 2y = -1,$
      $2x + 3y = 6$

## Synthesis

*Find an equation of a hyperbola satisfying the given conditions.*

**49.** Vertices: $(3, -8)$ and $(3, -2)$;
      asymptotes: $y = 3x - 14,\ y = -3x + 4$

**50.** Vertices: $(-9, 4)$ and $(-5, 4)$;
      asymptotes: $y = 3x + 25,\ y = -3x - 17$

*Use a graphing calculator to find the center, the vertices, and the asymptotes.*

**51.** $5x^2 - 3.5y^2 + 14.6x - 6.7y + 3.4 = 0$

**52.** $x^2 - y^2 - 2.046x - 4.088y - 4.228 = 0$

**53.** *Navigation.* Two radio transmitters positioned 300 mi apart along the shore send simultaneous signals to a ship that is 200 mi offshore, sailing parallel to the shoreline. The signal from transmitter $S$ reaches the ship 200 microseconds later than the signal from transmitter $T$. The signals travel at a speed of 186,000 miles per second, or 0.186 mile per microsecond. Find the equation of the hyperbola with foci $S$ and $T$ on which the ship is located. (*Hint:* For any point on the hyperbola, the absolute value of the difference of its distances from the foci is $2a$.)

## 7.4 Nonlinear Systems of Equations and Inequalities

- Solve a nonlinear system of equations.
- Use nonlinear systems of equations to solve applied problems.
- Graph nonlinear systems of inequalities.

The systems of equations that we have studied so far have been composed of linear equations. Now we consider systems of two equations in two variables in which at least one equation is not linear.

### Nonlinear Systems of Equations

The graphs of the equations in a nonlinear system of equations can have no point of intersection or one or more points of intersection. The coordinates of each point of intersection represent a solution of the system of equations. When no point of intersection exists, the system of equations has no real-number solution.

Solutions of nonlinear systems of equations can be found using the substitution method or the elimination method. The substitution method is preferable for a system consisting of one linear equation and one nonlinear equation. The elimination method is preferable in most, but not all, cases when both equations are nonlinear.

**EXAMPLE 1**   Solve the following system of equations:

$$x^2 + y^2 = 25, \quad \textbf{(1)} \qquad \text{The graph is a circle.}$$
$$3x - 4y = 0. \quad \textbf{(2)} \qquad \text{The graph is a line.}$$

### Algebraic Solution

We use the substitution method. First, we solve equation (2) for $x$:

$$x = \tfrac{4}{3}y. \quad \textbf{(3)} \qquad \text{We could have solved for } y \text{ instead.}$$

Next, we substitute $\tfrac{4}{3}y$ for $x$ in equation (1) and solve for $y$:

$$\left(\tfrac{4}{3}y\right)^2 + y^2 = 25$$
$$\tfrac{16}{9}y^2 + y^2 = 25$$
$$\tfrac{25}{9}y^2 = 25$$
$$y^2 = 9 \qquad \text{Multiplying by } \tfrac{9}{25}$$
$$y = \pm 3.$$

Now we substitute these numbers for $y$ in equation (3) and solve for $x$:

$$x = \tfrac{4}{3}(3) = 4, \qquad (4, 3) \text{ appears to be a solution.}$$
$$x = \tfrac{4}{3}(-3) = -4. \qquad (-4, -3) \text{ appears to be a solution.}$$

## Algebraic Solution (*continued*)

*Check:*  For $(4, 3)$:

$$\begin{array}{c|c}
\dfrac{x^2 + y^2 = 25}{4^2 + 3^2 \;?\; 25} & \dfrac{3x - 4y = 0}{3(4) - 4(3) \;?\; 0} \\
\begin{array}{c|c} 16 + 9 & \\ & 25 \end{array} \begin{array}{c} \\ 25 \end{array} \text{TRUE} & \begin{array}{c|c} 12 - 12 & \\ & 0 \end{array} \begin{array}{c} \\ 0 \end{array} \text{TRUE}
\end{array}$$

For $(-4, -3)$:

$$\begin{array}{c|c}
\dfrac{x^2 + y^2 = 25}{(-4)^2 + (-3)^2 \;?\; 25} & \dfrac{3x - 4y = 0}{3(-4) - 4(-3) \;?\; 0} \\
\begin{array}{c|c} 16 + 9 & \\ & 25 \end{array} \begin{array}{c} \\ 25 \end{array} \text{TRUE} & \begin{array}{c|c} -12 + 12 & \\ & 0 \end{array} \begin{array}{c} \\ 0 \end{array} \text{TRUE}
\end{array}$$

The pairs $(4, 3)$ and $(-4, -3)$ check, so they are the solutions.

## Graphical Solution

We graph both equations in the same viewing window. Note that there are two points of intersection. We can find their coordinates using the INTERSECT feature.

$y_1 = \sqrt{25 - x^2}, \; y_2 = -\sqrt{25 - x^2},$
$y_3 = \frac{3}{4}x$

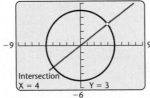

$y_1 = \sqrt{25 - x^2}, \; y_2 = -\sqrt{25 - x^2},$
$y_3 = \frac{3}{4}x$

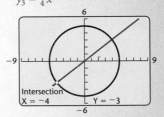

The solutions are $(4, 3)$ and $(-4, -3)$.

**Now Try Exercise 7.**

In the algebraic solution in Example 1, suppose that to find $x$ we had substituted 3 and $-3$ in equation (1) rather than equation (3). If $y = 3$, $y^2 = 9$, and if $y = -3$, $y^2 = 9$, so both substitutions can be performed at the same time:

$$\begin{aligned}
x^2 + y^2 &= 25 \qquad \textbf{(1)} \\
x^2 + (\pm 3)^2 &= 25 \\
x^2 + 9 &= 25 \\
x^2 &= 16 \\
x &= \pm 4.
\end{aligned}$$

Each $y$-value produces two values for $x$. Thus, if $y = 3$, $x = 4$ or $x = -4$, and if $y = -3$, $x = 4$ or $x = -4$. The possible solutions are $(4, 3)$, $(-4, 3)$, $(4, -3)$, and $(-4, -3)$. A check reveals that $(4, -3)$ and $(-4, 3)$ are not solutions of equation (2). Since a circle and a line can intersect in at most two points, it is clear that there can be at most two real-number solutions.

**EXAMPLE 2** Solve the following system of equations:

$$x + y = 5, \qquad \textbf{(1)} \qquad \text{The graph is a line.}$$
$$y = 3 - x^2. \qquad \textbf{(2)} \qquad \text{The graph is a parabola.}$$

## Algebraic Solution

We use the substitution method, substituting $3 - x^2$ for $y$ in equation (1):

$$x + 3 - x^2 = 5$$
$$-x^2 + x - 2 = 0 \qquad \text{Subtracting 5 and rearranging}$$
$$x^2 - x + 2 = 0. \qquad \text{Multiplying by } -1$$

Next, we use the quadratic formula:

$$x = \frac{-b \pm \sqrt{b^2 - 4ac}}{2a} = \frac{-(-1) \pm \sqrt{(-1)^2 - 4(1)(2)}}{2(1)}$$
$$= \frac{1 \pm \sqrt{1 - 8}}{2} = \frac{1 \pm \sqrt{-7}}{2} = \frac{1 \pm i\sqrt{7}}{2} = \frac{1}{2} \pm \frac{\sqrt{7}}{2}i.$$

Now, we substitute these values for $x$ in equation (1) and solve for $y$:

$$\frac{1}{2} + \frac{\sqrt{7}}{2}i + y = 5$$
$$y = 5 - \frac{1}{2} - \frac{\sqrt{7}}{2}i = \frac{9}{2} - \frac{\sqrt{7}}{2}i$$

and

$$\frac{1}{2} - \frac{\sqrt{7}}{2}i + y = 5$$
$$y = 5 - \frac{1}{2} + \frac{\sqrt{7}}{2}i = \frac{9}{2} + \frac{\sqrt{7}}{2}i.$$

The solutions are

$$\left(\frac{1}{2} + \frac{\sqrt{7}}{2}i, \frac{9}{2} - \frac{\sqrt{7}}{2}i\right) \quad \text{and} \quad \left(\frac{1}{2} - \frac{\sqrt{7}}{2}i, \frac{9}{2} + \frac{\sqrt{7}}{2}i\right).$$

There are no real-number solutions.

## Graphical Solution

We graph both equations in the same viewing window.

$y_1 = 5 - x, \quad y_2 = 3 - x^2$

Note that there are no points of intersection. This indicates that there are no real-number solutions. Algebra must be used, as at left, to find the imaginary-number solutions.

**Now Try Exercise 17.**

**EXAMPLE 3** Solve the following system of equations:

$$2x^2 + 5y^2 = 39, \quad (1) \qquad \text{The graph is an ellipse.}$$
$$3x^2 - y^2 = -1. \quad (2) \qquad \text{The graph is a hyperbola.}$$

## Algebraic Solution

We use the elimination method. First, we multiply equation (2) by 5 and add to eliminate the $y^2$-term:

$$\begin{aligned} 2x^2 + 5y^2 &= 39 \qquad \textbf{(1)} \\ \underline{15x^2 - 5y^2 = -5} \qquad & \text{Multiplying (2)} \\ & \text{by 5} \\ 17x^2 \qquad &= 34 \qquad \text{Adding} \\ x^2 &= 2 \\ x &= \pm\sqrt{2}. \end{aligned}$$

If $x = \sqrt{2}$, $x^2 = 2$, and if $x = -\sqrt{2}$, $x^2 = 2$. Thus substituting $\sqrt{2}$ or $-\sqrt{2}$ for $x$ in equation (2) gives us

$$\begin{aligned} 3(\pm\sqrt{2})^2 - y^2 &= -1 \\ 3 \cdot 2 - y^2 &= -1 \\ 6 - y^2 &= -1 \\ -y^2 &= -7 \\ y^2 &= 7 \\ y &= \pm\sqrt{7}. \end{aligned}$$

Each $x$-value produces two values for $y$. Thus, for $x = \sqrt{2}$, we have $y = \sqrt{7}$ or $y = -\sqrt{7}$, and for $x = -\sqrt{2}$, we have $y = \sqrt{7}$ or $y = -\sqrt{7}$. The possible solutions are $(\sqrt{2}, \sqrt{7})$, $(\sqrt{2}, -\sqrt{7})$, $(-\sqrt{2}, \sqrt{7})$, and $(-\sqrt{2}, -\sqrt{7})$. All four pairs check, so they are the solutions.

## Graphical Solution

We graph both equations in the same viewing window. There are four points of intersection. We can use the INTERSECT feature to find their coordinates.

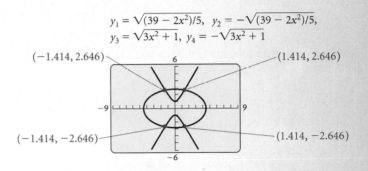

$$y_1 = \sqrt{(39 - 2x^2)/5}, \quad y_2 = -\sqrt{(39 - 2x^2)/5},$$
$$y_3 = \sqrt{3x^2 + 1}, \quad y_4 = -\sqrt{3x^2 + 1}$$

$(-1.414, 2.646)$     $(1.414, 2.646)$

$(-1.414, -2.646)$     $(1.414, -2.646)$

Note that the algebraic solution yields exact solutions, whereas the graphical solution yields decimal approximations of the solutions on most graphing calculators.

The solutions are approximately $(1.414, 2.646)$, $(1.414, -2.646)$, $(-1.414, 2.646)$, and $(-1.414, -2.646)$.

**Now Try Exercise 27.**

**EXAMPLE 4** Solve the following system of equations:

$$x^2 - 3y^2 = 6, \quad (1)$$
$$xy = 3. \quad (2)$$

## Algebraic Solution

We use the substitution method. First, we solve equation (2) for $y$:

$$xy = 3 \quad \textbf{(2)}$$

$$y = \frac{3}{x}. \quad \textbf{(3)} \qquad \text{Dividing by } x$$

Next, we substitute $3/x$ for $y$ in equation (1) and solve for $x$:

$$x^2 - 3\left(\frac{3}{x}\right)^2 = 6$$

$$x^2 - 3 \cdot \frac{9}{x^2} = 6$$

$$x^2 - \frac{27}{x^2} = 6$$

$$x^4 - 27 = 6x^2 \qquad \text{Multiplying by } x^2$$

$$x^4 - 6x^2 - 27 = 0$$

$$u^2 - 6u - 27 = 0 \qquad \text{Letting } u = x^2$$

$$(u - 9)(u + 3) = 0 \qquad \text{Factoring}$$

$$u = 9 \quad or \quad u = -3 \qquad \text{Principle of zero products}$$

$$x^2 = 9 \quad or \quad x^2 = -3 \qquad \text{Substituting } x^2 \text{ for } u$$

$$x = \pm 3 \quad or \quad x = \pm\sqrt{3}i.$$

Since $y = 3/x$,

when $x = 3$, $\qquad y = \frac{3}{3} = 1$;

when $x = -3$, $\qquad y = \frac{3}{-3} = -1$;

when $x = \sqrt{3}i$, $\quad y = \frac{3}{\sqrt{3}i} = \frac{3}{\sqrt{3}i} \cdot \frac{-\sqrt{3}i}{-\sqrt{3}i} = \sqrt{3}i$;

when $x = -\sqrt{3}i$, $\quad y = \frac{3}{-\sqrt{3}i} = \frac{3}{-\sqrt{3}i} \cdot \frac{\sqrt{3}i}{\sqrt{3}i} = \sqrt{3}i$.

The pairs $(3, 1)$, $(-3, -1)$, $(\sqrt{3}i, -\sqrt{3}i)$, and $(-\sqrt{3}i, \sqrt{3}i)$ check, so they are the solutions.

## Graphical Solution

We graph both equations in the same viewing window and find the coordinates of their points of intersection.

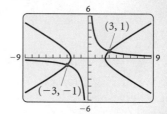

$$x^2 - 3y^2 = 6$$

$$y_1 = \sqrt{(x^2 - 6)/3}, \quad y_2 = -\sqrt{(x^2 - 6)/3},$$
$$y_3 = 3/x$$

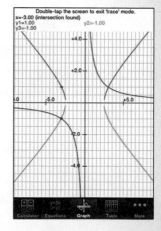

We also can use an app to find the coordinates of their points of intersection.

Again, note that the graphical methods yield only the real-number solutions of the system of equations. The algebraic method must be used in order to find *all* the solutions.

**Now Try Exercise 19.**

## ■ Modeling and Problem Solving

**EXAMPLE 5** *Dimensions of a Piece of Land.* For a student recreation building at Southport Community College, an architect wants to lay out a rectangular piece of land that has a perimeter of 204 m and an area of 2565 m². Find the dimensions of the piece of land.

*Solution*

1. **Familiarize.** We make a drawing and label it, letting $l =$ the length of the piece of land, in meters, and $w =$ the width, in meters.

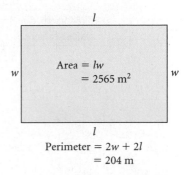

Perimeter $= 2w + 2l$
$= 204$ m

2. **Translate.** We now have the following:

Perimeter:  $2w + 2l = 204,$  **(1)**
Area:          $lw = 2565.$  **(2)**

3. **Carry out.** We solve the system of equations both algebraically and graphically.

### Algebraic Solution

We solve the system of equations

$$2w + 2l = 204, \quad \textbf{(1)}$$
$$lw = 2565. \quad \textbf{(2)}$$

Solving the second equation for $l$ gives us $l = 2565/w$. We then substitute $2565/w$ for $l$ in equation (1) and solve for $w$:

$$2w + 2\left(\frac{2565}{w}\right) = 204$$

$$2w^2 + 2(2565) = 204w \qquad \text{Multiplying by } w$$
$$2w^2 - 204w + 2(2565) = 0$$
$$w^2 - 102w + 2565 = 0 \qquad \text{Multiplying by } \tfrac{1}{2}$$
$$(w - 57)(w - 45) = 0$$
$$w = 57 \quad or \quad w = 45. \qquad \text{Principle of zero products}$$

If $w = 57$, then $l = 2565/w = 2565/57 = 45$. If $w = 45$, then $l = 2565/w = 2565/45 = 57$. Since length is generally considered to be longer than width, we have the solution $l = 57$ and $w = 45$, or $(57, 45)$.

### Graphical Solution

We replace $l$ with $x$ and $w$ with $y$, graph $y_1 = (204 - 2x)/2$ and $y_2 = 2565/x$, and find the point(s) of intersection of the graphs.

$$y_1 = (204 - 2x)/2, \ y_2 = 2565/x$$

As in the algebraic solution, we have two possible solutions: $(45, 57)$ and $(57, 45)$. Since length, $x$, is generally considered to be longer than width, $y$, we have the solution $(57, 45)$.

4. **Check.** If $l = 57$ and $w = 45$, the perimeter is $2 \cdot 45 + 2 \cdot 57$, or 204. The area is $57 \cdot 45$, or 2565. The numbers check.

5. **State.** The length of the piece of land is 57 m and the width is 45 m.

**Now Try Exercise 61.**

## ■ Nonlinear Systems of Inequalities

SYSTEMS OF INEQUALITIES

REVIEW SECTION **6.7.**

Recall that a solution of a system of inequalities is an ordered pair that is a solution of each inequality in the system. We graphed systems of linear inequalities in Section 6.7. Now we graph a system of nonlinear inequalities.

**EXAMPLE 6**   Graph the solution set of the system

$$x^2 + y^2 \le 25,$$
$$3x - 4y > 0.$$

**Solution**   We graph $x^2 + y^2 \le 25$ by first graphing the related equation of the circle $x^2 + y^2 = 25$. We use a solid curve since the inequality symbol is $\le$. Next, we choose $(0, 0)$ as a test point and find that it is a solution of $x^2 + y^2 \le 25$, so we shade the region that contains $(0, 0)$ using red. This is the region inside the circle. Now we graph the line $3x - 4y = 0$ using a dashed line since the inequality symbol is $>$. The point $(0, 0)$ is on the line, so we choose another test point, say, $(0, 2)$. We find that this point is not a solution of $3x - 4y > 0$, so we shade the half-plane that does not contain $(0, 2)$ using green. The solution set of the system of inequalities is the region shaded both red and green, or brown, including part of the circle $x^2 + y^2 = 25$.

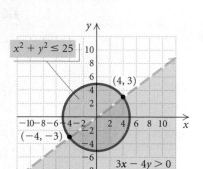

To find the points of intersection of the graphs of the related equations, we solve the system composed of those equations:

$$x^2 + y^2 = 25,$$
$$3x - 4y = 0.$$

In Example 1, we found that these points are $(4, 3)$ and $(-4, -3)$.

**Now Try Exercise 75.**

**EXAMPLE 7**   Use a graphing calculator to graph the system

$$y \le 4 - x^2,$$
$$x + y \ge 2.$$

**Solution**   We graph $y_1 = 4 - x^2$ and $y_2 = 2 - x$. Using the test point $(0, 0)$ for each inequality, we find that we should shade below $y_1$ and above $y_2$. We can find the points of intersection of the graphs of the related equations, $(-1, 3)$ and $(2, 0)$, using the INTERSECT feature.

$$y_1 = 4 - x^2, \ y_2 = 2 - x$$

**Now Try Exercise 81.**

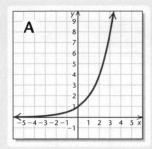

A

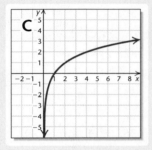

B

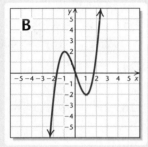

C

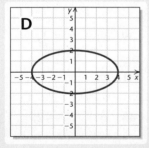

D

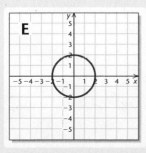

E

# Visualizing the Graph

Match the equation or system of equations with its graph.

**1.** $y = x^3 - 3x$

**2.** $y = x^2 + 2x - 3$

**3.** $y = \dfrac{x - 1}{x^2 - x - 2}$

**4.** $y = -3x + 2$

**5.** $\begin{array}{l} x + y = 3, \\ 2x + 5y = 3 \end{array}$

**6.** $\begin{array}{l} 9x^2 - 4y^2 = 36, \\ x^2 + y^2 = 9 \end{array}$

**7.** $5x^2 + 5y^2 = 20$

**8.** $4x^2 + 16y^2 = 64$

**9.** $y = \log_2 x$

**10.** $y = 2^x$

**Answers on page A-49**

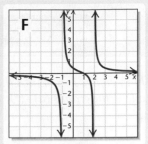

F

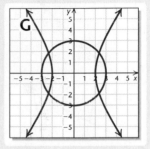

G

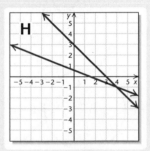

H

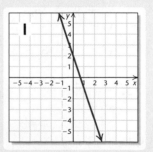

I

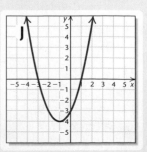

J

# 7.4 Exercise Set

*In Exercises 1–6, match the system of equations with one of the graphs (a)–(f), which follow.*

a)
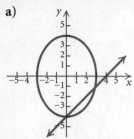

b)
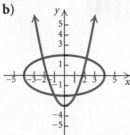

c)
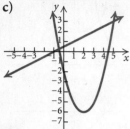

d)
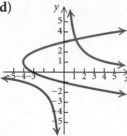

e)
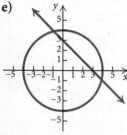

f)

**1.** $x^2 + y^2 = 16$,
$x + y = 3$

**2.** $16x^2 + 9y^2 = 144$,
$x - y = 4$

**3.** $y = x^2 - 4x - 2$,
$2y - x = 1$

**4.** $4x^2 - 9y^2 = 36$,
$x^2 + y^2 = 25$

**5.** $y = x^2 - 3$,
$x^2 + 4y^2 = 16$

**6.** $y^2 - 2y = x + 3$,
$xy = 4$

*Solve.*

**7.** $x^2 + y^2 = 25$,
$y - x = 1$

**8.** $x^2 + y^2 = 100$,
$y - x = 2$

**9.** $4x^2 + 9y^2 = 36$,
$3y + 2x = 6$

**10.** $9x^2 + 4y^2 = 36$,
$3x + 2y = 6$

**11.** $x^2 + y^2 = 25$,
$y^2 = x + 5$

**12.** $y = x^2$,
$x = y^2$

**13.** $x^2 + y^2 = 9$,
$x^2 - y^2 = 9$

**14.** $y^2 - 4x^2 = 4$,
$4x^2 + y^2 = 4$

**15.** $y^2 - x^2 = 9$,
$2x - 3 = y$

**16.** $x + y = -6$,
$xy = -7$

**17.** $y^2 = x + 3$,
$2y = x + 4$

**18.** $y = x^2$,
$3x = y + 2$

**19.** $x^2 + y^2 = 25$,
$xy = 12$

**20.** $x^2 - y^2 = 16$,
$x + y^2 = 4$

**21.** $x^2 + y^2 = 4$,
$16x^2 + 9y^2 = 144$

**22.** $x^2 + y^2 = 25$,
$25x^2 + 16y^2 = 400$

**23.** $x^2 + 4y^2 = 25$,
$x + 2y = 7$

**24.** $y^2 - x^2 = 16$,
$2x - y = 1$

**25.** $x^2 - xy + 3y^2 = 27$,
$x - y = 2$

**26.** $2y^2 + xy + x^2 = 7$,
$x - 2y = 5$

**27.** $x^2 + y^2 = 16$,
$y^2 - 2x^2 = 10$

**28.** $x^2 + y^2 = 14$,
$x^2 - y^2 = 4$

**29.** $x^2 + y^2 = 5$,
$xy = 2$

**30.** $x^2 + y^2 = 20$,
$xy = 8$

**31.** $3x + y = 7$,
$4x^2 + 5y = 56$

**32.** $2y^2 + xy = 5$,
$4y + x = 7$

**33.** $a + b = 7$,
$ab = 4$

**34.** $p + q = -4$,
$pq = -5$

**35.** $x^2 + y^2 = 13$,
$xy = 6$

**36.** $x^2 + 4y^2 = 20$,
$xy = 4$

**37.** $x^2 + y^2 + 6y + 5 = 0$,
$x^2 + y^2 - 2x - 8 = 0$

**38.** $2xy + 3y^2 = 7$,
$3xy - 2y^2 = 4$

**39.** $2a + b = 1$,
$b = 4 - a^2$

**40.** $4x^2 + 9y^2 = 36$,
$x + 3y = 3$

**41.** $a^2 + b^2 = 89$,
$a - b = 3$

**42.** $xy = 4$,
$x + y = 5$

**43.** $xy - y^2 = 2$,
$2xy - 3y^2 = 0$

**44.** $4a^2 - 25b^2 = 0$,
$2a^2 - 10b^2 = 3b + 4$

**45.** $m^2 - 3mn + n^2 + 1 = 0$,
$3m^2 - mn + 3n^2 = 13$

**46.** $ab - b^2 = -4$,
$ab - 2b^2 = -6$

**47.** $x^2 + y^2 = 5$,
$x - y = 8$

**48.** $4x^2 + 9y^2 = 36$,
$y - x = 8$

**49.** $a^2 + b^2 = 14$,
$ab = 3\sqrt{5}$

**50.** $x^2 + xy = 5$,
$2x^2 + xy = 2$

**51.** $x^2 + y^2 = 25$,
$9x^2 + 4y^2 = 36$

**52.** $x^2 + y^2 = 1$,
$9x^2 - 16y^2 = 144$

**53.** $5y^2 - x^2 = 1$,
$xy = 2$

**54.** $x^2 - 7y^2 = 6$,
$xy = 1$

*In Exercises 55–58, determine whether the statement is true or false.*

**55.** A nonlinear system of equations can have both real-number solutions and imaginary-number solutions.

**56.** If the graph of a nonlinear system of equations consists of a line and a parabola, the system has two real-number solutions.

**57.** If the graph of a nonlinear system of equations consists of a line and a circle, the system has at most two real-number solutions.

**58.** If the graph of a nonlinear system of equations consists of a line and an ellipse, it is possible for the system to have exactly one real-number solution.

**59.** *Picture Frame Dimensions.* Frank's Frame Shop is building a frame for a rectangular oil painting with a perimeter of 28 cm and a diagonal of 10 cm. Find the dimensions of the painting.

**60.** *Sign Dimensions.* Peden's Advertising is building a rectangular sign with an area of 2 yd$^2$ and a perimeter of 6 yd. Find the dimensions of the sign.

**61.** *Graphic Design.* Marcia Graham, owner of Graham's Graphics, is designing an advertising brochure for the Art League's spring show. Each page of the brochure is rectangular with an area of 20 in$^2$ and a perimeter of 18 in. Find the dimensions of the brochure.

**62.** *Landscaping.* Green Leaf Landscaping is planting a rectangular wildflower garden with a perimeter of 6 m and a diagonal of $\sqrt{5}$ m. Find the dimensions of the garden.

**63.** *Fencing.* It will take 210 yd of fencing to enclose a rectangular dog pen. The area of the pen is 2250 yd$^2$. What are the dimensions of the pen?

**64.** *Carpentry.* Ted Hansen of Hansen Woodworking Designs has been commissioned to make a rectangular tabletop with an area of $\sqrt{2}$ m$^2$ and a diagonal of $\sqrt{3}$ m for the Decorators' Show House. Find the dimensions of the tabletop.

**65.** *Banner Design.* A rectangular banner with an area of $\sqrt{3}$ m$^2$ is being designed to advertise an exhibit at the Davis Gallery. The length of a diagonal is 2 m. Find the dimensions of the banner.

**66.** *Investment.* Jenna made an investment for 1 year that earned $7.50 simple interest. If the principal had been $25 more and the interest rate 1% less, the interest would have been the same. Find the principal and the interest rate.

**67.** *Seed Test Plots.* The Burton Seed Company has two square test plots. The sum of their areas is 832 ft², and the difference of their areas is 320 ft². Find the length of a side of each plot.

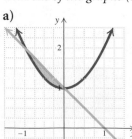

**68.** *Office Dimensions.* The diagonal of the floor of a rectangular office cubicle is 1 ft longer than the length of the cubicle and 3 ft longer than twice the width. Find the dimensions of the cubicle.

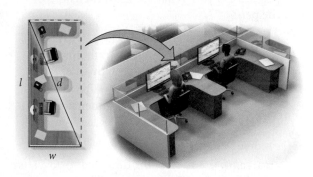

*In Exercises 69–74, match the system of inequalities with one of the graphs (a)–(f), which follow.*

**a)**

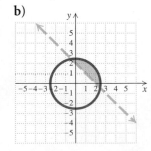

**b)**

**c)**

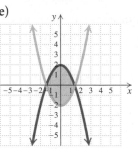

**d)**

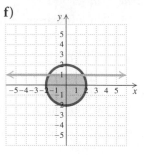

**e)**

**f)**

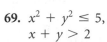

**69.** $x^2 + y^2 \leq 5,$
$\quad x + y > 2$

**70.** $y \leq 2 - x^2,$
$\quad y \geq x^2 - 2$

**71.** $y \geq x^2,$
$\quad y > x$

**72.** $x^2 + y^2 \leq 4,$
$\quad y \leq 1$

**73.** $y \geq x^2 + 1,$
$\quad x + y \leq 1$

**74.** $x^2 + y^2 \leq 9,$
$\quad y > x$

*Graph the system of inequalities. Then find the coordinates of the points of intersection of the graphs of the related equations.*

**75.** $x^2 + y^2 \leq 16,$
$\quad y < x$

**76.** $x^2 + y^2 \leq 10,$
$\quad y > x$

**77.** $x^2 \leq y,$
$\quad x + y \geq 2$

**78.** $x \geq y^2,$
$\quad x - y \leq 2$

**79.** $x^2 + y^2 \leq 25,$
$\quad x - y > 5$

**80.** $x^2 + y^2 \geq 9,$
$\quad x - y > 3$

**81.** $y \geq x^2 - 3,$
$\quad y \leq 2x$

**82.** $y \leq 3 \quad x^2,$
$\quad y \geq x + 1$

**83.** $y \geq x^2,$
$\quad y < x + 2$

**84.** $y \leq 1 - x^2,$
$\quad y > x - 1$

## Skill Maintenance

*Solve.*

**85.** $2^{3x} = 64$

**86.** $5^x = 27$

**87.** $\log_3 x = 4$

**88.** $\log(x - 3) + \log x = 1$

## Synthesis

**89.** Find an equation of the circle that passes through the points $(2, 4)$ and $(3, 3)$ and whose center is on the line $3x - y = 3$.

**90.** Find an equation of the circle that passes through the points $(2, 3)$, $(4, 5)$, and $(0, -3)$.

**91.** Find an equation of an ellipse centered at the origin that passes through the points $(1, \sqrt{3}/2)$ and $(\sqrt{3}, 1/2)$.

**92.** Find an equation of a hyperbola of the type

$$\frac{x^2}{b^2} - \frac{y^2}{a^2} = 1$$

that passes through the points $(-3, -3\sqrt{5}/2)$ and $(-3/2, 0)$.

**93.** Show that a hyperbola does not intersect its asymptotes. That is, solve the system of equations

$$\frac{x^2}{a^2} - \frac{y^2}{b^2} = 1,$$

$$y = \frac{b}{a}x \left( \text{or } y = -\frac{b}{a}x \right).$$

**94.** *Numerical Relationship.* Find two numbers whose product is 2 and the sum of whose reciprocals is $\frac{33}{8}$.

**95.** *Numerical Relationship.* The sum of two numbers is 1, and their product is 1. Find the sum of their cubes. There is a method to solve this problem that is easier than solving a nonlinear system of equations. Can you discover it?

**96.** *Box Dimensions.* Four squares with sides 5 in. long are cut from the corners of a rectangular metal sheet that has an area of 340 in². The edges are bent up to form an open box with a volume of 350 in³. Find the dimensions of the box.

*Solve.*

**97.** $x^3 + y^3 = 72,$
$x + y = 6$

**98.** $a + b = \dfrac{5}{6},$

$\dfrac{a}{b} + \dfrac{b}{a} = \dfrac{13}{6}$

**99.** $p^2 + q^2 = 13,$
$\dfrac{1}{pq} = -\dfrac{1}{6}$

**100.** $e^x - e^{x+y} = 0,$
$e^y - e^{x-y} = 0$

*Solve using a graphing calculator. Find all real solutions.*

**101.** $y - \ln x = 2,$
$y = x^2$

**102.** $y = \ln(x + 4),$
$x^2 + y^2 = 6$

**103.** $y = e^x,$
$x - y = -2$

**104.** $y - e^{-x} = 1,$
$y = 2x + 5$

**105.** $14.5x^2 - 13.5y^2 - 64.5 = 0,$
$5.5x - 6.3y - 12.3 = 0$

**106.** $2x + 2y = 1660,$
$xy = 35{,}325$

**107.** $0.319x^2 + 2688.7y^2 = 56{,}548,$
$0.306x^2 - 2688.7y^2 = 43{,}452$

**108.** $13.5xy + 15.6 = 0,$
$5.6x - 6.7y - 42.3 = 0$

# Chapter 7 Summary and Review

## STUDY GUIDE

| KEY TERMS AND CONCEPTS | EXAMPLES |
|---|---|

### SECTION 7.1: THE PARABOLA

**Standard Equation of a Parabola with Vertex (0, 0) and Vertical Axis of Symmetry**

The standard equation of a parabola with vertex $(0, 0)$ and directrix $y = -p$ is

$$x^2 = 4py.$$

The focus is $(0, p)$ and the $y$-axis is the axis of symmetry.

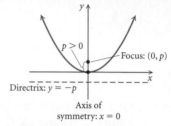

(When $p < 0$, the parabola opens down.)

---

**Standard Equation of a Parabola with Vertex (0, 0) and Horizontal Axis of Symmetry**

The standard equation of a parabola with vertex $(0, 0)$ and directrix $x = -p$ is

$$y^2 = 4px.$$

The focus is $(p, 0)$ and the $x$-axis is the axis of symmetry.

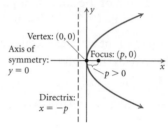

(When $p < 0$, the parabola opens to the left.)

---

**Standard Equation of a Parabola with Vertex ($h, k$) and Vertical Axis of Symmetry**

The standard equation of a parabola with vertex $(h, k)$ and vertical axis of symmetry is

$$(x - h)^2 = 4p(y - k),$$

where the vertex is $(h, k)$, the focus is $(h, k + p)$, and the directrix is $y = k - p$.

The parabola opens up if $p > 0$. It opens down if $p < 0$.

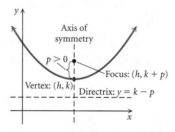

(When $p < 0$, the parabola opens down.)

See also Examples 1 and 3 on pp. 579 and 581.

**Standard Equation of a Parabola with Vertex ($h, k$) and Horizontal Axis of Symmetry**

The standard equation of a parabola with vertex ($h, k$) and horizontal axis of symmetry is

$$(y - k)^2 = 4p(x - h),$$

where the vertex is ($h, k$), the focus is ($h + p, k$), and the directrix is $x = h - p$.

The parabola opens to the right if $p > 0$. It opens to the left if $p < 0$.

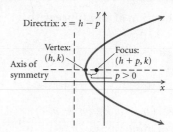

(When $p < 0$, the parabola opens to the left.)

See also Examples 2 and 4 on pp. 580 and 582.

---

### SECTION 7.2: THE CIRCLE AND THE ELLIPSE

**Standard Equation of a Circle**

The standard equation of a circle with center ($h, k$) and radius $r$ is

$$(x - h)^2 + (y - k)^2 = r^2.$$

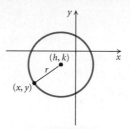

See also Example 1 on p. 586.

---

**Standard Equation of an Ellipse with Center at the Origin**

*Major Axis Horizontal*

$$\frac{x^2}{a^2} + \frac{y^2}{b^2} = 1, \quad a > b > 0$$

Vertices: $(-a, 0), (a, 0)$
$y$-intercepts: $(0, -b), (0, b)$
Foci: $(-c, 0), (c, 0)$, where $c^2 = a^2 - b^2$

*Major Axis Vertical*

$$\frac{x^2}{b^2} + \frac{y^2}{a^2} = 1, \quad a > b > 0$$

Vertices: $(0, -a), (0, a)$
$x$-intercepts: $(-b, 0), (b, 0)$
Foci: $(0, -c), (0, c)$, where $c^2 = a^2 - b^2$

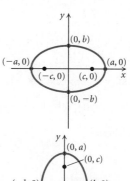

See also Examples 2 and 3 on pp. 588 and 589.

**Standard Equation of an Ellipse with Center at $(h, k)$**

*Major Axis Horizontal*

$$\frac{(x - h)^2}{a^2} + \frac{(y - k)^2}{b^2} = 1, \quad a > b > 0$$

Vertices: $(h - a, k), (h + a, k)$
Length of minor axis: $2b$
Foci: $(h - c, k), (h + c, k),$
   where $c^2 = a^2 - b^2$

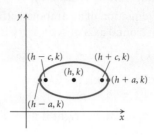

*Major Axis Vertical*

$$\frac{(x - h)^2}{b^2} + \frac{(y - k)^2}{a^2} = 1, \quad a > b > 0$$

Vertices: $(h, k - a), (h, k + a)$
Length of minor axis: $2b$
Foci: $(h, k - c), (h, k + c),$
   where $c^2 = a^2 - b^2$

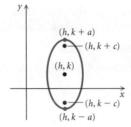

See also Example 4 on p. 590.

## SECTION 7.3: THE HYPERBOLA

**Standard Equation of a Hyperbola with Center at the Origin**

*Transverse Axis Horizontal*

$$\frac{x^2}{a^2} - \frac{y^2}{b^2} = 1$$

Vertices: $(-a, 0), (a, 0)$

Asymptotes: $y = -\frac{b}{a}x, y = \frac{b}{a}x$

Foci: $(-c, 0), (c, 0),$ where $c^2 = a^2 + b^2$

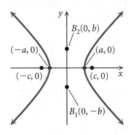

*Transverse Axis Vertical*

$$\frac{y^2}{a^2} - \frac{x^2}{b^2} = 1$$

Vertices: $(0, -a), (0, a)$

Asymptotes: $y = -\frac{a}{b}x, y = \frac{a}{b}x$

Foci: $(0, -c), (0, c),$ where $c^2 = a^2 + b^2$

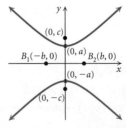

See also Examples 1 and 2 on p. 598.

**Standard Equation of a Hyperbola
with Center at $(h, k)$**

*Transverse Axis Horizontal*

$$\frac{(x - h)^2}{a^2} - \frac{(y - k)^2}{b^2} = 1$$

Vertices: $(h - a, k), (h + a, k)$

Asymptotes: $y - k = \dfrac{b}{a}(x - h),$

$$y - k = -\frac{b}{a}(x - h)$$

Foci: $(h - c, k), (h + c, k),$
where $c^2 = a^2 + b^2$

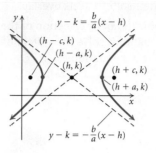

*Transverse Axis Vertical*

$$\frac{(y - k)^2}{a^2} - \frac{(x - h)^2}{b^2} = 1$$

Vertices: $(h, k - a), (h, k + a)$

Asymptotes: $y - k = \dfrac{a}{b}(x - h),$

$$y - k = -\frac{a}{b}(x - h)$$

Foci: $(h, k - c), (h, k + c),$
where $c^2 = a^2 + b^2$

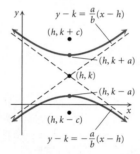

See also Example 3 on p. 600.

## SECTION 7.4: NONLINEAR SYSTEMS OF EQUATIONS AND INEQUALITIES

Substitution or elimination can be used to solve **systems of equations containing at least one nonlinear equation.**

Solve: $x^2 - y = 2,$  **(1)**   The graph is a parabola.

$x - y = -4.$  **(2)**   The graph is a line.

$x = y - 4$  Solving equation (2) for $x$

$(y - 4)^2 - y = 2$  Substituting for $x$ in equation (1)

$y^2 - 8y + 16 - y = 2$

$y^2 - 9y + 14 = 0$

$(y - 2)(y - 7) = 0$

$y - 2 = 0 \quad or \quad y - 7 = 0$

$y = 2 \quad or \quad\quad y = 7$

If $y = 2$, then $x = 2 - 4 = -2$.
If $y = 7$, then $x = 7 - 4 = 3$.

The pairs $(-2, 2)$ and $(3, 7)$ check, so they are the solutions.

To graph a **nonlinear system of inequalities**, graph each inequality in the system and then shade the region where their solution sets overlap.

To find the point(s) of intersection of the graphs of the related equations, solve the system of equations composed of those equations.

Graph: $x^2 - y \le 2$,
$\quad\quad\;\; x - y > -4$.

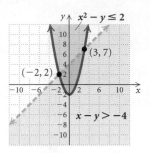

To find the points of intersection of the graphs of the related equations, we solve the system of equations

$$x^2 - y = 2,$$
$$x - y = -4.$$

We saw in the example above that these points are $(-2, 2)$ and $(3, 7)$.

## REVIEW EXERCISES

*Determine whether the statement is true or false.*

1. The graph of $x + y^2 = 1$ is a parabola that opens to the left. [7.1]

2. The graph of $\dfrac{(x-2)^2}{4} + \dfrac{(y+3)^2}{9} = 1$ is an ellipse with center $(-2, 3)$. [7.2]

3. The hyperbola $\dfrac{x^2}{5} - \dfrac{y^2}{10} = 1$ has a horizontal transverse axis. [7.3]

4. Every nonlinear system of equations has at least one real-number solution. [7.4]

5. The graph of a nonlinear system of equations shows all the solutions of the system of equations. [7.4]

*In Exercises 6–13, match the equation with one of the graphs (a)–(h), which follow.*

a)

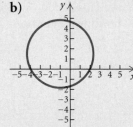

b)

c)

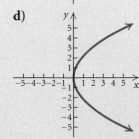

d)

**e)**

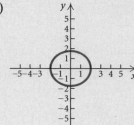

**f)**

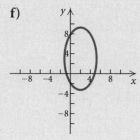

**g)**

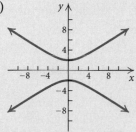

**h)**

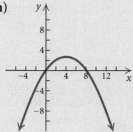

**6.** $y^2 = 5x$ [7.1]

**7.** $y^2 = 9 - x^2$ [7.2]

**8.** $3x^2 + 4y^2 = 12$ [7.2]

**9.** $9y^2 - 4x^2 = 36$ [7.3]

**10.** $x^2 + y^2 + 2x - 3y = 8$ [7.2]

**11.** $4x^2 + y^2 - 16x - 6y = 15$ [7.2]

**12.** $x^2 - 8x + 6y = 0$ [7.1]

**13.** $\dfrac{(x + 3)^2}{16} - \dfrac{(y - 1)^2}{25} = 1$ [7.3]

**14.** Find an equation of the parabola with directrix $y = \frac{3}{2}$ and focus $\left(0, -\frac{3}{2}\right)$. [7.1]

**15.** Find the focus, the vertex, and the directrix of the parabola given by $y^2 = -12x$. [7.1]

**16.** Find the vertex, the focus, and the directrix of the parabola given by
$$x^2 + 10x + 2y + 9 = 0. \quad [7.1]$$

**17.** Find the center, the vertices, and the foci of the ellipse given by
$$16x^2 + 25y^2 - 64x + 50y - 311 = 0.$$
Then draw the graph. [7.2]

**18.** Find an equation of the ellipse having vertices $(0, -4)$ and $(0, 4)$ with minor axis of length 6. [7.2]

**19.** Find the center, the vertices, the foci, and the asymptotes of the hyperbola given by
$$x^2 - 2y^2 + 4x + y - \tfrac{1}{8} = 0. \quad [7.3]$$

**20.** *Spotlight.*  A spotlight has a parabolic cross section that is 2 ft wide at the opening and 1.5 ft deep at the vertex. How far from the vertex is the focus? [7.1]

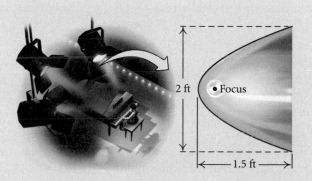

**Solve.** [7.4]

**21.** $x^2 - 16y = 0,$
$\quad x^2 - y^2 = 64$

**22.** $4x^2 + 4y^2 = 65,$
$\quad 6x^2 - 4y^2 = 25$

**23.** $x^2 - y^2 = 33,$
$\quad x + y = 11$

**24.** $x^2 - 2x + 2y^2 = 8,$
$\quad 2x + y = 6$

**25.** $x^2 - y = 3,$
$\quad 2x - y = 3$

**26.** $x^2 + y^2 = 25,$
$\quad x^2 - y^2 = 7$

**27.** $x^2 - y^2 = 3,$
$\quad y = x^2 - 3$

**28.** $x^2 + y^2 = 18,$
$\quad 2x + y = 3$

**29.** $x^2 + y^2 = 100,$
$\quad 2x^2 - 3y^2 = -120$

**30.** $x^2 + 2y^2 = 12,$
$\quad xy = 4$

**31.** *Numerical Relationship.*  The sum of two numbers is 11, and the sum of their squares is 65. Find the numbers. [7.4]

**32.** *Dimensions of a Rectangle.*  A rectangle has a perimeter of 38 m and an area of 84 m². What are the dimensions of the rectangle? [7.4]

**33.** *Numerical Relationship.*  Find two positive integers whose sum is 12 and the sum of whose reciprocals is $\frac{3}{8}$. [7.4]

**34.** *Perimeter.*  The perimeter of a square is 12 cm more than the perimeter of another square. The area of the first square exceeds the area of the other by 39 cm². Find the perimeter of each square. [7.4]

**35.** *Radius of a Circle.* The sum of the areas of two circles is $130\pi$ ft$^2$. The difference of the areas is $112\pi$ ft$^2$. Find the radius of each circle. [7.4]

*Graph the system of inequalities. Then find the coordinates of the points of intersection of the graphs of the related equations.* [7.4]

**36.** $y \leq 4 - x^2$,
$x - y \leq 2$

**37.** $x^2 + y^2 \leq 16$,
$x + y < 4$

**38.** $y \geq x^2 - 1$,
$y < 1$

**39.** $x^2 + y^2 \leq 9$,
$x \leq -1$

**40.** The vertex of the parabola $y^2 - 4y - 12x - 8 = 0$ is which of the following? [7.1]

**A.** $(1, -2)$      **B.** $(-1, 2)$

**C.** $(2, -1)$      **D.** $(-2, 1)$

**41.** Which of the following cannot be a number of solutions possible for a system of equations representing an ellipse and a straight line? [7.4]

**A.** 0      **B.** 1

**C.** 2      **D.** 4

**42.** The graph of $x^2 + 4y^2 = 4$ is which of the following? [7.2], [7.3]

**A.**
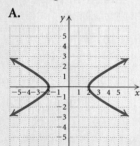

**B.**
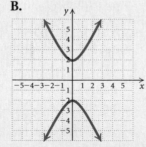

**C.**
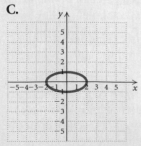

**D.**
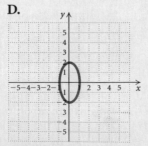

## Synthesis

**43.** Find two numbers whose product is 4 and the sum of whose reciprocals is $\frac{65}{56}$. [7.4]

**44.** Find an equation of the circle that passes through the points $(10, 7)$, $(-6, 7)$, and $(-8, 1)$. [7.2], [7.4]

**45.** Find an equation of the ellipse containing the point $(-1/2, 3\sqrt{3}/2)$ and with vertices $(0, -3)$ and $(0, 3)$. [7.2]

**46.** *Navigation.* Two radio transmitters positioned 400 mi apart along the shore send simultaneous signals to a ship that is 250 mi offshore, sailing parallel to the shoreline. The signal from transmitter $A$ reaches the ship 300 microseconds before the signal from transmitter $B$. The signals travel at a speed of 186,000 miles per second, or 0.186 mile per microsecond. Find the equation of the hyperbola with foci $A$ and $B$ on which the ship is located. (*Hint*: For any point on the hyperbola, the absolute value of the difference of its distances from the foci is $2a$.) [7.3]

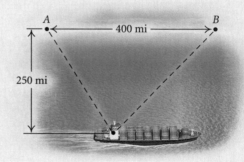

## Collaborative Discussion and Writing

**47.** Is a circle a special type of ellipse? Why or why not? [7.2]

**48.** How does the graph of a parabola differ from the graph of one branch of a hyperbola? [7.1], [7.3]

**49.** Are the asymptotes of a hyperbola part of the graph of the hyperbola? Why or why not? [7.3]

**50.** What would you say to a classmate who tells you that it is always possible to visualize all of the solutions of a nonlinear system of equations? [7.4]

## Chapter 7 Test

In Exercises 1–4, match the equation with one of the graphs (a)–(d), which follow.

a)

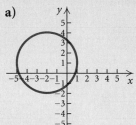

b)

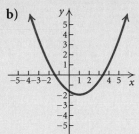

c)

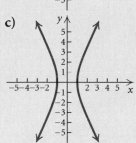

d)

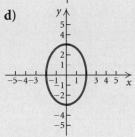

**1.** $4x^2 - y^2 = 4$

**2.** $x^2 - 2x - 3y = 5$

**3.** $x^2 + 4x + y^2 - 2y - 4 = 0$

**4.** $9x^2 + 4y^2 = 36$

*Find the vertex, the focus, and the directrix of the parabola. Then draw the graph.*

**5.** $x^2 = 12y$

**6.** $y^2 + 2y - 8x - 7 = 0$

**7.** Find an equation of the parabola with focus $(0, 2)$ and directrix $y = -2$.

**8.** Find the center and the radius of the circle given by $x^2 + y^2 + 2x - 6y - 15 = 0$. Then draw the graph.

*Find the center, the vertices, and the foci of the ellipse. Then draw the graph.*

**9.** $9x^2 + 16y^2 = 144$

**10.** $\dfrac{(x + 1)^2}{4} + \dfrac{(y - 2)^2}{9} = 1$

**11.** Find an equation of the ellipse having vertices $(0, -5)$ and $(0, 5)$ and with minor axis of length 4.

*Find the center, the vertices, the foci, and the asymptotes of the hyperbola. Then draw the graph.*

**12.** $4x^2 - y^2 = 4$

**13.** $\dfrac{(y - 2)^2}{4} - \dfrac{(x + 1)^2}{9} = 1$

**14.** Find the asymptotes of the hyperbola given by $2y^2 - x^2 = 18$.

**15.** *Satellite Dish.* A satellite dish has a parabolic cross section that is 18 in. wide at the opening and 6 in. deep at the vertex. How far from the vertex is the focus?

*Solve.*

**16.** $2x^2 - 3y^2 = -10,$
$x^2 + 2y^2 = 9$

**17.** $x^2 + y^2 = 13,$
$x + y = 1$

**18.** $x + y = 5,$
$xy = 6$

**19.** *Landscaping.* Leisurescape is planting a rectangular flower garden with a perimeter of 18 ft and a diagonal of $\sqrt{41}$ ft. Find the dimensions of the garden.

**20.** *Fencing.* It will take 210 ft of fencing to enclose a rectangular playground with an area of 2700 ft². Find the dimensions of the playground.

**21.** Graph the system of inequalities. Then find the coordinates of the points of intersection of the graphs of the related equations.

$$y \geq x^2 - 4,$$
$$y < 2x - 1$$

**22.** The graph of $(y - 1)^2 = 4(x + 1)$ is which of the following?

**A.**

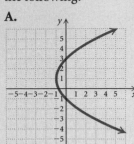

**B.**

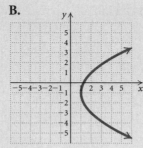

**C.**

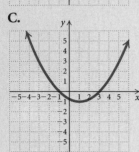

**D.**

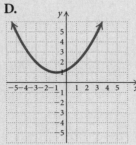

## Synthesis

**23.** Find an equation of the circle for which the endpoints of a diameter are $(1, 1)$ and $(5, -3)$.

# Sequences, Series, and Combinatorics

## APPLICATION

A formation of a marching band has 10 marchers in the first row, 12 in the second row, 14 in the third row, and so on, for 8 rows. How many marchers are in the last row? How many marchers are there altogether?

**This problems appears as Exercise 45 in Section 8.2.**

## 8.1

# Sequences and Series

- Find terms of sequences given the *n*th term.
- Look for a pattern in a sequence and try to determine a general term.
- Convert between sigma notation and other notation for a series.
- Construct the terms of a recursively defined sequence.

In this section, we discuss sets or lists of numbers, considered in order, and their sums.

### ▪ Sequences

Suppose that $1000 is invested at 6%, compounded annually. The amounts to which the account will grow after 1 year, 2 years, 3 years, 4 years, and so on, form the following sequence of numbers:

$$\begin{array}{cccc} (1) & (2) & (3) & (4) \\ \downarrow & \downarrow & \downarrow & \downarrow \end{array}$$

$1060.00, \quad $1123.60, \quad $1191.02, \quad $1262.48, \ldots .$

We can think of this as a function that pairs 1 with $1060.00, 2 with $1123.60, 3 with $1191.02, and so on. A **sequence** is thus a *function*, where the domain is a set of consecutive positive integers beginning with 1.

If we continue to compute the amounts of money in the account forever, we obtain an **infinite sequence** with function values

$1060.00, \quad $1123.60, \quad $1191.02, \quad $1262.48, \quad $1338.23, \quad $1418.52, \ldots .$

The dots "..." at the end indicate that the sequence goes on without stopping. If we stop after a certain number of years, we obtain a **finite sequence:**

$1060.00, \quad $1123.60, \quad $1191.02, \quad $1262.48.$

> ### SEQUENCES
>
> An **infinite sequence** is a function having for its domain the set of positive integers, $\{1, 2, 3, 4, 5, \ldots\}$.
>
> A **finite sequence** is a function having for its domain a set of positive integers, $\{1, 2, 3, 4, 5, \ldots, n\}$, for some positive integer $n$.

Consider the sequence given by the formula

$$a(n) = 2^n, \quad \text{or} \quad a_n = 2^n.$$

Some of the function values, also known as the **terms** of the sequence, are as follows:

$$a_1 = 2^1 = 2,$$
$$a_2 = 2^2 = 4,$$
$$a_3 = 2^3 = 8,$$
$$a_4 = 2^4 = 16,$$
$$a_5 = 2^5 = 32.$$

The first term of the sequence is denoted as $a_1$, the fifth term as $a_5$, and the $n$th term, or **general term**, as $a_n$. This sequence can also be denoted as

$$2, 4, 8, \quad , \quad \text{or as} \quad 2, 4, 8, \ldots, 2^n, \ldots.$$

**EXAMPLE 1** Find the first 4 terms and the 23rd term of the sequence whose general term is given by $a_n = (-1)^n n^2$.

***Solution*** We have $a_n = (-1)^n n^2$, so

$$a_1 = (-1)^1 \cdot 1^2 = -1,$$
$$a_2 = (-1)^2 \cdot 2^2 = 4,$$
$$a_3 = (-1)^3 \cdot 3^2 = -9,$$
$$a_4 = (-1)^4 \cdot 4^2 = 16,$$
$$a_{23} = (-1)^{23} \cdot 23^2 = -529.$$

We can also use a graphing calculator to find the desired terms of this sequence. We enter $y_1 = (-1)^x x^2$. We then set up a table in ASK mode and enter 1, 2, 3, 4, and 23 as values for $x$. ▸ **Now Try Exercise 1.**

| X | Y₁ |
|---|---|
| 1 | −1 |
| 2 | 4 |
| 3 | −9 |
| 4 | 16 |
| 23 | −529 |

X =

Note in Example 1 that the power $(-1)^n$ causes the signs of the terms to alternate between positive and negative, depending on whether $n$ is even or odd. This kind of sequence is called an **alternating sequence.**

GCM **EXAMPLE 2** Use a graphing calculator to find the first 5 terms of the sequence whose general term is given by $a_n = n/(n + 1)$.

***Solution*** We can use a table or the SEQ feature, as shown here. We select SEQ from the LIST OPS menu and enter the general term, the variable, and the numbers of the first and last terms desired. The calculator will write the terms horizontally as a list. The list can also be written in fraction notation. The first 5 terms of the sequence are

$$\frac{1}{2}, \quad \frac{2}{3}, \quad \frac{3}{4}, \quad \frac{4}{5}, \quad \text{and} \quad \frac{5}{6}.$$

seq(X/(X+1),X,1,5) ▸ Frac
$\{\frac{1}{2} \quad \frac{2}{3} \quad \frac{3}{4} \quad \frac{4}{5} \quad \frac{5}{6}\}$

We can graph a sequence just as we graph other functions. Consider the function given by $f(x) = x + 1$ and the sequence whose general term is given by $a_n = n + 1$. The graph of $f(x) = x + 1$ is shown on the left below. Since the domain of a sequence is a set of positive integers, the graph of a sequence is a set of points that are not connected. Thus if we use only positive integers for inputs of $f(x) = x + 1$, we have the graph of the sequence $a_n = n + 1$, as shown on the right below.

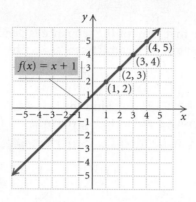
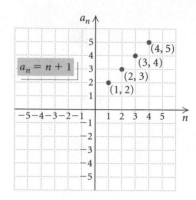

We can also use a graphing calculator to graph a sequence. Since we are graphing a set of unconnected points, we use DOT mode. We also select SEQUENCE mode. In this mode, the variable is $n$ and functions are named $u(n)$, $v(n)$, and $w(n)$ rather than $y_1$, $y_2$, and $y_3$. On many graphing calculators, **X,T,θ,n** is used to enter $n$ in SEQUENCE mode.

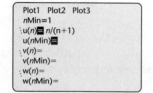

**EXAMPLE 3** Use a graphing calculator to graph the sequence whose general term is given by $a_n = n/(n + 1)$.

**Solution** With the calculator set in DOT mode and SEQUENCE mode, we enter $u(n) = n/(n + 1)$. All the function values will be positive numbers that are less than 1, so we choose the window $[0, 10, 0, 1]$ and we also choose $n\text{Min} = 1$, $n\text{Max} = 10$, PlotStart $= 1$, and PlotStep $= 1$.

> **Now Try Exercise 19.**

## ▪ Finding the General Term

When only the first few terms of a sequence are known, we do not know for sure what the general term is, but we might be able to make a prediction by looking for a pattern.

**EXAMPLE 4** For each of the following sequences, predict the general term.

**a)** $1, \sqrt{2}, \sqrt{3}, 2, \ldots$
**b)** $-1, 3, -9, 27, -81, \ldots$
**c)** $2, 4, 8, \ldots$

**Solution**

**a)** These are square roots of consecutive integers, so the general term might be $\sqrt{n}$.

**b)** These are powers of 3 with alternating signs, so the general term might be $(-1)^n 3^{n-1}$.

**c)** If we see the pattern of powers of 2, we will see 16 as the next term and guess $2^n$ for the general term. Then the sequence could be written with more terms as

$$2, 4, 8, 16, 32, 64, 128, \ldots .$$

If we see that we can get the second term by adding 2, the third term by adding 4, and the next term by adding 6, and so on, we will see 14 as the next term. A general term for the sequence is $n^2 - n + 2$, and the sequence can be written with more terms as

$$2, 4, 8, 14, 22, 32, 44, 58, \ldots .$$ **Now Try Exercise 23.**

Example 4(c) illustrates that, in fact, you can never be certain about the general term when only a few terms are given. The fewer the given terms, the greater the uncertainty.

## Sums and Series

**SERIES**

Given the infinite sequence

$$a_1, a_2, a_3, a_4, \ldots, a_n, \ldots,$$

the sum of the terms

$$a_1 + a_2 + a_3 + \cdots + a_n + \cdots$$

is called an **infinite series**. A **partial sum** is the sum of the first $n$ terms:

$$a_1 + a_2 + a_3 + \cdots + a_n.$$

A partial sum is also called a **finite series**, or **$n$th partial sum**, and is denoted $S_n$.

**EXAMPLE 5** For the sequence $-2, 4, -6, 8, -10, 12, -14, \ldots$, find each of the following.

**a)** $S_1$ **b)** $S_4$ **c)** $S_5$

*Solution*

**a)** $S_1 = -2$

**b)** $S_4 = -2 + 4 + (-6) + 8 = 4$

**c)** $S_5 = -2 + 4 + (-6) + 8 + (-10) = -6$

**Now Try Exercise 33.**

We can also use a graphing calculator to find partial sums of a sequence when a formula for the general term is known.

GCM

**EXAMPLE 6** Use a graphing calculator to find $S_1$, $S_2$, $S_3$, and $S_4$ for the sequence whose general term is given by $a_n = n^2 - 3$.

```
cumSum(seq(n² – 3,n,1,4))
            {-2 -1 5 18}
```

**Solution** We can use the CUMSUM feature from the LIST OPS menu. The calculator will write the partial sums as a list. (Note that the calculator can be set in either FUNCTION mode or SEQUENCE mode. Here we show SEQUENCE mode.)

We have $S_1 = -2$, $S_2 = -1$, $S_3 = 5$, and $S_4 = 18$.

### ▪ Sigma Notation

The Greek letter $\Sigma$ (sigma) can be used to denote a sum when the general term of a sequence is a formula. For example, the sum of the first four terms of the sequence $3, 5, 7, 9, \ldots, 2k + 1, \ldots$ can be named as follows, using what is called **sigma notation**, or **summation notation**:

$$\sum_{k=1}^{4} (2k + 1).$$

This is read "the sum as $k$ goes from 1 to 4 of $2k + 1$." The letter $k$ is called the **index of summation**. The index of summation might start at a number other than 1, and letters other than $k$ can be used.

GCM

**EXAMPLE 7** Find and evaluate each of the following sums.

a) $\displaystyle\sum_{k=1}^{5} k^3$     b) $\displaystyle\sum_{k=0}^{4} (-1)^k 5^k$     c) $\displaystyle\sum_{i=8}^{11} \left(2 + \frac{1}{i}\right)$

**Solution**

a) We replace $k$ with 1, 2, 3, 4, and 5. Then we add the results.

```
sum(seq(n³,n,1,5))
                225
```

$$\sum_{k=1}^{5} k^3 = 1^3 + 2^3 + 3^3 + 4^3 + 5^3$$

$$= 1 + 8 + 27 + 64 + 125 = 225$$

We can also combine the SUM from the LIST MATH submenu and the SEQ features on a graphing calculator to add the terms of this sequence.

b) $\displaystyle\sum_{k=0}^{4} (-1)^k 5^k = (-1)^0 5^0 + (-1)^1 5^1 + (-1)^2 5^2 + (-1)^3 5^3 + (-1)^4 5^4$

$$= 1 - 5 + 25 - 125 + 625 = 521$$

c) $\displaystyle\sum_{i=8}^{11} \left(2 + \frac{1}{i}\right) = \left(2 + \frac{1}{8}\right) + \left(2 + \frac{1}{9}\right) + \left(2 + \frac{1}{10}\right) + \left(2 + \frac{1}{11}\right)$

$$= 8\frac{1691}{3960}$$

**Now Try Exercise 37.**

**EXAMPLE 8** Write sigma notation for each sum.

a) $1 + 2 + 4 + 8 + 16 + 32 + 64$

b) $-2 + 4 - 6 + 8 - 10$

c) $x + \dfrac{x^2}{2} + \dfrac{x^3}{3} + \dfrac{x^4}{4} + \cdots$

*Solution*

**a)** $1 + 2 + 4 + 8 + 16 + 32 + 64$

This is the sum of powers of 2, beginning with $2^0$, or 1, and ending with $2^6$, or 64. Sigma notation is $\Sigma_{k=0}^{6} 2^k$.

**b)** $-2 + 4 - 6 + 8 - 10$

Disregarding the alternating signs, we see that this is the sum of the first 5 even integers. Note that $2k$ is a formula for the $k$th positive even integer, and $(-1)^k = -1$ when $k$ is odd and $(-1)^k = 1$ when $k$ is even. Thus the general term is $(-1)^k(2k)$. The sum begins with $k = 1$ and ends with $k = 5$, so sigma notation is $\Sigma_{k=1}^{5}(-1)^k(2k)$.

**c)** $x + \dfrac{x^2}{2} + \dfrac{x^3}{3} + \dfrac{x^4}{4} + \cdots$

The general term is $x^k/k$, beginning with $k = 1$. This is also an infinite series. We use the symbol $\infty$ for infinity and write the series using sigma notation: $\Sigma_{k=1}^{\infty}(x^k/k)$.

> **Now Try Exercise 55.**

## ■ Recursive Definitions

A sequence may be defined **recursively** or by using a **recursion formula**. Such a definition lists the first term, or the first few terms, and then describes how to determine the remaining terms from the given terms.

GCM **EXAMPLE 9** Find the first 5 terms of the sequence defined by

$$a_1 = 5, \qquad a_{n+1} = 2a_n - 3, \quad \text{for } n \geq 1.$$

*Solution* We have

$$a_1 = 5,$$

$$a_2 = 2a_1 - 3 = 2 \cdot 5 - 3 = 7,$$

$$a_3 = 2a_2 - 3 = 2 \cdot 7 - 3 = 11,$$

$$a_4 = 2a_3 - 3 = 2 \cdot 11 - 3 = 19,$$

$$a_5 = 2a_4 - 3 = 2 \cdot 19 - 3 = 35.$$

Many graphing calculators have the capability to work with recursively defined sequences when they are set in SEQUENCE mode. For this sequence, for instance, the function could be entered as $u(n) = 2 * u(n - 1) - 3$ with $u(nMin) = 5$. We can read the terms of the sequence from a table.

> **Now Try Exercise 65.**

```
Plot1  Plot2  Plot3
 nMin=1
\u(n)■2*u(n−1)−3
 u(nMin)■{5}
\v(n)=
 v(nMin)=
\w(n)=
 w(nMin)=
```

| $n$ | $u(n)$ |
|---|---|
| 1 | 5 |
| 2 | 7 |
| 3 | 11 |
| 4 | 19 |
| 5 | 35 |
| 6 | 67 |
| 7 | 131 |

$n = 1$

# 8.1 Exercise Set

*In each of the following, the nth term of a sequence is given. Find the first 4 terms, $a_{10}$, and $a_{15}$.*

**1.** $a_n = 4n - 1$

**2.** $a_n = (n - 1)(n - 2)(n - 3)$

**3.** $a_n = \dfrac{n}{n - 1}$, $n \geq 2$

**4.** $a_n = n^2 - 1$, $n \geq 3$

**5.** $a_n = \dfrac{n^2 - 1}{n^2 + 1}$

**6.** $a_n = \left(-\dfrac{1}{2}\right)^{n-1}$

**7.** $a_n = (-1)^n n^2$

**8.** $a_n = (-1)^{n-1}(3n - 5)$

**9.** $a_n = 5 + \dfrac{(-2)^{n+1}}{2^n}$

**10.** $a_n = \dfrac{2n - 1}{n^2 + 2n}$

*Find the indicated term of the given sequence.*

**11.** $a_n = 5n - 6$; $a_8$

**12.** $a_n = (3n - 4)(2n + 5)$; $a_7$

**13.** $a_n = (2n + 3)^2$; $a_6$

**14.** $a_n = (-1)^{n-1}(4.6n - 18.3)$; $a_{12}$

**15.** $a_n = 5n^2(4n - 100)$; $a_{11}$

**16.** $a_n = \left(1 + \dfrac{1}{n}\right)^2$; $a_{80}$

**17.** $a_n = \ln e^n$; $a_{67}$     **18.** $a_n = 2 - \dfrac{1000}{n}$; $a_{100}$

*Use a graphing calculator to construct a table of values and a graph for the first 10 terms of the sequence.*

**19.** $a_n = \left(1 + \dfrac{1}{n}\right)^n$

**20.** $a_n = \sqrt{n + 1} - \sqrt{n}$

**21.** $a_1 = 2$, $a_{n+1} = \sqrt{1 + \sqrt{a_n}}$

**22.** $a_1 = 2$, $a_{n+1} = \dfrac{1}{2}\left(a_n + \dfrac{2}{a_n}\right)$

*Predict the general term, or nth term, $a_n$, of the sequence. Answers may vary.*

**23.** $2, 4, 6, 8, 10, \ldots$     **24.** $3, 9, 27, 81, 243, \ldots$

**25.** $-2, 6, -18, 54, \ldots$

**26.** $-2, 3, 8, 13, 18, \ldots$

**27.** $\dfrac{2}{3}, \dfrac{3}{4}, \dfrac{4}{5}, \dfrac{5}{6}, \dfrac{6}{7}, \ldots$

**28.** $\sqrt{2}, 2, \sqrt{6}, 2\sqrt{2}, \sqrt{10}, \ldots$

**29.** $1 \cdot 2, 2 \cdot 3, 3 \cdot 4, 4 \cdot 5, \ldots$

**30.** $-1, -4, -7, -10, -13, \ldots$

**31.** $0, \log 10, \log 100, \log 1000, \ldots$

**32.** $\ln e^2, \ln e^3, \ln e^4, \ln e^5, \ldots$

*Find the indicated partial sums for the sequence.*

**33.** $1, 2, 3, 4, 5, 6, 7, \ldots$; $S_3$ and $S_7$

**34.** $1, -3, 5, -7, 9, -11, \ldots$; $S_2$ and $S_5$

**35.** $2, 4, 6, 8, \ldots$; $S_4$ and $S_5$

**36.** $1, \dfrac{1}{4}, \dfrac{1}{9}, \dfrac{1}{16}, \dfrac{1}{25}, \ldots$; $S_1$ and $S_5$

*Find and evaluate the sum.*

**37.** $\displaystyle\sum_{k=1}^{5} \dfrac{1}{2k}$     **38.** $\displaystyle\sum_{i=1}^{6} \dfrac{1}{2i + 1}$

**39.** $\displaystyle\sum_{i=0}^{6} 2^i$     **40.** $\displaystyle\sum_{k=4}^{7} \sqrt{2k - 1}$

**41.** $\displaystyle\sum_{k=7}^{10} \ln k$     **42.** $\displaystyle\sum_{k=1}^{4} \pi k$

**43.** $\displaystyle\sum_{k=1}^{8} \dfrac{k}{k + 1}$     **44.** $\displaystyle\sum_{i=1}^{5} \dfrac{i - 1}{i + 3}$

**45.** $\displaystyle\sum_{i=1}^{5} (-1)^i$     **46.** $\displaystyle\sum_{k=0}^{5} (-1)^{k+1}$

**47.** $\displaystyle\sum_{k=1}^{8} (-1)^{k+1} 3k$     **48.** $\displaystyle\sum_{k=0}^{7} (-1)^k 4^{k+1}$

**49.** $\displaystyle\sum_{k=0}^{6} \dfrac{2}{k^2 + 1}$     **50.** $\displaystyle\sum_{i=1}^{10} i(i + 1)$

**51.** $\displaystyle\sum_{k=0}^{5} (k^2 - 2k + 3)$     **52.** $\displaystyle\sum_{k=1}^{10} \dfrac{1}{k(k + 1)}$

**53.** $\displaystyle\sum_{i=0}^{10} \dfrac{2^i}{2^i + 1}$     **54.** $\displaystyle\sum_{k=0}^{3} (-2)^{2k}$

*Write sigma notation. Answers may vary.*

**55.** $5 + 10 + 15 + 20 + 25 + \cdots$

**56.** $7 + 14 + 21 + 28 + 35 + \cdots$

**57.** $2 - 4 + 8 - 16 + 32 - 64$

**58.** $3 + 6 + 9 + 12 + 15$

**59.** $-\dfrac{1}{2} + \dfrac{2}{3} - \dfrac{3}{4} + \dfrac{4}{5} - \dfrac{5}{6} + \dfrac{6}{7}$

**60.** $\dfrac{1}{1^2} + \dfrac{1}{2^2} + \dfrac{1}{3^2} + \dfrac{1}{4^2} + \dfrac{1}{5^2}$

**61.** $4 - 9 + 16 - 25 + \cdots + (-1)^n n^2$

**62.** $9 - 16 + 25 + \cdots + (-1)^{n+1} n^2$

**63.** $\dfrac{1}{1 \cdot 2} + \dfrac{1}{2 \cdot 3} + \dfrac{1}{3 \cdot 4} + \dfrac{1}{4 \cdot 5} + \cdots$

**64.** $\dfrac{1}{1 \cdot 2^2} + \dfrac{1}{2 \cdot 3^2} + \dfrac{1}{3 \cdot 4^2} + \dfrac{1}{4 \cdot 5^2} + \cdots$

*Find the first 4 terms of the recursively defined sequence.*

**65.** $a_1 = 4, \; a_{n+1} = 1 + \dfrac{1}{a_n}$

**66.** $a_1 = 256, \; a_{n+1} = \sqrt{a_n}$

**67.** $a_1 = 6561, \; a_{n+1} = (-1)^n \sqrt{a_n}$

**68.** $a_1 = e^Q, \; a_{n+1} = \ln a_n$

**69.** $a_1 = 2, \; a_2 = 3, \; a_{n+1} = a_n + a_{n-1}$

**70.** $a_1 = -10, \; a_2 = 8, \; a_{n+1} = a_n - a_{n-1}$

**71.** *Compound Interest.* Suppose that $1000 is invested at 6.2%, compounded annually. The value of the investment after $n$ years is given by the sequence model

$$a_n = \$1000(1.062)^n, \quad n = 1, 2, 3, \ldots.$$

**a)** Find the first 10 terms of the sequence.
**b)** Find the value of the investment after 20 years.

**72.** *Salvage Value.* The value of a post-hole digger is $5200. Its salvage value each year is 75% of its value the year before. Give a sequence that lists the salvage value of the post-hole digger for each year of a 10-year period.

**73.** *Wage Sequence.* Adahy is paid $9.80 per hour for working at Red Freight Limited. Each year he receives a $1.10 hourly raise. Give a sequence that lists Adahy's hourly wage over a 10-year period.

**74.** *Bacteria Growth.* Suppose that a single cell of bacteria divides into two every 15 min. Suppose that the same rate of division is maintained for 4 hr. Give a sequence that lists the number of cells after successive 15-min periods.

**75.** *Fibonacci Sequence: Rabbit Population Growth.* One of the most famous recursively defined sequences is the **Fibonacci sequence**. In 1202, the Italian mathematician Leonardo da Pisa, also called Fibonacci, proposed the following model for rabbit population growth. Suppose that every month each mature pair of rabbits in the population produces a new pair that begins reproducing after two months, and also suppose that no rabbits die. Beginning with one pair of newborn rabbits, the population can be modeled by the following recursively defined sequence:

$$a_1 = 1, \; a_2 = 1, \; a_n = a_{n-1} + a_{n-2}, \text{ for } n \geq 3,$$

where $a_n$ is the total number of pairs of rabbits in month $n$. Find the first 7 terms of the Fibonacci sequence.

**76.** *Weekly Earnings.* The table below lists the average weekly earnings of U.S. production workers in recent years.

| Year, $n$ | Average Weekly Earnings of U.S. Production Workers |
|---|---|
| 2004, 0 | $529.09 |
| 2005, 1 | 544.33 |
| 2006, 2 | 567.87 |
| 2007, 3 | 590.04 |
| 2008, 4 | 607.95 |
| 2009, 5 | 617.11 |

*Sources:* U.S. Bureau of Labor Statistics, U.S. Department of Labor

**a)** Use a graphing calculator to fit a linear sequence regression function

$$a_n = an + b$$

to the data, where $n$ is the number of years since 2004.

**b)** Estimate the average weekly earnings in 2010, in 2012, and in 2015.

**77.** *Vehicle Production.*   The table below lists world motor vehicle production in recent years.

| Year, $n$ | World Motor Vehicle Production (in thousands of units) |
|---|---|
| 2001, 0 | 57,705 |
| 2002, 1 | 59,587 |
| 2003, 2 | 61,562 |
| 2004, 3 | 65,654 |
| 2005, 4 | 67,892 |
| 2006, 5 | 70,992 |
| 2007, 6 | 74,647 |
| 2008, 7 | 67,602 |
| 2009, 8 | 59,096 |

*Sources*: Automotive New Data Center; R. L. Polk

**a)** Use a graphing calculator to fit a quadratic sequence regression function

$$a_n = an^2 + bn + c$$

to the data, where $n$ is the number of years since 2001.

**b)** Estimate the world motor vehicle production in 2010, in 2011, and in 2012.

## Skill Maintenance

*Solve.*

**78.** $3x - 2y = 3,$
$2x + 3y = -11$

**79.** *Patent Applications.*   In 2010, the United States, Japan, and China had a total of 89,348 patent applications. The number of applications in China was 3741 less than one-half the number of applications in Japan. The number of applications in the United States was 32,518 more than the number of applications in China. (*Source*: World Intellectual Property Organization) Find the number of patent applications in each country.

*Find the center and the radius of the circle with the given equation.*

**80.** $x^2 + y^2 - 6x + 4y = 3$

**81.** $x^2 + y^2 + 5x - 8y = 2$

## Synthesis

*Find the first 5 terms of the sequence, and then find $S_5$.*

**82.** $a_n = \dfrac{1}{2^n} \log 1000^n$      **83.** $a_n = i^n,\ i = \sqrt{-1}$

**84.** $a_n = \ln(1 \cdot 2 \cdot 3 \cdot \cdots \cdot n)$

*For each sequence, find a formula for $S_n$.*

**85.** $a_n = \ln n$      **86.** $a_n = \dfrac{1}{n} - \dfrac{1}{n+1}$

---

### 8.2   Arithmetic Sequences and Series

- For any arithmetic sequence, find the $n$th term when $n$ is given and $n$ when the $n$th term is given, and given two terms, find the common difference and construct the sequence.

- Find the sum of the first $n$ terms of an arithmetic sequence.

A sequence in which each term after the first term is found by adding the same number to the preceding term is an **arithmetic sequence**.

## ■ Arithmetic Sequences

The sequence 2, 5, 8, 11, 14, 17, . . . is arithmetic because adding 3 to any term produces the next term. In other words, the difference between any term and the preceding one is 3. Arithmetic sequences are also called *arithmetic progressions*.

> **ARITHMETIC SEQUENCE**
>
> A sequence is **arithmetic** if there exists a number $d$, called the **common difference**, such that $a_{n+1} = a_n + d$ for any integer $n \geq 1$.

**EXAMPLE 1** For each of the following arithmetic sequences, identify the first term, $a_1$, and the common difference, $d$.

**a)** 4, 9, 14, 19, 24, . . .

**b)** 34, 27, 20, 13, 6, −1, −8, . . .

**c)** $2, 2\frac{1}{2}, 3, 3\frac{1}{2}, 4, 4\frac{1}{2}, \ldots$

*Solution* The first term, $a_1$, is the first term listed. To find the common difference, $d$, we choose any term beyond the first and subtract the preceding term from it.

| SEQUENCE | FIRST TERM, $a_1$ | COMMON DIFFERENCE, $d$ |
|---|---|---|
| **a)** 4, 9, 14, 19, 24, . . . | 4 | 5 $(9 - 4 = 5)$ |
| **b)** 34, 27, 20, 13, 6, −1, −8, . . . | 34 | −7 $(27 - 34 = -7)$ |
| **c)** $2, 2\frac{1}{2}, 3, 3\frac{1}{2}, 4, 4\frac{1}{2}, \ldots$ | 2 | $\frac{1}{2}$ $\left(2\frac{1}{2} - 2 = \frac{1}{2}\right)$ |

We obtained the common difference by subtracting $a_1$ from $a_2$. Had we subtracted $a_2$ from $a_3$ or $a_3$ from $a_4$, we would have obtained the same values for $d$. Thus we can check by adding $d$ to each term in a sequence to see if we progress correctly to the next term.

*Check:*

**a)** $4 + 5 = 9$, $\quad 9 + 5 = 14$, $\quad 14 + 5 = 19$, $\quad 19 + 5 = 24$

**b)** $34 + (-7) = 27$, $\quad 27 + (-7) = 20$, $\quad 20 + (-7) = 13$,
$13 + (-7) = 6$, $\quad 6 + (-7) = -1$, $\quad -1 + (-7) = -8$

**c)** $2 + \frac{1}{2} = 2\frac{1}{2}$, $\quad 2\frac{1}{2} + \frac{1}{2} = 3$, $\quad 3 + \frac{1}{2} = 3\frac{1}{2}$, $\quad 3\frac{1}{2} + \frac{1}{2} = 4$,
$4 + \frac{1}{2} = 4\frac{1}{2}$

**Now Try Exercise 1.**

To find a formula for the general, or *n*th, term of any arithmetic sequence, we denote the common difference by $d$, write out the first few terms, and look for a pattern:

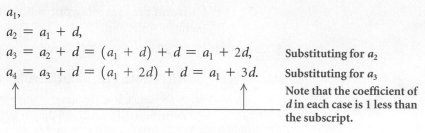

$a_1,$

$a_2 = a_1 + d,$

$a_3 = a_2 + d = (a_1 + d) + d = a_1 + 2d,$     Substituting for $a_2$

$a_4 = a_3 + d = (a_1 + 2d) + d = a_1 + 3d.$     Substituting for $a_3$

Note that the coefficient of $d$ in each case is 1 less than the subscript.

Generalizing, we obtain the following formula.

> ### *n*TH TERM OF AN ARITHMETIC SEQUENCE
> The **nth term** of an arithmetic sequence is given by
> $a_n = a_1 + (n - 1)d$, for any integer $n \geq 1$.

**EXAMPLE 2**    Find the 14th term of the arithmetic sequence

$$4, 7, 10, 13, \ldots.$$

**Solution**    We first note that $a_1 = 4$, $d = 7 - 4$, or 3, and $n = 14$. Then using the formula for the *n*th term, we obtain

$$a_n = a_1 + (n - 1)d$$
$$a_{14} = 4 + (14 - 1) \cdot 3 \quad \text{Substituting}$$
$$= 4 + 13 \cdot 3 = 4 + 39$$
$$= 43.$$

The 14th term is 43.

 **Now Try Exercise 9.** 

**EXAMPLE 3**    In the sequence of Example 2, which term is 301? That is, find $n$ if $a_n = 301$.

**Solution**    We substitute 301 for $a_n$, 4 for $a_1$, and 3 for $d$ in the formula for the *n*th term and solve for $n$:

$$a_n = a_1 + (n - 1)d$$
$$301 = 4 + (n - 1) \cdot 3 \quad \text{Substituting}$$
$$\left. \begin{array}{l} 301 = 4 + 3n - 3 \\ 301 = 3n + 1 \\ 300 = 3n \\ 100 = n. \end{array} \right\} \quad \text{Solving for } n$$

The term 301 is the 100th term of the sequence.

 **Now Try Exercise 15.** 

Given two terms and their places in an arithmetic sequence, we can construct the sequence.

**EXAMPLE 4** The 3rd term of an arithmetic sequence is 8, and the 16th term is 47. Find $a_1$ and $d$ and construct the sequence.

*Solution* We know that $a_3 = 8$ and $a_{16} = 47$. Thus we would need to add $d$ 13 times to get from 8 to 47. That is,

$$8 + 13d = 47. \qquad a_3 \text{ and } a_{16} \text{ are } 16 - 3, \text{ or } 13, \text{ terms apart.}$$

Solving $8 + 13d = 47$, we obtain

$$13d = 39$$
$$d = 3.$$

Since $a_3 = 8$, we subtract $d$ twice to get $a_1$. Thus,

$$a_1 = 8 - 2 \cdot 3 = 2. \qquad a_1 \text{ and } a_3 \text{ are } 3 - 1, \text{ or } 2, \text{ terms apart.}$$

The sequence is 2, 5, 8, 11, . . . . Note that we could also subtract $d$ 15 times from $a_{16}$ in order to find $a_1$.

Now Try Exercise 23.

In general, $d$ should be subtracted $n - 1$ times from $a_n$ in order to find $a_1$.

## Sum of the First *n* Terms of an Arithmetic Sequence

Consider the arithmetic sequence

$$3, 5, 7, 9, . . . .$$

When we add the first 4 terms of the sequence, we get $S_4$, which is

$$3 + 5 + 7 + 9, \quad \text{or} \quad 24.$$

This sum is called an **arithmetic series**. To find a formula for the sum of the first $n$ terms, $S_n$, of an arithmetic sequence, we first denote an arithmetic sequence, as follows:

> This term is two terms back from the last. If you add $d$ to this term, the result is the next-to-last term, $a_n - d$.

$$a_1, \quad (a_1 + d), \quad (a_1 + 2d), . . . , \quad \overbrace{(a_n - 2d)}, \quad \underbrace{(a_n - d)}, \quad a_n.$$

> This is the next-to-last term. If you add $d$ to this term, the result is $a_n$.

Then $S_n$ is given by

$$S_n = a_1 + (a_1 + d) + (a_1 + 2d) + \cdots + (a_n - 2d)$$
$$+ (a_n - d) + a_n. \qquad (1)$$

Reversing the order of the addition gives us

$$S_n = a_n + (a_n - d) + (a_n - 2d) + \cdots + (a_1 + 2d)$$
$$+ (a_1 + d) + a_1. \qquad (2)$$

If we add corresponding terms of each side of equations (1) and (2), we get

$$2S_n = [a_1 + a_n] + [(a_1 + d) + (a_n - d)] + [(a_1 + 2d) + (a_n - 2d)]$$
$$+ \cdots + [(a_n - 2d) + (a_1 + 2d)]$$
$$+ [(a_n - d) + (a_1 + d)] + [a_n + a_1].$$

In the expression for $2S_n$, there are $n$ expressions in square brackets. Each of these expressions is equivalent to $a_1 + a_n$. Thus the expression for $2S_n$ can be written in simplified form as

$$2S_n = [a_1 + a_n] + [a_1 + a_n] + [a_1 + a_n] + \cdots + [a_n + a_1]$$
$$+ [a_n + a_1] + [a_n + a_1].$$

Since $a_1 + a_n$ is being added $n$ times, it follows that

$$2S_n = n(a_1 + a_n),$$

from which we get the following formula.

---

### SUM OF THE FIRST $n$ TERMS

The sum of the first $n$ terms of an arithmetic sequence is given by

$$S_n = \frac{n}{2}(a_1 + a_n).$$

---

**EXAMPLE 5** Find the sum of the first 100 natural numbers.

*Solution* The sum is

$$1 + 2 + 3 + \cdots + 99 + 100.$$

This is the sum of the first 100 terms of the arithmetic sequence for which

$$a_1 = 1, \qquad a_n = 100, \quad \text{and} \quad n = 100.$$

Thus substituting into the formula

$$S_n = \frac{n}{2}(a_1 + a_n),$$

we get

$$S_{100} = \frac{100}{2}(1 + 100) = 50(101) = 5050.$$

The sum of the first 100 natural numbers is 5050.

**Now Try Exercise 27.**

**EXAMPLE 6**  Find the sum of the first 15 terms of the arithmetic sequence 4, 7, 10, 13, . . . .

**Solution**  Note that $a_1 = 4$, $d = 3$, and $n = 15$. Before using the formula

$$S_n = \frac{n}{2}(a_1 + a_n),$$

we find the last term, $a_{15}$:

$$a_{15} = 4 + (15 - 1)3 \qquad \text{Substituting into the formula } a_n = a_1 + (n - 1)d$$
$$= 4 + 14 \cdot 3 = 46.$$

Thus,

$$S_{15} = \frac{15}{2}(4 + 46) = \frac{15}{2}(50) = 375.$$

The sum of the first 15 terms is 375.

**Now Try Exercise 25.**

**EXAMPLE 7**  Find the sum: $\displaystyle\sum_{k=1}^{130} (4k + 5)$.

**Solution**  It is helpful to first write out a few terms:

$$9 + 13 + 17 + \cdots .$$

It appears that this is an arithmetic series coming from an arithmetic sequence with $a_1 = 9$, $d = 4$, and $n = 130$. Before using the formula

$$S_n = \frac{n}{2}(a_1 + a_n),$$

we find the last term, $a_{130}$:

$$a_{130} = 4 \cdot 130 + 5 \qquad \text{The } k\text{th term is } 4k + 5.$$
$$= 520 + 5$$
$$= 525.$$

Thus,

$$S_{130} = \frac{130}{2}(9 + 525) \qquad \text{Substituting into } S_n = \frac{n}{2}(a_1 + a_n)$$
$$= 34{,}710.$$

```
sum(seq(4X+5,X,1,130))
                  34710
```

This sum can also be found on a graphing calculator. It is not necessary to have the calculator set in SEQUENCE mode in order to do this.

**Now Try Exercise 33.**

## ■ Applications

The translation of some applications and problem-solving situations may involve arithmetic sequences or series. We consider some examples.

**EXAMPLE 8**  *Hourly Wages.*  Rachel accepts a job, starting with an hourly wage of $14.25, and is promised a raise of 15¢ per hour every 2 months for 5 years. At the end of 5 years, what will Rachel's hourly wage be?

*Solution*    It helps to first write down the hourly wage for several 2-month time periods:

Beginning:           $14.25,
After 2 months:   $14.40,
After 4 months:   $14.55,
and so on.

What appears is a sequence of numbers: 14.25, 14.40, 14.55, . . . . This sequence is arithmetic, because adding 0.15 each time gives us the next term.

We want to find the last term of an arithmetic sequence, so we use the formula $a_n = a_1 + (n - 1)d$. We know that $a_1 = 14.25$ and $d = 0.15$, but what is $n$? That is, how many terms are in the sequence? Each year there are 12/2, or 6 raises, since Rachel gets a raise every 2 months. There are 5 years, so the total number of raises will be $5 \cdot 6$, or 30. Thus there will be 31 terms: the original wage and 30 increased rates.

Substituting in the formula $a_n = a_1 + (n - 1)d$ gives us

$$a_{31} = 14.25 + (31 - 1) \cdot 0.15 = 18.75.$$

Thus, at the end of 5 years, Rachel's hourly wage will be $18.75.

**Now Try Exercise 41.**

The calculations in Example 8 could be done in a number of ways. There is often a variety of ways in which a problem can be solved. In this chapter, we concentrate on the use of sequences and series and their related formulas.

**EXAMPLE 9**    *Total in a Stack.*    A stack of telephone poles has 30 poles in the bottom row. There are 29 poles in the second row, 28 in the next row, and so on. How many poles are in the stack if there are 5 poles in the top row?

*Solution*    A picture will help in this case. The figure below shows the ends of the poles and the way in which they stack.

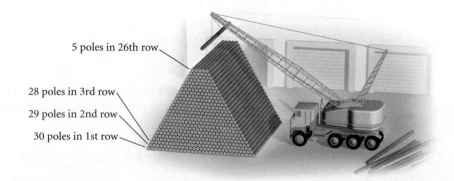

5 poles in 26th row

28 poles in 3rd row

29 poles in 2nd row

30 poles in 1st row

Since the number of poles goes from 30 in a row up to 5 in the top row, there must be 26 rows. We want the sum

$$30 + 29 + 28 + \cdots + 5.$$

Thus we have an arithmetic series. We use the formula

$$S_n = \frac{n}{2}(a_1 + a_n),$$

with $n = 26$, $a_1 = 30$, and $a_{26} = 5$.
Substituting, we get

$$S_{26} = \frac{26}{2}(30 + 5) = 455.$$

There are 455 poles in the stack.

**Now Try Exercise 39.**

---

## 8.2   Exercise Set

*Find the first term and the common difference.*

1. $3, 8, 13, 18, \ldots$

2. $\$1.08, \$1.16, \$1.24, \$1.32, \ldots$

3. $9, 5, 1, -3, \ldots$

4. $-8, -5, -2, 1, 4, \ldots$

5. $\frac{3}{2}, \frac{9}{4}, 3, \frac{15}{4}, \ldots$

6. $\frac{3}{5}, \frac{1}{10}, -\frac{2}{5}, \ldots$

7. $\$316, \$313, \$310, \$307, \ldots$

8. Find the 11th term of the arithmetic sequence $0.07, 0.12, 0.17, \ldots$.

9. Find the 12th term of the arithmetic sequence $2, 6, 10, \ldots$.

10. Find the 17th term of the arithmetic sequence $7, 4, 1, \ldots$.

11. Find the 14th term of the arithmetic sequence $3, \frac{7}{3}, \frac{5}{3}, \ldots$.

12. Find the 13th term of the arithmetic sequence $\$1200, \$964.32, \$728.64, \ldots$.

13. Find the 10th term of the arithmetic sequence $\$2345.78, \$2967.54, \$3589.30, \ldots$.

14. In the sequence of Exercise 8, what term is the number 1.67?

15. In the sequence of Exercise 9, what term is the number 106?

16. In the sequence of Exercise 10, what term is $-296$?

17. In the sequence of Exercise 11, what term is $-27$?

18. Find $a_{20}$ when $a_1 = 14$ and $d = -3$.

19. Find $a_1$ when $d = 4$ and $a_8 = 33$.

20. Find $d$ when $a_1 = 8$ and $a_{11} = 26$.

21. Find $n$ when $a_1 = 25$, $d = -14$, and $a_n = -507$.

22. In an arithmetic sequence, $a_{17} = -40$ and $a_{28} = -73$. Find $a_1$ and $d$. Write the first 5 terms of the sequence.

23. In an arithmetic sequence, $a_{17} = \frac{25}{3}$ and $a_{32} = \frac{95}{6}$. Find $a_1$ and $d$. Write the first 5 terms of the sequence.

24. Find the sum of the first 14 terms of the series $11 + 7 + 3 + \cdots$.

25. Find the sum of the first 20 terms of the series $5 + 8 + 11 + 14 + \cdots$.

26. Find the sum of the first 300 natural numbers.

27. Find the sum of the first 400 even natural numbers.

28. Find the sum of the odd numbers 1 to 199, inclusive.

29. Find the sum of the multiples of 7 from 7 to 98, inclusive.

30. Find the sum of all multiples of 4 that are between 14 and 523.

31. If an arithmetic series has $a_1 = 2$, $d = 5$, and $n = 20$, what is $S_n$?

32. If an arithmetic series has $a_1 = 7$, $d = -3$, and $n = 32$, what is $S_n$?

*Find the sum.*

**33.** $\sum_{k=1}^{40} (2k + 3)$

**34.** $\sum_{k=5}^{20} 8k$

**35.** $\sum_{k=0}^{19} \frac{k - 3}{4}$

**36.** $\sum_{k=2}^{50} (2000 - 3k)$

**37.** $\sum_{k=12}^{57} \frac{7 - 4k}{13}$

**38.** $\sum_{k=101}^{200} (1.14k - 2.8) - \sum_{k=1}^{5} \left( \frac{k + 4}{10} \right)$

**39.** *Stacking Poles.*   How many poles will be in a stack of telephone poles if there are 50 in the first layer, 49 in the second, and so on, with 6 in the top layer?

**40.** *Investment Return.*   Max, an investment counselor, sets up an investment situation for a client that will return $5000 the first year, $6125 the second year, $7250 the third year, and so on, for 25 years. How much is received from the investment altogether?

**41.** *Theater Seating.*   Theaters are often built with more seats per row as the rows move toward the back. Suppose that the first balcony of a theater has 28 seats in the first row, 32 in the second, 36 in the third, and so on, for 20 rows. How many seats are in the first balcony altogether?

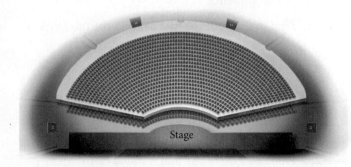

**42.** *Total Savings.*   If 10¢ is saved on October 1, 20¢ is saved on October 2, 30¢ on October 3, and so on, how much is saved during the 31 days of October?

**43.** *Parachutist Free Fall.*   When a parachutist jumps from an airplane, the distances, in feet, that the parachutist falls in each successive second before pulling the ripcord to release the parachute are as follows:

$$16, \ 48, \ 80, \ 112, \ 144, \ldots.$$

Is this sequence arithmetic? What is the common difference? What is the total distance fallen in 10 sec?

**44.** *Lightning Distance.*   The table below lists the distance in miles from lightning $d_n$ when thunder is heard $n$ seconds after lightning is seen. Is this sequence arithmetic? What is the common difference?

| $n$ (in seconds) | $d_n$ (in miles) |
|:---:|:---:|
| 5 | 1 |
| 6 | 1.2 |
| 7 | 1.4 |
| 8 | 1.6 |
| 9 | 1.8 |
| 10 | 2 |

**45.** *Band Formation.*   A formation of a marching band has 10 marchers in the first row, 12 in the second row, 14 in the third row, and so on, for 8 rows. How many marchers are in the last row? How many marchers are there altogether?

**46.** *Garden Plantings.* A gardener is making a planting in the shape of a trapezoid. It will have 35 plants in the first row, 31 in the second row, 27 in the third row, and so on. If the pattern is consistent, how many plants will there be in the last row? How many plants are there altogether?

**47.** *Raw Material Production.* In a manufacturing process, it took 3 units of raw materials to produce 1 unit of a product. The raw material needs thus formed the sequence

$$3, 6, 9, \ldots, 3n, \ldots.$$

Is this sequence arithmetic? What is the common difference?

## Skill Maintenance

*Solve.*

**48.** $7x - 2y = 4,$
$\phantom{aa}x + 3y = 17$

**49.** $2x + y + 3z = 12,$
$\phantom{aaa}x - 3y + 2z = 11,$
$\phantom{a}5x + 2y - 4z = -4$

**50.** Find the vertices and the foci of the ellipse with equation $9x^2 + 16y^2 = 144$.

**51.** Find an equation of the ellipse with vertices $(0, -5)$ and $(0, 5)$ and minor axis of length 4.

## Synthesis

**52.** *Straight-Line Depreciation.* A company buys an office machine for $5200 on January 1 of a given year. The machine is expected to last for 8 years, at the end of which time its **trade-in value**, or **salvage value**, will be $1100. If the company's accountant figures the decline in value to be the same each year, then its **book values**, or **salvage values**, after $t$ years, $0 \leq t \leq 8$, form an arithmetic sequence given by

$$a_t = C - t\left(\frac{C - S}{N}\right),$$

where $C$ is the original cost of the item ($5200), $N$ is the number of years of expected life (8), and $S$ is the salvage value ($1100).

**a)** Find the formula for $a_t$ for the straight-line depreciation of the office machine.

**b)** Find the salvage value after 0 year, 1 year, 2 years, 3 years, 4 years, 7 years, and 8 years.

**53.** Find a formula for the sum of the first $n$ odd natural numbers:

$$1 + 3 + 5 + \cdots + (2n - 1).$$

**54.** Find three numbers in an arithmetic sequence such that the sum of the first and third is 10 and the product of the first and second is 15.

**55.** Find the first term and the common difference for the arithmetic sequence for which

$$a_2 = 40 - 3q \quad \text{and} \quad a_4 = 10p + q.$$

*If $p$, $m$, and $q$ form an arithmetic sequence, it can be shown that $m = (p + q)/2$. The number $m$ is the **arithmetic mean,** or **average,** of $p$ and $q$. Given two numbers $p$ and $q$, if we find $k$ other numbers $m_1$, $m_2, \ldots, m_k$ such that*

$$p, m_1, m_2, \ldots, m_k, q$$

*forms an arithmetic sequence, we say that we have "inserted $k$ arithmetic means between $p$ and $q$."*

**56.** Insert three arithmetic means between $-3$ and 5.

**57.** Insert four arithmetic means between 4 and 13.

---

## 8.3    Geometric Sequences and Series

- Identify the common ratio of a geometric sequence, and find a given term and the sum of the first $n$ terms.
- Find the sum of an infinite geometric series, if it exists.

A sequence in which each term after the first term is found by multiplying the preceding term by the same number is a **geometric sequence**.

## ■ Geometric Sequences

Consider the sequence:

$$2, \ 6, \ 18, \ 54, \ 162, \dots.$$

Note that multiplying each term by 3 produces the next term. We call the number 3 the **common ratio** because it can be found by dividing any term by the preceding term. A geometric sequence is also called a *geometric progression*.

---

**GEOMETRIC SEQUENCE**

A sequence is **geometric** if there is a number $r$, called the **common ratio**, such that

$$\frac{a_{n+1}}{a_n} = r, \quad \text{or} \quad a_{n+1} = a_n r, \quad \text{for any integer } n \geq 1.$$

---

**EXAMPLE 1** For each of the following geometric sequences, identify the common ratio.

**a)** $3, 6, 12, 24, 48, \dots$
**b)** $1, \ -\dfrac{1}{2}, \ \dfrac{1}{4}, \ -\dfrac{1}{8}, \dots$

**c)** $\$5200, \$3900, \$2925, \$2193.75, \dots$
**d)** $\$1000, \$1060, \$1123.60, \dots$

*Solution*

| SEQUENCE | COMMON RATIO |
|---|---|
| **a)** $3, \ 6, \ 12, \ 24, \ 48, \dots$ | $2 \quad \left(\frac{6}{3} = 2, \frac{12}{6} = 2, \text{ and so on}\right)$ |
| **b)** $1, \ -\dfrac{1}{2}, \ \dfrac{1}{4}, \ -\dfrac{1}{8}, \dots$ | $-\dfrac{1}{2} \quad \left(\dfrac{-\frac{1}{2}}{1} = -\dfrac{1}{2}, \dfrac{\frac{1}{4}}{-\frac{1}{2}} = -\dfrac{1}{2}, \text{ and so on}\right)$ |
| **c)** $\$5200, \$3900, \$2925, \$2193.75, \dots$ | $0.75 \quad \left(\dfrac{\$3900}{\$5200} = 0.75, \dfrac{\$2925}{\$3900} = 0.75, \text{ and so on}\right)$ |
| **d)** $\$1000, \$1060, \$1123.60, \dots$ | $1.06 \quad \left(\dfrac{\$1060}{\$1000} = 1.06, \dfrac{\$1123.60}{\$1060} = 1.06, \text{ and so on}\right)$ |

**Now Try Exercise 1.**

We now find a formula for the general, or $n$th, term of a geometric sequence. If we let $a_1$ be the first term and $r$ the common ratio, then the first few terms are as follows:

$$a_1,$$
$$a_2 = a_1 r,$$
$$a_3 = a_2 r = (a_1 r)r = a_1 r^2, \qquad \text{Substituting } a_1 r \text{ for } a_2$$
$$a_4 = a_3 r = (a_1 r^2)r = a_1 r^3. \qquad \text{Substituting } a_1 r^2 \text{ for } a_3$$

Note that the exponent is 1 less than the subscript.

Generalizing, we obtain the following.

---

### *n*TH TERM OF A GEOMETRIC SEQUENCE

The *n*th **term** of a geometric sequence is given by

$$a_n = a_1 r^{n-1}, \quad \text{for any integer } n \geq 1.$$

---

**EXAMPLE 2** Find the 7th term of the geometric sequence 4, 20, 100, . . . .

*Solution* We first note that

$$a_1 = 4 \quad \text{and} \quad n = 7.$$

To find the common ratio, we can divide any term (other than the first) by the preceding term. Since the second term is 20 and the first is 4, we get

$$r = \frac{20}{4}, \quad \text{or} \quad 5.$$

Then using the formula $a_n = a_1 r^{n-1}$, we have

$$a_7 = 4 \cdot 5^{7-1} = 4 \cdot 5^6 = 4 \cdot 15{,}625 = 62{,}500.$$

Thus the 7th term is 62,500. **Now Try Exercise 11.**

**EXAMPLE 3** Find the 10th term of the geometric sequence 64, $-32$, 16, $-8$, . . . .

*Solution* We first note that

$$a_1 = 64, \qquad n = 10, \quad \text{and} \quad r = \frac{-32}{64}, \text{or} -\frac{1}{2}.$$

Then using the formula $a_n = a_1 r^{n-1}$, we have

$$a_{10} = 64 \cdot \left(-\frac{1}{2}\right)^{10-1} = 64 \cdot \left(-\frac{1}{2}\right)^9 = 2^6 \cdot \left(-\frac{1}{2^9}\right) = -\frac{1}{2^3} = -\frac{1}{8}.$$

Thus the 10th term is $-\frac{1}{8}$. **Now Try Exercise 15.**

### ● Sum of the First *n* Terms of a Geometric Sequence

Next, we develop a formula for the sum $S_n$ of the first $n$ terms of a geometric sequence:

$$a_1, \ a_1 r, \ a_1 r^2, \ a_1 r^3, \ \ldots, a_1 r^{n-1}, \ldots.$$

The associated **geometric series** is given by

$$S_n = a_1 + a_1 r + a_1 r^2 + a_1 r^3 + \cdots + a_1 r^{n-1}. \tag{1}$$

We want to find a formula for this sum. If we multiply by $r$ on both sides of equation (1), we have

$$rS_n = a_1 r + a_1 r^2 + a_1 r^3 + a_1 r^4 + \cdots + a_1 r^n. \tag{2}$$

Subtracting equation (2) from equation (1), we see that the differences of the terms shown in red are 0, leaving

$$S_n - rS_n = a_1 - a_1r^n,$$

or

$$S_n(1 - r) = a_1(1 - r^n). \qquad \text{Factoring}$$

Dividing by $1 - r$ on both sides gives us the following formula.

> **SUM OF THE FIRST $n$ TERMS**
>
> The sum of the first $n$ terms of a geometric sequence is given by
>
> $$S_n = \frac{a_1(1 - r^n)}{1 - r}, \quad \text{for any } r \neq 1.$$

**EXAMPLE 4**   Find the sum of the first 7 terms of the geometric sequence $3, 15, 75, 375, \ldots$.

*Solution*   We first note that

$$a_1 = 3, \quad n = 7, \quad \text{and} \quad r = \tfrac{15}{3}, \text{ or } 5.$$

Then using the formula

$$S_n = \frac{a_1(1 - r^n)}{1 - r},$$

we have

$$S_7 = \frac{3(1 - 5^7)}{1 - 5} = \frac{3(1 - 78{,}125)}{-4} = 58{,}593.$$

Thus the sum of the first 7 terms is 58,593.      **Now Try Exercise 23.**

**EXAMPLE 5**   Find the sum: $\displaystyle\sum_{k=1}^{11}(0.3)^k$.

*Solution*   This is a geometric series with $a_1 = 0.3$, $r = 0.3$, and $n = 11$. Thus,

$$S_{11} = \frac{0.3(1 - 0.3^{11})}{1 - 0.3} \approx 0.42857.$$

We can also find this sum using a graphing calculator set in either FUNCTION mode or SEQUENCE mode.      **Now Try Exercise 41.**

sum(seq(.3^X,X,1,11))
                .4285706694

■ **Infinite Geometric Series**

The sum of the terms of an infinite geometric sequence is an **infinite geometric series.** For some geometric sequences, $S_n$ gets close to a specific number as $n$ gets large. For example, consider the infinite series

$$\frac{1}{2} + \frac{1}{4} + \frac{1}{8} + \frac{1}{16} + \cdots + \frac{1}{2^n} + \cdots.$$

We can visualize $S_n$ by considering the area of a square. For $S_1$, we shade half the square. For $S_2$, we shade half the square plus half the remaining half, or $\frac{1}{4}$. For $S_3$, we shade the parts shaded in $S_2$ plus half the remaining part. We see that the values of $S_n$ will continue to get close to 1 (shading the complete square).

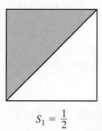

$S_1 = \frac{1}{2}$

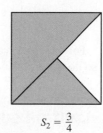

$S_2 = \frac{3}{4}$

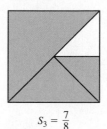

$S_3 = \frac{7}{8}$

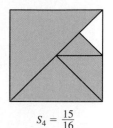

$S_4 = \frac{15}{16}$

We examine some partial sums. Note that each of the partial sums is less than 1, but $S_n$ gets very close to 1 as $n$ gets large.

| $n$ | $S_n$ |
|---|---|
| 1 | 0.5 |
| 5 | 0.96875 |
| 10 | 0.9990234375 |
| 20 | 0.9999990463 |
| 30 | 0.9999999991 |

```
sum(seq(1/2^X,X,1,20))
                .9999990463
sum(seq(1/2^X,X,1,30))
                .9999999991
```

We say that 1 is the **limit** of $S_n$ and also that 1 is the **sum of the infinite geometric sequence**. The sum of an infinite geometric sequence is denoted $S_\infty$. In this case, $S_\infty = 1$.

Some infinite sequences do not have sums. Consider the infinite geometric series

$$2 + 4 + 8 + 16 + \cdots + 2^n + \cdots.$$

We again examine some partial sums. Note that as $n$ gets large, $S_n$ gets large without bound. This sequence does not have a sum.

| $n$ | $S_n$ |
|---|---|
| 1 | 2 |
| 5 | 62 |
| 10 | 2,046 |
| 20 | 2,097,150 |
| 30 | 2,147,483,646 |

```
sum(seq(2^X,X,1,20))
                2097150
sum(seq(2^X,X,1,30))
             2147483646
```

It can be shown (but we will not do so here) that the sum of an infinite geometric series exists if and only if $|r| < 1$ (that is, the absolute value of the common ratio is less than 1).

To find a formula for the sum of an infinite geometric series, we first consider the sum of the first $n$ terms:

$$S_n = \frac{a_1(1 - r^n)}{1 - r} = \frac{a_1 - a_1 r^n}{1 - r}. \qquad \textbf{Using the distributive law}$$

For $|r| < 1$, values of $r^n$ get close to 0 as $n$ gets large. As $r^n$ gets close to 0, so does $a_1 r^n$. Thus, $S_n$ gets close to $a_1/(1 - r)$.

---

**LIMIT OR SUM OF AN INFINITE GEOMETRIC SERIES**

When $|r| < 1$, the limit or sum of an infinite geometric series is given by

$$S_\infty = \frac{a_1}{1 - r}.$$

---

**EXAMPLE 6** Determine whether each of the following infinite geometric series has a limit. If a limit exists, find it.

**a)** $1 + 3 + 9 + 27 + \cdots$ **b)** $-2 + 1 - \frac{1}{2} + \frac{1}{4} - \frac{1}{8} + \cdots$

*Solution*

**a)** Here $r = 3$, so $|r| = |3| = 3$. Since $|r| > 1$, the series *does not* have a limit.

**b)** Here $r = -\frac{1}{2}$, so $|r| = \left|-\frac{1}{2}\right| = \frac{1}{2}$. Since $|r| < 1$, the series *does* have a limit. We find the limit:

$$S_\infty = \frac{a_1}{1 - r} = \frac{-2}{1 - \left(-\frac{1}{2}\right)} = \frac{-2}{\frac{3}{2}} = -\frac{4}{3}.$$

**Now Try Exercises 33 and 37.**

**EXAMPLE 7** Find fraction notation for $0.78787878\ldots$, or $0.\overline{78}$.

*Solution* We can express this as

$$0.78 + 0.0078 + 0.000078 + \cdots.$$

Then we see that this is an infinite geometric series, where $a_1 = 0.78$ and $r = 0.01$. Since $|r| < 1$, this series has a limit:

$$S_\infty = \frac{a_1}{1 - r} = \frac{0.78}{1 - 0.01} = \frac{0.78}{0.99} = \frac{78}{99}, \quad \text{or} \quad \frac{26}{33}.$$

Thus fraction notation for $0.78787878\ldots$ is $\frac{26}{33}$. You can check this on your calculator.

**Now Try Exercise 51.**

## Applications

The translation of some applications and problem-solving situations may involve geometric sequences or series. Examples 9 and 10, in particular, show applications in business and economics.

**EXAMPLE 8** *A Daily Doubling Salary.* Suppose someone offered you a job for the month of September (30 days) under the following conditions. You will be paid $0.01 for the first day, $0.02 for the second, $0.04 for the third, and so on, doubling your previous day's salary each day. How much would you earn? (Would you take the job? Make a conjecture before reading further.)

**Solution** You earn $0.01 the first day, $0.01(2)$ the second day, $0.01(2)(2)$ the third day, and so on. The amount earned is the geometric series

$$\$0.01 + \$0.01(2) + \$0.01(2^2) + \$0.01(2^3) + \cdots + \$0.01(2^{29}),$$

where $a_1 = \$0.01$, $r = 2$, and $n = 30$. Using the formula

$$S_n = \frac{a_1(1 - r^n)}{1 - r},$$

we have

$$S_{30} = \frac{\$0.01(1 - 2^{30})}{1 - 2} = \$10{,}737{,}418.23.$$

The pay exceeds $10.7 million for the month.

> **Now Try Exercise 57.**

**EXAMPLE 9** *The Amount of an Annuity.* An **annuity** is a sequence of equal payments, made at equal time intervals, that earn interest. Fixed deposits in a savings account are an example of an annuity. Suppose that to save money to buy a car, Andrea deposits $2000 at the *end* of each of 5 years in an account that pays 4% interest, compounded annually. The total amount in the account at the end of 5 years is called the **amount of the annuity.** Find that amount.

**Solution** The time diagram below can help visualize the problem. Note that no deposit is made until the end of the first year.

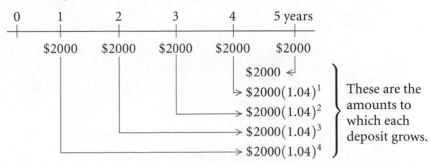

The amount of the annuity is the geometric series

$$\$2000 + \$2000(1.04)^1 + \$2000(1.04)^2 + \$2000(1.04)^3 + \$2000(1.04)^4,$$

where $a_1 = \$2000$, $n = 5$, and $r = 1.04$. Using the formula

$$S_n = \frac{a_1(1 - r^n)}{1 - r},$$

we have

$$S_5 = \frac{\$2000(1 - 1.04^5)}{1 - 1.04} \approx \$10,832.65.$$

The amount of the annuity is $10,832.65.　**Now Try Exercise 61.**

**EXAMPLE 10    *The Economic Multiplier.*    Large sporting events have a significant impact on the economy of the host city. Those attending the 2010 Super Bowl in Miami poured $333 million into the economy of south Florida (*Source*: Sport Management Research Institute). Assume that 60% of that amount is spent again in the area, and then 60% of that amount is spent again, and so on. This is known as the *economic multiplier effect*. Find the total effect on the economy.

*Solution*    The total economic effect is given by the infinite series

$$\$333,000,000 + \$333,000,000(0.6) + \$333,000,000(0.6)^2 + \cdots.$$

Since $|r| = |0.6| = 0.6 < 1$, the series has a sum. Using the formula for the sum of an infinite geometric series, we have

$$
\begin{aligned}
S_\infty &= \frac{a_1}{1 - r} \\
&= \frac{\$333,000,000}{1 - 0.6} \\
&= \$832,500,000.
\end{aligned}
$$

The total effect of the spending on the economy is $832,500,000.

**Now Try Exercise 67.**

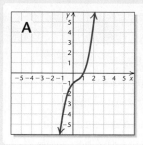

A

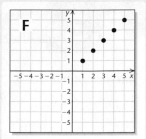

F

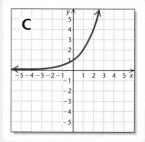

B

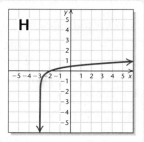

G

# Visualizing the Graph

Match the equation with its graph.

**1.** $(x - 1)^2 + (y + 2)^2 = 9$

**2.** $y = x^3 - x^2 + x - 1$

**3.** $f(x) = 2^x$

**4.** $f(x) = x$

**5.** $a_n = n$

**6.** $y = \log(x + 3)$

**7.** $f(x) = -(x - 2)^2 + 1$

**8.** $f(x) = (x - 2)^2 - 1$

**9.** $y = \dfrac{1}{x - 1}$

**10.** $y = -3x + 4$

Answers on page A-52

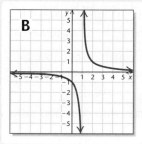

C

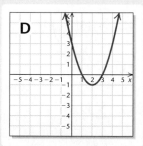

D

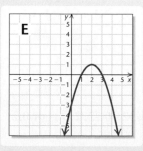

E

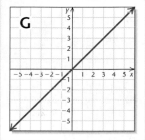

H

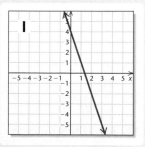

I

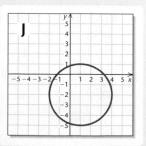

J

# 8.3 Exercise Set

*Find the common ratio.*

**1.** 2, 4, 8, 16, . . .

**2.** 18, −6, 2, −$\frac{2}{3}$, . . .

**3.** −1, 1, −1, 1, . . .

**4.** −8, −0.8, −0.08, −0.008, . . .

**5.** $\frac{2}{3}$, −$\frac{4}{3}$, $\frac{8}{3}$, −$\frac{16}{3}$, . . .

**6.** 75, 15, 3, $\frac{3}{5}$, . . .

**7.** 6.275, 0.6275, 0.06275, . . .

**8.** $\dfrac{1}{x}$, $\dfrac{1}{x^2}$, $\dfrac{1}{x^3}$, . . .

**9.** 5, $\dfrac{5a}{2}$, $\dfrac{5a^2}{4}$, $\dfrac{5a^3}{8}$, . . .

**10.** $780, $858, $943.80, $1038.18, . . .

*Find the indicated term of the geometric sequence.*

**11.** 2, 4, 8, 16, . . . ; the 7th term

**12.** 2, −10, 50, −250, . . . ; the 9th term

**13.** 2, 2$\sqrt{3}$, 6, . . . ; the 9th term

**14.** 1, −1, 1, −1, . . . ; the 57th term

**15.** $\frac{7}{625}$, −$\frac{7}{25}$, . . . ; the 23rd term

**16.** $1000, $1060, $1123.60, . . . ; the 5th term

*Find the nth, or general, term.*

**17.** 1, 3, 9, . . .         **18.** 25, 5, 1, . . .

**19.** 1, −1, 1, −1, . . .         **20.** −2, 4, −8, . . .

**21.** $\dfrac{1}{x}$, $\dfrac{1}{x^2}$, $\dfrac{1}{x^3}$, . . .         **22.** 5, $\dfrac{5a}{2}$, $\dfrac{5a^2}{4}$, $\dfrac{5a^3}{8}$, . . .

**23.** Find the sum of the first 7 terms of the geometric series

$$6 + 12 + 24 + \cdots.$$

**24.** Find the sum of the first 10 terms of the geometric series

$$16 − 8 + 4 − \cdots.$$

**25.** Find the sum of the first 9 terms of the geometric series

$$\tfrac{1}{18} − \tfrac{1}{6} + \tfrac{1}{2} − \cdots.$$

**26.** Find the sum of the geometric series

$$−8 + 4 + (−2) + \cdots + \left(−\tfrac{1}{32}\right).$$

*Determine whether the statement is true or false.*

**27.** The sequence 2, −2$\sqrt{2}$, 4, −4$\sqrt{2}$, 8, . . . is geometric.

**28.** The sequence with general term $3n$ is geometric.

**29.** The sequence with general term $2^n$ is geometric.

**30.** Multiplying a term of a geometric sequence by the common ratio produces the next term of the sequence.

**31.** An infinite geometric series with common ratio −0.75 has a sum.

**32.** Every infinite geometric series has a limit.

*Find the sum, if it exists.*

**33.** 4 + 2 + 1 + $\cdots$

**34.** 7 + 3 + $\frac{9}{7}$ + $\cdots$

**35.** 25 + 20 + 16 + $\cdots$

**36.** 100 − 10 + 1 − $\frac{1}{10}$ + $\cdots$

**37.** 8 + 40 + 200 + $\cdots$

**38.** −6 + 3 − $\frac{3}{2}$ + $\frac{3}{4}$ − $\cdots$

**39.** 0.6 + 0.06 + 0.006 + $\cdots$

**40.** $\displaystyle\sum_{k=0}^{10} 3^k$         **41.** $\displaystyle\sum_{k=1}^{11} 15\left(\frac{2}{3}\right)^k$

**42.** $\displaystyle\sum_{k=0}^{50} 200(1.08)^k$         **43.** $\displaystyle\sum_{k=1}^{\infty} \left(\frac{1}{2}\right)^{k-1}$

**44.** $\displaystyle\sum_{k=1}^{\infty} 2^k$         **45.** $\displaystyle\sum_{k=1}^{\infty} 12.5^k$

**46.** $\displaystyle\sum_{k=1}^{\infty} 400(1.0625)^k$         **47.** $\displaystyle\sum_{k=1}^{\infty} \$500(1.11)^{-k}$

**48.** $\displaystyle\sum_{k=1}^{\infty} \$1000(1.06)^{-k}$         **49.** $\displaystyle\sum_{k=1}^{\infty} 16(0.1)^{k-1}$

**50.** $\displaystyle\sum_{k=1}^{\infty} \frac{8}{3}\left(\frac{1}{2}\right)^{k-1}$

*Find fraction notation.*

**51.** 0.131313. . . , or 0.$\overline{13}$         **52.** 0.2222. . . , or 0.$\overline{2}$

**53.** $8.999\overline{9}$

**54.** $6.1616\overline{16}$

**55.** $3.4125\overline{125}$

**56.** $12.7809\overline{809}$

**57.** *Daily Doubling Salary.* Suppose someone offers you a job for the month of February (28 days) under the following conditions. You will be paid $0.01 the 1st day, $0.02 the 2nd, $0.04 the 3rd, and so on, doubling your previous day's salary each day. How much would you earn altogether?

**58.** *Bouncing Ping-Pong Ball.* A ping-pong ball is dropped from a height of 16 ft and always rebounds $\frac{1}{4}$ of the distance fallen.

**a)** How high does it rebound the 6th time?

**b)** Find the total sum of the rebound heights of the ball.

**59.** *Bungee Jumping.* A bungee jumper always rebounds 60% of the distance fallen. A bungee jump is made using a cord that stretches to 200 ft.

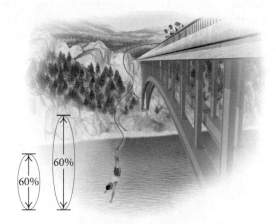

**a)** After jumping and then rebounding 9 times, how far has a bungee jumper traveled upward (the total rebound distance)?

**b)** About how far will a jumper have traveled upward (bounced) before coming to rest?

**60.** *Population Growth.* Hadleytown has a present population of 100,000, and the population is increasing by 3% each year.

**a)** What will the population be in 15 years?

**b)** How long will it take for the population to double?

**61.** *Amount of an Annuity.* To save for the down payment on a house, the Clines make a sequence of 10 yearly deposits of $3200 each in a savings account on which interest is compounded annually at 4.6%. Find the amount of the annuity.

**62.** *Amount of an Annuity.* To create a college fund, a parent makes a sequence of 18 yearly deposits of $1000 each in a savings account on which interest is compounded annually at 3.2%. Find the amount of the annuity.

**63.** *Doubling the Thickness of Paper.* A piece of paper is 0.01 in. thick. It is cut and stacked repeatedly in such a way that its thickness is doubled each time for 20 times. How thick is the result?

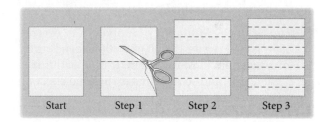

**64.** *Amount of an Annuity.* A sequence of yearly payments of $P$ dollars is invested at the end of each of $N$ years at interest rate $i$, compounded annually. The total amount in the account, or the amount of the annuity, is $V$.

**a)** Show that

$$V = \frac{P\left[(1 + i)^N - 1\right]}{i}.$$

**b)** Suppose that interest is compounded $n$ times per year and deposits are made every compounding period. Show that the formula for $V$ is then given by

$$V = \frac{P\left[\left(1 + \dfrac{i}{n}\right)^{nN} - 1\right]}{i/n}.$$

**65.** *Amount of an Annuity.* A sequence of payments of $300 is invested over 12 years at the end of each quarter at 5.1%, compounded quarterly. Find the amount of the annuity. Use the formula in Exercise 64(b).

**66.** *Amount of an Annuity.*   A sequence of yearly payments of $750 is invested at the end of each of 10 years at 4.75%, compounded annually. Find the amount of the annuity. Use the formula in Exercise 64(a).

**67.** *Advertising Effect.*   Great Grains Cereal Company is about to market a new low-carbohydrate cereal in a city of 5,000,000 people. They plan an advertising campaign that they think will induce 30% of the people to buy the product. They estimate that if those people like the product, they will induce $30\% \cdot 30\% \cdot 5{,}000{,}000$ more to buy the product, and those will induce $30\% \cdot 30\% \cdot 30\% \cdot 5{,}000{,}000$, and so on. In all, how many people will buy the product as a result of the advertising campaign? What percentage of the population is this?

**68.** *The Economic Multiplier.*   Suppose the government is making a $13,000,000,000 expenditure for educational improvement. If 85% of this is spent again, and so on, what is the total effect on the economy?

## Skill Maintenance

*For each pair of functions, find* $(f \circ g)(x)$ *and* $(g \circ f)(x)$.

**69.** $f(x) = x^2$, $g(x) = 4x + 5$

**70.** $f(x) = x - 1$, $g(x) = x^2 + x + 3$

*Solve.*

**71.** $5^x = 35$          **72.** $\log_2 x = -4$

## Synthesis

**73.** Prove that
$$\sqrt{3} - \sqrt{2}, \quad 4 - \sqrt{6}, \quad \text{and} \quad 6\sqrt{3} - 2\sqrt{2}$$
form a geometric sequence.

**74.** Assume that $a_1, a_2, a_3, \ldots$ is a geometric sequence. Prove that $\ln a_1, \ln a_2, \ln a_3, \ldots$ is an arithmetic sequence.

**75.** Consider the sequence
$$x + 3, \quad x + 7, \quad 4x - 2, \ldots.$$
   **a)** If the sequence is arithmetic, find $x$ and then determine each of the 3 terms and the 4th term.
   **b)** If the sequence is geometric, find $x$ and then determine each of the 3 terms and the 4th term.

**76.** Find the sum of the first $n$ terms of
$$1 + x + x^2 + \cdots.$$

**77.** Find the sum of the first $n$ terms of
$$x^2 - x^3 + x^4 - x^5 + \cdots.$$

**78.** The sides of a square are 16 cm long. A second square is inscribed by joining the midpoints of the sides, successively. In the second square, we repeat the process, inscribing a third square. If this process is continued indefinitely, what is the sum of all the areas of all the squares? (*Hint*: Use an infinite geometric series.)

## Mathematical Induction

**8.4**

- Prove infinite sequences of statements using mathematical induction.

In this section, we learn to prove a sequence of mathematical statements using a procedure called *mathematical induction.*

### ■ Proving Infinite Sequences of Statements

Infinite sequences of statements occur often in mathematics. In an infinite sequence of statements, there is a statement for each natural number. For example, consider the sequence of statements represented by the following:

"The sum of the first $n$ positive odd integers is $n^2$,"   or
$$1 + 3 + 5 + \cdots + (2n - 1) = n^2.$$

Let's think of this as $S(n)$, or $S_n$. Substituting natural numbers for $n$ gives a sequence of statements. We list the first four:

$S_1$:  $1 = 1^2$;
$S_2$:  $1 + 3 = 4 = 2^2$;
$S_3$:  $1 + 3 + 5 = 9 = 3^2$;
$S_4$:  $1 + 3 + 5 + 7 = 16 = 4^2$.

The fact that the statement is true for $n = 1, 2, 3,$ and 4 might tempt us to conclude that the statement is true for any natural number $n$, but we cannot be sure that this is the case. We can, however, use the principle of mathematical induction to prove that the statement is true for all natural numbers.

---

**THE PRINCIPLE OF MATHEMATICAL INDUCTION**

We can prove an infinite sequence of statements $S_n$ by showing the following.

1. *Basis step*:  $S_1$ is true.
2. *Induction step*:  For all natural numbers $k$, $S_k \rightarrow S_{k+1}$.

---

Mathematical induction is analogous to lining up a sequence of dominoes. The induction step tells us that if any one domino is knocked over, then the one next to it will be hit and knocked over. The basis step tells us that the first domino can indeed be knocked over. Note that in order for all dominoes to fall, *both* conditions must be satisfied.

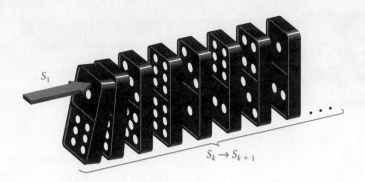

When you are learning to do proofs by mathematical induction, it is helpful to first write out $S_n$, $S_1$, $S_k$, and $S_{k+1}$. This helps to identify what is to be assumed and what is to be deduced.

**EXAMPLE 1** Prove: For every natural number $n$,

$$1 + 3 + 5 + \cdots + (2n - 1) = n^2.$$

**Proof.** We first list $S_n$, $S_1$, $S_k$, and $S_{k+1}$.

$S_n$:     $1 + 3 + 5 + \cdots + (2n - 1) = n^2$

$S_1$:     $1 = 1^2$

$S_k$:     $1 + 3 + 5 + \cdots + (2k - 1) = k^2$

$S_{k+1}$:     $1 + 3 + 5 + \cdots + (2k - 1) + \left[2(k + 1) - 1\right] = (k + 1)^2$

1. *Basis step*: $S_1$, as listed, is true since $1 = 1^2$, or $1 = 1$.

2. *Induction step*: We let $k$ be any natural number. We assume $S_k$ to be true and try to show that it implies that $S_{k+1}$ is true. Now $S_k$ is

   $$1 + 3 + 5 + \cdots + (2k - 1) = k^2.$$

   Starting with the left side of $S_{k+1}$ and substituting $k^2$ for $1 + 3 + 5 + \cdots + (2k - 1)$, we have

   $$\underbrace{1 + 3 + \cdots + (2k - 1)} + \left[2(k + 1) - 1\right]$$

   $$\begin{aligned}
   &= k^2 + \left[2(k + 1) - 1\right] &&\text{We assume } S_k \text{ is true.} \\
   &= k^2 + 2k + 2 - 1 \\
   &= k^2 + 2k + 1 \\
   &= (k + 1)^2.
   \end{aligned}$$

   We have shown that for all natural numbers $k$, $S_k \rightarrow S_{k+1}$. This completes the induction step. It and the basis step tell us that the proof is complete.

   **Now Try Exercise 5.**

**EXAMPLE 2** Prove: For every natural number $n$,

$$\frac{1}{2} + \frac{1}{4} + \frac{1}{8} + \cdots + \frac{1}{2^n} = \frac{2^n - 1}{2^n}.$$

**Proof.** We first list $S_n$, $S_1$, $S_k$, and $S_{k+1}$.

$S_n$: $\quad \dfrac{1}{2} + \dfrac{1}{4} + \dfrac{1}{8} + \cdots + \dfrac{1}{2^n} = \dfrac{2^n - 1}{2^n}$

$S_1$: $\quad \dfrac{1}{2^1} = \dfrac{2^1 - 1}{2^1}$

$S_k$: $\quad \dfrac{1}{2} + \dfrac{1}{4} + \dfrac{1}{8} + \cdots + \dfrac{1}{2^k} = \dfrac{2^k - 1}{2^k}$

$S_{k+1}$: $\quad \dfrac{1}{2} + \dfrac{1}{4} + \dfrac{1}{8} + \cdots + \dfrac{1}{2^k} + \dfrac{1}{2^{k+1}} = \dfrac{2^{k+1} - 1}{2^{k+1}}$

1. *Basis step*: We show $S_1$ to be true as follows:

$$\frac{2^1 - 1}{2^1} = \frac{2 - 1}{2} = \frac{1}{2}.$$

2. *Induction step*: We let $k$ be any natural number. We assume $S_k$ to be true and try to show that it implies that $S_{k+1}$ is true. Now $S_k$ is

$$\frac{1}{2} + \frac{1}{4} + \frac{1}{8} + \cdots + \frac{1}{2^k} = \frac{2^k - 1}{2^k}.$$

We start with the left side of $S_{k+1}$. Since we assume $S_k$ is true, we can substitute

$$\frac{2^k - 1}{2^k} \quad \text{for} \quad \frac{1}{2} + \frac{1}{4} + \cdots + \frac{1}{2^k}.$$

We have

$$\underbrace{\frac{1}{2} + \frac{1}{4} + \frac{1}{8} + \cdots + \frac{1}{2^k}} + \frac{1}{2^{k+1}}$$

$$= \frac{2^k - 1}{2^k} + \frac{1}{2^{k+1}} = \frac{2^k - 1}{2^k} \cdot \frac{2}{2} + \frac{1}{2^{k+1}}$$

$$= \frac{(2^k - 1) \cdot 2 + 1}{2^{k+1}}$$

$$= \frac{2^{k+1} - 2 + 1}{2^{k+1}}$$

$$= \frac{2^{k+1} - 1}{2^{k+1}}.$$

We have shown that for all natural numbers $k$, $S_k \to S_{k+1}$. This completes the induction step. It and the basis step tell us that the proof is complete.

**Now Try Exercise 15.**

**EXAMPLE 3** Prove: For every natural number $n$, $n < 2^n$.

**Proof.** We first list $S_n$, $S_1$, $S_k$, and $S_{k+1}$.

$$S_n: \quad n < 2^n$$
$$S_1: \quad 1 < 2^1$$
$$S_k: \quad k < 2^k$$
$$S_{k+1}: \quad k + 1 < 2^{k+1}$$

1. *Basis step*: $S_1$, as listed, is true since $2^1 = 2$ and $1 < 2$.

2. *Induction step*: We let $k$ be any natural number. We assume $S_k$ to be true and try to show that it implies that $S_{k+1}$ is true. Now

$$k < 2^k \qquad \text{This is } S_k.$$
$$2k < 2 \cdot 2^k \qquad \text{Multiplying by 2 on both sides}$$
$$2k < 2^{k+1} \qquad \text{Adding exponents on the right}$$
$$k + k < 2^{k+1}. \qquad \text{Rewriting } 2k \text{ as } k + k$$

Since $k$ is any natural number, we know that $1 \le k$. Thus,

$$k + 1 \le k + k. \qquad \text{Adding } k \text{ on both sides of } 1 \le k$$

Putting the results $k + 1 \le k + k$ and $k + k < 2^{k+1}$ together gives us

$$k + 1 < 2^{k+1}. \qquad \text{This is } S_{k+1}.$$

We have shown that for all natural numbers $k$, $S_k \rightarrow S_{k+1}$. This completes the induction step. It and the basis step tell us that the proof is complete.

<kbd>Now Try Exercise 11.</kbd>

# 8.4 Exercise Set

*List the first five statements in the sequence that can be obtained from each of the following. Determine whether each of the statements is true or false.*

**1.** $n^2 < n^3$

**2.** $n^2 - n + 41$ is prime. Find a value for $n$ for which the statement is false.

**3.** A polygon of $n$ sides has $[n(n-3)]/2$ diagonals.

**4.** The sum of the angles of a polygon of $n$ sides is $(n-2) \cdot 180°$.

*Use mathematical induction to prove each of the following.*

**5.** $2 + 4 + 6 + \cdots + 2n = n(n+1)$

**6.** $4 + 8 + 12 + \cdots + 4n = 2n(n+1)$

**7.** $1 + 5 + 9 + \cdots + (4n - 3) = n(2n - 1)$

**8.** $3 + 6 + 9 + \cdots + 3n = \dfrac{3n(n+1)}{2}$

**9.** $2 + 4 + 8 + \cdots + 2^n = 2(2^n - 1)$

**10.** $2 \le 2^n$      **11.** $n < n + 1$

**12.** $3^n < 3^{n+1}$      **13.** $2n \le 2^n$

**14.** $\dfrac{1}{1 \cdot 2} + \dfrac{1}{2 \cdot 3} + \cdots + \dfrac{1}{n(n+1)} = \dfrac{n}{n+1}$

**15.** $\dfrac{1}{1 \cdot 2 \cdot 3} + \dfrac{1}{2 \cdot 3 \cdot 4} + \dfrac{1}{3 \cdot 4 \cdot 5} + \cdots$
$\qquad + \dfrac{1}{n(n+1)(n+2)} = \dfrac{n(n+3)}{4(n+1)(n+2)}$

**16.** If $x$ is any real number greater than 1, then for any natural number $n$, $x \le x^n$.

*The following formulas can be used to find sums of powers of natural numbers. Use mathematical induction to prove each formula.*

**17.** $1 + 2 + 3 + \cdots + n = \dfrac{n(n + 1)}{2}$

**18.** $1^2 + 2^2 + 3^2 + \cdots + n^2 = \dfrac{n(n + 1)(2n + 1)}{6}$

**19.** $1^3 + 2^3 + 3^3 + \cdots + n^3 = \dfrac{n^2(n + 1)^2}{4}$

**20.** $1^4 + 2^4 + 3^4 + \cdots + n^4$
$= \dfrac{n(n + 1)(2n + 1)(3n^2 + 3n - 1)}{30}$

*Use mathematical induction to prove each of the following.*

**21.** $\displaystyle\sum_{i=1}^{n} i(i + 1) = \dfrac{n(n + 1)(n + 2)}{3}$

**22.** $\left(1 + \dfrac{1}{1}\right)\left(1 + \dfrac{1}{2}\right)\left(1 + \dfrac{1}{3}\right) \cdots \left(1 + \dfrac{1}{n}\right)$
$= n + 1$

**23.** The sum of $n$ terms of an arithmetic sequence:
$a_1 + (a_1 + d) + (a_1 + 2d) + \cdots + [a_1 + (n - 1)d]$
$= \dfrac{n}{2}[2a_1 + (n - 1)d]$

## Skill Maintenance

*Solve.*

**24.** $2x - 3y = 1,$
$3x - 4y = 3$

**25.** *Investment.* Martin received $104 in simple interest one year from three investments. Part is invested at 1.5%, part at 2%, and part at 3%. The amount invested at 2% is twice the amount invested at 1.5%. There is $400 more invested at 3% than at 2%. Find the amount invested at each rate.

## Synthesis

*Use mathematical induction to prove each of the following.*

**26.** The sum of $n$ terms of a geometric sequence:
$$a_1 + a_1 r + a_1 r^2 + \cdots + a_1 r^{n-1} = \dfrac{a_1 - a_1 r^n}{1 - r}$$

**27.** $x + y$ is a factor of $x^{2n} - y^{2n}$.

*Prove each of the following using mathematical induction. Do the basis step for $n = 2$.*

**28.** For every natural number $n \geq 2$,
$2n + 1 < 3^n$.

**29.** For every natural number $n \geq 2$,
$\log_a(b_1 b_2 \cdots b_n)$
$= \log_a b_1 + \log_a b_2 + \cdots + \log_a b_n$.

*Prove each of the following for any complex numbers $z_1, z_2, \ldots, z_n$, where $i^2 = -1$ and $\bar{z}$ is the conjugate of $z$. (See Section 3.1.)*

**30.** $\overline{z^n} = \bar{z}^n$

**31.** $\overline{z_1 + z_2 + \cdots + z_n} = \bar{z_1} + \bar{z_2} + \cdots + \bar{z_n}$

**32.** *The Tower of Hanoi Problem.* There are three pegs on a board. On one peg are $n$ disks, each smaller than the one on which it rests. The problem is to move this pile of disks to another peg. The final order must be the same, but you can move only one disk at a time and can never place a larger disk on a smaller one.

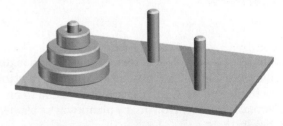

**a)** What is the *smallest* number of moves needed to move 3 disks? 4 disks? 2 disks? 1 disk?
**b)** Conjecture a formula for the *smallest* number of moves needed to move $n$ disks. Prove it by mathematical induction.

*Determine whether the statement is true or false.*

**1.** The general term of the sequence $1, -2, 3, -4, \ldots$ can be expressed as $a_n = n$. [8.1]

**2.** To find the common difference of an arithmetic sequence, choose any term except the first and then subtract the preceding term from it. [8.2]

**3.** The sequence $7, 3, -1, -5, \ldots$ is geometric. [8.2], [8.3]

**4.** If we can show that $S_k \rightarrow S_{k+1}$ for some natural number $k$, then we know that $S_n$ is true for all natural numbers $n$. [8.4]

*In each of the following, the nth term of a sequence is given. Find the first 4 terms, $a_9$, and $a_{14}$.*

**5.** $a_n = 3n + 5$ [8.1]

**6.** $a_n = (-1)^{n+1}(n - 1)$ [8.1]

*Predict the general term, or nth term, $a_n$, of the sequence. Answers may vary.*

**7.** $3, 6, 9, 12, 15, \ldots$ [8.1]

**8.** $-1, 4, -9, 16, -25, \ldots$ [8.1]

**9.** Find the partial sum $S_4$ for the sequence $1, \frac{1}{2}, \frac{1}{4}, \frac{1}{8}, \frac{1}{16} \ldots$ [8.1]

**10.** Find and evaluate the sum $\sum_{k=1}^{5} k(k + 1)$. [8.1]

**11.** Write sigma notation for the sum $-4 + 8 - 12 + 16 - 20 + \cdots$. [8.1]

**12.** Find the first 4 terms of the sequence defined by $a_1 = 2, a_{n+1} = 4a_n - 2$. [8.1]

**13.** Find the common difference of the arithmetic sequence $12, 7, 2, -3, \ldots$. [8.2]

**14.** Find the 10th term of the arithmetic sequence $4, 6, 8, 10, \ldots$. [8.2]

**15.** In the sequence in Exercise 14, what term is the number 44? [8.2]

**16.** Find the sum of the first 16 terms of the arithmetic series $6 + 11 + 16 + 21 + \cdots$. [8.2]

**17.** Find the common ratio of the geometric sequence $16, -8, 4, -2, 1, \ldots$. [8.3]

**18.** Find **(a)** the 8th term and **(b)** the sum of the first 10 terms of the geometric sequence $\frac{1}{16}, \frac{1}{8}, \frac{1}{4}, \frac{1}{2}, 1, \ldots$. [8.3]

*Find the sum, if it exists.*

**19.** $-8 + 4 - 2 + 1 - \cdots$ [8.3]

**20.** $\sum_{k=0}^{\infty} 5^k$ [8.3]

**21.** *Landscaping.* A landscaper is planting a triangular flower bed with 36 plants in the first row, 30 plants in the second row, 24 in the third row, and so on, for a total of 6 rows. How many plants will be planted in all? [8.2]

**22.** *Amount of an Annuity.* To save money for adding a bedroom to their home, at the end of each of 4 years the Davidsons deposit $1500 in an account that pays 4% interest, compounded annually. Find the total amount of the annuity. [8.3]

**23.** Prove: For every natural number $n$,
$$1 + 4 + 7 + \cdots + (3n - 2) = \tfrac{1}{2}n(3n - 1).$$
[8.4]

## Collaborative Discussion and Writing

**24.** The sum of the first $n$ terms of an arithmetic sequence can be given by

$$S_n = \frac{n}{2}\left[2a_1 + (n-1)d\right].$$

Compare this formula to

$$S_n = \frac{n}{2}(a_1 + a_n).$$

Discuss the reasons for the use of one formula over the other. [8.2]

**26.** Write a problem for a classmate to solve. Devise the problem so that a geometric series is involved and the solution is "The total amount in the bank is $900(1.08)^{40}$, or about \$19,552." [8.3]

**25.** It is said that as a young child, the mathematician Karl F. Gauss (1777–1855) was able to compute the sum $1 + 2 + 3 + \cdots + 100$ very quickly in his head to the amazement of a teacher. Explain how Gauss might have done this had he possessed some knowledge of arithmetic sequences and series. Then give a formula for the sum of the first $n$ natural numbers. [8.2]

**27.** Write an explanation of the idea behind mathematical induction for a fellow student. [8.4]

# Combinatorics: Permutations

## 8.5

- Evaluate factorial notation and permutation notation and solve related applied problems.

In order to study probability, it is first necessary to learn about **combinatorics**, the theory of counting.

## ■ Permutations

In this section, we will consider the part of combinatorics called *permutations*.

> The study of permutations involves *order* and *arrangements*.

**EXAMPLE 1** How many 3-letter code symbols can be formed with the letters A, B, C *without* repetition (that is, using each letter only once)?

**Solution** Consider placing the letters in these boxes.

We can select any of the 3 letters for the first letter in the symbol. Once this letter has been selected, the second must be selected from the

2 remaining letters. After this, the third letter is already determined, since only 1 possibility is left. That is, we can place any of the 3 letters in the first box, either of the remaining 2 letters in the second box, and the only remaining letter in the third box. The possibilities can be determined using a **tree diagram,** as shown below.

| TREE DIAGRAM | OUTCOMES | |
|---|---|---|

```
         B —— C       ABC
     A <
         C —— B       ACB        Each outcome
                                  represents one
         A —— C       BAC        permutation of the
     B <                         letters A, B, C.
         C —— A       BCA
         A —— B       CAB
     C <
         B —— A       CBA
```

We see that there are 6 possibilities. The set of all the possibilities is

{ABC, ACB, BAC, BCA, CAB, CBA}.

This is the set of all *permutations* of the letters A, B, C.

Suppose that we perform an experiment such as selecting letters (as in the preceding example), flipping a coin, or drawing a card. The results are called **outcomes.** An **event** is a set of outcomes. The following principle enables us to count actions that are combined to form an event.

---

**THE FUNDAMENTAL COUNTING PRINCIPLE**

Given a combined action, or *event*, in which the first action can be performed in $n_1$ ways, the second action can be performed in $n_2$ ways, and so on, the total number of ways in which the combined action can be performed is the product

$$n_1 \cdot n_2 \cdot n_3 \cdots \cdot n_k.$$

---

Thus, in Example 1, there are 3 choices for the first letter, 2 for the second letter, and 1 for the third letter, making a total of $3 \cdot 2 \cdot 1$, or 6 possibilities.

**EXAMPLE 2** How many 3-letter code symbols can be formed with the letters A, B, C, D, E *with* repetition (that is, allowing letters to be repeated)?

*Solution* Since repetition is allowed, there are 5 choices for the first letter, 5 choices for the second, and 5 for the third. Thus, by the fundamental counting principle, there are $5 \cdot 5 \cdot 5$, or 125 code symbols.

---

**PERMUTATION**

A **permutation** of a set of $n$ objects is an ordered arrangement of all $n$ objects.

---

We can use the fundamental counting principle to count the number of permutations of the objects in a set. Consider, for example, a set of 4 objects

$$\{A, B, C, D\}.$$

To find the number of ordered arrangements of the set, we select a first letter: There are 4 choices. Then we select a second letter: There are 3 choices. Then we select a third letter: There are 2 choices. Finally, there is 1 choice for the last selection. Thus, by the fundamental counting principle, there are $4 \cdot 3 \cdot 2 \cdot 1$, or 24, permutations of a set of 4 objects.

We can find a formula for the total number of permutations of all objects in a set of $n$ objects. We have $n$ choices for the first selection, $n - 1$ choices for the second, $n - 2$ for the third, and so on. For the $n$th selection, there is only 1 choice.

---

**THE TOTAL NUMBER OF PERMUTATIONS OF $n$ OBJECTS**

The total number of permutations of $n$ objects, denoted $_nP_n$, is given by

$$_nP_n = n(n - 1)(n - 2) \cdots 3 \cdot 2 \cdot 1.$$

---

**EXAMPLE 3** Find each of the following.

**a)** $_4P_4$                                     **b)** $_7P_7$

*Solution*

Start with 4.

**a)** $_4P_4 = \underbrace{4 \cdot 3 \cdot 2 \cdot 1}_{\text{4 factors}} = 24$

**b)** $_7P_7 = 7 \cdot 6 \cdot 5 \cdot 4 \cdot 3 \cdot 2 \cdot 1 = 5040$

We can also find the total number of permutations of $n$ objects using the $_nP_r$ operation from the MATH PRB (probability) menu on a graphing calculator.

**Now Try Exercise 1.**

**EXAMPLE 4** In how many ways can 9 packages be placed in 9 mailboxes, one package in a box?

*Solution* We have

$$_9P_9 = 9 \cdot 8 \cdot 7 \cdot 6 \cdot 5 \cdot 4 \cdot 3 \cdot 2 \cdot 1 = 362{,}880.$$

**Now Try Exercise 23.**

## ■ Factorial Notation

We will use products such as $7 \cdot 6 \cdot 5 \cdot 4 \cdot 3 \cdot 2 \cdot 1$ so often that it is convenient to adopt a notation for them. For the product

$$7 \cdot 6 \cdot 5 \cdot 4 \cdot 3 \cdot 2 \cdot 1,$$

we write 7!, read "7 factorial."

We now define factorial notation for natural numbers and for 0.

> **FACTORIAL NOTATION**
>
> For any natural number $n$,
>
> $$n! = n(n - 1)(n - 2) \cdots 3 \cdot 2 \cdot 1.$$
>
> For the number 0,
>
> $$0! = 1.$$

We define 0! as 1 so that certain formulas can be stated concisely and with a consistent pattern.

Here are some examples of factorial notation.

**GCM**

$$7! = 7 \cdot 6 \cdot 5 \cdot 4 \cdot 3 \cdot 2 \cdot 1 = 5040$$
$$6! = \quad 6 \cdot 5 \cdot 4 \cdot 3 \cdot 2 \cdot 1 = \quad 720$$
$$5! = \qquad 5 \cdot 4 \cdot 3 \cdot 2 \cdot 1 = \quad 120$$
$$4! = \qquad\quad 4 \cdot 3 \cdot 2 \cdot 1 = \qquad 24$$
$$3! = \qquad\qquad 3 \cdot 2 \cdot 1 = \qquad\quad 6$$
$$2! = \qquad\qquad\quad 2 \cdot 1 = \qquad\quad 2$$
$$1! = \qquad\qquad\qquad 1 = \qquad\quad 1$$
$$0! = \qquad\qquad\qquad 1 = \qquad\quad 1$$

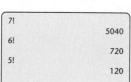

Factorial notation can also be evaluated using the ! operation from the MATH PRB (probability) menu, as shown at left.

We now see that the following statement is true.

> $$_nP_n = n!$$

We will often need to manipulate factorial notation. For example, note that

$$8! = 8 \cdot 7 \cdot 6 \cdot 5 \cdot 4 \cdot 3 \cdot 2 \cdot 1$$
$$= 8 \cdot (7 \cdot 6 \cdot 5 \cdot 4 \cdot 3 \cdot 2 \cdot 1) = 8 \cdot 7!.$$

Generalizing, we get the following.

> For any natural number $n$, $n! = n(n - 1)!$.

By using this result repeatedly, we can further manipulate factorial notation.

**EXAMPLE 5** Rewrite 7! with a factor of 5!.

*Solution* We have

$$7! = 7 \cdot 6! = 7 \cdot 6 \cdot 5!.$$

In general, we have the following.

> For any natural numbers $k$ and $n$, with $k < n$,
>
> $$n! = \underbrace{n(n-1)(n-2) \cdots [n-(k-1)]}_{k \text{ factors}} \cdot \underbrace{(n-k)!}_{n-k \text{ factors}}$$

## ■ Permutations of *n* Objects Taken *k* at a Time

Consider a set of 5 objects

$$\{A, B, C, D, E\}.$$

How many ordered arrangements can be formed using 3 objects from the set without repetition? Examples of such an arrangement are EBA, CAB, and BCD. There are 5 choices for the first object, 4 choices for the second, and 3 choices for the third. By the fundamental counting principle, there are

$$5 \cdot 4 \cdot 3,$$

or

60 *permutations* of a set of 5 objects taken 3 at a time.

Note that

$$5 \cdot 4 \cdot 3 = \frac{5 \cdot 4 \cdot 3 \cdot 2 \cdot 1}{2 \cdot 1}, \quad \text{or} \quad \frac{5!}{2!}.$$

> **PERMUTATION OF *n* OBJECTS TAKEN *k* AT A TIME**
>
> A **permutation** of a set of $n$ objects taken $k$ at a time is an ordered arrangement of $k$ objects taken from the set.

Consider a set of $n$ objects and the selection of an ordered arrangement of $k$ of them. There would be $n$ choices for the first object. Then there would remain $n - 1$ choices for the second, $n - 2$ choices for the third, and so on. We make $k$ choices in all, so there are $k$ factors in the product. By the fundamental counting principle, the total number of permutations is

$$\underbrace{n(n-1)(n-2) \cdots [n-(k-1)]}_{k \text{ factors}}.$$

We can express this in another way by multiplying by 1, as follows:

$$n(n-1)(n-2) \cdots [n-(k-1)] \cdot \frac{(n-k)!}{(n-k)!}$$

$$= \frac{n(n-1)(n-2) \cdots [n-(k-1)](n-k)!}{(n-k)!}$$

$$= \frac{n!}{(n-k)!}.$$

This gives us the following.

---

**THE NUMBER OF PERMUTATIONS OF *n* OBJECTS TAKEN *k* AT A TIME**

The number of permutations of a set of *n* objects taken *k* at a time, denoted $_nP_k$, is given by

$$_nP_k = \underbrace{n(n-1)(n-2) \cdots [n-(k-1)]}_{k \text{ factors}} \qquad (1)$$

$$= \frac{n!}{(n-k)!}. \qquad (2)$$

---

GCM **EXAMPLE 6** Compute $_8P_4$ using both forms of the formula.

*Solution* Using form (1), we have

The 8 tells where to start.

$$_8P_4 = \underbrace{8 \cdot 7 \cdot 6 \cdot 5}_{} = 1680.$$

The 4 tells how many factors.

Using form (2), we have

$$_8P_4 = \frac{8!}{(8-4)!}$$

$$= \frac{8!}{4!}$$

$$= \frac{8 \cdot 7 \cdot 6 \cdot 5 \cdot 4!}{4!} = \frac{8 \cdot 7 \cdot 6 \cdot 5 \cdot \cancel{4!}}{\cancel{4!}}$$

$$= 8 \cdot 7 \cdot 6 \cdot 5 = 1680.$$

```
8 nPr 4
              1680
```

We can also evaluate $_8P_4$ using the $_nP_r$ operation from the MATH PRB menu on a graphing calculator.

**Now Try Exercise 3.**

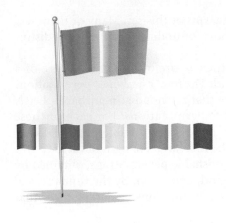

**EXAMPLE 7** *Flags of Nations.* The flags of many nations consist of three vertical stripes. For example, the flag of Ireland, shown here, has its first stripe green, second white, and third orange.

Suppose that the following 9 colors are available:

{black, yellow, red, blue, white, gold, orange, pink, purple}.

How many different flags of 3 colors can be made without repetition of colors in a flag? This assumes that the order in which the stripes appear is considered.

*Solution* We are determining the number of permutations of 9 objects taken 3 at a time. There is no repetition of colors. Using form (1), we get

$$_9P_3 = 9 \cdot 8 \cdot 7 = 504.$$ **Now Try Exercise 37(a).**

**EXAMPLE 8** *Batting Orders.* A baseball manager arranges the batting order as follows: The 4 infielders will bat first. Then the 3 outfielders, the catcher, and the pitcher will follow, not necessarily in that order. How many different batting orders are possible?

*Solution* The infielders can bat in $_4P_4$ different ways, the rest in $_5P_5$ different ways. Then by the fundamental counting principle, we have

$$_4P_4 \cdot {_5P_5} = 4! \cdot 5!, \quad \text{or} \quad 2880 \text{ possible batting orders.}$$ **Now Try Exercise 31.**

If we allow repetition, a situation like the following can occur.

**EXAMPLE 9** How many 5-letter code symbols can be formed with the letters A, B, C, and D if we allow a letter to occur more than once?

*Solution* We can select each of the 5 letters of the code in 4 ways. That is, we can select the first letter in 4 ways, the second in 4 ways, and so on. Thus there are $4^5$, or 1024 arrangements. **Now Try Exercise 37(b).**

> The number of distinct arrangements of $n$ objects taken $k$ at a time, allowing repetition, is $n^k$.

## ■ Permutations of Sets with Nondistinguishable Objects

Consider a set of 7 marbles, 4 of which are blue and 3 of which are red. When they are lined up, one red marble will look just like any other red marble. In this sense, we say that the red marbles are nondistinguishable and, similarly, the blue marbles are nondistinguishable.

We know that there are 7! permutations of this set. Many of them will look alike, however. We develop a formula for finding the number of distinguishable permutations.

Consider a set of $n$ objects in which $n_1$ are of one kind, $n_2$ are of a second kind, ..., and $n_k$ are of a $k$th kind. The total number of permutations of the set is $n!$, but this includes many that are nondistinguishable. Let $N$ be the total number of distinguishable permutations. For each of these $N$ permutations, there are $n_1!$ actual, nondistinguishable permutations, obtained by permuting the objects of the first kind. For each of these $N \cdot n_1!$ permutations, there are $n_2!$ nondistinguishable permutations, obtained by permuting the objects of the second kind, and so on. By the fundamental counting principle, the total number of permutations, including those that are nondistinguishable, is

$$N \cdot n_1! \cdot n_2! \cdots \cdots n_k!.$$

Then we have $N \cdot n_1! \cdot n_2! \cdots \cdots n_k! = n!$. Solving for $N$, we obtain

$$N = \frac{n!}{n_1! \cdot n_2! \cdots \cdots n_k!}.$$

Now, to finish our problem with the marbles, we have

$$N = \frac{7!}{4!\,3!}$$

$$= \frac{7 \cdot 6 \cdot 5 \cdot 4!}{4! \cdot 3 \cdot 2 \cdot 1} = \frac{7 \cdot 3 \cdot 2 \cdot 5 \cdot 4!}{4! \cdot 3 \cdot 2 \cdot 1}$$

$$= \frac{7 \cdot 5}{1}, \quad \text{or} \quad 35$$

distinguishable permutations of the marbles.

In general, we have the following.

For a set of $n$ objects in which $n_1$ are of one kind, $n_2$ are of another kind, ..., and $n_k$ are of a $k$th kind, the number of distinguishable permutations is

$$\frac{n!}{n_1! \cdot n_2! \cdots \cdots n_k!}.$$

**EXAMPLE 10** In how many distinguishable ways can the letters of the word CINCINNATI be arranged?

**Solution** There are 2 C's, 3 I's, 3 N's, 1 A, and 1 T for a total of 10 letters. Thus,

$$N = \frac{10!}{2! \cdot 3! \cdot 3! \cdot 1! \cdot 1!}, \quad \text{or} \quad 50,400.$$

The letters of the word CINCINNATI can be arranged in 50,400 distinguishable ways.

**Now Try Exercise 35.**

# 8.5   Exercise Set

*Evaluate.*

**1.** $_6P_6$

**2.** $_4P_3$

**3.** $_{10}P_7$

**4.** $_{10}P_3$

**5.** $5!$    **GCM**   **6.** $7!$

**7.** $0!$

**8.** $1!$

**9.** $\dfrac{9!}{5!}$   **GCM**

**10.** $\dfrac{9!}{4!}$

**11.** $(8-3)!$

**12.** $(8-5)!$

**13.** $\dfrac{10!}{7!\,3!}$

**14.** $\dfrac{7!}{(7-2)!}$

**15.** $_8P_0$

**16.** $_{13}P_1$

**17.** $_{52}P_4$

**18.** $_{52}P_5$

**19.** $_nP_3$

**20.** $_nP_2$

**21.** $_nP_1$

**22.** $_nP_0$

*In each of Exercises 23–41, give your answer using permutation notation, factorial notation, or other operations. Then evaluate.*

*How many permutations are there of the letters in each of the following words, if all the letters are used without repetition?*

**23.** CREDIT

**24.** FRUIT

**25.** EDUCATION

**26.** TOURISM

**27.** How many permutations are there of the letters of the word EDUCATION if the letters are taken 4 at a time?

**28.** How many permutations are there of the letters of the word TOURISM if the letters are taken 5 at a time?

**29.** How many 5-digit numbers can be formed using the digits 2, 4, 6, 8, and 9 without repetition? with repetition?

**30.** In how many ways can 7 athletes be arranged in a straight line?

**31.** *Program Planning.*   A program is planned to have 5 musical numbers and 4 speeches. In how many ways can this be done if a musical number and a speech are to alternate and a musical number is to come first?

**32.** A professor is going to grade her 24 students on a curve. She will give 3 A's, 5 B's, 9 C's, 4 D's, and 3 F's. In how many ways can she do this?

**33.** *Phone Numbers.*   How many 7-digit phone numbers can be formed with the digits 0, 1, 2, 3, 4, 5, 6, 7, 8, and 9, assuming that the first number cannot be 0 or 1? Accordingly, how many telephone numbers can there be within a given area code, before the area needs to be split with a new area code?

**34.** How many distinguishable code symbols can be formed from the letters of the word BUSINESS? BIOLOGY? MATHEMATICS?

**35.** Suppose the expression $a^2b^3c^4$ is rewritten without exponents. In how many distinguishable ways can this be done?

**36.** *Coin Arrangements.*   A penny, a nickel, a dime, and a quarter are arranged in a straight line.

**a)** Considering just the coins, in how many ways can they be lined up?

**b)** Considering the coins and heads and tails, in how many ways can they be lined up?

**37.** How many code symbols can be formed using 5 of the 6 letters A, B, C, D, E, F if the letters:

   **a)** are not repeated?
   **b)** can be repeated?
   **c)** are not repeated but must begin with D?
   **d)** are not repeated but must begin with DE?

**38.** *License Plates.* A state forms its license plates by first listing a number that corresponds to the county in which the owner of the car resides. (The names of the counties are alphabetized and the number is its location in that order.) Then the plate lists a letter of the alphabet, and this is followed by a number from 1 to 9999. How many such plates are possible if there are 80 counties?

**39.** *Zip Codes.* A U.S. postal zip code is a five-digit number.

   **a)** How many zip codes are possible if any of the digits 0 to 9 can be used?
   **b)** If each post office has its own zip code, how many possible post offices can there be?

**40.** *Zip-Plus-4 Codes.* A zip-plus-4 postal code uses a 9-digit number like 75247-5456. How many 9-digit zip-plus-4 postal codes are possible?

**41.** *Social Security Numbers.* A social security number is a 9-digit number like 243-47-0825.

   **a)** How many different social security numbers can there be?
   **b)** There are about 310 million people in the United States. Can each person have a unique social security number?

## Skill Maintenance

*Find the zero(s) of the function.*

**42.** $f(x) = 4x - 9$

**43.** $f(x) = x^2 + x - 6$

**44.** $f(x) = 2x^2 - 3x - 1$

**45.** $f(x) = x^3 - 4x^2 - 7x + 10$

## Synthesis

*Solve for n.*

**46.** $_nP_5 = 7 \cdot {_nP_4}$

**47.** $_nP_4 = 8 \cdot {_{n-1}P_3}$

**48.** $_nP_5 = 9 \cdot {_{n-1}P_4}$

**49.** $_nP_4 = 8 \cdot {_nP_3}$

**50.** Show that $n! = n(n-1)(n-2)(n-3)!$.

**51.** *Single-Elimination Tournaments.* In a single-elimination sports tournament consisting of $n$ teams, a team is eliminated when it loses one game. How many games are required to complete the tournament?

**52.** *Double-Elimination Tournaments.* In a double-elimination softball tournament consisting of $n$ teams, a team is eliminated when it loses two games. At most, how many games are required to complete the tournament?

## Combinatorics: Combinations

▪ Evaluate combination notation and solve related applied problems.

We now consider counting techniques in which order is not considered.

### ▪ Combinations

We sometimes make a selection from a set *without regard to order.* Such a selection is called a *combination.* If you play cards, for example, you know that in most situations the *order* in which you hold cards is not important. That is,

The hand

is "equivalent" to these hands.

Each hand contains the same combination of three cards.

**EXAMPLE 1** Find all the combinations of 3 letters taken from the set of 5 letters {A, B, C, D, E}.

*Solution* The combinations are

$$\{A, B, C\}, \quad \{A, B, D\},$$
$$\{A, B, E\}, \quad \{A, C, D\},$$
$$\{A, C, E\}, \quad \{A, D, E\},$$
$$\{B, C, D\}, \quad \{B, C, E\},$$
$$\{B, D, E\}, \quad \{C, D, E\}.$$

There are 10 combinations of the 5 letters taken 3 at a time.

When we find all the combinations from a set of 5 objects taken 3 at a time, we are finding all the 3-element subsets. When a set is named, the order of the elements is *not* considered. Thus,

$$\{A, C, B\} \quad \text{names the same set as} \quad \{A, B, C\}.$$

> ## COMBINATION; COMBINATION NOTATION
>
> A **combination** containing $k$ objects chosen from a set of $n$ objects, $k \leq n$, is denoted using **combination notation** $_nC_k$.

We want to derive a general formula for $_nC_k$ for any $k \leq n$. First, it is true that $_nC_n = 1$, because a set with $n$ objects has only 1 subset with $n$ objects, the set itself. Second, $_nC_1 = n$, because a set with $n$ objects has $n$ subsets with 1 element each. Finally, $_nC_0 = 1$, because a set with $n$ objects has only one subset with 0 elements, namely, the empty set $\varnothing$. To consider other possibilities, let's return to Example 1 and compare the number of combinations with the number of permutations.

| COMBINATIONS | | | PERMUTATIONS | | |
|---|---|---|---|---|---|

$_5C_3$ of these $\begin{cases} \{A, B, C\} \longrightarrow & ABC \quad BCA \quad CAB \quad CBA \quad BAC \quad ACB \\ \{A, B, D\} \longrightarrow & ABD \quad BDA \quad DAB \quad DBA \quad BAD \quad ADB \\ \{A, B, E\} \longrightarrow & ABE \quad BEA \quad EAB \quad EBA \quad BAE \quad AEB \\ \{A, C, D\} \longrightarrow & ACD \quad CDA \quad DAC \quad DCA \quad CAD \quad ADC \\ \{A, C, E\} \longrightarrow & ACE \quad CEA \quad EAC \quad ECA \quad CAE \quad AEC \\ \{A, D, E\} \longrightarrow & ADE \quad DEA \quad EAD \quad EDA \quad DAE \quad AED \\ \{B, C, D\} \longrightarrow & BCD \quad CDB \quad DBC \quad DCB \quad CBD \quad BDC \\ \{B, C, E\} \longrightarrow & BCE \quad CEB \quad EBC \quad ECB \quad CBE \quad BEC \\ \{B, D, E\} \longrightarrow & BDE \quad DEB \quad EBD \quad EDB \quad DBE \quad BED \\ \{C, D, E\} \longrightarrow & CDE \quad DEC \quad ECD \quad EDC \quad DCE \quad CED \end{cases}$ $3! \cdot {}_5C_3$ of these

Note that each combination of 3 objects yields 6, or 3!, permutations.

$$3! \cdot {}_5C_3 = 60 = {}_5P_3 = 5 \cdot 4 \cdot 3,$$

so

$$_5C_3 = \frac{{}_5P_3}{3!} = \frac{5 \cdot 4 \cdot 3}{3 \cdot 2 \cdot 1} = 10.$$

In general, the number of combinations of $n$ objects taken $k$ at a time, $_nC_k$, times the number of permutations of these objects, $k!$, must equal the number of permutations of $n$ objects taken $k$ at a time:

$$k! \cdot {}_nC_k = {}_nP_k$$

$$_nC_k = \frac{{}_nP_k}{k!} = \frac{1}{k!} \cdot {}_nP_k$$

$$= \frac{1}{k!} \cdot \frac{n!}{(n-k)!} = \frac{n!}{k!(n-k)!}.$$

**COMBINATIONS OF *n* OBJECTS TAKEN *k* AT A TIME**

The total number of combinations of *n* objects taken *k* at a time, denoted $_nC_k$, is given by

$$_nC_k = \frac{n!}{k!\,(n-k)!},$$  (1)

or

$$_nC_k = \frac{_nP_k}{k!} = \frac{n(n-1)(n-2)\,\cdots\,[n-(k-1)]}{k!}.$$  (2)

Another kind of notation for $_nC_k$ is **binomial coefficient notation.** The reason for such terminology will be seen later.

**BINOMIAL COEFFICIENT NOTATION**

$$\binom{n}{k} = {}_nC_k$$

You should be able to use either notation and either form of the formula.

GCM  **EXAMPLE 2**  Evaluate $\binom{7}{5}$, using forms (1) and (2).

*Solution*

**a)** By form (1),

$$\binom{7}{5} = \frac{7!}{5!\,(7-5)!} = \frac{7!}{5!\,2!}$$

$$= \frac{7 \cdot 6 \cdot 5!}{5! \cdot 2!} = \frac{7 \cdot 6 \cdot \cancel{5}!}{\cancel{5}! \cdot 2!} = \frac{7 \cdot 6}{2 \cdot 1} = 21.$$

**b)** By form (2),

The 7 tells where to start.

$$\binom{7}{5} = \frac{7 \cdot 6 \cdot 5 \cdot 4 \cdot 3}{5 \cdot 4 \cdot 3 \cdot 2 \cdot 1} = \frac{7 \cdot 6}{2 \cdot 1} = 21$$

The 5 tells how many factors there are in both the numerator and the denominator and where to start the denominator.

We can also find combinations using the $_nC_r$ operation from the MATH PRB (probability) menu on a graphing calculator.

```
7 nCr 5
                    21
```

Now Try Exercise 11.

Be sure to keep in mind that $\binom{n}{k}$ does not mean $n \div k$, or $n/k$.

**EXAMPLE 3**   Evaluate $\binom{n}{0}$ and $\binom{n}{2}$.

*Solution*   We use form (1) for the first expression and form (2) for the second. Then

$$\binom{n}{0} = \frac{n!}{0!(n-0)!} = \frac{n!}{1 \cdot n!} = 1,$$

using form (1), and

$$\binom{n}{2} = \frac{n(n-1)}{2!} = \frac{n(n-1)}{2}, \quad \text{or} \quad \frac{n^2 - n}{2},$$

using form (2).

**Now Try Exercise 19.**

Note that

$$\binom{7}{2} = \frac{7 \cdot 6}{2 \cdot 1} = 21.$$

Using the result of Example 2 gives us

$$\binom{7}{5} = \binom{7}{2}.$$

This says that the number of 5-element subsets of a set of 7 objects is the same as the number of 2-element subsets of a set of 7 objects. When 5 elements are chosen from a set, one also chooses *not* to include 2 elements. To see this, consider the set $\{A, B, C, D, E, F, G\}$:

$$\{B, G\}$$

$$\{A, B, C, D, E, F, G\}$$
$$\{A, C, D, E, F\}$$

Each time we form a subset with 5 elements, we leave behind a subset with 2 elements, and vice versa.

In general, we have the following. This result provides an alternative way to compute combinations.

**SUBSETS OF SIZE $k$ AND OF SIZE $n - k$**

$$\binom{n}{k} = \binom{n}{n-k} \quad \text{and} \quad {}_nC_k = {}_nC_{n-k}$$

The number of subsets of size $k$ of a set with $n$ objects is the same as the number of subsets of size $n - k$. The number of combinations of $n$ objects taken $k$ at a time is the same as the number of combinations of $n$ objects taken $n - k$ at a time.

We now solve problems involving combinations.

**EXAMPLE 4** *Michigan Lottery.* Run by the state of Michigan, Classic Lotto 47 is a twice-weekly lottery game with jackpots starting at $1 million. For a wager of $1, a player can choose 6 numbers from 1 through 47. If the numbers match those drawn by the state, the player wins. (*Source:* www.michigan.gov/lottery)

**a)** How many 6-number combinations are there?

**b)** Suppose it takes you 10 min to pick your numbers and buy a game ticket. How many tickets can you buy in 4 days?

**c)** How many people would you have to hire for 4 days to buy tickets with all the possible combinations and ensure that you win?

*Solution*

**a)** No order is implied here. You pick any 6 different numbers from 1 through 47. Thus the number of combinations is

$$_{47}C_6 = \binom{47}{6} = \frac{47!}{6!(47-6)!} = \frac{47!}{6!\,41!}$$
$$= \frac{47 \cdot 46 \cdot 45 \cdot 44 \cdot 43 \cdot 42}{6 \cdot 5 \cdot 4 \cdot 3 \cdot 2 \cdot 1} = 10{,}737{,}573.$$

**b)** First, we find the number of minutes in 4 days:

$$4 \text{ days} = 4 \text{ days} \cdot \frac{24 \text{ hr}}{1 \text{ day}} \cdot \frac{60 \text{ min}}{1 \text{ hr}} = 5760 \text{ min}.$$

Thus you could buy 5760/10, or 576 tickets in 4 days.

**c)** You would need to hire 10,737,573/576, or about 18,642 people, to buy tickets with all the possible combinations and ensure a win. (This presumes lottery tickets can be bought 24 hours a day.)

> **Now Try Exercise 23.**

**EXAMPLE 5** How many committees can be formed from a group of 5 governors and 7 senators if each committee consists of 3 governors and 4 senators?

*Solution* The 3 governors can be selected in $_5C_3$ ways and the 4 senators can be selected in $_7C_4$ ways. If we use the fundamental counting principle, it follows that the number of possible committees is

$$_5C_3 \cdot {}_7C_4 = \frac{5!}{3!\,2!} \cdot \frac{7!}{4!\,3!} = \frac{5 \cdot 4 \cdot 3!}{3! \cdot 2 \cdot 1} \cdot \frac{7 \cdot 6 \cdot 5 \cdot 4!}{4! \cdot 3 \cdot 2 \cdot 1}$$
$$= \frac{5 \cdot 2 \cdot 2 \cdot 3!}{3! \cdot 2 \cdot 1} \cdot \frac{7 \cdot 3 \cdot 2 \cdot 5 \cdot 4!}{4! \cdot 3 \cdot 2 \cdot 1}$$
$$= 10 \cdot 35$$
$$= 350.$$

> **Now Try Exercise 27.**

```
5 nCr 3*7 nCr 4
                        350
```

## CONNECTING THE CONCEPTS

### Permutations and Combinations

| PERMUTATIONS | COMBINATIONS |
|---|---|
| Permutations involve order and arrangements of objects. | Combinations do not involve order or arrangements of objects. |
| Given 5 books, we can arrange 3 of them on a shelf in $_5P_3$, or 60 ways. | Given 5 books, we can select 3 of them in $_5C_3$, or 10 ways. |
| Placing the books in different orders produces different arrangements. | The order in which the books are chosen does not matter. |

## 8.6 Exercise Set

*Evaluate.*

**1.** $_{13}C_2$

**2.** $_9C_6$

**3.** $\begin{pmatrix} 13 \\ 11 \end{pmatrix}$

**4.** $\begin{pmatrix} 9 \\ 3 \end{pmatrix}$

**5.** $\begin{pmatrix} 7 \\ 1 \end{pmatrix}$

**6.** $\begin{pmatrix} 8 \\ 8 \end{pmatrix}$

**7.** $\dfrac{_5P_3}{3!}$

**8.** $\dfrac{_{10}P_5}{5!}$

**9.** $\begin{pmatrix} 6 \\ 0 \end{pmatrix}$

**10.** $\begin{pmatrix} 6 \\ 1 \end{pmatrix}$

**11.** $\begin{pmatrix} 6 \\ 2 \end{pmatrix}$

**12.** $\begin{pmatrix} 6 \\ 3 \end{pmatrix}$

**13.** $\begin{pmatrix} 7 \\ 0 \end{pmatrix} + \begin{pmatrix} 7 \\ 1 \end{pmatrix} + \begin{pmatrix} 7 \\ 2 \end{pmatrix} + \begin{pmatrix} 7 \\ 3 \end{pmatrix} + \begin{pmatrix} 7 \\ 4 \end{pmatrix} + \begin{pmatrix} 7 \\ 5 \end{pmatrix} + \begin{pmatrix} 7 \\ 6 \end{pmatrix} + \begin{pmatrix} 7 \\ 7 \end{pmatrix}$

**14.** $\begin{pmatrix} 6 \\ 0 \end{pmatrix} + \begin{pmatrix} 6 \\ 1 \end{pmatrix} + \begin{pmatrix} 6 \\ 2 \end{pmatrix} + \begin{pmatrix} 6 \\ 3 \end{pmatrix} + \begin{pmatrix} 6 \\ 4 \end{pmatrix} + \begin{pmatrix} 6 \\ 5 \end{pmatrix} + \begin{pmatrix} 6 \\ 6 \end{pmatrix}$

**15.** $_{52}C_4$

**16.** $_{52}C_5$

**17.** $\begin{pmatrix} 27 \\ 11 \end{pmatrix}$

**18.** $\begin{pmatrix} 37 \\ 8 \end{pmatrix}$

**19.** $\begin{pmatrix} n \\ 1 \end{pmatrix}$

**20.** $\begin{pmatrix} n \\ 3 \end{pmatrix}$

**21.** $\begin{pmatrix} m \\ m \end{pmatrix}$

**22.** $\begin{pmatrix} t \\ 4 \end{pmatrix}$

*In each of the following exercises, give an expression for the answer using permutation notation, combination notation, factorial notation, or other operations. Then evaluate.*

**23.** *Fraternity Officers.* There are 23 students in a fraternity. How many sets of 4 officers can be selected?

**24.** *League Games.* How many games can be played in a 9-team sports league if each team plays all other teams once? twice?

**25.** *Test Options.* On a test, a student is to select 10 out of 13 questions. In how many ways can this be done?

**26.** *Senate Committees.* Suppose the Senate of the United States consists of 58 Republicans and 42 Democrats. How many committees can be formed consisting of 6 Republicans and 4 Democrats?

**27.** *Test Options.* Of the first 10 questions on a test, a student must answer 7. Of the second 5 questions, the student must answer 3. In how many ways can this be done?

**28.** *Lines and Triangles from Points.*   How many lines are determined by 8 points, no 3 of which are collinear? How many triangles are determined by the same points?

**29.** *Poker Hands.*   How many 5-card poker hands are possible with a 52-card deck?

**30.** *Bridge Hands.*   How many 13-card bridge hands are possible with a 52-card deck?

**31.** *Baskin-Robbins Ice Cream.*   Burt Baskin and Irv Robbins began making ice cream in 1945. Initially they developed 31 flavors—one for each day of the month. (*Source*: Baskin-Robbins)

  **a)** How many 2-dip cones are possible using the 31 original flavors if order of flavors is to be considered and no flavor is repeated?

  **b)** How many 2-dip cones are possible if order is to be considered and a flavor can be repeated?

  **c)** How many 2-dip cones are possible if order is not considered and no flavor is repeated?

## Skill Maintenance

*Solve.*

**32.** $3x - 7 = 5x + 10$

**33.** $2x^2 - x = 3$

**34.** $x^2 + 5x + 1 = 0$

**35.** $x^3 + 3x^2 - 10x = 24$

## Synthesis

**36.** *Full House.*   A full house in poker consists of three of a kind and a pair (two of a kind). How many full houses are there that consist of 3 aces and 2 queens? (See Section 8.8 for a description of a 52-card deck.)

**37.** *Flush.*   A flush in poker consists of a 5-card hand with all cards of the same suit. How many 5-card hands (flushes) are there that consist of all diamonds?

**38.** There are $n$ points on a circle. How many quadrilaterals can be inscribed with these points as vertices?

**39.** *League Games.*   How many games are played in a league with $n$ teams if each team plays each other team once? twice?

*Solve for n.*

**40.** $\dbinom{n}{3} = 2 \cdot \dbinom{n-1}{2}$     **41.** $\dbinom{n}{n-2} = 6$

**42.** Prove that

$$\dbinom{n}{k-1} + \dbinom{n}{k} = \dbinom{n+1}{k}$$

for any natural numbers $n$ and $k$, $k \leq n$.

## 8.7 The Binomial Theorem

- Expand a power of a binomial using Pascal's triangle or factorial notation.
- Find a specific term of a binomial expansion.
- Find the total number of subsets of a set of $n$ objects.

In this section, we consider ways of expanding a binomial $(a + b)^n$.

## ■ Binomial Expansions Using Pascal's Triangle

Consider the following expanded powers of $(a + b)^n$, where $a + b$ is any binomial and $n$ is a whole number. Look for patterns.

$$(a + b)^0 = 1$$
$$(a + b)^1 = a + b$$
$$(a + b)^2 = a^2 + 2ab + b^2$$
$$(a + b)^3 = a^3 + 3a^2b + 3ab^2 + b^3$$
$$(a + b)^4 = a^4 + 4a^3b + 6a^2b^2 + 4ab^3 + b^4$$
$$(a + b)^5 = a^5 + 5a^4b + 10a^3b^2 + 10a^2b^3 + 5ab^4 + b^5$$

Each expansion is a polynomial. There are some patterns to be noted.

1. There is one more term than the power of the exponent, $n$. That is, there are $n + 1$ terms in the expansion of $(a + b)^n$.

2. In each term, the sum of the exponents is $n$, the power to which the binomial is raised.

3. The exponents of $a$ start with $n$, the power of the binomial, and decrease to 0. The last term has no factor of $a$. The first term has no factor of $b$, so powers of $b$ start with 0 and increase to $n$.

4. The coefficients start at 1 and increase through certain values about "halfway" and then decrease through these same values back to 1.

Let's explore the coefficients further. Suppose that we want to find an expansion of $(a + b)^6$. The patterns we just noted indicate that there are 7 terms in the expansion:

$$a^6 + c_1a^5b + c_2a^4b^2 + c_3a^3b^3 + c_4a^2b^4 + c_5ab^5 + b^6.$$

How can we determine the value of each coefficient, $c_i$? We can do so in two ways. The first method involves writing the coefficients in a triangular array, as follows. This is known as **Pascal's triangle**:

$$
\begin{array}{ccccccccccccc}
(a+b)^0: & & & & & & 1 & & & & & & \\
(a+b)^1: & & & & & 1 & & 1 & & & & & \\
(a+b)^2: & & & & 1 & & 2 & & 1 & & & & \\
(a+b)^3: & & & 1 & & 3 & & 3 & & 1 & & & \\
(a+b)^4: & & 1 & & 4 & & 6 & & 4 & & 1 & & \\
(a+b)^5: & 1 & & 5 & & 10 & & 10 & & 5 & & 1 & \\
\end{array}
$$

There are many patterns in the triangle. Find as many as you can.

Perhaps you discovered a way to write the next row of numbers, given the numbers in the row above it. There are always 1's on the outside. Each remaining number is the sum of the two numbers above it. Let's try to find an expansion for $(a + b)^6$ by adding another row using the patterns we have discovered:

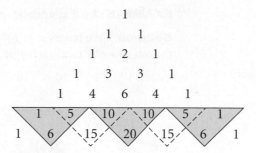

We see that in the last row

the 1st and last numbers are **1**;
the 2nd number is $1 + 5$, or **6**;
the 3rd number is $5 + 10$, or **15**;
the 4th number is $10 + 10$, or **20**;
the 5th number is $10 + 5$, or **15**; and
the 6th number is $5 + 1$, or **6**.

Thus the expansion for $(a + b)^6$ is

$$(a + b)^6 = 1a^6 + 6a^5b + 15a^4b^2 + 20a^3b^3 + 15a^2b^4 + 6ab^5 + 1b^6.$$

To find an expansion for $(a + b)^8$, we complete two more rows of Pascal's triangle:

```
                  1
               1     1
            1     2     1
         1     3     3     1
      1     4     6     4     1
   1     5    10    10     5     1
1     6    15    20    15     6     1
1  7    21    35    35    21     7     1
1  8    28    56    70    56    28     8     1
```

Thus the expansion of $(a + b)^8$ is

$$(a + b)^8 = a^8 + 8a^7b + 28a^6b^2 + 56a^5b^3 + 70a^4b^4 + 56a^3b^5$$
$$+ 28a^2b^6 + 8ab^7 + b^8.$$

We can generalize our results as follows.

**THE BINOMIAL THEOREM USING PASCAL'S TRIANGLE**

For any binomial $a + b$ and any natural number $n$,

$$(a + b)^n = c_0a^nb^0 + c_1a^{n-1}b^1 + c_2a^{n-2}b^2 + \cdots$$
$$+ c_{n-1}a^1b^{n-1} + c_na^0b^n,$$

where the numbers $c_0, c_1, c_2, \ldots, c_{n-1}, c_n$ are from the $(n + 1)$st row of Pascal's triangle.

**EXAMPLE 1**  Expand: $(u - v)^5$.

**Solution**  We have $(a + b)^n$, where $a = u$, $b = -v$, and $n = 5$. We use the 6th row of Pascal's triangle:

$$1 \quad 5 \quad 10 \quad 10 \quad 5 \quad 1$$

Then we have

$$\begin{aligned}
(u - v)^5 &= [u + (-v)]^5 \\
&= 1(u)^5 + 5(u)^4(-v)^1 + 10(u)^3(-v)^2 + 10(u)^2(-v)^3 \\
&\quad + 5(u)(-v)^4 + 1(-v)^5 \\
&= u^5 - 5u^4v + 10u^3v^2 - 10u^2v^3 + 5uv^4 - v^5.
\end{aligned}$$

Note that the signs of the terms alternate between $+$ and $-$. When the power of $-v$ is odd, the sign is $-$.

Now Try Exercise 5.

**EXAMPLE 2**  Expand: $\left(2t + \dfrac{3}{t}\right)^4$.

**Solution**  We have $(a + b)^n$, where $a = 2t$, $b = 3/t$, and $n = 4$. We use the 5th row of Pascal's triangle:

$$1 \quad 4 \quad 6 \quad 4 \quad 1$$

Then we have

$$\begin{aligned}
\left(2t + \frac{3}{t}\right)^4 &= 1(2t)^4 + 4(2t)^3\left(\frac{3}{t}\right)^1 + 6(2t)^2\left(\frac{3}{t}\right)^2 + 4(2t)^1\left(\frac{3}{t}\right)^3 + 1\left(\frac{3}{t}\right)^4 \\
&= 1(16t^4) + 4(8t^3)\left(\frac{3}{t}\right) + 6(4t^2)\left(\frac{9}{t^2}\right) + 4(2t)\left(\frac{27}{t^3}\right) + 1\left(\frac{81}{t^4}\right) \\
&= 16t^4 + 96t^2 + 216 + 216t^{-2} + 81t^{-4}.
\end{aligned}$$

Now Try Exercise 9.

## ▪ Binomial Expansion Using Combination Notation

Suppose that we want to find the expansion of $(a + b)^{11}$. The disadvantage in using Pascal's triangle is that we must compute all the preceding rows of the triangle to obtain the row needed for the expansion. The following method avoids this. It also enables us to find a specific term—say, the 8th term—without computing all the other terms of the expansion. This method is useful in such courses as finite mathematics, calculus, and statistics, and it uses the *binomial coefficient notation* $\dbinom{n}{k}$ developed in Section 8.6.

We can restate the binomial theorem as follows.

---

**THE BINOMIAL THEOREM USING COMBINATION NOTATION**

For any binomial $a + b$ and any natural number $n$,

$$(a + b)^n = \binom{n}{0}a^n b^0 + \binom{n}{1}a^{n-1}b^1 + \binom{n}{2}a^{n-2}b^2 + \cdots$$
$$+ \binom{n}{n-1}a^1 b^{n-1} + \binom{n}{n}a^0 b^n$$
$$= \sum_{k=0}^{n}\binom{n}{k}a^{n-k}b^k.$$

---

The binomial theorem can be proved by mathematical induction. (See Exercise 57.) This form shows why $\binom{n}{k}$ is called a *binomial coefficient*.

**EXAMPLE 3**   Expand: $(x^2 - 2y)^5$.

**Solution**   We have $(a + b)^n$, where $a = x^2$, $b = -2y$, and $n = 5$. Then using the binomial theorem, we have

$$(x^2 - 2y)^5 = \binom{5}{0}(x^2)^5 + \binom{5}{1}(x^2)^4(-2y) + \binom{5}{2}(x^2)^3(-2y)^2$$
$$+ \binom{5}{3}(x^2)^2(-2y)^3 + \binom{5}{4}x^2(-2y)^4 + \binom{5}{5}(-2y)^5$$
$$= \frac{5!}{0!\,5!}x^{10} + \frac{5!}{1!\,4!}x^8(-2y) + \frac{5!}{2!\,3!}x^6(4y^2) + \frac{5!}{3!\,2!}x^4(-8y^3)$$
$$+ \frac{5!}{4!\,1!}x^2(16y^4) + \frac{5!}{5!\,0!}(-32y^5)$$
$$= 1 \cdot x^{10} + 5x^8(-2y) + 10x^6(4y^2) + 10x^4(-8y^3)$$
$$+ 5x^2(16y^4) + 1 \cdot (-32y^5)$$
$$= x^{10} - 10x^8 y + 40x^6 y^2 - 80x^4 y^3 + 80x^2 y^4 - 32y^5.$$

> **Now Try Exercise 11.**

**EXAMPLE 4**   Expand: $\left(\dfrac{2}{x} + 3\sqrt{x}\right)^4$.

**Solution**   We have $(a + b)^n$, where $a = 2/x$, $b = 3\sqrt{x}$, and $n = 4$.

Then using the binomial theorem, we have

$$\left(\frac{2}{x} + 3\sqrt{x}\right)^4 = \binom{4}{0}\left(\frac{2}{x}\right)^4 + \binom{4}{1}\left(\frac{2}{x}\right)^3(3\sqrt{x}) + \binom{4}{2}\left(\frac{2}{x}\right)^2(3\sqrt{x})^2$$

$$+ \binom{4}{3}\left(\frac{2}{x}\right)(3\sqrt{x})^3 + \binom{4}{4}(3\sqrt{x})^4$$

$$= \frac{4!}{0!\,4!}\left(\frac{16}{x^4}\right) + \frac{4!}{1!\,3!}\left(\frac{8}{x^3}\right)(3x^{1/2})$$

$$+ \frac{4!}{2!\,2!}\left(\frac{4}{x^2}\right)(9x) + \frac{4!}{3!\,1!}\left(\frac{2}{x}\right)(27x^{3/2}) + \frac{4!}{4!\,0!}(81x^2)$$

$$= \frac{16}{x^4} + \frac{96}{x^{5/2}} + \frac{216}{x} + 216x^{1/2} + 81x^2.$$

> **Now Try Exercise 13.**

### ◾ Finding a Specific Term

Suppose that we want to determine only a particular term of an expansion. The method we have developed will allow us to find such a term without computing all the rows of Pascal's triangle or all the preceding coefficients.

Note that in the binomial theorem, $\binom{n}{0}a^n b^0$ gives us the 1st term, $\binom{n}{1}a^{n-1}b^1$ gives us the 2nd term, $\binom{n}{2}a^{n-2}b^2$ gives us the 3rd term, and so on. This can be generalized as follows.

---

**FINDING THE $(k + 1)$st TERM**

The $(k + 1)$st term of $(a + b)^n$ is $\binom{n}{k}a^{n-k}b^k$.

---

**EXAMPLE 5**   Find the 5th term in the expansion of $(2x - 5y)^6$.

***Solution***   First, we note that $5 = 4 + 1$. Thus, $k = 4$, $a = 2x$, $b = -5y$, and $n = 6$. Then the 5th term of the expansion is

$$\binom{6}{4}(2x)^{6-4}(-5y)^4, \quad \text{or} \quad \frac{6!}{4!\,2!}(2x)^2(-5y)^4, \quad \text{or} \quad 37{,}500x^2y^4.$$

> **Now Try Exercise 21.**

**EXAMPLE 6**   Find the 8th term in the expansion of $(3x - 2)^{10}$.

***Solution***   First, we note that $8 = 7 + 1$. Thus, $k = 7$, $a = 3x$, $b = -2$, and $n = 10$. Then the 8th term of the expansion is

$$\binom{10}{7}(3x)^{10-7}(-2)^7, \quad \text{or} \quad \frac{10!}{7!\,3!}(3x)^3(-2)^7, \quad \text{or} \quad -414{,}720x^3.$$

> **Now Try Exercise 27.**

## ■ Total Number of Subsets

Suppose that a set has $n$ objects. The number of subsets containing $k$ elements is $\binom{n}{k}$ by a result of Section 8.6. The total number of subsets of a set is the number of subsets with 0 elements, plus the number of subsets with 1 element, plus the number of subsets with 2 elements, and so on. The total number of subsets of a set with $n$ elements is

$$\binom{n}{0} + \binom{n}{1} + \binom{n}{2} + \cdots + \binom{n}{n}.$$

Now consider the expansion of $(1 + 1)^n$:

$$(1 + 1)^n = \binom{n}{0} \cdot 1^n + \binom{n}{1} \cdot 1^{n-1} \cdot 1^1 + \binom{n}{2} \cdot 1^{n-2} \cdot 1^2$$
$$+ \cdots + \binom{n}{n} \cdot 1^n$$
$$= \binom{n}{0} + \binom{n}{1} + \binom{n}{2} + \cdots + \binom{n}{n}.$$

Thus the total number of subsets is $(1 + 1)^n$, or $2^n$. We have proved the following.

---

**TOTAL NUMBER OF SUBSETS**

The total number of subsets of a set with $n$ elements is $2^n$.

---

**EXAMPLE 7** The set $\{A, B, C, D, E\}$ has how many subsets?

*Solution* The set has 5 elements, so the number of subsets is $2^5$, or 32.

    Now Try Exercise 31.

**EXAMPLE 8** Wendy's, a national restaurant chain, offers the following toppings for its hamburgers:

    {catsup, mustard, mayonnaise, tomato, lettuce, onions, pickle}.

How many different kinds of hamburgers can Wendy's serve, excluding size of hamburger or number of patties?

*Solution* The toppings on each hamburger are the elements of a subset of the set of all possible toppings, the empty set being a plain hamburger. The total number of possible hamburgers is

$$\binom{7}{0} + \binom{7}{1} + \binom{7}{2} + \cdots + \binom{7}{7} = 2^7 = 128.$$

Thus Wendy's serves hamburgers in 128 different ways.

    Now Try Exercise 33.

## 8.7 Exercise Set

*Expand.*

**1.** $(x + 5)^4$

**2.** $(x - 1)^4$

**3.** $(x - 3)^5$

**4.** $(x + 2)^9$

**5.** $(x - y)^5$

**6.** $(x + y)^8$

**7.** $(5x + 4y)^6$

**8.** $(2x - 3y)^5$

**9.** $\left(2t + \dfrac{1}{t}\right)^7$

**10.** $\left(3y - \dfrac{1}{y}\right)^4$

**11.** $(x^2 - 1)^5$

**12.** $(1 + 2q^3)^8$

**13.** $(\sqrt{5} + t)^6$

**14.** $(x - \sqrt{2})^6$

**15.** $\left(a - \dfrac{2}{a}\right)^9$

**16.** $(1 + 3)^n$

**17.** $(\sqrt{2} + 1)^6 - (\sqrt{2} - 1)^6$

**18.** $(1 - \sqrt{2})^4 + (1 + \sqrt{2})^4$

**19.** $(x^{-2} + x^2)^4$

**20.** $\left(\dfrac{1}{\sqrt{x}} - \sqrt{x}\right)^6$

*Find the indicated term of the binomial expansion.*

**21.** 3rd; $(a + b)^7$

**22.** 6th; $(x + y)^8$

**23.** 6th; $(x - y)^{10}$

**24.** 5th; $(p - 2q)^9$

**25.** 12th; $(a - 2)^{14}$

**26.** 11th; $(x - 3)^{12}$

**27.** 5th; $(2x^3 - \sqrt{y})^8$

**28.** 4th; $\left(\dfrac{1}{b^2} + \dfrac{b}{3}\right)^7$

**29.** Middle; $(2u - 3v^2)^{10}$

**30.** Middle two; $(\sqrt{x} + \sqrt{3})^5$

*Determine the number of subsets of each of the following.*

**31.** A set of 7 elements

**32.** A set of 6 members

**33.** The set of letters of the Greek alphabet, which contains 24 letters

**34.** The set of letters of the English alphabet, which contains 26 letters

**35.** What is the degree of $(x^5 + 3)^4$?

**36.** What is the degree of $(2 - 5x^3)^7$?

*Expand each of the following, where $i^2 = -1$.*

**37.** $(3 + i)^5$

**38.** $(1 + i)^6$

**39.** $(\sqrt{2} - i)^4$

**40.** $\left(\dfrac{\sqrt{3}}{2} - \dfrac{1}{2}i\right)^{11}$

**41.** Find a formula for $(a - b)^n$. Use sigma notation.

**42.** Expand and simplify:
$$\dfrac{(x + h)^{13} - x^{13}}{h}.$$

**43.** Expand and simplify:
$$\dfrac{(x + h)^n - x^n}{h}.$$
Use sigma notation.

### Skill Maintenance

*Given that $f(x) = x^2 + 1$ and $g(x) = 2x - 3$, find each of the following.*

**44.** $(f + g)(x)$

**45.** $(fg)(x)$

**46.** $(f \circ g)(x)$

**47.** $(g \circ f)(x)$

### Synthesis

*Solve for x.*

**48.** $\displaystyle\sum_{k=0}^{8} \binom{8}{k} x^{8-k} 3^k = 0$

**49.** $\displaystyle\sum_{k=0}^{4} \binom{4}{k} (-1)^k x^{4-k} 6^k = 81$

**50.** Find the ratio of the 4th term of
$$\left(p^2 - \dfrac{1}{2}\sqrt[3]{q}\right)^5$$
to the 3rd term.

**51.** Find the term of
$$\left(\sqrt[3]{x} - \dfrac{1}{\sqrt{x}}\right)^7$$
containing $1/x^{1/6}$.

**52.** *Money Combinations.* A money clip contains one each of the following bills: \$1, \$2, \$5, \$10, \$20, \$50, and \$100. How many different sums of money can be formed using the bills?

*Find the sum.*

**53.** $_{100}C_0 + {_{100}C_1} + \cdots + {_{100}C_{100}}$

**54.** $_{n}C_0 + {_{n}C_1} + \cdots + {_{n}C_n}$

*Simplify.*

**55.** $\displaystyle\sum_{k=0}^{23}\binom{23}{k}(\log_a x)^{23-k}(\log_a t)^k$

**56.** $\displaystyle\sum_{k=0}^{15}\binom{15}{k}i^{30-2k}$

**57.** Use mathematical induction and the property
$$\binom{n}{r-1} + \binom{n}{r} = \binom{n+1}{r}$$
to prove the binomial theorem.

# Probability

## 8.8

- Compute the probability of a simple event.

When a coin is tossed, we can reason that the chance, or likelihood, that it will fall heads is 1 out of 2, or the **probability** that it will fall heads is $\frac{1}{2}$. Of course, this does not mean that if a coin is tossed 10 times it will necessarily fall heads 5 times. If the coin is a "fair coin" and it is tossed a great many times, however, it will fall heads very nearly half of the time. Here we give an introduction to two kinds of probability, **experimental** and **theoretical**.

### ▪ Experimental Probability and Theoretical Probability

If we toss a coin a great number of times—say, 1000—and count the number of times it falls heads, we can determine the probability that it will fall heads. If it falls heads 503 times, we would calculate the probability of its falling heads to be

$$\frac{503}{1000}, \quad \text{or} \quad 0.503.$$

This is an **experimental** determination of probability. Such a determination of probability is discovered by the observation and study of data and is quite common and very useful. Here, for example, are some probabilities that have been determined *experimentally*:

**1.** The probability that a woman will get breast cancer in her lifetime is $\frac{1}{11}$.

**2.** If you kiss someone who has a cold, the probability of your catching a cold is 0.07.

**3.** A person who has just been released from prison has an 80% probability of returning.

If we consider a coin and reason that it is just as likely to fall heads as tails, we would calculate the probability that it will fall heads to be $\frac{1}{2}$. This is a **theoretical** determination of probability. Here are some other probabilities that have been determined *theoretically*, using mathematics:

1. If there are 30 people in a room, the probability that two of them have the same birthday (excluding year) is 0.706.

2. While on a trip, you meet someone and, after a period of conversation, discover that you have a common acquaintance. The typical reaction, "It's a small world!", is actually not appropriate, because the probability of such an occurrence is quite high—just over 22%.

In summary, experimental probabilities are determined by making observations and gathering data. Theoretical probabilities are determined by reasoning mathematically. Examples of experimental probability and theoretical probability like those above, especially those we do not expect, lead us to see the value of a study of probability. You might ask, "What is the *true* probability?" In fact, there is none. Experimentally, we can determine probabilities within certain limits. These may or may not agree with the probabilities that we obtain theoretically. There are situations in which it is much easier to determine one of these types of probabilities than the other. For example, it would be quite difficult to arrive at the probability of catching a cold using theoretical probability.

## ■ Computing Experimental Probabilities

We first consider experimental determination of probability. The basic principle we use in computing such probabilities is as follows.

> **PRINCIPLE *P* (EXPERIMENTAL)**
>
> Given an experiment in which $n$ observations are made, if a situation, or event, $E$ occurs $m$ times out of $n$ observations, then we say that the **experimental probability** of the event, $P(E)$, is given by
>
> $$P(E) = \frac{m}{n}.$$

**EXAMPLE 1**   *Television Ratings.*   There are an estimated 114,900,000 households in the United States that have at least one television. Each week, viewing information is collected and reported. One week 19,400,000 households tuned in to the 2010 World Cup soccer match between the United States and Ghana on ABC and Univision, and 9,700,000 households tuned in to the action series "NCIS" on CBS (*Source*: Nielsen Media Research).

What is the probability that a television household tuned in to the soccer match during the given week? to "NCIS"?

**Solution** The probability that a television household was tuned in to the soccer match is *P*, where

$$P = \frac{19,400,000}{114,900,000} \approx 0.169 \approx 16.9\%.$$

The probability that a television household was tuned in to "NCIS" is *P*, where

$$P = \frac{9,700,000}{114,900,000} \approx 0.084 \approx 8.4\%.$$

**Now Try Exercise 1.**

**EXAMPLE 2** *Sociological Survey.* The authors of this text conducted a survey to determine the number of people who are left-handed, right-handed, or both. The results are shown in the graph at left.

**a)** Determine the probability that a person is right-handed.

**b)** Determine the probability that a person is left-handed.

**c)** Determine the probability that a person is ambidextrous (uses both hands with equal ability).

**d)** There are 120 bowlers in most tournaments held by the Professional Bowlers Association. On the basis of the data in this experiment, how many of the bowlers would you expect to be left-handed?

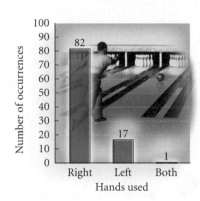

*Number of occurrences* — 82, 17, 1 — Right, Left, Both — Hands used

**Solution**

**a)** The number of people who are right-handed is 82, the number who are left-handed is 17, and the number who are ambidextrous is 1. The total number of observations is $82 + 17 + 1$, or 100. Thus the probability that a person is right-handed is *P*, where

$$P = \frac{82}{100}, \quad \text{or} \quad 0.82, \quad \text{or} \quad 82\%.$$

**b)** The probability that a person is left-handed is *P*, where

$$P = \frac{17}{100}, \quad \text{or} \quad 0.17, \quad \text{or} \quad 17\%.$$

c) The probability that a person is ambidextrous is $P$, where

$$P = \frac{1}{100}, \quad \text{or} \quad 0.01, \quad \text{or} \quad 1\%.$$

d) There are 120 bowlers, and from part (b) we can expect 17% to be left-handed. Since

$$17\% \text{ of } 120 = 0.17 \cdot 120 = 20.4,$$

we can expect that about 20 of the bowlers will be left-handed.

**Now Try Exercise 3.**

### ■ Theoretical Probability

Suppose that we perform an experiment such as flipping a coin, throwing a dart, drawing a card from a deck, or checking an item off an assembly line for quality. Each possible result of such an experiment is called an **outcome.** The set of all possible outcomes is called the **sample space.** An **event** is a set of outcomes, that is, a subset of the sample space.

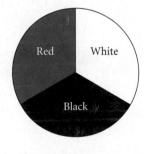

**EXAMPLE 3** *Dart Throwing.* Consider this dartboard. Assume that the experiment is "throwing a dart" and that the dart hits the board. Find each of the following.

**a)** The outcomes **b)** The sample space

*Solution*

**a)** The outcomes are *hitting black* (B), *hitting red* (R), and *hitting white* (W).

**b)** The sample space is $\{hitting\ black, hitting\ red, hitting\ white\}$, which can be simply stated as $\{B, R, W\}$.

**EXAMPLE 4** *Die Rolling.* A die (pl., dice) is a cube, with six faces, each containing a number of dots from 1 to 6.

Suppose a die is rolled. Find each of the following.

**a)** The outcomes **b)** The sample space

*Solution*

**a)** The outcomes are 1, 2, 3, 4, 5, 6.

**b)** The sample space is $\{1, 2, 3, 4, 5, 6\}$.

We denote the probability that an event $E$ occurs as $P(E)$. For example, "a coin falling heads" may be denoted $H$. Then $P(H)$ represents the probability of the coin falling heads. When all the outcomes of an experiment have the same probability of occurring, we say that they are *equally likely.*

To see the distinction between events that are equally likely and those that are not, consider the dartboards shown below.

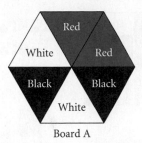

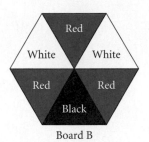

Board A                    Board B

For board A, the events *hitting black*, *hitting red*, and *hitting white* are equally likely, because the black, red, and white areas are the same. However, for board B the areas are not the same so these events are not equally likely.

---

### PRINCIPLE *P* (THEORETICAL)

If an event *E* can occur *m* ways out of *n* possible equally likely outcomes of a sample space *S*, then the **theoretical probability** of the event, $P(E)$, is given by

$$P(E) = \frac{m}{n}.$$

---

**EXAMPLE 5** Suppose that we select, without looking, one marble from a bag containing 3 red marbles and 4 green marbles. What is the probability of selecting a red marble?

*Solution* There are 7 equally likely ways of selecting any marble, and since the number of ways of getting a red marble is 3, we have

$$P(\text{selecting a red marble}) = \frac{3}{7}.$$

**Now Try Exercise 5(a).**

**EXAMPLE 6** What is the probability of rolling an even number on a die?

*Solution* The event is rolling an *even* number. It can occur 3 ways (rolling 2, 4, or 6). The number of equally likely outcomes is 6. By Principle *P*, we have

$$P(\text{even}) = \frac{3}{6}, \quad \text{or} \quad \frac{1}{2}.$$

**Now Try Exercise 7.**

We will use a number of examples related to a standard bridge deck of 52 cards. Such a deck is made up as shown in the following figure.

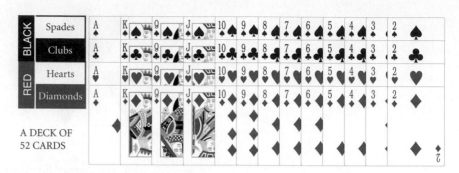

A DECK OF
52 CARDS

**EXAMPLE 7**    What is the probability of drawing an ace from a well-shuffled deck of cards?

*Solution*    There are 52 outcomes (the number of cards in the deck), they are equally likely (from a well-shuffled deck), and there are 4 ways to obtain an ace, so by Principle P, we have

$$P(\text{drawing an ace}) = \frac{4}{52}, \quad \text{or} \quad \frac{1}{13}.$$

**Now Try Exercise 9(a).**

The following are some results that follow from Principle P.

---

## PROBABILITY PROPERTIES

**a)** If an event $E$ cannot occur, then $P(E) = 0$.
**b)** If an event $E$ is certain to occur, then $P(E) = 1$.
**c)** The probability that an event $E$ will occur is a number from 0 to 1: $0 \leq P(E) \leq 1$.

---

For example, in coin tossing, the event that a coin will land on its edge has probability 0. The event that a coin falls either heads or tails has probability 1.

In the following examples, we use the combinatorics that we studied in Sections 8.5 and 8.6 to calculate theoretical probabilities.

**EXAMPLE 8**    Suppose that 2 cards are drawn from a well-shuffled deck of 52 cards. What is the probability that both of them are spades?

*Solution*    The number of ways $n$ of drawing 2 cards from a well-shuffled deck of 52 cards is $_{52}C_2$. Since 13 of the 52 cards are spades, the number of ways $m$ of drawing 2 spades is $_{13}C_2$. Thus,

| 13 nCr 2/52 nCr 2▶Frac |
|---|
| $\frac{1}{17}$ |

$$P(\text{drawing 2 spades}) = \frac{m}{n} = \frac{_{13}C_2}{_{52}C_2} = \frac{78}{1326} = \frac{1}{17}.$$

**Now Try Exercise 11.**

**EXAMPLE 9** Suppose that 3 people are selected at random from a group that consists of 6 men and 4 women. What is the probability that 1 man and 2 women are selected?

*Solution* The number of ways of selecting 3 people from a group of 10 is $_{10}C_3$. One man can be selected in $_6C_1$ ways, and 2 women can be selected in $_4C_2$ ways. By the fundamental counting principle, the number of ways of selecting 1 man and 2 women is $_6C_1 \cdot {}_4C_2$. Thus the probability that 1 man and 2 women are selected is

$$P = \frac{_6C_1 \cdot {}_4C_2}{_{10}C_3} = \frac{3}{10}.$$

**Now Try Exercise 13.**

**EXAMPLE 10** *Rolling Two Dice.* What is the probability of getting a total of 8 on a roll of a pair of dice?

*Solution* On each die, there are 6 possible outcomes. The outcomes are paired so there are $6 \cdot 6$, or 36, possible ways in which the two can fall. (Assuming that the dice are different—say, one red and one blue—can help in visualizing this.)

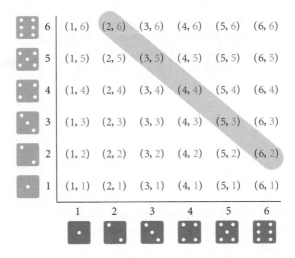

The pairs that total 8 are as shown in the figure above. There are 5 possible ways of getting a total of 8, so the probability is $\frac{5}{36}$.

**Now Try Exercise 19.**

## 8.8   Exercise Set

1. *Select a Number.* In a survey conducted by the authors, 100 people were polled and asked to select a number from 1 to 5. The results are shown in the table below.

| Number Chosen | 1 | 2 | 3 | 4 | 5 |
|---|---|---|---|---|---|
| Number Who Chose That Number | 18 | 24 | 23 | 23 | 12 |

   **a)** What is the probability that the number chosen is 1? 2? 3? 4? 5?

   **b)** What general conclusion might be made from the results of the experiment?

2. *Mason Dots®.* Made by the Tootsie Industries of Chicago, Illinois, Mason Dots® is a gumdrop candy. A box was opened by the authors and was found to contain the following number of gumdrops:

     Orange        9
     Lemon         8
     Strawberry    7
     Grape          6
     Lime           5
     Cherry        4

If we take one gumdrop out of the box, what is the probability of getting lemon? lime? orange? grape? strawberry? licorice?

3. *Junk Mail.* In experimental studies, the U.S. Postal Service has found that the probability that a piece of advertising is opened and read is 78%. A business sends out 15,000 pieces of advertising. How many of these can the company expect to be opened and read?

4. *Linguistics.* An experiment was conducted by the authors to determine the relative occurrence of various letters of the English alphabet. The front page of a newspaper was considered. In all, there were 9136 letters. The number of occurrences of each letter of the alphabet is listed in the table below.

| Letter | Number of Occurrences | Probability |
|---|---|---|
| A | 853 | 853/9136 ≈ 9.3% |
| B | 136 | |
| C | 273 | |
| D | 286 | |
| E | 1229 | |
| F | 173 | |
| G | 190 | |
| H | 399 | |
| I | 539 | |
| J | 21 | |
| K | 57 | |
| L | 417 | |
| M | 231 | |
| N | 597 | |
| O | 705 | |
| P | 238 | |
| Q | 4 | |
| R | 609 | |
| S | 745 | |
| T | 789 | |
| U | 240 | |
| V | 113 | |
| W | 127 | |
| X | 20 | |
| Y | 124 | |
| Z | 21 | 21/9136 ≈ 0.2% |

**a)** Complete the table of probabilities with the percentage, to the nearest tenth of a percent, of the occurrence of each letter.

**b)** What is the probability of a vowel occurring?

**c)** What is the probability of a consonant occurring?

**5.** *Marbles.* Suppose that we select, without looking, one marble from a bag containing 4 red marbles and 10 green marbles. What is the probability of selecting each of the following?

**a)** A red marble

**b)** A green marble

**c)** A purple marble

**d)** A red marble or a green marble

**6.** *Selecting Coins.* Suppose that we select, without looking, one coin from a bag containing 5 pennies, 3 dimes, and 7 quarters. What is the probability of selecting each of the following?

**a)** A dime

**b)** A quarter

**c)** A nickel

**d)** A penny, a dime, or a quarter

**7.** *Rolling a Die.* What is the probability of rolling a number less than 4 on a die?

**8.** *Rolling a Die.* What is the probability of rolling either a 1 or a 6 on a die?

**9.** *Drawing a Card.* Suppose that a card is drawn from a well-shuffled deck of 52 cards. What is the probability of drawing each of the following?

**a)** A queen

**b)** An ace or a 10

**c)** A heart

**d)** A black 6

**10.** *Drawing a Card.* Suppose that a card is drawn from a well-shuffled deck of 52 cards. What is the probability of drawing each of the following?

**a)** A 7

**b)** A jack or a king

**c)** A black ace

**d)** A red card

**11.** *Drawing Cards.* Suppose that 3 cards are drawn from a well-shuffled deck of 52 cards. What is the probability that they are all aces?

**12.** *Drawing Cards.* Suppose that 4 cards are drawn from a well-shuffled deck of 52 cards. What is the probability that they are all red?

**13.** *Production Unit.* The sales force of a business consists of 10 men and 10 women. A production unit of 4 people is set up at random. What is the probability that 2 men and 2 women are chosen?

**14.** *Coin Drawing.* A sack contains 7 dimes, 5 nickels, and 10 quarters. Eight coins are drawn at random. What is the probability of getting 4 dimes, 3 nickels, and 1 quarter?

*Five-Card Poker Hands.* Suppose that 5 cards are drawn from a deck of 52 cards. What is the probability of drawing each of the following?

**15.** 3 sevens and 2 kings

**16.** 5 aces

**17.** 5 spades

**18.** 4 aces and 1 five

**19.** *Tossing Three Coins.* Three coins are flipped. An outcome might be HTH.

**a)** Find the sample space.

What is the probability of getting each of the following?

**b)** Exactly one head

**c)** At most two tails

**d)** At least one head

**e)** Exactly two tails

*Roulette.* An American roulette wheel contains 38 slots numbered 00, 0, 1, 2, 3, . . . , 35, 36. Eighteen of the slots numbered 1–36 are colored red and 18 are colored black. The 00 and 0 slots are considered to be uncolored. The wheel is spun, and a ball is rolled around the rim until it falls into a slot. What is the probability that the ball falls in each of the following?

**20.** A red slot   **21.** A black slot

**22.** The 00 slot

**23.** The 0 slot

**24.** Either the 00 or the 0 slot

**25.** A red slot or a black slot

**26.** The number 24

**27.** An odd-numbered slot

**28.** *Dartboard.* The figure below shows a dartboard. A dart is thrown and hits the board. Find the probabilities

$$P(\text{red}), P(\text{green}), P(\text{blue}), P(\text{yellow}).$$

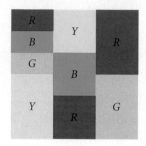

## Skill Maintenance

## Vocabulary Reinforcement

*In each of Exercises 29–36, fill in the blank with the correct term. Some of the given choices will be used more than once. Others will not be used.*

| | |
|---|---|
| range | zero |
| domain | *y*-intercept |
| function | one-to-one |
| an inverse function | rational |
| a composite function | permutation |
| direct variation | combination |
| inverse variation | arithmetic sequence |
| factor | geometric sequence |
| solution | |

**29.** A(n) _____ of a function is an input for which the output is 0.

**30.** A function is _____ if different inputs have different outputs.

**31.** A(n) _____ is a correspondence between a first set, called the _____, and a second set, called the _____, such that each member of the _____ corresponds to exactly one member of the _____.

**32.** The first coordinate of an *x*-intercept of a function is a(n) _____ of the function.

**33.** A selection made from a set without regard to order is a(n) _____.

**34.** If we have a function $f(x) = k/x$, where $k$ is a positive constant, we have _____.

**35.** For a polynomial function $f(x)$, if $f(c) = 0$, then $x - c$ is a(n) _____ of the polynomial.

**36.** We have $\dfrac{a_{n+1}}{a_n} = r$, for any integer $n \geq 1$, in a(n) _____.

## Synthesis

*Five-Card Poker Hands.* *Suppose that 5 cards are drawn from a deck of 52 cards. For the following exercises, give both a reasoned expression and an answer.*

**37.** *Two Pairs.* A hand with *two pairs* is a hand like Q-Q-3-3-A.

   **a)** How many are there?

   **b)** What is the probability of getting two pairs?

**38.** *Full House.* A *full house* consists of 3 of a kind and a pair such as Q-Q-Q-4-4.

   **a)** How many full houses are there?

   **b)** What is the probability of getting a full house?

**39.** *Three of a Kind.* A *three-of-a-kind* is a 5-card hand in which exactly 3 of the cards are of the same denomination and the other 2 are not a pair, such as Q-Q-Q-10-7.

   **a)** How many three-of-a-kind hands are there?

   **b)** What is the probability of getting three of a kind?

**40.** *Four of a Kind.* A *four-of-a-kind* is a 5-card hand in which 4 of the cards are of the same denomination, such as J-J-J-J-6, 7-7-7-7-A, or 2-2-2-2-5.

   **a)** How many four-of-a-kind hands are there?

   **b)** What is the probability of getting four of a kind?

## STUDY GUIDE

| KEY TERMS AND CONCEPTS | EXAMPLES |
|---|---|

### SECTION 8.1: SEQUENCES AND SERIES

An **infinite sequence** is a function having for its domain the set of positive integers $\{1, 2, 3, 4, 5, \ldots\}$.

A **finite sequence** is a function having for its domain a set of positive integers $\{1, 2, 3, 4, 5, \ldots, n\}$ for some positive integer $n$.

The first four terms of the sequence whose general term is given by $a_n = 3n + 2$ are

$$a_1 = 3 \cdot 1 + 2 = 5,$$
$$a_2 = 3 \cdot 2 + 2 = 8,$$
$$a_3 = 3 \cdot 3 + 2 = 11, \quad \text{and}$$
$$a_4 = 3 \cdot 4 + 2 = 14.$$

---

The sum of the terms of an infinite sequence is an **infinite series**. A **partial sum** is the sum of the first $n$ terms. It is also called a **finite series** or the **$n$th partial sum** and is denoted $S_n$.

For the sequence above, $S_4 = 5 + 8 + 11 + 14$, or 38. We can denote this sum using **sigma notation** as

$$\sum_{k=1}^{4}(3k + 2).$$

---

A sequence can be defined **recursively** by listing the first term, or the first few terms, and then using a **recursion formula** to determine the remaining terms from the given term.

The first four terms of the recursively defined sequence

$$a_1 = 3, \ a_{n+1} = (a_n - 1)^2$$

are

$$a_1 = 3,$$
$$a_2 = (a_1 - 1)^2 = (3 - 1)^2 = 4,$$
$$a_3 = (a_2 - 1)^2 = (4 - 1)^2 = 9, \quad \text{and}$$
$$a_4 = (a_3 - 1)^2 = (9 - 1)^2 = 64.$$

### SECTION 8.2: ARITHMETIC SEQUENCES AND SERIES

For an arithmetic sequence:

$$a_{n+1} = a_n + d;$$    **$d$ is the common difference.**

$$a_n = a_1 + (n - 1)d;$$    **The $n$th term**

$$S_n = \frac{n}{2}(a_1 + a_n).$$    **The sum of the first $n$ terms**

For the arithmetic sequence $5, 8, 11, 14, \ldots$ :

$$a_1 = 5;$$
$$d = 3 \quad (8 - 5 = 3, 11 - 8 = 3, \text{and so on});$$
$$a_6 = 5 + (6 - 1)3$$
$$= 5 + 15 = 20;$$
$$S_6 = \frac{6}{2}(5 + 20)$$
$$= 3(25) = 75.$$

## SECTION 8.3: GEOMETRIC SEQUENCES AND SERIES

For a geometric sequence:

$a_{n+1} = a_n r;$        $r$ is the common ratio.

$a_n = a_1 r^{n-1};$        The $n$th term

$S_n = \dfrac{a_1(1 - r^n)}{1 - r};$      The sum of the first $n$ terms

$S_\infty = \dfrac{a_1}{1 - r}, \quad |r| < 1.$    The limit, or sum, of an infinite geometric series

For the geometric sequence $12, -6, 3, -\frac{3}{2}, \ldots$ :

$a_1 = 12;$

$r = -\dfrac{1}{2} \quad \left( \dfrac{-6}{12} = -\dfrac{1}{2}, \dfrac{3}{-6} = -\dfrac{1}{2}, \text{and so on} \right);$

$a_6 = 12\left( -\dfrac{1}{2} \right)^{6-1}$

$\quad = 12\left( -\dfrac{1}{2^5} \right) = -\dfrac{3}{8};$

$S_6 = \dfrac{12\left[ 1 - \left( -\frac{1}{2} \right)^6 \right]}{1 - \left( -\frac{1}{2} \right)}$

$\quad = \dfrac{12\left( 1 - \frac{1}{64} \right)}{\frac{3}{2}} = \dfrac{63}{8};$

$|r| = \left| -\frac{1}{2} \right| = \frac{1}{2} < 1,$ so we have

$S_\infty = \dfrac{12}{1 - \left( -\frac{1}{2} \right)} = \dfrac{12}{\frac{3}{2}} = 8.$

## SECTION 8.4: MATHEMATICAL INDUCTION

**The Principle of Mathematical Induction**

We can prove an infinite sequence of statements $S_n$ by showing the following.

**(1)** *Basic step*: $S_1$ is true.

**(2)** *Induction step*: For all natural numbers $k$, $S_k \rightarrow S_{k+1}$.

See Examples 1–3 on pp. 658–660.

## SECTION 8.5: COMBINATORICS: PERMUTATIONS

**The Fundamental Counting Principle**

Given a combined action, or *event*, in which the first action can be performed in $n_1$ ways, the second action can be performed in $n_2$ ways, and so on, the total number of ways in which the combined action can be performed is the product

$\qquad n_1 \cdot n_2 \cdot n_3 \cdots \cdots n_k.$

The product $n(n - 1)(n - 2) \cdots 3 \cdot 2 \cdot 1$, for any natural number $n$, can also be written in **factorial notation** as $n!$. For the number 0, $0! = 1$.

The total number of permutations, or ordered arrangements, of $n$ objects, denoted $_nP_n$, is given by

$_nP_n = n(n - 1)(n - 2) \cdots 3 \cdot 2 \cdot 1, \quad \text{or} \quad n!.$

In how many ways can 7 books be arranged in a straight line?

We have

$\qquad _7P_7 = 7! = 7 \cdot 6 \cdot 5 \cdot 4 \cdot 3 \cdot 2 \cdot 1$
$\qquad\qquad = 5040.$

**The Number of Permutations of *n* Objects Taken *k* at a Time**

$$_nP_k = n(n-1)(n-2)\cdots[n-(k-1)] \qquad (1)$$

$$= \frac{n!}{(n-k)!} \qquad (2)$$

Compute: $_7P_4$.

Using form (1), we have

$$_7P_4 = 7 \cdot 6 \cdot 5 \cdot 4 = 840.$$

Using form (2), we have

$$_7P_4 = \frac{7!}{(7-4)!} = \frac{7!}{3!}$$

$$= \frac{7 \cdot 6 \cdot 5 \cdot 4 \cdot 3!}{3!} = 840.$$

---

The number of distinct arrangements of *n* objects taken *k* at a time, allowing repetition, is $n^k$.

The number of 4-number code symbols that can be formed with the numbers 5, 6, 7, 8, and 9, if we allow a number to occur more than once, is $5^4$, or 625.

---

For a set of *n* objects in which $n_1$ are of one kind, $n_2$ are of another kind, . . . , and $n_k$ are of a *k*th kind, the number of distinguishable permutations is

$$\frac{n!}{n_1! \cdot n_2! \cdot \cdots \cdot n_k!}.$$

Find the number of distinguishable code symbols that can be formed using the letters in the word MISSISSIPPI.

There are 1 M, 4 I's, 4 S's, and 2 P's, for a total of 11 letters, so we have

$$\frac{11!}{1!\,4!\,4!\,2!}, \quad \text{or} \quad 34{,}650.$$

---

## SECTION 8.6: COMBINATORICS: COMBINATIONS

**The Number of Combinations of *n* Objects Taken *k* at a Time**

$$_nC_k = \frac{n!}{k!\,(n-k)!} \qquad (1)$$

$$= \frac{_nP_k}{k!}$$

$$= \frac{n(n-1)(n-2)\cdots[n-(k-1)]}{k!} \qquad (2)$$

We can also use **binomial coefficient notation**:

$$\binom{n}{k} = {_nC_k}.$$

Compute: $\binom{6}{4}$, or $_6C_4$.

Using form (1), we have

$$\binom{6}{4} = \frac{6!}{4!\,(6-4)!} = \frac{6!}{4!\,2!}$$

$$= \frac{6 \cdot 5 \cdot 4!}{4!\,2!} = \frac{6 \cdot 5 \cdot 4!}{4! \cdot 2 \cdot 1} = 15.$$

Using form (2), we have

$$\binom{6}{4} = \frac{_6P_4}{4!} = \frac{6 \cdot 5 \cdot 4 \cdot 3}{4 \cdot 3 \cdot 2 \cdot 1} = 15.$$

**The Binomial Theorem Using Pascal's Triangle**

For any binomial $a + b$ and any natural number $n$,

$$(a + b)^n = c_0 a^n b^0 + c_1 a^{n-1} b^1 + c_2 a^{n-2} b^2$$
$$+ \cdots + c_{n-1} a^1 b^{n-1} + c_n a^0 b^n,$$

where the numbers $c_0, c_1, c_2, \ldots, c_{n-1}, c_n$ are from the $(n + 1)$st row of Pascal's triangle. (See Pascal's triangle on p. 680.)

Expand: $(x - 2)^4$.

We have $a = x$, $b = -2$, and $n = 4$. We use the fourth row of Pascal's triangle.

$$(x - 2)^4 = 1 \cdot x^4 + 4 \cdot x^3 (-2)^1$$
$$+ 6 \cdot x^2 (-2)^2 + 4 \cdot x^1 (-2)^3 + 1(-2)^4$$
$$= x^4 + 4x^3(-2) + 6x^2 \cdot 4 + 4x(-8) + 16$$
$$= x^4 - 8x^3 + 24x^2 - 32x + 16$$

---

**The Binomial Theorem Using Factorial Notation**

For any binomial $a + b$ and any natural number $n$,

$$(a + b)^n = \binom{n}{0} a^n b^0 + \binom{n}{1} a^{n-1} b^1$$
$$+ \binom{n}{2} a^{n-2} b^2 + \cdots$$
$$+ \binom{n}{n-1} a^1 b^{n-1} + \binom{n}{n} a^0 b^n$$
$$= \sum_{k=0}^{n} \binom{n}{k} a^{n-k} b^k.$$

Expand: $(x^2 + 3)^3$.

We have $a = x^2$, $b = 3$, and $n = 3$.

$$(x^2 + 3)^3 = \binom{3}{0}(x^2)^3 + \binom{3}{1}(x^2)^2(3)$$
$$+ \binom{3}{2}(x^2)3^2 + \binom{3}{3}3^3$$
$$= \frac{3!}{0!\,3!} x^6 + \frac{3!}{1!\,2!}(x^4)(3) + \frac{3!}{2!\,1!}(x^2)(9)$$
$$+ \frac{3!}{3!\,0!}(27)$$
$$= 1 \cdot x^6 + 3 \cdot 3x^4 + 3 \cdot 9x^2 + 1 \cdot 27$$
$$= x^6 + 9x^4 + 27x^2 + 27$$

---

The $(k + 1)$st term of $(a + b)^n$ is $\binom{n}{k} a^{n-k} b^k$.

The third term of $(x^2 + 3)^3$ is

$$\binom{3}{2}(x^2)^{3-2} \cdot 3^2 = 3 \cdot x^2 \cdot 9 = 27x^2. \qquad (k = 2)$$

---

The **total number of subsets** of a set with $n$ elements is $2^n$.

How many subsets does the set $\{W, X, Y, Z\}$ have?

The set has 4 elements, so we have

$$2^4, \quad \text{or} \quad 16.$$

---

**Principle $P$ (Experimental)**

Given an experiment in which $n$ observations are made, if a situation, or event, $E$ occurs $m$ times out of $n$ observations, then we say that the **experimental probability** of the event, $P(E)$, is given by

$$P(E) = \frac{m}{n}.$$

From a batch of 1000 gears, 35 were found to be defective. The probability that a defective gear is produced is

$$\frac{35}{1000} = 0.035, \quad \text{or} \quad 3.5\%.$$

**Principle P (Theoretical)**

If an event $E$ can occur $m$ ways out of $n$ possible equally likely outcomes of a sample space $S$, then the **theoretical probability** of the event, $P(E)$, is given by

$$P(E) = \frac{m}{n}.$$

What is the probability of drawing 2 red marbles and 1 green marble from a bag containing 5 red marbles, 6 green marbles, and 4 white marbles?

Number of ways of drawing 3 marbles from a bag of 15: $_{15}C_3$

Number of ways of drawing 2 red marbles from 5 red marbles: $_5C_2$

Number of ways of drawing 1 green marble from 6 green marbles: $_6C_1$

Probability that 2 red marbles and 1 green marble are drawn:

$$\frac{_5C_2 \cdot _6C_1}{_{15}C_3} = \frac{10 \cdot 6}{455} = \frac{12}{91}$$

# REVIEW EXERCISES

*Determine whether the statement is true or false.*

**1.** A sequence is a function.  [8.1]

**2.** An infinite geometric series with $r = -1$ has a limit.  [8.3]

**3.** Permutations involve order and arrangements of objects.  [8.5]

**4.** The total number of subsets of a set with $n$ elements is $n^2$.  [8.7]

**5.** Find the first 4 terms, $a_{11}$, and $a_{23}$:

$$a_n = (-1)^n \left( \frac{n^2}{n^4 + 1} \right). \ \ [8.1]$$

**6.** Predict the general, or $n$th, term. Answers may vary.

$$2, -5, 10, -17, 26, \ldots \ \ [8.1]$$

**7.** Find and evaluate:

$$\sum_{k=1}^{4} \frac{(-1)^{k+1}3^k}{3^k - 1}. \ \ [8.1]$$

**8.** Use a graphing calculator to construct a table of values and a graph for the first 10 terms of this sequence.

$$a_1 = 0.3, \quad a_{k+1} = 5a_k + 1 \ \ [8.1]$$

**9.** Write sigma notation. Answers may vary.

$$0 + 3 + 8 + 15 + 24 + 35 + 48 \ \ [8.1]$$

**10.** Find the 10th term of the arithmetic sequence

$$\tfrac{3}{4}, \tfrac{13}{12}, \tfrac{17}{12}, \ldots . \ \ [8.2]$$

**11.** Find the 6th term of the arithmetic sequence

$$a - b, a, a + b, \ldots . \ \ [8.2]$$

**12.** Find the sum of the first 18 terms of the arithmetic sequence

$$4, 7, 10, \ldots . \ \ [8.2]$$

**13.** Find the sum of the first 200 natural numbers.  [8.2]

**14.** The 1st term in an arithmetic sequence is 5, and the 17th term is 53. Find the 3rd term.  [8.2]

**15.** The common difference in an arithmetic sequence is 3. The 10th term is 23. Find the first term.  [8.2]

**16.** For a geometric sequence, $a_1 = -2$, $r = 2$, and $a_n = -64$. Find $n$ and $S_n$.  [8.3]

**17.** For a geometric sequence, $r = \tfrac{1}{2}$ and $S_5 = \tfrac{31}{2}$. Find $a_1$ and $a_5$.  [8.3]

*Find the sum of each infinite geometric series, if it exists.* [8.3]

**18.** $25 + 27.5 + 30.25 + 33.275 + \cdots$

**19.** $0.27 + 0.0027 + 0.000027 + \cdots$

**20.** $\frac{1}{2} - \frac{1}{6} + \frac{1}{18} - \cdots$

**21.** Find fraction notation for $2.\overline{43}$. [8.3]

**22.** Insert four arithmetic means between 5 and 9. [8.2]

**23.** *Bouncing Golfball.* A golfball is dropped from a height of 30 ft to the pavement. It always rebounds three-fourths of the distance that it drops. How far (up and down) will the ball have traveled when it hits the pavement for the 6th time? [8.3]

**24.** *The Amount of an Annuity.* To create a college fund, a parent makes a sequence of 18 yearly deposits of $2000 each in a savings account on which interest is compounded annually at 2.8%. Find the amount of the annuity. [8.3]

**25.** *Total Gift.* Suppose you receive 10¢ on the first day of the year, 12¢ on the 2nd day, 14¢ on the 3rd day, and so on.
   **a)** How much will you receive on the 365th day? [8.2]
   **b)** What is the sum of these 365 gifts? [8.2]

**26.** *The Economic Multiplier.* Suppose the government is making a $24,000,000,000 expenditure for travel to Mars. If 73% of this amount is spent again, and so on, what is the total effect on the economy? [8.3]

*Use mathematical induction to prove each of the following.* [8.4]

**27.** For every natural number $n$,
$$1 + 4 + 7 + \cdots + (3n - 2) = \frac{n(3n - 1)}{2}.$$

**28.** For every natural number $n$,
$$1 + 3 + 3^2 + \cdots + 3^{n-1} = \frac{3^n - 1}{2}.$$

**29.** For every natural number $n \geq 2$,
$$\left(1 - \frac{1}{2}\right)\left(1 - \frac{1}{3}\right) \cdots \left(1 - \frac{1}{n}\right) = \frac{1}{n}.$$

**30.** *Book Arrangements.* In how many ways can 6 books be arranged on a shelf? [8.5]

**31.** *Flag Displays.* If 9 different signal flags are available, how many different displays are possible using 4 flags in a row? [8.5]

**32.** *Prize Choices.* The winner of a contest can choose any 8 of 15 prizes. How many different sets of prizes can be chosen? [8.6]

**33.** *Fraternity–Sorority Names.* The Greek alphabet contains 24 letters. How many fraternity or sorority names can be formed using 3 different letters? [8.5]

**34.** *Letter Arrangements.* In how many distinguishable ways can the letters of the word TENNESSEE be arranged? [8.5]

**35.** *Floor Plans.* A manufacturer of houses has 1 floor plan but achieves variety by having 3 different roofs, 4 different ways of attaching the garage, and 3 different types of entrances. Find the number of different houses that can be produced. [8.5]

**36.** *Code Symbols.* How many code symbols can be formed using 5 out of 6 of the letters of G, H, I, J, K, L if the letters:
   **a)** cannot be repeated? [8.5]
   **b)** can be repeated? [8.5]
   **c)** cannot be repeated but must begin with K? [8.5]
   **d)** cannot be repeated but must end with IGH? [8.5]

**37.** Determine the number of subsets of a set containing 8 members. [8.7]

*Expand.* [8.7]

**38.** $(m + n)^7$

**39.** $(x - \sqrt{2})^5$

**40.** $(x^2 - 3y)^4$

**41.** $\left(a + \frac{1}{a}\right)^8$

**42.** $(1 + 5i)^6$, where $i^2 = -1$

**43.** Find the 4th term of $(a + x)^{12}$. [8.7]

**44.** Find the 12th term of $(2a - b)^{18}$. Do not multiply out the factorials. [8.7]

**45.** *Rolling Dice.* What is the probability of getting a 10 on a roll of a pair of dice? on a roll of 1 die? [8.8]

**46.** *Drawing a Card.* From a deck of 52 cards, 1 card is drawn at random. What is the probability that it is a club? [8.8]

**47.** *Drawing Three Cards.* From a deck of 52 cards, 3 are drawn at random without replacement. What is the probability that 2 are aces and 1 is a king? [8.8]

**48.** *Election Poll.* Three people were running for mayor in an election campaign. A poll was conducted to see which candidate was favored. During the polling, 86 favored candidate A, 97 favored B, and 23 favored C. Assuming that the poll is a valid indicator of the election results, what is the probability that the election will be won by A? B? C? [8.8]

**49.** *Consumption of American Cheese.* The table below lists the number of pounds of American cheese consumed per capita for selected years.

| Year, $n$ | American Cheese Consumed per Capita (in pounds) |
|---|---|
| 1930, 0 | 3.2 |
| 1950, 20 | 5.5 |
| 1970, 40 | 7.0 |
| 1990, 60 | 11.1 |
| 2008, 78 | 13.0 |

*Sources:* Economic Research Service; U.S. Department of Agriculture

  **a)** Find a linear sequence function $a_n = an + b$ that models the data. Let $n$ represent the number of years since 1930.

  **b)** Use the sequence found in part(a) to estimate the number of pounds of American cheese consumed per capita in 2010.

**50.** Which of the following is the 25th term of the arithmetic sequence 12, 10, 8, 6, . . . ? [8.2]

  **A.** $-38$         **B.** $-36$

  **C.** 32           **D.** 60

**51.** What is the probability of getting a total of 4 on a roll of a pair of dice? [8.8]

  **A.** $\frac{1}{12}$         **B.** $\frac{1}{9}$

  **C.** $\frac{1}{6}$         **D.** $\frac{5}{36}$

**52.** The graph of the sequence whose general term is $a_n = n - 1$ is which of the following? [8.1]

  **A.**

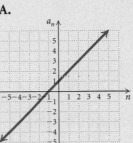

  **B.**

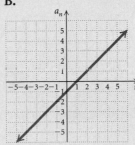

  **C.**

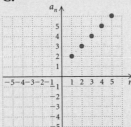

  **D.**

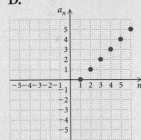

## Synthesis

**53.** Suppose that $a_1, a_2, \ldots, a_n$ and $b_1, b_2, \ldots, b_n$ are geometric sequences. Prove that $c_1, c_2, \ldots, c_n$ is a geometric sequence, where $c_n = a_n b_n$. [8.3]

**54.** Suppose that $a_1, a_2, \ldots, a_n$ is an arithmetic sequence. Is $b_1, b_2, \ldots, b_n$ an arithmetic sequence if:

  **a)** $b_n = |a_n|$? [8.2]     **b)** $b_n = a_n + 8$? [8.2]

  **c)** $b_n = 7a_n$? [8.2]     **d)** $b_n = \dfrac{1}{a_n}$? [8.2]

  **e)** $b_n = \log a_n$? [8.2]     **f)** $b_n = a_n^3$? [8.2]

**55.** The zeros of this polynomial function form an arithmetic sequence. Find them. [8.2]

$$f(x) = x^4 - 4x^3 - 4x^2 + 16x$$

**56.** Write the first 3 terms of the infinite geometric series with $r = -\frac{1}{3}$ and $S_\infty = \frac{3}{8}$. [8.3]

**57.** Simplify:

$$\sum_{k=0}^{10}(-1)^k\binom{10}{k}(\log x)^{10-k}(\log y)^k. \quad [8.6]$$

*Solve for n.* [8.6]

**58.** $\binom{n}{6} = 3 \cdot \binom{n-1}{5}$    **59.** $\binom{n}{n-1} = 36$

**60.** Solve for *a*:

$$\sum_{k=0}^{5}\binom{5}{k}9^{5-k}a^k = 0. \quad [8.7]$$

## Collaborative Discussion and Writing

**61.** How "long" is 15? Suppose you own 15 books and decide to make up all the possible arrangements of the books on a shelf. About how long, in years, would it take you if you were to make one arrangement per second? Write out the reasoning you used for this problem in the form of a paragraph. [8.5]

**62.** *Circular Arrangements.* In how many ways can the numbers on a clock face be arranged? See if you can derive a formula for the number of distinct circular arrangements of *n* objects. Explain your reasoning. [8.5]

**63.** Give an explanation that you might use with a fellow student to explain that

$$\binom{n}{k} = \binom{n}{n-k}. \quad [8.6]$$

**64.** Explain why a "combination" lock should really be called a "permutation" lock. [8.6]

**65.** Discuss the advantages and disadvantages of each method of finding a binomial expansion. Give examples of when you might use one method rather than the other. [8.7]

## Chapter 8 Test

**1.** For the sequence whose *n*th term is $a_n = (-1)^n(2n + 1)$, find $a_{21}$.

**2.** Find the first 5 terms of the sequence with general term

$$a_n = \frac{n+1}{n+2}.$$

**3.** Find and evaluate:

$$\sum_{k=1}^{4}(k^2 + 1).$$

**4.** Use a graphing calculator to construct a table of values and a graph for the first 10 terms of the sequence with general term

$$a_n = \frac{n+1}{n+2}.$$

*Write sigma notation. Answers may vary.*

**5.** $4 + 8 + 12 + 16 + 20 + 24$

**6.** $2 + 4 + 8 + 16 + 32 + \cdots$

**7.** Find the first 4 terms of the recursively defined sequence

$$a_1 = 3, \quad a_{n+1} = 2 + \frac{1}{a_n}.$$

**8.** Find the 15th term of the arithmetic sequence 2, 5, 8, . . . .

**9.** The 1st term of an arithmetic sequence is 8 and the 21st term is 108. Find the 7th term.

**10.** Find the sum of the first 20 terms of the series $17 + 13 + 9 + \cdots$ .

**11.** Find the sum: $\sum_{k=1}^{25}(2k + 1)$.

**12.** Find the 11th term of the geometric sequence $10, -5, \frac{5}{2}, -\frac{5}{4}, \ldots$ .

13. For a geometric sequence, $r = 0.2$ and $S_4 = 1248$. Find $a_1$.

*Find the sum, if it exists.*

14. $\displaystyle\sum_{k=1}^{8} 2^k$

15. $18 + 6 + 2 + \cdots$

16. Find fraction notation for $0.\overline{56}$.

17. *Salvage Value.* The value of an office machine is $10,000. Its salvage value each year is 80% of its value the year before. Give a sequence that lists the salvage value of the machine for each year of a 6-year period.

18. *Hourly Wage.* Tamika accepts a job, starting with an hourly wage of $8.50, and is promised a raise of 25¢ per hour every three months for 4 years. What will Tamika's hourly wage be at the end of the 4-year period?

19. *Amount of an Annuity.* To create a college fund, a parent makes a sequence of 18 equal yearly deposits of $2500 in a savings account on which interest is compounded annually at 5.6%. Find the amount of the annuity.

20. Use mathematical induction to prove that, for every natural number $n$,

$$2 + 5 + 8 + \cdots + (3n - 1) = \frac{n(3n + 1)}{2}.$$

*Evaluate.*

21. $_{15}P_6$

22. $_{21}C_{10}$

23. $\dbinom{n}{4}$

24. How many 4-digit numbers can be formed using the digits 1, 3, 5, 6, 7, and 9 without repetition?

25. How many code symbols can be formed using 4 of the 6 letters A, B, C, X, Y, Z if the letters:
    a) can be repeated?
    b) are not repeated and must begin with Z?

26. *Scuba Club Officers.* The Bay Woods Scuba Club has 28 members. How many sets of 4 officers can be selected from this group?

27. *Test Options.* On a test with 20 questions, a student must answer 8 of the first 12 questions and 4 of the last 8. In how many ways can this be done?

28. Expand: $(x + 1)^5$.

29. Find the 5th term of the binomial expansion $(x - y)^7$.

30. Determine the number of subsets of a set containing 9 members.

31. *Marbles.* Suppose we select, without looking, one marble from a bag containing 6 red marbles and 8 blue marbles. What is the probability of selecting a blue marble?

32. *Drawing Coins.* Ethan has 6 pennies, 5 dimes, and 4 quarters in his pocket. Six coins are drawn at random. What is the probability of getting 1 penny, 2 dimes, and 3 quarters?

33. The graph of the sequence whose general term is $a_n = 2n - 2$ is which of the following?

A.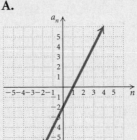

B.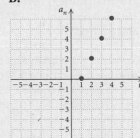

C.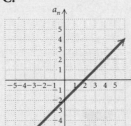

D.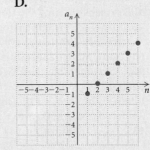

## Synthesis

34. Solve for $n$: $_nP_7 = 9 \cdot {_nP_6}$.

# Answers

## Chapter R

### Exercise Set R.1

1. $\frac{2}{3}, 6, -2.45, 18.\overline{4}, -11, \sqrt[3]{27}, 5\frac{1}{6}, -\frac{8}{7}, 0, \sqrt{16}$
3. $\sqrt{3}, \sqrt[6]{26}, 7.151551555\ldots, -\sqrt{35}, \sqrt[5]{3}$
5. $6, \sqrt[3]{27}, 0, \sqrt{16}$    7. $-11, 0$    9. $\frac{2}{3}, -2.45, 18.\overline{4}, 5\frac{1}{6}, -\frac{8}{7}$
11. $[-5, 5]$;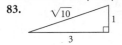
13. $(-3, -1]$;
15. $(-\infty, -2]$;
17. $(3.8, \infty)$;
19. $(7, \infty)$;    21. $(0, 5)$
23. $[-9, -4)$    25. $[x, x + h]$    27. $(p, \infty)$    29. True
31. False    33. True    35. False    37. False    39. True
41. True    43. True    45. False    47. Commutative
property of addition    49. Multiplicative identity property
51. Commutative property of multiplication
53. Commutative property of multiplication
55. Commutative property of addition
57. Multiplicative inverse property    59. 8.15    61. 295
63. $\sqrt{97}$    65. 0    67. $\frac{5}{4}$    69. 22    71. 6    73. 5.4
75. $\frac{21}{8}$    77. 7    79. Answers may vary; $0.124124412444\ldots$
81. Answers may vary; $-0.00999$
83.

### Exercise Set R.2

1. $\frac{1}{3^7}$    3. $\frac{y^4}{x^5}$    5. $\frac{t^6}{mn^{12}}$    7. 1    9. $z^7$    11. $5^2$, or 25

13. 1    15. $y^{-4}$, or $\frac{1}{y^4}$    17. $(x + 3)^2$    19. $3^6$, or 729

21. $6x^5$    23. $-15a^{-12}$, or $-\frac{15}{a^{12}}$    25. $-42x^{-1}y^{-4}$, or $-\frac{42}{xy^4}$

27. $432x^7$    29. $-200n^5$    31. $y^4$    33. $b^{-19}$, or $\frac{1}{b^{19}}$

35. $x^3y^{-3}$, or $\frac{x^3}{y^3}$    37. $8xy^{-5}$, or $\frac{8x}{y^5}$    39. $16x^8y^4$    41. $-32x^{15}$

43. $\frac{c^2d^4}{25}$    45. $432m^{-8}$, or $\frac{432}{m^8}$    47. $\frac{8x^{-9}y^{21}}{z^{-3}}$, or $\frac{8y^{21}z^3}{x^9}$

49. $2^{-5}a^{-20}b^{25}c^{-10}$, or $\frac{b^{25}}{32a^{20}c^{10}}$    51. $1.65 \times 10^7$

53. $4.37 \times 10^{-7}$    55. $2.346 \times 10^{11}$    57. $1.04 \times 10^{-3}$
59. $1.67 \times 10^{-27}$    61. 760,000    63. 0.000000109
65. 34,960,000,000    67. 0.0000000541    69. 231,900,000
71. $1.344 \times 10^6$    73. $2.21 \times 10^{-10}$    75. $8 \times 10^{-14}$
77. $2.5 \times 10^5$    79. About $6.737 \times 10^2$ pieces of trash
81. About $5.0667 \times 10^4$ people per square mile
83. $2.48136 \times 10^{13}$ mi    85. $1.332 \times 10^{14}$ disintegrations
87. 103    89. 2048    91. 5    93. $3647.28
95. $4704.84    97. $170,797.30    99. $419.67    101. $x^{8t}$
103. $t^{8x}$    105. $9x^{2a}y^{2b}$

### Exercise Set R.3

1. $7x^3, -4x^2, 8x, 5; 3$    3. $3a^4b, -7a^3b^3, 5ab, -2; 6$
5. $2ab^2 - 9a^2b + 6ab + 10$    7. $3x + 2y - 2z - 3$
9. $-2x^2 + 6x - 2$    11. $x^4 - 3x^3 - 4x^2 + 9x - 3$
13. $-21a^6$    15. $54x^5y^5$    17. $2a^4 - 2a^3b - a^2b + 4ab^2 - 3b^3$
19. $y^2 + 2y - 15$    21. $x^2 + 9x + 18$
23. $2a^2 + 13a + 15$    25. $4x^2 + 8xy + 3y^2$
27. $x^2 + 6x + 9$    29. $y^2 - 10y + 25$
31. $25x^2 - 30x + 9$    33. $4x^2 + 12xy + 9y^2$
35. $4x^4 - 12x^2y + 9y^2$    37. $n^2 - 36$    39. $9y^2 - 16$
41. $9x^2 - 4y^2$    43. $4x^2 + 12xy + 9y^2 - 16$    45. $x^4 - 1$
47. $a^{2n} - b^{2n}$    49. $a^{2n} + 2a^nb^n + b^{2n}$    51. $x^6 - 1$
53. $x^{a^2-b^2}$    55. $a^2 + b^2 + c^2 + 2ab + 2ac + 2bc$

### Exercise Set R.4

1. $3(x + 6)$    3. $2z^2(z - 4)$    5. $4(a^2 - 3a + 4)$
7. $(b - 2)(a + c)$    9. $(3x - 1)(x^2 + 6)$
11. $(y - 1)(y^2 + 2)$    13. $12(2x - 3)(x^2 + 3)$
15. $(x - 1)(x^2 - 5)$    17. $(a - 3)(a^2 - 2)$
19. $(w - 5)(w - 2)$    21. $(x + 1)(x + 5)$
23. $(t + 3)(t + 5)$    25. $(x + 3y)(x - 9y)$
27. $2(n - 12)(n + 2)$    29. $(y - 7)(y + 3)$
31. $y^2(y - 2)(y - 7)$    33. $2x(x + 3y)(x - 4y)$
35. $(2n - 7)(n + 8)$    37. $(3x + 2)(4x + 1)$
39. $(4x + 3)(x + 3)$    41. $(2y - 3)(y + 2)$
43. $(3a - 4b)(2a - 7b)$    45. $4(3a - 4)(a + 1)$
47. $(z + 9)(z - 9)$    49. $(4x + 3)(4x - 3)$
51. $6(x + y)(x - y)$    53. $4x(y^2 + z)(y^2 - z)$
55. $7p(q^2 + y^2)(q + y)(q - y)$    57. $(x + 6)^2$
59. $(3z - 2)^2$    61. $(1 - 4x)^2$    63. $a(a + 12)^2$
65. $4(p - q)^2$    67. $(x + 4)(x^2 - 4x + 16)$
69. $(m - 6)(m^2 + 6m + 36)$    71. $8(t + 1)(t^2 - t + 1)$

**73.** $3a^2(a - 2)(a^2 + 2a + 4)$    **75.** $(t^2 + 1)(t^4 - t^2 + 1)$
**77.** $3ab(6a - 5b)$    **79.** $(x - 4)(x^2 + 5)$
**81.** $8(x + 2)(x - 2)$    **83.** Prime    **85.** $(m + 3n)(m - 3n)$
**87.** $(x + 4)(x + 5)$    **89.** $(y - 5)(y - 1)$
**91.** $(2a + 1)(a + 4)$    **93.** $(3x - 1)(2x + 3)$
**95.** $(y - 9)^2$    **97.** $(3z - 4)^2$    **99.** $(xy - 7)^2$
**101.** $4a(x + 7)(x - 2)$    **103.** $3(z - 2)(z^2 + 2z + 4)$
**105.** $2ab(2a^2 + 3b^2)(4a^4 - 6a^2b^2 + 9b^4)$
**107.** $(y - 3)(y + 2)(y - 2)$    **109.** $(x - 1)(x^2 + 1)$
**111.** $5(m^2 + 2)(m^2 - 2)$    **113.** $2(x + 3)(x + 2)(x - 2)$
**115.** $(2c - d)^2$    **117.** $(m^3 + 10)(m^3 - 2)$
**119.** $p(1 - 4p)(1 + 4p + 16p^2)$    **121.** $(y^2 + 12)(y^2 - 7)$
**123.** $\left(y + \frac{4}{7}\right)\left(y - \frac{2}{7}\right)$    **125.** $\left(x + \frac{3}{2}\right)^2$    **127.** $\left(x - \frac{1}{2}\right)^2$
**129.** $h(3x^2 + 3xh + h^2)$    **131.** $(y + 4)(y - 7)$
**133.** $(x^n + 8)(x^n - 3)$    **135.** $(x + a)(x + b)$
**137.** $(5y^m + x^n - 1)(5y^m - x^n + 1)$
**139.** $y(y - 1)^2(y - 2)$

## Exercise Set R.5

**1.** 12    **3.** $-4$    **5.** 3    **7.** 10    **9.** 11    **11.** $-1$
**13.** $-12$    **15.** 2    **17.** $-1$    **19.** $\frac{18}{5}$    **21.** $-3$    **23.** 1
**25.** 0    **27.** $-\frac{1}{10}$    **29.** 5    **31.** $-\frac{3}{2}$    **33.** $\frac{20}{7}$    **35.** $-7, 4$
**37.** $-5, 0$    **39.** $-3$    **41.** 10    **43.** $-4, 8$    **45.** $-2, -\frac{2}{3}$
**47.** $-\frac{3}{4}, \frac{2}{3}$    **49.** $-\frac{4}{3}, \frac{7}{4}$    **51.** $-2, 7$    **53.** $-6, 6$    **55.** $-12, 12$
**57.** $-\sqrt{10}, \sqrt{10}$    **59.** $-\sqrt{3}, \sqrt{3}$    **61.** $b = \dfrac{2A}{h}$
**63.** $w = \dfrac{P - 2l}{2}$    **65.** $b_2 = \dfrac{2A - hb_1}{h}$, or $\dfrac{2A}{h} - b_1$
**67.** $\pi = \dfrac{3V}{4r^3}$    **69.** $C = \frac{5}{9}(F - 32)$    **71.** $A = \dfrac{C - By}{x}$
**73.** $h = \dfrac{p - l - 2w}{2}$    **75.** $y = \dfrac{2x - 6}{3}$    **77.** $b = \dfrac{a}{1 + cd}$
**79.** $x = \dfrac{z}{y - y^2}$    **81.** $\frac{23}{66}$    **83.** 8    **85.** $-\frac{6}{5}, -\frac{1}{4}, 0, \frac{2}{3}$
**87.** $-3, -2, 3$

## Exercise Set R.6

**1.** $\{x \mid x \text{ is a real number}\}$
**3.** $\{x \mid x \text{ is a real number } and \ x \neq 0 \ and \ x \neq 1\}$
**5.** $\{x \mid x \text{ is a real number } and \ x \neq -5 \ and \ x \neq 1\}$
**7.** $\{x \mid x \text{ is a real number } and \ x \neq -2 \ and \ x \neq 2 \ and \ x \neq -5\}$
**9.** $\dfrac{x + 2}{x - 2}$    **11.** $\dfrac{x - 3}{x}$    **13.** $\dfrac{2(y + 4)}{y - 1}$    **15.** $-\dfrac{1}{x + 8}$
**17.** 1    **19.** $\dfrac{(x - 5)(3x + 2)}{7x}$    **21.** $\dfrac{4x + 1}{3x - 2}$    **23.** $m + n$
**25.** $\dfrac{3(x - 4)}{2(x + 4)}$    **27.** $\dfrac{1}{x + y}$    **29.** $\dfrac{x - y - z}{x + y + z}$
**31.** $\dfrac{2}{x}$    **33.** 1    **35.** $\dfrac{7}{8z}$    **37.** $\dfrac{3x - 4}{(x + 2)(x - 2)}$
**39.** $\dfrac{-y + 10}{(y + 4)(y - 5)}$    **41.** $\dfrac{4x - 8y}{(x + y)(x - y)}$    **43.** $\dfrac{y - 2}{y - 1}$
**45.** $\dfrac{x + y}{2x - 3y}$    **47.** $\dfrac{3x - 4}{(x - 2)(x - 1)}$    **49.** $\dfrac{5a^2 + 10ab - 4b^2}{(a + b)(a - b)}$

**51.** $\dfrac{11x^2 - 18x + 8}{(2 + x)(2 - x)^2}$, or $\dfrac{11x^2 - 18x + 8}{(x + 2)(x - 2)^2}$    **53.** 0
**55.** $\dfrac{a}{a + b}$    **57.** $x - y$    **59.** $\dfrac{c^2 - 2c + 4}{c}$    **61.** $\dfrac{xy}{x - y}$
**63.** $\dfrac{a^2 - 1}{a^2 + 1}$    **65.** $\dfrac{3(x - 1)^2(x + 2)}{(x - 3)(x + 3)(-x + 10)}$    **67.** $\dfrac{1 + a}{1 - a}$
**69.** $\dfrac{b + a}{b - a}$    **71.** $2x + h$    **73.** $3x^2 + 3xh + h^2$    **75.** $x^5$
**77.** $\dfrac{(n + 1)(n + 2)(n + 3)}{2 \cdot 3}$    **79.** $\dfrac{x^3 + 2x^2 + 11x + 20}{2(x + 1)(2 + x)}$

## Exercise Set R.7

**1.** 21    **3.** $3|y|$    **5.** $|a - 2|$    **7.** $-3x$    **9.** $3x^2$    **11.** 2
**13.** $6\sqrt{5}$    **15.** $6\sqrt{2}$    **17.** $3\sqrt[3]{2}$    **19.** $8\sqrt{2}|c|d^2$
**21.** $2|x||y|\sqrt[4]{3x^2}$    **23.** $|x - 2|$    **25.** $5\sqrt{21}$    **27.** $4\sqrt{5}$
**29.** $2x^2y\sqrt{6}$    **31.** $3x\sqrt[3]{4y}$    **33.** $2(x + 4)\sqrt[3]{(x + 4)^2}$
**35.** $\dfrac{m^2n^3}{2}$    **37.** $\sqrt{5y}$    **39.** $\dfrac{1}{2x}$    **41.** $\dfrac{4a\sqrt[3]{a}}{3b}$    **43.** $\dfrac{x\sqrt{7x}}{6y^3}$
**45.** $17\sqrt{2}$    **47.** $4\sqrt{5}$    **49.** $-2x\sqrt{2} - 12\sqrt{5x}$
**51.** $-12$    **53.** $-9 - 5\sqrt{15}$    **55.** $27 - 10\sqrt{2}$
**57.** $11 - 2\sqrt{30}$    **59.** About 53.2 yd    **61.** (a) $h = \dfrac{a}{2}\sqrt{3}$;
(b) $A = \dfrac{a^2}{4}\sqrt{3}$    **63.** 8    **65.** $\dfrac{\sqrt{21}}{7}$    **67.** $\dfrac{\sqrt[3]{35}}{5}$
**69.** $\dfrac{2\sqrt[3]{6}}{3}$    **71.** $\sqrt{3} + 1$    **73.** $-\dfrac{\sqrt{6}}{6}$    **75.** $\dfrac{6\sqrt{m} + 6\sqrt{n}}{m - n}$
**77.** $\dfrac{10}{3\sqrt{2}}$    **79.** $\dfrac{2}{\sqrt[3]{20}}$    **81.** $\dfrac{11}{\sqrt{33}}$
**83.** $\dfrac{76}{27 + 3\sqrt{5} - 9\sqrt{3} - \sqrt{15}}$    **85.** $\dfrac{a - b}{3a\sqrt{a} - 3a\sqrt{b}}$
**87.** $\sqrt[6]{y^5}$    **89.** 8    **91.** $\frac{1}{5}$    **93.** $\dfrac{a\sqrt[4]{a}}{\sqrt[4]{b^3}}$, or $a\sqrt[4]{\dfrac{a}{b^3}}$
**95.** $mn^2\sqrt[3]{m^2n}$    **97.** $17^{3/5}$    **99.** $12^{4/5}$    **101.** $11^{1/6}$
**103.** $5^{5/6}$    **105.** 4    **107.** $8a^2$    **109.** $\dfrac{x^3}{3b^{-2}}$, or $\dfrac{x^3b^2}{3}$
**111.** $x\sqrt[3]{y}$    **113.** $n\sqrt[3]{mn^2}$    **115.** $a^{12}\sqrt{a^5} + a^2\sqrt[12]{a}$
**117.** $\sqrt[6]{288}$    **119.** $\sqrt[12]{x^{11}y^7}$    **121.** $a\sqrt[6]{a^5}$
**123.** $(a + x)\sqrt[12]{(a + x)^{11}}$    **125.** $\dfrac{(2 + x^2)\sqrt{1 + x^2}}{1 + x^2}$
**127.** $a^{a/2}$

## Review Exercises: Chapter R

**1.** True    **2.** False    **3.** True    **4.** True
**5.** $-7, 43, -\frac{4}{9}, 0, \sqrt[3]{64}, 4\frac{3}{4}, \frac{12}{7}, 102$    **6.** $43, 0, \sqrt[3]{64}, 102$
**7.** $-7, 43, 0, \sqrt[3]{64}, 102$    **8.** All of them    **9.** $43, \sqrt[3]{64}, 102$
**10.** $\sqrt{17}, 2.191191119\ldots, -\sqrt{2}, \sqrt[5]{5}$    **11.** $(-4, 7]$

**12.** 24   **13.** $\frac{7}{8}$   **14.** 10   **15.** 2   **16.** $-10$   **17.** 0.000083
**18.** 20,700,000   **19.** $4.05 \times 10^5$   **20.** $3.9 \times 10^{-7}$
**21.** $1.395 \times 10^3$   **22.** $7.8125 \times 10^{-22}$
**23.** $-12x^2y^{-4}$, or $\dfrac{-12x^2}{y^4}$   **24.** $8a^{-6}b^3c$, or $\dfrac{8b^3c}{a^6}$   **25.** 3

**26.** $-2$   **27.** $\dfrac{b}{a}$   **28.** $\dfrac{x+y}{xy}$   **29.** $-4$   **30.** $27 - 10\sqrt{2}$
**31.** $13\sqrt{5}$   **32.** $x^3 + t^3$   **33.** $10a^2 - 7ab - 12b^2$
**34.** $10x^2y + xy^2 + 2xy - 11$   **35.** $8x(4x^3 - 5y^3)$
**36.** $(y + 3)(y^2 - 2)$   **37.** $(x + 12)^2$
**38.** $x(9x - 1)(x + 4)$   **39.** $(3x - 5)^2$
**40.** $(2x - 1)(4x^2 + 2x + 1)$   **41.** $3(6x^2 - x + 2)$
**42.** $(x - 1)(2x + 3)(2x - 3)$   **43.** $6(x + 2)(x^2 - 2x + 4)$
**44.** $(ab - 3)(ab + 2)$   **45.** $(2x - 1)(x + 3)$   **46.** 7
**47.** $-1$   **48.** 3   **49.** $-\frac{1}{13}$   **50.** $-8$   **51.** $-4, 5$
**52.** $-6, \frac{1}{2}$   **53.** $-1, 3$   **54.** $-4, 4$   **55.** $-\sqrt{7}, \sqrt{7}$
**56.** $b = \dfrac{-2a + 10}{-5} = \dfrac{2a - 10}{5}$, or $\dfrac{2}{5}a - 2$   **57.** $x = \dfrac{3}{y-1}$

**58.** 3   **59.** $\dfrac{x - 5}{(x + 5)(x + 3)}$   **60.** $y^3\sqrt[6]{y}$   **61.** $\sqrt[3]{(a+b)^2}$

**62.** $b\sqrt[5]{b^2}$   **63.** $\dfrac{m^4n^2}{3}$   **64.** $\dfrac{23 - 9\sqrt{3}}{22}$

**65.** About 18.8 ft   **66.** B   **67.** C   **68.** \$553.67
**69.** \$606.92   **70.** $t^{2a} + 2 + t^{-2a}$
**71.** $a^{3n} - 3a^{2n}b^n + 3a^nb^{2n} - b^{3n}$
**72.** $(x^t - 7)(x^t + 4)$   **73.** $m^{3n}(m^n - 1)(m^{2n} + m^n + 1)$

## Test: Chapter R

**1.** [R.1] **(a)** 0, $\sqrt[3]{8}$, 29; **(b)** $\sqrt{12}$; **(c)** 0, $-5$;
**(d)** $6\frac{6}{7}$, $-\frac{13}{4}$, $-1.2$   **2.** [R.1] 17.6   **3.** [R.1] $\frac{15}{11}$   **4.** [R.1] 0
**5.** [R.1] $(-3, 6]$;

$\xleftarrow{\hspace{1em}}\overset{-3 \hspace{2em} 0 \hspace{3em} 6}{(\!-\!\!-\!\!-\!\!-\!\!-\!\!-\!\!-\!\!-\!\!]}\xrightarrow{\hspace{1em}}$

**6.** [R.1] 15   **7.** [R.2] $-5$   **8.** [R.2] $4.509 \times 10^6$
**9.** [R.2] 0.000086   **10.** [R.2] $7.5 \times 10^6$   **11.** [R.2] $x^{-3}$, or $\dfrac{1}{x^3}$

**12.** [R.2] $72y^{14}$   **13.** [R.2] $-15a^4b^{-1}$, or $-\dfrac{15a^4}{b}$

**14.** [R.3] $8xy^4 - 9xy^2 + 4x^2 + 2y - 7$
**15.** [R.3] $3y^2 - 2y - 8$   **16.** [R.3] $16x^2 - 24x + 9$
**17.** [R.6] $\dfrac{x - y}{xy}$   **18.** [R.7] $3\sqrt{5}$   **19.** [R.7] $2\sqrt[3]{7}$
**20.** [R.7] $21\sqrt{3}$   **21.** [R.7] $6\sqrt{5}$   **22.** [R.7] $4 + \sqrt{3}$
**23.** [R.4] $2(2x + 3)(2x - 3)$   **24.** [R.4] $(y + 3)(y - 6)$
**25.** [R.4] $(2n - 3)(n + 4)$   **26.** [R.4] $x(x + 5)^2$
**27.** [R.4] $(m - 2)(m^2 + 2m + 4)$   **28.** [R.5] 4
**29.** [R.5] $\frac{15}{4}$   **30.** [R.5] $-\frac{3}{2}, -1$   **31.** [R.5] $-\sqrt{11}, \sqrt{11}$
**32.** [R.5] $y = \dfrac{-8x + 24}{3}$, or $-\dfrac{8}{3}x + 8$   **33.** [R.6] $\dfrac{x - 5}{x - 2}$
**34.** [R.6] $\dfrac{x + 3}{(x + 1)(x + 5)}$   **35.** [R.7] $\dfrac{35 + 5\sqrt{3}}{46}$
**36.** [R.7] $\sqrt[8]{m^3}$   **37.** [R.7] $3^{5/6}$   **38.** [R.7] 13 ft
**39.** [R.3] $x^2 - 2xy + y^2 - 2x + 2y + 1$

# Chapter 1

## Visualizing the Graph

**1.** H   **2.** B   **3.** D   **4.** A   **5.** G   **6.** I   **7.** C
**8.** J   **9.** F   **10.** E

## Exercise Set 1.1

**1.** $A: (-5, 4)$; $B: (2, -2)$; $C: (0, -5)$;
$D: (3, 5)$; $E: (-5, -4)$; $F: (3, 0)$
**3.** 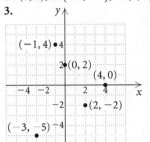   **5.**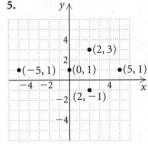

**7.** $(1970, 4.7\%)$, $(1980, 6.2\%)$, $(1990, 8.0\%)$,
$(2000, 10.4\%)$, $(2010, 12.8\%)$   **9.** Yes; no
**11.** Yes; no   **13.** No; yes   **15.** No; yes
**17.** $x$-intercept: $(-3, 0)$;   **19.** $x$-intercept: $(2, 0)$;
$y$-intercept: $(0, 5)$;   $y$-intercept: $(0, 4)$;

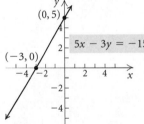   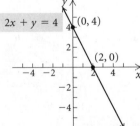

**21.** $x$-intercept: $(-4, 0)$;   **23.**
$y$-intercept: $(0, 3)$;

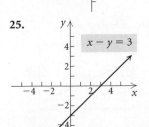   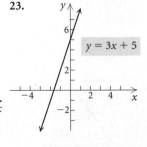

**25.**   **27.**

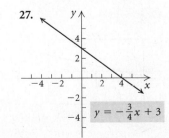

**29.**
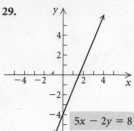
5x − 2y = 8

**31.**
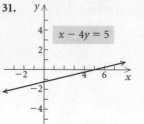
x − 4y = 5

**33.**
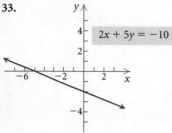
2x + 5y = −10

**35.**
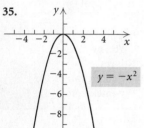
$y = -x^2$

**37.**
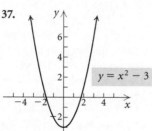
$y = x^2 - 3$

**39.**
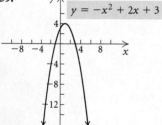
$y = -x^2 + 2x + 3$

**41.** (b)   **43.** (a)

**45.** $y = 2x + 1$
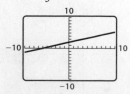

**47.** $4x + y = 7$
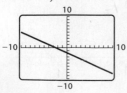

**49.** $y = \frac{1}{3}x + 2$

**51.** $2x + 3y = -5$

**53.** $y = x^2 + 6$
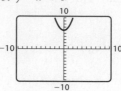

**55.** $y = 2 - x^2$
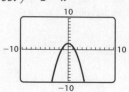

**57.** $y = x^2 + 4x - 2$
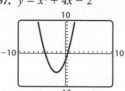

**59.** Standard window

**61.** $[-1, 1, -0.3, 0.3]$   **63.** $\sqrt{10}$, 3.162   **65.** 13
**67.** $\sqrt{45}$, 6.708   **69.** 16   **71.** $\frac{14}{3}$   **73.** $\sqrt{128.05}$, 11.316
**75.** $\sqrt{a^2 + b^2}$   **77.** 6.5   **79.** Yes   **81.** No
**83.** $(-4, -6)$   **85.** $\left(-\frac{1}{5}, \frac{1}{4}\right)$   **87.** $(4.95, -4.95)$
**89.** $\left(-6, \frac{13}{2}\right)$   **91.** $\left(-\frac{5}{12}, \frac{13}{40}\right)$
**93.**
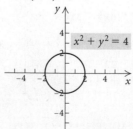

$\left(-\frac{1}{2}, \frac{3}{2}\right), \left(\frac{7}{2}, \frac{1}{2}\right), \left(\frac{5}{2}, \frac{9}{2}\right), \left(-\frac{3}{2}, \frac{11}{2}\right)$; no

**95.** $\left(\dfrac{\sqrt{7} + \sqrt{2}}{2}, -\dfrac{1}{2}\right)$   **97.** Square the window;
for example, use $[-12, 9, -4, 10]$.
**99.** $(x - 2)^2 + (y - 3)^2 = \frac{25}{9}$
**101.** $(x + 1)^2 + (y - 4)^2 = 25$
**103.** $(x - 2)^2 + (y - 1)^2 = 169$
**105.** $(x + 2)^2 + (y - 3)^2 = 4$
**107.** $(0, 0)$; 2;

$x^2 + y^2 = 4$

**109.** $(0, 3)$; 4;

$x^2 + (y - 3)^2 = 16$

**111.** $(1, 5)$; 6;

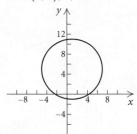

$$(x - 1)^2 + (y - 5)^2 = 36$$

**113.** $(-4, -5)$; 3;

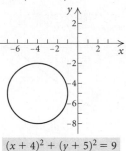

$$(x + 4)^2 + (y + 5)^2 = 9$$

**115.** $(x + 2)^2 + (y - 1)^2 = 3^2$
**117.** $(x - 5)^2 + (y + 5)^2 = 15^2$　**119.** Third
**121.** $\sqrt{h^2 + h + 2a - 2\sqrt{a^2 + ah}}$,
$\left(\dfrac{2a + h}{2}, \dfrac{\sqrt{a} + \sqrt{a + h}}{2}\right)$
**123.** $(x - 2)^2 + (y + 7)^2 = 36$　**125.** $(0, 4)$
**127. (a)** $(0, -3)$; **(b)** 5 ft　**129.** Yes　**131.** Yes
**133.** Let $P_1 = (x_1, y_1)$, $P_2 = (x_2, y_2)$, and
$M = \left(\dfrac{x_1 + x_2}{2}, \dfrac{y_1 + y_2}{2}\right)$. Let $d(AB)$ denote the distance
from point $A$ to point $B$.

$$d(P_1 M) = \sqrt{\left(\frac{x_1 + x_2}{2} - x_1\right)^2 + \left(\frac{y_1 + y_2}{2} - y_1\right)^2}$$
$$= \frac{1}{2}\sqrt{(x_2 - x_1)^2 + (y_2 - y_1)^2};$$
$$d(P_2 M) = \sqrt{\left(\frac{x_1 + x_2}{2} - x_2\right)^2 + \left(\frac{y_1 + y_2}{2} - y_2\right)^2}$$
$$= \frac{1}{2}\sqrt{(x_1 - x_2)^2 + (y_1 - y_2)^2}$$
$$= \frac{1}{2}\sqrt{(x_2 - x_1)^2 + (y_2 - y_1)^2} = d(P_1 M).$$

## Exercise Set 1.2

**1.** Yes　**3.** Yes　**5.** No　**7.** Yes　**9.** Yes　**11.** Yes
**13.** No　**15.** Function; domain: $\{2, 3, 4\}$; range: $\{10, 15, 20\}$
**17.** Not a function; domain: $\{-7, -2, 0\}$; range: $\{3, 1, 4, 7\}$
**19.** Function; domain: $\{-2, 0, 2, 4, -3\}$; range: $\{1\}$
**21. (a)** 1; **(b)** 6; **(c)** 22; **(d)** $3x^2 + 2x + 1$; **(e)** $3t^2 - 4t + 2$
**23. (a)** 8; **(b)** $-8$; **(c)** $-x^3$; **(d)** $27y^3$;
**(e)** $8 + 12h + 6h^2 + h^3$　**25. (a)** $\frac{1}{8}$; **(b)** 0; **(c)** does not exist;
**(d)** $\frac{81}{53}$, or approximately 1.5283; **(e)** $\dfrac{x + h - 4}{x + h + 3}$　**27.** 0; does

not exist; does not exist as a real number; $\dfrac{1}{\sqrt{3}}$, or $\dfrac{\sqrt{3}}{3}$

**29.** $g(-2.1) \approx -21.8$; $g(5.08) \approx -130.4$;
$g(10.003) \approx -468.3$

**31.**

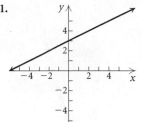

$$f(x) = \frac{1}{2}x + 3$$

**33.**

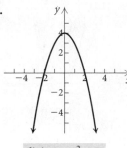

$$f(x) = -x^2 + 4$$

**35.**

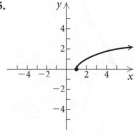

$$f(x) = \sqrt{x - 1}$$

**37.** $h(1) = -2$; $h(3) = 2$; $h(4) = 1$
**39.** $s(-4) = 3$; $s(-2) = 0$; $s(0) = -3$
**41.** $f(-1) = 2$; $f(0) = 0$; $f(1) = -2$
**43.** No　**45.** Yes　**47.** Yes　**49.** No
**51.** All real numbers, or $(-\infty, \infty)$
**53.** All real numbers, or $(-\infty, \infty)$
**55.** $\{x | x \neq 0\}$, or $(-\infty, 0) \cup (0, \infty)$
**57.** $\{x | x \neq 2\}$, or $(-\infty, 2) \cup (2, \infty)$
**59.** $\{x | x \neq -1 \text{ and } x \neq 5\}$, or $(-\infty, -1) \cup (-1, 5) \cup (5, \infty)$
**61.** All real numbers, or $(-\infty, \infty)$
**63.** $\{x | x \neq 0 \text{ and } x \neq 7\}$, or $(-\infty, 0) \cup (0, 7) \cup (7, \infty)$
**65.** All real numbers, or $(-\infty, \infty)$
**67.** Domain: $[0, 5]$; range: $[0, 3]$
**69.** Domain: $[-2\pi, 2\pi]$; range: $[-1, 1]$
**71.** Domain: $(-\infty, \infty)$; range: $\{-3\}$
**73.** Domain: $[-5, 3]$; range: $[-2, 2]$
**75.** Domain: $(-\infty, \infty)$; range: $[0, \infty)$
**77.** Domain: $(-\infty, \infty)$; range: $(-\infty, \infty)$
**79.** Domain: $(-\infty, 3) \cup (3, \infty)$; range: $(-\infty, 0) \cup (0, \infty)$
**81.** Domain: $(-\infty, \infty)$; range: $(-\infty, \infty)$
**83.** Domain: $(-\infty, 7]$; range: $[0, \infty)$
**85.** Domain: $(-\infty, \infty)$; range: $(-\infty, 3]$
**87. (a)** 2018: \$26.22; 2025: \$29.48; **(b)** about 63 years
after 1985, or in 2048　**89.** 645 m; 0 m
**90.** [1.1] $(-3, -2)$, yes; $(2, -3)$, no
**91.** [1.1] $(0, -7)$, no; $(8, 11)$, yes
**92.** [1.1] $\left(\frac{4}{5}, -2\right)$, yes; $\left(\frac{11}{5}, \frac{1}{10}\right)$, yes

**93.** [1.1]

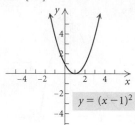

$y = (x-1)^2$

**94.** [1.1]

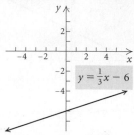

$y = \frac{1}{3}x - 6$

**95.** [1.1]

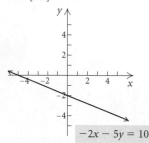

$-2x - 5y = 10$

**96.** [1.1]

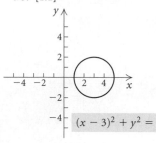

$(x-3)^2 + y^2 = 4$

**97.** $\left[-\frac{5}{2}, \infty\right)$   **99.** $[-6, -2) \cup (-2, 3) \cup (3, \infty)$
**101.** $f(x) = x, g(x) = x + 1$   **103.** $-7$

## Visualizing the Graph

**1.** E   **2.** D   **3.** A   **4.** J   **5.** C   **6.** F   **7.** H
**8.** G   **9.** B   **10.** I

## Exercise Set 1.3

**1. (a)** Yes;  **(b)** yes;  **(c)** yes   **3. (a)** Yes;  **(b)** no;  **(c)** no
**5.** $\frac{6}{5}$   **7.** $-\frac{3}{5}$   **9.** 0   **11.** $\frac{1}{5}$   **13.** Not defined   **15.** 0.3
**17.** 0   **19.** $-\frac{6}{5}$   **21.** $-\frac{1}{3}$   **23.** Not defined   **25.** $-2$
**27.** 5   **29.** 0   **31.** 1.3   **33.** Not defined   **35.** $-\frac{1}{2}$
**37.** $-1$   **39.** 0   **41.** The average rate of change in U.S.
fireworks revenue from 1998 to 2009 was about $47.3 million
per year.   **43.** The average rate of change in the population
of St. Louis, Missouri, over the 10-year period was $-2890$
people per year.   **45.** The average rate of change in the sales
of electric bikes in China from 2009 to 2012 was 1.3 million
bikes per year.   **47.** The average rate of change in the consumption of broccoli per capita from 1990 to 2008 was about
0.13 lb per year.   **49.** $\frac{3}{5}$; $(0, -7)$   **51.** Slope is not defined;
there is no $y$-intercept.   **53.** $-\frac{1}{2}$; $(0, 5)$   **55.** $-\frac{3}{2}$; $(0, 5)$
**57.** 0; $(0, -6)$   **59.** $\frac{4}{5}$; $\left(0, \frac{8}{5}\right)$   **61.** $\frac{1}{4}$; $\left(0, -\frac{1}{2}\right)$
**63.**

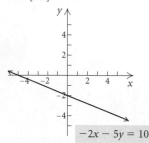

$y = -\frac{1}{2}x - 3$

**65.**

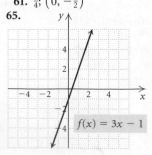

$f(x) = 3x - 1$

**67.**

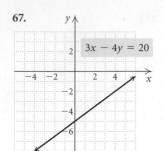

$3x - 4y = 20$

**69.**

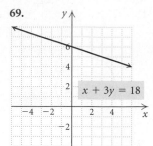

$x + 3y = 18$

**71. (a)** $y = \frac{1}{33}x + 1$

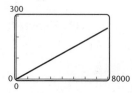

**(b)** 1 atm, 2 atm, $31\frac{10}{33}$ atm, $152\frac{17}{33}$ atm, $213\frac{4}{33}$ atm
**73. (a)** $\frac{11}{10}$. For each mile per hour faster that the
car travels, it takes $\frac{11}{10}$ ft longer to stop;
**(b)** $y = \frac{11}{10}x + \frac{1}{2}$

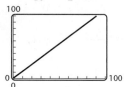

**(c)** 6 ft, 11.5 ft, 22.5 ft, 55.5 ft, 72 ft;  **(d)** $\{r \mid r > 0\}$,
or $(0, \infty)$. If $r$ is allowed to be 0, the function says
that a stopped car has a reaction distance of $\frac{1}{2}$ ft.
**75.** $C(t) = 89 + 114.99t$; $C(24) = \$2848.76$
**77.** $C(x) = 750 + 15x$; $C(32) = \$1230$
**79.** [1.2] $-\frac{5}{4}$   **80.** [1.2] 10   **81.** [1.2] 40
**82.** [1.2] $a^2 + 3a$   **83.** [1.2] $a^2 + 2ah + h^2 - 3a - 3h$
**85.** $2a + h$   **87.** False   **89.** $f(x) = x + b$

## Mid-Chapter Mixed Review: Chapter 1

**1.** False   **2.** True   **3.** False   **4.** $x$-intercept: $(5, 0)$;
$y$-intercept: $(0, -8)$   **5.** $\sqrt{605} = 11\sqrt{5} \approx 24.6$; $\left(-\frac{5}{2}, -4\right)$
**6.** $\sqrt{2} \approx 1.4$; $\left(-\frac{1}{4}, -\frac{3}{10}\right)$   **7.** $(x+5)^2 + (y-2)^2 = 169$
**8.** Center: $(3, -1)$; radius: 2
**9.**

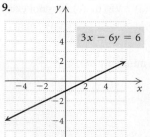

$3x - 6y = 6$

**10.**

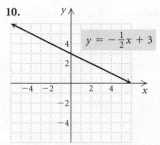

$y = -\frac{1}{2}x + 3$

**11.**

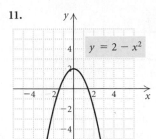

$y = 2 - x^2$

**12.**

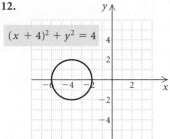

$(x + 4)^2 + y^2 = 4$

**13.** $f(-4) = -36; f(0) = 0; f(1) = -1$
**14.** $g(-6) = 0; g(0) = -2; g(3)$ is not defined
**15.** All real numbers, or $(-\infty, \infty)$
**16.** $\{x \mid x \neq -5\}$, or $(-\infty, -5) \cup (-5, \infty)$
**17.** $\{x \mid x \neq -3 \text{ and } x \neq 1\}$, or $(-\infty, -3) \cup (-3, 1) \cup (1, \infty)$

**18.**

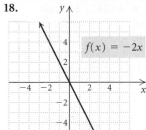

$f(x) = -2x$

**19.**

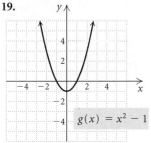

$g(x) = x^2 - 1$

**20.** Domain: $[-4, 3)$; range: $[-4, 5)$     **21.** Not defined
**22.** $-\frac{1}{4}$     **23.** 0     **24.** Slope: $-\frac{1}{9}$; $y$-intercept: $(0, 12)$
**25.** Slope: 0; $y$-intercept: $(0, -6)$     **26.** Slope is not defined; there is no $y$-intercept     **27.** Slope: $\frac{3}{16}$; $y$-intercept: $\left(0, \frac{1}{16}\right)$
**28.** The sign of the slope indicates the slant of a line. A line that slants up from left to right has positive slope because corresponding changes in $x$ and $y$ have the same sign. A line that slants down from left to right has negative slope, because corresponding changes in $x$ and $y$ have opposite signs. A horizontal line has zero slope, because there is no change in $y$ for a given change in $x$. The slope of a vertical line is not defined, because there is no change in $x$ for a given change in $y$ and division by 0 is not defined. The larger the absolute value of slope, the steeper the line. This is because a larger absolute value corresponds to a greater change in $y$, compared to the change in $x$, than a smaller absolute value.     **29.** A vertical line $(x = a)$ crosses the graph more than once. Thus, $x = a$ fails the vertical-line test.     **30.** The domain of a function is the set of all inputs of the function. The range is the set of all outputs. The range depends on the domain.     **31.** Let $A = (a, b)$ and $B = (c, d)$. The coordinates of a point $C$ one-half of the way from $A$ to $B$ are $\left(\dfrac{a + c}{2}, \dfrac{b + d}{2}\right)$. A point $D$ that is one-half of the way from $C$ to $B$ is $\frac{1}{2} + \frac{1}{2} \cdot \frac{1}{2}$, or $\frac{3}{4}$, of the way from $A$ to $B$. Its coordinates are $\left(\dfrac{\frac{a + c}{2} + c}{2}, \dfrac{\frac{b + d}{2} + d}{2}\right)$, or $\left(\dfrac{a + 3c}{4}, \dfrac{b + 3d}{4}\right)$. Then a point $E$ that is one-half of

the way from $D$ to $B$ is $\frac{3}{4} + \frac{1}{2} \cdot \frac{1}{4}$, or $\frac{7}{8}$, of the way from $A$ to $B$. Its coordinates are $\left(\dfrac{\frac{a + 3c}{4} + c}{2}, \dfrac{\frac{b + 3d}{4} + d}{2}\right)$, or $\left(\dfrac{a + 7c}{8}, \dfrac{b + 7d}{8}\right)$.

## Exercise Set 1.4

**1.** 4; $(0, -2)$; $y = 4x - 2$     **3.** $-1$, $(0, 0)$; $y = -x$
**5.** 0, $(0, -3)$; $y = -3$     **7.** $y = \frac{2}{9}x + 4$
**9.** $y = -4x - 7$     **11.** $y = -4.2x + \frac{3}{4}$
**13.** $y = \frac{2}{9}x + \frac{19}{3}$     **15.** $y = 8$     **17.** $y = -\frac{3}{5}x - \frac{17}{5}$
**19.** $y = -3x + 2$     **21.** $y = -\frac{1}{2}x + \frac{7}{2}$     **23.** $y = \frac{2}{3}x - 6$
**25.** $y = 7.3$     **27.** Horizontal: $y = -3$; vertical: $x = 0$
**29.** Horizontal: $y = -1$; vertical: $x = \frac{2}{11}$
**31.** $h(x) = -3x + 7; 1$     **33.** $f(x) = \frac{2}{5}x - 1; -1$
**35.** Perpendicular     **37.** Neither parallel nor perpendicular
**39.** Parallel     **41.** Perpendicular     **43.** $y = \frac{2}{7}x + \frac{29}{7}$; $y = -\frac{7}{2}x + \frac{31}{2}$     **45.** $y = -0.3x - 2.1$; $y = \frac{10}{3}x + \frac{70}{3}$
**47.** $y = -\frac{3}{4}x + \frac{1}{4}$; $y = \frac{4}{3}x - 6$     **49.** $x = 3$; $y = -3$
**51.** True     **53.** True     **55.** False     **57.** No     **59.** Yes
**61.** **(a)** Using $(2, 785)$ and $(8, 1858)$ gives us $y = 178.83x + 427.34$, where $x$ is the number of years after 2001 and $y$ is in millions;  **(b)** 2013: about 2573 million Internet users; 2018: about 3467 million Internet users     **63.** Using $(3, 24.9)$ and $(15, 56.9)$ gives us $y = 2.67x + 16.89$, where $x$ is the number of years after 1991 and $y$ is the total amount of U.S. expenditures on pets, in billions of dollars; 2005: $54.27 billion; 2015: $80.97 billion     **65.** Using $(2, 16.3)$ and $(5, 19.0)$ gives us $y = 0.9x + 14.5$, where $x$ is the number of years after 2005 and $y$ is in billions of dollars; $23.5 billion
**67.** **(a)** $y = 171.2606061x + 430.5272727$, where $x$ is the number of years after 2001 and $y$ is in millions; **(b)** about 2486 million; this value is 87 million lower than the value found in Exercise 61;  **(c)** $r \approx 0.9890$; the line fits the data fairly well.     **69.** **(a)** $y = 2.628571429x + 17.41428571$, where $x$ is the number of years after 1991 and $y$ is the total amount of U.S. expenditures on pets, in billions of dollars; **(b)** about $80.5 billion; this value is $0.47 billion less than the value found in Exercise 65; **(c)** $r \approx 0.9950$; the line fits the data well.     **71.** **(a)** $M = 0.2H + 156$; **(b)** 164, 169, 171, 173; **(c)** $r = 1$; the regression line fits the data perfectly and should be a good predictor.
**73.** [1.3] $-1$     **74.** [1.3] Not defined
**75.** [1.1] $(x + 7)^2 + (y + 1)^2 = \frac{81}{25}$
**76.** [1.1] $x^2 + (y - 3)^2 = 6.25$
**77.** $-7.75$     **79.** $y = -\frac{1}{2}x + 7$

## Exercise Set 1.5

**1.** 4     **3.** All real numbers, or $(-\infty, \infty)$     **5.** $-\frac{3}{4}$     **7.** $-9$
**9.** 6     **11.** No solution     **13.** $\frac{11}{5}$     **15.** $\frac{35}{6}$     **17.** 8
**19.** $-4$     **21.** 6     **23.** $-1$     **25.** $\frac{4}{5}$     **27.** $-\frac{3}{2}$     **29.** $-\frac{2}{3}$
**31.** $\frac{1}{2}$     **33.** About 105,000 students

**35.** 260,000 metric tons   **37.** 10,040 ft   **39.** Toyota Tundra: 5740 lb; Ford Mustang: 3605 lb; Smart For Two: 1805 lb   **41.** CBS: 11.4 million viewers; ABC: 9.7 million viewers; NBC: 8.0 million viewers   **43.** $1300   **45.** $9800   **47.** $8.50   **49.** 26°, 130°, 24°   **51.** Length: 93 m; width: 68 m   **53.** Length: 100 yd; width: 65 yd   **55.** 74.25 lb   **57.** 3 hr   **59.** 4.5 hr   **61.** 2.5 hr   **63.** $2400 at 3%; $2600 at 4%   **65.** YouTube.com: 114,254,701 visitors; amazon.com: 71,428,476 visitors   **67.** 709 ft   **69.** About 4,318,978 people   **71.** $-5$   **73.** $\frac{11}{2}$   **75.** 16   **77.** $-12$   **79.** 6   **81.** 20   **83.** 25   **85.** 15   **87.** (a) $(4, 0)$; (b) 4   **89.** (a) $(-2, 0)$; (b) $-2$   **91.** (a) $(-4, 0)$; (b) $-4$   **93.** [1.4] $y = -\frac{3}{4}x + \frac{13}{4}$   **94.** [1.4] $y = -\frac{3}{4}x + \frac{1}{4}$   **95.** [1.1] 13   **96.** [1.1] $\left(-1, \frac{1}{2}\right)$   **97.** [1.2] $f(-3) = \frac{1}{2}; f(0) = 0; f(3)$ does not exist.   **98.** [1.3] $m = 7$; $y$-intercept: $\left(0, -\frac{1}{2}\right)$   **99.** Yes   **101.** No   **103.** $-\frac{2}{3}$   **105.** No; the 6-oz cup costs about 6.4% more per ounce.   **107.** 11.25 mi

## Exercise Set 1.6

**1.** $\{x | x > 5\}$, or $(5, \infty)$;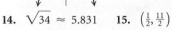

**3.** $\{x | x > 3\}$, or $(3, \infty)$;

**5.** $\{x | x \geq -3\}$, or $[-3, \infty)$;

**7.** $\left\{y | y \geq \frac{22}{13}\right\}$, or $\left[\frac{22}{13}, \infty\right)$;

**9.** $\{x | x > 6\}$, or $(6, \infty)$;

**11.** $\left\{x | x \geq -\frac{5}{12}\right\}$, or $\left[-\frac{5}{12}, \infty\right)$;

**13.** $\left\{x | x \leq \frac{15}{34}\right\}$, or $\left(-\infty, \frac{15}{34}\right]$;

**15.** $\{x | x < 1\}$, or $(-\infty, 1)$;

**17.** $\{x | x \geq 7\}$, or $[7, \infty)$   **19.** $\left\{x | x \leq \frac{1}{5}\right\}$, or $\left(-\infty, \frac{1}{5}\right]$   **21.** $\{x | x > -4\}$, or $(-4, \infty)$

**23.** $[-3, 3)$;

**25.** $[8, 10]$;

**27.** $[-7, -1]$;

**29.** $\left(-\frac{3}{2}, 2\right)$;

**31.** $(1, 5]$;

**33.** $\left(-\frac{11}{3}, \frac{13}{3}\right)$;

**35.** $(-\infty, -2] \cup (1, \infty)$;

**37.** $\left(-\infty, -\frac{7}{2}\right] \cup \left[\frac{1}{2}, \infty\right)$;

**39.** $(-\infty, 9.6) \cup (10.4, \infty)$;

**41.** $\left(-\infty, -\frac{57}{4}\right] \cup \left[-\frac{55}{4}, \infty\right)$;

**43.** Years after 2016   **45.** Less than 4 hr   **47.** $5000   **49.** $12,000   **51.** Sales greater than $18,000   **53.** [1.2] Function; domain; range; domain; exactly one; range   **54.** [1.1] Midpoint formula   **55.** [1.1] $x$-intercept   **56.** [1.3] Constant; identity   **57.** $\left(-\frac{1}{4}, \frac{5}{9}\right)$   **59.** $\left(-\frac{1}{8}, \frac{1}{2}\right)$

## Review Exercises: Chapter 1

**1.** True   **2.** True   **3.** False   **4.** False   **5.** True   **6.** False   **7.** Yes; no   **8.** Yes; no

**9.** $x$-intercept: $(3, 0)$; $y$-intercept: $(0, -2)$;

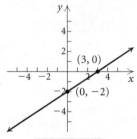

$2x - 3y = 6$

**10.** $x$-intercept: $(2, 0)$; $y$-intercept: $(0, 5)$;

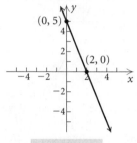

$10 - 5x = 2y$

**11.**

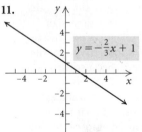

$y = -\frac{2}{3}x + 1$

**12.**

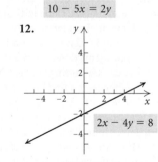

$2x - 4y = 8$

**13.**

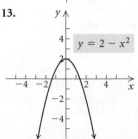

$y = 2 - x^2$

**14.** $\sqrt{34} \approx 5.831$   **15.** $\left(\frac{1}{2}, \frac{11}{2}\right)$

**16.** Center: $(-1, 3)$; radius: 3;

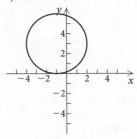

$$(x + 1)^2 + (y - 3)^2 = 9$$

**17.** $x^2 + (y + 4)^2 = \frac{9}{4}$    **18.** $(x + 2)^2 + (y - 6)^2 = 13$
**19.** $(x - 2)^2 + (y - 4)^2 = 26$    **20.** No    **21.** Yes
**22.** Not a function; domain: $\{3, 5, 7\}$; range: $\{1, 3, 5, 7\}$
**23.** Function; domain: $\{-2, 0, 1, 2, 7\}$; range: $\{-7, -4, -2, 2, 7\}$
**24.** (a) $-3$; (b) $9$; (c) $a^2 - 3a - 1$; (d) $x^2 + x - 3$
**25.** (a) $0$; (b) $\dfrac{x - 6}{x + 6}$; (c) does not exist; (d) $-\frac{5}{3}$
**26.** $f(2) = -1$; $f(-4) = -3$; $f(0) = -1$    **27.** No
**28.** Yes    **29.** No    **30.** Yes    **31.** All real numbers,
or $(-\infty, \infty)$    **32.** $\{x \mid x \neq 0\}$, or $(-\infty, 0) \cup (0, \infty)$
**33.** $\{x \mid x \neq 5 \text{ and } x \neq 1\}$, or $(-\infty, 1) \cup (1, 5) \cup (5, \infty)$
**34.** $\{x \mid x \neq -4 \text{ and } x \neq 4\}$, or $(-\infty, -4) \cup (-4, 4) \cup (4, \infty)$
**35.** Domain: $[-4, 4]$; range: $[0, 4]$    **36.** Domain: $(-\infty, \infty)$;
range: $[0, \infty)$    **37.** Domain: $(-\infty, \infty)$; range: $(-\infty, \infty)$
**38.** Domain: $(-\infty, \infty)$; range: $[0, \infty)$    **39.** (a) Yes; (b) no;
(c) no, strictly speaking, but data might be modeled by a
linear regression function.    **40.** (a) Yes; (b) yes; (c) yes
**41.** $\frac{5}{3}$    **42.** $0$    **43.** Not defined    **44.** The average rate
of change over the 20-year period was about $0.11 per year.
**45.** $m = -\frac{7}{11}$; $y$-intercept: $(0, -6)$    **46.** $m = -2$;
$y$-intercept: $(0, -7)$
**47.**

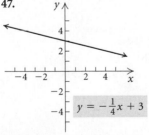

$$y = -\frac{1}{4}x + 3$$

**48.** $C(t) = 110 + 85t$; $1130    **49.** (a) $70°C$, $220°C$, $10,020°C$;
(b) $[0, 5600]$    **50.** $y = -\frac{2}{3}x - 4$    **51.** $y = 3x + 5$
**52.** $y = \frac{1}{3}x - \frac{1}{3}$    **53.** Horizontal: $y = \frac{2}{5}$; vertical: $x = -4$
**54.** $h(x) = 2x - 5$; $-5$    **55.** Parallel    **56.** Neither
**57.** Perpendicular    **58.** $y = -\frac{2}{3}x - \frac{1}{3}$    **59.** $y = \frac{3}{2}x - \frac{5}{2}$
**60.** (a) Using $(98.7, 194.8)$ and $(120.7, 238.6)$ gives us
$H(c) = 1.99c - 1.61$ and $H(118 \text{ cm}) = 233.21 \text{ cm}$;
(b) $y = 1.990731128x - 1.653424424$; $233.25 \text{ cm}$;
$r \approx 0.99999956$; the line fits the data extremely well.
**61.** $\frac{3}{2}$    **62.** $-6$    **63.** $-1$    **64.** $-21$    **65.** $\frac{95}{24}$
**66.** No solution    **67.** All real numbers, or $(-\infty, \infty)$
**68.** 2780 milligrams    **69.** $2300    **70.** 3.4 hr
**71.** 3    **72.** 4    **73.** 0.2, or $\frac{1}{5}$    **74.** 4

**75.** $(-\infty, 12)$;

**76.** $(-\infty, -4]$;

**77.** $\left[-\frac{4}{3}, \frac{4}{3}\right]$;

**78.** $\left(\frac{2}{5}, 2\right]$;

**79.** $\left(-\infty, -\frac{1}{2}\right) \cup (3, \infty)$;

**80.** $\left(-\infty, -\frac{5}{3}\right] \cup [1, \infty)$;

**81.** Years after 2013    **82.** Fahrenheit temperatures less than
$113°$    **83.** B    **84.** B    **85.** C    **86.** $\left(\frac{5}{2}, 0\right)$
**87.** $\{x \mid x < 0\}$, or $(-\infty, 0)$    **88.** $\{x \mid x \neq -3 \text{ and }$
$x \neq 0 \text{ and } x \neq 3\}$, or $(-\infty, -3) \cup (-3, 0) \cup (0, 3) \cup (3, \infty)$
**89.** Think of the slopes as $\dfrac{-3/5}{1}$ and $\dfrac{1/2}{1}$. The graph of $f(x)$
changes $\frac{3}{5}$ unit vertically for each unit of horizontal change,
whereas the graph of $g(x)$ changes $\frac{1}{2}$ unit vertically for each
unit of horizontal change. Since $\frac{3}{5} > \frac{1}{2}$, the graph of
$f(x) = -\frac{3}{5}x + 4$ is steeper than the graph of $g(x) = \frac{1}{2}x - 6$.
**90.** If an equation contains no fractions, using the addition
principle before using the multiplication principle eliminates
the need to add or subtract fractions.    **91.** The solution
set of a disjunction is a union of sets, so it is only possible
for a disjunction to have no solution when the solution set
of each inequality is the empty set.    **92.** The graph of
$f(x) = mx + b$, $m \neq 0$, is a straight line that is not
horizontal. The graph of such a line intersects the $x$-axis
exactly once. Thus the function has exactly one zero.
**93.** By definition, the notation $3 < x < 4$ indicates that
$3 < x$ and $x < 4$. The disjunction $x < 3$ or $x > 4$ cannot be
written $3 > x > 4$, or $4 < x < 3$, because it is not possible
for $x$ to be greater than 4 and less than 3.    **94.** A function is a
correspondence between two sets in which each member of the
first set corresponds to exactly one member of the second set.

## Test: Chapter 1

**1.** [1.1] Yes    **2.** [1.1] $x$-intercept: $(-2, 0)$; $y$-intercept: $(0, 5)$;

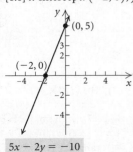

$$5x - 2y = -10$$

**3.** [1.1] $\sqrt{45} \approx 6.708$  **4.** [1.1] $\left(-3, \frac{9}{2}\right)$  **5.** [1.1] Center: $(-4, 5)$; radius: 6  **6.** [1.1] $(x + 1)^2 + (y - 2)^2 = 5$
**7.** [1.2] **(a)** Yes;  **(b)** $\{-4, 3, 1, 0\}$;  **(c)** $\{7, 0, 5\}$  **8.** [1.2] **(a)** 8;
**(b)** $2a^2 + 7a + 11$  **9.** [1.2] **(a)** Does not exist;  **(b)** 0
**10.** [1.2] 0  **11.** [1.2] **(a)** No;  **(b)** yes
**12.** [1.2] $\{x \mid x \neq 4\}$, or $(-\infty, 4) \cup (4, \infty)$
**13.** [1.2] All real numbers, or $(-\infty, \infty)$
**14.** [1.2] $\{x \mid -5 \leq x \leq 5\}$, or $[-5, 5]$
**15.** [1.2] **(a)**

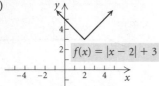

**(b)** $(-\infty, \infty)$;
**(c)** $[3, \infty)$

**16.** [1.3] Not defined  **17.** [1.3] $-\frac{11}{6}$  **18.** [1.3] 0
**19.** [1.3] The average rate of change in the percent of 12th graders who smoke from 1995 to 2008 was about $-0.8\%$ per year.
**20.** [1.3] Slope: $\frac{3}{2}$; $y$-intercept: $\left(0, \frac{5}{2}\right)$
**21.** [1.3] $C(t) = 80 + 49.95t$; $1278.80
**22.** [1.4] $y = -\frac{5}{8}x - 5$  **23.** [1.4] $y - 4 = -\frac{3}{4}(x - (-5))$,
or $y - (-2) = -\frac{3}{4}(x - 3)$, or $y = -\frac{3}{4}x + \frac{1}{4}$
**24.** [1.4] $x = -\frac{3}{8}$  **25.** [1.4] Perpendicular
**26.** [1.4] $y - 3 = -\frac{1}{2}(x + 1)$, or $y = -\frac{1}{2}x + \frac{5}{2}$
**27.** [1.4] $y - 3 = 2(x + 1)$, or $y = 2x + 5$
**28.** [1.4] **(a)** Using $(1, 12{,}485)$ and $(3, 11{,}788)$ gives us
$y = -348.5x + 12{,}833.5$, where $x$ is the number of years after 2005 and $y$ is the average number of miles per passenger car; 2010: 11,091 mi, 2013: 10,045.5 mi;
**(b)** $y = -234.7x + 12{,}623.8$; 2010: 11,450 mi, 2013: 10,746 mi; $r \approx -0.9036$  **29.** [1.5] $-1$  **30.** [1.5] All real numbers, or $(-\infty, \infty)$  **31.** [1.5] $-60$  **32.** [1.5] $\frac{21}{11}$
**33.** [1.5] Length: 60 m; width: 45 m
**34.** [1.5] $1.80  **35.** [1.5] $-3$
**36.** [1.6] $(-\infty, -3]$;
**37.** [1.6] $(-5, 3)$;
**38.** [1.6] $(-\infty, 2] \cup [4, \infty)$;
**39.** [1.6] More than 6 hr  **40.** [1.3] B  **41.** [1.2] $-2$

## Chapter 2

### Exercise Set 2.1

**1.** **(a)** $(-5, 1)$;  **(b)** $(3, 5)$;  **(c)** $(1, 3)$
**3.** **(a)** $(-3, -1)$, $(3, 5)$;  **(b)** $(1, 3)$;  **(c)** $(-5, -3)$
**5.** **(a)** $(-\infty, -8)$, $(-3, -2)$;  **(b)** $(-8, -6)$;
**(c)** $(-6, -3)$, $(-2, \infty)$  **7.** Domain: $[-5, 5]$; range: $[-3, 3]$
**9.** Domain: $[-5, -1] \cup [1, 5]$; range: $[-4, 6]$
**11.** Domain: $(-\infty, \infty)$; range: $(-\infty, 3]$  **13.** Relative maximum: 3.25 at $x = 2.5$; increasing: $(-\infty, 2.5)$; decreasing: $(2.5, \infty)$  **15.** Relative maximum: 2.370 at $x = -0.667$; relative minimum: 0 at $x = 2$;

increasing: $(-\infty, -0.667)$, $(2, \infty)$; decreasing: $(-0.667, 2)$
**17.** Increasing: $(0, \infty)$; decreasing: $(-\infty, 0)$; relative minimum: 0 at $x = 0$  **19.** Increasing: $(-\infty, 0)$; decreasing: $(0, \infty)$; relative maximum: 5 at $x = 0$
**21.** Increasing: $(3, \infty)$; decreasing: $(-\infty, 3)$; relative minimum: 1 at $x = 3$  **23.** Increasing: $(1, 3)$; decreasing: $(-\infty, 1)$, $(3, \infty)$; relative maximum: $-4$ at $x = 3$; relative minimum: $-8$ at $x = 1$  **25.** Increasing: $(-1.552, 0)$, $(1.552, 0)$; decreasing: $(-\infty, -1.552)$, $(0, 1.552)$; relative maximum: 4.07 at $x = 0$; relative minima: $-2.314$ at $x = -1.552$, $-2.314$ at $x = 1.552$
**27.** **(a)** $y = -x^2 + 300x + 6$   ;  **(b)** 22,506

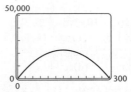

**(c)** When $150 thousand is spent on advertising, 22,506 baskets of three potted amaryllises will be sold.
**29.** Increasing: $(-1, 1)$; decreasing: $(-\infty, -1)$, $(1, \infty)$
**31.** Increasing: $(-1.414, 1.414)$; decreasing: $(-2, -1.414)$, $(1.414, 2)$  **33.** $A(x) = x(240 - x)$, or $240x - x^2$
**35.** $d(t) = \sqrt{(120t)^2 + 400^2}$  **37.** $A(w) = 10w - \frac{w^2}{2}$
**39.** $d(s) = \frac{14}{s}$  **41.** **(a)** $A(x) = x(30 - x)$, or $30x - x^2$;
**(b)** $\{x \mid 0 < x < 30\}$;  **(c)** 15 ft by 15 ft
**43.** **(a)** $V(x) = 8x(14 - 2x)$, or $112x - 16x^2$;
**(b)** $\{x \mid 0 < x < 7\}$;  **(c)** $y = 112x - 16x^2$

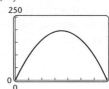

**45.** **(a)** $A(x) = x\sqrt{256 - x^2}$; **(b)** $\{x \mid 0 < x < 16\}$;
**(c)** $y = x\sqrt{256 - x^2}$     **(d)** 11.314 ft by 11.314 ft

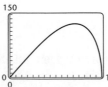

**47.** $g(-4) = 0$; $g(0) = 4$; $g(1) = 5$; $g(3) = 5$
**49.** $h(-5) = 1$; $h(0) = 1$; $h(1) = 3$; $h(4) = 6$
**51.**     **53.**

**55.**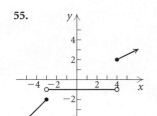

**57.**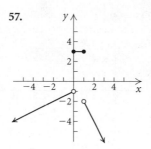

**59.**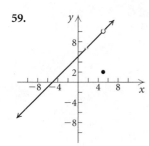

**61.**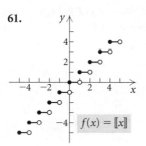

$f(x) = [\![x]\!]$

**63.**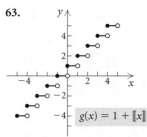

$g(x) = 1 + [\![x]\!]$

**65.** Domain: $(-\infty, \infty)$; range: $(-\infty, 0] \cup [3, \infty)$
**67.** Domain: $(-\infty, \infty)$; range: $[-1, \infty)$
**69.** Domain: $(-\infty, \infty)$; range: $(-\infty, -2] \cup \{-1\} \cup [2, \infty)$
**71.** Domain: $(-\infty, \infty)$; range: $\{-5, -2, 4\}$;

$$f(x) = \begin{cases} -2, & \text{for } x < 2, \\ -5, & \text{for } x = 2, \\ 4, & \text{for } x > 2 \end{cases}$$

**73.** Domain: $(-\infty, \infty)$; range: $(-\infty, -1] \cup [2, \infty)$;

$$g(x) = \begin{cases} x, & \text{for } x \leq -1, \\ 2, & \text{for } -1 < x < 2, \\ x, & \text{for } x \geq 2 \end{cases}$$

or

$$g(x) = \begin{cases} x, & \text{for } x \leq -1, \\ 2, & \text{for } -1 < x \leq 2, \\ x, & \text{for } x > 2 \end{cases}$$

**75.** Domain: $[-5, 3]$; range: $(-3, 5)$;

$$h(x) = \begin{cases} x + 8, & \text{for } -5 \leq x < -3, \\ 3, & \text{for } -3 \leq x \leq 1, \\ 3x - 6, & \text{for } 1 < x \leq 3 \end{cases}$$

**77.** [1.2] **(a)** 38; **(b)** 38; **(c)** $5a^2 - 7$; **(d)** $5a^2 - 7$
**78.** [1.2] **(a)** 22; **(b)** $-22$; **(c)** $4a^3 - 5a$;
**(d)** $-4a^3 + 5a$    **79.** [1.4] $y = -\frac{1}{8}x + \frac{7}{8}$
**80.** [1.4] Slope is $\frac{2}{9}$; $y$-intercept is $\left(0, \frac{1}{9}\right)$.
**81.** Increasing: $(-5, -2)$, $(4, \infty)$; decreasing: $(-\infty, -5)$, $(-2, 4)$; relative maximum: 560 at $x = -2$; relative minima: 425 at $x = -5$, $-304$ at $x = 4$

**83. (a)**    **(b)** $C(t) = 2([\![t]\!] + 1)$, $t > 0$

**85.** $\{x \mid -5 \leq x < -4 \text{ or } 5 \leq x < 6\}$
**87. (a)** $h(r) = \dfrac{30 - 5r}{3}$; **(b)** $V(r) = \pi r^2 \left(\dfrac{30 - 5r}{3}\right)$;

**(c)** $V(h) = \pi h \left(\dfrac{30 - 3h}{5}\right)^2$

## Exercise Set 2.2

**1.** 33    **3.** $-1$    **5.** Does not exist    **7.** 0    **9.** 1
**11.** Does not exist    **13.** 0    **15.** 5
**17. (a)** Domain of $f$, $g$, $f + g$, $f - g$, $fg$, and $ff$: $(-\infty, \infty)$;
domain of $f/g$: $\left(-\infty, \frac{3}{5}\right) \cup \left(\frac{3}{5}, \infty\right)$;
domain of $g/f$: $\left(-\infty, -\frac{3}{2}\right) \cup \left(-\frac{3}{2}, \infty\right)$;
**(b)** $(f + g)(x) = -3x + 6$; $(f - g)(x) = 7x$;
$(fg)(x) = -10x^2 - 9x + 9$; $(ff)(x) = 4x^2 + 12x + 9$;
$(f/g)(x) = \dfrac{2x + 3}{3 - 5x}$; $(g/f)(x) = \dfrac{3 - 5x}{2x + 3}$
**19. (a)** Domain of $f$: $(-\infty, \infty)$; domain of $g$: $[-4, \infty)$;
domain of $f + g$, $f - g$, and $fg$: $[-4, \infty)$;
domain of $ff$: $(-\infty, \infty)$; domain of $f/g$: $(-4, \infty)$;
domain of $g/f$: $[-4, 3) \cup (3, \infty)$;
**(b)** $(f + g)(x) = x - 3 + \sqrt{x + 4}$;
$(f - g)(x) = x - 3 - \sqrt{x + 4}$; $(fg)(x) = (x - 3)\sqrt{x + 4}$;
$(ff)(x) = x^2 - 6x + 9$; $(f/g)(x) = \dfrac{x - 3}{\sqrt{x + 4}}$;
$(g/f)(x) = \dfrac{\sqrt{x + 4}}{x - 3}$
**21. (a)** Domain of $f$, $g$, $f + g$, $f - g$, $fg$, and $ff$: $(-\infty, \infty)$;
domain of $f/g$: $(-\infty, 0) \cup (0, \infty)$;
domain of $g/f$: $\left(-\infty, \frac{1}{2}\right) \cup \left(\frac{1}{2}, \infty\right)$
**(b)** $(f + g)(x) = -2x^2 + 2x - 1$;
$(f - g)(x) = 2x^2 + 2x - 1$; $(fg)(x) = -4x^3 + 2x^2$;
$(ff)(x) = 4x^2 - 4x + 1$; $(f/g)(x) = \dfrac{2x - 1}{-2x^2}$;
$(g/f)(x) = \dfrac{-2x^2}{2x - 1}$
**23. (a)** Domain of $f$: $[3, \infty)$; domain of $g$: $[-3, \infty)$;
domain of $f + g$, $f - g$, $fg$, and $ff$: $[3, \infty)$;
domain of $f/g$: $[3, \infty)$; domain of $g/f$: $(3, \infty)$;
**(b)** $(f + g)(x) = \sqrt{x - 3} + \sqrt{x + 3}$;
$(f - g)(x) = \sqrt{x - 3} - \sqrt{x + 3}$; $(fg)(x) = \sqrt{x^2 - 9}$;
$(ff)(x) = |x - 3|$; $(f/g)(x) = \dfrac{\sqrt{x - 3}}{\sqrt{x + 3}}$; $(g/f)(x) = \dfrac{\sqrt{x + 3}}{\sqrt{x - 3}}$
**25. (a)** Domain of $f$, $g$, $f + g$, $f - g$, $fg$, and $ff$: $(-\infty, \infty)$;
domain of $f/g$: $(-\infty, 0) \cup (0, \infty)$;
domain of $g/f$: $(-\infty, -1) \cup (-1, \infty)$;

**(b)** $(f + g)(x) = x + 1 + |x|; (f - g)(x) = x + 1 - |x|;$
$(fg)(x) = (x + 1)|x|; (ff)(x) = x^2 + 2x + 1;$
$(f/g)(x) = \dfrac{x + 1}{|x|}; (g/f)(x) = \dfrac{|x|}{x + 1}$

**27. (a)** Domain of $f, g, f + g, f - g, fg,$ and $ff$: $(-\infty, \infty)$;
domain of $f/g$: $(-\infty, -3) \cup \left(-3, \frac{1}{2}\right) \cup \left(\frac{1}{2}, \infty\right)$;
domain of $g/f$: $(-\infty, 0) \cup (0, \infty)$;
**(b)** $(f + g)(x) = x^3 + 2x^2 + 5x - 3;$
$(f - g)(x) = x^3 - 2x^2 - 5x + 3;$
$(fg)(x) = 2x^5 + 5x^4 - 3x^3; (ff)(x) = x^6;$
$(f/g)(x) = \dfrac{x^3}{2x^2 + 5x - 3}; (g/f)(x) = \dfrac{2x^2 + 5x - 3}{x^3}$

**29. (a)** Domain of $f$: $(-\infty, -1) \cup (-1, \infty)$;
domain of $g$: $(-\infty, 6) \cup (6, \infty)$; domain of $f + g, f - g,$ and
$fg$: $(-\infty, -1) \cup (-1, 6) \cup (6, \infty)$;
domain of $ff$: $(-\infty, -1) \cup (-1, \infty)$;
domain of $f/g$ and $g/f$: $(-\infty, -1) \cup (-1, 6) \cup (6, \infty)$;

**(b)** $(f + g)(x) = \dfrac{4}{x + 1} + \dfrac{1}{6 - x};$

$(f - g)(x) = \dfrac{4}{x + 1} - \dfrac{1}{6 - x};$

$(fg)(x) = \dfrac{4}{(x + 1)(6 - x)}; (ff)(x) = \dfrac{16}{(x + 1)^2};$

$(f/g)(x) = \dfrac{4(6 - x)}{x + 1}; (g/f)(x) = \dfrac{x + 1}{4(6 - x)}$

**31. (a)** Domain of $f$: $(-\infty, 0) \cup (0, \infty)$;
domain of $g$: $(-\infty, \infty)$; domain of $f + g, f - g, fg,$ and $ff$:
$(-\infty, 0) \cup (0, \infty)$; domain of $f/g$: $(-\infty, 0) \cup (0, 3) \cup (3, \infty)$;
domain of $g/f$: $(-\infty, 0) \cup (0, \infty)$;

**(b)** $(f + g)(x) = \dfrac{1}{x} + x - 3; (f - g)(x) = \dfrac{1}{x} - x + 3;$

$(fg)(x) = 1 - \dfrac{3}{x}; (ff)(x) = \dfrac{1}{x^2}; (f/g)(x) = \dfrac{1}{x(x - 3)};$

$(g/f)(x) = x(x - 3)$

**33. (a)** Domain of $f$: $(-\infty, 2) \cup (2, \infty)$; domain of $g$: $[1, \infty)$;
domain of $f + g, f - g,$ and $fg$: $[1, 2) \cup (2, \infty)$; domain
of $ff$: $(-\infty, 2) \cup (2, \infty)$; domain of $f/g$: $(1, 2) \cup (2, \infty)$;
domain of $g/f$: $[1, 2) \cup (2, \infty)$;

**(b)** $(f + g)(x) = \dfrac{3}{x - 2} + \sqrt{x - 1};$

$(f - g)(x) = \dfrac{3}{x - 2} - \sqrt{x - 1};$

$(fg)(x) = \dfrac{3\sqrt{x - 1}}{x - 2}; (ff)(x) = \dfrac{9}{(x - 2)^2};$

$(f/g)(x) = \dfrac{3}{(x - 2)\sqrt{x - 1}}; (g/f)(x) = \dfrac{(x - 2)\sqrt{x - 1}}{3}$

**35.** Domain of $F$: $[2, 11]$; domain of $G$: $[1, 9]$;
domain of $F + G$: $[2, 9]$   **37.** $[2, 3) \cup (3, 9]$

**39.**

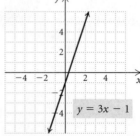

**41.** Domain of $F$: $[0, 9]$; domain of $G$: $[3, 10]$;
domain of $F + G$: $[3, 9]$   **43.** $[3, 6) \cup (6, 8) \cup (8, 9]$

**45.**

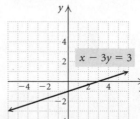

**47. (a)** $P(x) = -0.4x^2 + 57x - 13;$ **(b)** $R(100) = 2000;$
$C(100) = 313; P(100) = 1687;$ **(c)** Left to the student

**49.** 3   **51.** 6   **53.** $\frac{1}{3}$   **55.** $\dfrac{-1}{3x(x + h)},$ or $-\dfrac{1}{3x(x + h)}$

**57.** $\dfrac{1}{4x(x + h)}$   **59.** $2x + h$   **61.** $-2x - h$

**63.** $6x + 3h - 2$   **65.** $\dfrac{5|x + h| - 5|x|}{h}$

**67.** $3x^2 + 3xh + h^2$   **69.** $\dfrac{7}{(x + h + 3)(x + 3)}$

**71.** [1.1], [1.3]          **72.** [1.1], [1.3]

$y = 3x - 1$

$2x + y = 4$

**73.** [1.1], [1.3]          **74.** [1.1]

$x - 3y = 3$

$y = x^2 + 1$

**75.** $f(x) = \dfrac{1}{x+7}, g(x) = \dfrac{1}{x-3}$; answers may vary

**77.** $(-\infty, -1) \cup (-1, 1) \cup \left(1, \frac{7}{3}\right) \cup \left(\frac{7}{3}, 3\right) \cup (3, \infty)$

## Exercise Set 2.3

**1.** $-8$  **3.** $64$  **5.** $218$  **7.** $-80$
**9.** $-6$  **11.** $512$  **13.** $-32$  **15.** $x^9$
**17.** $(f \circ g)(x) = (g \circ f)(x) = x$;
domain of $f \circ g$ and $g \circ f$: $(-\infty, \infty)$
**19.** $(f \circ g)(x) = 3x^2 - 2x$; $(g \circ f)(x) = 3x^2 + 4x$;
domain of $f \circ g$ and $g \circ f$: $(-\infty, \infty)$
**21.** $(f \circ g)(x) = 16x^2 - 24x + 6$; $(g \circ f)(x) = 4x^2 - 15$;
domain of $f \circ g$ and $g \circ f$: $(-\infty, \infty)$
**23.** $(f \circ g)(x) = \dfrac{4x}{x-5}$; $(g \circ f)(x) = \dfrac{1-5x}{4}$;
domain of $f \circ g$: $(-\infty, 0) \cup (0, 5) \cup (5, \infty)$;
domain of $g \circ f$: $\left(-\infty, \frac{1}{5}\right) \cup \left(\frac{1}{5}, \infty\right)$
**25.** $(f \circ g)(x) = (g \circ f)(x) = x$;
domain of $f \circ g$ and $g \circ f$: $(-\infty, \infty)$
**27.** $(f \circ g)(x) = 2\sqrt{x} + 1$; $(g \circ f)(x) = \sqrt{2x+1}$;
domain of $f \circ g$: $[0, \infty)$; domain of $g \circ f$: $\left[-\frac{1}{2}, \infty\right)$
**29.** $(f \circ g)(x) = 20$; $(g \circ f)(x) = 0.05$;
domain of $f \circ g$ and $g \circ f$: $(-\infty, \infty)$
**31.** $(f \circ g)(x) = |x|$; $(g \circ f)(x) = x$;
domain of $f \circ g$: $(-\infty, \infty)$; domain of $g \circ f$: $[-5, \infty)$
**33.** $(f \circ g)(x) = 5 - x$; $(g \circ f)(x) = \sqrt{1-x^2}$; domain of
$f \circ g$: $(-\infty, 3]$; domain of $g \circ f$: $[-1, 1]$
**35.** $(f \circ g)(x) = (g \circ f)(x) = x$; domain of $f \circ g$:
$(-\infty, -1) \cup (-1, \infty)$; domain of $g \circ f$: $(-\infty, 0) \cup (0, \infty)$
**37.** $(f \circ g)(x) = x^3 - 2x^2 - 4x + 6$;
$(g \circ f)(x) = x^3 - 5x^2 + 3x + 8$;
domain of $f \circ g$ and $g \circ f$: $(-\infty, \infty)$
**39.** $f(x) = x^5$; $g(x) = 4 + 3x$
**41.** $f(x) = \dfrac{1}{x}$; $g(x) = (x-2)^4$
**43.** $f(x) = \dfrac{x-1}{x+1}$; $g(x) = x^3$
**45.** $f(x) = x^6$; $g(x) = \dfrac{2+x^3}{2-x^3}$
**47.** $f(x) = \sqrt{x}$; $g(x) = \dfrac{x-5}{x+2}$
**49.** $f(x) = x^3 - 5x^2 + 3x - 1$; $g(x) = x + 2$
**51.** (a) $r(t) = 3t$; (b) $A(r) = \pi r^2$; (c) $(A \circ r)(t) = 9\pi t^2$;
the function gives the area of the ripple in terms of time $t$.
**53.** $P(m) = 1.5m + 3$  **55.** [1.3] (c)  **56.** [1.3] None
**57.** [1.3] (b), (d), (f), and (h)  **58.** [1.3] (b)  **59.** [1.3] (a)
**60.** [1.3] (c) and (g)  **61.** [1.4] (c) and (g)  **62.** [1.4] (a)
and (f)  **63.** Only $(c \circ p)(a)$ makes sense. It represents the
cost of the grass seed required to seed a lawn with area $a$.

## Mid-Chapter Mixed Review: Chapter 2

**1.** True  **2.** False  **3.** True  **4.** (a) $(2, 4)$;
(b) $(-5, -3), (4, 5)$; (c) $(-3, -1)$  **5.** Relative maximum:
6.30 at $x = -1.29$; relative minimum: $-2.30$ at $x = 1.29$;

increasing: $(-\infty, -1.29), (1.29, \infty)$; decreasing: $(-1.29, 1.29)$
**6.** Domain: $[-5, -1] \cup [2, 5]$; range: $[-3, 5]$
**7.** $A(h) = \dfrac{h^2}{2} - h$  **8.** $-10; -8; 1; 3$
**9.**

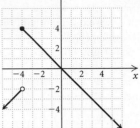

**10.** 1  **11.** $-4$  **12.** 5
**13.** Does not exist

**14.** (a) Domain of $f, g, f + g, f - g, fg$, and $ff$: $(-\infty, \infty)$;
domain of $f/g$: $(-\infty, -4) \cup (-4, \infty)$;
domain of $g/f$: $\left(-\infty, -\frac{5}{2}\right) \cup \left(-\frac{5}{2}, \infty\right)$;
(b) $(f + g)(x) = x + 1$; $(f - g)(x) = 3x + 9$;
$(fg)(x) = -2x^2 - 13x - 20$; $(ff)(x) = 4x^2 + 20x + 25$;
$(f/g)(x) = \dfrac{2x+5}{-x-4}$; $(g/f)(x) = \dfrac{-x-4}{2x+5}$
**15.** (a) Domain of $f$: $(-\infty, \infty)$; domain of $g, f + g, f - g$,
and $fg$: $[-2, \infty)$; domain of $ff$: $(-\infty, \infty)$; domain of
$f/g$: $(-2, \infty)$; domain of $g/f$: $[-2, 1) \cup (1, \infty)$;
(b) $(f + g)(x) = x - 1 + \sqrt{x+2}$;
$(f - g)(x) = x - 1 - \sqrt{x+2}$;
$(fg)(x) = (x - 1)\sqrt{x+2}$; $(ff)(x) = x^2 - 2x + 1$;
$(f/g)(x) = \dfrac{x-1}{\sqrt{x+2}}$; $(g/f)(x) = \dfrac{\sqrt{x+2}}{x-1}$
**16.** 4  **17.** $-2x - h$  **18.** 6  **19.** 28  **20.** $-24$
**21.** 102  **22.** $(f \circ g)(x) = 3x + 2$; $(g \circ f)(x) = 3x + 4$;
domain of $f \circ g$ and $g \circ f$: $(-\infty, \infty)$
**23.** $(f \circ g)(x) = 3\sqrt{x} + 2$; $(g \circ f)(x) = \sqrt{3x+2}$;
domain of $f \circ g$: $[0, \infty)$; domain of $g \circ f$: $\left[-\frac{2}{3}, \infty\right)$
**24.** The graph of $y = (h - g)(x)$ will be the same as the
graph of $y = h(x)$ shifted down $b$ units.  **25.** Under the
given conditions, $(f + g)(x)$ and $(f/g)(x)$ have different
domains if $g(x) = 0$ for one or more real numbers $x$.
**26.** If $f$ and $g$ are linear functions, then any real number
can be an input for each function. Thus the domain of
$f \circ g$ = the domain of $g \circ f = (-\infty, \infty)$.  **27.** This approach
is not valid. Consider Exercise 23 in Section 2.3, for example.
Since $(f \circ g)(x) = \dfrac{4x}{x-5}$, an examination of only this
composed function would lead to the incorrect conclusion
that the domain of $f \circ g$ is $(-\infty, 5) \cup (5, \infty)$. However, we
must also exclude from the domain of $f \circ g$ those values
of $x$ that are not in the domain of $g$. Thus the domain of
$f \circ g$ is $(-\infty, 0) \cup (0, 5) \cup (5, \infty)$.

## Exercise Set 2.4

**1.** $x$-axis, no; $y$-axis, yes; origin, no  **3.** $x$-axis, yes;
$y$-axis, no; origin, no  **5.** $x$-axis, no; $y$-axis, no; origin, yes

**7.** *x*-axis, no; *y*-axis, yes; origin, no    **9.** *x*-axis, no;
*y*-axis, no; origin, no    **11.** *x*-axis, no; *y*-axis, yes; origin, no
**13.** *x*-axis, no; *y*-axis, no; origin, yes    **15.** *x*-axis, no;
*y*-axis, no; origin, yes    **17.** *x*-axis, yes; *y*-axis, yes;
origin, yes    **19.** *x*-axis, no; *y*-axis, yes; origin, no
**21.** *x*-axis, yes; *y*-axis, yes; origin, yes    **23.** *x*-axis, no;
*y*-axis, no; origin, no    **25.** *x*-axis, no; *y*-axis, no; origin, yes
**27.** *x*-axis, $(-5, -6)$; *y*-axis, $(5, 6)$; origin, $(5, -6)$
**29.** *x*-axis, $(-10, 7)$; *y*-axis, $(10, -7)$; origin, $(10, 7)$
**31.** *x*-axis, $(0, 4)$; *y*-axis, $(0, -4)$; origin, $(0, 4)$
**33.** Even    **35.** Odd    **37.** Neither    **39.** Odd
**41.** Even    **43.** Odd    **45.** Neither    **47.** Even
**49.** [2.1]

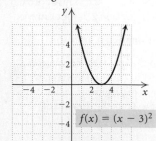

**50.** [1.5] Opening ceremonies: \$1100; closing ceremonies: \$775
**51.** Odd    **53.** *x*-axis, yes; *y*-axis, no; origin, no

**55.** $E(-x) = \dfrac{f(-x) + f(-(-x))}{2} = \dfrac{f(-x) + f(x)}{2} = E(x)$

**57. (a)** $E(x) + O(x) = \dfrac{f(x) + f(-x)}{2} + \dfrac{f(x) - f(-x)}{2} =$
$\dfrac{2f(x)}{2} = f(x)$; **(b)** $f(x) = \dfrac{-22x^2 + \sqrt{x} + \sqrt{-x} - 20}{2} +$
$\dfrac{8x^3 + \sqrt{x} - \sqrt{-x}}{2}$    **59.** True

## Visualizing the Graph

**1.** C    **2.** B    **3.** A    **4.** E    **5.** G    **6.** D    **7.** H
**8.** I    **9.** F

## Exercise Set 2.5

**1.** Start with the graph of $y = x^2$.    **3.** Start with the graph of $y = x$.
Shift it right 3 units.    Shift it down 3 units.

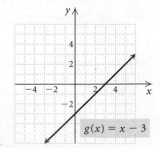

**5.** Start with the graph of $y = \sqrt{x}$. Reflect it across the *x*-axis.    **7.** Start with the graph of $y = \dfrac{1}{x}$. Shift it up 4 units.

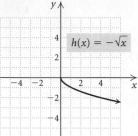
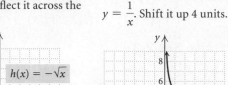

**9.** Start with the graph of $y = x$. Stretch it vertically by multiplying each *y*-coordinate by 3. Then reflect it across the *x*-axis and shift it up 3 units.

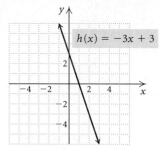

**11.** Start with the graph of $y = |x|$. Shrink it vertically by multiplying each *y*-coordinate by $\frac{1}{2}$. Then shift it down 2 units.

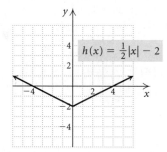

**13.** Start with the graph of $y = x^3$. Shift it right 2 units. Then reflect it across the *x*-axis.

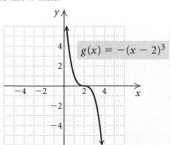

**15.** Start with the graph of $y = x^2$. Shift it left 1 unit. Then shift it down 1 unit.

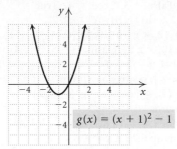

$$g(x) = (x + 1)^2 - 1$$

**17.** Start with the graph of $y = x^3$. Shrink it vertically by multiplying each $y$-coordinate by $\frac{1}{3}$. Then shift it up 2 units.

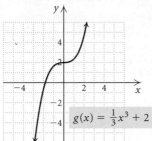

$$g(x) = \frac{1}{3}x^3 + 2$$

**19.** Start with the graph of $y = \sqrt{x}$. Shift it left 2 units.

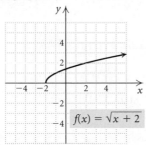

$$f(x) = \sqrt{x + 2}$$

**21.** Start with the graph of $y = \sqrt[3]{x}$. Shift it down 2 units.

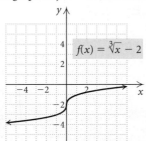

$$f(x) = \sqrt[3]{x} - 2$$

**23.** Start with the graph of $y = |x|$. Shrink it horizontally by multiplying each $x$-coordinate by $\frac{1}{3}$ (or dividing each $x$-coordinate by 3).

**25.** Start with the graph of $y = \frac{1}{x}$. Stretch it vertically by multiplying each $y$-coordinate by 2.

**27.** Start with the graph of $y = \sqrt{x}$. Stretch it vertically by multiplying each $y$-coordinate by 3. Then shift it down 5 units.

**29.** Start with the graph of $y = |x|$. Stretch it horizontally by multiplying each $x$-coordinate by 3. Then shift it down 4 units.

**31.** Start with the graph of $y = x^2$. Shift it right 5 units, shrink it vertically by multiplying each $y$-coordinate by $\frac{1}{4}$, and then reflect it across the $x$-axis.

**33.** Start with the graph of $y = \frac{1}{x}$. Shift it left 3 units, then up 2 units.

**35.** Start with the graph of $y = x^2$. Shift it right 3 units. Then reflect it across the $x$-axis and shift it up 5 units.

**37.** $(-12, 2)$    **39.** $(12, 4)$    **41.** $(-12, 2)$    **43.** $(-12, 16)$
**45.** B    **47.** A    **49.** $f(x) = -(x - 8)^2$

**51.** $f(x) = |x + 7| + 2$    **53.** $f(x) = \frac{1}{2x} - 3$

**55.** $f(x) = -(x - 3)^2 + 4$    **57.** $f(x) = \sqrt{-(x + 2)} - 1$

**59.**

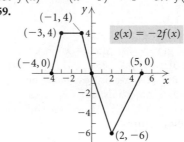

$$g(x) = -2f(x)$$

**61.**

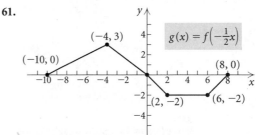

$$g(x) = f\left(-\frac{1}{2}x\right)$$

**63.**

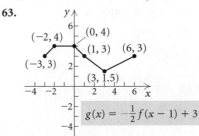

$$g(x) = -\frac{1}{2}f(x - 1) + 3$$

**65.**

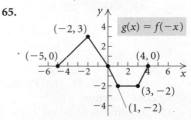

$$g(x) = f(-x)$$

**67.**

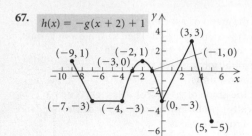

$h(x) = -g(x + 2) + 1$

$(-9, 1)$ $(-2, 1)$ $(3, 3)$
$(-3, 0)$ $(-1, 0)$
$(-7, -3)$ $(-4, -3)$ $(0, -3)$
$(5, -5)$

**69.**

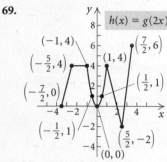

$h(x) = g(2x)$

$(-1, 4)$ $\left(\frac{7}{2}, 6\right)$
$\left(-\frac{5}{2}, 4\right)$ $(1, 4)$
$\left(-\frac{7}{2}, 0\right)$ $\left(\frac{1}{2}, 1\right)$
$\left(-\frac{1}{2}, 1\right)$ $\left(\frac{5}{2}, -2\right)$
$(0, 0)$

**71.** (f)  **73.** (f)  **75.** (d)  **77.** (c)
**79.** $f(-x) = 2(-x)^4 - 35(-x)^3 + 3(-x) - 5 =$
$2x^4 + 35x^3 - 3x - 5 = g(x)$  **81.** $g(x) = x^3 - 3x^2 + 2$
**83.** $k(x) = (x + 1)^3 - 3(x + 1)^2$  **85.** [2.4] $x$-axis, no;
$y$-axis, yes; origin, no  **86.** [2.4] $x$-axis, yes; $y$-axis, no;
origin, no  **87.** [2.4] $x$-axis, no; $y$-axis, no; origin, yes
**88.** [1.5] 1.5 million games  **89.** [1.5] About $83.3 billion
**90.** [1.5] About 68.3 million returns
**91.**     **93.**

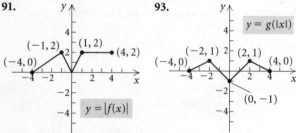

$(-1, 2)$ $(1, 2)$
$(-4, 0)$ $(4, 2)$
$y = |f(x)|$

$y = g(|x|)$
$(-2, 1)$ $(2, 1)$
$(-4, 0)$ $(4, 0)$
$(0, -1)$

**95.** Start with the graph of $g(x) = [\![x]\!]$. Shift it right $\frac{1}{2}$ unit.
Domain: all real numbers; range: all integers.
**97.** $(3, 8); (3, 6); \left(\frac{3}{2}, 4\right)$

## Exercise Set 2.6

**1.** $4.5; y = 4.5x$  **3.** $36; y = \dfrac{36}{x}$  **5.** $4; y = 4x$

**7.** $4; y = \dfrac{4}{x}$  **9.** $\dfrac{3}{8}; y = \dfrac{3}{8}x$  **11.** $0.54; y = \dfrac{0.54}{x}$  **13.** $1.41

**15.** About 686 kg  **17.** 90 g  **19.** 3.5 hr  **21.** $66\frac{2}{3}$ cm

**23.** 1.92 ft  **25.** $y = \dfrac{0.0015}{x^2}$  **27.** $y = 15x^2$  **29.** $y = xz$

**31.** $y = \frac{3}{10}xz^2$  **33.** $y = \dfrac{1}{5} \cdot \dfrac{xz}{wp}$, or $y = \dfrac{xz}{5wp}$  **35.** 2.5 m

**37.** 36 mph  **39.** About 103 earned runs  **41.** [1.4] Parallel
**42.** [1.5] Zero  **43.** [2.1] Relative minimum
**44.** [2.4] Odd function  **45.** [2.6] Inverse variation

**47.** $3.73; $4.50  **49.** $\dfrac{\pi}{4}$

## Review Exercises: Chapter 2

**1.** True  **2.** False  **3.** True  **4.** True
**5.** (a) $(-4, -2)$; (b) $(2, 5)$; (c) $(-2, 2)$
**6.** (a) $(-1, 0), (2, \infty)$; (b) $(0, 2)$; (c) $(-\infty, -1)$
**7.** Increasing: $(0, \infty)$; decreasing: $(-\infty, 0)$; relative
minimum: $-1$ at $x = 0$  **8.** Increasing: $(-\infty, 0)$;
decreasing: $(0, \infty)$; relative maximum: 2 at $x = 0$
**9.** Increasing: $(2, \infty)$; decreasing: $(-\infty, 2)$;
relative minimum: $-1$ at $x = 2$
**10.** Increasing: $(-\infty, 0.5)$; decreasing: $(0.5, \infty)$;
relative maximum: 6.25 at $x = 0.5$
**11.** Increasing: $(-\infty, -1.155), (1.155, \infty)$; decreasing:
$(-1.155, 1.155)$; relative maximum: 3.079 at $x = -1.155$;
relative minimum: $-3.079$ at $x = 1.155$
**12.** Increasing: $(-1.155, 1.155)$; decreasing:
$(-\infty, -1.155), (1.155, \infty)$; relative maximum: 1.540 at
$x = 1.155$; relative minimum: $-1.540$ at $x = -1.155$
**13.** $A(l) = l(10 - l)$, or $10l - l^2$  **14.** $A(x) = 2x\sqrt{4 - x^2}$

**15.** (a) $A(x) = x\left(33 - \dfrac{x}{2}\right)$, or $33x - \dfrac{x^2}{2}$;

(b) $\{x \mid 0 < x < 66\}$;
(c)     (d) 33 ft by 16.5 ft

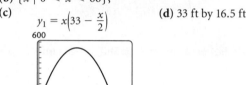

$y_1 = x\left(33 - \dfrac{x}{2}\right)$

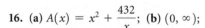

**16.** (a) $A(x) = x^2 + \dfrac{432}{x}$; (b) $(0, \infty)$;

(c) $x = 6$ in., height $= 3$ in.
**17.**     **18.**

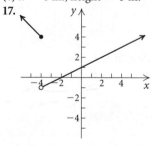

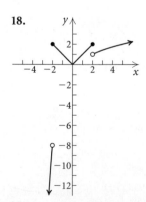

**19.**     **20.**

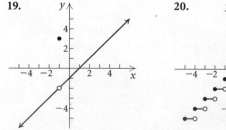

$f(x) = [\![x]\!]$

**21.**

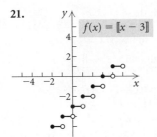

**22.** $f(-1) = 1; f(5) = 2; f(-2) = 2; f(-3) = -27$
**23.** $f(-2) = -3; f(-1) = 3; f(0) = -1; f(4) = 3$
**24.** 33    **25.** 0    **26.** Does not exist
**27. (a)** Domain of $f$: $(-\infty, 0) \cup (0, \infty)$; domain of $g$:
$(-\infty, \infty)$; domain of $f + g, f - g$, and $fg$: $(-\infty, 0) \cup (0, \infty)$;
domain of $f/g$: $(-\infty, 0) \cup \left(0, \frac{3}{2}\right) \cup \left(\frac{3}{2}, \infty\right)$

**(b)** $(f + g)(x) = \frac{4}{x^2} + 3 - 2x; (f - g)(x) = \frac{4}{x^2} - 3 + 2x;$

$(fg)(x) = \frac{12}{x^2} - \frac{8}{x}; (f/g)(x) = \frac{4}{x^2(3 - 2x)}$

**28. (a)** Domain of $f, g, f + g, f - g$, and $fg$: $(-\infty, \infty)$;
domain of $f/g$: $\left(-\infty, \frac{1}{2}\right) \cup \left(\frac{1}{2}, \infty\right)$;
**(b)** $(f + g)(x) = 3x^2 + 6x - 1; (f - g)(x) = 3x^2 + 2x + 1;$

$(fg)(x) = 6x^3 + 5x^2 - 4x; (f/g)(x) = \frac{3x^2 + 4x}{2x - 1}$

**29.** $P(x) = -0.5x^2 + 105x - 6$    **30.** 2    **31.** $-2x - h$

**32.** $\frac{-4}{x(x + h)}$, or $-\frac{4}{x(x + h)}$    **33.** 9    **34.** 5

**35.** 128    **36.** 580    **37.** 7    **38.** $-509$
**39.** $4x - 3$    **40.** $-24 + 27x^3 - 9x^6 + x^9$

**41. (a)** $(f \circ g)(x) = \frac{4}{(3 - 2x)^2}; (g \circ f)(x) = 3 - \frac{8}{x^2};$

**(b)** domain of $f \circ g$: $\left(-\infty, \frac{3}{2}\right) \cup \left(\frac{3}{2}, \infty\right)$; domain of
$g \circ f$: $(-\infty, 0) \cup (0, \infty)$
**42. (a)** $(f \circ g)(x) = 12x^2 - 4x - 1; (g \circ f)(x) = 6x^2 + 8x - 1;$
**(b)** domain of $f \circ g$ and $g \circ f$: $(-\infty, \infty)$
**43.** $f(x) = \sqrt{x}, g(x) = 5x + 2$; answers may vary.
**44.** $f(x) = 4x^2 + 9, g(x) = 5x - 1$; answers may vary.
**45.** $x$-axis, yes; $y$-axis, yes; origin, yes    **46.** $x$-axis, yes;
$y$-axis, yes; origin, yes    **47.** $x$-axis, no; $y$-axis, no; origin, no
**48.** $x$-axis, no; $y$-axis, yes; origin, no    **49.** $x$-axis, no;
$y$-axis, no; origin, yes    **50.** $x$-axis, no; $y$-axis, yes; origin, no
**51.** Even    **52.** Even    **53.** Odd    **54.** Even    **55.** Even
**56.** Neither    **57.** Odd    **58.** Even    **59.** Even
**60.** Odd    **61.** $f(x) = (x + 3)^2$
**62.** $f(x) = -\sqrt{x - 3} + 4$    **63.** $f(x) = 2|x - 3|$
**64.**

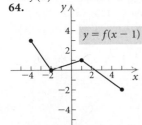

**65.**

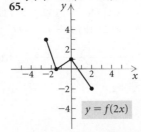

**66.**

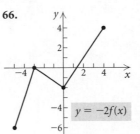

**67.**

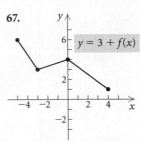

**68.** $y = 4x$    **69.** $y = \frac{2}{3}x$    **70.** $y = \frac{2500}{x}$    **71.** $y = \frac{54}{x}$

**72.** $y = \frac{48}{x^2}$    **73.** $y = \frac{1}{10} \cdot \frac{xz^2}{w}$    **74.** 20 min    **75.** 75

**76.** 500 watts    **77.** A    **78.** C    **79.** B
**80.** Let $f(x)$ and $g(x)$ be odd functions. Then by definition,
$f(-x) = -f(x)$, or $f(x) = -f(-x)$,
and $g(-x) = -g(x)$, or $g(x) = -g(-x)$. Thus,
$(f + g)(x) = f(x) + g(x) = -f(-x) + [-g(-x)] = -[f(-x) + g(-x)] = -(f + g)(-x)$ and $f + g$ is odd.
**81.** Reflect the graph of $y = f(x)$ across the $x$-axis
and then across the $y$-axis.    **82. (a)** $4x^3 - 2x + 9$;
**(b)** $4x^3 + 24x^2 + 46x + 35$; **(c)** $4x^3 - 2x + 42$.
**(a)** Adds 2 to each function value; **(b)** adds 2 to each input
before finding a function value; **(c)** adds the output for
2 to the output for $x$    **83.** In the graph of $y = f(cx)$,
the constant $c$ stretches or shrinks the graph of $y = f(x)$
horizontally. The constant $c$ in $y = cf(x)$ stretches or shrinks
the graph of $y = f(x)$ vertically. For $y = f(cx)$, the
$x$-coordinates of $y = f(x)$ are divided by $c$; for $y = cf(x)$,
the $y$-coordinates of $y = f(x)$ are multiplied by $c$.
**84.** The graph of $f(x) = 0$ is symmetric with respect
to the $x$-axis, the $y$-axis, and the origin. This function is
both even and odd.    **85.** If all the exponents are even
numbers, then $f(x)$ is an even function. If $a_0 = 0$ and all the
exponents are odd numbers, then $f(x)$ is an odd function.
**86.** Let $y(x) = kx^2$. Then $y(2x) = k(2x)^2 = k \cdot 4x^2 = 4 \cdot kx^2 = 4 \cdot y(x)$. Thus doubling $x$ causes $y$

to be quadrupled.    **87.** Let $y = k_1 x$ and $x = \frac{k_2}{z}$.

Then $y = k_1 \cdot \frac{k_2}{z}$, or $y = \frac{k_1 k_2}{z}$, so $y$ varies inversely as $z$.

## Test: Chapter 2

**1.** [2.1] **(a)** $(-5, -2)$; **(b)** $(2, 5)$; **(c)** $(-2, 2)$
**2.** [2.1] Increasing: $(-\infty, 0)$; decreasing: $(0, \infty)$;
relative maximum: 2 at $x = 0$    **3.** [2.1] Increasing:
$(-\infty, -2.667), (0, \infty)$; decreasing: $(-2.667, 0)$; relative
maximum: 9.481 at $x = -2.667$; relative minimum: 0
at $x = 0$    **4.** [2.1] $A(b) = \frac{1}{2}b(4b - 6)$, or $2b^2 - 3b$

**5.** [2.1]

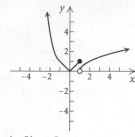

**6.** [2.1] $f\left(-\frac{7}{8}\right) = \frac{7}{8}; f(5) = 2; f(-4) = 16$    **7.** [2.2] 66
**8.** [2.2] 6    **9.** [2.2] $-1$    **10.** [2.2] 0    **11.** [1.2] $(-\infty, \infty)$
**12.** [1.2], [1.6] $[3, \infty)$    **13.** [2.2] $[3, \infty)$    **14.** [2.2] $[3, \infty)$
**15.** [2.2] $[3, \infty)$    **16.** [2.2] $(3, \infty)$
**17.** [2.2] $(f + g)(x) = x^2 + \sqrt{x - 3}$
**18.** [2.2] $(f - g)(x) = x^2 - \sqrt{x - 3}$
**19.** [2.2] $(fg)(x) = x^2\sqrt{x - 3}$
**20.** [2.2] $(f/g)(x) = \dfrac{x^2}{\sqrt{x - 3}}$    **21.** [2.2] $\frac{1}{2}$
**22.** [2.2] $4x + 2h - 1$    **23.** [2.3] 83    **24.** [2.3] 0
**25.** [2.3] 4    **26.** [2.3] $16x + 15$
**27.** [2.3] $(f \circ g)(x) = \sqrt{x^2 - 4}; (g \circ f)(x) = x - 4$
**28.** [2.3] Domain of $(f \circ g)(x) = (-\infty, -2] \cup [2, \infty)$;
domain of $(g \circ f)(x) = [5, \infty)$
**29.** [2.3] $f(x) = x^4; g(x) = 2x - 7$; answers may vary
**30.** [2.4] $x$-axis: no; $y$-axis: yes; origin: no    **31.** [2.4] Odd
**32.** [2.5] $f(x) = (x - 2)^2 - 1$
**33.** [2.5] $f(x) = -(x + 2)^2 + 3$
**34.** [2.5]

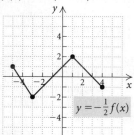

**35.** [2.6] $y = \dfrac{30}{x}$    **36.** [2.6] $y = 5x$    **37.** [2.6] $y = \dfrac{50xz^2}{w}$
**38.** [2.6] 50 ft    **39.** [2.5] C    **40.** [2.5] $(-1, 1)$

# Chapter 3

## Exercise Set 3.1

**1.** $\sqrt{3}i$    **3.** $5i$    **5.** $-\sqrt{33}i$    **7.** $-9i$    **9.** $7\sqrt{2}i$
**11.** $2 + 11i$    **13.** $5 - 12i$    **15.** $4 + 8i$    **17.** $-4 - 2i$
**19.** $5 + 9i$    **21.** $5 + 4i$    **23.** $5 + 7i$    **25.** $11 - 5i$
**27.** $-1 + 5i$    **29.** $2 - 12i$    **31.** $-12$    **33.** $-45$
**35.** $35 + 14i$    **37.** $6 + 16i$    **39.** $13 - i$    **41.** $-11 + 16i$
**43.** $-10 + 11i$    **45.** $-31 - 34i$    **47.** $-14 + 23i$

**49.** 41    **51.** 13    **53.** 74    **55.** $12 + 16i$
**57.** $-45 - 28i$    **59.** $-8 - 6i$    **61.** $2i$    **63.** $-7 + 24i$
**65.** $\frac{15}{146} + \frac{33}{146}i$    **67.** $\frac{10}{13} - \frac{15}{13}i$    **69.** $-\frac{14}{13} + \frac{5}{13}i$
**71.** $\frac{11}{25} - \frac{27}{25}i$    **73.** $\dfrac{-4\sqrt{3} + 10}{41} + \dfrac{5\sqrt{3} + 8}{41}i$
**75.** $-\frac{1}{2} + \frac{1}{2}i$    **77.** $-\frac{1}{2} - \frac{13}{2}i$    **79.** $-i$    **81.** $-i$
**83.** 1    **85.** $i$    **87.** 625    **89.** [1.4] $y = -2x + 1$
**90.** [2.2] All real numbers, or $(-\infty, \infty)$
**91.** [2.2] $\left(-\infty, -\frac{5}{3}\right) \cup \left(-\frac{5}{3}, \infty\right)$    **92.** [2.2] $x^2 - 3x - 1$
**93.** [2.2] $\frac{8}{11}$    **94.** [2.2] $2x + h - 3$    **95.** True    **97.** True
**99.** $a^2 + b^2$    **101.** $x^2 - 6x + 25$

## Exercise Set 3.2

**1.** $\frac{2}{3}, \frac{3}{2}$    **3.** $-2, 10$    **5.** $-1, \frac{2}{3}$    **7.** $-\sqrt{3}, \sqrt{3}$
**9.** $-\sqrt{7}, \sqrt{7}$    **11.** $-\sqrt{2}i, \sqrt{2}i$    **13.** $-4i, 4i$
**15.** $0, 3$    **17.** $-\frac{1}{3}, 0, 2$    **19.** $-1, -\frac{1}{7}, 1$
**21.** **(a)** $(-4, 0), (2, 0)$;    **(b)** $-4, 2$
**23.** **(a)** $(-1, 0), (3, 0)$;    **(b)** $-1, 3$
**25.** **(a)** $(-2, 0), (2, 0)$;    **(b)** $-2, 2$    **27.** **(a)** $(1, 0)$;    **(b)** 1
**29.** $-7, 1$    **31.** $4 \pm \sqrt{7}$    **33.** $-4 \pm 3i$    **35.** $-2, \frac{1}{3}$
**37.** $-3, 5$    **39.** $-1, \frac{2}{5}$    **41.** $\dfrac{5 \pm \sqrt{7}}{3}$    **43.** $-\frac{1}{2} \pm \dfrac{\sqrt{7}}{2}i$
**45.** $\dfrac{4 \pm \sqrt{31}}{5}$    **47.** $\frac{5}{6} \pm \dfrac{\sqrt{23}}{6}i$    **49.** $4 \pm \sqrt{11}$
**51.** $\dfrac{-1 \pm \sqrt{61}}{6}$    **53.** $\dfrac{5 \pm \sqrt{17}}{4}$    **55.** $-\frac{1}{5} \pm \frac{3}{5}i$
**57.** 144; two real    **59.** $-7$; two imaginary
**61.** 49; two real    **63.** 2, 6    **65.** 0.143, 6
**67.** $-0.151, 1.651$    **69.** $-0.637, 3.137$    **71.** $-5, -1$
**73.** $\dfrac{3 \pm \sqrt{21}}{2}$    **75.** $\dfrac{5 \pm \sqrt{21}}{2}$    **77.** $-1 \pm \sqrt{6}$
**79.** $\frac{1}{4} \pm \dfrac{\sqrt{31}}{4}i$    **81.** $\dfrac{1 \pm \sqrt{13}}{6}$    **83.** $\dfrac{1 \pm \sqrt{6}}{5}$
**85.** $\dfrac{-3 \pm \sqrt{57}}{8}$    **87.** $-1.535, 0.869$    **89.** $-0.347, 1.181$
**91.** $\pm 1, \pm \sqrt{2}$    **93.** $\pm \sqrt{2}, \pm \sqrt{5}i$    **95.** $\pm 1, \pm \sqrt{5}i$
**97.** 16    **99.** $-8, 64$    **101.** 1, 16    **103.** $\frac{5}{2}, 3$
**105.** $-\frac{3}{2}, -1, \frac{1}{2}, 1$    **107.** 1995    **109.** 2010
**111.** About 10.216 sec    **113.** Length: 4 ft; width: 3 ft
**115.** 4 and 9; $-9$ and $-4$    **117.** 2 cm    **119.** Length: 8 ft;
width: 6 ft    **121.** Linear    **123.** Quadratic    **125.** Linear
**127.** [1.2] $16.68 billion    **128.** [1.2] About 12 years after 2004,
or in 2016    **129.** [2.4] $x$-axis: yes; $y$-axis: yes; origin: yes
**130.** [2.4] $x$-axis: no; $y$-axis: yes; origin: no
**131.** [2.4] Odd    **132.** [2.4] Neither    **133.** **(a)** 2;    **(b)** $\frac{11}{2}$
**135.** **(a)** 2;    **(b)** $1 - i$    **137.** 1    **139.** $-\sqrt{7}, -\frac{3}{2}, 0, \frac{1}{3}, \sqrt{7}$
**141.** $\dfrac{-1 \pm \sqrt{1 + 4\sqrt{2}}}{2}$    **143.** $3 \pm \sqrt{5}$    **145.** 19
**147.** $-2 \pm \sqrt{2}, \frac{1}{2} \pm \dfrac{\sqrt{7}}{2}i$

## Visualizing the Graph

**1.** C  **2.** B  **3.** A  **4.** J  **5.** F  **6.** D  **7.** I
**8.** G  **9.** H  **10.** E

## Exercise Set 3.3

**1.** (a) $\left(-\frac{1}{2}, -\frac{9}{4}\right)$;  (b) $x = -\frac{1}{2}$;  (c) minimum: $-\frac{9}{4}$
**3.** (a) $(4, -4)$;  (b) $x = 4$;  (c) minimum: $-4$;
(d)

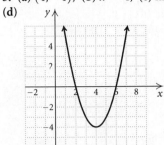

$$f(x) = x^2 - 8x + 12$$

**5.** (a) $\left(\frac{7}{2}, -\frac{1}{4}\right)$;  (b) $x = \frac{7}{2}$;  (c) minimum: $-\frac{1}{4}$;
(d)

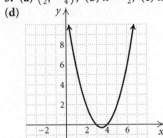

$$f(x) = x^2 - 7x + 12$$

**7.** (a) $(-2, 1)$;  (b) $x = -2$;  (c) minimum: 1;
(d)

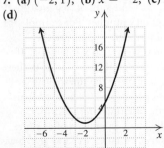

$$f(x) = x^2 + 4x + 5$$

**9.** (a) $(-4, -2)$;  (b) $x = -4$;  (c) minimum: $-2$;
(d)

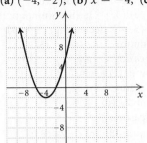

$$g(x) = \frac{x^2}{2} + 4x + 6$$

**11.** (a) $\left(-\frac{3}{2}, \frac{7}{2}\right)$;  (b) $x = -\frac{3}{2}$;  (c) minimum: $\frac{7}{2}$;
(d)

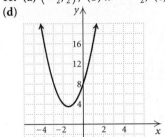

$$g(x) = 2x^2 + 6x + 8$$

**13.** (a) $(-3, 12)$;  (b) $x = -3$;  (c) maximum: 12;
(d)

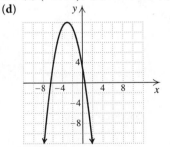

$$f(x) = -x^2 - 6x + 3$$

**15.** (a) $\left(\frac{1}{2}, \frac{3}{2}\right)$;  (b) $x = \frac{1}{2}$;  (c) maximum: $\frac{3}{2}$;
(d)

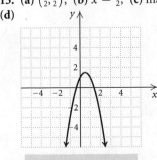

$$g(x) = -2x^2 + 2x + 1$$

**17.** (f)    **19.** (b)    **21.** (h)    **23.** (c)    **25.** True
**27.** False    **29.** True    **31. (a)** $(3, -4)$; **(b)** minimum: $-4$;
**(c)** $[-4, \infty)$; **(d)** increasing: $(3, \infty)$; decreasing: $(-\infty, 3)$
**33. (a)** $(-1, -18)$; **(b)** minimum: $-18$; **(c)** $[-18, \infty)$;
**(d)** increasing: $(-1, \infty)$; decreasing: $(-\infty, -1)$
**35. (a)** $\left(5, \frac{9}{2}\right)$; **(b)** maximum: $\frac{9}{2}$; **(c)** $\left(-\infty, \frac{9}{2}\right]$;
**(d)** increasing: $(-\infty, 5)$; decreasing: $(5, \infty)$
**37. (a)** $(-1, 2)$; **(b)** minimum: 2; **(c)** $[2, \infty)$;
**(d)** increasing: $(-1, \infty)$; decreasing: $(-\infty, -1)$
**39. (a)** $\left(-\frac{3}{2}, 18\right)$; **(b)** maximum: 18; **(c)** $\left(-\infty, 18\right]$;
**(d)** increasing: $\left(-\infty, -\frac{3}{2}\right)$; decreasing: $\left(-\frac{3}{2}, \infty\right)$
**41.** 0.625 sec; 12.25 ft    **43.** 3.75 sec; 305 ft
**45.** 4.5 in.    **47.** Base: 10 cm; height: 10 cm
**49.** 350 chairs    **51.** \$797; 40 units    **53.** 4800 yd$^2$
**55.** 350.6 ft    **57.** [2.2] 3    **58.** [2.2] $4x + 2h - 1$
**59.** [2.5]

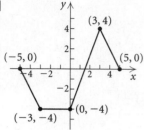

$$g(x) = -2f(x)$$

**60.** [2.5]    **61.** $-236.25$

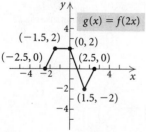

$$g(x) = f(2x)$$

**63.**

$$f(x) = (|x| - 5)^2 - 3$$

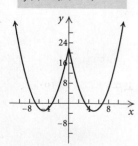

**65.** Pieces should be
$\dfrac{24\pi}{4 + \pi}$ in. and $\dfrac{96}{4 + \pi}$ in.

$x - 1 = \pm \dfrac{\sqrt{7}}{2}$; $x = 1 \pm \dfrac{\sqrt{7}}{2} = \dfrac{2 \pm \sqrt{7}}{2}$

**22. (a)** 29; two real; **(b)** $\dfrac{3 \pm \sqrt{29}}{2}$; $-1.193, 4.193$
**23. (a)** 0; one real; **(b)** $\frac{3}{2}$    **24. (a)** $-8$; two nonreal;
**(b)** $-\dfrac{1}{3} \pm \dfrac{\sqrt{2}}{3}i$    **25.** $\pm 1, \pm \sqrt{6}i$    **26.** $\frac{1}{4}, 4$
**27.** 5 and 7; $-7$ and $-5$    **28. (a)** $(3, -2)$;
**(b)** $x = 3$; **(c)** minimum: $-2$; **(d)** $[-2, \infty)$;
**(e)** increasing: $(3, \infty)$; decreasing: $(-\infty, 3)$;
**(f)**

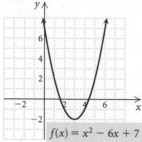

$$f(x) = x^2 - 6x + 7$$

**29. (a)** $(-1, -3)$; **(b)** $x = -1$; **(c)** maximum: $-3$;
**(d)** $(-\infty, -3]$; **(e)** increasing: $(-\infty, -1)$; decreasing: $(-1, \infty)$;
**(f)**

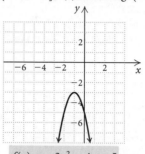

$$f(x) = -2x^2 - 4x - 5$$

**30.** Base: 8 in., height: 8 in.    **31.** The sum of two imaginary numbers is not always an imaginary number. For example, $(2 + i) + (3 - i) = 5$, a real number.    **32.** Use the discriminant. If $b^2 - 4ac < 0$, there are no $x$-intercepts. If $b^2 - 4ac = 0$, there is one $x$-intercept. If $b^2 - 4ac > 0$, there are two $x$-intercepts.    **33.** Completing the square was used in Section 3.2 to solve quadratic equations. It was used again in Section 3.3 to write quadratic functions in the form $f(x) = a(x - h)^2 + k$.    **34.** The $x$-intercepts of $g(x)$ are also $(x_1, 0)$ and $(x_2, 0)$. This is true because $f(x)$ and $g(x)$ have the same zeros. Consider $g(x) = 0$, or $-ax^2 - bx - c = 0$. Multiplying by $-1$ on both sides, we get an equivalent equation $ax^2 + bx + c = 0$, or $f(x) = 0$.

## Mid-Chapter Mixed Review: Chapter 3

**1.** True    **2.** False    **3.** True    **4.** False    **5.** $6i$    **6.** $\sqrt{5}i$
**7.** $-4i$    **8.** $4\sqrt{2}i$    **9.** $-1 + i$    **10.** $-7 + 5i$
**11.** $23 + 2i$    **12.** $-\frac{1}{29} - \frac{17}{29}i$    **13.** $i$    **14.** 1    **15.** $-i$
**16.** $-64$    **17.** $-4, 1$    **18.** $-2, -\frac{3}{2}$    **19.** $\pm \sqrt{6}$
**20.** $\pm 10i$    **21.** $4x^2 - 8x - 3 = 0$; $4x^2 - 8x = 3$;
$x^2 - 2x = \frac{3}{4}$; $x^2 - 2x + 1 = \frac{3}{4} + 1$; $(x - 1)^2 = \frac{7}{4}$;

## Exercise Set 3.4

**1.** $\frac{20}{9}$    **3.** 286    **5.** 6    **7.** 6    **9.** 4    **11.** 2, 3
**13.** $-1, 6$    **15.** $\frac{1}{2}, 5$    **17.** 7    **19.** No solution
**21.** $-\frac{69}{14}$    **23.** $-\frac{37}{18}$    **25.** 2    **27.** No solution
**29.** $\{x \,|\, x \text{ is a real number } and \ x \neq 0 \ and \ x \neq 6\}$
**31.** $\frac{5}{3}$    **33.** $\frac{9}{2}$    **35.** 3    **37.** $-4$    **39.** $-5$    **41.** $\pm \sqrt{2}$

**43.** No solution  **45.** 6  **47.** $-1$  **49.** $\frac{35}{2}$  **51.** $-98$
**53.** $-6$  **55.** 5  **57.** 7  **59.** 2  **61.** $-1, 2$  **63.** 7
**65.** 7  **67.** No solution  **69.** 1  **71.** 3, 7  **73.** 5

**75.** $-1$  **77.** $-8$  **79.** 81  **81.** $T_1 = \dfrac{P_1 V_1 T_2}{P_2 V_2}$

**83.** $C = \dfrac{1}{LW^2}$  **85.** $R_2 = \dfrac{RR_1}{R_1 - R}$

**87.** $P = \dfrac{A}{I^2 + 2I + 1}$, or $\dfrac{A}{(I + 1)^2}$  **89.** $p = \dfrac{Fm}{m - F}$

**91.** [1.5] $\frac{15}{2}$, or 7.5  **92.** [1.5] 3
**93.** [1.5] About 4975 fatalities
**94.** [1.5] Mall of America: 96 acres; Disneyland: 85 acres
**95.** $3 \pm 2\sqrt{2}$  **97.** $-1$  **99.** 0, 1

## Exercise Set 3.5

**1.** $-7, 7$  **3.** 0  **5.** $-\frac{5}{6}, \frac{5}{6}$  **7.** No solution  **9.** $-\frac{1}{3}, \frac{1}{3}$
**11.** $-3, 3$  **13.** $-3, 5$  **15.** $-8, 4$  **17.** $-1, -\frac{1}{3}$
**19.** $-24, 44$  **21.** $-2, 4$  **23.** $-13, 7$  **25.** $-\frac{4}{3}, \frac{2}{3}$
**27.** $-\frac{3}{4}, \frac{9}{4}$  **29.** $-13, 1$  **31.** 0, 1
**33.** $(-7, 7)$;

**35.** $[-2, 2]$;

**37.** $(-\infty, -4.5] \cup [4.5, \infty)$;

**39.** $(-\infty, -3) \cup (3, \infty)$;

**41.** $\left(-\frac{1}{3}, \frac{1}{3}\right)$;

**43.** $(-\infty, -3] \cup [3, \infty)$;

**45.** $(-17, 1)$;

**47.** $(-\infty, -17] \cup [1, \infty)$;

**49.** $\left(-\frac{1}{4}, \frac{3}{4}\right)$;

**51.** $[-6, 3]$;

**53.** $(-\infty, 4.9) \cup (5.1, \infty)$;

**55.** $\left(-\infty, -\frac{1}{2}\right] \cup \left[\frac{7}{2}, \infty\right)$;

**57.** $\left[-\frac{7}{3}, 1\right]$;

**59.** $(-\infty, -8) \cup (7, \infty)$;

**61.** No solution  **63.** $(-\infty, \infty)$  **65.** [1.1] $y$-intercept
**66.** [1.1] Distance formula  **67.** [1.2] Relation
**68.** [1.2] Function  **69.** [1.3] Horizontal lines
**70.** [1.4] Parallel  **71.** [2.1] Decreasing
**72.** [2.4] Symmetric with respect to the $y$-axis  **73.** $\left(-\infty, \frac{1}{2}\right)$
**75.** No solution  **77.** $\left(-\infty, -\frac{8}{3}\right) \cup (-2, \infty)$

## Review Exercises: Chapter 3

**1.** True  **2.** True  **3.** False  **4.** False  **5.** $-\frac{5}{2}, \frac{1}{3}$
**6.** $-5, 1$  **7.** $-2, \frac{4}{3}$  **8.** $-\sqrt{3}, \sqrt{3}$  **9.** $-\sqrt{10}i, \sqrt{10}i$
**10.** 1  **11.** $-5, 3$  **12.** $\dfrac{1 \pm \sqrt{41}}{4}$  **13.** $-\frac{1}{3} \pm \frac{2\sqrt{2}}{3}i$
**14.** $\frac{27}{7}$  **15.** $-\frac{1}{2}, \frac{9}{4}$  **16.** 0, 3  **17.** 5  **18.** 1, 7  **19.** $-8, 1$
**20.** $(-\infty, -3] \cup [3, \infty)$;

**21.** $\left(-\frac{14}{3}, 2\right)$;

**22.** $\left(-\frac{2}{3}, 1\right)$;

**23.** $(-\infty, -6] \cup [-2, \infty)$;

**24.** $P = \dfrac{MN}{M + N}$  **25.** $-2\sqrt{10}i$  **26.** $-4\sqrt{15}$
**27.** $-\frac{7}{8}$  **28.** $2 - i$  **29.** $1 - 4i$  **30.** $-18 - 26i$
**31.** $\frac{11}{10} + \frac{3}{10}i$  **32.** $-i$  **33.** $x^2 - 3x + \frac{9}{4} = 18 + \frac{9}{4}$;
$\left(x - \frac{3}{2}\right)^2 = \frac{81}{4}; x - \frac{3}{2} = \pm\frac{9}{2}; x = \frac{3}{2} \pm \frac{9}{2}; -3, 6$
**34.** $3x^2 - 12x = 6; x^2 - 4x = 2$;
$x^2 - 4x + 4 = 2 + 4; (x - 2)^2 = 6; x - 2 = \pm\sqrt{6}$;
$x = 2 \pm \sqrt{6}; 2 - \sqrt{6}, 2 + \sqrt{6}$  **35.** $-4, \frac{2}{3}$
**36.** $1 - 3i, 1 + 3i$  **37.** $-2, 5$  **38.** 1  **39.** $\pm\sqrt{\dfrac{3 \pm \sqrt{5}}{2}}$
**40.** $-\sqrt{3}, 0, \sqrt{3}$  **41.** $-2, -\frac{2}{3}, 3$  **42.** $-5, -2, 2$
**43.** (a) $\left(\frac{3}{8}, -\frac{7}{16}\right)$; (b) $x = \frac{3}{8}$; (c) maximum: $-\frac{7}{16}$;
(d) $\left(-\infty, -\frac{7}{16}\right]$;
(e)

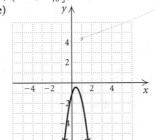

$f(x) = -4x^2 + 3x - 1$

**44.** (a) $(1, -2)$; (b) $x = 1$; (c) minimum: $-2$; (d) $[-2, \infty)$;
(e)

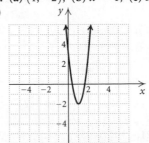

$f(x) = 5x^2 - 10x + 3$

**45.** (d)  **46.** (c)  **47.** (b)  **48.** (a)  **49.** 30 ft, 40 ft
**50.** Cassidy: 15 km/h; Logan: 8 km/h  **51.** $35 - 5\sqrt{33}$ ft,
or about 6.3 ft  **52.** 6 ft by 6 ft  **53.** $\dfrac{15 - \sqrt{115}}{2}$ cm,
or about 2.1 cm  **54.** B  **55.** B  **56.** A  **57.** 256
**58.** $4 \pm 3\sqrt[4]{3}$, or 0.052, 7.948  **59.** $-7, 9$
**60.** $-\frac{1}{4}, 2$  **61.** $-1$  **62.** 9%  **63.** $\pm 6$
**64.** The product of two imaginary numbers is not always an
imaginary number. For example, $i \cdot i = i^2 = -1$, a real
number.  **65.** No; consider the quadratic formula
$x = \dfrac{-b \pm \sqrt{b^2 - 4ac}}{2a}$. If $b^2 - 4ac = 0$, then $x = \dfrac{-b}{2a}$,
so there is one real zero. If $b^2 - 4ac > 0$, then $\sqrt{b^2 - 4ac}$ is
a real number and there are two real zeros. If $b^2 - 4ac < 0$,
then $\sqrt{b^2 - 4ac}$ is an imaginary number and there are two
imaginary zeros. Thus a quadratic function cannot have one
real zero and one imaginary zero.  **66.** You can conclude
that $|a_1| = |a_2|$ since these constants determine how wide
the parabolas are. Nothing can be concluded about the $h$'s
and the $k$'s.  **67.** When both sides of an equation are multiplied by the LCD, the resulting equation might not be equivalent to the original equation. One or more of the possible
solutions of the resulting equation might make a denominator of the original equation 0.  **68.** When both sides of an
equation are raised to an even power, the resulting equation
might not be equivalent to the original equation. For example, the solution set of $x = -2$ is $\{-2\}$, but the solution set of
$x^2 = (-2)^2$, or $x^2 = 4$, is $\{-2, 2\}$.  **69.** Absolute value is
nonnegative.  **70.** $|x| \geq 0 > p$ for any real number $x$.

## Test: Chapter 3

**1.** $[3.2] \frac{1}{2}, -5$  **2.** $[3.2] -\sqrt{6}, \sqrt{6}$  **3.** $[3.2] -2i, 2i$
**4.** $[3.2] -1, 3$  **5.** $[3.2] \dfrac{5 \pm \sqrt{13}}{2}$  **6.** $[3.2] \dfrac{3}{4} \pm \dfrac{\sqrt{23}}{4}i$
**7.** $[3.2]$ 16  **8.** $[3.4] -1, \frac{13}{6}$  **9.** $[3.4]$ 5  **10.** $[3.4]$ 5
**11.** $[3.5] -11, 3$  **12.** $[3.5] -\frac{1}{2}, 2$
**13.** $[3.5] [-7, 1]$;

**14.** $[3.5] (-2, 3)$;

**15.** $[3.5] (-\infty, -7) \cup (-3, \infty)$;

**16.** $[3.5] \left(-\infty, -\frac{2}{3}\right) \cup [4, \infty)$;

**17.** $[3.4] B = \dfrac{AC}{A - C}$  **18.** $[3.4] n = \dfrac{R^2}{3p}$
**19.** $[3.2] x^2 + 4x = 1; x^2 + 4x + 4 = 1 + 4; (x + 2)^2 = 5$;
$x + 2 = \pm\sqrt{5}; x = -2 \pm \sqrt{5}; -2 - \sqrt{5}, -2 + \sqrt{5}$
**20.** $[3.2]$ About 11.4 sec.  **21.** $[3.1] \sqrt{43}i$  **22.** $[3.1] -5i$
**23.** $[3.1] 3 - 5i$  **24.** $[3.1] 10 + 5i$  **25.** $[3.1] \frac{1}{10} - \frac{1}{5}i$
**26.** $[3.1] i$  **27.** $[3.2] -\frac{1}{4}, 3$  **28.** $[3.2] \dfrac{1 \pm \sqrt{57}}{4}$

**29.** $[3.3]$ **(a)** $(1, 9)$; **(b)** $x = 1$; **(c)** maximum: 9; **(d)** $(-\infty, 9]$;
**(e)**

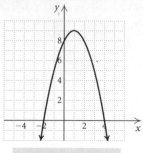

$$f(x) = -x^2 + 2x + 8$$

**30.** $[3.3]$ 20 ft by 40 ft  **31.** $[3.3]$ C  **32.** $[3.3], [3.4] -\frac{4}{9}$

# Chapter 4

## Exercise Set 4.1

**1.** $\frac{1}{2}x^3; \frac{1}{2}; 3$; cubic  **3.** $0.9x; 0.9; 1$; linear  **5.** $305x^4; 305$;
4; quartic  **7.** $x^4; 1; 4$; quartic  **9.** $4x^3; 4; 3$; cubic
**11.** (d)  **13.** (b)  **15.** (c)  **17.** (a)  **19.** (c)
**21.** (d)  **23.** Yes; no; no  **25.** No; yes; yes  **27.** $-3$,
multiplicity 2; 1, multiplicity 1  **29.** 4, multiplicity 3;
$-6$, multiplicity 1  **31.** $\pm 3$, each has multiplicity 3
**33.** 0, multiplicity 3; 1, multiplicity 2; $-4$, multiplicity 1
**35.** 3, multiplicity 2; $-4$, multiplicity 3; 0, multiplicity 4
**37.** $\pm\sqrt{3}, \pm 1$, each has multiplicity 1  **39.** $-3, \pm 1$,
each has multiplicity 1  **41.** $\pm 2, \frac{1}{2}$, each has multiplicity 1
**43.** $-1.532, -0.347, 1.879$  **45.** $-1.414, 0, 1.414$
**47.** $-1, 0, 1$  **49.** $-10.153, -1.871, -0.821, -0.303, 0.098$,
$0.535, 1.219, 3.297$  **51.** $-1.386$; relative maximum: 1.506
at $x = -0.632$, relative minimum: 0.494 at $x = 0.632$;
$(-\infty, \infty)$  **53.** $-1.249, 1.249$; relative minimum: $-3.8$ at
$x = 0$, no relative maxima; $[-3.8, \infty)$  **55.** $-1.697, 0$,
$1.856$; relative maximum: 11.012 at $x = 1.258$, relative
minimum: $-8.183$ at $x = -1.116$; $(-\infty, \infty)$  **57.** False
**59.** True  **61.** 1995: 98,856 births; 2005: 134,457 births
**63.** 26, 64, and 80  **65.** 5 sec  **67.** 2002: $159,556; 2005:
$214,783; 2008: $201,015; 2009: $175,752  **69.** 6.3%
**71.** (b)  **73.** (c)  **75.** (a)  **77.** (a) Quadratic:
$y = -342.2575758x^2 + 2687.051515x + 17,133.10909$,
$R^2 \approx 0.9388$; cubic: $y = -31.88461538x^3 +$
$88.18473193x^2 + 1217.170746x + 17,936.6014$,
$R^2 \approx 0.9783$; quartic: $y = 3.968531469x^4 -$
$103.3181818x^3 + 489.0064103x^2 + 502.8350816x +$
$18,108.04196, R^2 \approx 0.9815$; the $R^2$-value for the quartic function, 0.9815, is the closest to 1; thus the quartic function
is the best fit; (b) 8404 foreign adoptions
**79.** (a) Cubic: $y = -0.0029175919x^3 + 0.1195127076x^2 -$
$0.5287827297x + 3.087289783, R^2 \approx 0.9023$; quartic:
$y = -0.0001884254x^4 + 0.0099636973x^3 -$
$0.1550252422x^2 + 1.300895521x + 1.77274528$,

$R^2 \approx 0.9734$; the $R^2$-value for the quartic function, 0.9734, is the closest to 1; thus the quartic function is the better fit; **(b)** 1988: \$8.994 billion; 2002: \$19.862 billion; 2010: \$1.836 billion   **81.** [1.1] 5   **82.** [1.1] $6\sqrt{2}$   **83.** [1.1] Center: $(3, -5)$; radius: 7   **84.** [1.1] Center: $(-4, 3)$; radius: $2\sqrt{2}$   **85.** [1.6] $\{y | y \geq 3\}$, or $[3, \infty)$
**86.** [1.6] $\{x | x > \frac{5}{3}\}$, or $\left(\frac{5}{3}, \infty\right)$
**87.** [3.5] $\{x | x \leq -13 \ or \ x \geq 1\}$, or $(-\infty, -13] \cup [1, \infty)$
**88.** [3.5] $\{x | -\frac{11}{12} \leq x \leq \frac{5}{12}\}$, or $\left[-\frac{11}{12}, \frac{5}{12}\right]$   **89.** 16; $x^{16}$

## Visualizing the Graph

**1.** H   **2.** D   **3.** J   **4.** B   **5.** A   **6.** C
**7.** I   **8.** E   **9.** G   **10.** F

## Exercise Set 4.2

**1.** (a) 5; (b) 5; (c) 4   **3.** (a) 10; (b) 10; (c) 9
**5.** (a) 3; (b) 3; (c) 2   **7.** (d)   **9.** (f)   **11.** (b)

**13.**
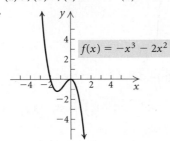
$$f(x) = -x^3 - 2x^2$$

**15.**
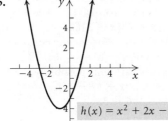
$$h(x) = x^2 + 2x - 3$$

**17.**
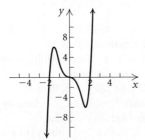
$$h(x) = x^5 - 4x^3$$

**19.**
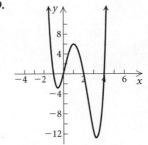
$$h(x) = x(x-4)(x+1)(x-2)$$

**21.**
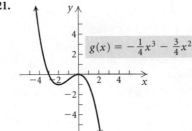
$$g(x) = -\frac{1}{4}x^3 - \frac{3}{4}x^2$$

**23.**
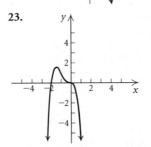
$$g(x) = -x^4 - 2x^3$$

**25.**
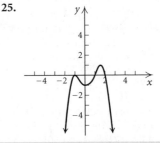
$$f(x) = -\frac{1}{2}(x-2)(x+1)^2(x-1)$$

**27.**
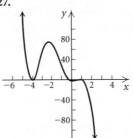
$$g(x) = -x(x-1)^2(x+4)^2$$

**29.**
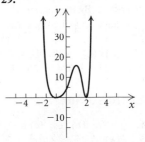
$$f(x) = (x-2)^2(x+1)^4$$

**31.**
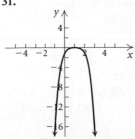
$$g(x) = -(x-1)^4$$

**33.**
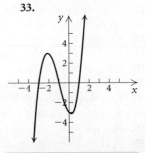
$$h(x) = x^3 + 3x^2 - x - 3$$

**35.**

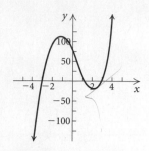

$f(x) = 6x^3 - 8x^2 - 54x + 72$

**37.**

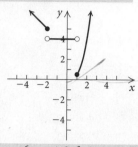

$$g(x) = \begin{cases} -x + 3, & \text{for } x \le -2, \\ 4, & \text{for } -2 < x < 1, \\ \frac{1}{2}x^3, & \text{for } x \ge 1 \end{cases}$$

**39.** $f(-5) = -18$ and $f(-4) = 7$. By the intermediate value theorem, since $f(-5)$ and $f(-4)$ have opposite signs, then $f(x)$ has a zero between $-5$ and $-4$.
**41.** $f(-3) = 22$ and $f(-2) = 5$. Both $f(-3)$ and $f(-2)$ are positive. We cannot use the intermediate value theorem to determine if there is a zero between $-3$ and $-2$.
**43.** $f(2) = 2$ and $f(3) = 57$. Both $f(2)$ and $f(3)$ are positive. We cannot use the intermediate value theorem to determine if there is a zero between 2 and 3.    **45.** $f(4) = -12$ and $f(5) = 4$. By the intermediate value theorem, since $f(4)$ and $f(5)$ have opposite signs, then $f(x)$ has a zero between 4 and 5.
**47.** [1.1] (d)    **48.** [1.3] (f)    **49.** [1.1] (e)    **50.** [1.1] (a)
**51.** [1.1] (b)    **52.** [1.3] (c)    **53.** [1.5] $\frac{9}{10}$
**54.** [4.1] $-3, 0, 4$    **55.** [3.2] $-\frac{5}{3}, \frac{11}{2}$    **56.** [1.5] $\frac{196}{25}$

## Exercise Set 4.3

**1.** (a) No; (b) yes; (c) no    **3.** (a) Yes; (b) no; (c) yes
**5.** $P(x) = (x + 2)(x^2 - 2x + 4) - 16$
**7.** $P(x) = (x + 9)(x^2 - 3x + 2) + 0$
**9.** $P(x) = (x + 2)(x^3 - 2x^2 + 2x - 4) + 11$
**11.** $Q(x) = 2x^3 + x^2 - 3x + 10$, $R(x) = -42$
**13.** $Q(x) = x^2 - 4x + 8$, $R(x) = -24$
**15.** $Q(x) = 3x^2 - 4x + 8$, $R(x) = -18$
**17.** $Q(x) = x^4 + 3x^3 + 10x^2 + 30x + 89$, $R(x) = 267$
**19.** $Q(x) = x^3 + x^2 + x + 1$, $R(x) = 0$
**21.** $Q(x) = 2x^3 + x^2 + \frac{7}{2}x + \frac{7}{4}$, $R(x) = -\frac{1}{8}$
**23.** $0; -60; 0$    **25.** $10; 80; 998$    **27.** $5,935,988; -772$
**29.** $0; 0; 65; 1 - 12\sqrt{2}$    **31.** Yes; no    **33.** Yes; yes
**35.** No; yes    **37.** No; no
**39.** $f(x) = (x - 1)(x + 2)(x + 3); 1, -2, -3$
**41.** $f(x) = (x - 2)(x - 5)(x + 1); 2, 5, -1$
**43.** $f(x) = (x - 2)(x - 3)(x + 4); 2, 3, -4$

**45.** $f(x) = (x - 3)^3(x + 2); 3, -2$
**47.** $f(x) = (x - 1)(x - 2)(x - 3)(x + 5); 1, 2, 3, -5$
**49.**                         **51.**

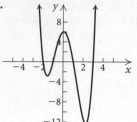    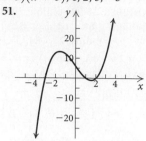

$f(x) = x^4 - x^3 - 7x^2 + x + 6$    $f(x) = x^3 - 7x + 6$

**53.**

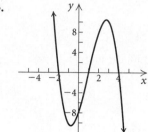

$f(x) = -x^3 + 3x^2 + 6x - 8$

**55.** [3.2] $\dfrac{5}{4} \pm \dfrac{\sqrt{71}}{4}i$    **56.** [3.2] $-1, \frac{3}{7}$    **57.** [3.2] $-5, 0$
**58.** [1.2] 10    **59.** [3.2] $-3, -2$
**60.** [1.4] $f(x) = 0.25x + 5.39$; 2005: \$6.64; 2015: \$9.14
**61.** [3.3] $b = 15$ in., $h = 15$ in.    **63.** (a) $x + 4, x + 3,$
$x - 2, x - 5$; (b) $P(x) = (x + 4)(x + 3)(x - 2)(x - 5)$;
(c) yes; two examples are $f(x) = c \cdot P(x)$ for any nonzero
constant $c$ and $g(x) = (x - a)P(x)$; (d) no    **65.** $\frac{14}{3}$
**67.** $0, -6$    **69.** Answers may vary. One possibility is
$P(x) = x^{15} - x^{14}$.    **71.** $x^2 + 2ix + (2 - 4i)$,
$R -6 - 2i$    **73.** $x - 3 + i, R\ 6 - 3i$

## Mid-Chapter Mixed Review: Chapter 4

**1.** False    **2.** True    **3.** True    **4.** False    **5.** 5; multiplicity 6    **6.** $-5, -\frac{1}{2}, 5$; each has multiplicity 1
**7.** $\pm 1, \pm \sqrt{2}$; each has multiplicity 1    **8.** 3, multiplicity 2;
$-4$, multiplicity 1    **9.** (d)    **10.** (a)    **11.** (b)
**12.** (c)    **13.** $f(-2) = -13$ and $f(0) = 3$. By the
intermediate value theorem, since $f(-2)$ and $f(0)$ have opposite signs, then $f(x)$ has a zero between $-2$ and 0.
**14.** $f\left(-\frac{1}{2}\right) = \frac{19}{8}$ and $f(1) = 2$. Both $f\left(-\frac{1}{2}\right)$ and $f(1)$ are
positive. We cannot use the intermediate value theorem to
determine if there is a zero between $-\frac{1}{2}$ and 1.
**15.** $P(x) = (x - 1)(x^3 - 5x^2 - 5x - 4) - 6$
**16.** $Q(x) = 3x^3 + 5x^2 + 12x + 18$, $R(x) = 42$
**17.** $Q(x) = x^4 - x^3 + x^2 - x + 1$, $R(x) = -6$
**18.** $-380$    **19.** $-15$    **20.** $20 - \sqrt{2}$
**21.** Yes; no    **22.** Yes; yes
**23.** $h(x) = (x - 1)(x - 8)(x + 7); 1, 8, -7$
**24.** $g(x) = (x + 1)(x - 2)(x - 4)(x + 3); -1, 2, 4, -3$

**25.** The range of a polynomial function with an odd degree is $(-\infty, \infty)$. The range of a polynomial function with an even degree is $[s, \infty)$ for some real number $s$ if $a_n > 0$ and is $(-\infty, s]$ for some real number $s$ if $a_n < 0$.     **26.** Since we can find $f(0)$ for any polynomial function $f(x)$, it is not possible for the graph of a polynomial function to have no $y$-intercept. It is possible for a polynomial function to have no $x$-intercepts. For instance, a function of the form $f(x) = x^2 + a$, $a > 0$, has no $x$-intercepts. There are other examples as well.
**27.** The zeros of a polynomial function are the first coordinates of the points at which the graph of the function crosses or is tangent to the $x$-axis.     **28.** For a polynomial $P(x)$ of degree $n$, when we have $P(x) = d(x) \cdot Q(x) + R(x)$, where the degree of $d(x)$ is 1, then the degree of $Q(x)$ must be $n - 1$.

## Exercise Set 4.4

**1.** $f(x) = x^3 - 6x^2 - x + 30$
**3.** $f(x) = x^3 + 3x^2 + 4x + 12$
**5.** $f(x) = x^3 - 3x^2 - 2x + 6$     **7.** $f(x) = x^3 - 6x - 4$
**9.** $f(x) = x^3 + 2x^2 + 29x + 148$
**11.** $f(x) = x^3 - \frac{5}{3}x^2 - \frac{2}{3}x$
**13.** $f(x) = x^5 + 2x^4 - 2x^2 - x$
**15.** $f(x) = x^4 + 3x^3 + 3x^2 + x$     **17.** $-\sqrt{3}$
**19.** $i, 2 + \sqrt{5}$     **21.** $-3i$     **23.** $-4 + 3i, 2 + \sqrt{3}$
**25.** $-\sqrt{5}, 4i$     **27.** $2 + i$     **29.** $-3 - 4i, 4 + \sqrt{5}$
**31.** $4 + i$     **33.** $f(x) = x^3 - 4x^2 + 6x - 4$
**35.** $f(x) = x^2 + 16$     **37.** $f(x) = x^3 - 5x^2 + 16x - 80$
**39.** $f(x) = x^4 - 2x^3 - 3x^2 + 10x - 10$
**41.** $f(x) = x^4 + 4x^2 - 45$     **43.** $-\sqrt{2}, \sqrt{2}$     **45.** $i, 2, 3$
**47.** $1 + 2i, 1 - 2i$     **49.** $\pm 1$     **51.** $\pm 1, \pm \frac{1}{2}, \pm 2, \pm 4, \pm 8$
**53.** $\pm 1, \pm 2, \pm \frac{1}{3}, \pm \frac{1}{5}, \pm \frac{2}{3}, \pm \frac{2}{5}, \pm \frac{1}{15}, \pm \frac{2}{15}$
**55.** (a) Rational: $-3$; other: $\pm\sqrt{2}$;
(b) $f(x) = (x + 3)(x + \sqrt{2})(x - \sqrt{2})$
**57.** (a) Rational: $\frac{1}{3}$; other: $\pm\sqrt{5}$;
(b) $f(x) = 3\left(x - \frac{1}{3}\right)(x + \sqrt{5})(x - \sqrt{5})$, or
$(3x - 1)(x + \sqrt{5})(x - \sqrt{5})$
**59.** (a) Rational: $-2, 1$; other: none;
(b) $f(x) = (x + 2)(x - 1)^2$     **61.** (a) Rational: $-\frac{3}{2}$;
other: $\pm 3i$; (b) $f(x) = 2\left(x + \frac{3}{2}\right)(x + 3i)(x - 3i)$, or
$(2x + 3)(x + 3i)(x - 3i)$
**63.** (a) Rational: $-\frac{1}{5}, 1$; other: $\pm 2i$;
(b) $f(x) = 5\left(x + \frac{1}{5}\right)(x - 1)(x + 2i)(x - 2i)$
**65.** (a) Rational: $-2, -1$; other: $3 \pm \sqrt{13}$;
(b) $f(x) = (x + 2)(x + 1)(x - 3 - \sqrt{13})(x - 3 + \sqrt{13})$
**67.** (a) Rational: 2; other: $1 \pm \sqrt{3}$;
(b) $f(x) = (x - 2)(x - 1 - \sqrt{3})(x - 1 + \sqrt{3})$
**69.** (a) Rational: $-2$; other: $1 \pm \sqrt{3}i$;
(b) $f(x) = (x + 2)(x - 1 - \sqrt{3}i)(x - 1 + \sqrt{3}i)$
**71.** (a) Rational: $\frac{1}{2}$; other: $\dfrac{1 \pm \sqrt{5}}{2}$;
(b) $f(x) = \frac{1}{3}\left(x - \frac{1}{2}\right)\left(x - \dfrac{1 + \sqrt{5}}{2}\right)\left(x - \dfrac{1 - \sqrt{5}}{2}\right)$
**73.** $1, -3$     **75.** No rational zeros     **77.** No rational zeros

**79.** $-2, 1, 2$     **81.** 3 or 1; 0     **83.** 0; 3 or 1     **85.** 2 or 0; 2 or 0     **87.** 1; 1     **89.** 1; 0     **91.** 2 or 0; 2 or 0
**93.** 3 or 1; 1     **95.** 1; 1
**97.**
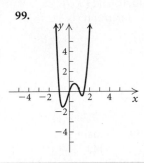

**99.**

$f(x) = 4x^3 + x^2 - 8x - 2$ $\qquad$ $f(x) = 2x^4 - 3x^3 - 2x^2 + 3x$

**101.** [3.3] (a) $(4, -6)$; (b) $x = 4$; (c) minimum: $-6$ at $x = 4$
**102.** [3.3] (a) $(1, -4)$; (b) $x = 1$; (c) minimum: $-4$ at $x = 1$
**103.** [1.5] 10     **104.** [3.2] $-3, 11$     **105.** [4.1] Cubic; $-x^3$; $-1$; 3; as $x \to \infty$, $g(x) \to -\infty$, and as $x \to -\infty$, $g(x) \to \infty$
**106.** [3.3] Quadratic; $-x^2$; $-1$; 2; as $x \to \infty$, $f(x) \to -\infty$, and as $x \to -\infty$, $f(x) \to -\infty$     **107.** [1.3], [4.1] Constant; $-\frac{4}{9}$; $-\frac{4}{9}$; 0 degree; for all $x$, $f(x) = -\frac{4}{9}$
**108.** [1.3], [4.1] Linear; $x$; 1; 1; as $x \to \infty$, $h(x) \to \infty$, and as $x \to -\infty$, $h(x) \to -\infty$     **109.** [4.1] Quartic; $x^4$; 1; 4; as $x \to \infty$, $g(x) \to \infty$, and as $x \to -\infty$, $g(x) \to \infty$
**110.** [4.1] Cubic; $x^3$; 1; 3; as $x \to \infty$, $h(x) \to \infty$, and as $x \to -\infty$, $h(x) \to -\infty$     **111.** (a) $-1, \frac{1}{2}, 3$; (b) $0, \frac{3}{2}, 4$;
(c) $-3, -\frac{3}{2}, 1$; (d) $-\frac{1}{2}, \frac{1}{4}, \frac{3}{2}$     **113.** $-8, -\frac{3}{2}, 4, 7, 15$

## Visualizing the Graph

**1.** A     **2.** C     **3.** D     **4.** H     **5.** G     **6.** F
**7.** B     **8.** I     **9.** J     **10.** E

## Exercise Set 4.5

**1.** $\{x | x \neq 2\}$, or $(-\infty, 2) \cup (2, \infty)$
**3.** $\{x | x \neq 1 \text{ and } x \neq 5\}$, or $(-\infty, 1) \cup (1, 5) \cup (5, \infty)$
**5.** $\{x | x \neq -5\}$, or $(-\infty, -5) \cup (-5, \infty)$     **7.** (d); $x = 2$, $x = -2, y = 0$     **9.** (e); $x = 2, x = -2, y = 0$
**11.** (c); $x = 2, x = -2, y = 8x$     **13.** $x = 0$     **15.** $x = 2$
**17.** $x = 4, x = -6$     **19.** $x = \frac{3}{2}, x = -1$     **21.** $y = \frac{3}{4}$
**23.** $y = 0$     **25.** No horizontal asymptote
**27.** $y = x + 1$     **29.** $y = x$     **31.** $y = x - 3$
**33.** Domain: $\qquad$ **35.** Domain:
$(-\infty, 0) \cup (0, \infty)$; no $\qquad$ $(-\infty, 0) \cup (0, \infty)$; no
$x$-intercepts, no $y$-intercept; $\qquad$ $x$-intercepts, no $y$-intercept;

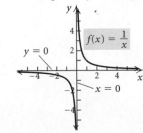

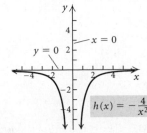

**37.** Domain: $(-\infty, -1) \cup$ $(-1, \infty)$; $x$-intercepts: $(1, 0)$ and $(3, 0)$, $y$-intercept: $(0, 3)$;

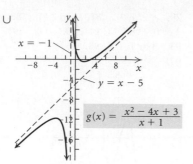

$$g(x) = \frac{x^2 - 4x + 3}{x + 1}$$

**39.** Domain: $(-\infty, 5) \cup (5, \infty)$; no $x$-intercepts, $y$-intercept: $\left(0, \frac{2}{5}\right)$;

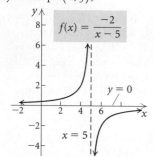

$$f(x) = \frac{-2}{x - 5}$$

**41.** Domain: $(-\infty, 0) \cup (0, \infty)$; $x$-intercept: $\left(-\frac{1}{2}, 0\right)$, no $y$-intercept;

$$f(x) = \frac{2x + 1}{x}$$

**43.** Domain: $(-\infty, -3) \cup$ $(-3, 3) \cup (3, \infty)$; no $x$-intercepts, $y$-intercept: $\left(0, -\frac{1}{3}\right)$;

$$f(x) = \frac{x + 3}{x^2 - 9}$$

**45.** Domain: $(-\infty, -3) \cup$ $(-3, 0) \cup (0, \infty)$; no $x$-intercepts, no $y$-intercept;

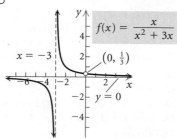

$$f(x) = \frac{x}{x^2 + 3x}$$

**47.** Domain: $(-\infty, 2) \cup (2, \infty)$; no $x$-intercepts, $y$-intercept: $\left(0, \frac{1}{4}\right)$;

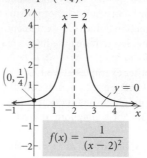

$$f(x) = \frac{1}{(x - 2)^2}$$

**49.** Domain: $(-\infty, -3) \cup (-3, -1) \cup$ $(-1, \infty)$; $x$-intercept: $(1, 0)$, $y$-intercept: $(0, -1)$;

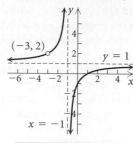

$$f(x) = \frac{x^2 + 2x - 3}{x^2 + 4x + 3}$$

**51.** Domain: $(-\infty, \infty)$; no $x$-intercepts, $y$-intercept: $\left(0, \frac{1}{3}\right)$;

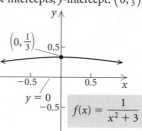

$$f(x) = \frac{1}{x^2 + 3}$$

**53.** Domain: $(-\infty, 2) \cup (2, \infty)$; $x$-intercept: $(-2, 0)$, $y$-intercept: $(0, 2)$;

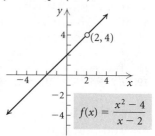

$$f(x) = \frac{x^2 - 4}{x - 2}$$

**55.** Domain: $(-\infty, -2) \cup (-2, \infty)$; $x$-intercept: $(1, 0)$, $y$-intercept: $\left(0, -\frac{1}{2}\right)$;

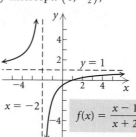

$$f(x) = \frac{x - 1}{x + 2}$$

**57.** Domain: $\left(-\infty, -\frac{1}{2}\right) \cup \left(-\frac{1}{2}, 0\right) \cup$ $(0, 3) \cup (3, \infty)$; $x$-intercept: $(-3, 0)$, no $y$-intercept;

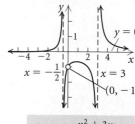

$$f(x) = \frac{x^2 + 3x}{2x^3 - 5x^2 - 3x}$$

**59.** Domain: $(-\infty, -1) \cup (-1, \infty)$; $x$-intercepts: $(-3, 0)$ and $(3, 0)$, $y$-intercept: $(0, -9)$;

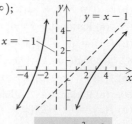

$$f(x) = \frac{x^2 - 9}{x + 1}$$

**61.** Domain: $(-\infty, \infty)$; $x$-intercepts: $(-2, 0)$ and $(1, 0)$, $y$-intercept: $(0, -2)$;

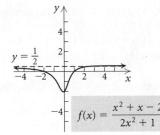

$$f(x) = \frac{x^2 + x - 2}{2x^2 + 1}$$

**63.** Domain: $(-\infty, 1) \cup (1, \infty)$; $x$-intercept: $\left(-\frac{2}{3}, 0\right)$, $y$-intercept: $(0, 2)$;

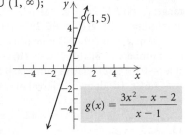

$$g(x) = \frac{3x^2 - x - 2}{x - 1}$$

**65.** Domain: $(-\infty, -1) \cup (-1, 3) \cup (3, \infty)$; $x$-intercept: $(1, 0)$, $y$-intercept: $\left(0, \frac{1}{3}\right)$;

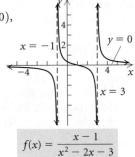

$$f(x) = \frac{x - 1}{x^2 - 2x - 3}$$

**67.** Domain: $(-\infty, -4) \cup (-4, 2) \cup (2, \infty)$; $x$-intercept: $\left(\frac{1}{3}, 0\right)$, $y$-intercept: $\left(0, \frac{1}{2}\right)$;

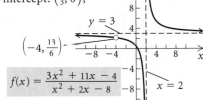

$$f(x) = \frac{3x^2 + 11x - 4}{x^2 + 2x - 8}$$

**69.** Domain: $(-\infty, -1) \cup (-1, \infty)$; $x$-intercept: $(3, 0)$, $y$-intercept: $(0, -3)$;

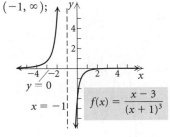

$$f(x) = \frac{x - 3}{(x + 1)^3}$$

**71.** Domain: $(-\infty, 0) \cup (0, \infty)$; $x$-intercept: $(-1, 0)$, no $y$-intercept;

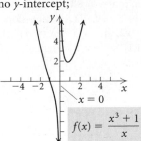

$$f(x) = \frac{x^3 + 1}{x}$$

**73.** Domain: $(-\infty, -2) \cup (-2, 7) \cup (7, \infty)$; $x$-intercepts: $(-5, 0)$, $(0, 0)$, and $(3, 0)$, $y$-intercept: $(0, 0)$;

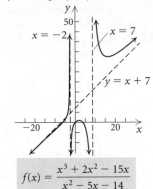

$$f(x) = \frac{x^3 + 2x^2 - 15x}{x^2 - 5x - 14}$$

**75.** Domain: $(-\infty, \infty)$; $x$-intercept: $(0, 0)$, $y$-intercept: $(0, 0)$;

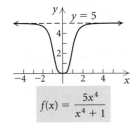

$$f(x) = \frac{5x^4}{x^4 + 1}$$

**77.** Domain: $(-\infty, -1) \cup (-1, 2) \cup (2, \infty)$; $x$-intercept: $(0, 0)$, $y$-intercept: $(0, 0)$;

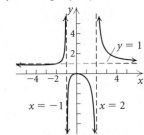

$$f(x) = \frac{x^2}{x^2 - x - 2}$$

**79.** $f(x) = \dfrac{1}{x^2 - x - 20}$

**81.** $f(x) = \dfrac{3x^2 + 6x}{2x^2 - 2x - 40}$, or $f(x) = \dfrac{3x^2 + 12x + 12}{2x^2 - 2x - 40}$

**83.** (a) $N(t)$

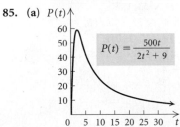

$$N(t) = \frac{0.8t + 1000}{5t + 4}, t \geq 15$$

$N(t) \to 0.16$ as $t \to \infty$; (b) The medication never completely disappears from the body; a trace amount remains.

**85.** (a) $P(t)$

$$P(t) = \frac{500t}{2t^2 + 9}$$

**(b)** $P(0) = 0$; $P(1) = 45,455$; $P(3) = 55,556$; $P(8) = 29,197$;
**(c)** $P(t) \to 0$ as $t \to \infty$; **(d)** In time, no one lives in the senior community. **(e)** 58,926 at $t \approx 2.12$ months
**87.** [1.2] Domain, range, domain, range **88.** [1.3] Slope
**89.** [1.3] Slope–intercept equation **90.** [1.4] Point–slope
equation **91.** [1.1] $x$-intercept **92.** [2.4] $f(-x) = -f(x)$
**93.** [1.3] Vertical lines **94.** [1.1] Midpoint formula
**95.** [1.1] $y$-intercept **97.** $y = x^3 + 4$
**99.**

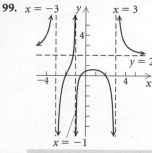

$$f(x) = \frac{2x^3 + x^2 - 8x - 4}{x^3 + x^2 - 9x - 9}$$

## Exercise Set 4.6

**1.** $-5, 3$ **3.** $[-5, 3]$ **5.** $(-\infty, -5] \cup [3, \infty)$
**7.** $(-\infty, -4) \cup (2, \infty)$ **9.** $(-\infty, -4) \cup [2, \infty)$
**11.** $0$ **13.** $(-5, 0] \cup (1, \infty)$
**15.** $(-\infty, -5) \cup (0, 1)$ **17.** $(-\infty, -3) \cup (0, 3)$
**19.** $(-3, 0) \cup (3, \infty)$ **21.** $(-\infty, -5) \cup (-3, 2)$
**23.** $(-2, 0] \cup (2, \infty)$ **25.** $(-4, 1)$
**27.** $(-\infty, -2) \cup (1, \infty)$ **29.** $(-\infty, -1] \cup [3, \infty)$
**31.** $(-\infty, -5) \cup (5, \infty)$ **33.** $(-\infty, -2] \cup [2, \infty)$
**35.** $(-\infty, 3) \cup (3, \infty)$ **37.** $\varnothing$ **39.** $\left(-\infty, -\frac{5}{4}\right] \cup [0, 3]$
**41.** $[-3, -1] \cup [1, \infty)$ **43.** $(-\infty, -2) \cup (1, 3)$
**45.** $\left[-\sqrt{2}, -1\right] \cup \left[\sqrt{2}, \infty\right)$ **47.** $(-\infty, -1] \cup \left[\frac{3}{2}, 2\right]$
**49.** $(-\infty, 5]$ **51.** $(-\infty, -1.680) \cup (2.154, 5.526)$
**53.** $-4; (-4, \infty)$ **55.** $-\frac{5}{2}; \left(-\frac{5}{2}, \infty\right)$ **57.** $0, 4$;
$(-\infty, 0] \cup (4, \infty)$ **59.** $-3, -\frac{1}{5}, 1; \left(-3, -\frac{1}{5}\right] \cup (1, \infty)$
**61.** $2, \frac{46}{11}, 5; \left(2, \frac{46}{11}\right) \cup (5, \infty)$ **63.** $2, \frac{7}{2}; \left(2, \frac{7}{2}\right]$
**65.** $1 - \sqrt{2}, 0, 1 + \sqrt{2}; (1 - \sqrt{2}, 0) \cup (1 + \sqrt{2}, \infty)$
**67.** $-3, 1, 3, \frac{11}{3}; (-\infty, -3) \cup (1, 3) \cup \left[\frac{11}{3}, \infty\right)$
**69.** $0; (-\infty, \infty)$ **71.** $-3, \dfrac{1 - \sqrt{61}}{6}, -\dfrac{1}{2}, 0, \dfrac{1 + \sqrt{61}}{6}$;
$\left(-3, \dfrac{1 - \sqrt{61}}{6}\right) \cup \left(-\dfrac{1}{2}, 0\right) \cup \left(\dfrac{1 + \sqrt{61}}{6}, \infty\right)$
**73.** $-1, 0, \frac{2}{7}, \frac{7}{2}; (-1, 0) \cup \left(\frac{2}{7}, \frac{7}{2}\right)$
**75.** $-6 - \sqrt{33}, -5, -6 + \sqrt{33}, 1, 5$;
$\left[-6 - \sqrt{33}, -5\right) \cup \left[-6 + \sqrt{33}, 1\right) \cup (5, \infty)$
**77.** $(0.408, 2.449)$ **79.** **(a)** $(10, 200)$;
**(b)** $(0, 10) \cup (200, \infty)$ **81.** $\{n | 9 \le n \le 23\}$
**83.** [1.1] $(x + 2)^2 + (y - 4)^2 = 9$
**84.** [1.1] $x^2 + (y + 3)^2 = \frac{49}{16}$ **85.** [3.3] **(a)** $\left(\frac{3}{4}, -\frac{55}{8}\right)$;
**(b)** maximum: $-\frac{55}{8}$ when $x = \frac{3}{4}$; **(c)** $\left(-\infty, -\frac{55}{8}\right]$

**86.** [3.3] **(a)** $(5, -23)$; **(b)** minimum: $-23$ when $x = 5$;
**(c)** $[-23, \infty)$ **87.** $\left[-\sqrt{5}, \sqrt{5}\right]$ **89.** $\left[-\frac{3}{2}, \frac{3}{2}\right]$
**91.** $\left(-\infty, -\frac{1}{4}\right) \cup \left(\frac{1}{2}, \infty\right)$ **93.** $x^2 + x - 12 < 0$; answers
may vary **95.** $(-\infty, -3) \cup (7, \infty)$

## Review Exercises: Chapter 4

**1.** True **2.** True **3.** False **4.** False **5.** False
**6.** **(a)** $-2.637, 1.137$; **(b)** relative maximum: 7.125 at
$x = -0.75$; **(c)** none; **(d)** domain: all real numbers;
range: $(-\infty, 7.125]$ **7.** **(a)** $-3, -1.414, 1.414$; **(b)** relative
maximum: 2.303 at $x = -2.291$; **(c)** relative minimum:
$-6.303$ at $x = 0.291$; **(d)** domain: all real numbers; range: all
real numbers **8.** **(a)** 0, 1, 2; **(b)** relative maximum: 0.202
at $x = 0.610$; **(c)** relative minima: 0 at $x = 0$, $-0.620$ at
$x = 1.640$; **(d)** domain: all real numbers; range: $[-0.620, \infty)$
**9.** $0.45x^4$; $0.45$; 4; quartic **10.** $-25$; $-25$; 0; constant
**11.** $-0.5x$; $-0.5$; 1; linear **12.** $\frac{1}{3}x^3$; $\frac{1}{3}$; 3; cubic
**13.** As $x \to \infty$, $f(x) \to -\infty$, and as $x \to -\infty$, $f(x) \to -\infty$.
**14.** As $x \to \infty$, $f(x) \to \infty$, and as $x \to -\infty$, $f(x) \to -\infty$.
**15.** $\frac{2}{3}$, multiplicity 1; $-2$, multiplicity 3; 5, multiplicity 2
**16.** $\pm 1, \pm 5$, each has multiplicity 1 **17.** $\pm 3, -4$, each
has multiplicity 1 **18.** **(a)** 4%; **(b)** 5%
**19.** **(a)** Linear: $f(x) = 0.5408695652x - 30.30434783$;
quadratic: $f(x) = 0.0030322581x^2 - 0.5764516129x + 57.53225806$; cubic: $f(x) = 0.0000247619x^3 - 0.0112857143x^2 + 2.002380952x - 82.14285714$;
**(b)** the cubic function; **(c)** 298, 498

**20.**

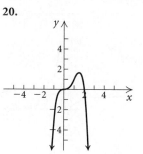

$f(x) = -x^4 + 2x^3$

**21.**

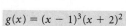

$g(x) = (x - 1)^3(x + 2)^2$

**22.**

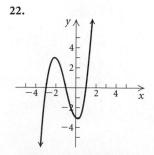

$h(x) = x^3 + 3x^2 - x - 3$

**23.**

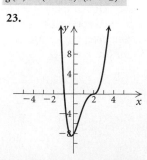

$f(x) = x^4 - 5x^3 + 6x^2 + 4x - 8$

**24.**

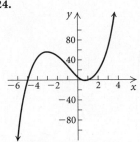

$$g(x) = 2x^3 + 7x^2 - 14x + 5$$

**25.** $f(1) = -4$ and $f(2) = 3$. Since $f(1)$ and $f(2)$ have opposite signs, $f(x)$ has a zero between 1 and 2.

**26.** $f(-1) = -3.5$ and $f(1) = -0.5$. Since $f(-1)$ and $f(1)$ have the same sign, the intermediate value theorem does not allow us to determine whether there is a zero between $-1$ and 1.   **27.** $Q(x) = 6x^2 + 16x + 52$, $R(x) = 155$; $P(x) = (x - 3)(6x^2 + 16x + 52) + 155$

**28.** $Q(x) = x^3 - 3x^2 + 3x - 2$, $R(x) = 7$; $P(x) = (x + 1)(x^3 - 3x^2 + 3x - 2) + 7$

**29.** $x^2 + 7x + 22$, R 120   **30.** $x^3 + x^2 + x + 1$, R 0

**31.** $x^4 - x^3 + x^2 - x - 1$, R 1   **32.** 36   **33.** 0

**34.** $-141,220$   **35.** Yes, no   **36.** No, yes   **37.** Yes, no

**38.** No, yes   **39.** $f(x) = (x - 1)^2(x + 4)$; $-4, 1$

**40.** $f(x) = (x - 2)(x + 3)^2$; $-3, 2$

**41.** $f(x) = (x - 2)^2(x - 5)(x + 5)$; $-5, 2, 5$

**42.** $f(x) = (x - 1)(x + 1)(x - \sqrt{2})(x + \sqrt{2})$; $-\sqrt{2}, -1, 1, \sqrt{2}$   **43.** $f(x) = x^3 + 3x^2 - 6x - 8$

**44.** $f(x) = x^3 + x^2 - 4x + 6$

**45.** $f(x) = x^3 - \frac{5}{2}x^2 + \frac{1}{2}$, or $2x^3 - 5x^2 + 1$

**46.** $f(x) = x^4 + \frac{29}{2}x^3 + \frac{135}{2}x^2 + \frac{175}{2}x - \frac{125}{2}$, or $2x^4 + 29x^3 + 135x^2 + 175x - 125$

**47.** $f(x) = x^5 + 4x^4 - 3x^3 - 18x^2$   **48.** $-\sqrt{5}, 4 - i$

**49.** $1 - \sqrt{3}, \sqrt{3}$   **50.** $\sqrt{2}$   **51.** $f(x) = x^2 - 11$

**52.** $f(x) = x^3 - 6x^2 + x - 6$

**53.** $f(x) = x^4 - 5x^3 + 4x^2 + 2x - 8$

**54.** $f(x) = x^4 - x^2 - 20$   **55.** $f(x) = x^3 + \frac{8}{3}x^2 - x$, or $3x^3 + 8x^2 - 3x$

**56.** $\pm\frac{1}{4}, \pm\frac{1}{2}, \pm\frac{3}{4}, \pm 1, \pm\frac{3}{2}, \pm 2, \pm 3, \pm 4, \pm 6, \pm 12$

**57.** $\pm\frac{1}{3}, \pm 1$   **58.** $\pm 1, \pm 2, \pm 3, \pm 4, \pm 6, \pm 8, \pm 12, \pm 24$

**59.** **(a)** Rational: $0, -2, \frac{1}{3}, 3$; other: none;
**(b)** $f(x) = 3x(x - \frac{1}{3})(x + 2)^2(x - 3)$, or $x(3x - 1)(x + 2)^2(x - 3)$

**60.** **(a)** Rational: 2; other: $\pm\sqrt{3}$;
**(b)** $f(x) = (x - 2)(x + \sqrt{3})(x - \sqrt{3})$

**61.** **(a)** Rational: $-1, 1$; other: $3 \pm i$;
**(b)** $f(x) = (x + 1)(x - 1)(x - 3 - i)(x - 3 + i)$

**62.** **(a)** Rational: $-5$; other: $1 \pm \sqrt{2}$;
**(b)** $f(x) = (x + 5)(x - 1 - \sqrt{2})(x - 1 + \sqrt{2})$

**63.** **(a)** Rational: $\frac{2}{3}, 1$; other: none;
**(b)** $f(x) = 3(x - \frac{2}{3})(x - 1)^2$, or $(3x - 2)(x - 1)^2$

**64.** **(a)** Rational: 2; other: $1 \pm \sqrt{5}$;
**(b)** $f(x) = (x - 2)^3(x - 1 + \sqrt{5})(x - 1 - \sqrt{5})$

**65.** **(a)** Rational: $-4, 0, 3, 4$; other: none;
**(b)** $f(x) = x^2(x + 4)^2(x - 3)(x - 4)$

**66.** **(a)** Rational: $\frac{5}{2}, 1$; other: none;
**(b)** $f(x) = 2(x - \frac{5}{2})(x - 1)^4$   **67.** 3 or 1; 0

**68.** 4 or 2 or 0; 2 or 0   **69.** 3 or 1; 0

**70.** Domain: $(-\infty, -2) \cup (-2, \infty)$; $x$-intercepts: $(-\sqrt{5}, 0)$ and $(\sqrt{5}, 0)$, $y$-intercept: $(0, -\frac{5}{2})$

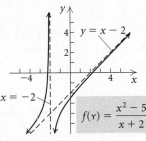

**71.** Domain: $(-\infty, 2) \cup (2, \infty)$; $x$-intercepts: none, $y$-intercept: $(0, \frac{5}{4})$

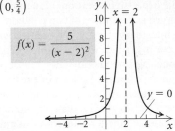

**72.** Domain: $(-\infty, -4) \cup (-4, 5) \cup (5, \infty)$; $x$-intercepts: $(-3, 0)$ and $(2, 0)$, $y$-intercept: $(0, \frac{3}{10})$

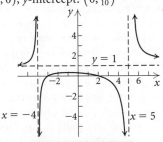

$$f(x) = \frac{x^2 + x - 6}{x^2 - x - 20}$$

**73.** Domain: $(-\infty, -3) \cup (-3, 5) \cup (5, \infty)$; $x$-intercept: $(2, 0)$, $y$-intercept: $(0, \frac{2}{15})$

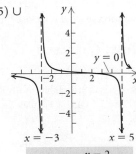

$$f(x) = \frac{x - 2}{x^2 - 2x - 15}$$

**74.** $f(x) = \dfrac{1}{x^2 - x - 6}$   **75.** $f(x) = \dfrac{4x^2 + 12x}{x^2 - x - 6}$

**76.** **(a)** $N(t) \to 0.0875$ as $t \to \infty$;   **(b)** The medication never completely disappears from the body; a trace amount remains.

**77.** $(-3, 3)$    **78.** $\left(-\infty, -\frac{1}{2}\right) \cup (2, \infty)$
**79.** $[-4, 1] \cup [2, \infty)$    **80.** $\left(-\infty, -\frac{14}{3}\right) \cup (-3, \infty)$
**81.** (a) 7 sec after launch; (b) $(2, 3)$
**82.** $\left[\dfrac{5 - \sqrt{15}}{2}, \dfrac{5 + \sqrt{15}}{2}\right]$
**83.** A    **84.** C    **85.** B
**86.** $\left(-\infty, -1 - \sqrt{6}\right] \cup \left[-1 + \sqrt{6}, \infty\right)$
**87.** $\left(-\infty, -\frac{1}{2}\right) \cup \left(\frac{1}{2}, \infty\right)$
**88.** $1 + i, 1 - i, i, -i$    **89.** $(-\infty, 2)$
**90.** $(x - 1)\left(x + \dfrac{1}{2} - \dfrac{\sqrt{3}}{2}i\right)\left(x + \dfrac{1}{2} + \dfrac{\sqrt{3}}{2}i\right)$
**91.** 7    **92.** $-4$    **93.** $(-\infty, -5] \cup [2, \infty)$
**94.** $(-\infty, 1.1] \cup [2, \infty)$    **95.** $\left(-1, \frac{3}{7}\right)$
**96.** A polynomial function is a function that can be defined by a polynomial expression. A rational function is a function that can be defined as a quotient of two polynomials.
**97.** No; since imaginary zeros of polynomials with rational coefficients occur in conjugate pairs, a third-degree polynomial with rational coefficients can have at most two imaginary zeros. Thus there must be at least one real zero.
**98.** Vertical asymptotes occur at any $x$-values that make the denominator zero and do not also make the numerator zero. The graph of a rational function does not cross any vertical asymptotes. Horizontal asymptotes occur when the degree of the numerator is less than or equal to the degree of the denominator. Oblique asymptotes occur when the degree of the numerator is 1 greater than the degree of the denominator. Graphs of rational functions may cross horizontal or oblique asymptotes.    **99.** If $P(x)$ is an even function, then $P(-x) = P(x)$ and thus $P(-x)$ has the same number of sign changes as $P(x)$. Hence, $P(x)$ has one negative real zero also.    **100.** A horizontal asymptote occurs when the degree of the numerator of a rational function is less than or equal to the degree of the denominator. An oblique asymptote occurs when the degree of the numerator is 1 greater than the degree of the denominator. Thus a rational function cannot have both a horizontal asymptote and an oblique asymptote.    **101.** A quadratic inequality $ax^2 + bx + c \leq 0, a > 0$, or $ax^2 + bx + c \geq 0, a < 0$, has a solution set that is a closed interval.

## Test: Chapter 4

**1.** [4.1] $-x^4$; $-1$; 4; quartic    **2.** [4.1] $-4.7x$; $-4.7$; 1; linear
**3.** [4.1] $0, \frac{5}{3}$, each has multiplicity 1; 3, multiplicity 2; $-1$, multiplicity 3    **4.** [4.1] 1930: 11.3%; 1990: 7.9%; 2000: 10.5%
**5.** [4.2]

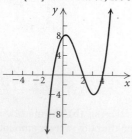

$$f(x) = x^3 - 5x^2 + 2x + 8$$

**6.** [4.2]

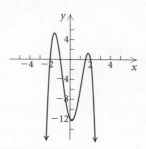

$$f(x) = -2x^4 + x^3 + 11x^2 - 4x - 12$$

**7.** [4.2] $f(0) = 3$ and $f(2) = -17$. Since $f(0)$ and $f(2)$ have opposite signs, $f(x)$ has a zero between 0 and 2.
**8.** [4.2] $g(-2) = 5$ and $g(-1) = 1$. Both $g(-2)$ and $g(-1)$ are positive. We cannot use the intermediate value theorem to determine if there is a zero between $-2$ and $-1$.    **9.** [4.3] $Q(x) = x^3 + 4x^2 + 4x + 6, R(x) = 1$; $P(x) = (x - 1)(x^3 + 4x^2 + 4x + 6) + 1$
**10.** [4.3] $3x^2 + 15x + 63$, R 322    **11.** [4.3] $-115$
**12.** [4.3] Yes    **13.** [4.4] $f(x) = x^4 - 27x^2 - 54x$
**14.** [4.4] $-\sqrt{3}, 2 + i$
**15.** [4.4] $f(x) = x^3 + 10x^2 + 9x + 90$
**16.** [4.4] $f(x) = x^5 - 2x^4 - x^3 + 6x^2 - 6x$
**17.** [4.4] $\pm 1, \pm 2, \pm 3, \pm 4, \pm 6, \pm 12, \pm \frac{1}{2}, \pm \frac{3}{2}$
**18.** [4.4] $\pm \frac{1}{10}, \pm \frac{1}{5}, \pm \frac{1}{2}, \pm 1, \pm \frac{5}{2}, \pm 5$
**19.** [4.4] (a) Rational: $-1$; other: $\pm \sqrt{5}$;
(b) $f(x) = (x + 1)(x - \sqrt{5})(x + \sqrt{5})$
**20.** [4.4] (a) Rational: $-\frac{1}{2}, 1, 2, 3$; other: none;
(b) $f(x) = 2\left(x + \frac{1}{2}\right)(x - 1)(x - 2)(x - 3)$, or $(2x + 1)(x - 1)(x - 2)(x - 3)$    **21.** [4.4] (a) Rational: $-4$; other: $\pm 2i$; (b) $f(x) = (x - 2i)(x + 2i)(x + 4)$
**22.** [4.4] (a) Rational: $\frac{2}{3}, 1$; other: none;
(b) $f(x) = 3\left(x - \frac{2}{3}\right)(x - 1)^3$, or $(3x - 2)(x - 1)^3$
**23.** [4.4] 2 or 0; 2 or 0
**24.** [4.5] Domain: $(-\infty, 3) \cup (3, \infty)$; $x$-intercepts: none, $y$-intercept: $\left(0, \frac{2}{9}\right)$;

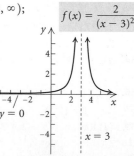

$$f(x) = \frac{2}{(x - 3)^2}$$

**25.** [4.5] Domain: $(-\infty, -1) \cup (-1, 4) \cup (4, \infty)$; $x$-intercept: $(-3, 0)$, $y$-intercept: $\left(0, -\frac{3}{4}\right)$;

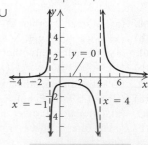

$$f(x) = \frac{x + 3}{x^2 - 3x - 4}$$

**26.** [4.5] Answers may vary; $f(x) = \dfrac{x + 4}{x^2 - x - 2}$

**27.** [4.6] $\left(-\infty, -\frac{1}{2}\right) \cup (3, \infty)$

**28.** [4.1], [4.6] $(-\infty, 4) \cup \left[\frac{13}{2}, \infty\right)$

**29.** (a) [4.1] 6 sec; (b) [4.1], [4.6] $(1, 3)$

**30.** [4.2] D     **31.** [4.1], [4.6] $(-\infty, -4] \cup [3, \infty)$

# Chapter 5

## Exercise Set 5.1

**1.** $\{(8, 7), (8, -2), (-4, 3), (-8, 8)\}$     **3.** $\{(-1, -1), (4, -3)\}$

**5.** $x = 4y - 5$     **7.** $y^3x = -5$     **9.** $y = x^2 - 2x$

**11.**
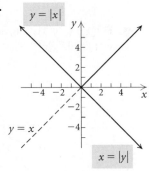

**13.**
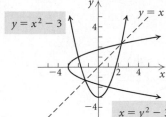

**15.**
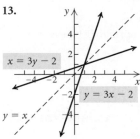

**17.** Assume $f(a) = f(b)$ for any numbers $a$ and $b$ in the domain of $f$. Since $f(a) = \frac{1}{3}a - 6$ and $f(b) = \frac{1}{3}b - 6$, we have

$$\frac{1}{3}a - 6 = \frac{1}{3}b - 6$$
$$\frac{1}{3}a = \frac{1}{3}b \qquad \text{Adding 6}$$
$$a = b. \qquad \text{Multiplying by 3}$$

Thus, if $f(a) = f(b)$, then $a = b$ and $f$ is one-to-one.

**19.** Assume $f(a) = f(b)$ for any numbers $a$ and $b$ in the domain of $f$. Since $f(a) = a^3 + \frac{1}{2}$ and $f(b) = b^3 + \frac{1}{2}$, we have

$$a^3 + \frac{1}{2} = b^3 + \frac{1}{2}$$
$$a^3 = b^3 \qquad \text{Subtracting } \frac{1}{2}$$
$$a = b. \qquad \text{Taking the cube root}$$

Thus, if $f(a) = f(b)$, then $a = b$ and $f$ is one-to-one.

**21.** Find two numbers $a$ and $b$ for which $a \neq b$ and $g(a) = g(b)$. Two such numbers are $-2$ and $2$, because $g(-2) = g(2) = -3$. Thus, $g$ is not one-to-one.

**23.** Find two numbers $a$ and $b$ for which $a \neq b$ and $g(a) = g(b)$. Two such numbers are $-1$ and $1$, because $g(-1) = g(1) = 0$. Thus, $g$ is not one-to-one.

**25.** Yes     **27.** No     **29.** No     **31.** Yes     **33.** Yes
**35.** No     **37.** No     **39.** Yes     **41.** No     **43.** No

**45.** $y_1 = 0.8x + 1.7$,     Domain and range of both $f$
$y_2 = \dfrac{x - 1.7}{0.8}$     and $f^{-1}$: $(-\infty, \infty)$

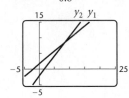

**47.** $y_1 = \dfrac{1}{2}x - 4$,     Domain and range of both $f$
$y_2 = 2x + 8$     and $f^{-1}$: $(-\infty, \infty)$

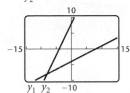

**49.** $y_1 = \sqrt{x - 3}$,     Domain of $f$: $[3, \infty)$,
$y_2 = x^2 + 3, x \geq 0$     range of $f$: $[0, \infty)$;
domain of $f^{-1}$: $[0, \infty)$,
range of $f^{-1}$: $[3, \infty)$

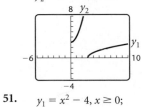

**51.** $y_1 = x^2 - 4, x \geq 0$;     Domain of $f$: $[0, \infty)$,
$y_2 = \sqrt{4 + x}$     range of $f$: $[-4, \infty)$;
domain of $f^{-1}$: $[-4, \infty)$,
range of $f^{-1}$: $[0, \infty)$

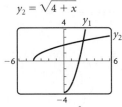

**53.** $y_1 = (3x - 9)^3$,     Domain and range of
$y_2 = \dfrac{\sqrt[3]{x} + 9}{3}$     both $f$ and $f^{-1}$: $(-\infty, \infty)$

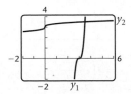

**55.** (a) Yes;  (b) $f^{-1}(x) = x - 4$

**57.** (a) Yes;  (b) $f^{-1}(x) = \dfrac{x + 1}{2}$

**59.** (a) Yes;  (b) $f^{-1}(x) = \dfrac{4}{x} - 7$

**61. (a)** Yes; **(b)** $f^{-1}(x) = \dfrac{3x + 4}{x - 1}$

**63. (a)** Yes; **(b)** $f^{-1}(x) = \sqrt[3]{x + 1}$

**65. (a)** No

**67. (a)** Yes; **(b)** $f^{-1}(x) = \sqrt{\dfrac{x + 2}{5}}$

**69. (a)** Yes; **(b)** $f^{-1}(x) = x^2 - 1, x \geq 0$

**71.** $\frac{1}{3}x$    **73.** $-x$    **75.** $x^3 + 5$

**77.**

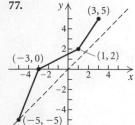

**79.**

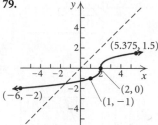

**81.**

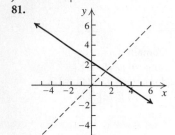

**83.** $f^{-1}(f(x)) = f^{-1}\left(\frac{7}{8}x\right) = \frac{8}{7} \cdot \frac{7}{8}x = x;$
$f(f^{-1}(x)) = f\left(\frac{8}{7}x\right) = \frac{7}{8} \cdot \frac{8}{7}x = x$

**85.** $f^{-1}(f(x)) = f^{-1}\left(\dfrac{1 - x}{x}\right) = \dfrac{1}{\dfrac{1 - x}{x} + 1} =$

$\dfrac{1}{\dfrac{1 - x + x}{x}} = \dfrac{1}{\dfrac{1}{x}} = 1 \cdot \dfrac{x}{1} = x; f(f^{-1}(x)) = f\left(\dfrac{1}{x + 1}\right) =$

$\dfrac{1 - \dfrac{1}{x + 1}}{\dfrac{1}{x + 1}} = \dfrac{\dfrac{x + 1 - 1}{x + 1}}{\dfrac{1}{x + 1}} = \dfrac{x}{x + 1} \cdot \dfrac{x + 1}{1} = x$

**87.** $f^{-1}(f(x)) = f^{-1}\left(\dfrac{2}{5}x + 1\right) = \dfrac{5\left(\dfrac{2}{5}x + 1\right) - 5}{2} =$

$\dfrac{2x + 5 - 5}{2} = \dfrac{2x}{2} = x; f(f^{-1}(x)) = f\left(\dfrac{5x - 5}{2}\right) =$

$\dfrac{2}{5}\left(\dfrac{5x - 5}{2}\right) + 1 = x - 1 + 1 = x$

**89.** $f^{-1}(x) = \frac{1}{5}x + \frac{3}{5}$; domain and range of both $f$ and $f^{-1}$: $(-\infty, \infty)$

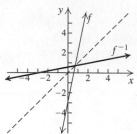

**91.** $f^{-1}(x) = \dfrac{2}{x}$; domain and range of both $f$ and $f^{-1}$:
$(-\infty, 0) \cup (0, \infty)$

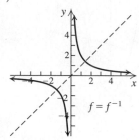

**93.** $f^{-1}(x) = \sqrt[3]{3x + 6}$; domain of $f$ and $f^{-1}$: $(-\infty, \infty)$;
range of $f$ and $f^{-1}$: $(-\infty, \infty)$

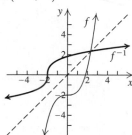

**95.** $f^{-1}(x) = \dfrac{3x + 1}{x - 1}$; domain of $f$: $(-\infty, 3) \cup (3, \infty)$;
range of $f$: $(-\infty, 1) \cup (1, \infty)$;
domain of $f^{-1}$: $(-\infty, 1) \cup (1, \infty)$;
range of $f^{-1}$: $(-\infty, 3) \cup (3, \infty)$;

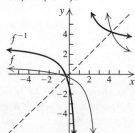

**97.** $5; a$    **99.** $C^{-1}(x) = \dfrac{60}{x - 2}$; $C^{-1}(x)$ represents the number
of people in the group, where $x$ is the cost per person, in dollars.
**101. (a)** $T(-13°) = -25°, T(86°) = 30°$;
**(b)** $T^{-1}(x) = \frac{9}{5}x + 32$; $T^{-1}(x)$ represents the Fahrenheit temperature when the Celsius temperature is $x$.

**103.** [3.3] (b), (d), (f), (h)   **104.** [3.3] (a), (c), (e), (g)
**105.** [3.3] (a)   **106.** [3.3] (d)   **107.** [3.3] (f)
**108.** [3.3] (a), (b), (c), (d)   **109.** Yes   **111.** $f(x) = x^2 - 3$,
$x \geq 0$; $f^{-1}(x) = \sqrt{x + 3}, x \geq -3$   **113.** Answers may
vary. $f(x) = 3/x, f(x) = 1 - x, f(x) = x$

## Exercise Set 5.2

**1.** 54.5982   **3.** 0.0856   **5.** (f)   **7.** (e)   **9.** (a)
**11.**

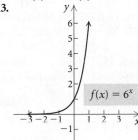

$f(x) = 3^x$

**13.**

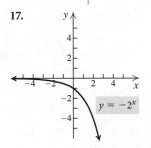

$f(x) = 6^x$

**15.**

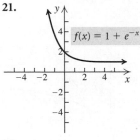

$f(x) = \left(\frac{1}{4}\right)^x$

**17.**

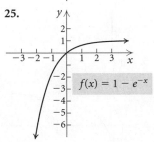

$y = -2^x$

**19.**

$f(x) = -0.25^x + 4$

**21.**

$f(x) = 1 + e^{-x}$

**23.**

$y = \frac{1}{4}e^x$

**25.**

$f(x) = 1 - e^{-x}$

**27.** Shift the graph of $y = 2^x$ left 1 unit.

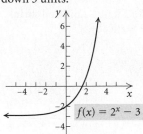

$f(x) = 2^{x+1}$

**29.** Shift the graph of $y = 2^x$ down 3 units.

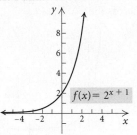

$f(x) = 2^x - 3$

**31.** Shift the graph of $y = 2^x$ left 1 unit, reflect it across the $y$-axis, and shift it up 2 units.

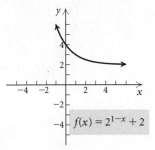

$f(x) = 2^{1-x} + 2$

**33.** Reflect the graph of $y = 3^x$ across the $y$-axis and then across the $x$-axis and then shift it up 4 units.

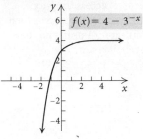

$f(x) = 4 - 3^{-x}$

**35.** Shift the graph of $y = \left(\frac{3}{2}\right)^x$ right 1 unit.

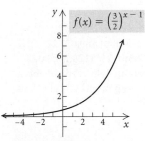

$f(x) = \left(\frac{3}{2}\right)^{x-1}$

**37.** Shift the graph of $y = 2^x$ left 3 units and then down 5 units.

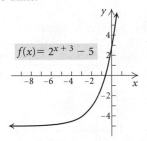

$f(x) = 2^{x+3} - 5$

**39.** Shift the graph of $y = 2^x$ right 1 unit, stretch it vertically, and shift it up 1 unit.

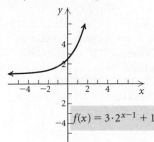

$f(x) = 3 \cdot 2^{x-1} + 1$

**41.** Shrink the graph of $y = e^x$ horizontally.

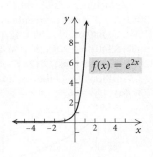

$f(x) = e^{2x}$

**43.** Reflect the graph of $y = e^x$ across the $x$-axis, shift it up 1 unit, and shrink it vertically.

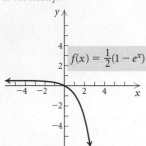

$f(x) = \frac{1}{2}(1 - e^x)$

**45.** Shift the graph of $y = e^x$ left 1 unit and then reflect it across the $y$-axis.

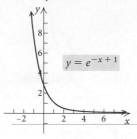

$y = e^{-x+1}$

**47.** Reflect the graph of $y = e^x$ across the $y$-axis, then across the $x$-axis, then shift it up 1 unit, and then stretch it vertically.

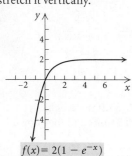

$f(x) = 2(1 - e^{-x})$

**49.**

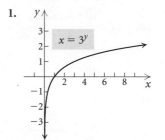

$$f(x) = \begin{cases} e^{-x} - 4, & \text{for } x < -2, \\ x + 3, & \text{for } -2 \le x < 1, \\ x^2, & \text{for } x \ge 1 \end{cases}$$

**51. (a)** $A(t) = 82{,}000(1.01125)^{4t}$;
**(b)** $y = 82{,}000(1.01125)^{4x}$; **(c)** \$82,000; \$89,677.22; \$102,561.54; \$128,278.90;

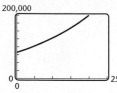

**(d)** about 4.43 years, or about 4 years, 5 months, and 5 days

**53.** \$4930.86 **55.** \$3247.30 **57.** \$153,610.15
**59.** \$76,305.59 **61.** \$26,086.69 **63.** 2000: 29,272 servicemembers; 2008: 68,436 servicemembers; 2013: 116,362 servicemembers **65.** 1950: 2900 pages; 1990: 24,689 pages; 2000: 42,172 pages **67.** 2003: \$278,574; 2007: \$540,460; 2013: \$1,460,486 **69.** 1999: 122,370 tons; 2007: 380,099 tons; 2012: 771,855 tons **71.** 2007: about 25.5 million; 2014: about 42.4 million
**73. (a)** $y = 56{,}395(0.85)^x$; **(b)** \$56,395; \$50,756; \$41,112; \$29,971; \$19,664; **(c)** in 6 years

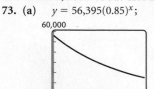

**75. (a)** $y = 100(1 - e^{-0.04x})$ **(b)** about 63%;
**(c)** after 58 days

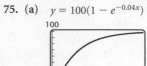

**77.** (c) **79.** (a) **81.** (1) **83.** (g) **85.** (i)
**87.** (k) **89.** (m) **91.** (1.481, 4.090)
**93.** $(-0.402, -1.662)$, $(1.051, 2.722)$ **95.** 4.448
**97.** $(0, \infty)$ **99.** 2.294, 3.228 **101.** [3.1] $31 - 22i$
**102.** [3.1] $\frac{1}{2} - \frac{1}{2}i$ **103.** [3.2] $\left(-\frac{1}{2}, 0\right)$, $(7, 0)$; $-\frac{1}{2}, 7$
**104.** [4.3] $(1, 0)$; 1 **105.** [4.1] $(-1, 0)$, $(0, 0)$, $(1, 0)$; $-1, 0, 1$
**106.** [4.1] $(-4, 0)$, $(0, 0)$, $(3, 0)$; $-4, 0, 3$ **107.** [4.1] $-8, 0, 2$
**108.** [3.2] $\dfrac{5 \pm \sqrt{97}}{6}$ **109.** $\pi^7$; $70^{80}$
**111. (a)** $y = e^{-x^2}$ **(b)** none; **(c)** relative maximum: 1 at $x = 0$

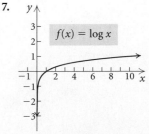

**113.** $f(x) = \dfrac{2e^x(e^h - 1)}{h}$

### Visualizing the Graph

**1.** J **2.** F **3.** H **4.** B **5.** E **6.** A **7.** C
**8.** I **9.** D **10.** G

### Exercise Set 5.3

**1.**

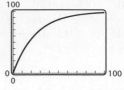

$x = 3^y$

**3.**

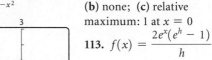

$x = \left(\frac{1}{2}\right)^y$

**5.**

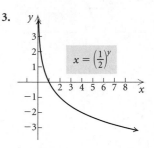

$y = \log_3 x$

**7.**

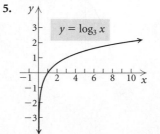

$f(x) = \log x$

**9.** 4 **11.** 3 **13.** $-3$ **15.** $-2$ **17.** 0 **19.** 1
**21.** 4 **23.** $\frac{1}{4}$ **25.** $-7$ **27.** $\frac{1}{2}$ **29.** $\frac{3}{4}$ **31.** 0 **33.** $\frac{1}{2}$
**35.** $\log_{10} 1000 = 3$, or $\log 1000 = 3$ **37.** $\log_8 2 = \frac{1}{3}$

**39.** $\log_e t = 3$, or $\ln t = 3$
**41.** $\log_e 7.3891 = 2$, or $\ln 7.3891 = 2$   **43.** $\log_p 3 = k$
**45.** $5^1 = 5$   **47.** $10^{-2} = 0.01$   **49.** $e^{3.4012} = 30$
**51.** $a^{-x} = M$   **53.** $a^x = T^3$   **55.** $0.4771$   **57.** $2.7259$
**59.** $-0.2441$   **61.** Does not exist   **63.** $0.6931$   **65.** $6.6962$
**67.** Does not exist   **69.** $3.3219$   **71.** $-0.2614$
**73.** $0.7384$   **75.** $2.2619$   **77.** $0.5880$
**79.** $y_1 = 3^x$,

$y_2 = \dfrac{\log x}{\log 3}$

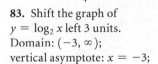

**81.** $y_1 = \log x$,

$y_2 = 10^x$

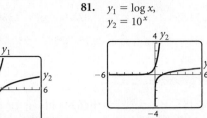

**83.** Shift the graph of $y = \log_2 x$ left 3 units. Domain: $(-3, \infty)$; vertical asymptote: $x = -3$;

$y = \dfrac{\log (x + 3)}{\log 2}$

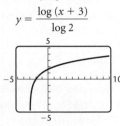

**85.** Shift the graph of $y = \log_3 x$ down 1 unit. Domain: $(0, \infty)$; vertical asymptote: $x = 0$;

$y = \dfrac{\log x}{\log 3} - 1$

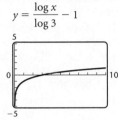

**87.** Stretch the graph of $y = \ln x$ vertically. Domain: $(0, \infty)$; vertical asymptote: $x = 0$;

$y = 4 \ln x$

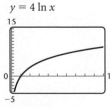

**89.** Reflect the graph of $y = \ln x$ across the $x$-axis and shift it up 2 units. Domain: $(0, \infty)$; vertical asymptote: $x = 0$;

$y = 2 - \ln x$

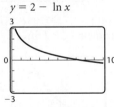

**91.** Shift the graph of $y = \log x$ right 1 unit, shrink it vertically, and shift it down 2 units. Domain: $(1, \infty)$; vertical asymptote: $x = 1$.

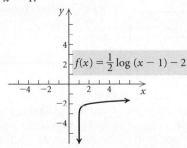

$f(x) = \dfrac{1}{2} \log (x - 1) - 2$

**93.**

$$g(x) = \begin{cases} 5, & \text{for } x \le 0, \\ \log x + 1, & \text{for } x > 0 \end{cases}$$

**95.** **(a)** 2.4 ft/sec;
**(b)** 3.0 ft/sec; **(c)** 2.3 ft/sec;
**(d)** 2.8 ft/sec; **(e)** 2.4 ft/sec;
**(f)** 2.2 ft/sec; **(g)** 2.5 ft/sec;
**(h)** 2.1 ft/sec   **97.** **(a)** 7.7;
**(b)** 9.5; **(c)** 6.6; **(d)** 7.4;
**(e)** 8.0; **(f)** 7.9; **(g)** 9.1; **(h)** 9.3
**99.** **(a)** $10^{-7}$; **(b)** $4.0 \times 10^{-6}$;
**(c)** $6.3 \times 10^{-4}$; **(d)** $1.6 \times 10^{-5}$

**101.** **(a)** 140 decibels; **(b)** 115 decibels; **(c)** 90 decibels;
**(d)** 65 decibels; **(e)** 100 decibels; **(f)** 194 decibels
**102.** [1.4] $m = \frac{3}{10}$; $y$-intercept: $\left(0, -\frac{7}{5}\right)$   **103.** [1.4] $m = 0$;
$y$-intercept: $(0, 6)$   **104.** [1.4] Slope is not defined; no
$y$-intercept   **105.** [4.3] $-280$   **106.** [4.3] $-4$   **107.** [4.4]
$f(x) = x^3 - 7x$   **108.** [4.4] $f(x) = x^3 - x^2 + 16x - 16$
**109.** 3   **111.** $(0, \infty)$   **113.** $(-\infty, 0) \cup (0, \infty)$
**115.** $\left(-\frac{5}{2}, -2\right)$   **117.** (d)   **119.** (b)
**121.** **(a)** $y = x \ln x$

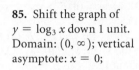

**(b)** 1; **(c)** relative
minimum: $-0.368$ at
$x = 0.368$

**123.** **(a)** $y = \dfrac{\ln x}{x^2}$

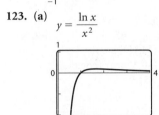

**(b)** 1; **(c)** relative
maximum: 0.184 at
$x = 1.649$

**125.** $(1.250, 0.891)$

## Mid-Chapter Mixed Review: Chapter 5

**1.** False   **2.** True   **3.** False   **4.** Yes; $f^{-1}(x) = -\dfrac{2}{x}$
**5.** No   **6.** Yes; $f^{-1}(x) = \dfrac{5}{x} + 2$
**7.** $(f^{-1} \circ f)(x) = f^{-1}(\sqrt{x - 5}) = (\sqrt{x - 5})^2 + 5 =$
$x - 5 + 5 = x$; $(f \circ f^{-1})(x) = f(x^2 + 5) =$
$\sqrt{(x^2 + 5) - 5} = \sqrt{x^2} = x$   **8.** $f^{-1}(x) = \sqrt[3]{x} - 2$;
domain and range of both $f$ and $f^{-1}$: $(-\infty, \infty)$   **9.** (d)
**10.** (h)   **11.** (c)   **12.** (g)   **13.** (b)   **14.** (f)   **15.** (e)
**16.** (a)   **17.** \$4185.57   **18.** 0   **19.** $-\frac{4}{5}$   **20.** $-2$
**21.** 2   **22.** 0   **23.** $-4$   **24.** 0   **25.** 3   **26.** $\frac{1}{4}$   **27.** 1
**28.** $\log_e 0.0025 = -6$, or $\ln 0.0025 = -6$
**29.** $10^r = T$   **30.** 2.7268

**31.** 2.0115   **32.** For an even function $f$, $f(x) = f(-x)$, so we have $f(x) = f(-x)$ but $x \neq -x$ (for $x \neq 0$). Thus, $f$ is not one-to-one and hence it does not have an inverse.
**33.** The most interest will be earned the eighth year, because the principal is greatest during that year.
**34.** In $f(x) = x^3$, the variable $x$ is the base. The range of $f$ is $(-\infty, \infty)$. In $g(x) = 3^x$, the variable $x$ is the exponent. The range of $g$ is $(0, \infty)$. The graph of $f$ does not have an asymptote. The graph of $g$ has an asymptote $y = 0$.
**35.** If $\log b < 0$, then $0 < b < 1$.

## Exercise Set 5.4

**1.** $\log_3 81 + \log_3 27 = 4 + 3 = 7$
**3.** $\log_5 5 + \log_5 125 = 1 + 3 = 4$
**5.** $\log_t 8 + \log_t Y$   **7.** $\ln x + \ln y$   **9.** $3 \log_b t$
**11.** $8 \log y$   **13.** $-6 \log_c K$   **15.** $\frac{1}{3} \ln 4$
**17.** $\log_t M - \log_t 8$   **19.** $\log x - \log y$   **21.** $\ln r - \ln s$
**23.** $\log_a 6 + \log_a x + 5 \log_a y + 4 \log_a z$
**25.** $2 \log_b p + 5 \log_b q - 4 \log_b m - 9$
**27.** $\ln 2 - \ln 3 - 3 \ln x - \ln y$
**29.** $\frac{3}{2} \log r + \frac{1}{2} \log t$   **31.** $3 \log_a x - \frac{5}{2} \log_a p - 4 \log_a q$
**33.** $2 \log_a m + 3 \log_a n - \frac{3}{4} - \frac{5}{4} \log_a b$   **35.** $\log_a 150$
**37.** $\log 100 = 2$   **39.** $\log m^3 \sqrt{n}$
**41.** $\log_a x^{-5/2} y^4$, or $\log_a \dfrac{y^4}{x^{5/2}}$   **43.** $\ln x$   **45.** $\ln (x - 2)$
**47.** $\log \dfrac{x - 7}{x - 2}$   **49.** $\ln \dfrac{x}{(x^2 - 25)^3}$   **51.** $\ln \dfrac{2^{11/5} x^9}{y^8}$
**53.** $-0.74$   **55.** 1.991   **57.** 0.356   **59.** 4.827
**61.** $-1.792$   **63.** 0.099   **65.** 3   **67.** $|x - 4|$   **69.** $4x$
**71.** $w$   **73.** $8t$   **75.** $\frac{1}{2}$   **77.** [4.1] Quartic
**78.** [5.2] Exponential   **79.** [1.4] Linear (constant)
**80.** [5.2] Exponential   **81.** [4.5] Rational
**82.** [5.3] Logarithmic   **83.** [4.1] Cubic
**84.** [4.5] Rational   **85.** [1.4] Linear
**86.** [3.3] Quadratic   **87.** 4   **89.** $\log_a (x^3 - y^3)$
**91.** $\frac{1}{2} \log_a (x - y) - \frac{1}{2} \log_a (x + y)$   **93.** 7   **95.** True
**97.** True   **99.** True   **101.** $-2$   **103.** 3
**105.** $e^{-xy} = \dfrac{a}{b}$
**107.** $\log_a \left( \dfrac{x + \sqrt{x^2 - 5}}{5} \cdot \dfrac{x - \sqrt{x^2 - 5}}{x - \sqrt{x^2 - 5}} \right)$
$= \log_a \dfrac{5}{5(x - \sqrt{x^2 - 5})} = \log_a \dfrac{1}{x - \sqrt{x^2 - 5}}$
$= \log_a (x - \sqrt{x^2 - 5})^{-1} = -\log_a (x - \sqrt{x^2 - 5})$

## Exercise Set 5.5

**1.** 4   **3.** $\frac{3}{2}$   **5.** 5.044   **7.** $\frac{5}{2}$   **9.** $-3, \frac{1}{2}$   **11.** 0.959
**13.** 0   **15.** 0   **17.** 6.908   **19.** 84.191   **21.** $-1.710$
**23.** 2.844   **25.** $-1.567, 1.567$   **27.** 1.869
**29.** $-1.518, 0.825$   **31.** 625   **33.** 0.0001   **35.** $e$
**37.** $-\frac{1}{3}$   **39.** $\frac{22}{3}$   **41.** 10   **43.** 4   **45.** $\frac{1}{63}$   **47.** 2
**49.** $\frac{2}{5}$   **51.** 5   **53.** $\frac{21}{8}$   **55.** $\frac{8}{7}$   **57.** 6   **59.** 6.192
**61.** 0   **63.** $-1.911, 4.222$   **65.** 0.621   **67.** $-10, 0.366$
**69.** 7.062   **71.** 2.444   **73.** $(4.093, 0.786)$

**75.** $(7.586, 6.684)$   **77.** [3.3] (a) $(0, -6)$; (b) $x = 0$;
(c) minimum: $-6$ when $x = 0$   **78.** [3.3] (a) $(3, 1)$;
(b) $x = 3$; (c) maximum: 1 when $x = 3$
**79.** [3.3] (a) $(-1, -5)$; (b) $x = -1$; (c) maximum: $-5$
when $x = -1$   **80.** [3.3] (a) $(2, 4)$; (b) $x = 2$;
(c) minimum: 4 when $x = 2$   **81.** $\dfrac{\ln 2}{2}$, or 0.347   **83.** 1, $e^4$
or 1, 54.598   **85.** $\frac{1}{3}$, 27   **87.** 1, $e^2$ or 1, 7.389   **89.** 0, $\dfrac{\ln 2}{\ln 5}$, or
0, 0.431   **91.** $e^{-2}$, $e^2$ or 0.135, 7.389   **93.** $\frac{7}{4}$   **95.** $a = \frac{2}{3} b$

## Exercise Set 5.6

**1.** (a) $P(t) = 5.9 e^{0.0176t}$; (b) 6.6 million; (c) about 9.7 years
after 2010; (d) 39.4 years   **3.** (a) 0.98%; (b) 1.51%;
(c) 21.6 years; (d) 57.8 years; (e) 0.28%; (f) 29.9 years;
(g) 49.5 years; (h) 0.66%; (i) 2.04%; (j) 87.7 years
**5.** In about 151 years   **7.** (a) $P(t) = 10,000 e^{0.054t}$;
(b) \$10,554.85; \$11,140.48; \$13,099.64; \$17,160.07;
(c) about 12.8 years   **9.** About 12,320 years
**11.** (a) 22.4% per minute; (b) 3.1% per year; (c) 60.3 days;
(d) 10.7 years; (e) 2.4% per year; (f) 1.0% per year;
(g) 0.0029% per year   **13.** (a) $k \approx 0.1268$;
$C(t) = 4.85 e^{-0.1268t}$; (b) 2015: \$2.27; 2018: \$1.55;
(c) in 2017   **15.** (a) $k \approx 0.1002$; $C(x) = 28 e^{0.1002x}$;
(b) 2013: 764; 2017: 1141; (c) about 40 years after 1980, or in 2020
**17.** (a)
$$y = \frac{3500}{1 + 19.9 e^{-0.6x}}$$
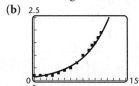
(b) 167; (c) 500; 1758; 3007; 3449; 3495; (d) as $t \to \infty$, $N(t) \to 3500$; the number approaches 3500 but never actually reaches it.
**19.** 46.7°F   **21.** 59.6°F
**23.** (d)   **25.** (a)   **27.** (e)
**29.** (a) $y = 0.1377082721(1.023820625)^x$; $r \approx 0.9824$, the function is a good fit;
(b) 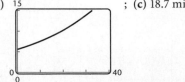   ; (c) 2007: 1.7%; 2015: 2.1%; 2020: 2.3%
**31.** (a) $y = 5.600477432(1.030576807)^x$;
(b) ; (c) 18.7 million homes
**33.** (a) $y = 2079.930585(1.079796876)^x$;
(b) 2012: \$5226; 2015: \$6579; (c) about 16 years after 2000, or in 2016   **35.** [1.6] Multiplication principle for inequalities   **36.** [5.4] Product rule   **37.** [3.2] Principle of zero products   **38.** [3.2] Principle of square roots
**39.** [5.4] Power rule   **40.** [1.5] Multiplication principle for equations   **41.** \$14,182.70   **43.** (a) 24.7%; 1.5%; 0.09%; $(3.98 \times 10^{-29})$%; (b) 0.00008%   **45.** Linear

## Review Exercises: Chapter 5

**1.** True    **2.** False    **3.** False    **4.** True    **5.** False
**6.** True    **7.** $\{(-2.7, 1.3), (-3, 8), (3, -5), (-3, 6), (-5, 7)\}$
**8. (a)** $x = -2y + 3$;  **(b)** $x = 3y^2 + 2y - 1$;
**(c)** $0.8y^3 - 5.4x^2 = 3y$    **9.** No    **10.** No    **11.** Yes
**12.** Yes    **13. (a)** Yes;  **(b)** $f^{-1}(x) = \dfrac{-x + 2}{3}$

**14. (a)** Yes;  **(b)** $f^{-1}(x) = \dfrac{x + 2}{x - 1}$
**15. (a)** Yes;  **(b)** $f^{-1}(x) = x^2 + 6, x \geq 0$
**16. (a)** Yes;  **(b)** $f^{-1}(x) = \sqrt[3]{x + 8}$    **17. (a)** No
**18. (a)** Yes;  **(b)** $f^{-1}(x) = \ln x$
**19.** $(f^{-1} \circ f)(x) =$
$f^{-1}(f(x)) = f^{-1}(6x - 5) = \dfrac{6x - 5 + 5}{6} = \dfrac{6x}{6} = x$;
$(f \circ f^{-1})(x) =$
$f(f^{-1}(x)) = f\left(\dfrac{x + 5}{6}\right) = 6\left(\dfrac{x + 5}{6}\right) - 5 = x + 5 - 5 = x$

**20.** $(f^{-1} \circ f)(x) =$
$f^{-1}(f(x)) = f^{-1}\left(\dfrac{x + 1}{x}\right) = \dfrac{1}{\dfrac{x + 1}{x} - 1} =$

$\dfrac{1}{\dfrac{x + 1 - x}{x}} = \dfrac{1}{\dfrac{1}{x}} = x; (f \circ f^{-1})(x) =$

$f(f^{-1}(x)) = f\left(\dfrac{1}{x - 1}\right) =$

$\dfrac{\dfrac{1}{x - 1} + 1}{\dfrac{1}{x - 1}} = \dfrac{\dfrac{1 + x - 1}{x - 1}}{\dfrac{1}{x - 1}} = \dfrac{x}{x - 1} \cdot \dfrac{x - 1}{1} = x$

**21.** $f^{-1}(x) = \dfrac{2 - x}{5}$; domain and range of both $f$ and $f^{-1}$:
$(-\infty, \infty)$;

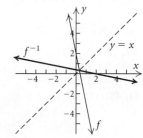

**22.** $f^{-1}(x) = \dfrac{-2x - 3}{x - 1}$;
domain of $f$: $(-\infty, -2) \cup (-2, \infty)$;
range of $f$: $(-\infty, 1) \cup (1, \infty)$;
domain of $f^{-1}$: $(-\infty, 1) \cup (1, \infty)$;
range of $f^{-1}$: $(-\infty, -2) \cup (-2, \infty)$

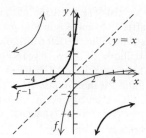

**23.** 657    **24.** $a$
**25.**

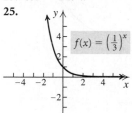

**26.**

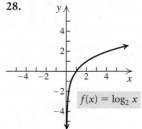

**27.**

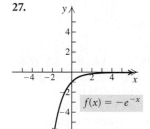

**28.**

**29.**

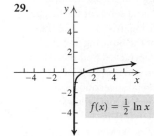

**30.**

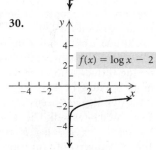

**31.** (c)    **32.** (a)    **33.** (b)    **34.** (f)    **35.** (e)
**36.** (d)    **37.** 3    **38.** 5    **39.** 1    **40.** 0    **41.** $\frac{1}{4}$
**42.** $\frac{1}{2}$    **43.** 0    **44.** 1    **45.** $\frac{1}{3}$    **46.** $-2$    **47.** $4^2 = x$
**48.** $a^k = Q$    **49.** $\log_4 \frac{1}{64} = -3$    **50.** $\ln 80 = x$, or
$\log_e 80 = x$    **51.** 1.0414    **52.** $-0.6308$    **53.** 1.0986
**54.** $-3.6119$    **55.** Does not exist    **56.** Does not exist
**57.** 1.9746    **58.** 0.5283    **59.** $\log_b \dfrac{x^3 \sqrt{z}}{y^4}$, or $\log_b \dfrac{x^3 z^{1/2}}{y^4}$
**60.** $\ln (x^2 - 4)$    **61.** $\frac{1}{4} \ln w + \frac{1}{2} \ln r$    **62.** $\frac{2}{3} \log M - \frac{1}{3} \log N$
**63.** 0.477    **64.** 1.699    **65.** $-0.699$    **66.** 0.233
**67.** $-5k$    **68.** $-6t$    **69.** 16    **70.** $\frac{1}{5}$    **71.** 4.382    **72.** 2
**73.** $\frac{1}{2}$    **74.** 5    **75.** 4    **76.** 9    **77.** 1    **78.** 3.912
**79. (a)** $A(t) = 30{,}000(1.0105)^{4t}$;  **(b)** $30{,}000; \$38{,}547.20$;
$\$49{,}529.56; \$63{,}640.87$    **80.** $\$65.6$ million; $\$363.1$ million
**81.** 8.1 years    **82.** 2.7%    **83.** About 2623 years    **84.** 5.6
**85.** 6.3    **86.** 30 decibels    **87. (a)** 2.2 ft/sec;  **(b)** 8,553,143

**88.** (a) $k \approx 0.1432$; (b) $S(t) = 0.035e^{0.1432t}$, where $t$ is the number of years after 1940 and $S$ is in billions of dollars; (c) about $1.256 billion; about $92.166 billion; about $1616 billion, or $1.616 trillion; (d) about 76 years after 1940, or in 2016 **89.** (a) $P(t) = 6.188e^{0.01985t}$, where $t$ is the number of years since 2011 and $P$ is in millions; (b) 2013: 6.439 million; 2015: 6.699 million; (c) 24 years after 2011; (d) 34.9 years **90.** (a) $C(x) = 4.687002171(1.047188864)^x$, where $x$ is the number of years since 1958; (b)

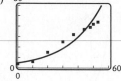 ;

(c) 1970: 8¢; 2000: 33¢; 2015: 65¢
**91.** No **92.** (a) $y = 5e^{-x}\ln x$

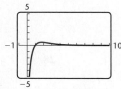

(b) relative maximum: 0.486 at $x = 1.763$; no relative minimum **93.** D **94.** A **95.** D **96.** B **97.** $\frac{1}{64}, 64$
**98.** 1 **99.** 16 **100.** $(1, \infty)$ **101.** Reflect the graph of $f(x) = \ln x$ across the line $y = x$ to obtain the graph of $h(x) = e^x$. Then shift this graph right 2 units to obtain the graph of $g(x) = e^{x-2}$. **102.** Measure the atmospheric pressure $P$ at the top of the building. Substitute that value in the equation $P = 14.7e^{-0.00005a}$, and solve for the height, or altitude, $a$ of the top of the building. Also measure the atmospheric pressure at the base of the building and solve for the altitude of the base. Then subtract to find the height of the building. **103.** $\log_a ab^3 \neq (\log_a a)(\log_a b^3)$. If the first step had been correct, then the second step would be as well. The correct procedure follows: $\log_a ab^3 = \log_a a + \log_a b^3 = 1 + 3\log_a b$.
**104.** The inverse of a function $f(x)$ is written $f^{-1}(x)$, whereas $[f(x)]^{-1}$ means $\frac{1}{f(x)}$.

## Test: Chapter 5

**1.** [5.1] $\{(5, -2), (3, 4), (-1, 0), (-3, -6)\}$ **2.** [5.1] No
**3.** [5.1] Yes **4.** [5.1] (a) Yes; (b) $f^{-1}(x) = \sqrt[3]{x - 1}$
**5.** [5.1] (a) Yes; (b) $f^{-1}(x) = 1 - x$
**6.** [5.1] (a) Yes; (b) $f^{-1}(x) = \dfrac{2x}{1 + x}$ **7.** [5.1] (a) No

**8.** [5.1] $(f^{-1} \circ f)(x) =$
$f^{-1}(f(x)) = f^{-1}(-4x + 3) = \dfrac{3 - (-4x + 3)}{4} =$
$\dfrac{4x}{4} = x$; $(f \circ f^{-1})(x) =$
$f(f^{-1}(x)) = f\left(\dfrac{3 - x}{4}\right) = -4\left(\dfrac{3 - x}{4}\right) + 3 =$
$-3 + x + 3 = x$

**9.** [5.1] $f^{-1}(x) = \dfrac{4x + 1}{x}$; domain of $f$: $(-\infty, 4) \cup (4, \infty)$; range of $f$: $(-\infty, 0) \cup (0, \infty)$; domain of $f^{-1}$: $(-\infty, 0) \cup (0, \infty)$; range of $f^{-1}$: $(-\infty, 4) \cup (4, \infty)$;

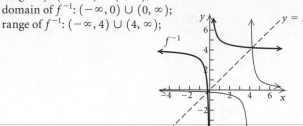

**10.** [5.2] **11.** [5.3]

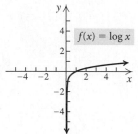 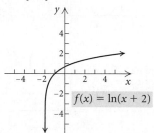

**12.** [5.2] **13.** [5.3]

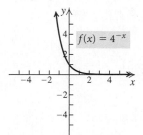 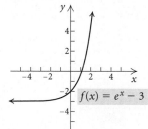

**14.** [5.3] $-5$ **15.** [5.3] 1 **16.** [5.3] 0 **17.** [5.3] $\frac{1}{5}$
**18.** [5.3] $x = e^4$ **19.** [5.3] $x = \log_3 5.4$ **20.** [5.3] 2.7726
**21.** [5.3] $-0.5331$ **22.** [5.3] 1.2851 **23.** [5.4] $\log_a \dfrac{x^2\sqrt{z}}{y}$, or $\log_a \dfrac{x^2 z^{1/2}}{y}$ **24.** [5.4] $\frac{2}{5}\ln x + \frac{1}{5}\ln y$ **25.** [5.4] 2.322
**26.** [5.4] $-4t$ **27.** [5.5] $\frac{1}{2}$ **28.** [5.5] 1 **29.** [5.5] 1
**30.** [5.5] 4.174 **31.** [5.3] 6.6 **32.** [5.6] 0.0154
**33.** [5.6] (a) 4.5%; (b) $P(t) = 1000e^{0.045t}$; (c) $1433.33; (d) 15.4 years **34.** [5.2] C **35.** [5.5] $\frac{27}{8}$

# Chapter 6

## Visualizing the Graph

**1.** C **2.** G **3.** D **4.** J **5.** A **6.** F **7.** I
**8.** B **9.** H **10.** E

## Exercise Set 6.1

**1.** (c)  **3.** (f)  **5.** (b)  **7.** $(-1, 3)$  **9.** $(-1, 1)$
**11.** No solution  **13.** $(-2, 4)$  **15.** Infinitely many
solutions; $\left(x, \dfrac{x-1}{2}\right)$, or $(2y + 1, y)$  **17.** $(5, 4)$
**19.** $(1, -3)$  **21.** $(2, -2)$  **23.** No solution
**25.** $\left(\frac{39}{11}, -\frac{1}{11}\right)$  **27.** $\left(\frac{1}{2}, \frac{3}{4}\right)$  **29.** Infinitely many solutions;
$(x, 3x - 5)$ or $\left(\frac{1}{3}y + \frac{5}{3}, y\right)$  **31.** $(1, 3)$; consistent,
independent  **33.** $(-4, -2)$; consistent, independent
**35.** Infinitely many solutions; $(4y + 2, y)$ or $\left(x, \frac{1}{4}x - \frac{1}{2}\right)$;
consistent, dependent  **37.** No solution; inconsistent,
independent  **39.** $(1, 1)$; consistent, independent
**41.** $(-3, 0)$; consistent, independent  **43.** $(10, 8)$;
consistent, independent  **45.** True  **47.** False  **49.** True
**51.** Knee replacements: 3.5 million; hip replacements:
0.572 million, or 572,000  **53.** Cell-phone providers:
37,477 complaints; cable/satellite TV providers: 32,616 complaints  **55.** Ohio: 58,590 Amish; Wisconsin: 15,360 Amish
**57.** Pecan: 503 calories; lemon meringue: 362 calories
**59.** Standard: 76 packages; express: 44 packages
**61.** 4%: \$6000; 5%: \$9000  **63.** 6 lb of French roast, 4 lb
of Kenyan  **65.** 1.5 servings of spaghetti, 2 servings of lettuce
**67.** Boat: 20 km/h; stream: 3 km/h  **69.** 2 hr
**71.** $(15, \$100)$  **73.** 140  **75.** 6000
**77.** (a) $r(x) = 0.2308826185x + 44.61085102$;
$p(x) = 0.6288961625x + 26.81925282$;  (b) 45 years after 1995
**79.** [1.5] About 838.3 million travelers  **80.** [1.5] About
\$44.5 billion  **81.** [3.2] $-2, 6$  **82.** [3.2] $-1, 5$
**83.** [1.2] 120  **84.** [3.2] 1, 3  **85.** 4 km  **87.** First train:
36 km/h; second train: 54 km/h  **89.** $A = \frac{1}{10}, B = -\frac{7}{10}$
**91.** City: 294 mi; highway: 153 mi

## Exercise Set 6.2

**1.** $(3, -2, 1)$  **3.** $(-3, 2, 1)$  **5.** $\left(2, \frac{1}{2}, -2\right)$
**7.** No solution  **9.** Infinitely many solutions;
$\left(\dfrac{11y + 19}{5}, y, \dfrac{9y + 11}{5}\right)$
**11.** $\left(\frac{1}{2}, \frac{2}{3}, -\frac{5}{6}\right)$  **13.** $(-1, 4, 3)$  **15.** $(1, -2, 4, -1)$
**17.** North America: 6 sites; Europe: 13 sites; Asia: 2 sites
**19.** China: 59 billion lb; United States: 10 billion lb; Turkey:
5 billion lb  **21.** Brewed coffee: 80 mg; Red Bull: 80 mg;
Mountain Dew: 37 mg  **23.** Fish: 183 million; cats: 94 million; dogs: 78 million  **25.** *The Dark Knight*: \$158 million;
*Spider-Man 3*: \$151 million; *The Twilight Saga: New Moon*:
\$143 million  **27.** $1\frac{1}{4}$ servings of beef, 1 baked potato,
$\frac{3}{4}$ serving of strawberries  **29.** 3%: \$1300; 4%: \$900; 6%:
\$2800  **31.** Orange juice: \$2.40; bagel: \$2.75; coffee: \$3.00
**33.** (a) $f(x) = \frac{5}{52}x^2 - \frac{87}{52}x + 16$;  (b) 10.5%
**35.** (a) $f(x) = -\frac{37}{84}x^2 + \frac{353}{84}x + 431$;  (b) 2000: about
440 acres, 2010: about 411 acres

**37.** (a) $f(x) = 0.143707483x^2 - 0.6921768707x +$
$5.882482993$;  (b) 2003: 5.3%, 2007: 6.0%, 2009: 8.1%
**39.** [1.4] Perpendicular  **40.** [4.1] The leading-term test
**41.** [1.2] A vertical line  **42.** [5.1] A one-to-one function
**43.** [4.5] A rational function  **44.** [2.5] Inverse variation
**45.** [4.5] A vertical asymptote  **46.** [4.5] A horizontal
asymptote  **47.** $\left(-\frac{1}{2}, -1, -\frac{1}{3}\right)$  **49.** $3x + 4y + 2z = 12$
**51.** $y = -4x^3 + 5x^2 - 3x + 1$

## Exercise Set 6.3

**1.** $3 \times 2$  **3.** $1 \times 4$  **5.** $3 \times 3$  **7.** $\begin{bmatrix} 2 & -1 & | & 7 \\ 1 & 4 & | & 5 \end{bmatrix}$

**9.** $\begin{bmatrix} 1 & -2 & 3 & | & 12 \\ 2 & 0 & -4 & | & 8 \\ 0 & 3 & 1 & | & 7 \end{bmatrix}$

**11.** $3x - 5y = 1,$
   $x + 4y = -2$
**13.** $2x + y - 4z = 12,$
   $3x \quad\ + 5z = -1,$
   $x - y + z = 2$
**15.** $\left(\frac{3}{2}, \frac{5}{2}\right)$  **17.** $\left(-\frac{63}{29}, -\frac{114}{29}\right)$  **19.** $\left(-1, \frac{5}{2}\right)$
**21.** $(0, 3)$  **23.** No solution  **25.** Infinitely
many solutions; $(3y - 2, y)$  **27.** $(-1, 2, -2)$
**29.** $\left(\frac{3}{2}, -4, 3\right)$  **31.** $(-1, 6, 3)$
**33.** Infinitely many solutions; $\left(\frac{1}{2}z + \frac{1}{2}, -\frac{1}{2}z - \frac{1}{2}, z\right)$
**35.** Infinitely many solutions; $(r - 2, -2r + 3, r)$
**37.** No solution  **39.** $(1, -3, -2, -1)$  **41.** 15 stamps
**43.** 2010: \$2 million; 2011: \$3 million; 2012: \$6 million
**45.** [5.2] Exponential  **46.** [1.3] Linear  **47.** [4.5] Rational
**48.** [4.1] Quartic  **49.** [5.3] Logarithmic  **50.** [4.1] Cubic
**51.** [1.3] Linear  **52.** [3.2] Quadratic
**53.** $y = 3x^2 + \frac{5}{2}x - \frac{15}{2}$
**55.** $\begin{bmatrix} 1 & 5 \\ 0 & 1 \end{bmatrix}, \begin{bmatrix} 1 & 0 \\ 0 & 1 \end{bmatrix}$  **57.** $\left(-\frac{4}{3}, -\frac{1}{3}, 1\right)$
**59.** Infinitely many solutions; $\left(-\frac{14}{13}z - 1, \frac{3}{13}z - 2, z\right)$
**61.** $(-3, 3)$

## Exercise Set 6.4

**1.** $x = -3, y = 5$  **3.** $x = -1, y = 1$
**5.** $\begin{bmatrix} -2 & 7 \\ 6 & 2 \end{bmatrix}$  **7.** $\begin{bmatrix} 1 & 3 \\ 2 & 6 \end{bmatrix}$  **9.** $\begin{bmatrix} 9 & 9 \\ -3 & -3 \end{bmatrix}$
**11.** $\begin{bmatrix} 11 & 13 \\ 5 & 3 \end{bmatrix}$  **13.** $\begin{bmatrix} -4 & 3 \\ -2 & -4 \end{bmatrix}$  **15.** $\begin{bmatrix} 17 & 9 \\ -2 & 1 \end{bmatrix}$
**17.** $\begin{bmatrix} 0 & 0 \\ 0 & 0 \end{bmatrix}$  **19.** $\begin{bmatrix} 1 & 2 \\ 4 & 3 \end{bmatrix}$  **21.** $\begin{bmatrix} 1 \\ 40 \end{bmatrix}$
**23.** $\begin{bmatrix} -10 & 28 \\ 14 & -26 \\ 0 & -6 \end{bmatrix}$  **25.** Not defined

**27.** $\begin{bmatrix} 3 & 16 & 3 \\ 0 & -32 & 0 \\ -6 & 4 & 5 \end{bmatrix}$

**29. (a)** $\begin{bmatrix} 40 & 20 & 30 \end{bmatrix}$; **(b)** $\begin{bmatrix} 44 & 22 & 33 \end{bmatrix}$; **(c)** $\begin{bmatrix} 84 & 42 & 63 \end{bmatrix}$; the total amount of each type of produce ordered for both weeks
**31. (a)** $\mathbf{C} = \begin{bmatrix} 140 & 27 & 3 & 13 & 64 \end{bmatrix}$,
$\mathbf{P} = \begin{bmatrix} 180 & 4 & 11 & 24 & 662 \end{bmatrix}$, $\mathbf{B} = \begin{bmatrix} 50 & 5 & 1 & 82 & 20 \end{bmatrix}$;
**(b)** $\begin{bmatrix} 650 & 50 & 28 & 307 & 1448 \end{bmatrix}$, the total nutritional value of
a meal of 1 3-oz serving of chicken, 1 cup of potato salad, and
3 broccoli spears

**33. (a)** $\begin{bmatrix} 1.50 & 0.15 & 0.26 & 0.23 & 0.64 \\ 1.55 & 0.14 & 0.24 & 0.21 & 0.75 \\ 1.62 & 0.22 & 0.31 & 0.28 & 0.53 \\ 1.70 & 0.20 & 0.29 & 0.33 & 0.68 \end{bmatrix}$;

**(b)** $\begin{bmatrix} 65 & 48 & 93 & 57 \end{bmatrix}$;
**(c)** $\begin{bmatrix} 419.46 & 48.33 & 73.78 & 69.88 & 165.65 \end{bmatrix}$;
**(d)** the total cost, in dollars, for each item for the day's meals

**35. (a)** $\begin{bmatrix} 900 & 500 \\ 450 & 1000 \\ 600 & 700 \end{bmatrix}$; **(b)** $\begin{bmatrix} 5 & 8 & 4 \end{bmatrix}$; **(c)** $\begin{bmatrix} 10{,}500 & 13{,}300 \end{bmatrix}$;

**(d)** the total profit from each distributor
**37. (a)** $\begin{bmatrix} 20 & 25 & 15 \end{bmatrix}$; **(b)** $\mathbf{CM} = \begin{bmatrix} 38{,}250 & 45{,}500 \end{bmatrix}$

**39.** $\begin{bmatrix} 2 & -3 \\ 1 & 5 \end{bmatrix} \begin{bmatrix} x \\ y \end{bmatrix} = \begin{bmatrix} 7 \\ -6 \end{bmatrix}$

**41.** $\begin{bmatrix} 1 & 1 & -2 \\ 3 & -1 & 1 \\ 2 & 5 & -3 \end{bmatrix} \begin{bmatrix} x \\ y \\ z \end{bmatrix} = \begin{bmatrix} 6 \\ 7 \\ 8 \end{bmatrix}$

**43.** $\begin{bmatrix} 3 & -2 & 4 \\ 2 & 1 & -5 \end{bmatrix} \begin{bmatrix} x \\ y \\ z \end{bmatrix} = \begin{bmatrix} 17 \\ 13 \end{bmatrix}$

**45.** $\begin{bmatrix} -4 & 1 & -1 & 2 \\ 1 & 2 & -1 & -1 \\ -1 & 1 & 4 & -3 \\ 2 & 3 & 5 & -7 \end{bmatrix} \begin{bmatrix} w \\ x \\ y \\ z \end{bmatrix} = \begin{bmatrix} 12 \\ 0 \\ 1 \\ 9 \end{bmatrix}$

**47.** [3.3] **(a)** $\left( \frac{1}{2}, -\frac{25}{4} \right)$;
**(b)** $x = \frac{1}{2}$;
**(c)** minimum: $-\frac{25}{4}$;
**(d)**

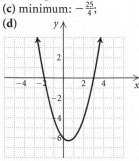

$f(x) = x^2 - x - 6$

**48.** [3.3] **(a)** $\left( \frac{5}{4}, -\frac{49}{8} \right)$;
**(b)** $x = \frac{5}{4}$;
**(c)** minimum: $-\frac{49}{8}$;
**(d)**

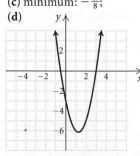

$f(x) = 2x^2 - 5x - 3$

**49.** [3.3] **(a)** $\left( -\frac{3}{2}, \frac{17}{4} \right)$;
**(b)** $x = -\frac{3}{2}$;
**(c)** maximum: $\frac{17}{4}$;
**(d)**

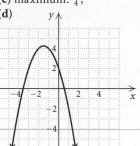

$f(x) = -x^2 - 3x + 2$

**50.** [3.3] **(a)** $\left( \frac{2}{3}, \frac{16}{3} \right)$;
**(b)** $x = \frac{2}{3}$;
**(c)** maximum: $\frac{16}{3}$;
**(d)**

$f(x) = -3x^2 + 4x + 4$

**51.** $(\mathbf{A} + \mathbf{B})(\mathbf{A} - \mathbf{B}) = \begin{bmatrix} -2 & 1 \\ 2 & -1 \end{bmatrix}$; $\mathbf{A}^2 - \mathbf{B}^2 = \begin{bmatrix} 0 & 3 \\ 0 & -3 \end{bmatrix}$

**53.** $(\mathbf{A} + \mathbf{B})(\mathbf{A} - \mathbf{B}) = \begin{bmatrix} -2 & 1 \\ 2 & -1 \end{bmatrix}$
$= \mathbf{A}^2 + \mathbf{BA} - \mathbf{AB} - \mathbf{B}^2$

## Mid-Chapter Mixed Review: Chapter 6

**1.** False    **2.** True    **3.** True    **4.** False    **5.** $(-3, 2)$
**6.** No solution    **7.** $(1, -2)$    **8.** Infinitely many solutions;
$\left( x, \frac{x-1}{3} \right)$ or $(3y + 1, y)$    **9.** $\left( \frac{1}{3}, -\frac{1}{6}, \frac{4}{3} \right)$
**10.** Under 10 lb: 60 packages; 10 lb up to 15 lb: 70 packages;
15 lb or more: 20 packages    **11.** $(4, -3)$    **12.** $(-3, 2, -1)$

**13.** $\begin{bmatrix} 1 & 5 \\ 6 & 1 \end{bmatrix}$   **14.** $\begin{bmatrix} -5 & 7 \\ -4 & -7 \end{bmatrix}$   **15.** $\begin{bmatrix} -8 & 12 & 0 \\ 4 & -4 & 8 \\ -12 & 16 & 4 \end{bmatrix}$

**16.** $\begin{bmatrix} 0 & 16 \\ 13 & -1 \end{bmatrix}$   **17.** $\begin{bmatrix} -7 & 21 \\ -6 & 18 \end{bmatrix}$   **18.** $\begin{bmatrix} 24 & 26 \\ -12 & -13 \end{bmatrix}$

**19.** $\begin{bmatrix} 20 & 16 & -10 \\ -10 & -8 & 5 \end{bmatrix}$   **20.** Not defined

**21.** $\begin{bmatrix} 2 & -1 & 3 \\ 1 & 2 & -1 \\ 3 & -4 & 2 \end{bmatrix} \begin{bmatrix} x \\ y \\ z \end{bmatrix} = \begin{bmatrix} 7 \\ 3 \\ 5 \end{bmatrix}$

**22.** When a variable is not alone on one side of an equation or when solving for a variable is difficult or produces an expression containing fractions, the elimination method is preferable to the substitution method.    **23.** Add a nonzero multiple of one equation to a nonzero multiple of the other equation, where the multiples are not opposites.    **24.** See Example 4 in Section 6.3.    **25.** No; see Exercise 17 in Section 6.4, for example.

## Exercise Set 6.5

**1.** Yes    **3.** No    **5.** $\begin{bmatrix} -3 & 2 \\ 5 & -3 \end{bmatrix}$    **7.** Does not exist

**9.** $\begin{bmatrix} \frac{3}{8} & -\frac{1}{4} & \frac{1}{8} \\ -\frac{1}{8} & \frac{3}{4} & -\frac{3}{8} \\ -\frac{1}{4} & \frac{1}{2} & \frac{1}{4} \end{bmatrix}$    **11.** Does not exist

**13.** $\begin{bmatrix} 0.4 & -0.6 \\ 0.2 & -0.8 \end{bmatrix}$    **15.** $\begin{bmatrix} -1 & -1 & -6 \\ 1 & 0 & 2 \\ 0 & 1 & 3 \end{bmatrix}$

**17.** $\begin{bmatrix} 1 & 1 & 2 \\ 1 & 1 & 1 \\ 2 & 3 & 4 \end{bmatrix}$    **19.** Does not exist

**21.** $\begin{bmatrix} 1 & -2 & 3 & 8 \\ 0 & 1 & -3 & 1 \\ 0 & 0 & 1 & -2 \\ 0 & 0 & 0 & -1 \end{bmatrix}$

**23.** $\begin{bmatrix} 0.25 & 0.25 & 1.25 & -0.25 \\ 0.5 & 1.25 & 1.75 & -1 \\ -0.25 & -0.25 & -0.75 & 0.75 \\ 0.25 & 0.5 & 0.75 & -0.5 \end{bmatrix}$

**25.** $(-23, 83)$    **27.** $(-1, 5, 1)$    **29.** $(2, -2)$    **31.** $(0, 2)$
**33.** $(3, -3, -2)$    **35.** $(-1, 0, 1)$    **37.** $(1, -1, 0, 1)$
**39.** Sausages: 50; hot dogs: 95    **41.** Topsoil: $239; mulch: $179; pea gravel: $222    **43.** [4.3] $-48$    **44.** [4.3] 194

**45.** [3.2] $\dfrac{-1 \pm \sqrt{57}}{4}$    **46.** [3.4] $-3, -2$    **47.** [3.4] 4

**48.** [3.2], [3.4] 9    **49.** [4.3] $(x + 2)(x - 1)(x - 4)$
**50.** [4.3] $(x + 5)(x + 1)(x - 1)(x - 3)$

**51.** $\mathbf{A}^{-1}$ exists if and only if $x \neq 0$. $\mathbf{A}^{-1} = \begin{bmatrix} \frac{1}{x} \end{bmatrix}$

**53.** $\mathbf{A}^{-1}$ exists if and only if $xyz \neq 0$. $\mathbf{A}^{-1} = \begin{bmatrix} 0 & 0 & \frac{1}{z} \\ 0 & \frac{1}{y} & 0 \\ \frac{1}{x} & 0 & 0 \end{bmatrix}$

## Exercise Set 6.6

**1.** $-14$    **3.** $-2$    **5.** $-11$    **7.** $x^3 - 4x$
**9.** $M_{11} = 6, M_{32} = -9, M_{22} = -29$
**11.** $A_{11} = 6, A_{32} = 9, A_{22} = -29$    **13.** $-10$
**15.** $-10$    **17.** $-10$    **19.** $M_{41} = -14$,
$M_{33} = 20$    **21.** $A_{24} = 15, A_{43} = 30$    **23.** 110
**25.** $-109$    **27.** $-x^4 + x^2 - 5x$    **29.** $\left(-\frac{25}{2}, -\frac{11}{2}\right)$
**31.** $(3, 1)$    **33.** $\left(\frac{1}{2}, -\frac{1}{3}\right)$    **35.** $(1, 1)$    **37.** $\left(\frac{3}{2}, \frac{13}{14}, \frac{33}{14}\right)$
**39.** $(3, -2, 1)$    **41.** $(1, 3, -2)$    **43.** $\left(\frac{1}{2}, \frac{2}{3}, -\frac{5}{6}\right)$

**45.** [5.1] $f^{-1}(x) = \dfrac{x - 2}{3}$    **46.** [5.1] Not one-to-one
**47.** [5.1] Not one-to-one    **48.** [5.1] $f^{-1}(x) = (x - 1)^3$
**49.** [3.1] $5 - 3i$    **50.** [3.1] $6 - 2i$    **51.** [3.1] $10 - 10i$
**52.** [3.1] $\frac{9}{25} + \frac{13}{25}i$    **53.** $3, -2$    **55.** 4

**57.** Answers may vary. $\begin{vmatrix} a & b \\ -b & a \end{vmatrix}$

**59.** Answers may vary. $\begin{vmatrix} 2\pi r & 2\pi r \\ -h & r \end{vmatrix}$

## Exercise Set 6.7

**1.** (f)    **3.** (h)    **5.** (g)    **7.** (b)
**9.**

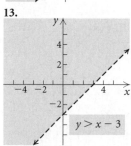

$y > 2x$

**11.**

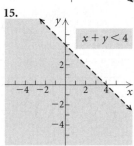

$y + x \geq 0$

**13.**

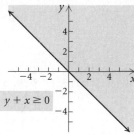

$y > x - 3$

**15.**

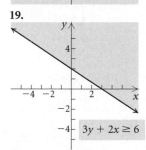

$x + y < 4$

**17.**

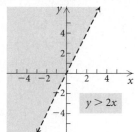

$3x - 2y \leq 6$

**19.**

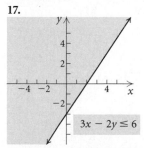

$3y + 2x \geq 6$

**21.**

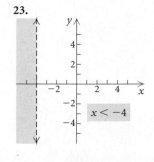

$3x - 2 \leq 5x + y$

**23.**

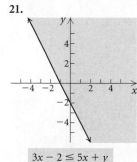

$x < -4$

**25.**

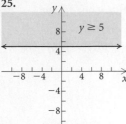

$y \geq 5$

**27.**

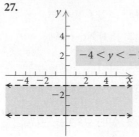

$-4 < y < -1$

**55.**

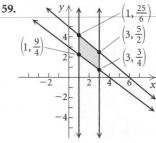

(0, 4)    (6, 4)    (4, 2)

**57.**

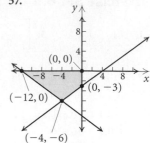

(0, 0)    (0, −3)    (−12, 0)    (−4, −6)

**29.**

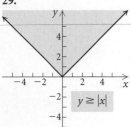

$y \geq |x|$

**31.** (f)    **33.** (a)    **35.** (b)

**59.**

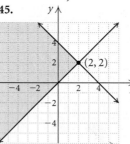

$\left(1, \frac{25}{6}\right)$    $\left(3, \frac{5}{2}\right)$    $\left(1, \frac{9}{4}\right)$    $\left(3, \frac{3}{4}\right)$

**37.** $y \leq -x + 4,$      **39.** $x < 2,$
$\quad\ y \leq 3x$             $\quad\ y > -1$

**41.** $y \leq -x + 3,$
$\quad\ y \leq x + 1,$
$\quad\ x \geq 0,$
$\quad\ y \geq 0$

**61.** Maximum: 179 when $x = 7$ and $y = 0$; minimum: 48 when $x = 0$ and $y = 4$    **63.** Maximum: 216 when $x = 0$ and $y = 6$; minimum: 0 when $x = 0$ and $y = 0$

**65.** Maximum income of \$130 is achieved when 100 of each type of biscuit are made.    **67.** Maximum profit of \$11,000 is achieved by producing 100 units of lumber and 300 units of plywood.    **69.** Minimum cost of \$$36\frac{12}{13}$ is achieved by using $1\frac{11}{13}$ sacks of soybean meal and $1\frac{11}{13}$ sacks of oats.    **71.** Maximum income of \$3110 is achieved when \$22,000 is invested in corporate bonds and \$18,000 is invested in municipal bonds.    **73.** Minimum cost of \$460 thousand is achieved using 30 $P_1$'s and 10 $P_2$'s.    **75.** Maximum profit per day of \$192 is achieved when 2 knit suits and 4 worsted suits are made.    **77.** Minimum weekly cost of \$19.05 is achieved when 1.5 lb of meat and 3 lb of cheese are used.

**79.** Maximum total number of 800 is achieved when there are 550 of A and 250 of B.

**81.** [1.6] $\{x \mid -7 \leq x < 2\}$, or $[-7, 2)$
**82.** [3.5] $\{x \mid x \leq 1 \ or \ x \geq 5\}$, or $(-\infty, 1] \cup [5, \infty)$
**83.** [4.6] $\{x \mid -1 \leq x \leq 3\}$, or $[-1, 3]$
**84.** [4.6] $\{x \mid -3 < x < -2\}$, or $(-3, -2)$

**43.**

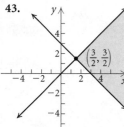

$\left(\frac{3}{2}, \frac{3}{2}\right)$

**45.**

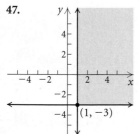

(2, 2)

**47.**

(1, −3)

**49.**

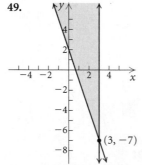

(3, −7)

**85.**

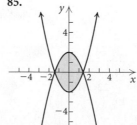

**87.**

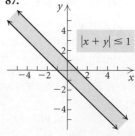

$|x + y| \leq 1$

**51.**

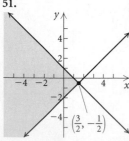

$\left(\frac{3}{2}, -\frac{1}{2}\right)$

**53.**

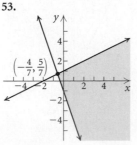

$\left(-\frac{4}{7}, \frac{5}{7}\right)$

**89.**

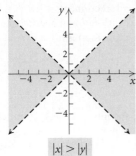

$|x| > |y|$

**91.** Maximum income of $11,400 is achieved by making 95 chairs and 0 sofas.

## Exercise Set 6.8

**1.** $\dfrac{2}{x-3} - \dfrac{1}{x+2}$   **3.** $\dfrac{5}{2x-1} - \dfrac{4}{3x-1}$

**5.** $\dfrac{2}{x+2} - \dfrac{3}{x-2} + \dfrac{4}{x+1}$

**7.** $-\dfrac{3}{(x+2)^2} - \dfrac{1}{x+2} + \dfrac{1}{x-1}$   **9.** $\dfrac{3}{x-1} - \dfrac{4}{2x-1}$

**11.** $x - 2 + \dfrac{\frac{17}{16}}{x+1} - \dfrac{\frac{11}{4}}{(x+1)^2} - \dfrac{\frac{17}{16}}{x-3}$

**13.** $\dfrac{3x+5}{x^2+2} - \dfrac{4}{x-1}$   **15.** $\dfrac{3}{2x-1} - \dfrac{2}{x+2} + \dfrac{10}{(x+2)^2}$

**17.** $3x + 1 + \dfrac{2}{2x-1} + \dfrac{3}{x+1}$

**19.** $-\dfrac{1}{x-3} + \dfrac{3x}{x^2+2x-5}$

**21.** $\dfrac{5}{3x+5} - \dfrac{3}{x+1} + \dfrac{4}{(x+1)^2}$   **23.** $\dfrac{8}{4x-5} + \dfrac{3}{3x+2}$

**25.** $\dfrac{2x-5}{3x^2+1} - \dfrac{2}{x-2}$   **27.** [4.1], [4.3], [4.4] $-1, \pm 3i$

**28.** [4.1], [4.3], [4.4] $3, \pm i$   **29.** [4.4] $-2, \dfrac{1 \pm \sqrt{5}}{2}$

**30.** [4.4] $-2, 3, \pm i$   **31.** [4.1], [4.3], [4.4] $-3, -1 \pm \sqrt{2}$

**33.** $-\dfrac{\frac{1}{2a^2}x}{x^2+a^2} + \dfrac{\frac{1}{4a^2}}{x-a} + \dfrac{\frac{1}{4a^2}}{x+a}$

**35.** $-\dfrac{3}{25(\ln x + 2)} + \dfrac{3}{25(\ln x - 3)} + \dfrac{7}{5(\ln x - 3)^2}$

## Review Exercises: Chapter 6

**1.** True   **2.** False   **3.** True   **4.** False   **5.** (a)
**6.** (e)   **7.** (h)   **8.** (d)   **9.** (b)   **10.** (g)
**11.** (c)   **12.** (f)   **13.** $(-2, -2)$   **14.** $(-5, 4)$
**15.** No solution   **16.** Infinitely many solutions;
$(-y-2, y)$, or $(x, -x-2)$   **17.** $(3, -1, -2)$   **18.** No
solution   **19.** $(-5, 13, 8, 2)$   **20.** Consistent: 13, 14, 16,
17, 19; the others are inconsistent.   **21.** Dependent: 16; the
others are independent.   **22.** $(1, 2)$   **23.** $(-3, 4, -2)$

**24.** Infinitely many solutions; $\left(\dfrac{z}{2}, -\dfrac{z}{2}, z\right)$

**25.** $(-4, 1, -2, 3)$   **26.** Nickels: 31; dimes: 44
**27.** 3%: $1600; 3.5%: $3400   **28.** 1 bagel, $\frac{1}{2}$ serving of
cream cheese, 2 bananas   **29.** 75, 69, 82
**30. (a)** $f(x) = -8x^2 + 16x + 40$;
**(b)** 16 thousand trademarks

**31.** $\begin{bmatrix} 0 & -1 & 6 \\ 3 & 1 & -2 \\ -2 & 1 & -2 \end{bmatrix}$   **32.** $\begin{bmatrix} -3 & 3 & 0 \\ -6 & -9 & 6 \\ 6 & 0 & -3 \end{bmatrix}$

**33.** $\begin{bmatrix} -1 & 1 & 0 \\ -2 & -3 & 2 \\ 2 & 0 & -1 \end{bmatrix}$   **34.** $\begin{bmatrix} -2 & 2 & 6 \\ 1 & -8 & 18 \\ 2 & 1 & -15 \end{bmatrix}$

**35.** Not defined   **36.** $\begin{bmatrix} 2 & -1 & -6 \\ 1 & 5 & -2 \\ -2 & -1 & 4 \end{bmatrix}$

**37.** $\begin{bmatrix} -13 & 1 & 6 \\ -3 & -7 & 4 \\ 8 & 3 & -5 \end{bmatrix}$   **38.** $\begin{bmatrix} -2 & -1 & 18 \\ 5 & -3 & -2 \\ -2 & 3 & -8 \end{bmatrix}$

**39. (a)** $\begin{bmatrix} 0.98 & 0.23 & 0.30 & 0.28 & 0.45 \\ 1.03 & 0.19 & 0.27 & 0.34 & 0.41 \\ 1.01 & 0.21 & 0.35 & 0.31 & 0.39 \\ 0.99 & 0.25 & 0.29 & 0.33 & 0.42 \end{bmatrix}$;

**(b)** $\begin{bmatrix} 32 & 19 & 43 & 38 \end{bmatrix}$;
**(c)** $\begin{bmatrix} 131.98 & 29.50 & 40.80 & 41.29 & 54.92 \end{bmatrix}$;
**(d)** the total cost, in dollars, for each item for the day's meals

**40.** $\begin{bmatrix} -\frac{1}{2} & 0 \\ \frac{1}{6} & \frac{1}{3} \end{bmatrix}$   **41.** $\begin{bmatrix} 0 & 0 & \frac{1}{4} \\ 0 & -\frac{1}{2} & 0 \\ \frac{1}{3} & 0 & 0 \end{bmatrix}$

**42.** $\begin{bmatrix} 1 & 0 & 0 & 0 \\ 0 & \frac{1}{9} & \frac{5}{18} & 0 \\ 0 & \frac{1}{9} & \frac{2}{9} & 0 \\ 0 & 0 & 0 & 1 \end{bmatrix}$

**43.** $\begin{bmatrix} 3 & -2 & 4 \\ 1 & 5 & -3 \\ 2 & -3 & 7 \end{bmatrix} \begin{bmatrix} x \\ y \\ z \end{bmatrix} = \begin{bmatrix} 13 \\ 7 \\ -8 \end{bmatrix}$   **44.** $(-8, 7)$

**45.** $(1, -2, 5)$   **46.** $(2, -1, 1, -3)$   **47.** 10
**48.** $-18$   **49.** $-6$   **50.** $-1$   **51.** $(3, -2)$
**52.** $(-1, 5)$   **53.** $\left(\frac{3}{2}, \frac{13}{14}, \frac{33}{14}\right)$   **54.** $(2, -1, 3)$

**55.**   **56.**

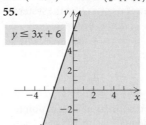

**57.**

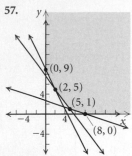

(0, 9)
(2, 5)
(5, 1)
(8, 0)

**58.** Minimum: 52 when $x = 2$ and $y = 4$; maximum: 92 when $x = 2$ and $y = 8$    **59.** Maximum score of 96 is achieved when 0 group A questions and 8 group B questions are answered.

**60.** $\dfrac{5}{x + 1} - \dfrac{5}{x + 2} - \dfrac{5}{(x + 2)^2}$    **61.** $\dfrac{2}{2x - 3} - \dfrac{5}{x + 4}$

**62.** C    **63.** A    **64.** B

**65.** 4%:$10,000; 5%: $12,000; $5\frac{1}{2}$%: $18,000

**66.** $\left(\frac{5}{18}, \frac{1}{7}\right)$    **67.** $\left(1, \frac{1}{2}, \frac{1}{3}\right)$

**68.**

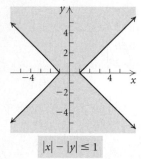

$|x| - |y| \le 1$

**69.**

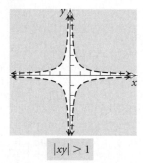

$|xy| > 1$

**70.** The solution of the equation $2x + 5 = 3x - 7$ is the first coordinate of the point of intersection of the graphs of $y_1 = 2x + 5$ and $y_2 = 3x - 7$. The solution of the system of equations $y = 2x + 5$, $y = 3x - 7$ is the ordered pair that is the point of intersection of the two lines.    **71.** In general, $(\mathbf{AB})^2 \ne \mathbf{A}^2\mathbf{B}^2$. $(\mathbf{AB})^2 = \mathbf{ABAB}$ and $\mathbf{A}^2\mathbf{B}^2 = \mathbf{AABB}$. Since matrix multiplication is not commutative, $\mathbf{BA} \ne \mathbf{AB}$, so $(\mathbf{AB})^2 \ne \mathbf{A}^2\mathbf{B}^2$.

**72.** If $\begin{vmatrix} a_1 & b_1 \\ a_2 & b_2 \end{vmatrix} = 0$, then $a_1 = ka_2$ and $b_1 = kb_2$ for some number $k$. This means that the equations $a_1x + b_1y = c_1$ and $a_2x + b_2y = c_2$ are dependent if $c_1 = kc_2$, or the system is inconsistent if $c_1 \ne kc_2$.    **73.** If $a_1x + b_1y = c_1$ and $a_2x + b_2y = c_2$ are parallel lines, then $a_1 = ka_2$, $b_1 = kb_2$, and $c_1 \ne kc_2$, for some number $k$.

Then $\begin{vmatrix} a_1 & b_1 \\ a_2 & b_2 \end{vmatrix} = 0$, $\begin{vmatrix} c_1 & b_1 \\ c_2 & b_2 \end{vmatrix} \ne 0$, and $\begin{vmatrix} a_1 & c_1 \\ a_2 & c_2 \end{vmatrix} \ne 0$.

**74.** The graph of a linear equation consists of a set of points on a line. The graph of a linear inequality consists of the set of points in a half-plane and might also include the points on the line that is the boundary of the half-plane.

**75.** The denominator of the second fraction, $x^2 - 5x + 6$, can be factored into linear factors with rational coefficients: $(x - 3)(x - 2)$. Thus the given expression is not a partial fraction decomposition.

## Test: Chapter 6

**1.** [6.1] $(-3, 5)$; consistent, independent    **2.** [6.1] Infinitely many solutions; $(x, 2x - 3)$ or $\left(\dfrac{y + 3}{2}, y\right)$; consistent, dependent

**3.** [6.1] No solution; inconsistent, independent

**4.** [6.1] $(1, -2)$; consistent, independent

**5.** [6.2] $(-1, 3, 2)$    **6.** [6.1] Student: 342 tickets; nonstudent: 408 tickets    **7.** [6.2] Latonna: 120 orders; Cole: 104 orders; Sam: 128 orders

**8.** [6.4] $\begin{bmatrix} -2 & -3 \\ -3 & 4 \end{bmatrix}$    **9.** [6.4] Not defined

**10.** [6.4] $\begin{bmatrix} -7 & -13 \\ 5 & -1 \end{bmatrix}$    **11.** [6.4] Not defined

**12.** [6.4] $\begin{bmatrix} 2 & -2 & 6 \\ -4 & 10 & 4 \end{bmatrix}$    **13.** [6.5] $\begin{bmatrix} 0 & -1 \\ -\frac{1}{4} & -\frac{3}{4} \end{bmatrix}$

**14.** [6.4] **(a)** $\begin{bmatrix} 0.95 & 0.40 & 0.39 \\ 1.10 & 0.35 & 0.41 \\ 1.05 & 0.39 & 0.36 \end{bmatrix}$; **(b)** $\begin{bmatrix} 26 & 18 & 23 \end{bmatrix}$;

**(c)** $\begin{bmatrix} 68.65 & 25.67 & 25.80 \end{bmatrix}$; **(d)** the total cost, in dollars, for each type of menu item served on the given day

**15.** [6.4] $\begin{bmatrix} 3 & -4 & 2 \\ 2 & 3 & 1 \\ 1 & -5 & -3 \end{bmatrix}\begin{bmatrix} x \\ y \\ z \end{bmatrix} = \begin{bmatrix} -8 \\ 7 \\ 3 \end{bmatrix}$

**16.** [6.5] $(-2, 1, 1)$    **17.** [6.6] 61    **18.** [6.6] $-33$

**19.** [6.6] $\left(-\frac{1}{2}, \frac{3}{4}\right)$    **20.** [6.7]

$3x + 4y \le -12$

**21.** [6.7] Maximum: 15 when $x = 3$ and $y = 3$; minimum: 2 when $x = 1$ and $y = 0$    **22.** [6.7] Maximum profit of $1475 occurs when 25 coffeecakes and 75 cheesecakes are prepared.    **23.** [6.8] $-\dfrac{2}{x - 1} + \dfrac{5}{x + 3}$    **24.** [6.7] D

**25.** [6.2] $A = 1$, $B = -3$, $C = 2$

# Chapter 7

## Exercise Set 7.1

**1.** (f)   **3.** (b)   **5.** (d)

**7.** $V: (0, 0); F: (0, 5);$
$D: y = -5$

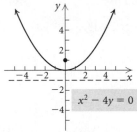
$x^2 = 20y$

**9.** $V: (0, 0); F: \left(-\frac{3}{2}, 0\right);$
$D: x = \frac{3}{2}$

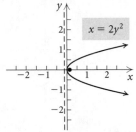
$y^2 = -6x$

**11.** $V: (0, 0); F: (0, 1);$
$D: y = -1$

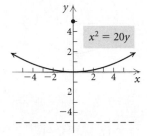
$x^2 - 4y = 0$

**13.** $V: (0, 0); F: \left(\frac{1}{8}, 0\right);$
$D: x = -\frac{1}{8}$

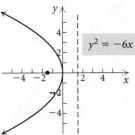
$x = 2y^2$

**15.** $y^2 = -12x$   **17.** $y^2 = 49x$   **19.** $x^2 = -4\pi y$
**21.** $(y - 2)^2 = 14\left(x + \frac{1}{2}\right)$
**23.** $V: (-2, 1); F: \left(-2, -\frac{1}{2}\right);$
$D: y = \frac{5}{2}$

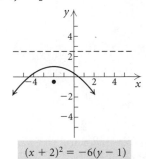
$(x + 2)^2 = -6(y - 1)$

**25.** $V: (-1, -3);$
$F: \left(-1, -\frac{7}{2}\right); D: y = -\frac{5}{2}$

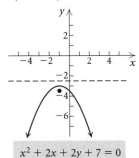
$x^2 + 2x + 2y + 7 = 0$

**27.** $V: (0, -2); F: \left(0, -1\frac{3}{4}\right);$
$D: y = -2\frac{1}{4}$

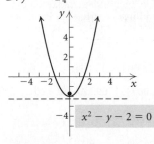
$x^2 - y - 2 = 0$

**29.** $V: (-2, -1);$
$F: \left(-2, -\frac{3}{4}\right); D: y = -1\frac{1}{4}$

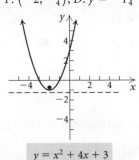
$y = x^2 + 4x + 3$

**31.** $V: \left(5\frac{3}{4}, \frac{1}{2}\right); F: \left(6, \frac{1}{2}\right); D: x = 5\frac{1}{2}$

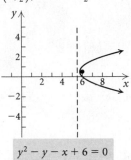
$y^2 - y - x + 6 = 0$

**33.** (a) $y^2 = 16x$; (b) $3\frac{33}{64}$ ft   **35.** About 11.75 in.   **37.** [1.1] (h)
**38.** [1.1], [1.4] (d)   **39.** [1.3] (a), (b), (f), (g)   **40.** [1.3] (b)
**41.** [1.4] (b)   **42.** [1.1] (f)   **43.** [1.4] (a) and (g)
**44.** [1.4] (a) and (h); (g) and (h); (b) and (c)   **45.** $(x + 1)^2 =$
$-4(y - 2)$   **47.** $V: (0.867, 0.348); F: (0.867, -0.190);$
$D: y = 0.887$   **49.** 10 ft, 11.6 ft, 16.4 ft, 24.4 ft, 35.6 ft, 50 ft

## Exercise Set 7.2

**1.** (b)   **3.** (d)   **5.** (a)
**7.** $(7, -2); 8$

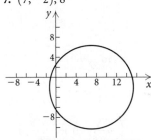
$x^2 + y^2 - 14x + 4y = 11$

**9.** $(-3, 1); 4$

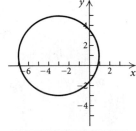
$x^2 + y^2 + 6x - 2y = 6$

**11.** $(-2, 3); 5$

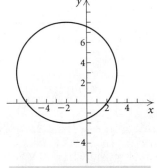
$x^2 + y^2 + 4x - 6y - 12 = 0$

**13.** $(3, 4); 3$

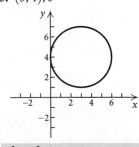
$x^2 + y^2 - 6x - 8y + 16 = 0$

**15.** $(-3, 5)$; $\sqrt{34}$

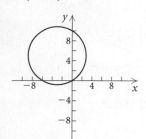

$$x^2 + y^2 + 6x - 10y = 0$$

**17.** $\left(\dfrac{9}{2}, -2\right)$; $\dfrac{5\sqrt{5}}{2}$

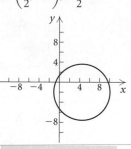

$$x^2 + y^2 - 9x = 7 - 4y$$

**39.** $C$: $(-3, 5)$; $V$: $(-3, 11)$, $(-3, -1)$; $F$: $(-3, 5 + \sqrt{11})$, $(-3, 5 - \sqrt{11})$

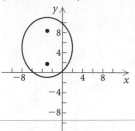

$$\dfrac{(x + 3)^2}{25} + \dfrac{(y - 5)^2}{36} = 1$$

**41.** $C$: $(-2, 1)$; $V$: $(-10, 1)$, $(6, 1)$; $F$: $(-6, 1)$, $(2, 1)$

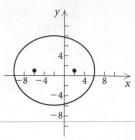

$$3(x + 2)^2 + 4(y - 1)^2 = 192$$

**19.** (c)    **21.** (d)

**23.** $V$: $(2, 0)$, $(-2, 0)$; $F$: $(\sqrt{3}, 0)$, $(-\sqrt{3}, 0)$

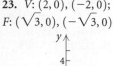

$$\dfrac{x^2}{4} + \dfrac{y^2}{1} = 1$$

**25.** $V$: $(0, 4)$, $(0, -4)$; $F$: $(0, \sqrt{7})$, $(0, -\sqrt{7})$

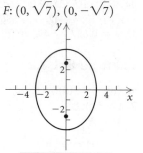

$$16x^2 + 9y^2 = 144$$

**43.** $C$: $(2, -1)$; $V$: $(-1, -1)$, $(5, -1)$; $F$: $(2 + \sqrt{5}, -1)$, $(2 - \sqrt{5}, -1)$

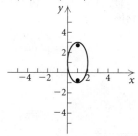

$$4x^2 + 9y^2 - 16x + 18y - 11 = 0$$

**45.** $C$: $(1, 1)$; $V$: $(1, 3)$, $(1, -1)$; $F$: $(1, 1 + \sqrt{3})$, $(1, 1 - \sqrt{3})$

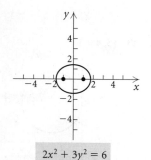

$$4x^2 + y^2 - 8x - 2y + 1 = 0$$

**27.** $V$: $(-\sqrt{3}, 0)$, $(\sqrt{3}, 0)$; $F$: $(-1, 0)$, $(1, 0)$

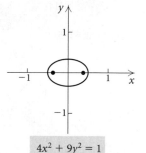

$$2x^2 + 3y^2 = 6$$

**29.** $V$: $\left(-\dfrac{1}{2}, 0\right)$, $\left(\dfrac{1}{2}, 0\right)$; $F$: $\left(-\dfrac{\sqrt{5}}{6}, 0\right)$, $\left(\dfrac{\sqrt{5}}{6}, 0\right)$

$$4x^2 + 9y^2 = 1$$

**31.** $\dfrac{x^2}{49} + \dfrac{y^2}{40} = 1$    **33.** $\dfrac{x^2}{25} + \dfrac{y^2}{64} = 1$    **35.** $\dfrac{x^2}{9} + \dfrac{y^2}{5} = 1$

**37.** $C$: $(1, 2)$; $V$: $(4, 2)$, $(-2, 2)$; $F$: $(1 + \sqrt{5}, 2)$, $(1 - \sqrt{5}, 2)$

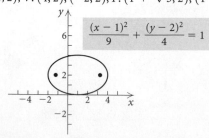

$$\dfrac{(x - 1)^2}{9} + \dfrac{(y - 2)^2}{4} = 1$$

**47.** Example 2; $\dfrac{3}{5} < \dfrac{\sqrt{12}}{4}$    **49.** $\dfrac{x^2}{15} + \dfrac{y^2}{16} = 1$

**51.** $\dfrac{x^2}{2500} + \dfrac{y^2}{144} = 1$    **53.** $2 \times 10^6$ mi

**55.** [1.1] Midpoint    **56.** [1.5] Zero    **57.** [1.1] $y$-intercept
**58.** [3.2] Two different real-number solutions
**59.** [4.3] Remainder    **60.** [7.2] Ellipse    **61.** [7.1] Parabola
**62.** [7.2] Circle    **63.** $\dfrac{(x - 3)^2}{4} + \dfrac{(y - 1)^2}{25} = 1$
**65.** $\dfrac{x^2}{9} + \dfrac{y^2}{484/5} = 1$    **67.** $C$: $(2.003, -1.005)$;
$V$: $(-1.017, -1.005)$, $(5.023, -1.005)$    **69.** About 9.1 ft

## Mid-Chapter Mixed Review: Chapter 7

**1.** True    **2.** False    **3.** False    **4.** True    **5.** (c)    **6.** (h)
**7.** (d)    **8.** (a)    **9.** (b)    **10.** (f)    **11.** (g)    **12.** (e)

**13.** $V$: $(0, 0)$; $F$: $(3, 0)$; $D$: $x = -3$

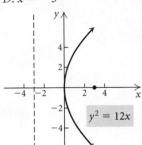

$y^2 = 12x$

**14.** $V$: $(3, 2)$; $F$: $(3, 3)$; $D$: $y = 1$

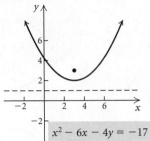

$x^2 - 6x - 4y = -17$

**15.** $x^2 = 4(y - 2)$    **16.** $(y - 6)^2 = -12(x + 1)$

**17.** $(-2, 4)$; 5    **18.** $(3, -1)$; 4

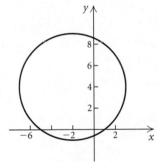

$x^2 + y^2 + 4x - 8y = 5$

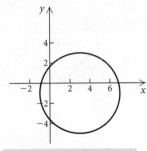

$x^2 + y^2 - 6x + 2y - 6 = 0$

**19.** $V$: $(0, -3)$, $(0, 3)$; $F$: $(0, -2\sqrt{2})$, $(0, 2\sqrt{2})$

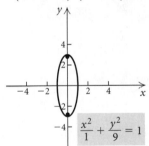

$\dfrac{x^2}{1} + \dfrac{y^2}{9} = 1$

**20.** $V$: $(-\sqrt{6}, 0)$, $(\sqrt{6}, 0)$; $F$: $(-\sqrt{2}, 0)$, $(\sqrt{2}, 0)$

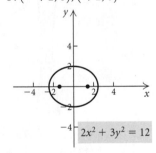

$2x^2 + 3y^2 = 12$

**21.** $V$: $(-2, -1)$, $(6, -1)$; $F$: $(2 - 2\sqrt{3}, -1)$, $(2 + 2\sqrt{3}, -1)$

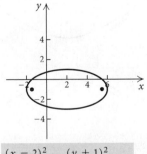

$\dfrac{(x - 2)^2}{16} + \dfrac{(y + 1)^2}{4} = 1$

**22.** $V$: $(1, -6)$, $(1, 4)$; $F$: $(1, -1 - \sqrt{21})$, $(1, -1 + \sqrt{21})$

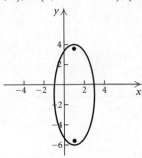

$25x^2 + 4y^2 - 50x + 8y = 71$

**23.** $\dfrac{x^2}{25} + \dfrac{y^2}{21} = 1$    **24.** $\dfrac{x^2}{4} + \dfrac{y^2}{9} = 1$    **25.** $\dfrac{x^2}{16} + \dfrac{y^2}{7} = 1$

**26.** No; parabolas with a horizontal axis of symmetry fail the vertical-line test.    **27.** See p. 579 in the text.    **28.** Circles and ellipses are not functions.    **29.** No; the center of an ellipse is not part of the graph of the ellipse. Its coordinates do not satisfy the equation of the ellipse.

**Exercise Set 7.3**

**1.** (b)    **3.** (c)    **5.** (a)    **7.** $\dfrac{y^2}{9} - \dfrac{x^2}{16} = 1$

**9.** $\dfrac{x^2}{4} - \dfrac{y^2}{9} = 1$

**11.** $C$: $(0, 0)$; $V$: $(2, 0)$, $(-2, 0)$; $F$: $(2\sqrt{2}, 0)$, $(-2\sqrt{2}, 0)$; $A$: $y = x$, $y = -x$

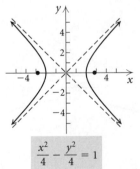

$\dfrac{x^2}{4} - \dfrac{y^2}{4} = 1$

**13.** $C$: $(2, -5)$; $V$: $(-1, -5)$, $(5, -5)$; $F$: $(2 - \sqrt{10}, -5)$, $(2 + \sqrt{10}, -5)$; $A$: $y = -\dfrac{x}{3} - \dfrac{13}{3}$, $y = \dfrac{x}{3} - \dfrac{17}{3}$

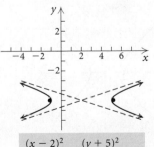

$\dfrac{(x - 2)^2}{9} - \dfrac{(y + 5)^2}{1} = 1$

**15.** $C: (-1, -3)$; $V: (-1, -1), (-1, -5)$;
$F: (-1, -3 + 2\sqrt{5}), (-1, -3 - 2\sqrt{5})$;
$A: y = \frac{1}{2}x - \frac{5}{2}, y = -\frac{1}{2}x - \frac{7}{2}$

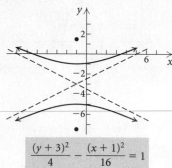

$$\frac{(y+3)^2}{4} - \frac{(x+1)^2}{16} = 1$$

**17.** $C: (0, 0)$; $V: (-2, 0), (2, 0)$; $F: (-\sqrt{5}, 0), (\sqrt{5}, 0)$;
$A: y = -\frac{1}{2}x, y = \frac{1}{2}x$

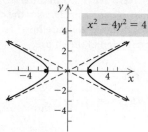

$x^2 - 4y^2 = 4$

**19.** $C: (0, 0)$; $V: (0, -3), (0, 3)$; $F: (0, -3\sqrt{10}), (0, 3\sqrt{10})$;
$A: y = \frac{1}{3}x, y = -\frac{1}{3}x$

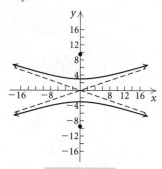

$9y^2 - x^2 = 81$

**21.** $C: (0, 0)$; $V: (-\sqrt{2}, 0), (\sqrt{2}, 0)$; $F: (-2, 0), (2, 0)$;
$A: y = x, y = -x$

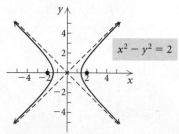

$x^2 - y^2 = 2$

**23.** $C: (0, 0)$; $V: \left(0, -\frac{1}{2}\right), \left(0, \frac{1}{2}\right)$; $F: \left(0, -\frac{\sqrt{2}}{2}\right),$
$\left(0, \frac{\sqrt{2}}{2}\right)$; $A: y = x, y = -x$

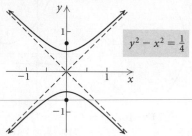

$y^2 - x^2 = \frac{1}{4}$

**25.** $C: (1, -2)$; $V: (0, -2), (2, -2)$; $F: (1 - \sqrt{2}, -2),$
$(1 + \sqrt{2}, -2)$; $A: y = -x - 1, y = x - 3$

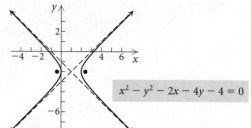

$x^2 - y^2 - 2x - 4y - 4 = 0$

**27.** $C: \left(\frac{1}{3}, 3\right)$; $V: \left(-\frac{2}{3}, 3\right), \left(\frac{4}{3}, 3\right)$; $F: \left(\frac{1}{3} - \sqrt{37}, 3\right),$
$\left(\frac{1}{3} + \sqrt{37}, 3\right)$; $A: y = 6x + 1, y = -6x + 5$

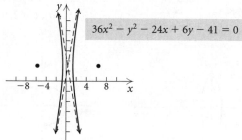

$36x^2 - y^2 - 24x + 6y - 41 = 0$

**29.** $C: (3, 1)$; $V: (3, 3), (3, -1)$; $F: (3, 1 + \sqrt{13}),$
$(3, 1 - \sqrt{13})$; $A: y = \frac{2}{3}x - 1, y = -\frac{2}{3}x + 3$

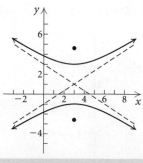

$9y^2 - 4x^2 - 18y + 24x - 63 = 0$

**31.** $C: (1, -2)$; $V: (2, -2), (0, -2)$; $F: (1 + \sqrt{2}, -2)$, $(1 - \sqrt{2}, -2)$; $A: y = x - 3, y = -x - 1$

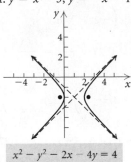

$$x^2 - y^2 - 2x - 4y = 4$$

**33.** $C: (-3, 4)$; $V: (-3, 10), (-3, -2)$; $F: (-3, 4 + 6\sqrt{2})$, $(-3, 4 - 6\sqrt{2})$; $A: y = x + 7, y = -x + 1$

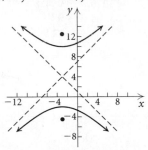

$$y^2 - x^2 - 6x - 8y - 29 = 0$$

**35.** Example 3; $\dfrac{\sqrt{5}}{1} > \dfrac{5}{4}$    **37.** $\dfrac{x^2}{9} - \dfrac{(y - 7)^2}{16} = 1$

**39.** Top: 275 ft; base: 453 ft    **41.** [5.1] **(a)** Yes;

**(b)** $f^{-1}(x) = \dfrac{x + 3}{2}$    **42.** [5.1] **(a)** Yes;

**(b)** $f^{-1}(x) = \sqrt[3]{x - 2}$    **43.** [5.1] **(a)** Yes;

**(b)** $f^{-1}(x) = \dfrac{5}{x} + 1$, or $\dfrac{5 + x}{x}$    **44.** [5.1] **(a)** Yes;

**(b)** $f^{-1}(x) = x^2 - 4, x \geq 0$    **45.** [6.1], [6.3], [6.5], [6.6]
$(6, -1)$    **46.** [6.1], [6.3], [6.5], [6.6] $(1, -1)$    **47.** [6.1],
[6.3], [6.5], [6.6] $(2, -1)$    **48.** [6.1], [6.3], [6.5], [6.6] $(-3, 4)$

**49.** $\dfrac{(y + 5)^2}{9} - (x - 3)^2 = 1$

**51.** $C: (-1.460, -0.957)$; $V: (-2.360, -0.957)$,
$(-0.560, -0.957)$; $A: y = -1.20x - 2.70, y = 1.20x + 0.79$

**53.** $\dfrac{x^2}{345.96} - \dfrac{y^2}{22,154.04} = 1$

## Visualizing the Graph

**1.** B    **2.** J    **3.** F    **4.** I    **5.** H    **6.** G    **7.** E
**8.** D    **9.** C    **10.** A

## Exercise Set 7.4

**1.** (e)    **3.** (c)    **5.** (b)    **7.** $(-4, -3), (3, 4)$
**9.** $(0, 2), (3, 0)$    **11.** $(-5, 0), (4, 3), (4, -3)$
**13.** $(3, 0), (-3, 0)$    **15.** $(0, -3), (4, 5)$    **17.** $(-2, 1)$
**19.** $(3, 4), (-3, -4), (4, 3), (-4, -3)$

**21.** $\left( \dfrac{6\sqrt{21}}{7}, \dfrac{4\sqrt{35}}{7} i \right), \left( \dfrac{6\sqrt{21}}{7}, -\dfrac{4\sqrt{35}}{7} i \right),$
$\left( -\dfrac{6\sqrt{21}}{7}, \dfrac{4\sqrt{35}}{7} i \right), \left( -\dfrac{6\sqrt{21}}{7}, -\dfrac{4\sqrt{35}}{7} i \right)$

**23.** $(3, 2), \left( 4, \frac{3}{2} \right)$

**25.** $\left( \dfrac{5 + \sqrt{70}}{3}, \dfrac{-1 + \sqrt{70}}{3} \right), \left( \dfrac{5 - \sqrt{70}}{3}, \dfrac{-1 - \sqrt{70}}{3} \right)$

**27.** $(\sqrt{2}, \sqrt{14}), (-\sqrt{2}, \sqrt{14}), (\sqrt{2}, -\sqrt{14}),$
$(-\sqrt{2}, -\sqrt{14})$    **29.** $(1, 2), (-1, -2), (2, 1), (-2, -1)$

**31.** $\left( \dfrac{15 + \sqrt{561}}{8}, \dfrac{11 - 3\sqrt{561}}{8} \right),$
$\left( \dfrac{15 - \sqrt{561}}{8}, \dfrac{11 + 3\sqrt{561}}{8} \right)$

**33.** $\left( \dfrac{7 - \sqrt{33}}{2}, \dfrac{7 + \sqrt{33}}{2} \right), \left( \dfrac{7 + \sqrt{33}}{2}, \dfrac{7 - \sqrt{33}}{2} \right)$

**35.** $(3, 2), (-3, -2), (2, 3), (-2, -3)$

**37.** $\left( \dfrac{5 - 9\sqrt{15}}{20}, \dfrac{-45 + 3\sqrt{15}}{20} \right),$
$\left( \dfrac{5 + 9\sqrt{15}}{20}, \dfrac{-45 - 3\sqrt{15}}{20} \right)$    **39.** $(3, -5), (-1, 3)$

**41.** $(8, 5), (-5, -8)$    **43.** $(3, 2), (-3, -2)$
**45.** $(2, 1), (-2, -1), (1, 2), (-1, -2)$

**47.** $\left( 4 + \dfrac{3\sqrt{6}}{2} i, -4 + \dfrac{3\sqrt{6}}{2} i \right), \left( 4 - \dfrac{3\sqrt{6}}{2} i, -4 - \dfrac{3\sqrt{6}}{2} i \right)$

**49.** $(3, \sqrt{5}), (-3, -\sqrt{5}), (\sqrt{5}, 3), (-\sqrt{5}, -3)$

**51.** $\left( \dfrac{8\sqrt{5}}{5} i, \dfrac{3\sqrt{105}}{5} \right), \left( \dfrac{8\sqrt{5}}{5} i, -\dfrac{3\sqrt{105}}{5} \right),$
$\left( -\dfrac{8\sqrt{5}}{5} i, \dfrac{3\sqrt{105}}{5} \right), \left( -\dfrac{8\sqrt{5}}{5} i, -\dfrac{3\sqrt{105}}{5} \right)$

**53.** $(2, 1), (-2, -1), \left( -\sqrt{5}i, \dfrac{2\sqrt{5}i}{5} \right), \left( \sqrt{5}i, -\dfrac{2\sqrt{5}i}{5} \right)$

**55.** True    **57.** True    **59.** 6 cm by 8 cm    **61.** 4 in. by 5 in.
**63.** 30 yd by 75 yd    **65.** Length: $\sqrt{3}$ m; width: 1 m
**67.** 16 ft, 24 ft    **69.** (b)    **71.** (d)    **73.** (a)
**75.**

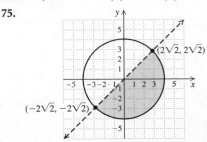

**77.**

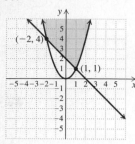

**79.**

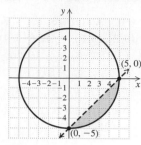

**81.**

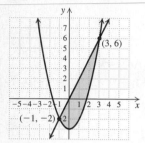

**83.**

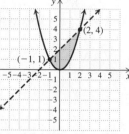

**85.** [5.5] 2    **86.** [5.5] 2.048    **87.** [5.5] 81    **88.** [5.5] 5

**89.** $(x - 2)^2 + (y - 3)^2 = 1$    **91.** $\dfrac{x^2}{4} + y^2 = 1$

**93.** There is no number $x$ such that $\dfrac{x^2}{a^2} - \dfrac{\left(\dfrac{b}{a}x\right)^2}{b^2} = 1$,

because the left side simplifies to $\dfrac{x^2}{a^2} - \dfrac{x^2}{a^2}$, which is 0.

**95.** Factor: $x^3 + y^3 = (x + y)(x^2 - xy + y^2)$.
We know that $x + y = 1$, so $(x + y)^2 = x^2 + 2xy + y^2 = 1$,
or $x^2 + y^2 = 1 - 2xy$. We also know that $xy = 1$, so
$x^2 + y^2 = 1 - 2 \cdot 1 = -1$. Then $x^3 + y^3 = 1 \cdot (-1 - 1) = -2$.    **97.** $(2, 4), (4, 2)$
**99.** $(3, -2), (-3, 2), (2, -3), (-2, 3)$    **101.** $(1.564, 2.448)$,
$(0.138, 0.019)$    **103.** $(1.146, 3.146), (-1.841, 0.159)$
**105.** $(2.112, -0.109), (-13.041, -13.337)$    **107.** $(400, 1.431)$,
$(-400, 1.431), (400, -1.431), (-400, -1.431)$

## Review Exercises: Chapter 7

**1.** True    **2.** False    **3.** True    **4.** False    **5.** False
**6.** (d)    **7.** (a)    **8.** (e)    **9.** (g)    **10.** (b)    **11.** (f)
**12.** (h)    **13.** (c)    **14.** $x^2 = -6y$    **15.** $F: (-3, 0)$;
$V: (0, 0); D: x = 3$    **16.** $V: (-5, 8); F: \left(-5, \frac{15}{2}\right); D: y = \frac{17}{2}$
**17.** $C: (2, -1); V: (-3, -1), (7, -1); F: (-1, -1), (5, -1)$

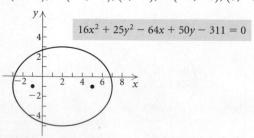

**18.** $\dfrac{x^2}{9} + \dfrac{y^2}{16} = 1$    **19.** $C: \left(-2, \frac{1}{4}\right); V: \left(0, \frac{1}{4}\right), \left(-4, \frac{1}{4}\right)$;

$F: \left(-2 + \sqrt{6}, \frac{1}{4}\right), \left(-2 - \sqrt{6}, \frac{1}{4}\right)$;

$A: y - \dfrac{1}{4} = \dfrac{\sqrt{2}}{2}(x + 2), y - \dfrac{1}{4} = -\dfrac{\sqrt{2}}{2}(x + 2)$

**20.** 0.167 ft    **21.** $(-8\sqrt{2}, 8), (8\sqrt{2}, 8)$

**22.** $\left(3, \dfrac{\sqrt{29}}{2}\right), \left(-3, \dfrac{\sqrt{29}}{2}\right), \left(3, -\dfrac{\sqrt{29}}{2}\right), \left(-3, -\dfrac{\sqrt{29}}{2}\right)$

**23.** $(7, 4)$    **24.** $(2, 2), \left(\frac{32}{9}, -\frac{10}{9}\right)$    **25.** $(0, -3), (2, 1)$
**26.** $(4, 3), (4, -3), (-4, 3), (-4, -3)$    **27.** $(-\sqrt{3}, 0)$,
$(\sqrt{3}, 0), (-2, 1), (2, 1)$    **28.** $\left(-\frac{3}{5}, \frac{21}{5}\right), (3, -3)$
**29.** $(6, 8), (6, -8), (-6, 8), (-6, -8)$
**30.** $(2, 2), (-2, -2), (2\sqrt{2}, \sqrt{2}), (-2\sqrt{2}, -\sqrt{2})$
**31.** 7, 4    **32.** 7 m by 12 m    **33.** 4, 8
**34.** 32 cm, 20 cm    **35.** 11 ft, 3 ft
**36.**

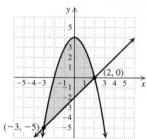

**37.**

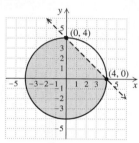

**38.**

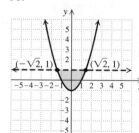

**39.**

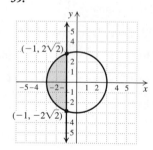

**40.** B    **41.** D    **42.** C
**43.** $\frac{8}{7}, \frac{7}{2}$    **44.** $(x - 2)^2 + (y - 1)^2 = 100$
**45.** $x^2 + \dfrac{y^2}{9} = 1$    **46.** $\dfrac{x^2}{778.41} - \dfrac{y^2}{39{,}221.59} = 1$
**47.** The equation of a circle can be written as

$$\dfrac{(x - h)^2}{a^2} + \dfrac{(y - k)^2}{b^2} = 1,$$

where $a = b = r$, the radius of the circle. In an ellipse, $a > b$,
so a circle is not a special type of ellipse.    **48.** See the
figure on p. 578 of the text.    **49.** No; the asymptotes
of a hyperbola are not part of the graph of the hyperbola.
The coordinates of points on the asymptotes do not satisfy
the equation of the hyperbola.    **50.** Although we can
always visualize the real-number solutions, we cannot
visualize the imaginary-number solutions.

## Test: Chapter 7

**1.** [7.3] (c) **2.** [7.1] (b) **3.** [7.2] (a) **4.** [7.2] (d)

**5.** [7.1] $V: (0, 0)$; $F: (0, 3)$; $D: y = -3$

**6.** [7.1] $V: (-1, -1)$; $F: (1, -1)$; $D: x = -3$

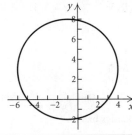

$x^2 = 12y$

$y^2 + 2y - 8x - 7 = 0$

**7.** [7.1] $x^2 = 8y$

**8.** [7.2] Center: $(-1, 3)$; radius: 5

**9.** [7.2] $C: (0, 0)$; $V: (-4, 0)$, $(4, 0)$; $F: (-\sqrt{7}, 0), (\sqrt{7}, 0)$

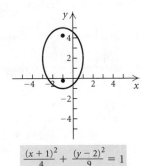

$x^2 + y^2 + 2x - 6y - 15 = 0$

$9x^2 + 16y^2 = 144$

**10.** [7.2] $C: (-1, 2)$; $V: (-1, -1), (-1, 5)$; $F: (-1, 2 - \sqrt{5})$, $(-1, 2 + \sqrt{5})$

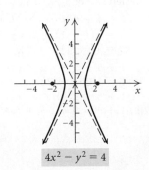

$\frac{(x + 1)^2}{4} + \frac{(y - 2)^2}{9} = 1$

**11.** [7.2] $\frac{x^2}{4} + \frac{y^2}{25} = 1$

**12.** [7.3] $C: (0, 0)$; $V: (-1, 0), (1, 0)$; $F: (-\sqrt{5}, 0), (\sqrt{5}, 0)$; $A: y = -2x, y = 2x$

$4x^2 - y^2 = 4$

**13.** [7.3] $C: (-1, 2)$; $V: (-1, 0), (-1, 4)$; $F: (-1, 2 - \sqrt{13})$, $(-1, 2 + \sqrt{13})$; $A: y = -\frac{2}{3}x + \frac{4}{3}, y = \frac{2}{3}x + \frac{8}{3}$

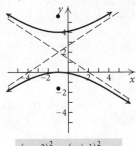

$\frac{(y - 2)^2}{4} - \frac{(x + 1)^2}{9} = 1$

**14.** [7.3] $y = \frac{\sqrt{2}}{2}x, y = -\frac{\sqrt{2}}{2}x$ **15.** [7.1] $\frac{27}{8}$ in.

**16.** [7.4] $(1, 2), (1, -2), (-1, 2), (-1, -2)$

**17.** [7.4] $(3, -2), (-2, 3)$ **18.** [7.4] $(2, 3), (3, 2)$

**19.** [7.4] 5 ft by 4 ft **20.** [7.4] 60 ft by 45 ft

**21.** [7.4]

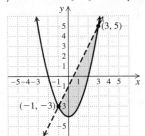

**22.** [7.1] A **23.** [7.2] $(x - 3)^2 + (y + 1)^2 = 8$

## Chapter 8

### Exercise Set 8.1

**1.** 3, 7, 11, 15; 39; 59 **3.** 2, $\frac{3}{2}, \frac{4}{3}, \frac{5}{4}; \frac{10}{9}; \frac{15}{14}$

**5.** 0, $\frac{3}{5}, \frac{4}{5}, \frac{15}{17}; \frac{99}{101}; \frac{112}{113}$ **7.** $-1, 4, -9, 16; 100; -225$

**9.** 7, 3, 7, 3; 3; 7 **11.** 34 **13.** 225 **15.** $-33,880$ **17.** 67

**19.**

| $n$ | $u_n$ |
|-----|-------|
| 1 | 2 |
| 2 | 2.25 |
| 3 | 2.3704 |
| 4 | 2.4414 |
| 5 | 2.4883 |
| 6 | 2.5216 |
| 7 | 2.5465 |
| 8 | 2.5658 |
| 9 | 2.5812 |
| 10 | 2.5937 |

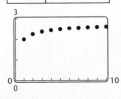

**21.**

| $n$ | $u_n$ |
|-----|-------|
| 1 | 2 |
| 2 | 1.5538 |
| 3 | 1.4988 |
| 4 | 1.4914 |
| 5 | 1.4904 |
| 6 | 1.4902 |
| 7 | 1.4902 |
| 8 | 1.4902 |
| 9 | 1.4902 |
| 10 | 1.4902 |

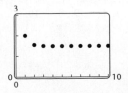

**23.** $2n$    **25.** $(-1)^n \cdot 2 \cdot 3^{n-1}$    **27.** $\dfrac{n+1}{n+2}$

**29.** $n(n+1)$    **31.** $\log 10^{n-1}$, or $n-1$    **33.** 6; 28

**35.** 20; 30    **37.** $\frac{1}{2} + \frac{1}{4} + \frac{1}{6} + \frac{1}{8} + \frac{1}{10} = \frac{137}{120}$

**39.** $1 + 2 + 4 + 8 + 16 + 32 + 64 = 127$

**41.** $\ln 7 + \ln 8 + \ln 9 + \ln 10 = \ln(7 \cdot 8 \cdot 9 \cdot 10) = \ln 5040 \approx 8.5252$

**43.** $\frac{1}{2} + \frac{2}{3} + \frac{3}{4} + \frac{4}{5} + \frac{5}{6} + \frac{6}{7} + \frac{7}{8} + \frac{8}{9} = \frac{15{,}551}{2520}$

**45.** $-1 + 1 - 1 + 1 - 1 = -1$

**47.** $3 - 6 + 9 - 12 + 15 - 18 + 21 - 24 = -12$

**49.** $2 + 1 + \frac{2}{5} + \frac{1}{5} + \frac{2}{17} + \frac{1}{13} + \frac{2}{37} = \frac{157{,}351}{40{,}885}$

**51.** $3 + 2 + 3 + 6 + 11 + 18 = 43$

**53.** $\frac{1}{2} + \frac{2}{3} + \frac{4}{5} + \frac{8}{9} + \frac{16}{17} + \frac{32}{33} + \frac{64}{65} + \frac{128}{129} + \frac{256}{257} + \frac{512}{513} + \frac{1024}{1025} \approx 9.736$    **55.** $\displaystyle\sum_{k=1}^{\infty} 5k$    **57.** $\displaystyle\sum_{k=1}^{6} (-1)^{k+1} 2^k$

**59.** $\displaystyle\sum_{k=1}^{6} (-1)^k \frac{k}{k+1}$    **61.** $\displaystyle\sum_{k=2}^{n} (-1)^k k^2$    **63.** $\displaystyle\sum_{k=1}^{\infty} \frac{1}{k(k+1)}$

**65.** $4, 1\frac{1}{4}, 1\frac{4}{5}, 1\frac{5}{9}$    **67.** $6561, -81, 9i, -3\sqrt{i}$    **69.** $2, 3, 5, 8$

**71. (a)** 1062, 1127.84, 1197.77, 1272.03, 1350.90, 1434.65, 1523.60, 1618.07, 1718.39, 1824.93; **(b)** $3330.35

**73.** $9.80, $10.90, $12.00, $13.10, $14.20, $15.30, $16.40, $17.50, $18.60, $19.70    **75.** 1, 1, 2, 3, 5, 8, 13

**77. (a)** $a_n = -659.8950216n^2 + 6297.77684n + 54{,}737.29091$; **(b)** 2010: 57,966,000 vehicles, 2011: 51,726,000 vehicles, 2012: 44,166,000 vehicles    **78.** [6.1], [6.3], [6.5], [6.6] $(-1, -3)$    **79.** [6.2], [6.3], [6.5], [6.6] United States: 44,855; Japan: 32,156; China: 12,337

**80.** [7.2] $(3, -2); 4$    **81.** [7.2] $\left(-\frac{5}{2}, 4\right); \dfrac{\sqrt{97}}{2}$

**83.** $i, -1, -i, 1, i; i$    **85.** $\ln(1 \cdot 2 \cdot 3 \cdot \ \cdots \ \cdot n)$

## Exercise Set 8.2

**1.** $a_1 = 3, d = 5$    **3.** $a_1 = 9, d = -4$

**5.** $a_1 = \frac{3}{2}, d = \frac{3}{4}$    **7.** $a_1 = $316, d = -$3$

**9.** $a_{12} = 46$    **11.** $a_{14} = -\frac{17}{3}$    **13.** $a_{10} = $7941.62$

**15.** 27th    **17.** 46th    **19.** $a_1 = 5$    **21.** $n = 39$

**23.** $a_1 = \frac{1}{3}; d = \frac{1}{2}; \frac{1}{3}, \frac{5}{6}, \frac{4}{3}, \frac{11}{6}, \frac{7}{3}$    **25.** 670    **27.** 160,400

**29.** 735    **31.** 990    **33.** 1760    **35.** $\frac{65}{2}$    **37.** $-\frac{6026}{13}$

**39.** 1260 poles    **41.** 1320 seats    **43.** Yes; 32; 1600 ft

**45.** 24 marchers; 136 marchers    **47.** Yes; 3

**48.** [6.1], [6.3], [6.5], [6.6] $(2, 5)$

**49.** [6.2], [6.3], [6.5], [6.6] $(2, -1, 3)$

**50.** [7.2] $(-4, 0), (4, 0); (-\sqrt{7}, 0), (\sqrt{7}, 0)$

**51.** [7.2] $\dfrac{x^2}{4} + \dfrac{y^2}{25} = 1$    **53.** $n^2$    **55.** $a_1 = 60 - 5p - 5q$; $d = 5p + 2q - 20$    **57.** $5\frac{4}{5}, 7\frac{3}{5}, 9\frac{2}{5}, 11\frac{1}{5}$

## Visualizing the Graph

**1.** J    **2.** A    **3.** C    **4.** G    **5.** F    **6.** H    **7.** E
**8.** D    **9.** B    **10.** I

## Exercise Set 8.3

**1.** 2    **3.** $-1$    **5.** $-2$    **7.** 0.1    **9.** $\dfrac{a}{2}$    **11.** 128

**13.** 162    **15.** $7(5)^{40}$    **17.** $3^{n-1}$    **19.** $(-1)^{n-1}$    **21.** $\dfrac{1}{x^n}$

**23.** 762    **25.** $\frac{4921}{18}$    **27.** True    **29.** True    **31.** True

**33.** 8    **35.** 125    **37.** Does not exist    **39.** $\frac{2}{3}$    **41.** $29\frac{38{,}569}{59{,}049}$

**43.** 2    **45.** Does not exist    **47.** $4545.\overline{45}$    **49.** $\frac{160}{9}$

**51.** $\frac{13}{99}$    **53.** 9    **55.** $\frac{34{,}091}{9990}$    **57.** $2,684,354.55

**59. (a)** About 297 ft; **(b)** 300 ft    **61.** $39,505.71

**63.** 10,485.76 in.    **65.** $19,694.01    **67.** 2,142,857; 42.9%

**69.** [2.3] $(f \circ g)(x) = 16x^2 + 40x + 25; (g \circ f)(x) = 4x^2 + 5$
**70.** [2.3] $(f \circ g)(x) = x^2 + x + 2; (g \circ f)(x) = x^2 - x + 3$

**71.** [5.5] 2.209    **72.** [5.5] $\frac{1}{16}$

**73.** $(4 - \sqrt{6})/(\sqrt{3} - \sqrt{2}) = 2\sqrt{3} + \sqrt{2}$, $(6\sqrt{3} - 2\sqrt{2})/(4 - \sqrt{6}) = 2\sqrt{3} + \sqrt{2}$; there exists a common ratio, $2\sqrt{3} + \sqrt{2}$; thus the sequence is geometric.

**75. (a)** $\frac{13}{3}, \frac{22}{3}, \frac{34}{3}, \frac{46}{3}, \frac{58}{3}$; **(b)** $-\frac{11}{3}, -\frac{2}{3}, \frac{10}{3}, -\frac{50}{3}, \frac{250}{3}$ or 5; 8, 12, 18, 27

**77.** $S_n = \dfrac{x^2(1 - (-x)^n)}{x+1}$

## Exercise Set 8.4

**1.** $1^2 < 1^3$, false; $2^2 < 2^3$, true; $3^2 < 3^3$, true; $4^2 < 4^3$, true; $5^2 < 5^3$, true    **3.** A polygon of 3 sides has $\dfrac{3(3-3)}{2}$ diagonals. True; A polygon of 4 sides has $\dfrac{4(4-3)}{2}$ diagonals. True; A polygon of 5 sides has $\dfrac{5(5-3)}{2}$ diagonals. True; A polygon of 6 sides has $\dfrac{6(6-3)}{2}$ diagonals. True; A polygon of 7 sides has $\dfrac{7(7-3)}{2}$ diagonals. True.

**5.** $S_n$:    $2 + 4 + 6 + \cdots + 2n = n(n+1)$
$S_1$:    $2 = 1(1+1)$
$S_k$:    $2 + 4 + 6 + \cdots + 2k = k(k+1)$
$S_{k+1}$:    $2 + 4 + 6 + \cdots + 2k + 2(k+1)$
        $= (k+1)(k+2)$

(1) *Basis step*: $S_1$ true by substitution.
(2) *Induction step*: Assume $S_k$. Deduce $S_{k+1}$. Starting with the left side of $S_{k+1}$, we have

$$2 + 4 + 6 + \cdots + 2k + 2(k+1)$$
$$= k(k+1) + 2(k+1) \qquad \textbf{By } S_k$$
$$= (k+1)(k+2). \qquad \textbf{Factoring}$$

**7.** $S_n$:    $1 + 5 + 9 + \cdots + (4n-3) = n(2n-1)$
$S_1$:    $1 = 1(2 \cdot 1 - 1)$
$S_k$:    $1 + 5 + 9 + \cdots + (4k-3) = k(2k-1)$
$S_{k+1}$:    $1 + 5 + 9 + \cdots + (4k-3) + [4(k+1) - 3]$
        $= (k+1)[2(k+1) - 1]$
        $= (k+1)(2k+1)$

(1) *Basis step*: $S_1$ true by substitution.

(2) *Induction step*: Assume $S_k$. Deduce $S_{k+1}$. Starting with the left side of $S_{k+1}$, we have

$$1 + 5 + 9 + \cdots + (4k - 3) + [4(k + 1) - 3]$$
$$= k(2k - 1) + [4(k + 1) - 3] \quad \textbf{By } S_k$$
$$= 2k^2 - k + 4k + 4 - 3$$
$$= 2k^2 + 3k + 1$$
$$= (k + 1)(2k + 1).$$

**9.** $S_n$:  $\quad 2 + 4 + 8 + \cdots + 2^n = 2(2^n - 1)$

$S_1$:  $\quad 2 = 2(2 - 1)$

$S_k$:  $\quad 2 + 4 + 8 + \cdots + 2^k = 2(2^k - 1)$

$S_{k+1}$:  $\quad 2 + 4 + 8 + \cdots + 2^k + 2^{k+1} = 2(2^{k+1} - 1)$

(1) *Basis step*: $S_1$ is true by substitution.

(2) *Induction step*: Assume $S_k$. Deduce $S_{k+1}$. Starting with the left side of $S_{k+1}$, we have

$$\underbrace{2 + 4 + 8 + \cdots + 2^k} + 2^{k+1}$$
$$= 2(2^k - 1) + 2^{k+1} \quad \textbf{By } S_k$$
$$= 2^{k+1} - 2 + 2^{k+1}$$
$$= 2 \cdot 2^{k+1} - 2$$
$$= 2(2^{k+1} - 1).$$

**11.** $S_n$:  $\quad n < n + 1$

$S_1$:  $\quad 1 < 1 + 1$

$S_k$:  $\quad k < k + 1$

$S_{k+1}$:  $\quad k + 1 < (k + 1) + 1$

(1) *Basis step*: Since $1 < 1 + 1$, $S_1$ is true.

(2) *Induction step*: Assume $S_k$. Deduce $S_{k+1}$. Now

$$k < k + 1 \qquad \textbf{By } S_k$$
$$k + 1 < k + 1 + 1 \qquad \textbf{Adding 1}$$
$$k + 1 < k + 2. \qquad \textbf{Simplifying}$$

**13.** $S_n$:  $\quad 2n \leq 2^n$

$S_1$:  $2 \cdot 1 \leq 2^1$

$S_k$:  $\quad 2k \leq 2^k$

$S_{k+1}$:  $2(k + 1) \leq 2^{k+1}$

(1) *Basis step*: Since $2 = 2$, $S_1$ is true.

(2) *Induction step*: Let $k$ be any natural number. Assume $S_k$. Deduce $S_{k+1}$.

$$2k \leq 2^k \qquad \textbf{By } S_k$$
$$2 \cdot 2k \leq 2 \cdot 2^k \qquad \textbf{Multiplying by 2}$$
$$4k \leq 2^{k+1}$$

Since $1 \leq k$, $k + 1 \leq k + k$, or $k + 1 \leq 2k$.
Then $2(k + 1) \leq 4k$. $\qquad$ **Multiplying by 2**
Thus, $2(k + 1) \leq 4k \leq 2^{k+1}$, so $2(k + 1) \leq 2^{k+1}$.

**15.**

$S_n$:  $\dfrac{1}{1 \cdot 2 \cdot 3} + \dfrac{1}{2 \cdot 3 \cdot 4} + \dfrac{1}{3 \cdot 4 \cdot 5} + \cdots$

$\qquad + \dfrac{1}{n(n + 1)(n + 2)} = \dfrac{n(n + 3)}{4(n + 1)(n + 2)}$

$S_1$:  $\dfrac{1}{1 \cdot 2 \cdot 3} = \dfrac{1(1 + 3)}{4(1 + 1)(1 + 2)}$

$S_k$:  $\dfrac{1}{1 \cdot 2 \cdot 3} + \dfrac{1}{2 \cdot 3 \cdot 4} + \cdots + \dfrac{1}{k(k + 1)(k + 2)}$
$$= \dfrac{k(k + 3)}{4(k + 1)(k + 2)}$$

$S_{k+1}$:  $\dfrac{1}{1 \cdot 2 \cdot 3} + \dfrac{1}{2 \cdot 3 \cdot 4} + \cdots + \dfrac{1}{k(k + 1)(k + 2)}$
$$+ \dfrac{1}{(k + 1)(k + 2)(k + 3)}$$
$$= \dfrac{(k + 1)(k + 1 + 3)}{4(k + 1 + 1)(k + 1 + 2)} = \dfrac{(k + 1)(k + 4)}{4(k + 2)(k + 3)}$$

(1) *Basis step*: Since $\dfrac{1}{1 \cdot 2 \cdot 3} = \dfrac{1}{6}$ and

$\dfrac{1(1 + 3)}{4(1 + 1)(1 + 2)} = \dfrac{1 \cdot 4}{4 \cdot 2 \cdot 3} = \dfrac{1}{6}$, $S_1$ is true.

(2) *Induction step*: Assume $S_k$. Deduce $S_{k+1}$.

Add $\dfrac{1}{(k + 1)(k + 2)(k + 3)}$ on both sides of $S_k$ and simplify the right side. Only the right side is shown here.

$$\dfrac{k(k + 3)}{4(k + 1)(k + 2)} + \dfrac{1}{(k + 1)(k + 2)(k + 3)}$$
$$= \dfrac{k(k + 3)(k + 3) + 4}{4(k + 1)(k + 2)(k + 3)}$$
$$= \dfrac{k^3 + 6k^2 + 9k + 4}{4(k + 1)(k + 2)(k + 3)}$$
$$= \dfrac{(k + 1)^2(k + 4)}{4(k + 1)(k + 2)(k + 3)}$$
$$= \dfrac{(k + 1)(k + 4)}{4(k + 2)(k + 3)}$$

**17.**

$S_n$:  $\quad 1 + 2 + 3 + \cdots + n = \dfrac{n(n + 1)}{2}$

$S_1$:  $\quad 1 = \dfrac{1(1 + 1)}{2}$

$S_k$:  $\quad 1 + 2 + 3 + \cdots + k = \dfrac{k(k + 1)}{2}$

$S_{k+1}$:  $\quad 1 + 2 + 3 + \cdots + k + (k + 1) = \dfrac{(k + 1)(k + 2)}{2}$

(1) *Basis step*: $S_1$ true by substitution.

(2) *Induction step*: Assume $S_k$. Deduce $S_{k+1}$. Starting with the left side of $S_{k+1}$, we have

$$\underbrace{1 + 2 + 3 + \cdots + k} + (k + 1)$$
$$= \dfrac{k(k + 1)}{2} + (k + 1) \qquad \textbf{By } S_k$$
$$= \dfrac{k(k + 1) + 2(k + 1)}{2} \qquad \textbf{Adding}$$
$$= \dfrac{(k + 1)(k + 2)}{2}. \qquad \textbf{Factoring}$$

**19.** $S_n$:  $\quad 1^3 + 2^3 + 3^3 + \cdots + n^3 = \dfrac{n^2(n + 1)^2}{4}$

$S_1$:  $\quad 1^3 = \dfrac{1^2(1 + 1)^2}{4}$

$S_k$:　　$1^3 + 2^3 + 3^3 + \cdots + k^3 = \dfrac{k^2(k+1)^2}{4}$

$S_{k+1}$:　　$1^3 + 2^3 + 3^3 + \cdots + k^3 + (k+1)^3$
$$= \dfrac{(k+1)^2[(k+1)+1]^2}{4}$$

(1) *Basis step*: $S_1$: $1^3 = \dfrac{1^2(1+1)^2}{4} = 1$. True.

(2) *Induction step*: Assume $S_k$. Deduce $S_{k+1}$.

$$1^3 + 2^3 + \cdots + k^3 = \dfrac{k^2(k+1)^2}{4} \qquad S_k$$

$$1^3 + 2^3 + \cdots + k^3 + (k+1)^3 = \dfrac{k^2(k+1)^2}{4} + (k+1)^3$$

**Adding $(k+1)^3$**

$$= \dfrac{k^2(k+1)^2 + 4(k+1)^3}{4}$$

$$= \dfrac{(k+1)^2}{4}[k^2 + 4(k+1)]$$

$$= \dfrac{(k+1)^2}{4}(k^2 + 4k + 4)$$

$$= \dfrac{(k+1)^2(k+2)^2}{4}$$

**21.**

$S_n$:　$2 + 6 + 12 + \cdots + n(n+1) = \dfrac{n(n+1)(n+2)}{3}$

$S_1$:　$1(1+1) = \dfrac{1(1+1)(1+2)}{3}$

$S_k$:　$2 + 6 + 12 + \cdots + k(k+1) = \dfrac{k(k+1)(k+2)}{3}$

$S_{k+1}$:

$2 + 6 + 12 + \cdots + k(k+1) + (k+1)[(k+1)+1]$
$$= \dfrac{(k+1)[(k+1)+1][(k+1)+2]}{3}$$

(1) *Basis step*: $S_1$: $1(1+1) = \dfrac{1(1+1)(1+2)}{3}$. True.

(2) *Induction step*: Assume $S_k$:

$2 + 6 + 12 + \cdots + k(k+1) = \dfrac{k(k+1)(k+2)}{3}$.

Then $2 + 6 + 12 + \cdots + k(k+1) + (k+1)(k+1+1)$

$$= \dfrac{k(k+1)(k+2)}{3} + (k+1)(k+2)$$

$$= \dfrac{k(k+1)(k+2) + 3(k+1)(k+2)}{3}$$

$$= \dfrac{(k+1)(k+2)(k+3)}{3}$$

$$= \dfrac{(k+1)(k+1+1)(k+1+2)}{3}.$$

**23.** $S_n$:　$a_1 + (a_1 + d) + (a_1 + 2d) + \cdots +$
$$[a_1 + (n-1)d] = \dfrac{n}{2}[2a_1 + (n-1)d]$$

$S_1$:　$a_1 = \dfrac{1}{2}[2a_1 + (1-1)d]$

$S_k$:　$a_1 + (a_1 + d) + (a_1 + 2d) + \cdots +$
$$[a_1 + (k-1)d] = \dfrac{k}{2}[2a_1 + (k-1)d]$$

$S_{k+1}$:　$a_1 + (a_1 + d) + (a_1 + 2d) + \cdots +$
$[a_1 + (k-1)d] + [a_1 + ((k+1)-1)d]$
$$= \dfrac{k+1}{2}[2a_1 + ((k+1)-1)d]$$

(1) *Basis step*: Since $\frac{1}{2}[2a_1 + (1-1)d] = \frac{1}{2} \cdot 2a_1 = a_1$, $S_1$ is true.

(2) *Induction step*: Assume $S_k$. Deduce $S_{k+1}$. Starting with the left side of $S_{k+1}$, we have

$$\underbrace{a_1 + (a_1 + d) + \cdots + [a_1 + (k-1)d]}_{= \dfrac{k}{2}[2a_1 + (k-1)d]} + [a_1 + kd]$$
$$+ [a_1 + kd]$$

**By $S_k$**

$$= \dfrac{k[2a_1 + (k-1)d]}{2 \cdot} + \dfrac{2[a_1 + kd]}{2}$$

$$= \dfrac{2ka_1 + k(k-1)d + 2a_1 + 2kd}{2}$$

$$= \dfrac{2a_1(k+1) + k(k-1)d + 2kd}{2}$$

$$= \dfrac{2a_1(k+1) + (k-1+2)kd}{2}$$

$$= \dfrac{2a_1(k+1) + (k+1)kd}{2} = \dfrac{k+1}{2}[2a_1 + kd].$$

**24.** [6.1], [6.3], [6.5], [6.6] $(5, 3)$

**25.** [6.2], [6.3], [6.5], [6.6] 1.5%: \$800; 2%: \$1600; 3%: \$2000

**27.** $S_n$:　$x + y$ is a factor of $x^{2n} - y^{2n}$.

$S_1$:　$x + y$ is a factor of $x^2 - y^2$.

$S_k$:　$x + y$ is a factor of $x^{2k} - y^{2k}$.

$S_{k+1}$:　$x + y$ is a factor of $x^{2(k+1)} - y^{2(k+1)}$.

(1) *Basis step*: $S_1$: $x + y$ is a factor of $x^2 - y^2$. True.
　　$S_2$: $x + y$ is a factor of $x^4 - y^4$. True.

(2) *Induction step*: Assume $S_{k-1}$: $x + y$ is a factor of $x^{2(k-1)} - y^{2(k-1)}$. Then $x^{2(k-1)} - y^{2(k-1)} = (x+y)Q(x)$ for some polynomial $Q(x)$.

Assume $S_k$: $x + y$ is a factor of $x^{2k} - y^{2k}$. Then $x^{2k} - y^{2k} = (x+y)P(x)$ for some polynomial $P(x)$.

$x^{2(k+1)} - y^{2(k+1)}$
$$= (x^{2k} - y^{2k})(x^2 + y^2) - (x^{2(k-1)} - y^{2(k-1)})(x^2y^2)$$
$$= (x+y)P(x)(x^2 + y^2) - (x+y)Q(x)(x^2y^2)$$
$$= (x+y)[P(x)(x^2 + y^2) - Q(x)(x^2y^2)]$$

so $x + y$ is a factor of $x^{2(k+1)} - y^{2(k+1)}$.

**29.** $S_2$:　$\log_a(b_1 b_2) = \log_a b_1 + \log_a b_2$

$S_k$:　$\log_a(b_1 b_2 \cdots b_k) = \log_a b_1 + \log_a b_2 + \cdots + \log_a b_k$

$S_{k+1}$:　$\log_a(b_1 b_2 \cdots b_{k+1}) = \log_a b_1 + \log_a b_2 + \cdots + \log_a b_{k+1}$

(1) *Basis step*: $S_2$ is true by the properties of logarithms.

(2) *Induction step*: Let $k$ be a natural number $k \geq 2$. Assume $S_k$. Deduce $S_{k+1}$.

$\log_a(b_1 b_2 \cdots b_{k+1})$    **Left side of $S_{k+1}$**
$= \log_a(b_1 b_2 \cdots b_k) + \log_a b_{k+1}$    **By $S_2$**
$= \log_a b_1 + \log_a b_2 + \cdots + \log_a b_k + \log_a b_{k+1}$

**31.** $S_2$:  $\overline{z_1 + z_2} = \bar{z}_1 + \bar{z}_2$:
$\overline{(a+bi)+(c+di)} = \overline{(a+c)+(b+d)i}$
$= (a+c) - (b+d)i$
$\overline{(a+bi)} + \overline{(c+di)} = a - bi + c - di$
$= (a+c) - (b+d)i.$
$S_k$:  $\overline{z_1 + z_2 + \cdots + z_k} = \bar{z}_1 + \bar{z}_2 + \cdots + \bar{z}_k.$
$\overline{(z_1 + z_2 + \cdots + z_k) + z_{k+1}}$
$= \overline{(z_1 + z_2 + \cdots + z_k)} + \overline{z_{k+1}}$    **By $S_2$**
$= \bar{z}_1 + \bar{z}_2 + \cdots + \bar{z}_k + \bar{z}_{k+1}$    **By $S_k$**

## Mid-Chapter Mixed Review: Chapter 8

**1.** False    **2.** True    **3.** False    **4.** False    **5.** 8, 11, 14, 17;
32; 47    **6.** 0, $-1$, 2, $-3$; 8; $-13$    **7.** $a_n = 3n$
**8.** $a_n = (-1)^n n^2$    **9.** $1\frac{7}{8}$, or $\frac{15}{8}$    **10.** $2 + 6 + 12 +$
$20 + 30 = 70$    **11.** $\sum_{k=1}^{\infty} (-1)^k 4k$    **12.** 2, 6, 22, 86
**13.** $-5$    **14.** 22    **15.** 21    **16.** 696    **17.** $-\frac{1}{2}$
**18.** (a) 8; (b) $\frac{1023}{16}$, or 63.9375    **19.** $-\frac{16}{3}$    **20.** Does not exist
**21.** 126 plants    **22.** \$6369.70
**23.** $S_n$:    $1 + 4 + 7 + \cdots + (3n - 2) = \frac{1}{2}n(3n - 1)$
$S_1$:    $3 \cdot 1 - 2 = \frac{1}{2} \cdot 1(3 \cdot 1 - 1)$
$S_k$:    $1 + 4 + 7 + \cdots + (3k - 2) = \frac{1}{2}k(3k - 1)$
$S_{k+1}$:    $1 + 4 + 7 + \cdots + (3k - 2) + [3(k + 1) - 2]$
$= \frac{1}{2}(k + 1)[3(k + 1) - 1]$
$= \frac{1}{2}(k + 1)(3k + 2)$

(1) *Basis step*: $S_1$: $3 \cdot 1 - 2 = \frac{1}{2} \cdot 1(3 \cdot 1 - 1)$.    **True**
(2) *Induction step*: Assume $S_k$:
$1 + 4 + 7 + \cdots + (3k - 2) = \frac{1}{2}k(3k - 1)$.
Then $1 + 4 + 7 + \cdots + (3k - 2) + [3(k + 1) - 2]$
$= \frac{1}{2}k(3k - 1) + [3(k + 1) - 2]$
$= \frac{3}{2}k^2 - \frac{1}{2}k + 3k + 1$
$= \frac{3}{2}k^2 + \frac{5}{2}k + 1$
$= \frac{1}{2}(3k^2 + 5k + 2)$
$= \frac{1}{2}(k + 1)(3k + 2).$
**24.** The first formula can be derived from the second by substituting $a_1 + (n - 1)d$ for $a_n$. When the first and last terms of the sum are known, the second formula is the better one to use. If the last term is not known, the first formula allows us to compute the sum in one step without first finding $a_n$.
**25.** $1 + 2 + 3 + \cdots + 100$
$= (1 + 100) + (2 + 99) + (3 + 98) + \cdots +$
$(50 + 51)$
$= \underbrace{101 + 101 + 101 + \cdots + 101}_{\text{50 addends of 101}}$
$= 50 \cdot 101$
$= 5050$
A formula for the first $n$ natural numbers is $\frac{n}{2}(1 + n)$.

**26.** Answers may vary. One possibility is given. Casey invests \$900 at 8% interest, compounded annually. How much will be in the account at the end of 40 years?    **27.** We can prove an infinite sequence of statements $S_n$ by showing that a basis statement $S_1$ is true and then that for all natural numbers $k$, if $S_k$ is true, then $S_{k+1}$ is true.

## Exercise Set 8.5

**1.** 720    **3.** 604,800    **5.** 120    **7.** 1    **9.** 3024    **11.** 120
**13.** 120    **15.** 1    **17.** 6,497,400    **19.** $n(n - 1)(n - 2)$
**21.** $n$    **23.** $6! = 720$    **25.** $9! = 362,880$    **27.** $_9P_4 = 3024$
**29.** $_5P_5 = 120$; $5^5 = 3125$    **31.** $_5P_5 \cdot {}_4P_4 = 2880$
**33.** $8 \cdot 10^6 = 8,000,000$; 8 million    **35.** $\frac{9!}{2!3!4!} = 1260$
**37.** (a) $_6P_5 = 720$; (b) $6^5 = 7776$; (c) $1 \cdot {}_5P_4 = 120$;
(d) $1 \cdot 1 \cdot {}_4P_3 = 24$    **39.** (a) $10^5$, or 100,000; (b) 100,000
**41.** (a) $10^9 = 1,000,000,000$; (b) yes    **42.** [1.5] $\frac{9}{4}$, or 2.25
**43.** [3.2] $-3, 2$    **44.** [3.2] $\frac{3 \pm \sqrt{17}}{4}$
**45.** [4.4] $-2, 1, 5$    **47.** 8    **49.** 11    **51.** $n - 1$

## Exercise Set 8.6

**1.** 78    **3.** 78    **5.** 7    **7.** 10    **9.** 1    **11.** 15
**13.** 128    **15.** 270,725    **17.** 13,037,895    **19.** $n$    **21.** 1
**23.** $_{23}C_4 = 8855$    **25.** $_{13}C_{10} = 286$
**27.** $\binom{10}{7} \cdot \binom{5}{3} = 1200$    **29.** $\binom{52}{5} = 2,598,960$
**31.** (a) $_{31}P_2 = 930$; (b) $31^2 = 961$; (c) $_{31}C_2 = 465$
**32.** [1.5] $-\frac{17}{2}$    **33.** [3.2] $-1, \frac{3}{2}$    **34.** [3.2] $\frac{-5 \pm \sqrt{21}}{2}$
**35.** [4.4] $-4, -2, 3$    **37.** $\binom{13}{5} = 1287$
**39.** $\binom{n}{2}$; $2\binom{n}{2}$    **41.** 4

## Exercise Set 8.7

**1.** $x^4 + 20x^3 + 150x^2 + 500x + 625$
**3.** $x^5 - 15x^4 + 90x^3 - 270x^2 + 405x - 243$
**5.** $x^5 - 5x^4 y + 10x^3 y^2 - 10x^2 y^3 + 5xy^4 - y^5$
**7.** $15,625x^6 + 75,000x^5 y + 150,000x^4 y^2 +$
$160,000x^3 y^3 + 96,000x^2 y^4 + 30,720xy^5 + 4096y^6$
**9.** $128t^7 + 448t^5 + 672t^3 + 560t + 280t^{-1} + 84t^{-3} +$
$14t^{-5} + t^{-7}$    **11.** $x^{10} - 5x^8 + 10x^6 - 10x^4 + 5x^2 - 1$
**13.** $125 + 150\sqrt{5}t + 375t^2 + 100\sqrt{5}t^3 + 75t^4 +$
$6\sqrt{5}t^5 + t^6$
**15.** $a^9 - 18a^7 + 144a^5 - 672a^3 + 2016a - 4032a^{-1} +$
$5376a^{-3} - 4608a^{-5} + 2304a^{-7} - 512a^{-9}$    **17.** $140\sqrt{2}$
**19.** $x^{-8} + 4x^{-4} + 6 + 4x^4 + x^8$    **21.** $21a^5 b^2$
**23.** $-252x^5 y^5$    **25.** $-745,472a^3$    **27.** $1120x^{12}y^2$
**29.** $-1,959,552u^5 v^{10}$    **31.** $2^7$, or 128    **33.** $2^{24}$, or
16,777,216    **35.** 20    **37.** $-12 + 316i$

**39.** $-7 - 4\sqrt{2}i$    **41.** $\sum_{k=0}^{n}\binom{n}{k}(-1)^k a^{n-k}b^k$, or

$\sum_{k=0}^{n}\binom{n}{k}a^{n-k}(-b)^k$

**43.** $\sum_{k=1}^{n}\binom{n}{k}x^{n-k}h^{k-1}$    **44.** [2.2] $x^2 + 2x - 2$

**45.** [2.2] $2x^3 - 3x^2 + 2x - 3$    **46.** [2.3] $4x^2 - 12x + 10$

**47.** [2.3] $2x^2 - 1$    **49.** $3, 9, 6 \pm 3i$    **51.** $-\dfrac{35}{x^{1/6}}$

**53.** $2^{100}$    **55.** $\left[\log_a(xt)\right]^{23}$

**57.** (1) *Basis step*: Since $a + b = (a + b)^1$, $S_1$ is true.

(2) *Induction step*: Let $S_k$ be the statement of the binomial theorem with $n$ replaced by $k$. Multiply both sides of $S_k$ by $(a + b)$ to obtain

$(a + b)^{k+1}$

$= \left[a^k + \cdots + \binom{k}{r-1}a^{k-(r-1)}b^{r-1} \right.$

$\left. + \binom{k}{r}a^{k-r}b^r + \cdots + b^k\right](a + b)$

$= a^{k+1} + \cdots + \left[\binom{k}{r-1} + \binom{k}{r}\right]a^{(k+1)-r}b^r$

$\qquad + \cdots + b^{k+1}$

$= a^{k+1} + \cdots + \binom{k+1}{r}a^{(k+1)-r}b^r + \cdots + b^{k+1}.$

This proves $S_{k+1}$, assuming $S_k$. Hence $S_n$ is true for $n = 1, 2, 3, \ldots$

## Exercise Set 8.8

**1.** (a) 0.18, 0.24, 0.23, 0.23, 0.12;  (b) Opinions may vary, but it seems that people tend not to pick the first or last numbers.
**3.** 11,700 pieces    **5.** (a) $\frac{2}{7}$; (b) $\frac{5}{7}$; (c) 0; (d) 1    **7.** $\frac{1}{2}$
**9.** (a) $\frac{1}{13}$; (b) $\frac{2}{13}$; (c) $\frac{1}{4}$; (d) $\frac{1}{26}$    **11.** $\frac{1}{5525}$    **13.** $\frac{135}{323}$
**15.** $\frac{1}{108,290}$    **17.** $\frac{33}{66,640}$    **19.** (a) HHH, HHT, HTH, HTT, THH, THT, TTH, TTT; (b) $\frac{3}{8}$; (c) $\frac{7}{8}$; (d) $\frac{7}{8}$; (e) $\frac{3}{8}$    **21.** $\frac{9}{19}$
**23.** $\frac{1}{38}$    **25.** $\frac{18}{19}$    **27.** $\frac{9}{19}$    **29.** [1.5] Zero
**30.** [5.1] One-to-one    **31.** [1.2] Function; domain; range; domain; range    **32.** [1.5] Zero    **33.** [8.6] Combination
**34.** [2.5] Inverse variation    **35.** [4.3] Factor
**36.** [8.3] Geometric sequence
**37.** (a) $4 \cdot \left(\frac{13}{5}\right) - 4 - 36 = 5108$; (b) 0.0475
**39.** (a) $13 \cdot \binom{4}{3} \cdot \binom{48}{2} - 3744$, or 54,912;

(b) $\dfrac{54{,}912}{\binom{52}{5}} \approx 0.0211$

## Review Exercises:Chapter 8

**1.** True    **2.** False    **3.** True    **4.** False
**5.** $-\frac{1}{2}, \frac{4}{17}, -\frac{9}{82}, \frac{16}{257}; -\frac{121}{14,642}; -\frac{529}{279,842}$    **6.** $(-1)^{n+1}(n^2 + 1)$
**7.** $\frac{3}{2} - \frac{9}{8} + \frac{27}{26} - \frac{81}{80} = \frac{417}{1040}$

**8.**

| $n$ | $u_n$ |
|----|----------|
| 1 | 0.3 |
| 2 | 2.5 |
| 3 | 13.5 |
| 4 | 68.5 |
| 5 | 343.5 |
| 6 | 1718.5 |
| 7 | 8593.5 |
| 8 | 42968.5 |
| 9 | 214843.5 |
| 10 | 1074218.5 |

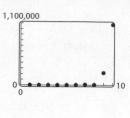

**9.** $\sum_{k=1}^{7}(k^2 - 1)$    **10.** $\frac{15}{4}$    **11.** $a + 4b$    **12.** 531
**13.** 20,100    **14.** 11    **15.** $-4$    **16.** $n = 6, S_n = S_6 = -126$
**17.** $a_1 = 8, a_5 = \frac{1}{2}$    **18.** Does not exist    **19.** $\frac{3}{11}$
**20.** $\frac{3}{8}$    **21.** $\frac{241}{99}$    **22.** $5\frac{4}{5}, 6\frac{3}{5}, 7\frac{2}{5}, 8\frac{1}{5}$    **23.** 167.3 ft
**24.** \$45,993.04    **25.** (a) \$7.38;  (b) \$1365.10
**26.** \$88,888,888,889

**27.** $S_n$:    $1 + 4 + 7 + \cdots + (3n - 2) = \dfrac{n(3n - 1)}{2}$

$S_1$:    $1 = \dfrac{1(3 \cdot 1 - 1)}{2}$

$S_k$:    $1 + 4 + 7 + \cdots + (3k - 2) = \dfrac{k(3k - 1)}{2}$

$S_{k+1}$:    $1 + 4 + 7 + \cdots + (3k - 2) + \left[3(k + 1) - 2\right]$
$\qquad = 1 + 4 + 7 + \cdots + (3k - 2) + (3k + 1)$
$\qquad = \dfrac{(k + 1)(3k + 2)}{2}$

(1) *Basis step*: $\dfrac{1(3 \cdot 1 - 1)}{2} = \dfrac{2}{2} = 1$ is true.

(2) *Induction step*: Assume $S_k$. Add $(3k + 1)$ on both sides.
$1 + 4 + 7 + \cdots + (3k - 2) + (3k + 1)$

$= \dfrac{k(3k - 1)}{2} + (3k + 1)$

$= \dfrac{k(3k - 1)}{2} + \dfrac{2(3k + 1)}{2}$

$= \dfrac{3k^2 - k + 6k + 2}{2}$

$= \dfrac{3k^2 + 5k + 2}{2}$

$= \dfrac{(k + 1)(3k + 2)}{2}$

**28.** $S_n$:    $1 + 3 + 3^2 + \cdots + 3^{n-1} = \dfrac{3^n - 1}{2}$

$S_1$:    $1 = \dfrac{3^1 - 1}{2}$

$S_k$: $\quad 1 + 3 + 3^2 + \cdots + 3^{k-1} = \dfrac{3^k - 1}{2}$

$S_{k+1}$: $\quad 1 + 3 + 3^2 + \cdots + 3^{(k+1)-1} = \dfrac{3^{k+1} - 1}{2}$

(1) *Basis step*: $\dfrac{3^1 - 1}{2} = \dfrac{2}{2} = 1$ is true.

(2) *Induction step*: Assume $S_k$. Add $3^k$ on both sides.

$\quad 1 + 3 + \cdots + 3^{k-1} + 3^k$

$\qquad = \dfrac{3^k - 1}{2} + 3^k = \dfrac{3^k - 1}{2} + 3^k \cdot \dfrac{2}{2}$

$\qquad = \dfrac{3 \cdot 3^k - 1}{2} = \dfrac{3^{k+1} - 1}{2}$

**29.** $S_n$: $\left(1 - \dfrac{1}{2}\right)\left(1 - \dfrac{1}{3}\right)\cdots\left(1 - \dfrac{1}{n}\right) = \dfrac{1}{n}$

$S_2$: $\left(1 - \dfrac{1}{2}\right) = \dfrac{1}{2}$

$S_k$: $\left(1 - \dfrac{1}{2}\right)\left(1 - \dfrac{1}{3}\right)\cdots\left(1 - \dfrac{1}{k}\right) = \dfrac{1}{k}$

$S_{k+1}$: $\left(1 - \dfrac{1}{2}\right)\left(1 - \dfrac{1}{3}\right)\cdots\left(1 - \dfrac{1}{k}\right)\left(1 - \dfrac{1}{k+1}\right)$

$\qquad = \dfrac{1}{k+1}$.

(1) *Basis step*: $S_2$ is true by substitution.

(2) *Induction step*: Assume $S_k$. Deduce $S_{k+1}$. Starting with the left side of $S_{k+1}$, we have

$\underbrace{\left(1 - \dfrac{1}{2}\right)\left(1 - \dfrac{1}{3}\right)\cdots\left(1 - \dfrac{1}{k}\right)}\left(1 - \dfrac{1}{k+1}\right)$

$\qquad = \dfrac{1}{k} \cdot \left(1 - \dfrac{1}{k+1}\right) \qquad$ **By $S_k$**

$\qquad = \dfrac{1}{k} \cdot \left(\dfrac{k+1-1}{k+1}\right)$

$\qquad = \dfrac{1}{k} \cdot \dfrac{k}{k+1}$

$\qquad = \dfrac{1}{k+1}$. $\qquad$ **Simplifying**

**30.** $6! = 720$ $\quad$ **31.** $9 \cdot 8 \cdot 7 \cdot 6 = 3024$

**32.** $\dbinom{15}{8} = 6435$ $\quad$ **33.** $24 \cdot 23 \cdot 22 = 12{,}144$

**34.** $\dfrac{9!}{1!\,4!\,2!\,2!} = 3780$ $\quad$ **35.** $3 \cdot 4 \cdot 3 = 36$

**36.** (a) $_6P_5 = 720$; (b) $6^5 = 7776$; (c) $1 \cdot {}_5P_4 = 120$;
(d) $_3P_2 \cdot 1 \cdot 1 \cdot 1 = 6$ $\quad$ **37.** $2^8$, or $256$
**38.** $m^7 + 7m^6n + 21m^5n^2 + 35m^4n^3 + 35m^3n^4 +$
$21m^2n^5 + 7mn^6 + n^7$
**39.** $x^5 - 5\sqrt{2}x^4 + 20x^3 - 20\sqrt{2}x^2 + 20x - 4\sqrt{2}$
**40.** $x^8 - 12x^6y + 54x^4y^2 - 108x^2y^3 + 81y^4$
**41.** $a^8 + 8a^6 + 28a^4 + 56a^2 + 70 + 56a^{-2} +$
$28a^{-4} + 8a^{-6} + a^{-8}$ $\quad$ **42.** $-6624 + 16{,}280i$ $\quad$ **43.** $220a^9x^3$

**44.** $-\dbinom{18}{11}128a^7b^{11} = -4{,}073{,}472a^7b^{11}$ $\quad$ **45.** $\frac{1}{12}; 0$

**46.** $\frac{1}{4}$ $\quad$ **47.** $\frac{6}{5525}$ $\quad$ **48.** $\frac{86}{206} \approx 0.42, \frac{97}{206} \approx 0.47, \frac{23}{206} \approx 0.11$

**49.** (a) $a_n = 0.1285179017n + 2.870691091$;
(b) about 13.2 lb of American cheese $\quad$ **50.** B

**51.** A $\quad$ **52.** D $\quad$ **53.** $\dfrac{a_{k+1}}{a_k} = r_1, \dfrac{b_{k+1}}{b_k} = r_2$, so
$\dfrac{a_{k+1}b_{k+1}}{a_kb_k} = r_1r_2$, a constant. $\quad$ **54.** (a) No (unless $a_n$ is all
positive or all negative); (b) yes; (c) yes; (d) no (unless $a_n$ is
constant); (e) no (unless $a_n$ is constant); (f) no (unless $a_n$ is
constant) $\quad$ **55.** $-2, 0, 2, 4$ $\quad$ **56.** $\frac{1}{2}, -\frac{1}{6}, \frac{1}{18}$

**57.** $\left(\log\dfrac{x}{y}\right)^{10}$ $\quad$ **58.** $18$ $\quad$ **59.** $36$ $\quad$ **60.** $-9$

**61.** Put the following in the form of a paragraph. First find
the number of seconds in a year (365 days):

$365 \text{ days} \cdot \dfrac{24 \text{ hr}}{1 \text{ day}} \cdot \dfrac{60 \text{ min}}{1 \text{ hr}} \cdot \dfrac{60 \text{ sec}}{1 \text{ min}} = 31{,}536{,}000 \text{ sec.}$

The number of arrangements possible is 15!.

The time is $\dfrac{15!}{31{,}536{,}000} \approx 41{,}466$ years. $\quad$ **62.** For each
circular arrangement of the numbers on a clock face, there
are 12 distinguishable ordered arrangements on a line. The
number of arrangements of 12 objects on a line is $_{12}P_{12}$,
or 12!. Thus the number of circular permutations is
$\dfrac{_{12}P_{12}}{12} = \dfrac{12!}{12} = 11! = 39{,}916{,}800$. In general, for each circular
arrangement of $n$ objects, there are $n$ distinguishable ordered
arrangements on a line. The total number of arrangements of
$n$ objects on a line is $_nP_n$, or $n!$. Thus the number of circular
permutations is $\dfrac{n!}{n} = \dfrac{n(n-1)!}{n} = (n-1)!$.

**63.** Choosing $k$ objects from a set of $n$ objects is equivalent to
not choosing the other $n - k$ objects. $\quad$ **64.** Order is considered in a combination lock. $\quad$ **65.** In expanding $(a + b)^n$,
it would probably be better to use Pascal's triangle when $n$ is
relatively small. When $n$ is large, and many rows of Pascal's
triangle must be computed to get to the $(n + 1)$st row, it
would probably be better to use factorial notation. In addition, factorial notation allows us to write a particular term of
the expansion more efficiently than Pascal's triangle does.

## Test: Chapter 8

**1.** [8.1] $-43$ $\quad$ **2.** [8.1] $\frac{2}{3}, \frac{3}{4}, \frac{4}{5}, \frac{5}{6}, \frac{6}{7}$
**3.** [8.1] $2 + 5 + 10 + 17 = 34$
**4.** [8.1]

| $n$ | $u_n$ |
|---|---|
| 1 | 0.66667 |
| 2 | 0.75 |
| 3 | 0.8 |
| 4 | 0.83333 |
| 5 | 0.85714 |
| 6 | 0.875 |
| 7 | 0.88889 |
| 8 | 0.9 |
| 9 | 0.90909 |
| 10 | 0.91667 |

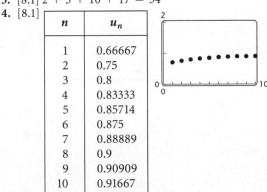

**5.** [8.1] $\sum_{k=1}^{6} 4k$    **6.** [8.1] $\sum_{k=1}^{\infty} 2^k$    **7.** [8.1] $3, 2\frac{1}{3}, 2\frac{3}{7}, 2\frac{7}{17}$

**8.** [8.2] 44    **9.** [8.2] 38    **10.** [8.2] $-420$

**11.** [8.2] 675    **12.** [8.3] $\frac{5}{512}$    **13.** [8.3] 1000

**14.** [8.3] 510    **15.** [8.3] 27    **16.** [8.3] $\frac{56}{99}$

**17.** [8.1] $10,000, $8000, $6400, $5120, $4096,

$3276.80    **18.** [8.2] $12.50    **19.** [8.3] $74,399.77

**20.** [8.4]

$S_n$:   $2 + 5 + 8 + \cdots + (3n - 1) = \dfrac{n(3n + 1)}{2}$

$S_1$:   $2 = \dfrac{1(3 \cdot 1 + 1)}{2}$

$S_k$:   $2 + 5 + 8 + \cdots + (3k - 1) = \dfrac{k(3k + 1)}{2}$

$S_{k+1}$:  $2 + 5 + 8 + \cdots + (3k - 1) + [3(k + 1) - 1]$
$$= \dfrac{(k + 1)[3(k + 1) + 1]}{2}$$

**(1)** *Basis step:* $\dfrac{1(3 \cdot 1 + 1)}{2} = \dfrac{1 \cdot 4}{2} = 2$, so $S_1$ is true.

**(2)** *Induction step:*

$\underbrace{2 + 5 + 8 + \cdots + (3k - 1)} + [3(k + 1) - 1]$

$= \dfrac{k(3k + 1)}{2} + [3k + 3 - 1]$   **By $S_k$**

$= \dfrac{3k^2}{2} + \dfrac{k}{2} + 3k + 2$

$= \dfrac{3k^2}{2} + \dfrac{7k}{2} + 2$

$= \dfrac{3k^2 + 7k + 4}{2}$

$= \dfrac{(k + 1)(3k + 4)}{2}$

$= \dfrac{(k + 1)[3(k + 1) + 1]}{2}$

**21.** [8.5] 3,603,600    **22.** [8.6] 352,716

**23.** [8.6] $\dfrac{n(n - 1)(n - 2)(n - 3)}{24}$    **24.** [8.5] $_6P_4 = 360$

**25.** [8.5] **(a)** $6^4 = 1296$; **(b)** $1 \cdot {}_5P_3 = 60$

**26.** [8.6] $_{28}C_4 = 20,475$    **27.** [8.6] $_{12}C_8 \cdot {}_8C_4 = 34,650$

**28.** [8.7] $x^5 + 5x^4 + 10x^3 + 10x^2 + 5x + 1$

**29.** [8.7] $35x^3y^4$    **30.** [8.7] $2^9 = 512$    **31.** [8.8] $\frac{4}{7}$

**32.** [8.8] $\frac{48}{1001}$    **33.** [8.1] B    **34.** [8.5] 15

# Index

## A

Abscissa, 59
Absolute value, 6, 53
  equations with, 283, 292
  inequalities with, 284, 292
  and roots, 45, 53
Absolute value function, 199
Absorption, coefficient of, 470
*ac*-method of factoring, 26
Addition
  associative property, 5, 53
  commutative property, 5, 53
  of complex numbers, 238, 287
  of exponents, 10, 53
  of functions, 175, 223
  of logarithms, 435, 441, 474
  of matrices, 519, 521, 566
  of polynomials, 18
  of radical expressions, 46
  of rational expressions, 38
Addition principle
  for equations, 31, 53, 121, 149
  for inequalities, 139, 150
Addition properties
  for matrices, 521
  for real numbers, 5, 53
Additive identity
  for matrices, 520, 521
  for real numbers, 5, 53
Additive inverse
  for matrices, 520, 521
  for real numbers, 5, 53
Algebra, fundamental theorem of, 335, 380
Algebra of functions, 175, 223
  and domains, 177, 223
Alternating sequence, 629
Amount of an annuity, 651, 655
Annuity, 651, 655
Appcylon, 63
Applications, *see Index of Applications*
Arithmetic mean, 645
Arithmetic progressions, *see* Arithmetic
    sequences
Arithmetic sequences, 636, 637, 697
  common difference, 637, 697
  *n*th term, 638, 697
  sums, 640, 697
Arithmetic series, 639
ASK mode, 123
Associative properties
  for matrices, 521
  for real numbers, 5, 53
Astronomical unit, 12
Asymptotes
  of exponential functions, 406, 420, 472
  horizontal, 349–352, 354, 382

  of a hyperbola, 597
  of logarithmic functions, 420, 473
  oblique, 353, 354, 383
  slant, 353, 354, 383
  vertical, 347–349, 354, 382
AU (astronomical unit), 12
Augmented matrix, 511, 532
AUTO mode, 63
Average, 645
Average rate of change, 92, 96, 147, 178
Axes, 58. *See also* Axis.
  of an ellipse, 587
  of a hyperbola, 596, 597
Axis
  conjugate, 597
  major, 587
  minor, 587
  of symmetry, 260, 290, 578
  transverse, 596
  *x*-, 59
  *y*-, 59

## B

Back-substitution, 501
Bar graph, 58
Base
  change of, 424, 474
  converting base *b* to base *e*, 462
  of an exponential function, 405, 472
  in exponential notation, 9
  of a logarithmic function, 418, 419, 473
Base *e* logarithm, 423, 472
Base–exponent property, 443, 475
Base-10 logarithm, 422, 473
Basis step, 657, 698
Beer–Lambert law, 470
Bel, 432
Binomial, *see* Binomials
Binomial coefficient notation, 675, 699
Binomial expansion
  using combination notation, 683, 700
  $(k + 1)$st term, 684, 700
  using Pascal's triangle, 680, 700
Binomial theorem, 681, 683, 700
Binomials, 18
  product of, 20, 53
Book value, 645
Break-even point, 498

## C

Calculator. *See also* Graphing
    calculator.
  logarithms on, 422
Canceling, 36
Carbon dating, 459, 460
Carry out, 124

Cartesian coordinate system, 58, 59
Center
  of circle, 67, 145, 586
  of ellipse, 587
  of hyperbola, 596
Change-of-base formula, 424, 474
Checking the solution, 31, 32, 124, 276, 280,
    291, 292
Circle APP, 587
Circle graph, 58
Circles, 67, 586
  center, 67, 145, 586
  equation, standard, 67, 145, 586, 619
  graph, 67, 68, 586, 619
  radius, 67, 145
  unit, 73
Circular arrangements, 703
Clearing fractions, 275, 291
Closed interval, 4
Coefficient, 17, 298, 375
  of absorption, 470
  binomial, 675, 699
  of determination, 308
  of linear correlation, 116
  leading, 17, 298, 375
  of a polynomial, 17
  of a polynomial function, 298, 375
Coefficient matrix, 511
Cofactor, 539
Collecting like (or similar) terms, 18
Column matrix, 525
Columns of a matrix, 511
Combinations, 673–675, 699
Combinatorics, 663, 673. *See also*
    Combinations; Permutations.
Combined variation, 216, 229
Combining like (or similar) terms, 18
Common denominator, 38
Common difference, 637, 697
Common factors, 22
Common logarithms, 422, 473
Common ratio, 646, 698
Commutative properties
  for matrices, 521
  for real numbers, 5, 53
Completing the square, 246–248, 289
Complex conjugates, 240, 287
  as zeros, 336
Complex numbers, 237, 287
  addition, 238, 287
  conjugates of, 240, 287
  division of, 241, 287
  imaginary, 237
  imaginary part, 237, 287
  multiplication of, 238, 239, 287
  powers of *i*, 239
  pure imaginary, 237

**I-1**

# Index of Applications

# Geometry

## Plane Geometry

**Rectangle**
Area: $A = lw$
Perimeter: $P = 2l + 2w$

**Square**
Area: $A = s^2$
Perimeter: $P = 4s$

**Triangle**
Area: $A = \frac{1}{2}bh$

**Sum of Angle Measures**
$A + B + C = 180°$

**Right Triangle**
Pythagorean theorem
(equation):
$a^2 + b^2 = c^2$

**Parallelogram**
Area: $A = bh$

**Trapezoid**
Area: $A = \frac{1}{2}h(a + b)$

**Circle**
Area: $A = \pi r^2$
Circumference:
$C = \pi d = 2\pi r$

## Solid Geometry

**Rectangular Solid**
Volume: $V = lwh$

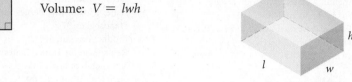

**Cube**
Volume: $V = s^3$

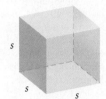

**Right Circular Cylinder**
Volume: $V = \pi r^2 h$
Lateral surface area:
$L = 2\pi rh$
Total surface area:
$S = 2\pi rh + 2\pi r^2$

**Right Circular Cone**
Volume: $V = \frac{1}{3}\pi r^2 h$
Lateral surface area:
$L = \pi rs$
Total surface area:
$S = \pi r^2 + \pi rs$
Slant height:
$s = \sqrt{r^2 + h^2}$

**Sphere**
Volume: $V = \frac{4}{3}\pi r^3$
Surface area: $S = 4\pi r^2$

# Trigonometry

## Trigonometric Functions

**Acute Angles**

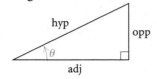

**Any Angle**

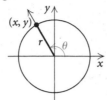

**Real Numbers**

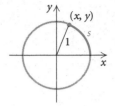

$$\sin\theta = \frac{\text{opp}}{\text{hyp}}, \quad \csc\theta = \frac{\text{hyp}}{\text{opp}},$$

$$\cos\theta = \frac{\text{adj}}{\text{hyp}}, \quad \sec\theta = \frac{\text{hyp}}{\text{adj}},$$

$$\tan\theta = \frac{\text{opp}}{\text{adj}}, \quad \cot\theta = \frac{\text{adj}}{\text{opp}}$$

$$\sin\theta = \frac{y}{r}, \quad \csc\theta = \frac{r}{y},$$

$$\cos\theta = \frac{x}{r}, \quad \sec\theta = \frac{r}{x},$$

$$\tan\theta = \frac{y}{x}, \quad \cot\theta = \frac{x}{y}$$

$$\sin s = y, \quad \csc s = \frac{1}{y},$$

$$\cos s = x, \quad \sec s = \frac{1}{x},$$

$$\tan s = \frac{y}{x}, \quad \cot s = \frac{x}{y}$$

## Basic Trigonometric Identities

$$\sin(-x) = -\sin x,$$
$$\cos(-x) = \cos x,$$
$$\tan(-x) = -\tan x,$$

$$\tan x = \frac{\sin x}{\cos x},$$

$$\cot x = \frac{\cos x}{\sin x},$$

$$\csc x = \frac{1}{\sin x},$$

$$\sec x = \frac{1}{\cos x},$$

$$\cot x = \frac{1}{\tan x}$$

## Pythagorean Identities

$$\sin^2 x + \cos^2 x = 1,$$
$$1 + \cot^2 x = \csc^2 x,$$
$$1 + \tan^2 x = \sec^2 x$$

## Identities Involving $\pi/2$

$$\sin(\pi/2 - x) = \cos x,$$
$$\cos(\pi/2 - x) = \sin x, \quad \sin(x \pm \pi/2) = \pm\cos x,$$
$$\tan(\pi/2 - x) = \cot x, \quad \cos(x \pm \pi/2) = \mp\sin x$$

## Sum and Difference Identities

$$\sin(u \pm v) = \sin u\cos v \pm \cos u\sin v,$$
$$\cos(u \pm v) = \cos u\cos v \mp \sin u\sin v,$$
$$\tan(u \pm v) = \frac{\tan u \pm \tan v}{1 \mp \tan u\tan v}$$

## Double-Angle Identities

$$\sin 2x = 2\sin x\cos x,$$
$$\cos 2x = \cos^2 x - \sin^2 x$$
$$= 1 - 2\sin^2 x$$
$$= 2\cos^2 x - 1,$$
$$\tan 2x = \frac{2\tan x}{1 - \tan^2 x}$$

## Half-Angle Identities

$$\sin\frac{x}{2} = \pm\sqrt{\frac{1 - \cos x}{2}}, \quad \cos\frac{x}{2} = \pm\sqrt{\frac{1 + \cos x}{2}},$$

$$\tan\frac{x}{2} = \pm\sqrt{\frac{1 - \cos x}{1 + \cos x}} = \frac{\sin x}{1 + \cos x} = \frac{1 - \cos x}{\sin x}$$

*(Trigonometry continued)*

2318

# Trigonometry  *(continued)*

## The Law of Sines

In any $\triangle ABC$,

$$\frac{a}{\sin A} = \frac{b}{\sin B} = \frac{c}{\sin C}.$$

## The Law of Cosines

In any $\triangle ABC$,

$$a^2 = b^2 + c^2 - 2bc \cos A,$$
$$b^2 = a^2 + c^2 - 2ac \cos B,$$
$$c^2 = a^2 + b^2 - 2ab \cos C.$$

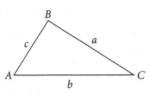

## Trigonometric Function Values of Special Angles

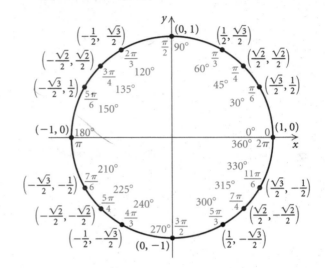

## Graphs of Trigonometric Functions

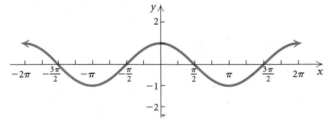

The sine function: $f(x) = \sin x$

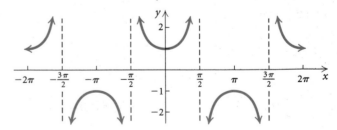

The cosecant function: $f(x) = \csc x$

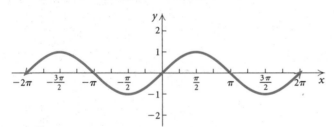

The cosine function: $f(x) = \cos x$

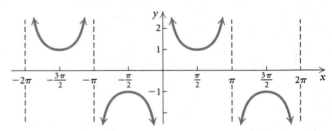

The secant function: $f(x) = \sec x$

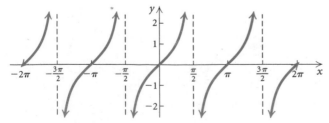

The tangent function: $f(x) = \tan x$

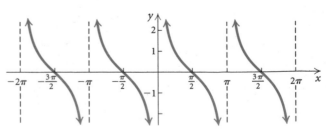

The cotangent function: $f(x) = \cot x$